FIFTH EDITION

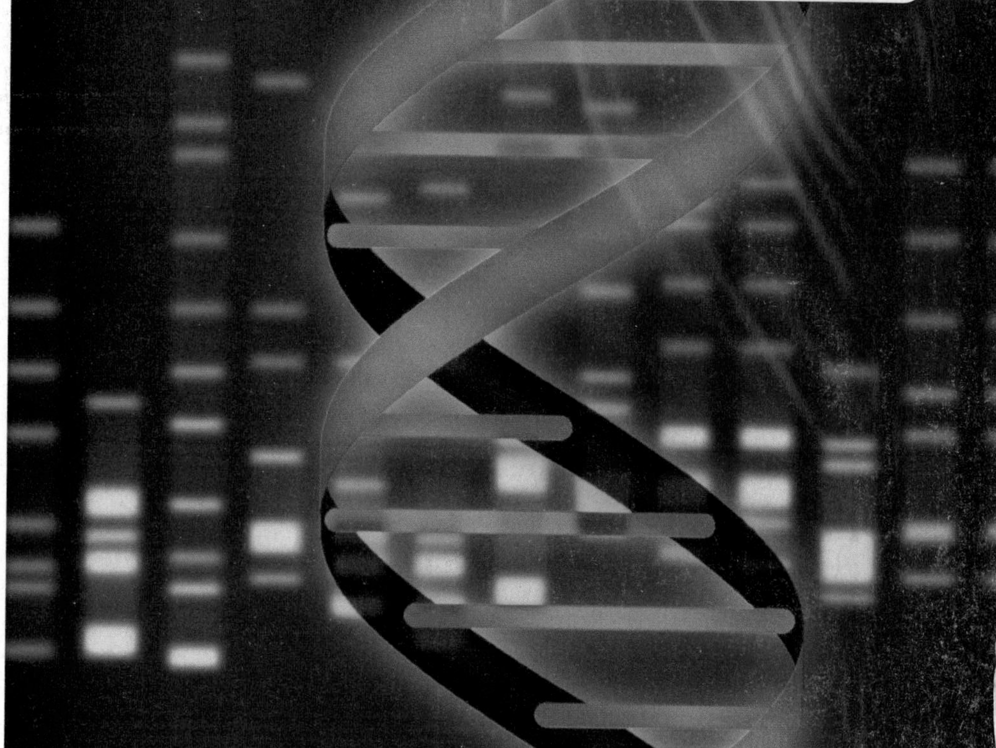

Polytene chromosomes from the salivary gland of the fruit fly *Drosophila melanogaster*, subjected to comparative immunofluorescent analysis.

ESSENTIALS OF GENETICS

William S. Klug
The College of New Jersey

Michael R. Cummings
University of Illinois at Chicago

With contributions by
Jon Herron, University of Washington
Charlotte Spencer, University of Alberta
Sara Ward, Colorado State University

Upper Saddle River, New Jersey 07458

Library of Congress Cataloging-in-Publication Data
Klug, William S.
Essentials of genetics / by William S. Klug and Michael R. Cummings.
p. cm.
Includes bibliographical references and index.
ISBN 0-13-143510-8 (alk. paper)
1. Genetics. I. Cummings, Michael R. II. Title.

QH430.K576 2005
576.5—dc22

2003066437

Executive Editor: Gary Carlson
Editor in Chief, Science: John Challice
Executive Managing Editor: Kathleen Schiaparelli
Assistant Managing Editor: Beth Sweeten
Vice President ESM Production and Manufacturing: David W. Riccardi
Manufacturing Manager: Trudy Pisciotti
Manufacturing Buyer: Alan Fischer
Creative Director: Carole Anson
Art Director: Jonathan Boylan
Director of Creative Service: Paul Belfanti
Managing Editor, AV Production & Management: Patty Burns
AV Art Editor: Jessica Einsig
Artwork: Imagineering
Media Editor: Andrew Stull
Assistant Managing Editor, Science Media: Nicole Bush
Assistant Managing Editor, Science Supplements: Becca Richter
Senior Marketing Manager: Shari Meffert
Project Manager: Chrissy Dudonis
Editorial Assistant: Susan Ziegler
Interior Designer: Anne Flanagan
Cover Designer: Tom Nery
Cover Photo: Laguna Design/Photo Researchers, Inc.
Photo Research: Lynda Sykes
Photo Editor: Debbie Hewitson
Production Services/Composition: Preparé Inc.

Published by Pearson Education, Inc.
Pearson Prentice Hall
Pearson Education, Inc.
Upper Saddle River, NJ 07458

Printed in the United States of America
10 9 8 7 6 5 4 3

ISBN 0-13-143510-8

Pearson Education Ltd., *London*
Pearson Education Australia Pty., Limited, *Sydney*
Pearson Education Singapore, Pte. Ltd.
Pearson Education North Asia Ltd., *Hong Kong*
Pearson Education Canada, Ltd., *Toronto*
Pearson Educación de Mexico, S.A. de C.V.
Pearson Education—Japan, *Tokyo*
Pearson Education Malaysia, Pte. Ltd.

DEDICATION

For geneticists, it is much about siblings and offspring. Ours are the best and very special.

To Sallie and Mike, and to Cindy, Braden, and Dori

To Mark and Kathleen, and to Brendan and Kerry

WSK
MRC

ABOUT THE AUTHORS

WILLIAM S. KLUG is currently a Professor of Biology at The College of New Jersey (formerly Trenton State College) in Ewing, New Jersey. He served as Chair of the Biology Department for 17 years, a position to which he was first elected in 1974. He received his B.A. degree in Biology from Wabash College in Crawfordsville, Indiana, and his Ph.D. from Northwestern University in Evanston, Illinois. Prior to coming to The College of New Jersey, he was on the faculty of Wabash College as an Assistant Professor, where he first taught genetics, as well as general biology and electron microscopy. His research interests have involved ultrastructural and molecular genetic studies of oogenesis in *Drosophila*. He has taught the genetics course as well as the senior capstone seminar course in human and molecular genetics to undergraduate biology majors for each of the last 35 years. He was the recent recipient of the first annual teaching award given at The College of New Jersey granted to the faculty member who most challenges students to achieve high standards.

MICHAEL R. CUMMINGS is currently an Associate Professor in the Department of Biological Sciences and in the Department of Molecular Genetics at the University of Illinois at Chicago. He has also served on the faculty at Northwestern University and Florida State University. He received his B.A. from St. Mary's College in Winona, Minnesota, and his M.S. and Ph.D. from Northwestern University in Evanston, Illinois. He has also written textbooks in human genetics and general biology for nonmajors. His research interests center on the molecular organization and physical mapping of human acrocentric chromosomes. At the undergraduate level, he teaches courses in Mendelian genetics, human genetics, and general biology for nonmajors. He has received numerous teaching awards given by the university and by student organizations.

BRIEF CONTENTS

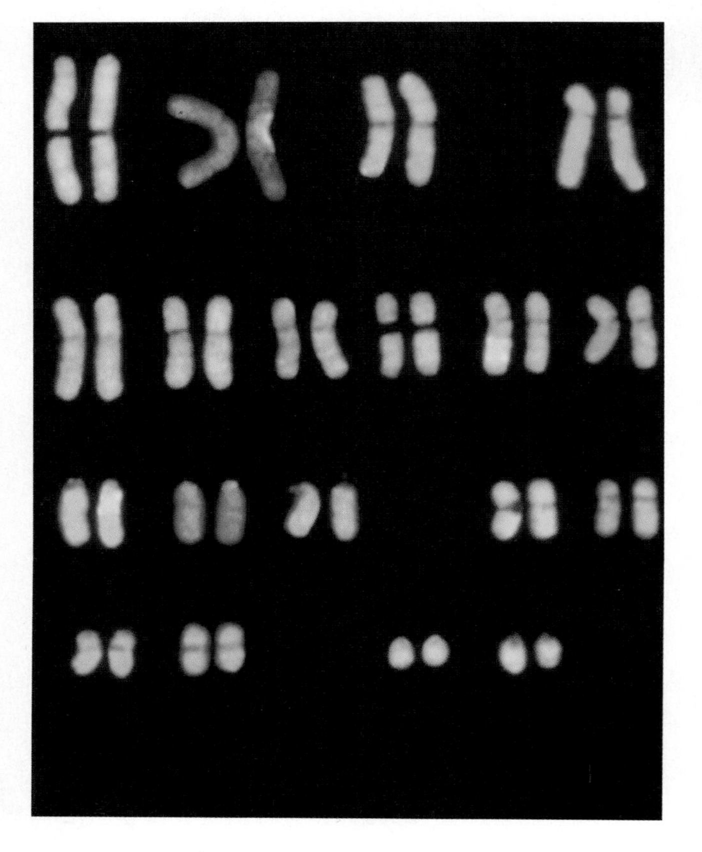

DETAILED CONTENTS

CHAPTER 12

Chromosome Structure and DNA Sequence Organization 253

CHAPTER 13

The Genetic Code and Transcription 272

CHAPTER 14

Translation and Proteins 295

CHAPTER 15

Gene Mutation, DNA Repair, and Transposable Elements 322

CHAPTER 24

Conservation Genetics 552

PREFACE

Essentials of Genetics is written for courses requiring a text that is shorter and more basic than its more comprehensive companion, *Concepts of Genetics*. While coverage is thorough, current, and of high quality, *Essentials* is written to be more accessible to biology majors early in their undergraduate careers, as well as to students majoring in agriculture, chemistry, engineering, forestry, psychology, or wildlife management. Because the text is shorter than many other books, *Essentials of Genetics* will also be more manageable in one-quarter and one-semester courses.

Goals

Although *Essentials of Genetics* is almost 300 pages shorter than its companion volume, our goals during revision are the same for both books. Specifically, we seek to

- Emphasize concepts rather than excessive detail.
- Write clearly and directly to students in order to provide understandable explanations of complex analytical topics.
- Establish careful organization within and between chapters.
- Maintain constant emphasis on scientific analysis as the means to illlustrate the nature of scientific discovery.
- Propagate the rich history of genetics that so beautifully illustrates how information is acquired and extended within the discipline as it develops and grows.
- Emphasize problem solving, thereby guiding students to think analytically and to apply and extend their knowledge of genetics.
- Provide the most modern and up-to-date coverage of this exciting field.
- Include whole chapters that provide comprehensive coverage of topics at the cutting edge of genetics.
- Create inviting, engaging, and pedagogically useful full-color figures enhanced by equally helpful photographs to support concept development.
- Provide outstanding On-line Media Tutorials where students are guided in their understanding of important concepts by working through the best animations, tutorial exercises, and self-assessment tools available.

These goals serve as the cornerstones of *Essentials of Genetics*. This pedagogic foundation allows the book to accommodate courses with many different approaches and lecture formats. Chapters are written to be as independent of one another as possible, allowing instructors to utilize them in various sequences. We believe that the varied approaches embodied in these goals work together to provide students with optimal support for their study of genetics.

Features New to the Fifth Edition

This edition has several new features that continue to make *Essentials of Genetics* the most distinctive and pedagogically useful textbook available to students of genetics, including:

- **How Do We Know What We Know?** This is a new feature found at the end of the introduction in each chapter. Based on a modern approach to teaching biology referred to as "Science as a way of Knowing," this section asks

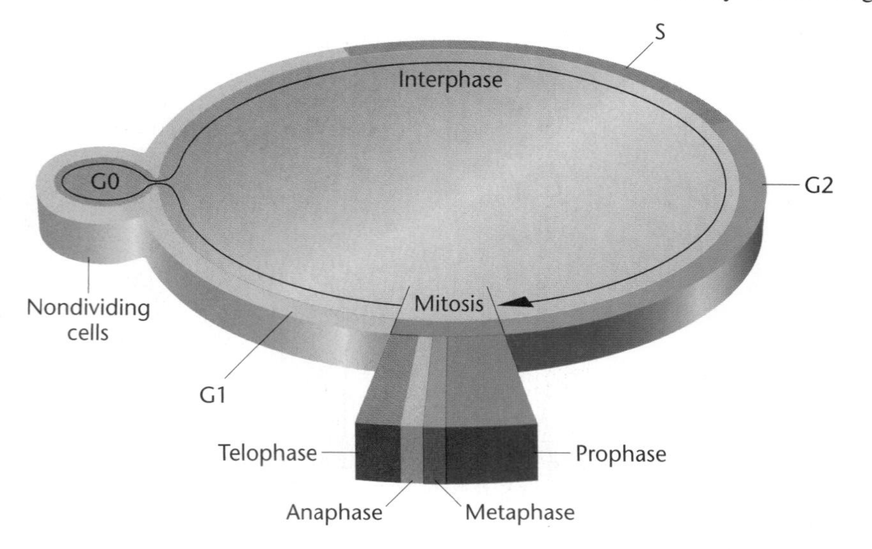

the student to be aware of the most important issues that have framed our thinking and to constantly ask themselves how we discovered this information. We believe that this approach, which has always been pursued in the text of each chapter but is now formalized, will enhance students' understanding of the topics covered in each chapter. The ideas behind this approach are explained in Chapter 1, and the execution begins in Chapter 2.

- **New Problems** Over 300 new problems, many based on research data derived from the literature of genetics, have been added to chapters throughout the text. Students will find in excess of 30 Problems and Discussion Questions at the end of almost all of the chapters. As in past editions, solutions appear in the *Student Handbook and Solutions Manual.*
- **Extranuclear Inheritance** Formal coverage, absent from recent editions, has been returned to this edition. This topic is covered in Chapter 4, "Modification of Mendelian Ratios", and Chapter 12, "Chromosome Structure and DNA Sequence Organization."
- **New Genetics, Technology, and Society Essays** Six new topics have been added to this edition.

  A Question of Gender: Sex Selection in Humans (Chapter 5) looks at the future of this controversial topic.

  Why Is There No Effective HIV Vaccine? (Chapter 16) explores the immunological aspects of AIDS.

  Footprints of a Killer (Chapter 18) examines our attempts to genetically identify anthrax strains.

  Gene Therapy—Two Steps Forward or Two Steps Back (Chapter 19) examines why this approach has yet to reach its full potential in curing genetic diseases.

  Stem Cell Wars (Chapter 20) examines the issues surrounding the use of these undifferentiated cells.

  Tracking Our Genetic Footprints out of Africa (Chapter 22) explores human origins.

- **Revised Genetics, Technology, and Society Essays** In addition to the new essays, five existing essays have been updated.

  The Twists and Turns of the Helical Revolution (Chapter 10)

  Antisense Oligonucleotides: Attacking the Messenger (Chapter 13)

  Mad Cows and Heresies: The Prion Story (Chapter 14)

  Beyond Dolly: The Cloning of Humans (Chapter 17)

  Gene Pools and Endangered Species: The Plight of the Florida Panther (Chapter 24)

- **Conversion of Major Headings to a Narrative Style** To better summarize the main point of each major section in the book, we have converted the headings to narrative statements rather than descriptive phrases.

Continued Emphasis on Concepts

As in its companion volume, *Essentials of Genetics* continues to emphasize the conceptual framework of genetics. Our experience with this approach shows that students more easily comprehend and take with them to succeeding courses the most important ideas in genetics as well as an analytic view of biological problems. To aid students in identifying conceptual aspects of a major topic, each chapter begins with Chapter Concepts, the set of narrative descriptions of each major section within the chapter. Then, each chapter ends with a Chapter Summary, which enumerates the five to ten key points that have been covered. These two features help to ensure that students focus on concepts and are not distracted by the many, albeit important, details of genetics. Specific examples and carefully designed figures support this approach throughout the book.

Insights and Solutions

Genetics, more than any other discipline within biology, requires problem solving and analytical thinking. At the end of many chapters we include what has become an extremely popular and successful section called Insights and Solutions. In this section we stress:

- Problem solving
- Quantitative analysis

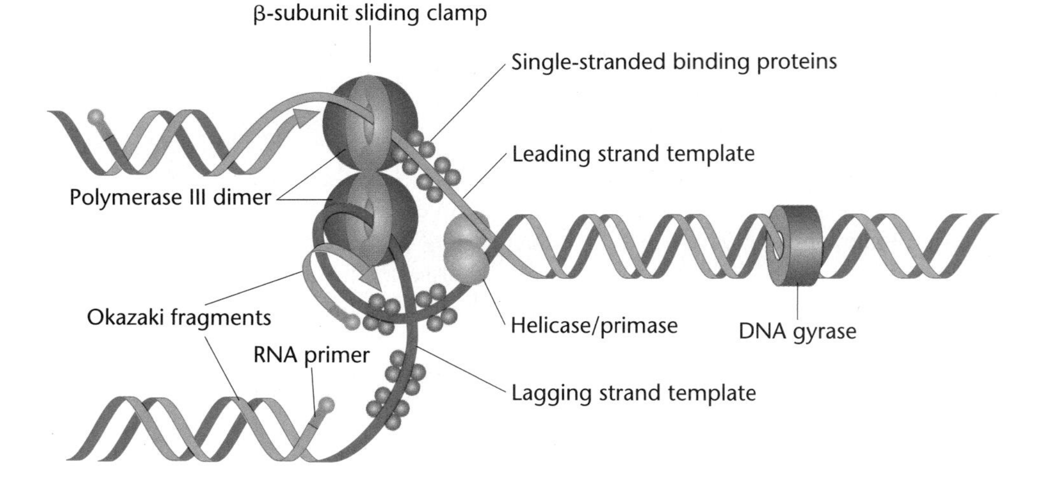

- Analytical thinking
- Experimental rationale

Problems or questions are posed, and detailed solutions or answers are provided. This feature primes students as they move on to the Problems and Discussion Questions that conclude each chapter.

Problems and Discussion Questions

Each chapter ends with an extensive collection of problems and discussion questions that optimize the opportunities for student growth in the important areas of problem solving and analytical thinking. Various levels of difficulty are presented, with the most challenging problems located at the end of each section. Brief answers to half of the problems are in Appendix A. The *Student Handbook and Solutions Manual* is available for faculty who wish to expose their students to detailed answers to all problems and questions. As mentioned above, we have greatly expanded the number of these problems, particularly those that are more challenging and those involving data analysis, so almost every chapter has at least 30 entries.

Acknowledgments

All comprehensive texts are dependent upon the valuable input provided by many colleagues. While we take full responsibility for any errors in this book, we gratefully acknowledge the help provided by those individuals who reviewed or otherwise contributed to the content and pedagogy of this and previous editions.

In particular, we thank Sarah Ward at Colorado State University for creating and revising Chapter 24, "Conservation Genetics." We are grateful for the valuable contributions of Charlotte Spencer at the Cross Cancer Institute at the University of Alberta, who wrote and or revised most of the Genetics, Technology, and Society essays. We thank Mark Shotwell at Slippery Rock University for his contribution of several other essays. We also much appreciate the careful attention and dedication provided by Arlene Larson at the University of Colorado, Denver, and Laura Runyen-Janecky at the University of Richmond, as they proofread the entire final manuscript. And, as always, it has been a pleasure working with Harry Nickla at Creighton U. who has served many roles during the production of this text. Finally, we are pleased to have received input from many genetics colleagues during their reviews of *Essentials of Genetics*:

Robert W. Adkinson, Louisiana State University
Nancy Bachman, SUNY College at Oneonta
Hank W. Bass, Florida State University
Andrew Bohonak, San Diego State University
Janice Bossart, The College of New Jersey
Paul Bottino, University of Maryland
James Bricker, The College of New Jersey
Hugh Britten, University of South Dakota
Aaron Cassill, University of Texas, San Antonio
Michael John Christoffers, North Dakota State University
Mary Anne Clark, Texas Wesleyan University
Jimmy D. Clark, University of Kentucky
Gregory P. Copenhaver, University of North Carolina at Chapel Hill
Jason Curtis, Purdue University North Central
Stephen J. D'Surney, University of Mississippi
Robert Duronio, University of North Carolina at Chapel Hill
Scott Erdman, Syracuse University
Asim Esen, Virginia Tech
Nancy H. Ferguson, Clemson University
Gail Fraizer, Kent State University
Kim Gaither, Oklahoma Christian University
David Galbreath, McMaster University
Derek J. Girman, Sonoma State University
Elliott Goldstein, Arizona State University
Mark L. Hammond, Campbell University
Jon Herron, University of Washington
Karen Hicks, Kenyon College
Kent Holingser, University of Connecticut
Mike Hoopman, The College of New Jersey
David Hoppe, University of Minnesota, Morris
John A. Hunt, University of Hawaii, Honolulu
Cheryl Jorcyk, Boise State University
David Kass, Eastern Michigan University
Beth A. Krueger, Monroe Community College
M. A. Lachance, University of Western Ontario
Arlene Larson, University of Colorado at Denver
Hsiu-Ping Liu, Southwest Missouri State University
Paul F. Lurquin, Washington State University
Sally Mackenzie, University of Nebraska
Terry C. Matthews, Millikan University
Cynthia Moore, Washington University
Janet Morrison, The College of New Jersey
Michelle A. Murphy, University of Notre Dame
Marcia O'Connell, The College of New Jersey
Dan Panaccione, West Virginia University
Todd Rimkus, Marymount University
Laura Runyen-Janecky, University of Richmond
Tom Savage, Oregon State University
Gerald Schlink, Missouri Southern State College
Randy Scholl, Ohio State University
Malcolm Schug, University of North Carolina, Greensboro
Ralph Seelke, University of Wisconsin, Superior
Gurel S. Sidhu, California State University
Theresa Spradling, University of Northern Iowa
Mark Sturtevant, Northern Arizona University
Christine Tachibana, University of Washington
Daniel Wang, University of Miami
R. C. Woodruff, Bowling Green State University
Marie Wooten, Auburn University

For the Student

Online Media Tutorials

The most sophisticated learning and tutorial package available for students of genetics, these on-line tutorials address the concepts that students find most difficult. Each of the 44 tutorials is composed of several animations and interactive exercises, along with a post-tutorial quiz of self-grading questions to reinforce the important concepts. Found on the Student Web Site, the tutorials offer timely and relevant support for students. Look for this icon in the margin of this page to identify media tutorials that support concepts presented in the textbook.

Student Handbook and Solutions Manual

Harry Nickla, Creighton University
(0-13-143524-8)
Completely reviewed and checked for accuracy, this valuable handbook provides a detailed step-by-step solution or extended discussion for every problem in the text in a chapter-by-chapter format. The handbook also contains extra study problems and a thorough review of the concepts and vocabulary.

The* New York Times *Themes of the Times: Genetics and Molecular Biology

Coordinated by Harry Nickla, Creighton University
This exciting newspaper-format supplement brings together recent genetics and molecular biology articles from the pages of the highly respected *New York Times*. This free supplement , available through your local representative, encourages students to make the connections between genetics concepts and the latest research and breakthroughs in the field

Research Navigator
www.researchnavigator.com
Prentice Hall's new *Research Navigator*™ gives your students access to the most current information available for a wide array of subjects via EBSCO's ContentSelect™ Academic Journal Database, *The New York Times* Search by Subject Archive, "Best of the Web" Link Library, and information on the latest news and current events. This valuable tool helps students find the most useful articles and journals, cite sources, and write effective papers for research assignments.

For the Instructor

Instructor's Resource Center on CD-ROM

For adopters of *Essentials, Fifth Edition*, the Instructor's Resource Center on CD-ROM (IRC) contains

- Presentation art PowerPoints for each chapter
- Personal Response System PowerPoints for each chapter
- Image files (all illustrations, all tables, and many photographs)
- Instructor's Manual
- Test item file for each chapter (easily customized)
- 44 Web Tutorials exploring the most challenging material in genetics
- ~200 Animations and Interactive Exercises presenting key concepts in genetics
- Glossary of genetic terms
- Link to the Student Web Site for OneKey

Instructors will be able to coordinate lecture presentations with text content, knowing that students will be studying using the same animated tutorials, images, and review material.

Instructor's Resource Manual with Tests

Harry Nickla, Creighton University
(0-13-143511-6)
This manual with tests contains over 1000 questions and problems an instructor can use to prepare exams. The manual also provides optional course sequences, a guide to audiovisual supplements, and a section on searching the Web.

TestGen-EQ
(0-13-143514-0)
This text-specific testing program is networkable for administering tests. It also allows you to edit existing test items or add your own questions to create a nearly unlimited number of quizzes and tests.

Transparencies
(0-13-143513-2)
Two hundred figures from the text are included in the transparency package: 150 four-color transparencies from the text plus 50 transparency masters. The font size of the labels has been increased for easy viewing from the back of the classroom.

MEDIA TUTORIAL

OneKey

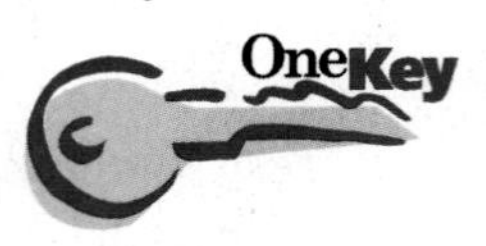

OneKey offers the best teaching and learning resources all in one place. Conveniently organized by textbook chapter, these compiled resources help you save time and help your students reinforce and apply what they have learned in class. The student media program in OneKey includes web tutorials, self-grading quizzes, problems and discussion, and much more.

Course Management Tools

If your campus is currently an adopter of either the WebCT or Blackboard course management environment, you can use our pre-formatted cartridges to help you get a jump-start building your course.

WebCT ***Blackboard***

Both of these options include all the on-line study opportunities and multimedia resources that accompany *Essentials of Genetics, Fifth Edition*. Extensive on-line support and information is provided to help you more effectively build, customize, and manage your on-line course.

CHAPTER 1

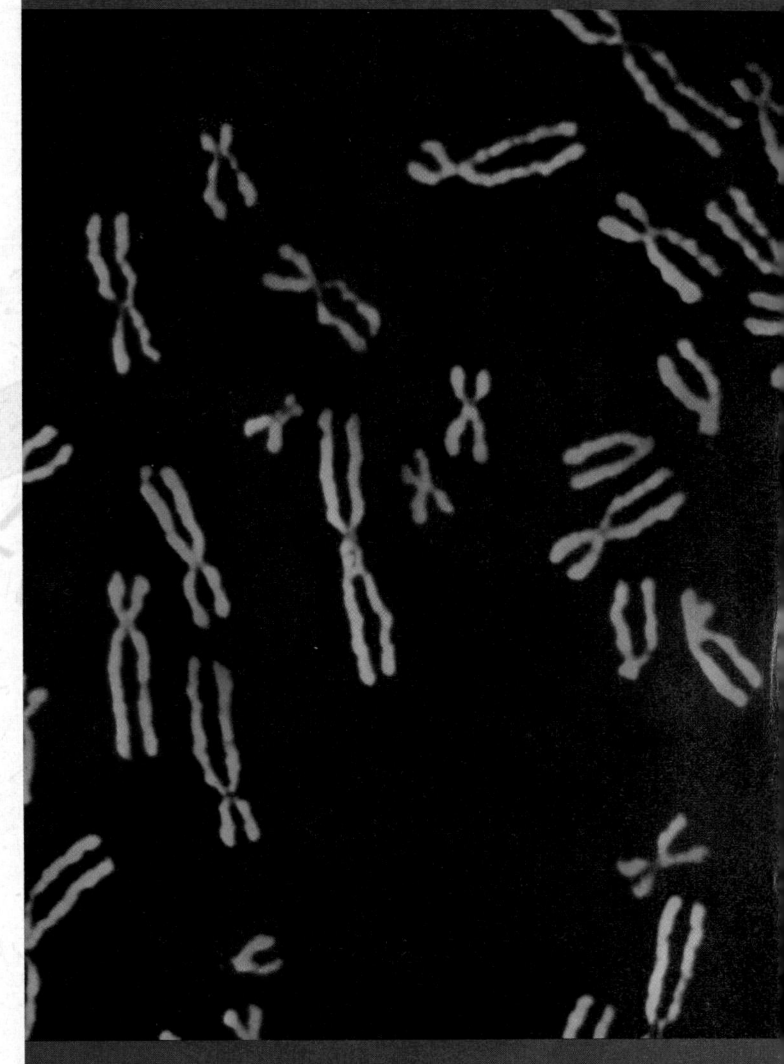

Human metaphase chromosomes, each composed of two sister chromatids joined at a common centromere. *(Omikron/Science Source/Photo Researchers, Inc.)*

An Introduction to Genetics

CHAPTER CONCEPTS

1.1 Genetics Has a Rich and Interesting History
Prehistoric Times: Domesticated Animals and Cultivated Plants
The Greek Influence: Hippocrates and Aristotle
1600–1850: The Dawn of Modern Biology
Charles Darwin and Evolution

1.2 Nucleic Acids and Proteins Serve as the Molecular Basis of Genetics
The Trinity of Molecular Genetics
The Structure of Nucleic Acids
The Genetic Code and RNA Triplets
Proteins and Biological Function

1.3 Genetics Has Been Investigated Using Many Different Approaches

1.4 Genetics Has a Profound Impact on Society
Eugenics and Euphenics
Genetic Advances in Agriculture and Medicine

Throughout history, genetics has had a profound effect on humankind. As knowledge in the discipline has grown, many different issues have arisen that have led to controversies at the interface between science and society. We begin this text with one instance where current advances in genetics directly affect all members of an entire country. This example captures the flavor of how genetic technology may impact on society, and provides a glimpse of how the future might be envisioned as further advances are made.

In early 1999, a controversy was growing among the 270,000 residents of the remote island nation of Iceland. As the new millennium approached, following months of fierce and sometimes heated debate, Iceland's parliament passed a law that gave a local biotechnology company, deCODE Genetics, the exclusive rights to develop a large database containing detailed DNA profiles of every resident of the country. This genetic information would be correlated with the existing genealogical and medical records of each person and marketed to researchers around the world over the next decade.

This is not a passage drawn from Aldous Huxley's *Brave New World* but rather an example of the current interface between genetics and society as we begin a new century. It is also the subject of a number of projects in several other countries. The largest is the "Biobank" effort launched in Great Britain in 2003. There, a huge database of 500,000 Britons will be compiled with the hope of ultimately defining the role of genetics and environment in the origin of many human diseases. In the United States, a similar effort involving 40,000 individuals is already under way in Marshfield, Wisconsin.

In Iceland, there are many reasons why a possible invasion of genetic privacy is about to occur to the residents of this small remote country. The people of Iceland represent a unique case of genetic uniformity seldom seen or accessible to scientific investigation. Indeed, for a variety of reasons, almost all Icelanders share a close genetic resemblance not only to one another but to their Viking ancestors who settled this country over 1000 years ago. Thus, geneticists believe that the Icelandic population is a tremendous asset in studying the link between genes and disease. Because of the state-supported health-care system, medical records exist for all residents as far back as 1900. Genealogical information is available for every living person, as well as for over 500,000 of the estimated 750,000 individuals who have ever lived in Iceland.

On the flip side are issues of privacy, consent, and commercialism—issues that are at the heart of most dilemmas and controversies arising from applications of newly acquired genetic technology. The central questions currently being asked throughout the global scientific community include what will actually be done with the vast genetic information that is acquired, and how does society factor into these decisions. For example, how will knowledge of the complete nucleotide sequence of the human genome be used? More than at any other time in the history of science, addressing the ethical questions surrounding an emerging technology is clearly as important as the information that is gained from that technology.

As you launch your study of the discipline of genetics, try to remain sensitive to issues like those just described. There has never been a more exciting time to be immersed in the study of this science, but never has the need for caution been so apparent. This text will enable you to achieve a thorough understanding of modern-day genetics. Along the way, enjoy your studies, but take your responsibility as a novice geneticist very seriously.

HOW DO WE KNOW WHAT WE KNOW?

IN THE EARLY 1980S, A MOVEMENT was under way to define a new approach for teaching Biology. John A. Moore, representing a committee of the American Society of Zoologists and cosponsored by the Genetics Society of America, suggested that we move away from the focus on facts, terms, and details as the traditional way to convey information. Instead, Moore and others proposed the "Science As A Way Of Knowing" approach to conveying the essence of Biology. This pedagogical shift suggested that we emphasize the conceptual framework of the topics covered, that we address the ideas that have shaped our knowledge, and that we pursue the data supporting our understanding of living systems. In essence, Moore and others were suggesting that we continually ask "How do we know what we know?" The book that you are now reading is the product of 11 previous editions of Genetics texts in which we have applied Moore's advice as we shaped each chapter. In this edition, we have decided to formalize our approach by writing a short section after the introduction of each chapter that we call "How Do We Know What We Know." We shall pose a few fundamental, important questions that will be addressed in the chapter, directing you, a student of genetics, to the issues that have framed our thinking. As you proceed through the chapter, be aware of these questions and attempt to answer each one. We believe that your understanding of the material will be greatly enhanced if you utilize this approach, which begins in Chapter 2. ■

1.1 Genetics Has a Rich and Interesting History

Genetic processes are fundamental to the comprehension of life itself, and the discipline of genetics sits at the center of the field of biology. Genetic information directs cellular function, determines an organism's external appearance, and serves as the link between generations in every species. As such, knowledge of genetics is essential to the thorough understanding of other disciplines of biology, including molecular biology, cell biology, physiology, evolution, ecology, systematics, and behavior. Genetics thus unifies biology and serves as its "core."

It is thus not surprising that genetics has a long, rich history. In the chapters that follow, we will discuss the nature of chromosomes, the way in which genetic information is transmitted from one generation to the next, and how this information is stored, altered, expressed, and regulated. The most significant scientific findings—serving as the foundation for our discussions—were obtained in the nineteenth century. As the twentieth century dawned, certain discoveries began to clarify our understanding of the physical basis of living organisms and their relationship to one another. Several related ideas were gaining acceptance and were particularly significant: (1) Matter is composed of atoms; (2) cells are the fundamental units of living organisms; (3) nuclei somehow serve as the "life force" of cells; and (4) chromosomes housed within nuclei somehow play an important role in heredity. When these ideas were correlated with the newly rediscovered genetic findings of Gregor Mendel and integrated with Darwin's theory of natural selection and the origin of species, a more complete picture of life at the level of the individual and of the population emerged. The era of modern-day biology was built on this foundation.

But what of the many important ideas and hypotheses that served as forerunners of nineteenth-century thought? In the following short discussion, we consider some of these ideas, several of which can be traced back well over 1000 years!

Prehistoric Times: Domesticated Animals and Cultivated Plants

We don't know when people first recognized the existence of heredity, but various types of archeological evidence (e.g., primitive art, preserved bones and skulls, and dried seeds) provide insights. Such evidence documents the successful domestication of animals and cultivation of plants thousands of years ago. These ancient endeavors represent the artificial selection of genetic variants within populations. For example, between 8000 and 1000 B.C. horses, camels, oxen, and various breeds of dogs (derived from the wolf family) were domesticated, and selective breeding soon followed. Cultivation of many plants, including maize, wheat, rice, and the date palm, began around 5000 B.C. Remains of maize dating to this period have been recovered in caves in the Tehuacan Valley of Mexico. Assyrian art depicts artificial pollination of the date palm, thought to have originated in Babylonia (Figure 1–1). Such cultivation very likely led to the types of date palms that are found in that region today, where over 400 varieties exist in just four oases in the Sahara Desert. They differ from one another in various traits such as fruit taste.

Prehistoric evidence of cultivated plants and domesticated animals documents our ancient ancestors' successful attempts to manipulate the genetic composition of useful species. There is little doubt that people soon learned that desirable and undesirable traits are passed to successive generations, and more desirable varieties of animals and plants could be bred. Human awareness of heredity was thus apparent during prehistoric times.

FIGURE 1–1 Relief carving depicting artificial pollination of date palms (800 B.C.). *(The Metropolitan Museum of Art, Gift of John D. Rockefeller, Jr., 1932. (32.143.3) Photograph © 1983, The Metropolitan Museum of Art)*

The Greek Influence: Hippocrates and Aristotle

Few, if any, significant ideas were put forward to explain heredity during prehistoric times; however, during the Golden Age of Greek culture, philosophers directed much more attention to this subject as it relates to the origin of humans, and of reproduction and heredity in particular. This is evident in the writings of the Hippocratic school of medicine (500–400 B.C.), and subsequently of the philosopher and naturalist Aristotle (384–322 B.C.).

For example, the Hippocratic treatise *On the Seed* argued that active "humors" resided in various parts of the body, serving as the bearers of hereditary traits. Drawn from various parts of the male body to the semen, they could be healthy or diseased, the latter condition accounting for the appearance of newborns with congenital disorders or deformities. It was also believed that these humors could be altered in individuals before they were passed on to offspring, explaining how newborns could "inherit" traits that their parents had "acquired" because of their environment.

Aristotle (Figure 1–2), who studied under Plato for some 20 years, was more critical and more expansive than Hippocrates in his analysis of human origins and heredity. Aristotle proposed that the generative power of male semen resided in a "vital heat" contained within it. This vital heat had the capacity to produce offspring of the same "form" (i.e., basic structure and capacities) as the parent. Aristotle believed that it cooked and shaped the menstrual blood produced by the female,

FIGURE 1–2 Aristotle describes the animals Alexander has sent him. (A fresco in The Main Hall of The Assemblée Nationale in Paris. Painting by Eugene Delacroix.) *(Eugene Delacroix, frescos from the spandrels of the main hall of the Assemblée Nationale, Paris, France. © Photograph by Erich Lessing/Art Resource, NY)*

which was the "physical substance" that gave rise to an offspring. The embryo developed not because it already contained the parts in miniature (as some Hippocratics had thought) but because of the shaping power of the vital heat. These ideas constituted only one part of the Aristotelian philosophy of order in the living world.

Although the ideas of Hippocrates and Aristotle sound primitive and naive today, we should recall that prior to the 1800s neither sperm nor eggs had been observed in mammals. Thus, the Greek philosophers' ideas were worthy ones in their time and for centuries to come.

1600–1850: The Dawn of Modern Biology

During the ensuing 1900 years (from 300 B.C. to A.D. 1600), the theoretical understanding of genetics was not extended by any new and significant ideas. However, in Roman times, plant grafting and animal breeding were common. By the Middle Ages, naturalists, well aware of the impact of heredity on organisms they studied, were faced with reconciling their findings with current religious beliefs. The theories of Hippocrates and Aristotle still prevailed and when applied to humans, no doubt conflicted with the prevailing religious doctrines.

Between 1600 and 1850, major strides were made that provided much greater insight into the biological basis of life, setting the scene for the revolutionary work and principles presented by Charles Darwin and Gregor Mendel. In the 1600s, the English anatomist William Harvey (1578–1657) wrote a treatise on reproduction and development patterned after Aristotle's work. He is credited with the earliest statement of the theory of **epigenesis**: An organism is derived from substances present in the egg that differentiate into adult structures during embryonic development. Epigenesis holds that structures such as body organs are not initially present in the early embryo, but instead are formed de novo (anew).

The theory of epigenesis directly conflicted with the theory of **preformation**, first put forward in the seventeenth century, which stated that sex cells contain a complete miniature adult, perfect in every form, called a **homunculus** (Figure 1–3). Preformation was popular well into the eighteenth century. However, work by the embryologist Kasper Wolff (1733–1794) and others clearly disproved this theory, thus favoring epigenesis. Wolff was convinced that several structures, such as the alimentary canal, were not present in the earliest embryos he studied, but were instead formed later during development.

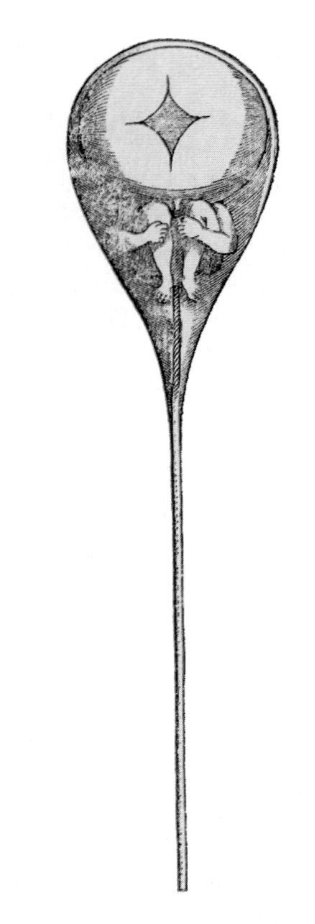

FIGURE 1–3 Depiction of the "homunculus," a sperm containing a miniature adult, perfect in proportion and fully formed. *(Hartsoeker,* N. Essay de dioptrique, *Paris, 1694, p. 230. National Library of Medicine)*

During this same period, other significant chemical and biological discoveries affected future scientific thinking. In 1808, John Dalton expounded his **atomic theory**, which stated that all matter is composed of small invisible units called atoms. Around 1830, Matthias Schleiden and Theodor Schwann, using improved microscopes, proposed the **cell theory**, stating that all organisms are composed of basic visible units called cells, which are derived from similar preexisting structures. The idea of **spontaneous generation**, the creation of living organisms from nonliving components, had clearly been disproved by this time, and living organisms were considered to be derived from preexisting organisms and to consist of cells made up of atoms.

Another prevailing notion prevalent in the nineteenth century, the **fixity of species**, was influential. According to this doctrine, animal and plant groups have remained unchanged in form since the moment of their appearance on earth. This doctrine was popularized by several people and embraced particularly by those who also adhered to a belief in special creation, including the Swedish physician and plant taxonomist Carolus Linnaeus (1707–1778), who is better known for devising the binomial system of species classification.

The influence of the fixity of species tenet is illustrated in the work of a German plant breeder, Joseph Gottlieb Kolreuter (1733–1806), who worked with tobacco plants. He crossbred two groups and derived a new hybrid form, which he then converted back to one of the parental types by repeated backcrosses. In other breeding experiments using carnations, he clearly observed segregation of traits, which was to become one of Mendel's principles of genetics. These results seemed to contradict the idea of "fixed species" that do not change with time. While Kolreuter was puzzled by these outcomes, because of his belief in both special creation and the fixity of species, he failed to recognize the real significance of his findings.

Charles Darwin and Evolution

With the preceding information as background, we conclude our coverage of the historical context of genetics with a brief discussion of the work of Charles Darwin, who published the book-length statement of his evolutionary theory, *The Origin of Species*, in 1859. Darwin's many geological, geographical, and biological observations convinced him that existing species arose by descent with modification from other ancestral species. Greatly influenced by his famous voyage on the HMS *Beagle* (1831–1836), Darwin's thinking culminated in his formulation of the theory of **natural selection**, a theory that attempted to explain the causes of evolutionary change. Formulated and proposed at the same time (but independently) by Alfred Russel Wallace, natural selection was based on the observation that populations tend to consist of more offspring than the environment can support, leading to a struggle for survival among them. Those organisms with heritable traits that allow them to adapt to their environment are better able to survive and reproduce than those with less adaptive traits. Over a long period of time, slight but advantageous variations will accumulate. If a population bearing these inherited variations becomes reproductively isolated, a new species may result.

The primary gap in Darwin's theory was a lack of understanding of the genetic basis of variation and inheritance, a gap that left his theory open to reasonable criticism well into the twentieth century. Aware of this weakness, Darwin published a second book in 1868, *Variations in Animals and Plants Under Domestication*, in which he attempted to provide a more definitive explanation of how heritable variation arises gradually over time. Even though Darwin never understood the genetic basis for inherited variation, his ideas concerning evolution may be the most influential theory ever put forth in the history of biology. His ability to distill his extensive observations and synthesize his ideas into a cohesive hypothesis describing the origin of the diversity of organisms populating the earth constituted a major achievement in the history of science.

As Darwin's work ensued, Gregor Johann Mendel (Figure 1–4) conducted experiments between 1856 and 1863. He published a paper in 1866, in which he demonstrated a number of statistical patterns underlying inheritance and developed a theory involving hereditary factors in the germ cells to explain these patterns. His research

FIGURE 1–4 Gregor Johann Mendel, who in 1866 put forward the major postulates of transmission genetics as a result of experiments with the garden pea. *(Archiv/Photo Researchers, Inc.)*

was virtually ignored until it was partially duplicated and then cited by Karl Correns, Hugo de Vries, and Eric Tschermak around 1900, and subsequently Mendel's classic paper was championed by William Bateson.

By the early part of the twentieth century, chromosomes were discovered and support for the epigenetic interpretation of development had grown considerably. It gradually became clear that heredity and development were dependent upon genetic information residing in genes contained in chromosomes, which were then contributed to each individual by gametes—the so-called **chromosomal theory of inheritance**. The gap in Darwin's theory had narrowed considerably, and Mendel's research has continued to serve as the foundation of genetics.

1.2 Nucleic Acids and Proteins Serve as the Molecular Basis of Genetics

As we proceed through the first part of this text, it will be useful for you to have a basic understanding of the molecular basis for genetic function, which serves as the underpinnings of more classical information. This will enable you to better understand Mendel's work as well as that which followed. It is indeed remarkable that so much was learned in the first half of the twentieth century without the knowledge of molecular genetics, knowledge we take for granted today.

The Trinity of Molecular Genetics

The way in which genes control inherited variation is best understood in terms of three molecules, sometimes referred to as the trinity of molecular genetics: **DNA**, **RNA**, and **protein**. The nucleic acid DNA (deoxyribonucleic acid) serves as the genetic material in all living organisms, as well as in most viruses. DNA is organized into genes and stores genetic information. As part of the chromosomes, the information contained in genes can be transmitted faithfully by parents through gametes to their offspring. For the gene's DNA to subsequently influence an inherited trait, the stored genetic information in the DNA must first be transferred to a closely related nucleic acid, RNA (ribonucleic acid). In eukaryotic organisms, RNA carries the genetic information out of the nucleus, where chromosomes reside, into the cytoplasm of the cell. There, the information in RNA is translated into proteins, which serve as the end products of most all genes. It is the diverse functions of proteins that determine the biochemical identity of cells and ultimately determine the expression of inherited traits. The process of storage and expression of the genetic information upon which life on earth is based may be summarized as

DNA makes RNA that makes proteins.

The process of transferring information from DNA to RNA is called **transcription**. The subsequent conversion of the genetic information contained in RNA into a protein is called **translation**.

The Structure of Nucleic Acids

Two depictions of the structure and components of DNA are shown in Figure 1–5. The molecule exists in cells as a long, coiled ladderlike structure described as a double helix. Each strand of the helix consists of a linear polymer made up of genetic building blocks called **nucleotides**, of which there are four types. Nucleotides vary, depending upon which of four nitrogenous bases is part of the molecule—A (adenine), G (guanine), T (thymine), or C (cytosine). These comprise the genetic alphabet, and various combinations specify the components of proteins. James Watson and Francis Crick, who proposed the double helix model in 1953, had made one of the great discoveries of the twentieth century. A critical component of their model was that the two strands of the helix are exact complements of one another, so the rungs of the ladder always consist of either A=T or G≡C base pairs. This **complementarity** between adenine and thymine nitrogenous base pairs and between guanine and cytosine nitrogenous base pairs—attracted to one another by **hydrogen bonds**—is extremely important to genetic function. Complementarity serves as the basis for both the replication of DNA and for the transcription of DNA into RNA. During both processes, DNA strands serve as templates for the synthesis of the complementary molecule under the direction of the appropriate enzyme.

RNA is chemically very similar to DNA. However, it demonstrates a small variation in its component sugar (ribose versus deoxyribose) and also contains the nitrogenous base uracil in place of thymine. In addition, in contrast to the double helix of DNA, RNA is generally single-stranded. Importantly, it can form complementary structures with a strand of DNA, where uracil replaces thymine in the base pair with adenine. As noted earlier, this complementarity is the basis for transcription of the chemical information in DNA into RNA (Figure 1–6).

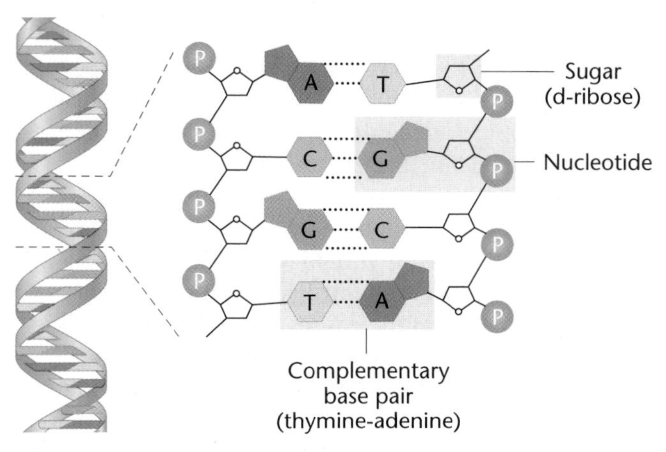

FIGURE 1–5 Summary of the structure of DNA, illustrating on the left the nature of the double helix, and on the right the chemical components making up both strands.

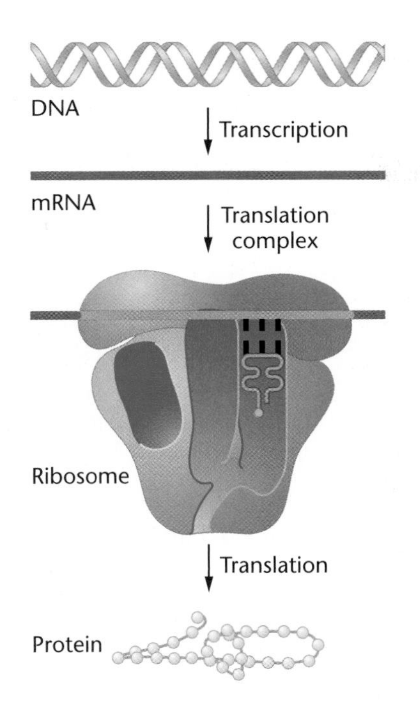

FIGURE 1–6 Depiction of genetic expression involving transcription of DNA into mRNA and the translation of mRNA on a ribosome into a protein.

The Genetic Code and RNA Triplets

Once an RNA molecule complementary to one strand of a gene's DNA is transcribed, the RNA behaves as a messenger that directs the synthesis of proteins. This is accomplished during the association of this RNA molecule (called messenger RNA, or mRNA) with a complex cellular structure called the **ribosome**. The process by which proteins are synthesized under the direction of mRNA, as mentioned earlier, is called translation. Ribosomes serve as nonspecific workbenches for protein synthesis.

Proteins, the end product of genes, are linear polymers made up of amino acids, of which there are 20 different types in living organisms. A major question is how information present in mRNA is encoded to direct the insertion of specific amino acids into protein chains as they are synthesized. The genetic code consists of a linear series of triplet nucleotides present in mRNA molecules. Each triplet reflects, through complementarity, the information stored in DNA and specifies the insertion of a specific amino acid as the mRNA is translated into the growing protein chain (Figure 1–6). A key discovery in how this is accomplished involved the identification of a series of adapter molecules called **transfer RNA (tRNA)**. Within the ribosome, tRNAs adapt the information encoded in the mRNA triplets to the specific amino acid during translation.

So we see that DNA makes RNA that makes protein, and with great specificity. Using an alphabet of only four letters (A, T, C, and G), a language exists that directs the synthesis of highly specific proteins that collectively serve as the basis for all biological function.

Proteins and Biological Function

Proteins are the end products of genetic expression. They are the molecules responsible for imparting many properties that we attribute to the living process. The potential for achieving the diverse nature of biological function rests with fact that the instructions used to construct proteins consist of 20 words (amino acids) that combine to create sentences that can be thousands of words long (Figure 1–7).

If we consider a protein chain just 100 amino acids in length, and at each position there can be any one of

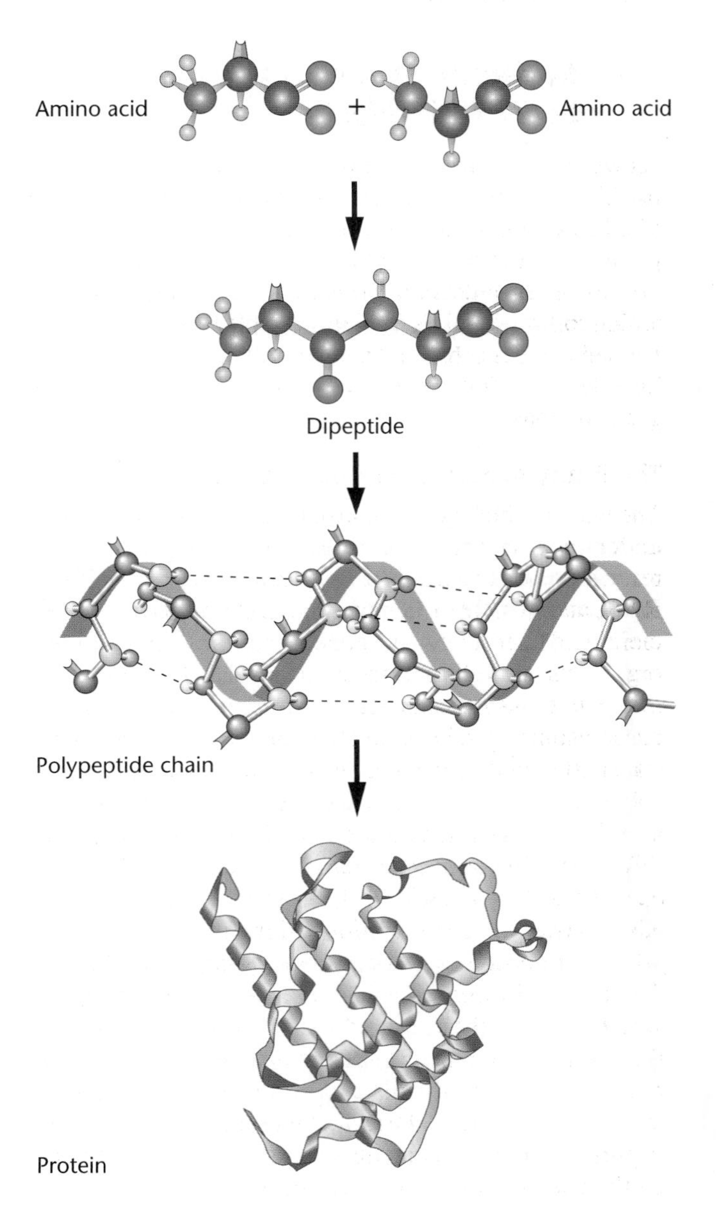

FIGURE 1–7 Depiction of three steps leading to the formation of a protein. Initially, two amino acids are joined to form a dipeptide. As synthesis continues, a longer polypeptide chain is formed, which often coils into a right-handed alpha helical structure. This polypeptide then folds into a three-dimensional conformation specific to the protein's function.

20 amino acids, then the number of different molecules—each with a unique sequence—is equal to

$$20^{100}$$

Since 20^{10} exceeds 5×10^{12}, or over 5 trillion, imagine how large this number is! Obviously, evolution has seized on a class of molecules with the potential for enormous structural diversity as they serve as the mainstay in biological systems.

The main category of proteins includes **enzymes**. These molecules are biological catalysts that enable biochemical reactions to proceed at rates that sustain life. For example, by lowering the energy of activation in reactions, metabolism is able to proceed under the direction of enzymes at body temperature (37°C) in humans, where in the absence of enzymes, these chemical reactions would proceed at rates thousands of times more slowly. As a result, enzymes, each under the control of one or more specific genes, are capable of directing both the anabolism (the synthesis) and catabolism (the breakdown) of all organic molecules in the cell, including carbohydrates, lipids, nucleic acids, and proteins themselves.

There are countless proteins other than enzymes that are critical components of cells and organisms. These include such diverse examples as **hemoglobin**, the oxygen-binding pigment in red blood cells; **insulin**, the pancreatic hormone; **collagen**, the connective tissue molecule; **keratin**, the structural molecule in hair; **histones**, the proteins integral to chromosome structure in eukaryotes; **actin** and **myosin**, the contractile muscle proteins; and **immunoglobulins**, the antibody molecules of the immune system. Specific proteins are also critical components of all membranes and serve as molecules that regulate genetic expression. The potential for such diverse functions rests with the enormous variation of three-dimensional conformation that can be achieved by proteins, and the final conformation of a protein is the direct result of the unique linear sequence of amino acids constituting the molecule. To come full circle, this sequence is dictated by the stored information in the DNA of a gene that is transferred to RNA, which then directs the synthesis of a protein. DNA makes RNA that makes protein.

1.3 Genetics Has Been Investigated Using Many Different Approaches

It will be useful in your study of genetics to know something about the various research approaches that have advanced our knowledge of the field. Investigations have involved viruses, bacteria, and a wide variety of plants and animals and have spanned all levels of biological organization, from molecules to populations. Although some overlap exists, most researchers have used one of four basic approaches.

The classical investigative approach is the study of **transmission genetics**, in which the patterns of inheritance of traits are examined. Experiments are designed so that the transmission of traits from parents to offspring can be analyzed through several generations. Patterns of inheritance are sought that will provide insights into genetic principles. The first significant experimentation to have a major impact on understanding heredity was performed by Gregor Mendel in the middle of the nineteenth century, and his work serves as the foundation of transmission genetics. In human studies, where designed matings are neither possible nor desirable, **pedigree analysis** is often useful. As shown in Figure 1–8, patterns of inheritance are traced through several generations, leading to inferences concerning the mode of inheritance of the trait under investigation.

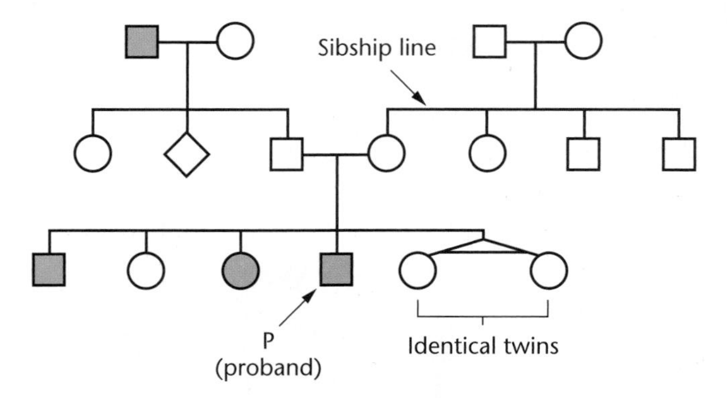

FIGURE 1–8 A representative human pedigree tracing a genetic characteristic through three generations.

The second approach involves **cytogenetics**—the study of chromosomes. The earliest studies used light microscopy. The initial discovery of chromosome behavior during mitosis and meiosis early in the twentieth century was a critical event in the history of genetics because of the important role these observations played in the rediscovery and acceptance of Mendelian principles. The light microscope continues to be useful in the investigation of chromosome structure and abnormalities and is instrumental in preparing a **karyotype**, an illustration of the chromosomes arranged in a standard sequence (Figure 1–9). With the advent of electron microscopy, the repertoire of investigative approaches in genetics has grown. In high-resolution microscopy, genetic molecules and their behavior during gene expression can be seen directly.

The third general approach, **molecular genetic analysis**, began in the 1940s and has had the greatest impact on the recent growth of genetic knowledge. Although experiments initially relied on bacteria and the viruses that infect them, extensive information is now available concerning the nature, expression, replication, and regulation of the genetic information in eukaryotes as well. The precise nucleotide sequence has been determined for many genes cloned in the laboratory. Recombinant DNA studies (Figure 1–10), in which genes from another organism are inserted into bacterial or viral DNA and then cloned, are the basis of a far-reaching research technology used in

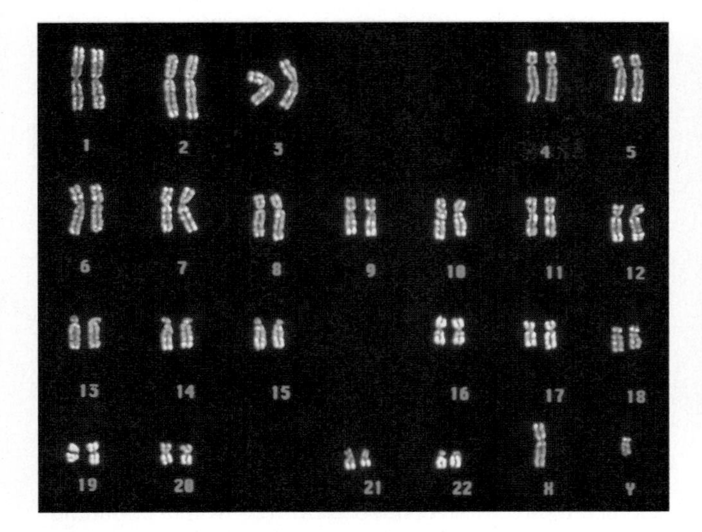

FIGURE 1–9 The human male karyotype. *(Sovereign/Phototake)*

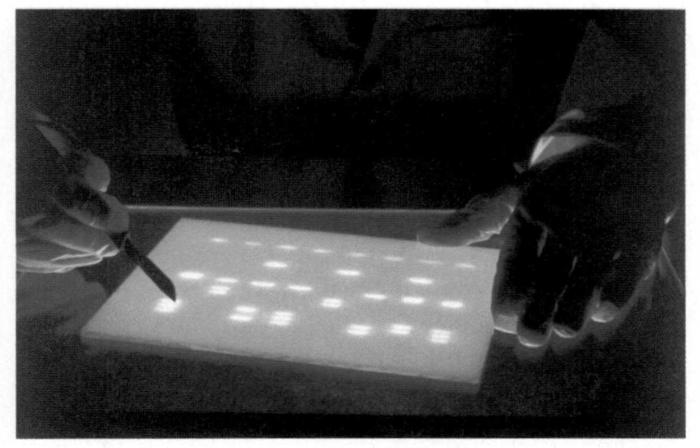

FIGURE 1–10 DNA fragments shown under ultraviolet light. The bands were produced using recombinant DNA technology. *(Matt Meadows/Peter Arnold, Inc.)*

molecular genetic investigations. Building on this approach, the new field of DNA biotechnology identifies, sequences, clones, and manipulates genes. Furthermore, using the most recent technology, it is now possible to probe gene function in extreme detail. Such molecular and biochemical analysis has created the potential for gene therapy and has profound implications for medicine, agriculture, and bioethics.

Perhaps the most striking achievement thus far in the history of biotechnology occurred in 1996 at the Roslin Institute in Scotland, when the world's most famous lamb, Dolly, was born (Figure 1–11). The first animal ever to be cloned from an adult somatic cell, Dolly was the result of the research of Ian Wilmut, who fused the nucleus of a frozen udder cell taken from a six-year-old sheep with an enucleated oocyte of another sheep. Following implantation into a surrogate mother, complete embryonic and fetal development was accomplished under the direction of the genetic material of the udder cell. While the implications of this event are far-reaching and raise numerous ethical concerns, the ultimate goal of Wilmut's research is to use cloned animals as models to study human disease and to produce therapeutic drugs beneficial to humans. Since this success, other organisms, including mice, cows, a cat, and most recently, a horse, have been successfully cloned.

The final investigative approach involves the study of **population genetics**. In these investigations, scientists attempt to define how and why certain genetic variation (Figure 1–12) is maintained in populations, while other variation diminishes or is lost with time. Such information is critical to the understanding of evolutionary processes. Population genetics also allows us to predict gene frequencies in future generations.

These varied approaches have transformed a subject that was only poorly understood in 1900 into one of the most advanced scientific disciplines today. As a result, the impact of genetics on society has been, and will continue to be, immense.

FIGURE 1–11 Dolly, a Finn Dorset sheep cloned from the genetic material of an adult mammary cell, shown next to her firstborn lamb, Bonnie. *(Photo courtesy of Roslin Institute)*

FIGURE 1–12 Genetic variation exhibited in the skin of corn snakes. The normal variety displays orange and black markings. *(Zig Leszczynski/Animals Animals/Earth Scenes)*

1.4 Genetics Has a Profound Impact on Society

In addition to acquiring information for the sake of extending knowledge in any discipline of science—an experimental approach called **basic research**—scientists conduct investigations to solve problems facing society or simply to improve the well-being of members of our society, an approach called **applied research**. Both types of genetic research have combined to enhance the quality of our lives and provide a more thorough understanding of life processes. As we shall see throughout this text, there are few aspects of our lives that genetics fails to touch.

Eugenics and Euphenics

There is always the danger that scientific findings will be used to formulate policies and/or actions that are unjust or even tragic. Here, we will review such a case, which began near the end of the nineteenth century. At that time, Darwin's theory of natural selection had strongly influenced some people's thinking concerning the human condition. Our story recounts the first attempt to apply genetic knowledge for the improvement of human existence. Championed in England by Francis Galton, the general approach is called **eugenics**, a term he coined in 1883.

Galton, a cousin of Charles Darwin, believed that many human characteristics are inherited and could be subject to artificial selection if human matings were controlled. *Positive eugenics* encouraged parents displaying favorable characteristics to have large families. Superior intelligence, intellectual achievement, and artistic talent are examples. *Negative eugenics*, on the other hand, attempted to restrict reproduction in people displaying unfavorable characteristics. For example, low intelligence, mental retardation, and criminal behavior were discouraged.

In the United States, the eugenics movement was a significant social force that led to state and federal laws requiring the sterilization of those considered to be "genetically inferior." Over half of the states passed such laws, commencing in 1907 with Indiana. Sterilization was mandated for "imbeciles, idiots, convicted rapists, and habitual criminals." Not without significant legal controversy, the issue rose to the U.S. Supreme Court, where in the 1927 case of Buck vs. Bell, Justice Oliver Wendell Holmes wrote in favor of upholding these laws:

> It is better for all the world, if instead of waiting to execute degenerate offspring for crime,
> or to let them starve for their imbecility, society can prevent those who are manifestly
> unfit from continuing their kind ... Three generations of imbeciles are enough.*

* See *The Sterilization of Carrie Buck* by J. D. Smith and K. R. Nelson, in the Selected Readings section.

By 1931, involuntary sterilization also applied to "sexual perverts, drug fiends, drunkards, and epileptics." Often, individuals were just deemed "feebleminded," a term applied to countless people displaying a plethora of physical characteristics or unacceptable behavior. Sterilization programs continued in the United States into the 1940s.

Immigration to the United States from certain areas of Europe and from Asia was also restricted to prevent the influx of what were regarded as genetically inferior people. In addition to the violation of individual human rights, such policies were seriously flawed by an inadequate understanding of the genetic basis of various characteristics. The formulation of eugenic policies was premised on the mistaken notions that "superior" and "inferior" traits are totally under genetic control and that genes deemed unfavorable could be removed from a population by sterilizing individuals expressing those traits. The potential impact of environment and the genetic theory underlying population genetics were largely ignored as eugenics policies were developed.

In Nazi Germany in the 1930s, the concept of achieving a superior, racially pure group was an extension of the eugenics movement. Initially applied to those considered socially and physically defective, the underlying rationale of negative eugenics was soon applied to entire ethnic groups, including Jews and Gypsies. Fueled by various forms of racial prejudice, Adolf Hitler and the Nazi regime took eugenics to a horrific extreme by instituting policies aimed at the extinction of these "impure" human populations. This deplorable disregard for human life had been preceded by incremental policies involving forced sterilization and mercy killings. The eugenics movement, based on scientifically invalid premises, culminated in the mass murder we call the Holocaust.

Even before the Nazi Party rose to power in 1933, English and American geneticists started distancing themselves from the eugenics movement. They were concerned about the validity of the premises underlying the movement and the evidence supporting these premises. Thus, many geneticists chose not to study human genetics for fear of being grouped with those who advocated eugenics.

However, since the end of World War II, tremendous strides have been made in human genetics research. A new term, **euphenics**, has replaced eugenics. Euphenics refers to medical and/or genetic intervention designed to reduce the impact of defective genotypes on individuals. The use of insulin by diabetics and the dietary control of newborn phenylketonurics are long-standing examples. Today, "genetic surgery" to replace defective genes looms on the horizon. Furthermore, social policies now have a solid genetic foundation upon which they may be based. Nevertheless, caution is still required to ensure that our expanded knowledge of human genetics does not obscure the role played by the environment in determining an individual's phenotype.

Genetic Advances in Agriculture and Medicine

As a result of research in genetics, major benefits have accrued to society in the fields of agriculture and medicine. Although cultivation of plants and domestication of animals had begun long before, the rediscovery of Mendel's work in the early twentieth century spurred scientists to apply genetic principles to these human endeavors. The use of selective breeding and hybridization techniques has had the most significant impact in agriculture.

Plants have been improved in four major ways: (1) enhanced potential for more vigorous growth and increased yields, including the unique genetic phenomenon of hybrid vigor (heterosis); (2) increased resistance to natural predators and pests, including insects and disease-causing microorganisms; (3) production of hybrids exhibiting a combination of superior traits derived from two different strains or even two different species (Figure 1–13); and (4) selection of genetic variants with desirable qualities such as increased protein value, increased content of limiting amino acids (essential in the human diet), or smaller plant size, which reduces vulnerability to adverse weather conditions.

Over the past five decades, genetic manipulation has been applied to such crops as barley, beans, corn, oats, rice, rye, and wheat. It is estimated that in the United States the use of improved genetic strains has led to a threefold increase in crop yield per acre. In Mexico, where corn is the staple crop, a significant increase in protein content and yield has occurred. A team of researchers led by Norman Borlaug has developed varieties of wheat that incorporate favorable genes from other strains found in various parts of the world, revolutionizing wheat production in Mexico and other underdeveloped countries. Because of this effort, which led to the well-publicized "Green Revolution," Borlaug received the Nobel Peace Prize in 1970. There is little question that this application of genetics has contributed to the well-being of our own species by improving the quality of nutrition worldwide.

Applied research in genetics has also resulted in the development of superior breeds of livestock (Figure 1–14). Selective breeding has produced chickens that grow faster, yield more high-quality meat per chicken, and lay greater numbers of larger eggs. In larger animals, including pigs and cows, the use of artificial insemination has been particularly important; sperm samples from a single male with superior genetic traits can now be used to fertilize thousands of females located anywhere in the world.

Equivalent strides have been made in medicine as a result of advances in genetics, particularly since 1950. Numerous disorders in humans have been discovered to result from either a single mutation or a specific chromosomal abnormality (Figure 1–15). For example, the genetic basis of disorders such as sickle-cell anemia, erythroblastosis fetalis, cystic fibrosis, hemophilia, muscular dystrophy, Tay-Sachs disease, Down syndrome, and many metabolic disorders is now well documented and often understood at the molecular level.

The genes responsible for human diseases such as cystic fibrosis, Duchenne muscular dystrophy, and Huntington disease have been identified, isolated, cloned, and studied. Physicians hope that such research will pave the way for **gene therapy**, whereby genetic disorders are treated by

FIGURE 1–13 *Triticale*, a hybrid grain derived from wheat and rye, produced as a result of applied genetic breeding experiments. *(Grant Heilman/Grant Heilman Photography, Inc.)*

FIGURE 1–14 The effects of breeding and selection, as illustrated by the production of this Vietnamese potbellied pig. *(Renee Lynn/Photo Researchers, Inc.)*

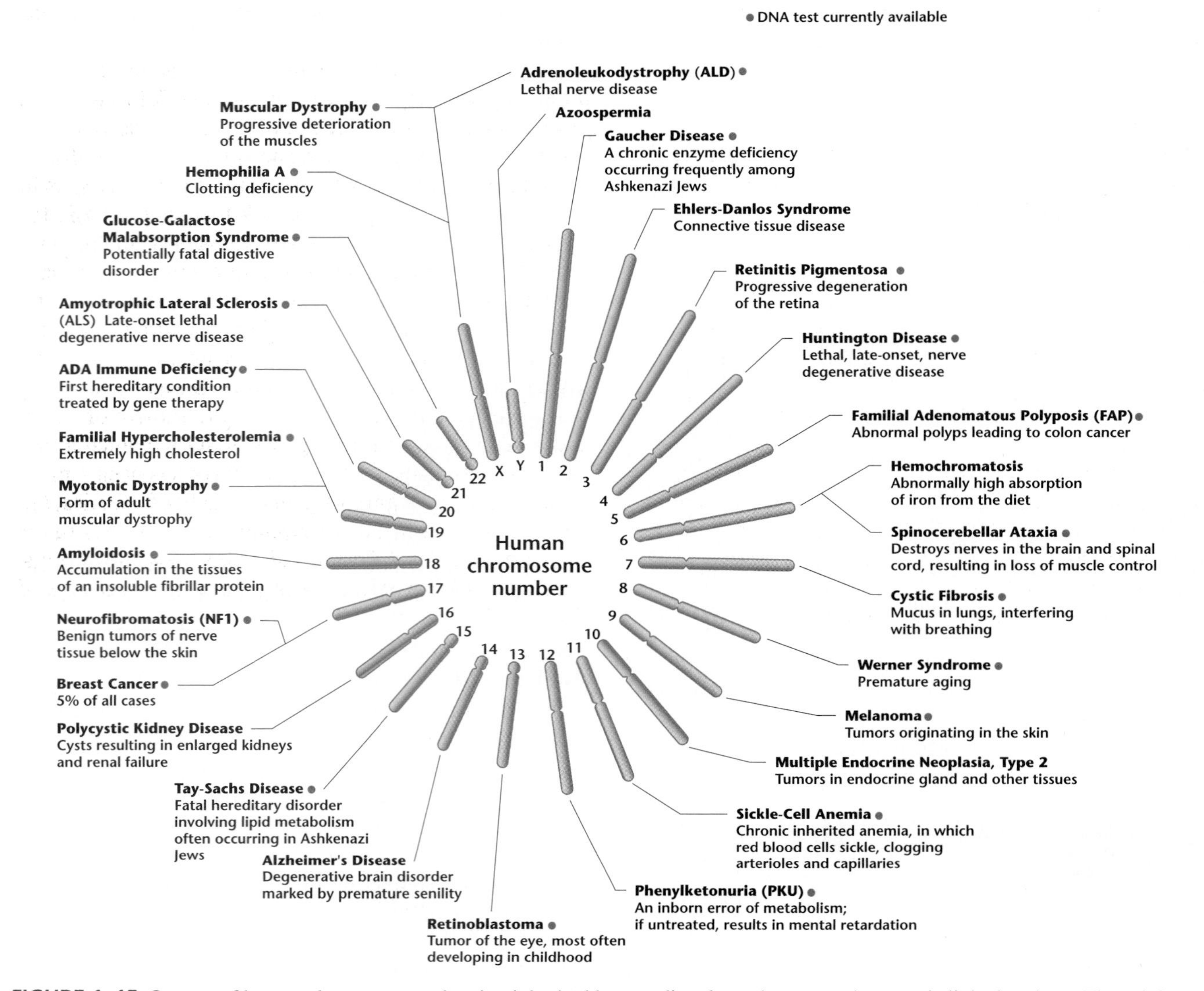

FIGURE 1–15 One set of human chromosomes showing inherited human disorders whose genetic cause is linked to them. Those inherited conditions that can be diagnosed using DNA analysis are indicated by a red dot (•)

inserting normal copies of genes into the cells of afflicted individuals.

The importance of acquiring knowledge of inherited disorders is underscored by the estimate that more than 10 million children and adults in the United States suffer from some form of genetic affliction, and that every child bearing couple has an approximately 3% risk of having a child with some form of serious genetic anomaly.

Additionally, it is now apparent that all forms of cancer have a genetic basis. Although cancer is not usually an inherited disease, it is very clear that *cancer is a genetic disorder at the somatic cell level.* Most cancers are derived from somatic cells that have undergone some type of genetic change; malignant tumors are then derived from the genetically altered cell. However, in some cases a genetic predisposition to cancer also exists.

Recognition of the molecular basis of human genetic disorders and cancer is providing the impetus for the development of methods for detection and treatment. Genetic counseling gives couples objective information on which they can base informed decisions about childbearing. In the case of cancer, recent genetic discoveries have already led to more effective early detection and more efficient treatment.

Applied research in genetics is also providing other medical benefits. Advances in immunogenetics have made compatible blood transfusions as well as organ transplants a reality. In conjunction with immunosuppressive drugs, the number of successful transplant operations involving human organs, including the heart, liver, pancreas, and kidney, is increasing annually.

The most recent advances in human genetics have been dependent upon the application of DNA biotechnology. First developed in the 1970s, recombinant DNA techniques paved the way for manipulating and cloning a variety of genes, including those that encode many medically important molecules, such as insulin, blood-clotting factors, growth hormone, and interferon. Human genes are isolated and spliced into vectors and transferred to host cells that serve as "production centers" for the synthesis of these proteins.

Recombinant DNA techniques are being extended considerably. The DNA of any organism of interest is routinely manipulated in the laboratory; this has had a major impact on plants and animals, which have been altered genetically in countless ways. Such **genetically modified organisms** (known as **GMOs**), the source of potentially great improvements in both agriculture and animal breeding, are also the source of great debate. The "Genetics, Technology, and Society" Essay in this chapter (see page 14) contrasts both sides of the argument involving the use of these organisms.

Perhaps the most far-reaching use of DNA biotechnology involves the Human Genome Project, in which the entire genetic complement (the **genome**) of numerous species, including our own, has been sequenced. The genomic sequence of several bacterial species, as well as that of yeast, the fruit fly, the mustard plant (*Arabidopsis*), rice, a nematode (*Caenorhabditis elegans*), and the mouse has now been deciphered.

With the sequence of the human genome in hand, one promising application involves the new field known as **pharmacogenetics**. Because of their genetic makeup, some individuals are very responsive to specific drugs, while others are resistant to the same drug. Researchers in this field are working to establish the genetic basis for varied responses. Genetic scans are now available on a limited basis, and in a few years they will be commonly used to determine the predisposition to genetic disorders, as well as in the design of individualized drug-treatment profiles to effectively treat these diseases.

Although other scientific disciplines are also expanding in knowledge, none has paralleled the growth of information that is occurring in genetics. As we noted at the outset of this chapter, while there never has been a more exciting time to be immersed in the study of genetics, the potential impact of this discipline on society has never been more profound. By the end of this book, we are confident you will agree that the present truly is the "Age of Genetics."

The Frankenfood Debates: Genetically Modified Foods

Until recently, North Americans paid little attention to the nature or extent of genetic engineering of agricultural products. We have assumed that genetically engineered, or genetically modified (GM), foods are equivalent to nonengineered foods and that they pose no particular threats to human or environmental health. However, this attitude is gradually changing as we learn more about the controversies surrounding GM foods in Europe and as more countries ban the importation of GM seeds, crops, and foods.

Genetic engineering of plants involves recombinant DNA techniques. Scientists clone a gene of interest from a plant or animal, attach the gene to a suitable promoter sequence, and introduce the gene into plant cells that are growing in tissue culture. After the plant cells have successfully integrated this new transgene, the cultured cells are grown into whole plants. In this way, every cell of the new GM plant contains the new gene. The GM plants are then extensively tested for growth and expression of the transgene.

GM crops burst onto the agricultural scene in the mid-1990s. By 1999, half of U.S. soybeans, 40% of U.S. corn, and half of Canada's canola crop were grown from genetically engineered seeds. Other common GM crops include cotton, potatoes, and tomatoes. It is estimated that over 60% of all processed foods in North America contain ingredients derived from GM plants. Several other GM crops are in the pipeline for approval and marketing. Virtually all GM crops currently in use have been developed and are marketed by large international agrochemical companies—companies that hold patents on the seeds and the technology to generate the seeds.

At present, the two genetic modifications found in the majority of GM crops are *Bt* pest resistance and glyphosate herbicide resistance. The *Bt* gene is obtained from a bacterium (*Bacillus thuringiensis*) and encodes a toxin that kills certain insects. This toxin does not appear to affect other animals and plants. Plants that make their own *Bt* toxin do not require spraying for these insects. In theory, insecticide use will decrease, yields will increase, and farmers' profits will rise. Glyphosate (known commercially as RoundUp™) is a herbicide that kills all plants except those with natural or engineered resistance. Use of glyphosate-resistant crops should allow farmers to kill weeds that compete with the GM crop, thereby increasing yields and profits.

Proponents of GM crops argue that genetic engineering will reduce the use of pesticides and herbicides, improve the quality of foods, and increase yields sufficiently to feed the world's hungry population. Critics are less optimistic and even label GM foods "Frankenfoods." The most frequent criticism of GM foods is a perceived threat to human health. Consumer polls suggest that over half of North Americans worry about the safety of GM foods. In Europe, the figure is closer to 90%. The government of the United Kingdom banned all commercial GM crops until the year 2003. Many British and European supermarket chains are voluntarily removing GM foods from their stores. The European Parliament is considering legislation to make agrochemical companies financially responsible for any ill effects from GM food consumption. Unfortunately, there has been little animal research—and no human research—on the long-term health effects of any GM food. So far, animal studies appear to show no ill effects of GM foods; however, each GM crop is different, and few have been tested.

Environmental criticisms of GM crops are numerous. Many scientists argue that the continuous presence of *Bt* toxin in *Bt*-engineered crops will keep insect pests under constant pressure to evolve resistance. Once insects are resistant, the advantages of growing *Bt*-engineered crops will vanish, requiring farmers to resort to using other, more toxic, pesticides. Recent data suggest that *Bt*-expressing plants leak the *Bt* toxin through their roots into the soil, with unknown effects on soil ecology. And, *Bt*-expressing plants and their pollen may be toxic to beneficial insects, as suggested by recent studies of monarch butterflies and *Bt*-corn pollen.

Another criticism is that glyphosate resistance may be transferred to wild relatives of the glyphosate-resistant plants through cross-pollination, creating "superweeds" that cannot be killed with glyphosate. Similarly, engineered traits such as virus or fungus resistance may be transferred to weeds, with unpredictable effects. These scenarios are not impossible, as genetic exchange between cultivated and wild plants has been documented for decades. Interestingly, herbicide-resistant canola has already become a common weed in Canadian wheat fields. If wild and weedy plants acquire glyphosate resistance, it will be necessary to use higher levels of glyphosate, or other, more toxic, herbicides to control weeds. If these resistance scenarios are correct, the benefits of *Bt*-engineered and glyphosate-engineered crops will be short-lived.

Critics also argue that GM crops will do little to feed the world's hungry. They contend that the millions of people who are malnourished are too poor to afford patented seed, the petrochemicals required to produce the crop, or GM foods. They feel that GM technologies favor wealthier farmers, large monocultures, and exports. In a similar vein, critics of GM crops worry that an increasingly large proportion of agricultural biotechnology will be controlled by a small number of global corporations, leading to knowledge that is proprietary and patented. This may lead to a kind of "bioserfdom" of the world's farmers. Understandably, biotechnology companies want to protect their investments in genetic engineering by patenting the technology and preventing farmers from saving seeds for replanting. To ensure that farmers cannot replant GM seeds, some companies are developing "terminator technologies" that render GM seeds sterile—although these restrictions counter the argument that GM crops will feed the world's hungry. In addition, the concept that life can be patented and privatized is a profound moral issue for some people.

The major problem with the Frankenfood debates is that there is very little solid information about either the benefits or perils of this new technology. GM foods have been developed and marketed so quickly that little research has been done on the long-term medical or ecological effects of each genetically engineered crop. If this new genetic technology is to deliver benefits as promised, we must take time to address the scientific, ethical, and political questions that surround it.

References

Chrispeels, M. J. 2000. Biotechnology and the poor. *Plant Physiol.* 124: 3–6.

Ferber, D. 1999. Risks and benefits: GM crops in the crosshairs. *Science* 286: 1662–66.

Serageldin, I. 1999. Biotechnology and food security in the 21st century. *Science* 285: 387–89.

Web Site

Living in a GM World, (online articles about GM crops and biotechnology). *New Scientist*. Available from World Wide Web: ***http://www.nsplus.com/gm/gm.jsp***

Chapter Summary

1. The history of genetics, which emerged as a fundamental discipline of biology early in the twentieth century, dates to prehistoric times.
2. Molecular genetics, based on the concept that DNA makes RNA that makes protein, serves as the underpinnings of the more classical work referred to as transmission genetics.
3. The four investigative approaches most often used in the study of genetics are transmission genetic studies, cytogenetic analyses, molecular experimentation, and inquiries into the genetic structure of populations.
4. Genetic research is either basic or applied. Basic genetic research extends our knowledge of the discipline; the objective of applied genetics research is to solve specific problems affecting the quality of our lives and society in general.
5. Eugenics, the application of knowledge of genetics for the improvement of human existence, has a long and controversial history. Euphenics, genetic intervention designed to ameliorate the impact of harmful genotypes on individuals, represents the modern eugenic approach.
6. Genetic research has had a highly positive impact on many facets of agriculture and medicine.
7. DNA biotechnology is greatly expanding our research capability. It has also had a profound impact on the elucidation of inherited diseases, has enabled the mass production of medically important gene products, and will serve as the foundation on which gene therapy is developed.

Key Terms

actin, 8
applied research, 9
atomic theory, 5
basic research, 9
cell theory, 5
chromosomal theory of inheritance, 6
collagen, 8
complementarity, 6
cytogenetics, 8
DNA (deoxyribonucleic acid), 6
enzymes, 8
epigenesis, 4
eugenics, 9
euphenics, 9
fixity of species, 5
gene therapy, 12
genetically modified organisms (GMOs), 13
genome, 13
hemoglobin, 8
histone, 8
homunculus, 4
hydrogen bonds, 6
immunoglobulins, 8
insulin, 8
karyotype, 8
keratin, 8
molecular genetic analysis, 8
myosin, 8
natural selection, 5
nucleotides, 6
pedigree analysis, 8
pharmacogenetics, 13
population genetics, 8
preformation, 4
protein, 6
ribosome, 7
RNA (ribonucleic acid), 6
spontaneous generation, 5
transcription, 6
transfer RNA (tRNA), 7
translation, 6
transmission genetics, 8

Problems and Discussion Questions

1. Describe and contrast the ideas of Hippocrates and Aristotle relating to the genetic basis of life.
2. Define and contrast epigenesis and preformationism.
3. Which ideas and doctrines that preceded Darwin were central to his thinking?
4. Describe Darwin's and Wallace's theories of natural selection. What information was missing from their theories. That is, what gap remained?
5. Describe the trinity of molecular genetics and how it serves as the basis of this discipline.
6. Describe the four major investigative approaches used in studying genetics.
7. Contrast basic and applied research.
8. Norman Borlaug received the Nobel Peace Prize for his work in genetics. Why do you think he was awarded this prize?
9. Contrast positive and negative eugenics. Which of these categories includes the approach called euphenics? Define this approach.
10. Describe several examples of how genetic research has been applied to agriculture and medicine.

Selected Readings

Allen, G. E. 1996. Science misapplied: The eugenics age revisited. *Technol. Rev.* 99: 23–31.

Anderson, W. F., and Dircumakos, E. G. 1981. Genetic engineering in mammalian cells. *Sci. Am.* (July) 245: 106–21.

Borlaug, N. E. 1983. Contributions of conventional plant breeding to food production. *Science* 219: 689–93.

Bowler, P. J. 1989. *The Mendelian Revolution: The Emergence of Hereditarian Concepts in Modern Science and Society*. London: Athione.

Cocking, E. C., Davey, M. R., Pental, D., and Power, J. B. 1981. Aspects of plant genetic manipulation. *Nature* 293: 265–70.

Day, P. R. 1977. Plant genetics: Increasing crop yield. *Science* 197: 1334–39.

Dunn, L. C. 1965. *A Short History of Genetics*. New York: McGraw-Hill.

Gardner, E. J. 1972. *History of Biology*. 3rd ed. New York: Macmillan.

Garver, K. L., and Garver, B. 1991. Eugenics: Past, present, and future. *Am. J. Hum. Genet.* 49: 1109–18.

Gasser, C. S., and Fraley, R. T. 1989. Genetically engineering plants for crop improvement. *Science* 244: 1293–99.

———. 1992. Transgenic crops. *Sci. Am.* (June) 266: 62–69.

Horgan, J. 1993. Eugenics revisited. *Sci. Am.* (June) 268: 123–31.

Jones, S. 2000. *"The Origin of Species"—Updated*. New York: Random House.

Keller, E. F. 2000. *The Century of the Gene*. Cambridge: Harvard Univ. Press.

Kerr, A., and Shakespeare, T. 2002. *Genetic Politics: From Eugenics to Genome*. Cheltenham, U.K.: New Clarion Press.

King, R. C., and Stansfield, W. D. 2002. *A Dictionary of Genetics*. 6th ed. New York: Oxford Univ. Press.

Kolata, G. 1998. *Clone—The Road to Dolly, and the Path Ahead*. New York: William Morrow.

Moore, J. A. 1985. Science as a way of knowing: III. Genetics. *Amer. Zool.* 25: 1–165.

Olby, R. C. 1985. *Origins of Mendelism*. 2nd ed. London: Constable.

Pilnick, A. 2002. *Genetics and Society: An Introduction*. Cheltenham, U.K.: Open Univ. Press.

Silver, L. 1998. *Remaking Eden: Cloning and Beyond in a Brave New World*. New York: Avon Books.

Smith, K. D., and Nelsen, K. R. 1989. *The Sterilization of Carrie Buck*. Far Hills: New Horizon Press.

Stubbe, H. 1972. *History of Genetics: From Prehistoric Times to the Rediscovery of Mendel* (transl. by T. R. W. Waters). Cambridge: MIT Press.

Torrey, J. G. 1985. The development of plant biotechnology. *Am. Sci.* 73: 354–63.

Vasil, I. K. 1990. The realities and challenges of plant biotechnology. *Bio/Technology* 8: 296–301.

Weinberg, R. A. 1985. The molecules of life. *Sci. Am.* (Oct.) 253: 48–57.

CHAPTER 2

Mitosis and Meiosis

Chromosomes in the prometaphase stage of mitosis, from a cell in the flower of *Haemanthus*. *(Dr. Andrew S. Bajer, University of Oregon)*

CHAPTER CONCEPTS

In every living thing, there exists a substance referred to as the genetic material. Except in certain viruses, this material is composed of the nucleic acid, DNA. A molecule of DNA is organized into units called genes, the products of which direct the metabolic activities of cells. DNA, with its array of genes, is organized into structures called chromosomes, which serve as vehicles for transmitting genetic information. The manner in which chromosomes are transmitted from one generation of cells to the next and from organisms to their descendants must be exceedingly precise. In this chapter, we consider exactly how genetic continuity is maintained between cells and organisms.

Two major processes are involved in eukaryotes: **mitosis** and **meiosis**. Although the mechanisms of the two processes are similar in many ways, the outcomes are quite different. Mitosis leads to the production of two cells, each with the same number of chromosomes as the parent cell. Meiosis, on the other hand, reduces the genetic content and the number of chromosomes to precisely half. This reduction is essential if sexual reproduction is to occur without doubling the amount of genetic material at each generation. Strictly speaking, mitosis is that portion of the cell cycle during which the hereditary components are equally divided into daughter cells. Meiosis is part of a special type of cell division that leads to the production of sex cells: **gametes** or **spores**. This process is an essential step in the transmission of genetic information from an organism to its offspring.

Normally, chromosomes are visible only during mitosis and meiosis. When cells are not undergoing division, the genetic material making up chromosomes unfolds and uncoils into a diffuse network within the nucleus, generally referred to as chromatin. We will briefly review the structure of cells, emphasizing those components that are of particular significance to genetic function. We then devote the remainder of the chapter to the behavior of chromosomes during cell division.

HOW DO WE KNOW WHAT WE KNOW?

IN THIS CHAPTER, WE WILL FOCUS ON how chromosomes are distributed during cell division, both in dividing somatic cells (mitosis) and in gamete-forming cells (meiosis). A key concept in genetics is the presence of chromosomes as homologous pairs. As you study this topic, you should try to answer several fundamental questions:

1. How do we know that chromosomes exist in homologous pairs?
2. How did we learn that DNA replication occurs during interphase, not early in mitosis?
3. How do we know about the various stages of mitosis?
4. How do we know that the meiosis varies from mitosis?
5. How do we know that mitotic chromosomes are derived from chromatin? ■

2.1 Cell Structure Is Closely Tied to Genetic Function

Before describing mitosis and meiosis, a brief review of the structure of cells will be helpful. Many components, such as the nucleolus, ribosome, and centriole, are involved directly or indirectly with genetic processes. Other components, like the mitochondria and chloroplasts, contain their own unique genetic information. It will also be useful to compare the structural differences between the prokaryotic bacterial cell and the eukaryotic cell. Variation in the structure and function of cells is dependent upon specific genetic expression by each cell type.

Before 1940, our knowledge of cell structure was limited to what we could see with the light microscope. Around 1940, the transmission electron microscope was in its early stages of development, and by 1950, many details of cell ultrastructure had emerged. Under the electron microscope, cells were seen as highly organized, precise structures. A new world of whorled membranes, organelles, microtubules, granules, and filaments was revealed. These discoveries revolutionized thinking in the entire field of biology, but we will be concerned only with those aspects of cell structure that relate to genetic study. The typical animal cell shown in Figure 2–1 illustrates most of the structures we will discuss.

All cells are surrounded by a **plasma membrane**, an outer covering that defines the cell boundary and delimits the cell from its immediate external environment. This membrane is not passive, but instead actively controls the movement of materials into and out of the cell. In addition to this membrane, plant cells have an outer covering called the **cell wall** whose major component is a polysaccharide called cellulose.

Many, if not most, animal cells have a covering over the plasma membrane, referred to as the **cell coat**. Consisting of glycoproteins and polysaccharides, its chemical composition differs from comparable structures in either plants or bacteria. The cell coat, among other functions, provides biochemical identity at the surface of cells, and these forms of cellular identity are under genetic control. For example, various antigenic determinants such as the **AB** and **MN antigens** are found on the surface of red blood cells. In other cells, **histocompatibility antigens**, which elicit an immune response during tissue and organ transplants, are present. A variety of **receptor molecules** are also important components at the surface of cells. These are recognition sites that transfer specific chemical signals across the cell membrane into the cell.

The presence of a nucleus and other membranous organelles characterizes eukaryotic cells. The **nucleus** houses the genetic material, DNA, which is complexed with an array of acidic and basic proteins into thin fibers. During nondivisional phases of the cell cycle, these fibers are uncoiled and dispersed into **chromatin**. During mitosis and meiosis, chromatin fibers coil and condense into

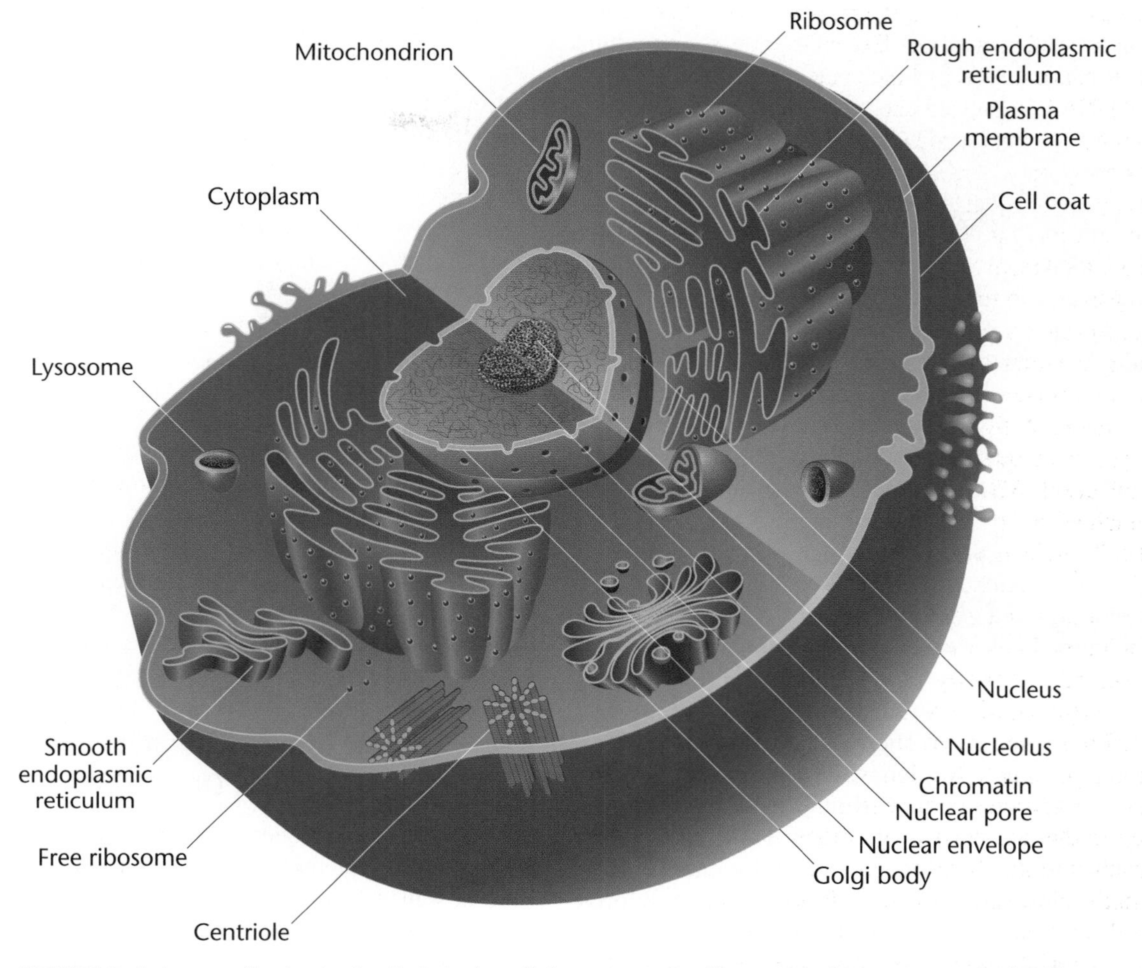

FIGURE 2–1 A generalized animal cell. Only the cellular components discussed in the text are emphasized here.

MEDIA TUTORIAL Parts of the cell

structures called **chromosomes**. Also present in the nucleus is the **nucleolus**, an amorphous component where ribosomal RNA (rRNA) is synthesized and where the initial stages of ribosomal assembly occur. The areas of DNA that encode rRNA are collectively referred to as the **nucleolus organizer region**, or the **NOR**.

Prokaryotes lack a nuclear envelope and membraneous organelles. In bacteria such as *Escherichia coli*, the genetic material is present as a long, circular DNA molecule compacted into the **nucleoid** area. Part of the DNA may be attached to the cell membrane, but in general the nucleoid constitutes a large area throughout the cell. Although the DNA is compacted, it does not undergo the extensive coiling characteristic of the stages of mitosis where, in eukaryotes, chromosomes become visible. Nor is the DNA in these organisms associated as extensively with proteins as is eukaryotic DNA. Figure 2–2, in which two bacteria are forming during cell division, illustrates the nucleoid regions that house the bacterial chromosome. Prokaryotic cells do not have a distinct nucleolus, but they do contain genes that specify rRNA molecules.

The remainder of the eukaryotic cell enclosed by the plasma membrane, excluding the nucleus, is composed of **cytoplasm** and all associated cellular organelles. Cytoplasm is a nonparticulate, colloidal material referred to as the cytosol, which surrounds and encompasses the cellular organelles. Beyond these components, an extensive system of tubules and filaments comprising the cytoskeleton provides a lattice of support structures within the cytoplasm. Consisting primarily of tubulin-derived microtubules and actin-derived microfilaments, this structural framework maintains cell shape, facilitates cell mobility, and anchors the various organelles.

One such organelle, the membranous **endoplasmic reticulum (ER)**, compartmentalizes the cytoplasm, greatly increasing the surface area available for biochemical synthesis. The ER may appear smooth, in which case it serves as the site for synthesizing fatty acids and phospholipids, or it may appear rough because it is studded with ribosomes. Ribosomes serve as sites where genetic information contained in messenger RNA (mRNA) is translated into proteins.

Three other cytoplasmic structures are very important in the eukaryotic cell's activities: mitochondria, chloroplasts, and centrioles. **Mitochondria** are found in both animal and plant cells and are the sites of the oxidative

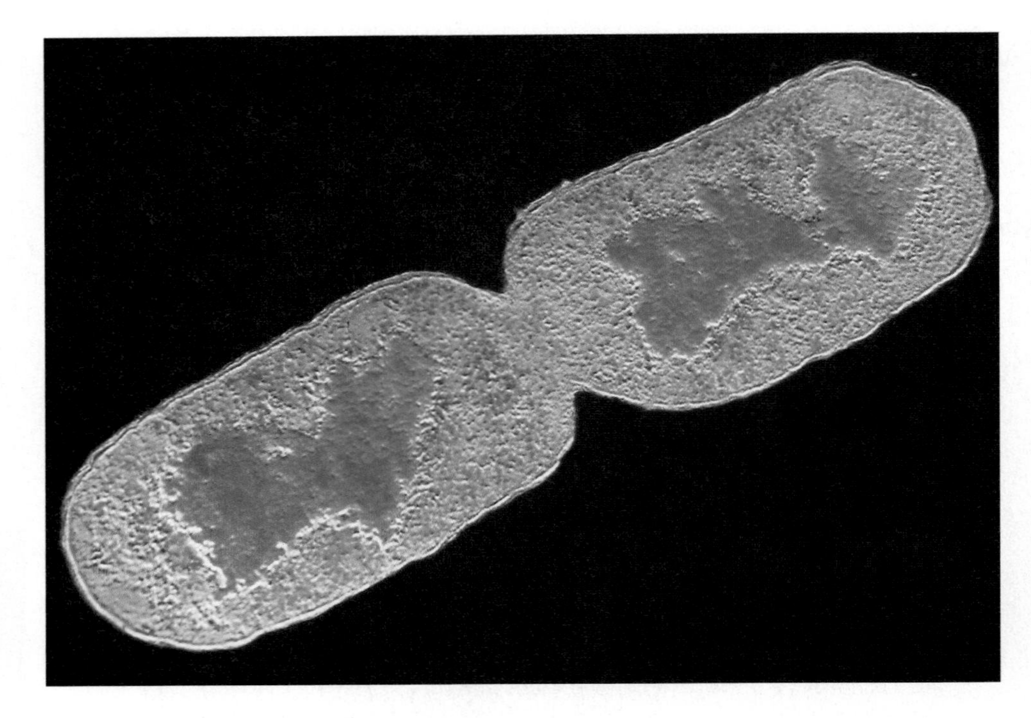

FIGURE 2–2 Color-enhanced electron micrograph of *E. coli* undergoing cell division. Particularly prominent are the two chromosomal areas (shown in red), called nucleoids, that have been partitioned into the daughter cells. *(CNRI/Science Photo Library/Photo Researchers, Inc.)*

phases of cell respiration. These chemical reactions generate large amounts of adenosine triphosphate (ATP), an energy-rich molecule. **Chloroplasts** are found in plants, algae, and some protozoans. This organelle is associated with photosynthesis, the major energy-trapping process on earth. Both mitochondria and chloroplasts contain a type of DNA distinct from that found in the nucleus. Furthermore, these organelles can duplicate themselves and transcribe and translate their genetic information. It is interesting to note that the genetic machinery of mitochondria and chloroplasts closely resembles that of prokaryotic cells. This and other observations have led to the proposal that these organelles were once primitive free-living organisms that established a symbiotic relationship with a primitive eukaryotic cell. This theory, which describes the evolutionary origin of these organelles, is called the **endosymbiont hypothesis**.

Animal cells and some plant cells also contain a pair of complex structures called the **centrioles**. These cytoplasmic bodies, located in a specialized region called the centrosome, are associated with the organization of spindle fibers that function in mitosis and meiosis. In some organisms, the centriole is derived from another structure, the basal body, which is associated with the formation of cilia and flagella. Over the years, many reports have suggested that centrioles and basal bodies contain DNA, which could be involved in the replication of these structures. Currently, this is thought not to be the case.

The organization of **spindle fibers** by the centrioles occurs during the early phases of mitosis and meiosis. These fibers play an important role in the movement of chromosomes as they separate during cell division. They are composed of arrays of microtubules consisting of polymers of polypeptide subunits of the protein tubulin.

2.2 Chromosomes Exist in Homologous Pairs in Diploid Organisms

As we discuss the processes of mitosis and meiosis, it is important that you understand the concept of homologous chromosomes. Such an understanding will also be of critical importance in our future discussions of Mendelian genetics. Chromosomes are most easily visualized during mitosis. When they are examined carefully, distinctive lengths and shapes are apparent. Each chromosome contains a condensed or constricted region called the **centromere**, which establishes the general appearance of each chromosome. Figure 2–3 shows chromosomes with centromere placements at different points along their lengths. Extending from either side of the centromere are the arms of the chromosome. Depending on the position of the centromere, different arm ratios are produced. As Figure 2–3 illustrates, chromosomes are classified as **metacentric**, **submetacentric**, **acrocentric**, or **telocentric** on the basis of the centromere location. The shorter arm, by convention, is shown above the centromere and is called the **p arm** (p, for "petite"). The longer arm is shown below the centromere and is called the **q arm** (because q is the next letter in the alphabet).

When studying mitosis, several observations are of particular relevance. First, all somatic cells derived from members of the same species contain an identical number of chromosomes. In most cases, this represents the **diploid number (2*n*)**. When the lengths and centromere placements of the chromosomes are examined, a second general feature is apparent: Nearly all chromosomes exist in pairs with regard to these two criteria, and the members of each pair are called **homologous chromosomes**. So for each chromosome exhibiting a specific length and centromere placement, another exists with identical features.

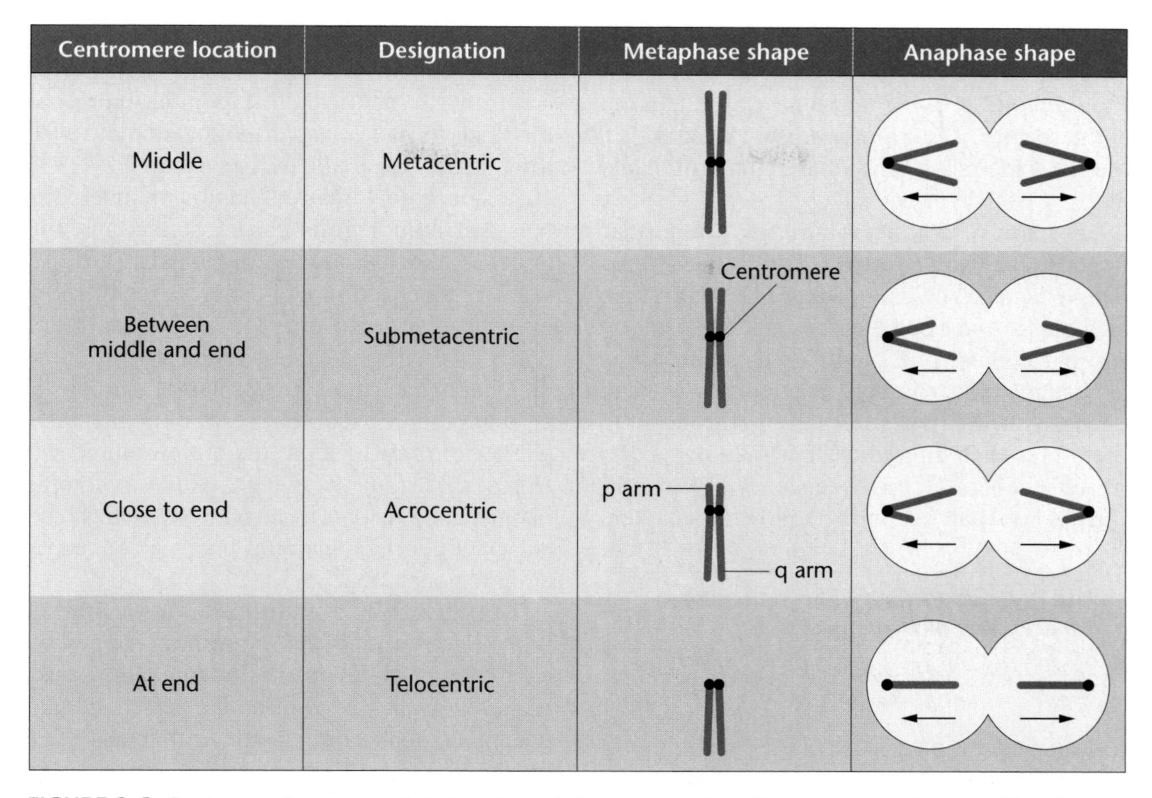

Centromere location
Designation
Metaphase shape
Anaphase shape
Middle
Metacentric
Between middle and end
Submetacentric
Centromere
Close to end
Acrocentric
p arm
q arm
At end
Telocentric

FIGURE 2–3 Centromere locations and designations of chromosomes based on centromere location. Note that the shape of the chromosome during anaphase is determined by the position of the centromere.

There are exceptions to this rule. Bacteria and viruses have only one chromosome, and organisms such as yeasts and molds and certain plants such as bryophytes (mosses), spend the predominant phase of the life cycle in the haploid stage. That is, they contain only one member of each homologous pair of chromosomes during most of their lives.

Figure 2–4 illustrates the physical appearance of different pairs of homologous chromosomes. There, the human mitotic chromosomes have been photographed, cut out of the print, and matched up, creating a **karyotype**. As you can see, humans have a $2n$ number of 46, which on close examination exhibit a diversity of sizes and centromere placements. Note also that each of the 46 chromosomes is clearly a double structure consisting of two parallel **sister chromatids** connected by a common centromere. Had these chromosomes been allowed to continue dividing, the sister chromatids, which are replicas of one another, would have separated into two new cells as division continued.

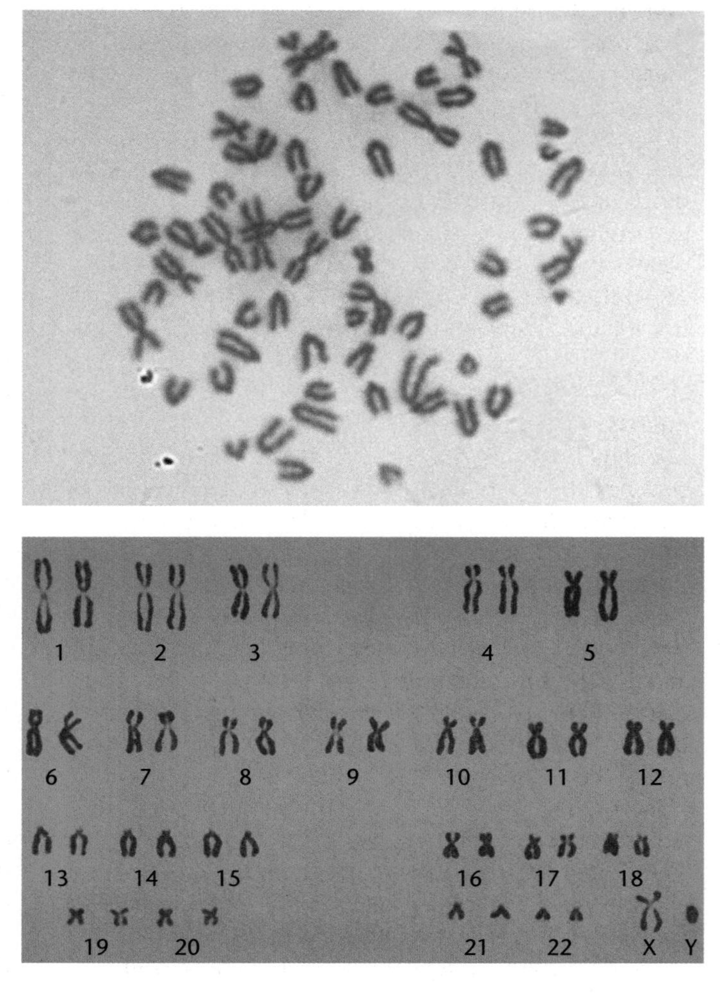

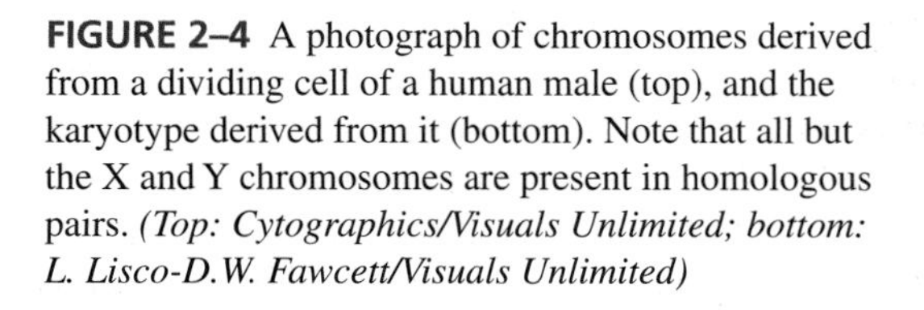

FIGURE 2–4 A photograph of chromosomes derived from a dividing cell of a human male (top), and the karyotype derived from it (bottom). Note that all but the X and Y chromosomes are present in homologous pairs. *(Top: Cytographics/Visuals Unlimited; bottom: L. Lisco-D.W. Fawcett/Visuals Unlimited)*

The haploid number (n) of chromosomes is equal to one-half the diploid number. Collectively, the total set of genes contained in a haploid set of chromosomes constitutes the **genome** of the species. The examples listed in Table 2.1 demonstrate the wide range of n values found in plants and animals.

Homologous pairs of chromosomes have important genetic similarities. They contain identical gene sites along their lengths, each called a **locus** (pl., loci). Thus, they are identical in their genetic potential. In sexually reproducing organisms, one member of each pair is derived from the maternal parent (through the ovum) and one is derived from the paternal parent (through the sperm). Therefore, each diploid organism contains two copies of each gene as a consequence of **biparental inheritance**. As we shall see in the following chapters on transmission genetics, the members of each pair of genes, while influencing the same characteristic or trait, need not be identical. In a population of members of the same species, many different alternative forms of the same gene, called **alleles**, can exist.

The conceptual issues of haploid number, diploid number, and homologous chromosomes are important in understanding the process of meiosis. During the formation of gametes or spores, meiosis converts the diploid number of chromosomes to the haploid number. As a result, haploid gametes or spores contain precisely one member of each homologous pair of chromosomes—that is, one complete haploid set. Following fusion of two gametes in fertilization, the diploid number is reestablished; that is, the zygote contains two complete sets of haploid chromosomes, one set from each parent. The constancy of genetic material is thus maintained from generation to generation.

There is one important exception to the concept of homologous pairs of chromosomes. In many species, one pair, the **sex-determining chromosomes**, is often not homologous in size, centromere placement, arm ratio, or genetic content. For example, in humans, females carry two homologous X chromosomes, while males carry one Y chromosome and one X chromosome (Figure 2–4). These X and Y chromosomes are not strictly homologous. The Y is considerably smaller and lacks most of the gene sites contained on the X. Nevertheless, in meiosis they behave as homologs so that gametes produced by males receive either one X or one Y chromosome.

TABLE 2.1 The Haploid Number of Chromosomes for a Variety of Organisms

Common Name	*Scientific Name*	*Haploid Number*
Black bread mold	*Aspergillus nidulans*	8
Broad bean	*Vicia faba*	6
Cat	*Felis domesticus*	19
Cattle	*Bos taurus*	30
Chicken	*Gallus domesticus*	39
Chimpanzee	*Pan troglodytes*	24
Corn	*Zea mays*	10
Cotton	*Gossypium hirsutum*	26
Dog	*Canis familiaris*	39
Evening primrose	*Oenothera biennis*	7
Frog	*Rana pipiens*	13
Fruitfly	*Drosophila melanogaster*	4
Garden onion	*Allium cepa*	8
Garden pea	*Pisum sativum*	7
Grasshopper	*Melanoplus differentialis*	12
Green alga	*Chlamydomonas reinhardi*	18
Horse	*Equus caballus*	32
Housefly	*Musca domestica*	6
House mouse	*Mus musculus*	20
Human	*Homo sapiens*	23
Jimson weed	*Datura stramonium*	12
Mosquito	*Culex pipiens*	3
Mustard plant	*Arabidopsis thaliana*	5
Pink bread mold	*Neurospora crassa*	7
Potato	*Solanum tuberosum*	24
Rhesus monkey	*Macaca mulatta*	21
Roundworm	*Caenorhabditis elegans*	6
Silkworm	*Bombyx mori*	28
Slime mold	*Dictyostelium discoidium*	7
Snapdragon	*Antirrhinum majus*	8
Tobacco	*Nicotiana tabacum*	24
Tomato	*Lycopersicon esculentum*	12
Water fly	*Nymphaea alba*	80
Wheat	*Triticum aestivum*	21
Yeast	*Saccharomyces cerevisiae*	16
Zebrafish	*Danio rerio*	25

2.3 Mitosis Partitions Chromosomes into Dividing Cells

The process of mitosis is critical to all eukaryotic organisms. In some single-celled organisms, such as protozoans and some fungi and algae, mitosis (as a part of cell division) provides the basis for asexual reproduction. Multicellular diploid organisms begin life as single-celled fertilized eggs called **zygotes**. The mitotic activity of the zygote and the subsequent daughter cells is the foundation for the development and growth of the organism. In adult organisms, mitotic activity is prominent in wound healing and other forms of cell replacement in certain tissues. For example, the epidermal skin cells of humans are continuously sloughed off and replaced. Cell division also results in the continuous production of reticulocytes (immature red blood cells) that eventually shed their nuclei and replenish the supply of red blood cells in vertebrates. In abnormal situations, somatic cells may lose control of cell division and form a tumor.

The genetic material is partitioned into daughter cells during nuclear division or **karyokinesis**. This process is quite complex and requires great precision. The chromosomes must first be exactly replicated and then accurately partitioned. The end result is the production of two daughter nuclei, each with a chromosome composition identical to that of the parent cell.

Karyokinesis is followed by cytoplasmic division, or **cytokinesis**. The less complex division of the cytoplasm requires a mechanism that partitions the volume into two parts, then encloses both new cells within a distinct plasma membrane. Cytoplasmic organelles either replicate themselves, arise from existing membrane structures, or are synthesized *de novo* (anew) in each cell. The subsequent proliferation of these structures is a reasonable and adequate mechanism for reconstituting the cytoplasm in daughter cells.

Following cell division, the initial size of each new daughter cell is approximately one-half the size of the parent cell. However, the nucleus of each new cell is not appreciably smaller than the nucleus of the original cell. Quantitative measurements of DNA confirm that there is an amount of genetic material in the daughter nuclei equivalent to that in the parent cell.

Interphase and the Cell Cycle

Many cells undergo a continuous alternation between division and nondivision. The events that occur from the completion of one division until the beginning of the next division constitute the **cell cycle** (Figure 2–5). We will consider the initial **interphase** stage of the cycle as the interval between divisions. It was once thought that the biochemical activity during interphase was devoted solely to the cell's growth and its normal function. However, we now know that another biochemical step critical to the ensuing mitosis occurs during interphase: *replication of the DNA of each chromosome*. This period during which DNA is synthesized occurs before the cell enters mitosis and is called the **S phase**. The initiation and completion of synthesis can be detected by monitoring the incorporation of radioactive precursors into DNA.

Investigations of this nature show two periods during interphase when no DNA synthesis occurs, one before and one after S phase. These are designated **G1 (gap I)** and **G2 (gap II)**, respectively. During both of these intervals, as well as during S, intensive metabolic activity, cell growth, and cell differentiation occur. By the end of G2, the volume of the cell has roughly doubled, DNA has been replicated, and mitosis (M) is initiated. Following mitosis, continuously dividing cells then repeat this cycle (G1, S, G2, M) over and over, as shown in Figure 2–5.

Much is known about the cell cycle based on *in vitro* (in glass) studies. When grown in culture, many cell types in different organisms traverse the complete cycle in about 16 hours. The actual process of mitosis occupies only a small part of the overall cycle, often less than an hour. The length of the S and G2 phase of interphase are fairly consistent among different cell types. Most variation is seen in the length of time spent in the G1 stage. Figure 2–6 shows the length of these intervals in a typical cell.

G1 is of great interest in the study of cell proliferation and its control. At a point late in G1, all cells follow one of two paths. They either withdraw from the cycle, become quiescent, and enter the **G0 stage** (see Figure 2–5), or they become committed to initiating DNA synthesis and completing the cycle. Cells that enter G0 remain viable and metabolically active but are nonproliferative. Cancer cells apparently avoid entering G0 or pass through it very quickly. Other cells enter G0 and never reenter the cell cycle. Still others remain in G0, but they can be stimulated to return to G1 and thereby reenter the cycle.

Cytologically, interphase is characterized by the absence of visible chromosomes. Instead, the nucleus is filled with chromatin fibers that are formed as the chromosomes are uncoiled and dispersed after the previous mitosis (Figure 2–7a). Once G1, S, and G2 are completed, mitosis is initiated. Mitosis is a dynamic period of vigorous and continual activity. For discussion purposes, the entire process is subdivided into discrete stages, and specific events are assigned to each one.

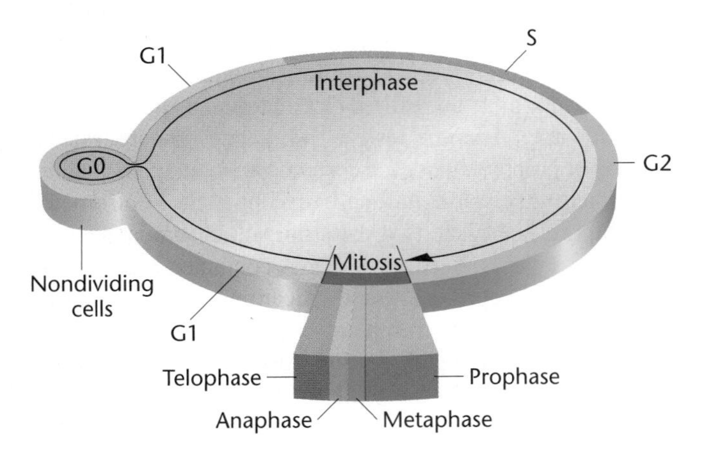

FIGURE 2–5 The phases of the cell cycle. Following mitosis (M), cells enter the G1 stage of interphase, initiating a new cycle. Cells may become nondividing (G0) or continue through G1, where they become committed to begin DNA synthesis (S) and complete the cycle (G2 and M). Following mitosis, two daughter cells are produced and the cycle begins anew for each cell.

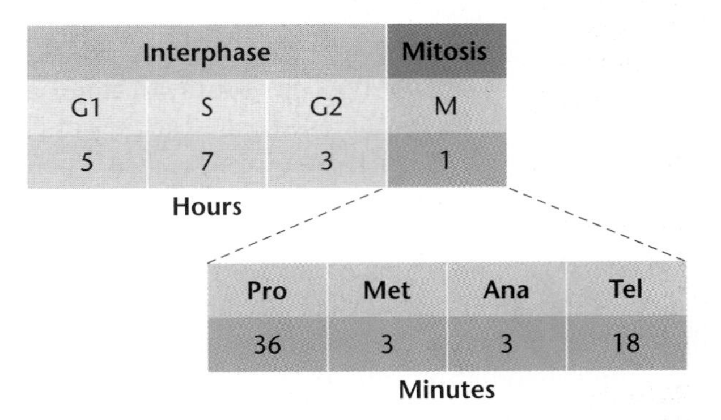

Interphase			Mitosis
G1	S	G2	M
5	7	3	1

Hours

Pro	Met	Ana	Tel
36	3	3	18

Minutes

FIGURE 2–6 The length of time spent in each phase of one complete cell cycle of a human cell in culture. Actual times vary according to cell types and conditions.

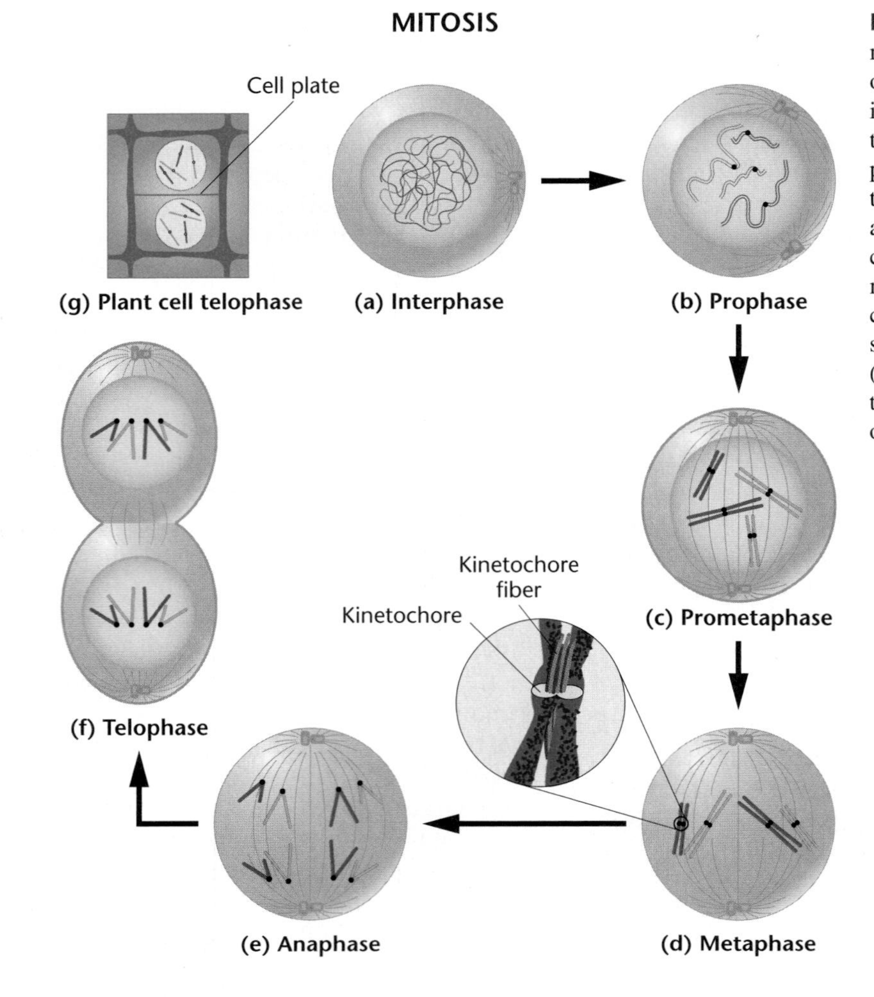

FIGURE 2–7 Mitosis in an animal cell with a diploid number of $2n = 4$. The events occurring in each stage are described in the text. Of the two homologous pairs of chromosomes, one contains longer, metacentric arms and the other, shorter, submetacentric arms. The maternal chromosome and paternal chromosome of each pair are shown in different colors. Part (g) illustrates the formation of the cell plate and lack of centrioles in a plant cell.

MEDIA TUTORIAL Mitosis and the cell cycle

These stages, in order of occurrence, are **prophase**, **prometaphase**, **metaphase**, **anaphase**, and **telophase**. They are diagrammed in Figure 2–7, and a photograph of each stage is shown in Figure 2–8.

Prophase

Often, over half of mitosis is spent in prophase [Figure 2–7(b) and 2–8(b)], a stage characterized by several significant activities. One of the early events in prophase of all animal cells involves the migration of two pairs of centrioles to opposite ends of the cell. These structures are found just outside the nuclear envelope in an area of differentiated cytoplasm called the **centrosome**. It is believed that each pair of centrioles consists of one mature unit and a smaller, newly formed centriole.

The direction of migration of the centrioles is such that two poles are established at opposite ends of the cell. Following their migration, the centrioles are responsible for organizing cytoplasmic microtubules into a series of spindle fibers that are formed and run between these poles to create an axis along which chromosomal separation occurs. Interestingly, the cells of most plants (there are a few exceptions), fungi, and certain algae seem to lack centrioles. Spindle fibers are nevertheless apparent during mitosis. Thus, centrioles are not universally responsible for the organization of spindle fibers.

As the centrioles migrate, the nuclear envelope begins to break down and gradually disappears. In a similar fashion, the nucleolus disintegrates within the nucleus. While these events are taking place, the diffuse chromatin fibers condense, continuing until distinct threadlike structures, the chromosomes, become visible. It becomes apparent near the end of prophase that each chromosome is actually a double structure split longitudinally except at the centromere (see Figure 2–4). The two parts of each chromosome are called chromatids. Because the DNA contained in each pair of chromatids represents the duplication of a single chromosome, these sister chromatids are genetically identical. Therefore, they are called sister chromatids. In humans, with a diploid number of 46, a cytological preparation of late prophase reveals 46 chromosomes randomly distributed in the area formerly occupied by the nucleus.

Prometaphase and Metaphase

The distinguishing event of the ensuing stages is the migration of each chromosome, led by the centromeric region, to the equatorial plane. The equatorial plane, or **metaphase plate**, is the midline region of the cell, a

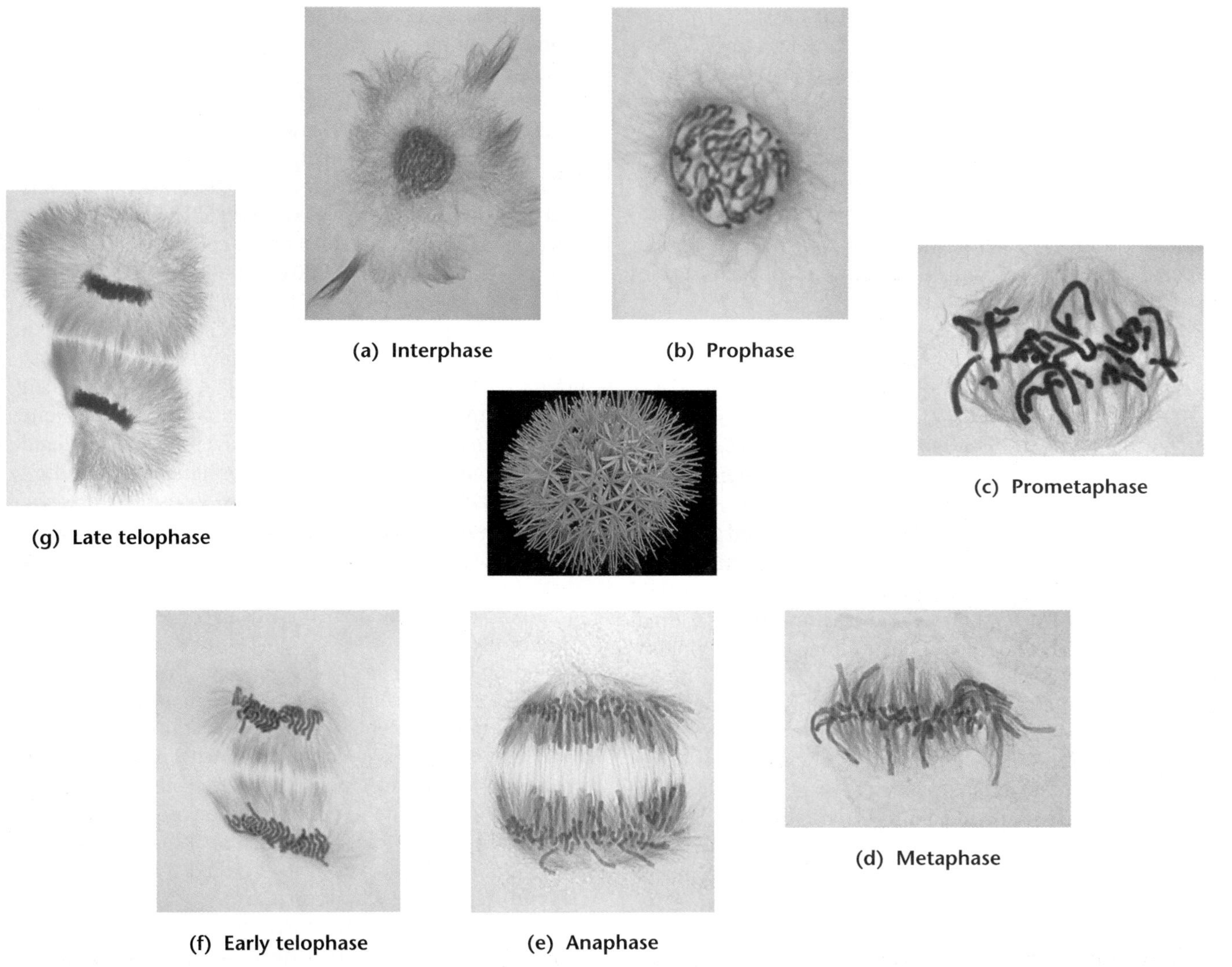

FIGURE 2–8 Light micrographs illustrating the stages of mitosis depicted in Figure 2–7. These stages are derived from the flower of *Haemanthus*, shown in the center. *(Dr. Andrew S. Bajer, University of Oregon)*

plane that lies perpendicular to the axis established by the spindle fibers. In some descriptions, the term **prometaphase** refers to the period of chromosome movement [Figure 2–7(c) and 2–8(c)], and **metaphase** is applied strictly to the chromosome configuration after migration.

Migration is made possible by the binding of spindle fibers to a structure associated with the centromere of each chromosome called the **kinetochore**. This structure, consisting of multilayered plates of proteins, forms on opposite sides of each centromere, intimately associating with the two sister chromatids of each chromosome. Once attached to microtubules making up the spindle fibers, the sister chromatids are now ready to be pulled to opposite poles during the ensuing anaphase stage.

At the completion of metaphase, each centromere is aligned at the metaphase plate, with the chromosome arms extending outward in a random array. This configuration is shown in Figures 2–7(d) and 2–8(d).

Anaphase

Events critical to chromosome distribution during mitosis occur during *anaphase*, the shortest stage of mitosis. During this phase, sister chromatids of each chromosome *disjoin* (separate) from each other and migrate to opposite ends of the cell. For complete disjunction to occur, each centromeric region must be split in two. This event signals the initiation of anaphase. Once it occurs, each chromatid is referred to as a **daughter chromosome**.

Movement of daughter chromosomes to the opposite poles of the cell is dependent upon the centromere–spindle fiber attachment. Recent investigations reveal that chromosome migration results from the activity of a series of specific proteins, generally called motor proteins. These proteins use the energy generated by the hydrolysis of ATP, and their activity is said to constitute **molecular motors** in the cell. These motors act at several positions within the dividing cell, but all of them are involved in the activity of microtubules and ultimately serve to propel the chromosomes to opposite ends of the cell. The centromeres of each chromosome *appear* to lead the way during migration, with the chromosome arms trailing behind. The location of the centromere determines the shape of the chromosome during separation, as you saw in Figure 2–3.

The steps that occur during anaphase are critical in providing each subsequent daughter cell with an identical set of chromosomes. In human cells, there would now be 46 chromosomes at each pole, one from each original sister pair. Figures 2–7(e) and 2–8(e) show anaphase just prior to its completion.

Telophase

Telophase is the final stage of mitosis and is depicted in Figures 2–7(f) and (g) and 2–8(f) and (g). At its beginning, there are two complete sets of chromosomes, one set at each pole. The most significant event is **cytokinesis**, the division or partitioning of the cytoplasm. Cytokinesis is essential if two new cells are to be produced from one cell. The mechanism differs greatly in plant and animal cells. In plant cells, a **cell plate** is synthesized and laid down across the region of the metaphase plate. Animal cells, however, undergo a constriction of the cytoplasm in the same way a loop of string might be tightened around the middle of a balloon. The end result is the same: Two distinct cells are formed.

It is not surprising that the process of cytokinesis varies among cells of different organisms. Plant cells, which are more regularly shaped and are structurally rigid, require a mechanism for depositing new cell wall material around the plasma membrane. The cell plate, laid down during telophase, becomes the **middle lamella**. Subsequently, the primary and secondary layers of the cell wall are deposited between the cell membrane and middle lamella on both sides of the boundary between the two daughter cells. In animals, complete constriction of the cell membrane produces a **cell furrow** characteristic of newly divided cells.

Other events necessary for the transition from mitosis to interphase are initiated during late telophase. They are generally a reversal of the events that occurred during prophase. In each new cell, the chromosomes begin to uncoil and become diffuse chromatin once again while the nuclear envelope re-forms around them. The spindle fibers disappear, and the nucleolus gradually re-forms and becomes visible in the nucleus during early interphase. At the completion of telophase, the cell enters interphase.

2.4 The Cell Cycle Is Genetically Regulated

The cell cycle, including mitosis, is fundamentally the same in all eukaryotic organisms. The similarity of the events leading to cell duplication in many diverse organisms suggests that the cell cycle is governed by a genetic program that has been conserved throughout evolution and is therefore genetically regulated. Elucidation of this genetic program provides information basic to our understanding of the nature of living organisms. Furthermore, because disruption of this regulation can lead to the uncontrolled cell division that characterizes malignant tumors, interest in how genes regulate the cell cycle is keen.

A mammoth research effort over the past decade has paid high dividends, and we now have knowledge of many genes involved in the control of the cell cycle. This work was recognized when the 2001 Nobel Prize was awarded to Lee Hartwell, Paul Nurse, and Tim Hunt (see the inside of the front cover). As with other studies of genetic input into essential biological processes, investigation is focusing on the discovery of mutations that interrupt the cell cycle. Many mutations that exert their effect at various stages of the cell cycle are now known. First discovered in yeast but now evident in all organisms, including humans, such mutations were originally designated as ***cdc* mutations** (**cell *d*ivision cycle mutations**). The study of these mutations has established that during the cell cycle, at least three major **checkpoints** exist, where the cell is monitored or "checked" before it can proceed to the next stage of the cycle.

The products of many of the genes that regulate checkpoints are enzymes called ***cdc* kinases** that can add phosphates to other proteins. They serve as "master control" molecules that work in conjunction with proteins called **cyclins**. These kinases phosphorylate cyclins and influence their activity at the cell-cycle checkpoints, thus regulating the cell cycle. When a *cdc* kinase works in conjunction with a cyclin, it is called a **Cdk protein** (**Cyclin-dependent kinase protein**).

Figure 2–9 identifies the three checkpoints within the cell cycle. First is the **G1/S checkpoint**, which monitors the cell size following the previous mitosis and determines if the DNA has been damaged. If the cell has not achieved an adequate size, or if the DNA has been damaged, further progress through the cycle is arrested until these conditions are "corrected." If both conditions are initially "normal," then the checkpoint is traversed and the cell proceeds to the S phase of the cycle.

The second important checkpoint is the **G2/M checkpoint**, where physiological conditions in the cell are monitored prior to entering mitosis. If DNA replication or repair to any DNA damage has not been completed, the cell cycle is arrested until these processes are completed.

The final checkpoint occurs during mitosis and is called the **M checkpoint**. Here, both the successful formation of the spindle-fiber system and the attachment of spindle fibers to the kinetochores associated with the centromeres are monitored. If spindle fibers are not properly formed or attachment is inadequate, mitosis is arrested.

The importance of cell-cycle control and checkpoints can be demonstrated by considering what happens when the regulatory system is impaired. If, for example, a cell that has incurred damage to its DNA is allowed to proceed through the cell cycle, the damage may lead to

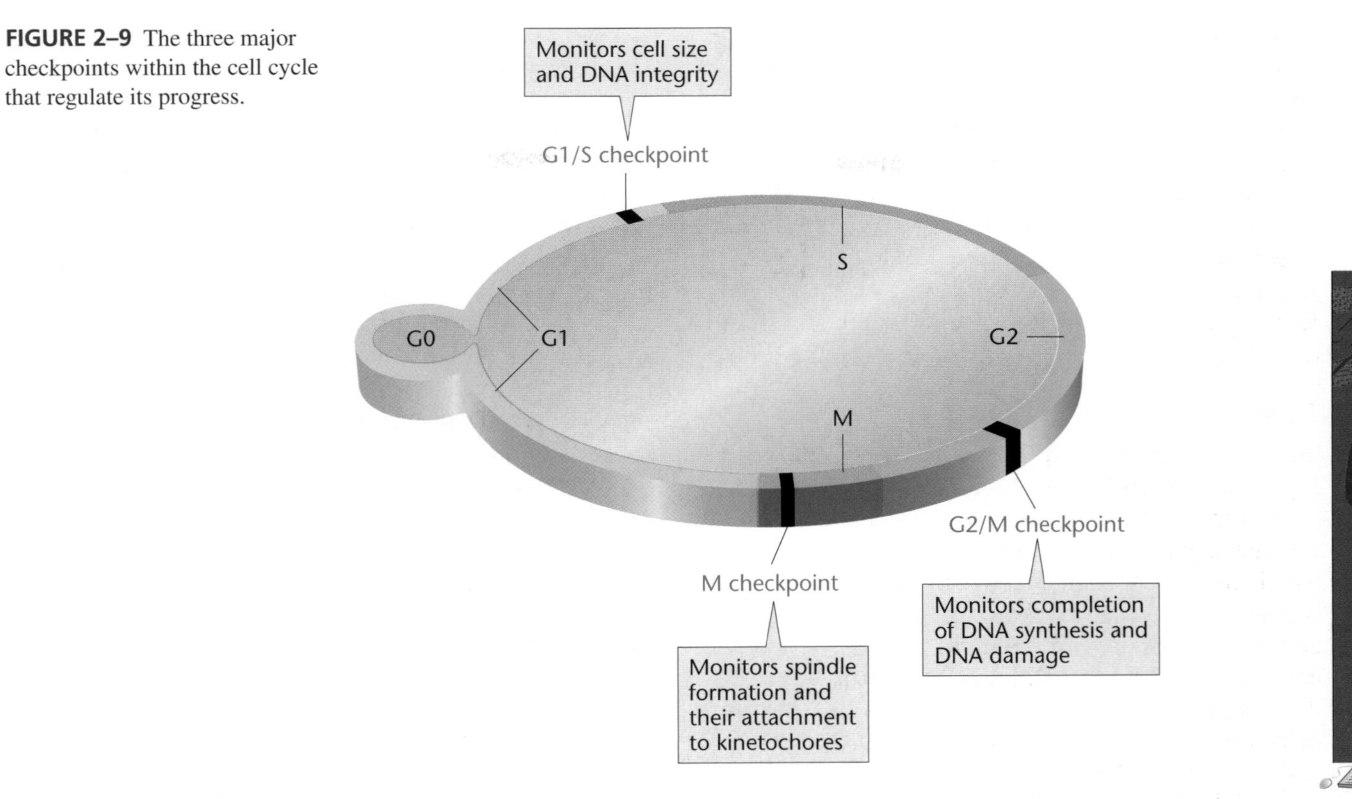

FIGURE 2–9 The three major checkpoints within the cell cycle that regulate its progress.

MEDIA TUTORIAL Cell cycle regulation

uncontrolled cell division—the definition of a cancerous cell. As we saw earlier, such a damaged cell would normally be arrested at either the G1/S or the G2/M checkpoint.

An interesting related finding involves the protein product of the ***p53* gene** in humans and its involvement during scrutiny at the G1/S checkpoint. This protein functions during the regulation of **apoptosis**, the genetic process whereby programmed cell death occurs. When the normal *p53* gene product is present, a proliferative cell that has incurred severe damage to its DNA is targeted for programmed cell death at the G1/S checkpoint and thus effectively removed from the cell population. However, if the *p53* gene has mutated, resulting in abnormal function of its gene product, the damaged cell may proceed through the checkpoint and continue to proliferate in an uncontrolled manner. In fact, a high percentage of human cancers have been found to contain mutations in the *p53* gene, including colon, breast, lung, and bladder malignancies. In the language of cancer genetics, *p53* is referred to as a tumor-suppressor gene. The role of this and other genes involved in the cell cycle has sparked great interest in cancer research.

2.5 Meiosis Reduces the Chromosome Number from Diploid to Haploid in Germ Cells and Spores

The process of meiosis, unlike mitosis, reduces the amount of genetic material to half. Whereas in diploids, mitosis produces daughter cells with a full diploid complement, meiosis produces gametes or spores with only one haploid set of chromosomes. During sexual reproduction, gametes then combine at fertilization to reconstitute the diploid complement found in parental cells. Figure 2–10 compares the two processes by following two pairs of homologous chromosomes.

Meiosis must be highly specific since, by definition, haploid gametes or spores contain precisely one member of each homologous pair of chromosomes. If it is successfully completed, meiosis ensures genetic continuity from generation to generation.

The process of sexual reproduction also ensures genetic variety among members of a species. As you study meiosis, you will see that this process results in gametes with many unique combinations of maternally and paternally derived chromosomes among the haploid complement. With such a tremendous genetic variation among the gametes, a large number of chromosome combinations are possible at fertilization. Furthermore, the meiotic event referred to as **crossing over** results in genetic exchange between members of each homologous pair of chromosomes. This creates intact chromosomes that are mosaics of the maternal and paternal homologs from which they arise, further enhancing the potential genetic variation in gametes and the offspring derived from them. Sexual reproduction therefore reshuffles the genetic material, producing offspring that often differ greatly from either parent. Thus, meiosis is the major form of genetic recombination within species.

An Overview of Meiosis

In the preceding discussion, we established what might be considered the goals of meiosis. Before we consider

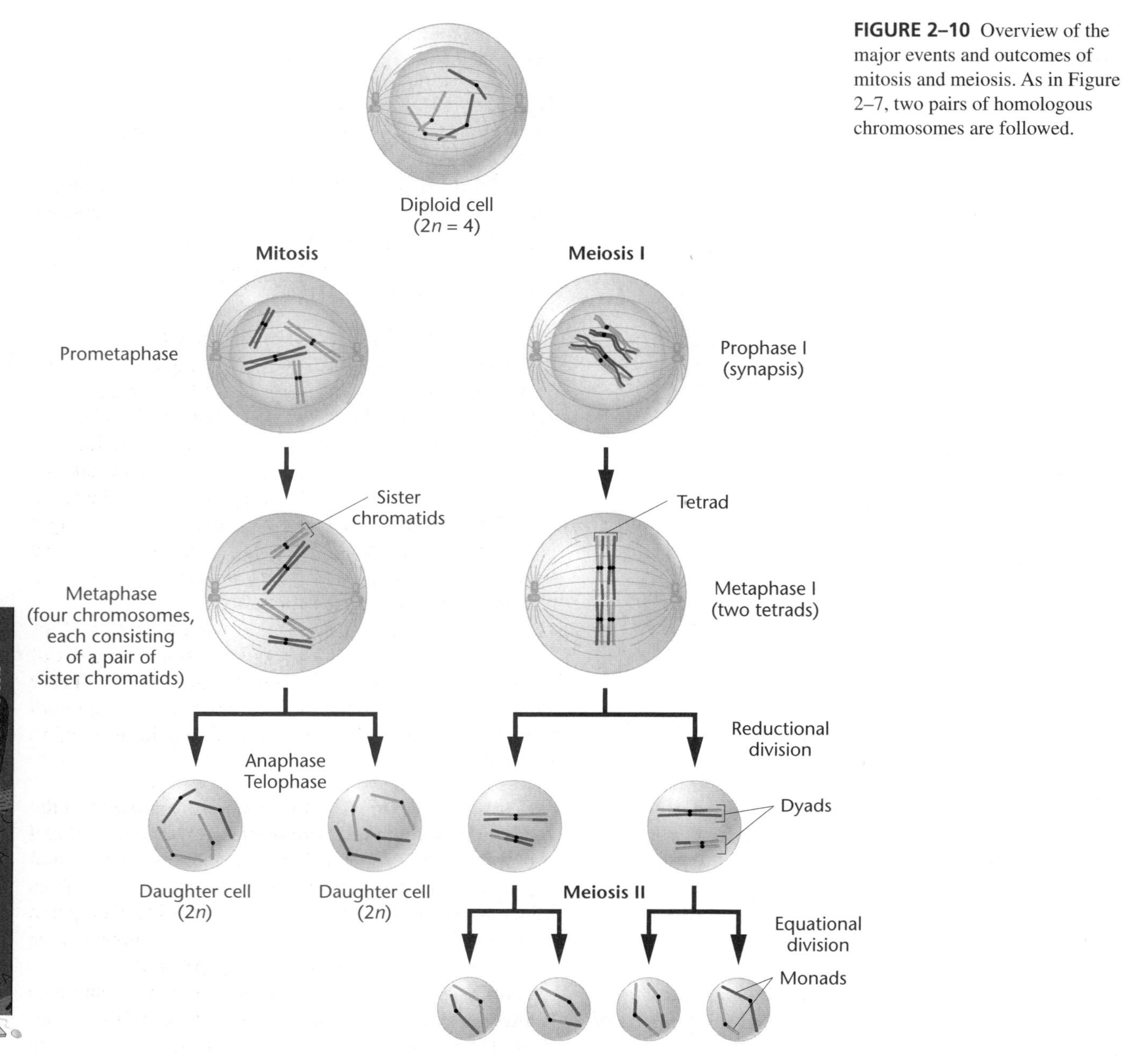

FIGURE 2–10 Overview of the major events and outcomes of mitosis and meiosis. As in Figure 2–7, two pairs of homologous chromosomes are followed.

MEDIA TUTORIAL Overview of meiosis

the phases of this process systematically, we will briefly examine how diploid cells give rise to haploid gametes or spores. You should refer to the meiosis I side of Figure 2–10 during the following discussion.

You have seen that in mitosis each paternally and maternally derived member of any homologous pair of chromosomes behaves autonomously during division. By contrast, early in meiosis, homologous chromosomes form pairs; that is, they synapse. Each synapsed structure is initially called a **bivalent**, which eventually gives rise to a unit, the **tetrad**, consisting of four chromatids. The presence of four chromatids demonstrates that *both* homologs (making up the bivalent) have, in fact, duplicated. Therefore, in order to achieve haploidy, two divisions are necessary.

The first division occurs in meiosis I and is described as a **reductional division** (because the number of centromeres, each representing one chromosome, is *reduced* by one-half). Components of each tetrad—representing the two homologs—separate, yielding two **dyads**. Each dyad is composed of two sister chromatids joined at a common centromere. The second division occurs during meiosis II and is described as an **equational division** (because the number of centromeres remains *equal*). Here each dyad splits into two **monads** of one chromosome each. Thus, the two divisions potentially produce four haploid cells.

The First Meiotic Division: Prophase I

We turn now to a more detailed account of meiosis. As in mitosis, meiosis is a continuous process. We name the parts of each stage of division only to facilitate discussion. From a genetic standpoint, three events characterize the initial stage, *prophase I* (Figure 2–11). First, as in mitosis, the chromatin present in interphase condenses and coils into visible chromosomes. Second, unlike mitosis,

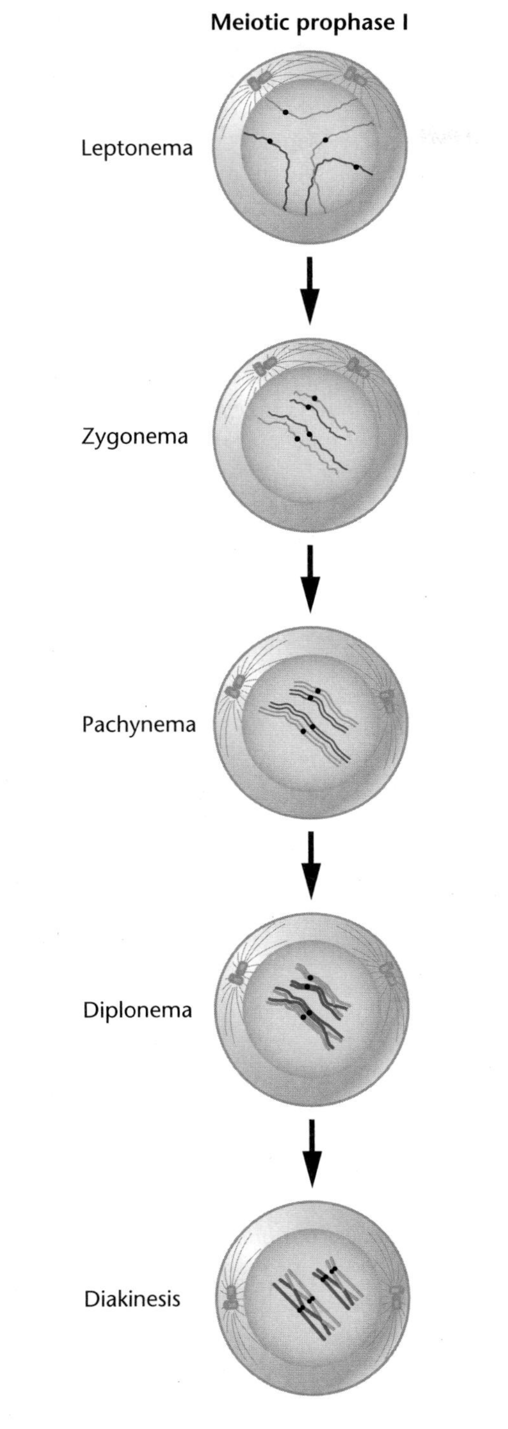

FIGURE 2–11 The substages of meiotic prophase I for the chromosomes depicted in Figure 2–10.

members of each homologous pair of chromosomes undergo **synapsis**. Third, crossing over occurs between synapsed homologs. Because of the complexity of these genetic events, prophase I is further divided into five substages: leptonema, zygonema, pachynema, diplonema,* and diakinesis. As we discuss these substages, be aware that even though it is not immediately apparent in the earliest phases of meiosis, the DNA of chromosomes has already been replicated during the prior interphase.

*These are the noun forms of these substages. The adjective forms (leptotene, zygotene, pachytene, and diplotene) are also used.

Leptonema During the **leptotene stage**, the interphase chromatin material begins to condense, and the chromosomes, although still extended, become visible. Along each chromosome are **chromomeres**, localized condensations that resemble beads on a string. Recent evidence suggests that a process called homology search, which precedes and is essential to the initial pairing of homologs, begins during leptonema.

Zygonema The chromosomes continue to shorten and thicken during the **zygotene stage**. During the **homology search**, homologous chromosomes undergo initial alignment with one another. This so-called rough pairing is complete by the end of zygonema. In yeast, homologs are separated by about 300 nm, and near the end of zygonema, structures called "lateral elements" are visible between paired homologs. As meiosis proceeds, the overall length of the lateral elements increases and a more extensive ultrastructural component, the **synaptonemal complex**, begins to form between the homologs.

At the completion of zygonema, the paired homologs are referred to as bivalents. Although both members of each bivalent have already replicated their DNA, it is not yet visually apparent that each member is a double structure. The number of bivalents in each species is equal to the haploid (n) number.

Pachynema In the transition from the zygotene to the **pachytene stage**, the chromosomes continue to coil and shorten, and further development of the synaptonemal complex occurs between the two members of each bivalent. This leads to synapsis, a more intimate pairing. Compared to the rough-pairing characteristic of yeast zygonema, the homologs are now separated by only 100 nm.

During pachynema, each homolog is first evident as a double structure, providing visual evidence of the earlier replication of the DNA of each chromosome. Thus, each bivalent contains four chromatids. As in mitosis, replicates are called sister chromatids, while chromatids from maternal and paternal members of a homologous pair are called nonsister chromatids. The four-membered structure is a tetrad, and each tetrad contains two pairs of sister chromatids.

Diplonema During the ensuing **diplotene stage**, it is even more apparent that each tetrad consists of two pairs of sister chromatids. Within each tetrad, each pair of sister chromatids begins to separate. However, one or more areas remain in contact where chromatids are intertwined. Each such area, called a **chiasma** (pl., chiasmata), is thought to represent a point where nonsister chromatids have undergone genetic exchange through the process of crossing over. Although the physical exchange between chromosome areas occurred during the previous pachytene stage, the result of crossing over is visible only when the duplicated chromosomes begin to separate. Crossing over is an important source of genetic variability, and new combinations of genetic material are formed during this process.

Diakinesis The final stage of prophase I is **diakinesis**. The chromosomes pull farther apart, but the nonsister chromatids remain loosely associated via the chiasmata. As separation proceeds, the chiasmata move toward the ends of the tetrad. This process of **terminalization** begins in late diplonema and is completed during diakinesis. During this final substage period, the nucleolus and nuclear envelope break down, and the two centromeres of each tetrad attach to the recently formed spindle fibers. By the completion of prophase I, the centromeres of each tetrad structure are present on the metaphase plate of the cell.

Metaphase, Anaphase, and Telophase I

The remainder of the meiotic process is depicted in Figure 2–12. After meiotic prophase I, steps similar to those of mitosis occur. In the first division, **metaphase I**, the chromosomes have maximally shortened and thickened. The terminal chiasmata of each tetrad are visible and appear to be the only factor holding the nonsister chromatids together. Each tetrad interacts with spindle fibers, facilitating movement to the metaphase plate. The alignment of each tetrad prior to the first anaphase is random. Half of each tetrad is pulled randomly to one or the other pole, and the other half then moves to the opposite pole.

During the stages of meiosis I, a single centromere holds each pair of sister chromatids together. It does *not* divide. At **anaphase I**, one-half of each tetrad (the dyad) is pulled toward each pole of the dividing cell. This separation process is the physical basis of **disjunction**, the separation of chromosomes from one another. Occasionally, errors in meiosis occur and separation is not achieved. The term **nondisjunction** describes such an error. At the completion of a normal anaphase I, a series of dyads equal to the haploid number is present at each pole.

If crossing over had not occurred in the first meiotic prophase, each dyad at each pole would consist solely of either paternal or maternal chromatids. However, the exchanges produced by crossing over create mosaic chromatids of paternal and maternal origin.

In many organisms, **telophase I** reveals a nuclear membrane forming around the dyads. Next, the nucleus enters into a short interphase period. If interphase occurs, the chromosomes do not replicate since they already consist of two chromatids. In other organisms, the cells go directly from anaphase I to meiosis II. In general, meiotic telophase is much shorter than the corresponding stage in mitosis.

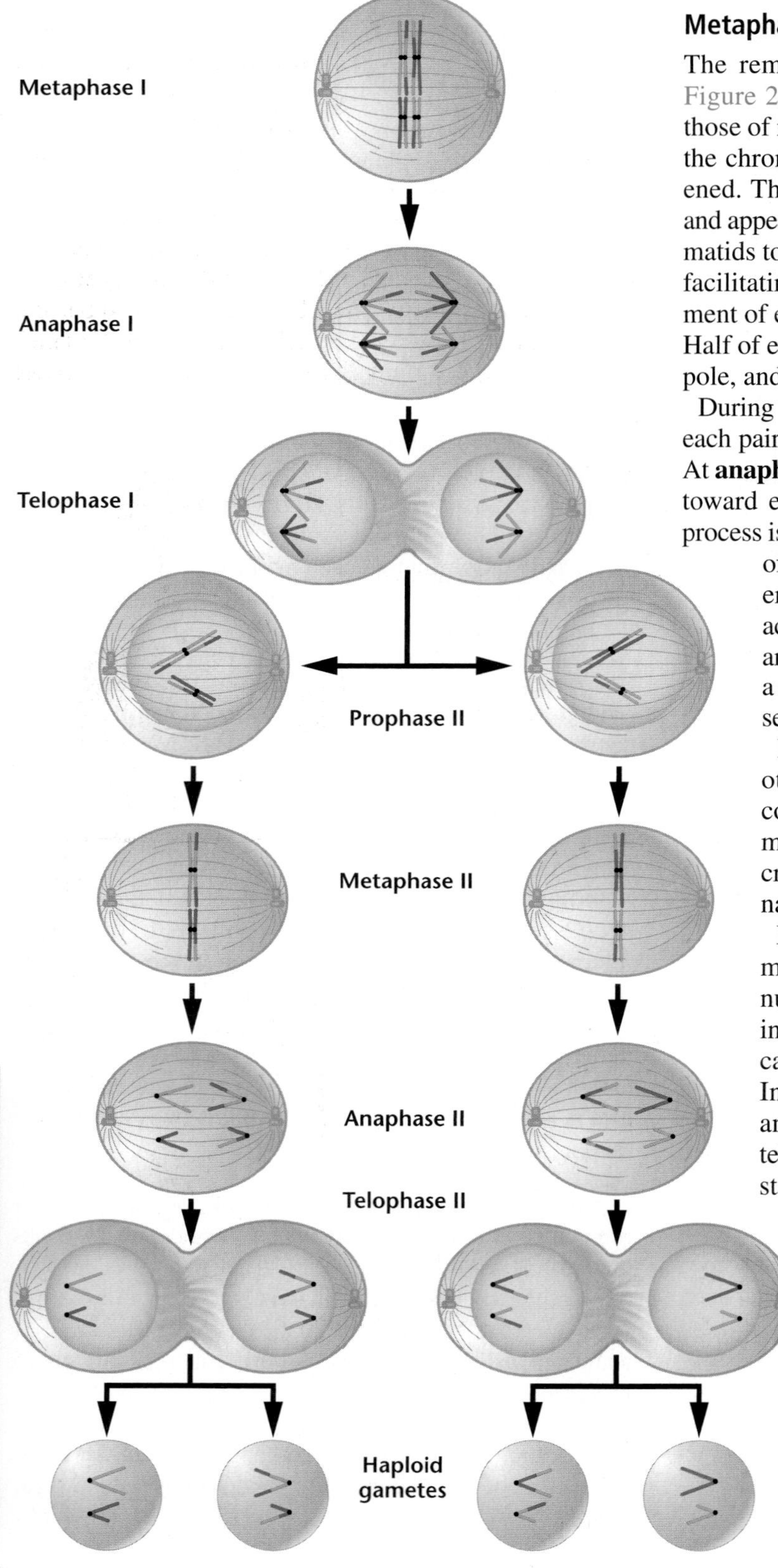

FIGURE 2–12 The major events during meiosis in an animal with a diploid number of $2n = 4$, beginning with metaphase I. Note that the combination of chromosomes in the cells produced following telophase II depends on the random alignment of each tetrad and dyad on the metaphase plate during metaphase I and metaphase II. Several other combinations (not shown) can also be formed.

MEDIA TUTORIAL Meiosis

The Second Meiotic Division

A second division, **meiosis II**, is essential if each gamete or spore is to receive only one chromatid from each original tetrad. The stages characterizing meiosis II are shown in the bottom half of Figure 2–12. During **prophase II**, each dyad is composed of one pair of sister chromatids attached by a common centromere. During **metaphase II**, the centromeres are positioned on the metaphase plate. When they divide, **anaphase II** is initiated and the sister chromatids of each dyad are pulled to opposite poles. Because the number of dyads is equal to the haploid number, **telophase II** reveals one member of each pair of homologous chromosomes at each pole. Each chromosome is now a monad. Following cytokinesis in telophase II, four haploid gametes may result from a single meiotic event. At the conclusion of meiosis II, not only has the haploid state been achieved but if crossing over has occurred, each monad is a combination of maternal and paternal genetic information. As a result, the offspring produced by any gamete receives a mixture of genetic information originally present in his or her grandparents. Meiosis thus significantly increases the level of genetic variation in each ensuing generation.

2.6 The Development of Gametes Varies During Spermatogenesis and Oogenesis

Although events that occur during the meiotic divisions are similar in all cells that participate in gametogenesis in most animal species, there are certain differences between the production of a male gamete (spermatogenesis) and a female gamete (oogenesis). Figure 2–13 summarizes these processes.

Spermatogenesis takes place in the testes, the male reproductive organs. The process begins with the expanded growth of an undifferentiated diploid germ cell called a **spermatogonium**. This cell enlarges to become a **primary spermatocyte**, which undergoes the first meiotic division. The products of this division, called

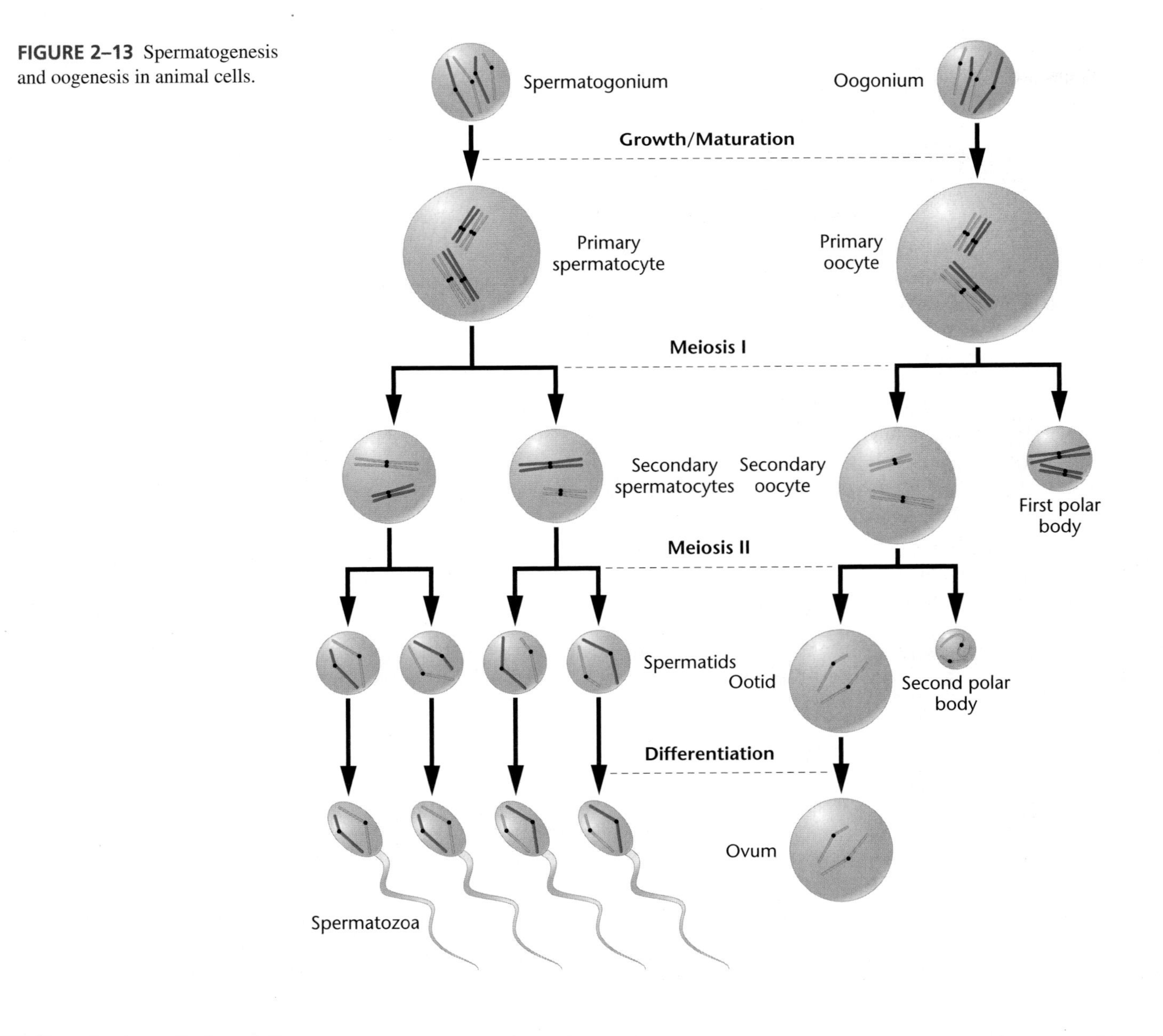

FIGURE 2–13 Spermatogenesis and oogenesis in animal cells.

secondary spermatocytes, contain a haploid number of dyads. The secondary spermatocytes then undergo meiosis II, and each of these cells produces two haploid **spermatids**. Spermatids go through a series of developmental changes, **spermiogenesis**, and become highly specialized, motile **spermatozoa** or **sperm**. All sperm cells produced during spermatogenesis receive equal amounts of genetic material and cytoplasm.

Spermatogenesis may be continuous or may occur periodically in mature male animals; its onset is determined by the species' reproductive cycle. Animals that reproduce year-round produce sperm continuously, whereas those whose breeding period is confined to a particular season produce sperm only during that time.

In animal **oogenesis**, the formation of **ova** (sing., ovum), or eggs, takes place in the ovaries, the female reproductive organs. The daughter cells resulting from the two meiotic divisions receive equal amounts of genetic material, but they do *not* receive equal amounts of cytoplasm. Instead, during each division almost all the cytoplasm of the **primary oocyte**, which is derived from the **oogonium**, is concentrated in one of the two daughter cells. This concentration of cytoplasm is necessary because a major function of the mature ovum is to nourish the developing embryo after fertilization.

During anaphase I in oogenesis, the tetrads of the primary oocyte separate, and the dyads move toward opposite poles. During telophase I, the dyads at one pole are pinched off with very little surrounding cytoplasm to form the **first polar body**. The first polar body may or may not divide again to produce two small haploid cells. The other daughter cell produced by this first meiotic division contains most of the cytoplasm and is called the **secondary oocyte**. The mature ovum will be produced from the secondary oocyte during the second meiotic division. During this division, the cytoplasm of the secondary oocyte again divides unequally, producing an **ootid** and a **second polar body**. The ootid then differentiates into the mature ovum.

Unlike the divisions of spermatogenesis, the two meiotic divisions of oogenesis may not be continuous. In some animal species, the two divisions may directly follow each other. In others, including humans, the first division of all oocytes begins in the embryonic ovary, but arrests in prophase I. Many years later, meiosis resumes in each oocyte just prior to its ovulation. The second division is completed only after fertilization.

2.7 Meiosis Is Critical to the Successful Sexual Reproduction of All Diploid Organisms

The process of meiosis is critical to the successful sexual reproduction of all diploid organisms. It is the mechanism by which the diploid amount of genetic information is reduced to the haploid amount. In animals, meiosis leads to the formation of gametes; while in plants, haploid spores are produced, which in turn leads to the formation of haploid gametes.

Each diploid organism contains its genetic information in the form of homologous pairs of chromosomes. Each pair consists of one member derived from the maternal parent and one from the paternal parent. Following meiosis, haploid cells potentially contain either the paternal or maternal representative of each homologous pair of chromosomes. However, the process of crossing over, which occurs in the first meiotic prophase, reshuffles the genetic information. Crossing over occurs between the maternal and paternal members of each homologous pair, which then assort independently into gametes. This results in the great amounts of genetic variation in gametes.

It is important to touch briefly on the significant role that meiosis plays in the life cycles of fungi and plants. In many fungi, the predominant stage of the life cycle consists of haploid vegetative cells. They arise through meiosis and proliferate by mitotic cell division. In multicellular plants, the life cycle alternates between the diploid **sporophyte stage** and the haploid **gametophyte stage** (Figure 2–14). While one or the other predominates in different plant groups during this "alternation of generations," the processes of meiosis and fertilization constitute the "bridge" between the sporophyte and gametophyte generations. Therefore, meiosis is an essential component of the life cycle of plants.

2.8 Electron Microscopy Has Revealed the Cytological Nature of Mitotic and Meiotic Chromosomes

Thus far in this chapter, we have focused on mitotic and meiotic chromosomes, emphasizing their behavior during cell division and gamete formation. An interesting question is why chromosomes are invisible during interphase but present during the various stages of mitosis and meiosis. Studies using electron microscopy clearly show why this is.

Recall that during interphase, only dispersed chromatin fibers are present in the nucleus (Figure 2–15a). Once mitosis begins, however, the fibers coil and fold, condensing into typical mitotic chromosomes (Figure 2–15b). If the fibers comprising the mitotic chromosome are loosened, the areas of greatest spreading reveal individual fibers similar to those seen in interphase chromatin (Figure 2–15c). Very few fiber ends seem to be present, and in some cases, none can be seen. Instead, individual fibers always seem to loop back into the interior where they are twisted and coiled around one another, forming the regular pattern of the mitotic chromosome. Starting in late telophase of mitosis and continuing during G1 of interphase, chromosomes unwind to form the long fibers characteristic of chromatin, which consist of DNA and associated proteins, particularly proteins called histones. It is in this physical arrangement that DNA can most efficiently function during transcription and replication.

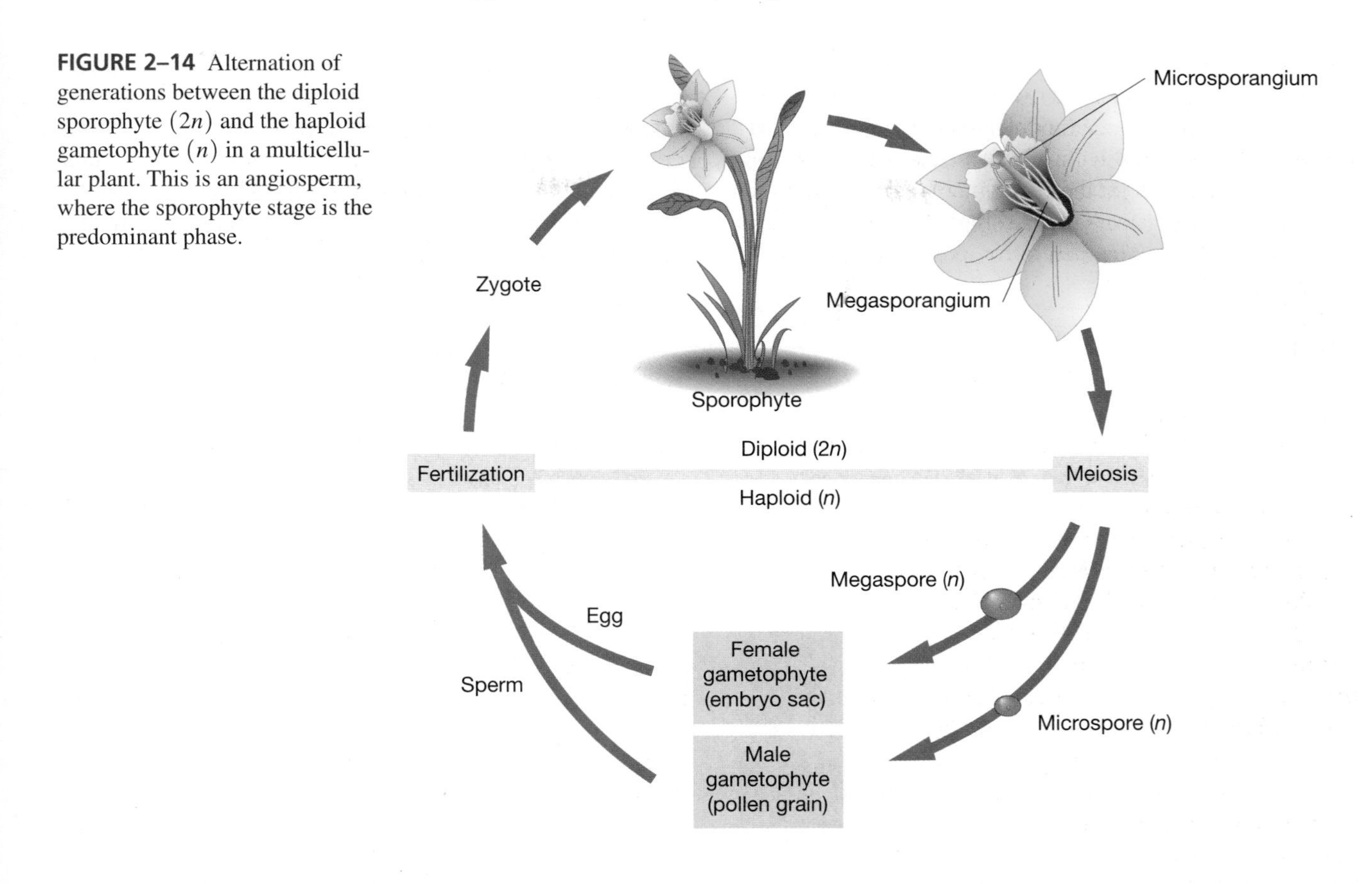

FIGURE 2–14 Alternation of generations between the diploid sporophyte $(2n)$ and the haploid gametophyte (n) in a multicellular plant. This is an angiosperm, where the sporophyte stage is the predominant phase.

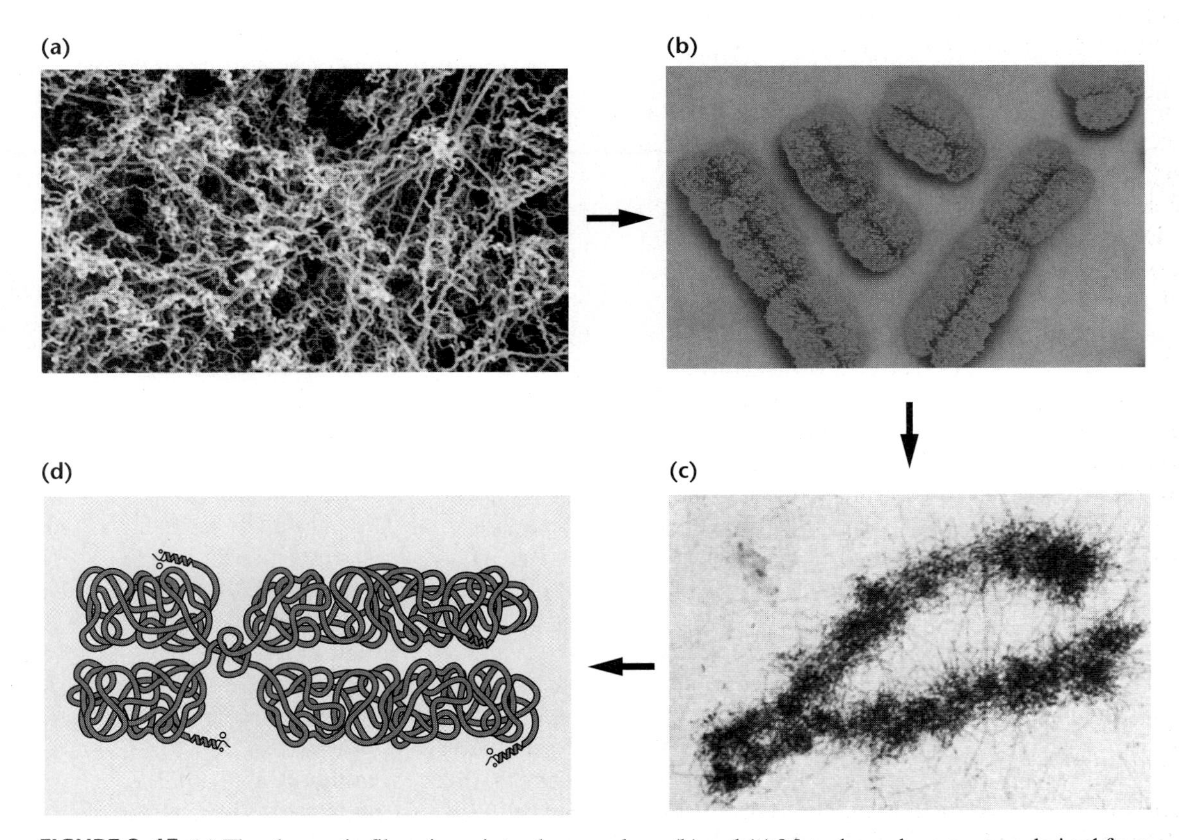

FIGURE 2–15 (a) The chromatin fibers in an interphase nucleus. (b) and (c) Metaphase chromosomes derived from chromatin during mitosis. Part (d) is a diagram of the mitotic chromosome and its various components, showing how chromatin is condensed into it. Parts (a) and (c) are transmission electron micrographs; part (b) is a scanning electron micrograph. *(Biophoto Associates/Photo Researchers, Inc.)*

Electron microscopic observations of mitotic chromosomes in varying states of coiling led Ernest DuPraw to postulate the **folded-fiber model** shown in Figure 2–15(d). During metaphase, each chromosome consists of two sister chromatids joined at the centromeric region. Each arm of the chromatid appears to be a single fiber wound much like a skein of yarn. The fiber is composed of tightly coiled double-stranded DNA and protein. An orderly coiling-twisting-condensing process appears to be involved in the transition of the interphase chromatin to the more condensed, mitotic chromosomes. Geneticists believe that during the transition from interphase to prophase, a 5000-fold contraction occurs in the length of DNA within the chromatin fiber!

Chapter Summary

1. The structure of cells is elaborate and complex. Many components of cells are involved directly or indirectly with genetic processes.
2. In diploid organisms, chromosomes exist in homologous pairs. Each pair shares the same size, centromere placement, and gene sites. One member of each pair is derived from the maternal parent and one is derived from the paternal parent.
3. Mitosis and meiosis are mechanisms by which cells distribute the genetic information contained in their chromosomes to progeny cells in a precise, orderly fashion.
4. Mitosis, or nuclear division, is part of the cell cycle and is the basis of cellular reproduction. Daughter cells are produced that are genetically identical to their progenitor cell.
5. Mitosis may be subdivided into discrete stages: prophase, prometaphase, metaphase, anaphase, and telophase. Condensation of chromatin into chromosome structures occurs during prophase. During prometaphase, chromosomes appear as double structures, each composed of a pair of sister chromatids. In metaphase, chromosomes line up on the metaphase plate of the cell. During anaphase, sister chromatids of each chromosome are pulled apart and directed toward opposite poles. Daughter-cell formation is completed at telophase and is characterized by cytokinesis, the division of the cytoplasm.
6. The cell cycle is characteristic of all eukaryotes and is tightly regulated at three checkpoints: G1/S, G2/M, and M.
7. Meiosis, the underlying basis of sexual reproduction, results in the conversion of a diploid cell into a haploid gamete or spore. As a result of chromosome duplication and two subsequent divisions, each haploid cell receives one member of each homologous pair of chromosomes.
8. A major difference in meiosis exists between males and females. Spermatogenesis partitions the cytoplasmic volume equally and produces four haploid sperm cells. Oogenesis, on the other hand, accumulates the cytoplasm in one egg cell and reduces the other haploid sets of genetic material to polar bodies. The extra cytoplasm contributes to zygote development after fertilization.
9. Meiosis results in extensive genetic variation by virtue of the exchange during crossing over between maternal and paternal chromatids and their random segregation into gametes. In addition, meiosis plays an important role in the life cycles of fungi and plants, serving as the bridge between alternating generations.
10. Mitotic chromosomes are produced as a result of the coiling and condensation of the chromatin fibers that are characteristic of interphase.

Key Terms

AB antigens, 18
acrocentric, 20
allele, 22
anaphase, 24
anaphase I, 30
anaphase II, 31
apoptosis, 27
biparental inheritance, 20
bivalent, 28
cdc kinase, 26
cdc mutation (*c*ell *d*ivision *c*ycle mutation), 26
Cdk protein (Cyclin-dependent kinase protein), 26
cell coat, 18
cell cycle, 23
cell furrow, 26
cell plate, 26
cell wall, 18
centriole, 20
centromere, 24
centrosome, 20, 24
checkpoint, 26
chiasma, 29
chloroplast, 20
chromatid, 21
chromatin, 18
chromomere, 29
chromosome, 19
crossing over, 27
cyclin, 26
cytokinesis, 23, 26
cytoplasm, 19
daughter chromosome, 25
diakinesis, 30
diploid number ($2n$), 20
diplotene stage, 29
disjunction, 30
dyad, 28
endoplasmic reticulum (ER), 19
endosymbiont hypothesis, 20
equational division, 28
first polar body, 32
folded-fiber model, 34
gamete, 18
gametophyte stage, 32
genome, 22
G1 (gap I), 23
G1/S checkpoint, 26
G2 (gap II), 23
G2/M checkpoint, 26
G0 stage, 23
histocompatibility antigen, 18
homologous chromosome, 20
homology search, 29
interphase, 23
karyokinesis, 22
karyotype, 21
kinetochore, 25
leptotene stage, 29
locus, 22
M checkpoint, 26
meiosis, 18
meiosis II, 31
metacentric, 20
metaphase, 24–25
metaphase I, 30
metaphase II, 31
metaphase plate, 24

INSIGHTS AND SOLUTIONS

With this initial appearance of "Insights and Solutions," it is appropriate to describe its value to you as a student. This section precedes the "Problems and Discussion Questions" in each chapter; it provides sample problems and solutions that demonstrate approaches useful in genetic analysis. The insights you gain will help you arrive at correct solutions to ensuing problems.

1. In an organism with a diploid number of $2n = 6$, how many individual chromosomal structures will align on the metaphase plate during (a) mitosis, (b) meiosis I, and (c) meiosis II? Describe each configuration.

Solution: (a) In mitosis, where homologous chromosomes do not synapse, there will be six double structures, each consisting of a pair of sister chromatids. The number of structures is equivalent to the diploid number.

(b) In meiosis I, the homologs have synapsed, reducing the number of structures to three. Each is a tetrad and consists of two pairs of sister chromatids.

(c) In meiosis II, the same number of structures exist (three), but in this case they are dyads. Each dyad is a pair of sister chromatids. When crossing over has occurred, each chromatid may contain parts of one of its nonsister chromatids obtained during exchange in prophase I.

2. Draw all possible alignment configurations that can occur during metaphase I for the chromosomes shown in Figure 2–12.

Solution: As shown in the diagram below, four cases are possible when $n = 2$.

3. Describe the composition of a meiotic tetrad as it exists during prophase I, assuming no crossover event has occurred. What impact would a single crossover event have on this structure?

Solution: Such a tetrad contains four chromatids, existing as two pairs. Members of each pair are sister chromatids. They are held together by a common centromere. Members of one pair are maternally derived, whereas members of the other are paternally derived. Maternal and paternal members are nonsister chromatids. A single crossover event has the effect of exchanging a portion of a maternal *and* a paternal chromatid, leading to a chiasma, where the two chromatids overlap physically in the tetrad.

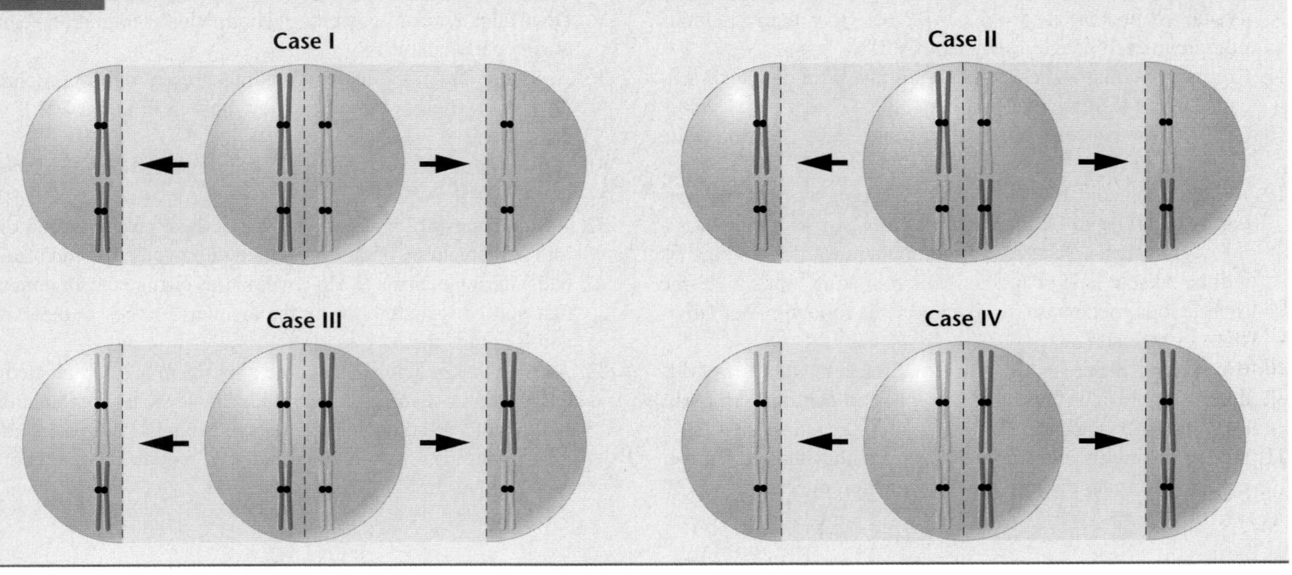

blems and Discussion Questions

1. What role do the following cellular components play in the storage, expression, or transmission of genetic information: (a) chromatin, (b) nucleolus, (c) ribosome, (d) mitochondrion, (e) centriole, (f) centromere?
2. Discuss the concepts of homologous chromosomes, diploidy, and haploidy. What characteristics are shared between two homologous chromosomes?
3. If two chromosomes of a species are the same length and have similar centromere placements yet are not homologous, what is different about them?
4. Describe the events that characterize each stage of mitosis.
5. If an organism has a diploid number of 16, how many chromatids are visible at the end of mitotic prophase? How many chromosomes are moving to each pole during anaphase of mitosis?
6. How are chromosomes named on the basis of their centromere placement?
7. Contrast telophase in plant and animal mitosis.
8. Describe the phases of the cell cycle and the events that characterize each phase.
9. What checkpoints occur in the cell cycle? What is the role of each checkpoint?
10. Describe the role and significance of the *p53* gene in humans. How does the process of apoptosis relate to your answer to the preceding question?
11. Examine Figure 2–13, which shows oogenesis in animal cells. Will the genotype of the second polar body (derived from meiosis II) always be identical to that of the ootid? Why or why not?
12. Contrast the end results of meiosis with those of mitosis.
13. Define and discuss these terms: (a) synapsis, (b) bivalent, (c) chiasmata, (d) crossing over, (e) chromomere, (f) sister chromatids, (g) tetrad, (h) dyad, (i) monad.
14. Contrast the genetic content and the origin of sister versus nonsister chromatids during their earliest appearance in prophase I of meiosis. How might the genetic content of these change by the time tetrads have aligned at the metaphase plate during metaphase I?
15. Given the end results of the two types of division, why is it necessary for homologs to pair during meiosis and not desirable for them to pair during mitosis?
16. An organism has a diploid number of 16 in a primary oocyte. (a) How many tetrads are present in prophase I? (b) How many dyads are present in prophase II? (c) How many monads migrate to each pole during anaphase II?
17. Contrast spermatogenesis and oogenesis. What is the significance of the formation of polar bodies?
18. Explain why meiosis leads to significant genetic variation while mitosis does not.
19. A diploid cell contains three pairs of homologous chromosomes designated C1 and C2, M1 and M2, and S1 and S2; no crossing over occurs. What possible combinations of chromosomes will be present in (a) daughter cells following mitosis, (b) the first meiotic metaphase, (c) haploid cells following both divisions of meiosis?
20. Considering the preceding problem, predict the number of different haploid cells that will occur if a fourth chromosome pair (W1 and W2) is added.
21. During oogenesis in an animal species with a haploid number $n = 6$, one dyad undergoes nondisjunction during meiosis II. After the second meiotic division, the dyad ends up intact in the ovum. How many chromosomes are present in (a) the mature ovum and (b) the second polar body? (c) Following fertilization by a normal sperm, what chromosome condition is created?
22. What is the probability that, in an organism with a haploid number of 10, a sperm will be formed which contains all 10 chromosomes whose centromeres were derived from maternal homologs?
23. During the first meiotic prophase, (a) when does crossing over occur; (b) when does synapsis occur; (c) during which stage are the chromosomes least condensed; and (d) when are chiasmata first visible?
24. Describe the role of meiosis in the life cycle of a plant.
25. Contrast the chromatin fiber with the mitotic chromosome. How are the two structures related?
26. Describe the folded-fiber model of the mitotic chromosome.
27. You are given a metaphase chromosome preparation (a slide) from an unknown organism that contains 12 chromosomes. Two are clearly smaller than the rest, appearing identical in length and centromere placement. Describe all that you can about these two chromosomes.

For Questions 28–33, consider a diploid cell that contains three pairs of chromosomes designated *AA*, *BB*, and *CC*. Each pair contains a maternal and a paternal member (e.g., A^m and A^p, etc.). Using these designations, demonstrate your understanding of mitosis and meiosis by drawing chromatid combinations as requested. Be sure to indicate when chromatids are paired as a result of replication and/or synapsis. You may wish to use a large piece of brown manila wrapping paper or a large cut-up paper bag and work with another student as you deal with these problems. Such cooperative learning may be a useful approach as you solve problems throughout the text.

28. In mitosis, what chromatid combination(s) will be present during metaphase? What combination(s) will be present at each pole at the completion of anaphase?
29. During meiosis I, assuming no crossing over, what chromatid combination(s) will be present at the completion of prophase? Draw all possible alignments of chromatids as migration begins during early anaphase.
30. Are there any possible combinations present during prophase of meiosis II other than those you drew in Problem 29? If so, draw them.
31. Draw all possible combinations of chromatids during anaphase in meiosis II.
32. Assume that during meiosis I, none of the *C* chromosomes disjoin at metaphase, but they separate into dyads (instead of monads) during meiosis II. How would this change the alignments that you constructed during the anaphase stages in meiosis I and II? Draw them.
33. Assume that each resultant gamete (problem 32) participated in fertilization with a normal haploid gamete. What combinations will result? What percentage of zygotes will be diploid, containing one paternal and one maternal member of each chromosome pair?

Selected Readings

Alberts, B., et al. 2002. *Molecular Biology of the Cell*. 4th ed. New York: Garland.

Brachet, J., and Mirsky, A. E. 1961. *The Cell: Meiosis and Mitosis, Vol. 3*. Orlando: Academic Press.

DuPraw, E. J. 1970. *DNA and Chromosomes*. New York: Holt, Rinehart & Winston.

Glover, D. M., Gonzalez, C., and Raff, J. W. 1993. The centrosome. *Sci. Am.* (June) 268: 62–68.

Golomb, H. M., and Bahr, G. F. 1971. Scanning electron microscopic observations of surface structures of isolated human chromosomes. *Science* 171: 1024–26.

Hartwell, L. H., and Karstan, M. B. 1994. Cell cycle control and cancer. *Science* 266: 1821–28.

Hartwell, L. H., and Weinert, T. A. 1989. Checkpoint controls that ensure the order of cell cycle events. *Science* 246: 629–34.

Kleckner, N. 1996. Meiosis: How could it work? *Proc. Natl. Acad. Sci.* 93: 8167–74.

Mazia, D. 1961. How cells divide. *Sci. Am.* (Jan.) 205: 101–20.

———. 1974. The cell cycle. *Sci. Am.* (Jan.) 235: 54–64.

McIntosh, J. R., and McDonald, K. L. 1989. The mitotic spindle. *Sci. Am.* (Oct.) 261: 48–56.

Murray, A. W., and Kirschner, M. 1993. *The Cell Cycle: An Introduction*. New York: Oxford Univ. Press.

Prescott, D. M., and Flexer, A. S. 1986. *Cancer, the Misguided Cell*. 2nd ed. Sunderland, MA: Sinauer.

Swanson, C. P., Merz, T., and Young, W. J. 1981. *Cytogenetics, the Chromosome in Division, Inheritance, and Evolution*, 2nd ed. Englewood Cliffs, NJ: Prentice-Hall.

Westergard, M., and von Wettstein, D. 1972. The synaptinemal complex. *Annu. Rev. Genet.* 6: 71–110.

Wheatley, D. N. 1982. *The Centriole: A Central Enigma of Cell Biology*. New York: Elsevier/North-Holland Biomedical.

APTER 3

Wrinkled and round garden peas, the phenotypic traits in one of Mendel's monohybrid crosses. (*Mike Dunton/ VictorySeeds.com.*)

Mendelian Genetics

CHAPTER CONCEPTS

Although inheritance of biological traits has been recognized for thousands of years, the first significant insights into the mechanisms involved occurred about 135 years ago. In 1866, Gregor Johann Mendel published the results of a series of experiments that would lay the foundation for the formal discipline of genetics. Mendel's work went largely unnoticed until the turn of the century, but in the ensuing years the concept of the gene as a distinct hereditary unit was established. The ways in which genes, as members of chromosomes, are transmitted to offspring and control traits were clarified. Research has continued unabated throughout the twentieth century—indeed, studies in genetics, most recently at the molecular level, have remained continually at the forefront of biological research since the early 1900s.

When Mendel began his studies of inheritance using *Pisum sativum*, the garden pea, chromosomes and the role and mechanism of meiosis were totally unknown. Nevertheless, he determined that discrete **units of inheritance** exist and predicted their behavior during the formation of gametes. Subsequent investigators, with access to cytological data, were able to relate their observations of chromosome behavior during meiosis to Mendel's principles of inheritance. Once this correlation was made, Mendel's postulates were accepted as the basis for the study of what is known as **transmission genetics**.

HOW DO WE KNOW WHAT WE KNOW?

IN THIS CHAPTER, WE FOCUS ON how Mendel was able to derive the essential postulates that explain inheritance. As you study this topic, you should try to answer several fundamental questions:

1. How did Mendel know that unit factors existed as fundamental genetic components if he couldn't directly observe them?
2. How do we know that an organism expressing a recessive trait is homozygous or heterozygous?
3. In genetic data, how do we know that deviation from the ideal ratio is due to chance rather than another independent factor?
4. How do we know how a trait is inherited in humans? ■

3.1 Mendel Used a Model Experimental Approach to Study Patterns of Inheritance

Johann Mendel was born in 1822 to a peasant family in the central European village of Heinzendorf. An excellent student in high school, he studied philosophy for several years afterward, and in 1843 was admitted to the Augustinian Monastery of St. Thomas in Brno, now part of the Czech Republic, taking the name of Gregor. In 1849, he was relieved of pastoral duties and accepted a teaching appointment that lasted several years. From 1851 to 1853, he attended the University of Vienna, where he studied physics and botany. He returned to Brno in 1854, where he taught physics and natural science for the next 16 years. Mendel received support from the monastery for his studies and research throughout his life.

In 1856, Mendel performed his first set of hybridization experiments with the garden pea. The research phase of his career lasted until 1868, when he was elected abbot of the monastery. Although he retained his interest in genetics, his new responsibilities demanded most of his time. In 1884, Mendel died of a kidney disorder. The local newspaper paid him the following tribute:

> "His death deprives the poor of a benefactor, and mankind at large of a man of the noblest character, one who was a warm friend, a promoter of the natural sciences, and an exemplary priest."

Mendel first reported the results of some simple genetic crosses between certain strains of the garden pea in 1865. Although his was not the first attempt to provide experimental evidence pertaining to inheritance, Mendel's success where others failed can be attributed, at least in part, to his elegant model of experimental design and analysis.

Mendel showed remarkable insight into the methodology necessary for good experimental biology. He chose an organism that is easy to grow and hybridize artificially: the pea plant is self-fertilizing in nature, but is easy to crossbreed experimentally. It reproduces well and grows to maturity in a single season. Mendel followed seven visible features (unit characters), each represented by two contrasting forms, or **traits** (Figure 3–1). For the character stem height, for example, he experimented with the traits *tall* and *dwarf*. He selected six other visibly contrasting pairs of traits involving seed shape and color, pod shape and color, and pod and flower arrangement. From local seed merchants, Mendel obtained true-breeding strains—those in which each trait appeared unchanged generation after generation in self-fertilizing plants.

There were several reasons for Mendel's success. In addition to his choice of a suitable organism, he restricted his examination to one or very few pairs of contrasting traits in each experiment. He also kept accurate quantitative records, a necessity in genetic experiments. From the analysis of his data, Mendel derived certain postulates that became principles of transmission genetics.

The results of Mendel's experiments were unappreciated until the turn of the century, well after his death. However, once Mendel's publications were rediscovered by geneticists investigating the function and behavior of chromosomes, the implications of his postulates were immediately apparent. He had discovered the basis for the transmission of hereditary traits!

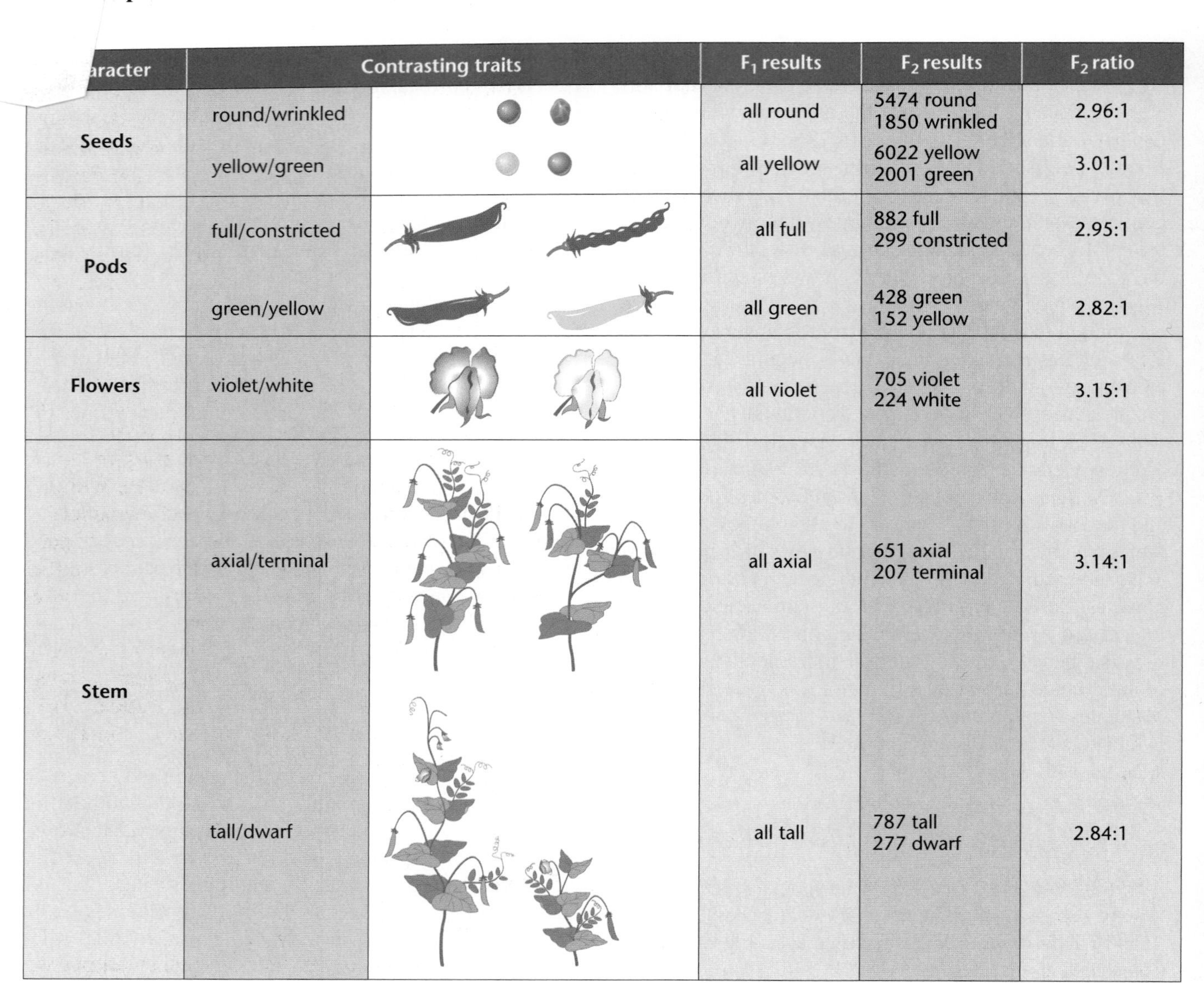

aracter	Contrasting traits	F_1 results	F_2 results	F_2 ratio
Seeds	round/wrinkled	all round	5474 round 1850 wrinkled	2.96:1
	yellow/green	all yellow	6022 yellow 2001 green	3.01:1
Pods	full/constricted	all full	882 full 299 constricted	2.95:1
	green/yellow	all green	428 green 152 yellow	2.82:1
Flowers	violet/white	all violet	705 violet 224 white	3.15:1
Stem	axial/terminal	all axial	651 axial 207 terminal	3.14:1
	tall/dwarf	all tall	787 tall 277 dwarf	2.84:1

FIGURE 3–1 Seven pairs of contrasting traits and the results of Mendel's seven monohybrid crosses of the garden pea (*Pisum sativum*). In each case, pollen derived from plants exhibiting one trait was used to fertilize the ova of plants exhibiting the other trait. In the F_1 generation, one of the two traits was exhibited by all plants. The contrasting trait reappeared in approximately 1/4 of the F_2 plants.

3.2 The Monohybrid Cross Reveals How One Trait Is Transmitted from Generation to Generation

Mendel's simplest crosses involved only one pair of contrasting traits. Each such breeding experiment is a **monohybrid cross**, which is made by mating individuals from two parent strains, each exhibiting one of the two contrasting forms of the character under study. Initially, we examine the first generation of offspring of such a cross, and then we consider the results of **selfing**, the offspring of self-fertilizing individuals from this first generation. The original parents are the $\mathbf{P_1}$, or **parental**, **generation**, their offspring are the $\mathbf{F_1}$, or **first filial**, **generation**, and the individuals resulting from the selfed F_1 generation are the $\mathbf{F_2}$, or **second filial**, **generation**. We can, of course, continue to follow subsequent generations.

The cross between true-breeding pea plants with tall stems and dwarf stems is representative of Mendel's monohybrid crosses. *Tall* and *dwarf* are contrasting traits of the character of stem height. Unless tall or dwarf plants are crossed together or with another strain, they will undergo self-fertilization and breed true, producing their respective trait generation after generation. However, when Mendel crossed tall plants with dwarf plants, the resulting F_1 generation consisted only of tall plants. When members of the F_1 generation were selfed, Mendel observed that 787 of 1064 F_2 plants were tall, while the remaining 277 were dwarf. Note that in this cross (Figure 3–1) the dwarf trait disappears in the F_1, only to reappear in the F_2 generation.

Genetic data are usually expressed and analyzed as ratios. In this particular example, many identical P_1 crosses were made, and many F_1 plants—all tall—were produced. Of the 1064 F_2 offspring, 787 were tall and 277 were dwarf—a ratio of 2.84:1.0, or about 3:1.

Mendel made similar crosses between pea plants exhibiting other pairs of contrasting traits, the results of these crosses are shown in Figure 3–1. In every case, the outcome was similar to the tall/dwarf cross just described. All F_1 offspring were identical to one of the parents, but in the F_2 offspring, an approximate ratio of 3:1 was obtained. That is, three-fourths looked like the F_1 plants, while one-fourth exhibited the contrasting trait, which had disappeared in the F_1 generation.

We will point out one further aspect of Mendel's monohybrid crosses. In each cross, the F_1 and F_2 patterns of inheritance were similar regardless of which P_1 plant served as the source of pollen (sperm) and which served as the source of the ovum (egg). The crosses could be made either way—pollination of dwarf plants by tall plants or vice versa. These are called **reciprocal crosses**. Therefore, the results of Mendel's monohybrid crosses were not sex-dependent.

To explain these results, Mendel proposed the existence of particular unit factors for each trait. He suggested that these factors serve as the basic units of heredity and are passed unchanged from generation to generation, determining the various traits expressed by each individual plant. Using these general ideas, Mendel proceeded to hypothesize precisely how unit factors could account for the results of the monohybrid crosses.

Mendel's First Three Postulates

Using the consistent pattern of results in the monohybrid crosses, Mendel derived the following three postulates or principles of inheritance.

1. **Unit Factors in Pairs**
   Genetic characters are controlled by unit factors that exist in pairs in individual organisms.
   In the monohybrid cross involving tall and dwarf stems, a specific unit factor exists for each trait. Because the factors occur in pairs, three combinations are possible: two factors for tallness, two factors for dwarfness, or one factor for each trait. Every individual contains one of these three combinations, which determines stem height.

2. **Dominance/Recessiveness**
   When two unlike unit factors responsible for a single character are present in a single individual, one unit factor is dominant to the other, which is said to be recessive.
   In each monohybrid cross, the trait expressed in the F_1 generation is controlled by the **dominant** unit factor. The trait not expressed is controlled by the **recessive** unit factor. Note that this dominance/recessiveness relationship pertains only when unlike unit factors are present in pairs. The terms *dominant* and *recessive* are also used to designate traits. In this case, tall stems are said to be dominant over the recessive dwarf stems.

3. **Segregation**
   During the formation of gametes, the paired unit factors separate or segregate randomly so that each gamete receives one or the other with equal likelihood.
   If an individual contains a pair of like unit factors (e.g., both specific for tall), then all gametes receive one tall unit factor. If an individual contains unlike unit factors (e.g., one for tall and one for dwarf), then each gamete has a 50% probability of receiving either the tall or the dwarf unit factor.

These postulates provide a suitable explanation for the results of the monohybrid crosses. Let's use the tall/dwarf cross to illustrate. Mendel reasoned that P_1 tall plants contain identical paired unit factors, as do the P_1 dwarf plants. The gametes of tall plants all receive one tall unit factor as a result of segregation. Likewise, the gametes of dwarf plants all receive one dwarf unit factor. Following fertilization, all F_1 plants receive one unit factor from each parent: a tall factor from one and a dwarf factor from the other, reestablishing the paired relationship—but because tall is dominant to dwarf, all F_1 plants are tall.

When F_1 plants form gametes, the postulate of segregation demands that each gamete randomly receives either the tall or the dwarf unit factor. Following random fertilization events during F_1 selfing, four F_2 combinations result in equal frequency:

1. tall/tall
2. tall/dwarf
3. dwarf/tall
4. dwarf/dwarf

Combinations (1) and (4) result in tall and dwarf plants, respectively. According to the postulate of dominance/recessiveness, combinations (2) and (3) both yield tall plants. Therefore, the F_2 is predicted to consist of 3/4 tall and 1/4 dwarf, or a ratio of 3:1. This is approximately what Mendel observed in the cross between tall and dwarf plants. A similar pattern was observed in each of the other monohybrid crosses (see Figure 3–1).

Modern Genetic Terminology

To illustrate the monohybrid cross and Mendel's first three postulates, we must first introduce several new terms as well as a symbol convention for the unit factors.

Traits such as tall or dwarf are visible expressions of the information contained in unit factors. The physical appearance of a trait is the **phenotype** of the individual. Mendel's unit factors represent units of inheritance called **genes** by modern geneticists. For any given character, such as plant height, the phenotype is determined by alternative forms of a single gene called **alleles**. For example, the unit factors representing tall and dwarf are alleles determining the height of the pea plant.

The convention we will use is to choose the first letter of the recessive trait to symbolize the character in ıon—the lowercase italic letter designates the allele ‿ the recessive trait, and the uppercase italic letter designates the allele for the dominant trait. Thus, for Mendel's pea plants, we use *d* for the dwarf allele and *D* for the tall allele. When alleles are written in pairs to represent the two unit factors present in any individual (*DD*, *Dd*, or *dd*), these symbols are called the **genotype**. This term reflects the genetic makeup of an individual, whether it is haploid or diploid. By reading the genotype, we know the phenotype of the individual: *DD* and *Dd* are tall, and *dd* is dwarf. When both alleles are the same (*DD* or *dd*), the individual is *homozygous* or a **homozygote**; when the alleles are different (*Dd*), we use the term *heterozygous* or **heterozygote**. These symbols and terms are used in Figure 3–2 to illustrate the monohybrid cross.

Because he operated without the hindsight that modern geneticists enjoy, Mendel's analytical reasoning must be considered a truly outstanding scientific achievement. On the basis of rather simple but precisely executed breeding experiments, he not only proposed that discrete particulate units of heredity exist, he also explained how they are transmitted from one generation to the next.

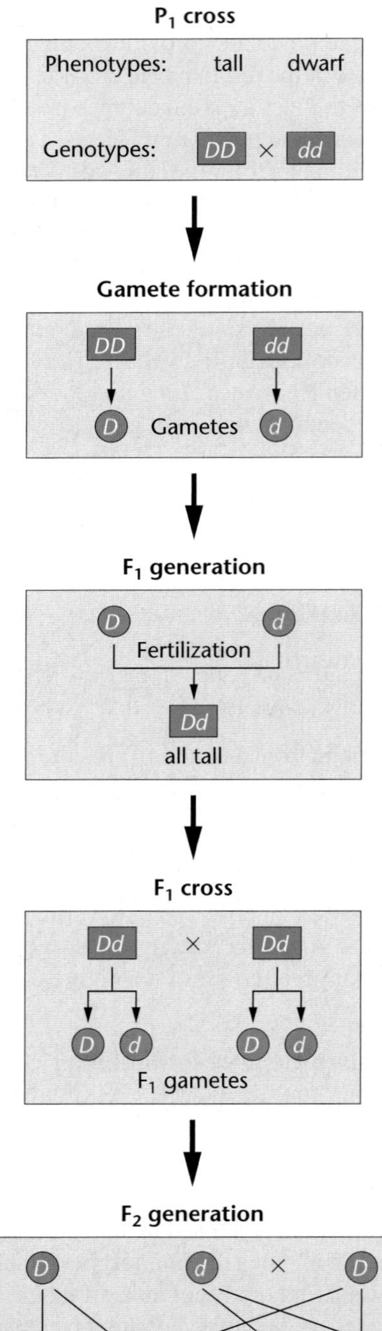

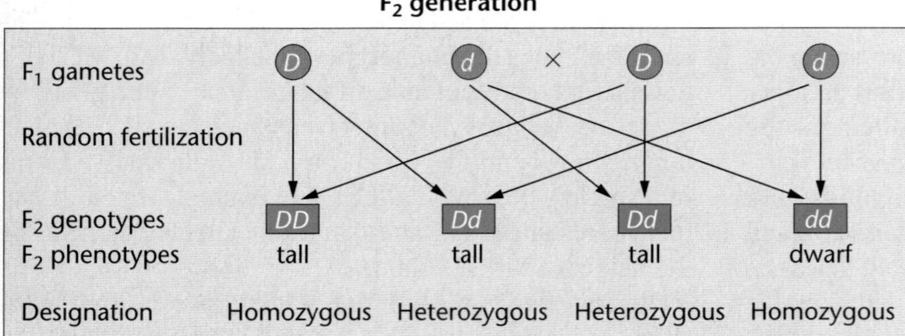

FIGURE 3–2 The monohybrid cross between tall (*D*) and dwarf (*d*) pea plants. Individuals are shown in rectangles, and gametes in circles.

MEDIA TUTORIAL Monohybrid cross and punnett squares

Punnett Squares

The genotypes and phenotypes resulting from the recombination of gametes during fertilization can be easily visualized by constructing a **Punnett square**, named after the person who first devised this approach, Reginald C. Punnett. Figure 3–3 demonstrates this method of analysis for our $F_1 \times F_1$ monohybrid cross. Each of the possible gametes is assigned to a column or a row; the vertical column represents those of the female parent, and the horizontal row represents those of the male parent. After putting the gametes into the rows and columns, the new generation is predicted by combining the male and female gametic information for each combination and entering the resulting genotypes in the boxes. This process thus lists all possible random fertilization events. The genotypes and phenotypes of all potential offspring are ascertained by reading the entries in the boxes.

The Punnett square method is particularly useful when you are first learning about genetics and how to solve problems. Note the ease with which the 3:1 phenotypic ratio and the 1:2:1 genotypic ratio is derived in the F_2 generation in Figure 3–3.

The Test Cross: One Character

Tall plants produced in the F_2 generation are predicted to be either the *DD* or *Dd* genotype. You might ask if there is a way to distinguish the genotype. Mendel devised a rather simple method that is still used today in breeding plants and animals: the **test cross**. The organism expressing the dominant phenotype, but of unknown genotype, is crossed to a known homozygous recessive individual. For

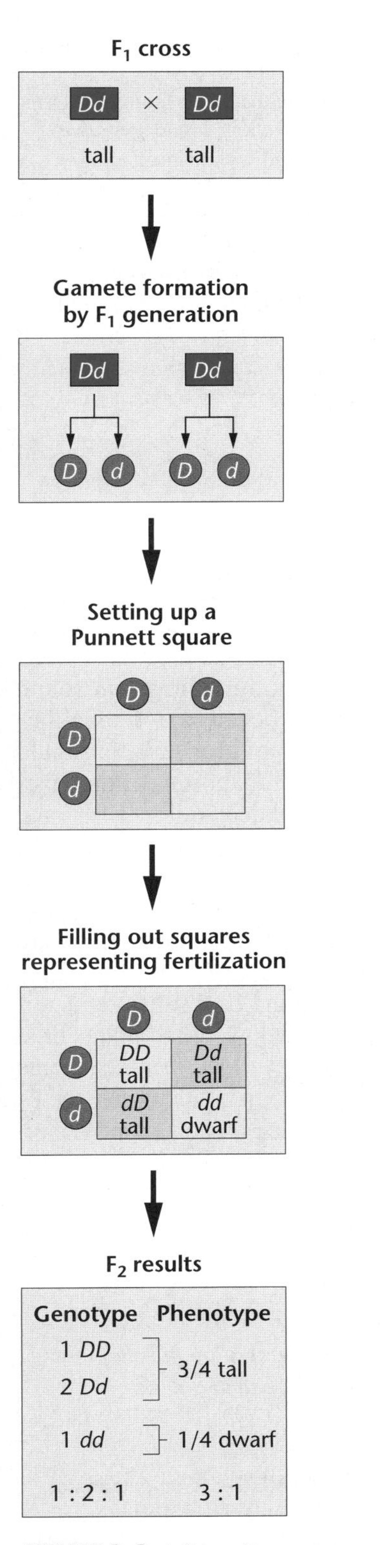

FIGURE 3–3 A Punnett square generating the F_2 ratio of the $F_1 \times F_1$ cross shown in Figure 3–2.

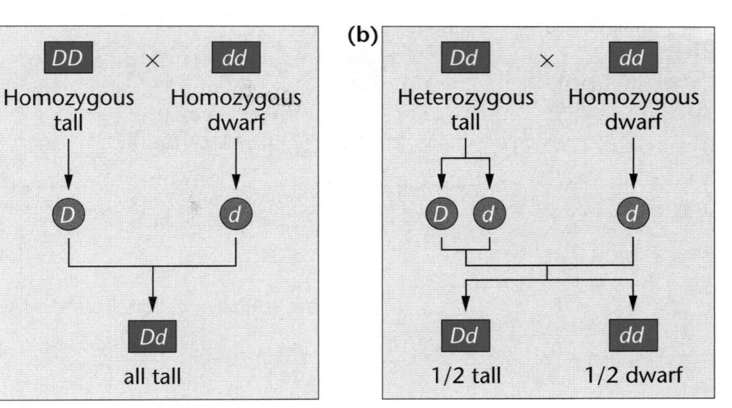

FIGURE 3–4 Test cross of a single character. In (a), the tall parent is homozygous, but in (b), the tall parent is heterozygous. The genotype of each tall P_1 plant can be determined by examining the offspring when each is crossed to a homozygous recessive dwarf plant.

example, as shown in Figure 3–4(a), if a tall plant of genotype *DD* is test-crossed to a dwarf plant, which must have the *dd* genotype, all offspring will be tall phenotypically and *Dd* genotypically. However, as shown in Figure 3–4(b), if a tall plant is *Dd* and it is crossed to a dwarf plant (*dd*), then one-half of the offspring will be tall (*Dd*) and the other half will be dwarf (*dd*). Therefore, a 1:1 tall/dwarf ratio demonstrates the heterozygous nature of the tall plant of unknown genotype. The test cross reinforced Mendel's conclusion that separate unit factors control traits.

3.3 Mendel's Dihybrid Cross Generated a Unique F_2 Ratio

As a natural extension of the monohybrid cross, Mendel also designed experiments in which he examined two characters simultaneously. Such a cross, involving two pairs of contrasting traits, is a **dihybrid cross**, or *two-factor cross*. For example, if pea plants having yellow seeds that are round are bred with those having green seeds that are wrinkled, the results shown in Figure 3–5 will occur: The F_1 offspring will all be yellow and round. It is therefore apparent that yellow is dominant to green, and that round is dominant to wrinkled. When the F_1 individuals are selfed, approximately 9/16 of the F_2 plants express yellow and round, 3/16 express yellow and wrinkled, 3/16 express green and round, and 1/16 express green and wrinkled.

A variation of this cross is also shown in Figure 3–5. Instead of crossing one P_1 parent with both dominant traits (yellow, round) and one with both recessive traits (green, wrinkled), plants with yellow, wrinkled seeds are crossed with those with green, round seeds. In spite of the change in the P_1 phenotypes, both the F_1 and F_2 results remain unchanged. It will become clear in the next section why this is so.

Mendel's Fourth Postulate: Independent Assortment

We can most easily understand the results of a dihybrid cross if we consider it theoretically as consisting of two monohybrid crosses conducted separately. Think of the two sets of traits as inherited independently of each other;

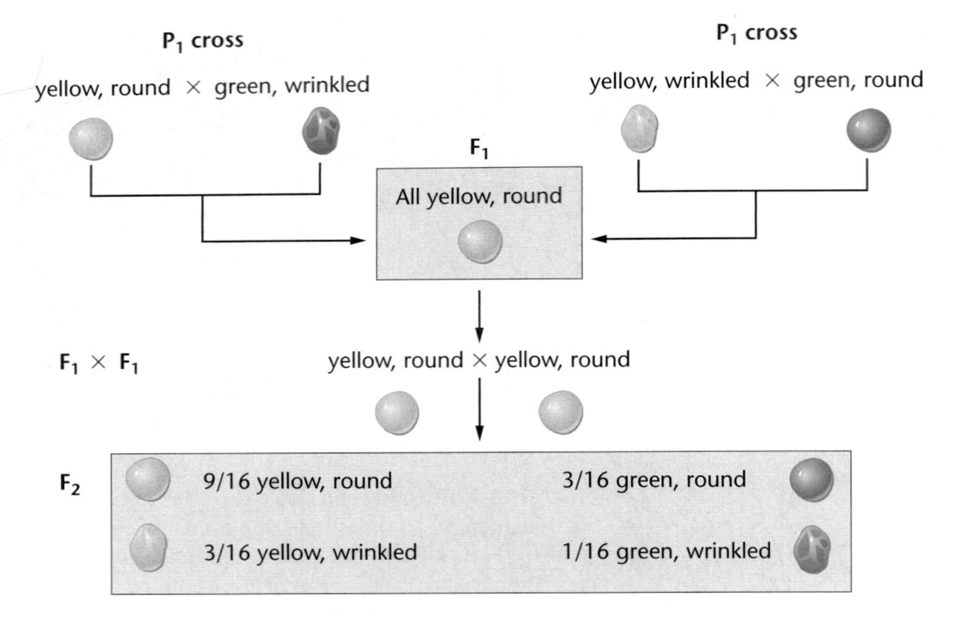

FIGURE 3–5 F_1 and F_2 results of Mendel's dihybrid crosses, where the plants on the top left with yellow, round seeds are crossed with plants having green, wrinkled seeds, and the plants on the top right with yellow, wrinkled seeds are crossed with plants having green, round seeds.

that is, the chance of any plant having yellow or green seeds is not at all influenced by the chance that this plant will have round or wrinkled seeds. Thus, because yellow is dominant to green, all F_1 plants in the first theoretical cross would have yellow seeds. In the second theoretical cross, all F_1 plants would have round seeds because round is dominant to wrinkled. When Mendel examined the F_1 plants of the dihybrid cross, all were yellow and round, as we just predicted.

The predicted F_2 results of the first cross are 3/4 yellow and 1/4 green. Similarly, the second cross would yield 3/4 round and 1/4 wrinkled. Figure 3–5 shows that in the dihybrid cross, 12/16 F_2 plants are yellow while 4/16 are green, exhibiting the expected 3:1 (3/4 : 1/4) ratio. Similarly, 12/16 F_2 plants have round seeds while 4/16 have wrinkled seeds, again revealing the 3:1 ratio.

It is evident that the two pairs of contrasting traits are inherited independently, so we can predict the frequencies of all possible F_2 phenotypes by applying the **product law** of probabilities: *When two independent events occur simultaneously, the combined probability of the two outcomes is equal to the product of their individual probabilities of occurrence.* For example, the probability of an F_2 plant's having yellow *and* round seeds is (3/4)(3/4), or 9/16, because 3/4 of all F_2 plants should be yellow and 3/4 of all F_2 plants should be round. In a like manner, the probabilities of the other three F_2 phenotypes can be calculated: Yellow (3/4) and wrinkled (1/4) are predicted to be present together 3/16 of the time; green (1/4) and round (3/4) are predicted 3/16 of the time; and green (1/4) and wrinkled (1/4) are predicted 1/16 of the time. These calculations are shown in Figure 3–6.

It is now apparent why the F_1 and F_2 results are identical whether the initial cross is yellow, round plants bred with green, wrinkled plants, or if yellow, wrinkled plants are bred with green, round plants. In both crosses, the F_1 genotype of all plants is identical. Each plant is heterozygous for both gene pairs. As a result, the F_2 generation is also identical in both crosses.

On the basis of similar results in numerous dihybrid crosses, Mendel proposed a fourth postulate called **independent assortment**:

During gamete formation, segregating pairs of unit factors assort independently of each other.

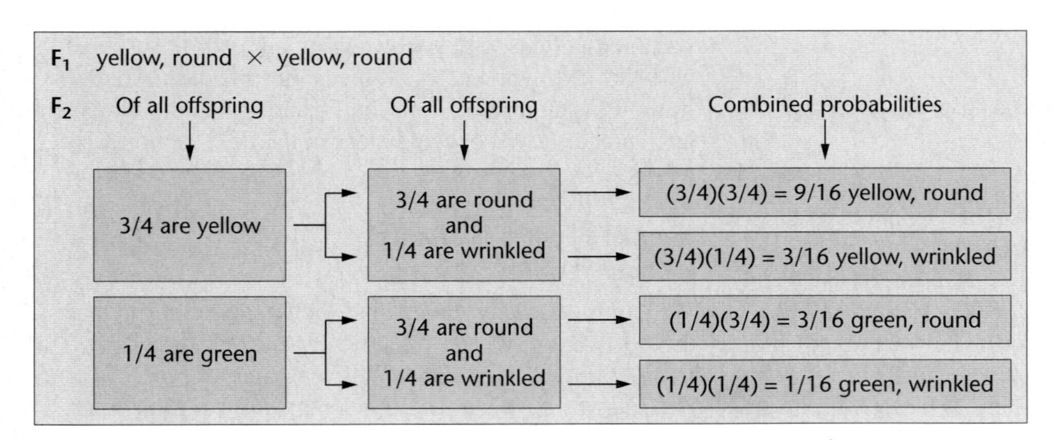

FIGURE 3–6 Computation of the combined probabilities of each F_2 phenotype for two independently inherited characters. The probability of each plant's being yellow or green is independent of the probability of its bearing round or wrinkled seeds.

This postulate stipulates that segregation of any pair of unit factors occurs independently of all others. As a result of random segregation, each gamete receives one member of every pair of unit factors. For one pair, whichever unit factor is received does not influence the outcome of segregation of any other pair. Thus, according to the postulate of independent assortment, all possible combinations of gametes are formed in equal frequency.

The Punnett square in Figure 3–7 shows how independent assortment works in the formation of the F_2 generation. Examine the formation of gametes by the F_1 plants; segregation prescribes that every gamete receives either a

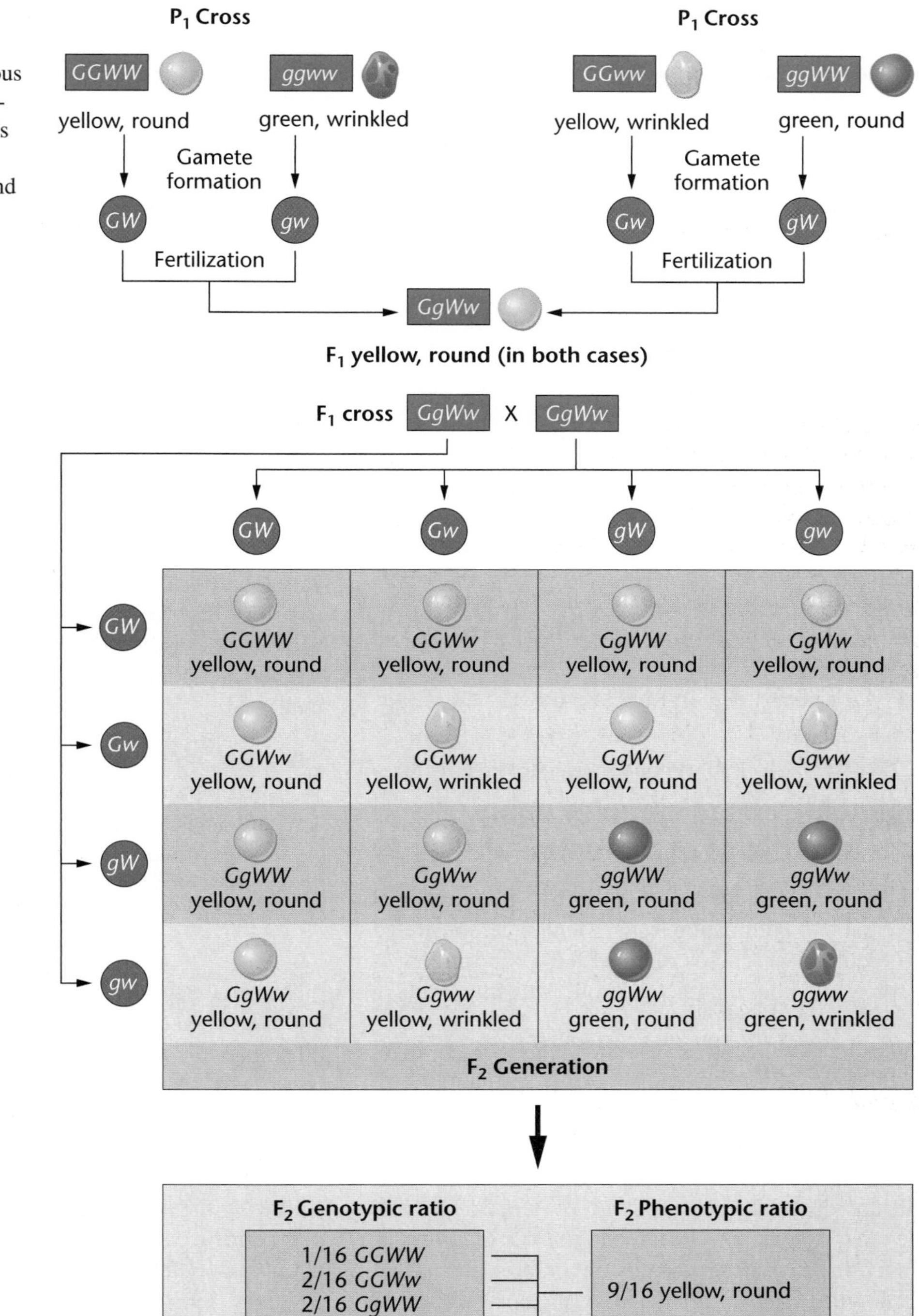

FIGURE 3–7 Analysis of the dihybrid crosses shown in Figure 3–5. The F_1 heterozygous plants are self-fertilized to produce an F_2 generation, which is computed using a Punnett square. Both the phenotypic and genotypic F_2 ratios are shown.

MEDIA TUTORIAL Indipendent assortment

G or *g* allele, and a *W* or *w* allele. Independent assortment stipulates that all four combinations (*GW*, *Gw*, *gW*, and *gw*) will be formed with equal probabilities.

In every $F_1 \times F_1$ fertilization event, each zygote has an equal probability of receiving one of the four combinations from each parent. If many offspring are produced, 9/16 have yellow, round seeds, 3/16 have yellow, wrinkled seeds, 3/16 have green, round seeds, and 1/16 have green, wrinkled seeds, yielding what is designated as **Mendel's 9:3:3:1 dihybrid ratio**. This is an ideal ratio based on probability events involving segregation, independent assortment, and random fertilization. Because of deviation due strictly to chance, particularly if small numbers of offspring are produced, actual results seldom match the ideal ratio.

The Test Cross: Two Characters

The test cross can also be applied to individuals that express two dominant traits but whose genotypes are unknown. For example, the expression of the yellow, round seed phenotype in the F_2 generation just described may result from the *GGWW*, *GGWw*, *GgWW*, and *GgWw* genotypes. If an F_2 yellow, round plant is crossed with a homozygous recessive green, wrinkled plant (*ggww*), analysis of the offspring will indicate the actual genotype of that yellow, round plant. Each of the above genotypes results in a different set of gametes, and in a testcross, a different set of phenotypes in the resulting offspring. You should work out the results of each of these four crosses to be sure you understand this concept.

3.4 The Trihybrid Cross Demonstrates That Mendel's Principles Apply to Inheritance of Multiple Traits

Thus far, we have considered inheritance by individuals of up to two pairs of contrasting traits. Mendel demonstrated that the identical processes of segregation and independent assortment apply to three pairs of contrasting traits in what is called a **trihybrid cross**, or *three-factor cross*.

Although a trihybrid cross is somewhat more complex than a dihybrid cross, its results are easily calculated if the principles of segregation and independent assortment are followed. For example, consider the cross shown in Figure 3–8, where the gene pairs of theoretical contrasting traits are represented by the symbols *A*, *a*, *B*, *b*, *C*, and *c*. In the cross between *AABBCC* and *aabbcc* individuals, all F_1 individuals are heterozygous for all three gene pairs. Their genotype, *AaBbCc*, results in the phenotypic expression of the dominant *A*, *B*, and *C* traits. When F_1 individuals serve as parents, each produces eight different gametes in equal frequencies. At this point, we could construct a Punnett square with 64 separate boxes and read out the phenotypes—but such a method is cumbersome in a cross involving so many factors. Therefore another method has been devised to calculate the predicted ratio.

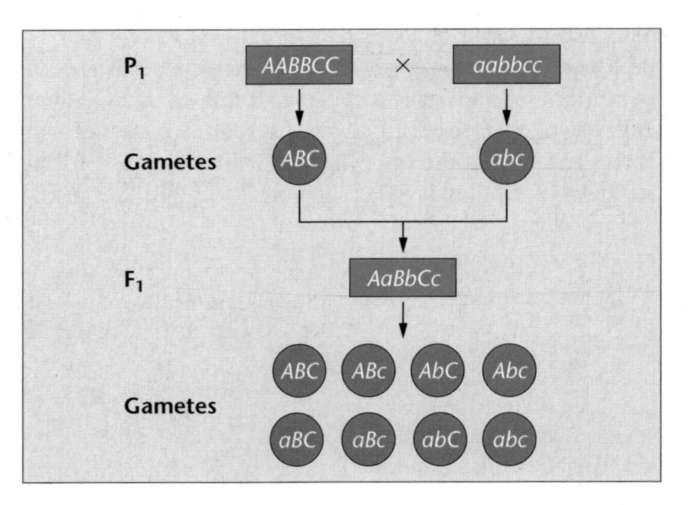

FIGURE 3–8 Formation of P_1 and F_1 gametes in a trihybrid cross.

The Forked-Line Method

It is much less difficult to consider each contrasting pair of traits separately and then to combine these results by using the **forked-line method**, first shown in Figure 3–6. This method, also called a **branch diagram**, relies on the simple application of the laws of probability established for the dihybrid cross. Each gene pair is assumed to behave independently during gamete formation.

When the monohybrid cross $AA \times aa$ is made, we know that:

1. All F_1 individuals have the genotype *Aa* and express the phenotype represented by the *A* allele, which is called the A phenotype in the following discussion.

2. The F_2 generation consists of individuals with either the A phenotype or the a phenotype in the ratio of 3:1.

The same generalizations can be made for the $BB \times bb$ and $CC \times cc$ crosses. Thus, in the F_2 generation, 3/4 of all organisms express phenotype A, 3/4 express B, and 3/4 express C. Similarly, 1/4 of all organisms express phenotype a, 1/4 express b, and 1/4 express c. The proportions of organisms that express each phenotypic combination can be predicted by assuming that fertilization, following the independent assortment of these three gene pairs during gamete formation, is a random process—we simply apply the product law of probabilities once again. Figure 3–9 uses the forked-line method to calculate the phenotypic proportions of the F_2 generation. They fall into the trihybrid ratio of 27:9:9:9:3:3:3:1. The same method can be used to solve crosses involving any number of gene pairs, *provided that all gene pairs assort independently of each other*. We shall see later that this is not always the case. However, it appeared to be true for all of Mendel's characters.

Generation of F_2 trihybrid phenotypes

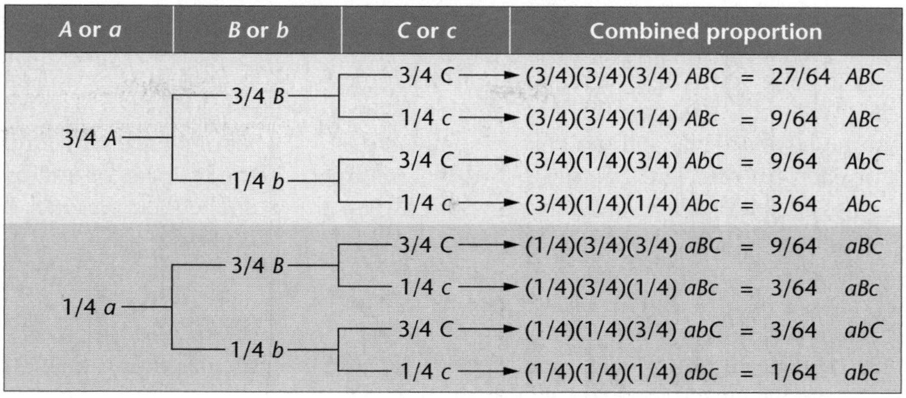

A or a	B or b	C or c	Combined proportion
3/4 *A*	3/4 *B*	3/4 *C*	(3/4)(3/4)(3/4) *ABC* = 27/64 *ABC*
		1/4 *c*	(3/4)(3/4)(1/4) *ABc* = 9/64 *ABc*
	1/4 *b*	3/4 *C*	(3/4)(1/4)(3/4) *AbC* = 9/64 *AbC*
		1/4 *c*	(3/4)(1/4)(1/4) *Abc* = 3/64 *Abc*
1/4 *a*	3/4 *B*	3/4 *C*	(1/4)(3/4)(3/4) *aBC* = 9/64 *aBC*
		1/4 *c*	(1/4)(3/4)(1/4) *aBc* = 3/64 *aBc*
	1/4 *b*	3/4 *C*	(1/4)(1/4)(3/4) *abC* = 3/64 *abC*
		1/4 *c*	(1/4)(1/4)(1/4) *abc* = 1/64 *abc*

FIGURE 3–9 Generation of the F_2 trihybrid phenotypic ratio, using the forked-line method. This method is based on the expected probability of occurrence of each phenotype.

3.5 Mendel's Work Was Rediscovered in the Early Twentieth Century

Mendel's work, initiated in 1856, was presented to the Brünn Society of Natural Science in 1865 and published the following year. However, while his findings were often cited and discussed, their significance went unappreciated for about 35 years. Many reasons have been suggested to explain why the significance of his research was not immediately recognized.

Mendel's adherence to mathematical analysis of probability events was an unusual approach in those days for biological studies. Perhaps his approach seemed foreign to his contemporaries. More important, his conclusions did not fit well with existing theories on the cause of variation among organisms. The source of natural variation intrigued students of evolutionary theory. These individuals, stimulated by the proposal developed by Charles Darwin and Alfred Russel Wallace, believed in **continuous variation**, whereby offspring were a *blend* of their parents' phenotypes. As we mentioned earlier, Mendel theorized that variation was due to discrete or particulate units, resulting in **discontinuous variation**. For example, Mendel proposed that the F_2 offspring of a dihybrid cross are expressing traits produced by new combinations of previously existing unit factors. As a result, Mendel's theories did not fit well with the evolutionists' preconceptions about causes of variation.

In the latter part of the nineteenth century, a remarkable observation set the scene for the rebirth of Mendel's work: Walther Flemming's discovery of chromosomes in the nuclei of salamander cells. In 1879, Flemming described the behavior of these threadlike structures during cell division. As a result of his findings and the work of many other cytologists, the presence of discrete units within the nucleus soon became an integral part of ideas about inheritance. It was this mind-set that prompted scientists to reexamine Mendel's findings.

In the early twentieth century, research led to renewed interest in Mendel's work. Hybridization experiments similar to Mendel's were performed independently by three botanists: Hugo de Vries, Karl Correns, and Erich Tschermak. De Vries's work demonstrated the principle of segregation in his experiments with several plant species. Apparently, he searched the existing literature and found that Mendel's work anticipated his own conclusions! Correns and Tschermak also reached conclusions similar to those of Mendel.

In 1902, two cytologists, Walter Sutton and Theodor Boveri, independently published papers linking their discoveries of the behavior of chromosomes during meiosis to the Mendelian principles of segregation and independent assortment. They pointed out that the separation of chromosomes during meiosis could serve as the cytological basis of these two postulates. Although they thought Mendel's unit factors were probably chromosomes rather than genes on chromosomes, their findings also reestablished the importance of Mendel's work, which became the basis of ensuing genetic investigations. Sutton and Boveri are credited with initiating the **chromosomal theory of inheritance**, which was developed during the next two decades.

Unit Factors, Genes, and Homologous Chromosomes

Because the correlation between Sutton's and Boveri's observations and Mendelian principles serves as the foundation for the modern interpretation of transmission genetics, we will examine this correlation in some depth before moving on to other topics.

As we know, each species possesses a specific number of chromosomes in each somatic cell nucleus (except in gametes). For diploid organisms, this number is called the **diploid number** (**2*n***) and is characteristic of that species. During the formation of gametes, this number is precisely halved (n), and when two gametes combine during fertilization, the diploid number is reestablished. During meiosis, however, the chromosome number is not reduced in a random manner. It was apparent to early cytologists that the diploid number of chromosomes is composed of homologous pairs identifiable by their morphological appearance and behavior. The gametes contain one member of each pair—thus the chromosome complement of a gamete is quite specific, and the number of chromosomes in each gamete is equal to the haploid number.

With this basic information, we can see the correlation between the behavior of unit factors and chromosomes and genes. Figure 3–10 shows three of Mendel's postulates (in the left column) and the chromosomal explanation of each (in the right column). Unit factors are really genes located on homologous pairs of chromosomes (Figure 3–10a). Members of each pair of homologs separate, or segregate, during gamete formation (Figure 3–10b). Two different alignments are possible, both of which are shown.

To illustrate the principle of independent assortment, we must distinguish between members of any given homologous pair of chromosomes. One member of each pair comes from the **maternal parent**, while the other member comes from the **paternal parent**. (We represent the different parental origins with different colors). As shown in Figure 3–10(c), following independent segregation of each pair of homologs, each gamete receives one member from each pair of chromosomes. All possible combinations are formed with equal probability. If we add the symbols used in Mendel's dihybrid cross (*G*, *g* and *W*, *w*) to the diagram, we can see why equal numbers of the four types of gametes are formed. The independent behavior of Mendel's pairs of unit factors (*G* and *W* in this example) is due to their presence on separate pairs of homologous chromosomes.

Observations of the phenotypic diversity of living organisms make it logical to assume that there are many more genes than chromosomes. Therefore, each homolog must carry genetic information for more than one trait. The currently accepted concept is that a chromosome is composed of a large number of linearly ordered, information-containing *genes*. Mendel's unit factors (which determine tall or dwarf stems, for example) actually constitute a pair of genes located on one pair of homologous chromosomes. The location on a given chromosome where any particu-

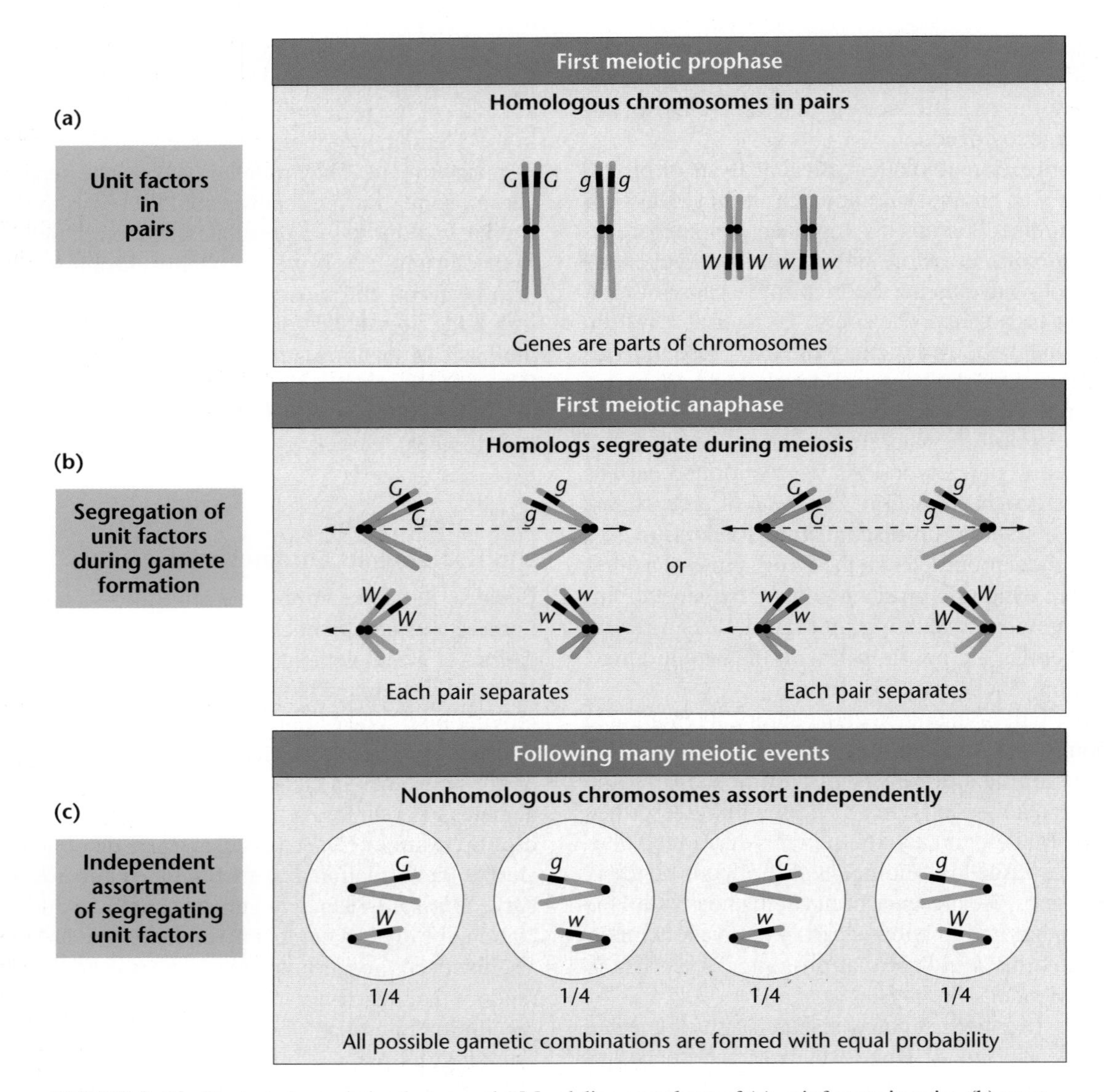

FIGURE 3–10 Illustrated correlation between the Mendelian postulates of (a) unit factors in pairs, (b) segregation, and (c) independent assortment, showing the presence of genes located on homologous chromosomes and their behavior during meiosis.

lar gene occurs is called its **locus** (pl., loci). The different forms taken by a given gene, called *alleles* (G or g), contain slightly different genetic information that determines the same character (seed color in this case). Although we have examined only genes with two alternative alleles, most genes have more than two allelic forms. We conclude this section by reviewing the criteria necessary to classify two chromosomes as a homologous pair:

1. During mitosis and meiosis, when chromosomes are visible as distinct figures, both members of a homologous pair are the same size and exhibit identical centromere locations.
2. During early stages of meiosis, homologous chromosomes form pairs, or synapse.
3. Although not generally microscopically visible, homologs contain identical, linearly ordered gene loci.

3.6 Independent Assortment Leads to Extensive Genetic Variation

One major consequence of independent assortment is the production by an individual of genetically dissimilar gametes. Genetic variation results because the two members of any homologous pair of chromosomes are rarely, if ever, genetically identical. Therefore, because independent assortment leads to the production of all possible chromosome combinations, extensive genetic diversity results.

We have seen that the number of possible gametes, each with different chromosome compositions, is 2^n, where n equals the haploid number. Thus, if a species has haploid number of $n = 4$, then $2^4 = 16$ different gamete combinations can be formed as a result of independent assortment. Although this number is not high, consider the human species, where $n = 23$. If 2^{23} is calculated, we find that in excess of 8×10^6, or over 8 million, different types of gametes are represented. Because fertilization represents an event involving only one of approximately 8×10^6 possible gametes from each of two parents, each offspring represents only one of $(8 \times 10^6)^2$, or 64×10^{12} potential genetic combinations! No wonder that, except for identical twins, each member of the human species demonstrates a distinctive appearance and individuality—this number of combinations is far greater than the number of humans who have ever lived on earth! Genetic variation resulting from independent assortment has been extremely important to the process of evolution in all organisms.

3.7 Laws of Probability Help to Explain Genetic Events

Recall that genetic ratios are expressed as probabilities—for example, 3/4 tall : 1/4 dwarf. These values predict the outcome of each fertilization event, such that the probability of each zygote having the genetic potential for becoming tall is 3/4, while the potential for becoming dwarf is 1/4. Probabilities range from 0.0, when an event is *certain not to occur*, to 1.0, when an event is *certain to occur*. When two or more events occur independently but at the same time, we can calculate the probability of possible outcomes when they occur together. This is accomplished by applying the *product law*—the probability of two or more events occurring simultaneously is equal to the product of their individual probabilities. Two or more events are independent of one another if the outcome of each one does not affect the outcome of any of the others under consideration.

To illustrate the product law, consider the possible results if you toss a penny (P) and a nickel (N) at the same time and examine all combinations of heads (H) and tails (T) that can occur. There are four possible outcomes:

$$(P_H : N_H) = (1/2)(1/2) = 1/4$$
$$(P_T : N_H) = (1/2)(1/2) = 1/4$$
$$(P_H : N_T) = (1/2)(1/2) = 1/4$$
$$(P_T : N_T) = (1/2)(1/2) = 1/4$$

How Mendel's Peas Become Wrinkled: A Molecular Explanation

Only recently, well over a hundred years after Mendel used wrinkled peas in his groundbreaking hybridization experiments, have we come to find out how the *wrinkled* gene makes peas wrinkled. The wild-type allele of the gene encodes a protein called **starch-branching enzyme (*SBEI*)**. This enzyme catalyzes the formation of highly branched starch molecules as the seed matures.

Wrinkled peas, which result from the homozygous presence of the mutant form of the gene, lack the activity of this enzyme. The production of branch points is inhibited during the synthesis of starch within the seed, which in turn leads to the accumulation of more sucrose and a higher water content while the seed develops. Osmotic pressure inside rises, causing the seed to lose water internally, and ultimately results in the wrinkled appearance of the seed at maturation. In contrast, developing seeds that bear at least one copy of the normal gene (being either homozygous or heterozygous for the dominant allele) synthesize starch and reach an osmotic balance that minimizes the loss of water. The end result is a smooth-textured outer coat.

The *SBEI* gene has been cloned and analyzed, and study of the *SBEI* gene provides greater insight into the relationship between genotypes and phenotypes. Interestingly, the mutant gene contains a foreign sequence of some 800 base pairs that disrupts the normal coding sequence. This foreign segment closely resembles other such sequences, called **transposable elements**; these sequences have the ability to move from place to place in the genome of organisms. Transposable elements have been found in maize (corn), parsley, and snapdragons, fruit flies, among many other organisms.

MEDIA TUTORIAL Probability

The probability of obtaining a head or a tail in the toss of either coin is 1/2 and is unrelated to the outcome of the toss of the other coin. Thus, all four possible combinations are predicted to occur with equal probability.

If we want to calculate the probability where the possible outcomes of two events are independent of one another but can be accomplished in more than one way, we apply the **sum law**. For example, what is the probability of tossing our penny and nickel and obtaining one head and one tail? In such a case, we don't care whether it is the penny or the nickel that comes up heads, provided the other coin has the alternative outcome. As we saw above, there are two ways in which the desired outcome can be accomplished, each with a probability of 1/4. Thus, according to the sum law, the overall probability is equal to

$$(1/4) + (1/4) = 1/2.$$

One-half of all coin tosses are predicted to yield the desired outcome.

These simple probability laws will be useful throughout our discussions of transmission genetics and for solving genetics problems. In fact, we already applied the product law when we used the forked-line method to calculate the phenotypic results of Mendel's dihybrid and trihybrid crosses. When we wish to know the results of a cross, we need only calculate the probability of each possible outcome. The results of this calculation then allow us to predict the proportion of offspring expressing each phenotype or each genotype.

An important point to remember when you deal with probability is that predictions of possible outcomes are based on large sample sizes. If we predict that 9/16 of the offspring of a dihybrid cross will express both dominant traits, it is very unlikely that, in a small sample, exactly 9 of every 16 offspring will express this phenotype. Instead, our prediction is that, of a large number of offspring, approximately 9/16 of them will do so. The deviation from the predicted ratio in smaller sample sizes is attributed to chance, a subject we examine in our discussion of statistics in the next section. As you shall see, the impact of deviation due strictly to chance diminishes as the sample size increases.

3.8 Chi-Square Analysis Evaluates the Influence of Chance on Genetic Data

Mendel's 3:1 monohybrid and 9:3:3:1 dihybrid ratios are hypothetical predictions based on the following assumptions: (1) Each allele is dominant or recessive; (2) segregation is operative; (3) independent assortment occurs; and (4) fertilization is random. The last three assumptions are influenced by chance events and are therefore subject to random fluctuation. This concept of **chance deviation** is most easily illustrated by tossing a single coin numerous times and recording the number of heads and tails observed. In each toss, there is a probability of 1/2 that a head will occur and a probability of 1/2 that a tail will occur for each coin toss. Therefore, the expected ratio of many tosses is 1:1. If a coin is tossed 1000 times, usually *about* 500 heads and 500 tails will be observed. Any reasonable fluctuation from this hypothetical ratio (e.g., 486 heads and 514 tails) is attributed to chance.

As the total number of tosses is reduced, the impact of chance deviation increases. For example, if a coin is tossed only four times, you wouldn't be too surprised if all four tosses result in only heads or only tails. For 1000 tosses, however, 1000 heads or 1000 tails would be most unexpected. In fact, you might believe that such a result would be impossible. Actually, all heads or all tails in 1000 tosses can be predicted to occur with a probability of $(1/2)^{1000}$. Since $(1/2)^{20}$ is equivalent to less than one in a million times, an event occurring with a probability of $(1/2)^{1000}$ is virtually impossible. Two major points are significant before we consider *chi-square analysis*:

1. The outcomes of segregation, independent assortment, and fertilization, like coin tossing, are subject to random fluctuations from their predicted occurrences as a result of chance deviation.
2. As the sample size increases, the average deviation from the expected results decreases. Therefore, a larger sample size diminishes the impact of chance deviation on the final outcome.

Chi-Square Calculations and Their Interpretation

In genetics, the ability to evaluate observed deviation is a crucial skill. When we assume that data will fit a given ratio such as 1:1, 3:1, or 9:3:3:1, we establish what is called the **null hypothesis** (H_0). It is so named because the hypothesis assumes that *no real difference* exists between *measured values* (or ratio) and *predicted values* (or ratio). The apparent difference can be attributed purely to chance. The null hypothesis is evaluated using statistical analysis. On this basis, the null hypothesis may either (1) be rejected or (2) fail to be rejected. If it is rejected, the observed deviation from the expected result is not attributed to chance alone. The null hypothesis and the underlying assumptions leading to it must be reexamined. If the null hypothesis fails to be rejected, any observed deviations are attributed to chance.

One of the simplest statistical tests devised to assess the null hypothesis is **chi-square (χ^2) analysis**. This test takes into account the observed deviation in each component of an expected ratio as well as the sample size and reduces them to a single numerical value. The value for χ^2 is then used to estimate how frequently the observed deviation can be expected to occur strictly as a result of chance. The formula for chi-square analysis is

$$\chi^2 = \Sigma \frac{(o - e)^2}{e}$$

where o is the observed value for a given category, e is the expected value for that category, and Σ (the Greek letter sigma) represents the sum of the calculated values for each category of the ratio. Because $(o - e)$ is the deviation (d) in each case, the equation reduces to

$$\chi^2 = \Sigma \frac{d^2}{e}$$

Table 3.1(*a*) shows a χ^2 calculation for the F_2 results of a hypothetical monohybrid cross. To analyze these data, you work from left to right, calculating and entering the appropriate numbers in each column. Regardless of whether the deviation d is positive or negative, d^2 always becomes positive after the number is squared. In Table 3.1(*b*) the F_2 results of a hypothetical dihybrid cross are analyzed—be sure that you understand how each number was calculated in the dihybrid example.

The final step in chi-square analysis is to interpret the χ^2 value. To do so, you must initially determine the value of the **degrees of freedom** (***df***), which is equal to $n - 1$, where n is the number of different categories into which each datum point may fall. For the 3:1 ratio, $n = 2$, so $df = 2 - 1 = 1$. For the 9:3:3:1 ratio, $n = 4$ and $df = 3$. Degrees of freedom must be taken into account because the greater the number of categories, the more deviation is expected as a result of chance.

Once you have determined the degrees of freedom, we can interpret the χ^2 value in terms of a corresponding **probability value** (***p***). Since this calculation is complex, we usually take the p value from a standard table or graph. Figure 3–11 shows a wide range of χ^2 and p values for various degrees of freedom in both a graph and a table. Let's use the graph to determine the p value. The caption for Figure 3–11(b) explains how to use the table.

To determine p, execute the following steps:

1. Locate the χ^2 value on the abscissa (the horizontal or x-axis).
2. Draw a vertical line from this point up to the angled line on the graph representing the appropriate *df*.
3. Extend a horizontal line from this point to the left until it intersects the ordinate (the vertical or y-axis).
4. Estimate, by interpolation, the corresponding p value.

We used these steps for the monohybrid cross in Table 3.1(*a*) to estimate the p value of 0.48 shown in Figure 3–11(a). For the dihybrid cross, try this method to see if you can determine the p value. Since the χ^2 value is 4.16 and $df = 3$, an approximate p value is 0.26. Checking this result in the table confirms that p values for both the monohybrid and dihybrid crosses are between 0.20 and 0.50.

So far, we have been concerned only with finding p. The most important aspect of chi-square analysis is understanding the meaning of the p value. Let's use the example of the dihybrid cross in Table 3.1(*b*) $(p = 0.26)$. In these discussions, it is simplest to think of the p value as a percentage (e.g., 0.26 = 26%). In our example, the p value indicates that if we repeat the same experiment many times, 26% of the trials would be expected to exhibit chance deviation as great or greater than that seen in the initial trial. Conversely, 74% of the trials would show less deviation than initially observed as a result of chance. Thus, the p value reveals that a hypothesis (the 9:3:3:1 ratio in this case) is never proved or disproved absolutely. Instead, a relative standard is set that enables us to either *reject* or *fail to reject* the null hypothesis—this standard is most often a p value of 0.05. When applied to chi-square analysis, a p value less than 0.05 means that the observed deviation in the set of results will be obtained by chance alone less than 5% of the time. Such a p value indicates that the difference between the observed and predicted results is substantial and thus enables us to reject the null hypothesis.

TABLE 3.1 Chi-Square Analysis

(a) Monohybrid Cross

Expected Ratio	*Observed* (o)	*Expected* (e)	*Deviation* $(o - e)$	*Deviation*^{2} (d^2)	d^2/e
3/4	740	3/4(1000) = 750	740 − 750 = −10	$(-10)^2 = 100$	100/750 = 0.13
1/4	260	1/4(1000) = 250	260 − 250 = +10	$(+10)^2 = 100$	100/250 = 0.40
	Total = 1000				$\chi^2 = 0.53$
					$p = 0.48$

(b) Dihybrid Cross

Expected Ratio	(o)	(e)	$(o - e)$	(d^2)	(d^2/e)
9/16	587	567	+20	400	0.71
3/16	197	189	+8	64	0.34
3/16	168	189	−21	441	2.33
1/16	56	63	−7	49	0.78
	Total = 1008				$\chi^2 = 4.16$
					$p = 0.26$

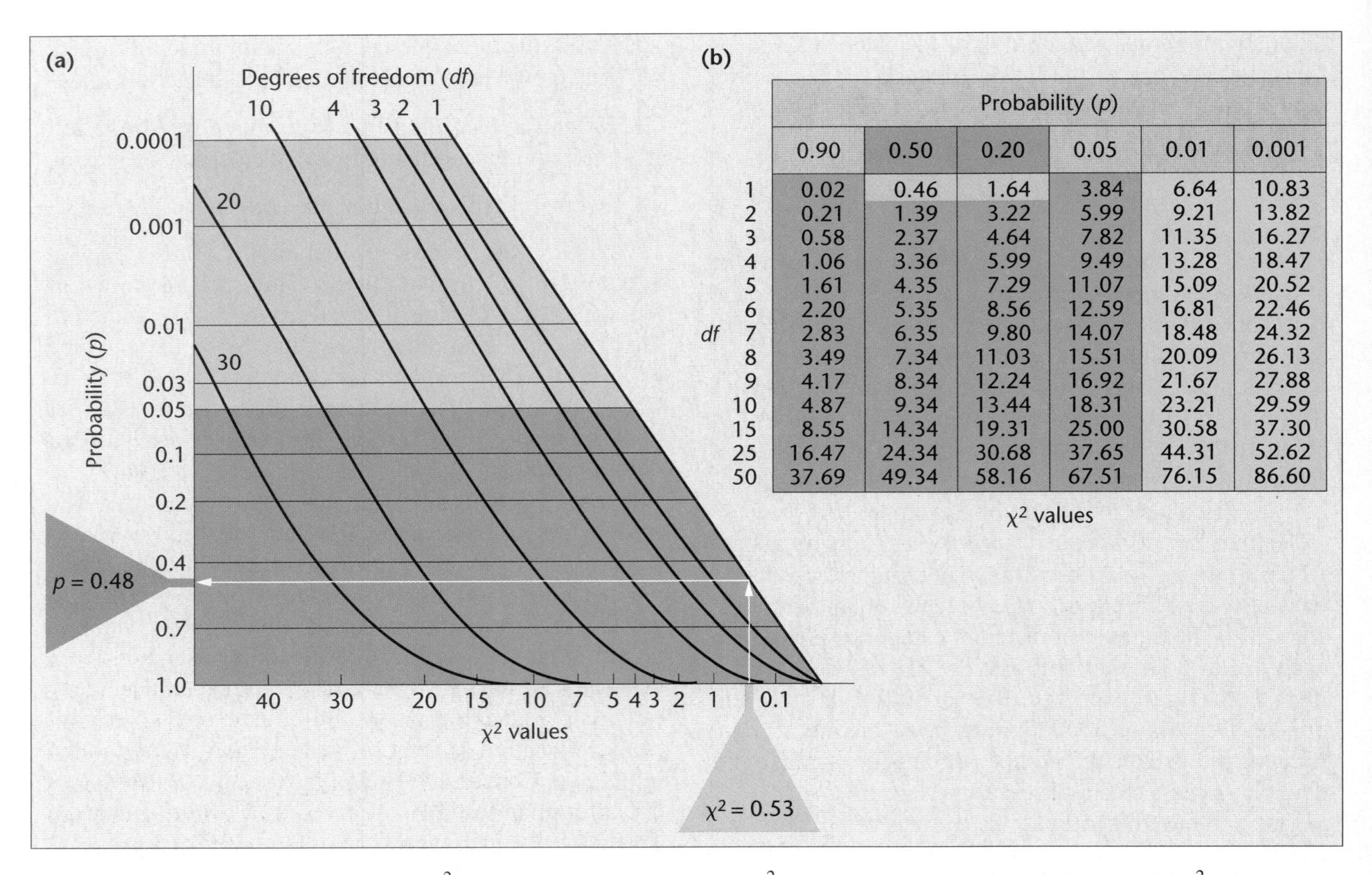

df	0.90	0.50	0.20	0.05	0.01	0.001
1	0.02	0.46	1.64	3.84	6.64	10.83
2	0.21	1.39	3.22	5.99	9.21	13.82
3	0.58	2.37	4.64	7.82	11.35	16.27
4	1.06	3.36	5.99	9.49	13.28	18.47
5	1.61	4.35	7.29	11.07	15.09	20.52
6	2.20	5.35	8.56	12.59	16.81	22.46
7	2.83	6.35	9.80	14.07	18.48	24.32
8	3.49	7.34	11.03	15.51	20.09	26.13
9	4.17	8.34	12.24	16.92	21.67	27.88
10	4.87	9.34	13.44	18.31	23.21	29.59
15	8.55	14.34	19.31	25.00	30.58	37.30
25	16.47	24.34	30.68	37.65	44.31	52.62
50	37.69	49.34	58.16	67.51	76.15	86.60

Probability (p) (columns); χ^2 values (cells)

FIGURE 3–11 (a) Graph for converting χ^2 values to p values. (b) Table of χ^2 values for selected values of df and p. The χ^2 values greater than that shown at $p = 0.05$ (darker blue areas) justify failure to reject the null hypothesis. Values less than those at $p = 0.05$ (lighter blue areas) justify rejecting the null hypothesis. For example, using the table in part (b), where $\chi^2 = 0.53$ for 1 degree of freedom, the corresponding p value is between 0.20 and 0.50. The graph in (a) gives a more precise p value of 0.48 by interpolation. Thus, we fail to reject the null hypothesis.

On the other hand, p values of 0.05 or greater (0.05 to 1.0) indicate that the observed deviation will be obtained by chance alone 5% or more of the time. The conclusion is not to reject the null hypothesis. Thus, for the p value of 0.26, assessing the hypothesis that independent assortment accounts for the results fails to be rejected. Therefore, the observed deviation can be reasonably attributed to chance.

A final note is relevant here for the case where the null hypothesis is rejected, that is, where $p \leq 0.05$. Suppose we are testing the null hypothesis that the data represented a 9:3:3:1 ratio, indicative of independent assortment. If the null hypothesis is rejected, what are alternative interpretations of the data? Researchers will reassess the assumptions that underlie the null hypothesis. In our example, we assumed that segregation operates faithfully for both gene pairs. We also assumed that fertilization is random and that the viability of all gametes is equal irrespective of genotype—that is, that all gametes are equally likely to participate in fertilization. Finally, following fertilization, we assumed that all preadult stages and adult offspring are equally viable regardless of their genotype. If any of these assumptions is incorrect, the original hypothesis is not necessarily invalid.

An example will clarify this: Suppose our null hypothesis is that a dihybrid cross between fruit flies will result in 3/16 mutant wingless fly zygotes. However, not as many of the mutant embryos may survive their preadult development or as young adults, compared to flies whose genotype gives rise to wings. As a result, when the data are gathered, there are fewer than 3/16 wingless flies. Rejection of the null hypothesis alone is not cause for us to disregard the validity of the postulates of segregation and independent assortment, because other factors are operative.

3.9 Pedigrees Reveal Patterns of Inheritance in Humans

In the crosses discussed so far, one of the two traits for each character has been dominant to the other. Based on this observation, two significant questions arise:

1. Does the expression of all genes occur in this fashion?
2. Is it possible to ascertain the mode of inheritance of genes in organisms such as humans, where designed crosses and the production of large numbers of offspring are not practical?

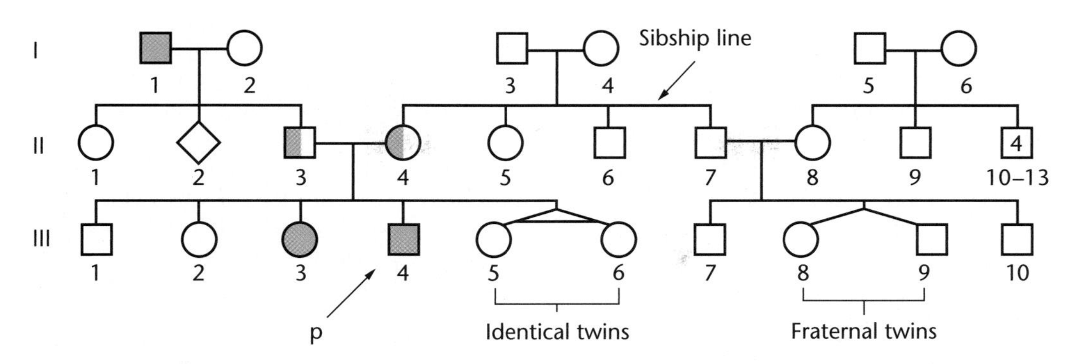

FIGURE 3–12 A representative pedigree for a single characteristic through three generations.

The answer to the first question is no. Many modes of inheritance exist that modify the monohybrid and dihybrid ratios observed by Mendel.

The answer to the second question is yes. The pattern of inheritance of a specific phenotype can be studied, even in humans. The simplest way to study this pattern is to construct a family tree that shows the phenotype of the trait in question for each member. Such a family tree is called a **pedigree**. By analyzing the pedigree, we may be able to predict how the gene controlling the trait is inherited. If many similar pedigrees for the same trait are found, our prediction is strengthened.

Figure 3–12 shows the conventions we use to construct pedigrees. Circles represent females, and squares designate males. Parents are connected by a horizontal line with vertical lines leading to their offspring. All such offspring are called **sibs** and are connected by a horizontal **sibship line**. Sibs are placed from left to right according to birth order and are labeled with Arabic numerals. Each generation is indicated by a Roman numeral. If the sex is unknown, a diamond is used (see II-2). If a pedigree traces only a single trait, as Figure 3–12 does, the circles, squares, and diamonds are shaded if the phenotype being considered is expressed. Individuals who fail to express a recessive trait, when known with certainty to be heterozygous, have only the left half of their square or circle shaded (see II-3 and II-4).

In the special case of twins, diagonal lines are drawn from the vertical line and connected to the sibship line. For **monozygotic (identical) twins**, the diagonal lines are linked by a horizontal line (III-5 and 6). **Dizygotic (fraternal) twins** lack this connecting line (III-8 and 9). A number within a symbol (II-10–13) represents numerous sibs of the same or unknown phenotypes. The individual whose phenotype drew the attention of a physician or geneticist is called the **proband** and is indicated by an arrow connected to the designation **p** (III-4).

The pedigree shown in Figure 3–12 traces the pattern of inheritance of the human trait **albinism** (the inability to produce the pigment melanin). By analyzing the pedigree, we can see that albinism is a recessive trait. Here we can see that the male parent of the first generation (I-1) is affected, and since none of his offspring show the disorder, we can conclude that the unaffected female parent (I-2) was a homozygous normal individual. Had she been heterozygous, one-half of the offspring would be expected to exhibit albinism. However, such a small sample (three offspring) prevents any certainty in the matter.

An unaffected second generation is characteristic of a rare recessive trait. If albinism were inherited as a dominant trait, individual II-3 would have to express the disorder in order to pass it to his offspring (III-3 and III-4). He does not. Inspection of the offspring constituting the third generation (row III) provides further support for the hypothesis that albinism is a recessive trait. If so, parents II-3 and II-4 must both be heterozygous, and approximately one-fourth of their offspring should be affected—two of the six offspring do show albinism. Note that this deviation from the expected ratio is not unexpected in crosses with few offspring.

Pedigree analysis of many traits has been an extremely valuable research technique in human genetic studies. However, this approach does not usually provide the certainty in drawing conclusions that is afforded by designed crosses yielding large numbers of offspring. Nevertheless, when many independent pedigrees of the same trait or disorder are analyzed, consistent conclusions can often be drawn. Table 3.2 lists numerous human traits and classifies them according to their recessive or dominant expression. We shall see in Chapter 4 that the genes controlling some of these traits are located on the sex-determining chromosomes.

TABLE 3.2 Representative Recessive and Dominant Human Traits

Recessive Traits	*Dominant Traits*
Albinism	Achondroplasia
Alkaptonuria	Brachydactyly
Ataxia telangiectasia	Congenital stationary night blindness
Color blindness	Ehler-Danlos syndrome
Cystic fibrosis	Hypotrichosis
Duchenne muscular dystrophy	Huntington disease
Galactosemia	Hypercholesterolemia
Hemophilia	Marfan syndrome
Lesch-Nyhan syndrome	Neurofibromatosis
Phenylketonuria	Phenylthiocarbamide tasting (PTC)
Sickle-cell anemia	Porphyria
Tay-Sachs disease	Widow's peak

Chapter Summary

1. Over a century ago, Mendel studied inheritance patterns in the garden pea and established the principles of transmission genetics.
2. Mendel's postulates help describe the basis for the inheritance of phenotypic expression. He showed that unit factors, later called alleles, exist in pairs and exhibit a dominant/recessive relationship in determining the expression of traits.
3. Mendel postulated that unit factors must segregate during gamete formation, such that each gamete receives only one of the two factors with equal probability.
4. Mendel's postulate of independent assortment states that each pair of unit factors segregates independently of other such pairs. As a result, all possible combinations of gametes are formed with equal probability.
5. The discovery of chromosomes in the late 1800s, along with subsequent studies of their behavior during meiosis, led to the rebirth of Mendel's work, linking the behavior of his unit factors to that of chromosomes during meiosis.
6. The Punnett square and the forked-line methods are used to predict the probabilities of phenotypes (and genotypes) from crosses involving two or more gene pairs.
7. Genetic ratios are expressed as probabilities. Thus, deriving outcomes of genetic crosses requires an understanding of the laws of probability.
8. Statistical analysis is used to test the validity of experimental outcomes. In genetics, variations from the expected ratios due to chance deviations can be anticipated.
9. Chi-square analysis allows us to assess the null hypothesis, which states that there is no real difference between the expected and observed values. As such, it tests the probability of whether observed variations can be attributed to chance deviation.
10. Pedigree analysis is a method for studying the inheritance pattern of human traits over several generations. It frequently provides the basis for determining the mode of inheritance of human characteristics and disorders.

Key Terms

albinism, 53
allele, 41
branch diagram, 46
chance deviation, 50
chi-square (χ^2) analysis, 50
chromosomal theory of inheritance, 47
continuous variation, 47
degrees of freedom (df), 51
dihybrid cross, 43
discontinuous variation, 47
diploid number ($2n$), 47
dizygotic twins, 53
dominant, 41
F_1 generation (first filial), 40
forked-line method, 46
fraternal twins, 53
F_2 generation (second filial), 40
gene, 41
genotype, 42
heterozygote, 42
homozygote, 42
identical twins, 53
independent assortment, 44
locus, 49
maternal parent, 48
Mendel's 9:3:3:1 dihybrid ratio, 46
monohybrid cross, 40
monozygotic twins, 53
null hypothesis, 50
paternal parent, 48
pedigree, 53
phenotype, 41
P_1 generation (parental), 40
probability value, 51
proband, 53
product law, 44, 49
Punnett square, 42
recessive, 41
reciprocal cross, 41
segregation, 41
self-fertilizing, 40
sib, 53
sibship line, 53
starch-branching enzyme (*SBEI*), 49
sum law, 50
testcross, 42
traits, 39
transmission genetics, 39
transposable elements, 49
trihybrid cross, 46
unit factors, 41
unit of inheritance, 39

INSIGHTS AND SOLUTIONS

As a student, you will be asked to demonstrate your knowledge of transmission genetics by solving genetics problems. Success at this task represents not only comprehension of theory but its application to more practical genetic situations. Most students find problem solving in genetics to be challenging but rewarding. This section will provide you with basic insights into the reasoning essential to this process.

Genetics problems are in many ways similar to word problems in algebra. The approach to solving them is identical: (1) Analyze the problem carefully; (2) translate words into symbols, first defining each one; and (3) choose and apply a specific technique to solve the problem. The first two steps are critical. The third step is largely mechanical.

The simplest problems state all necessary information about the P_1 generation and ask you to find the expected ratios of the F_1 and F_2 genotypes and/or phenotypes. Always follow these steps when you encounter this type of problem:

1. Determine insofar as possible the genotypes of the individuals in the P_1 generation.

2. Determine what gametes may be formed by the P_1 parents.

3. Recombine gametes by the Punnett square or the forked-line methods, or if the situation is very simple, by inspection. Read the F_1 phenotypes.

4. Repeat the process to obtain information about the F_2 generation.

Determining the genotypes from the given information requires that you understand the basic theory of transmission genetics. Consider this problem: *A recessive mutant allele, black, causes a very dark body in Drosophila (a fruit fly) when homozygous. The wild-type (normal) color is gray. What F_1 phenotypic ratio* *is predicted when a black female is crossed with a gray male whose father was black?*

To work out this problem, you must understand dominance and recessiveness, as well as the principle of segregation. Furthermore, you must use the information about the male parent's father. Here is one way to solve this problem:

1. The female parent is black, so she must be homozygous for the mutant allele (bb).

2. The male parent is gray; therefore, he must have at least one dominant allele (B). His father was black (bb), and he received one of the chromosomes bearing these alleles, so the male parent must be heterozygous (Bb).

With this information, the problem is simple:

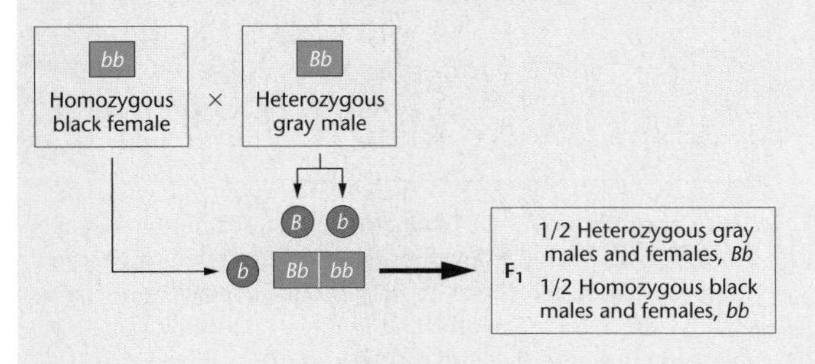

Apply this approach to the following problems.

1. In Mendel's work, he found that full pods are dominant over constricted pods while round seeds are dominant over wrinkled seeds. One of his crosses was between full, round plants and constricted, wrinkled plants. From this cross, he obtained an F_1 generation that was all full and round. In the F_2 generation, Mendel obtained his classic 9:3:3:1 ratio. Using this information, determine the expected F_1 and F_2 results of a cross between homozygous constricted, round plants and full, wrinkled plants.

Solution: Define gene symbols for each pair of contrasting traits. Use the lowercase first letter of the recessive traits to designate those phenotypes and the uppercase first letter to designate the dominant traits. Thus, C and c indicate full and constricted, and W and w indicate round and wrinkled phenotypes, respectively.

Determine the genotypes of the P_1 generation, form the gametes, reconstitute the F_1 generation, and read off the phenotype(s):

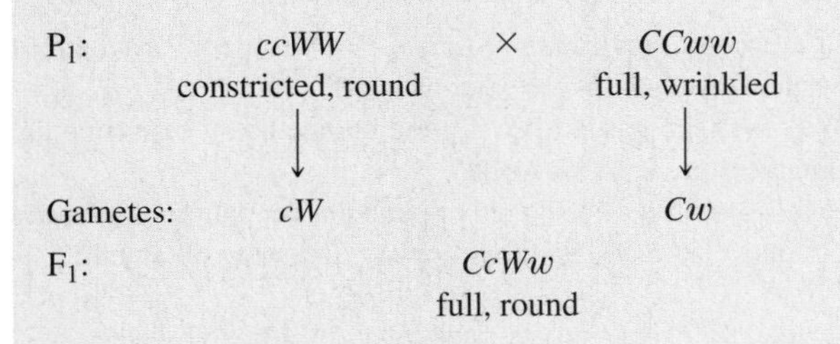

You can immediately see that the F_1 generation expresses both dominant phenotypes and is heterozygous for both gene pairs. Thus, you expect that the F_2 generation will yield the classic Mendelian ratio of 9:3:3:1. Let's work it out anyway just to confirm this, using the forked-line method. Both gene pairs are heterozygous and can be expected to assort independently, so we can predict the F_2 outcomes from each gene pair separately and then proceed with the forked-line method.

Every F_2 offspring is subject to the following probabilities:

$Cc \times Cc$ ↓ : CC, Cc, cC } full; cc constricted

$Ww \times Ww$ ↓ : WW, Ww, wW } round; ww wrinkled

The forked-line method then confirms the 9:3:3:1 phenotypic ratio. Remember that this represents proportions of 9/16 : 3/16 : 3/16 : 1/16. Note that we are applying the product law as we compute the final probabilities:

3/4 full
- 3/4 round $\xrightarrow{(3/4)(3/4)}$ 9/16 full, round
- 1/4 wrinkled $\xrightarrow{(3/4)(1/4)}$ 3/16 full, wrinkled

1/4 constricted
- 3/4 round $\xrightarrow{(1/4)(3/4)}$ 3/16 constricted, round
- 1/4 wrinkled $\xrightarrow{(1/4)(1/4)}$ 1/16 constricted, wrinkled

2. In another cross involving parent plants of unknown genotype and phenotype, the following offspring were obtained.

F_1: 3/8 full, round
3/8 full, wrinkled
1/8 constricted, round
1/8 constricted, wrinkled

Determine the genotypes and phenotypes of the parents.

Solution: This problem is more difficult and requires keener insight because you must work backward. The best approach is to consider the outcomes of pod shape separately from those of seed texture.

Of all the plants, $3/8 + 3/8 = 3/4$ are full and $1/8 + 1/8 = 1/4$ are constricted. Of the various genotypic combinations that can serve as parents, which combination will give rise to a ratio of 3/4 : 1/4? This ratio is identical to Mendel's monohybrid F_2 results, and we can propose that both unknown parents share the same genetic characteristic as the monohybrid F_1 parents; they must both be heterozygous for the genes controlling pod shape and thus are Cc.

Before we accept this hypothesis, let's consider the possible genotypic combinations that control seed texture. If we consider this characteristic alone, we see that the traits are expressed in a ratio of $3/8 + 1/8 = 1/2$ round: $3/8 + 1/8 = 1/2$ wrinkled. To generate such a ratio, the parents cannot both be heterozygous, or their offspring would yield a 3/4 : 1/4 phenotypic ratio. They cannot both be homozygous, or all of their offspring would express a single phenotype. Thus, we are left with testing the hypothesis that one parent is homozygous and one is heterozygous for the alleles controlling texture. The potential case of $WW \times Ww$ does not work, since it yields only a single phenotype. This leaves us with the potential case of $Ww \times ww$. Offspring in such a mating will yield 1/2 Ww (round) : 1/2 ww (wrinkled), exactly the outcome we are seeking.

Now, let's combine the hypotheses and predict the outcome of the cross. In our solution, we use a dash (–) to indicate that the second allele may be either dominant or recessive, since we are only predicting phenotypes.

$3/4\ C-$
- $1/2\ Ww \longrightarrow 3/8\ C{-}Ww$ full, round
- $1/2\ ww \longrightarrow 3/8\ C{-}ww$ full, wrinkled

$1/4\ cc$
- $1/2\ Ww \longrightarrow 1/8\ ccWw$ constricted, round
- $1/2\ ww \longrightarrow 1/8\ ccww$ constricted, wrinkled

As you can see, this cross produces offspring according to our initial information, and we have solved the problem. Note that in this solution, we used genotypes in the forked-line method, in contrast to the use of phenotypes in the earlier solution.

3. Determine the probability that a plant of genotype *CcWw* will be produced from parental plants with the genotypes *CcWw* and *Ccww*.

Solution: The two gene pairs demonstrate straightforward dominance and recessiveness and assort independently during gamete formation. We need only calculate the individual probabilities of obtaining the two separate outcomes (*Cc* and *Ww*) and apply the product law to calculate the final probability:

$$Cc \times Cc \longrightarrow 1/4\ CC : 1/2\ Cc : 1/4\ cc$$

$$Ww \times ww \longrightarrow 1/2\ Ww : 1/2\ ww$$

$$p = (1/2\ Cc)(1/2\ Ww) = 1/4\ CcWw$$

4. In the laboratory, a genetics student crossed flies that had normal, long wings with flies expressing the *dumpy* mutation (truncated wings), which she believed was a recessive trait. In the F_1 generation, all flies had long wings. The following results were obtained in the F_2 generation:

792 long-winged flies

208 dumpy-winged flies

The student tested the hypothesis that the dumpy wing is inherited as a recessive trait, using chi-square analysis of the F_2 data.

(a) What ratio was hypothesized?

(b) Did the analysis support the hypothesis?

(c) What do the data suggest about the *dumpy* mutation?

Solution:

(a) The student hypothesized that the F_2 data (792 : 208) fit Mendel's 3 : 1 monohybrid ratio for recessive genes.

(b) The initial step in χ^2 analysis is to calculate the expected results (e) if the ratio is 3 : 1. Then we can compute deviation $o - e$ (d) and the remaining numbers.

Ratio	*o*	*e*	*d* (*o* − *e*)	d^2	d^2/e
3/4	792	750	42	1764	2.35
1/4	208	250	−42	1764	7.06
Total =	1000				

$$\chi^2 = \Sigma \frac{d^2}{e}$$
$$= 2.35 + 7.06$$
$$= 9.41$$

We consult Figure 3–11 to determine the probability (p) and determine whether the deviations can be attributed to chance. There are two possible outcomes (n), so the degrees of freedom (df) $= n - 1$ or 1. The table in Figure 3–11(b) shows that p is a value between 0.01 and 0.001; the graph in Figure 3–11(a) gives an estimate of about 0.001. Since $p < 0.05$, we reject the null hypothesis. The data do not fit a 3:1 ratio.

(c) When we accepted Mendel's 3:1 ratio as a valid expression of the monohybrid cross, numerous assumptions were made. Examining our underlying assumptions may explain why the null hypothesis was rejected. We assumed that all genotypes are equally viable—that genotypes yielding long wings are equally likely to survive from fertilization through adulthood as the genotype yielding dumpy wings. Further study may reveal that dumpy-winged flies are somewhat less viable than normal flies. As a result, we would expect less than 1/4 of the total offspring to express dumpy wings. This observation is borne out in the data, although we have not proven that this is true.

Problems and Discussion Questions

When working out genetics problems in this and succeeding chapters, always assume that members of the P_1 generation are homozygous, unless the information given, or the data, indicates otherwise.

1. In a cross between a black and a white guinea pig, all members of the F_1 generation are black. The F_2 generation is made up of approximately 3/4 black and 1/4 white guinea pigs. Diagram this cross, and show the genotypes and phenotypes.
2. Albinism in humans is inherited as a simple recessive trait. Determine the genotypes of the parents and offspring for the following families. When two alternative genotypes are possible, list both. (a) Two nonalbino (normal) parents have five children, four normal and one albino. (b) A normal male and an albino female have six children, all normal.
3. In a problem involving albinism (see Problem 2), which of Mendel's postulates are demonstrated?
4. Why was the garden pea a good choice as an experimental organism in Mendel's work?
5. Pigeons exhibit a checkered or plain feather pattern. In a series of controlled matings, the following data were obtained:

	F_1 Progeny	
P_1 Cross	*Checkered*	*Plain*
(a) checkered × checkered	36	0
(b) checkered × plain	38	0
(c) plain × plain	0	35

How are the checkered and plain patterns inherited? Predict the results of the $F_1 \times F_1$ mating from cross (b).

(Left photo: Joyce Photographics/Photo Researchers, Inc.; Right photo: R. J. Erwin/Photo Researchers, Inc.)

6. Mendel crossed peas having round seeds and yellow cotyledons with peas having wrinkled seeds and green cotyledons. All the F_1 plants had round seeds with yellow cotyledons. Diagram this cross through the F_2 generation, using both the Punnett square and forked-line methods.
7. Determine the genotypes of the parental plants by analyzing the phenotypes of the offspring from these crosses:

Parental Plants	*Offspring*
(a) round, yellow × round, yellow	3/4 round, yellow 1/4 wrinkled, yellow
(b) round, yellow × wrinkled, yellow	6/16 wrinkled, yellow 2/16 wrinkled, green 6/16 round, yellow 2/16 round, green
(c) round, yellow × wrinkled green	1/4 round, yellow 1/4 round, green 1/4 wrinkled, yellow 1/4 wrinkled, green

8. Are any of the crosses in Problem 7 testcrosses? If so, which one(s)?
9. Which of Mendel's postulates can be demonstrated in the crosses of Problem 7, but not in those in Problems 1 and 5? State this postulate.
10. Correlate Mendel's four postulates with what is now known about homologous chromosomes, genes, alleles, and the process of meiosis.
11. What is the basis for homology among chromosomes?
12. Distinguish between homozygosity and heterozygosity.
13. In *Drosophila*, gray body color is dominant over ebony body color, while long wings are dominant over vestigial wings. Work the following crosses through the F_2 generation, and determine the genotypic and phenotypic ratios for each generation. Assume that the P_1 individuals are homozygous:
(a) gray, long × ebony, vestigial;
(b) gray, vestigial × ebony, long;
(c) gray, long × gray, vestigial.
14. How many different types of gametes can be formed by individuals of the following genotypes? What are they in each case? (a) *AaBb*, (b) *AaBB*, (c) *AaBbCc*, (d) *AaBBcc*, (e) *AaBbcc*, and (f) *AaBbCcDdEe*?
15. Using the forked-line method, determine the genotypic and phenotypic ratios of these trihybrid crosses:
(a) *AaBbCc* × *AaBBCC*, (b) *AaBBCc* × *aaBBCc*, and (c) *AaBbCc* × *AaBbCc*.
16. Mendel crossed peas with round, green seeds with peas having wrinkled, yellow seeds. All F_1 plants had seeds that were round and yellow. Predict the results of testcrossing these F_1 plants.
17. Shown are F_2 results of two of Mendel's monohybrid crosses. State a null hypothesis that you will test using chi-square analysis. Calculate the χ^2 value and determine the p value for both crosses, then interpret the p values. Which cross shows a greater amount of deviation?

(a) Full pods	882
Constricted pods	299
(b) Violet flowers	705
White flowers	224

18. In one of Mendel's dihybrid crosses, he observed 315 round, yellow; 108 round, green; 101 wrinkled, yellow; and 32 wrinkled, green F_2 plants. Analyze these data using chi-square analysis to see whether (a) they fit a 9:3:3:1 ratio; (b) the round, wrinkled traits fit a 3:1 ratio; or (c) the yellow, green traits fit a 3:1 ratio.
19. A geneticist, in assessing data that fell into two phenotypic classes, observed values of 250:150. He decided to perform chi-square analysis using two different null hypotheses: (a) The data fit a 3:1 ratio; and (b) the data fit a 1:1 ratio. Calculate the χ^2 values for each hypothesis. What can you conclude about each hypothesis?
20. The basis for rejecting any null hypothesis is arbitrary. The researcher can set more or less stringent standards by deciding to raise or lower the critical p value. Would the use of a standard of $p = 0.10$ be more or less stringent in failing to reject the null hypothesis? Explain.
21. Consider three independently assorting gene pairs, A/a, B/b, and C/c, where each demonstrates typical dominance (*A*–, *B*–, *C*–), and recessiveness (*aa*, *bb*, *cc*). What is the probability of obtaining an offspring that is *AABbCc* from parents that are *AaBbCC* and *AABbCc*?
22. What is the probability of obtaining a triply recessive individual from the parents shown in Problem 21?
23. Of all offspring of the parents in Problem 21, what proportion will express all three dominant traits?
24. For the following pedigree, predict the mode of inheritance and the resulting genotypes of each individual. Assume that the alleles *A* and *a* control the expression of the trait.

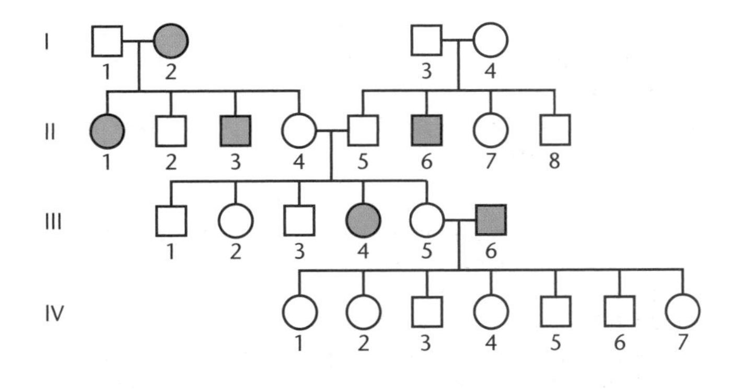

25. Which of Mendel's postulates are demonstrated by the pedigree in Problem 24? List and define these postulates.

26. The following pedigree follows the inheritance of myopia (nearsightedness) in humans. Predict whether the disorder is inherited as a dominant or a recessive trait. Based on your prediction, indicate the most probable genotype for each individual.

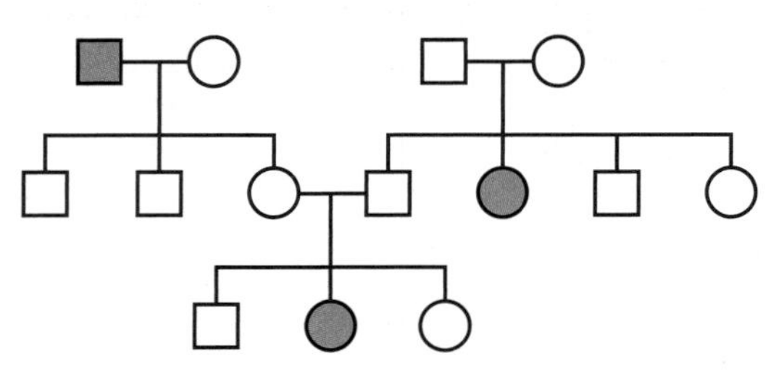

27. Draw all possible conclusions concerning the mode of inheritance of the trait expressed in each of the following limited pedigrees. (Each case is based on a different trait.)

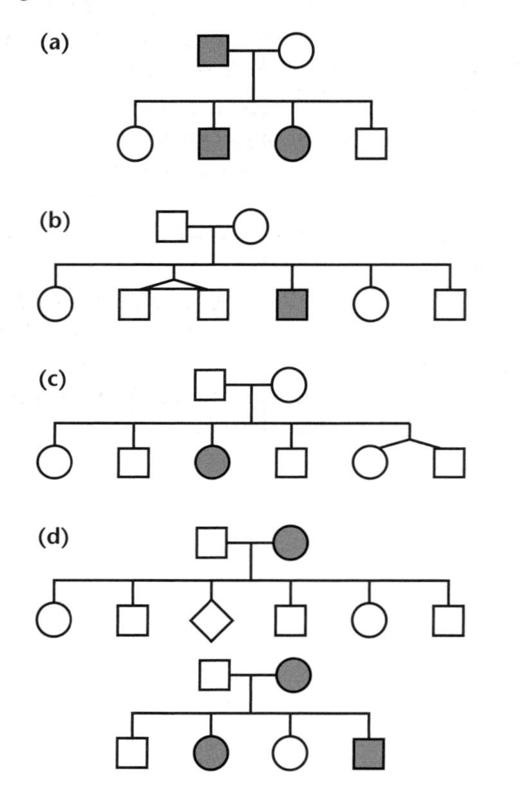

28. Two true-breeding pea plants are crossed. One parent is round, terminal, violet, constricted; while the other expresses the contrasting phenotypes of wrinkled, axial, white, full. The four pairs of contrasting traits are controlled by four genes, each located on a separate chromosome. In the F_1 generation, only round, axial, violet, and full are expressed. In the F_2 generation, all possible combinations of these traits are expressed in ratios consistent with Mendelian inheritance.
   (a) What conclusion can you draw about the inheritance of these traits based on the F_1 results?
   (b) Which phenotype appears most frequently in the F_2 results? Write a mathematical expression that predicts the frequency of occurrence of this phenotype.
   (c) Which F_2 phenotype is expected to occur least frequently? Write a mathematical expression that predicts this frequency.
   (d) How often is either P_1 phenotype likely to occur in the F_2 generation?
   (e) If the F_1 plant is test-crossed, how many different phenotypes will be produced, and how does this number compare to the number of different phenotypes in the F_2 generation discussed in part (b)?
29. Tay-Sachs disease (TSD) is an inborn error of metabolism that results in death, usually by the age of two years. You are a genetic counselor, and you interview a phenotypically normal couple who consult you because the man had a female first cousin (on his father's side) who died from TSD, and the woman had a maternal uncle with TSD. There are no other known cases in either family, and none of the matings were/are between related individuals. Assume that this trait is rare in this population.
   (a) Using standard pedigree symbols, draw a pedigree of these individuals' families, showing the relevant individuals.
   (b) The couple asks you to calculate the probability that they both are heterozygous for the TSD allele.
   (c) They also want to know the probability that neither of them is heterozygous.
   (d) They also ask you for the probability that one of them is heterozygous but the other is not.
   [*Hint:* The answers to (b), (c), and (d) should add up to 1.0.]
30. The wild-type (normal) fruit fly, *Drosophila melanogaster*, has straight wings and long bristles. Mutant strains have been isolated with either curled wings or short bristles. The genes representing these two mutant traits are located on separate chromosomes. Carefully examine the data from the five crosses below. (a) For each mutation, determine whether it is dominant or recessive. In each case, identify which crosses support your answer; and (b) define gene symbols, and determine the genotypes of the parents for each cross.

	Cross	*Number of Progeny*			
		Straight Wings, Long Bristles	*Straight Wings, Short Bristles*	*Curled Wings, Long Bristles*	*Curled Wings, Short Bristles*
1	straight, short × straight, short	30	90	10	30
2	straight, long × straight, long	120	0	40	0
3	curled, long × straight, short	40	40	40	40
4	straight, short × straight, short	40	120	0	0
5	curled, short × straight, short	20	60	20	60

31. To assess Mendel's law of segregation using tomatoes, a true-breeding tall variety (*SS*) is crossed with a true-breeding short variety (*ss*). The heterozygous tall plants (*Ss*) were crossed to produce the two sets of F_2 data shown below:

Set I	*Set II*
30 tall	300 tall
5 short	50 short

(a) Using chi-square analysis, analyze the results for both data sets. Calculate χ^2 values, and estimate the p values in both cases.

(b) From the analysis in part (a), what can you conclude about the importance of generating large data sets in experimental settings?

Selected Readings

Carlson, E. A. 1987. *The Gene: A Critical History*. 2nd ed. Philadelphia: Saunders.

Dunn, L. C. 1965. *A Short History of Genetics*. New York: McGraw-Hill.

Henig, R. M. 2001. *The Monk in the Garden: The Lost and Found Genius of Gregor Mendel, the Father of Genetics*. Boston: Houghton Mifflin.

Miller, J. A. 1984. Mendel's peas: A matter of genius or guile? *Science News* 125: 108–09.

Olby, R. C. 1985. *Origins of Mendelism*. 2nd ed. London: Constable.

Orel, V. 1996. *Gregor Mendel. The First Geneticist*. New York: Oxford University Press.

Peters, J., ed. 1959. *Classic Papers in Genetics*. Englewood Cliffs: Prentice-Hall.

Sokal, R. R., and Rohlf, F. J. 1987. *Introduction to Biostatistics*. 2nd ed. New York: W. H. Freeman.

Soudek, D. 1984. Gregor Mendel and the people around him. *Am. J. Hum. Genet.* 36: 495–98.

Stern, C. 1950. *The Birth of Genetics*. (Supplement to *Genetics* 35.)

Stern, C., and Sherwood, E. 1966. *The Origin of Genetics: A Mendel Source Book*. San Francisco: W. H. Freeman.

Stubbe, H. 1972. *History of Genetics: From Prehistoric Times to Rediscovery of Mendel's Laws*. Cambridge: MIT Press.

Sturtevant, A. H. 1965. *A History of Genetics*. New York: Harper & Row.

Tschermak-Seysenegg, E. 1951. The rediscovery of Mendel's work. *J. Hered.* 42: 163–72.

Voeller, B. R., ed. 1968. *The Chromosome Theory of Inheritance: Classical Papers in Development and Heredity*. New York: Appleton-Century-Crofts.

Welling, F. 1991. Historical study: Johann Gregor Mendel 1822—1884. *Am. J. Med. Genet.* 40: 1–25.

CHAPTER 4

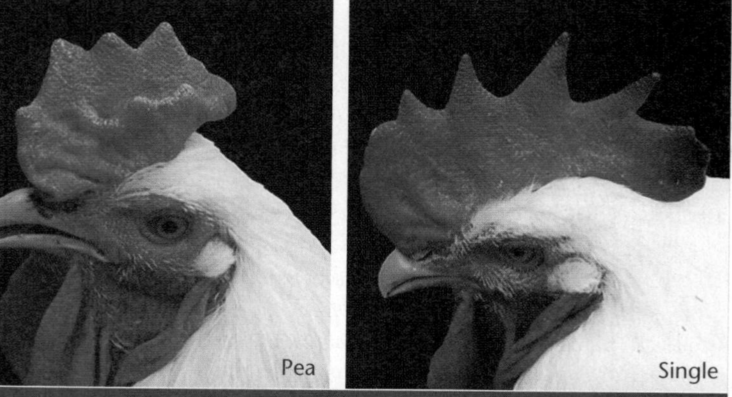

Inherited variation in comb shape of chickens, controlled by two pairs of genes. *(Top photos, Dr. Ralph Somes. Bottom photos, J. James Bitgood/University of Wisconsin, Animal Sciences Dept.)*

Modification of Mendelian Ratios

CHAPTER CONCEPTS

4.1 Alleles Alter Phenotypes in Different Ways

4.2 Geneticists Use a Variety of Symbols for Alleles

4.3 Neither Allele Is Dominant in Incomplete, or Partial, Dominance

4.4 In Codominance, the Influence of Both Alleles in a Heterozygote Is Clearly Evident

4.5 Multiple Alleles of a Gene May Exist in a Population

The ABO Blood Group
The Bombay Phenotype
The *white* Locus in *Drosophila*

4.6 Lethal Alleles Represent Essential Genes

4.7 Combinations of Two Gene Pairs with Two Modes of Inheritance Modify the 9:3:3:1 Ratio

4.8 Phenotypes Are Often Affected by More than One Gene

Epistasis
Novel Phenotypes
Other Modified Dihybrid Ratios

4.9 Complementation Analysis Can Determine if Two Mutations Causing a Similar Phenotype Are Alleles

4.10 X-Linkage Describes Genes on the X Chromosome

X-Linkage in *Drosophila*
X-Linkage in Humans

4.11 In Sex-Limited and Sex-Influenced Inheritance, an Individual's Sex Influences the Phenotype

4.12 Phenotypic Expression Is Not Always a Direct Reflection of the Genotype

Penetrance and Expressivity
Temperature Effects
Onset of Genetic Expression
Genetic Anticipation
Genomic Imprinting

4.13 Extranuclear Inheritance Modifies Mendelian Patterns

Organelle Heredity: DNA in Chloroplasts and Mitochondria
Chloroplasts: Variegation in Four-o'clock Plants
Mitochondrial Mutations: *poky* in *Neurospora* and *petite* in *Saccharomyces*
Mitochondrial Mutations: Human Genetic Disorders
Maternal Effect: *Limnaea* Coiling

In Chapter 3, we discussed the simplest principles of transmission genetics. We saw that genes are present on homologous chromosomes, and that these chromosomes segregate from each other and assort independently with other segregating chromosomes during gamete formation. These two postulates are the fundamental principles of gene transmission from parent to offspring. However, when gene expression does not adhere to a simple dominant/recessive mode or when more than one pair of genes influences the expression of a single character, the classic 3:1 and 9:3:3:1 ratios are usually modified. Although more complex modes of inheritance result, the fundamental principles set down by Mendel still hold true in these situations.

In this chapter we will restrict our initial discussion to the inheritance of traits that are under the control of only one set of genes. In diploid organisms, which have homologous pairs of chromosomes, two copies of each gene influence such traits. The copies need not be identical because alternative forms of genes (alleles) occur within populations. How alleles influence phenotypes is our primary focus. We will then consider how a single phenotype can be controlled by more than one set of genes, a situation sometimes described as gene interaction.

Thus far, we have restricted our discussion to chromosomes other than the X and Y pair, or **autosomes**. In this chapter, we will also examine cases where genes are present on the X chromosome, illustrating X-linkage. We will also discuss internal and external influences on phenotypic expression. The expression of a given phenotype depends an organism's overall environment. We conclude with a consideration of extranuclear inheritance–cases where DNA within organelles influences an organism's phenotype.

HOW DO WE KNOW WHAT WE KNOW?

IN THIS CHAPTER, WE WILL FOCUS ON how genes control phenotypes in ways that modify simple Mendelian inheritance patterns. As you study this topic, you should try to answer several fundamental questions:

1. What experimental approach did early geneticists use to explain inheritance patterns that did not fit typical Mendelian ratios?
2. How did geneticists establish that inheritance of some phenotypic characteristics involved the interaction of two or more gene pairs?
3. How do we know how many genes are involved in the inheritance of a trait?
4. What observations enabled us to distinguish between various modes of qualitative inheritance?
5. How do we know that specific genes are located on the sex-determining chromosomes (i. e., that X-linked inheritance exists)?
6. How was extranuclear inheritance discovered? ■

4.1 Alleles Alter Phenotypes in Different Ways

After Mendel's work was rediscovered in the early 1900s, researchers focused on the many ways in which genes influence an individual's phenotype. Each type of inheritance was more thoroughly investigated when observations of genetic data did not conform precisely to the expected Mendelian ratios, and hypotheses that modified and extended the Mendelian principles were proposed and tested with specifically designed crosses. The explanations were in accord with the principle that a phenotype is under the control of one or more genes located at specific loci on one or more pairs of homologous chromosomes.

To understand the various modes of inheritance, we must first examine the potential function of alleles. Alleles are alternative forms of the same gene. The allele that occurs most frequently in a population, the one that we arbitrarily designate as normal, is called the **wild-type allele** and is usually dominant (the allele for tall plants in the garden pea, for example). Its product is therefore functional in the cell. Wild-type alleles are responsible for the corresponding wild-type phenotype and are the standards against which all mutations at a particular locus are compared.

A mutant allele contains modified genetic information and often specifies an altered gene product. For example, in human populations, there are many known alleles of the gene that encode the β chain of human hemoglobin. All such alleles store information necessary for the synthesis of the β-chain polypeptide, but each allele specifies a slightly different form of the same molecule. Once the allele's product has been manufactured, the product may or may not have its function altered.

Mutation is the source of new alleles. A new allele often leads to a change in the phenotype. A new phenotype is the result of a change in the functional activity of the cellular product encoded by that gene. Usually, the alteration or mutation is expressed as a loss of the specific wild-type function. For example, if a gene is responsible ultimately for the synthesis of a specific enzyme, a mutation in the gene may change the conformation of this enzyme and eliminate its affinity for the substrate. Conversely, another organism may have a different mutation in the same gene—a different allele—and the resulting enzyme may demonstrate a reduced or increased affinity for binding the substrate, or it may not have its affinity altered at all. Thus, a mutation may reduce, enhance, leave unchanged, or result in the total loss of the functional capacity of the enzyme.

Although phenotypic traits can be affected by a single mutation, traits are often influenced by many gene products. In the case of enzymatic reactions, most are part of complex metabolic pathways. Therefore, phenotypic traits are often under the control of more than one gene and the allelic forms of each gene involved. In each of the many crosses discussed in the next few chapters, only one or a relatively few gene pairs are involved. Keep in mind that, in each cross, all genes that are *not* under consideration are assumed to have no effect on the inheritance patterns described.

4.2 Geneticists Use a Variety of Symbols for Alleles

We have previously symbolized alleles for very simple Mendelian traits where the initial letter of the name of a recessive trait, lowercased and italic, denotes the recessive allele. The same letter in uppercase refers to the dominant allele. Thus, for tall and dwarf, where dwarf is recessive, D and d represent the alleles responsible for these respective traits. Mendel used upper- and lowercase letters such as these to symbolize his unit factors.

Another useful system was developed in genetic studies of the fruit fly *Drosophila melanogaster* to discriminate between wild-type and mutant traits. This system uses the initial letter, or a combination of two or three letters, of the name of the mutant trait. If the trait is recessive, lowercase is used; if it is dominant, uppercase is used. The contrasting wild-type trait is denoted by the same letter, but with a superscript +. For example, *ebony* is a recessive body color mutation in *Drosophila*. The normal wild-type body color is gray. Using this system, *ebony* is denoted by the symbol e, while gray is denoted by e^+. The responsible locus can be occupied by either the wild-type allele (e^+) or the mutant allele (e). A diploid fly may thus exhibit one of three possible genotypes:

e^+/e^+	gray homozygote (wild type)
e^+/e	gray heterozygote (wild type)
e/e	ebony homozygote (mutant)

The slash between the letters indicates that the two allele designations represent the same locus on two homologous chromosomes. If we instead consider a dominant wing mutation such as *Wrinkled* (*Wr*) wing in *Drosophila*, the three possible designations are Wr^+/Wr^+, Wr^+/Wr, and Wr/Wr. The latter two genotypes express the wrinkled-wing phenotype.

One advantage of this system is that further abbreviation can be used when convenient: The wild-type allele may simply be denoted by the + symbol. Using *ebony* as an example, the designations of the three possible genotypes become

$+/+$	gray homozygote (wild type)
$+/e$	gray heterozygote (wild type)
e/e	ebony homozygote (mutant)

Another variation is utilized when no dominance exists between alleles. We simply use uppercase italic letters and superscripts to denote alternative alleles (e.g., R^1 and R^2, L^M and L^N, I^A and I^B). Their use will become apparent later in this chapter.

Although we have adopted a standard convention for assigning genetic symbols, many diverse systems of genetic nomenclature are used to identify genes in various organisms. Usually, the symbol selected reflects the function of the gene, or even a disorder caused by a mutant gene. For example, the *cdk* gene, involved in cell-cycle regulation in yeast, refers to *c*yclin *d*ependent *k*inase genes; and in bacteria, leu^- refers to a mutation that interrupts the biosynthesis of the amino acid leucine, where the wild-type gene is designated leu^+. The symbol *dnaA* represents a bacterial gene involved in DNA replication (and DnaA is the protein made by that gene). In humans, capital letters are used to name genes: *BRCA1* represents a gene associated with susceptibility to *br*east *ca*ncer. Although these different systems may seem complex, they are useful ways to symbolize genes.

4.3 Neither Allele Is Dominant in Incomplete, or Partial, Dominance

A cross between parents with contrasting traits may generate offspring with an intermediate phenotype. For example, if plants such as four-o'clocks or snapdragons with red flowers are crossed with white-flowered plants, the offspring have pink flowers. Some red pigment is produced in the F_1 intermediate-colored pink flowers. Therefore neither red nor white flower color is dominant. This situation is known as **incomplete**, or **partial**, **dominance**.

If this phenotype is under the control of a single gene and two alleles where neither is dominant, the results of the F_1(pink) × F_1(pink) cross can be predicted. The resulting F_2 generation shown in Figure 4–1 confirms the hypothesis that only one pair of alleles determines these phenotypes. The genotypic ratio (1:2:1) of the F_2 generation is identical to that of Mendel's monohybrid cross. However, because neither allele is dominant, the phenotypic ratio is identical to the genotypic ratio. Note that because neither allele is recessive, we have chosen not to use upper- and lowercase letters as symbols. Instead, we denoted the red and white alleles as R^1 and R^2. We could have used W^1 and W^2 or still other designations such as C^W and C^R, where C indicates "color" and the W and R superscripts indicate white and red.

Clear-cut cases of incomplete dominance, which result in intermediate expression of the overt phenotype, are relatively rare. However, even when complete dominance seems apparent, careful examination of the gene product, rather than the phenotype, often reveals an intermediate level of gene expression. An example is the human biochemical disorder *Tay-Sachs disease*, in which homozygous recessive individuals are severely affected with a fatal lipid storage disorder, and neonates die during their first one to three years of life. There is almost no activity of the enzyme **hexosaminidase** in afflicted individuals, an enzyme normally involved in lipid metabolism. Heterozygotes, with only a single copy of the mutant gene, are phenotypically normal, but express only about 50% of the enzyme activity found in homozygous normal individuals. Fortunately, this level of enzyme activity is adequate to

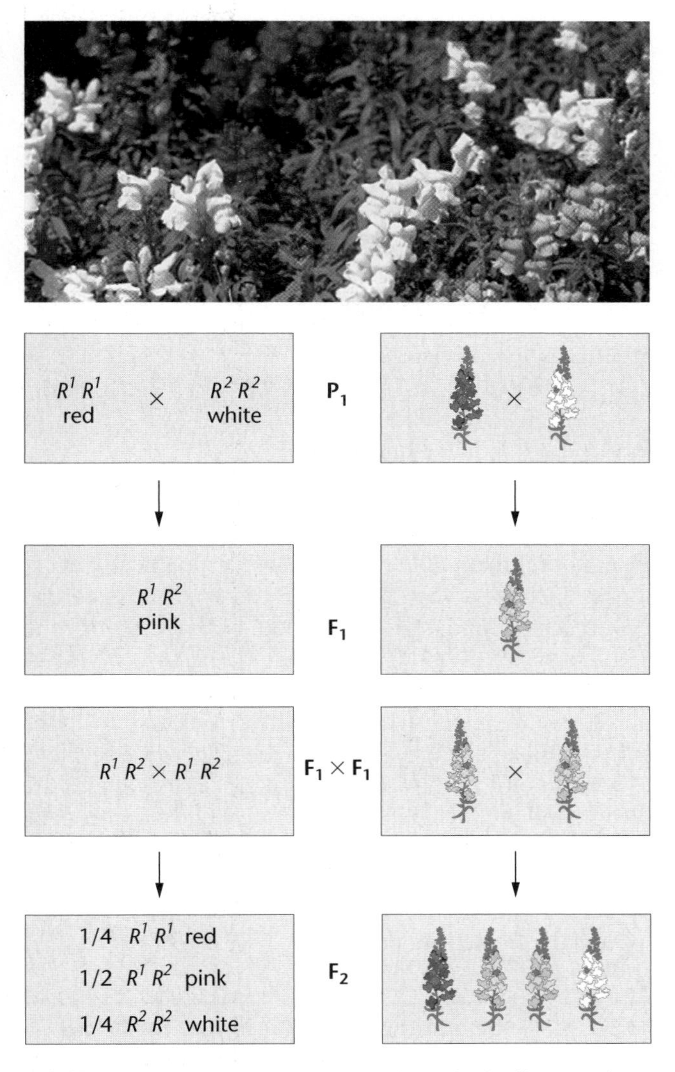

FIGURE 4–1 Incomplete dominance shown in the flower color of snapdragons. *(Photo: John D. Cunningham/Visuals Unlimited)*

achieve normal biochemical function—a situation not uncommon in enzyme disorders.

4.4 In Codominance, the Influence of Both Alleles in a Heterozygote Is Clearly Evident

If two alleles of a single gene are responsible for producing two distinct, detectable gene products, a situation different from incomplete dominance or dominance/recessiveness arises. In this case, the joint expression of both alleles in a heterozygote is called **codominance**. The **MN blood group** in humans illustrates this phenomenon and is characterized by an antigen called a glycoprotein, found on the surface of red blood cells. In the human population, two forms of this glycoprotein exist, designated M and N; an individual may exhibit either one or both of them.

The MN system is under the control of an autosomal locus found on chromosome 4 and two alleles designated L^M and L^N. Humans are diploid, so three combinations are possible, each resulting in a distinct blood type:

Genotype	*Phenotype*
$L^M L^M$	M
$L^M L^N$	MN
$L^N L^N$	N

As predicted, a mating between two MN parents may produce children of all three blood types:

$$L^M L^N \times L^M L^N$$
$$\downarrow$$
$$1/4 \quad L^M L^M = \text{M}$$
$$1/2 \quad L^M L^N = \text{MN}$$
$$1/4 \quad L^N L^N = \text{N}$$

Once again the genotypic ratio, 1:2:1, is upheld.

Codominant inheritance is characterized by distinct expression of the gene products of both alleles. This characteristic distinguishes it from incomplete dominance, where heterozygotes express an intermediate, blended phenotype.

MEDIA TUTORIAL Extensions of mendelian inheritance

4.5 Multiple Alleles of a Gene May Exist in a Population

The information stored in any gene is extensive, and mutations can modify this information in many ways. Each change produces a different allele. Therefore, for any specific gene, the number of alleles within members of a population need not be restricted to two. When three or more alleles of the same gene are found, **multiple alleles** are present that create a unique mode of inheritance. It is important to realize that *multiple alleles can be studied only in populations*. An individual diploid organism has, at most, two homologous gene loci that may be occupied by different alleles of the same gene. However, among many members of a species, numerous alternative forms of the same gene can exist.

The ABO Blood Group

The simplest case of multiple alleles is that in which three alternative alleles of one gene exist. This situation is illustrated by the **ABO blood group** in humans, discovered by Karl Landsteiner in the early 1900s. The ABO system, like the MN blood group, is characterized by the presence of antigens on the surface of red blood cells. The A and B antigens are distinct from MN antigens and are under the control of a different gene, located on chromosome 9. As in the MN system, one combination of alleles in the ABO system exhibits a codominant mode of inheritance.

When individuals are tested using antisera that contain antibodies against the A or B antigen, four phenotypes are revealed. Each individual has either the A antigen

(A phenotype), the B antigen (B phenotype), the A and B antigens (AB phenotype), or neither antigen (O phenotype). In 1924, it was hypothesized that these phenotypes were inherited as the result of three alleles of a single gene. This hypothesis was based on studies of the blood types of many different families.

Although different designations can be used, we use the symbols I^A, I^B, and I^O to distinguish these three alleles; the *I* stands for *isoagglutinogen*, another term for antigen. If we assume that the I^A and I^B alleles are responsible for the production of their respective A and B antigens and that I^O is an allele that does not produce any detectable A or B antigens, we can list the various genotypic possibilities and assign the appropriate phenotype to each:

Genotype	*Antigen*	*Phenotype*
I^AI^A	A	A
I^AI^O	A	
I^BI^B	B	B
I^BI^O	B	
I^AI^B	A, B	AB
I^OI^O	Neither	O

In these assignments the I^A and I^B alleles are dominant to the I^O allele, but are codominant to each other. Our knowledge of human blood types has several practical applications, the most important of which are compatible blood transfusions and organ transplantations.

The Bombay Phenotype

The biochemical basis of the ABO blood type system has been carefully worked out. The A and B antigens are actually carbohydrate groups (sugars) that are bound to lipid molecules (fatty acids) protruding from the membrane of the red blood cell. The specificity of the A and B antigens is based on the terminal sugar of the carbohydrate group. Both the A and B antigens are derived from a precursor molecule called the **H substance**, to which one or two terminal sugars are added.

In extremely rare instances, first recognized in a woman in Bombay in 1952, the H substance is incompletely formed. As a result, it is an inadequate substrate for the enzyme that normally adds the terminal sugar. This condition results in blood type O and is called the **Bombay phenotype**. Research has revealed that it is due to a rare recessive mutation, *h*, at a locus separate from that controlling the A and B antigens. Thus, even though individuals may have the I^A and/or I^B alleles, if they display the *hh* genotype, neither the A nor B antigen can be added to the cell surface. This information helped explain why the woman in Bombay was blood typed as O even though one of her parents was type AB (thus she should not have been type O), and why she was able to pass the I^B allele to her children (Figure 4–2).

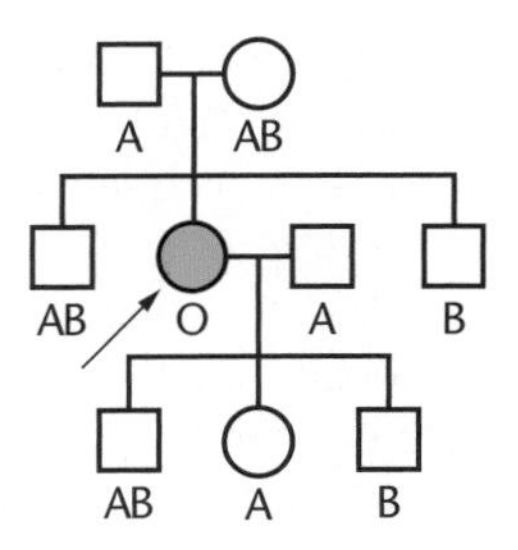

FIGURE 4–2 A partial pedigree of a woman with the Bombay phenotype. Functionally, her ABO blood group behaves as type O. Genetically, she is type B.

The *white* Locus in *Drosophila*

Many other phenotypes in plants and animals are known to be controlled by multiple allelic inheritance. In *Drosophila*, many alleles are known at practically every locus. The recessive mutation that causes white eyes, discovered by Thomas H. Morgan and Calvin Bridges in 1912, is one of over 100 alleles that can occupy this locus. In this allelic series, eye colors range from complete absence of pigment in the *white* allele, to deep ruby in the *white-satsuma* allele, to orange in the *white-apricot* allele, to a buff color in the *white-buff* allele. These alleles are designated w, w^{sat}, w^a, and w^{bf} respectively. In each case, the total amount of pigment in these mutant eyes is reduced to less than 20% of that found in the brick-red, wild-type eye. Table 4.1 lists these and other *white* alleles and their color phenotypes.

TABLE 4.1 Some of the Alleles Present at the *White* Locus of *Drosophila Melanogaster* and Their Eye-Color Phenotype

Allele	*Name*	*Eye Color*
w	*white*	pure white
w^a	*white-apricot*	yellowish orange
w^{bf}	*white-buff*	light buff
w^{bl}	*white-blood*	yellowish ruby
w^{cf}	*white-coffee*	deep ruby
w^e	*white-eosin*	yellowish pink
w^{mo}	*white-mottled orange*	light mottled orange
w^{sat}	*white-satsuma*	deep ruby
w^{sp}	*white-spotted*	fine grain, yellow mottling
w^t	*white-tinged*	light pink

4.6 Lethal Alleles Represent Essential Genes

Many gene products are essential to an organism's survival. Mutations resulting in the synthesis of a gene product that is nonfunctional can sometimes be tolerated in the heterozygous state; that is, one wild-type allele may be sufficient to produce enough of the essential product to allow survival. However, such a mutation behaves as a recessive **lethal allele**, and homozygous recessive individuals will not survive. The time of death will depend on when the product is essential. In mammals, for example, this might occur during development, early childhood, or even during adulthood.

In some cases, the allele responsible for a lethal effect when homozygous may also result in a distinctive mutant phenotype when present heterozygously. It is behaving as a recessive lethal allele but is dominant with respect to the phenotype. For example, a mutation that causes a yellow coat in mice was discovered in the early part of this century. The yellow coat varies from the normal agouti (wild-type) coat phenotype, as shown in Figure 4–3. Crosses between the various combinations of the two strains yield unusual results:

Crosses
A: agouti × agouti → all agouti
B: yellow × yellow → 2/3 yellow : 1/3 agouti
C: agouti × yellow → 1/2 yellow : 1/2 agouti

These results are explained on the basis of a single pair of alleles. With regard to coat color, the mutant *yellow* allele (A^Y) is dominant to the wild-type *agouti* allele (A), so heterozygous mice will have yellow coats. However, the *yellow* allele is also a homozygous recessive lethal. When present in two copies, the mice die before birth. Thus, there are no homozygous yellow mice. The genetic basis for these three crosses is shown in Figure 4–3.

Other mutant genes are known to behave as *dominant* lethal alleles, where the presence of just one copy of the allele results in the death of the individual. In humans, a disorder called **Huntington disease** (previously referred to as Huntington's chorea) is due to a dominant autosomal allele H, where the onset of the disease in heterozygotes (Hh) is delayed, usually well into adulthood. Affected individuals then undergo gradual nervous and motor degeneration until they die. This lethal disorder is particularly tragic because it has such a late onset, typically at about age 40. By that time, the affected individual may have produced a family, and each of their children has a 50% probability of inheriting the lethal allele, transmitting the allele to his or her offspring, and eventually developing the disorder. The American folk singer and composer Woody Guthrie (father of modern-day folk singer Arlo Guthrie) died from this disease at age 39.

Dominant lethal alleles are rarely observed. For these alleles to exist in a population, the affected individuals must reproduce before the lethal allele is expressed, as can occur in Huntington disease. If all affected individuals die

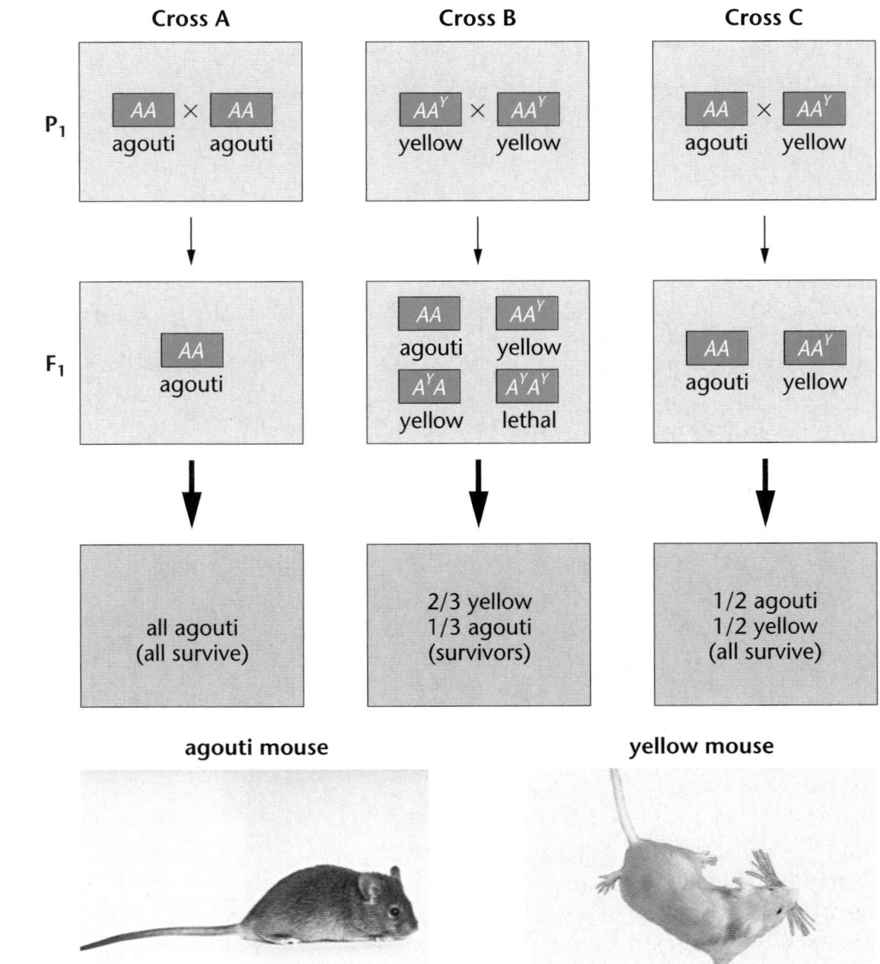

FIGURE 4–3 Inheritance patterns in three crosses involving the wild-type agouti allele (A) and the mutant yellow allele (A^Y) in mice. Note that the mutant allele behaves as a homozygous lethal allele, and the genotype $A^Y A^Y$ does not survive. *(Left photo: Courtesy of Stanton K. Short [The Jackson Laboratory, Bar Harbor, ME]; Right photo: Tom Cerniglio/Oak Ridge National Laboratory)*

before reaching reproductive age, the mutant gene will not be passed to future generations, and the mutation will disappear from the population unless it arises again as a result of a new mutation.

4.7 Combinations of Two Gene Pairs with Two Modes of Inheritance Modify the 9:3:3:1 Ratio

Each example discussed so far modifies Mendel's 3:1 F_2 monohybrid ratio. Therefore, combining any two of these modes of inheritance in a dihybrid cross will likewise modify the classical 9:3:3:1 ratio. Having established the foundation for the modes of inheritance of incomplete dominance, codominance, multiple alleles, and lethal alleles, we can now deal with the situation of two modes of inheritance occurring simultaneously. Mendel's principle of independent assortment applies to these situations, provided that the genes controlling each character are not linked on the same chromosome.

Consider, for example, a mating that occurs between two humans who are both heterozygous for the autosomal recessive gene that causes albinism and who are both of blood type AB. What is the probability of a particular phenotypic combination occurring in each of their children? Albinism is inherited in the simple Mendelian fashion, and the blood types are determined by the series of three multiple alleles, I^A, I^B, and I^O. The solution to this problem is diagrammed in Figure 4–4, using the forked-line method. This dihybrid cross does not yield the classical four phenotypes in a 9:3:3:1 ratio. Instead, six phenotypes occur in a 3:6:3:1:2:1 ratio, establishing the expected probability for each phenotype. This is just one of the many variants of modified ratios that are possible when different modes of inheritance are combined.

4.8 Phenotypes Are Often Affected by More than One Gene

Soon after Mendel's work was rediscovered, experimentation revealed that individual characteristics displaying discrete phenotypes are often under the control of more than one gene. This was a significant discovery because it revealed that genetic influence on the pheno-

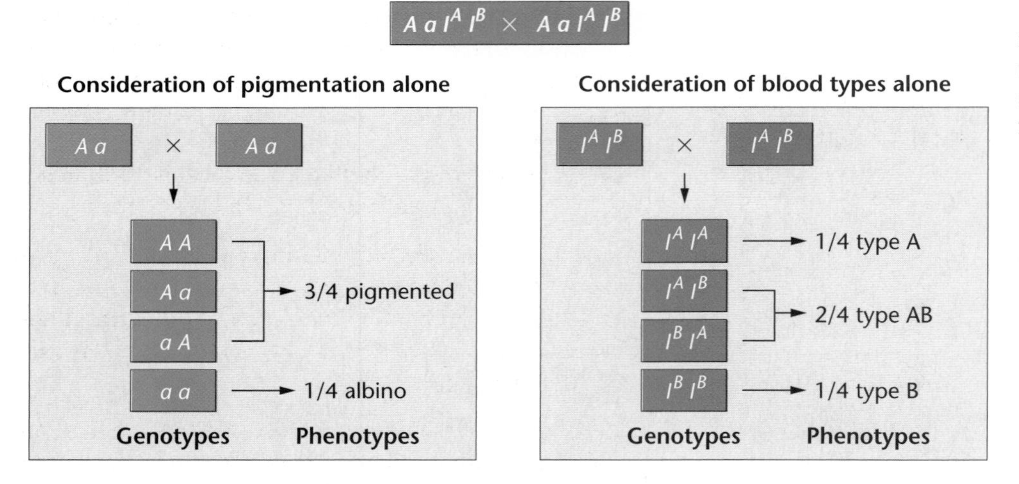

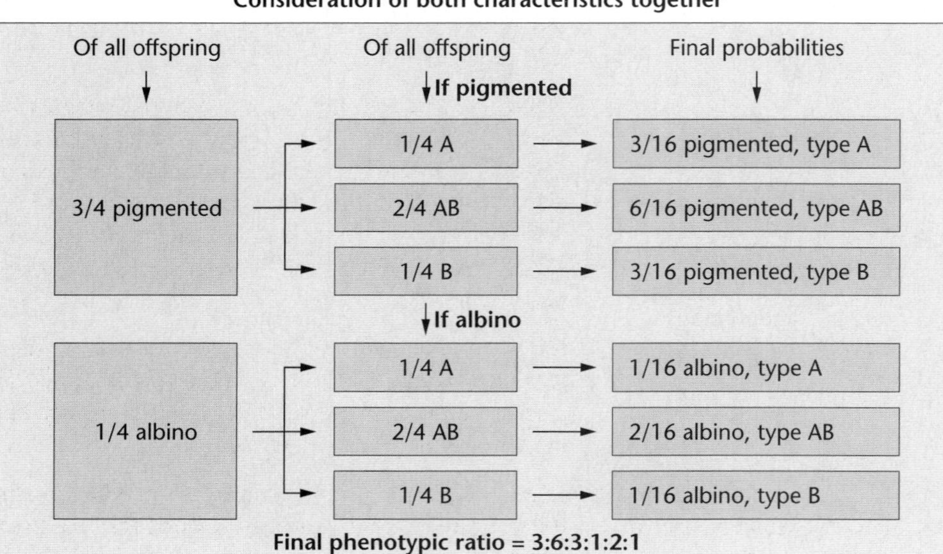

FIGURE 4–4 Calculation of the mating probabilities involving the ABO blood type and albinism in humans, using the forked-line method.

type is often much more complex than envisioned by Mendel. Instead of single genes controlling the development of individual parts of the plant or animal body, it soon became clear that phenotypic characters can be influenced by the interactions of many different genes and their products.

The term **gene interaction** is often used to describe the idea that several genes influence a particular characteristic. This does not mean, however, that two or more genes, or their products, necessarily interact directly with one another to influence a particular phenotype. Rather, the cellular function of numerous gene products contributes to the development of a common phenotype. For example, the development of an organ such as the compound eye of an insect is exceedingly complex and leads to a structure with multiple phenotypic manifestations—such as specific size, shape, texture, and color. The development of the eye is a complex cascade of developmental events leading to its formation. This process exemplifies the developmental concept of **epigenesis**, whereby each step of development increases the complexity of this sensory organ and is under the control and influence of one or more genes.

Epistasis

Some of the best examples of gene interaction are those that reveal the phenomenon of **epistasis** (Greek for "stoppage"). Epistasis occurs when the expression of one gene or gene pair masks or modifies the expression of another gene or gene pair. Sometimes the genes involved control the expression of the same general phenotypic characteristic in an antagonistic manner, as when masking occurs. In other cases, however, the genes involved exert their influence on one another in a complementary, or cooperative, fashion.

For example, the homozygous presence of a recessive allele prevents or overrides the expression of other alleles at a second locus (or several other loci). In this case, the alleles at the first locus are said to be *epistatic* to those at the second locus, and the alleles at the second locus are *hypostatic* to those at the first locus. In another example, a single dominant allele at the first locus influences the expression of the alleles at a second gene locus. In a third example, two gene pairs complement one another such that at least one dominant allele at each locus is required to express a particular phenotype.

The Bombay phenotype discussed earlier is an example of a homozygous recessive condition at one locus masking the expression of a second locus (see Figure 4–2). There, we had established that the homozygous condition (*hh*) masks the expression of the I^A and I^B alleles. Only individuals containing at least one *H* allele (designated *H–*) can form the A or B antigens. As a result, individuals whose genotypes include the I^A or I^B allele and who are also *hh* express the type O phenotype, regardless of their potential to make either antigen. An example of the outcome of matings between individuals heterozygous at both loci is shown in Figure 4–5. If many individuals of the genotype I^AI^BHh have children, the phenotypic ratio of 3 A : 6 AB : 3 B : 4 O is expected in their offspring.

It is important to note the following points when examining this cross and the predicted phenotypic ratio:

1. A key distinction exists in this cross compared to the modified dihybrid cross shown in Figure 4–4: *only one characteristic—blood type—is being followed.* In the modified dihybrid cross of Figure 4–4, blood type *and* skin pigmentation are followed as separate phenotypic characteristics.

2. Even though only a single character was followed, the phenotypic ratio is expressed in sixteenths. If we knew nothing about the H substance and the genes controlling it, we could still be confident that a second gene pair, other than that controlling the A and B antigens, is involved in the phenotypic expression. *When studying a single character, a ratio that is expressed in 16 parts (e.g., 3:6:3:4) suggests that two gene pairs are "interacting" during the expression of the phenotype under consideration.*

The study of gene interaction reveals inheritance patterns that modify the classical Mendelian dihybrid F_2 ratio (9:3:3:1) in other ways as well. In these examples, epistasis combines one or more of the four phenotypic categories in various ways. The generation of these four groups is reviewed in Figure 4–6, along with several modified ratios.

As we discuss these and other examples, we will make several assumptions and adopt certain conventions:

1. In each case, distinct phenotypic classes are produced, each clearly discernible from all others. Such traits illustrate discontinuous variation, where phenotypic categories are discrete and qualitatively different from one another.

2. The genes considered in each cross are not linked and therefore assort independently of one another during gamete formation. To allow you to easily compare the results of different crosses, we designated alleles as *A*, *a* and *B*, *b* in each case.

3. When we assume that complete dominance exists between the alleles of any gene pair, such that *AA* and *Aa*, or *BB* and *Bb* are equivalent in their genetic effects, we used the designations *A–* or *B–* for both combinations, where the dash (–) indicates that either allele may be present, without consequence to the phenotype.

4. All P_1 crosses involve homozygous individuals (e.g., $AABB \times aabb$, $AAbb \times aaBB$, or $aaBB \times AAbb$). Therefore, each F_1 generation consists of only heterozygotes of genotype *AaBb*.

5. In each example, the F_2 generation produced from these heterozygous parents is our main focus of analysis. When two genes are involved (as in

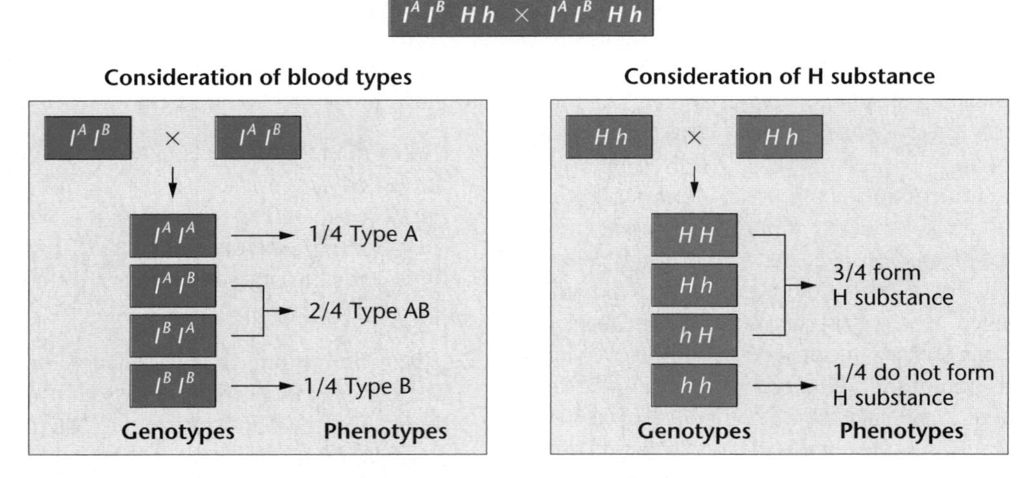

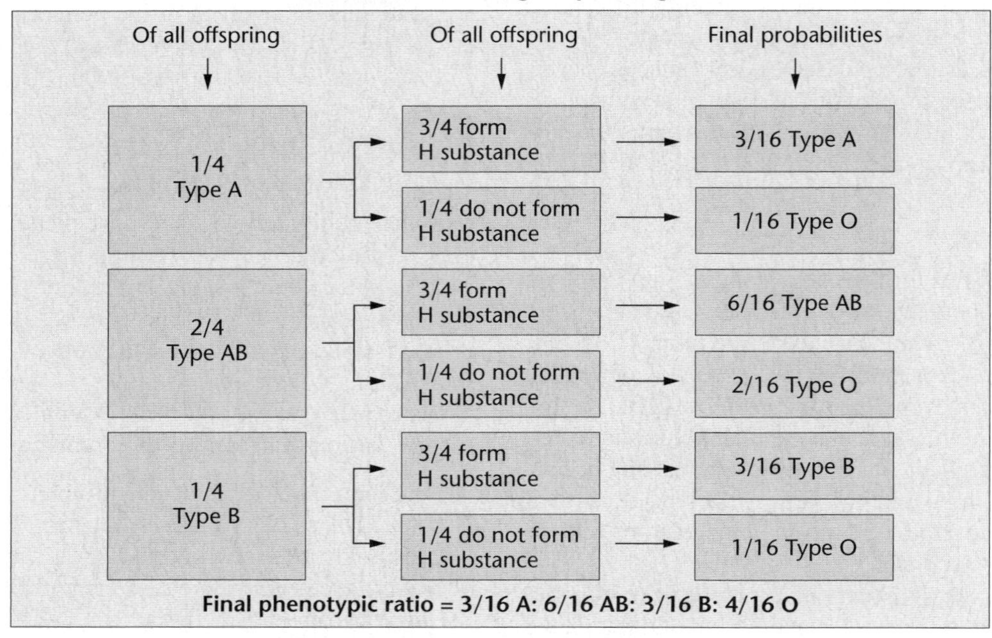

FIGURE 4–5 The outcome of a mating between individuals who are heterozygous at two genes determining their ABO blood type. Final phenotypes are calculated by considering both genes separately and then combining the results using the forked-line method.

Figure 4–6), the F_2 genotypes fall into four categories: 9/16 *A–B–*, 3/16 *A–bb*, 3/16 *aaB–*, and 1/16 *aabb*. Because of dominance, all genotypes in each category are equivalent in their effect on the phenotype.

Case 1 is the inheritance of coat color in mice (Figure 4–7). Normal wild-type coat color is agouti, a grayish pattern formed by alternating bands of pigment on each hair. Agouti is dominant to black (non-agouti) hair, which is caused by a recessive mutation, *a*. Thus, *A–* results in agouti, while *aa* yields black coat color. When it is homozygous, a recessive mutation, *b*, at a separate locus, eliminates pigmentation altogether, yielding albino mice (*bb*), regardless of the genotype at the other locus. The presence of at least one *B* allele allows pigmentation to occur in much the same way that the *H* allele in humans allows the expression of the ABO blood types. In a cross between agouti (*AABB*) and albino (*aabb*), members of the F_1 are all *AaBb* and have agouti coat color. In the F_2 progeny of a cross between two F_1 heterozygotes, the following genotypes and phenotypes are observed:

$$F_1: AaBb \times AaBb$$
$$\downarrow$$

F_2 Ratio	*Genotype*	*Phenotype*	*Final Phenotypic Ratio*
9/16	*A–B–*	agouti	9/16 agouti
3/16	*A–bb*	albino	4/16 albino
3/16	*aaB–*	black	3/16 black
1/16	*aabb*	albino	

We can envision gene interaction yielding the observed 9:3:4 F_2 ratio as a two-step process:

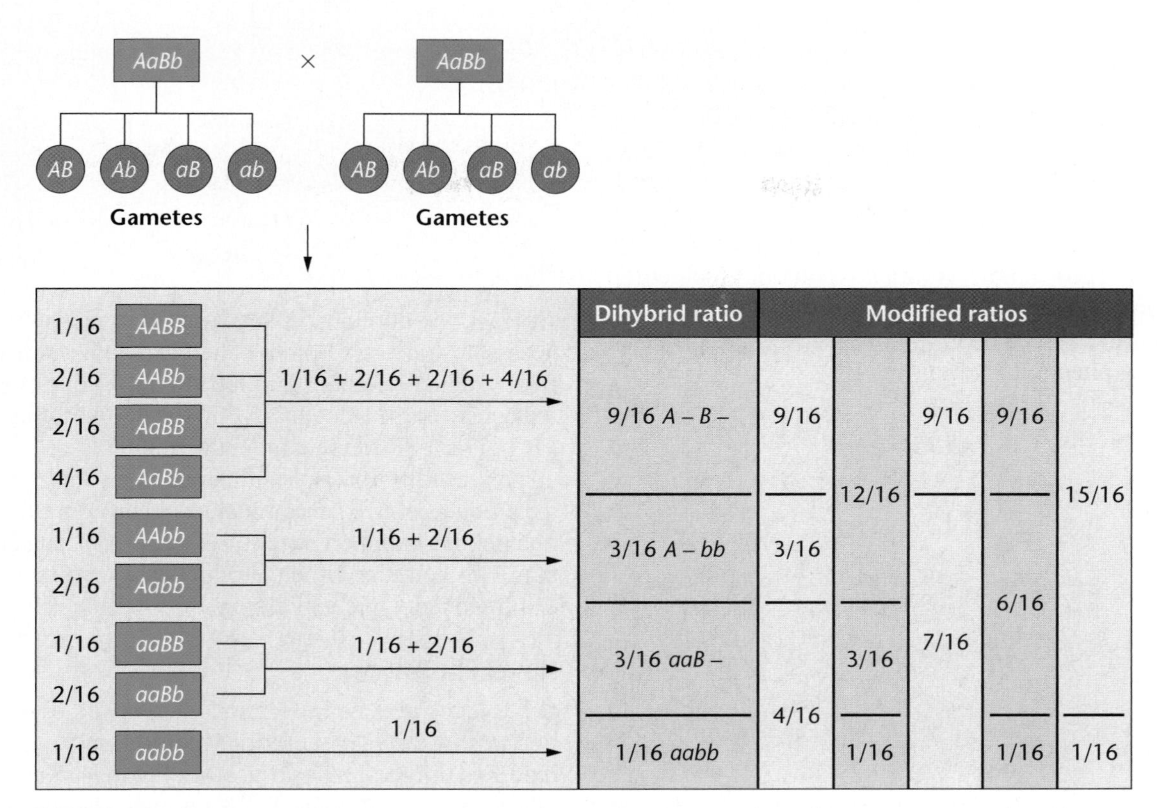

FIGURE 4–6 Generation of the various modified dihybrid ratios from the nine unique genotypes produced in a cross between individuals who are heterozygous at two genes.

	Gene *B*		Gene *A*	
Precursor	↓	Black	↓	Agouti
Molecule	⟶	Pigment	⟶	Pattern
(colorless)	*B*–		*A*–	

In the presence of a *B* allele, black pigment can be made from a colorless substance. In the presence of an *A* allele, the black pigment is deposited during the development of hair in a pattern that produces the agouti phenotype. If the *aa* genotype occurs, all of the hair remains black. If the *bb* genotype occurs, no black pigment is produced, regardless of the presence of the *A* or *a* alleles, and the mouse is albino. Therefore, the *bb* genotype masks or suppresses the expression of the *A* gene, thus demonstrating epistasis.

Case	Organism	Character	F_2 Phenotypes 9/16	3/16	3/16	1/16	Modified ratio
1	Mouse	Coat color	agouti	albino	black	albino	9:3:4
2	Squash	Color	white		yellow	green	12:3:1
3	Pea	Flower color	purple	white			9:7
4	Squash	Fruit shape	disc	sphere		long	9:6:1
5	Chicken	Color	white		colored	white	13:3
6	Mouse	Color	white-spotted	white	colored	white-spotted	10:3:3
7	Shepherd's purse	Seed capsule	triangular			ovoid	15:1
8	Flour beetle	Color	red / sooty / red / sooty	black	jet	black	6:3:3:4

FIGURE 4–7 The basis of modified dihybrid F_2 phenotypic ratios, resulting from crosses between doubly heterozygous F_1 individuals. The four groupings of the F_2 genotypes shown in Figure 4–6 and across the top of this figure are combined in various ways to produce these ratios.

A second type of epistasis occurs when a dominant allele at one genetic locus masks the expression of the alleles at a second locus. For instance, case 2 of Figure 4–7 deals with the inheritance of fruit color in summer squash. Here, the dominant allele *A* results in white fruit color regardless of the genotype at a second locus, *B*. In the absence of the dominant *A* allele (the *aa* genotype), *BB* or *Bb* results in yellow color, while *bb* results in green color. Therefore, if two white-colored double heterozygotes (*AaBb*) are crossed, this type of epistasis generates an interesting phenotypic ratio:

$$F_1: AaBb \times AaBb$$

$$\downarrow$$

F_2 Ratio	Genotype	Phenotype	Final Phenotypic Ratio
9/16	*A–B–*	white	12/16 white
3/16	*A–bb*	white	
3/16	*aaB–*	yellow	3/16 yellow
1/16	*aabb*	green	1/16 green

Of the offspring, 9/16 are *A–B–* and are thus white. The 3/16 bearing the genotypes *A–bb* are also white. Finally, 3/16 are yellow (*aaB–*), while 1/16 are green (*aabb*); and we obtain the modified ratio of 12:3:1.

Our third type of gene interaction (case 3 of Figure 4–7) was first discovered by William Bateson and Reginald Punnett (of Punnett-square fame). It is demonstrated in a cross between two true-breeding strains of white-flowered sweet peas. Unexpectedly, the results of this cross yield all purple F_1 plants, and the F_2 plants occur in a ratio of 9/16 purple to 7/16 white. The proposed explanation suggests that the presence of at least one dominant allele of each of two gene pairs is essential for flowers to be purple. All other genotype combinations yield white flowers because the homozygous condition of *either* recessive allele masks the expression of the dominant allele at the other locus. The cross is shown as follows:

P_1: *AAbb* × *aaBB*
white white

↓

F_1: All *AaBb* (purple)

↓

F_2 Ratio	Genotype	Phenotype	Final Phenotypic Ratio
9/16	*A–B–*	purple	9/16 purple
3/16	*A–bb*	white	
3/16	*aaB–*	white	7/16 white
1/16	*aabb*	white	

We can now see how two gene pairs might yield such results:

	Gene *A*		Gene *B*	
Precursor	↓	Intermediate	↓	Final
Substance	⟶	Product	⟶	Product
(colorless)	*A–*	(colorless)	*B–*	(purple)

At least one dominant allele from each pair of genes is necessary to ensure both biochemical conversions to the final product, yielding purple flowers. In our cross, this will occur in 9/16 of the F_2 offspring. All other plants (7/16) have flowers that remain white.

The preceding examples illustrate in a simple way how the products of two genes "interact" to influence the development of a common phenotype. In other instances, more than two genes and their products are involved in controlling phenotypic expression.

Novel Phenotypes

Other cases of gene interaction yield novel, or new, phenotypes in the F_2 generation, in addition to producing modified dihybrid ratios. Case 4 in Figure 4–7 depicts the inheritance of fruit shape in the summer squash *Cucurbita pepo*. When plants with disc-shaped fruit (*AABB*) are crossed to plants with long fruit (*aabb*), the F_1 generation all have disc fruit. However, in the F_2 progeny, fruit with a novel shape—sphere—appear, along with fruit exhibiting the parental phenotypes. A variety of fruit shapes are shown in Figure 4–8.

The F_2 generation, with a modified 9:6:1 ratio, is generated as follows:

F_1: *AaBb* × *AaBb*
disc ↓ disc

F_2 Ratio	Genotype	Phenotype	Final Phenotypic Ratio
9/16	*A–B–*	disc	9/16 disc
3/16	*A–bb*	sphere	6/16 sphere
3/16	*aaB–*	sphere	
1/16	*aabb*	long	1/16 long

In this example of gene interaction, both gene pairs influence fruit shape equally. A dominant allele at either locus ensures a sphere-shaped fruit. In the absence of dominant alleles, the fruit is long. However, if both dominant alleles (*A* and *B*) are present, the fruit displays a flattened, disc shape.

Other Modified Dihybrid Ratios

The remaining cases (5–8) in Figure 4–7 show additional modifications of the dihybrid ratio and provide still other examples of gene interactions. However, all eight cases have two things in common. First, we have not violated the principles of segregation and independent assortment to

FIGURE 4–8 Summer squash exhibiting the fruit-shape phenotypes disc (white), long (orange gooseneck), and sphere (bottom left). *(Photo: Irene Vandermolen/Animals Animals/Earth Scenes)*

explain the inheritance pattern of each case. Therefore, the added complexity of inheritance in these examples does not detract from the validity of Mendel's conclusions. Second, the F_2 phenotypic ratio in each example has been expressed in sixteenths. When similar observations are made in crosses where the inheritance pattern is unknown, it suggests to geneticists that two gene pairs are controlling the observed phenotypes. You should make the same inference in your analysis of genetics problems.

4.9 Complementation Analysis Can Determine if Two Mutations Causing a Similar Phenotype Are Alleles

An interesting situation arises when two mutations, both of which produce a similar phenotype, are isolated independently. Suppose that two investigators, one in a genetics laboratory in the United States and the other in a genetics laboratory in Canada, independently isolate and establish a true-breeding strain of wingless *Drosophila* and demonstrate that each is due to a recessive mutation. We might assume that both strains contain mutations in the same gene. However, since we know that many genes are involved in the formation of wings, mutations in any one of them might inhibit wing formation during development. The experimental approach called **complementation analysis** allows us to determine whether two such mutations are in the same gene—that is, whether they are alleles, or whether they represent mutations in separate genes.

Our analysis seeks to answer this simple question: *Are two mutations that yield similar phenotypes present in the same gene or in two different genes?* To find the answer, we cross the two mutant strains and analyze the F_1 generation. There are two alternative outcomes and interpretations of this cross shown in Figure 4–9. We discuss both cases, using the designations m^{usa} for the mutation isolated in the United States and m^{can} for the mutation

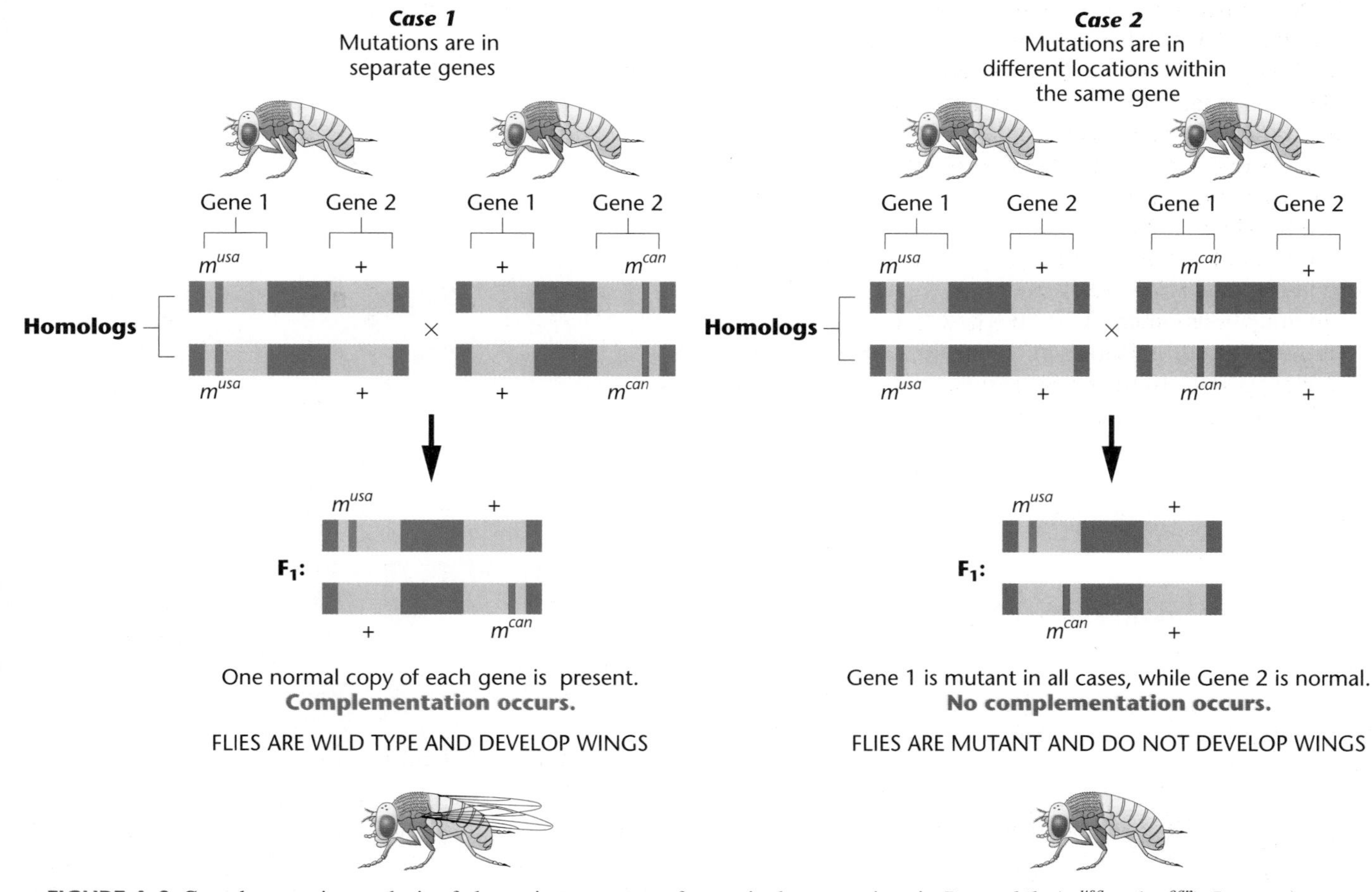

FIGURE 4–9 Complementation analysis of alternative outcomes of two wingless mutations in *Drosophila* (m^{usa} and m^{can}). In case 1, the mutations are not alleles of the same gene, while in case 2, the mutations are alleles of the same gene.

isolated in Canada. Now we will see if they are alleles of the same gene, or not.

Case 1. *All offspring develop normal wings.*

Interpretation: The two recessive mutations are in separate genes and are not alleles of one another. Following the cross, all F_1 flies are heterozygous for both genes. *Complementation* is said to occur. Since each mutation is in a separate gene and each F_1 fly is heterozygous at both loci, the normal products of both genes are produced (by the one normal copy of each gene), and wings develop.

Case 2. *All offspring fail to develop wings.*

Interpretation: The two mutations affect the same gene and are alleles of one another. Complementation does *not* occur. Since the two mutations affect the same gene, the F_1 flies are homozygous for the two mutant alleles (the m^{usa} allele and the m^{can} allele). No normal product of the gene is produced, and in the absence of this essential product, wings do not form.

Complementation analysis, as originally devised by the *Drosophila* geneticist Edward B. Lewis, is often called the **cis-trans test**. Borrowed from nomenclature used in organic chemistry, **cis** (alongside one another) refers to the case where the two mutations are on the same homolog. Likewise, trans (opposite one another) refers to the case where the mutations are on separate homologs. As shown in Figure 4–9, the trans configuration is critical in determining whether the two mutations are alleles of the same gene or not. If we determine that the two mutations are indeed alleles, their presence as heterozygotes in the cis configuration serves as an important control in complementation analyses. In this configuration, flies will develop wings.

Complementation analysis may be used to screen any number of individual mutations that result in the same phenotype. Such an analysis may reveal that only a single gene is involved or that two or more genes are involved. All mutations determined to be present in any single gene are said to fall into the same **complementation group**, and they will complement mutations in all other groups.

When large numbers of mutations affecting the same trait are available and studied using complementation analysis, it is possible to predict the total number of genes involved in the determination of that trait.

MEDIA TUTORIAL X-linked inheritance

4.10 X-Linkage Describes Genes on the X Chromosome

In many animal and some plant species, one of the sexes contains a pair of unlike chromosomes that are involved in sex determination. In many cases, these are designated as the X and Y. For example, in both *Drosophila* and humans, males contain an X and a Y chromosome, whereas females contain two X chromosomes. While the Y chromosome must contain a region of pairing homology with the X chromosome if the two are to synapse and segregate during meiosis, much of the remainder of the Y chromosome in humans and other species is considered to be relatively inert genetically. Thus, it lacks most genes that are present on the X chromosome. As a result, genes present on the X chromosome exhibit unique patterns of inheritance in comparison with autosomal genes. The term **X-linkage** is used to describe these situations.

In the discussion below, we will focus on inheritance patterns resulting from genes present on the X but absent from the Y chromosome—this situation results in a modification of Mendelian ratios, the central theme of this chapter.

X-Linkage in *Drosophila*

One of the first cases of X-linkage was documented by Thomas H. Morgan around 1920 during his studies of the *white* mutation in the eyes of *Drosophila* (Figure 4–10). The normal wild-type red eye color is dominant to white. We will use this case to illustrate **X-linkage**.

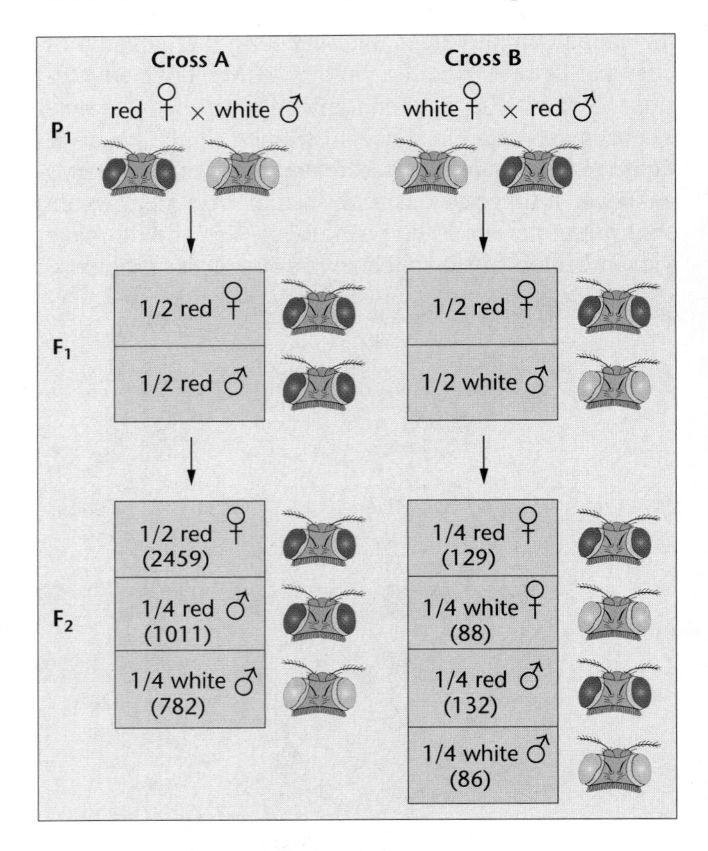

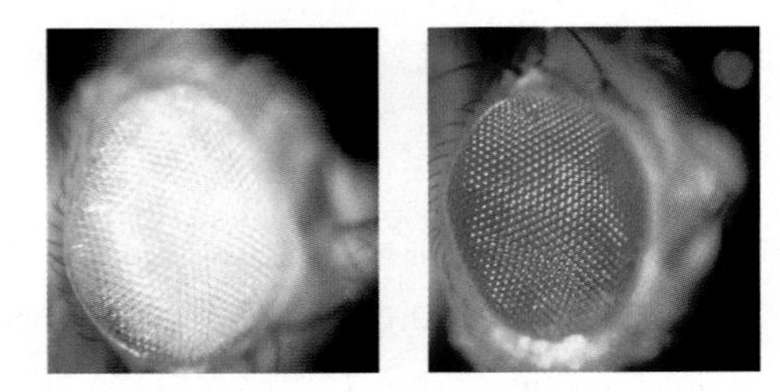

FIGURE 4–10 The F_1 and F_2 results of T. H. Morgan's reciprocal crosses involving the X-linked *white* mutation in *Drosophila melanogaster*. The actual F_2 data are shown in parentheses. The photographs show white eyes and the brick-red wild-type eye color. *(Photos: Carolina Biological Supply Co./Phototake NYC)*

Morgan's work established that the inheritance pattern of the white-eye trait is clearly related to the sex of the parent carrying the mutant allele. Unlike the outcome of the typical monohybrid cross, reciprocal crosses between white- and red-eyed flies did not yield identical results. In contrast, in all of Mendel's monohybrid crosses, F_1 and F_2 data were very similar regardless of which P_1 parent exhibited the recessive mutant trait. Morgan's analysis led to the conclusion that the *white* locus is present on the X chromosome rather than on one of the autosomes. As such, both the gene and the trait are said to be X-linked.

Results of reciprocal crosses between white-eyed and red-eyed flies are shown in Figure 4–10. The obvious differences in phenotypic ratios in both the F_1 and F_2 generations are dependent on whether or not the P_1 white-eyed parent was male or female.

Morgan was able to correlate these observations with the difference found in the sex chromosome composition between male and female *Drosophila*. He hypothesized that the recessive allele for white eyes is found on the X chromosome, but its corresponding locus is absent from the Y chromosome. Females thus have two available gene sites, one on each X chromosome, while males have only one available gene site on their single X chromosome.

Morgan's interpretation of X-linked inheritance, shown in Figure 4–11, provides a suitable theoretical explanation for his results. Since the Y chromosome lacks homology with most genes on the X chromosome, whatever alleles are present on the X chromosome of the males will be expressed directly in their phenotype. Males cannot be homozygous or heterozygous for X-linked genes, and this condition is referred to as being **hemizygous**.

One result of X-linkage is the **crisscross pattern of inheritance**, whereby phenotypic traits controlled by recessive X-linked genes are passed from homozygous mothers to all sons. This pattern occurs because females exhibiting a recessive trait carry the mutant allele on both X chromosomes. Because male offspring receive one of their mother's two X chromosomes and are hemizygous for all alleles present on that X, all sons will express the same recessive X-linked traits as their mother.

Morgan's work has taken on great historical significance. By 1910, the correlation between Mendel's work and the behavior of chromosomes during meiosis had provided the basis for the **chromosome theory of inheritance**. Work involving the X chromosome is considered to be the first solid experimental evidence in support of this theory. In the ensuing two decades, these findings inspired further research, which provided indisputable evidence in support of this theory.

X-Linkage in Humans

In humans, many genes and the respective traits controlled by them are recognized as being linked to the X chromosome (see Table 4.2). These X-linked traits can be easily identified in a pedigree because of the crisscross pattern of inheritance. A pedigree for one form of human

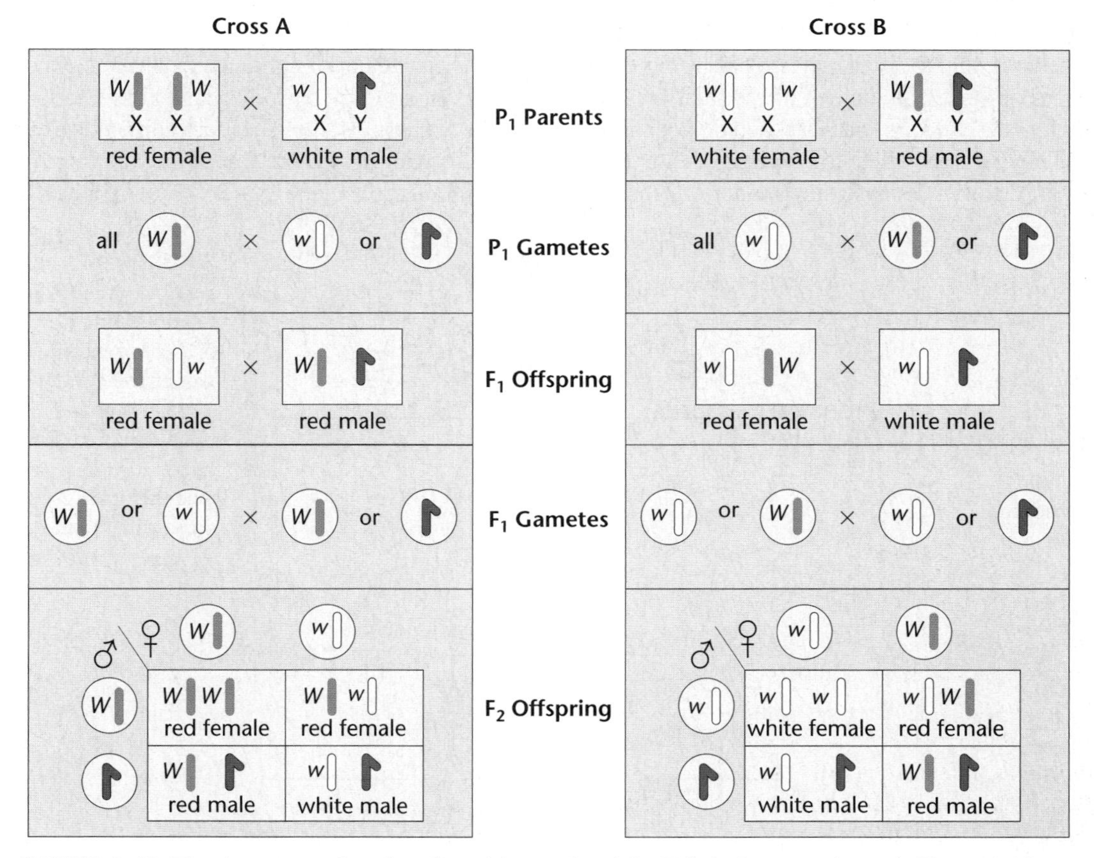

FIGURE 4–11 The chromosomal explanation of the results of the X-linked crosses shown in Figure 4–10.

TABLE 4.2 Human X-Linked Traits

Condition	*Characteristics*
Color blindness, deutan type	Insensitivity to green light.
Color blindness, protan type	Insensitivity to red light.
Fabry disease	Deficiency of galactosidase A; heart and kidney defects, early death.
G-6-PD deficiency	Deficiency of glucose-6-phosphate dehydrogenase, severe anemic reaction following intake of primaquines in drugs and certain foods, including fava beans.
Hemophilia A	Classical form of clotting deficiency; absence of clotting factor VIII.
Hemophilia B	Christmas disease; absence of clotting factor IX.
Hunter syndrome	Mucopolysaccharide storage disease resulting from iduronate sulfatase enzyme deficiency; short stature, clawlike fingers, coarse facial features, slow mental deterioration, and deafness.
Ichthyosis	Deficiency of steroid sulfatase enzyme; scaly dry skin, particularly on extremities.
Lesch-Nyhan syndrome	Deficiency of hypoxanthine-guanine phosphoribosyl transferase enzyme (HGPRT) leading to motor and mental retardation, self-mutilation, and early death.
Duchenne muscular dystrophy	Progressive, life-shortening disorder characterized by muscle degeneration and weakness; sometimes associated with mental retardation; absence of the protein dystrophin.

color blindness is shown in Figure 4–12. The mother in generation I passes the trait to all her sons but to none of her daughters. If the offspring in generation II marry normal individuals, the color-blind sons will produce all normal male and female offspring (III-1, 2, and 3); the normal-visioned daughters will produce normal-visioned female offspring (III-4, 6, and 7), as well as color-blind (III-8) and normal-visioned (III-5) male offspring.

The way in which X-linked genes are transmitted causes unusual circumstances associated with recessive X-linked disorders, in comparison to recessive autosomal disorders. For example, if an X-linked disorder debilitates or is lethal to the affected individual prior to reproductive maturation, the disorder occurs exclusively in males. This is so because the only sources of the lethal allele in the population are in heterozygous females who are "carriers" and do not express the disorder. They pass the allele to one-half of their sons, who develop the disorder because they are hemizygous but rarely, if ever, reproduce. Heterozygous females also pass the allele to one-half of their daughters, who become carriers but do not develop the disorder. An example of such an X-linked disorder is Duchenne muscular dystrophy. The disease has an onset prior to age 6 and is often lethal prior to age 20. It normally occurs only in males.

4.11 In Sex-Limited and Sex-Influenced Inheritance, an Individual's Sex Influences the Phenotype

In other instances, inheritance patterns may be affected by the sex of an individual, although not necessarily by genes on the X chromosome. There are numerous examples in different organisms where the sex of the individual plays a determining role in the expression of certain phenotypes. In

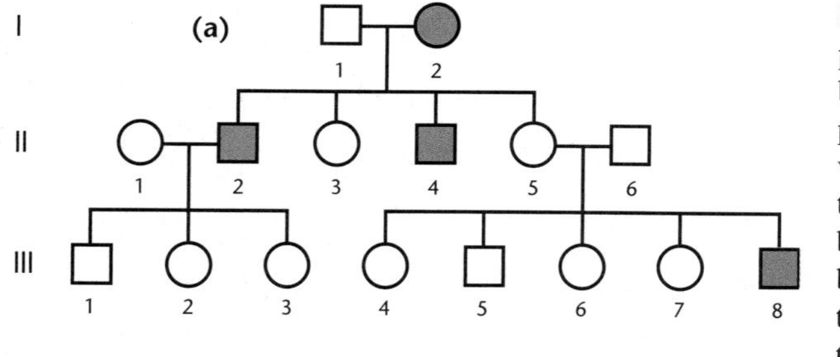

Symbols
c = color blindness
C = normal vision
↱ = Y chromosome

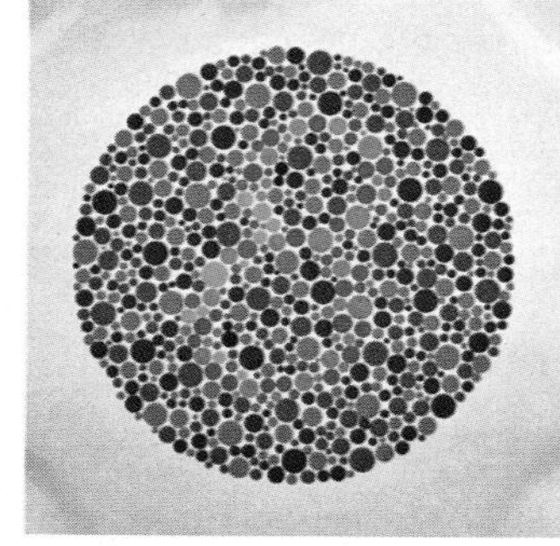

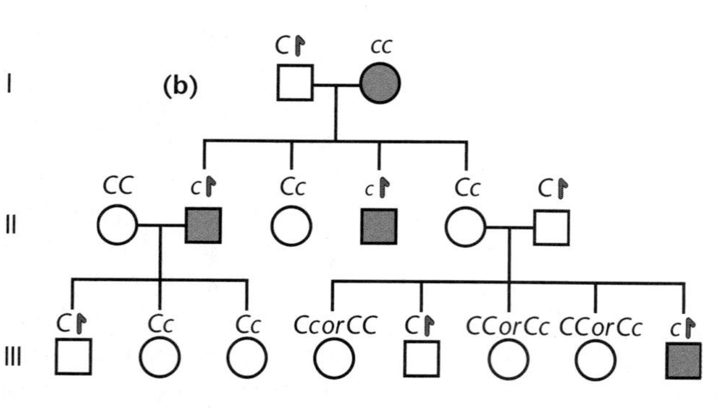

FIGURE 4–12 (a) A human pedigree of the X-linked color blindness trait. (b) The most probable genotype of each individual in the pedigree. The photograph is of an Ishihara color blindness chart. Red-green color-blind individuals see a 3 rather than the figure 8 visualized by those with normal color vision. *(Photo: Mary Teresa Giancoli)*

some cases, the expression of a specific phenotype is absolutely limited to one sex; in others, the sex of an individual influences the expression of a phenotype that is not limited to one sex or the other. This distinction differentiates **sex-limited inheritance** from **sex-influenced inheritance**.

In domestic fowl, tail and neck plumage is often distinctly different in males and females (Figure 4–13), demonstrating sex-limited inheritance. Cock feathering is longer, more curved, and pointed, while hen feathering is shorter and less curved. Inheritance of these feather phenotypes is controlled by a single pair of autosomal alleles whose expression is modified by the individual's sex hormones.

As shown in the following chart, hen feathering is due to a dominant allele, *H*, but regardless of the homozygous presence of the recessive *h* allele, all females remain hen-feathered. Only in males does the *hh* genotype result in cock feathering.

Genotype	*Phenotype*	
	Females	*Males*
HH	Hen-feathered	Hen-feathered
Hh	Hen-feathered	Hen-feathered
hh	Hen-feathered	Cock-feathered

In certain breeds of fowl, the hen-feathering or cock-feathering allele has become fixed in the population. In the Leghorn breed, all individuals are of the *hh* genotype; as a result, males always differ from females in their plumage. Sebright bantams are all *HH*, resulting in no sexual distinction in feathering phenotypes.

Another example of sex-limited inheritance involves the autosomal genes responsible for milk yield in dairy cattle. Regardless of the overall genotype that influences the quantity of milk production, those genes are obviously expressed only in females.

Cases of sex-influenced inheritance include pattern baldness in humans, horn formation in certain breeds of sheep

FIGURE 4–13 Hen feathering (left) and cock feathering (right) in domestic fowl. Note that the hen's feathers are shorter and less curved. *(Photo: Hans Reinhard/Bruce Coleman, Inc.)*

FIGURE 4–14 Pattern baldness, a sex-influenced autosomal trait in humans. *(Photo: Debra P. Hershkowitz/Bruce Coleman, Inc.)*

(e.g., Dorset Horn sheep), and certain coat-color patterns in cattle. In such cases, autosomal genes are responsible for the contrasting phenotypes displayed by both males and females, but the expression of these genes is dependent on the hormonal constitution of the individual. Thus, the heterozygous genotype exhibits one phenotype in one sex and the contrasting one in the other. For example, **pattern baldness** in humans, where the hair is very thin on the top of the head (Figure 4–14), is inherited in this way:

Genotype	*Phenotype*	
	Females	*Males*
BB	Bald	Bald
Bb	Not bald	Bald
bb	Not bald	Not bald

Females can display pattern baldness, but this phenotype is much more prevalent in males. When females do inherit the *BB* genotype, the phenotype is less pronounced than in males and is expressed later in life.

4.12 Phenotypic Expression Is Not Always a Direct Reflection of the Genotype

We now focus on **phenotypic expression**. In previous discussions, we assumed that the genotype of an organism is always directly expressed in its phenotype. For example, pea plants homozygous for the recessive *d* allele (*dd*) will always be dwarf. We discussed gene expression as though the genes operate in a closed system in which the presence or absence of functional products directly determines the collective phenotype of an individual. The situation is actually much more complex. Most gene products function within the internal milieu of the cell, and cells interact with one another in various ways. Further, the organism exists under diverse environmental influences. Thus, gene expression and the resultant phenotype are often modified through the interaction between an individual's particular genotype and the internal and external environment. Here, we deal with several important variables that are known to modify gene expression.

Penetrance and Expressivity

Some mutant genotypes are always expressed as a distinct phenotype, whereas others produce a proportion of individuals whose phenotypes cannot be distinguished from normal (wild type). The degree of expression of a particular trait can be studied quantitatively by determining the *penetrance* and *expressivity* of the genotype under investigation. The percentage of individuals that show at least some degree of expression of a mutant genotype defines the **penetrance** of the mutation. For example, the phenotypic expression of many mutant alleles in *Drosophila* can overlap with wild type. If 15% of mutant flies show the wild-type appearance, the mutant gene is said to have a penetrance of 85%.

By contrast, **expressivity** reflects the *range of expression* of the mutant genotype. Flies homozygous for the recessive mutant *eyeless* gene yield phenotypes that range from the presence of normal eyes to a partial reduction in size to the complete absence of one or both eyes (Figure 4–15). Although the average reduction of eye size is one-fourth to one-half, expressivity ranges from complete loss of both eyes to completely normal eyes.

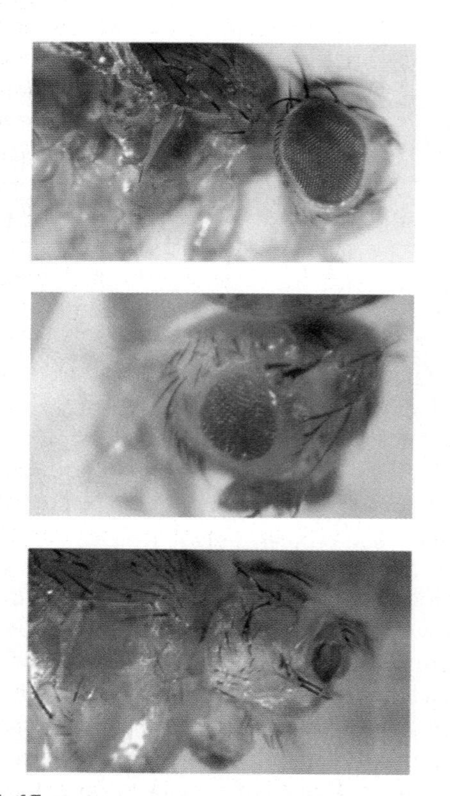

FIGURE 4–15 Variable expressivity, as shown in flies homozygous for the *eyeless* mutation in *Drosophila*. Gradations in phenotype range from wild type to partial reduction to eyeless. *(Top and bottom photos: Tanya Wolff, Washington University School of Medicine. Middle photo: Joel C. Eisenberg, Ph.D., Dept. of Biochemistry, St. Louis University Medical Center)*

Examples such as the expression of the *eyeless* gene provide the basis for experiments to determine the causes of phenotypic variation. If a laboratory environment is held constant and extensive phenotypic variation is still observed, other genes may be influencing or modifying the *eyeless* phenotype. On the other hand, if the genetic background is not the cause of the phenotypic variation, environmental factors such as temperature, humidity, and nutrition may be involved. In the case of the *eyeless* phenotype, experiments have shown that both genetic background and environmental factors influence its expression.

Temperature Effects

Chemical activity depends on the kinetic energy of the reacting substances, which in turn depends on the surrounding temperature. We can thus expect temperature to influence phenotypes. One example is the evening primrose, which produces red flowers when grown at 23°C and white flowers when grown at 18°C. An even more striking example is seen in Siamese cats and Himalayan rabbits, which exhibit dark fur in certain body regions where the body temperature is slightly cooler, particularly the nose, ears, and paws (Figure 4–16). In these animals, it appears that the enzyme responsible for pigment production is functional at the lower temperatures present in the extremities, but it loses its catalytic function at the slightly higher temperatures found throughout the rest of the body.

Mutations whose expression is affected by temperature are examples of **conditional mutations**. They are called **temperature-sensitive mutations**. Examples are known in viruses and a variety of organisms, including bacteria, fungi, and *Drosophila*. In extreme cases, an organism carrying a mutant allele may express a mutant phenotype when grown at one temperature, but express the wild-type phenotype when reared at another temperature. This type of temperature effect is useful in studying mutations that interrupt essential processes during development and are thus normally detrimental to the organism. For example, if bacterial viruses are cultured under *permissive conditions* of 25°C, the gene product is functional, infection proceeds normally, and new viruses are produced; but if bacterial viruses carrying temperature-sensitive mutations infect bacteria cultured at 42°C (the *restrictive condition*), infection progresses up to the point where the essential gene product is required (e.g., for viral assembly) and then arrests. The use of temperature-sensitive mutations, which can be induced and isolated, has added immensely to the study of viral genetics.

Onset of Genetic Expression

The age at which an organism expresses a gene corresponds to the normal sequence of growth and development. In humans, the prenatal, infant, preadult, and adult

FIGURE 4–16 (a) A Himalayan rabbit. (b) A Siamese cat. Both species show dark fur color on the snout, ears, and paws. The patches are due to the temperature-sensitive allele responsible for pigment production. *(Left Jane Burton/Bruce Coleman Inc., right Dr. William S. Klug.)*

phases require different genetic information. As a result, many severe inherited disorders are not manifested until after birth. For example, **Tay-Sachs disease**, inherited as an autosomal recessive, is a lethal lipid metabolism disease involving an abnormal enzyme, hexosaminidase A. Newborns appear to be phenotypically normal for the first few months. Then developmental retardation, paralysis, and blindness ensue, and most affected children die around the age of three.

The **Lesch-Nyhan syndrome**, inherited as an X-linked recessive disease, is characterized by abnormal nucleic acid metabolism (biochemical salvage of nitrogenous purine bases), leading to the accumulation of uric acid in blood and tissues, mental retardation, palsy, and self-mutilation of the lips and fingers. The disorder is due to a mutation in the gene encoding hypoxanthine-guanine phosphoribosyl transferase (HGPRT). Newborns are normal for six to eight months prior to the onset of the first symptoms.

Still another example involves **Duchenne muscular dystrophy (DMD)**, an X-linked recessive disorder associated with progressive muscular wasting. It is not usually diagnosed until the child is three to five years old. Even with modern medical intervention, the disease is often fatal in the early twenties.

Perhaps the most age-variable of all inherited human disorders is **Huntington disease**. Inherited as an autosomal dominant, Huntington disease affects the frontal lobes of the cerebral cortex, where progressive cell death occurs over a period of more than a decade. Brain deterioration is accompanied by spastic uncontrolled movements, intellectual and emotional deterioration, and ultimately death. Onset of this disease has been reported at all ages, but it most frequently occurs between ages 30 and 50, with a mean onset age of 38 years.

These conditions support the concept that the critical expression of normal genes varies throughout the life cycle of all organisms, including humans. Gene products may play more essential roles at certain life stages, and it is likely that the internal physiological environment of an organism changes with age.

Genetic Anticipation

Interest in studying the genetic onset of phenotypic expression has intensified with the discovery of heritable disorders that exhibit a progressively earlier age of onset and an increased severity of the disorder in each successive generation. This phenomenon is called **genetic anticipation**.

Myotonic dystrophy (DM), the most common type of adult muscular dystrophy, clearly illustrates genetic anticipation. Individuals afflicted with this autosomal dominant disorder exhibit extreme variation in the severity of symptoms. Mildly affected individuals develop cataracts as adults but have little or no muscular weakness. Severely affected individuals demonstrate more extensive myopathy and may be mentally retarded. In its most extreme form, the disease is fatal just after birth. A great deal of excitement was generated in 1989 when C. J. Howeler and colleagues confirmed the correlation of increased severity with earlier onset. They studied 61 parent-child pairs, and in 60 cases, age of onset was earlier in the child than in his or her affected parent.

In 1992, an explanation was put forward for both the molecular cause of the mutation responsible for DM, as well as the basis of genetic anticipation. Interestingly, a particular region of the DM gene is repeated a variable number of times and is unstable. Normal individuals average about five copies of this region, minimally affected individuals have about 50 copies, and severely affected individuals have over 1000 copies. The most remarkable observation was that in successive generations, the size of the repeated segment increases. Although it is not yet clear how this expansion in size affects onset and phenotypic expression, the correlation is extremely strong. Several other inherited human disorders, including the fragile-X syndrome, Kennedy disease, and Huntington disease, also reveal an association between the size of specific regions of the responsible gene and disease severity.

Genomic Imprinting

There are some cases involving phenotypic expression where the phenotype depends on the parental origin of the chromosome carrying a particular gene, a phenomenon called **genomic** (or **parental**) **imprinting**. Certain chromosomal regions and the genes contained within them somehow retain a memory, or an "imprint," of their parental origin in some species, influencing whether specific genes are expressed or remain genetically silent—that is, they are not expressed.

The imprinting step is thought to occur before or during gamete formation, leading to differentially marked genes (or chromosome regions) in sperm-forming versus egg-forming tissues. The process differs from mutation because the imprint can be reversed in succeeding generations as genes pass from mother to son to granddaughter, and so on.

The first example of genomic imprinting was discovered in 1991, when three specific mouse genes were shown to undergo imprinting. One is the gene encoding insulin-like growth factor II (*Igf2*). A mouse that carries two normal wild-type alleles of this gene is normal in size, whereas a mouse that carries two mutant alleles lacks a growth factor and is dwarf. The size of a heterozygous mouse (one allele normal and one mutant; Figure 4–17) depends on the parental origin of the wild-type allele. The mouse will be normal in size if the wild-type allele comes from the father, but will be dwarf if the wild-type allele comes from the mother. From this, we can deduce that the normal *Igf2* gene is imprinted during egg production but functions normally when it has passed through sperm-producing tissue in males.

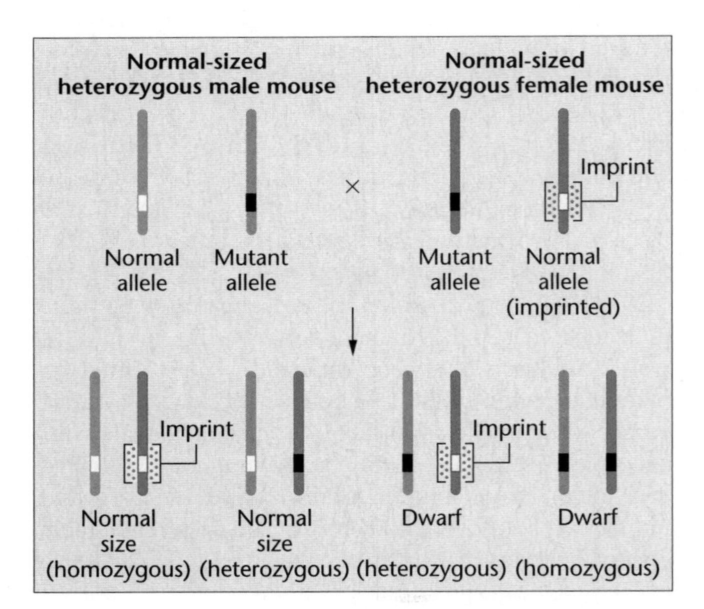

FIGURE 4–17 The effect of imprinting on the mouse *Igf2* gene, which produces dwarf mice in the homozygous condition. Heterozygotes that receive an imprinted normal allele from their mother are dwarf.

In humans, two distinct genetic disorders are thought to be caused by differential imprinting of the same region of chromosome 15. In both cases, the disorders result from the deletion of an identical region (15q11–15) in one member of the chromosome-15 pair. The first disorder, **Prader-Willi syndrome (PWS)**, results when the chromosome bearing the deletion is inherited from the father. If this chromosome is inherited from the mother, a different disorder, **Angelman syndrome (AS)**, results.

These two conditions exhibit different phenotypes. PWS entails mental retardation, a severe eating disorder marked by an uncontrollable appetite, obesity, and diabetes. AS also exhibits mental retardation, but involuntary muscle contractions (chorea) and seizures accompany the disorder. We can conclude that the involved region of chromosome 15 is imprinted differently in male and female gametes and that an undeleted maternal and paternal region are required for normal development.

Researchers are investigating how a region of a chromosome may be imprinted. Current thinking suggests that methylation of certain nitrogenous bases found in DNA is involved in the imprinting mechanism. Such an explanation is in keeping with the knowledge that methylation of cytosine in DNA inhibits gene activity. Whatever the cause, this phenomenon is a fascinating topic under intense investigation.

4.13 Extranuclear Inheritance Modifies Mendelian Patterns

Throughout the history of genetics, occasional reports have challenged the basic tenet of Mendelian transmission genetics—that the phenotype is transmitted by nuclear genes located on chromosomes of both parents. In this final section of the chapter, we consider several examples where inheritance patterns vary from patterns predicted by the traditional biparental inheritance of nuclear genes upon which Mendelian genetics is based. In the following cases, we will focus on two situations where transmission of genetic information is extranuclear. In our first case, an organism's phenotype is affected by the expression of genes contained in the DNA of mitochondria or chloroplasts rather than the nucleus. In our second case, an organism's phenotype is determined by genetic information expressed in the gamete of one parent—usually the mother—such that, following fertilization, the zygote is influenced by gene products from only one of the parents during development. As a result, the phenotype is influenced by genetic information transmitted through the cytoplasm rather than the nucleus.

Initially, such observations met with skepticism. However, increasing knowledge of molecular genetics and the discovery of DNA in mitochondria and chloroplasts caused **extranuclear inheritance** to be recognized as an important aspect of genetics.

Organelle Heredity: DNA in Chloroplasts and Mitochondria

Let us examine examples of inheritance patterns related to chloroplast and mitochondrial function. Before DNA was discovered in these organelles, the exact mechanism of transmission of the traits was not clear, except that their inheritance appeared to be linked to something in the cytoplasm rather than to genes in the nucleus. Furthermore, transmission was most often from the maternal parent through the ooplasm, causing the results of reciprocal crosses to vary. Such an extranuclear pattern of inheritance is now appropriately called **organelle heredity**.

Analysis of the inheritance patterns resulting from mutant alleles in chloroplasts and mitochondria has been difficult for two reasons. First, the function of these organelles is dependent upon gene products from both nuclear and organelle DNA, making the discovery of the genetic origin of mutations affecting organelle function difficult. Second, many mitochondria and chloroplasts are contributed to each progeny. Thus, if only one or a few of the organelles contain a mutant gene in a cell with a population of mostly normal mitochondria, the corresponding mutant phenotype may not be revealed. This condition of **heteroplasmy** may lead to normal cells since the organelles lacking the mutation provide the basis of wild-type function. Analysis is therefore much more complex than for Mendelian characters.

Chloroplasts: Variegation in Four-o'clock Plants

In 1908, Carl Correns (one of the rediscoverers of Mendel's work) provided the earliest example of inheritance linked to chloroplast transmission. Correns discovered a variant of the four-o'clock plant, *Mirabilis jalapa*, that had branches with either white, green, or variegated white-and-green leaves. The white areas in variegated leaves and the completely white leaves lack chlorophyll that provides the green color to normal leaves. Chlorophyll is the light-absorbing pigment made within chloroplasts.

Correns was curious about how inheritance of this phenotypic trait occurred. As shown in Figure 4–18, inheritance in all possible combinations of crosses is strictly determined by the phenotype of the ovule source. For example, if the seeds (representing the progeny) were derived from ovules on branches with green leaves, all progeny plants bore only green leaves, regardless of the phenotype of the source of pollen. Correns concluded that inheritance was transmitted through the cytoplasm of the maternal parent because the pollen, which contributes little or no cytoplasm to the zygote, had no apparent influence on the progeny phenotypes.

Since leaf coloration is related to the chloroplast, genetic information contained either in that organelle or somehow present in the cytoplasm and influencing the chloroplast must be responsible for the inheritance pattern. It now seems certain that the genetic "defect" that eliminates the green chlorophyll in the white patches on leaves is a mutation in the DNA housed in the chloroplast.

	Location of Ovule		
Source of Pollen	*White branch*	*Green branch*	*Variegated branch*
White branch	White	Green	White, green, or variegated
Green branch	White	Green	White, green, or variegated
Variegated branch	White	Green	White, green, or variegated

FIGURE 4–18 Offspring from crosses between flowers from various branches of four-o'clock plants. The photograph illustrates the variation in flower color displayed by four-o'clocks as well as variegation, seen in the pink flowers. *(Grant Heilman Photography, Inc.)*

Mitochondrial Mutations: *poky* in *Neurospora* and *petite* in *Saccharomyces*

Mutations affecting mitochondrial function have been discovered and studied, revealing that they too contain a distinctive genetic system. As with chloroplasts, mitochondrial mutations are transmitted through the cytoplasm. In our current discussion, we will emphasize the link between mitochondrial mutations and the resultant extranuclear inheritance patterns.

In 1952, Mary B. Mitchell and Hershel K. Mitchell studied the bread mold *Neurospora crassa*. They discovered a slow-growing mutant strain and named it *poky*. Slow growth is associated with impaired mitochondrial function, specifically in relation to certain cytochromes essential for electron transport. Results of genetic crosses between wild-type and *poky* strains suggest that *poky* is an extranuclear trait inherited through the cytoplasm. If the female parent is *poky* and the male parent is wild type, all progeny colonies are *poky*. The reciprocal cross, where *poky* is transmitted by the male parent, produces normal wild-type colonies.

Another extensive study of mitochondrial mutations has been performed with the yeast *Saccharomyces cerevisiae*. The first such mutation, described by Boris Ephrussi and his coworkers in 1956, was named *petite* because of the small size of the yeast colonies (Figure 4–19). Many independent *petite* mutations have since been discovered and studied, and all have a common characteristic—a deficiency in cellular respiration involving abnormal electron transport. This organism is a facultative anaerobe and can grow by fermenting glucose through glycolysis; thus, it may survive the loss of mitochondrial function by generating energy anaerobically.

The complex genetics of *petite* mutations has revealed that a small proportion are the result of nuclear mutations. They exhibit Mendelian inheritance and illustrate that mitochondria function depends on both nuclear and organellar gene products. The majority of them demonstrate cytoplasmic transmission, indicating mutations in the DNA of the mitochondria.

Mitochondrial Mutations: Human Genetic Disorders

As with other organisms, several criteria must be met in order for a human disorder to be attributable to genetically altered mitochondria:

1. Inheritance must exhibit a maternal rather than Mendelian pattern.
2. The disorder must reflect a deficiency in the bioenergetic function of the organelle.
3. There must be a specific genetic mutation in one or more of the mitochondrial genes.

Several disorders in humans are known to demonstrate these characteristics. For example, **myoclonic epilepsy and ragged-red fiber disease (MERRF)** demonstrates a pattern of inheritance consistent with maternal transmission. Only the offspring of affected mothers inherit this disorder, while the offspring of affected fathers are normal. Individuals with this rare disorder express ataxia (lack of muscular coordination), deafness, dementia, and epileptic seizures. The disease is named for the presence of "ragged-red" skeletal-muscle fibers that exhibit blotchy red patches resulting from the proliferation of aberrant mitochondria (Figure 4–20). Brain function, which has a high energy demand, is also affected in this disorder, leading to the neurological symptoms described above.

The mutation that causes MERRF has now been identified and is in a mitochondrial gene whose altered product interferes with the capacity for translation of proteins within the organelle. This, in turn, leads to the various manifestations of the disorder.

The cells of MERRF individuals exhibit heteroplasmy, containing a mixture of normal and abnormal mitochondria. Different patients display different proportions of the two, and even different tissues from the same patient exhibit various levels of abnormal mitochondria. Were it not for heteroplasmy, the mutation would very likely be lethal, testifying to the essential nature of mitochondrial

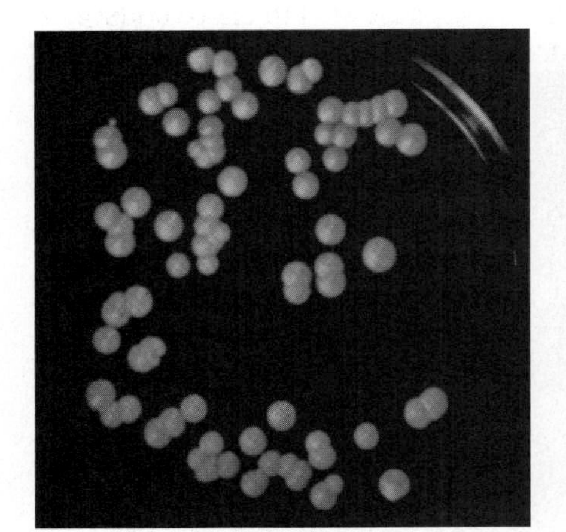

Normal colonies

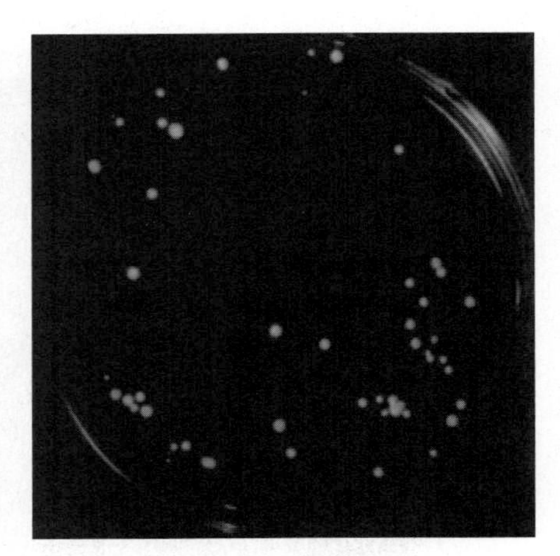

Petite colonies

FIGURE 4–19 Photos comparing normal versus *petite* colonies of the yeast *Saccharomyces cerevisiae*. *(Photo: Dr. Ronald A. Butow, Departement of Molecular Biology and Oncology, University of Texas Southern Medical Center.)*

(a)

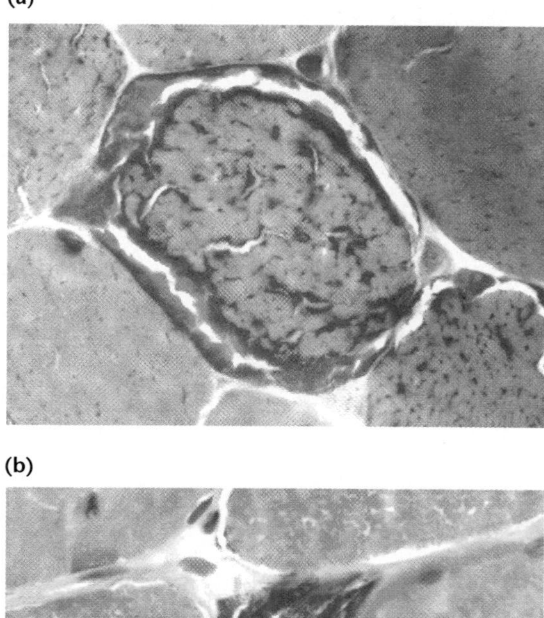

(b)

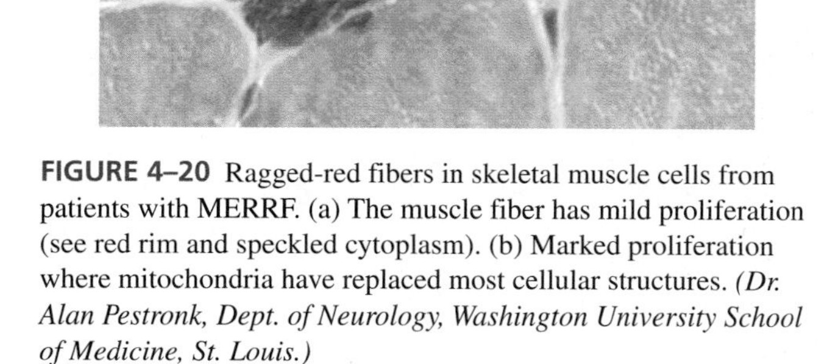

FIGURE 4–20 Ragged-red fibers in skeletal muscle cells from patients with MERRF. (a) The muscle fiber has mild proliferation (see red rim and speckled cytoplasm). (b) Marked proliferation where mitochondria have replaced most cellular structures. *(Dr. Alan Pestronk, Dept. of Neurology, Washington University School of Medicine, St. Louis.)*

function and its reliance on the genes encoded by DNA within the organelle.

A second human disorder, **Leber's hereditary optic neuropathy (LHON)**, also exhibits maternal inheritance as well as mitochondrial DNA lesions. The disease is characterized by sudden bilateral blindness. The average age of vision loss is 27, but onset is quite variable. Four mutations have been identified, all of which disrupt normal oxidative phosphorylation. Over 50% of cases are due to a mutation at a specific position in the mitochondrial gene encoding a subunit of NADH dehydrogenase, an enzyme essential to cell respiration. This mutation is transmitted maternally through the mitochondria to all offspring. It is interesting to note that in many instances of LHON, there is no family history; a significant number of cases are "sporadic," resulting from newly arisen mutations.

Maternal Effect: *Limnaea* Coiling

We conclude our discussion of extranuclear inheritance by considering a case that does not involve organelle heredity. **Maternal effect**, also referred to as *maternal influence*, is a situation where an offspring's phenotype for a particular trait is under the control of gene products present in the egg. The nuclear genes of the female gamete are transcribed, and the gene products (either proteins or yet untranslated mRNAs) accumulate in the egg cytoplasm. After fertilization, these products are distributed among newly formed cells and influence the patterns or traits established during early development. This interesting case of snails will illustrate the influence of the maternal genome on particular traits.

In the snail *Limnaea peregra*, some strains have left-handed, or sinistrally, coiled shells (*dd*), while others have right-handed, or dextrally, coiled shells (*DD* or *Dd*). These snails are hermaphroditic and may undergo either cross- or self-fertilization, providing a variety of types of matings.

Figure 4–21 illustrates the results of reciprocal crosses between true-breeding snails. These crosses yield different outcomes, even though both are between sinistral and dextral organisms and produce all heterozygous offspring. Examination of the progeny reveals that their phenotypes depend on the genotypes of the female parents. If we adopt that conclusion as a working hypothesis, we can test it by examining the offspring in subsequent generations of self-fertilization events. In each case, the hypothesis is upheld. Maternal parents that are *DD* or *Dd* produce only dextrally coiled progeny. Maternal parents that are *dd* produce only sinistrally coiled progeny. The coiling pattern of the progeny is determined by the genotype of the parent producing the egg, *regardless of the phenotype of that parent.*

Investigation of the developmental events in *Limnaea* reveals that the orientation of the spindle in the first cleavage division after fertilization determines the direction of coiling. Spindle orientation appears to be controlled by maternal genes acting on the developing eggs in the ovary. The orientation of the spindle, in turn, influences cell divisions following fertilization and establishes the permanent adult coiling pattern. The dextral allele (*D*) produces an active gene product that causes right-handed coiling. If ooplasm from dextral eggs is injected into uncleaved sinistral eggs, they cleave in a dextral pattern. However, in the converse experiment, sinistral ooplasm has no effect when injected into dextral eggs. Apparently the sinistral allele is the result of a classic recessive mutation that encodes an inactive gene product.

We can conclude, therefore, that females that are either *DD* or *Dd* produce oocytes that synthesize the *D* gene product, which is stored in the ooplasm. Even if the oocyte contains only the *d* allele following meiosis and is fertilized by a *d*-bearing sperm, the resulting *dd* snail will be dextrally coiled.

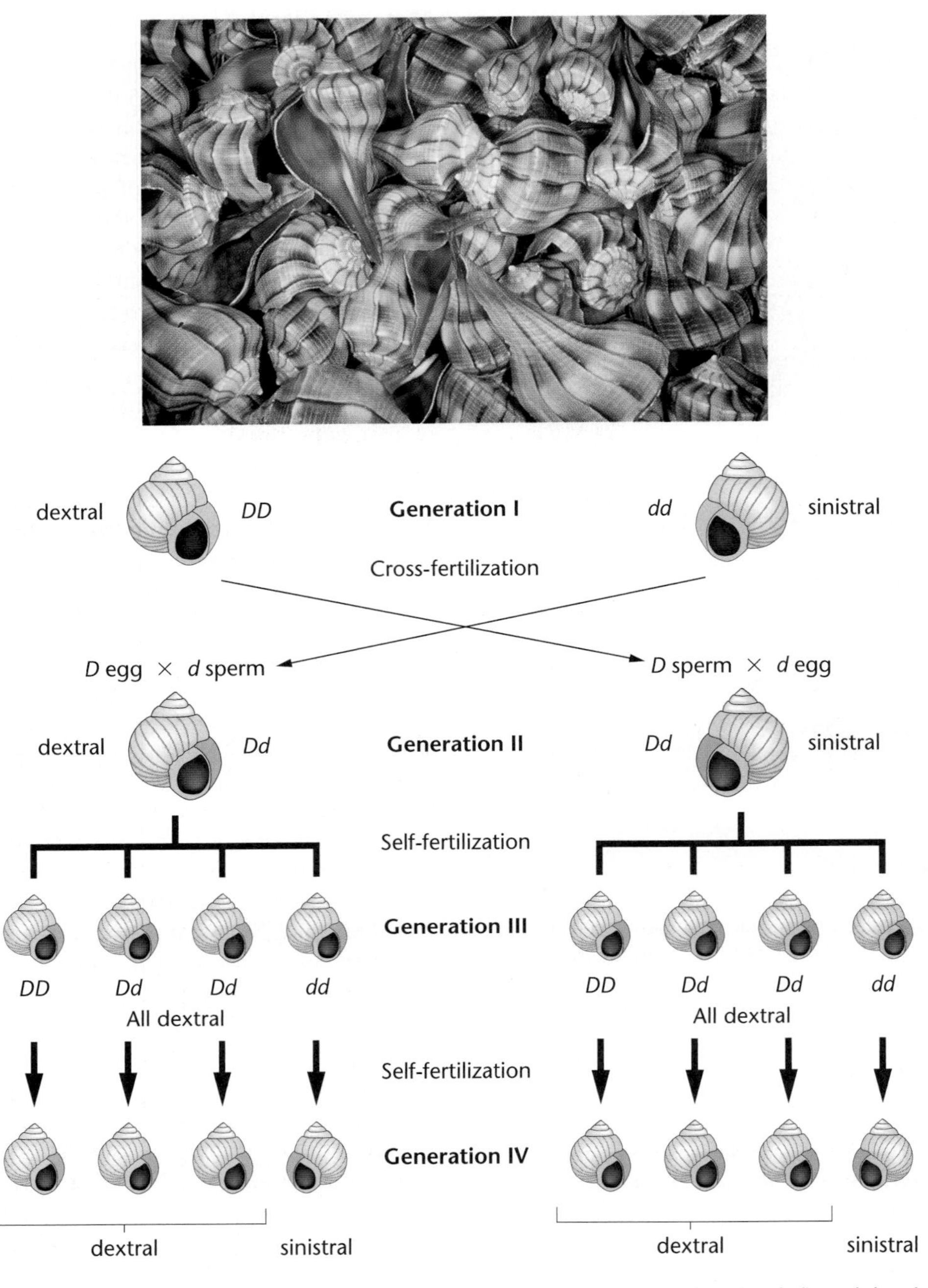

FIGURE 4–21 Inheritance of coiling in the snail *Limnaea peregra*. Coiling is either dextral (right-handed) or sinistral (left-handed). A maternal effect is evident in generations II and III, where the genotype of the maternal parent, rather than the offspring's own genotype, controls the phenotype of the offspring. The photograph shows a mixture of dextral and sinistral coiled snails. *(Robert & Linda Mitchell Photography.)*

GENETICS, TECHNOLOGY, AND SOCIETY

Improving the Genetic Fate of Purebred Dogs

Nothing is quite so heartbreaking for a dog lover as watching one's dog slowly go blind, standing by helplessly as the dog struggles to adapt to a life of perpetual darkness. That is what happens in progressive retinal atrophy (PRA), an inherited disorder first described in Gordon setters in 1911. Since that time, PRA has been found in many other breeds of dogs, including Irish setters, border collies, Norwegian elkhounds, toy poodles, miniature schnauzers, cocker spaniels, and Siberian huskies.

The products of many genes are required for the development and maintenance of a healthy retina, and a defect in any of them has the potential to cause retinal dysfunction. Decades of research have led to the identification of four genes (*rcd1*, *rcd2*, *erd*, and *prcd*) and more are likely to be discovered. Different genes are mutated in different breeds—for example, *rcd1* in Irish setters and *prcd* in Labrador retrievers. In all dogs examined thus far, PRA shows a recessive pattern of inheritance.

Whichever mutation is responsible, PRA is more common in particular pure breeds than in mixed breeds. The development of distinct breeds of dogs has involved the intensive selection for desirable attributes, such as a particular size, shape, color, or behavior. Most desirable characteristics are determined by recessive alleles, and the fastest way to increase the homozygosity of these alleles and establish the characteristics in a population is to mate close relatives, which are likely to carry the same alleles. In the practice known as linebreeding, for example, dogs may be mated to a cousin or a grandparent.

Unfortunately, the generations of inbreeding that have established favorable characteristics in purebreds have also increased the homozygosity of harmful recessive alleles, resulting in a variety of inherited diseases. Many breeds are plagued with inherited hip dysplasia, although it is particularly prevalent in German shepherds. Deafness and kidney disorders are common genetic maladies in Dalmatians.

Over 300 genetic diseases have been described in purebred dogs, and most breeds of dogs are affected by one or more. It is possible that fully 25% of purebred dogs are afflicted with some type of genetic ailment. Inbreeding is not the cause of genetic disease, but there is general agreement that improper breeding practices increase the frequency of disease.

Fortunately, advances in canine genetics are providing new tools for breeding healthy dogs. Since 1995, a genetic test has been available to identify mutations in the *rcd1* gene, which is responsible for the form of PRA that affects Irish setters. It is used to identify heterozygous carriers of *rcd1* mutations—dogs that show no symptoms of PRA but, if mated with other carriers, pass on the trait to about 25% of their offspring. Eliminating PRA carriers from breeding programs could theoretically eradicate this condition from Irish setters in just a few generations.

It took many years to identify *rcd1* and the other genes responsible for PRA. In the future, the isolation of genes underlying canine inherited disease should be quicker, thanks to the Dog Genome Project, a collaborative effort involving scientists at the University of California, the University of Oregon, the Fred Hutchinson Cancer Research Center in Seattle, and other research centers. Their goal is to create a complete genetic map of the 39 chromosomes in the dog. The preliminary stages were completed in late 1997, paving the way for the development of a more comprehensive map. Identifying the genes that cause inherited disease will enable the creation of genetic tests for diagnosis before symptoms develop, as well as for ensuring that breeding animals are free of harmful recessive alleles.

The Dog Genome Project may also benefit humans beyond the reduction in disease in their canine companions. Since about 85% of the genes in the dog genome have equivalents in humans, identifying a disease-causing dog gene may speed up isolation of the corresponding gene in humans. For example, some forms of PRA in dogs appear to be equivalent to retinitis pigmentosum (RP) in humans, which afflicts about 1.5 million people worldwide. Despite much research, RP remains poorly understood, and current treatments only retard its progress. Understanding the genetic basis of PRA in dogs may lead to breakthroughs in the diagnosis and treatment of RP, potentially saving the sight of thousands of people every year. By contributing to the cure of human diseases, dogs may prove to be "man's best friend" in an entirely new way.

References

Berson, E. L. 1996. Retinitis pigmentosum: Unfolding its mystery. *Proc. Natl. Acad. Sci. USA* 93: 4526–28.

Ray, K., Baldwin V., Acland G., and Aquirre, G. 1995. Molecular diagnostic tests for ascertainment of genotype at the rod cone dysplasia (*rcd1* locus) in Irish setters. *Curr. Eye Res.* 14: 243.

Smith, C. A. 1994. New hope for overcoming canine inherited disease. *Am. J. Vet. Med. Assoc.* 204: 41–46.

Chapter Summary

1. Since Mendel's work was rediscovered, the study of transmission genetics has expanded to include many alternative modes of inheritance. In many cases, phenotypes can be influenced by two or more genes.
2. Incomplete, or partial, dominance is exhibited when an intermediate phenotypic expression of a trait occurs in an organism that is heterozygous for two alleles.
3. Codominance is exhibited when distinctive expression of two alleles occurs in a heterozygous organism.
4. The concept of multiple alleles applies to populations, since a diploid organism may host only two alleles at any given locus. However, within a population many alternative alleles of the same gene can occur.
5. Lethal mutations usually result in the inactivation or lack of synthesis of gene products that are essential during an organism's development—these mutations can be recessive or dominant. Some lethal genes, such as the one that causes Huntington disease, are not expressed until adulthood.
6. Mendel's classic F_2 ratio is often modified in instances where gene interaction controls phenotypic variation.
7. Epistasis may occur when two or more genes influence a single characteristic. Usually, the expression of one of the genes masks the expression of the other gene or genes.
8. Complementation analysis determines whether independently isolated mutations producing similar phenotypes are alleles of one another or whether they represent separate genes.
9. Genes located on the X chromosome display a unique mode of inheritance called X-linkage.
10. Sex-limited and sex-influenced inheritance occur when the sex of the organism affects the phenotype controlled by a gene located on an autosome.
11. Phenotypic expression is not always the direct reflection of the genotype. Penetrance measures the percentage of organisms in a given population that exhibit evidence of the corresponding mutant phenotype. Expressivity, on the other hand, measures the range of phenotypic expression of a given genotype.
12. The time of onset of gene expression in organisms varies when the need for certain gene products occurs at different periods during development, growth, and aging.
13. Genetic anticipation is a phenomenon where the onset of phenotypic expression occurs earlier and becomes more severe in each ensuing generation.
14. Genomic imprinting is a process whereby a region of either the paternal or maternal chromosome is modified (marked or imprinted), thereby affecting phenotypic expression. Expression therefore depends on which parent contributes a mutant allele.
15. Patterns of inheritance sometimes vary from that expected during the biparental transmission of nuclear genes. In such instances, phenotypes most often appear to result from extranuclear genetic information transmitted through the egg.
16. Organelle heredity is based on the genotypes of chloroplast and mitochondrial DNA as these organelles are transmitted to offspring. Chloroplast mutations affect the photosynthetic capabilities of plants, whereas mitochondrial mutations affect cells highly dependent on energy generated through cellular respiration. The resulting mutants display phenotypes related to the loss of function of these organelles.
17. Patterns of maternal effect result when nuclear gene products controlled by the maternal genotype of the egg influence early development.

Key Terms

ABO blood group, 63
Angelman syndrome (AS), 78
autosome, 61
Bombay phenotype, 64
chromosome theory of inheritance, 73
cis-trans test, 72
codominance, 63
color blindness, 74
complementation analysis, 71
complementation group, 72
conditional mutation, 76
crisscross pattern of inheritance, 73
Duchenne muscular dystrophy (DMD), 77
epigenesis, 67
epistasis, 67
expressivity, 76
extranuclear inheritance, 79
gene interaction, 67
genomic imprinting, 78
genetic anticipation, 77
H substance, 64
hemizygous, 73
heteroplasmy, 79
hexosaminidase, 62
Huntington disease, 65, 77
incomplete dominance, 62
Leber hereditary optic neuropathy (LHON), 81
Lesch-Nyhan syndrome, 77
lethal allele, 64
maternal effect, 81
MN blood group, 63
multiple alleles, 63
myotonic dystrophy (DM), 77
myoclonic epilepsy and ragged-red fiber disease (MERRF), 80
organelle heredity, 79
parental imprinting, 78
partial dominance, 62
pattern baldness, 75
penetrance, 76
phenotypic expression, 75
Prader-Willi syndrome (PWS), 78
sex-influenced inheritance, 75
sex-limited inheritance, 75
Tay-Sachs disease, 77
temperature-sensitive mutation, 76
wild-type allele, 61
X-linkage, 72

INSIGHTS AND SOLUTIONS

Genetic problems take on added complexity if they involve two independent characters and multiple alleles, incomplete dominance, or epistasis. The most difficult types of problems are those that pioneering geneticists faced during laboratory or field studies. They had to determine the mode of inheritance by working backward from the observations of offspring to parents of unknown genotype.

1. Consider the problem of comb shape inheritance in chickens, where walnut, pea, rose, and single are the observed distinct phenotypes. (See the opening photograph in this chapter on page 60.) How is comb shape inherited, and what are the genotypes of the P_1 generation of each cross? Use the following data:

Cross 1:	single × single	→	all single
Cross 2:	walnut × walnut	→	all walnut
Cross 3:	rose × pea	→	all walnut
Cross 4:	F_1 × F_1 of Cross 3		
	walnut × walnut	→	93 walnut
			28 rose
			32 pea
			10 single

Solution: At first glance, this problem appears quite difficult. However, applying a systematic approach and breaking the analysis into steps usually simplifies it. Our approach involves two steps. Analyze the data carefully for any useful information. Once you identify something that is clearly helpful, follow an empirical approach—that is, formulate a hypothesis and, in a sense, test it against the given data. Look for a pattern of inheritance that is consistent with all cases.

This problem gives two immediately useful facts. First, in cross 1, P_1 singles breed true. Second, while P_1 walnut breeds true in cross 2, a walnut phenotype is also produced in cross 3 between rose and pea. When these F_1 walnuts are crossed in cross 4, all four comb shapes are produced in a ratio that approximates 9:3:3:1. This observation immediately suggests a cross involving two gene pairs, because the resulting data display the same ratio as in Mendel's dihybrid crosses. Since only one trait is involved (comb shape), epistasis may be occurring. This could serve as your working hypothesis, and you must now propose how the two gene pairs "interact" to produce each phenotype.

If you call the allele pairs *A*, *a* and *B*, *b*, you can predict that because walnut represents 9/16 in cross 4, *A–B–* will produce walnut. You might also hypothesize that in cross 2, the genotypes are *AABB* × *AABB*, where walnut bred true. (Recall that *A–* and *B–* mean *AA* or *Aa* and *BB* or *Bb*, respectively.)

The phenotype representing 1/16 of the offspring of cross 4 is single, therefore you could predict that this phenotype is the result of the *aabb* genotype. This is consistent with cross 1.

Now you have only to determine the genotypes for rose and pea. The most logical prediction is that at least one dominant *A* or *B* allele combined with the double recessive condition of the other allele pair accounts for these phenotypes. For example,

A–bb → rose
aaB– → pea

If *AAbb* (rose) is crossed with *aaBB* (pea) in cross 3, all offspring will be *AaBb* (walnut). This is consistent with the data, and you must now look at only cross 4. We predict these walnut genotypes to be *AaBb* (as above), and from the cross *AaBb* (walnut) × *AaBb* (walnut) we expect

9/16 *A–B–* (walnut)
3/16 *A–bb* (rose)
3/16 *aaB–* (pea)
1/16 *aabb* (single)

Our prediction is consistent with the information given. The initial hypothesis of the epistatic interaction of two gene pairs proves consistent throughout, and the problem is solved.

This problem demonstrates the need for a basic theoretical knowledge of transmission genetics. Then, you can search for appropriate clues that will enable you to proceed in a stepwise fashion toward a solution. Mastering problem solving requires practice, but can give you a great deal of satisfaction. Apply this general approach to the following problems.

2. Flower color in radishes may be red, purple, or white. The edible portion of the radish may be long or oval. When only flower color is studied, crossing red with white yields all purple. If these F_1 purple plants are interbred, no dominance is evident and the F_2 generation consists of 1/4 red : 1/2 purple : 1/4 white. Regarding radish shape, long is dominant to oval in a normal Mendelian fashion.

(a) Determine the F_1 and F_2 phenotypes from a cross between a true-breeding red, long radish and a white, oval radish.

(Hint: Be sure to define all gene symbols initially.*)*

Solution: This is a modified dihybrid cross in which the gene pair controlling color exhibits incomplete dominance; shape is controlled conventionally. We establish the gene symbols:

RR = red   *O–* = long
Rr = purple   *oo* = oval
rr = white

Our crosses yield

P_1: *RROO* × *rroo*
(red, long) (white, oval)
F_1: all *RrOo* (purple, long)
F_1 × F_1: *RrOo* × *RrOo*

F_2:	1/4 *RR*	3/4 *O–*	3/16 *RRO–*	red, long
		1/4 *oo*	1/16 *RRoo*	red, oval
	2/4 *Rr*	3/4 *O–*	6/16 *RrO–*	purple, long
		1/4 *oo*	2/16 *Rroo*	purple, oval
	1/4 *rr*	3/4 *O–*	3/16 *rrO–*	white, long
		1/4 *oo*	1/16 *rroo*	white, oval

Note that we used the forked-line method to generate the F_2 results. We considered the outcome of crossing F_1 parents for the color genes (*Rr* × *Rr*); then we considered the outcome of shape (*Oo* × *Oo*).

(b) A red, oval radish is crossed with a plant of unknown genotype and phenotype, yielding 103 red, long : 101 red, oval : 98 purple, long : 100 purple, oval offspring. Determine the genotype and phenotype of the unknown plant.

Solution: Since the two characters are inherited independently, we consider them separately. The data indicate a 1/4 : 1/4 : 1/4 : 1/4 ratio. For color we see that

P_1: red × ??? (unknown)
F_1: 204 red (1/2)
198 purple (1/2)

The red parent must be *RR*. The unknown parent must have a genotype of *Rr* to produce these results. It is thus purple.

Now for shape:

P_1: oval × ??? (unknown)
F_1: 201 long (1/2)
201 oval (1/2)

Let's consider the oval and long characters. Because the oval plant must be *oo*, the unknown plant must have a genotype of *Oo* to produce these results. So it is long. The unknown plant is thus *RrOo* purple, long.

3. In humans, red-green color blindness is inherited as an X-linked recessive trait. A woman with normal vision whose father is color-blind marries a male who has normal vision. Predict the color vision of their male and female offspring.

Solution: The female is heterozygous because she inherited an X chromosome with the mutant allele from her father. Her husband is normal. Therefore, the parental genotypes are *Cc* × *C*↾ (↾ is the Y chromosome). All female offspring are normal (*CC* or *Cc*). One-half of the male children will be color-blind (*c*↾) and the other half will have normal vision (*C*↾).

Problems and Discussion Questions

1. In Shorthorn cattle, coat color may be red, white, or roan. Roan is an intermediate phenotype expressed as a mixture of red and white hairs. The following data are obtained from various crosses:

red × red	⟶	all red
white × white	⟶	all white
red × white	⟶	all roan
roan × roan	⟶	1/4 red : 1/2 roan : 1/4 white

How is coat color inherited? What are the genotypes of parents and offspring for each cross?

2. Contrast incomplete dominance and codominance.
3. With regard to the ABO blood types in humans, determine the genotypes of the male parent and female parent:

Male parent: blood type B whose mother was type O
Female parent: blood type A whose father was type B

Predict the blood types of the offspring that this couple may have and the expected ratio of each.

4. In foxes, two alleles of a single gene, *P* and *p*, may result in lethality (*PP*), platinum coat (*Pp*), or silver coat (*pp*). What ratio is obtained when platinum foxes are interbred? Is the *P* allele behaving dominantly or recessively in causing (a) lethality; (b) platinum coat color?
5. Three gene pairs located on separate autosomes determine flower color and shape as well as plant height. The first pair exhibits incomplete dominance, where color can be red, pink (the heterozygote), or white. The second pair leads to the dominant personate or recessive peloric flower shape, while the third gene pair produces either the dominant tall trait or the recessive dwarf trait. Homozygous plants that are red, personate, and tall are crossed with those that are white, peloric, and dwarf. Determine the F_1 genotype(s) and phenotype(s). If the F_1 plants are interbred, what proportion of the offspring will exhibit the same phenotype as the F_1 plants?

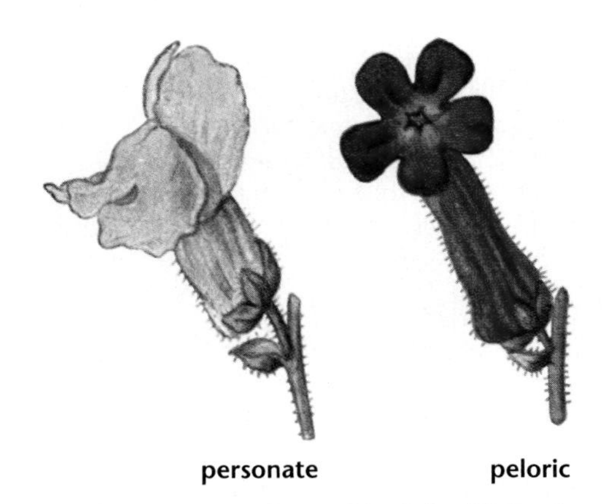

(Drawings of flowers: Erwin Bauer, The Scientific Basics of Plant Cultivation: A Textbook for Farmers, Gardeners, and Foresters. Gebruder Borntraeger, Berlin, 1924.)

6. As in the plants of Problem 5, color may be red, white, or pink; and flower shape may be personate or peloric. Determine the P_1 and F_1 genotypes for the following crosses:

(a) red, peloric × white, personate	⟶	F_1: all pink, personate
(b) red, personate × white, peloric	⟶	F_1: all pink, personate
(c) pink, personate × red, peloric	⟶	F_1: 1/4 red, personate; 1/4 red, peloric; 1/4 pink, personate; 1/4 pink, peloric
(d) pink, personate × white, peloric	⟶	F_1: 1/4 white, personate; 1/4 white, peloric; 1/4 pink, personate; 1/4 pink, peloric

What phenotype ratios would result from crossing the F_1 of (a) with the F_1 of (b)?

7. In some plants, a red pigment, cyanidin, is synthesized from a colorless precursor. The addition of a hydroxyl group (—OH) to the cyanidin molecule causes it to become purple. In a cross between two randomly selected purple plants, the following results are obtained:

94 purple : 31 red : 43 colorless

How many genes are involved in determining these flower colors? Which genotypic combinations produce which phenotypes? Diagram the purple × purple cross.

8. The following genotypes of two independently assorting autosomal genes determine coat color in rats:

A–B– (gray); *A–bb* (yellow); *aaB–* (black); *aabb* (cream)

A third gene pair on a separate autosome determines whether any color will be produced. The *CC* and *Cc* genotypes allow color according to the expression of the *A* and *B* alleles. However, the *cc* genotype results in albino rats regardless of the *A* and *B* alleles present. Determine the F_1 phenotypic ratio of the following crosses: (a) *AAbbCC* × *aaBBcc*; (b) *AaBBCC* × *AABbcc*; (c) *AaBbCc* × *AaBbcc*.

9. Given the inheritance pattern of coat color in rats described in Problem 8, predict the genotype and phenotype of the parents that produced the following F_1 offspring: (a) 9/16 gray : 3/16 yellow : 3/16 black : 1/16 cream; (b) 9/16 gray : 3/16 yellow : 4/16 albino; (c) 27/64 gray : 16/64 albino : 9/64 yellow : 9/64 black : 3/64 cream.

10. A husband and wife have normal vision, although both of their fathers are red-green color-blind, inherited as an X-linked recessive condition. What is the probability that their first child will be (a) a normal son, (b) a normal daughter, (c) a color-blind son, (d) a color-blind daughter?

11. In humans, the ABO blood type is under the control of autosomal multiple alleles. Red-green color blindness is a recessive X-linked trait. If two parents who are both type A and have normal vision produce a son who is color-blind and type O, what is the probability that their next child will be a female who has normal vision and is type O?

12. In spotted cattle, the colored regions may be mahogany or red. If a red female and a mahogany male, both derived from separate true-breeding lines, are mated and the cross is carried to an F_2 generation, the following results are obtained:

F_1: 1/2 mahogany males : 1/2 red females
F_2: 3/8 mahogany males : 1/8 red males :
1/8 mahogany females : 3/8 red females

When the reciprocal of the initial cross is performed (mahogany female and red male), identical results are obtained. Explain these results by postulating how the color is genetically determined, and diagram the crosses.

13. In cats, yellow coat color is determined by the *b* allele, and black coat color is determined by the *B* allele. The heterozygous condition results in a coat pattern known as tortoiseshell. These genes are X-linked. What kinds of offspring would be expected from a cross of a black male and a tortoiseshell female? What are the chances of getting a tortoiseshell male?

14. In *Drosophila*, an X-linked recessive mutation, *scalloped* (*sd*), causes irregular wing margins. Diagram the F_1 and F_2 results if
(a) A scalloped female is crossed with a normal male.
(b) A scalloped male is crossed with a normal female.
Compare these results to those that would be obtained if the *scalloped* gene is autosomal.

15. Another recessive mutation in *Drosophila*, *ebony* (*e*), is on an autosome (chromosome 3) and causes darkening of the body compared with wild-type flies. What phenotypic F_1 and F_2 male and female ratios will result if a scalloped-winged female with normal body color is crossed with a normal-winged ebony male? Work this problem by both the Punnett square method and the forked-line method.

16. While *vermilion* is X-linked in *Drosophila* and causes eye color to be bright red, *brown* is an autosomal recessive mutation that causes the eye to be brown. Flies carrying both mutations lose all pigmentation and are white-eyed. Predict the F_1 and F_2 results of the following crosses:
(a) vermilion females × brown males
(b) brown females × vermilion males
(c) white females × wild males

17. In pigs, coat color may be sandy, red, or white. A geneticist spent several years mating true-breeding pigs of all different color combinations, even obtaining true-breeding lines from different parts of the country. For crosses 1 and 4 below, she encountered a major problem: her computer crashed and she lost the F_2 data. She nevertheless persevered, and using the limited data shown here, was able to predict the mode of inheritance and the number of genes involved, and to assign genotypes to each coat color. Attempt to duplicate her analysis, based on the available data generated from the crosses shown.

Cross	P_1	F_1	F_2
1	Sandy × sandy	All red	Data lost
2	Red × sandy	All red	3/4 red : 1/4 sandy
3	Dandy × white	All sandy	3/4 sandy : 1/4 white
4	White × red	All red	Data lost

When you have formulated a hypothesis to explain the mode of inheritance and assigned genotypes to the respective coat colors, predict the outcomes of the F_2 generations where the data were lost.

18. A geneticist from an alien planet that prohibits genetic research brought with him two true-breeding lines of frogs. One frog line croaks by *uttering* "rib-it rib-it" and has purple eyes. The other frog line croaks by *muttering* "knee-deep knee-deep" and has green eyes. He mated the two frog lines, producing F_1 frogs that were all utterers with blue eyes. A large F_2 generation then yielded the following ratio:

27/64 blue, utterer
12/64 green, utterer
9/64 blue, mutterer
9/64 purple, utterer
4/64 green, mutterer
3/64 purple, mutterer

(a) How many total gene pairs are involved in the inheritance of both eye color and croaking?
(b) Of these, how many control eye color, and how many control croaking?

(c) Assign gene symbols for all phenotypes, and indicate the genotypes of the P_1, F_1, and F_2 frogs.

(d) After many years, the frog geneticist isolated true-breeding lines of all six F_2 phenotypes. Indicate the F_1 and F_2 phenotypic ratios of a cross between a blue, mutterer and a purple, utterer.

19. In another cross, the frog geneticist from Problem 18 mated two purple, utterers, with the results shown here. What were the genotypes of the parents?

9/16 purple, utterers
3/16 purple, mutterers
3/16 green, utterers
1/16 green, mutterers

20. In cattle, coats may be solid white, solid black, or black-and-white spotted. When true-breeding solid whites are mated with true-breeding solid blacks, the F_1 generation consists of all solid white individuals. After many $F_1 \times F_1$ matings, the following ratio was observed in the F_2 generation:

12/16 solid white
3/16 black-and-white spotted
1/16 solid black

Explain the mode of inheritance governing coat color by determining how many gene pairs are involved and which genotypes yield which phenotypes. Is it possible to isolate a true-breeding strain of black-and-white spotted cattle? If so, what genotype would they have? If not, explain why not.

21. In the guinea pig, a locus controlling coat color may be occupied by any of four alleles: C (*full color*), c^k (sepia), c^d (cream), or c^a (albino). A progressive order of dominance exists among these alleles when they are present heterozygously: $C > c^k > c^d > c^a$. Determine the genotype of each individual, and predict the phenotypic ratios of the offspring in the following crosses:

(a) sepia × cream, where both had an albino parent

(b) sepia × cream, where the sepia individual had an albino parent and the cream individual had two sepia parents

(c) sepia × cream, where the sepia individual had two full-color parents and the cream individual had two sepia parents

(d) sepia × cream, where the sepia individual had two full-color parents and the cream individual had two full-color parents

22. In parakeets, two autosomal genes (located on different chromosomes) control the production of feather pigment. Gene B controls the production of blue pigment, and gene Y controls the production of yellow pigment. Known recessive mutations in each gene result in the loss of pigment synthesis. Two green parakeets are mated and produce green, blue, yellow, and albino progeny.

(a) Based on this information, explain the pattern of inheritance. Be sure to include the genotypes of the green parents and all four phenotypic classes, as well as the fraction of total progeny that each phenotypic class represents in your answer.

(b) The parental (green) parakeets are the progeny of a cross between two true-breeding strains. What two types of crosses between true-breeding strains could have produced the green parents? Indicate the genotypes and phenotypes for each cross.

(Photo: Robert Pearcy/Animals Animals/Earth Scenes)

23. Consider the following three pedigrees (all involve the same human trait):

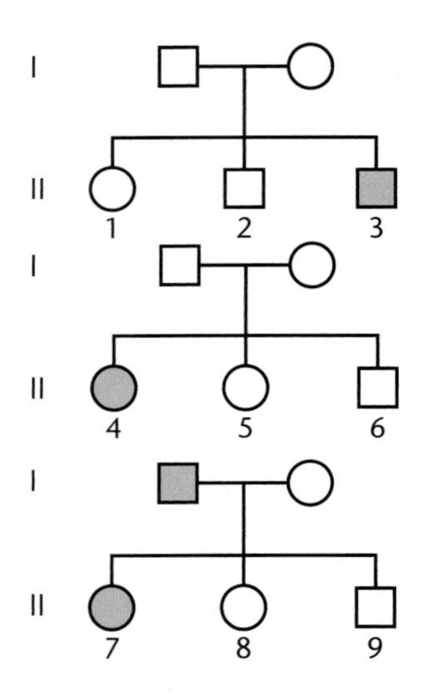

(a) Which sets of conditions, if any, can be excluded?

dominant and X-linked
dominant and autosomal
recessive and X-linked
recessive and autosomal

(b) For any set of conditions that you excluded, indicate the *single individual* in generation II (1–9) that was instrumental in your decision to exclude that condition. If none were excluded, answer "none apply."

(c) Given your conclusions in parts (a) and (b), indicate the *genotype* of individuals II-1, II-6, and II-9. If more than one possibility applies, list all possibilities. Use the symbols A and a for the genotypes.

24. Three autosomal recessive mutations in *Drosophila*, all with tan eye color ($r1$, $r2$, and $r3$) are independently isolated and subjected to complementation analysis. Of the results shown here, which, if any, are alleles of one another? Predict the results of the cross that is not shown—that is, $r2 \times r3$.

Cross 1:	$r1 \times r2 \longrightarrow$	F_1: all wild-type eyes
Cross 2:	$r1 \times r3 \longrightarrow$	F_1: all tan eyes

25. Labrador retrievers may be black, brown, or golden in color. While each color may breed true, many different outcomes occur if numerous litters are examined from a variety of matings, where the parents are not necessarily true-breeding. The results below show some of the possibilities. Propose a mode of inheritance that is consistent with these data, and indicate the corresponding genotypes of the parents in each mating. Indicate as well the genotypes of dogs that breed true for each color.

(Photo: William H. Mullins/Photo Researchers, Inc.)

(a) black × brown	⟶	all black
(b) black × brown	⟶	1/2 black 1/2 brown
(c) black × brown	⟶	3/4 black 1/4 golden
(d) black × golden	⟶	all black
(e) black × golden	⟶	4/8 golden 3/8 black 1/8 brown
(f) black × golden	⟶	2/4 golden 1/4 black 1/4 brown
(g) brown × brown	⟶	3/4 brown 1/4 golden
(h) black × black	⟶	9/16 black 4/16 golden 3/16 brown

26. In rabbits, a series of multiple alleles controls coat color in the following way: C is dominant to all other alleles and causes full color. The chinchilla phenotype is due to the c^{ch} allele, which is dominant to all alleles other than C. The c^h allele, dominant only to c^a (albino), results in the Himalayan coat color. Thus, the order of dominance is $C > c^{ch} > c^h > c^a$. For each of the following three cases, the phenotypes of the P_1 generations of two crosses are shown, as well as the phenotype of one member of the F_1 generation.

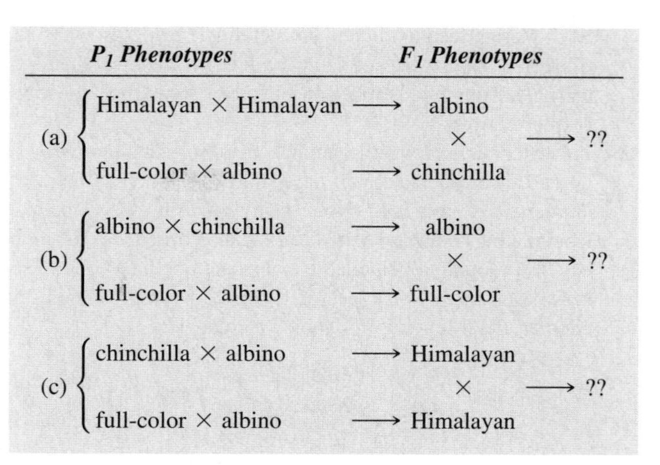

	P_1 Phenotypes		F_1 Phenotypes		
(a)	Himalayan × Himalayan	⟶	albino		
			×	⟶	??
	full-color × albino	⟶	chinchilla		
(b)	albino × chinchilla	⟶	albino		
			×	⟶	??
	full-color × albino	⟶	full-color		
(c)	chinchilla × albino	⟶	Himalayan		
			×	⟶	??
	full-color × albino	⟶	Himalayan		

Determine the genotypes of the P_1 generation and the F_1 offspring for each case, and predict the results of making each cross between F_1 individuals as shown.

27. Horses can be cremello (a light cream color), chestnut (a reddish brown color), or palomino (a golden color with white in the horse's tail and mane). Of these phenotypes, only palominos never breed true.

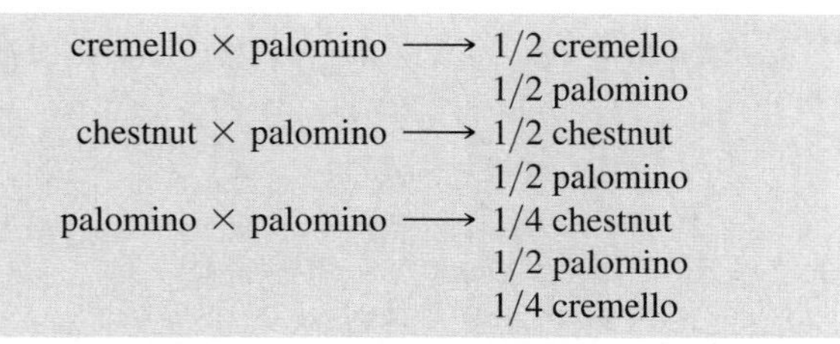

cremello × palomino	⟶	1/2 cremello 1/2 palomino
chestnut × palomino	⟶	1/2 chestnut 1/2 palomino
palomino × palomino	⟶	1/4 chestnut 1/2 palomino 1/4 cremello

(a) From these results, determine the mode of inheritance by assigning gene symbols and indicating which genotypes yield which phenotypes.
(b) Predict the F_1 and F_2 results of many initial matings between cremello and chestnut horses.

28. Pigment in the mouse is produced only when the C allele is present. Individuals of the cc genotype have no color. If color is present, it may be determined by the A and a alleles. AA or Aa results in agouti color, while aa results in black coats.

(a) What F_1 and F_2 genotypic and phenotypic ratios are obtained from a cross between $AACC$ and $aacc$ mice?
(b) In three crosses shown below between agouti females whose genotypes were unknown and males of the $aacc$ genotype, what are the genotypes of the female parents for each of the following phenotypic ratios?

(1)	(2)	(3)
8 agouti 8 colorless	9 agouti 10 black	4 agouti 5 black 10 colorless

29. Five human matings numbered 1–5 are shown in the following table. Included are both maternal and paternal phenotypes for ABO, MN, and Rh blood-group antigen status (Rh^+ is a dominant trait):

Parental Phenotypes								Offspring		
(1) A,	M,	Rh^-	×	A,	N,	Rh^-		(a) A,	N,	Rh^-
(2) B,	M,	Rh^-	×	B,	M,	Rh^+		(b) O,	N,	Rh^+
(3) O,	N,	Rh^+	×	B,	N,	Rh^+		(c) O,	MN,	Rh^-
(4) AB,	M,	Rh^+	×	O,	N,	Rh^+		(d) B,	M,	Rh^+
(5) AB,	MN,	Rh^-	×	AB,	MN,	Rh^-		(e) B,	MN,	Rh^+

Each mating resulted in one of the five offspring shown to the right (a–e). Match each offspring with one correct set of parents, using each parental set only once. Is there more than one set of correct answers?

30. In Dexter and Kerry cattle, animals may be polled (hornless) or horned. The Dexter animals have short legs, whereas the Kerry animals have long legs. When many offspring were obtained from matings between polled Kerrys and horned Dexters, one-half were found to be polled Dexters and one-half polled Kerrys. When these two types of F_1 cattle were bred with one another, the following F_2 data were obtained:

3/8	polled Dexters
3/8	polled Kerrys
1/8	horned Dexters
1/8	horned Kerrys

A geneticist was puzzled by these data and interviewed farmers who had bred these cattle for decades. She learned that Kerrys were true-breeding. Dexters, on the other hand, were not true-breeding and never produced as many offspring as Kerrys. Provide a genetic explanation for these observations.

Kerry cow

Dexter bull

(Top photo, Nigel J.H. Smith/Animals Animals/Earth Scenes. Bottom photo, Francois Gohier/Photo Researchers, Inc.)

31. What genetic criteria distinguish a case of extranuclear inheritance from (a) a case of Mendelian autosomal inheritance; (b) a case of X-linked inheritance?

32. In *Limnaea* (see Section 4.13), what results would you expect in a cross between a *Dd* dextrally coiled and a *Dd* sinistrally coiled snail, assuming cross-fertilization occurs as shown in Figure 4–21? What results would occur if the *Dd* dextral produced only eggs and the *Dd* sinistral produced only sperm?

33. In a cross of *Limnaea*, the snail contributing the eggs was dextral, but of unknown genotype. Both the genotype and the phenotype of the other snail are unknown. All F_1 offspring exhibited dextral coiling. Ten of the F_1 snails were allowed to undergo self-fertilization. One-half produced only dextrally coiled offspring, whereas the other half produced only sinistrally coiled offspring. What were the genotypes of the original parents?

34. The specification of the anterior–posterior axis in *Drosophila* embryos is initially controlled by various gene products that are synthesized and stored in the mature egg following oogenesis. Mutations in these genes result in abnormalities of the axis during embryogenesis, illustrating maternal effect. How do such mutations vary from those involved in organelle heredity that illustrate extranuclear inheritance? Devise a set of parallel crosses and expected outcomes involving mutant genes that contrast maternal effect and organelle heredity.

35. The maternal-effect *Drosophila* mutation, *bicoid* (*bcd*), is recessive. In the absence of the bicoid protein product, embryogenesis is not completed. Consider a cross between a female heterozygous for the *bicoid* mutation (bcd^+/bcd^-) and a male homozygous for the mutation (bcd^-/bcd^-). (a) How is it possible for a male homozygous for the mutation to exist? (b) Predict the outcome (normal versus failed embryogenesis) in the F_1 and F_2 generations of this cross.

Selected Readings

Bartolomei, M. S., and Tilghman, S. M. 1997. Genomic imprinting in mammals. *Annu. Rev. Genet.* 31: 493–525.

Brink, R. A., ed. 1967. *Heritage from Mendel.* Madison: University of Wisconsin Press.

Bultman, S. J., Michaud, E. J., and Woychik, R. P. 1992. Molecular characterization of the mouse *agouti* locus. *Cell* 71: 1195–1204.

Cattanach, B. M., and Jones, J. 1994. Genetic imprinting in the mouse: Implications for gene regulation. *J. Inherit. Metab. Dis.* 17: 403–20.

Carlson, E. A. 1987. *The Gene: A Critical History.* 2nd ed. Philadelphia: Saunders.

Choman, A. 1998. The myoclonic epilepsy and ragged-red fiber mutation provides new insights into human mitochondrial function and genetics. *Am. J. Hum. Genet.* 62: 745–51.

Drayna, D., and White, R. 1985. The genetic linkage map of the human X chromosome. *Science* 230: 753–58.

Dunn, L. C. 1966. *A Short History of Genetics.* New York: McGraw-Hill.

Feil, R., and Kelsey, G. 1997. Genomic imprinting: A chromatin connection. *Am. J. Hum. Genet.* 61: 1213–19.

Foster, M. 1965. Mammalian pigment genetics. *Adv. Genet.* 13: 311–39.

Freeman, G., and Lundelius, J.W. 1982. The developmental genetics of dextrality and sinistrality in the gastropod *Lymnaea Peregra. Wilhelm Roux Arch.* 191: 69–83.

Grant, V. 1975. *Genetics of Flowering Plants.* New York: Columbia University Press.

Harper, P. S., et al. 1992. Anticipation in myotonic dystrophy: New light on an old problem. *Am. J. Hum. Genet.* 51: 10–16.

Howeler, C. J., et al. 1989. Anticipation in myotonic dystrophy: Fact or fiction? *Brain* 112: 779–97.

Mahedevan, M., et al. 1992. Myotonic dystrophy mutation: An unstable CTG repeat in the 3′ untranslated region of the gene. *Science* 255: 1253–58.

McKusick, V. A. 1962 On the X chromosome of man. *Quart. Rev. Biol.* 37: 69–175.

Morgan, T. H. 1910. Sex-limited inheritance in *Drosophila. Science* 32: 120–22.

Nüsslein-Volhard, C. 1996. Gradients that organize embryo development. *Sci. Am.* (Aug.) 275: 54–61.

Peters, J. A., ed. 1959. *Classic Papers in Genetics.* Englewood Cliffs: Prentice-Hall.

Phillips, P. C. 1998. The language of gene interaction. *Genetics* 149: 1167–71.

Race, R. R., and Sanger, R. 1975. *Blood Groups in Man.* 6th ed. Oxford: Blackwell.

Sapienza, C. 1990. Parental imprinting of genes. *Sci. Am.* (Oct.) 363: 52–60.

Siracusa, L. D. 1994. The *agouti* gene: Turned on to yellow. *Cell* 10: 423–28.

Sturtevant, A. H. 1923. Inheritance of the direction of coiling in *Limnaea. Science* 58: 269–70.

Vogel, F., and Motulsky, A. G. 1997. *Human Genetics: Problems and Approaches.* 3rd ed. New York: Springer-Verlag.

Wallace, D. C. 1997. Mitochondrial DNA in aging and disease. *Sci. Am.* (Aug.) 277: 40–59.

——— 1999. Mitochondrial diseases in man and mouse. *Science* 283: 1482–88.

Wallace, D. C., et al. 1988. Familial mitochondrial encephalomyopathy (MERRF): Genetic, pathophysiological and biochemical characterization of a mitochondrial DNA disease. *Cell* 55: 601–10.

Watkins, M. W. 1966. Blood group substances. *Science* 152: 172–81.

Yoshida, A. 1982. Biochemical genetics of the human blood group ABO system. *Am. J. Hum. Genet.* 34: 1–14.

CHAPTER 5

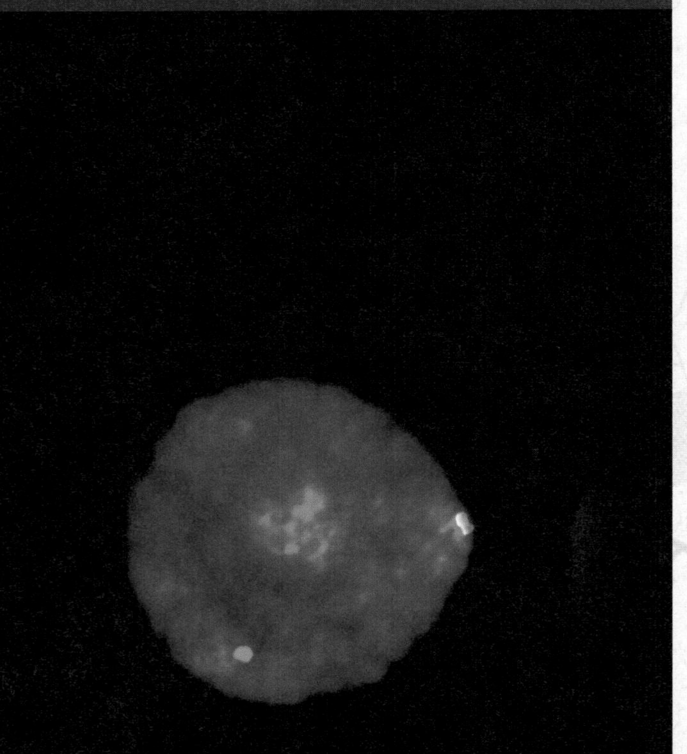

Demonstration of the X and Y chromosomes (the blue and the pink dots, respectively) in mammalian fetal cell using fluorescent *in situ* hybridization (FISH). *(James King-Holmes/Science Photo Library/Photo Researchers, Inc.)*

Sex Determination and Sex Chromosomes

CHAPTER CONCEPTS

In the biological world, a wide range of reproductive modes and life cycles are recognized. Asexual organisms exist where no evidence of sexual reproduction is evident, while other species alternate between short periods of sexual reproduction and prolonged periods of asexual reproduction. In most diploid eukaryotes, however, sexual reproduction is the only natural mechanism that results in new members of a species. Orderly transmission of genetic units from parents to offspring, and thus any phenotypic variability, relies on the processes of segregation and independent assortment that occur during meiosis. Meiosis produces haploid gametes so that, following fertilization, the resulting offspring maintain the diploid number of chromosomes characteristic of their species. Thus, meiosis ensures genetic constancy within members of the same species.

These events, which are involved in the perpetuation of all sexually reproducing organisms, ultimately depend on an efficient union of gametes during fertilization. In turn, successful mating between organisms, the basis for fertilization, relies on some form of sexual differentiation in organisms. Although not overtly apparent, this differentiation occurs as low on the evolutionary scale as bacteria and single-celled eukaryotic algae. In evolutionarily higher forms of life, the differentiation of the sexes is more evident as phenotypic dimorphism in the males and females of each species. The shield and spear (♂), the ancient symbols of iron and Mars; and the mirror (♀), the symbol of copper and Venus, represent the maleness and femaleness acquired by individuals.

Dissimilar, or **heteromorphic chromosomes**, such as the XY pair, often characterize one sex or the other, resulting in their label as **sex chromosomes**. Nevertheless, it is genes, rather than chromosomes, that ultimately serve as the underlying basis of sex determination. As we will see, some of these genes are present on sex chromosomes, but others are autosomal. Extensive investigation has revealed a wide variation in sex chromosome systems—even in closely related organisms—suggesting that mechanisms controlling sex determination have undergone rapid evolution many times.

In this chapter, we review several representative modes of sexual differentiation by examining the life cycles of three organisms often studied in genetics: the green alga, *Chlamydomonas*; the maize plant, *Zea mays*; and the nematode or roundworm, *Caenorhabditis elegans* (most often referred to as *C. elegans*). These organisms contrast the different roles that sexual differentiation plays in the lives of diverse organisms. Then, we delve more deeply into what is known about the genetic basis for the determination of sexual differences, with a particular emphasis on two organisms: our own species, representing mammals, and *Drosophila*, on which pioneering sex-determining studies were performed.

HOW DO WE KNOW WHAT WE KNOW?

IN THIS CHAPTER, WE WILL FOCUS ON how sex is determined in a variety of organisms and several other sex-related phenomena. As you study this topic, you should try to answer several fundamental questions:

1. How do we know whether or not a heteromorphic chromosome such as the Y chromosome plays a crucial role in the determination of sex?
2. How do we know that *Drosophila* utilizes a different sex determination mechanism from mammals, even though it has the same sex-chromosome compositions in males and females?
3. How do we know how mammals, including humans, solve the "dosage problem" caused by the presence of an X and Y chromosome in one sex and two X chromosomes in the other sex?
4. How did we learn that although the sex ratio at birth in humans favors males slightly, the sex ratio at conception vastly favors them?
5. When the sex of an organism (such as certain reptiles) is determined environmentally, how do we know which environmental factors dictate the sex of these organisms? ■

5.1 Life Cycles Depend on Sexual Differentiation

In multicellular organisms, it is important to distinguish between primary sexual differentiation, which involves only the gonads where gametes are produced, and secondary sexual differentiation, which involves the overall appearance of the organism, including clear differences in such organs as mammary glands and external genitalia. In plants and animals, the terms **unisexual**, **dioecious**, and **gonochoric** are equivalent; they refer to an individual containing only male *or* only female reproductive organs. Conversely, the terms **bisexual**, **monoecious**, and **hermaphroditic** refer to individuals containing both male *and* female reproductive organs, a common occurrence in both the plant and animal kingdoms. These organisms can produce both eggs and sperm. The term **intersex** is usually reserved for individuals of intermediate sexual differentiation, who are most often sterile.

Chlamydomonas

The life cycle of the green alga *Chlamydomonas* (Figure 5–1) is representative of organisms that exhibit only infrequent periods of sexual reproduction. Such organisms spend most of their life cycle in the haploid phase, asexually producing daughter cells by mitotic divisions. However, under unfavorable nutrient conditions,

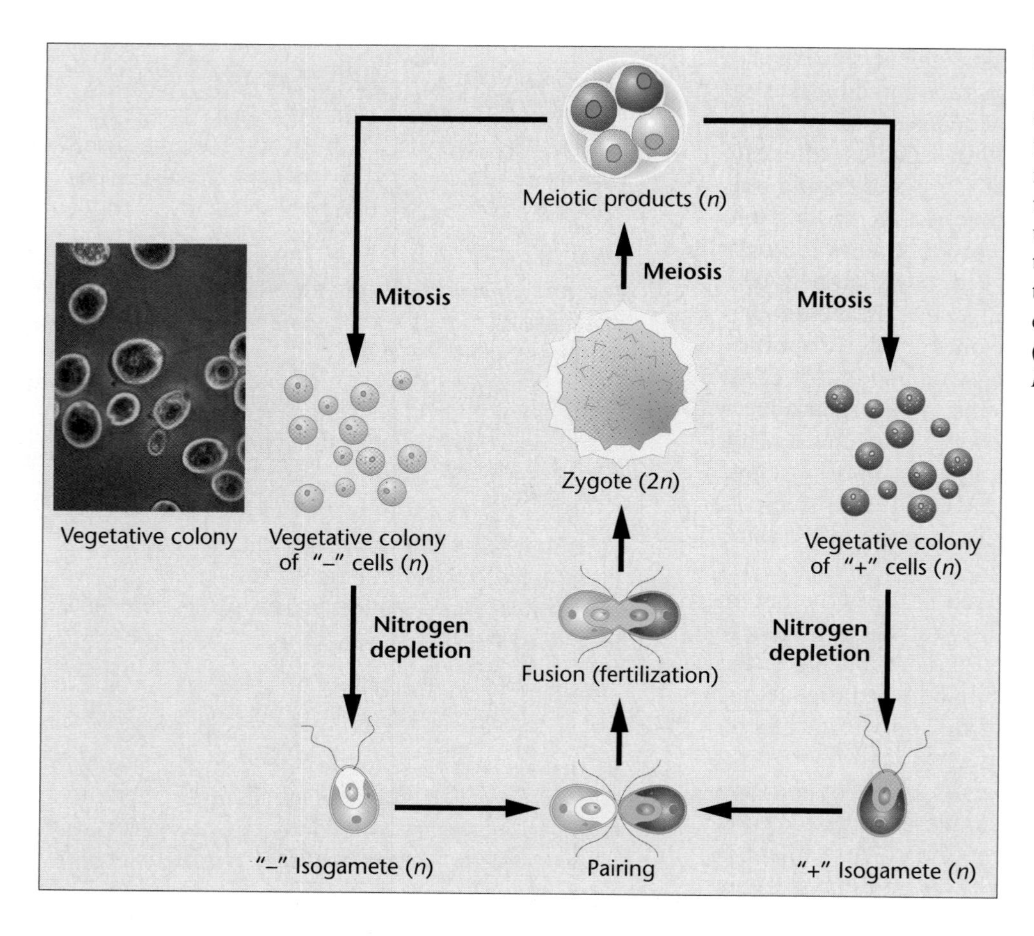

FIGURE 5–1 The life cycle of *Chlamydomonas*. Unfavorable conditions stimulate the formation of isogametes of opposite mating type that may fuse in fertilization. The resulting zygote undergoes meiosis, producing two haploid cells of each mating type. The photograph shows vegetative cells of this green alga. *(Photo: Biophoto Assoc./Photo Researchers, Inc.)*

such as nitrogen depletion, certain daughter cells function as gametes. Following fertilization, a diploid zygote, which can withstand the unfavorable environment, is formed. When conditions become more suitable, meiosis ensues and haploid vegetative cells are again produced.

In such species, there is little visible difference between the haploid vegetative cells that reproduce asexually and the haploid gametes that are involved in sexual reproduction. The two gametes that fuse during mating are morphologically indistinguishable and are called **isogametes**. Species producing them are said to be **isogamous**.

In 1954, Ruth Sager and Sam Granik demonstrated that gametes in *Chlamydomonas* can be subdivided into two mating types. Working with clones derived from single haploid cells, they showed that cells from a given clone mate with cells from some but not all other clones. When they tested mating abilities of large numbers of clones, all could be placed into one of two mating categories, either mt^+ or mt^-. "Plus" cells mate only with "minus" cells, and vice versa. Following fertilization and meiosis, the four haploid cells (zoospores) produced were found to consist of two plus types and two minus types.

Further experimentation established that plus and minus cells differ chemically. When extracts are prepared from cloned *Chlamydomonas* cells (or their flagella), and then added to cells of the opposite mating type, clumping or agglutination occurs. No such agglutination occurs if the extracts are added to cells of the mating type from which it was derived. These observations demonstrate that despite the morphological similarities between isogametes, they differentiate chemically. Therefore, in this alga, a primitive means of sex differentiation exists even though there is no morphological indication that such differentiation has occurred.

Maize (*Zea mays*)

Life cycles of many plants alternate between the haploid gametophyte stage and the diploid sporophyte stage (see Figure 2–14). The processes of meiosis and fertilization link the two phases during the life cycle. *Zea mays,* or maize, familiar to you as corn, exemplifies a monoecious seed plant where the sporophyte phase and the morphological structures representing this stage predominate during the life cycle. Both male and female structures are present on the adult plant. Thus, sex determination occurs differently in different tissues of the same organism, as shown in the life cycle of this plant (Figure 5–2). The *stamens* (which collectively constitute the tassel), produce diploid microspore mother cells, each of which undergoes meiosis and gives rise to four haploid microspores. Each haploid microspore in turn develops into a mature male microgametophyte—the pollen grain—which contains two sperm nuclei.

Equivalent female diploid cells, known as megaspore mother cells, exist in the *pistil* of the sporophyte. Following meiosis, only one of the four haploid megaspores survives.

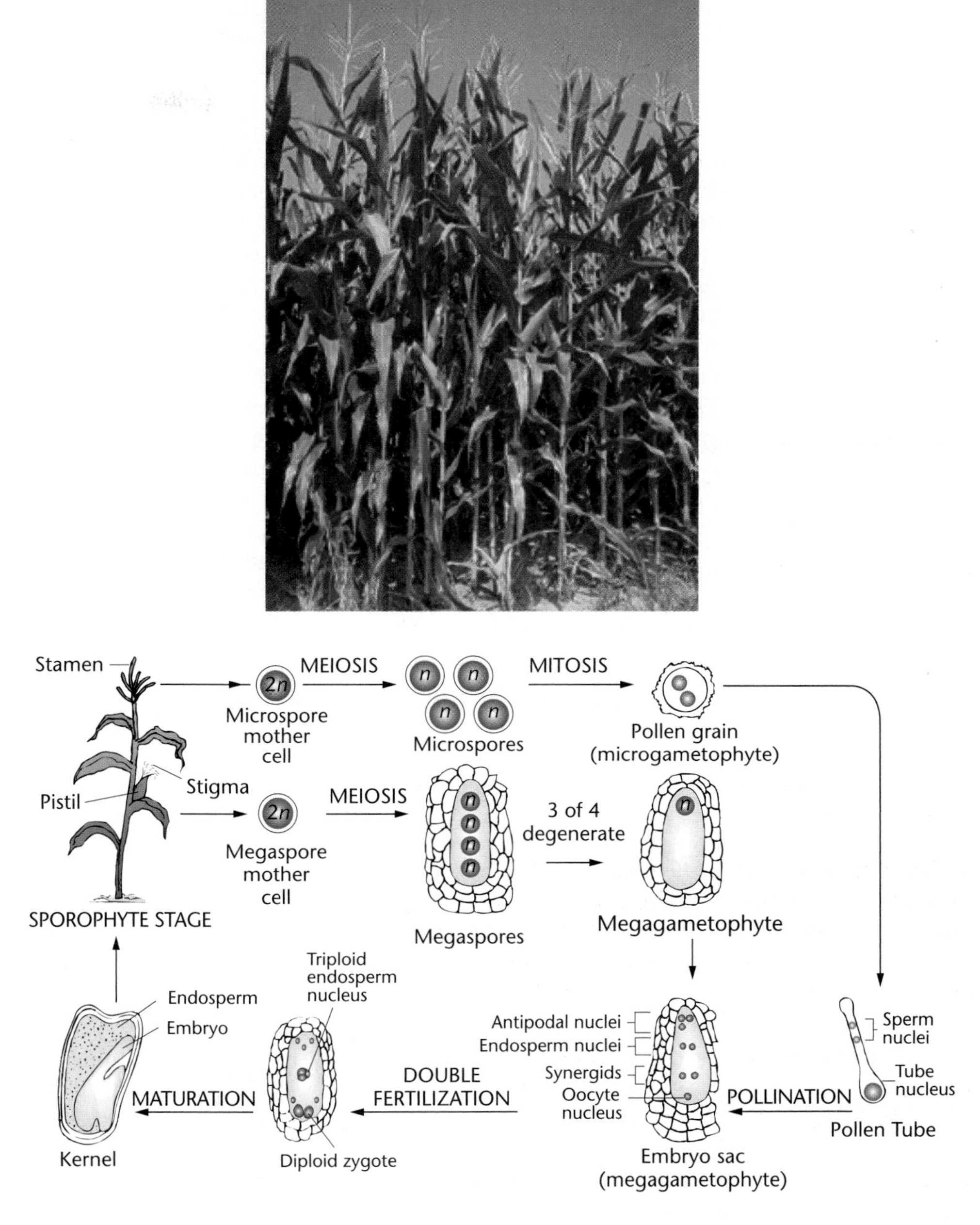

FIGURE 5–2 The life cycle of maize (*Zea mays*). The diploid sporophyte bears stamens and pistils that give rise to haploid microspores and megaspores, which develop into the pollen grain and the embryo sac that ultimately house the sperm and oocyte, respectively. Following fertilization, the embryo develops within the kernel and is nourished by the endosperm. Germination of the kernel gives rise to a new sporophyte (the mature corn plant), and the cycle repeats itself. *(Photo: Bill Beatty/Visuals Unlimited)*

It usually divides mitotically three times, producing a total of eight haploid nuclei enclosed in the embryo sac. Two of these nuclei unite near the center of the embryo sac, becoming the endosperm nuclei. At the micropyle end of the sac where the sperm enters, three nuclei remain: the oocyte nucleus and two synergids. The other three antipodal nuclei cluster at the opposite end of the embryo sac.

Pollination occurs when pollen grains make contact with the silks (or stigma) of the pistil and develop extensive pollen tubes that grow toward the embryo sac. When contact is made at the micropyle, the two sperm nuclei enter the embryo sac. One sperm nucleus unites with the haploid oocyte nucleus, and the other sperm nucleus unites with two endosperm nuclei. This process, known as double fertilization, results in the diploid zygote nucleus and the triploid endosperm nucleus, respectively. Each ear of corn can contain as many as 1000 of these structures, each of which develops into a single kernel. Each kernel, if allowed to germinate, gives rise to a new plant, the sporophyte.

The mechanism of sex determination and differentiation in a monoecious plant like *Zea mays*, where the tissues that form both male and female gametes are of the same genetic constitution, was difficult to comprehend at first. However, the discovery of a large number of mutant genes that disrupt normal tassel and pistil formation supports the concept that normal products of these genes play an important role in sex determination by affecting the differentiation of male or female tissue in several ways.

For example, mutant genes that cause sex reversal provide valuable information. When homozygous, all mutations classified as *tassel seed* (*ts*) interfere with tassel production and induce the formation of female structures. Thus, a single gene can cause a normally monoecious plant to become functionally female. On the other hand, the recessive mutations *silkless* (*sk*) and *barren stalk* (*ba*) interfere with the development of the pistil, resulting in plants with only functional male reproductive organs.

Data gathered from studies of these and other mutants suggest that the products of many wild-type alleles of these genes interact in controlling sex determination. During development, certain cells are "determined" to become male or female structures. Following sexual differentiation into either male or female structures, male or female gametes are produced.

Caenorhabditis elegans

The roundworm *Caenorhabditis elegans* has become a popular organism in genetic studies (Figure 5–3a), particularly during the investigation of the genetic control of development. Its usefulness is based on the fact that the adult consists of only about 1000 cells, the precise lineage of which can be traced back to specific embryonic origins. Among many interesting mutant phenotypes that have been studied, behavioral modifications are a favorite topic of inquiry.

There are two sexual phenotypes in these worms: males, which have only testes, and hermaphrodites, which contain both testes and ovaries. During larval development of hermaphrodites, testes form that produce sperm, which is stored. Ovaries are also produced, but oogenesis does not occur until the adult stage is reached several days later. The eggs that are then produced are fertilized by the stored sperm in the process of self-fertilization.

The outcome of this process is quite interesting (Figure 5–3b). The vast majority of organisms that result, like the parental worm, are hermaphrodites; less than 1% of the offspring are males. As adults, they can mate with hermaphrodites, producing about one-half male and one-half hermaphrodite offspring.

The genetic signal that determines maleness rather than hermaphroditic development is provided by genes located on both the X chromosome and autosomes. *C. elegans* lacks a Y chromosome altogether—hermaphrodites have two X chromosomes, while males have only one X chromosome. It is believed that the ratio of X chromosomes to the number of sets of autosomes ultimately determines the sex of these worms. A ratio of 1.0 results in hermaphrodites and a ratio of 0.5 results in males. The absence of a heteromorphic Y chromosome is not uncommon in organisms.

(a)

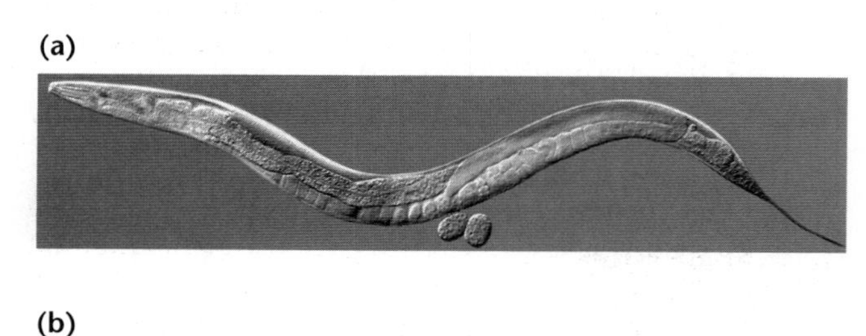

(b)

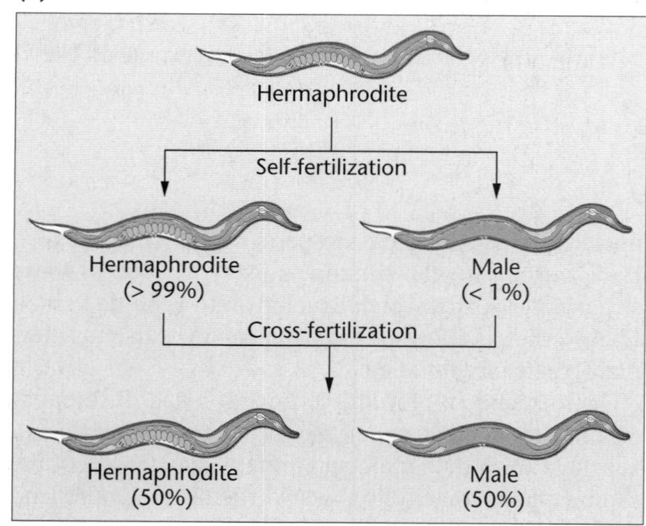

FIGURE 5–3 (a) Photomicrograph of an hermaphroditic nematode, *C. elegans*. (b) The outcomes of self-fertilization in a hermaphrodite, and a mating of a hermaphrodite and a male worm. *(Photo: Dr. Maria Gallegos, University of California, San Francisco)*

5.2 X and Y Chromosomes Were First Linked to Sex Determination Early in the Twentieth Century

How sex is determined has long intrigued geneticists. In 1891, H. Henking identified a nuclear structure in the sperm of certain insects, which he labeled the X-body. Several years later, Clarance McClung showed that some grasshopper sperm contain an unusual genetic structure, which he called a heterochromosome, but the remainder of the sperm lack such a structure. He mistakenly associated the presence of the heterochromosome with the production of male progeny. In 1906, Edmund B. Wilson clarified the findings of Henking and McClung when he demonstrated that female somatic cells in the insect *Protenor* contain 14 chromosomes, including two X chromosomes. During oogenesis, an even reduction occurs, producing gametes with seven chromosomes, including one X chromosome. Male somatic cells, on the other hand, contain only 13 chromosomes, including one X chromosome. During spermatogenesis, gametes are produced containing either six chromosomes, without an X, or seven chromosomes, one of which is an X. Fertilization by X-bearing sperm results in female offspring, and fertilization by X-deficient sperm results in male offspring (Figure 5–4a).

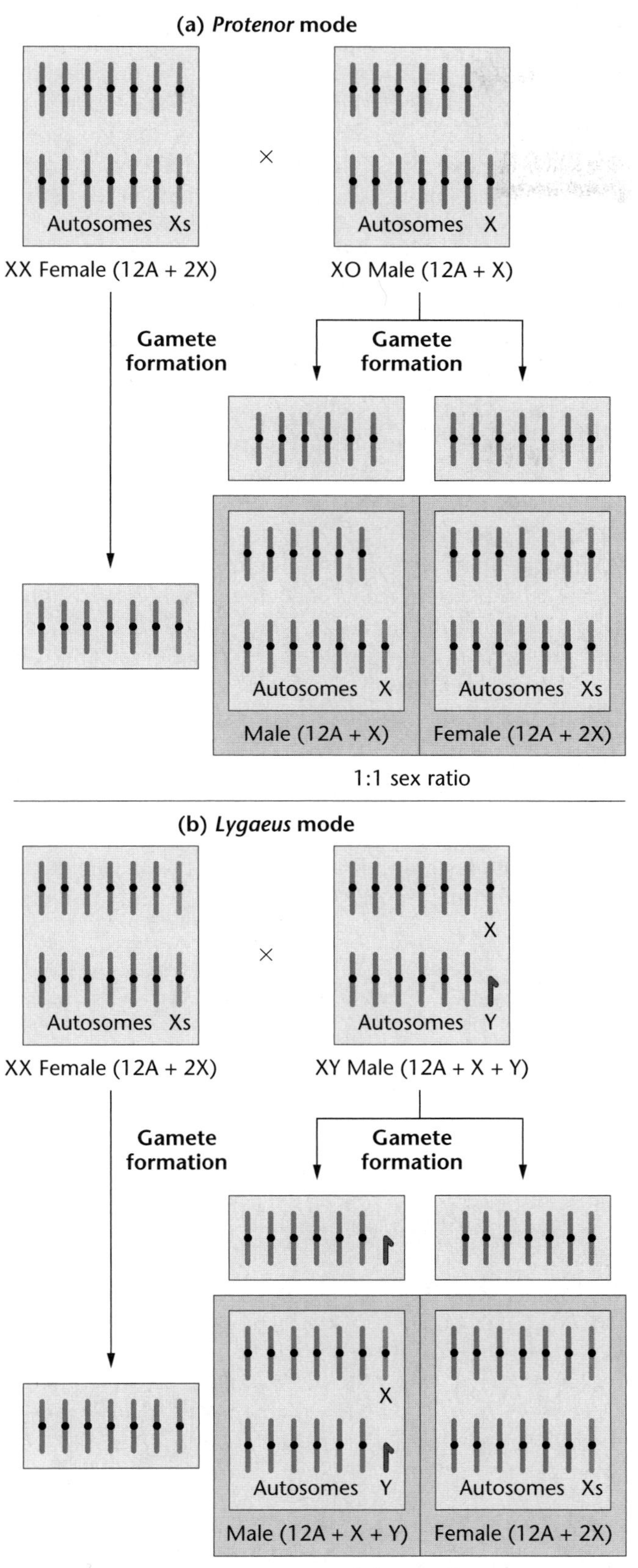

The presence or absence of the X chromosome in male gametes provides an efficient mechanism for sex determination in this species and also produces a 1:1 sex ratio in the resulting offspring. This mechanism, now called the **XX/XO** or ***Protenor* mode of sex determination**, depends on the random distribution of the X chromosome into one-half of the male gametes during segregation. As we saw earlier, *C. elegans* exhibits this system of sex determination.

Wilson also experimented with the hemipteran insect *Lygaeus turicus*, in which both sexes have 14 chromosomes. Twelve of these are autosomes (A). In addition, the females have two X chromosomes, while the males have only a single X and a smaller heterochromosome labeled the **Y chromosome**. Females in this species produce only gametes of the (6A + X) constitution, but males produce two types of gametes in equal proportions, (6A + X) and (6A + Y). Therefore, following random fertilization, equal numbers of male and female progeny are produced with distinct chromosome complements. This is called the ***Lygaeus*** or **XX/XY mode of sex determination** (Figure 5–4b).

In *Protenor* and *Lygaeus* insects, males produce unlike gametes. As a result, they are described as the **heterogametic sex**, and in effect, their gametes ultimately determine the sex of the progeny in those species. In such cases, the female, who has like sex chromosomes, is the **homogametic sex**, producing uniform gametes with regard to chromosome numbers and types.

The male is not always the heterogametic sex. In other organisms, the female produces unlike gametes, exhibiting either the *Protenor* (XX/XO) or *Lygaeus* (XX/XY) mode of sex determination. Examples include moths and butterflies, most birds, some fish, reptiles, amphibians, and at least one species of plants (*Fragaria orientalis*). To immediately distinguish situations in which the female is the heterogametic sex, some workers use the **ZZ/ZW** notation, where ZW is the heterogamous female, instead of the XX/XY notation.

The situation with fowl (chickens) demonstrates the difficulty in establishing which sex is heterogametic and whether the *Protenor* or *Lygaeus* mode is operative. While genetic evidence supported the hypothesis that the female is the heterogametic sex, the cytological identification of the sex chromosome was not accomplished until 1961 because of the large number of chromosomes (78) characteristic of chickens. When the sex chromosomes were finally identified, the female was shown to contain an unlike chromosome pair, including a heteromorphic chromosome (the W chromosome). Thus, in fowl, the female is indeed heterogametic and is characterized by the *Lygaeus* type of sex determination.

FIGURE 5–4 (a) The *Protenor* mode of sex determination, where the heterogametic sex (the male in this example) is XO and produces gametes with or without the X chromosome. (b) The *Lygaeus* mode of sex determination, where the heterogametic sex (again, the male in this example) is XY and produces gametes with either an X or a Y chromosome. In both cases, the chromosome composition of the offspring determines its sex.

5.3 The Y Chromosome Determines Maleness in Humans

The first attempt to understand sex determination in our own species occurred almost 100 years ago and involved the examination of chromosomes present in dividing cells. Efforts were made to accurately determine the diploid chromosome number of humans, but because of the relatively large number of chromosomes, this proved to be quite difficult. In 1912, H. von Winiwarter counted 47 chromosomes in a spermatogonial metaphase preparation. It was believed that the sex-determining mechanism in humans was based on the presence of an extra chromosome in females, who were thought to have 48 chromosomes. However, in the 1920s, Theophilus Painter observed between 45 and 48 chromosomes in cells of testicular tissue and also discovered the small Y chromosome, which we now know occurs only in males. In his original paper, Painter favored 46 as the diploid number in humans, but he later concluded incorrectly that 48 was the chromosome number in both males and females.

For 30 years, this number was accepted. Then, in 1956, Joe Hin Tjio and Albert Levan discovered a better way to prepare chromosomes. The improved technique led to a strikingly clear demonstration of metaphase stages showing that 46 is indeed the human diploid number. Later that same year, C. E. Ford and John L. Hamerton, also working with testicular tissue, confirmed this finding. The familiar karyotype of humans (Figure 5–5) is based on Tjio and Levan's technique.

Within the normal 23 pairs of human chromosomes, one pair was shown to vary in configuration in males and females. These two chromosomes were designated the X and Y sex chromosomes. The human female has two X chromosomes, and the human male has one X and one Y chromosome.

We might believe that this observation is sufficient to conclude that the Y chromosome determines maleness. However, several other interpretations are possible. The Y could play no role in sex determination; the presence of two X chromosomes could cause femaleness; or maleness could result from the lack of a second X chromosome. The evidence that clarified which explanation was correct emerged in the study of variations in the human sex chromosome composition. As such investigations reveal, the Y chromosome does indeed determine maleness in humans.

Klinefelter and Turner Syndromes

About 1940, scientists identified two human abnormalities characterized by aberrant sexual development, **Klinefelter syndrome (47,XXY)** and **Turner syndrome (45,X)**.* Individuals with Klinefelter syndrome have genitalia and internal ducts that are usually male, but their testes are rudimentary and fail to produce sperm. They are generally tall and have long arms and legs and large hands and feet. Although some masculine development does occur, feminine sexual development is not entirely suppressed. Slight enlargement of the breasts (gynecomastia) is common, and the hips are often rounded. This ambiguous sexual development can lead to abnormal social development. Intelligence is often below the normal range as well.

In Turner syndrome, the affected individual has female external genitalia and internal ducts, but the ovaries are rudimentary. Other characteristic abnormalities include short stature (usually under 5 ft), skin flaps on the back of the neck, and underdeveloped breasts. A broad, shieldlike chest is sometimes noted. Intelligence is usually normal.

* Although the possessive form of the names of most syndromes (eponyms) is sometimes used (e.g., Klinefelter's), the current preference is to use the nonpossessive form, which we have adopted for all human syndromes and diseases.

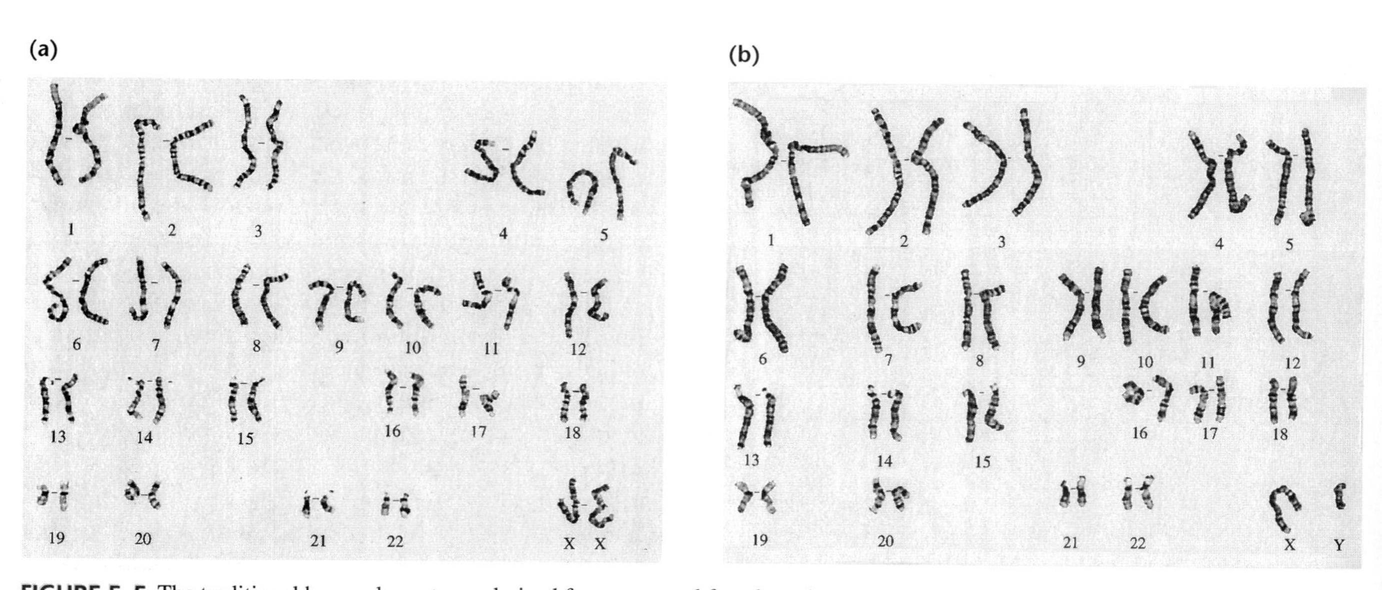

FIGURE 5–5 The traditional human karyotypes derived from a normal female and a normal male. Each contains 22 pairs of autosomes and two sex chromosomes. The female (a) contains two identical X chromosomes, while the male (b) contains one X and one Y chromosome (the Y is often referred to as a heterochromosome). *(Courtesy of the Greenwood Genetic Center, Greenwood, SC)*

In 1959, the karyotypes of individuals with these syndromes were determined to be abnormal with respect to the sex chromosomes. Individuals with Klinefelter syndrome have more than one X chromosome. Most often they have an XXY complement in addition to 44 autosomes (Figure 5–6a) and people with this karyotype are designated 47,XXY. Individuals with Turner syndrome most often have only 45 chromosomes, including just a single X chromosome, thus they are designated 45,X (Figure 5–6b). Note the convention used in designating these chromosome compositions. The number states the total number of chromosomes present, and the information after the comma indicates the deviation from the normal diploid content. Both conditions result from **nondisjunction**, the failure of the chromosomes to segregate properly during meiosis (see Figure 7–1).

The Klinefelter and Turner karyotypes and their corresponding sexual phenotypes allow us to conclude that the Y chromosome determines maleness in humans. In its absence, the sex of the individual is female, even if only a single X chromosome is present. The presence of the Y chromosome in the individual with Klinefelter syndrome is sufficient to determine maleness, even though its expression is not complete. Similarly, in the absence of a Y chromosome, as in the case of individuals with Turner syndrome, no masculinization occurs.

Klinefelter syndrome occurs in about 2 of every 1000 male births. The karyotypes 48,XXXY, 48,XXYY, 49,XXXXY, and 49,XXXYY are similar phenotypically to 47,XXY, but manifestations are often more severe in individuals with a greater number of X chromosomes.

Turner syndrome can also result from karyotypes other than 45,X, including individuals called *mosaics* whose somatic cells display two different genetic cell lines, each exhibiting a different karyotype. Such cell lines result from a mitotic error during early development, the most common chromosome combinations being 45,X/46,XY and 45,X/46,XX. Thus, an embryo that began life with a normal karyotype can give rise to an individual whose cells show a mixture of karyotypes and who exhibits this syndrome.

Turner syndrome is observed in about 1 in 2000 female births, a frequency much lower than that for Klinefelter syndrome. One explanation for this difference is the observation that a substantial majority of 45,X fetuses die *in utero* and are aborted spontaneously. Thus, a similar frequency of the two syndromes may occur at conception.

47,XXX Syndrome

The presence of three X chromosomes along with a normal set of autosomes (**47,XXX**) results in female differentiation. This syndrome, which occurs in about 1 of 1200 female births, is highly variable in expression. Frequently, 47,XXX women are perfectly normal. In other cases, underdeveloped secondary sex characteristics, sterility, and mental retardation can occur. In rare instances, 48,XXXX and 49,XXXXX karyotypes have been reported. The syndromes associated with these karyotypes are similar to, but more pronounced than, the 47,XXX. Thus, in many cases, the presence of additional X chromosomes appears to disrupt the delicate balance of genetic information essential to normal female development.

47,XYY Condition

Another human condition involving the sex chromosomes, **47,XYY**, has also been intensively investigated. Studies of this condition, where the only deviation from diploidy is the presence of an additional Y chromosome in an otherwise normal male karyotype, were initiated in 1965 by Patricia Jacobs. She discovered that 9 of 315 males in a Scottish maximum security prison had the 47,XYY karyotype. These males were significantly above average in height and had been incarcerated as a result of antisocial (nonviolent) criminal acts. Of the nine males studied, seven were of

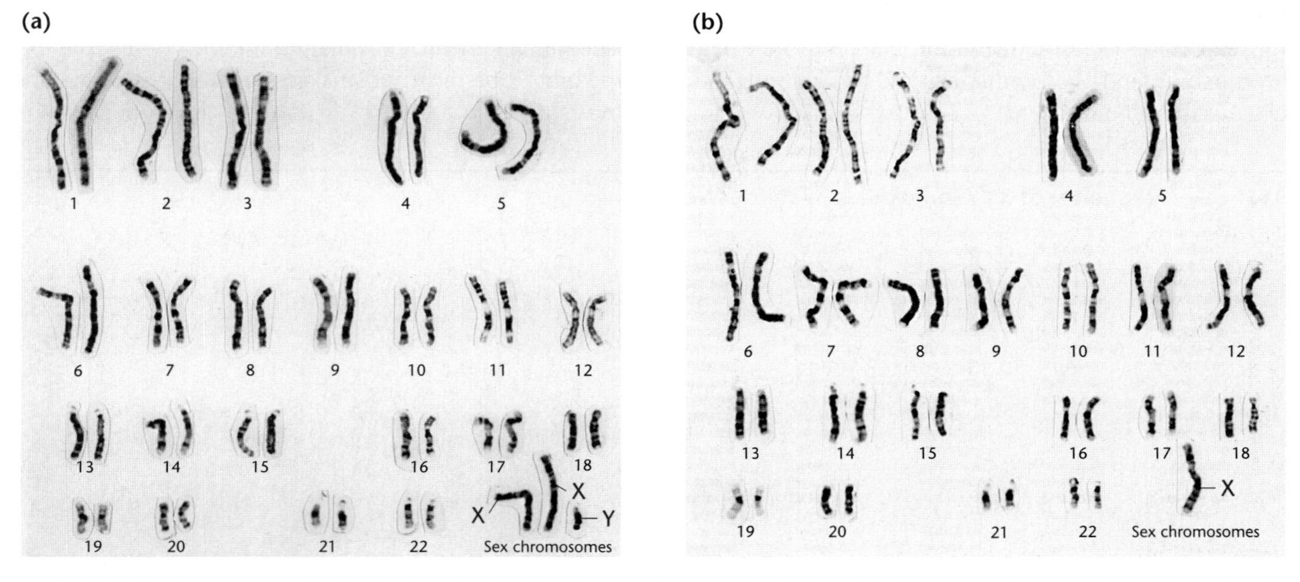

FIGURE 5–6 The karyotypes of individuals with (a) Klinefelter syndrome (47,XXY), and (b) Turner syndrome (45,X). *(Catherine G. Palmer, Dept. of Medical Genetics/Indiana University-Indianapolis)*

subnormal intelligence, and all suffered personality disorders. Several other studies produced similar findings.

The possible correlation between this chromosome composition and criminal behavior piqued considerable interest and extensive investigations of the phenotype and frequency of the 47,XYY condition in both criminal and noncriminal populations ensued. Above-average height (usually over 6 ft) and subnormal intelligence have been generally substantiated, and the frequency of males displaying this karyotype is indeed higher in penal and mental institutions compared with unincarcerated males (see Table 5.1). A particularly relevant question involves the characteristics displayed by XYY males who are not incarcerated. The only nearly constant association is that such individuals are over 6 feet tall.

A study that addressed this issue was initiated to identify 47,XYY individuals at birth and to follow their behavioral patterns during preadult and adult development. By 1974, the two investigators, Stanley Walzer and Park Gerald, had identified about 20 XYY newborns in 15,000 births at Boston Hospital for Women. However, they soon came under great pressure to abandon their research. Those opposed to the study argued that the investigation could not be justified and might cause great harm to those individuals who displayed this karyotype. The opponents argued that (1) no association between the additional Y chromosome and abnormal behavior had been previously established in the population at large, and (2) "labeling" these individuals in the study might become a self-fulfilling prophecy. That is, as a result of participation in the study, parents, relatives, and friends might treat individuals identified as 47,XYY differently, ultimately producing the expected antisocial behavior. Despite the support of a government funding agency and the faculty at Harvard Medical School, Walzer and Gerald abandoned the investigation in 1975.

More recently, it has become clear that many XYY males do not exhibit any form of antisocial behavior and lead normal lives. Therefore, we must conclude that there is a high, but not constant correlation between the extra Y chromosome and the predisposition of these males to behavioral problems.

Sexual Differentiation in Humans

Once researchers established that, in humans, it is the Y chromosome that houses genetic information necessary for maleness, they attempted to pinpoint a specific gene or genes capable of providing the "signal" responsible for sex determination. Before we delve into this topic, it is useful to consider how sexual differentiation occurs, in order to better comprehend how humans develop into sexually dimorphic males and females. During early development, every human embryo undergoes a period when it is potentially hermaphroditic. By the fifth week of gestation, gonadal primordia arise as a pair of ridges associated with each embryonic kidney. Primordial germ cells migrate to these ridges, where an outer cortex and inner medulla form. The cortex is capable of developing into an ovary, while the inner medulla may develop into a testis. In addition, two sets of undifferentiated male (Wolffian) and female (Mullerian) ducts exist in each embryo.

If the cells of the genital ridge have the XY constitution, development of the medullary region into a testis is initiated around the seventh week. However, in the absence of the Y chromosome, no male development occurs, and the cortex of the genital ridge subsequently forms ovarian tissue. Parallel development of the appropriate male or female duct system then occurs, and the other duct system degenerates. Substantial evidence indicates that in males, once testes differentiation is initiated, the embryonic testicular tissue secretes two hormones that are essential for continued male sexual differentiation.

In the absence of male development, as the twelfth week of fetal development approaches, the oogonia within the ovaries begin meiosis and primary oocytes can be detected. By the 25th week of gestation, all oocytes become arrested in meiosis and remain dormant until puberty is reached some 10–15 years later. In males, on the other hand, primary spermatocytes are not produced until puberty is reached.

The Y Chromosome and Male Development

The human Y chromosome, unlike the X, has long been thought to be mostly blank genetically. It is now known that this is not true, even though the Y chromosome

TABLE 5.1 Frequency of XYY Individuals in Various Settings

			XYY	
Setting	*Restriction*	*Number Studied*	*Number*	*Frequency (%)*
Control population	Newborns	28,366	29	0.10
Mental-penal	No height restriction	4239	82	1.93
Penal	No height restriction	5805	26	0.44
Mental	No height restriction	2562	8	0.31
Mental-penal	Height restriction	1048	48	4.61
Penal	Height restriction	1683	31	1.84
Mental	Height restriction	649	9	1.38

Source: Compiled from data presented in Hook, 1973, Tables 1–8. © 1973 by the American Association for the Advancement of Science.

contains far fewer genes than does the X. Current analysis has revealed numerous genes and regions with potential genetic function, some with and some without homologous counterparts on the X chromosome. For example, present on both ends of the Y chromosome are the so-called **pseudoautosomal regions** (**PARs**) that share homology with regions on the X chromosome and which synapse and recombine with it during meiosis. The presence of such a pairing region is critical to segregation of the X and Y chromosomes during male gametogenesis. The remainder of the chromosome about 95% of it, does not synapse or recombine with the X chromosome. As a result, it was originally referred to as the *nonrecombining region of the Y* (*NRY*). More recently, researchers have designated this region as the **male-specific region of the Y** (**MSY**). As you will see, some portions of the MSY share homology with genes on the X chromosome, and some do not.

The human Y chromosome is diagrammed in Figure 5–7. The MSY is divided about equally between *euchromatic* regions that contain functional genes and *heterochromatic* regions that lack genes. Within euchromatin, adjacent to the PAR of the short arm of the Y chromosome, is a critical gene that controls male sexual development, called the ***sex-determining region Y*** (***SRY***). In humans, the absence of a Y chromosome almost always leads to female development, thus this gene is absent from the X chromosome. *SRY* encodes a gene product that somehow triggers the undifferentiated gonadal tissue of the embryo to form testes. This product is called the **testis-determining factor** (**TDF**). *SRY* (or a closely related version) is present in all mammals thus far examined and is indicative of its essential function throughout this diverse group of animals.

Our ability to identify the presence or absence of DNA sequences in rare individuals whose expected sex chromosome composition does not correspond to their sexual phenotype has provided evidence that *SRY* is the gene responsible for male sex determination. For example, there are human males who have two X and no Y chromosomes. Often, attached to one of their X chromosomes is the region of the Y that contains *SRY*. There are also females who have one X and one Y chromosome. Their Y is almost always missing the *SRY* gene. These observations argue strongly in favor of the role of *SRY* in providing the primary signal for male development.

Further support of this conclusion involves an experiment using **transgenic mice**. These animals are produced from fertilized eggs injected with foreign DNA that is subsequently incorporated into the genetic composition of the developing embryo. In normal mice, a chromosome region designated *Sry* has been identified that is comparable to *SRY* in humans. When DNA containing only mouse *Sry* is injected into normal XX mouse eggs, most of the offspring develop into males.

The question of how the product of this gene triggers the embryonic gonadal tissue to develop into testes rather than ovaries is under extensive investigation. In humans, other autosomal genes are believed to be part of a cascade of genetic expression initiated by *SRY*. Examples include the *SOX9* gene and *WT1* (on chromosome 11), originally identified as an oncogene associated with Wilms tumor, which affects the kidney and gonads. Another, *SF1*, is involved in the regulation of enzymes affecting steroid metabolism. In mice, this gene is initially active in both the male and female bisexual genital ridge, persisting until the point in development when testis formation is apparent. At that time, its expression persists in males, but is extinguished in females. The link between these various genes and sex determination brings us closer to a complete understanding of how males and females arise in humans.

Some very recent findings by David Page and his many colleagues have now provided a reasonably complete picture of the MSY region of the human Y chromosome. This work, completed in 2003, is based on information gained through the Human Genome Project, where the DNA of all chromosomes has now been sequenced. Page has spearheaded the detailed study of the Y chromosome for the past several decades.

The MSY consists of about 23 million base pairs (23 Mb) and can be divided into three regions. The first region is the *X-transposed region*. It contains about 15% of the MSY and was originally derived from the X chromosome during human evolution (about 3 million–4 million years ago). The X-transposed region is 99% identical to region Xq21 of the modern human X chromosome. Two genes, both with X-chromosome homologs, are present in this region.

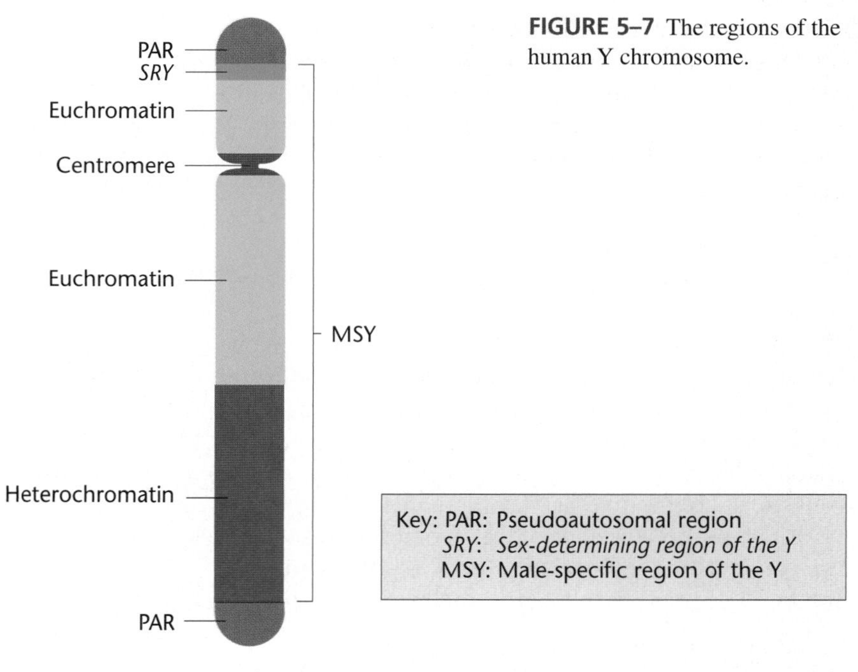

FIGURE 5–7 The regions of the human Y chromosome.

The second area is designated the *X-degenerative region*. Containing about 20% of the MSY, this region contains DNA sequences that are even more distantly related to those present on the X chromosome. The X-degenerative region contains 27 single-copy genes or *pseudogenes* (genes whose sequence has degenerated sufficiently during evolution to render them nonfunctional). As with the genes present in the X-transposed region, all share some homology with counterparts on the X chromosome. These 27 genetic units contain 14 that are capable of being transcribed, and each is present as a single copy. One of these is the *SRY* gene discussed above. Other X-degenerative genes that encode protein products are expressed ubiquitously in all tissues in the body, but *SRY* is expressed only in the testes.

The third area, the *ampliconic region*, contains about 30% of the MSY, including most of the genes closely associated with testes development. These genes lack counterparts on the X chromosome and their expression is limited to the testes. There are 60 transcription units divided among 9 gene families in this region, most represented by multiple copies. Members of each family have nearly identical ($> 98\%$) DNA sequences. Each repeat unit is an **amplicon** and is contained within seven segments scattered across the euchromatic regions present on both the short and long arms of the Y chromosome. Genes in the ampliconic region encode proteins specific to the development and function of the testes, and the products of many of these genes are directly related to fertility in males. It is currently believed that a great deal of male sterility in our population can be linked to mutations in these genes.

This recent work has provided a comprehensive picture of the genetic information present on this unique chromosome. This information clearly refutes the so-called "wasteland" theory, prevalent only 20 years ago, that depicted the human Y chromosome as almost devoid of genetic information other than a gene or two that caused maleness. The knowledge we have gained provides the basis for a much clearer picture of how maleness is determined. Additionally, this information provides important clues as to origin of the Y chromosome during human evolution.

5.4 The Ratio of Males to Females in Humans Is Not 1.0

The presence of heteromorphic sex chromosomes in one sex of a species but not the other provides a potential mechanism for producing equal proportions of male and female offspring. This potential is premised on the segregation of the X and Y (or Z and W) chromosomes during meiosis, such that one-half of the gametes of the heterogametic sex receive one of the chromosomes and one-half receive the other one. As we just learned in Section 5.3, small pseudoautosomal regions of pairing homology do exist at both ends of the X and the Y chromosomes in humans. Provided that both types of gametes are equally successful in fertilization and that the two sexes are equally viable during development, a 1 : 1 ratio of male and female offspring should result.

Given the potential for producing equal numbers of both sexes, the actual proportion of male to female offspring has been investigated and is referred to as the **sex ratio**. We can assess it in two ways. The **primary sex ratio** reflects the proportion of males to females conceived in a population. The **secondary sex ratio** reflects the proportion of each sex that is born. The secondary sex ratio is much easier to determine, but has the disadvantage of not accounting for any disproportionate embryonic or fetal mortality.

When the secondary sex ratio in the human population was determined in 1969 using worldwide census data, it did not equal 1.0. For example, in the Caucasian population in the United States, the secondary ratio was a little less than 1.06, indicating that about 106 males are born for each 100 females. In 1995, this ratio dropped to slightly less than 1.05. In the African-American population in the United States, the ratio was 1.025. In other countries, the excess of male births was even greater than is reflected in these values. For example, in Korea, the secondary sex ratio was 1.15.

Despite these ratios, it is possible that the *primary sex ratio* is 1.0, and that it is altered between conception and birth. For the secondary ratio to exceed 1.0, prenatal female mortality would have to be greater than prenatal male mortality. However, this hypothesis has been examined and shown to be false. In fact, just the opposite occurs. In a Carnegie Institute study, reported in 1948, the sex of approximately 6000 embryos and fetuses recovered from miscarriages and abortions was determined, and fetal mortality was actually higher in males. On the basis of the data derived from that study, the primary sex ratio in U. S. Caucasians was estimated to be 1.079. More recent data have estimated that this figure is much higher—between 1.20 and 1.60, suggesting that many more males than females are conceived in the human population.

It is not clear why such a radical departure from the expected primary sex ratio of 1.0 occurs. To come up with a suitable explanation, we must examine the assumptions on which the theoretical ratio is based:

1. Because of segregation, males produce equal numbers of X- and Y-bearing sperm.
2. Each type of sperm has equivalent viability and motility in the female reproductive tract.
3. The egg surface is equally receptive to both X- and Y-bearing sperm.

While no direct experimental evidence contradicts any of these assumptions, the human Y chromosome is smaller than the X chromosome and therefore has less mass. Thus, it has been speculated that Y-bearing sperm are more motile than X-bearing sperm. If this is true, then the probability of a fertilization event leading to a male zygote is increased, providing one possible explanation for the observed primary ratio.

5.5 Dosage Compensation Prevents Excessive Expression of X-Linked Genes in Humans and Other Mammals

The presence of two X chromosomes in normal human females and only one X chromosome in normal human males is unique compared with the equal numbers of autosomes present in the cells of both sexes. On theoretical grounds alone, it is possible to speculate that this disparity should create a "genetic dosage" problem between males and females for all X-linked genes. There is the potential for females to produce twice as much of each gene product for all X-linked genes. The additional X chromosomes in both males and females exhibiting the various syndromes discussed earlier should compound this dosage problem even more. In this section, we describe research findings on X-linked gene expression that demonstrate a genetic mechanism allowing for **dosage compensation**.

Barr Bodies

Murray L. Barr and Ewart G. Bertram's experiments with female cats, and Keith Moore and Barr's subsequent study with humans, demonstrate a genetic mechanism in mammals that compensates for X-chromosome dosage disparities. Barr and Bertram observed a darkly staining body in interphase nerve cells of female cats that was absent in similar cells of males. In humans, this body can be easily demonstrated in female cells derived from the buccal mucosa or in fibroblasts, but not in similar male cells (Figure 5–8). This highly condensed structure, about 1 μm in diameter, lies against the nuclear envelope of interphase cells. It stains positively in the Feulgen reaction for DNA.

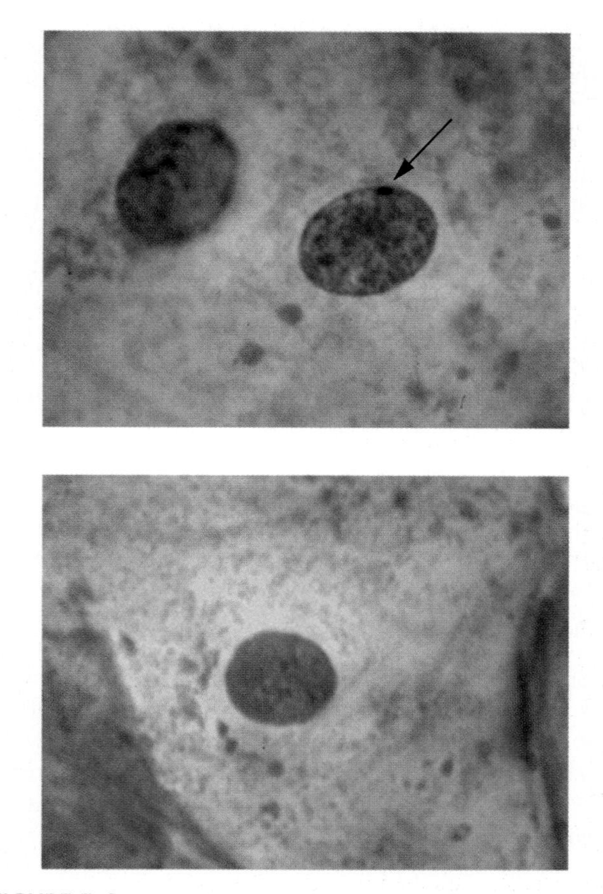

FIGURE 5–8 Photomicrographs comparing cheek epithelial cell nuclei from a male that fails to reveal Barr bodies (bottom) with a female that demonstrates Barr bodies (indicated by the arrow in the top image). This structure, also called a sex chromatin body, represents an inactivated X chromosome. *(Stuart Kenter Associates)*

Current experimental evidence demonstrates that this structure, called a **sex chromatin body** or simply a **Barr body**, is an inactivated X chromosome. Susumo Ohno was the first to suggest that the Barr body arises from one of the two X chromosomes. This hypothesis is attractive because it provides a mechanism for dosage compensation. If one of the two X chromosomes is inactive in the cells of females, the dosage of genetic information that can be expressed in males and females is equivalent. Convincing but indirect evidence for this hypothesis comes from the study of the sex chromosome syndromes described earlier in this chapter. Regardless of how many X chromosomes exist, all but one of them appear to be inactivated and can be seen as Barr bodies. For example, no Barr body is seen in Turner 45,X females; one is seen in Klinefelter 47,XXY males; two in 47,XXX females; three in 48,XXXX females; and so on (Figure 5–9). Therefore, the number of Barr bodies follows an $N - 1$ rule, where N is the total number of X chromosomes present.

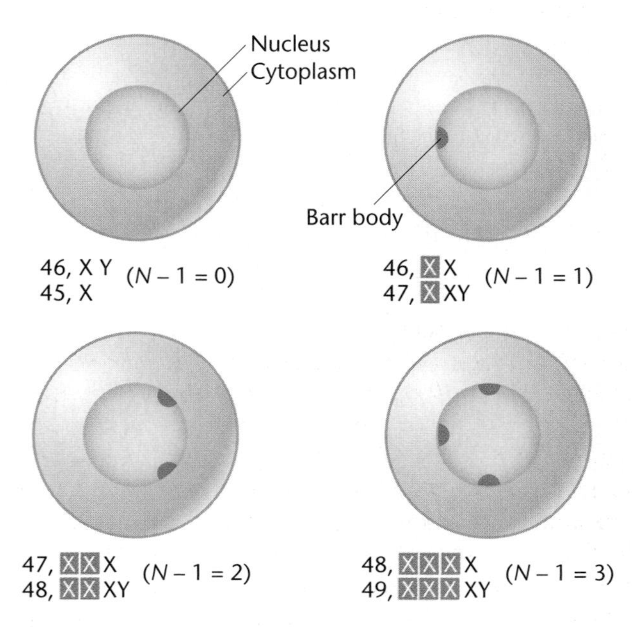

FIGURE 5–9 Occurrence of Barr bodies in various human karyotypes, where all X chromosomes except one $(N - 1)$ are inactivated.

Although this mechanism of inactivating all but one X chromosome increases our understanding of dosage compensation, it further complicates our perception of other matters. Because one of the two X chromosomes is inactivated in normal human females, why then is the Turner 45,X individual not entirely normal? Why aren't females with the triplo-X and tetra-X karyotypes (47,XXX and 48,XXXX, respectively) normal? Further, in Klinefelter syndrome (47,XXY), X-chromosome inactivation effectively renders such individuals 46,XY. Why aren't these males unaffected by the additional X chromosome in their nuclei?

One possible explanation is that chromosome inactivation does not normally occur in the very early developmental stages of those cells destined to form gonadal tissues. Another possible explanation is that not all of each X chromosome forming a Barr body is inactivated. If either hypothesis is correct, excessive expression of certain X-linked genes might still occur despite apparent inactivation of additional X chromosomes.

The Lyon Hypothesis

In mammalian females, one X chromosome is of maternal origin and the other is of paternal origin. Which one is inactivated? Is the inactivation random? Is the same chromosome inactive in all somatic cells? In 1961, Mary Lyon and Liane Russell independently proposed a hypothesis that answers these questions. They postulated that the inactivation of X chromosomes occurs randomly in somatic cells at a point early in embryonic development and that once inactivation has occurred, all progeny cells have the same X chromosome inactivated.

This explanation, which has come to be called the **Lyon hypothesis**, was initially based upon observations of female mice heterozygous for X-linked coat color genes. The pigmentation of these heterozygous females was mottled, with large patches expressing the color allele on one X chromosome and other patches expressing the allele on the other X chromosome. Indeed, such a phenotypic pattern would result if different X chromosomes were inactive in adjacent patches of cells. Similar mosaic patterns occur in the black and yellow-orange patches of female tortoiseshell and calico cats (Figure 5–10). Such X-linked coat color patterns do not occur in male cats because all their cells contain the single maternal X chromosome and are therefore hemizygous for only one X-linked coat color allele.

The most direct evidence in support of the Lyon hypothesis comes from studies of gene expression in clones of human fibroblast cells. Individual cells are isolated following biopsy and cultured *in vitro*. If each culture is derived from a single cell, it is called a **clone**. The synthesis of the enzyme *glucose-6-phosphate dehydrogenase (G6PD)* is controlled by an X-linked gene. Numerous mutant alleles of this gene have been detected, and their gene products can be differentiated from the wild-type enzyme by their migration pattern in an electrophoretic field.

Fibroblasts have been taken from females heterozygous for different allelic forms of G6PD and studied. The Lyon hypothesis predicts that if inactivation of an X chromosome occurs randomly early in development and is permanent in all progeny cells, such a female should show two types of clones, each showing only one electrophoretic form of G6PD, in approximately equal proportions.

In 1963, Ronald Davidson and his colleagues performed an experiment involving 14 clones from a single heterozygous female. Seven showed only one form of the enzyme, and seven showed only the other form. What was most important was that none of the 14 clones showed both forms of the enzyme. Studies of G6PD in humans thus provide strong support for the random inactivation of either the maternal or paternal X chromosome.

The Lyon hypothesis is generally accepted as valid; in fact, the inactivation of an X chromosome into a Barr body is sometimes referred to as *lyonization*. One extension of the hypothesis is that mammalian females are mosaics for all heterozygous X-linked alleles—some areas of the body express only the maternally derived alleles, and others express only the paternally derived alleles. Two especially interesting examples involve **red-green color blindness** and

(a) (b)

FIGURE 5–10 (a) A calico cat, where the random distribution of orange and black patches demonstrates the Lyon hypothesis. The white patches are due to another gene (*S*), which distinguish calico cats from tortoiseshell cats (b), which lack the white patches. *(Left photo: W. Layer/Okapia/Photo Researchers, Inc. Right photo: Walter Chandoha Photography)*

anhidrotic ectodermal dysplasia, both X-linked recessive disorders. In the former case, hemizygous males are fully color-blind in all retinal cells. However, heterozygous females display mosaic retinas with patches of defective color perception and surrounding areas with normal color perception. Males hemizygous for anhidrotic ectodermal dysplasia show an absence of teeth, sparse hair growth, and lack of sweat glands. The skin of females heterozygous for this disorder reveals random patterns of tissue with and without sweat glands. In both examples, random inactivation of one or the other X chromosome early in the development of heterozygous females leads to these occurrences.

The Mechanism of Inactivation

The least understood aspect of the Lyon hypothesis is the mechanism of chromosome inactivation in mammals. How are almost all genes of an entire chromosome inactivated? Recent investigations are beginning to clarify this issue. A single region of the human X chromosome, called the **X-inactivation center (Xic)**, is the major control unit. Genetic expression of this region, located on the proximal end of the p arm in humans, occurs only on the X chromosome that is inactivated. The constant association of expression of Xic and X chromosome inactivation supports the conclusion that this region is an important genetic component in the inactivation process.

The Xic is about 1 Mb (10^6 base pairs) in length and contains four genes. One of these, ***X-inactive specific transcript* (*XIST*)**, is now believed to represent the critical gene. Several interesting observations have been made regarding the RNA that is transcribed from it, with much of the underlying work having been done by using the equivalent gene in the mouse (*Xist*). First, the RNA product is quite large and lacks what is called an extended **open reading frame (ORF)**. An ORF includes the information necessary for translation of the RNA product into a protein. Thus, the RNA is not translated but instead serves a structural role in the nucleus, presumably in the mechanism of chromosome inactivation. This finding has led to the belief that the RNA products of *XIST* and *Xist* spread over and coat the X chromosome bearing the gene that produced it, creating some sort of molecular "cage" that entraps it, leading to its inactivation. Inactivation is therefore said to be cis-acting.

Second, transcription of *Xist* occurs initially at low levels on all X chromosomes. As the inactivation process begins, however, transcription continues and is enhanced only on the X chromosome(s) that becomes inactivated. In 1996, a research group led by Graeme Penny provided convincing evidence that transcription of *Xist* is the critical event in chromosome inactivation. These researchers were able to introduce a targeted deletion (7 kb) into this gene. As a result, the chromosome bearing this mutation lost its ability to become inactivated.

However, several intriguing questions remain regarding inactivation. In cells with more than two chromosomes, what sort of "counting" mechanism designates all but one X chromosome to be inactivated? What "blocks" the Xic of the active chromosome, preventing transcription of *Xist*? How is inactivation of the same X chromosome or chromosomes subsequently maintained in progeny cells, as the Lyon hypothesis calls for? The inactivation signal must somehow remain stable as cells proceed through mitosis. Whatever the answers to these questions, we have taken an exciting step toward understanding how dosage compensation is accomplished in mammals.

5.6 The Ratio of X Chromosomes to Sets of Autosomes Determines Sex in *Drosophila*

Because males and females in *Drosophila melanogaster* (and other *Drosophila* species) have the same general sex chromosome composition as humans (males are XY and females are XX), we might assume that the Y chromosome also causes maleness in these flies. However, the elegant work of Calvin Bridges in 1916 showed this is not true. He studied flies with quite varied chromosome compositions, leading him to conclude that the Y chromosome is not involved in sex determination in this organism. Instead, Bridges proposed that both the X chromosomes and autosomes together play a critical role in sex determination. Recall that in the nematode, *C. elegans*, which lacks a Y chromosome, the sex chromosomes and autosomes are also critical to sex determination.

The most telling observation that differentiates the mechanism operating in *Drosophila* from that in humans involves the XXY and XO sex chromosome compositions. Contrary to what was later discovered in humans, Bridges found that the XXY flies are normal females, and the XO flies are sterile males. The presence of the Y chromosome in the XXY flies did not cause maleness, and its absence in the XO flies did not produce femaleness. From these data, he concluded that the Y chromosome in *Drosophila* lacks male-determining factors, but since the XO males are sterile, it does contain genetic information essential to male fertility.

Bridges was able to clarify the mode of sex determination in *Drosophila* by studying the progeny of triploid females ($3n$), which have three copies each of the haploid complement of chromosomes. *Drosophila* has a haploid number of four, thereby displaying three pairs of autosomes in addition to its pair of sex chromosomes. Triploid females apparently originate from rare diploid eggs fertilized by normal haploid sperm. Triploid females have heavy-set bodies, coarse bristles, and coarse eyes, and they can be fertile. Because of the odd number of each chromosome (3), during meiosis a wide range of chromosome complements is distributed into gametes that give rise to offspring with a variety of abnormal chromosome constitutions. A correlation among the sexual morphology, chromosome composition, and Bridges' interpretation is shown in Figure 5–11.

FIGURE 5–11 Chromosome compositions, the ratios of X chromosomes to sets of autosomes, and the resultant sexual morphology in *Drosophila melanogaster*. The normal diploid male chromosome composition is shown as a reference on the left (XY/2A).

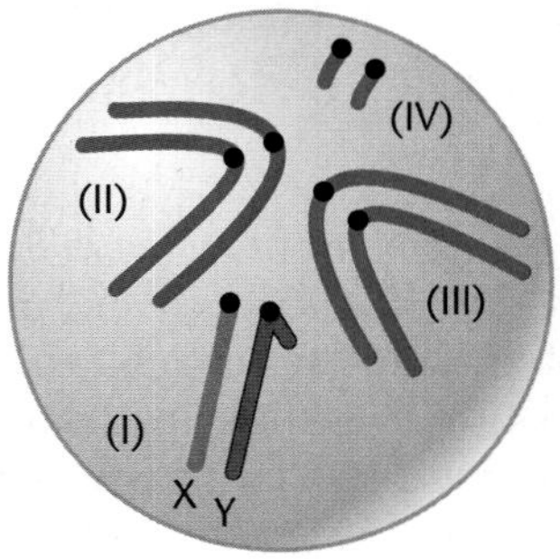

Normal diploid male

(II) (IV) (III) (I) X Y

2 sets of autosomes + X Y

Chromosome composition	Chromosome formulation	Ratio of X chromosomes to autosome sets	Sexual morphology
	3X/2A	1.5	Metafemale
	3X/3A	1.0	Female
	2X/2A	1.0	Female
	3X/4A	0.75	Intersex
	2X/3A	0.67	Intersex
	X/2A	0.50	Male
	XY/2A	0.50	Male
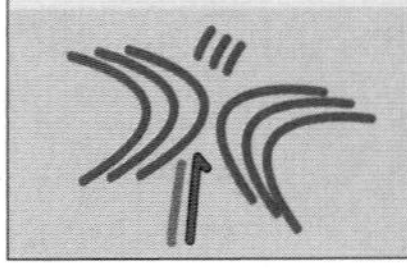	XY/3A	0.33	Metamale

Bridges realized that the critical factor in determining sex is the ratio of X chromosomes to the number of haploid sets of autosomes (A) present. Normal (2X:2A) and triploid (3X:3A) females each have a ratio equal to 1.0, and both are fertile. As the ratio exceeds unity (3X:2A, or 1.5, for example), what was once called a superfemale is produced. Because this female is rather weak and infertile and has lowered viability, this type is now more appropriately called a **metafemale**.

Normal (XY:2A) and sterile (XO:2A) males each have a ratio of 1:2 or 0.5. When the ratio decreases to 1:3, or 0.33, as in the case of an XY:3A male, infertile **metamales** result. Other flies recovered by Bridges in these studies contained an X:A ratio intermediate between 0.5 and 1.0. These flies were generally larger, and they exhibited a variety of morphological abnormalities and rudimentary bisexual gonads and genitalia. They were invariably sterile and expressed both male and female morphology, thus being designated as **intersexes**.

Bridges' results indicate that in *Drosophila*, factors that cause a fly to develop into a male are not localized on the sex chromosomes but are instead found on the autosomes. Some female-determining factors, however, are localized on the X chromosomes. Thus, with respect to primary sex determination, male gametes containing one of each autosome plus a Y chromosome result in male offspring not because of the presence of the Y but because of the lack of a second X chromosome. This mode of sex determination is explained by the **genic balance theory**. Bridges proposed that a threshold for maleness is reached when the X:A ratio is 1:2 (X:2A), but that the presence of an additional X chromosome (XX:2A) alters this balance and results in female differentiation.

Numerous mutant genes have been identified that are involved in sex determination in *Drosophila*. The recessive autosomal gene *transformer* (*tra*), discovered over 50 years ago by Alfred H. Sturtevant, clearly demonstrates that a single autosomal gene can have a profound impact on sex determination. Females homozygous for *tra* are transformed into sterile males, but homozygous males are unaffected.

More recently, another gene, *Sex-lethal* (*Sxl*), has been shown to play a critical role and serves as a "master switch" in sex determination. Activation of the X-linked *Sxl* gene, which relies on a ratio of X chromosomes to sets of autosomes that equals 1.0, is essential to female development. In the absence of activation, resulting, for example, from an X:A ratio of 0.5, male development occurs. It is interesting to note that mutations that inactivate the *Sxl* gene kill female embryos, but have no effect on male embryos, consistent with the role of the gene. While it is not yet exactly clear how this ratio influences the *Sxl* locus, we do have some insights into the question. The *Sxl* locus is part of a hierarchy of gene expression and exerts control over still other genes, including *tra* and *dsx* (*doublesex*), as well as others. The wild-type allele of *tra* is activated by the product of *Sxl* only in females, which in turn influences the expression of *dsx*. Depending on how the initial RNA transcript of *dsx* is processed (spliced), the resultant dsx protein activates either male- or female-specific genes required for sexual differentiation. Each step in this regulatory cascade requires a form of processing called **RNA splicing**, in which portions of the RNA are removed and the remaining fragments "spliced" back together prior to translation into a protein. In the case of the *Sxl* gene, its transcript can be spliced in several different ways, a phenomenon called **alternative splicing**. Two different RNA transcripts are produced in females and males, respectively. In potential females, the transcript is active and initiates a cascade of regulatory gene expression, ultimately leading to female differentiation. In potential males, the transcript is inactive, leading to different gene activity, whereby male differentiation occurs.

5.7 Temperature Variation Controls Sex Determination in Reptiles

We conclude this chapter by discussing several cases involving reptiles where the environment—specifically temperature—has a profound influence on sex determination. The investigations leading to this information may well come closer to revealing the true nature of the primary basis of sex determination than any finding previously discussed.

In many species of reptiles, sex is predetermined at conception by sex-chromosome composition, as is the case in many of the organisms already considered in this chapter. For example, in many snakes, including vipers, a ZZ/ZW mode is in effect, where the female is the heterogamous sex. However, in boas and pythons, it is impossible to distinguish one sex chromosome from the other in either sex. In lizards, both the XX/XY and ZZ/ZW systems are found, depending on the species. However, in other reptilian species, including all crocodiles, most turtles, and some lizards, sex determination is achieved according to the incubation temperature of eggs during a critical period of embryonic development.

Three distinct patterns of temperature-dependent sex determination emerge (cases I–III in Figure 5–12). In case I, low temperatures yield 100% males and high temperatures yield 100% females; just the opposite occurs in case II. In case III, low *and* high temperatures yield 100% females, while intermediate temperatures yield various proportions of males. The third pattern is seen in various species of crocodiles, turtles, and lizards, although some members of these groups are known to exhibit the first two patterns.

Two observations are noteworthy: First, under certain temperatures in all three patterns, both male and female offspring result; second, the pivotal temperature (T_p) is fairly narrow, usually less than 5°C and sometimes only 1°C. The central question raised by these observations is: What metabolic or physiological parameters are being affected by temperature that lead to the differentiation of one sex or the other?

The answer is thought to involve steroids (mainly estrogens) and the enzymes involved in their synthesis. Studies clearly demonstrate that the effects of temperature on estrogens, androgens, and inhibitors of the enzymes controlling their synthesis are involved in the sexual

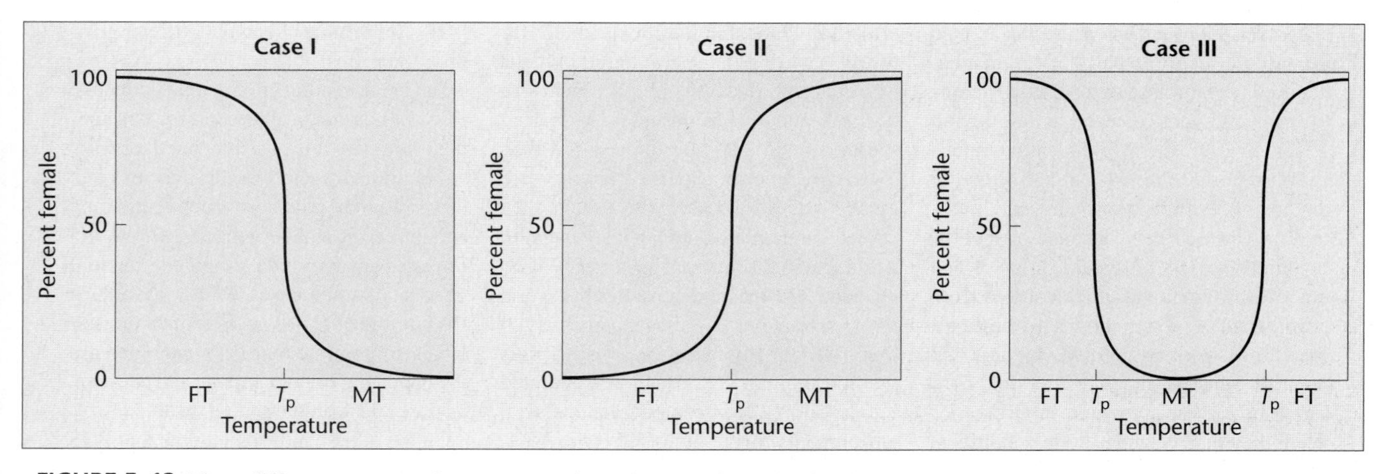

FIGURE 5–12 Three different patterns of temperature-dependent sex determination in reptiles. The relative pivotal temperature, T_p, is crucial to sex determination during a critical point in embryonic development; FT = female-determining temperature and MT = male-determining temperature.

differentiation of ovaries and testes. One enzyme in particular, *aromatase*, converts androgens (male hormones such as testosterone) to estrogens (female hormones such as estradiol). The activity of this enzyme is correlated with the pathway followed during gonadal differentiation activity and is high in developing ovaries and low in developing testes. Researchers in this field, including Claude Pieau and colleagues, have proposed that a thermosensitive factor mediates the transcription of the reptilian aromatase gene that leads to temperature-dependent sex determination. Several other genes are likely to be involved in this mediation.

The involvement of sex steroids in gonadal differentiation has also been documented in birds, fish, and amphibians. Thus, sex-determining mechanisms involving estrogens seem to be characteristic of nonmammalian vertebrates. The regulation of such a system, while temperature dependent in many reptiles, appears to be controlled by sex chromosomes (XX/XY or ZZ/ZW) in many of these other organisms.

GENETICS, TECHNOLOGY, AND SOCIETY

A Question of Gender: Sex Selection in Humans

The desire to choose a baby's gender is as pervasive as human nature itself. Throughout time, people have resorted to varied and sometimes bizarre methods to achieve the preferred gender of their offspring. In medieval Europe, prospective parents would place a hammer under the bed to help them conceive a boy, or a pair of scissors to conceive a girl. Equally effective were practices based on the ancient belief that semen from the right testicle created male offspring and that from the left testicle created females. Men in ancient Greece would lie on their right side during intercourse in order to conceive a boy. Up until the 18th century, European men would tie off (or remove) their left testicle to increase the chances of getting a male heir.

In some cultures, efforts to control the sex of offspring has a darker side—female infanticide. In ancient Greece, the murder of female infants was so common that the male:female ratio in some areas approached 4:1. Some societies, even to present times, practice female infanticide. In some parts of rural India, hundreds of families are reported to admitting to this practice, even as late as the 1990s. In 1997, the World Health Organization reported population data showing that about 50 million women were "missing" in China, likely because of institutionalized neglect of female children. The practice of female infanticide arises from poverty and age-old traditions. In these cultures, sons work and provide income and security, whereas daughters not only contribute no income but require large dowries when they marry. Under these conditions, it is easy to see why females are held in low esteem.

In recent times, sex-specific abortion has replaced much of the traditional female infanticide. Amniocentesis and ultrasound techniques have become lucrative businesses that provide prenatal sex determination. Studies in India estimate that hundreds of thousands of fetuses are aborted each year because they are female. As a result of sex-selective abortion, the male:female ratio in India was 1000:927 in 1991. In some Northern states, the ratio is as high as 1000:600. Although sex determination and selective abortion of female fetuses was outlawed in India and China in the mid-1990s, the practice is thought to continue.

In Western industrial countries, advances in genetics and reproductive technology offer parents ways to select their children's gender prior to implantation. Following *in vitro* fertilization, embryos can be biopsied and assessed for gender. Only sex-selected embryos are then implanted. The simplest and least invasive method for sex selection is preconception gender selection (PGS), which involves separating X- and Y-chromosome bearing spermatozoa. The only effective PGS method devised so far involves sorting the sperm based on their DNA content. Because of the different size of the X and Y chromosomes, X-bearing sperm contain 2.8–3.0% more DNA than Y-bearing sperm. Sperm samples are treated with a fluorescent DNA stain, then passed single file through a laser beam in a fluorescence-activated cell sorter (FACS) machine. The machine separates the sperm into two fractions based on the intensity of their DNA fluorescence. Using this method, human sperm can be separated into X- and Y-chromosome fractions, with enrichments of about 85% and 75%, respectively. The sorted sperm are then used for standard intrauterine insemination. The Genetics & IVF Institute (Fairfax, Virginia) is presently using this PGS technique in an FDA-approved clinical trial. As of January, 2002, 419 human pregnancies have resulted from the method. The company reports an approximately 80–90% success rate in producing the desired gender.

The emerging PGS methods raise a number of legal and ethical issues. Some feel that prospective parents have the legal right to use sex-selection techniques as part of their fundamental procreative liberty. Others believe that this liberty does not extend to custom-designing a child to the parents' specifications. Proponents state that the benefits far outweigh any dangers to offspring or society. The medical use of PGS is a clear case of benefit. People at risk for transmitting X-linked diseases such as hemophilia or Duchenne muscular dystrophy can now enhance their chance of conceiving a female child who will not express the disease. As there are over 500 known X-linked diseases and they are expressed in about 1 in 1000 live births, PGS could greatly reduce suffering for many families.

The greatest number of people undertake PGS for non-medical reasons, to "balance" their families. It is possible that the ability to intentionally select the desired sex of an offspring may reduce overpopulation and economic burdens for families who would repeatedly reproduce to get the desired gender. In some cases, PGS may reduce the number of abortions of female fetuses. It is also possible that PGS may increase the happiness of both parents and children, as the child would be more "wanted."

On the other hand, some argue that PGS serves neither the individual nor the common good. It is argued that PGS is inherently sexist, based on the concept of superiority of one sex over another, and leads to an increase in linking a child's worth to gender. Some fear that large-scale PGS will reinforce sex discrimination and lead to sex-ratio imbalances. Others feel that sexism and discrimination are not caused by sex ratios and would be better addressed through education and economic equality measures for men and women. The experience so far in Western countries suggests that sex-ratio imbalances would not result from PGS. Over half of couples in the United States who use PGS request female offspring. However, the consequences of widespread PGS in some Asian countries may be more problematic. Both India and China already have sex-ratio imbalances, which contribute to some socially undesirable side effects, such as the presence of millions of adult men who are unable to marry.

Some critics of PGS argue that this technology may contribute to social and economic inequality if it will be available only to those who can afford it. Other critics fear that our approval of PGS will open the door to accepting other genetic manipulations of children for socially acceptable characteristics such as skin color. It is difficult to predict the full effects that PGS will bring to the world. But the gender-selection genie is now out of the bottle and will be unwilling to step back in.

References

Sills, E. S., Kirman, I., Thatcher, S.S.III, and Palermo, G.D. 1998. Sex-selection of human spermatozoa: Evolution of current techniques and applications. *Arch. Gynecol. Obstet.* 261: 109–115.

Robertson, J. A. 2001. Preconception gender selection. *Am. J. Bioethics* 1: 2–9.

Web Site

Microsort technique, Genetics & IVF Institute, Fairfax, Virginia.
http://www.microsort.net
Female Infanticide, Gendercide Watch.
http://www.gendercide.org/case_infanticide.html

Chapter Summary

1. In sexually reproducing organisms, meiosis, which both creates genetic variability and ensures genetic constancy, is essential to fertilization. Fertilization ultimately relies on some form of sexual differentiation, which is achieved by a variety of sex-determining mechanisms.
2. The genetic basis of sexual differentiation is often related to different chromosome compositions in the two sexes. The heterogametic sex either lacks one chromosome or contains a unique heteromorphic chromosome, usually referred to as the Y chromosome.
3. In humans, the study of individuals with altered sex chromosome compositions has established that the Y chromosome is responsible for male differentiation. The absence of the Y chromosome leads to female differentiation. Similar studies in *Drosophila* have excluded the Y chromosome in such a role, instead demonstrating that a balance between the number of X chromosomes and sets of autosomes is the critical factor.
4. The primary sex ratio in humans substantially favors males at conception. During embryonic and fetal development, male mortality is higher than that of females. The secondary sex ratio at birth still favors males by a small margin.
5. Dosage compensation mechanisms limit the expression of X-linked genes in females, who have two X chromosomes, as compared to males who have only one X. In mammals, compensation is achieved by the inactivation of either the maternal or paternal X chromosome early in development. This process results in the formation of Barr bodies in female somatic cells.
6. The Lyon hypothesis states that early in development, inactivation is random between the maternal and paternal X chromosomes. All subsequent progeny cells inactivate the same X as their progenitor cell. Mammalian females thus develop as genetic mosaics with respect to their expression of heterozygous X-linked alleles.
7. In many reptiles, the incubation temperature at a critical time during embryogenesis is often responsible for sex determination. Temperature influences the activity of enzymes involved in the metabolism of steroids related to sexual differentiation.

Key Terms

alternative splicing, 107
amplicon, 102
anhidrotic ectodermal dysplasia, 105
Barr body, 103
bisexual, 93
clone, 104
dioecious, 93
dosage compensation, 103
47,XXX, 99
47,XYY, 99
genic balance theory, 106
gonochoric, 93
hermaphroditic, 93
heterogametic sex (ZZ/ZW), 97
heteromorphic chromosomes, 93
homogametic sex, 97
intersex, 106
isogametes, 94
isogamous, 94
Klinefelter syndrome (47,XXY), 98
Lygaeus mode of sex determination (XX/XY), 97
Lyon hypothesis, 104
lyonization, 104
male-specific region of the Y (MSY), 101
metafemale, 106
metamale, 106
monoecious, 93
open reading frame (ORF), 105
primary sex ratio, 102
Protenor mode of sex determination (XX/XO), 97
pseudoautosomal regions (PARs), 101
RNA splicing, 107
red-green color blindness, 104
secondary sex ratio, 102
sex chromatin body, 103
sex chromosomes, 93
sex ratio, 102

sex-determining region Y (*SRY*), 101
testis-determining factor (TDF), 101
transgenic mice, 101
Turner syndrome (45,X), 98
unisexual, 93
X-inactivation center (XIC), 105
X-inactive specific transcript (*XIST*), 105
Y chromosome, 97

INSIGHTS AND SOLUTIONS

1. In *Drosophila*, the X chromosomes can attach to one another ($\widehat{XX}$) such that they always segregate together. Some flies contain both an attached X chromosome and a Y chromosome. (a) What sex would such a fly be? Explain why this is so. (b) Given the answer in part (a), predict the sex of the offspring in a cross between this fly and a normal one of the opposite sex. (c) If the offspring of part (b) are allowed to interbreed, what will be the outcome?

Solution: (a) The fly will be a female. The ratio of X chromosomes to sets of autosomes will be 1.0, leading to normal female development. The Y chromosome has no influence on sex determination in *Drosophila*.

(b) All flies will have two sets of autosomes, but each offspring will have one of the following sex chromosome compositions:

(1) $\widehat{XX}X$ → a metafemale with 3 X's (a trisomic)

(2) $\widehat{XX}Y$ → a female like her mother

(3) XY → a normal male

(4) YY → no development occurs

(c) A true-breeding stock will be created that maintains the attached-X females generation after generation.

2. The Xg cell-surface antigen is coded for by a gene located on the X chromosome. No equivalent gene exists on the Y chromosome. Two codominant alleles of this gene have been identified: *Xg1* and *Xg2*. A woman of genotype *Xg2*/*Xg2* marries a man of genotype *Xg1*/Y and they produce a son with Klinefelter syndrome of genotype *Xg1*/*Xg2*/Y. Using proper genetic terminology, briefly explain how this individual was generated. In which parent and in which meiotic division did the mistake occur?

Solution: Because the son with Klinefelter syndrome is *Xg1*/*Xg2*/Y, he must have received both the *Xg1* allele and the Y chromosome from his father. Therefore, nondisjunction must have occurred during meiosis I in the father.

Problems and Discussion Questions

1. As related to sex determination, what is meant by (a) homomorphic and heteromorphic chromosomes; (b) isogamous and heterogamous organisms?
2. Contrast the life cycle of a plant such as *Zea mays* with an animal such as *C. elegans*.
3. Discuss the role of sexual differentiation in the life cycles of *Chlamydomonas*, *Zea mays*, and *C. elegans*.
4. Distinguish between the concepts of sexual differentiation and sex determination.
5. Contrast the *Protenor* and *Lygaeus* modes of sex determination.
6. Describe the major difference between sex determination in *Drosophila* and in humans.
7. What specific observations (evidence) support the conclusions you have drawn about sex determination in *Drosophila* and humans?
8. Describe how nondisjunction in human female gametes can give rise to Klinefelter and Turner syndrome offspring following fertilization by a normal male gamete.
9. An insect species is discovered in which the heterogametic sex is unknown. An X-linked recessive mutation for *reduced wing* (*rw*) is discovered. Contrast the F_1 and F_2 generations from a cross between a female with reduced wings and a male with normal-sized wings when (a) the female is the heterogametic sex; (b) the male is the heterogametic sex.
10. Based on your answers in Problem 9, is it possible to distinguish between the *Protenor* and *Lygaeus* mode of sex determination based on the outcome of these crosses?
11. When cows have twin calves of unlike sex (fraternal twins), the female twin is usually sterile and has masculinized reproductive organs. This calf is referred to as a freemartin. In cows, twins may share a common placenta and thus fetal circulation. Predict why a freemartin develops.
12. An attached-X female fly, $\widehat{XX}Y$ (see the "Insights and Solutions" box), expresses the recessive X-linked *white* eye phenotype. It is crossed to a male fly that expresses the X-linked recessive miniature wing phenotype. Determine the outcome of this cross regarding the sex, eye color, and wing size of the offspring.
13. Assume that rarely, the attached X chromosomes in female gametes become unattached. Based on the parental phenotypes in Problem 12, what outcomes in the F_1 generation would indicate that this has occurred during female meiosis?
14. It has been suggested that any male-determining genes contained on the Y chromosome in humans cannot be located in the limited region that synapses with the X chromosome during meiosis. What might be the outcome if such genes were located in this region?
15. What is a Barr body, and where is it found in a cell?
16. Indicate the expected number of Barr bodies in interphase cells of the following individuals: Klinefelter syndrome; Turner syndrome; and karyotypes 47,XYY, 47,XXX, and 48,XXXX.
17. Define the Lyon hypothesis.
18. Can the Lyon hypothesis be tested in a human female who is homozygous for one allele of the X-linked *G6PD* gene? Why, or why not?

19. Predict the potential effect of the Lyon hypothesis on the retina of a human female heterozygous for the X-linked red-green color-blindness trait.
20. Cat breeders are aware that kittens expressing the X-linked calico coat pattern and tortoiseshell pattern are almost invariably females. Why?
21. What does the apparent need for dosage compensation mechanisms suggest about the expression of genetic information in normal diploid individuals?
22. The marine echiurid worm *Bonellia viridis* is an extreme example of the environment's influence on sex determination. Undifferentiated larvae either remain free-swimming and differentiate into females or they settle on the proboscis of an adult female and become males. If larvae that have been on a female proboscis for a short period are removed and placed in seawater, they develop as intersexes. If larvae are forced to develop in an aquarium where pieces of proboscises have been placed, they develop into males. Contrast this mode of sexual differentiation with that of mammals. Suggest further experimentation to elucidate the mechanism of sex determination in *B. viridis*.
23. How do we know that the primary sex ratio in humans is as high as 1.40 to 1.60?
24. Devise as many hypotheses as you can that might explain why so many more human male conceptions than human female conceptions occur.
25. In mice, the *Sry* gene (see Section 5.3) is located on the Y chromosome very close to one of the pseudoautosomal regions that pairs with the X chromosome during male meiosis. Given this information, propose a model to explain the generation of unusual males who have two X chromosomes (with an *Sry*-containing piece of the Y chromosome attached to one X chromosome).
26. The genes encoding the red and green color-detecting proteins of the human eye are located next to one another on the X chromosome and probably arose during evolution from a common ancestral pigment gene. The two proteins demonstrate 76% homology in their amino acid sequences. A normal-visioned woman with one copy of each gene on each of her two X chromosomes has a red color-blind son who was shown to contain one copy of the green-detecting gene and no copies of the red-detecting gene. Devise an explanation at the chromosomal level (during meiosis) that explains these observations.
27. The X-linked dominant mutation in the mouse, *Testicular feminization* (*Tfm*), eliminates the normal response to the testicular hormone testosterone during sexual differentiation. An XY mouse bearing the *Tfm* allele on the X chromosome develops testes, but no further male differentiation occurs—the external genitalia of such an animal are female. From this information, what might you conclude about the role of the *Tfm* gene product and the X and Y chromosomes in sex determination and sexual differentiation in mammals? Can you devise an experiment, assuming you can "genetically engineer" the chromosomes of mice, to test and confirm your explanation?
28. Campomelic dysplasia (CMD1) is a congenital human syndrome, featuring malformation of bone and cartilage. It is caused by an autosomal dominant mutation of a gene located on chromosome 17. Consider the following observations in sequence, and in each case, draw whatever appropriate conclusions are warranted.
    (a) Of those with the syndrome who are karyotypically 46,XY, approximately 75% are sex reversed, exhibiting a wide range of female characteristics.
    (b) The nonmutant form of the gene, called *SOX9*, is expressed in the developing gonad of the XY male, but not the XX female.
    (c) The *SOX9* gene shares 71% amino acid coding sequence homology with the Y-linked *SRY* gene.
    (d) CMD1 patients who exhibit a 46,XX karyotype develop as females, with no gonadal abnormalities.
29. In the wasp, *Bracon hebetor*, a form of parthenogenesis (where unfertilized eggs initiate development) resulting in haploid organisms is not uncommon. All haploids are males. When offspring arise from fertilization, females almost invariably result. P. W. Whiting has shown that an X-linked gene with nine multiple alleles (X_a, X_b, etc.) controls sex determination. Any homozygous or hemizygous condition results in males and any heterozygous condition results in females. If an X_a/X_b female mates with an X_a male and lays 50% fertilized and 50% unfertilized eggs, what proportion of male and female offspring will result?
30. Shown below are two graphs that plot the percentage of males occurring against the atmospheric temperature during the early development of fertilized eggs in (a) snapping turtles and (b) most lizards. Interpret these data as they relate to the effect of temperature on sex determination.

(a) Snapping turtles

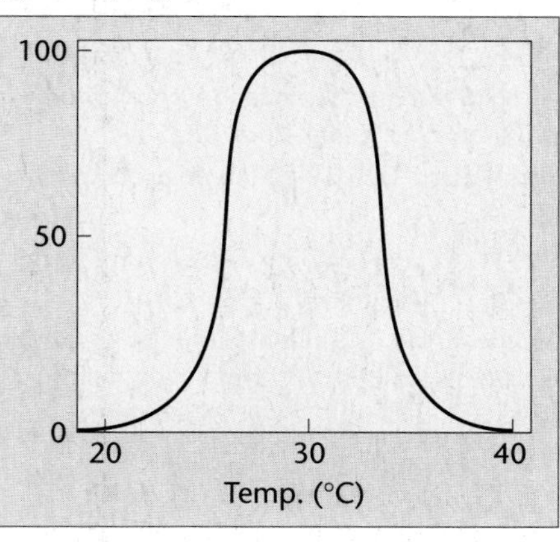

(b) Most lizards

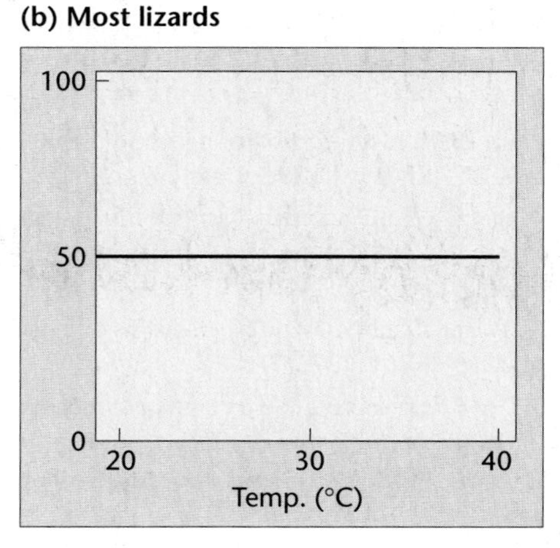

31. CC (Carbon Copy), the first cloned cat, was created from an ovarian cell taken from her genetic donor, Rainbow. The diploid nucleus from the cell was extracted and then injected into an enucleated egg. The resulting zygote was then allowed to divide in a petri dish and the cloned embryo was implanted in the uterus of a surrogate mother cat, who gave birth to CC. Rainbow is a calico cat. CC's surrogate mother is a tabby. Geneticists were very

interested in the outcome of cloning a calico cat, because they were not certain if the cat would have patches of orange and black, just orange, or just black. Taking into account the Lyon hypothesis, explain the basis of the uncertainty.

32. Let's assume hypothetically that Carbon Copy from Problem 31 is indeed a calico with black and orange patches, along with the patches of white characterizing a calico cat. Would you expect CC to appear identical to Rainbow? Explain why or why not.
33. When Carbon Copy was born (see Problem 31), she had black patches and white patches, but completely lacked any orange patches. The knowledgeable students of genetics were not surprized at this outcome. Starting with the somatic ovarian cell used as the source of the nucleus in the cloning process, explain how this outcome occurred.

CC (Carbon Copy), the first cloned cat, shown as a kitten along with her surrogate mother. *(Associated Press/Courtesy of Texas A & M University.)*

Selected Readings

Amory, J. K. et al. 2000. Klinefelter's syndrome. *Lancet* 356: 333–35.

Avner, P., and Heard, E. 2001. X-chromosome inactivation: Counting, choice and initiation. *Nat. Rev. Gen.* 2: 59–67.

Barr, M. L. 1966. The significance of sex chromatin. *Int. Rev. Cytol.* 19: 35–39.

Burgoyne, P. S. 1998. The mammalian Y chromosome: A new perspective. *Bioessays* 20: 363–66.

Carrel, L., and Willard, H. F. 1998. Counting on *Xist*. *Nature Genetics* 19: 211–12.

Court-Brown, W. M. 1968. Males with an XYY sex chromosome complement. *J. Med. Genet.* 5: 341–59.

Davidson, R. et al. 1963. Demonstration of two populations of cells in human females heterozygous for glucose-6-phosphate dehydrogenase variants. *Proc. Natl. Acad. Sci. USA* 50: 481–85.

Dellaporta, S. L., and Calderon-Urrea, A. 1994. The sex determination process in maize. *Science* 266: 1501–05.

Erickson, J. D. 1976. The secondary sex ratio of the United States, 1969–71: Association with race, parental ages, birth order, paternal education and legitimacy. *Ann. Hum. Genet.* (London) 40: 205–12.

Freije, D. et al. 1992. Identification of a second pseudoautosomal region near the Xq and Yq telomeres. *Science* 258: 1784–87.

Gorman, M. et al. 1993. Regulation of sex-specific binding of maleness dosage compensation protein to the male X chromosome in *Drosophila*. *Cell* 72: 39–49.

Haseltine, F. P., and Ohno S. 1981. Mechanisms of gonadal differentiation. *Science* 211: 1272–78.

Hodgkin, J. 1990. Sex determination compared in *Drosophila* and *Caenorhabditis*. *Nature* 344: 721–28.

Hook, E. B. 1973. Behavioral implications of the humans XYY genotype. *Science* 179: 139–50.

Irish, E. E. 1996. Regulation of sex determination in maize. *BioEssays* 18: 363–69.

Jacobs, P. A. et al. 1974. A cytogenetic survey of 11,680 newborn infants. *Ann. Hum. Genet.* 37: 359–76.

Jegalian, K., and Lahn, B.T. 2001. Why the Y is so weird. *Sci. Am.* (Feb.) 284: 56–61.

Koopman, P. et al. 1991. Male development of chromosomally female mice transgenic for *Sry*. *Nature* 351: 117–21.

Lahn, B. T. and Page, D. C. 1997. Functional coherence of the human Y chromosome. *Science* 278: 675–80.

Lahn, B.T. et al. 2001. The human Y chromosome, in the light of evolution. *Nature Reviews—Genetics* 2: 207–16.

Lucchesi, J. 1983. The relationship between gene dosage, gene expression, and sex in *Drosophila*. *Dev. Genet.* 3: 275–82.

Lyon, M. F. 1961. Gene action in the X chromosome of the mouse (*Mus musculus L.*). *Nature* 190: 372–73.

———. 1972. X-chromosome inactivation and developmental patterns in mammals. *Biol. Rev.* 47: 1–35.

———. 1988. X-chromosome inactivation and the location and expression of X-linked genes. *Am. J. Hum. Genet.* 42: 8–16.

———. 1998. X-Chromosome inactivation spreads itself: Effects in autosomes. *Am. J. Hum. Genet.* 63: 17–19.

Marin, I., and Baker, S. S. 1998. The evolutionary dynamics of sex determination. *Science* 281: 1990–94.

Marshall Graves, J. A. 1998. Interaction between *SRY* and *SOX* genes in mammalian sex determination. *BioEssays* 20: 264–69.

McMillen, M. M. 1979. Differential mortality by sex in fetal and neonatal deaths. *Science* 204: 89–91.

Penny, G. D. et al. 1996. Requirement for *Xist* in X chromosome inactivation. *Nature* 379: 131–37.

Pieau, C. 1996. Temperature variation and sex determination in reptiles. *BioEssays* 18: 19–26.

Reddy, K. S., and Sulcova, V. 1998. Pathogenetics of 45,X/46,XY gonadal mosaicism. *Cytogenet. Cell Genet.* 82: 52–57.

Skaletsky, H. et al. 2003. The male-specific region of the human Y chromosome is a mosaic of discrete sequence classes. *Nature* 423: 825–37.

Schafer, A. J. 1996. Sex determination in humans. *BioEssays* 18: 955–63.

Vainio. S. et al. 1999. Female development in mammals is regulated by Wnt-4 signalling. *Nature* 397: 405–09.

Westergaard, M. 1958. The mechanism of sex determination in dioecious flowering plants. *Adv. Genet.* 9: 217–81.

Whiting, P. W. 1939. Multiple alleles in sex determination in *Habrobracon*. *J. Morphology* 66: 323–55.

Witkin, H. A. et al. 1976. Criminality in XYY and XXY men. *Science* 193: 547–55.

Quantitative Genetics

A field of pumpkins, where size is a quantitative trait under the control of additive alleles. *(Ed Reschke/Peter Arnold, Inc.)*

CHAPTER CONCEPTS

6.1 Continuous Variation Characterizes the Inheritance of Quantitative Traits
- The Multiple-Factor Hypothesis
- Additive Alleles: The Basis of Continuous Variation
- Calculating the Number of Genes
- The Significance of Polygenic Inheritance

6.2 The Study of Polygenic Traits Relies on Statistical Analysis
- The Mean
- Variance
- Standard Deviation
- Standard Error of the Mean
- Analysis of a Quantitative Character

6.3 Heritability Is a Measure of the Genetic Contribution to Phenotypic Variability
- Broad-Sense Heritability
- Narrow-Sense Heritability
- Artificial Selection
- Twin Studies in Humans

6.4 Quantitative Trait Loci Can Be Mapped

We have thus far discussed numerous examples of gene interaction as modifications of Mendelian ratios. In each case, the resultant phenotypic variation was classified into distinct traits. Pea plants are tall or dwarf; squash shape is spherical, disc-shaped, or elongated; and fruit fly eye color is red or white. These phenotypes exemplify *discontinuous* variation, in which discrete phenotypic categories exist. Many other traits in a population demonstrate considerably more variation and are not as easily categorized into distinct classes. Such phenotypes are thus said to demonstrate *continuous* variation.

Traits exhibiting continuous variation are most often controlled by two or more genes that provide an additive component to the phenotype that can be quantified. In this chapter, we examine patterns of inheritance and outline some statistical techniques used to study such traits. These patterns illustrate **quantitative**, or **polygenic, inheritance**. In addition, we consider how geneticists assess the relative importance of genetic versus environmental factors as they contribute to phenotypic variation, and we discuss an approach used to localize these "quantitative" genes within the genome.

HOW DO WE KNOW WHAT WE KNOW?

IN THIS CHAPTER, WE WILL FOCUS ON traits that exhibit quantitative phenotypic variation and are under the genetic control of alleles whose influence is additive in nature. As you study this topic, you should try to answer several fundamental questions:

1. What observations made it apparent to early geneticists that quantitative traits must be under a different mode of inheritance than the more qualitative traits studied by Mendel?
2. What findings led geneticists to postulate the multiple-factor hypothesis that invoked the idea of additive alleles to explain inheritance patterns?
3. How do we know how many genes control a quantitative trait?
4. How do we assess environmental factors to determine if they impact the phenotype of a quantitatively inherited trait? ■

6.1 Continuous Variation Characterizes the Inheritance of Quantitative Traits

Near the end of the nineteenth century, scientists studied traits that exhibited a continuous gradation of phenotypes. For example, Sir Francis Galton investigated the diameter of sweet peas. When plants with large peas were crossed to those with small peas, the F_1 plants contained peas that were all of an intermediate diameter. When the F_2 generation was examined, peas were of many sizes: as large or as small as the original parents and many sizes in between. Galton's study showed a pattern of inheritance encountered by other investigators, including at least one cross made by Mendel. The F_1 generations were intermediate blends of the parental phenotypes, and the F_2 generation exhibited more or less continuous variation of phenotypic expression. In each case, the traits under investigation behaved in a quantitative fashion expressed as size, height, weight, color, and so on.

Not surprisingly, these traits were difficult to study, and their mode of inheritance was not clarified until early in the twentieth century. Because these results were exceptions to the patterns observed by Mendel in most of his crosses, they failed to support his hypotheses and no doubt delayed the acceptance of his work. Nevertheless, the genetic explanation of continuous variation serves as the foundation for our current understanding of the field of genetics called polygenic, or quantitative, inheritance.

The Multiple-Factor Hypothesis

One of the first cases of continuous phenotypic variation was encountered by Josef Gottlieb Kölreuter when he crossed tall and dwarf tobacco plants. The plants of the F_1 generation were all of intermediate height. When the F_2 generation was examined, individuals showed continuous variation in height, ranging from tall to dwarf (like the original parents), with many heights in between. A critical observation involved the distribution of phenotypes in the second generation: The majority of the F_2 plants were intermediate like the F_1 plants, but only a few were as tall or as dwarf as the P_1 parents. These distributions are shown in Figure 6–1. Note that the F_2 data demonstrate a normal distribution, as evidenced by the bell-shaped curve in the bottom histogram.

At the beginning of the twentieth century, geneticists noted that many characters in different species had similar patterns of inheritance, such as height and stature in humans, seed size in the broad bean, grain color in wheat, and kernel number and ear length in corn. In each case, offspring in the succeeding generation seemed to be a blend of their parents' characteristics.

The issue of whether continuous variation could be accounted for in Mendelian terms caused considerable controversy in the early 1900s. William Bateson and Gudney Yule, who adhered to the Mendelian explanation of inheritance, suggested that a large number of factors or genes could account for the observed patterns. This proposal, called the **multiple-factor hypothesis**, implied that many factors or genes contribute to the phenotype in a *cumulative* or *quantitative* way. However, other geneticists argued that Mendel's unit factors could not account for the blending of parental phenotypes characteristic of these patterns of inheritance and were thus skeptical of these ideas.

By 1920, the conclusions of several critical sets of experiments largely resolved the controversy and demonstrated that Mendelian factors could account for continuous variation. In one experiment, Edward M. East crossed two

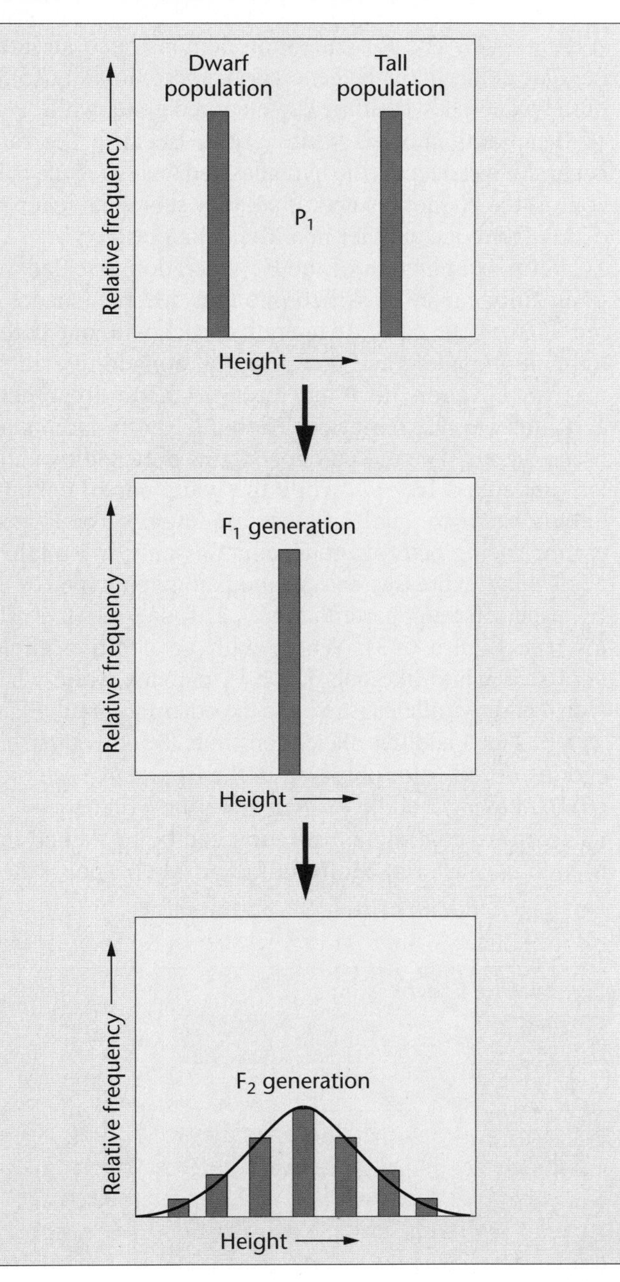

FIGURE 6–1 Histograms showing the relative frequency of individuals expressing various height phenotypes derived from Kölreuter's cross between dwarf and tall tobacco plants carried to the F_2 generation. The photograph shows a tobacco plant. *(Photo: Bildarchiv Okapia/Photo Researchers, Inc.)*

strains of the tobacco plant *Nicotiana longiflora*. The fused inner petals of the flower, or corollas, of strain A were decidedly shorter than the corollas of strain B. With only minor variation, each strain was true-breeding. Thus, the differences between them were clearly under genetic control.

When plants from the two strains were crossed, the F_1, F_2, and selected F_3 data demonstrated a distinct pattern (Figure 6–2). The F_1 generation displayed corollas that were intermediate in length compared with the P_1 varieties, and showed only minor variability among individuals. While corolla lengths of the P_1 plants were about 40 mm and 94 mm, the F_1 generation contained plants with corollas that were all about 64 mm. In the F_2 generation, lengths varied much more, ranging from 52 to 82 mm. The majority of individuals resembled their F_1 parents (64 mm), and as the deviation from this length increased, fewer and fewer plants were observed. When the data were plotted graphically (frequency versus length), a bell-shaped curve resulted.

East further experimented with this population by selecting F_2 plants of various corolla lengths and allowing them to produce separate F_3 generations. Several are shown in Figure 6–2. In each case, a bell-shaped distribution was observed, with most individuals similar in height to the selected F_2 parents, but with considerable variation around this value.

East's experiments demonstrated that although the variation in corolla length seemed continuous, experimental crosses resulted in the segregation of distinct phenotypic classes as observed in the three independent F_3 categories. This key finding was the basis for the multiple-factor hypothesis, which explains how traits can deviate considerably in their expression.

Additive Alleles: The Basis of Continuous Variation

The multiple-factor hypothesis, suggested by the observations of East and others, embodies the following major points:

1. Characters that exhibit continuous variation can usually be quantified (by measuring, weighing, counting, etc.).

2. Two or more pairs of genes, located throughout the genome, account for the hereditary influence on the phenotype in an *additive way*. Because many genes can be involved, inheritance of this type is often called *polygenic*.

3. Each gene locus may be occupied by either an **additive allele**, which contributes a set amount to the phenotype, or by a **nonadditive allele**, which does not contribute quantitatively to the phenotype.

4. The total effect on the phenotype of each additive allele, while small, is approximately equivalent to all other additive alleles at other gene sites.

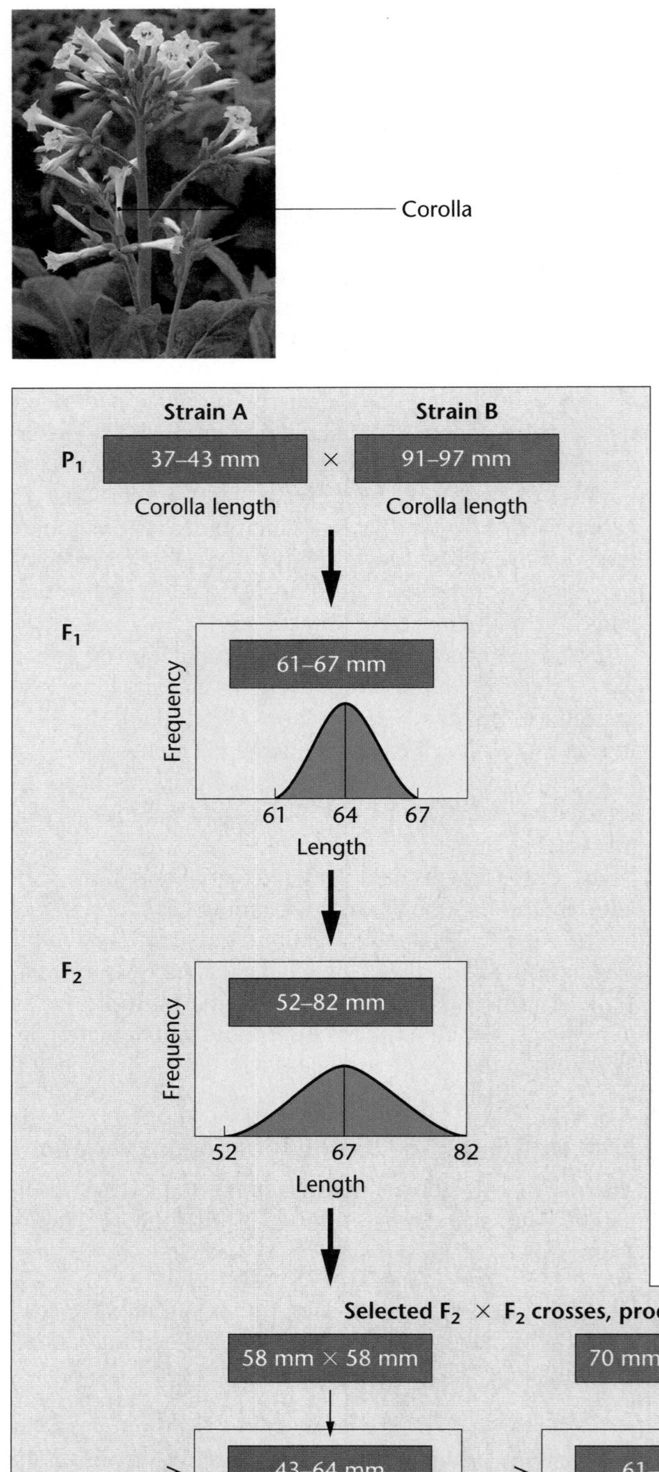

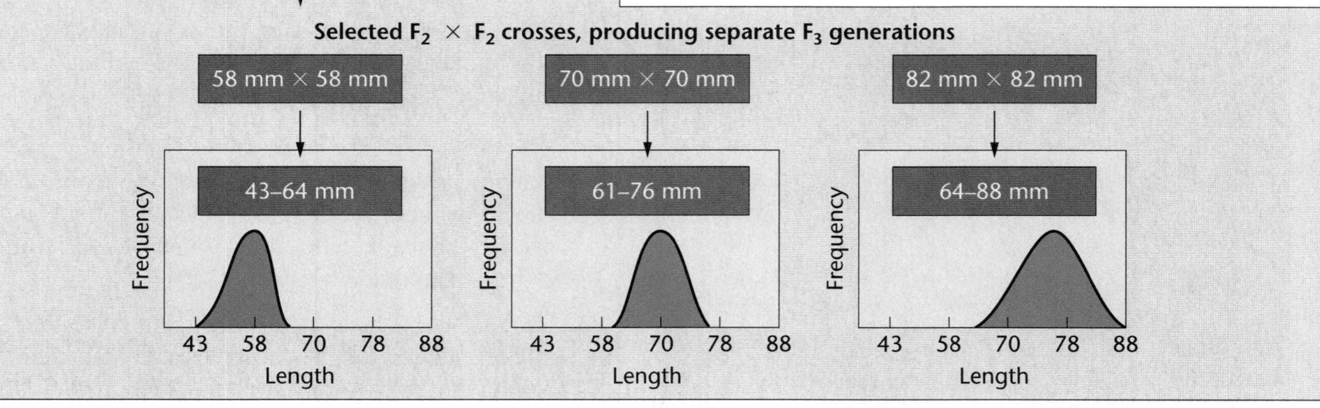

FIGURE 6–2 The F_1, F_2, and selected F_3 results of East's cross between two strains of *Nicotiana* with different corolla lengths. Plants of strain A vary from 37 to 43 mm, while plants of strain B vary from 91 to 97 mm. The photograph shows the flower and corolla of a tobacco plant. *(Photo: Norm Thomas/Photo Researchers, Inc.)*

5. Together, the genes controlling a single character produce substantial phenotypic variation.
6. Analysis of polygenic traits requires the study of large numbers of progeny from a population of organisms.

These points center around the concept that additive alleles at numerous loci control quantitative traits. To illustrate this, let's examine Herman Nilsson-Ehle's experiments involving grain color in wheat performed early in the twentieth century. In one set of experiments, wheat with red grain was crossed to wheat with white grain (Figure 6–3). The F_1 generation demonstrated an intermediate color. In the F_2 generation, approximately 15/16 of the plants showed some degree of red grain, while 1/16 of the plants showed white grain. Because the ratio occurred in sixteenths, he hypothesized that two gene pairs control the phenotype and, if so, they segregate independently from one another in a Mendelian fashion.

Careful examination of the F_2 generation revealed that grain color can be classified into four different shades of red. If two gene pairs are operative, each with one potential additive allele and one potential nonadditive allele, we can envision how the multiple-factor hypothesis accounts for this variation. In the P_1, both parents are homozygous. The red parent contains only additive alleles (uppercase letters), while the white parent contains only nonadditive alleles (lowercase letters). The F_1 generation, being heterozygous, contains only two additive alleles and expresses an intermediate phenotype. In the F_2, each offspring has either 4, 3, 2, 1, or 0 additive alleles (see Figure 6–3). Wheat with no additive alleles (1/16) is white like one of the P_1 parents, while wheat with 4 additive alleles is red like the other P_1 parent. Plants with 3, 2, or 1 additive alleles constitute the other three categories of red color observed in the F_2 plants, with most (6/16) having 2 additive alleles like the F_1 plants.

Therefore, continuous variation can be explained in a Mendelian fashion. Multiple-factor inheritance, where

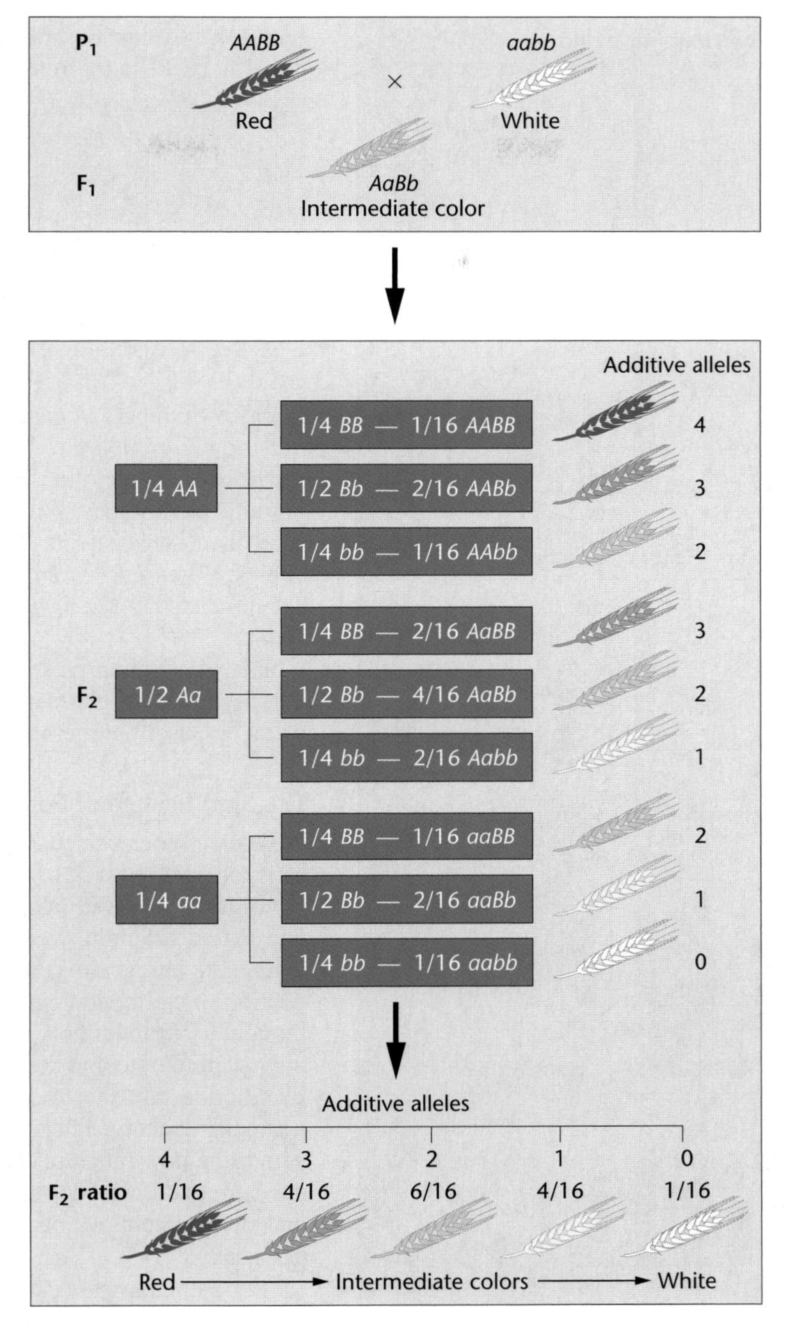

FIGURE 6–3 How the multiple-factor hypothesis accounts for the 1:4:6:4:1 phenotypic ratio of grain color when all alleles designated by uppercase letters are additive and contribute an equal amount of pigment to the phenotype.

additive alleles influence the phenotype in a quantitative manner, results in such variation. As we saw in Nilsson-Ehle's initial cross, if two gene pairs are involved, only five F_2 phenotypic categories, in a 1:4:6:4:1 ratio, are expected. There is no reason why three, four, or more gene pairs cannot function in controlling various phenotypes. As greater numbers of gene pairs become involved, the number of classes increases and results in more complex ratios. The number of phenotypes and the expected ratios of crosses involving up to five gene pairs are shown in Figure 6–4.

Calculating the Number of Genes

When additive effects control polygenic traits, it is of interest to determine the number of genes involved. If the ratio of F_2 individuals resembling *either* of the two most extreme phenotypes (the parental phenotypes) can be determined, then the number of gene pairs involved (n) can be calculated using the following simple formula:

$$\frac{1}{4^n} = \text{proportion of } F_2 \text{ individuals expressing either extreme phenotype}$$

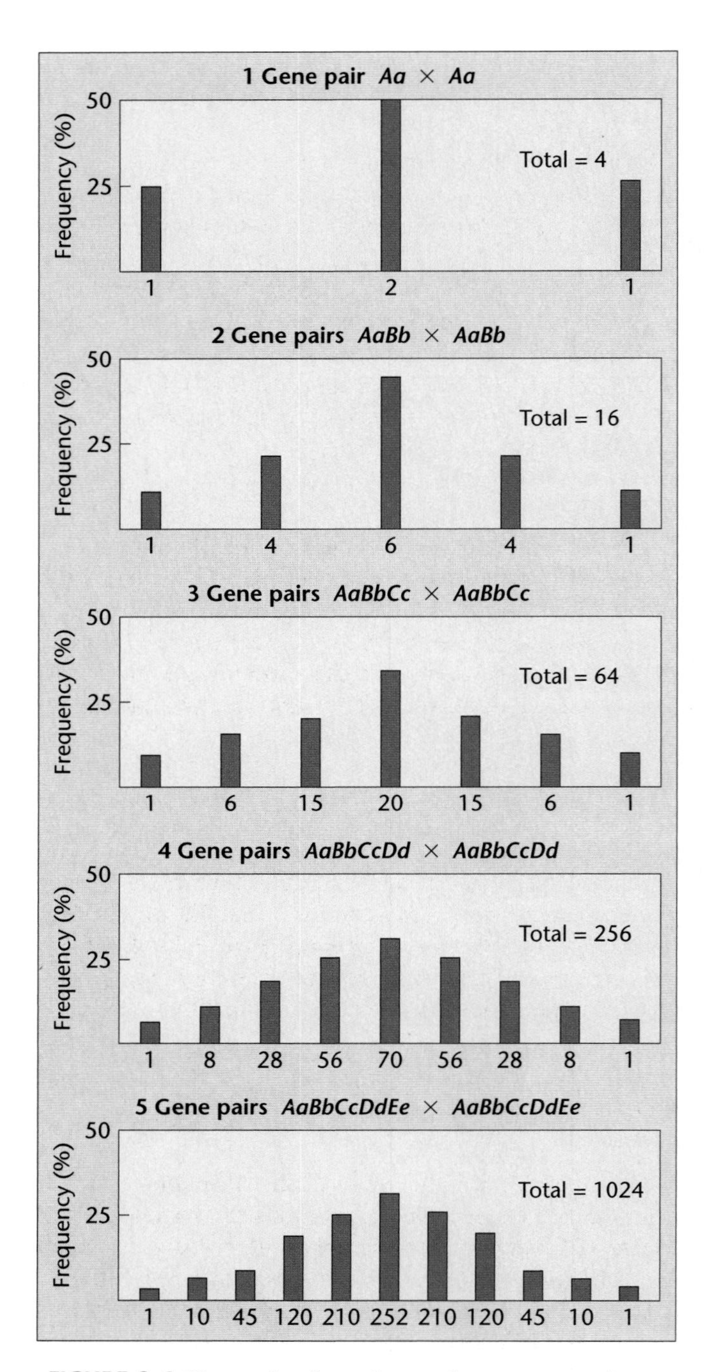

FIGURE 6–4 The results of crossing two heterozygotes when polygenic inheritance is operative with one to five gene pairs. Each histogram bar indicates a distinct phenotypic class from one extreme (at the left) to the other extreme (at the right). Each phenotype results from a different number of additive alleles.

In our previous example, the P_1 phenotypes represent these two extremes. In Figure 6–3, 1/16 of the F_2s are either red *or* white like the P_1 classes; this ratio can be substituted on the right side of the equation to solve for n:

$$\frac{1}{4^n} = \frac{1}{16}$$

$$\frac{1}{4^2} = \frac{1}{16}$$

$$n = 2$$

TABLE 6.1 Determination of the Number of Gene Pairs (*n*) Involved in Polygenic Crosses

n	*Individuals Expressing an Extreme Phenotype*	*Distinct F_2 Phenotypic Classes*
1	1/4	3
2	1/16	5
3	1/64	7
4	1/256	9
5	1/1024	11

For low numbers of gene pairs, it is sometimes easier to use the $(2n + 1)$ rule. If n equals the number of gene pairs, $2n + 1$ determines the total number of categories of possible phenotypes. When $n = 2$, $2n + 1 = 5$. That is, each phenotypic category can have 4, 3, 2, 1, or 0 additive alleles. When $n = 3$, $2n + 1 = 7$, and each phenotypic category can have 6, 5, 4, 3, 2, 1, or 0 additive alleles, and so on.

Table 6.1 summarizes information useful in determining the number of genes involved in polygenic inheritance, using either of the methods described above.

The Significance of Polygenic Inheritance

Polygenic inheritance is a significant concept because it appears to serve as the mode of inheritance for a vast number of traits involved in animal breeding and agriculture. For example, height, weight, and physical stature in animals, size and grain yield in crops, beef and milk production in cattle, and egg production in chickens are all thought to be under polygenic control. In most cases, it is important to note that the genotype, which is fixed at fertilization, establishes the potential range within which a particular phenotype falls. However, environmental factors determine how much of the potential will be realized. In the crosses described thus far, we have assumed an optimal environment, which minimizes variation from external sources.

Still other examples can be drawn from human genetic studies. Skin pigmentation, intelligence, various forms of behavior, obesity, and even the predisposition to certain diseases are all thought to be under the control of numerous genes. The latter two conditions, obesity and predisposition to disease (e.g., coronary heart disease) are good examples of **complex** or **multifactorial traits**. Unlike cases where continuous variation is observed, no clear Mendelian pattern of inheritance is observable in these diseases. However, both conditions run in families such that sons and daughters of affected parents are much more likely to be affected than are children of unaffected parents. Many genes are now known to be involved, but the environment substantially influences the ultimate expression of these traits. In such examples, when the combination of genetic and environmental factors are both substantial influences, genetic analysis is particularly difficult.

6.2 The Study of Polygenic Traits Relies on Statistical Analysis

Analysis of a polygenic trait most often involves quantitative measurements, usually from many offspring generated from many crosses. The outcome can be expressed as a frequency diagram that often demonstrates a normal (bell-shaped) distribution (Figure 6–5). While it is hoped that each series of crosses is representative of the population at large, variation in samples due strictly to chance can influence the data gathered. To assess the validity of the experimental data, statistical techniques are used. Such techniques were first devised by Galton early in the 20th century to assess the inheritance of traits exhibiting continuous variation. Galton's efforts served as this initial basis of this field of study, now called **biometry**.

Statistical analysis serves three purposes:

1. Data can be mathematically reduced to provide a descriptive summary of the sample.
2. Data from a small but random sample can be used to infer information about groups larger than those from which the original data were obtained (statistical inference).
3. Two or more sets of experimental data can be compared to determine whether they represent significantly different populations of measurements.

Several statistical methods are useful in the analysis of traits that exhibit a normal distribution, including the mean, variance, standard deviation, and standard error of the mean.

The Mean

The distributions of the two sets of phenotypic measurements graphed in Figure 6–6 cluster around a central value. This clustering is called the central tendency, one measurement of which is the **mean** ($\overline{X}$). The mean is simply the arithmetic average of a set of measurements or data and is calculated as

$$\overline{X} = \frac{\Sigma X_i}{n}$$

where $\overline{X}$ is the mean, ΣX_i represents the sum of all individual values in the sample, and n is the number of individual values.

Although the mean provides a descriptive summary of the sample, it is of limited value. As shown in Figure 6–6, a symmetrical distribution of values in the sample may, in one case, cluster near the mean, yet in another set of values may be distributed widely around it. These contrasting conditions represent different types of variation within each sample called the **frequency distribution**. Whether due to chance or to one or more experimental variables, such variation creates the need for methods to describe sample measurements statistically.

Variance

As seen in Figure 6–6, the range and distribution of values on either side of the mean determines the shape of the distribution curve. The degree to which values within this distribution diverge from the mean is called the **variance** (s^2) and is used to estimate the variation present in an infinitely large population. The variance for a sample is calculated as

$$s^2 = \frac{\Sigma(X_i - \overline{X})^2}{n - 1}$$

where the sum (Σ) of the squared differences between each measured value (X_i) and the mean ($\overline{X}$) is divided by one less than the total sample size ($n - 1$). To avoid the numerous subtraction functions necessary to calculate s^2 for a large sample, we convert the equation to its algebraic equivalent:

$$s^2 = \frac{\Sigma X_i^2 - n\overline{X}^2}{n - 1}$$

The variance is a valuable measure of sample variability. As noted earlier, two distributions can have identical means ($\overline{X}$) yet vary considerably in their frequency distribution around the mean. The variance represents the average squared deviation of the measurements from the mean. Estimation of variance is particularly valuable in

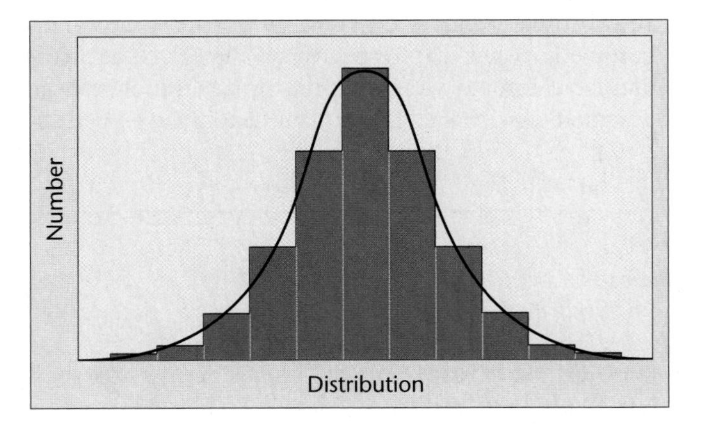

FIGURE 6–5 A normal frequency distribution characterized by a bell-shaped curve.

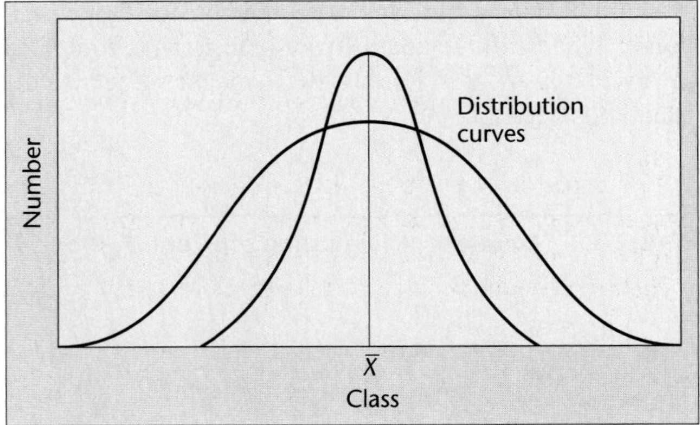

FIGURE 6–6 Two normal frequency distributions with the same mean but different amounts of variation.

TABLE 6.2 Sample Inclusion for Various s Values

Multiples of s	Sample Included (%)
$\overline{X} \pm 1s$	68.3
$\overline{X} \pm 1.96s$	95.0
$\overline{X} \pm 2s$	95.5
$\overline{X} \pm 3s$	99.7

determining the degree of genetic control of traits when the immediate environment also influences the phenotype.

Standard Deviation

The variance is a squared value. Therefore its unit of measurement is also squared (cm^2, mg^2, etc.). To express variation around the mean in the original units of measurement, we find the square root of the variance, or **standard deviation** (s):

$$s = \sqrt{s^2}$$

Table 6.2 shows what percentage of the individual values within a normal distribution is included with different multiples of the standard deviation. The mean plus or minus one standard deviation ($\overline{X} \pm 1s$) includes 68% of all values in the sample. Over 95% of all values are found within two standard deviations ($\overline{X} \pm 2s$). As such, the standard deviation is an important descriptive tool. Furthermore, s can be interpreted as a probability. The $\overline{X} \pm 1s$ indicates a 68% probability that a measured value picked at random will fall within that range.

Standard Error of the Mean

To estimate how much the means of other similar samples drawn from the same population might vary, we calculate the **standard error of the mean** ($S_{\overline{X}}$):

$$S_{\overline{X}} = \frac{s}{\sqrt{n}}$$

where s is the standard deviation, and $\sqrt{n}$ is the square root of the sample size. The standard error of the mean measures the accuracy of the sample mean—that is, the variation of sample means in replications of the experiment. Because the standard error of the mean is computed by dividing s by $\sqrt{n}$, it is always a smaller value than the standard deviation.

Analysis of a Quantitative Character

To illustrate how biometric methods are used in quantitative analysis, we consider a simple example involving fruit weight in tomatoes. Let's assume that fruit weight is a quantitative character and that one highly inbred strain (one that is highly homozygous) produces tomatoes averaging 18 oz in weight and another highly inbred strain produces fruit averaging 6 oz in weight. These two varieties are crossed and produce an F_1 generation with weights ranging from 10 oz to 14 oz The F_2 population contains individuals that produce fruit ranging from 6 oz to 18 oz The results characterizing both generations are shown in Table 6.3.

The mean value for fruit weight in the F_1 generation is calculated as

$$\overline{X} = \frac{\sum X_i}{n} = \frac{626}{52} = 12.04$$

Similarly, the mean value for fruit weight in the F_2 generation is calculated as

$$\overline{X} = \frac{\sum X_i}{n} = \frac{872}{72} = 12.11$$

Average fruit weight is 12.04 oz in the F_1 generation and 12.11 oz in the F_2 generation. Although these mean values are similar, it is apparent from the frequency distributions (Table 6.3) that more variation is present in the F_2 generation. Fruit weight ranges from 6 oz to 18 oz in the F_2 generation, but only from 10 oz to 14 oz in the F_1 generation.

To quantify the amount of variation present in each generation, we calculate the variance, s^2 (Table 6.4). As noted earlier, the sample variance is the sum of the squared differences between each value and the mean, divided by one less than the total number of observations. However, in the case where a number of observations (f) are grouped into representative classes (x), the variance is calculated according to the formula

$$s^2 = \frac{n\sum f(x^2) - (\sum fx)^2}{n(n-1)}$$

As shown in Table 6.4, the variance is 1.29 for the F_1 generation, and 4.27 for the F_2 generation. When converted to the standard deviation $\left(s = \sqrt{s^2}\right)$ the values become 1.13 and 2.07, respectively. Therefore, the distribution of tomato weight in the F_1 generation can be described as 12.04 ± 1.13 oz, and that in the F_2 genera-

TABLE 6.3 Frequency Distribution of F_1 and F_2 Progeny

		Weight (oz)												
		6	*7*	*8*	*9*	*10*	*11*	*12*	*13*	*14*	*15*	*16*	*17*	*18*
Number of	F_1					4	14	16	12	6				
Individuals	F_2	1	1	2	0	9	13	17	14	7	4	3	0	1

TABLE 6.4 Calculation of Variance, s^2

	F_1				F_2		
x	f	$f(x)$	$f(x)^2$	x	f	$f(x)$	$f(x)^2$
6				6	1	6	36
7				7	1	7	49
8				8	2	16	128
9				9	0	0	0
10	4	40	400	10	9	90	900
11	14	154	1694	11	13	143	1573
12	16	192	2304	12	17	204	2448
13	12	156	2028	13	14	182	2366
14	6	84	1176	14	7	98	1372
15				15	4	60	900
16				16	3	48	768
17				17	0	0	0
18				18	1	18	324

F_1: $n = 52 \quad \sum f(x) = 626 \quad \sum f(x)^2 = 7602$

$$s^2 = \frac{52 \times 7602 - (626)^2}{52(52 - 1)} = \frac{395{,}304 - 391{,}876}{2652} = 1.29$$

F_2: $n = 72 \quad \sum f(x) = 872 \quad \sum f(x)^2 = 10{,}864$

$$s^2 = \frac{72 \times 10{,}864 - (872)^2}{72(72 - 1)} = \frac{782{,}208 - 760{,}684}{5112} = 4.27$$

tion can be described as 12.11 ± 2.07 oz. This analysis indicates that the mean fruit weight of F_1 is nearly identical to that of F_2, but the F_2 shows greater variability in the distribution of weights than does F_1.

Observations about the inheritance of fruit weight in crosses between these two strains of tomatoes meet the expectations for polygenic traits. In this example, if we assume that each parental strain is homozygous for the additive or nonadditive alleles that control fruit weight, we can estimate the number of gene pairs involved in controlling fruit weight in these two strains of tomatoes. Since $1/72$ of the F_2 offspring have a phenotype that overlaps one of the parental strains (72 total F_2 offspring, one weighs 6 oz, and one weighs 18 oz; see Table 6.3), using the formula $1/4^n = 1/72$ shows that n is between 3 and 4, indicative of the number of genes that control fruit weight in these tomato strains. If this experiment were repeated many times with similar results, our confidence in this conclusion would be bolstered.

6.3 Heritability Is a Measure of the Genetic Contribution to Phenotypic Variability

Having just introduced several ways in which quantitative or continuous variation is measured and characterized in populations, we now consider how to assess the extent to which genetic factors contribute to such phenotypic variation. Often, much of the variation can be attributed to genetic factors, with the total environment having less influence. In other cases, the environment may have a greater influence on phenotypic variation within a population. The following discussion considers how geneticists attempt to define the impact of heredity versus environment on phenotypic variation.

Broad-Sense Heritability

If a trait can be measured quantitatively, experiments on many plants and animals can test the origin of variation. One approach uses inbred strains containing individuals of a relatively homogeneous (highly homozygous) genetic background. Experiments are then designed to test the effects of the range of prevailing environmental conditions on phenotypic variability. Variation observed *between* different inbred strains reared in a constant environment is due predominantly to genetic factors. Variation observed *among* members of the same inbred strain reared under different environmental conditions is due to nongenetic factors, which are generally categorized as "environmental."

The relative importance of genetic versus environmental factors can be formally assessed by examining the **heritability index (H^2)**, which we calculate using an analysis of variance among individuals of a known genetic relationship. (In the ensuing discussion, we replace the term s^2 for variance with the term V.) This is an important approach for investigating organisms with long generation times. Also called **broad-sense heritability**, $\boldsymbol{H^2}$ measures the degree to which **phenotypic variance** $(\boldsymbol{V_P})$ is due to variation in genetic factors for a single population under the limits of environmental variation during the study. Note that the H^2 value *does not* determine the proportion of the total phenotype attributed to genetic factors, but it *does* estimate the proportion of observed variation in the phenotype attributed to genetic factors, in comparison to environmental factors.

Phenotypic variance is due to the sum of three components: environmental variance (V_E), genetic variance (V_G), and variance resulting from the interaction of genetics and environment (V_{GE}). Therefore, phenotypic variance (V_P) is expressed as

$$V_P = V_E + V_G + V_{GE}$$

Because V_{GE} is often negligible, it is usually omitted. Therefore, the simpler equation is generally used:

$$V_P = V_E + V_G$$

Broad-sense heritability expresses that proportion of variance due to the genetic component:

$$H^2 = \frac{V_G}{V_P}$$

An H^2 value approaching 1.0 indicates that environmental conditions have little impact on phenotypic variance in the population studied. An H^2 value close to 0 indicates that the environment is almost solely responsible for the observed phenotypic variation.

It is not possible to obtain an absolute H^2 value for any given character. If measured in a different population under a greater or lesser degree of environmental variability, H^2 might be different for that character. For that reason, broad-sense heritability estimates are most useful in highly inbred strains or genetic clones such as artificially selected plant populations that are asexually propagated by cuttings. Furthermore, broad-sense heritability estimates are not very accurate in estimating the selection potential of quantitative traits since H^2 calculations take into account all forms of genetic variation, not simply additive genetic effects. Therefore, another type of calculation, narrow-sense heritability, has been devised that is useful for additive effects.

Narrow-Sense Heritability

Information regarding heritability is most useful in animal and plant breeding as a measure of potential response to selection. In this case, a different estimate of heritability must be used, based on a subcomponent of V_G referred to as *additive variance* (V_A):

$$V_G = V_A + V_D + V_I$$

Here, V_A represents the additive variance that results from the average effect of additive components of genes; *dominance variance*, V_D, is the deviation from additive components that results when phenotypic expression in heterozygotes is not precisely intermediate between the two homozygotes; and *interactive variance*, V_I, is the deviation from additive components that occurs when two or more loci behave epistatically. Interactive variance is not associated with the average effect of V_A and is often negligible. Thus, it is often excluded from calculations.

When V_G is partitioned into V_A and V_D, a new assessment of heritability, $\boldsymbol{h^2}$, or **narrow-sense heritability**, can be calculated. Thus, h^2 values are useful in assessing selection potential in randomly breeding animal and plant populations:

$$h^2 = \frac{V_A}{V_P}$$

Because $(V_P = V_E + V_G)$ and $(V_G = V_A + V_D)$, we obtain

$$h^2 = \frac{V_A}{V_E + V_A + V_D}$$

As we shall subsequently see, h^2 values are useful for predicting the phenotypes of offspring during selection procedures. The closer a value is to 1.0, the greater our ability to make an accurate prediction based on our knowledge of the parental phenotypes.

Artificial Selection

The process of selecting a specific group of organisms from an initially heterogeneous population for future breeding purposes is called **artificial selection**. A relatively high h^2 value predicts that selection will likely succeed in altering a population. As you can imagine based on our preceding discussion, measuring the components necessary to calculate h^2 is a complex task. A simplified approach involves measuring the central tendencies (the means) of a trait from (1) a parental population exhibiting a bell-shaped distribution (M), (2) a "selected" segment of the parental population that expresses the most desirable quantitative phenotypes (M_1), and (3) the offspring (M_2) resulting from interbreeding the selected M_1 group. When this is accomplished, the following relationship of the three means and h^2 exists:

$$M_2 = M + h^2(M_2 - M)$$

Solving this equation for h^2 gives us

$$h^2 = \frac{M_2 - M}{M_1 - M}$$

Once these relationships are established, the equation can be further simplified by defining $M_2 - M$ as the *response* (R) and $M_1 - M$ as the *selection differential* (S), where h^2 reflects the ratio of the response observed to the total response possible. Thus,

$$h^2 = \frac{R}{S}$$

Now let's assess one case of selection. Suppose we measure the diameter of corn kernels in a population where the mean diameter (M) is much larger than desirable (20 mm), and from that population we select a group with the smallest diameters, for which the mean (M_1) is 10 mm. Plants that yielded this selected population are then interbred, and the progeny kernels yield a mean (M_2) of 13 mm; we then calculate h^2 to estimate the potential for artificial selection on kernel size:

$$h^2 = \frac{13 - 20}{10 - 20}$$

$$= \frac{-7}{-10}$$

$$= 0.70$$

On the basis of this calculation, we can conclude that the selection potential for kernel size is relatively high.

The longest-running artificial selection experiment is still being conducted at the State Agricultural Laboratory in Illinois. Since 1896, corn has been selected for both high and low oil content. After 76 generations, selection continues to increase oil content (Figure 6–7). As selection has progressed, heritability of increased oil content has declined (see parenthetical values at generations 9, 25, 52, and 76 in Figure 6–7). The process will continue until all individuals in the population contain a uniform genotype for the additive alleles responsible for oil content. At that point, heritability will be reduced to zero and the response to artificial selection will cease. Examination of the selection for low oil content shows that heritability is approaching this point.

Similar calculations involving artificial selection for other traits in a variety of organisms have established whether selection is an effective approach for obtaining populations exhibiting desirable phenotypes. Table 6.5 lists estimates of narrow heritability for a variety of traits in various organisms. These proportions are expressed as percentages. As you can see, heritability varies considerably among traits.

In general, heritability is low for traits that are essential to an organism's survival, primarily because the genetic component has been largely optimized during evolution. Egg production, litter size, and conception rate are examples where such physiological limitations on selection have already been established. Nevertheless, traits that are less critical to survival, such as body weight, tail length, and wing length, show higher heritability values. Narrow-sense heritability estimates are more valuable when they are based on data collected in many populations and environments and where a clear

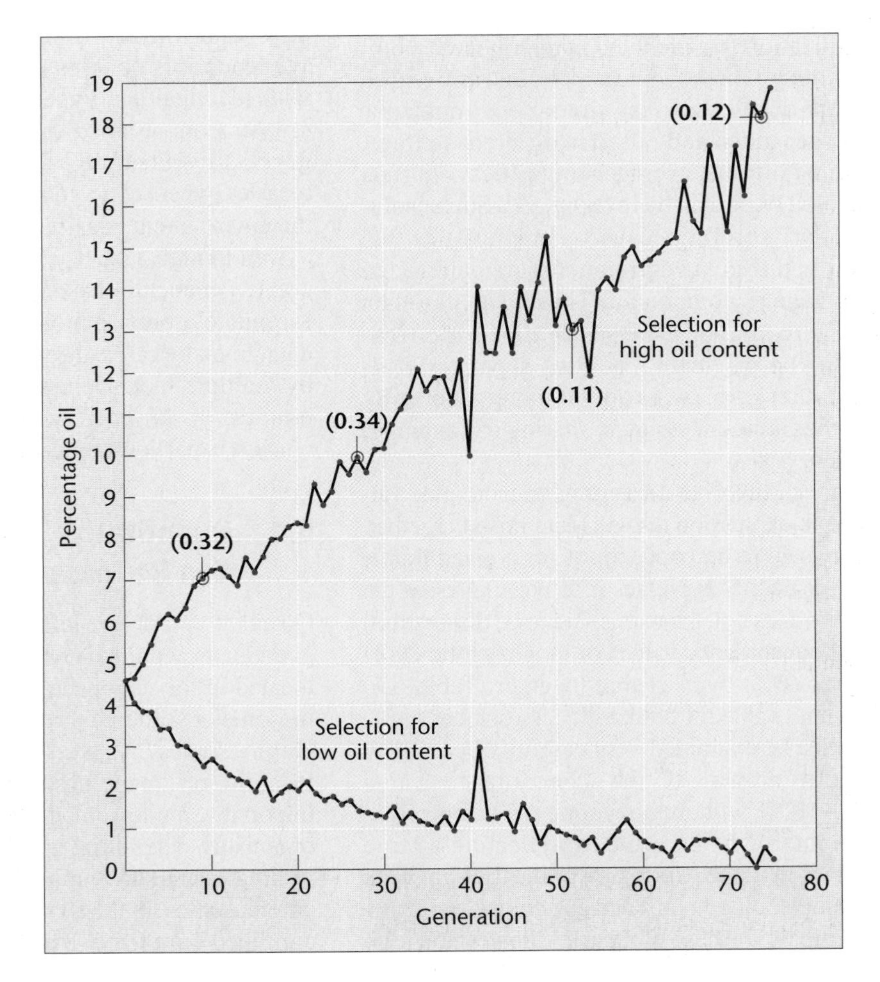

FIGURE 6–7 Response of corn selected for high and low oil content over 76 generations. The numbers in parentheses along the line for the high oil content indicate the calculation of heritability at these points in the continuing experiment.

TABLE 6.5 Estimates of Heritability for Traits in Different Organisms

Trait	Heritability (%) (h^2)
Mice	
Tail length	60
Body weight	37
Litter size	15
Chickens	
Body weight	50
Egg production	20
Egg hatchability	15
Cattle	
Birthweight	51
Milk yield	44
Conception rate	3

trend is established. Based on such estimates, selection techniques have led to vast improvements in the quality of animal and plant products.

TABLE 6.6 A Comparison of Concordance of Various Traits Between Monozygotic (MZ) and Dizygotic (DZ) Twins

	Concordance (%)	
Trait	MZ	DZ
Blood types	100	66
Eye color	99	28
Mental retardation	97	37
Measles	95	87
Hair color	89	22
Handedness	79	77
Idiopathic epilepsy	72	15
Schizophrenia	69	10
Diabetes	65	18
Identical allergy	59	5
Cleft lip	42	5
Clubfoot	32	3
Mammary cancer	6	3

Twin Studies in Humans

Traditional heritability studies are not possible in humans, for obvious reasons. However, human twins are very useful subjects for studying the heredity-versus-environment question. **Monozygotic (identical) twins**, derived from the division and splitting of a single egg following fertilization, are identical in their genetic compositions. Although most identical twins are reared together and are exposed to very similar environments, some pairs are separated and raised in different settings. Thus, for particular traits, average similarities or differences in separated twins can be investigated. Such analyses are particularly useful because characteristics that remain similar in different environments are believed to have a strong genetic component. These data can then be compared with similar analyses of **dizygotic (fraternal) twins**, who originate from two separate fertilization events. Dizygotic twins are no more genetically similar than other nontwin siblings, sharing (on average) one-half of their genes.

Another approach involves measuring concordance values of phenotypic expression in twin pairs raised together. These twins are said to be **concordant** for a given trait if both express it or neither expresses it. If one expresses the trait and the other does not, the pair is said to be **discordant**. Comparison of concordance values of monozygotic (MZ) versus dizygotic (DZ) twins reared together (Table 6.6) shows the potential value for heritability assessment.

These data must be examined very carefully before any conclusions are drawn. If the concordance value approaches 90–100% with monozygotic twins, we might be inclined to interpret this value as indicating a large genetic contribution to the expression of the trait. In some cases—for example, blood types and eye color—we know this is indeed true. In the case of measles, however, a high concordance value merely indicates that the trait is almost always induced by a factor in the environment—in this case, a virus.

It is more meaningful to compare the *difference* between the concordance values of monozygotic and dizygotic twins. If these values are significantly higher for monozygotic twins than for dizygotic twins, a strong genetic component may be involved in the determination of the trait. We reach this conclusion because monozygotic twins, with identical genotypes, would be expected to show a greater concordance than genetically related, but not genetically identical, dizygotic twins. In the case of measles, where concordance is high in both types of twins, the environment is assumed to contribute significantly.

Even though a particular trait may demonstrate considerable genetically based variation, it is often difficult to formulate a precise mode of inheritance using available data. In many cases, the trait is considered to be controlled by multiple-factor inheritance. However, when the environment is also exerting a partial influence, such a conclusion is particularly difficult to prove.

6.4 Quantitative Trait Loci Can Be Mapped

Quantitative traits are influenced by numerous genes. Thus, geneticists would like to know where these genes are located in the genome. Are they linked on a single chromosome or scattered throughout the genome among many chromosomes? As we shall see in subsequent chapters, locating, or "mapping," genes within the genome is an important step toward establishing the genetic identity of organisms. The initial approach is to localize genes controlling quantitative traits on a particular chromosome or chromosomes. In the context of this discussion, these genes are called **quantitative trait loci (QTLs)**, and some rather ingenious methods have been devised to map these genes.

As an example of this type of analysis, we consider the phenotypic trait in *Drosophila* of resistance to the insecticide DDT, which has been shown to be under polygenic control. To find the loci responsible, strains selected for resistance to DDT and strains selected for sensitivity to DDT were crossed to flies carrying dominant genes that serve as markers on each of *Drosophila*'s four chromosomes. Following a variety of crosses, offspring were produced that contained many different combinations of marker chromosomes and chromosomes from either resistant or sensitive strains. As shown in Figure 6–8, flies with various chromosome combinations were then tested for resistance (survival) when exposed to DDT. Results indicate that each chromosome in *Drosophila* contains genes that contribute to resistance. In other words, the loci bearing the genes that control DDT resistance are scattered throughout the genome.

We are now able to map the position of these genes more specifically along each chromosome in *Drosophila* because of the presence of molecular markers on each chromosome. The locations of QTLs are determined relative to the known positions of these markers. These markers, called **restriction fragment length polymorphisms (RFLPs)**, represent specific nuclease cleavage sites found along chromosomes. This approach enumerates and maps the loci responsible for quantitative traits.

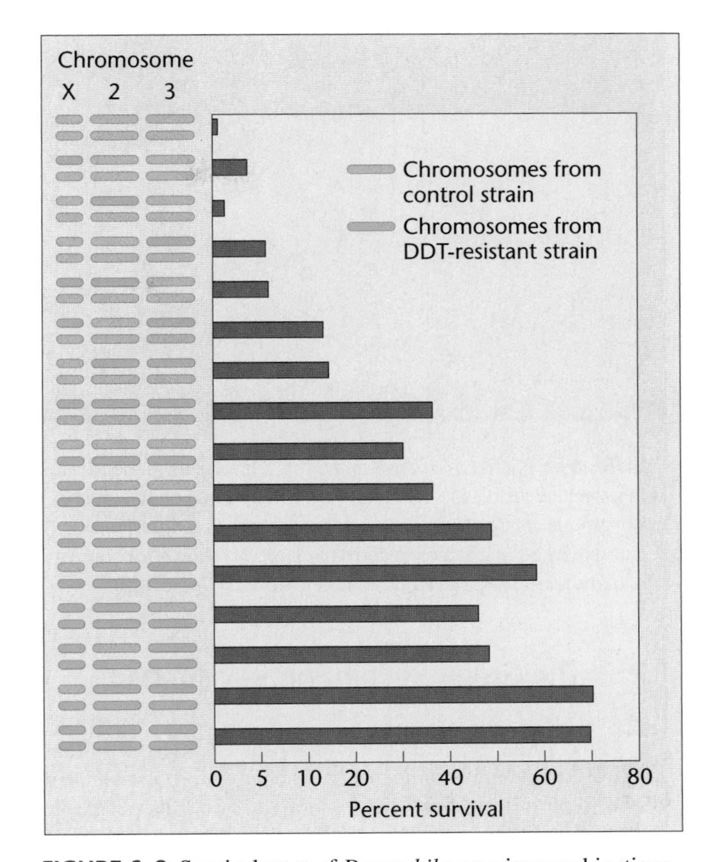

FIGURE 6–8 Survival rates of *Drosophila* carrying combinations of chromosomes from DDT-resistant and DDT-sensitive (control) strains when exposed to DDT. The results indicate that DDT resistance is polygenic, with genes on each of the major chromosomes making a major contribution. (Chromosome 4 carries only a few *Drosophila* genes and was omitted from this analysis).

RFLP markers are now available for many organisms of agricultural importance, making systematic mapping of QTLs possible. For example, hundreds of RFLP markers have been located in the tomato. They are spaced along all 12 chromosomes of this organism. Analysis is performed by crossing plants with extreme but opposite phenotypes and following the crosses through several generations. When both a marker and a phenotypic trait of interest are expressed together, they are said to *cosegregate*. Consistent cosegregation establishes the presence of a QTL at or near the RFLP marker along the chromosome. Whenever both an RFLP marker and a QTL responsible for the trait under investigation are closely linked along a single chromosome, they are more likely to demonstrate an association throughout a pedigree than if they are not closely linked. When numerous QTLs are located, a genetic map is created for the involved genes.

RFLP analysis has resulted in extensive mapping of QTLs in the tomato, *Lycopersicon esculentum*. Many loci have been identified that are responsible for fruit weight, soluble solid content, and acidity. These loci are distributed on all 12 chromosomes representing the haploid genome of this plant. Several chromosomes contain loci for all three traits. Note that the RFLP method initially allows the identification of small chromosomal regions, not individual genes, although each region may house only a single gene.

Fruit weight in the tomato has been the focus of a highly successful research effort conducted by Steven Tanksley and his colleagues. Currently, 28 QTLs responsible for phenotypic variation in fruit weight have been identified. One of these, *fw2.2*, has now been isolated, cloned, and transferred between plants, with most interesting results. While the cultivated tomato can weigh up to 1000 grams, the progenitor of the modern tomato is thought to have weighed only a few grams. Two distinct alleles of *fw2.2* are now recognized as a result of RFLP mapping studies—one allele is present in all of the wild small-fruited varieties of tomatoes investigated; the other allele is present in all domesticated large-fruited varieties. When the cloned allele from small-fruited varieties is transferred to a plant that normally produces large tomatoes, a remarkable result is achieved: The transformed plant produces fruits that are reduced in weight by about 30% (Figure 6–9). In the varieties employed in this study, this reduction averaged 17 grams, a significant phenotypic change caused by the action of a single gene.

Thus, for the first time, the genetic basis of quantitative variation is available for meaningful investigation. Tanksley's research group has now established that the *fw2.2* locus encodes the gene, *ORFX*, that is involved in the control of the number of carpels. Carpels are ovule-bearing units within the ovary of the flower that, following fertilization, develop into a fruit. The number of carpel units influences fruit size. That the allele for small fruit is partially dominant to the large fruit allele suggests that

FIGURE 6–9 Phenotypic effect of the *fw2.2* transgene in the tomato. When the allele causing small fruit is transferred to a plant that normally produces large fruit (marked +), the fruit is reduced in size. A fruit (marked −), which is normally larger, is shown for comparison. *(Courtesy of Dr. Steven D. Tanksley, Cornell University)*

the genetic alteration between the two alleles is involved in the regulation of floral development, ultimately determining carpel number. This is in contrast to a genetic change that alters the sequence and structure of a protein that somehow imparts weight to the fruit. Of added interest to these findings is the observation that the *ORFX* gene is structurally similar to a human oncogene in the *ras* family implicated in malignancy.

Mapping QTLs and defining the function of genes present in these regions in agriculturally important plants will greatly enhance programs designed to improve their yield. Knowledge gained from the study of so-called "quantitative genes" in plants will no doubt pave the way for investigations of similar genes in animals, including our own species. Since such loci are thought to be critical components of what we have previously been referred to as complex traits, our knowledge of how genes actually control phenotypes will be greatly extended.

GENETICS, TECHNOLOGY, AND SOCIETY

The Green Revolution Revisited

Of the greater than 6 billion people now living on Earth, about 750 million don't have enough to eat. And despite efforts to limit population growth, an additional 1 million people are expected to go hungry each year for the next several decades.

Will we be able to solve this problem? The past gives us some reasons to be optimistic. In the 1950s and 1960s, in the face of looming population increases, botanists around the world set about increasing the production of crop plants, including the three most important grains, wheat, rice, and maize. These efforts became known as the Green Revolution. The approach was three-pronged: (1) to increase the use of fertilizers, pesticides, and irrigation water, (2) to bring more land under cultivation, and (3) to develop improved varieties of crop plants by intensive plant breeding. While highly successful, in recent years, the rate of increase in grain yields has slowed. If food production is to keep pace with the projected increase in the world's population, plant breeders will have to depend more and more on the genetic improvement of crop plants to provide higher yields. But is this possible? Are we approaching the theoretical limits of yield in important crop plants? Recent work with rice suggests that the answer to this question is a resounding no.

Rice ranks third in worldwide production, just behind wheat and maize. About 2 billion people, fully one-third of Earth's population, depend on rice for their basic nourishment. The majority of the world's rice crop is grown and consumed in Asia, but it is also a dietary staple in Africa and Central America. The Green Revolution for rice began in 1960 with the establishment of the International Rice Research Institute (IRRI), headquartered at Los Baños, Philippines. The goal was to breed rice with improved disease resistance and higher yield. Breeders were almost too successful: The first high-yield varieties were so top-heavy with grain that they tended to fall over. (Plant breeders call this "lodging.") To reduce lodging, IRRI breeders crossed a high-yield line with a native dwarf variety to create semidwarf lines, which were introduced to farmers in 1966. Due in large part to the adoption of the semidwarf lines, the world production of rice doubled over the next 25 years.

Rice breeders cannot afford to rest on their laurels, however, as the yield of modern rice varieties has not improved much in recent years. Predictions suggest that a 70% increase in the annual rice harvest may be necessary to keep pace with anticipated population growth during the next 30 years. Breeders are now looking to wild rice varieties for further crop improvement. Leading the way are Susan McCouch, Steven Tanksley, and their coworkers at Cornell University. To test the hypothesis that wild rice species carry genes that will improve the yield of cultivated varieties, they crossed cultivated rice (*Oryza sativa*) with a low-yield wild ancestor species (*Oryza rufipogon*) and then successively backcrossed the interspecific hybrid to cultivated rice for three generations. In theory, this would create lines whose genomes were about 95% from *O. sativa* and 5% from *O. rufipogon*. When testing these backcrossed lines for grain yield, they found that several of them outproduced cultivated rice by as much as 30%. These results strikingly demonstrated that even though wild rice relatives have low yields and appear to be inferior to cultivated rice, they still carry genes that will increase the yield of elite rice varieties. It is now up to breeders to exploit the wild relatives of rice.

But introducing favorable genes from a wild relative into a cultivated variety by conventional breeding is a long and involved process, often requiring a decade or more of crossing, selection, backcrossing, and more selection. Future improvements in cultivated species must be quicker if crop yields are to keep up with population growth. Fortunately, modern gene-mapping techniques are now leading to the identification of quantitative trait loci (QTLs) that control complex traits such as yield and disease resistance. This makes possible a more direct approach to crop improvement, which has been termed the *advanced backcross QTL method.*

In this method, a cultivated variety is crossed with a wild relative, just as *O. sativa* was interbred with *O. rufipogon*. The hybrid is then backcrossed to the cultivated variety to generate lines that contain only a small fraction of the "wild" genome. The backcross lines with the best qualities (e.g., highest yield and most disease resistance) are selected, and the "wild QTLs" responsible for the superior performance are identified using a detailed molecular linkage map. Once beneficial QTLs are discovered by this method, they may be introduced into other cultivated varieties.

For this strategy to succeed, it is essential that wild relatives of crops be preserved as a storehouse of potentially useful genes. Efforts were begun in the 1970s to protect the existing plant relatives of many crops in their natural habitats and to preserve them in seed banks. As the work of McCouch and Tanksley and others has shown, it is not possible to predict which wild varieties may be needed decades or even centuries from now to contribute their beneficial alleles to cultivated varieties. To prevent the loss of superior genes, the widest possible spectrum of wild species must be preserved, even those that have no obvious favorable characteristics.

Almost 60 years ago, the great Russian plant geneticist N. I. Vavilov suggested that wild relatives of crop plants could be the source of genes to improve agriculture. In this century, Vavilov's vision may finally be realized, as genes from long-neglected wild relatives of crops, identified by new molecular methods, spark a revitalized Green Revolution.

References

Mann, C. 1997. Reseeding the green revolution. *Science* 277: 1038–43.

Ronald, P. C. 1997. Making rice disease-resistant. *Sci. Am.* (Nov.) 277: 98–105.

Tanksley, S. D., and McCouch, S. R. 1997. Seed banks and molecular maps: Unlocking genetic potential from the wild. *Science* 277: 1063–66.

Xiao, J. et al., 1996. Genes from wild rice improve yield. *Nature* 384: 223–24.

Chapter Summary

1. Continuous variation is exhibited in crosses involving traits under polygenic control. Such traits are quantitative in nature and are inherited as a result of the cumulative impact of additive alleles.
2. Polygenic characteristics can be analyzed using statistical methods, which include the mean, the variance, the standard deviation, and the standard error of the mean. Such statistical analysis is descriptive and can be used to make inferences about a population or to compare and contrast sets of data.
3. For many phenotypic characteristics, it is difficult to ascertain when variation is due to genetic or to environmental factors. Heritability, the estimate of the relative importance of genetic versus nongenetic factors in determining phenotypic variation in populations, can be calculated for many characters and is especially useful in selective breeding of commercially valuable plants and animals.
4. Studies involving twins are aimed at resolving the question of heredity versus environment in human traits. The degree of concordance of a trait is compared in monozygotic (identical) and dizygotic (fraternal) twins raised together or apart.
5. Loci bearing genes that control a quantitative trait are called quantitative trait loci (QTLs). Using either genetic or molecular markers, the location and distribution within the genome of QTLs can be ascertained.

Key Terms

additive allele, 115
artificial selection, 122
biometry, 119
broad-sense heritability, 122
complex traits, 118
concordance, 124
discordant, 124
dizygotic (fraternal) twins, 124
frequency distribution, 119
heritability index, 121
mean, 119
monozygotic (identical) twins, 124
multiple-factor hypothesis, 114
multifactorial traits, 118
narrow-sense heritability, 122
nonadditive allele, 115
phenotypic variance, 122
polygenic inheritance, 114
quantitative inheritance, 114
quantitative trait loci (QTLs), 124
restriction fragment length polymorphisms (RFLPs), 125
selection differential, 122
standard deviation, 120
standard error of the mean, 120
variance, 119

INSIGHTS AND SOLUTIONS

1. In a plant, height varies from 6 to 36 cm. When 6- and 36-cm plants are crossed, all F_1 plants are 21 cm. In the F_2 generation, a continuous range of heights is observed. Most are around 21 cm, and 3 of 200 plants are as short as the 6-cm P_1 parent. (a) What mode of inheritance is demonstrated, and how many gene pairs are involved? (b) How much does each additive allele contribute to height? (c) List all genotypes that give rise to 31-cm plants.

Solution: (a) Polygenic inheritance, where a continuous trait is involved and alleles contribute additively to the phenotype, is demonstrated. The 3/200 ratio of F_2 plants is the key to determining the number of gene pairs. This reduces to a proportion of 1/66.7, very close to 1/64. Using the formula $1/4^n = 1/64$ (where 1/64 is equal to the proportion of F_2 phenotypes as extreme as either P_1 parent), $n = 3$. Therefore, three gene pairs are involved.

(b) The variation between the two extreme phenotypes is

$$36 - 6 = 30 \text{ cm}$$

Because there are six potential additive alleles (*AABBCC*), each contributes

$$\frac{30}{6} = 5 \text{ cm}$$

to the base height of 6 cm, which results when no additive alleles (*aabbcc*) are part of the genotype.

(c) All genotypes that include 5 additive alleles will be $(5 \text{ alleles} \times 5 \text{ cm}) + 6\text{-cm base height} = 31 \text{ cm}$.

The genotypes *AABBCc*, *AABbCC*, and *AaBBCC* will result in 31-cm plants.

2. The results recorded for ear length in corn are shown in the table below. Calculate the mean values for ear length for each parental strain and for the F_1 plants.

Solution: The mean values are calculated as follows:

$$\bar{X} = \frac{\Sigma X_i}{n} \qquad \text{Parent A: } \bar{X} = \frac{\Sigma X_i}{n} = \frac{378}{57} = 6.63$$

$$\text{Parent B: } \bar{X} = \frac{\Sigma X_i}{n} = \frac{1697}{101} = 16.80$$

$$F_1\text{: } \bar{X} = \frac{\Sigma X_i}{n} = \frac{836}{69} = 12.12$$

3. Compare the mean of the F_1 generation in Problem 2 with that of each parental strain. What does this tell you about the type of gene action involved?

Solution: The F_1 mean (12.11) is almost midway between the parental means of 6.63 and 16.80. This indicates that the genes in question may be additive in effect.

4. The mean and variance of corolla length in two highly inbred strains of *Nicotiana* and their progeny are shown in the table below. One parent (P_1) has a short corolla, and the other parent (P_2) has a long corolla. Calculate the broad-sense heritability (H^2) of corolla length in this plant.

Strain	*Mean (mm)*	*Variance (mm)*
P_1 short	40.47	3.12
P_2 long	93.75	3.87
$F_1(P_1 \times P_2)$	63.90	4.74
$F_2(F_1 \times F_1)$	68.72	47.70

Solution: The formula for estimating broad-sense heritability is $H^2 = V_G/V_P$.

The main issue in this problem is obtaining some estimate of two components of phenotypic variation: genetic and environmental factors. Recall that V_P is the combination of genetic and environmental variance. Because the two parental strains are true-breeding, they are assumed to be homozygous and the variances of 3.12 and 3.87 are considered to result from environmental influences. The average of these two values is 3.50. The F_1 generation is also genetically homogeneous and gives us an additional estimate of the environmental factors. By averaging with the parents,

$$\frac{3.50 + 4.74}{2} = 4.12$$

we obtain a relatively good idea of environmental impact on the phenotype. The phenotypic variance in the F_2 is the sum of the genetic (V_G) and environmental (V_E) components. We have estimated the environmental input as 4.12, so 47.70 – 4.12 gives us an estimate of 43.58 for V_G. Heritability then becomes 43.58/47.70, or 0.91. This value indicates that about 91% of the variation in corolla length is due to genetic influences.

Problem 2: Ear Length Distribution (cm)

	5	*6*	*7*	*8*	*9*	*10*	*11*	*12*	*13*	*14*	*15*	*16*	*17*	*18*	*19*	*20*	*21*
Parent A	4	21	24	8													
Parent B									3	11	12	15	26	15	10	7	2
F_1					1	12	12	14	17	9	4						

Problems and Discussion Questions

1. Distinguish between discontinuous and continuous variation. Which type relates to inheritance of a quantitative nature?
2. Define and discuss (a) polygenes, (b) additive alleles, (c) multiple-factor hypothesis, (d) monozygotic versus dizygotic twins, (e) concordance versus discordance, and (f) heritability.
3. Weight of the nut of a commercially valuable species was determined to be under the control of two genes. Each locus can be occupied by either an additive or a nonadditive allele, and the effect of the additive alleles is approximately equal. In a population study, two true-breeding strains were isolated where the weights were 10.0 g and 13.2 g, respectively. Assuming these strains represent the upper and lower levels of nut weight, predict
   (a) the weight of the nuts in F_1 plants when the two strains are crossed, and
   (b) the range and distribution of phenotypes in the F_2 plants.
4. Of the F_2 phenotypes predicted in Problem 3, which of them could never be isolated as true-breeding strains?
5. A dark red strain and a white strain of wheat are crossed and produce an intermediate, medium red F_1. When the F_1 plants are interbred, an F_2 generation is produced in a ratio of 1 dark red: 4 medium-dark red: 6 medium red: 4 light red: 1 white. Further crosses reveal that the dark red and white F_2 plants are true-breeding.
   (a) Based on the ratio of offspring in the F_2, how many genes are involved in the production of color?
   (b) How many additive alleles, are needed to produce each possible phenotype?
   (c) Assign symbols to these alleles and list possible genotypes that give rise to the medium red and light red phenotypes.
   (d) Predict the outcome of the F_1 and F_2 generations in a cross between a true-breeding medium red plant and a white plant.
6. Height in humans depends on the additive action of genes. Assume that this trait is controlled by the four loci R, S, T, and U and that environmental effects are negligible. Instead of additive versus nonadditive alleles, assume that additive and partially additive alleles exist. Additive alleles contribute two units and partially additive alleles contribute one unit to height.
   (a) Can two individuals of moderate height produce offspring that are much taller or shorter than either parent? If so, how?
   (b) If an individual with the minimum height specified by these genes marries an individual of intermediate or moderate height, will any of their children be taller than the tall parent? Why or why not?
7. An inbred strain of plants has a mean height of 24 cm. A second strain of the same species from a different geographical region also has a mean height of 24 cm. When plants from the two strains are crossed, the F_1 plants are the same height as the parent plants. However, the F_2 generation shows a wide range of heights; the majority are like the P_1 and F_1 plants, but approximately 4 of 1000 are only 12 cm high, and about 4 of 1000 are 36 cm high.
   (a) What mode of inheritance is occurring here?
   (b) How many gene pairs are involved?
   (c) How much does each gene contribute to plant height?
   (d) Indicate one possible set of genotypes for the original P_1 and F_1 plants that could account for these results.
   (e) Indicate three possible genotypes that could account for 18-cm F_2 plants and three that could account for 33-cm F_2 plants.
8. Pig-parents Erma and Harvey were a compatible barnyard pair, but a curious sight. Harvey's tail was only 6 cm, while Erma's was 30 cm. Their F_1 piglet offspring all grew 18-cm tails. When inbred, an F_2 generation resulted in many piglets (Erma and Harvey's grandpigs) whose tails ranged in 4-cm intervals from 6 to 30 cm (6, 10, 14, 18, 22, 26, 30). Most had 18-cm tails, while 1/64 had 6-cm tails and 1/64 had 30-cm tails. (a) Explain how tail length is inherited, by describing the mode of inheritance, indicating how many gene pairs are at work, and designating the genotypes of Harvey, Erma, and their 18-cm-tailed offspring. (b) If one of the 18-cm F_1 pigs is mated with a 6-cm F_2 pig, what phenotypic ratio would be predicted if many offspring resulted? Diagram the cross.
9. In a plant where the seeds are valuable as food, two varieties are true-breeding. In one variety, the seeds all weigh about 2 g. In the other, the seeds are quite large and weigh about 12 g. When crossed, the offspring plants produce seeds that are all about 7 g. When an F_2 generation is produced, seeds of 2, 3, 4, 5, 6, 7, 8, 9, 10, 11, and 12 g were produced. Explain the inheritance by indicating the mode of inheritance, how many genes are involved; and indicate all genotypes that will yield seeds that are 3 g. In the F_2 above, what proportion of the offspring will be 12 g?
10. A 3" plant is crossed to a 27" plant, and the F_1 generation consists of only 15" plants. The F_2 generation exhibits a "normal distribution" of plants that are 3, 5, 7, 9, 11, 13, 15, 17, 19, 21, 23, 25, and 27" tall. Explain how height is inherited. What would change in the above cross if an F_1 plant was test-crossed instead of mating the F_1s?
11. Discuss the use of monozygotic and dizygotic twins reared together and apart in assessing the genetic component responsible for phenotypic variation in humans.
12. In the following table, average differences in height and weight of monozygotic twins (reared together and apart), dizygotic twins, and siblings are compared. Draw as many conclusions as you can concerning the effects of genetics and the environment in influencing these human traits.

Trait	*Monozygotic Reared Together*	*Monozygotic Reared Apart*	*Dizygotic Reared Together*	*Siblings Reared Together*
Height (cm)	1.7	1.8	4.4	4.5
Weight (kg)	1.9	4.5	4.5	4.7

13. List as many human traits as you can that are likely to be under the control of a polygenic mode of inheritance.
14. Corn plants from a test plot are measured, and the distribution of heights at 10-cm intervals is recorded:

Height (cm)	*Plants (no.)*
100	20
110	60
120	90
130	130
140	180
150	120
160	70
170	50
180	40

Calculate (a) the mean height, (b) the variance, (c) the standard deviation, and (d) the standard error of the mean. Plot a rough graph of plant height versus frequency. Do the values represent a normal distribution? Based on your calculations, how would you assess the variation within this population?

15. Phenotypic variation in the diameter of fruit of a commercially viable plant was analyzed with the following results:

$V_A = 1.7$
$V_D = 1.2$
$V_I = 1.3$
$V_E = 1.3$

Calculate the broad-sense heritability and the narrow-sense heritability for this trait.

16. Contrast broad-sense heritability (H^2) and narrow-sense heritability (h^2). To what type of population is each calculation applicable?
17. Considering the commercially viable fruit assessed in Problem 15, predict how accessible it is to artificial selection. Which measure of heritability will be the most useful in assessing the potential for artificial selection?
18. The mean and variance of plant height of two highly inbred strains (P_1 and P_2) and their progeny (F_1 and F_2) are shown in the table below. Calculate the broad-sense heritability (H^2) of plant height in this species.

Strain	*Mean (cm)*	*Variance*
P_1	34.2	4.2
P_2	55.3	3.8
F_1	44.2	5.6
F_2	46.3	10.3

19. In a hypothetical study, vitamin A content and cholesterol content of eggs from a large population of chickens is investigated. The variances (V) are calculated as shown in the table below. (a) Calculate the narrow-sense heritability (h^2) for both traits. (b) Which trait, if either, is likely to respond to selection?

	Trait	
Variance	*Vitamin A*	*Cholesterol*
V_P	123.5	862.0
V_E	96.2	484.6
V_A	12.0	192.1
V_D	15.3	185.3

20. In an assessment of learning in *Drosophila*, flies can be trained to avoid certain olfactory cues. In one population, a mean of 8.5 trials was required. A subgroup of this parental population that was trained most quickly (mean = 6.0 trials) was interbred and their progeny examined. These flies demonstrated a mean training value of 7.5 trials. Calculate and interpret narrow-sense heritability for olfactory learning in *Drosophila*.
21. A population of tomato plants with mean fruit weight of 60 g and an h^2 value of 0.3 is studied. Predict the results of artificial selection (the mean weight of the progeny) if tomato plants with an average fruit weight of 80 g are selected from the original population and interbred.
22. A mutant strain of *Drosophila* is isolated and shown to be resistant to an experimental insecticide, whereas normal (wild-type) flies are sensitive to the chemical. Following a cross between resistant flies and sensitive flies, isolated populations are derived that have various combinations of chromosomes from the two strains. Each population is tested for resistance as shown in the data below. Analyze the data and draw any appropriate conclusion about which chromosomes contain a gene responsible for inheritance of resistance to the insecticide.

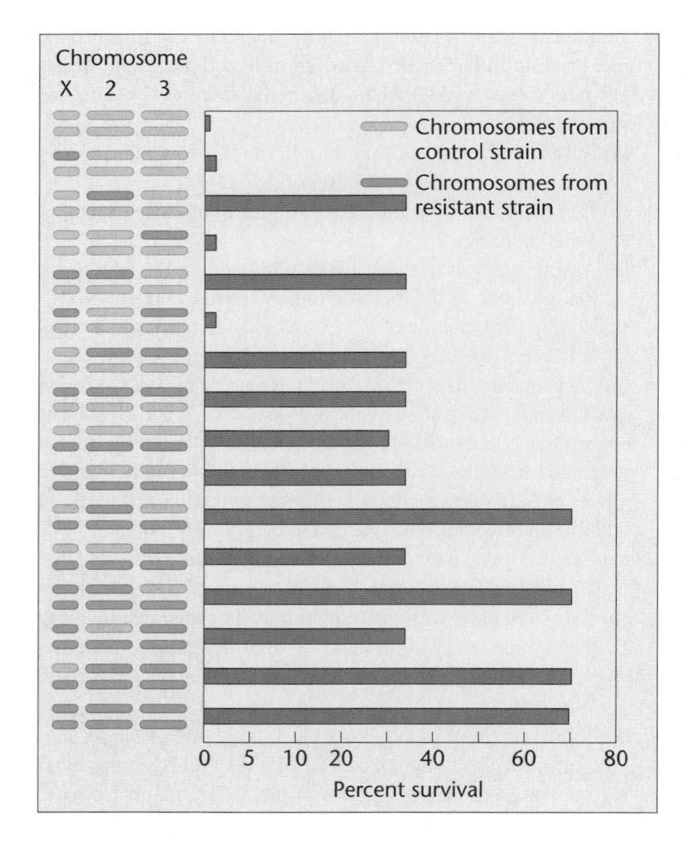

23. The locations of quantitative trait loci (QTLs) are often mapped relative to a variety of molecular markers. Restriction fragment length polymorphisms (RFLPs) are sites along chromosomes where a particular nucleotide sequence can be cleaved by a particular restriction enzyme. When both a marker and a phenotype of interest are expressed together through several generations, they are said to cosegregate. Explain why such markers are particularly useful and sometimes necessary, for mapping genes in some organisms but not others.

24. Horst Wilkens (Wilkens, H. 1988. *Ecol. Biol.* 23: 271–367) investigated blind cavefish, comparing them with members of a sibling species with normal vision that are found in a lake. (We will call them cavefish and lakefish.) Wilkens found that cavefish eyes are about seven times smaller than lakefish eyes. F_1 hybrids have eyes of intermediate size. These data as well as the $F_1 \times F_1$ cross and those from backcrosses ($F_1 \times$ cavefish and $F_1 \times$ lakefish) are depicted here:

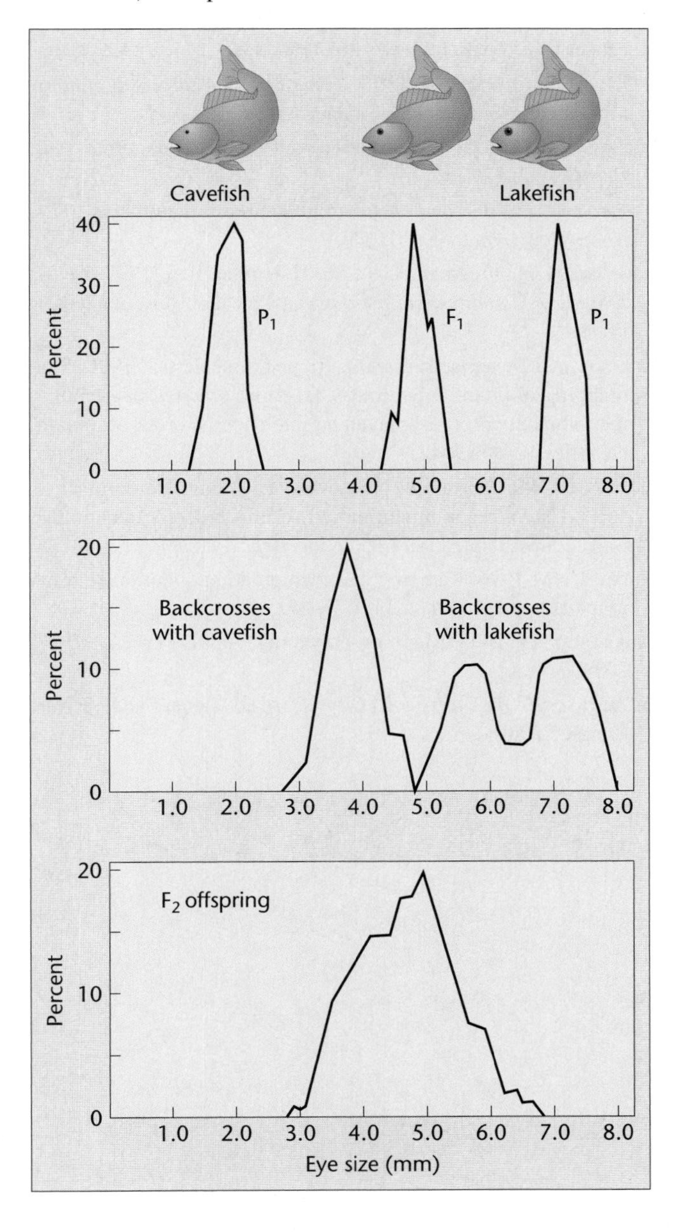

Based upon Wilkens' results, (a) what possible explanation concerning the inheritance of eye size seems most feasible for the F_1 and F_2 results? (b) Is your explanation supported by the results of the F_1 backcross with cavefish? Explain. (c) Is your explanation supported by the results of the F_1 backcross with lakefish? Explain. (d) Wilkens examined about 1000 F_2 progeny and estimated that 6–7 genes are involved in determining eye size. Is the sample size adequate to justify this conclusion? Propose an experimental protocol to test the hypothesis.

25. In a cross between a strain of large guinea pigs and a strain of small guinea pigs, the F_1s are phenotypically uniform with an average size about intermediate between that of the two parental strains. Among 1014 F_2 individuals, 3 are about the same size as the small parental strain and 5 are about the same size as the large parental strain. How many gene pairs are involved in the inheritance of size in these strains of guinea pigs?

26. Type A1B brachydactyly (short middle phalanges) is a genetically determined trait that maps to the short arm of chromosome 5 in humans. If you classify individuals as either having or not having brachydactyly, the trait appears to follow a single locus, incompletely dominant pattern of inheritance. However, if one examines the fingers and toes of affected individuals, one sees a range of expression from extremely short to only slightly short. What might cause such variation in the expression of brachylactyly?

27. In a series of crosses between two true-breeding strains of peaches, Strain A (38 g) $\times$ Strain B (22 g) produces an F_1 with 30-g peaches. The F_2 fruit mass ranges from 38 to 22 g at intervals of 2 g. (a) Using these data, determine the number of polygenic loci involved in the inheritance of peach mass. (b) Using gene symbols of your choice, give the genotypes of the parents and the F_1.

28. Students in a genetics laboratory began an experiment in an attempt to increase heat tolerance in two strains of *Drosophila melanogaster*. One strain was trapped from the wild six weeks before the experiment was to begin; the other was obtained from a *Drosophila* repository at a university laboratory. In which strain would you expect to see the most rapid and extensive response to heat-tolerance selection?

29. Consider a true-breeding plant, *AABBCC*, crossed with another true-breeding plant, *aabbcc*, whose resulting offspring are *AaBbCc*. If you cross the F_1s, where independent assortment is operational, the expected fraction of offspring in each phenotypic class is given by the expression $N!/[M!(N - M!)]$, where N is the total number of alleles (six in this example) and M is the number of uppercase alleles. In a cross of *AaBbCc* $\times$ *AaBbCc*, what proportion of the offspring would be expected to contain two uppercase alleles?

30. Canine hip dysplasia is a quantitative trait that continues to affect most large breeds of dogs in spite of approximately 40 years of effort to reduce the impact of this condition. Breeders and veterinarians rely on radiographic and universal registries to enhance the development of breeding schemes to reduce its incidence. Recent data (Wood and Lakhani, 2003. *Vet. Rec.* 152: 69–72) indicate that there is a "month-of-birth" effect on hip dysplasia in Labrador retrievers and Gordon setters, whereby the frequency and extent of expression of this disorder varies depending on the time of year dogs are born. Speculate on how breeders attempt to "select" out this disorder and what the month-of-birth phenomenon indicates about the expression of polygenic traits?

Selected Readings

Brink, R., ed. 1967. *Heritage from Mendel*. Madison: Univ. of Wisconsin Press.

Browman, K. W. 2001. Review of statistical methods of QTL mapping in experimental crosses. *Lab Animal* 30: 44–52.

Crow, J. F. 1993. Francis Galton: Count and measure, measure and count. *Genetics* 135: 1.

Cunningham, P. 1991. The genetics of thoroughbred horses. *Sci. Am.* (May) 264: 92–98.

Dudley, J. W. 1977. 76 generations of selection for oil and protein percentage in maize. *In* Pollack, E., Kempthorne, O. and Bailey T. (eds.), *Proceedings of the International Conference on Quantitative Genetics*, pp. 459–73. Ames: Iowa State Univ. Press.

East, E.M. 1916. Studies on size inheritance in *Nicotiana*. *Genetics* 1: 164–76.

Falconer, D. 1989. *Introduction to Quantitative Genetics*. 3rd ed. Harlow: Longman.

Farber, S. 1980. *Identical Twins Reared Apart*. New York: Basic Books.

Feldman, M. W., and Lewontin, R. C. 1975. The heritability hangup. *Science* 190: 1163–66.

Fowler, C., and Mooney, P. 1990. *Shattering: Food, Politics, and the Loss of Genetic Diversity*. Tucson: Univ. of Arizona Press.

Frary, A., et al. 2000. *fw2.2*: A quantitative trait locus key to the evolution of tomato fruit size. *Science* 289: 85–88.

Haley, C. 1991. Use of DNA fingerprints for the detection of major genes for quantitative traits in domestic species. *Anim. Genet.* 22: 259–77.

———. 1996. Livestock QTLs: Bringing home the bacon. *Trends Genet.* 11: 488–90.

Lander, E., and Botstein, D. 1989. Mapping mendelian factors underlying quantitative traits using RFLP linkage maps. *Genetics* 121: 185–99.

Lander, E., and Schork, N. 1994. Genetic dissection of complex traits. *Science* 265: 2037–48.

Lewontin, R. C. 1974. The analysis of variance and the analysis of causes. *Am. J. Hum. Genet.* 26: 400–11.

Lynch, M., and Walsh, B. 1998. *Genetics and Analysis of Quantitative Traits*. Sunderland: Sinauer Assoc.

Macay, T. F. C. 2001. Quantitative trait loci in *Drosophila*. *Nat. Rev. Gen.* 2: 11–19.

———. 2001. The genetic architecture of quantitative traits. *Annu. Rev. Genet.* 35: 303–39.

Newman, H. H., Freeman, F. N., and Holzinger, K. T. 1937. *Twins: A Study of Heredity and Environment.* Chicago: Univ. of Chicago Press.

Paterson, A., Deverna, J., Lanini, B., and Tanksley, S. 1990. Fine mapping of quantitative traits loci using selected overlapping recombinant chromosomes in an interspecific cross of tomato. *Genetics* 124: 735–42.

Plomin, R., McClearn, G., Gora-Maslak, G. and Neiderhiser, J. 1991. Use of recombinant inbred strains to detect quantitative trait loci associated with behavior. *Behav. Genet.* 21: 99–116.

Stuber, C. W. 1996. Mapping and manipulating quantitative traits in maize. *Trends Genet.* 11: 477–87.

Tanksley, S. D. 1993. Mapping polygenes. *Annu. Rev. Genet.* 27: 205–33.

Zar, J. H. 1996. *Biostatistical Analysis*. 3rd ed. Upper Saddle River: Prentice-Hall.

CHAPTER 7

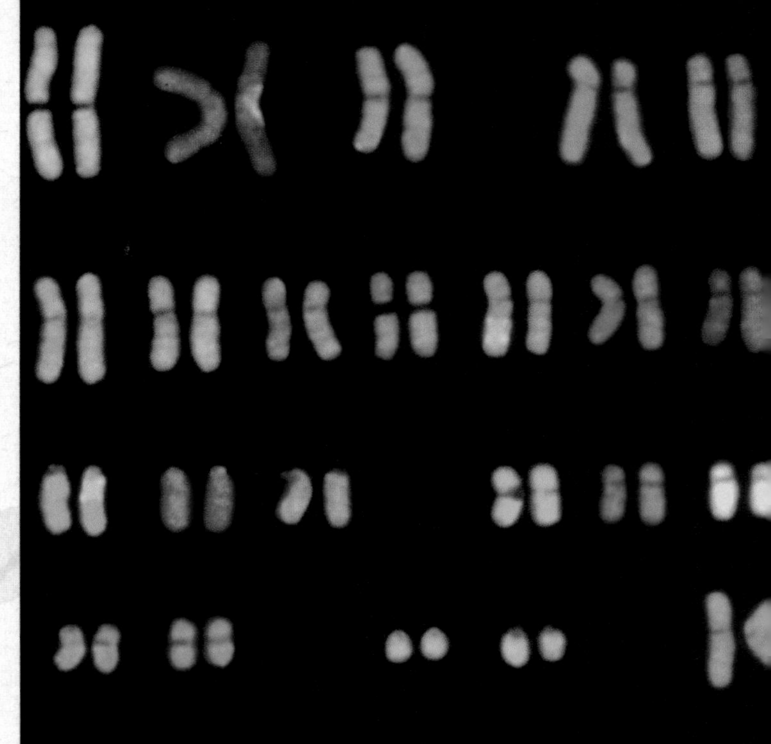

Spectral karyotyping of human chromosomes utilizing differentially labeled "painting" probes. *(Evelin Schrock, Stan du Manoir, and Tom Reid, National Institutes of Health)*

Chromosome Mutations: Variation in Number and Arrangement

CHAPTER CONCEPTS

In previous chapters, we have emphasized how mutations and the resulting alleles affect an organism's phenotype and how traits are passed from parents to offspring according to Mendelian principles. In this chapter, we look at phenotypic variation that results from more substantial changes than alterations of individual genes—modifications at the level of the chromosome.

Although most members of diploid species normally contain precisely two haploid chromosome sets, many known cases vary from this pattern. Modifications include a change in the total number of chromosomes, the deletion or duplication of genes or segments of a chromosome, and rearrangements of the genetic material either within or among chromosomes. Taken together, such changes are called **chromosome mutations** or **chromosome aberrations**, to distinguish them from gene mutations. Because the chromosome is the unit of genetic transmission, according to Mendelian laws, chromosome aberrations are passed to offspring in a predictable manner, resulting in many unique genetic outcomes.

Because the genetic component of an organism is delicately balanced, even minor alterations of either content or location of genetic information within the genome can result in some form of phenotypic variation. More substantial changes may be lethal, particularly in animals. Throughout this chapter, we consider many types of chromosomal aberrations, the phenotypic consequences for the organism that harbors an aberration, and the impact of the aberration on the offspring of an affected individual. We will also discuss the role of chromosome aberrations in the evolutionary process.

HOW DO WE KNOW WHAT WE KNOW?

IN THIS CHAPTER, WE WILL FOCUS ON mutations at the chromosomal level resulting from a change in number or structure of chromosomes. As you study this topic, you should try to answer several fundamental questions:

1. How do we know that changes in chromosome number or structure result in specific mutant phenotypes?
2. How did we learn how synapsis during meiosis occurs when chromosome mutations exist?
3. How do we know that the extra chromosome causing Down syndrome is usually maternal in origin?
4. In *Drosophila*, how do we know that the mutant *Bar* eye phenotype is due to a duplication?
5. How do we know that gene duplication is a source of new genes during evolution? ■

7.1 Specific Terminology Describes Variations in Chromosome Number

Variation in chromosome number ranges from the addition or loss of one or more chromosomes to the addition of one or more haploid sets of chromosomes. Before we embark on our discussion, it is useful to clarify the terminology that describes such changes. In the general condition known as **aneuploidy**, an organism gains or loses one or more chromosomes, but not a complete set. The loss of a single chromosome from an otherwise diploid genome is called *monosomy*. The gain of one chromosome results in *trisomy*. These changes are contrasted with the condition of **euploidy**, where complete haploid sets of chromosomes are present. If more than two sets are present, the term **polyploidy** applies. Organisms with three sets are specifically *triploid*; those with four sets are *tetraploid*, and so on. Table 7.1 provides an organizational framework for you to follow as we discuss each of these categories of aneuploid and euploid variation and the subsets within them.

7.2 Variation in the Number of Chromosomes Results from Nondisjunction

As we consider cases that include the gain or loss of chromosomes, it is useful to examine how such aberrations originate. For instance, how do the syndromes arise where the number of sex-determining chromosomes in humans is altered, as described in Chapter 5? As you may recall, the gain (47,XXY) or loss (45,X) of an X chromosome from an otherwise diploid genome affects the phenotype, resulting in **Klinefelter syndrome** or **Turner syndrome**, respectively (see Figure 5–6). Human females may contain extra

TABLE 7.1 Terminology for Variation in Chromosome Numbers

Term	*Explanation*
Aneuploidy	$2n \pm x$ chromosomes
Monosomy	$2n - 1$
Trisomy	$2n + 1$
Tetrasomy, pentasomy, etc.	$2n + 2$, $2n + 3$, etc.
Euploidy	Multiples of n
Diploidy	$2n$
Polyploidy	$3n$, $4n$, $5n$, ...
Triploidy	$3n$
Tetraploidy, pentaploidy, etc.	$4n$, $5n$, etc.
Autopolyploidy	Multiples of the same genome
Allopolyploidy (amphidiploidy)	Multiples of different genomes

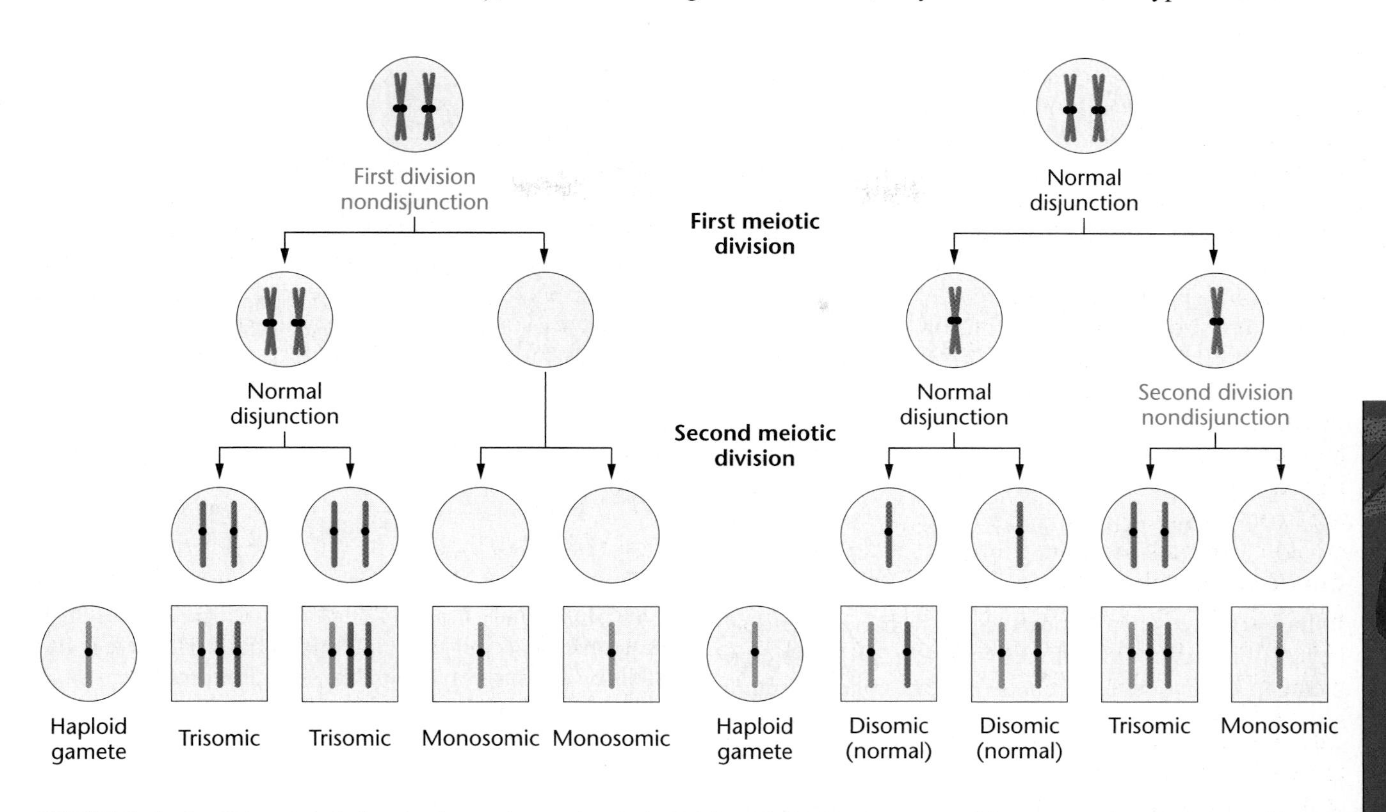

FIGURE 7–1 Nondisjunction during the first and second meiotic divisions. In both cases, some of the gametes that are formed either contain two members of a specific chromosome or lack that chromosome. After fertilization by a gamete with a normal haploid content, monosomic, disomic (normal), or trisomic zygotes are produced.

MEDIA TUTORIAL Chromosome separation in meiosis: disjunction

X chromosomes (e.g., 47,XXX, 48,XXXX), and some males contain an extra Y chromosome (47,XYY).

Chromosomal variation originates as the result of an error during meiosis, a phenomenon referred to as **nondisjunction**, whereby paired homologs fail to disjoin during segregation. This process disrupts the normal distribution of chromosomes into gametes. The results of nondisjunction during meiosis I and meiosis II for a single chromosome of a diploid organism are shown in Figure 7–1. As you can see, for the affected chromosome, abnormal gametes can form that contain either two members or none at all. Fertilizing these with a normal haploid gamete produces a zygote with either three members (trisomy) or only one member (monosomy) of this chromosome. As we shall see, nondisjunction leads to a variety of aneuploid conditons in humans and other organisms.

7.3 Monosomy, the Loss of a Single Chromosome, May Have Severe Phenotypic Effects

We turn now to a consideration of variations in the number of autosomes and the genetic consequence of such changes. The most common examples of aneuploidy, where an organism has a chromosome number other than an exact multiple of the haploid set, are cases in which a single chromosome is either added to, or lost from, a normal diploid set. The loss of one chromosome produces a $2n - 1$ complement called **monosomy**.

Although monosomy for the X chromosome occurs in humans, as we have seen in 45,X Turner syndrome, monosomy for any of the autosomes is not usually tolerated in humans or other animals. In *Drosophila*, flies that are monosomic for the very small chromosome 4—a condition referred to as **Haplo-IV**—develop more slowly, exhibit reduced body size, and have impaired viability. Monosomy for the larger chromosomes 2 and 3 is apparently lethal, because such flies have never been recovered.

The failure of monosomic individuals to survive is at first quite puzzling, since at least a single copy of every gene is present in the remaining homolog. However, if just one of those genes is represented by a lethal allele, the unpaired chromosome condition leads to the death of the organism. This occurs because monosomy unmasks recessive lethals that are tolerated in heterozygotes carrying the corresponding wild-type alleles.

Aneuploidy is better tolerated in the plant kingdom. Monosomy for autosomal chromosomes has been observed in maize; tobacco; the evening primrose, *Oenothera*; and the jimson weed, *Datura*, among many other plants. Nevertheless, such monosomic plants are usually less viable than their diploid derivatives. Haploid pollen grains, which undergo extensive development before participating in fertilization, are particularly sensitive to the lack of one chromosome and are seldom viable.

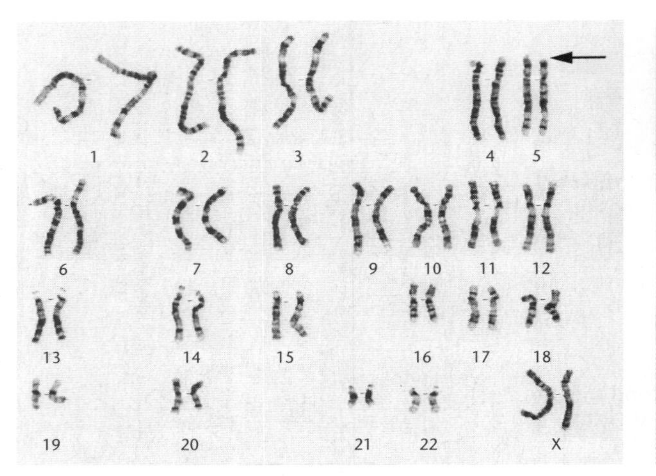

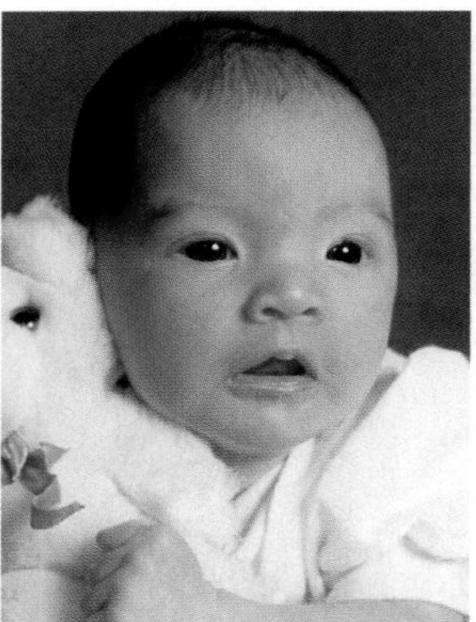

FIGURE 7–2 A representative karyotype and a photograph of a child exhibiting cri-du-chat syndrome (46,5p–). In the karyotype, the arrow identifies the absence of a small piece of the short arm of one member of the chromosome 5 homologs. *(Left photo: Courtesy of the Greenwood Genetic Center, Greenwood, SC; Right photo: Five P Minus Society)*

Cri-du-chat Syndrome

In humans, autosomal monosomy has not been reported beyond birth. Individuals with such chromosome complements are undoubtedly conceived, but none apparently survive embryonic and fetal development. There are, however, examples of survivors where only part of one chromosome is lost, sometimes referred to as **segmental deletions**. One case was first reported by Jérôme Le Jeune in 1963 when he described the clinical symptoms of the **cri-du-chat** (cry of the cat) **syndrome**. This syndrome is associated with the loss of part of the short arm of chromosome 5 (Figure 7–2). Thus, the genetic constitution may be designated as **46,5p −**, meaning that the individual has all 46 chromosomes but that some of the *p* arm (the petite arm) of one member of the chromosome 5 pair is missing.

Infants with this syndrome exhibit anatomic malformations, including gastrointestinal and cardiac complications, and they are often mentally retarded. Abnormal development of the glottis and larynx is also characteristic of individuals with this syndrome. As a result, the infant has an unusual cry, one that is similar to the meowing of a cat, giving the syndrome its name.

Since 1963, hundreds of cases of cri-du-chat syndrome have been reported worldwide. An incidence of 1 in 50,000 live births has been estimated. The length of the short arm that is deleted varies somewhat; longer deletions appear to have a greater impact on the physical, psychomotor, and mental skill levels of those children who survive. Although the effects of the syndrome are severe, many individuals achieve a moderate level of social development. Those who receive home care and early special schooling are ambulatory, develop self-care skills, and learn to communicate verbally.

7.4 Trisomy Involves the Addition of a Chromosome to a Diploid Genome

In general, the effects of trisomy $2n + 1$ parallel those of monosomy. However, the addition of an extra chromosome produces somewhat more viable individuals in both animal and plant species than does the loss of a chromosome. In animals, this is often true, provided that the chromosome involved is relatively small. However, the addition of a large autosome to the diploid complement in both *Drosophila* and humans has severe effects and is usually lethal during development.

In plants, trisomic individuals are viable, but their phenotype may be altered. A classical example involves the jimson weed, *Datura*, whose diploid number is 24. Twelve primary trisomic conditions are possible, and examples of each one have been recovered. Each trisomy alters the phenotype of the plant's capsule sufficiently to produce a unique phenotype. These capsule phenotypes were first thought to be caused by mutations in one or more genes.

Still another example is seen in the rice plant (*Oryza sativa*), which has a haploid number of 12. Trisomic strains for each chromosome have been isolated and studied—the plants of 11 strains can be distinguished from one another and from wild-type plants. Trisomics for the longer chromosomes are the most distinctive, and the plants grow more slowly. This is in keeping with the belief that larger chromosomes cause greater genetic imbalance than smaller ones. Leaf structure, foliage, stems, grain morphology, and plant height also vary among the various trisomies.

Down Syndrome

The only human autosomal trisomy in which a significant number of individuals survive longer than a year past birth was discovered in 1866 by Langdon Down. The condition is now known to result from trisomy of chromosome 21, one of the G group* (Figure 7–3), and is called **Down syndrome** or simply **trisomy 21** (designated **47,21 +**). This trisomy is found in approximately 1 infant in every 800 live births.

The overt phenotype of these individuals is so similar that they bear a striking resemblance to one another. They display a prominent epicanthic fold in the corner of the eye

*On the basis of size and centromere placement, human autosomal chromosomes are divided into seven groups: A (1–3), B (4–5), C (6–12), D (13–15), E (16–18), F (19–20), and G (21–22).

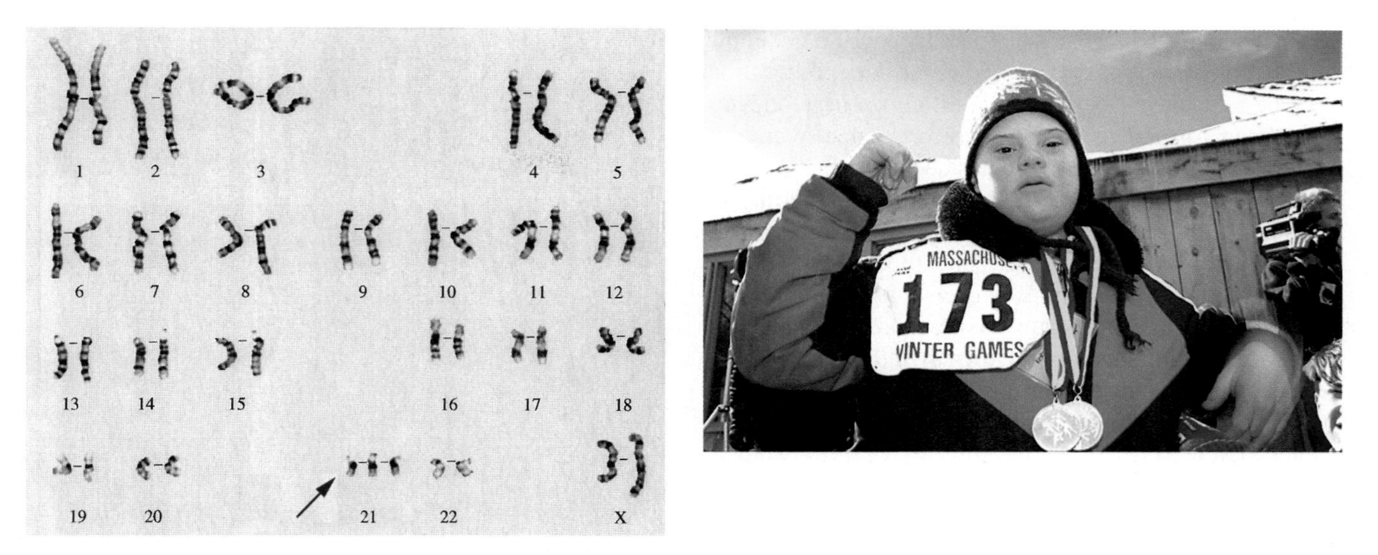

FIGURE 7–3 The karyotype and a photograph of a child with Down syndrome. In the karyotype, three members of the G-group chromosome 21 are present, creating the 47,21+ condition. *(Left photo: Courtesy of the Greenwood Genetic Center, Greenwood, SC; Right photo: William McCoy/Rainbow)*

and are characteristically short. They may have round heads with flat faces, a protruding furrowed tongue, which causes the mouth to remain partially open, and short, broad hands with fingers showing characteristic palm and fingerprint patterns. Physical, psychomotor, and mental development is retarded, and poor muscle tone is characteristic. Their life expectancy is shortened, although individuals are known to survive into their fifties.

Children afflicted with Down syndrome are prone to respiratory disease and heart malformations, and they show an incidence of leukemia approximately 20 times higher than that of the normal population. However, careful medical scrutiny and treatment throughout their lives has extended their survival significantly. A striking observation is that death of older Down syndrome adults is frequently due to Alzheimer's disease, although the onset of this disease occurs at a much earlier age than it does in the normal population.

Typical of other conditions referred to as a syndrome, there are many phenotypic characteristics that may be present, but any single affected individual usually expresses only a subset of these. In the case of Down syndrome, there are 12–14 such characteristics, but each individual, on average, expresses only 6–8 of them.

The most characteristic origin of this trisomic condition is through nondisjunction of chromosome 21 during meiosis. Failure of paired homologs to disjoin during anaphase I or failure of chromatids to disjoin during anaphase II can result in male or female gametes with the $n + 1$ chromosome composition. Following fertilization with a normal gamete, the trisomic condition is created. Chromosome analysis has shown that while the additional chromosome can be derived from either the mother or father, the ovum is the source in 95% of the cases.

Before the development of techniques that distinguish paternal from maternal homologs, this conclusion was supported by other indirect evidence derived from studies of the age of mothers giving birth to Down syndrome infants. Figure 7–4 shows the relationship between maternal age and the incidence of Down syndrome newborns. While the frequency is about 1 in 1000 at maternal age 30, a 10-fold increase to a frequency of 1 in 100 is noted at age 40. The frequency increases still further to about 1 in 50 at age 45. In spite of these statistics, it is important to point out that, overall, more than half of affected births occur to women who are under 35 years of age, primarily because there are many more pregnancies in this group of women.

While the nondisjunctional event that produces Down syndrome seems more likely to occur during oogenesis

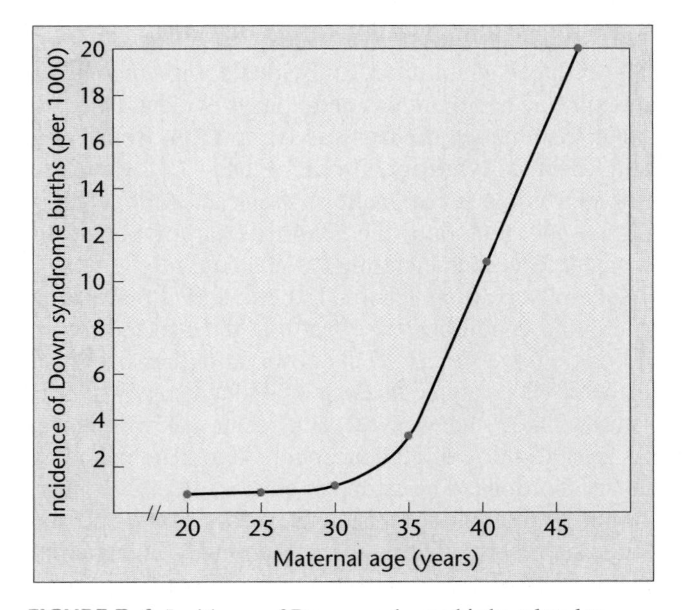

FIGURE 7–4 Incidence of Down syndrome births related to maternal age.

in women between the ages of 35 and 45, we do not know with certainty why this is so. However, one observation may be relevant. In human females, all primary oocytes have been formed by birth. Therefore, once ovulation begins, each succeeding ovum has been arrested in meiosis for about a month longer than the one preceding it. Women 30 or 40 years old produce ova that are significantly older and arrested longer than those they ovulated 10 or 20 years previously. However, it is not yet known whether ovum age is the cause of the increased incidence of nondisjunction leading to Down syndrome.

These statistics pose a serious problem for the woman who becomes pregnant late in her reproductive years. Genetic counseling early in the pregnancy serves two purposes. First, it informs the parents about the probability that their child will be affected and educates them about Down syndrome. Although some individuals with Down syndrome must be institutionalized, others benefit greatly from special education programs and can be cared for at home. Further, these children are noted for their affectionate, loving natures. Second, a genetic counselor may recommend a prenatal diagnostic technique such as **amniocentesis** or **chorionic villus sampling (CVS)**. These techniques require the removal and culture of fetal cells. The karyotype of the fetus is then determined by cytogenetic analysis. If the fetus is diagnosed as having Down syndrome, a therapeutic abortion is one option the parents may consider.

Because Down syndrome appears to be caused by a random error—nondisjunction of chromosome 21 during maternal or paternal meiosis—the disorder is not expected to be inherited. Nevertheless, Down syndrome occasionally runs in families. This condition, familial Down syndrome, involves a *translocation* of chromosome 21, another type of chromosomal aberration, which we will discuss in Section 7.10.

Viability in Human Aneuploid Conditions

The reduced viability of individuals with recognized monosomic and trisomic conditions is evident. Only two other trisomies in humans survive to term. Both **Patau** and **Edwards syndromes (47,13+** and **47,18+**, respectively) result in severe malformations and early lethality. Figure 7–5 illustrates the abnormal karyotype and the many defects characterizing Patau infants.

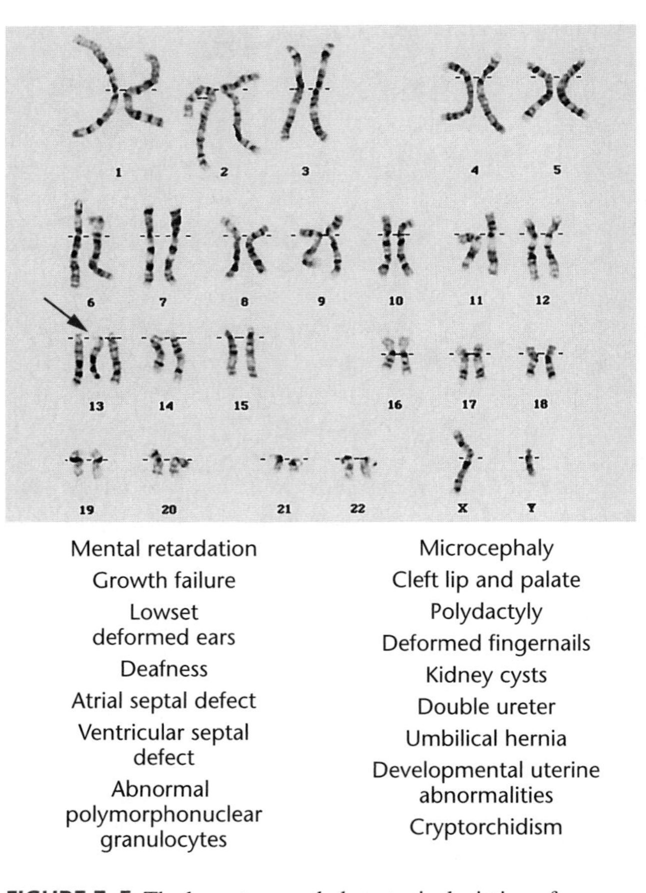

FIGURE 7–5 The karyotype and phenotypic depiction of an infant with Patau syndrome, where three members of the D-group chromosome 13 are present, creating the 47,13+ condition. (*David D. Weaver, M.D., Indiana University*)

Such observations lead us to believe that many other aneuploid conditions arise, but that the affected fetuses do not survive to term. This observation has been confirmed by karyotypic analysis of spontaneously aborted fetuses. These studies reveal some rather striking statistics. At least 15–20% of all conceptions terminate in spontaneous abortion (some estimates are considerably higher). About 30% of all spontaneously aborted fetuses demonstrate some form of chromosomal anomaly, and approximately 90% of all chromosomal anomalies are terminated prior to birth as a result of spontaneous abortion.

A large percentage of fetuses demonstrating chromosomal abnormalities are aneuploids. The aneuploid with highest incidence among abortuses is the 45,X condition, which produces an infant with Turner syndrome if the fetus survives to term.

An extensive review of this subject by David H. Carr also reveals that a significant percentage of aborted fetuses are trisomic for one of the chromosome groups. Trisomies for every human chromosome have been recovered. Monosomies are seldom found, however, even though nondisjunction should produce $n - 1$ gametes with a frequency equal to $n + 1$ gametes. This finding suggests that gametes lacking a single chromosome are functionally impaired to a serious degree or that the embryo dies so early in its development that recovery occurs infrequently. Various forms of polyploidy and other miscellaneous chromosomal anomalies were also found in Carr's study.

These observations support the hypothesis that normal embryonic development requires a precise diploid complement of chromosomes to maintain the delicate equilibrium in the expression of genetic information. The prenatal mortality of most aneuploids provides a barrier against the introduction of these genetic anomalies into the human population.

7.5 Polyploidy, in Which More than Two Haploid Sets of Chromosomes Are Present, Is Prevalent in Plants

The term *polyploidy* describes instances in which more than two multiples of the haploid chromosome set are found. The naming of polyploids is based on the number of sets of chromosomes found: A triploid has $3n$ chromosomes; a tetraploid has $4n$; a pentaploid, $5n$; and so forth. Several general statements can be made about polyploidy. This condition is relatively infrequent in many animal species, but is well known in lizards, amphibians, and fish, and is much more common in plant species. Odd numbers of chromosome sets are not usually reliably maintained from generation to generation, because a polyploid organism with an uneven number of homologs often does not produce genetically balanced gametes. For this reason, triploids, pentaploids, and so on, are not usually found in plant species that depend solely on sexual reproduction for propagation.

Polyploidy originates in two ways: (1) The addition of one or more extra sets of chromosomes, identical to the normal haploid complement of the same species, resulting in **autopolyploidy**; or (2) the combination of chromosome sets from different species occurring as a consequence of hybridization, resulting in **allopolyploidy** (from the Greek word *allo*, meaning "other" or "different"). The distinction between auto- and allopolyploidy is based on the genetic origin of the extra chromosome sets, as shown in Figure 7–6.

In our discussion of polyploidy, we use the following symbols to clarify the origin of additional chromosome sets. For example, if A represents the haploid set of chromosomes of any organism, then

$$A = a_1 + a_2 + a_3 + a_4 + \cdots + a_n$$

where a_1, a_2, and so on, are individual chromosomes and n is the haploid number. A normal diploid organism is represented simply as AA.

Autopolyploidy

In autopolyploidy, each additional set of chromosomes is identical to the parent species. Therefore, triploids are represented as AAA, tetraploids are $AAAA$, and so forth.

Autotriploids arise in several ways. A failure of all chromosomes to segregate during meiotic divisions can produce a diploid gamete. If such a gamete is fertilized by a haploid gamete, a zygote with three sets of chromosomes is produced. Or, rarely, two sperm may fertilize an ovum, resulting in a triploid zygote. Triploids are also produced under experimental conditions by crossing diploids with tetraploids. Diploid organisms produce gametes with n chromosomes, while tetraploids produce $2n$ gametes. Upon fertilization, the desired triploid is produced.

Because they have an even number of chromosomes, **autotetraploids** $(4n)$ are theoretically more likely to be found in nature than are autotriploids. Unlike triploids, which often produce genetically unbalanced gametes with odd numbers of chromosomes, tetraploids are more likely to produce balanced gametes when involved in sexual reproduction.

How polyploidy arises naturally is of great interest to geneticists. In theory, if chromosomes have replicated, but the parent cell never divides and instead reenters interphase, the chromosome number will be doubled. That this very likely occurs is supported by the observation that tetraploid cells can be produced experimentally from diploid cells. This is accomplished by applying cold or heat shock to meiotic cells or by applying colchicine to somatic cells undergoing mitosis. **Colchicine**, an alkaloid derived from the autumn crocus, interferes with spindle formation, and thus replicated chromosomes cannot separate at anaphase and do not migrate to the poles. When colchicine is removed, the cell can reenter interphase. When the paired sister chromatids separate and uncoil, the nucleus contains twice the diploid number of chromosomes and is therefore $4n$. This process is shown in Figure 7–7.

In general, autopolyploids are larger than their diploid relatives. This increase seems to be due to larger cell size

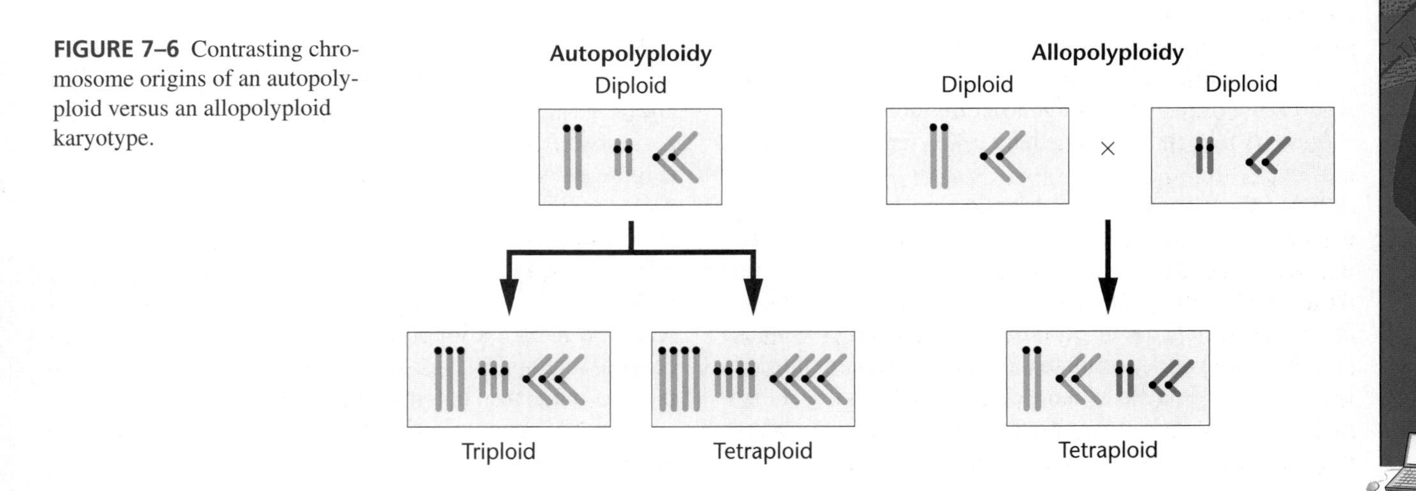

FIGURE 7–6 Contrasting chromosome origins of an autopolyploid versus an allopolyploid karyotype.

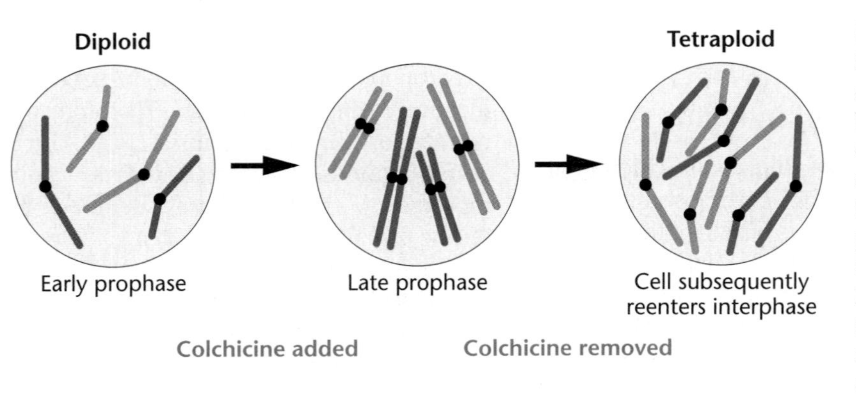

FIGURE 7–7 The potential involvement of colchicine in doubling the chromosome number. Two pairs of homologous chromosomes are shown. While each chromosome had replicated its DNA earlier during interphase, the chromosomes do not appear as double structures until late prophase. When anaphase fails to occur normally, the chromosome number doubles if the cell reenters interphase.

rather than greater cell number. Although autopolyploids do not contain new or unique information compared with their diploid relatives, the flower and fruit of plants are often increased in size, making such varieties of greater horticultural or commercial value. Economically important triploid plants include several potato species of the genus *Solanum*, Winesap apples, commercial bananas, seedless watermelons, and the cultivated tiger lily *Lilium tigrinum*. These plants are propagated asexually. Diploid bananas contain hard seeds, but the commercial, triploid, "seedless" variety has edible seeds. Tetraploid alfalfa, coffee, peanuts, and McIntosh apples are also of economic value because they are either larger or grow more vigorously than do their diploid or triploid counterparts. The commercial strawberry is an octoploid.

We have long been curious how cells with increased ploidy values, where no new genes are present, express different phenotypes from their diploid counterparts. Our current ability to examine gene expression using modern biotechnology has provided some interesting insights. For example, Gerald Fink and his colleagues have been able to create strains of the yeast *Saccaromyces cerevisiae* with one, two, three, or four copies of the genome. Thus, each strain contains identical genes (they are said to be isogenic) but different ploidy values. These scientists then examined the expression levels of all genes during the entire cell cycle of the organism. Using the rather stringent standards of a 10-fold increase or decrease of gene expression, Fink and coworkers proceeded to identify 10 cases where, as ploidy increased, gene expression was increased at least 10-fold and 7 cases where it was reduced by a similar level.

One of these genes provides insights into how polyploid cells are larger than their haploid or diploid counterparts. In polyploid yeast, two G1 cyclins, Cln1, and Pcl1 are repressed as ploidy increases, while the size of the yeast cells increases. This is explained based on the observation that G1 cyclins facilitate the cell's movement through G1, which is delayed when expression of these genes is repressed. The cell stays in the G1 phase longer and, on average, grows to a larger size before it moves beyond the G1 stage of the cell cycle. Yeast cells also show different morphology as ploidy increases. Several of the other genes, repressed as ploidy increases, have been linked to cytoskeletal dynamics, which accounts for the morphological changes.

Allopolyploidy

Polyploidy can also result from hybridizing two closely related species. If a haploid ovum from a species with chromosome sets *AA* is fertilized by sperm from a species with sets *BB*, the resulting hybrid is *AB*, where $A = a_1, a_2, a_3, \ldots, a_n$ and $B = b_1, b_2, b_3, \ldots, b_n$. The hybrid plant may be sterile because of its inability to produce viable gametes. Most often, this occurs when some or all of the *a* and *b* chromosomes are not homologous and therefore cannot synapse in meiosis. As a result, unbalanced genetic conditions result. If, however, the new *AB* genetic combination undergoes a natural or induced chromosomal doubling, two copies of all *a* chromosomes and two copies of all *b* chromosomes will be present, and they will pair during meiosis. As a result, a fertile *AABB* tetraploid is produced. These events are shown in Figure 7–8. Since this polyploid contains the equivalent of four haploid genomes derived from separate species, such an organism is called an **allotetraploid**. When both original species are known, an equivalent term, **amphidiploid**, is preferred in describing the allotetraploid.

Amphidiploid plants are often found in nature. Their reproductive success is based on their potential for forming balanced gametes. Since two homologs of each specific chromosome are present, meiosis occurs normally (Figure 7–8) and fertilization successfully propagates the plant sexually. This discussion assumes the simplest situation, where none of the chromosomes in set *A* are homologous to those in set *B*. In amphidiploids formed from closely related species, some homology between *a* and *b* chromosomes is likely. Allopolyploids are rare in most animals because mating behavior is most often species-specific, and thus the initial step in hybridization is unlikely to occur.

A classical example of amphidiploidy in plants is the cultivated species of American cotton, *Gossypium* (Figure 7–9). This species has 26 pairs of chromosomes:

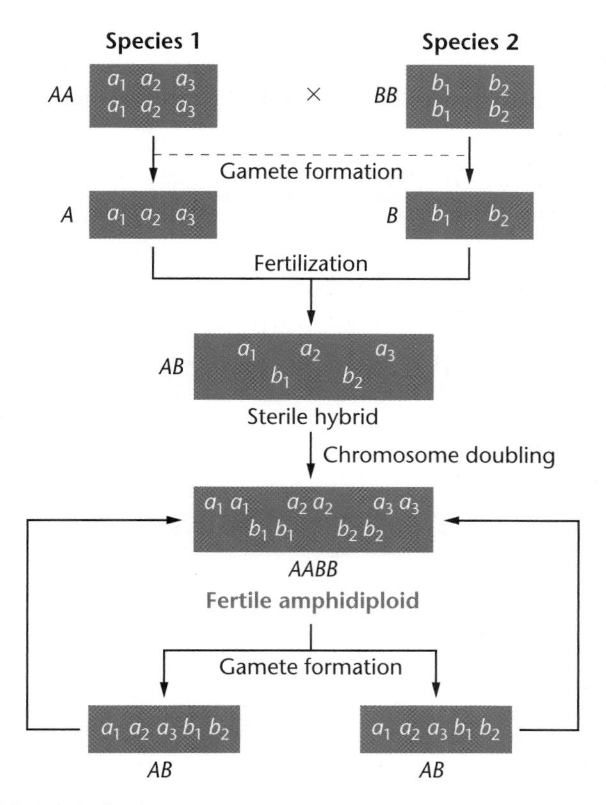

FIGURE 7–8 The origin and propagation of an amphidiploid. Species 1 contains genome *A* consisting of three distinct chromosomes, a_1, a_2, and a_3. Species 2 contains genome *B* consisting of two distinct chromosomes, b_1 and b_2. Following fertilization between members of the two species and chromosome doubling, a fertile amphidiploid containing two complete diploid genomes (*AABB*) is formed.

13 are large and 13 are much smaller. When it was discovered that Old World cotton had only 13 pairs of large chromosomes, allopolyploidy was suspected. After an examination of wild American cotton revealed 13 pairs of small chromosomes, this speculation was strengthened. J. O. Beasley reconstructed the origin of cultivated cotton experimentally by crossing the Old World strain with the wild American strain, then treating the hybrid with colchicine to double the chromosome number. The result of these treatments was a fertile amphidiploid variety of cotton. It contained 26 pairs of chromosomes as well as characteristics similar to the cultivated variety.

Amphidiploids often exhibit traits of both parental species. An interesting example, but one with no practical economic importance, is that of the hybrid formed between the radish *Raphanus sativus* and the cabbage *Brassica oleracea*. Both species have a haploid number $n = 9$. The initial hybrid consists of nine *Raphanus* and nine *Brassica* chromosomes (9R + 9B). While hybrids are almost always sterile, some fertile amphidiploids (18R + 18B) have been produced. Unfortunately, the root of this plant is more like the cabbage and its shoot more like the radish—had the converse occurred, the hybrid might have been of economic importance.

FIGURE 7–9 The pods of the amphidiploid form of *Gossypium*, the cultivated American cotton plant. *(Ken Wagner/Phototake NYC)*

A much more successful commercial hybridization uses the grasses wheat and rye. Wheat (genus *Triticum*) has a basic haploid genome of seven chromosomes. In addition to normal diploids ($2n = 14$), cultivated autopolyploids exist, including tetraploid ($4n = 28$) and hexaploid ($6n = 42$) species. Rye (genus *Secale*) also has a genome consisting of seven chromosomes. The only cultivated species is the diploid plant ($2n = 14$).

Using the technique outlined in Figure 7–8, geneticists have produced various hybrids. When tetraploid wheat is crossed with diploid rye and the F_1 plants are treated with colchicine, a hexaploid variety ($6n = 42$) is obtained; the hybrid, designated *Triticale* represents a new genus. Fertile hybrid varieties derived from various wheat and rye species can be crossed or backcrossed. These crosses have created many variations of the genus *Triticale*. The hybrid plants demonstrate characteristics of both wheat and rye. For example, certain hybrids combine the high protein content of wheat with rye's high content of the amino acid lysine. (The lysine content is low in wheat and thus is a limiting nutritional factor.) Wheat is considered to be a high-yielding grain, whereas rye is noted for its versatility of growth in unfavorable environments. *Triticale* species, which combine both traits, have the potential of significantly increasing grain production. Programs designed to improve crops through hybridization have long been under way in several underdeveloped countries.

7.6 Variation Occurs in the Structure and Arrangement of Chromosomes

The second general class of chromosome aberrations includes structural changes that delete, add, or rearrange substantial portions of one or more chromosomes. Included in this broad category are deletions and duplications of genes or part of a chromosome and rearrangements of genetic material in which a chromosome segment is

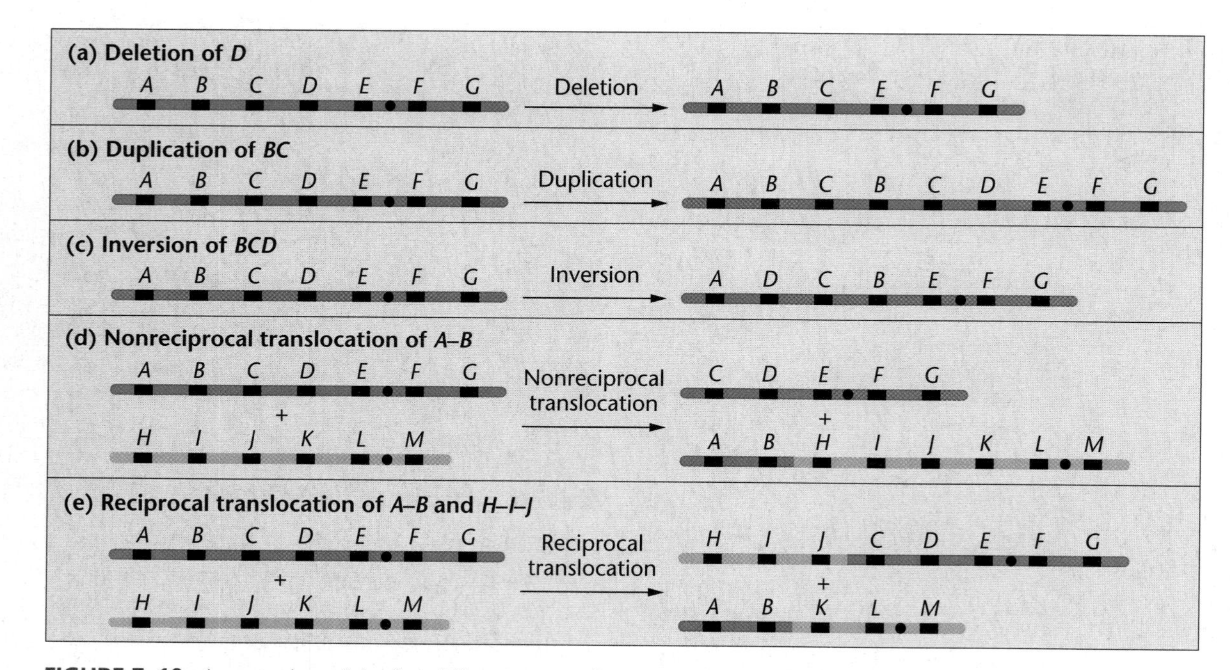

FIGURE 7–10 An overview of the five different types of rearrangement of chromosome segments.

inverted, exchanged with a segment of a nonhomologous chromosome, or merely transferred to another chromosome. Exchanges and transfers are called translocations, in which the location of a gene is altered within the genome. These types of chromosome alterations are illustrated in Figure 7–10.

In most instances, these structural changes are due to one or more breaks along the axis of a chromosome, followed by either the loss or rearrangement of genetic material. Chromosomes can break spontaneously, but the rate of breakage may increase in cells exposed to chemicals or radiation. Although the actual ends of chromosomes, known as telomeres, do not readily fuse with newly created ends of "broken" chromosomes or with other telomeres, the ends produced at points of breakage are "sticky" and can rejoin other broken ends. If breakage and rejoining does not reestablish the original relationship and if the alteration occurs in germ plasm, the gametes will contain the structural rearrangement, which is heritable.

If the aberration is found in one homolog, but not the other, the individual is said to be heterozygous for the aberration. In such cases, unusual but characteristic pairing configurations are formed during meiotic synapsis. These patterns are useful in identifying the type of change that has occurred. If no loss or gain of genetic material occurs, individuals bearing the aberration "heterozygously" are likely to be unaffected phenotypically. However, the unusual pairing arrangements often lead to gametes that are duplicated or deficient for some chromosomal regions. When this occurs, the offspring of "carriers" of certain aberrations often have an increased probability of demonstrating phenotypic changes.

7.7 A Deletion Is a Missing Region of a Chromosome

When a chromosome breaks in one or more places and a portion of it is lost, the missing piece is called a **deletion** (or a **deficiency**). The deletion can occur either near one end or within the interior of the chromosome. These are **terminal** and **intercalary deletions**, respectively (Figure 7–11a and b). The portion of the chromosome that retains the centromere region is usually maintained when the cell divides, whereas the segment without the centromere is eventually lost in progeny cells following mitosis or meiosis. For synapsis to occur between a chromosome with a large intercalary deletion and a normal homolog, the unpaired region of the normal homolog must "buckle out" into a **deletion**, or **compensation**, **loop** (Figure 7–11c).

As noted in our discussion of the cri-du-chat syndrome, where only part of the short arm of chromosome 5 is lost, deletion of a portion of a chromosome need not be very great before the effects become severe. If even more genetic information is lost as a result of a deletion, the aberration is often lethal, and these chromosome mutations never become available for study.

7.8 A Duplication Is a Repeated Segment of a Chromosome

When any part of the genetic material—a single locus or a large piece of a chromosome—is present more than once in the genome, it is called a duplication. As in deletions, pairing in heterozygotes can produce a compensation loop.

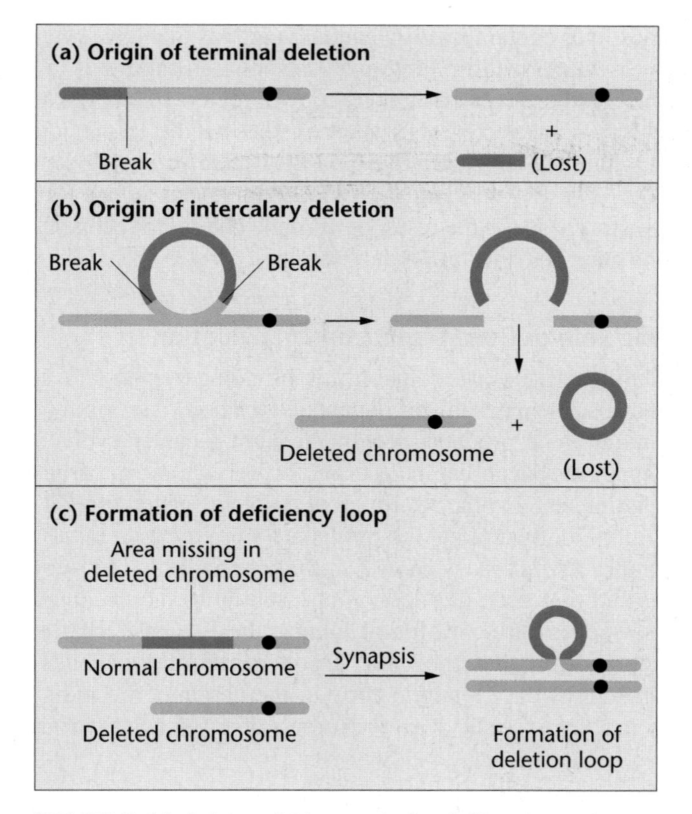

FIGURE 7–11 Origins of (a) a terminal and (b) an intercalary deletion. In (c), pairing occurs between a normal chromosome and one with an intercalary deletion by looping out the undeleted portion to form a deletion (or compensation) loop.

Duplications may arise as the result of unequal crossing over between synapsed chromosomes during meiosis (Figure 7–12) or through a replication error prior to meiosis. In the former case, both a duplication and a deletion are produced.

We consider three interesting aspects of duplications. First, they may result in gene redundancy. Second, as with deletions, duplications may produce phenotypic variation. Third, according to one convincing theory, duplications have also been an important source of genetic variability during evolution.

Gene Redundancy and Amplification—Ribosomal RNA Genes

Although many gene products are not needed in every cell of an organism, other gene products are known to be essential components of all cells. For example, ribosomal RNA must be present in abundance to support protein synthesis. The more metabolically active a cell is, the higher the demand for this molecule. We might hypothesize that a single copy of the gene encoding rRNA is inadequate in many cells. Studies using the technique of molecular hybridization, which enables us to determine the percentage of the genome that codes for specific RNA sequences, show that our hypothesis is correct. Indeed, multiple copies of genes code for rRNA. Such DNA is called **rDNA**, and the general phenomenon is called **gene redundancy**. For example, in the common intestinal bacterium *Escherichia coli* (*E. coli*), about 0.7% of the haploid genome consists of rDNA—this is the equivalent of 7 copies of the gene. In *Drosophila melanogaster*, 0.3% of the haploid genome, equivalent to 130 copies, consists of rDNA. Although the presence of multiple copies of the same gene is not restricted to those coding for rRNA, we will focus on them in this section.

In some cells, particularly oocytes, even the normal redundancy of rDNA is insufficient to provide adequate amounts of rRNA needed to construct ribosomes. Oocytes store abundant nutrients, including huge quantities of ribosomes, for use by the embryo during early development. More ribosomes are included in oocytes than in any other cell type. By considering how the amphibian *Xenopus laevis* acquires this abundance of ribosomes, we shall see a second way in which the amount of rRNA is increased. This phenomenon is called **gene amplification.**

The genes that code for rRNA are located in an area of the chromosome known as the **nucleolar organizer region** (**NOR**). The NOR is intimately associated with the nucleolus, which is a processing center for ribosome production. Molecular hybridization analysis has shown that each NOR in the frog *Xenopus* contains the equivalent of 400 redundant gene copies coding for rRNA. Even this number of genes is apparently inadequate to synthesize the vast amount of ribosomes that must accumulate in the amphibian oocyte to support development following fertilization.

To further amplify the number of rRNA genes, the rDNA is selectively replicated, and each new set of genes is released from its template. Because each new copy is equivalent to an NOR, multiple small nucleoli are formed around each NOR in the oocyte. As many as 1500 of these "micronucleoli" have been observed in a single oocyte. If we multiply the number of micronucleoli (1500) by the number of gene copies in each NOR (400), we see that amplification in *Xenopus* oocytes can result in over half a million gene copies! If each copy is transcribed only 20

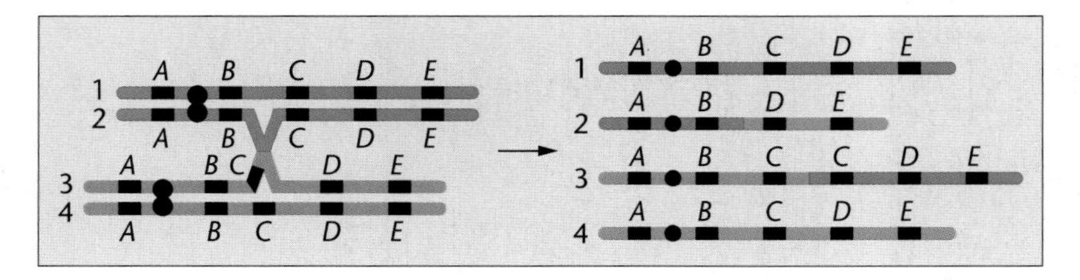

FIGURE 7–12 The origin of duplicated and deficient regions of chromosomes as a result of unequal crossing over. The tetrad on the left is mispaired during synapsis. A single crossover between chromatids 2 and 3 results in the deficient (chromosome 2) and duplicated (chromosome 3) chromosomal regions shown on the right. The two chromosomes uninvolved in the crossover event remain normal in gene sequence and content.

times during the maturation of the oocyte, in theory, sufficient copies of rRNA are produced and well over 12 million ribosomes will result.

The *Bar* Mutation in *Drosophila*

Duplications can cause phenotypic variation that might at first appear to be caused by a simple gene mutation. The *Bar* eye phenotype in *Drosophila* (Figure 7–13) is classic example. Instead of the normal oval eye shape, *Bar*-eyed flies have narrow, slitlike eyes. This phenotype is inherited in the same way as a dominant X-linked mutation.

In the early 1920s, Alfred H. Sturtevant and Thomas H. Morgan discovered and investigated this "mutation." Normal wild-type females (B^+/B^+) have about 800 facets in each eye. Heterozygous females (B/B^+) have about 350 facets, while homozygous females (B/B) average only about 70 facets. Females were occasionally recovered with even fewer facets and were designated as *double Bar* (B^D/B^+).

About 10 years later, Calvin Bridges and Herman J. Muller compared the polytene X chromosome banding pattern of the *Bar* fly with that of the wild-type fly. These chromosomes contain specific banding patterns that have been well categorized into regions. As shown in Figure 7–13, their studies revealed that one copy of region 16A of the X chromosome is present in wild-type flies but that this region was duplicated in *Bar* flies and triplicated in *double Bar* flies. These observations provided evidence that the *Bar* phenotype is not the result of a simple chemical change in the gene but is instead a duplication.

The Role of Gene Duplication in Evolution

During the study of evolution, it is intriguing to speculate on the possible mechanisms of genetic variation. The origin of unique gene products present in more recently evolved organisms but absent in ancestral forms is a topic of particular interest. In other words, how do "new" genes arise?

In 1970, Susumo Ohno published a provocative monograph, *Evolution by Gene Duplication*, in which he suggested that gene duplication is essential to the origin of new genes during evolution. Ohno's thesis is based on the supposition that the gene products of unique genes, present as only a single copy in the genome, are indispensable to the survival of members of any species during

(a) Genotypes and Phenotypes

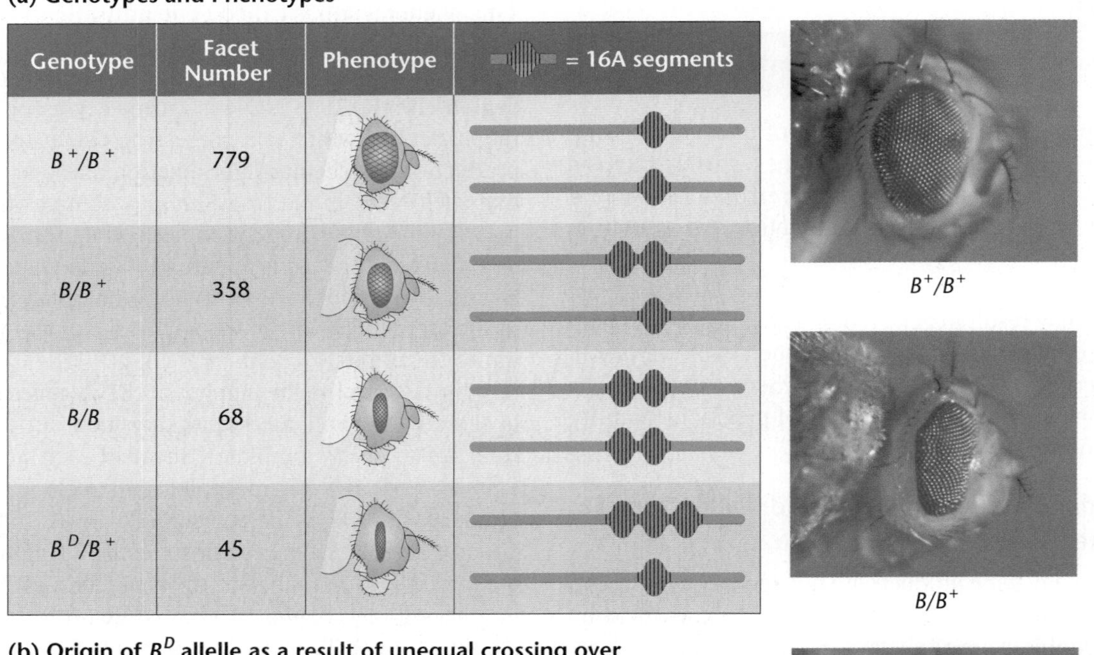

(b) Origin of B^D allele as a result of unequal crossing over

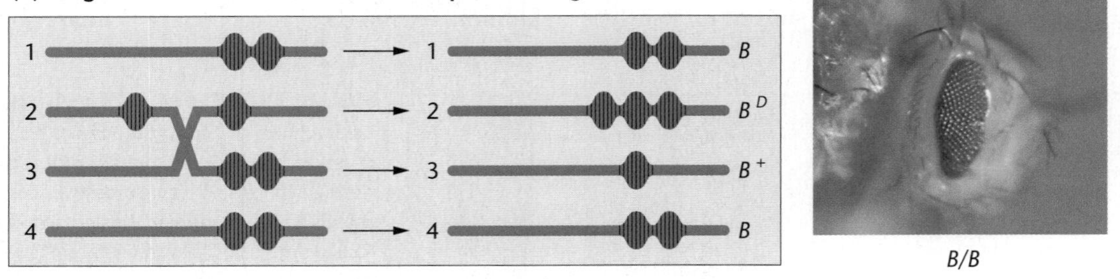

FIGURE 7–13 The duplication genotypes and resultant *Bar* eye phenotypes in *Drosophila*. Photographs show two *Bar* eye phenotypes and the wild type (B^+/B^+). *(Photos: Mary Lilly/Carnegie Institution of Washington)*

evolution. Therefore, unique genes are not free to accumulate mutations sufficient to alter their primary function and give rise to new genes.

However, if an essential gene is duplicated in the germ line, major mutational changes in this extra copy will be tolerated in future generations because the original gene provides the genetic information for its essential function. The duplicated copy will be free to acquire many mutational changes over extended periods of time. Over short intervals, the new genetic information may be of no practical advantage. However, over long evolutionary periods, the duplicated gene may change sufficiently so that its product assumes a divergent role in the cell. The new function may impart an "adaptive" advantage to organisms, enhancing their fitness. Ohno has outlined a mechanism through which sustained genetic variability may have originated.

Ohno's thesis is supported by the discovery of genes that have a substantial amount of their DNA sequence in common, but whose gene products are distinct. For example, trypsin and chymotrypsin fit this description, as do myoglobin and hemoglobin. The DNA sequence is so similar (homologous) in each case that we may conclude that members of each pair of genes arose from a common ancestral gene through duplication. During evolution, the related genes diverged sufficiently that their products became unique.

Other support includes the presence of **gene families**, groups of contiguous genes whose products perform the same function. Again, members of a family show DNA-sequence homology sufficient to conclude that they share a common origin. Examples are the various types of human hemoglobin polypeptide chains, as well as the immunologically important T-cell receptors and antigens encoded by the major histocompatibility complex.

Many recent findings derived from our ability to sequence entire genomes lend support to the idea that gene duplication has been a common feature of evolutionary progression. For example, Jurg Spring has compared a large number of genes in *Drosophila* and their counterparts in humans. In 50 genes studied, the fruit fly has only one copy, while there are multiple copies present in the human genome. There are many other investigations that have provided similar findings.

A new debate has begun concerning a second aspect of Ohno's thesis—that major evolutionary jumps, such as the transition from invertebrates to vertebrates, may have involved the duplication of entire genomes. Ohno has suggested that this might have occurred several times during the course of evolution. While it seems clear that genes, and even segments of chromosomes, have been duplicated, there is not yet any compelling evidence to convince evolutionary biologists that this was clearly the case. However, from the standpoint of gene expression, duplicating all genes proportionately seems more likely to be tolerated than duplicating just a portion of them. Whatever may be the case, it is interesting to examine evidence based on new technology that tests a hypothesis that was put forward over 40 years ago, long before that technology was available.

7.9 Inversions Rearrange the Linear Gene Sequence

The **inversion**, another class of structural variation, is a type of chromosomal aberration in which a segment of a chromosome is turned around 180° within a chromosome. An inversion does not involve a loss of genetic information, but simply rearranges the linear gene sequence. An inversion requires breaks at two points along the length of the chromosome and subsequent reinsertion of the inverted segment. Figure 7–14 illustrates how an inversion might arise. By forming a chromosomal loop prior to breakage, the newly created "sticky ends" are brought close together and rejoined.

The inverted segment may be short or quite long and may or may not include the centromere. If the centromere is not part of the rearranged chromosome segment, it is a **paracentric inversion**. If the centromere is part of the inverted segment, it is described as a **pericentric inversion**, which is the type shown in Figure 7–14.

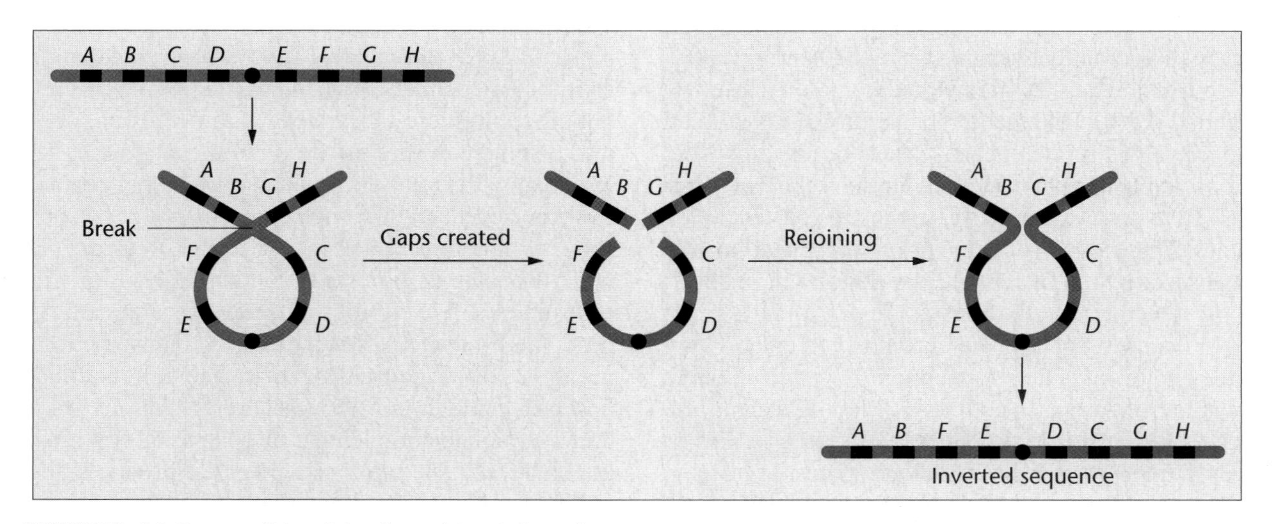

FIGURE 7–14 One possible origin of a pericentric inversion.

Although inversions appear to have a minimal impact on the individuals bearing them, their consequences are of great interest to geneticists. Organisms that are heterozygous for inversions may produce aberrant gametes that have a major impact on their offspring.

Consequences of Inversions During Gamete Formation

If only one member of a homologous pair of chromosomes has an inverted segment, normal linear synapsis during meiosis is not possible. Organisms with one inverted chromosome and one noninverted homolog are called **inversion heterozygotes**. Pairing between two such chromosomes in meiosis is accomplished only if they form an **inversion loop** (Figure 7–15).

If crossing over does not occur within the inverted segment of the inversion loop, the homologs will segregate, which results in two normal and two inverted chromatids that are distributed into gametes. However, if crossing over does occur within the inversion loop, abnormal chromatids are produced. The effect of a single exchange event within a paracentric inversion is diagrammed in Figure 7–15(a).

As in any meiotic tetrad, a single crossover between nonsister chromatids produces two parental chromatids and two recombinant chromatids. When the crossover occurs within a paracentric inversion, however, one recombinant **dicentric chromatid** (two centromeres), and one recombinant **acentric chromatid** (lacking a centromere) are produced. Both contain duplications and deletions of chromosome segments as well. During anaphase, an acentric chromatid moves randomly to one pole or the other or may be lost, while a dicentric chromatid is pulled in two directions. This polarized movement produces dicentric bridges that are cytologically recognizable. A dicentric chromatid usually breaks at some point so that part of the chromatid goes into one gamete and part into another gamete during the reduction divisions. Therefore, gametes containing either recombinant chromatid are deficient in genetic material. When such a gamete participates in fertilization, the zygote most often develops abnormally, if at all.

A similar chromosomal imbalance is produced as a result of a crossover event between a chromatid bearing a pericentric inversion and its noninverted homolog, as shown in Figure 7–15(b). The recombinant chromatids that are directly involved in the exchange have duplications and deletions. In plants, gametes receiving such aberrant chromatids fail to develop normally, leading to aborted pollen or ovules. Thus, lethality occurs prior to fertilization, and inviable seeds result. In animals, the gametes have developed prior to the meiotic error, so fertilization is more likely to occur in spite of the chromosome error. However, the end result is the production of inviable embryos following fertilization. In both cases, viability is reduced.

Since viable offspring do not result in either plants or animals, it *appears* as if the inversion suppresses crossing over, because offspring bearing crossover gametes are not recovered. Actually, in inversion heterozygotes, the inversion has the effect of *suppressing the recovery of crossover products* when chromosome exchange occurs within the inverted region. If crossing over always occurred within a paracentric or pericentric inversion, 50% of the gametes would be ineffective. The viability of the resulting zygotes is therefore greatly diminished. Furthermore, up to one-half of the viable gametes have the inverted chromosome, and the inversion will be perpetuated within the species. The cycle will be repeated continuously during meiosis in future generations.

7.10 Translocations Alter the Location of Chromosomal Segments in the Genome

Translocation, as the name implies, is the movement of a chromosomal segment to a new location in the genome. Reciprocal translocation, for example, involves the exchange of segments between two nonhomologous chromosomes. The least complex way for this event to occur is for two nonhomologous chromosome arms to come close to each other so that an exchange is facilitated. Figure 7–16(a) shows a simple reciprocal translocation in which only two breaks are required. If the exchange includes internal chromosome segments, four breaks are required, two on each chromosome.

The genetic consequences of reciprocal translocations are, in several instances, similar to those of inversions. For example, genetic information is not lost or gained. Rather, there is only a rearrangement of genetic material. The presence of a translocation does not, therefore, directly alter the viability of individuals bearing it.

Homologs that are heterozygous for a reciprocal translocation undergo unorthodox synapsis during meiosis. As shown in Figure 7–16(b), pairing results in a crosslike configuration. As with inversions, genetically unbalanced gametes are also produced as a result of this unusual alignment during meiosis. In the case of translocations, however, aberrant gametes are not necessarily the result of crossing over. To see how unbalanced gametes are produced, focus on the homologous centromeres in Figure 7–16(b) and Figure 7–16(c). According to the principle of independent assortment, the chromosome containing centromere 1 migrates randomly toward one pole of the spindle during the first meiotic anaphase; it travels along with *either* the chromosome having centromere 3 *or* the chromosome having centromere 4. The chromosome with centromere 2 moves to the other pole along with *either* the chromosome containing centromere 3 *or* centromere 4. This results in four potential meiotic products. The 1,4 combination contains chromosomes that are not involved in the translocation. The 2,3 combination, however, contains translocated chromosomes. These contain a complete complement of genetic information and are balanced. The other two potential products, the 1,3 and 2,4 combinations, contain chromosomes displaying duplicated and deleted segments.

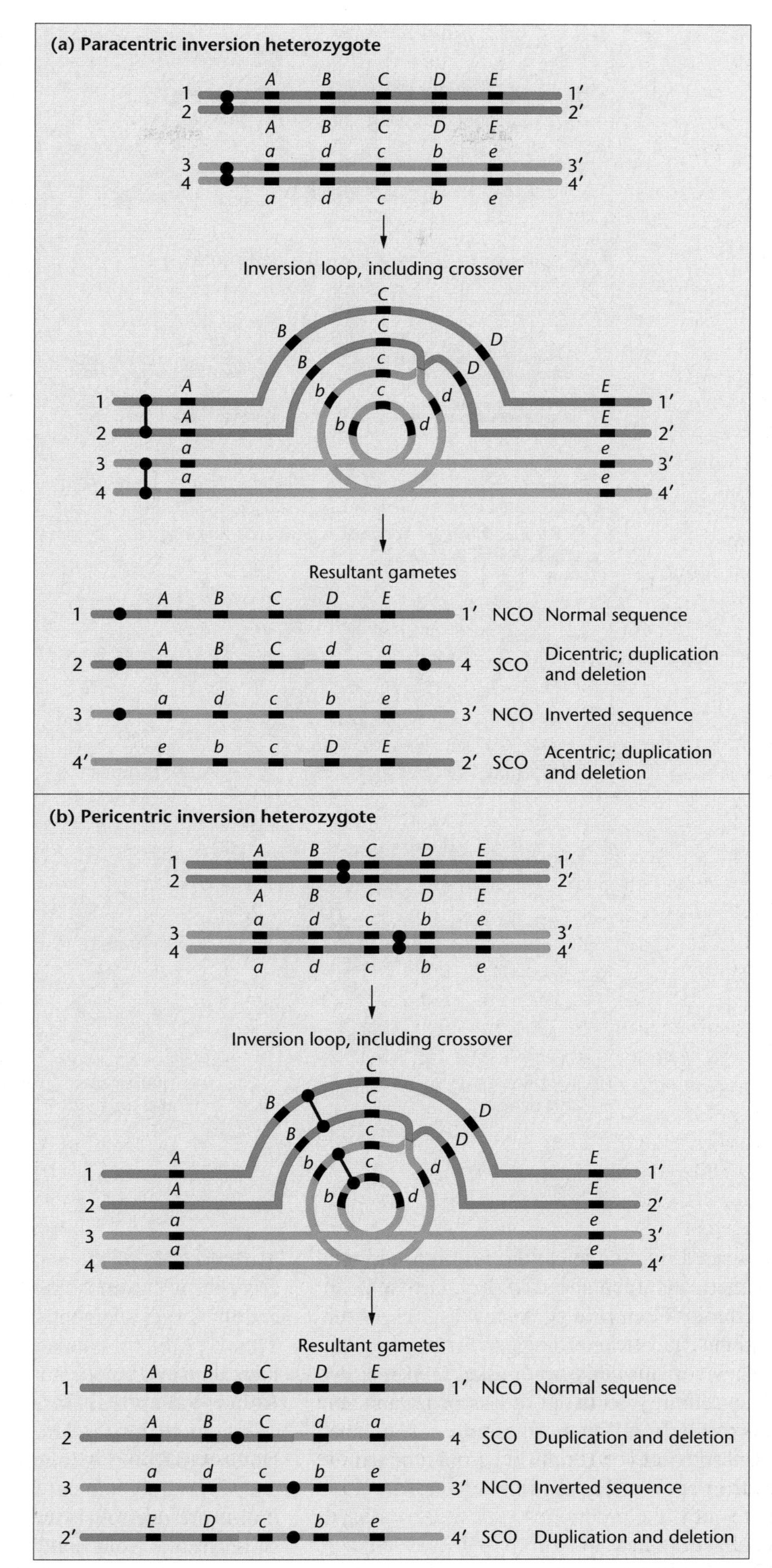

FIGURE 7–15 The effects of a single crossover within an inversion loop in (a) a paracentric inversion heterozygote, where two altered chromosomes are produced, one acentric and one dicentric. Both chromosomes also contain duplicated and deficient regions. The effects of this crossover in (b), a pericentric inversion heterozygote, where two altered chromosomes are produced, both with duplicated and deficient regions.

MEDIA TUTORIAL Aberrations

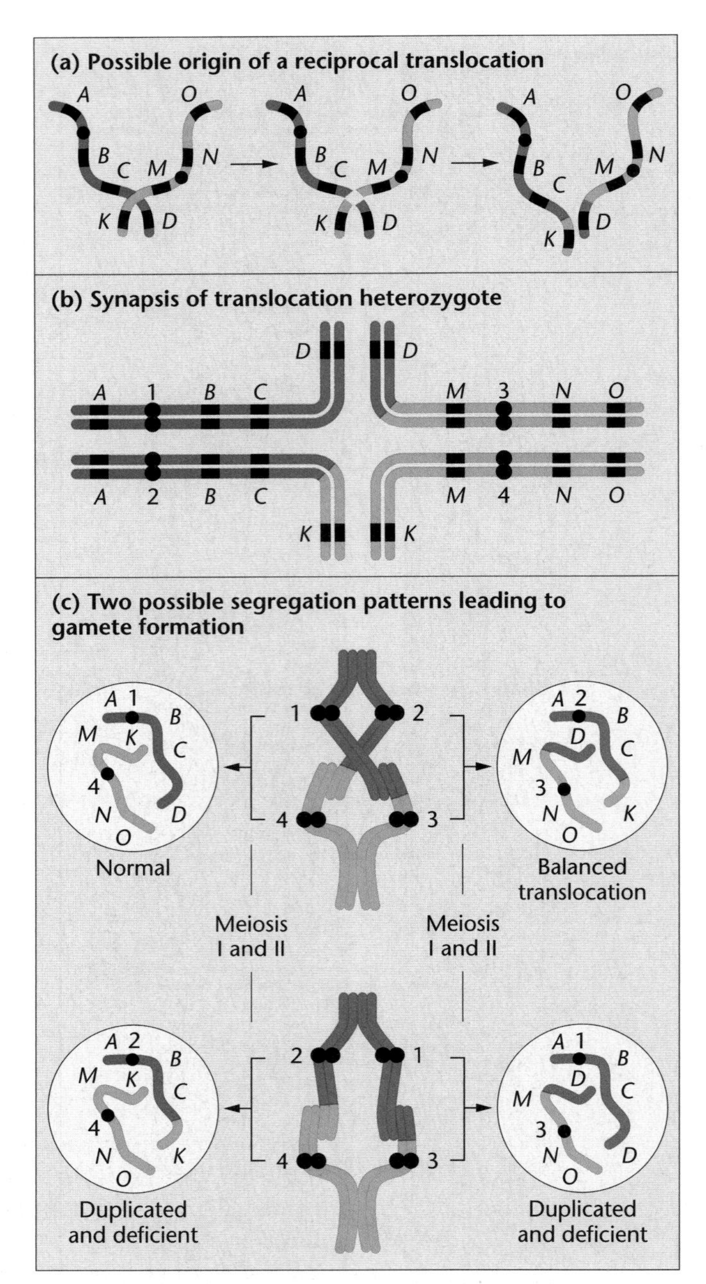

FIGURE 7–16 (a) Possible origin of a reciprocal translocation. (b) Synaptic configuration formed during meiosis in an individual that is heterozygous for the translocation. (c) Two possible segregation patterns, one of which leads to a normal and a balanced gamete (called alternate segregation) and one that leads to gametes containing duplications and deficiencies (called adjacent segregation).

When incorporated into gametes, the resultant meiotic products are genetically unbalanced. If they participate in fertilization, lethality often results. As few as 50% of the progeny of parents that are heterozygous for a reciprocal translocation survive. This condition, called *semisterility*, has an impact on the reproductive fitness of organisms, thus playing a role in evolution. Furthermore, in humans, such an unbalanced condition results in partial monosomy or trisomy, leading to a variety of birth defects.

Translocations in Humans: Familial Down Syndrome

Research conducted since 1959 has revealed numerous translocations in members of the human population. One common type of translocation involves breaks at the extreme ends of the short arms of two nonhomologous acrocentric chromosomes. These small segments are lost, and the larger segments fuse at their centromeric region. This type of translocation produces a new, large submetacentric or metacentric chromosome, often called a **Robertsonian translocation**.

One such translocation accounts for cases in which Down syndrome is inherited or familial. Earlier in this chapter, we pointed out that most instances of Down syndrome are due to trisomy 21. This chromosome composition results from nondisjunction during meiosis in one parent. Trisomy accounts for over 95% of all cases of Down syndrome. In such instances, the chance of the same parents producing a second affected child is extremely low. However, in the remaining families with a Down

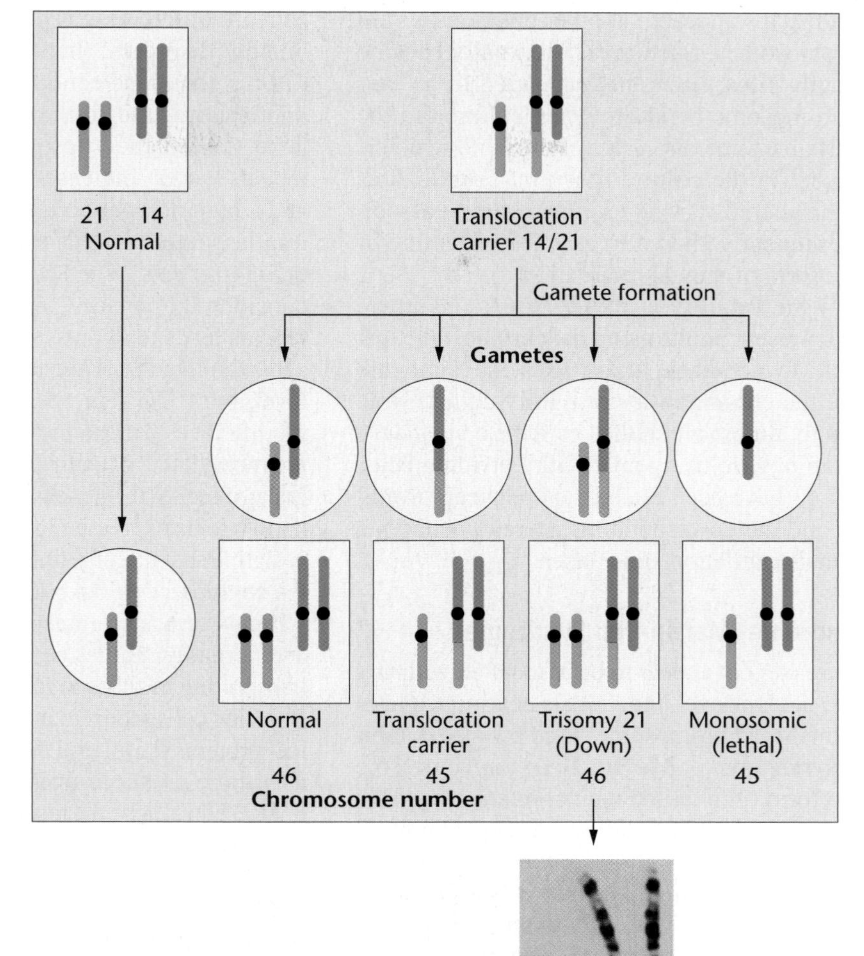

FIGURE 7–17 Chromosomal involvement and translocation in familial Down syndrome. The photograph shows the relevant chromosomes from a trisomy 21 offspring produced by a translocation carrier parent. *(Photo: Dr. Jorge Yunis)*

child, the syndrome occurs in a much higher frequency over several generations.

Cytogenetic studies of the parents and their offspring from these unusual cases explain the cause of **familial Down syndrome**. Analysis reveals that one of the parents contains a 14/21, D/G translocation (Figure 7–17). That is, one parent has the majority of the G-group chromosome 21 translocated to one end of the D-group chromosome 14. This individual is phenotypically normal even though he or she has only 45 chromosomes. During meiosis, one-fourth of the individual's gametes have two copies of chromosome 21: a normal chromosome and a second copy translocated to chromosome 14. When such a gamete is fertilized by a standard haploid gamete, the resulting zygote has 46 chromosomes but 3 copies of chromosome 21. These individuals exhibit Down syndrome. Other potential surviving offspring contain either the standard diploid genome (without a translocation) or the balanced translocation like the parent. Both cases result in normal individuals. Knowledge of translocations has allowed geneticists to resolve the seeming paradox of an inherited trisomic phenotype in an individual with an apparent diploid number of chromosomes.

It is interesting to note that the "carrier", who has 45 chromosomes and exhibits a normal phenotype, does not contain the *complete* diploid amount of genetic material. A small region is lost from both chromosomes 14 and 21 during the translocation event. This occurs because the ends of both chromosomes have broken off prior to their fusion. These specific regions are known to be two of many chromosomal locations housing multiple copies of the genes encoding rRNA, the major component of ribosomes. Despite the loss of up to 20% of these genes, the carrier is unaffected.

7.11 Fragile Sites in Humans Are Susceptible to Chromosome Breakage

We conclude this chapter with a brief discussion of the results of an intriguing discovery made around 1970 during observations of metaphase chromosomes prepared following human cell culture. In cells derived from certain individuals, a specific area along one of the chromosomes failed to stain, giving the appearance of a gap. In other individuals whose chromosomes displayed such morphology, the gaps appeared at other positions within the set of chromosomes. Such areas eventually became known as **fragile sites**, since they appeared to be susceptible to chromosome breakage when cultured in the absence of certain chemicals such as folic acid, which is normally present in the culture medium. Fragile sites were at first considered curiosities, until a strong association was subsequently shown to exist between one of the sites and a form of mental retardation.

The cause of the fragility at these sites is unknown. Because they represent points along the chromosome that are susceptible to breakage, these sites may indicate regions where the chromatin is not tightly coiled. Note that even though almost all studies of fragile sites have been carried out *in vitro* using mitotically dividing cells, clear associations have been established between several of these sites and the corresponding altered phenotype, including mental retardation and cancer.

Fragile X Syndrome (Martin-Bell Syndrome)

Most fragile sites do not appear to be associated with any clinical syndrome. However, individuals bearing a folate-sensitive site on the X chromosome (Figure 7–18) exhibit the **fragile X syndrome** (or **Martin-Bell syndrome**), the most common form of inherited mental retardation. This syndrome affects about 1 in 4000 males and 1 in 8000 females. Females carrying only one fragile X chromosome can be mentally retarded because it is a dominant trait. Fortunately, the trait is not fully expressed, as only about 30% of fragile X females are retarded, whereas about 80% of fragile X males are mentally retarded. In addition to mental retardation, affected males have characteristic long, narrow faces with protruding chins, enlarged ears, and increased testicular size.

A gene that spans the fragile site may be responsible for this syndrome. This gene, known as *FMR-1*, is one of a growing number of genes that have been discovered in which a sequence of three nucleotides is repeated many times, expanding the size of the gene. This phenomenon, called **trinucleotide repeats**, is also recognized in other human disorders, including Huntington disease. In *FMR-1*, the trinucleotide sequence CGG is repeated in an untranslated area adjacent to the coding sequence of the gene (called the "upstream" region). The number of repeats varies immensely within the human population, and a high number correlates directly with expression of fragile X syndrome. Normal individuals have between 6 and 54 repeats, whereas those with 55–230 repeats are considered "carriers" of the disorder. More than 230 repeats leads to expression of the syndrome.

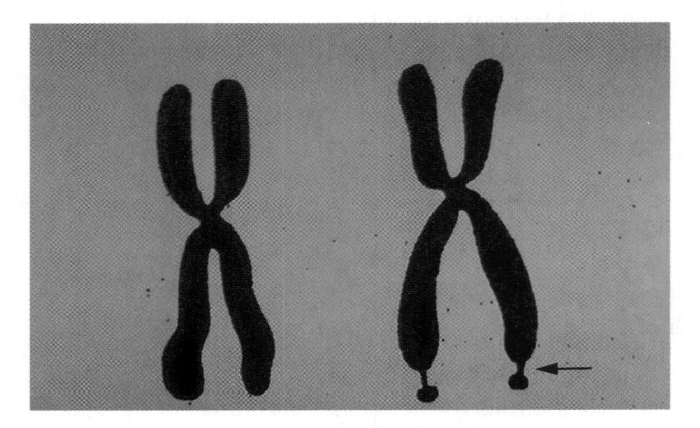

FIGURE 7–18 A normal human X chromosome (left) contrasted with a fragile X chromosome (right). The "gap" region (near the bottom of the chromosome) is associated with the fragile X syndrome. *(Science VU/Visuals Unlimited)*

It is thought that when the number of repeats reaches this level, the CGG regions of the gene become chemically modified so that the bases within and around the repeat are methylated, causing inactivation of the gene. The normal product of the gene is an RNA-binding protein, FMRP, known to be expressed in the brain. Evidence is now accumulating that directly links the absence of the protein with the cognitive defects associated with the syndrome.

The protein is prominent in cells of the developing brain and normally shuttles between the nucleus and cytoplasm, transporting mRNAs to ribosomal complexes. Its absence in dendritic cells seems to prevent the translation of other critical proteins during the development of the brain, presumably those produced from mRNAs that are transported by FMRP. Ultimately, the result is impairment of synaptic activity related to learning and memory. Molecular analysis has shown that FMRP is very selective, binding specifically to mRNAs that have a high guanine content. These RNAs form unique secondary structures called **G-quartets**, for which the protein has a strong binding affinity. The fact that possession of such quartets is important to recognition by FMRP becomes evident by examining cases where the formation of these stacked quartets is inhibited (using Li^+). In such cases, FMRP binding to these RNAs is abolished.

The *Drosophila* homolog (*dfrx*) of the human *FRM-1* gene has now been identified and used as a model for experimental investigation. The normal protein product of the fruit fly displays chemical properties that parallel its human counterpart. Mutations have been generated and studied that either overexpress the gene or eliminate its expression altogether. The latter case is designated a null mutation, where the gene has been "knocked out" using a molecular procedure now common in DNA biotechnology. The results of these studies show that the normal gene functions during the formation of synapses and that abnormal synaptic function accompanies the loss of *dfrx* expression. Knockout mice lacking this gene also display a similar phenotype. Ultimately, then, a closely related version of this gene is conserved in humans, *Drosophila*, and mice. In all three species, it has been shown to be essential to synaptic maturation and the subsequent establishment of proper cell communication in the brain. In humans, the neurological features of the syndrome are linked to the absence of its genetic expression.

From a genetic standpoint, perhaps the most interesting aspect of fragile X syndrome is the instability of the CGG repeats. An individual with 6–54 repeats transmits a gene containing the same number to his or her offspring. However, those with 55–230 repeats, while not at risk to develop the syndrome, may transmit to their offspring a gene with an increased number of repeats. The number of repeats continues to increase in future generations, demonstrating the phenomenon known as **genetic anticipation**. Once the threshold of 230 repeats is exceeded, expression of the malady becomes more severe in each successive generation as the number of trinucleotide repeats increases.

While the mechanism that leads to the trinucleotide expansion has not yet been established, several factors are known that influence the instability. Most significant is the observation that expansion from the carrier status (55–230 repeats) to the syndrome status (over 230 repeats) occurs during the transmission of the gene by the maternal parent, but not by the paternal parent. Furthermore, several reports suggest that male offspring are more likely to receive the increased repeat size leading to the syndrome than are female offspring. Obviously, we have much to learn about the genetic basis of instability and expansion of DNA sequences.

Fragile Sites and Cancer

A second link between a fragile site and a human disorder was reported in 1996 by Carlo Croce, Kay Huebner, and their colleagues. They demonstrated an association between an autosomal fragile site and cancer and showed that the gene *FHIT* (standing for *f*ragile *hi*stidine *t*riad), located within a well-defined fragile site on chromosome 3, is often altered in cells taken from the tumors of individuals with lung cancer. A variety of mutations were found in cells derived from the tumors where the DNA had apparently been broken and incorrectly re-fused, resulting in deletions within the gene. In most cases, these mutations caused the *FHIT* gene to become inactivated.

This gene is part of the fragile region of the autosome designated *FRA3B*, which has been linked to other cancers, including the esophagus, colon, and stomach. The nature of the genetic alterations found in cancer cells suggests that the *FHIT* gene, because it is within a fragile region, may be highly susceptible to induced breaks in DNA. If these breaks are incorrectly repaired, cancer-specific chromosome alterations may occur. Thus, this region of the chromosome appears to be particularly sensitive to carcinogen-induced damage, creating a susceptibility to cancer. It will be important to determine experimentally whether molecular polymorphism exists at this and other fragile sites within the human population, causing some individuals to be more susceptible to the effects of carcinogens than others.

Chapter Summary

1. Investigations into the uniqueness of each organism's chromosomal constitution is enhancing our understanding of genetic variation. Alterations of the precise diploid content of chromosomes are called chromosomal aberrations or chromosomal mutations.
2. Deviations from the expected chromosomal number, or mutations in the structure of the chromosome, are inherited in predictable Mendelian fashion; they often result in inviable organisms or substantial changes in the phenotype.
3. Aneuploidy is the gain or loss of one or more chromosomes from the diploid complement, resulting in conditions of monosomy, trisomy, tetrasomy, and so on. Studies of monosomic and trisomic disorders are increasing our understanding of the delicate genetic balance that enables normal development.
4. When complete sets of chromosomes are added to the diploid genome, polyploidy occurs. These sets can have identical or diverse genetic origin, creating either autopolyploidy or allopolyploidy, respectively.
5. Large segments of the chromosome can be modified by deletions or duplications. Deletions can produce serious conditions such as the cri-du-chat syndrome in humans, whereas duplications can be particularly important as a source of redundant or new genes.
6. Inversions and translocations, while altering the gene order along chromosomes, initially cause little or no loss of genetic information or deleterious effects. However, heterozygous combinations may cause genetically abnormal gametes following meiosis, with lethality often ensuing.
7. Fragile sites in human mitotic chromosomes have sparked research interest because one such site on the X chromosome is associated with the most common form of inherited mental retardation. Another fragile site, located on chromosome 3, has been linked to lung cancer.

Key Terms

acentric chromatid, 146
allopolyploidy, 139
allotetraploid, 140
amniocentesis, 138
amphidiploid, 140
aneuploidy, 134
autopolyploidy, 139
autotetraploid, 139
chorionic villus sampling (CVS), 138
chromosome aberration, 134
chromosome mutation, 134
colchicine, 139
compensation loop, 142
cri-du-chat syndrome (46,5p−), 136
deficiency, 142
deletion, 142
deletion loop, 142
dicentric chromatid, 146
Down syndrome (47,21+), 136
Edwards syndrome (47,18+), 138
euploidy, 134
familial Down syndrome, 149
fragile site, 150
fragile X syndrome, 150

INSIGHTS AND SOLUTIONS

1. In a cross using maize that involves three genes, *a*, *b*, and *c*, a heterozygote (*abc*/+++) is test-crossed to *abc*/*abc*. Even though the three genes are separated along the chromosome, thus predicting that crossover gametes and the resultant phenotype should be observed, only two phenotypes are recovered: *abc* and +++. Additionally, the cross produced significantly fewer viable plants than expected. Can you propose why no other phenotypes were recovered and why the viability was reduced?

Solution: One of the two chromosomes may contain an inversion that overlaps all three genes, effectively precluding the recovery of any "crossover" offspring. If this is a paracentric inversion and the genes are clearly separated (assuring that a significant number of crossovers occurs between them), then numerous acentric and dicentric chromosomes will form, resulting in the observed reduction in viability.

2. A male *Drosophila* from a wild-type stock is discovered to have only seven chromosomes, whereas normally $2n = 8$. Close examination reveals that one member of chromosome IV (the smallest chromosome) is attached to (translocated to) the distal end of chromosome II and is missing its centromere, thus accounting for the reduction in chromosome number. (a) Diagram all members of chromosomes II and IV during synapsis in meiosis I.

Solution:

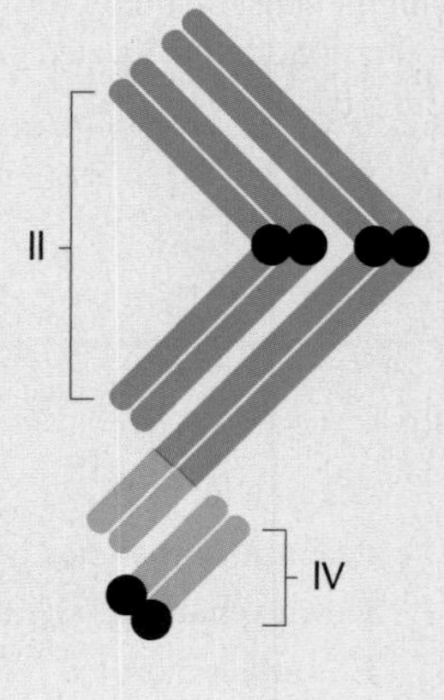

(b) If this male mates with a female with a normal chromosome composition who is homozygous for the recessive chromosome IV mutation *eyeless* (*ey*), what chromosome compositions will occur in the offspring regarding chromosomes II and IV?

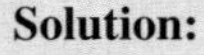
Solution:

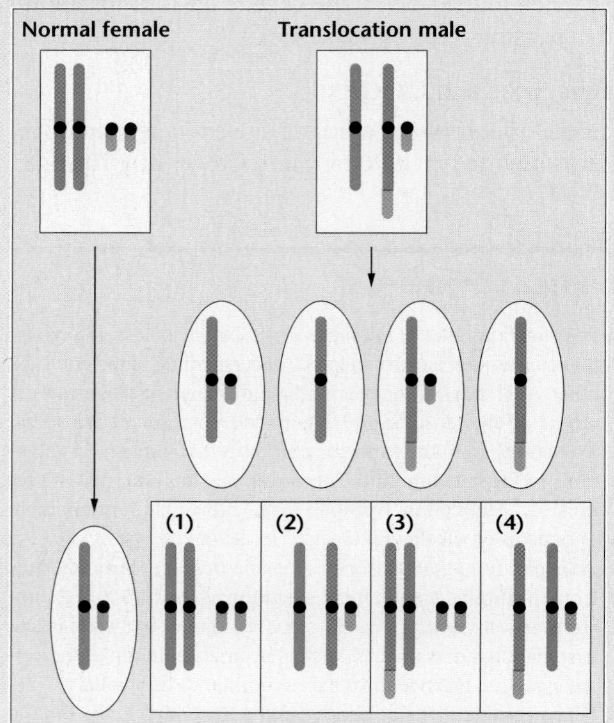

(c) Referring to the diagram in the solution to part (b), what phenotypic ratio will result regarding the presence of eyes, assuming all abnormal chromosome compositions survive?

Solution:

(1) normal (heterozygous)

(2) eyeless (monosomic, contains chromosome IV from mother)

(3) normal (heterozygous)

(4) normal (heterozygous)

The final ratio is 3/4 normal: 1/4 eyeless.

Problems and Discussion Questions

1. For a species with a diploid number of 18, indicate how many chromosomes will be present in the somatic nuclei of individuals that are haploid, triploid, tetraploid, trisomic, and monosomic.
2. Define these pairs of terms, and distinguish between them.

   aneuploidy/euploidy
   monosomy/trisomy
   Patau syndrome/Edwards syndrome
   autopolyploidy/allopolyploidy
   autotetraploid/amphidiploid
   paracentric inversion/pericentric inversion

3. Contrast the relative survival times of individuals with Down, Patau, and Edwards syndromes. Speculate as to why such differences exist.
4. What evidence suggests that Down syndrome is more often the result of nondisjunction during oogenesis rather than during spermatogenesis?
5. What evidence indicates that humans with aneuploid karyotypes occur at conception but are usually inviable?
6. Contrast the fertility of an allotetraploid with an autotriploid and an autotetraploid.
7. When two plants belonging to the same genus but different species are crossed, the F_1 hybrid is more viable and has more ornate flowers. Unfortunately, this hybrid is sterile and can only be propagated by vegetative cuttings. Explain the sterility of the hybrid. How might a horticulturist attempt to reverse its sterility?
8. Describe the origin of cultivated American cotton.
9. Predict how the synaptic configurations of homologous pairs of chromosomes might appear when one member is normal and the other member has sustained a deletion or duplication.
10. Inversions are said to "suppress crossing over." Is this terminology technically correct? If not, restate the description accurately.
11. Contrast the genetic composition of gametes derived from tetrads of inversion heterozygotes where crossing over occurs within a paracentric and a pericentric inversion.
12. Discuss Ohno's hypothesis on the role of gene duplication in the process of evolution.
13. What roles have inversions and translocations played in the evolutionary process?
14. A human female with Turner syndrome also expresses the X-linked trait hemophilia, as did her father. Which of her parents underwent nondisjunction during meiosis, giving rise to the gamete responsible for the syndrome?
15. The primrose, *Primula kewensis*, has 36 chromosomes that are similar in appearance to the chromosomes in two related species, *P. floribunda* ($2n = 18$) and *P. verticillata* ($2n = 18$). How could *P. kewensis* arise from these species? How would you describe *P. kewensis* in genetic terms?
16. Certain varieties of chrysanthemums contain 18, 36, 54, 72, and 90 chromosomes; all are multiples of a basic set of 9 chromosomes. How would you describe these varieties genetically? What feature is shared by the karyotypes of each variety? A variety with 27 chromosomes has been discovered, but it is sterile. Why?
17. *Drosophila* may be monosomic for chromosome 4 yet remain fertile. Contrast the F_1 and F_2 results of the following crosses involving the recessive chromosome 4 trait, *bent* bristles:
    (a) monosomic IV, bent bristles × diploid, normal bristles
    (b) monosomic IV, normal bristles × diploid, bent bristles
18. Mendelian ratios are modified in crosses involving autotetraploids. Assume that one plant expresses the dominant trait green seeds and is homozygous (*WWWW*). This plant is crossed to one with white seeds that is also homozygous (*wwww*). If only one dominant allele is sufficient to produce green seeds, predict the F_1 and F_2 results of such a cross. Assume that synapsis between chromosome pairs is random during meiosis.
19. Having correctly established the F_2 ratio in Problem 18, predict the F_2 ratio of a "dihybrid" cross involving two independently assorting characteristics (e.g., $P_1 = WWWWAAAA \times wwwwaaaa$).
20. In a cross between two varieties of corn, $gl_1gl_1\ Ws_3Ws_3$(egg parent) × $Gl_1Gl_1\ ws_3ws_3$ (pollen parent), a triploid offspring was produced with the genetic constitution $Gl_1Gl_1gl_1Ws_3ws_3ws_3$. From which parent, egg or pollen, did the $2n$ gamete originate? Is another explanation possible? Explain.
21. What is the effect of a rare double crossover (a) within a pericentric inversion present heterozygously, (b) within a paracentric inversion present heterozygously?
22. The outcome of a single crossover between nonsister chromatids in the inversion loop of an inversion heterozygote varies depending on whether the inversion is of the paracentric or pericentric type. What differences are expected?
23. A couple planning their family are aware that through the past three generations on the husband's side a substantial number of stillbirths have occurred and several malformed babies were born who died early in childhood. The wife has studied genetics and urges her husband to visit a genetic counseling clinic, where a complete karyotype-banding analysis is performed. Although the tests show that he has a normal complement of 46 chromosomes, banding analysis reveals that one member of the chromosome 1 pair (in group A) contains an inversion covering 70% of its length. The homolog of chromosome 1 and all other chromosomes show the normal banding sequence. (a) How would you explain the high incidence of past stillbirths? (b) What can you predict about the probability of abnormality/normality of their future children? (c) Would you advise the woman that she will have to bring each pregnancy to term to determine whether the fetus is normal? If not, what else can you suggest?
24. Briefly compare the fate of gametes in plants and animals carrying aberrant chromatids or unbalanced chromosomal complements.
25. In a *Drosophila* cross, a female heterozygous for the autosomally linked genes *a*, *b*, *c*, *d*, and *e* (*abcde*/+++++) is testcrossed to a male homozygous for all recessive alleles. Even though the distance between each of these loci is at least three map units, only four phenotypes are recovered, yielding the following data:

Phenotype	*No. of Flies*
+ + + + +	440
a b c d e	460
+ + + + *e*	48
a b c d +	52
Total =	1000

Why are many expected crossover phenotypes missing? Can any of these loci be mapped from the data given here? If so, determine map distances.

26. A woman who sought genetic counseling is found to be heterozygous for a chromosomal rearrangement between the second and third chromosomes. Her chromosomes, compared to those in a normal karyotype, are diagrammed here:

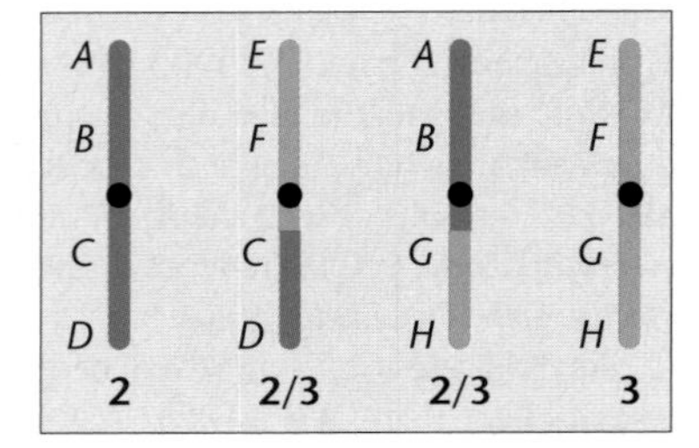

(a) What kind of chromosomal aberration is shown?
(b) Using a drawing, demonstrate how these chromosomes would pair during meiosis. Be sure to label the different segments of the chromosomes.
(c) This woman is phenotypically normal. Does this surprise you? Why or why not? Under what circumstances might you expect a phenotypic effect of such a rearrangement?

27. The woman in Problem 26 has had two miscarriages. She has come to you, an established genetic counselor, with these questions: (a) Is there a genetic explanation of her frequent miscarriages? (b) Should she abandon her attempts to have a child of her own? (c) If not, what is the chance that she could have a normal child? Provide an informed response to her concerns.

28. In a recent cytogenetic study on 1021 cases of Down syndrome, 46 were the result of translocations, the most frequent of which was symbolized as t(14;21). What does this symbol represent, and how many chromosomes would you expect to be present in t(14;21) Down syndrome individuals?

29. A boy with Klinefelter syndrome is born to a mother who is phenotypically normal and a father who has the X-linked skin condition called anhidrotic ectodermal dysplasia. The mother's skin is completely normal with no signs of the skin abnormality. In contrast, her son has patches of normal skin and patches of abnormal skin. (a) Which parent contributed the abnormal gamete? (b) Using the appropriate genetic terminology, describe the meiotic mistake that occurred. Be sure to indicate in which division the mistake occurred. (c) Using the appropriate genetic terminology, explain the son's skin phenotype.

30. To investigate the origin of nondisjunction, 200 human oocytes that had failed to be fertilized during *in vitro* fertilization procedures were subsequently examined. (Angel, R. 1997. *Am. J. Hum. Genet.* 61:23–32). These oocytes had completed meiosis I and were arrested in metaphase II (MII). The majority (67%) had a normal MII-metaphase complement, showing 23 chromosomes, each consisting of 2 sister chromatids joined at a common centromere. The remaining oocytes all had abnormal chromosome compositions. Surprisingly, when trisomy was considered, none of the abnormal oocytes had 24 chromosomes. (a) Interpret these results in regard to the origin of trisomy, as it relates to nondisjunction, and when it occurs. Why are the results surprising? (b) A large number of the abnormal oocytes contained 22 1/2 chromosomes; that is, 22 chromosomes plus a single chromatid representing the 1/2 chromosome. What chromosome compositions will result in the zygote if such oocytes proceed through meiosis and are fertilized by normal sperm? (c) How could the complement of 22 1/2 chromosomes arise? Provide a drawing that includes several pairs of MII chromosomes. (d) Do your answers support or dispute the generally accepted theory regarding nondisjunction and trisomy, as outlined in Figure 7–1?

31. In a human genetic study, a family with five phenotypically normal children was investigated. Two were "homozygous" for a Robertsonian translocation between chromosomes 19 and 20 (they contained two identical copies of the fused chromosome). These children have only 44 chromosomes but a complete genetic complement. Three of the children were "heterozygous" for the translocation and contained 45 chromosomes, with 1 translocated chromosome plus a normal copy of both chromosomes 19 and 20. The two other pregnancies resulted in stillbirths. It was later discovered that the parents were first cousins. Based on this information, determine the chromosome compositions of the parents. What led to the stillbirths? Why was the discovery that the parents were first cousins a key piece of information in understanding the genetics of this family?

Selected Readings

Antonarakis, S. E. 1998. Ten years of genomics, chromosome 21, and Down syndrome. *Genomics* 51: 1–16.

Ashley-Koch, A. E. et al. 1997. Examination of factors associated with instability of the *FMR1* CGG repeat. *Am. J. Hum. Genet.* 63: 776–85.

Beasley, J. O. 1942. Meiotic chromosome behavior in species, species hybrids, haploids, and induced polyploids of *Gossypium. Genetics* 27: 25–54.

Blakeslee, A. F. 1934. New jimson weeds from old chromosomes. *J. Hered.* 25: 80–108.

Borgaonker, D. S. 1989. *Chromosome Variation in Man: A Catalogue of Chromosomal Variants and Anomalies*. 5th ed. New York: Alan R. Liss.

Boue, A. 1985. Cytogenetics of pregnancy wastage. *Adv. Hum. Genet.* 14: 1–58.

Burgio, G. R. et al. eds. 1981. *Trisomy 21*. New York: Springer-Verlag.

Carr, D. H. 1971. Genetic basis of abortion. *Ann. Rev. Genet.* 5: 65–80.

Croce, C. M. 1996. The *FHIT* gene at 3p14.2 is abnormal in lung cancer. *Cell* 85: 17–26.

Cummings, M. R. 2003. *Human Heredity: Principles and Issues.* 6th ed. Pacific Grove: Brooks/Cole.

DeArce, M. A., and Kearns, A. 1984. The fragile X syndrome: The patients and their chromosomes. *J. Med. Genet.* 21: 84–91.

Feldman, M., and Sears, E. R. 1981. The wild gene resources of wheat. *Sci. Am.* (Jan.) 244: 102–12.

Galitski, T. et al. 1999. Ploidy regulation of gene expression. *Science* 285: 251–54.

Gersh, M. et al. 1995. Evidence for a distinct region causing a cat-like cry in patients with 5p deletions. *Am. J. Hum. Genet.* 56: 1404–10.

Gupta, P. K., and Priyadarshan, P. M. 1982. *Triticale*: Present status and future prospects. *Adv. Genet.* 21: 256–346.

Hassold, T. J. et al. 1980. Effect of maternal age on autosomal trisomies. *Ann. Hum. Genet.* (London) 44: 29–36.

Hassold, T. J., and Hunt, P. 2001. To err (meiotically) is human: The genesis of human aneuploidy. *Nat. Rev. Gen.* 2: 280–91.

Hassold, T., and Jacobs, P. A. 1984. Trisomy in man. *Annu. Rev. Genet.* 18: 69–98.

Hecht, F. 1988. Enigmatic fragile sites on human chromosomes. *Trends Genet.* 4: 121–22.

Henikoff, S. 1994. A reconsideration of the mechanism of position effect. *Genetics* 138: 1–5.

Hulse, J. H., and Spurgeon, D. 1974. *Triticale. Sci. Am.* (Aug.) 231: 72–81.

Jacobs, P. A. et al. 1974. A cytogenetic survey of 11,680 newborn infants. *Ann. Hum. Genet.* 37: 359–76.

Kaiser, P. 1984. Pericentric inversions: Problems and significance for clinical genetics. *Hum. Genet.* 68: 1–47.

Kaytor, M. D. and Orr, H. 2001. RNA targets of the fragile X protein. *Cell* 107: 555–57.

Khush, G. S. 1973. *Cytogenetics of Aneuploids.* Orlando: Academic Press.

Khush, G. S. et al. 1984. Primary trisomics of rice: Origin, morphology, cytology and use in linkage mapping. *Genetics.* 107: 141–63.

Lewis, E. B. 1950. The phenomenon of position effect. *Adv. Genet.* 3: 73–115.

Lewis, W. H., ed. 1980. *Polyploidy: Biological Relevance.* New York: Plenum Press.

Lupski, J. R., Roth, J. R., and Weinstock, G. M. 1996. Chromosomal duplications in bacteria, fruit flies, and humans. *Am. J. Hum. Genet.* 58: 21–26.

Lynch, M., and Conery, J. S. 2000. The evolutionary fate and consequences of duplicate genes. *Science* 290: 1151–54.

Madan, K. 1995. Paracentric inversions: A review. *Hum. Genet.* 96: 503–515.

Mantell, S. H., Mathews, J. A., and McKee, R. A. 1985. *Principles of Plant Biotechnology: An Introduction to Genetic Engineering in Plants.* Oxford: Blackwell.

Obe, G., and Basler, A. 1987. *Cytogenetics: Basic and Applied Aspects.* New York: Springer-Verlag.

Ohno, S. 1970. *Evolution by Gene Duplication.* New York: Springer-Verlag.

Oostra, B. A., and Verkerk, A. J. 1992. The fragile X syndrome: Isolation of the *FMR-1* gene and characterization of the fragile X mutation. *Chromosoma* 101: 381–87.

Page, S. L., and Shaffer, L. G. 1997. Nonhomologous Robertsonian translocations form predominantly during female meiosis. *Nat. Gen.* 15: 231–32.

Patterson, D. 1987. The causes of Down syndrome. *Sci. Am.* (Aug.) 257: 52–61.

Schimke, R. T., ed. 1982. *Gene Amplification.* Cold Spring Harbor: Cold Spring Harbor Laboratory Press.

Shepard, J. F. 1982. The regeneration of potato plants from protoplasts. *Sci. Am.* (May) 246: 154–166.

Shepard, J. et al. 1983. Genetic transfer in plants through interspecific protoplast fusion. *Science* 21: 683–88.

Simmonds, N. W., ed. 1976. *Evolution of Crop Plants.* London: Longman.

Stebbins, G. L. 1966. Chromosome variation and evolution. *Science* 152: 1463–69.

Strickberger, M. W. 2000. *Evolution.* 3rd ed. Boston: Jones and Bartlett.

Sutherland, G. 1985. The enigma of the fragile X chromosome. *Trends Genet.* 1: 108–11.

Taylor, A. I. 1968. Autosomal trisomy syndromes: A detailed study of 27 cases of Edwards syndrome and 27 cases of Patau syndrome. *J. Med. Genet.* 5: 227–52.

Therman, E. and Susman, B. 1995. *Human Chromosomes.* 3rd ed. New York: Springer-Verlag.

Tjio, J. H., and Levan, A. 1956. The chromosome number of man. *Hereditas* 42: 1–6.

Wilkins, L. E., Brown, J. X., and Wolf, B. 1980. Psychomotor development in 65 home-reared children with cri-du-chat syndrome. *J. Pediatr.* 97: 401–5.

Yunis, J. J., ed. 1977. *New Chromosomal Syndromes.* Orlando: Academic Press.

CHAPTER 8

Chiasmata between synapsed homologs during the first meiotic prophase. *(B. John/Cabisco/Visuals Unlimited)*

Linkage and Chromosome Mapping in Eukaryotes

CHAPTER CONCEPTS

Walter Sutton, along with Theodor Boveri, was instrumental in uniting the fields of cytology and genetics. As early as 1903, Sutton pointed out the likelihood that there must be many more "unit factors" than chromosomes in most organisms. Soon thereafter, genetics investigations revealed that certain genes segregate as if they were somehow joined or linked together. Further investigations showed that such genes are part of the same chromosome and are indeed transmitted as a single unit. We now know that most chromosomes contain a very large number of genes. Those that are part of the same chromosome are said to be *linked* and to demonstrate *linkage* in genetic crosses.

Because the chromosome, not the gene, is the unit of transmission during meiosis, linked genes are not free to undergo independent assortment. Instead, the alleles at all loci of one chromosome should, in theory, be transmitted as a unit during gamete formation. However, in many instances this does not occur. During the first meiotic prophase, when homologs are paired or synapsed, a reciprocal exchange of chromosome segments can take place. This **crossing over** results in the reshuffling, or recombination, of the alleles between homologs and always occurs during the tetrad stage.

The degree of crossing over between any two loci on a single chromosome is proportional to the distance between them. Therefore, depending on which loci are being studied, the percentage of recombinant gametes varies. This correlation allows us to construct chromosome maps, which give the relative locations of genes on chromosomes.

In this chapter, we will discuss linkage, crossing over, and chromosome mapping in more detail. We will conclude by entertaining the rather intriguing question of why Mendel, who studied seven genes in an organism with seven chromosomes, did not encounter linkage. Or did he?

HOW DO WE KNOW WHAT WE KNOW?

IN THIS CHAPTER, WE WILL FOCUS ON GENES located on the same eukaryotic chromosome, a concept called linkage. We are particularly interested in how such linked genes are transmitted to offspring and how the results of such transmission enable us to map them. As you study this topic, you should try to answer several fundamental questions:

1. How do we know that specific genes are linked on a single chromosome, in contrast to being located on separate chromosomes?
2. How was it determined that linked genes on homologous chromosomes are recombined during meiosis?
3. How do we know the sequence of, and intergenic distance between, genes found on the same chromosome?
4. In organisms where designed matings do not occur (such as humans), how do we know that genes are linked, and how do we map them?
5. How do we know that crossing over results from a physical exchange between chromatids? ■

8.1 Genes Linked on the Same Chromosome Segregate Together

A simplified overview of the major theme of this chapter is given in Figure 8–1, which contrasts the meiotic consequences of (a) independent assortment, (b) linkage *without* crossing over, and (c) linkage *with* crossing over. In Figure 8–1(a), we see the results of independent assortment of two pairs of chromosomes, each containing one heterozygous gene pair. No linkage is exhibited. When a large number of meiotic events are observed, four genetically different gametes are formed in equal proportions, and each contains a different combination of alleles of the two genes.

Now let's compare these results with what occurs if the same genes are linked on the same chromosome. If no crossing over occurs between the two genes (Figure 8–1b), only two genetically different gametes are formed. Each gamete receives the alleles present on one homolog or the other, which is transmitted intact as the result of segregation. This case demonstrates **complete linkage**, which produces only **parental** or **noncrossover gametes**. The two parental gametes are formed in equal proportions. Though complete linkage between two genes seldom occurs, it is useful to consider the theoretical consequences of this concept.

Figure 8–1(c) shows the results of crossing over between two linked genes. As you can see, this crossover involves only two nonsister chromatids of the four chromatids present in the tetrad. This exchange generates two new allele combinations, called **recombinant** or **crossover gametes**. The two chromatids not involved in the exchange result in noncrossover gametes, like those in Figure 8–1(b). The frequency with which crossing over occurs between any two linked genes is generally proportional to the distance separating the respective loci along the chromosome. In theory, two randomly selected genes can be so close to each other that crossover events are too infrequent to easily detect. As shown in Figure 8–1(b), this complete linkage produces only parental gametes. On the other hand, if a small but distinct distance separates two genes, few recombinant and many parental gametes will be formed. As the distance between the two genes increases, the proportion of recombinant gametes increases and that of the parental gametes decreases.

As we will discuss later in this chapter, when the loci of two linked genes are far apart, the number of recombinant gametes approaches, but does not exceed, 50%. If 50% recombinants occurs, the result is a 1:1:1:1 ratio of the four types (two parental and two recombinant gametes). In this case, transmission of two linked genes is indistinguishable from that of two unlinked, independently assorting genes. That is, the proportion of the four possible genotypes is identical, as shown in Figures 8–1(a) and (c).

The Linkage Ratio

If complete linkage exists between two genes because of their close proximity, and organisms heterozygous at both loci are mated, a unique F_2 phenotypic ratio results, which

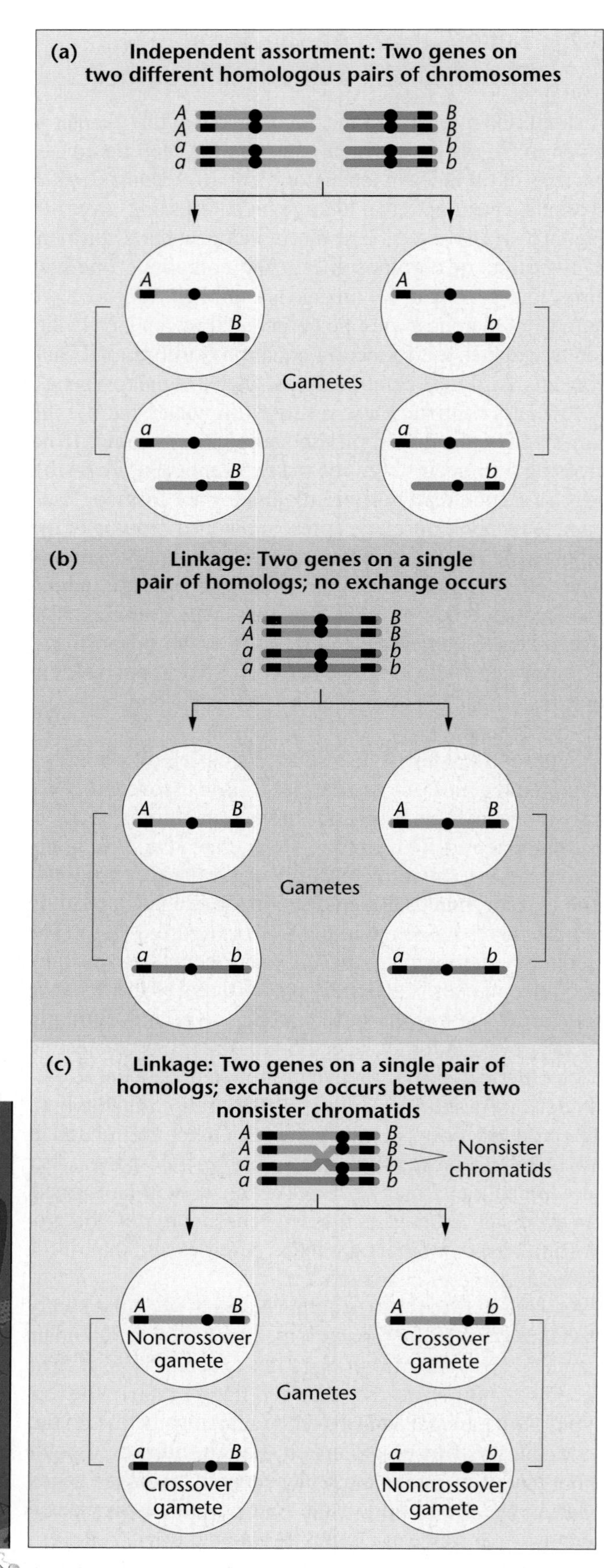

FIGURE 8–1 Results of gamete formation where two heterozygous genes are (a) on two different pairs of chromosomes; (b) on the same pair of homologs, but where no exchange occurs between them; and (c) on the same pair of homologs, where an exchange occurs between two nonsister chromatids.

we designate the **linkage ratio**. To illustrate this ratio, let's consider a cross involving the closely linked, recessive, mutant genes *brown* eye (*bw*) and *heavy* wing vein (*hv*) in *Drosophila melanogaster* (Figure 8–2). The normal, wild-type alleles bw^+ and hv^+ are both dominant and result in red eyes and thin wing veins, respectively.

In this cross, flies with mutant brown eyes and normal thin wing veins are mated to flies with normal red eyes and mutant heavy wing veins. In more concise terms, brown-eyed flies are crossed with heavy-veined flies. If we extend the system of genetic symbols established in Chapter 4, linked genes are represented by placing their allele designations above and below a single or double horizontal line. Those above the line are located at loci on one homolog, and those below are located at the homologous loci on the other homolog. Thus, we represent the P_1 generation as follows:

$$P_1: \quad \frac{bw\ hv^+}{bw\ hv^+} \times \frac{bw^+\ hv}{bw^+\ hv}$$

brown, thin   red, heavy

These genes are located on an autosome, so no distinction between males and females is necessary.

In the F_1 generation, each fly receives one chromosome of each pair from each parent. All flies are heterozygous for both gene pairs and exhibit the dominant traits of red eyes and thin wing veins:

$$F_1: \quad \frac{bw\ \ hv^+}{bw^+\ hv}$$

red, thin

As shown in Figure 8–2(a), when the F_1 generation is interbred, each F_1 individual forms only parental gametes because of complete linkage. After fertilization, the F_2 generation is produced in a 1:2:1 phenotypic and genotypic ratio. One-fourth of this generation shows brown eyes and thin wing veins; one-half shows both wild-type traits, namely, red eyes and thin wing veins; and one-fourth shows red eyes and heavy wing veins. In more concise terms, the ratio is 1 brown : 2 wild : 1 heavy. Such a ratio is characteristic of complete linkage. Complete linkage is usually observed only when genes are very close together and the number of progeny is relatively small.

Figure 8–2(b) also gives the results of a test cross with the F_1 flies. Such a cross produces a 1:1 ratio of brown, thin and red, heavy flies. Had the genes controlling these traits been incompletely linked or located on separate autosomes, the test cross would have produced four phenotypes rather than two.

When large numbers of mutant genes present in any given species are investigated, genes located on the same chromosome show evidence of linkage to one another. As a result, **linkage groups** can be established, one for each chromosome. In theory, the number of linkage groups should correspond to the haploid number of chromosomes. In diploid organisms in which large numbers of

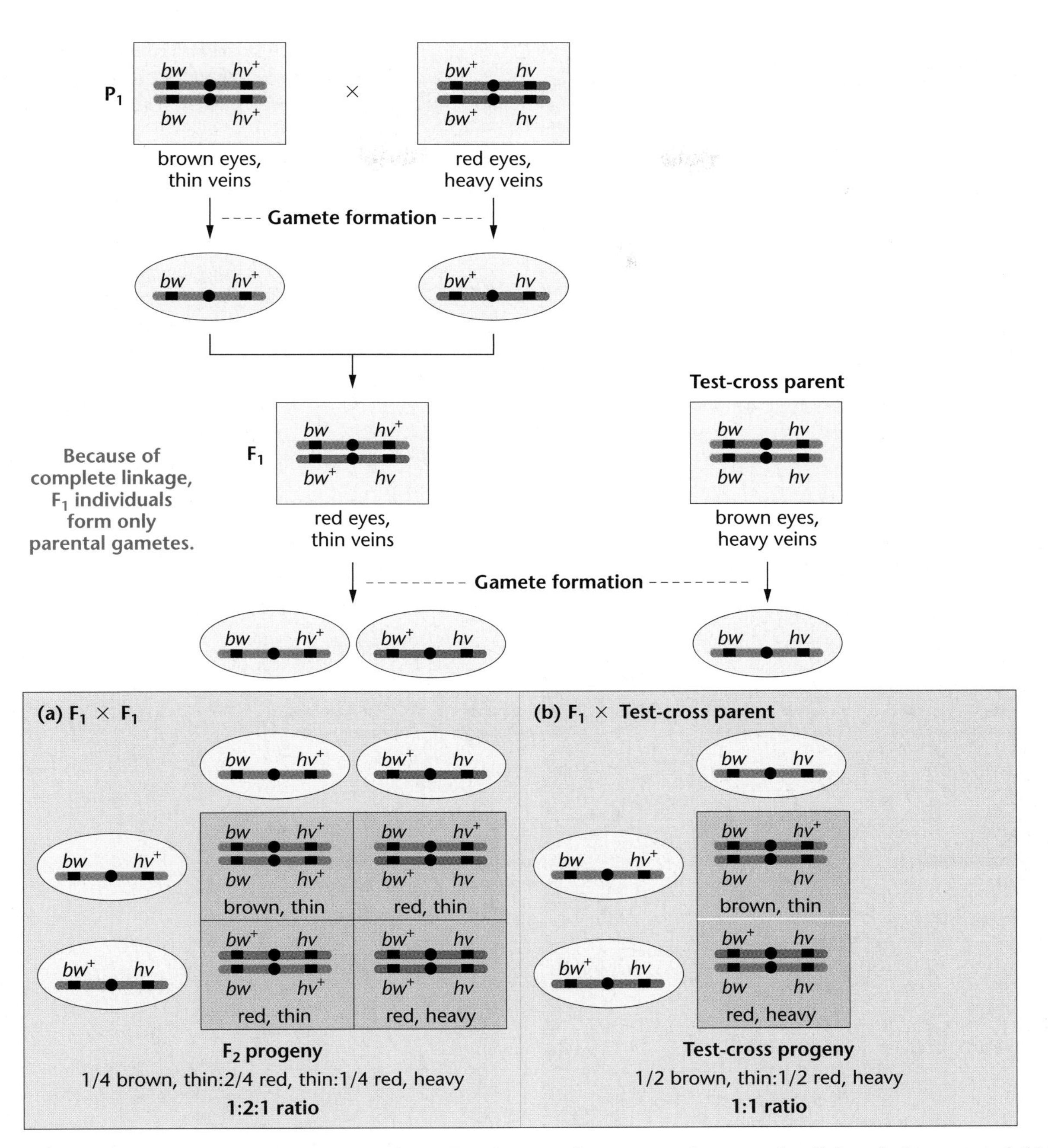

FIGURE 8–2 Results of a cross involving two genes located on the same chromosome where complete linkage is demonstrated. (a) The F_2 results of the cross. (b) The results of a test cross involving the F_1 progeny.

mutant genes are available for genetic study, this correlation has been confirmed.

8.2 Crossing Over Serves as the Basis of Determining the Distance Between Genes During Mapping

It is highly improbable that two randomly selected genes linked on the same chromosome will be so close to one another along the chromosome that they demonstrate complete linkage. Instead, crosses involving two such genes almost always produce a percentage of offspring resulting from recombinant gametes. This percentage is variable and depends on the distance between the two genes along the chromosome. This phenomenon was first explained in 1911 by two *Drosophila* geneticists, Thomas H. Morgan and his undergraduate student, Alfred H. Sturtevant.

Morgan and Crossing Over

In his studies, Morgan investigated numerous *Drosophila* mutations located on the X chromosome. When he analyzed crosses involving only one trait, he deduced the mode of X-linked inheritance. However, when he crossed two X-linked genes, his results were puzzling. For example, as shown in cross A of Figure 8–3, he crossed female flies expressing the mutant *yellow* body (y) and *white* eyes (w) alleles with wild-type males (gray bodies and red eyes). The F_1 females were wild type, while the F_1 males

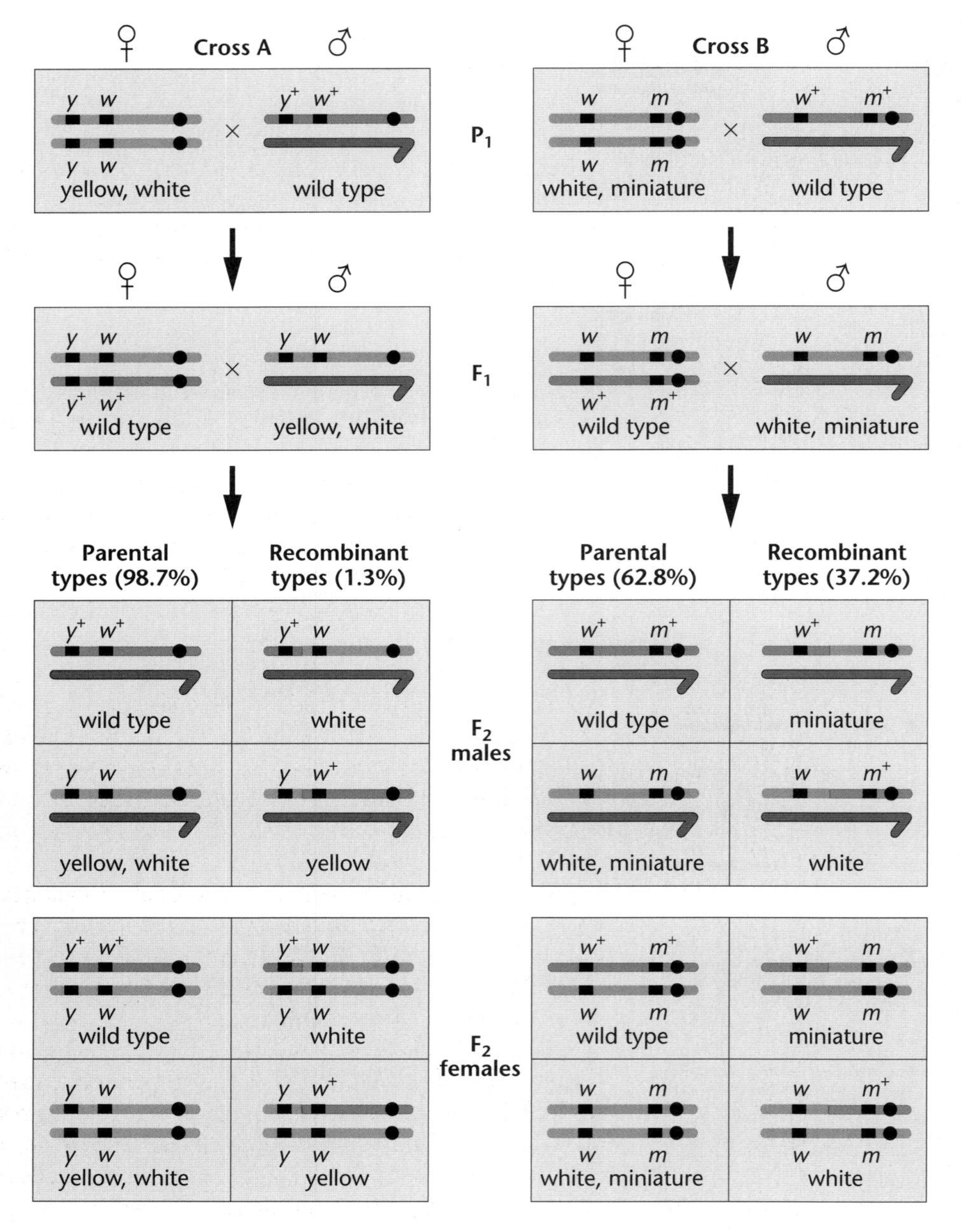

FIGURE 8–3 The F_1 and F_2 results of crosses involving the *yellow* body (y), *white* eye (w) mutations (Cross A), and the *white* eye, *miniature* wing (m) mutations (Cross B). In cross A, 1.3% of the F_2 flies (males and females) demonstrate recombinant phenotypes, which express either *white* or *yellow*. In cross B, 37.2% of the F_2 flies (males and females) demonstrate recombinant phenotypes, which express either *miniature* or *white*.

MEDIA TUTORIAL Linked genes and mapping

expressed both mutant traits. In the F_2, 98.7% of the offspring showed the parental phenotypes—either yellow-bodied, white-eyed flies or wild-type flies (gray-bodied, red-eyed). The remaining 1.3% of the flies were either yellow-bodied with red eyes or gray-bodied with white eyes. It was as if the genes had somehow separated from each other during gamete formation in the F_1 flies.

When Morgan crossed other X-linked genes, the results were even more puzzling (cross B of Figure 8–3). The same basic pattern was observed, but the proportion of F_2 phenotypes differed. For example, in a cross involving the mutant *white* eye, *miniature* wing (m) alleles, only 62.8% of the F_2 showed the parental phenotypes, while 37.2% of the offspring appeared as if the mutant genes had separated during gamete formation.

Morgan was faced with two questions: (1) What was the source of gene separation, and (2) why did the frequency of the apparent separation vary depending on the genes being studied? The answer he proposed for the first question was based on his knowledge of earlier cytological observations made by F. A. Janssens and others. Janssens had observed that synapsed homologous chromosomes in meiosis wrapped around each other, creating **chiasmata** (sing., *chiasma*) where points of overlap are evident—Morgan proposed that chiasmata could represent points of genetic exchange.

In the crosses shown in Figure 8–3, Morgan postulated that if an exchange occurs between the mutant genes on the two X chromosomes of the F_1 females, it leads to the observed results. He suggested that such exchanges led to 1.3% recombinant gametes in the *yellow–white* cross and 37.2% in the *white–miniature* cross. On the basis of this and other experiments, Morgan concluded that linked genes exist in a linear order along the chromosome and that a variable amount of exchange occurs between any two genes.

In answer to the second question, Morgan proposed that two genes located relatively close to each other along a chromosome are less likely to have a chiasma form between them than if the two genes are farther apart on the chromosome. Therefore, the closer two genes are, the less likely a genetic exchange will occur between them. Morgan was the first to propose the term "crossing over" to describe the physical exchange leading to recombination.

Sturtevant and Mapping

Morgan's student, Alfred H. Sturtevant, was the first to realize that his mentor's proposal could be used to map the sequence of linked genes. According to Sturtevant,

> "In a conversation with Morgan ... I suddenly realized that the variations in strength of linkage, already attributed by Morgan to differences in the spatial separation of the genes, offered the possibility of determining sequences in the linear dimension of a chromosome. I went home and spent most of the night (to the neglect of my undergraduate homework) in producing the first chromosomal map ..."

Sturtevant compiled data on recombination between the genes represented by the *yellow, white,* and *miniature* mutants initially studied by Morgan, and observed the following frequencies of crossing over between each pair of these three genes:

(1)	*yellow, white*	0.5%
(2)	*white, miniature*	34.5%
(3)	*yellow, miniature*	35.4%

Because the sum of (1) and (2) approximately equals (3), Sturtevant suggested that the recombination frequencies between linked genes are additive. On this basis, he predicted that the order of the genes on the X chromosome is *yellow–white–miniature*. In arriving at this conclusion, he reasoned as follows: The *yellow* and *white* genes are apparently close to each other because the recombination frequency is low. However, both of these genes are quite far apart from *miniature* gene because the *white, miniature* and *yellow, miniature* combinations show large recombination frequencies. Because *miniature* shows more recombination with *yellow* than with *white* (35.4% versus 34.5%), it follows that *white* is located between the other two genes, not outside of them.

Sturtevant knew from Morgan's work that the frequency of exchange could be used as an estimate of the distance between two genes or loci along the chromosome. He constructed a **chromosome map** of the three genes on the X chromosome, setting 1 map unit (mu) equal to 1% recombination between two genes.* In the preceding example, the distance between *yellow* and *white* is thus 0.5 mu, and the distance between *yellow* and *miniature* is 35.4 mu. It follows that the distance between *white* and *miniature* should be $35.4 - 0.5 = 34.9$ mu. This estimate is close to the actual frequency of recombination between *white* and *miniature* (34.5%). The map for these three genes is shown in Figure 8–4.

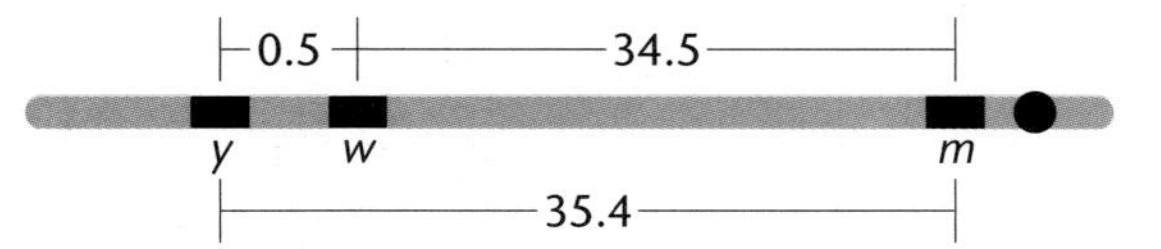

FIGURE 8–4 A map of the *yellow* body (*y*), *white* eye (*w*), and *miniature* wing (*m*) genes on the X chromosome of *Drosophila melanogaster*. Each number represents the percentage of recombinant offspring produced in one of three crosses, each involving two different genes.

In addition to these three genes, Sturtevant considered two other genes on the X chromosome and produced a more extensive map that included all five genes. He and a colleague, Calvin Bridges, soon began a search for autosomal linkage in *Drosophila*. By 1923, they had clearly shown that linkage and crossing over are not restricted to X-linked genes but can also be demonstrated with autosomes. During this work, they made another interesting observation. Crossing over in *Drosophila* was shown to occur only in females. The fact that no crossing over occurs in males made genetic mapping much less complex to analyze in *Drosophila*. However, crossing over does occur in both sexes in most other organisms.

Although many refinements in chromosome mapping have been developed since Sturtevant's initial work, his basic principles are considered to be correct. These principles are used to produce detailed chromosome maps of organisms for which large numbers of linked mutant genes are known. Sturtevant's findings are also historically significant to the broader field of genetics. In 1910, the **chromosomal theory of inheritance** was still widely disputed—even Morgan was skeptical of this theory before he conducted his experiments. Research has now firmly established that chromosomes contain genes in a linear order and that these genes are the equivalent of Mendel's unit factors.

Single Crossovers

Why should the relative distance between two loci influence the amount of recombination and crossing over observed between them? During meiosis, a limited number of crossover events occurs in each tetrad. These recombinant events occur randomly along the length of the tetrad. Therefore, the closer two loci reside along the axis of the chromosome, the less likely any single crossover event will occur between them. The same reasoning suggests that the farther apart two linked loci, the more likely a random crossover event will occur between them.

In Figure 8–5(a), a single crossover occurs between two nonsister chromatids, but not between the two loci; therefore, the crossover is not detected because no recombinant gametes are produced. In Figure 8–5(b), where two loci are quite far apart, crossover does occur between them, yielding recombinant gametes.

*In honor of Morgan's work, 1 map unit is now referred to as a centimorgan (cM).

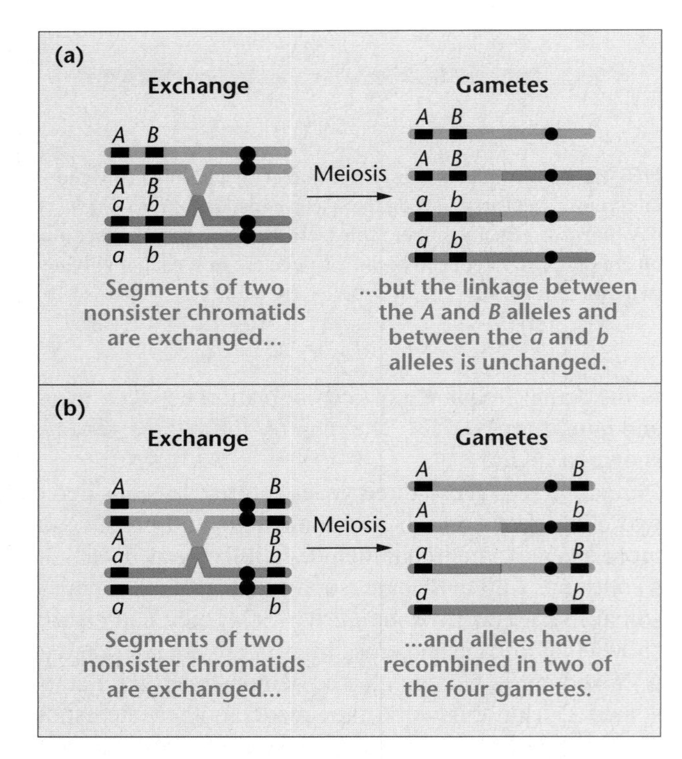

FIGURE 8–5 Two examples of a single crossover between two nonsister chromatids and the gametes subsequently produced. In (a), the exchange does not alter the linkage arrangement between the alleles of the two genes, only parental gametes form, and the exchange goes undetected. In (b), the exchange separates the alleles and results in recombinant gametes, which are detectable.

When a single crossover occurs between two nonsister chromatids, the other two chromatids of the tetrad are not involved in this exchange and enter the gamete unchanged. Even if a single crossover occurs 100% of the time between two linked genes, recombination is subsequently observed in only 50% of the potential gametes formed. This concept is diagrammed in Figure 8–6. Theoretically, if we consider only single exchanges and observe 20% recombinant gametes, crossing over actually occurred in 40% of the tetrads. Under these conditions, the general rule is that the percentage of tetrads involved in an exchange between two genes is twice the percentage of recombinant gametes produced. Therefore, the theoretical limit of recombination due to crossing over is 50%.

When 2 linked genes are more than 50 mu apart, a crossover can theoretically be expected to occur between them in 100% of the tetrads. If this prediction were achieved, each tetrad would yield equal proportions of the four gametes shown in Figure 8–6, just as if the genes were on different chromosomes and assorting independently. However, this theoretical limit is seldom achieved.

8.3 Determining the Gene Sequence During Mapping Relies on the Analysis of Multiple Crossovers

The study of single crossovers between two linked genes provides the basis of determining the *distance* between them. However, when many linked genes are studied, their *sequence* along the chromosome is more difficult to determine. Fortunately, the discovery that multiple exchanges occur between the chromatids of a tetrad has facilitated the process of producing more extensive chromosome maps. As we shall see next, when three or more linked genes are investigated simultaneously, it is possible to determine first the sequence of and then the distances between genes.

Multiple Crossovers

It is possible that in a single tetrad, two, three, or more exchanges will occur between nonsister chromatids as a result of several crossover events. Double exchanges of genetic material result from **double crossovers (DCOs)**, as shown in Figure 8–7. For a double exchange to be studied, three gene pairs must be investigated, each heterozygous for two alleles. Before we determine the frequency of recombination among all three loci, let's review some simple probability calculations.

As we have seen, the probability of a single exchange occurring between the *A* and *B* or the *B* and *C* genes relates directly to the distance between the respective loci. The closer *A* is to *B* and *B* is to *C*, the less likely a single exchange will occur between either of the two sets of loci. In the case of a double crossover, two separate and independent events or exchanges must occur simultaneously. The mathematical probability of two independent events occurring simultaneously is equal to the product of the individual probabilities (the product law).

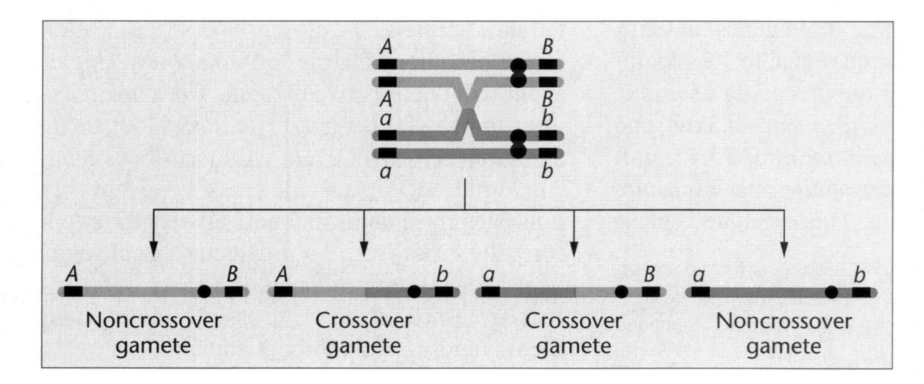

FIGURE 8–6 The consequences of a single exchange between two nonsister chromatids occurring in the tetrad stage. Two noncrossover (parental) and two crossover (recombinant) gametes are produced.

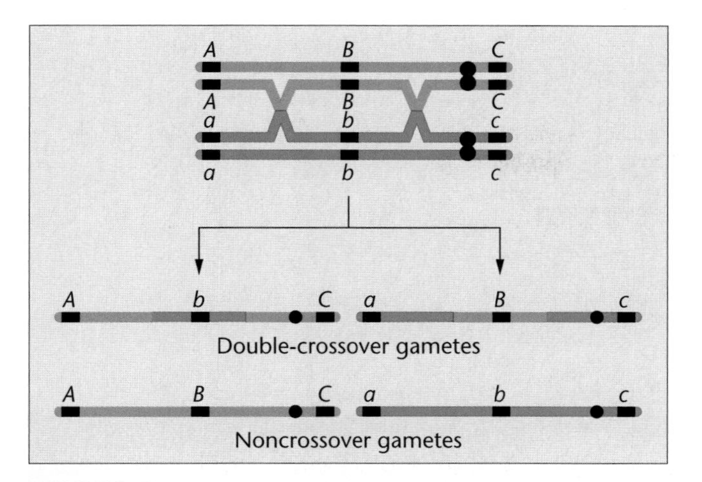

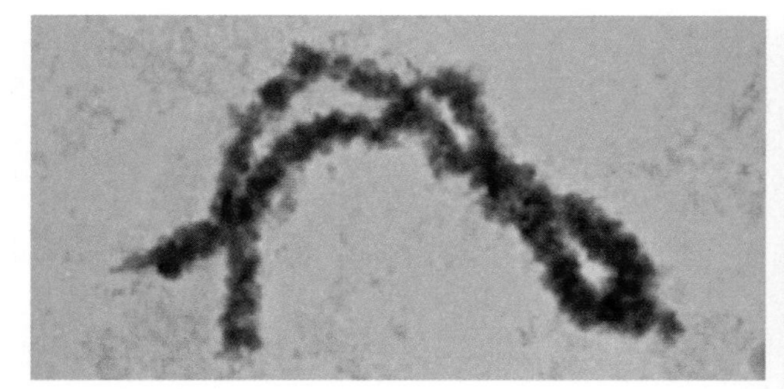

MEDIA TUTORIAL Three-point mapping

FIGURE 8–7 Consequences of a double exchange between two nonsister chromatids. Because the exchanges involve only two chromatids, two noncrossover gametes and two double-crossover gametes are produced. The photograph shows several chiasmata found in a tetrad isolated during the first meiotic prophase. *(© 2002 Clare A. Hasenkampf/Biological Photo Service)*

Suppose that crossover gametes resulting from single exchanges are recovered 20% of the time $(p = 0.20)$ between *A* and *B*, and 30% of the time $(p = 0.30)$ between *B* and *C*. The probability of recovering a double-crossover gamete arising from two exchanges (between *A* and *B*, and between *B* and *C*) is predicted to be $(0.20)(0.30) = 0.06$, or 6%. It is apparent from this calculation that the frequency of double-crossover gametes is always expected to be much lower than that of either single-crossover class of gametes.

If three genes are relatively close together along one chromosome, the expected frequency of double-crossover gametes is extremely low. For example, suppose the *A–B* distance in Figure 8–7 is 3 mu and the *B–C* distance is 2 mu. The expected double-crossover frequency is $(0.03)(0.02) = 0.0006$, or 0.06%. This translates to only 6 events in 10,000. Thus, in a mapping experiment where closely linked genes are involved, very large numbers of offspring are required to detect double-crossover events. In this example, it is unlikely that a double crossover will be observed even if 1000 offspring are examined. Thus, it is evident that if four or five genes are being mapped, even fewer triple and quadruple crossovers can be expected to occur.

Three-Point Mapping in *Drosophila*

The information in the preceding section enables us to map three or more linked genes in a single cross. To illustrate the mapping process in its entirety, we examine two situations involving three linked genes in two quite different organisms.

To execute a successful mapping cross, three criteria must be met:

1. The genotype of the organism producing the crossover gametes must be heterozygous at all loci under consideration.

2. The cross must be constructed so that genotypes of all gametes can be determined accurately by observing the phenotypes of the resulting offspring. This is necessary because the gametes and their genotypes can never be observed directly. To overcome this problem, each phenotypic class must reflect the genotype of the gametes of the parents producing it.

3. A sufficient number of offspring must be produced in the mapping experiment to recover a representative sample of all crossover classes.

These criteria are met in the three-point mapping cross from *D. melanogaster* shown in Figure 8–8. In this cross, three X-linked recessive mutant genes—*yellow* body color (*y*), *white* eye color (*w*), and *echinus* eye shape (*ec*)—are considered. To diagram the cross, we must assume some theoretical sequence, even though we do not yet know if it is correct. In Figure 8–8, we initially assume the sequence of the three genes to be *y–w–ec*. If this is incorrect, our analysis will demonstrate this and reveal the correct sequence.

In the P_1 generation, males hemizygous for all three wild-type alleles are crossed to females that are homozygous for all three recessive mutant alleles. Therefore, the P_1 males are wild type with respect to body color, eye color, and eye shape. They are said to have a *wild-type phenotype*. The females, on the other hand, exhibit the three mutant traits—yellow body color, white eyes, and echinus eye shape.

This cross produces an F_1 generation consisting of females that are heterozygous at all three loci and males that, because of the Y chromosome, are hemizygous for the three mutant alleles. Phenotypically, all F_1 females are wild type, while all F_1 males are yellow, white, and echinus. The genotype of the F_1 females fulfills the first criterion for mapping the three linked genes; that is, it is heterozygous at the three loci and can serve as the source of recombinant gametes generated by crossing over. Note that because of the genotypes of the P_1 parents, all three mutant alleles are on one homolog and all three wild-type alleles are on the other homolog. *Other arrangements are possible.* For example, the heterozygous F_1 female might have the *y* and *ec* mutant alleles on one homolog and the *w* allele on the other. This would occur if, in the P_1 cross, one parent was yellow, echinus and the other parent was white.

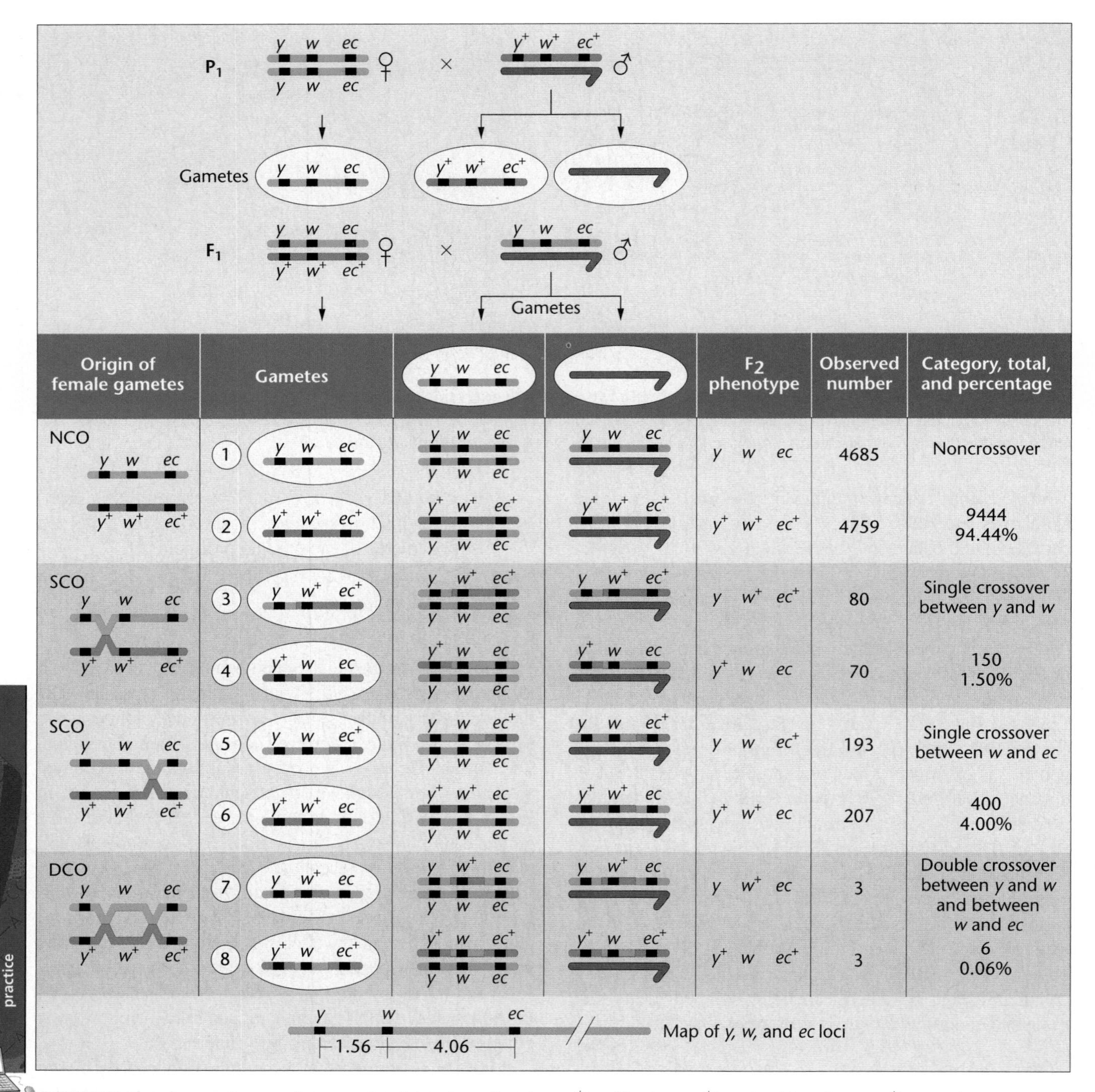

MEDIA TUTORIAL Three-point mapping practice

FIGURE 8–8 A three-point mapping cross involving the *yellow* (y or y^+), *white* (w or w^+), and *echinus* (ec or ec^+) genes in *D. melanogaster*. NCO, SCO, and DCO refer to noncrossover, single-crossover, and double-crossover groups, respectively. Centromeres are not included on the chromosomes, and only two nonsister chromatids are shown initially.

In our cross, the second criterion is met by virtue of the gametes formed by the F_1 males. Every gamete contains either an X chromosome bearing the three mutant alleles or a Y chromosome, which is genetically inert for the three loci being considered. Whichever type participates in fertilization, the genotype of the gamete produced by the F_1 female will be expressed phenotypically in the F_2 male and female offspring derived from it. Thus, all F_1 noncrossover and crossover gametes can be detected by observing the F_2 phenotypes.

With these two criteria met, we can now construct a chromosome map from the crosses shown in Figure 8–8. First, we determine which F_2 phenotypes correspond to the various noncrossover and crossover categories. To determine the noncrossover F_2 phenotypes, we must identify those derived from the parental gametes formed by the F_1 female. Each such gamete contains an X chromosome *unaffected by crossing over*. As a result of segregation, approximately equal proportions of the two types of gametes and, subsequently, the F_2 phenotypes, are produced. Because they derive from a heterozygote, the genotypes of the two parental gametes and the phenotypes of the two F_2 phenotypes complement one another. For example, if one is wild type, the other is completely mutant. This is the case in the cross being considered. In other situations, if one chromosome shows one mutant allele, the second chromosome shows the other two mutant alleles, and so on. They are therefore called **reciprocal classes** of gametes and phenotypes.

The two noncrossover phenotypes are most easily recognized because *they exist in the greatest proportion.* Figure 8–8 shows that gametes 1 and 2 are present in the greatest numbers. Therefore, flies that express yellow, white, and echinus phenotypes and flies that are normal (or wild type) for all three characters constitute the noncrossover category and represent 94.44% of the F_2 offspring.

The second category that can be easily detected is represented by the double-crossover phenotypes. Because of their low probability of occurrence, *they must be present in the least numbers*. Remember that this group represents two independent but simultaneous single-crossover events. Two reciprocal phenotypes can be identified: gamete 7, which shows the mutant traits yellow, echinus but normal eye color; and gamete 8, which shows the mutant trait white but normal body color and eye shape. Together these double-crossover phenotypes constitute only 0.06% of the F_2 offspring.

The remaining four phenotypic classes represent two categories resulting from single crossovers. Gametes 3 and 4, reciprocal phenotypes produced by single-crossover events occurring between the yellow and white loci, are equal to 1.50% of the F_2 offspring; and gametes 5 and 6, constituting 4.00% of the F_2 offspring, represent the reciprocal phenotypes resulting from single-crossover events occurring between the *white* and *echinus* loci.

The map distances separating the three loci can now be calculated. The distance between *y* and *w* or between *w* and *ec* is equal to the percentage of all detectable exchanges occurring between them. For any two genes under consideration, this includes all appropriate single crossovers as well as all double crossovers. The latter are included because they represent two simultaneous single crossovers. For the *y* and *w* genes, this includes gametes 3, 4, 7, and 8, totaling 1.50% + 0.06%, or 1.56 mu. Similarly, the distance between *w* and *ec* is equal to the percentage of offspring resulting from an exchange between these two loci: gametes 5, 6, 7, and 8, totaling 4.00% + 0.06%, or 4.06 mu. The map of these three loci on the X chromosome is shown at the bottom of Figure 8–8.

Determining the Gene Sequence

In the preceding example, the sequence (or order) of the three genes along the chromosome was assumed to be *y–w–ec*. Our analysis shows this sequence to be consistent with the data. However, in most mapping experiments the gene sequence is not known, and this constitutes another variable in the analysis. In our example, had the gene sequence been unknown, it could have been determined using a straightforward method.

This method is based on the fact that there are only three possible arrangements, each containing one of the three genes between the other two:

(I) *w–y–ec* (*y* in the middle)
(II) *y–ec–w* (*ec* in the middle)
(III) *y–w–ec* (*w* in the middle)

Use the following steps during your analysis to determine the gene order:

1. Assuming any one of the three orders, first determine the *arrangement of alleles* along each homolog of the heterozygous parent giving rise to noncrossover and crossover gametes (the F_1 female in our example).
2. Determine whether a double-crossover event occurring within that arrangement will produce the *observed double-crossover phenotypes*. Remember that these phenotypes occur least frequently and are easily identified.
3. If this order does not produce the predicted phenotypes, try each of the other two orders. One really does work!

These steps are shown in Figure 8–9, using our *y–w–ec* cross. The three possible arrangements are labeled I, II, and III, as shown above.

1. Assuming that *y* is between *w* and *ec*, arrangement I of alleles along the homologs of the F_1 heterozygote is

$$\text{(I)}\ \frac{w \quad y \quad ec}{w^+ \quad y^+ \quad ec^+}$$

We know this because of the way in which the P_1 generation was crossed: The P_1 female contributes an X chromosome bearing the *w*, *y*, and *ec* alleles, while the P_1 male contributes an X chromosome bearing the w^+, y^+, and ec^+ alleles.

2. A double crossover within that arrangement yields the following gametes:

$$\underline{w \quad y^+ \quad ec} \quad \text{and} \quad \underline{w^+ \quad y \quad ec^+}$$

Following fertilization, if *y* is in the middle, the F_2 double-crossover phenotypes will correspond to these gametic genotypes, yielding offspring that express the white, echinus phenotype and offspring that express the yellow phenotype. Instead, however, determination of the actual double-crossover phenotypes reveals them to be yellow, echinus flies and white flies. Therefore, our assumed order is incorrect.

3. If we consider arrangement II with the ec/ec^+ alleles in the middle or arrangement III with the w/w^+ alleles in the middle:

$$\text{(II)}\ \frac{y \quad ec \quad w}{y^+ \quad ec^+ \quad w^+} \quad \text{or} \quad \text{(III)}\ \frac{y \quad w \quad ec}{y^+ \quad w^+ \quad ec^+}$$

we see that arrangement II again provides *predicted* double-crossover phenotypes that *do not* correspond

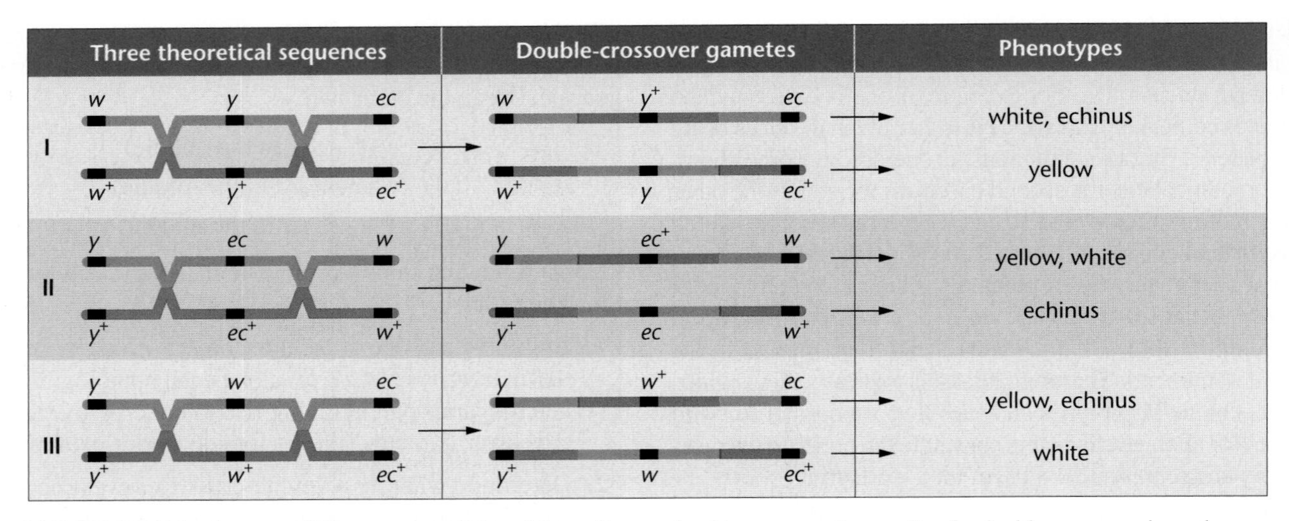

FIGURE 8–9 The three possible sequences of the *white*, *yellow*, and *echinus* genes, the results of a double crossover in each case, and the resulting phenotypes produced in a test cross. For simplicity, the two noncrossover chromatids of each tetrad are omitted.

to the *actual* (observed) double-crossover phenotypes. The predicted phenotypes are yellow, white flies and echinus flies in the F_2 generation. Therefore, this order is also incorrect. However, arrangement III produces the observed phenotypes—yellow, echinus flies and white flies. Therefore, this arrangement, with the *w* gene in the middle, is correct.

To summarize, this method is rather straightforward: First determine the arrangement of alleles on the homologs of the heterozygote yielding the crossover gametes by locating the reciprocal noncrossover phenotypes. Then, test each of three possible orders to determine which yields the observed double-crossover phenotypes—the one that does so represents the correct order.

A Mapping Problem in Maize

Having established the basic principles of chromosome mapping, we now consider a related problem in maize (corn), in which the gene sequence and interlocus distances are unknown.

This analysis differs from the preceding example in two ways. First, the previous mapping cross involved X-linked genes. Here, we consider autosomal genes. Second, in the discussion of this cross we have changed our use of symbols, as first suggested in Chapter 4. Instead of using the gene symbols and superscripts (e.g., bm^+, v^+, and pr^+), we simply use + to denote each wild-type allele. This system is easier to manipulate, but requires a better understanding of mapping procedures.

When we look at three autosomally linked genes in maize, the experimental cross must still meet the same three criteria we established for the X-linked genes in *Drosophila*: (1) One parent must be heterozygous for all traits under consideration; (2) the gametic genotypes produced by the heterozygote must be apparent from observing the phenotypes of the offspring; and (3) a sufficient sample size must be available for complete analysis.

In maize, the recessive mutant genes *brown* midrib (*bm*), *virescent* seedling (*v*), and *purple* aleurone (*pr*) are linked on chromosome 5. Assume that a female plant is known to be heterozygous for all three traits, but we do not know (1) the arrangement of the mutant alleles on the maternal and paternal homologs of this heterozygote, (2) the sequence of genes, or (3) the map distances between the genes. What genotype must the male plant have to allow successful mapping? To meet the second criterion, the male must be homozygous for all three recessive mutant alleles. Otherwise, offspring of this cross showing a given phenotype might represent more than one genotype, making accurate mapping impossible.

Figure 8–10 diagrams this cross. As shown, we know neither the arrangement of alleles nor the sequence of loci in the heterozygous female. Several possibilities are shown, but we have yet to determine which is correct. We don't know the sequence in the test-cross male parent either, and so we must designate it randomly. Note that we have initially placed *v* in the middle. This may or may not be correct.

The offspring are arranged in groups of two for each pair of reciprocal phenotypic classes. The two members of each reciprocal class are derived from no crossing over (NCO), one of two possible single-crossover events (SCO), or a double crossover (DCO).

To solve this problem, refer to Figures 8–10 and 8–11 as you consider the following questions.

1. *What is the correct heterozygous arrangement of alleles in the female parent?*
   Determine the two noncrossover classes, those that occur with the highest frequency. In this case, they are $\underline{+\ v\ bm}$ and $\underline{pr\ +\ +}$. Therefore, the alleles on the homologs of the female parent must be arranged as shown in Figure 8–11(a). These homologs segregate into gametes, unaffected by any recombination event. Any other arrangement of alleles will not yield the observed noncrossover classes. (Remember that $\underline{+\ v\ bm}$ is equivalent to $\underline{pr^+\ v\ bm}$, and that $\underline{pr\ +\ +}$ is equivalent to $\underline{pr\ v^+\ bm^+}$).

2. *What is the correct sequence of genes?*
   We know that the arrangement of alleles is

   $$\frac{+ \quad v \quad bm}{pr \quad + \quad +}$$

   But is the gene sequence correct? That is, will a double-crossover event yield the observed double-crossover phenotypes after fertilization? Observation shows that it will not (Figure 8–11b). Now try the other two orders (Figures 8–11c and d) *maintaining the same arrangement:*

   $$\frac{+ \quad bm \quad v}{pr \quad + \quad +} \quad \text{or} \quad \frac{v \quad + \quad bm}{+ \quad pr \quad +}$$

   Only the order on the right yields the observed double-crossover gametes (Figure 8–11d). Therefore, the *pr* gene is in the middle. From this point on, work the problem using this arrangement and sequence, with the *pr* locus in the middle.

3. *What is the distance between each pair of genes?*
   Having established the sequence of loci as *v–pr–bm*, we can determine the distance between *v* and *pr* and between *pr* and *bm*. Remember that the map distance between two genes is calculated on the basis of all detectable recombination events occurring between them. This includes both single- and double-crossover events.

Figure 8–11(e) shows that the phenotypes *v pr +* and *+ + bm* result from single crossovers between the *v* and *pr* loci, accounting for 14.5% of the offspring. By adding the percentage of double crossovers (7.8%) to the number obtained for single crossovers, the total distance between the *v* and *pr* loci is calculated to be 22.3 mu.

Figure 8–11(f) shows that the phenotypes *v + +* and *+ pr bm* result from single crossovers between the *pr* and *bm* loci, totaling 35.6%. Added to the double crossovers (7.8%), the distance between *pr* and *bm* is calculated to be 43.4 mu. The final map for all three genes in this example is shown in Figure 8–11(g).

FIGURE 8–10 (a) Some possible allele arrangements and gene sequences in a heterozygous female. The data from a three-point mapping cross depicted in (b), where the female is test-crossed, determine which combination of arrangement and sequence is correct (see Figure 8–11d).

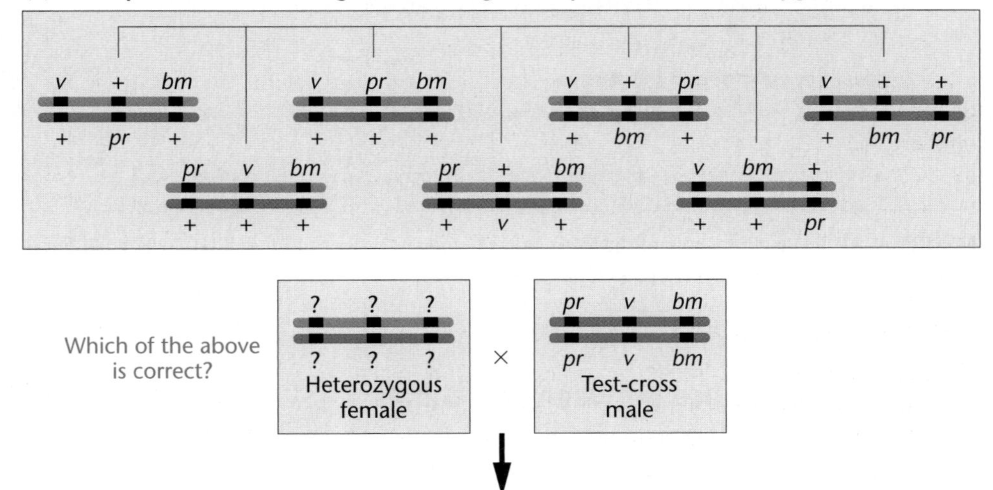

(b) Actual results of mapping cross *

Phenotypes of offspring	Number	Total and percentage	Exchange classification
+ v bm	230	467	Noncrossover
pr + +	237	42.1%	(NCO)
+ + bm	82	161	Single crossover
pr v +	79	14.5%	(SCO)
+ v +	200	395	Single crossover
pr + bm	195	35.6%	(SCO)
pr v bm	44	86	Double crossover
+ + +	42	7.8%	(DCO)

*The sequence *pr – v – bm* may or may not be correct

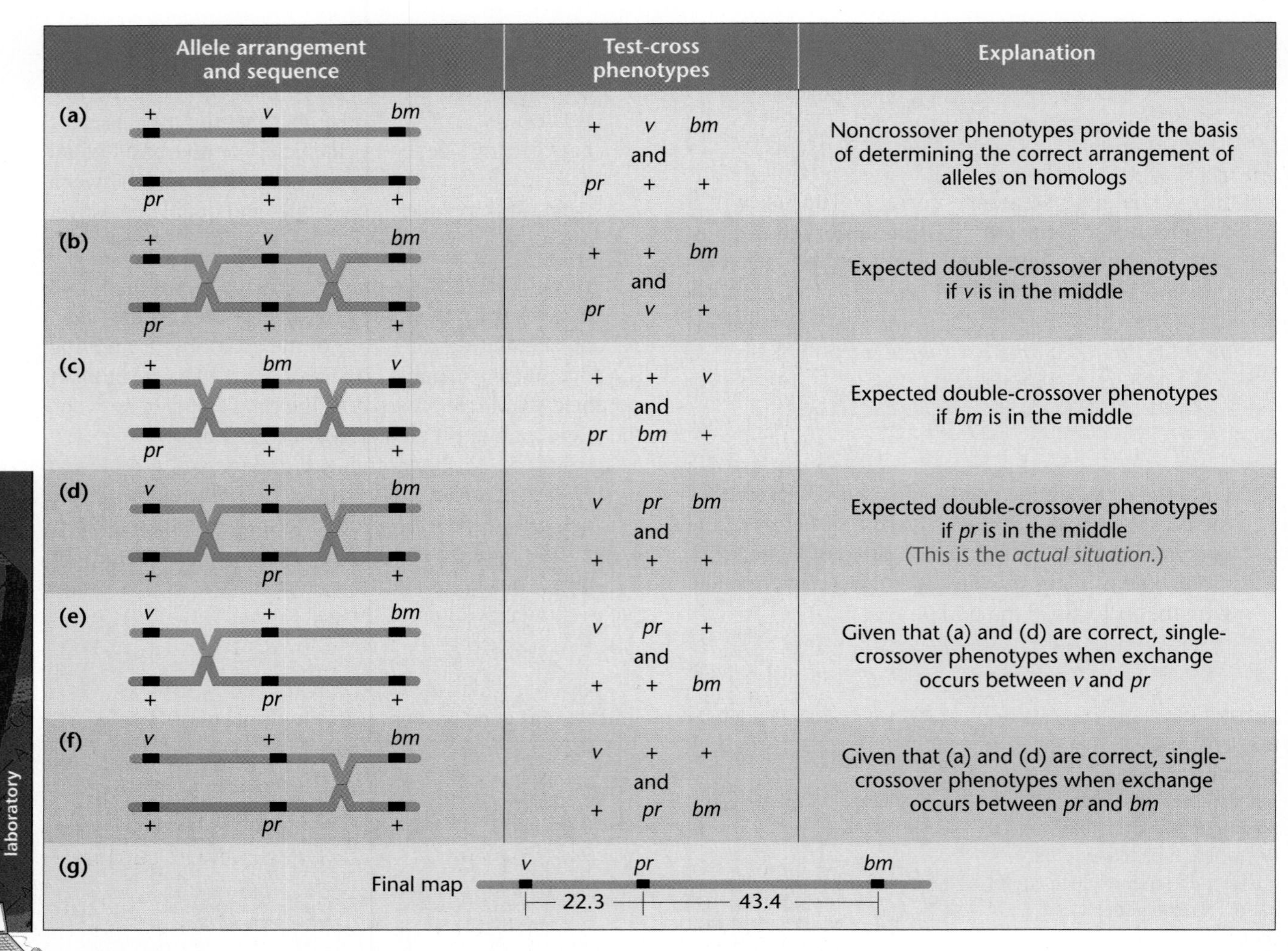

Allele arrangement and sequence
Test-cross phenotypes
Explanation
(a)
+ v bm
pr + +
+ v bm
and
pr + +
Noncrossover phenotypes provide the basis of determining the correct arrangement of alleles on homologs
(b)
+ v bm
pr + +
+ + bm
and
pr v +
Expected double-crossover phenotypes if v is in the middle
(c)
+ bm v
pr + +
+ + v
and
pr bm +
Expected double-crossover phenotypes if bm is in the middle
(d)
v + bm
+ pr +
v pr bm
and
+ + +
Expected double-crossover phenotypes if pr is in the middle
(This is the actual situation.)
(e)
v + bm
+ pr +
v pr +
and
+ + bm
Given that (a) and (d) are correct, single-crossover phenotypes when exchange occurs between v and pr
(f)
v + bm
+ pr +
v + +
and
+ pr bm
Given that (a) and (d) are correct, single-crossover phenotypes when exchange occurs between pr and bm
(g)
Final map
v pr bm
22.3
43.4

MEDIA TUTORIAL Virtual crossover laboratory

FIGURE 8–11 Producing a map of the three genes in the test cross of Figure 8–10, where neither the arrangement of alleles nor the sequence of genes in the heterozygous female parent is known.

8.4 As the Distance Between Two Genes Increases, Mapping Estimates Become More Inaccurate

So far, we have assumed that crossover frequencies are directly proportional to the distance between any two loci along the chromosome. However, it is not always possible to detect all crossover events. A case in point is a double exchange that occurs between the two loci in question. As shown in Figure 8–12(a), if a double exchange occurs, the original arrangement of alleles on each nonsister homolog is recovered. Therefore, even though crossing over has occurred, it is impossible to detect. This phenomenon is true for all even-numbered exchanges between two loci.

Furthermore, as a result of complications posed by multiple-strand exchanges, mapping determinations usually underestimate the actual distance between two genes. The farther apart two genes are, the greater the probability that undetected crossovers will occur. While the discrepancy is minimal for two genes relatively close together, the degree of inaccuracy increases as the distance increases, as shown in the graph of recombination frequency versus map distance in Figure 8–12(b). The most accurate maps are constructed from experiments whose genes are relatively close together.

Interference and the Coefficient of Coincidence

As shown in our maize example, we can predict the expected frequency of multiple exchanges, such as double crossovers, once the distance between genes is established. For example, in the maize cross, the distance between *v* and *pr* is 22.3 mu, and the distance between *pr* and *bm* is 43.4 mu. If the two single crossovers that make up a double crossover occur independently of one another, we can calculate the expected frequency of double crossovers (DCO_{exp}):

$$DCO_{exp} = (0.223) \times (0.434) = 0.097 = 9.7\%$$

Often in mapping experiments, the observed DCO frequency is less than the expected number of DCOs. In the maize cross, for example, only 7.8% DCOs are observed when 9.7% are expected. **Interference** (I) (when a crossover event in one region of the chromosome inhibits a second event in nearby regions), causes this reduction.

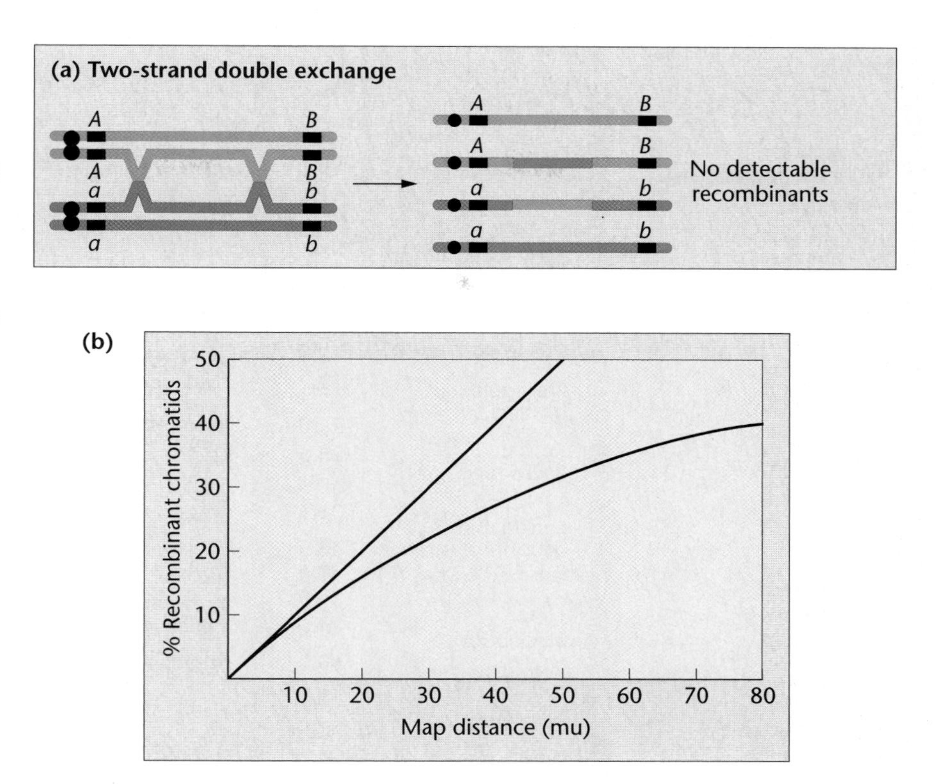

FIGURE 8–12 (a) A double crossover is undetected because no rearrangement of alleles occurs. (b) The theoretical and actual percentage of recombinant chromatids versus map distance. The straight line shows the theoretical relationship if a direct correlation between recombination and map distance exists. The curved line is the actual relationship derived from studies of *Drosophila*, *Neurospora*, and *Zea mays*.

To quantify the disparities that result from interference, we calculate the **coefficient of coincidence** (C):

$$C = \frac{\text{Observed DCO}}{\text{Expected DCO}}$$

In the maize cross, we have

$$C = \frac{0.078}{0.097} = 0.804$$

Once we have found C, we can quantify interference using the simple equation

$$I = 1 - C$$

In the maize cross, we have

$$I = 1.000 - 0.804 = 0.196$$

If interference is complete and no double crossovers occur, then $I = 1.0$. If fewer DCOs than expected occur, I is a positive number and positive interference has occurred. If more DCOs than expected occur, I is a negative number and negative interference has occurred. In the maize example, I is a positive number (0.196), indicating that 19.6% fewer double crossovers occurred than expected.

Positive interference is most often observed in eukaryotic systems. In general, the closer genes are to one another along the chromosome, the more positive interference occurs. In fact, interference in *Drosophila* is often complete within a distance of 10 mu, and no multiple crossovers are recovered. This observation suggests that physical constraints preventing the formation of closely aligned chiasmata cause interference. This interpretation is consistent with the finding that interference decreases as the genes in question are located farther apart. In the maize cross in Figures 8–10 and 8–11, the three genes are relatively far apart and 80% of the expected double crossovers are observed.

8.5 *Drosophila* Genes Have Been Extensively Mapped

In organisms such as *Drosophila*, maize, and the mouse, where large numbers of mutations have been discovered and where mapping crosses are possible, extensive chromosome maps have been constructed. Figure 8–13 shows partial maps for the four chromosomes (I, II, III, and IV) of *Drosophila*. Virtually every morphological feature of the fruit fly has been subjected to mutations. The locus of each mutant gene is first localized to one of the four chromosomes (or linkage groups) and then mapped in relation to all other genes present on that chromosome. Based on cytological evidence, the relative lengths of these genetic maps correlate roughly with the relative physical lengths of these chromosomes.

8.6 Lod Score Analysis and Somatic Cell Hybridization Were Historically Important in Creating Human Chromosome Maps

In humans, where neither designed matings nor large numbers of offspring are available, the earliest linkage studies were based on pedigree analysis. Attempts were made to establish whether a trait was X-linked or autosomal. For

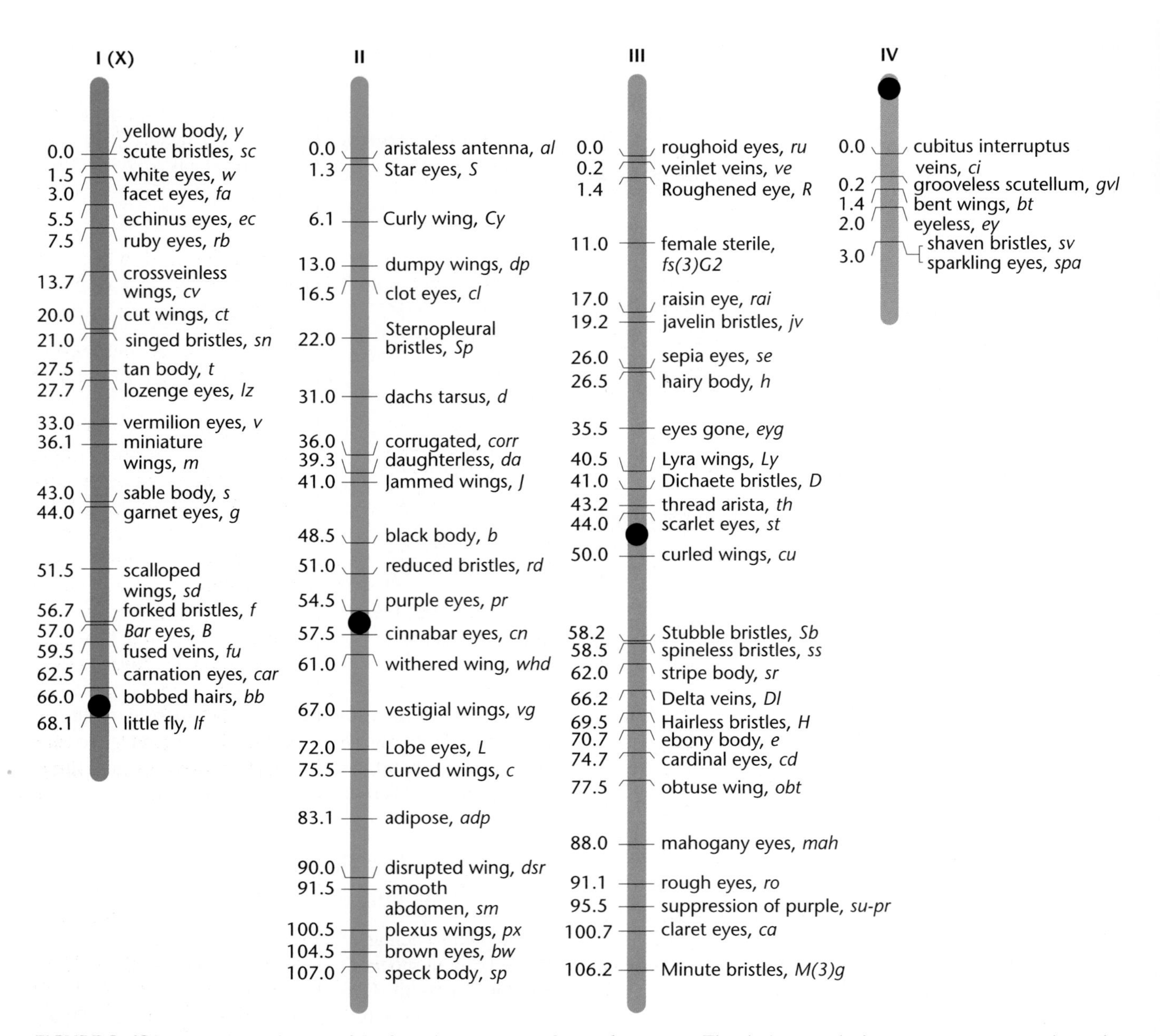

FIGURE 8–13 A partial genetic map of the four chromosomes of *D. melanogaster*. The circle on each chromosome represents the position of the centromere. Chromosome I is the X chromosome.

autosomal traits, geneticists tried to distinguish whether pairs of traits demonstrate linkage or independent assortment. In this way, it was hoped that a human gene map could be created.

The difficulty arises, however, when two genes of interest are separated on a chromosome such that recombinant gametes are formed, obscuring linkage in a pedigree. In such cases, the demonstration of linkage is enhanced by an approach that relies on probability, called the **lod score method**. First devised by J. B. S. Haldane and C. A. Smith in 1947, and refined by Newton Morton in 1955, the lod score (*l*og of the *od*ds favoring linkage) assesses the probability that a particular pedigree involving two traits reflects linkage or not. First, the probability is calculated that family data (pedigrees) concerning two or more traits conform to the transmission of traits without linkage. Then the probability that the identical family data following these same traits result from linkage with a specified recombination frequency is calculated. The ratio of these probabilities expresses the "odds" for and against linkage.

Accuracy using the lod score method is limited by the extent of the family data, but nevertheless represents an important advance in assigning human genes to specific chromosomes and constructing preliminary human chromosome maps. The initial results were discouraging because of the method's inherent limitations and the relatively high haploid number of human chromosomes (23), and by 1960, almost no linkage or mapping information had become available.

In the 1960s, a new technique, **somatic cell hybridization**, proved to be an immense aid in assigning human genes to their respective chromosomes. This technique, first discovered by Georges Barsky, relies on the fact that two cells in culture can be induced to fuse into a single hybrid cell. Barsky used two mouse-cell lines, but it soon became evident that cells from different organisms could also be fused. When fusion occurs, an initial cell type called a **heterokaryon** is produced. The hybrid cell contains two nuclei in a common cytoplasm. Using the proper techniques, it is possible to fuse human and mouse cells, for example, and isolate the hybrids from the parental cells.

As the heterokaryons are cultured *in vitro*, two interesting changes occur. The nuclei eventually fuse, creating a **synkaryon**. Then, as culturing is continued for many generations, chromosomes from one of the two parental species are gradually lost. In the case of the human–mouse hybrid, human chromosomes are lost randomly until eventually the synkaryon has a full complement of mouse chromosomes and only a few human chromosomes. As we shall see, it is the preferential loss of human chromosomes (rather than mouse chromosomes) that makes possible the assignment of human genes to the chromosomes on which they reside.

The experimental rationale is straightforward. If a specific human gene product is synthesized in a synkaryon containing one to three human chromosomes, then the gene responsible for that product must reside on one of the human chromosomes remaining in the hybrid cell. On the other hand, if the human gene product is not synthesized in a synkaryon, the gene responsible is not present on those human chromosomes that remain in this hybrid cell. Ideally, a panel of 23 hybrid-cell lines, each with just one unique human chromosome, would allow the immediate assignment of any human gene for which the product could be characterized.

In practice, a panel of cell lines, each with several remaining human chromosomes, is most often used. The correlation of the presence or absence of each chromosome with the presence or absence of each gene product is called **synteny testing**. Consider, for example, the hypothetical data provided in Figure 8–14, where four gene products (A, B, C, and D) are tested in relationship to eight human chromosomes. Let's analyze the gene that produces product *A*.

1. Product A is not produced by cell line 23, but chromosomes 1, 2, 3, and 4 are present in cell line 23. Therefore, we rule out the presence of gene *A* on those four chromosomes and conclude that it must be on chromosome 5, 6, 7, or 8.
2. Product A is produced by cell line 34, which contains chromosomes 5 and 6 but not 7 and 8. Therefore, gene *A* is on chromosome 5 or 6, but cannot be on 7 or 8 because they are absent even though product A is produced.
3. Product A is also produced by cell line 41, which contains chromosome 5 but not chromosome 6. Using similar reasoning, gene *A* must be on chromosome 5.

Using a similar approach, gene *B* can be assigned to chromosome 3. Perform this analysis for yourself to demonstrate that this is correct.

Gene *C* presents a unique situation. The data indicate that it is not present on chromosomes 1–7. While it might be on chromosome 8, no direct evidence supports this conclusion, and other panels are needed. We leave gene *D* for you to analyze—on what chromosome does it reside?

Using this technique, literally hundreds of human genes have been assigned to one chromosome or another. Some of the assignments shown in Figure 8–15 were either derived or confirmed with the use of somatic cell hybridization techniques. To map genes for which the products have yet to be discovered, researchers have had to rely on other approaches. For example, by using recombinant DNA technology in conjunction with pedigree analysis, it has been possible to assign the genes responsible for **Huntington disease**, **cystic fibrosis**, and **neurofibromatosis** to their respective chromosomes, 4, 7, and 17.

8.7 Linkage and Mapping Studies Can Be Performed in Haploid Organisms

Many of the single-celled eukaryotes are haploid during the vegetative stages of their life cycle. The alga *Chlamydomonas* and the mold *Neurospora* demonstrate this genetic condition. But these organisms do form reproductive cells that fuse during fertilization, producing a diploid zygote. However, this structure soon undergoes meiosis, resulting in haploid vegetative cells that then propagate by mitotic divisions. In genetic studies, small haploid organisms have several important advantages compared with diploid eukaryotes. They can be cultured and manipulated in genetic crosses much more

Hybrid cell lines	Human chromosomes present								Gene products expressed			
	1	2	3	4	5	6	7	8	A	B	C	D
23	●	●	●	●					–	+	–	+
34	●	●			●	●			+	–	–	+
41	●		●		●		●		+	+	–	+

FIGURE 8–14 A hypothetical grid of data used in synteny testing to assign genes to their appropriate human chromosomes. Three somatic hybrid-cell lines, designated 23, 34, and 41, have each been scored for the presence or absence of human chromosomes 1–8, as well as for their ability to produce the hypothetical human gene products A, B, C, and D.

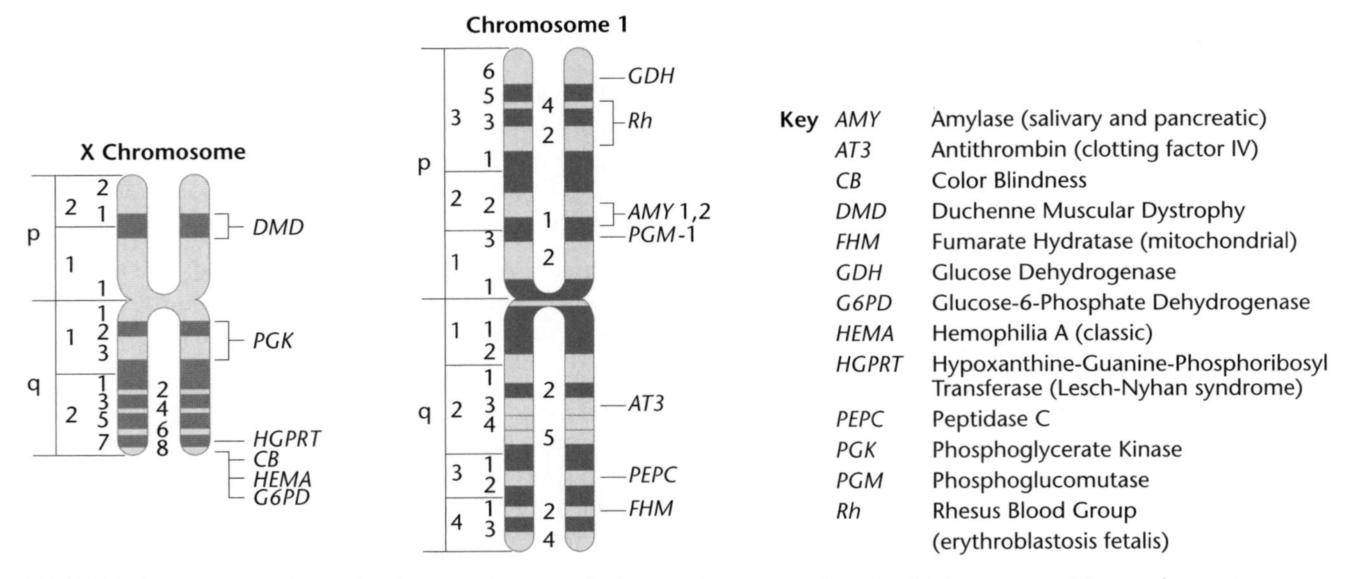

FIGURE 8–15 Representative regional gene assignments for human chromosome 1 and the X chromosome. Many assignments were initially derived using somatic cell hybridization techniques.

easily. In addition, a haploid organism contains only a single allele of each gene, which is expressed directly in the phenotype. This greatly simplifies genetic analysis. As a result, organisms such as *Chlamydomonas* and *Neurospora* serve as the subjects of research investigations in many areas of genetics, including linkage and mapping studies.

In order to perform genetics experiments with such organisms, crosses are made, and following fertilization, the meiotic structures are isolated. Because all four meiotic products give rise to spores in each structure, these structures are called **tetrads**. *The term "tetrad" has a different meaning here than earlier when it was used to describe a precise chromatid configuration in meiosis.* Individual tetrads are isolated, and the resultant cells are grown and analyzed separately from those of other tetrads. In the results we are about to describe, the data reflect the proportion of tetrads that show one combination of genotypes, the proportion that show another combination, and so on. Such experimentation is called **tetrad analysis**.

Gene-to-Centromere Mapping

When a single gene in *Neurospora* is analyzed (Figure 8–16), the data can be used to calculate the map distance between the gene and the centromere. This process is sometimes referred to as **mapping the centromere**. It is accomplished by experimentally determining the frequency of recombination using tetrad data. Note that once the four meiotic products of the tetrad are formed, a mitotic division occurs, resulting in eight ordered products (ascospores). If no crossover event occurs between the gene under study and the centromere, the pattern of ascospores contained within an ascus (pl., asci) appears as shown in Figure 8–16(a), *aaaa*++++.*

This pattern represents **first-division segregation** because the two alleles are separated during the first meiotic division. However, a crossover event will alter this pattern, as shown in Figure 8–16(b), *aa*++*aa*++, and 8–16(c), ++*aaaa*++. Two other recombinant patterns occur, but are not shown: ++*aa*++*aa* and *aa*++++*aa*. These depend on the chromatid orientation during the second meiotic division. These four patterns reflect **second-division segregation** because the two alleles are not separated until the second meiotic division. Usually, the ordered tetrad data are condensed to reflect the genotypes of the identical ascospore pairs, and six combinations are possible.

First-Division Segregation	***Second-Division Segregation***	
aa++	*a*+*a*+	+*aa*+
++*aa*	+*a*+*a*	*a*++*a*

To calculate the distance between the gene and the centromere, a large number of asci resulting from a controlled cross are counted. We then use these data to calculate the distance (*d*):

$$d = \frac{1/2\ \text{(second-division segregant asci)}}{\text{total asci scored}}$$

The distance (*d*) reflects the percentage of recombination and is only half the number of second-division segregant asci. This is because crossing over in each occurs in only two of the four chromatids during meiosis.

To illustrate, we use *a* for albino and + for wild type in *Neurospora*. In crosses between the two genetic types, suppose we observe 65 first-division segregants, and 70 second-division segregants. The distance between *a* and the centromere is

*The pattern (++++*aaaa*) can also be formed. However, it is indistinguishable from (*aaaa*++++).

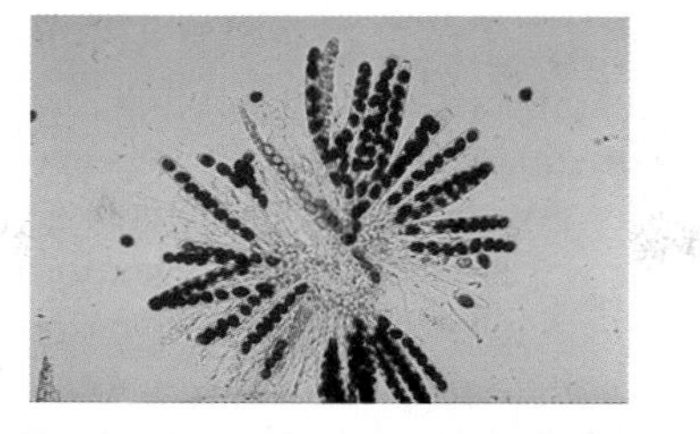

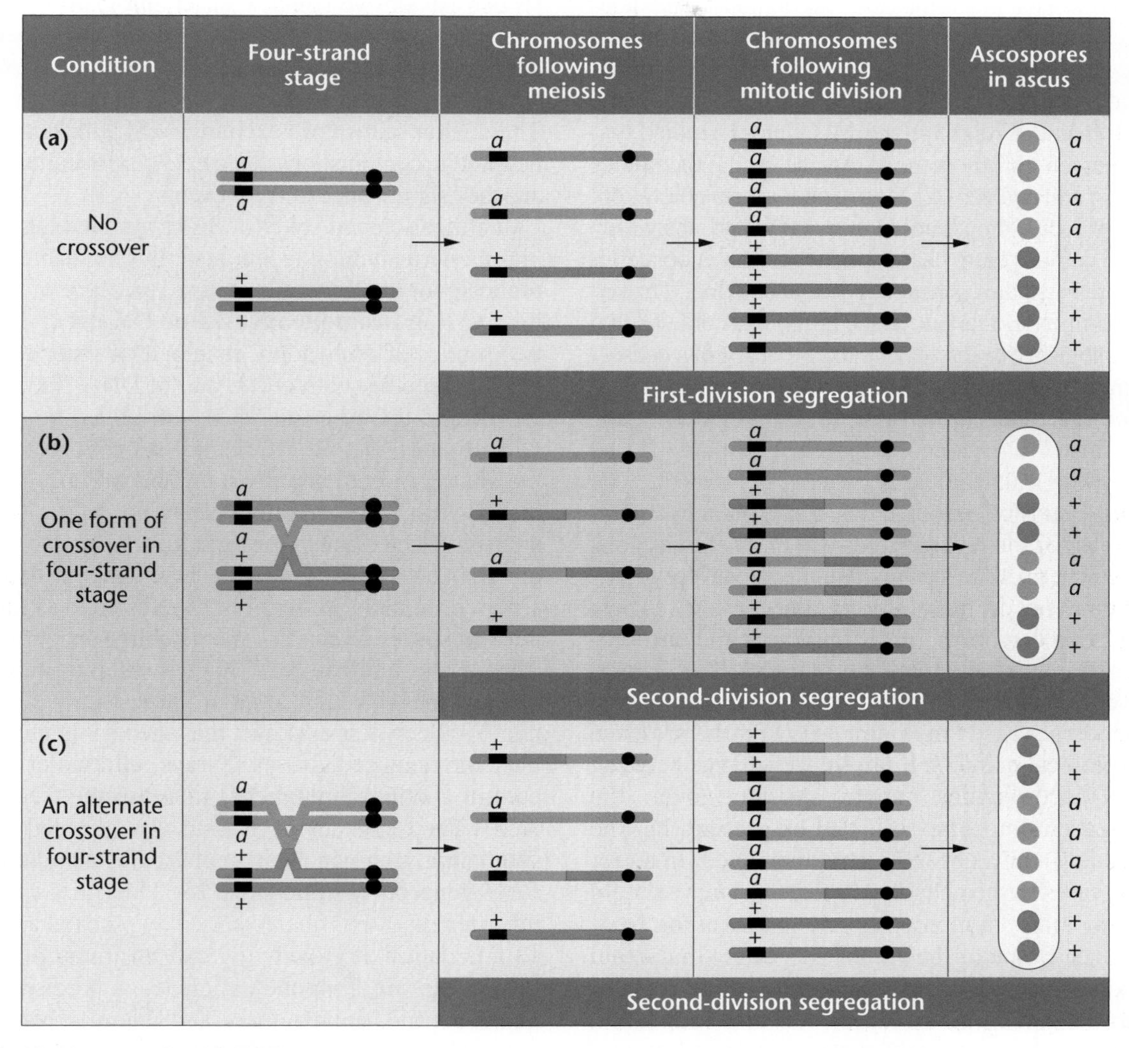

FIGURE 8–16 Three ways in which different ascospore patterns can be generated in *Neurospora*. Analysis of these patterns is the basis of gene-to-centromere mapping. The photograph shows a variety of ascospore arrangements within *Neurospora* asci. *(Photo: James W. Richardson/-Visuals Unlimited)*

$$d = \frac{(1/2)(70)}{135} = 0.259$$

or about 26 mu.

As the distance increases to 50 mu, all asci should reflect second-division segregation. However, numerous factors prevent this. As in diploid organisms, mapping accuracy based on crossover events is greatest when the gene and centromere are relatively close together.

We can also analyze haploid organisms to distinguish between linkage and independent assortment of two genes—mapping distances between gene loci are calculated once linkage is established. As a result, detailed maps of organisms such as *Neurospora* and *Chlamydomonas* are now available.

8.8 Other Aspects of Genetic Exchange

Careful analysis of crossing over during gamete formation allows us to construct chromosome maps in both diploid and haploid organisms. However, we should not lose sight of the real biological significance of crossing over, which is to generate genetic variation in gametes and, subsequently, in the offspring derived from the resultant eggs and sperm. Because of the critical role of crossing over in generating variation, the study of genetic exchange is a key topic for study in genetics. But many important questions remain. For example, does crossing over involve an actual exchange of chromosome arms? Does exchange occur between paired sister chromatids during mitosis? We shall briefly consider possible answers to these questions.

Cytological Evidence for Crossing Over

Once genetic mapping was understood, it was of great interest to investigate the relationship between chiasmata observed in meiotic prophase I and crossing over. For example, are chiasmata visible manifestations of crossover events? If so, then crossing over in higher organisms appears to result from an actual physical exchange between homologous chromosomes. That this is the case was demonstrated independently in the 1930s by Harriet Creighton and Barbara McClintock in *Zea mays*, and by Curt Stern in *Drosophila*.

Since the experiments are similar, we will consider only the work with maize. Creighton and McClintock studied two linked genes on chromosome 9. At one locus, the alleles *colorless* (*c*) and *colored* (*C*) control endosperm coloration. At the other locus, the alleles *starchy* (*Wx*) and *waxy* (*wx*) control the carbohydrate characteristics of the endosperm. The maize plant studied is heterozygous at both loci. The key to this experiment is that one of the homologs contains two unique cytological markers. The markers consist of a densely stained knob at one end of the chromosome and a translocated piece of another chromosome (8) at the other end. The arrangements of alleles and cytological markers can be detected cytologically and are shown in Figure 8–17.

Creighton and McClintock crossed this plant to a plant homozygous for the *colored* allele (*c*) and heterozygous for the endosperm alleles. They obtained a variety of different phenotypes in the offspring, but they were most interested in a crossover result involving the chromosome with the unique cytological markers. They examined the chromosomes of this plant with the colorless, waxy phenotype (case I in Figure 8–17) for the presence of the cytological markers. If physical exchange between homologs accompanies genetic crossing over, the translocated chromosome will still be present, but the knob will not—this is exactly what happened. In a second plant (case II), the phenotype colored, starchy should result from either nonrecombinant gametes or from crossing over. Some of the plants then ought to contain chromosomes with the dense knob but not the translocated chromosome. This condition was also found, and the conclusion that a physical exchange takes place was again supported. Along with Stern's findings with *Drosophila*, this work clearly established that crossing over has a cytological basis.

Sister Chromatid Exchanges

Knowing that crossing over occurs between synapsed homologs in meiosis, we might ask whether a similar physical exchange occurs between homologs during mitosis. While homologous chromosomes do not usually pair up or synapse in somatic cells (*Drosophila* is an exception), each individual chromosome in prophase and metaphase of mitosis consists of two identical sister chromatids, joined at a common centromere. Surprisingly, several experimental approaches have demonstrated that reciprocal exchanges similar to crossing over occur between sister chromatids. These **sister chromatid exchanges (SCEs)** do not produce new allelic combinations, but evidence is accumulating that attaches significance to these events.

Identification and study of SCEs are facilitated by several modern staining techniques. In one technique, cells replicate for two generations in the presence of the thymidine analog **bromodeoxyuridine (BUdR)**.* Following two rounds of replication, each pair of sister chromatids has one member with one strand of DNA "labeled" with BUdR and the other member with both strands labeled with BUdR. Using a differential stain, chromatids with the analog in both strands stain less brightly than chromatids with BUdR in only one strand. As a result, SCEs are readily detectable if they occur. In Figure 8–18, numerous instances of SCE events are clearly evident. These sister chromatids are sometimes referred to as **harlequin chromosomes** because of the patchlike appearance.

While the significance of SCEs is still uncertain, several observations have led to great interest in this phenomenon. We know, for example, that agents that induce chromosome damage (viruses, X-rays, ultraviolet light, and certain chemical mutagens) increase the frequency of SCEs. The frequency of SCEs is also elevated in **Bloom syndrome**, a human disorder caused by a mutation in the *BLM* gene on chromosome 15. This rare, recessively inherited disease is characterized by prenatal and postnatal retardation of growth, a great sensitivity of the facial skin to the sun, immune deficiency, a predisposition to malignant and benign tumors, and abnormal behavior patterns. The chromosomes from cultured leukocytes, bone

*The abbreviation BrdU is also used to denote bromodeoxyuridine.

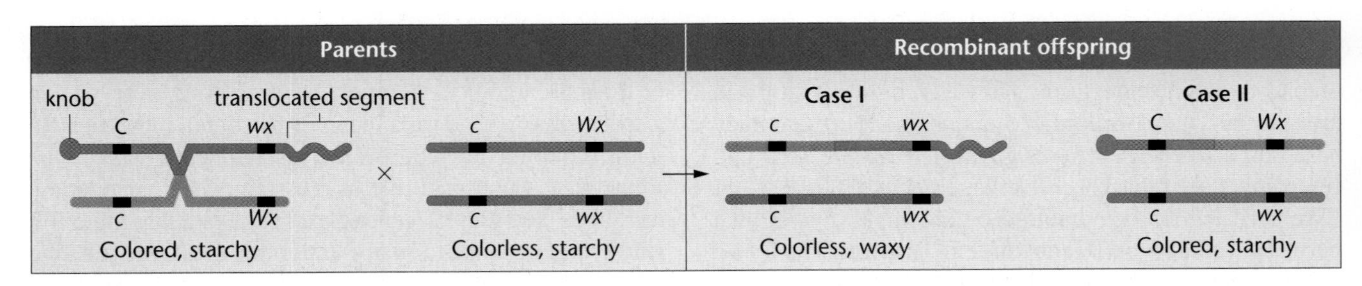

FIGURE 8–17 The phenotypes and chromosome compositions of parents and recombinant offspring in Creighton and McClintock's experiment in maize. The knob and translocated segment are the cytological markers that established that crossing over involves an actual exchange of chromosome arms.

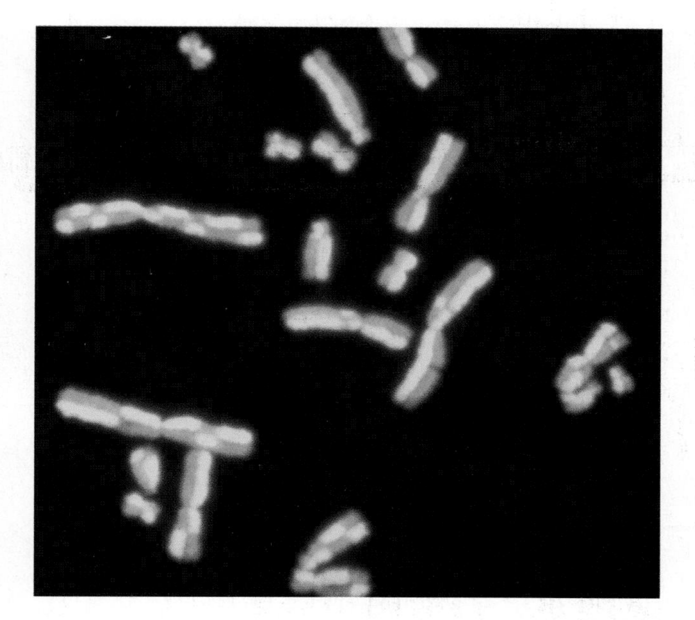

FIGURE 8–18 Light micrograph of sister chromatid exchanges (SCEs) in mitotic chromosomes. Sometimes called harlequin chromosomes because of their patchlike appearance, chromatids with the thymidine analog BUdR in both DNA strands fluoresce *less* brightly than do those with the analog in only one strand. These chromosomes were stained with 33258-Hoechst reagent and acridine orange and then viewed using fluorescence microscopy. *(Dr. Sheldon Wolff & Judy Bodycote/Laboratory of Radiology and Environmental Health, University of California, San Francisco)*

marrow cells, and fibroblasts derived from homozygotes are very fragile and unstable compared to those of homozygous and heterozygous normal individuals. Increased breaks and rearrangements between nonhomologous chromosomes are observed in addition to excessive amounts of sister chromatid exchanges. Recent work by James German and colleagues suggests that the *BLM* gene encodes an enzyme called DNA helicase, which is best known for its role in DNA replication.

8.9 Did Mendel Encounter Linkage?

We conclude this chapter by examining a modern-day interpretation of the experiments that form the cornerstone of transmission genetics—Mendel's crosses with garden peas.

Some observers believe that Mendel had extremely good fortune in his classic experiments. He did not encounter apparent linkage relationships between the seven mutant characters in any of his crosses. Had Mendel obtained highly variable data characteristic of linkage and crossing over, these unorthodox ratios might have hindered his successful analysis and interpretation.

The article by Stig Blixt, reprinted in its entirety in the following box, demonstrates the inadequacy of this hypothesis. As we shall see, some of Mendel's genes were indeed linked. We leave it to Stig Blixt to enlighten you as to why Mendel did not detect linkage.

Why Didn't Gregor Mendel Find Linkage?

It is quite often said that Mendel was very fortunate not to run into the complication of linkage during his experiments. He used seven genes, and the pea has only seven chromosomes. Some have said that had he taken just one more, he would have had problems. This, however, is a gross oversimplification. The actual situation, most probably, is shown in Table 8.1. This shows that Mendel worked with three genes in chromosome 4, two genes in chromosome 1, and one gene in each of chromosomes 5 and 7. It seems at first glance that, out of the 21 dihybrid combinations Mendel theoretically could have studied, no fewer than four (that is, *a–i*, *v–fa*, *v–le*, *fa–le*) ought to have resulted in linkages. However, as found in hundreds of crosses and shown by the genetic map of the pea, *a* and *i* in chromosome 1 are so distantly located on the chromosome that no linkage is normally detected. The same is true for *v* and *le* on the one hand, and *fa* on the other, in chromosome 4. This leaves *v–le*, which ought to have shown linkage.

Mendel, however, seems not to have published this particular combination and thus, presumably, never made the appropriate cross to obtain both genes segregating simultaneously. It is therefore not so astonishing that Mendel did not run into the complication of linkage, although he did not avoid it by choosing one gene from each chromosome.

Stig Blixt
Weibullsholm Plant Breeding Institute, Landskrona, Sweden, and Centro Energia Nucleate na Agricultura, Piracicaba, SP, Brazil.

TABLE 1 Relationship Between Modern Genetic Terminology and Character Pairs Used by Mendel

Character Pair Used by Mendel	*Alleles in Modern Terminology*	*Located in Chromosome*
Seed color, yellow–green	*I–i*	1
Seed coat and flowers, colored–white	*A–a*	1
Mature pods, smooth expanded–wrinkled indented	*V–v*	4
Inflorescences, from leaf axis–umbellate in top of plant	*Fa–fa*	4
Plant height, 0.5–1 m	*Le–le*	4
Unripe pods, green–yellow	*Gp–gp*	5
Mature seeds, smooth–wrinkled	*R–r*	7

Chapter Summary

1. Genes located on the same chromosome are said to be linked. Alleles located on the same homolog, therefore, can be transmitted together during gamete formation. However, crossing over between homologs during meiosis results in the reshuffling of alleles and thereby contributes to genetic variability within gametes.
2. Early in the 20th century century, geneticists realized that crossing over provides an experimental basis for mapping the location of linked genes relative to one another along the chromosome.
3. Somatic cell hybridization techniques have made possible linkage and mapping analysis of human genes.
4. Mapping studies may also be performed with haploid organisms such as *Chlamydomonas* and *Neurospora*.
5. Cytological investigations of both maize and *Drosophila* reveal that crossing over involves a physical exchange of segments between nonsister chromatids.
6. An exchange of genetic material between sister chromatids can occur during mitosis as well. These events are referred to as sister chromatid exchanges (SCEs). An elevated frequency of such events is seen in the human disorder Bloom syndrome.
7. Evidence now suggests that several of the genes studied by Mendel are, in fact, linked. However, in such cases, the genes are sufficiently far apart to prevent the detection of linkage.

Key Terms

Bloom syndrome, 174
bromodeoxyuridine (BUdR), 174
chiasmata, 160
chromosomal theory of inheritance, 161
chromosome map, 161
coefficient of coincidence, 169
complete linkage, 157
crossing over, 157
crossover gametes, 157
cystic fibrosis, 171
double crossover (DCO), 162
first-division segregation, 172
harlequin chromosomes, 174
heterokaryon, 171
Huntington disease, 171
interference, 168
linkage group, 158
linkage ratio, 158
lod score method, 170
mapping the centromere, 172
neurofibromatosis, 171
noncrossover gametes, 157
parental gametes, 157
reciprocal classes, 165
recombinant gametes, 157
second-division segregation, 172
sister chromatid exchanges (SCEs), 174
somatic cell hybridization, 170
synkaryon, 171
synteny testing, 171
tetrad, 172
tetrad analysis, 172

INSIGHTS AND SOLUTIONS

1. In rabbits, black color (B) is dominant to brown (b), while full color (C) is dominant to *chinchilla* (c^{ch}). The genes controlling these traits are linked. Rabbits that are heterozygous for both traits and express black, full color are crossed to rabbits that express brown, chinchilla with the following results:

31	brown, chinchilla
34	black, full
16	brown, full
19	black, chinchilla

Determine the arrangement of alleles in the heterozygous parents and the map distance between the two genes.

Solution: This is a two-point map problem, where the two reciprocal phenotypes most prevalent are the noncrossovers. The less frequent reciprocal phenotypes arise from a single crossover. The arrangement of alleles is derived from the noncrossover phenotypes because they enter gametes intact.

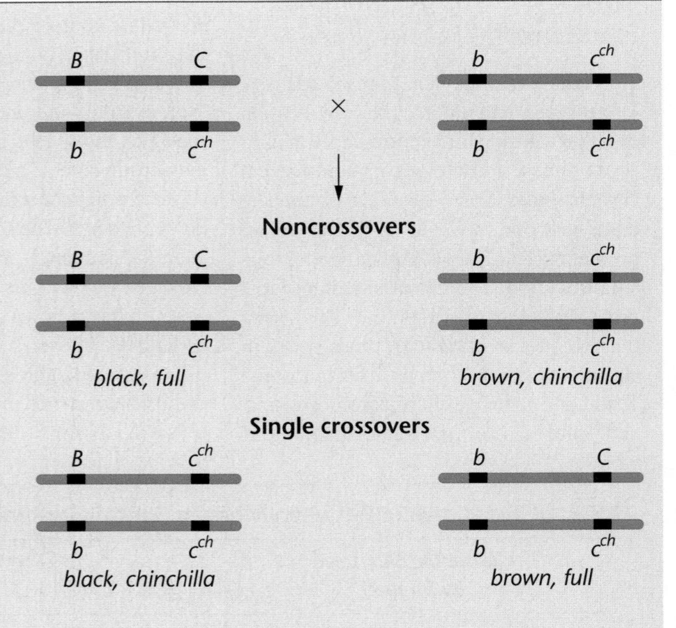

The single crossovers give rise to 35/100 offspring (35%). Therefore, the distance between the two genes is 35 mu.

2. In *Drosophila*, *Lyra* (*Ly*) and *Stubble* (*Sb*) are dominant mutations located at locus 40 and 58, respectively, on chromosome III. A recessive mutation with bright red eyes is discovered and shown also to be located on chromosome III. A map is obtained by crossing a female who is heterozygous for all three mutations to a male that is homozygous for the *bright-red* mutation (which we will call *br*), and the data in the table are generated. Determine the location of the *br* mutation on chromosome III.

Phenotype			*Number*
(1) *Ly*	*Sb*	*br*	404
(2) +	+	+	422
(3) *Ly*	+	+	18
(4) +	*Sb*	*br*	16
(5) *Ly*	+	*br*	75
(6) +	*Sb*	+	59
(7) *Ly*	*Sb*	+	4
(8) +	+	*br*	2
		Total =	1000

Solution: First, determine the arrangement of the alleles on the homologs of the heterozygous crossover parent (the female in this case). To do this, locate the most frequent reciprocal phenotypes, which arise from the noncrossover gametes—these are phenotypes (1) and (2). Each phenotype represents the arrangement of alleles on one of the homologs. Therefore, the arrangement is

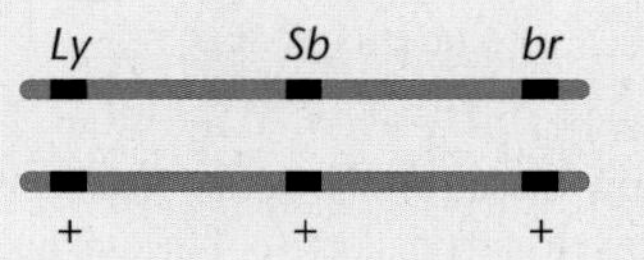

Second, find the correct sequence of the three loci along the chromosome. This is done by determining which sequence yields the observed double-crossover phenotypes, which are the least frequent reciprocal phenotypes (7 and 8). If the sequence is correct as written, then the double crossover depicted here,

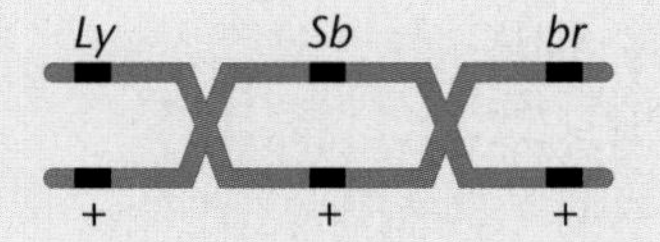

will yield *Ly* + *br* and + *Sb* + as phenotypes. Inspection shows that these categories (5 and 6) are actually single crossovers, not double crossovers. Therefore, the sequence is incorrect, as written. Only two other sequences are possible: The *br* gene is either to the left of *Ly* (case A), or it is between *Ly* and *Sb* (case B).

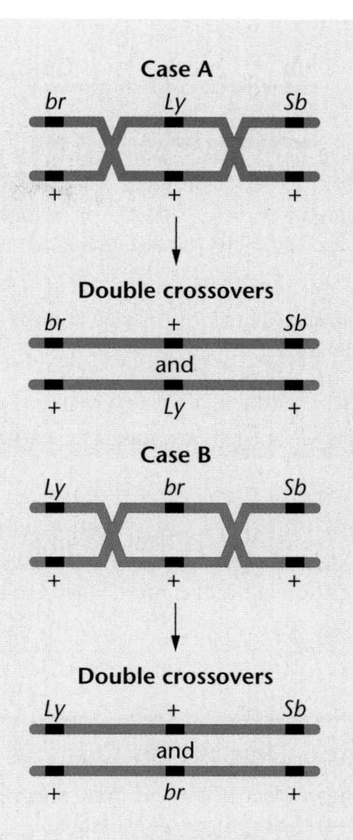

Comparison with the actual data shows that case B is correct. The double-crossover gametes yield flies that express *Ly* and *Sb* but not *br*, or express *br* but not *Ly* and *Sb*. Therefore, the correct arrangement and sequence is shown below.

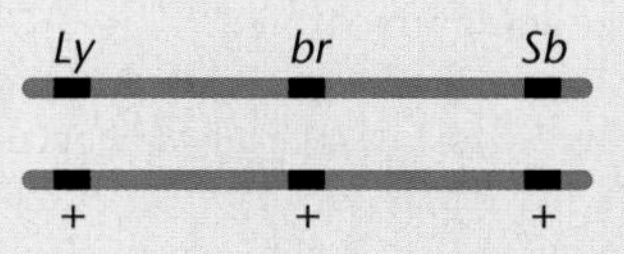

Once this sequence is found, determine the location of *br* relative to *Ly* and *Sb*. A single crossover between *Ly* and *br*, as shown here,

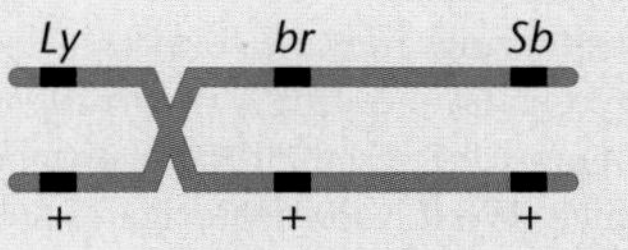

yields flies that are *Ly* + + and + *br Sb* (phenotypes 3 and 4). Therefore, the distance between the *Ly* and *br* loci is equal to

$$18 + 16 + 4 + 2/1000 = 40/1000 = 0.04 = 4 \text{ mu}$$

Remember to add the double crossovers because they represent two single crossovers occurring simultaneously. You need to know the frequency of all crossovers between *Ly* and *br*, so they must be included.

Similarly, the distance between the *br* and *Sb* loci is derived mainly from single crossovers between them.

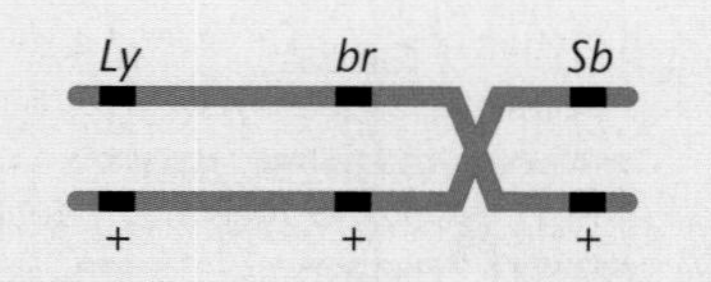

This event yields *Ly* + *br* + and + + *Sb* phenotypes (phenotypes 5 and 6). Therefore, the distance equals

$(75 + 59 + 4 + 2)/1000 = 140/1000 = 0.14 = 14$ mu.

The final map shows that *br* is located at locus 44, since *Lyra* and *Stubble* are known.

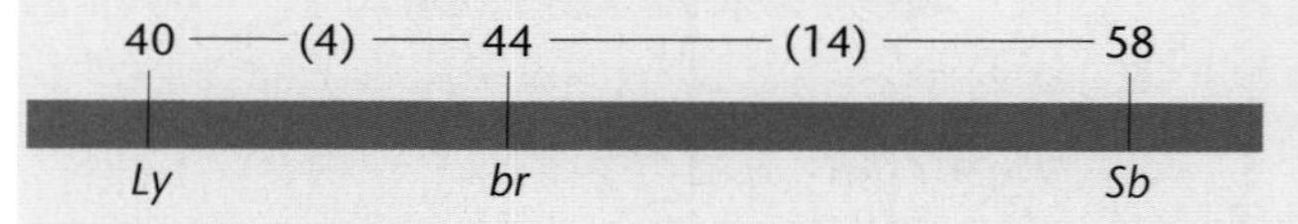

3. Refer to Figure 8–13, and predict what gene (which we called *br*) was discovered on chromosome III in Problem 2. Suggest an experimental cross that could confirm your prediction.

Solution: Inspection of Figure 8–13 reveals that the mutation *scarlet* (*st*) is present at locus 44.0, so it is reasonable to hypothesize that the *bright-red* eye mutation is an allele at the *scarlet* locus.

To test this hypothesis, you could perform complementation analysis (see Chapter 4) by crossing females expressing the bright red mutation with known *scarlet* males. If the two mutations are alleles, no complementation will occur and all progeny will reveal a bright red mutant eye phenotype. If complementation occurs, all progeny will express normal brick-red (wild-type) eyes, since the bright red mutation and *scarlet* are at different loci (they are probably very close together). In such a case, all progeny will be heterozygous at both the *bright red* eye and the *scarlet* loci and will not express either mutation because they are both recessive. This type of complementation analysis is called an *allelism test*.

Problems and Discussion Questions

1. What is the significance of crossing over (which leads to genetic recombination) to the process of evolution?
2. Describe the cytological observation that suggests that crossing over occurs during the first meiotic prophase.
3. Why does more crossing over occur between two distantly linked genes than between two genes that are very close together on the same chromosome?
4. Why is a 50% recovery of single-crossover products the upper limit, even when crossing over *always* occurs between two linked genes?
5. Why are double-crossover events expected in lower frequency than single-crossover events?
6. What is the proposed basis for positive interference?
7. What three essential criteria must be met in order to execute a successful mapping cross?
8. The genes *dumpy* wings (*dp*), *clot* eyes (*cl*), and *apterous* wings (*ap*) are linked on chromosome II of *Drosophila*. In a series of two-point mapping crosses, the genetic distances shown below were determined. What is the sequence of the three genes?

dp–ap	42
dp–cl	3
ap–cl	39

9. Consider two hypothetical recessive autosomal genes *a* and *b*, where a heterozygote is test-crossed to a double-homozygous mutant. Predict the phenotypic ratios under the following conditions:
   1. *a* and *b* are located on separate autosomes.
   2. *a* and *b* are linked on the same autosome, but are so far apart that a crossover always occurs between them.
   3. *a* and *b* are linked on the same autosome, but are so close together that a crossover almost never occurs.
   4. *a* and *b* are linked on the same autosome about 10 mu apart.
10. Colored aleurone in the kernels of corn is due to the dominant allele *R*. The recessive allele *r*, when homozygous, produces colorless aleurone. The plant color (not kernel color) is controlled by another gene with two alleles, *Y* and *y*. The dominant *Y* allele results in green color, whereas the homozygous presence of the recessive *y* allele causes the plant to appear yellow. In a test cross between a plant of unknown genotype and phenotype and a plant that is homozygous recessive for both traits, the following progeny were obtained:

colored, green	88
colored, yellow	12
colorless, green	8
colorless, yellow	92

Explain how these results were obtained by determining the exact genotype and phenotype of the unknown plant, including the precise association of the two genes on the homologs (i. e., the arrangement).

11. In the cross shown here, involving two linked genes, *ebony* (*e*) and *claret* (*ca*), in *Drosophila,* where crossing over does not occur in males, offspring were produced in a (2 + : 1 *ca* : 1 *e*) phenotypic ratio:

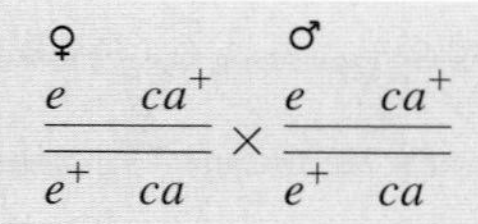

These genes are 30 mu apart on chromosome III. What did crossing over in the female contribute to these phenotypes?

12. With two pairs of genes involved (*P*, *p* and *Z*, *z*), a test cross (to *ppzz*) with an organism of unknown genotype indicated that the gametes were produced in these proportions: *PZ* = 42.4%; *Pz* = 6.9%; *pZ* = 7.1%; and *pz* = 43.6%. Draw all possible conclusions from these data.
13. In a series of two-point map crosses involving five genes located on chromosome II in *Drosophila,* the following recombinant (single-crossover) frequencies were observed:

pr–adp	29
pr–vg	13
pr–c	21
pr–b	6

adp–b	35
adp–c	8
adp–vg	16
vg–b	19
vg–c	8
c–b	27

(a) If the *adp* gene is present near the end of chromosome II (locus 83), construct a map of these genes. (b) In another set of experiments, a sixth gene (*d*) was tested against *b* and *pr*, and the results were *d–b* = 17% and *d–pr* = 23%. Predict the results of two-point maps between *d* and *c*, *d* and *vg*, and *d* and *adp*.

14. Two different female *Drosophila* were isolated, each heterozygous for the autosomally linked genes *black* body (*b*), *dachs* tarsus (*d*), and *curved* wings (*c*). These genes are in the order *d–b–c*, with *b* closer to *d* than to *c*. Shown below is the genotypic arrangement for each female, along with the various gametes formed by both. Identify which categories are noncrossovers (NCO), single crossovers (SCO), and double crossovers (DCO) in each case. Then, indicate the relative frequency in which each will be produced.

Female A		Female B	
d b + / *+ + c*		*d + +* / *+ b c*	
↓	Gamete formation	↓	
(1) *d b c*	(5) *d + +*	(1) *d b +*	(5) *d b c*
(2) *+ + +*	(6) *+ b c*	(2) *+ + c*	(6) *+ + +*
(3) *+ + c*	(7) *d + c*	(3) *d + c*	(7) *d + +*
(4) *d b +*	(8) *+ b +*	(4) *+ b +*	(8) *+ b c*

15. In *Drosophila*, a cross was made between females expressing the three X-linked recessive traits, *scute* bristles (*sc*), *sable* body (*s*), and *vermilion* eyes (*v*) and wild-type males. All females were wild type in the F_1, while all males expressed all three mutant traits. The cross was carried to the F_2 generation and 1000 offspring were counted, with the results shown in the table. No determination of sex was made in the F_2 data. (a) Using proper nomenclature, determine the genotypes of the P_1 and F_1 parents. (b) Determine the sequence of the three genes and the map distance between them. (c) Are there more or fewer double crossovers than expected? (d) Calculate the coefficient of coincidence; does this represent positive or negative interference?

Phenotype	Offspring
sc s v	314
+ + +	280
+ s v	150
sc + +	156
sc + v	46
+ s +	30
sc s +	10
+ + v	14

16. A cross in *Drosophila* involved the recessive, X-linked genes *yellow* body (*y*), *white* eyes (*w*), and *cut* wings (*ct*). A yellow-bodied, white-eyed female with normal wings was crossed to a male whose eyes and body were normal, but whose wings were cut. The F_1 females were wild type for all three traits, while the F_1 males expressed the yellow-body, white-eye traits. The cross was carried to F_2 progeny, and only male offspring were tallied. On the basis of the data shown here, a genetic map was constructed. (a) Diagram the genotypes of the F_1 parents. (b) Construct a map, assuming that *w* is at locus 1.5 on the X chromosome. (c) Were any double-crossover offspring expected? (d) Could the F_2 female offspring be used to construct the map? Why or why not?

Phenotype	Male Offspring
y + ct	9
+ w +	6
y w ct	90
+ + +	95
+ + ct	424
y w +	376
y + +	0
+ w ct	0

17. In *Drosophila*, *Dichaete* (*D*) is a mutation on chromosome III with a dominant effect on wing shape. It is lethal when homozygous. The genes *ebony* body (*e*) and *pink* eye (*p*) are recessive mutations on chromosome III. Flies from a Dichaete stock were crossed to homozygous ebony, pink flies, and the F_1 progeny with a Dichaete phenotype were backcrossed to the ebony, pink homozygotes. Using the results of this backcross shown in the table, (a) diagram the cross, showing the genotypes of the parents and offspring of both crosses. (b) What is the sequence and interlocus distance between these three genes?

Phenotype	Number
Dichaete	401
ebony, pink	389
Dichaete, ebony	84
pink	96
Dichaete, pink	2
ebony	3
Dichaete, ebony, pink	12
wild type	13

18. *Drosophila* females homozygous for the third chromosomal genes *pink* eye (*p*) and *ebony* body (*e*) were crossed with males homozygous for the second chromosomal gene *dumpy* wings (*dp*). Because these genes are recessive, all offspring were wild type (normal). F_1 females were test-crossed to triply recessive males. If we assume that the two linked genes (*p* and *e*) are 20 mu apart, predict the results of this cross. If the reciprocal cross were made (F_1 males—where no crossing over occurs—with triply recessive females), how would the results vary, if at all?

19. In *Drosophila*, the two mutations *Stubble* bristles (*Sb*) and *curled* wings (*cu*) are linked on chromosome III. *Sb* is a dominant gene that is lethal in a homozygous state, and *cu* is a recessive gene. If a female of the genotype

$$\frac{Sb \quad cu}{+ \quad +}$$

is to be mated to detect recombinants among her offspring, what male genotype would you choose as her mate?

20. In *Drosophila*, a heterozygous female for the X-linked recessive traits *a*, *b*, and *c* was crossed to a male that phenotypically expressed *a*, *b*, and *c*. The offspring occurred in the phenotypic

ratios in the table below, and no other phenotypes were observed. (a) What is the genotypic arrangement of the alleles of these genes on the X chromosome of the female? (b) Determine the correct sequence, and construct a map of these genes on the X chromosome. (c) What progeny phenotypes are missing, and why?

+	*b*	*c*	460
a	+	+	450
a	*b*	*c*	32
+	+	+	38
a	+	*c*	11
+	*b*	+	9

21. Are sister chromatid exchanges effective in producing genetic variability in an individual? In the offspring of individuals?
22. What conclusion can be drawn from the observations that in male *Drosophila,* no crossing over occurs and that during meiosis, synaptonemal complexes (ultrastructural components found between synapsed homologs in meiosis) are not seen in males, but are observed in females, where crossing over occurs?
23. An organism of the genotype *AaBbCc* was test-crossed to a triply recessive organism (*aabbcc*). The genotypes of the progeny are in the table below.

AaBbCc	20	*AaBbcc*	20
aabbCc	20	*aabbcc*	20
AabbCc	5	*Aabbcc*	5
aaBbCc	5	*aaBbcc*	5

(a) Assuming simple dominance and recessiveness in each gene pair, if these three genes were all assorting independently, how many genotypic and phenotypic classes would result in the offspring, and in what proportion?
(b) Answer part (a) again, assuming the three genes are so tightly linked on a single chromosome that no crossover gametes were recovered in the sample of offspring.
(c) What can you conclude from the *actual* data about the location of the three genes in relation to one another?

24. Based on our discussion of the potential inaccuracy of mapping (see Figure 8–12), would you revise your answer to Problem 23(c)? If so, how?
25. In a plant, fruit color is either red or yellow, and fruit shape is either oval or long. Red and oval are the dominant traits. Two plants, both heterozygous for these traits, were test-crossed, with the results shown below. Determine the location of the genes relative to one another and the genotypes of the two parental plants.

	Progeny	
Phenotype	**Plant A**	**Plant B**
red, long	46	4
yellow, oval	44	6
red, oval	5	43
yellow, long	5	47
	100	100

26. In a plant heterozygous for two gene pairs (*Ab/aB*), where the two loci are linked and 25 mu apart, two such individuals were crossed together. Assuming that crossing over occurs during the formation of both male and female gametes and that the *A* and *B* alleles are dominant, determine the phenotypic ratio of the offspring.
27. In a cross in *Neurospora* involving two alleles, *B* and *b,* the tetrad patterns in the next table were observed. Calculate the distance between the gene and the centromere.

Tetrad Pattern	Number
BBbb	36
bbBB	44
BbBb	4
bBbB	6
BbbB	3
bBBb	7

28. In Creighton and McClintock's experiment demonstrating that crossing over involves physical exchange between chromosomes (see Section 8.8), explain the importance of the cytological markers (the translocated segment and the chromosome knob) in the experimental rationale.
29. A number of human–mouse somatic cell hybrid clones were examined for the expression of specific human genes and the presence of human chromosomes—the results are summarized in this table. Assign each gene to the chromosome upon which it is located.

	Hybrid-Cell Clone					
	A	**B**	**C**	**D**	**E**	**F**
Genes (expressed or not)						
ENO1 (enolase-1)	−	+	−	+	+	−
MDH1 (malate dehydrogenase-1)	+	+	−	+	−	+
PEPS (peptidase S)	+	−	+	−	−	−
PGM1 (phosphoglucomutase-1)	−	+	−	+	+	−
Chromosomes (present or absent)						
1	−	+	−	+	+	−
2	+	+	−	+	−	+
3	+	+	−	−	+	−
4	+	−	+	−	−	−
5	−	+	+	+	+	+

30. A female of genotype

$$\frac{a \quad b \quad c}{+ \quad + \quad +}$$

produces 100 meiotic tetrads. Of these, 68 show no crossover events. Of the remaining 32, 20 show a crossover between *a* and *b*, 10 show a crossover between *b* and *c*, and 2 show a double crossover between *a* and *b* and between *b* and *c*. Of the 400 gametes produced, how many of each of the 8 different genotypes will be produced? Assuming the order *a–b–c* and the allele arrangement shown above, what is the map distance between these loci?
31. *D. melanogaster* has one pair of sex chromosomes (XX or XY) and three autosomes (chromosomes II, III, and IV). A genetics student discovered a male fly with very short (*sh*) legs. Using this male, the student was able to establish a pure-breeding stock of this mutant and found that it was recessive. She then incorporated the mutant into a stock containing the recessive gene *black* (*b*, body color, located on chromosome II) and the recessive gene *pink* (*p*, eye color, located on chromosome III). A female from the homozygous black, pink, short stock was then mated to a wild-type male. The F_1 males of this cross were all wild type and were

then backcrossed to the homozygous *b, p, sh* females. The F_2 results appeared as shown in the table that follows, and no other phenotypes were observed. (a) Based on these results, the student was able to assign *sh* to a linkage group (a chromosome). Determine which chromosome, and include step-by-step reasoning. (b) The student repeated the experiment, making the reciprocal cross: F_1 females backcrossed to homozygous *b, p, sh* males. She observed that 85% of the offspring fell into the given classes, but that 15% of the offspring were equally divided among $b{+}p$, $b{+}{+}$, $+shp$, and $+sh+$ phenotypic males and females. How can these results be explained, and what information can be derived from these data?

Phenotype	*Female*	*Male*
wild	63	59
pink*	58	65
black, short	55	51
black, pink, short	69	60

*Pink indicates that the other two traits are wild type (normal). Similarly, black, short offspring are wild type for eye color.

32. In *Drosophila*, a female fly is heterozygous for three mutations, *Bar* eyes (*B*), *miniature* wings (*m*), and *ebony* body (*e*). (Note that *Bar* is a dominant mutation.) The fly is crossed to a male with normal eyes, miniature wings, and ebony body. The results of the cross are shown below. Interpret the results of this cross. If you conclude that linkage is involved between any of the genes, determine the map distance(s) between them.

miniature	111	Bar	117
wild	29	Bar, miniature	26
Bar, ebony	101	ebony	35
Bar, miniature, ebony	31	miniature, ebony	115

Selected Readings

Allen, G. E. 1978. *Thomas Hunt Morgan: The Man and His Science.* Princeton: Princeton Univ. Press.

Blixt, S. 1975. Why didn't Gregor Mendel find linkage? *Nature* 256: 206.

Chaganti, R., Schonberg, S., and German, J. 1974. A manyfold increase in sister chromatid exchange in Bloom syndrome lymphocytes. *Proc. Natl. Acad. Sci.* 71: 4508–12.

Creighton, H. S., and McClintock, B. 1931. A correlation of cytological and genetical crossing over in *Zea mays*. *Proc. Natl. Acad. Sci.* 17: 492–97.

Douglas, L., and Novitski, E. 1977. What chance did Mendel's experiments give him of noticing linkage? *Heredity* 38: 253–57.

Ellis, N. A. et al. 1995. The Bloom's syndrome gene product is homologous to RecQ helicases. *Cell* 83: 655–66.

Ephrussi, B., and Weiss, M. C. 1969. Hybrid somatic cells. *Sci. Am.* (Apr) 220: 26–35.

Latt, S. A. 1981. Sister chromatid exchange formation. *Annu. Rev. Genet.* 15: 11–56.

Lindsley, D. L., and Grell, E. H. 1972. *Genetic Variations of Drosophila melanogaster*. Washington, DC: Carnegie Institute of Washington.

Morgan, T. H. 1911. An attempt to analyze the constitution of the chromosomes on the basis of sex-linked inheritance in *Drosophila*. *J. Exp. Zool.* 11: 365–414.

Morton, N. E. 1955. Sequential test for the detection of linkage. *Am. J. Hum. Genet.* 7: 277–318.

————. 1995. LODs—Past and present. *Genetics* 140: 7–12.

Neuffer, M. G., Jones, L., and Zober, M. 1968. *The mutants of maize*. Madison: Crop Sci. Soc. of America.

Perkins, D. 1962. Crossing over and interference in a multiply marked chromosome arm of *Neurospora*. *Genetics* 47: 1253–74.

Ruddle, F. H., and Kucherlapati, R. S. 1974. Hybrid cells and human genes. *Sci. Am.* (July) 231: 36–49.

Stahl, F. W. 1979. *Genetic Recombination.* New York: W. H. Freeman.

Stern, C. 1936. Somatic crossing over and segregation in *Drosophila Melanogaster*. *Genetics* 21: 625–31.

Sturtevant, A. H. 1913. The linear arrangement of six sex-linked factors in *Drosophila*, as shown by their mode of association. *J. Exp. Zool.* 14: 43–59.

————. 1965. *A History of Genetics*. New York: Harper & Row.

Voeller, B. R., ed. 1968. *The Chromosome Theory of Inheritance: Classical Papers in Development and Heredity*. New York: Appleton-Century-Croft.

Wolff, S., ed. 1982. *Sister Chromatid Exchange*. New York: Wiley-Interscience.

CHAPTER 9

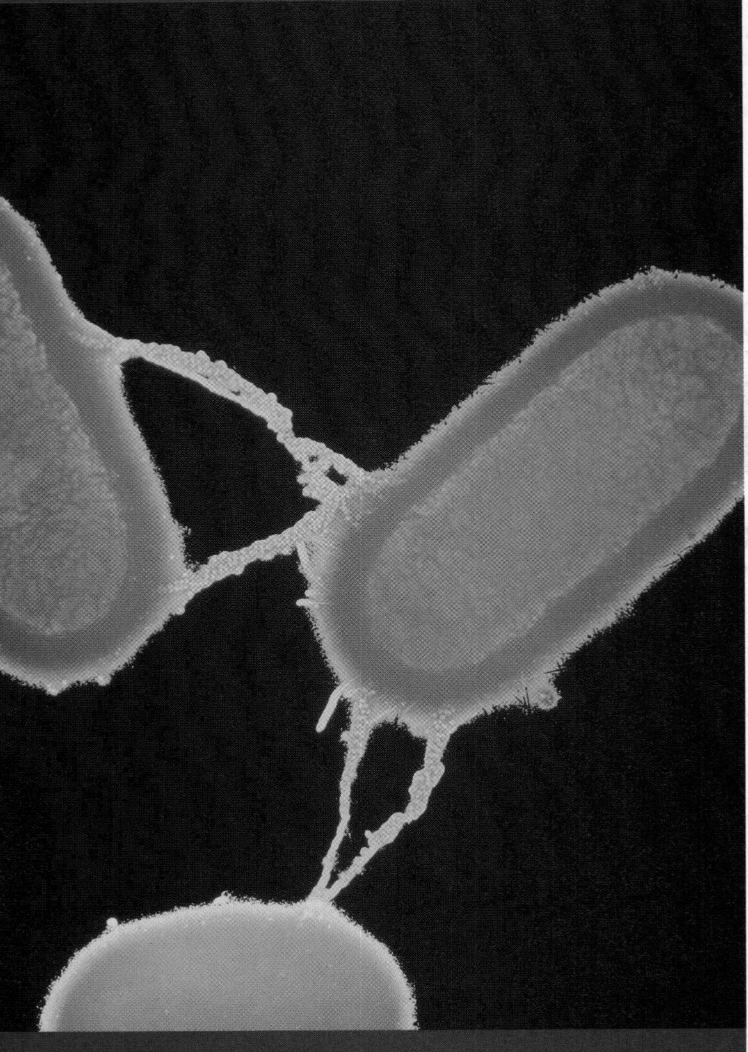

Transmission electron micrograph of conjugating *E. coli*. (*Dr. L. Caro/Science Photo Library/Photo Researchers, Inc.*)

Mapping in Bacteria and Bacteriophages

CHAPTER CONCEPTS

In this chapter, we shift from the consideration of transmission of genetic information in eukaryotes to a discussion of **bacteria** (prokaryotes) and **bacteriophages**, viruses that use bacteria as their hosts. The study of bacteria and bacteriophages has been essential to the accumulation of knowledge in many areas of genetic study. For example, much of what we know about molecular genetics, recombinational phenomena, and gene structure was initially derived from experimental work with these organisms. Furthermore, our extensive knowledge of bacteria and their resident plasmids has led to their widespread use in DNA cloning and other recombinant DNA studies.

Bacteria and their viruses are especially useful research organisms in genetics for several reasons. They have extremely short reproductive cycles—literally hundreds of generations, giving rise to billions of genetically identical bacteria or phages, can be produced in short periods of time. Furthermore, they can be studied in pure cultures. That is, a single species or mutant strain of bacteria or one type of virus can be isolated and investigated independently of other similar organisms.

In this chapter, we focus on genetic recombination and chromosome mapping. Complex processes have evolved in bacteria and bacteriophages that facilitate genetic recombination within populations. As we shall see, these processes are the basis for the chromosome mapping analysis that forms the cornerstone of molecular genetic investigations of bacteria and the viruses that invade them.

HOW DO WE KNOW WHAT WE KNOW?

IN THIS CHAPTER, WE WILL FOCUS ON GENETIC systems present in bacteria and the viruses that use bacteria as hosts (bacteriophages). In particular, we will discuss mechanisms by which bacteria and their phages undergo genetic recombination, the basis of chromosome mapping. As you study these topics, you should try the answer several fundamental questions:

1. How do we know that bacteria and bacteriophages display mutant phenotypes?
2. How do we know that bacteria undergo genetic recombination?
3. How do we know that conjugation involves cell contact?
4. How did we learn that the mechanism of genetic recombination differs between bacteria and bacteriophages?
5. How do we know that bacteriophages recombine genetic material through transduction? ■

9.1 Bacteria Mutate Spontaneously and Grow at an Exponential Rate

It has long been known that pure cultures of bacteria give rise to cells that exhibit heritable variation, particularly with respect to growth under unique environmental conditions. Prior to 1943, the source of this variation was hotly debated. The majority of bacteriologists believed that environmental factors induced changes in certain bacteria that led to their adaptation to the new conditions. For example, strains of *Escherichia coli* are known to be sensitive to infection by the bacteriophage T1. Infection by this bacteriophage leads to the virus reproducing at the expense of the bacterial cell, from which new phages are released as the host cell is disrupted, or lysed. If a plate of *E. coli* is uniformly sprayed with T1, almost all cells are lysed. Rare *E. coli* cells, however, survive infection and are not lysed. If these cells are isolated and established in pure culture, all of the descendants are resistant to T1 infection. The **adaptation hypothesis**, put forth to explain this type of observation, implies that the interaction of the phage and bacterium is essential to the acquisition of immunity. In other words, the phage "induces" resistance in the bacteria.

On the other hand, the existence of **spontaneous mutations**, which occur in the presence or the absence of bacteriophage T1, suggested an alternative model to explain the origin of resistance in *E. coli*. In 1943, Salvador Luria and Max Delbruck presented the first convincing evidence that bacteria, like eukaryotic organisms, are capable of spontaneous mutation. This experiment, referred to as the **fluctuation test**, marks the initiation of modern bacterial genetic study. Spontaneous mutation is now considered the primary source of genetic variation in bacteria.

Mutant cells that arise spontaneously in otherwise pure cultures can be isolated and established independently from the parent strain by using established selection techniques. As a result, mutations for almost any desired characteristic can now be induced and isolated. Because bacteria and viruses usually contain only a single chromosome and are therefore haploid, all mutations are expressed directly in the descendants of mutant cells, adding to the ease with which these microorganisms can be studied.

Bacteria are grown in a liquid culture medium or in a petri dish on a semisolid agar surface. If the nutrient components of the growth medium are simple and consist only of an organic carbon source (such as a glucose or lactose) and a variety of ions, including Na^+, K^+, Mg^{2+}, Ca^{2+}, and NH^{4+}, present as inorganic salts, it is called **minimal medium**. To grow on such a medium, a bacterium must be able to synthesize all essential organic compounds (e.g., amino acids, purines, pyrimidines, sugars, vitamins, and fatty acids). A bacterium that can accomplish this remarkable biosynthetic

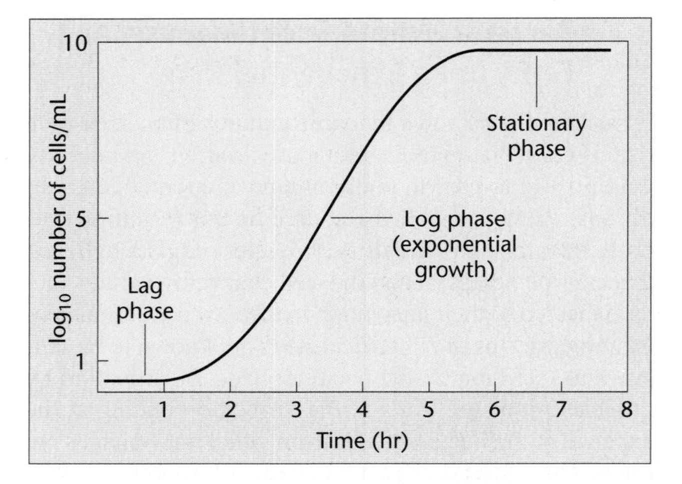

FIGURE 9–1 Typical bacterial population growth curve showing the initial lag phase, the subsequent log phase where exponential growth occurs, and the stationary phase that occurs when nutrients are exhausted.

feat—one that we ourselves cannot duplicate—is a **prototroph**. It is said to be wild type for all growth requirements. On the other hand, if a bacterium loses, through mutation, the ability to synthesize one or more organic components, it is an **auxotroph**. For example, if it loses the ability to make histidine, then this amino acid must be added as a supplement to the minimal medium for growth to occur. The resulting bacterium is designated as a his^- auxotroph, in contrast to its prototrophic his^+ counterpart.

To study mutant bacteria quantitatively, an inoculum of bacteria is placed in liquid culture medium. A graph of the characteristic growth pattern is shown in Figure 9–1. Initially, during the **lag phase**, growth is slow. Then, a period of rapid growth, called the **logarithmic (log) phase**, ensues. During this phase, cells divide continually with a fixed time interval between cell divisions, resulting in exponential growth. When a cell density of about 10^9 cells/mL is reached, nutrients and oxygen become limiting and the cells cease dividing; at this point, the cells enter the **stationary phase**. The doubling time during the log phase can be as short as 20 minutes. Thus, an initial inoculum of a few thousand cells easily achieves maximum cell density during an overnight incubation.

Cells grown in liquid medium can be quantified by transferring them to semisolid medium in a petri dish. Following incubation and many divisions, each cell gives rise to a visible colony on the surface of the medium. If the number of colonies is too great to count, then a series of successive dilutions (a technique called serial dilution) of the original liquid culture is made and plated, until the colony number is reduced to the point where it can be counted (Figure 9–2). This technique allows the number of bacteria present in the original culture to be calculated.

As an example, let's assume that the three dishes in Figure 9–2 represent serial dilutions of 10^{-3}, 10^{-4}, and 10^{-5}, (from left to right). We need only select the dish in which the number of colonies can be counted accurately. Because each colony arose from a single bacterium, the number of colonies multiplied by the dilution factor represents the number of bacteria in each milliliter of the initial inoculum used to start the serial dilutions. In Figure 9–2, the rightmost dish has 15 colonies. The dilution factor for a 10^{-5} dilution is 10^5. Therefore, the initial number of bacteria is 15×10^5 per mL.

9.2 Conjugation Is One Means of Genetic Recombination in Bacteria

Development of techniques that allowed the identification and study of bacterial mutations led to detailed investigations of the arrangement of genes on the bacterial chromosome. Such studies began in 1946 when Joshua Lederberg and Edward Tatum showed that bacteria undergo **conjugation**, a parasexual process in which the genetic information from one bacterium is transferred to and recombined with that of another bacterium (see the chapter opening photograph). Like meiotic crossing over in eukaryotes, genetic recombination in bacteria enabled the development of methodology for chromosome mapping. Note that the term **genetic recombination**, as applied to bacteria and bacteriophages, leads to the *replacement* of one or more genes present in one strain with those from a genetically distinct strain. While this is somewhat different from our use of genetic recombination in eukaryotes, where the term describes crossing over that results in *reciprocal exchange events*, the overall effect is the same:

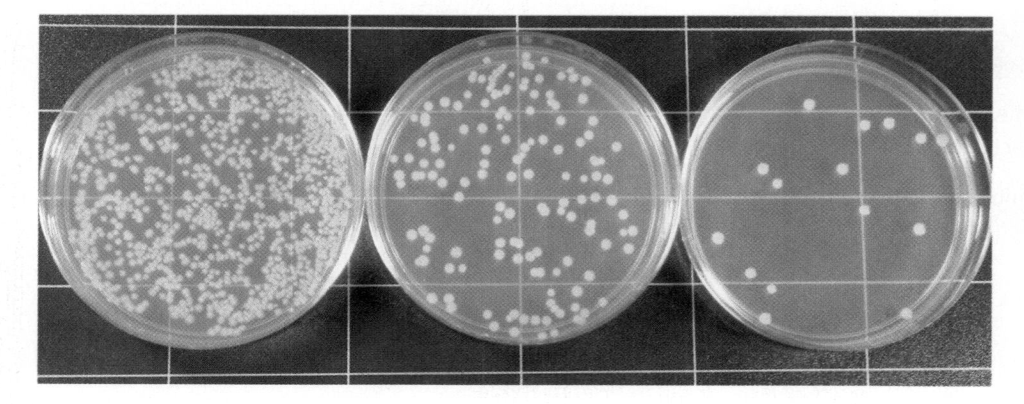

FIGURE 9–2 Results of the serial dilution technique and subsequent culture of bacteria. Each dilution varies by a factor of 10. Each colony is derived from a single bacterial cell. *(Michael G. Gabridge/Visuals Unlimited)*

Genetic information is transferred from one chromosome to another, resulting in an altered genotype. Two other phenomena that result in the transfer of genetic information from one bacterium to another, *transformation* and *transduction*, have helped us determine the arrangement of genes on the bacterial chromosome. We will discuss these processes in later sections of this chapter.

Lederberg and Tatum's initial experiments were performed with two multiple-auxotroph strains of *E. coli* K12. Strain A required methionine and biotin in order to grow, while strain B required threonine, leucine, and thiamine (Figure 9–3). Therefore neither strain would grow on minimal medium. The two strains were first grown separately in supplemented media, and cells from both were mixed and grown together for several more generations and then plated on minimal medium. Any bacterial cells that grew on minimal medium were prototrophs. It is highly improbable that any of the cells that contained two or three mutant genes underwent spontaneous mutation simultaneously at two or three independent locations, leading to wild-type cells. Therefore, the researchers assumed that any prototrophs recovered arose as a result of some form of genetic exchange and recombination between the two mutant strains.

In this experiment, prototrophs were recovered at a rate of $1/10^7$ or (10^{-7}) cells plated. The controls for this experiment involved separate plating of cells from strains A and B on minimal medium. No prototrophs were recovered. Based on these observations, Lederberg and Tatum proposed that genetic exchange had occurred.

F^+ and F^- Bacteria

The findings of Lederberg and Tatum were soon followed by numerous experiments that elucidated the genetic basis of conjugation. It quickly became evident that different strains of bacteria are involved in a unidirectional transfer of genetic material. When cells serve as donors of parts

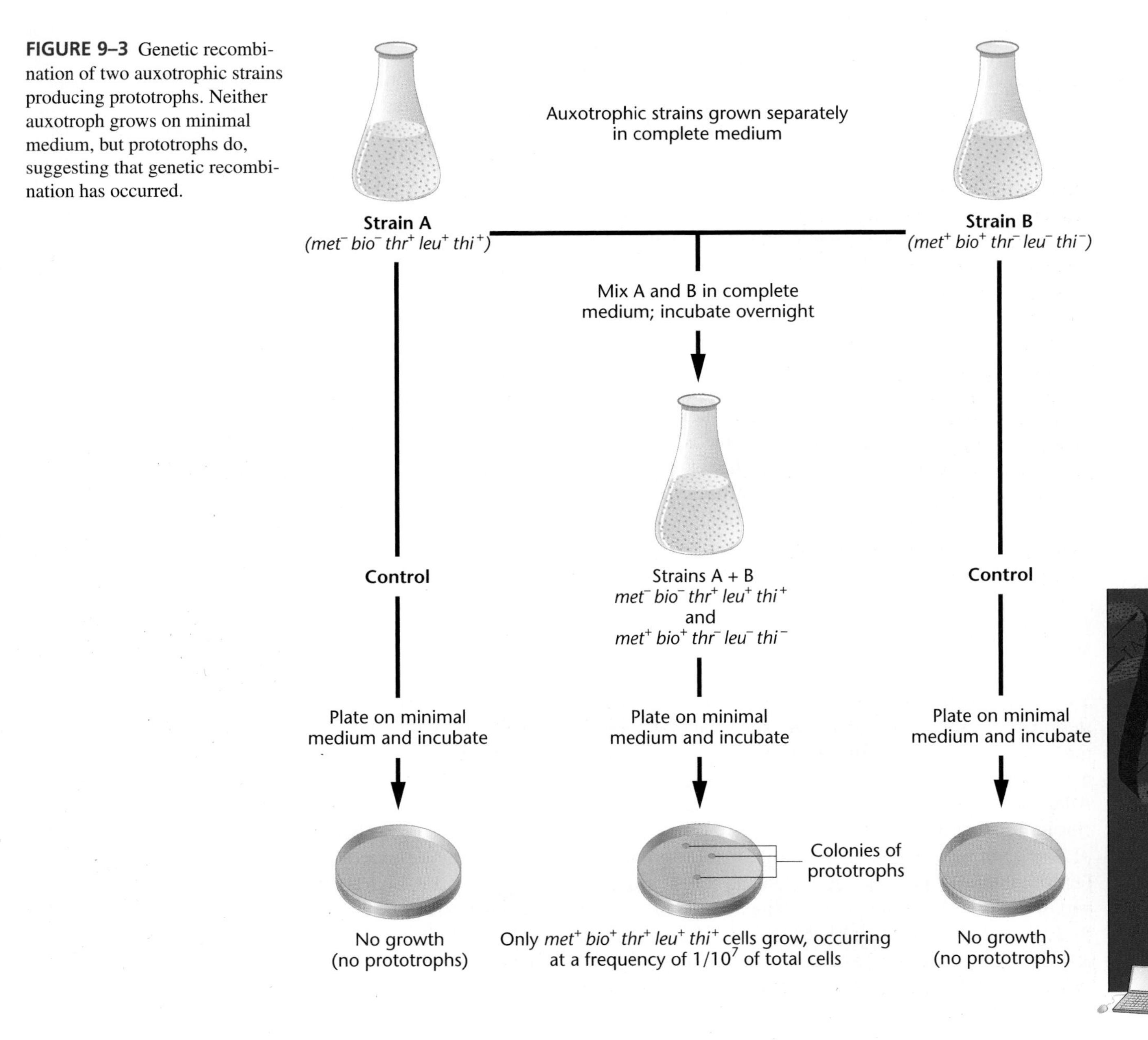

FIGURE 9–3 Genetic recombination of two auxotrophic strains producing prototrophs. Neither auxotroph grows on minimal medium, but prototrophs do, suggesting that genetic recombination has occurred.

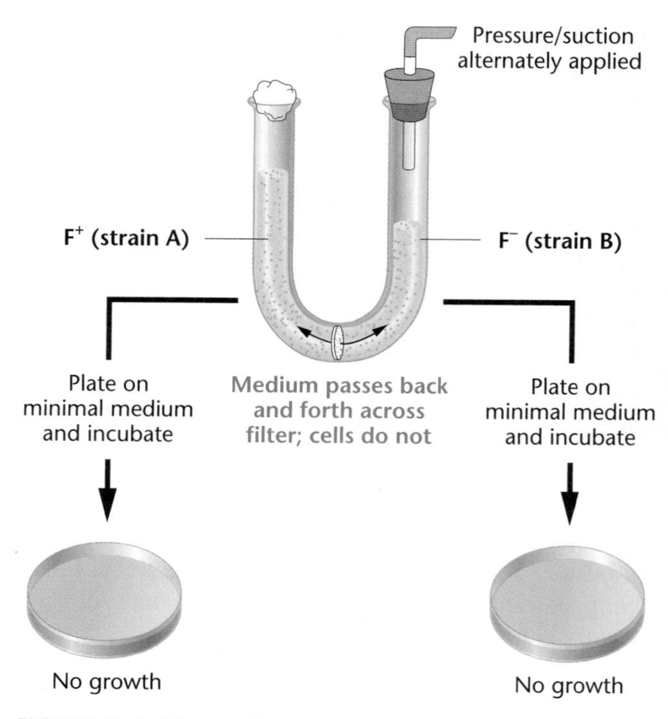

FIGURE 9–4 When strain A and B auxotrophs are grown in a common medium but separated by a filter, as in this Davis U-tube apparatus, no genetic recombination occurs and no prototrophs are produced.

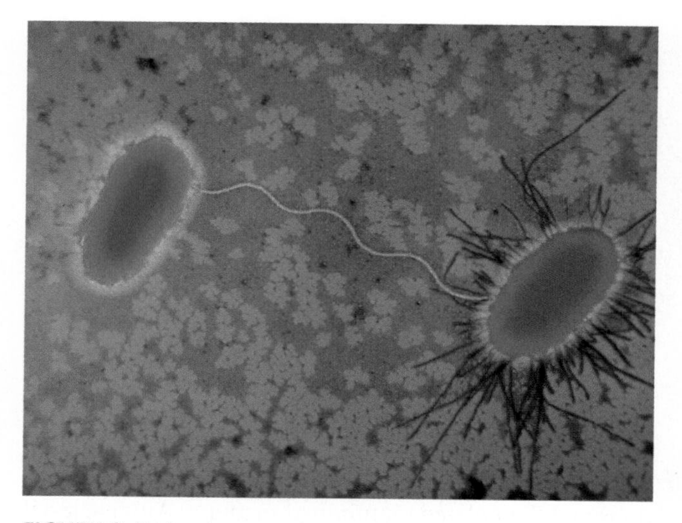

FIGURE 9–5 An electron micrograph of conjugation between an F^+ *E. coli* cell and an F^- cell. The sex pilus linking them is clearly visible. *(Dennis Kunkel/Phototake NYC)*

of their chromosomes, they are designated as $\mathbf{F^+}$ **cells** (F for "fertility"). Recipient bacteria receive the donor chromosome material (now known to be DNA), and recombine it with part of their own chromosome. They are designated as $\mathbf{F^-}$ **cells**.

Experimentation subsequently established that cell contact is essential for chromosome transfer to occur. Support for this concept was provided by Bernard Davis, who designed the Davis U-tube for growing F^+ and F^- cells shown in Figure 9–4. At the base of the tube is a sintered glass filter with a pore size that allows passage of the liquid medium, but is too small to allow the passage of bacteria. The F^+ cells are placed on one side of the filter, and F^- cells on the other side. The medium is moved back and forth across the filter so the cells share a common medium during bacterial incubation. When Davis plated samples from both sides of the tube on minimal medium, no prototrophs were found, and he logically concluded that *physical contact between cells of the two strains is essential to genetic recombination.* We now know that this physical interaction is the initial step in the process of conjugation established by a structure called the **F pilus** (or **sex pilus**; pl, pili). Bacteria often have many pili, which are tubular extensions of the cell. After contact is initiated between mating pairs (Figure 9–5), chromosome transfer begins.

Later evidence established that F^+ cells contain a **fertility factor** (**F factor**) that confers the ability to donate part of their chromosome during conjugation. Experiments by Joshua and Esther Lederberg and by William Hayes and Luca Cavalli-Sforza show that certain conditions eliminate the F factor in otherwise fertile cells. However, if these "infertile" cells are then grown with fertile donor cells, the F factor is regained.

The conclusion that the F factor is a mobile element is further supported by the observation that, following conjugation and genetic recombination, recipient cells always become F^+. Thus, in addition to the *rare* cases of gene transfer from the bacterial chromosome (genetic recombination), the F factor itself is passed to *all* recipient cells. On this basis, the initial crosses of Lederberg and Tatum (see Figure 9–3) is as follows:

Strain A		***Strain B***
$\mathbf{F^+}$	×	$\mathbf{F^-}$
Donor		**Recipient**

Isolation of the F factor confirmed these conclusions. Like the bacterial chromosome, though distinct from it, the F factor has been shown to consist of a circular, double-stranded DNA molecule, equivalent to about 2% of the bacterial chromosome (about 100,000 nucleotide pairs). There are 19 genes contained within the F factor, whose products are involved in the transfer of genetic information, including those essential to the formation of the sex pilus.

Geneticists believe that transfer of the F factor during conjugation involves separation of the two strands of its double helix and the movement of one of the two strands into the recipient cell. Both strands, one moving across the conjugation tube and one remaining in the donor cell, are replicated. The result is that both the donor *and* the recipient cells become F^+. This process is diagrammed in Figure 9–6.

To summarize, an *E. coli* cell may or may not contain the F factor. When this factor is present, the cell is able to form a sex pilus and potentially serves as a donor of genetic information. During conjugation, a copy of the F factor is almost always transferred from the F^+ cell to the F^- recipient, converting the recipient to the F^+ state. The question remained as to exactly why such a low proportion of cells involved in these matings (10^{-7}) also results in genetic recombination. The answer awaited further experimentation.

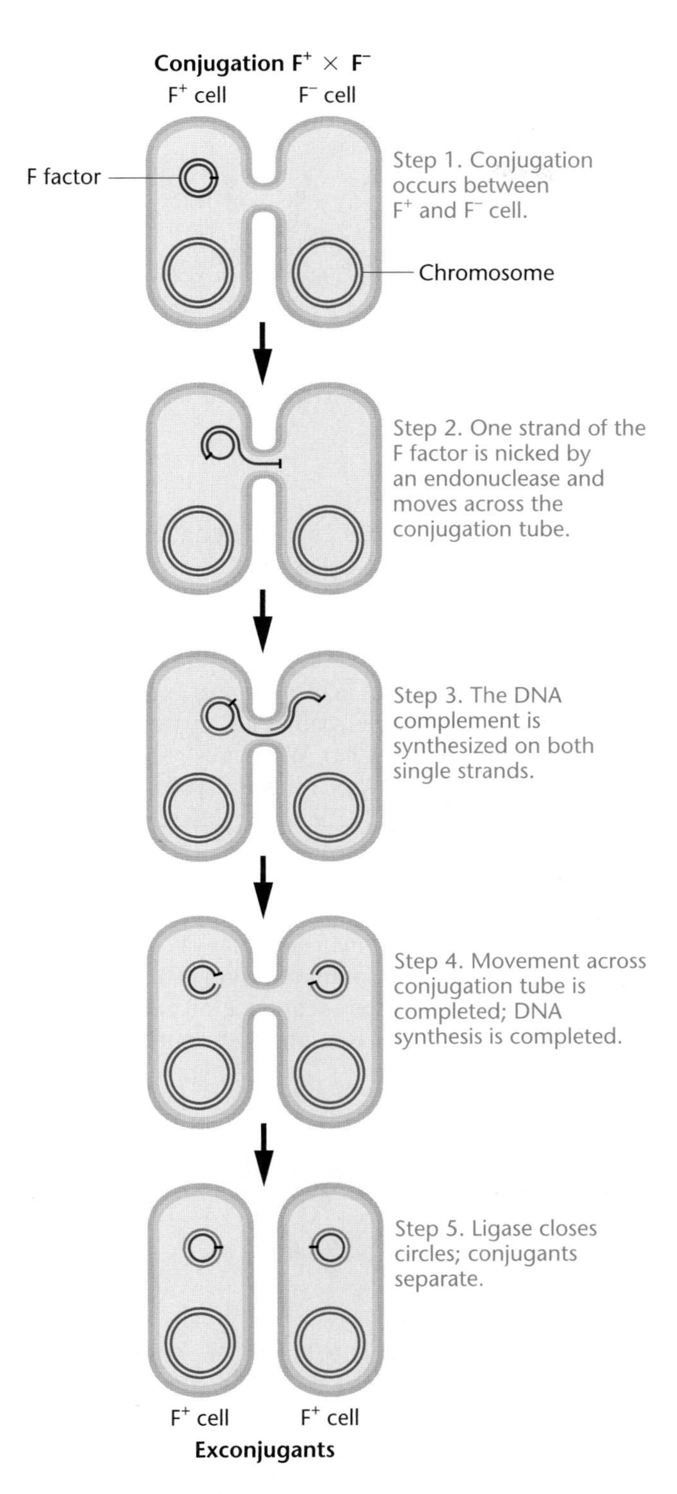

Conjugation F⁺ × F⁻
F⁺ cell
F⁻ cell
F factor
Step 1. Conjugation occurs between F⁺ and F⁻ cell.
Chromosome
Step 2. One strand of the F factor is nicked by an endonuclease and moves across the conjugation tube.
Step 3. The DNA complement is synthesized on both single strands.
Step 4. Movement across conjugation tube is completed; DNA synthesis is completed.
Step 5. Ligase closes circles; conjugants separate.
F⁺ cell
F⁺ cell
Exconjugants

FIGURE 9–6 An $F^+ \times F^-$ mating demonstrating how the recipient F^- cell converts to F^+. During conjugation, the DNA of the F factor replicates with one new copy entering the recipient cell, converting it to F^+. The black bars added to the F factors follow their clockwise rotation during replication.

As you soon shall see, the F factor is in reality an autonomous genetic unit called a *plasmid*. However, in our historical coverage of its discovery, we will continue to refer to it as a factor.

Hfr Bacteria and Chromosome Mapping

Subsequent discoveries not only clarified how genetic recombination occurs but also defined a mechanism by which the *E. coli* chromosome could be mapped. Let's address chromosome mapping first.

In 1950, Cavalli-Sforza treated an F^+ strain of *E. coli* K12 with nitrogen mustard, a chemical known to induce mutations. From these treated cells, he recovered a genetically altered strain of donor bacteria that underwent recombination at a rate of $1/10^4$ or (10^{-4})—1000 times more frequently than the original F^+ strains. In 1953, Hayes isolated another strain that demonstrated an elevated frequency. Both strains were designated **Hfr**, for **high-frequency recombination**. Because Hfr cells behave as donors, they are a special class of F^+ cells.

Another important difference was noted between Hfr strains and the original F^+ strains. If the donor is from an Hfr strain, recipient cells, while sometimes displaying genetic recombination, never become Hfr; that is, they remain F^-. In comparison, then,

$$F^+ \times F^- \rightarrow F^+ \qquad \text{(low rate of recombination)}$$

$$\text{Hfr} \times F^- \rightarrow F^- \qquad \text{(higher rate of recombination)}$$

Perhaps the most significant characteristic of Hfr strains is the *nature of recombination*. In any given strain, certain genes are more frequently recombined than others, and some not at all. This *nonrandom* pattern was shown to vary between Hfr strains. While these results were puzzling, Hayes interpreted them to mean that some physiological alteration of the F factor had occurred, resulting in the production of Hfr strains of *E. coli*.

In the mid-1950s, experimentation by Ellie Wollman and François Jacob elucidated the difference between Hfr and F^+ and showed how Hfr strains allow genetic mapping of the *E. coli* chromosome. In their experiments, Hfr and F^- strains with suitable marker genes were mixed, and recombination of specific genes was assayed at different times. To accomplish this, a culture containing a mixture of an Hfr and an F^- strain was first incubated and samples were removed at various intervals and placed in a blender. The shear forces in the blender separated conjugating bacteria so the transfer of the chromosome was terminated. The cells were then assayed for genetic recombination.

This process, called the **interrupted mating technique**, demonstrated that specific genes of a given Hfr strain were transferred and recombined sooner than others. The graph in Figure 9–7 illustrates this point. During the first eight minutes after the two strains were mixed, no genetic recombination was detected. At about 10 minutes, recombination of the *azi* gene was detected, but no transfer of the ton^s, lac^+, or gal^+, genes was noted. By 15 minutes, 50% of the recombinants were azi^+, and 15% were ton^s; but none was lac^+ or gal^+. Within 20 minutes, the lac^+ was found among the recombinants; and within 25 minutes, gal^+ was also being transferred. Wollman and Jacob had demonstrated *an ordered transfer of genes* that correlated with the length of time conjugation proceeded.

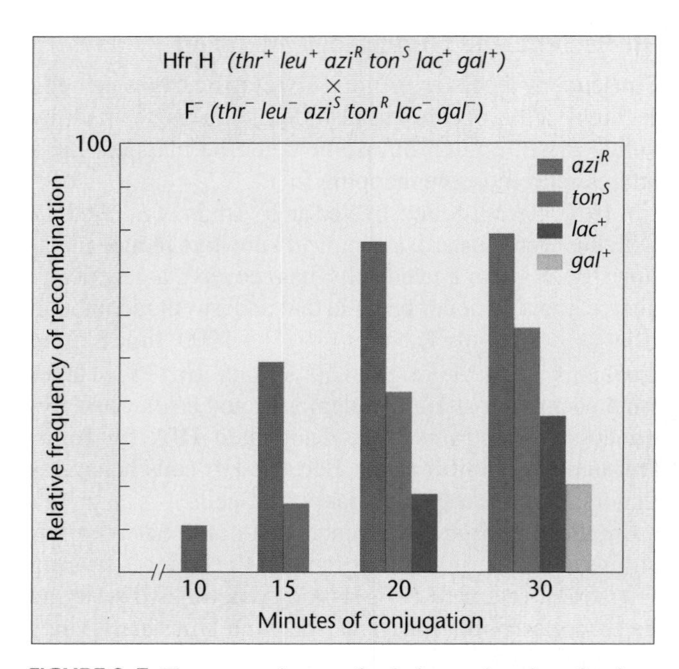

FIGURE 9–7 The progressive transfer during conjugation of various genes from a specific Hfr strain of *E. coli* to an F^- strain. Certain genes (*azi* and *ton*) transfer more quickly than others and recombine more frequently. Others (*lac* and *gal*) take longer to transfer and recombine with a lower frequency.

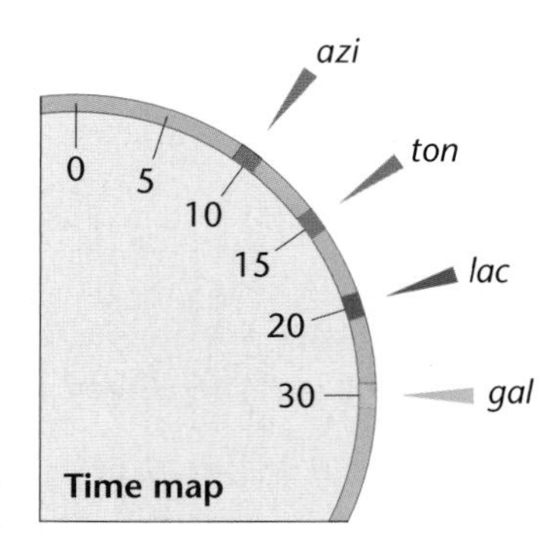

FIGURE 9–8 A time map of the genes studied in the experiment depicted in Figure 9–7.

It appeared that the chromosome of the Hfr bacterium was transferred linearly and that the gene order and distance between genes, as measured in minutes, could be predicted from such experiments (Figure 9–8). This information served as the basis for the first genetic map of the *E. coli* chromosome. Minutes in bacterial mapping are equivalent to map units in eukaryotes.

Wollman and Jacob then repeated the same type of experiment with other Hfr strains, obtaining similar results with one important difference. While genes were always transferred linearly with time, as in their original experiment, the order in which genes entered seemed to vary from Hfr strain to Hfr strain (see Figure 9–9a). When they reexamined the entry rate of genes, and thus the genetic maps for each strain, a definite pattern emerged. The major difference between each strain was simply the point of origin (*O*) and the direction in which entry proceeded from that point (Figure 9–9b).

To explain these results, Wollman and Jacob postulated that the *E. coli* chromosome is circular (a closed circle, with no free ends). If the point of origin (*O*) varies from strain to strain, a different sequence of genes will be transferred in each case. But what determines *O*? They

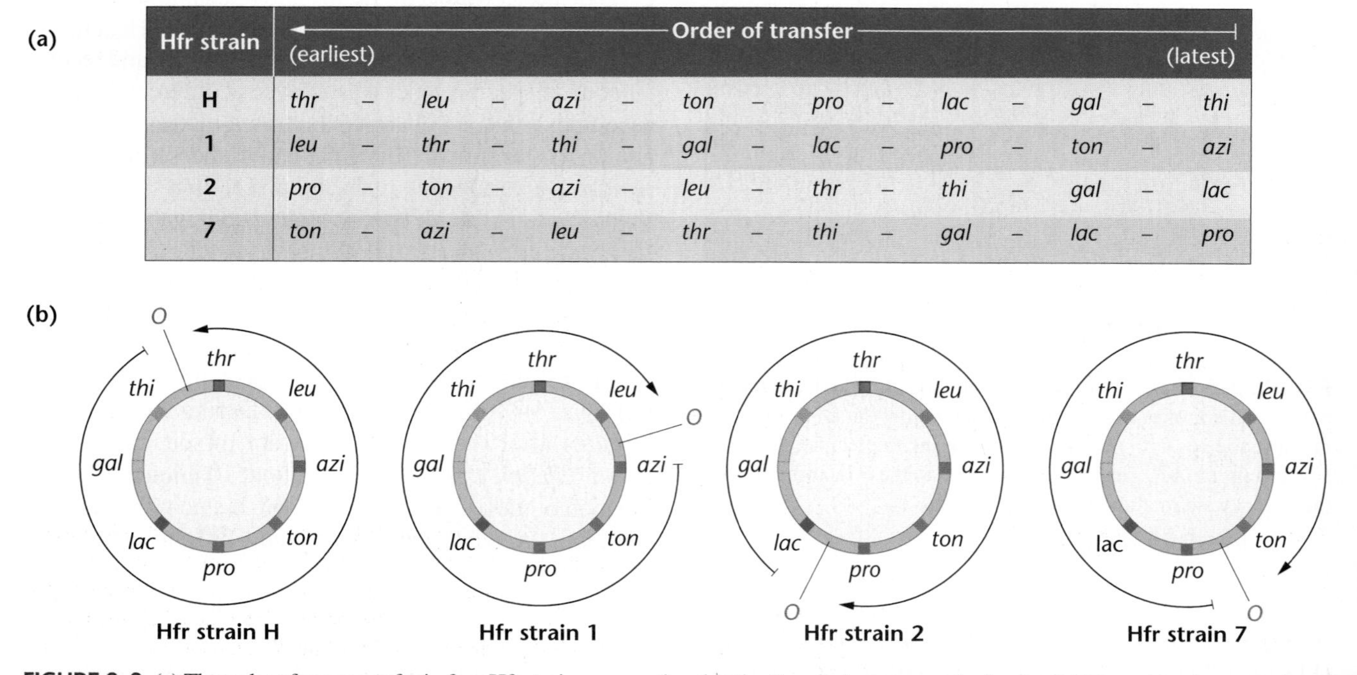

(a)

Hfr strain	Order of transfer (earliest)							(latest)
H	*thr* –	*leu* –	*azi* –	*ton* –	*pro* –	*lac* –	*gal* –	*thi*
1	*leu* –	*thr* –	*thi* –	*gal* –	*lac* –	*pro* –	*ton* –	*azi*
2	*pro* –	*ton* –	*azi* –	*leu* –	*thr* –	*thi* –	*gal* –	*lac*
7	*ton* –	*azi* –	*leu* –	*thr* –	*thi* –	*gal* –	*lac* –	*pro*

FIGURE 9–9 (a) The order of gene transfer in four Hfr strains, suggesting that the *E. coli* chromosome is circular. (b) The point where transfer originates (*O*) is identified in each strain. Note that transfer proceeds in either direction, depending on the strain. The origin is determined by the point of integration into the chromosome of the F factor, and the direction of transfer is determined by the orientation of the F factor as it integrates.

proposed that in various Hfr strains, the F factor integrates into the chromosome at different points, and its position determines the site of *O*. A case of integration is shown in step 1 of Figure 9–10. During conjugation between an Hfr and an F^- cell, the position of the F factor determines the initial point of transfer (steps 2 and 3). Those genes adjacent to *O* are transferred first, and *the F factor becomes the last part that can be transferred* (step 4). However, conjugation rarely, if ever, lasts long enough to allow the entire chromosome to pass across the conjugation tube (step 5). This proposal explains why recipient cells, when mated with Hfr cells, remain F^-.

Figure 9–10 also depicts the way in which the two strands making up a DNA molecule behave during transfer, allowing for the entry of one strand of DNA into the recipient (see step 3). Following replication, the entering DNA now has the potential to recombine with its homologous region of the host chromosome. The DNA strand that remains in the donor also undergoes replication.

Use of the interrupted mating technique with different Hfr strains enabled researchers to map the entire *E. coli* chromosome. Mapped in time units, strain *E. coli* K12 is 100 minutes long—over 900 genes have been placed on the map.

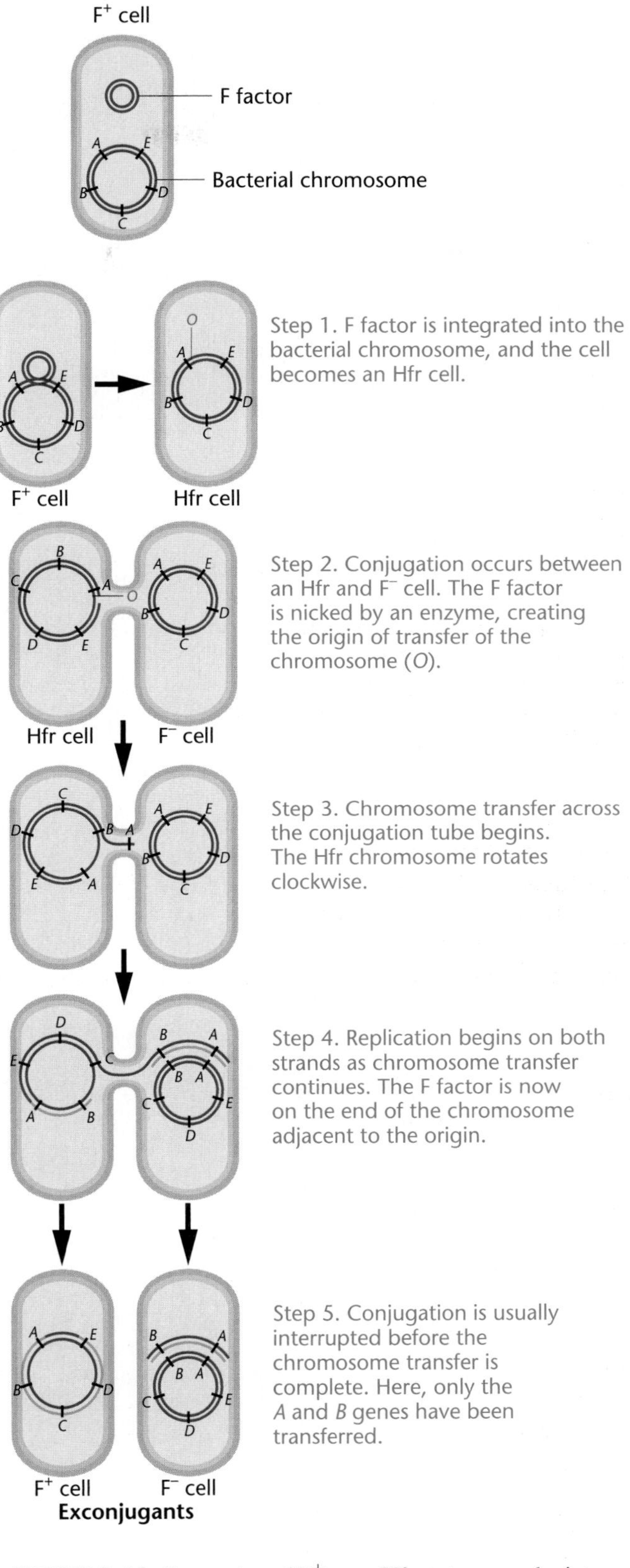

F^+ cell
F factor
Bacterial chromosome
A
E
B
D
C
Step 1. F factor is integrated into the bacterial chromosome, and the cell becomes an Hfr cell.
F^+ cell
Hfr cell
O
Step 2. Conjugation occurs between an Hfr and F^- cell. The F factor is nicked by an enzyme, creating the origin of transfer of the chromosome (*O*).
Hfr cell
F^- cell
Step 3. Chromosome transfer across the conjugation tube begins. The Hfr chromosome rotates clockwise.
Step 4. Replication begins on both strands as chromosome transfer continues. The F factor is now on the end of the chromosome adjacent to the origin.
Step 5. Conjugation is usually interrupted before the chromosome transfer is complete. Here, only the *A* and *B* genes have been transferred.
F^+ cell
F^- cell
Exconjugants

FIGURE 9–10 Conversion of F^+ to an Hfr state occurs by integrating the F factor into the bacterial chromosome. The point of integration determines the origin (*O*) of transfer. During conjugation, an enzyme nicks the F factor, now integrated into the host chromosome, initiating transfer of the chromosome at that point. Conjugation is usually interrupted prior to complete transfer. Above, only the *A* and *B* genes are transferred to the F^- cell, which may recombine with the host chromosome.

Recombination in $F^+ \times F^-$ Matings: A Reexamination

The preceding experiment helped geneticists better understand how genetic recombination occurs during $F^+ \times F^-$ matings. Recall that recombination occurs much less frequently than in Hfr $\times$ F^- matings and that random gene transfer is involved. The current belief is that when F^+ and F^- cells are mixed, conjugation occurs readily and each F^- cell involved in conjugation with an F^+ cell receives a copy of the F factor, *but no genetic recombination occurs*. However, at an extremely low frequency in a population of F^+ cells, the F factor integrates spontaneously from the cytoplasm to a random point in the bacterial chromosome, converting the F^+ cell to the Hfr state as we saw in Figure 9–10. Therefore, in $F^+ \times F^-$ matings, the extremely low frequency of genetic recombination (10^{-7}) is attributed to the rare, newly formed Hfr cells, which then undergo conjugation with F^- cells. Because the point of integration of the F factor is random, which gene or genes are transferred by any newly formed Hfr donor will also appear to be random within the larger F^+/F^- population. The recipient bacterium will appear as a recombinant, but will remain F^-. If it subsequently undergoes conjugation with an F^+ cell, it will then be converted to F^+.

The F′ State and Merozygotes

In 1959, during experiments with Hfr strains of *E. coli*, Edward Adelberg discovered that the F factor could lose its integrated status, causing the cell to revert to the F^+ state (Figure 9–11, step 1). When this occurs, the F factor frequently carries several adjacent bacterial genes along with it (step 2). Adelberg labeled this condition **F′** to distinguish

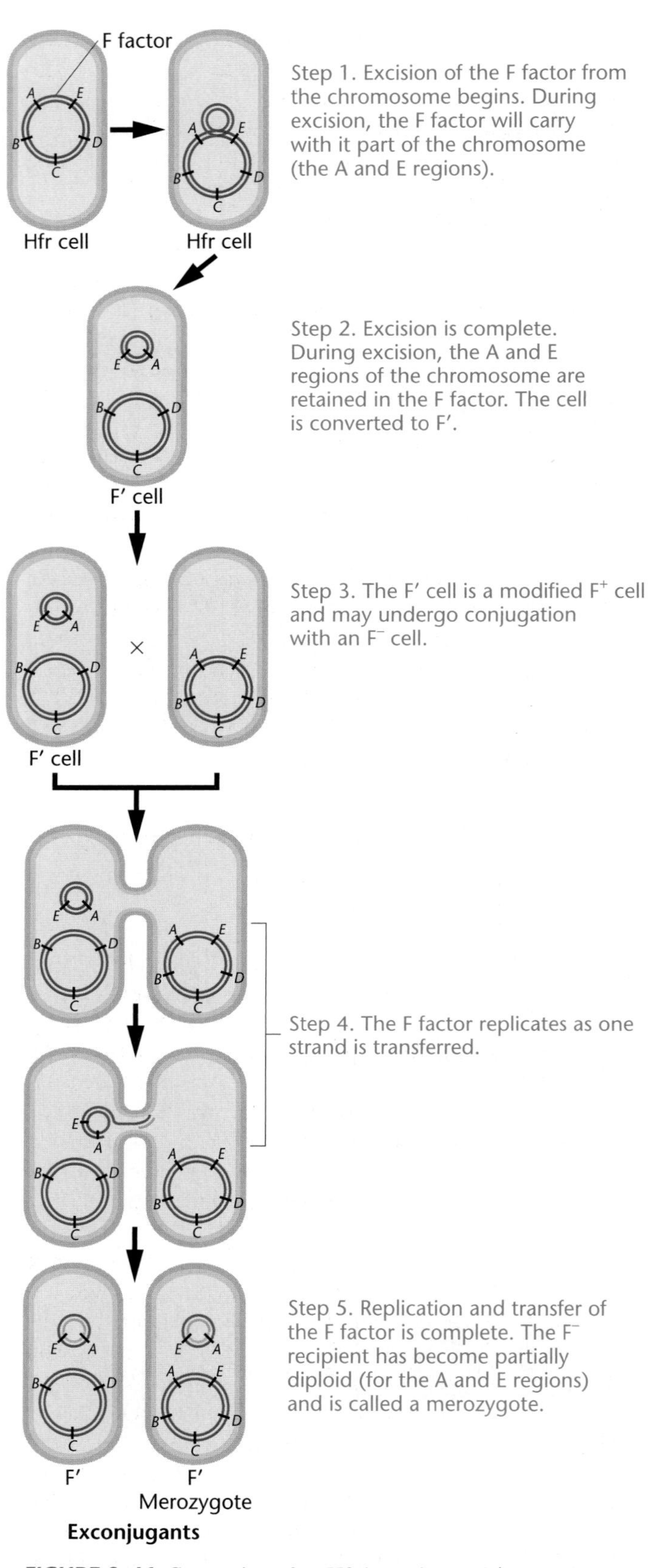

F factor
Hfr cell
Hfr cell
Step 1. Excision of the F factor from the chromosome begins. During excision, the F factor will carry with it part of the chromosome (the A and E regions).
Step 2. Excision is complete. During excision, the A and E regions of the chromosome are retained in the F factor. The cell is converted to F′.
F′ cell
Step 3. The F′ cell is a modified F+ cell and may undergo conjugation with an F− cell.
F′ cell
Step 4. The F factor replicates as one strand is transferred.
Step 5. Replication and transfer of the F factor is complete. The F− recipient has become partially diploid (for the A and E regions) and is called a merozygote.
F′
F′
Merozygote
Exconjugants

FIGURE 9–11 Conversion of an Hfr bacterium to F′ and its subsequent mating with an F^- cell. The conversion occurs when the F factor loses its integrated status. During excision from the chromosome, it carries with it one or more chromosomal genes (*A* and *E*). Following conjugation with an F^- cell, the recipient cell becomes partially diploid and is called a merozygote; it also behaves as an F^+ donor cell.

it from F^+ and Hfr. F′, like Hfr, is thus another special case of F^+, but this conversion is from Hfr to F′.

The presence of bacterial genes within a cytoplasmic F factor creates an interesting situation. An F′ bacterium behaves like an F^+ cell by initiating conjugation with F^- cells (Figure 9–11, step 3). When this occurs, the F factor, containing chromosomal genes, is transferred to the F^- cell (step 4). As a result, whatever chromosomal genes are part of the F factor are now present as duplicates in the recipient cell (step 5) because the recipient still has a complete chromosome. This creates a partially diploid cell called a **merozygote**. Pure cultures of F′ merozygotes can be established. They have been extremely useful in the study of bacterial genetics, particularly in genetic regulation.

9.3 Rec Proteins Are Essential to Bacterial Recombination

Once researchers established that a unidirectional transfer of DNA occurs between bacteria, they became interested in determining how the actual recombination event occurs in the recipient cell. Just how does the donor DNA replace the comparable region in the recipient chromosome? As with many systems, the biochemical mechanism by which recombination occurs was deciphered through genetic studies. Major insights were gained as a result of isolating a group of mutations representing ***rec* genes**.

The first relevant observation involved a series of mutant genes labeled *recA*, *recB*, *recC*, and *recD*. The first mutant gene, *recA*, diminished genetic recombination in bacteria 1000-fold, nearly eliminating it altogether; the other *rec* mutations reduced recombination by about 100 times. Clearly, the normal wild-type products of these genes must play some essential role in the process of recombination.

Researchers looked for, and subsequently isolated, several functional gene products present in normal cells but missing in mutant cells and showed that they play a role in genetic recombination. The first is called the **RecA protein**.* The second is a more complex protein called the **RecBCD protein**, an enzyme consisting of polypeptide subunits encoded by three other *rec* genes. The roles of these proteins have now been elucidated *in vitro*. This genetic research has considerably extended our knowledge of the process of recombination and underscores the value of isolating mutations, establishing their phenotypes and determining the biological role of the normal, wild-type gene. We will examine the molecular role of Rec proteins in recombination in Chapter 11 once we have thoroughly explored the topic of DNA structure.

*Note that the names of bacterial genes use lowercase letters and are italicized, while the corresponding gene products begin with capital letters and are not italicized, as illustrated by the *recA* gene and RecA protein.

9.4 F Factors Are Plasmids

In the preceding sections, we examined the extrachromosomal heredity unit called the F factor. When it exists autonomously in the bacterial cytoplasm, the F factor is composed of a double-stranded closed circle of DNA (Figure 9–12a). These characteristics place the F factor in the more general category of genetic structures called **plasmids**. These structures contain one or more genes, and often, quite a few. Their replication depends upon the same enzymes that replicate the chromosome of the host cell, and they are distributed to daughter cells along with the host chromosome during cell division.

Plasmids are generally classified according to the genetic information specified by their DNA. The F factor plasmid confers fertility and contains the genes essential for sex pilus formation, upon which genetic recombination depends. Other examples of plasmids include the R and Col plasmids.

Most **R plasmids** consist of two components: the **resistance transfer factor** (**RTF**) and one or more **r-determinants** (Figure 9–12b). The RTF encodes genetic information essential to transferring the plasmid between bacteria, and the r-determinants are genes that confer resistance to antibiotics. While RTFs are similar in a variety of plasmids from different bacterial species, r-determinants are specific for resistance to one class of antibiotic and vary widely. Resistance to tetracycline, streptomycin, ampicillin, sulfonamide, kanamycin, and chloramphenicol are most frequently encountered. Sometimes these occur in a single plasmid, conferring multiple resistance to several antibiotics (Figure 9–12b). Bacteria bearing these plasmids are of great medical significance not only because of their multiple resistance but because of the ease with which the plasmids can be transferred to other bacteria. Sometimes, a bacterial cell contains r-determinant plasmids but no RTF—the cell is resistant but cannot transfer the genetic information for resistance to recipient cells. The most commonly studied plasmids, however, contain the RTF as well as one or more r-determinants.

The **Col plasmid**, ColE1, (derived from *E. coli*), is clearly distinct from R plasmids. It encodes one or more proteins that are highly toxic to bacterial strains that do not harbor the same plasmid. These proteins, called **colicins**, can kill neighboring bacteria, and bacteria that carry the plasmid are said to be colicinogenic. Present in 10–20 copies per cell, a gene in the Col plasmid encodes an immunity protein that protects the host cell from the toxin. Unlike an R plasmid, the Col plasmid is not usually transmissible to other cells.

Interest in plasmids has increased dramatically because of their role in the genetic technology known as recombinant DNA research. Specific genes from any source can be inserted into a plasmid, which may then be inserted into a bacterial cell. As the altered cell replicates its DNA and undergoes division, the foreign gene is also replicated, thus cloning the foreign genes.

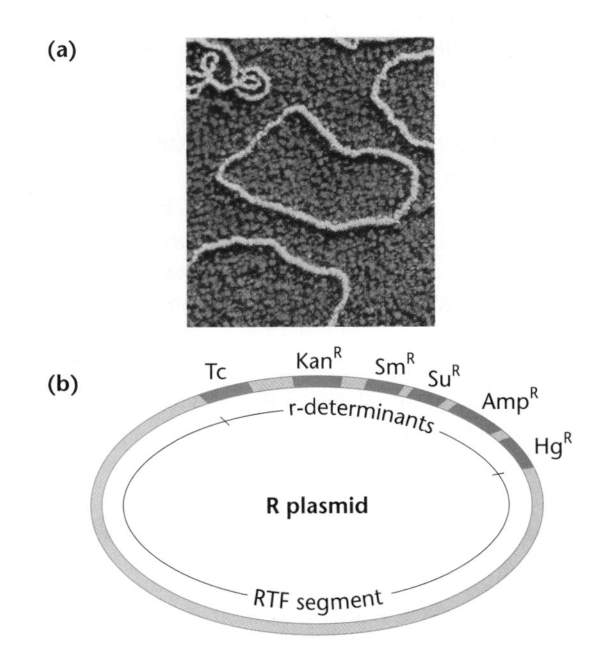

FIGURE 9–12 (a) Electron micrograph of a plasmid isolated from *E. coli*. (b) An R plasmid containing resistance transfer factors (RTFs) and multiple r-determinants (Tc, tetracycline; Kan, kanamycin; Sm, streptomycin; Su, sulfonamide; Amp, ampicillin; and Hg, mercury). *(Photo: K.G. Murti/Visuals Unlimited)*

9.5 Transformation Is Another Process Leading to Genetic Recombination in Bacteria

Transformation provides another mechanism for recombining genetic information in some bacteria. Small pieces of extracellular DNA are taken up by a living bacterium, ultimately leading to a stable genetic change in the recipient cell. We discuss transformation in this chapter because in those bacterial species where it occurs, the process can be used to map bacterial genes, though in a more limited way than conjugation. Transformation has played a central role in experiments proving that DNA is the genetic material.

The process of transformation (Figure 9–13) consists of numerous steps divided into two categories: (1) entry of DNA into a recipient cell, and (2) recombination of the donor DNA with its homologous region in the recipient chromosome. In a population of bacterial cells, only those in the particular physiological state of **competence** take up DNA. Entry is thought to occur at a limited number of receptor sites on the surface of the bacterial cell (Figure 9–13, step 1). Passage into the cell is an active process that requires energy and specific transport molecules. This model is supported by the fact that substances that inhibit energy production or protein synthesis in the recipient cell also inhibit the transformation process.

During entry, one of the two strands of the double helix is digested by nucleases, leaving only a single strand to participate in transformation (Figure 9–13, steps 2 and 3). The surviving strand of DNA then aligns with its complementary

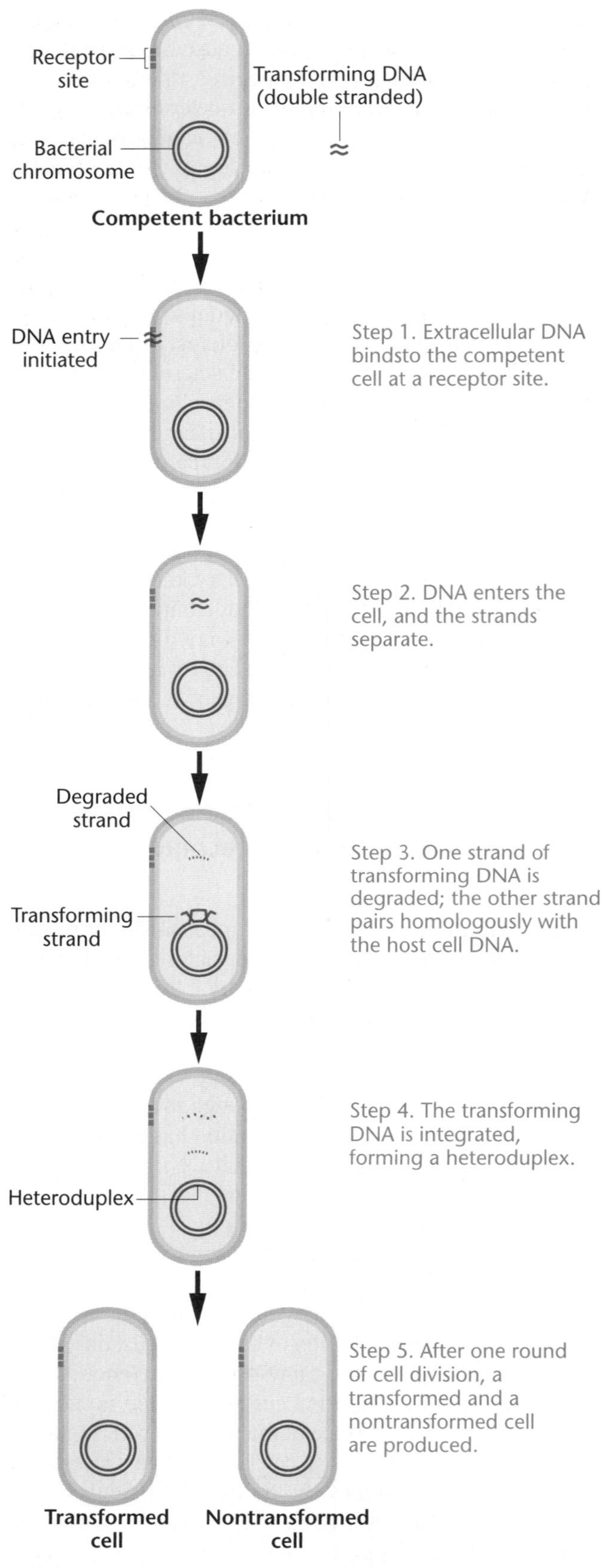

Receptor site
Transforming DNA (double stranded)
Bacterial chromosome
Competent bacterium
DNA entry initiated
Step 1. Extracellular DNA bindsto the competent cell at a receptor site.
Step 2. DNA enters the cell, and the strands separate.
Degraded strand
Transforming strand
Step 3. One strand of transforming DNA is degraded; the other strand pairs homologously with the host cell DNA.
Heteroduplex
Step 4. The transforming DNA is integrated, forming a heteroduplex.
Step 5. After one round of cell division, a transformed and a nontransformed cell are produced.
Transformed cell
Nontransformed cell

FIGURE 9–13 Proposed steps for transforming a bacterial cell by exogenous DNA. Only one of the two entering DNA strands is involved in the transformation event, which is completed following cell division.

region of the bacterial chromosome. In a process involving several enzymes, the segment replaces its counterpart in the chromosome, which is excised and degraded (step 4).

For recombination to be detected, the transforming DNA must be derived from a different strain of bacteria that bears some genetic variation, such as a mutation. Once it is integrated into the chromosome, the recombinant region contains one host strand (present originally) and one mutant strand. Because these strands are from different sources, this helical region is referred to as a **heteroduplex**. Following one round of DNA replication, one chromosome is restored to its original configuration, and the other contains the mutant gene. Following cell division, one untransformed cell (nonmutant) and one transformed cell (mutant) are produced (step 5).

Transformation and Linked Genes

For DNA to be effective in transformation, it must include between 10,000 and 20,000 nucleotide pairs, a length equal to about 1/200 of the *E. coli* chromosome—this size is sufficient to encode several genes. Genes adjacent to or very close to one another on the bacterial chromosome can be carried on a single segment of this size. Because of this, a single event can result in the **cotransformation** of several genes simultaneously. Genes that are close enough to each other to be cotransformed are *linked*. In contrast to *linkage* in eukaryotes, which indicates all genes on a single chromosome, linkage here refers to the close proximity of genes.

If two genes are not linked, simultaneous transformation occurs only as a result of two independent events involving two distinct segments of DNA. As in double crossovers in eukaryotes, the probability of two independent events occurring simultaneously is equal to the product of the individual probabilities. Thus, the frequency of two unlinked genes being transformed simultaneously is much lower than if they are linked.

Studies have shown that a variety of bacteria readily undergo transformation (e.g., *Hemophilus influenzae*, *Bacillus subtilis*, *Shigella paradysenteriae*, *Diplococcus pneumoniae,* and *E. coli*). Under certain conditions, the relative distances between linked genes can be determined from transformation data. While analysis is more complex, such data are interpreted in a manner analogous to chromosome mapping in eukaryotes.

9.6 Bacteriophages Are Bacterial Viruses

Bacteriophages, or **phages** as they are commonly known, are viruses that have bacteria as their hosts. During their reproduction, phages can be involved in still another mode of bacterial genetic recombination called transduction. To understand this process, we must consider the genetics of bacteriophages, which themselves undergo recombination.

A great deal of genetic research has been done using bacteriophages as a model system, making them a worthy subject of discussion. In this section, we will first examine the structure and life cycle of one type of bacteriophage. We then discuss how these phages are studied during their infection of bacteria. Finally, we contrast two possible modes of behavior once the initial phage infection occurs. This information is background for our discussion of transduction and bacteriophage recombination.

Phage T4: Structure and Life Cycle

Bacteriophage T4 is one of a group of related bacterial viruses referred to as T-even phages. It exhibits the intricate structure shown in Figure 9–14. The phage T4's genetic material (DNA) is contained within an icosahedral (a polyhedron with 20 faces) protein coat, making up the head of the virus. The DNA is sufficient in quantity to encode more than 150 average-sized genes. The head is connected to a tail that contains a collar and a contractile sheath surrounding a central core. Tail fibers, which protrude from the tail, contain binding sites in their tips that specifically recognize unique areas of the outer surface of the cell wall of the bacterial host, *E. coli*.

The life cycle of phage T4 (Figure 9–15) is initiated when the virus binds by adsorption to the bacterial host cell. Then, an ATP-driven contraction of the tail sheath causes the central core to penetrate the cell wall. The DNA in the head is extruded, and it moves across the cell membrane into the bacterial cytoplasm. Within minutes, all bacterial DNA, RNA, and protein synthesis in the host cell is inhibited and synthesis of viral molecules begins. At the same time, degradation of the host DNA is initiated.

A period of intensive viral gene activity characterizes infection. Initially, phage DNA replication occurs, leading to a pool of viral DNA molecules. Then, the components of the head, tail, and tail fibers are synthesized.

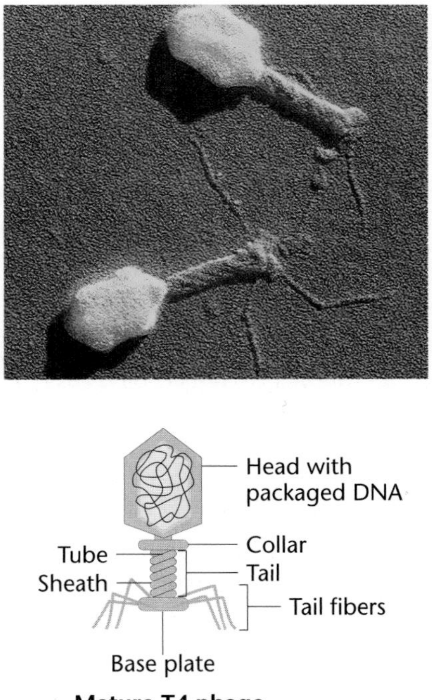

Head with packaged DNA
Collar
Tail
Tail fibers
Tube
Sheath
Base plate
Mature T4 phage

FIGURE 9–14 The structure of bacteriophage T4 includes an icosahedral head filled with DNA, a tail consisting of a collar, tube, sheath, base plate, and tail fibers. During assembly, the tail components are added to the head and then tail fibers are added. *(Photo: M. Wurtz/Biozentrum, University of Basel/Science Photo Library/Photo Researchers, Inc.)*

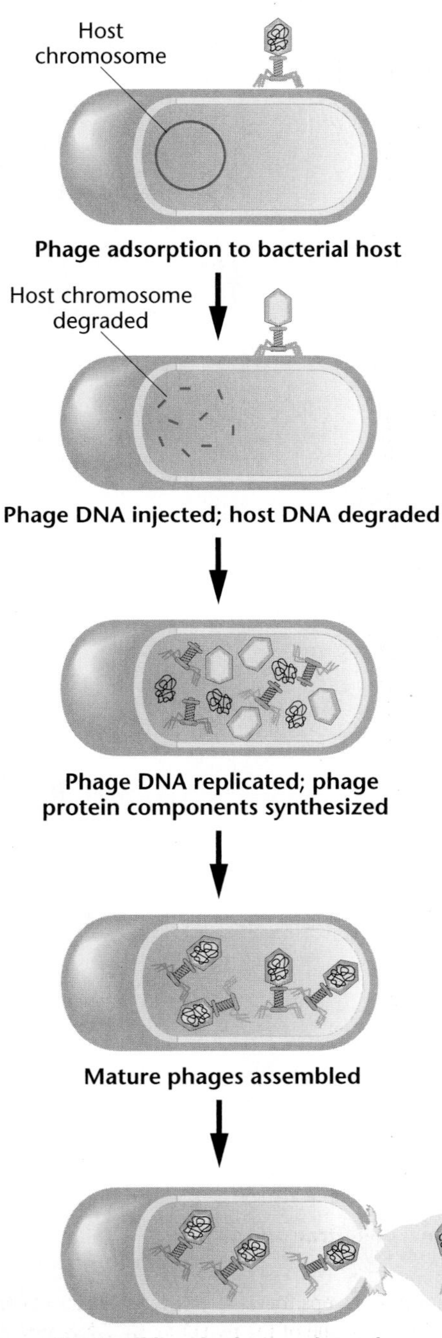
Host chromosome
Phage adsorption to bacterial host
Host chromosome degraded
Phage DNA injected; host DNA degraded
Phage DNA replicated; phage protein components synthesized
Mature phages assembled
Host cell lysed; phage released

FIGURE 9–15 Life cycle of bacteriophage T4.

MEDIA TUTORIAL Phage genetics

The assembly of mature viruses is a complex process that has been well studied by William Wood, Robert Edgar, and others. Three sequential pathways occur: (1) DNA packaging as the viral heads are assembled, (2) tail assembly, and (3) tail fiber asssembly. Once DNA is packaged into the head, it combines with the tail components, to which tail fibers are added. Total construction is a combination of self-assembly and enzyme-directed processes.

When approximately 200 new viruses have been constructed, the bacterial cell is ruptured by the action of the enzyme lysozyme (a phage gene product) and the mature phages are released from the host cell. The new phages infect other available bacterial cells, and the process repeats over and over again.

The Plaque Assay

Bacteriophages and other viruses have played a critical role in our understanding of molecular genetics. During infection of bacteria, enormous quantities of bacteriophages can be obtained for investigation. Often, over 10^{10} viruses are produced per milliliter of culture medium. Many genetic studies rely on the ability to quantify the number of phages produced following infection under specific culture conditions. The **plaque assay** is a routinely used technique, which is invaluable in mutational and recombinational studies of bacteriophages.

This assay is shown in Figure 9–16, where actual plaque morphology is also shown. A serial dilution of the original virally infected bacterial culture is performed first. Then, a 0.1-mL sample (an *aliquot*) from a dilution is

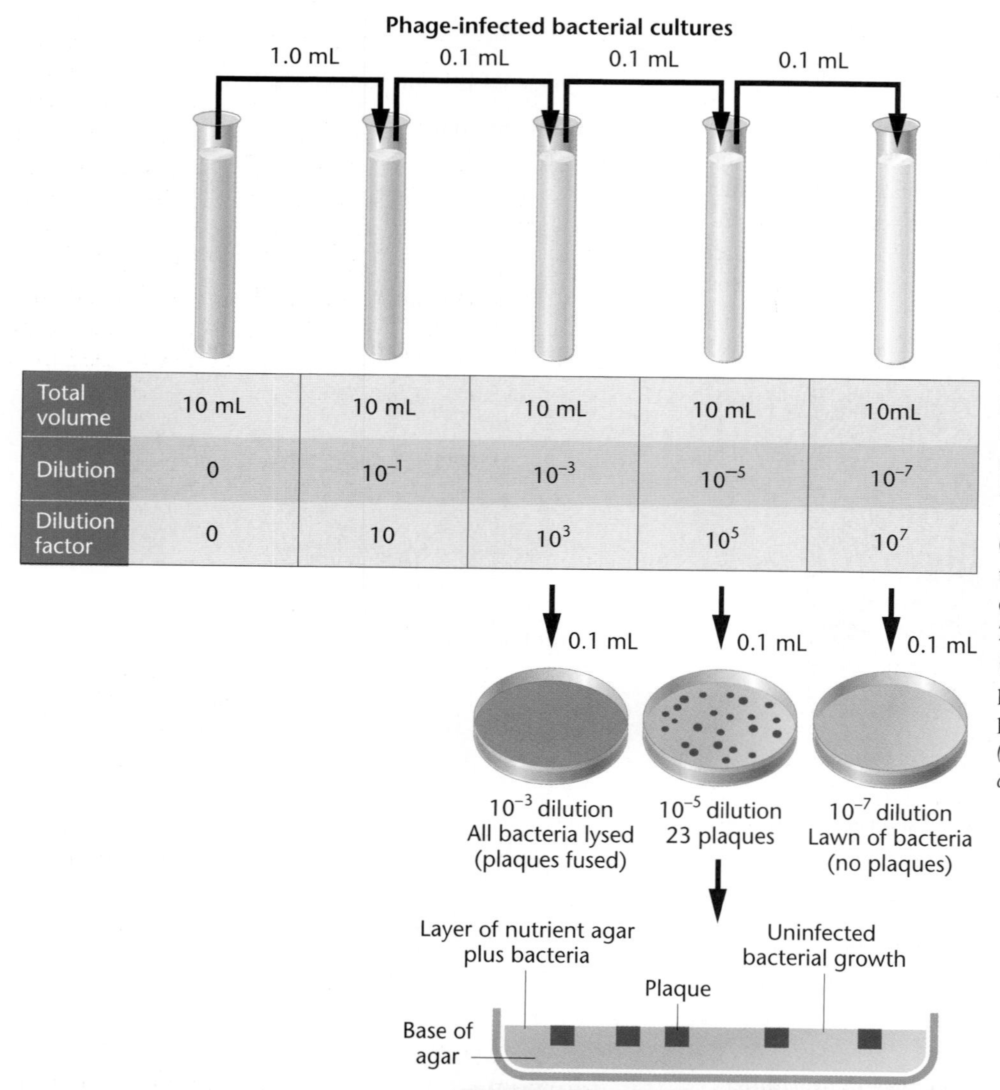

FIGURE 9–16 A plaque assay for bacteriophage analysis. Serial dilutions of a bacterial culture infected with bacteriophages are first made. Then three of the dilutions (10^{-3}, 10^{-5}, and 10^{-7}) are analyzed using the plaque assay technique. Each plaque represents the initial infection of one bacterial cell by one bacteriophage. In the 10^{-3} dilution, so many phages are present that all bacteria are lysed. In the 10^{-5} dilution, 23 plaques are produced. In the 10^{-7} dilution, the dilution factor is so great that no phages are present in the 0.1-mL sample, and thus no plaques form. From the 0.1-mL sample of the 10^{-5} dilution, the original bacteriophage density is calculated to be $23 \times 10 \times 10^5$ phages/mL (23×10^6, or 2.3×10^7). The photograph shows phage T2 plaques on lawns of *E. coli*. *(Photo: Bruce Iverson/Photomicrography)*

added to melted nutrient agar (about 3 mL) into which a few drops of a healthy bacterial culture have been added. The solution is then poured evenly over a base of solid nutrient agar in a petri dish and allowed to solidify before incubation. A clear area called a **plaque** occurs wherever a single virus initially infected one bacterium in the culture (the lawn) that has grown up during incubation. The plaque represents clones of the single infecting bacteriophage, created as reproduction cycles are repeated. If the dilution factor is too low, the plaques are plentiful and they will fuse, lysing the entire lawn—which has occurred in the 10^{-3} dilution of Figure 9–16. On the other hand, if the dilution factor is increased, plaques can be counted and the density of viruses in the initial culture can be estimated,

$$(\text{plaque number/mL}) \times (\text{dilution factor})$$

Using the results shown in Figure 9–16, 23 phage plaques are derived from the 0.1-mL aliquot of the 10^{-5} dilution. Therefore, we estimate that there are 230 phages/mL *at this dilution* (since the initial aliquot was 0.1 mL). The initial phage density in the undiluted sample, where 23 plaques are observed from 0.1 mL, may be calculated as

$$\text{initial phage density} = (230/\text{mL}) \times (10^5) = 230 \times 10^5/\text{mL}$$

Because this figure is derived from the 10^{-5} dilution, we can also estimate that there will be only 0.23 phage/0.1 mL in the 10^{-7} dilution. Thus, when 0.1 mL from this tube is assayed, it is predicted that no phage particles will be present. This possibility is borne out in Figure 9–16, where an intact lawn of bacteria lacking any plaques is depicted. The dilution factor is simply too great.

Lysogeny

The relationship between a virus and a bacterium does not always result in viral reproduction and lysis. As early as the 1920s, it was known that a virus can enter a bacterial cell and establish a symbiotic relationship with it. The precise molecular basis of this symbiosis is now well-understood. Upon entry, the viral DNA is integrated into the bacterial chromosome instead of replicating in the bacterial cytoplasm, a step that characterizes the developmental stage referred to as **lysogeny**. Subsequently, each time the bacterial chromosome is replicated, the viral DNA is also replicated and passed to daughter bacterial cells following division. No new viruses are produced, and no lysis of the bacterial cell occurs. However, under certain stimuli, such as chemical or ultraviolet light treatment, the viral DNA loses its integrated status and initiates replication, phage reproduction, and lysis of the bacterium.

Several terms are used to describe this relationship. The viral DNA that integrates into the bacterial chromosome is a **prophage**. Viruses that either lyse the cell or behave as a prophage are **temperate phages**. Those that only lyse the cell are referred to as **virulent phages**. A bacterium harboring a prophage is **lysogenic**; that is, it is capable of being lysed as a result of induced viral reproduction. The viral DNA, which can replicate either in the bacterial cytoplasm or become integrated into the bacterial chromosome, is thus classified as an **episome**.

9.7 Transduction Is Virus-Mediated Bacterial DNA Transfer

In 1952, Norton Zinder and Joshua Lederberg were investigating possible recombination in the bacterium *Salmonella typhimurium*. Although they recovered prototrophs from mixed cultures of two different auxotrophic strains, investigation revealed that recombination was occurring in a manner different from that attributable to the presence of an F factor, as in *E. coli*. What they had discovered was a process of bacterial recombination mediated by bacteriophages and now called **transduction**.

The Lederberg-Zinder Experiment

Lederberg and Zinder mixed the *Salmonella* auxotrophic strains LA-22 and LA-2 together, and when the mixture was plated on minimal medium, they recovered prototrophic cells. The LA-22 strain was unable to synthesize the amino acids phenylalanine and tryptophan (phe^-, trp^-), and LA-2 could not synthesize the amino acids methionine and histidine (met^-, his^-). Prototrophs (phe^+, trp^+, met^+, his^+) were recovered at a rate of about $1/10^5$ (10^{-5}) cells.

Although these observations at first suggested that the recombination involved was the type observed earlier in conjugative strains of *E. coli*, experiments using the Davis U-tube soon showed otherwise (Figure 9–17). The two aux-

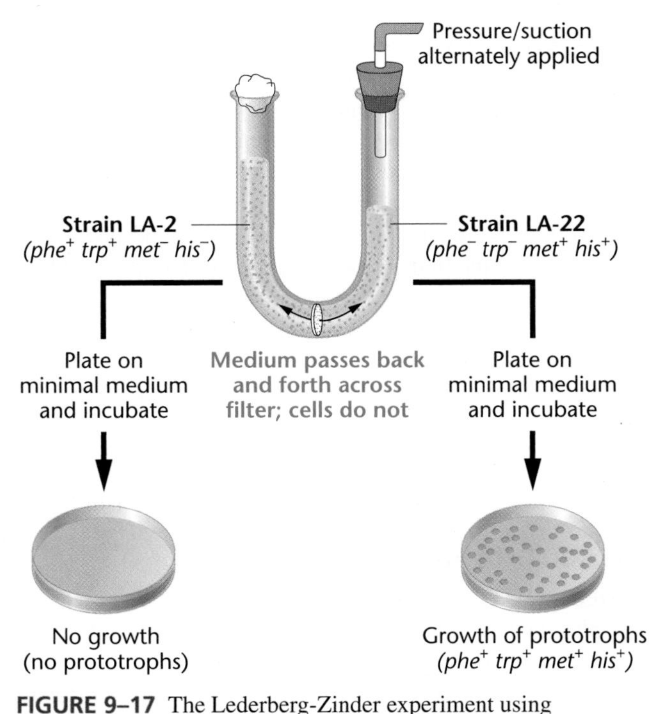

FIGURE 9–17 The Lederberg-Zinder experiment using *Salmonella*. After placing two auxotrophic strains on opposite sides of a Davis U-tube, Lederberg and Zinder recovered prototrophs from the side with the LA-22 strain, but not from the side containing the LA-2 strain.

otrophic strains were separated by a sintered glass filter, thus preventing cell contact but allowing growth to occur in a common medium. Surprisingly, when samples were removed from both sides of the filter and plated independently on minimal medium, prototrophs were recovered only from the side of the tube containing LA-22 bacteria. Recall that if conjugation were responsible, the conditions in the Davis U-tube would be expected to prevent recombination altogether (see Figure 9–4).

Since LA-2 cells appeared to be the source of the new genetic information (phe^+ and trp^+), how that information crossed the filter from the LA-2 cells to the LA-22 cells, allowing recombination to occur, was a mystery. The unknown source was designated simply as a **filterable agent (FA)**.

Three observations were used to identify the FA:

1. The FA was produced by the LA-2 cells only when they were grown in association with LA-22 cells. If LA-2 cells were grown independently and that culture medium was then added to LA-22 cells, recombination did not occur. Therefore, LA-22 cells play some role in the production of FA by LA-2 cells and do so only when they share a common growth medium.
2. The addition of DNase, which enzymatically digests DNA, did not render the FA ineffective. Therefore, the FA is not naked DNA, ruling out transformation.
3. The FA could not pass across the filter of the Davis U-tube when the pore size was reduced below the size of bacteriophages.

Aided by these observations and aware that temperate phages can lysogenize *Salmonella*, researchers proposed that the genetic recombination event is mediated by bacteriophage P22, present initially as a prophage in the chromosome of the LA-22 *Salmonella* cells. They hypothesized that P22 prophages rarely enter the vegetative or lytic phase, reproduce, and are released by the LA-22 cells. Such P22 phages, being much smaller than a bacterium, then cross the filter of the U-tube and subsequently infect and lyse some of the LA-2 cells. In the process of lysis of LA-2, these P22 phages occasionally package a region of the LA-2 chromosome in their heads. If this region contains the phe^+ and trp^+ genes and the phages subsequently pass back across the filter and infect LA-22 cells, these newly lysogenized cells will behave as prototrophs. This process of transduction, whereby bacterial recombination is mediated by bacteriophage P22, is diagrammed in Figure 9–18.

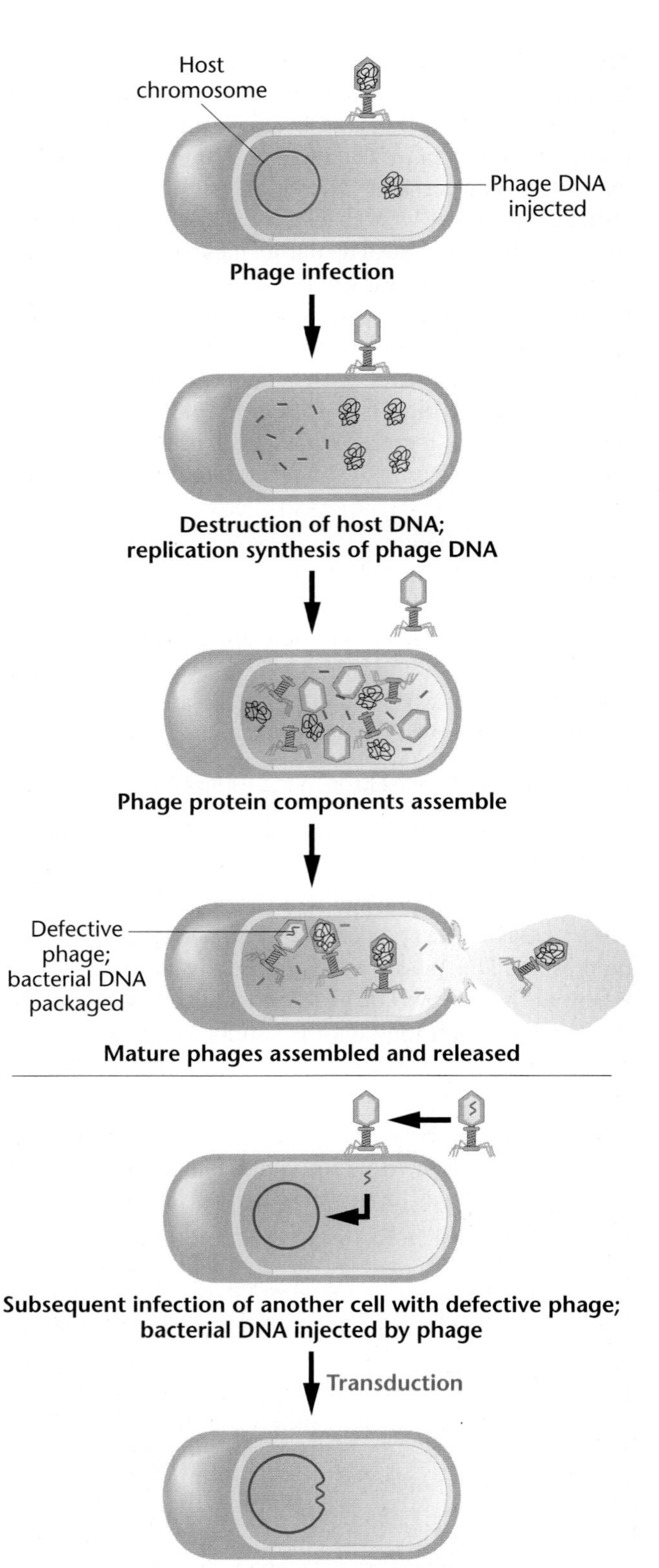
Host chromosome
Phage DNA injected
Phage infection
Destruction of host DNA; replication synthesis of phage DNA
Phage protein components assemble
Defective phage; bacterial DNA packaged
Mature phages assembled and released
Subsequent infection of another cell with defective phage; bacterial DNA injected by phage
Transduction

Integration of bacterial DNA into recipient chromosome

FIGURE 9–18 Generalized transduction.

The Nature of Transduction

Further studies have revealed the existence of transducing phages in other species of bacteria. For example, *E. coli* can be transduced by phages P1 and λ, and *B. subtilis* and *Pseudomonas aeruginosa* can be transduced by the phages SPO1 and F116, respectively. The details of several different modes of transduction have also been established. Even though the initial discovery of transduction involved a temperate phage and a lysogenized bacterium, the same process can occur during the normal lytic cycle. Sometimes a small piece of bacterial DNA is packaged *along with* the viral chromosome so that the transducing phage contains both viral and bacterial DNA. In such cases, only a few bacterial genes are present in the transducing phage. However, when *only* bacterial DNA is packaged, regions as large as 1% of the bacterial chromosome can become enclosed in the viral head. In either case, the ability to infect a host cell is unrelated to the type of DNA in the phage head, making transduction possible.

When bacterial rather than viral DNA is injected into the bacterium, it either remains in the cytoplasm or recombines with the homologous region of the bacterial chromosome. If the bacterial DNA remains in the cytoplasm, it does not replicate but is transmitted to one progeny cell following each division. When this happens, only a single cell, partially diploid for the transduced genes, is produced—a phenomenon called *abortive transduction*. If the bacterial DNA recombines with its homologous region of the bacterial chromosome, *complete transduction* occurs, where the transduced genes are replicated as part of the chromosome and passed to all daughter cells.

Both abortive and complete transduction are subclasses of the broader category of *generalized transduction*, which is characterized by the random nature of DNA fragments and genes that are transduced. Each fragment of the bacterial chromosome has a finite but small chance of being packaged in the phage head. Most cases of generalized transduction are of the abortive type; some data suggest that complete transduction occurs 10–20 times less frequently.

Transduction and Mapping

Like transformation, generalized transduction was used in linkage and mapping studies of the bacterial chromosome. The fragment of bacterial DNA involved in a transduction event is large enough to include numerous genes. As a result, two genes that closely align (are linked) on the bacterial chromosome can be simultaneously transduced, a process called **cotransduction**. Two genes that are not close enough to one another along the chromosome to be included on a single DNA fragment require two independent events to be transduced into a single cell. Since this occurs with a much lower probability than cotransduction, linkage can be determined.

By concentrating on two or three linked genes, transduction studies can also determine the precise order of these genes. The closer linked genes are to each other, the greater the frequency of cotransduction. Mapping studies involving three closely aligned genes can thus be executed, and the analysis of such an experiment is predicated on the same rationale underlying other mapping techniques.

9.8 Bacteriophages Undergo Intergenic Recombination

Around 1947, several research teams demonstrated that genetic recombination can be detected in bacteriophages. This led to the discovery that gene mapping can be performed in these viruses. Such studies relied on finding numerous phage mutations that could be visualized or assayed. As in bacteria and eukaryotes, these mutations allow genes to be identified and followed in mapping experiments. Thus, before considering recombination and mapping in these bacterial viruses, we briefly introduce several of the mutations that were studied.

Bacteriophage Mutations

Phage mutations often affect the morphology of the plaques formed following lysis of bacterial cells. For example, in 1946, Alfred Hershey observed unusual T2 plaques on plates of *E. coli* strain B. Normal T2 plaques are small and have a clear center surrounded by a diffuse (nearly invisible) halo, but the unusual plaques were larger and possessed a distinctive outer perimeter (Figure 9–19). When the viruses were isolated from these plaques and replated on *E. coli* B cells, the resulting plaque appearance was identical. Thus, the plaque phenotype was an inherited trait resulting from the reproduction of mutant phages. Hershey named the mutant *rapid lysis* (r) because the plaques were larger, apparently resulting from a more rapid or more efficient life

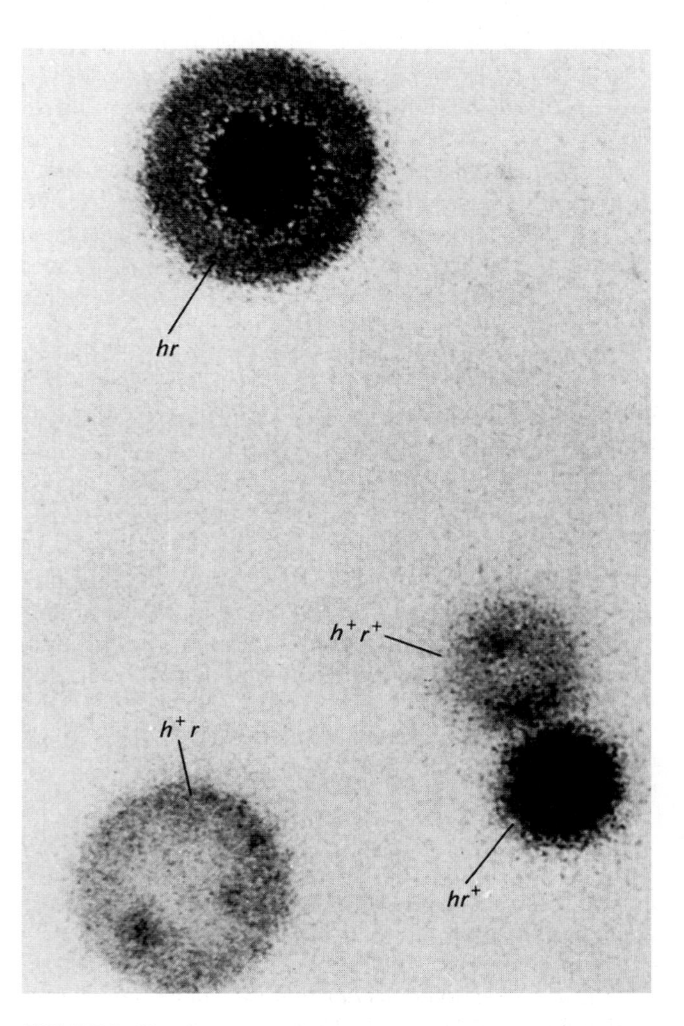

FIGURE 9–19 Plaque morphology phenotypes observed following simultaneous infection of *E. coli* by two strains of phage T2, h^+r and hr^+. In addition to the parental genotypes, recombinant plaques hr and h^+r^+ were recovered. *(From Hershey & Chase, 1951)*

TABLE 9.1 Some Mutant Types of T-even Phages

Name	*Description*
minute	Small plaques
turbid	Turbid plaques on *E. coli* B
star	Irregular plaques
UV-sensitive	Alters UV sensitivity
acriflavin-resistant	Forms plaques on acriflavin agar
osmotic shock	Withstands rapid dilution into distilled water
lysozyme	Does not produce lysozyme
amber	Grows in *E. coli* K12, but not B
temperature-sensitive	Grows at 25°C, but not at 42°C

TABLE 9.2 Results of a Cross Involving the *h* and *r* Genes in Phage T2 ($hr^+ \times h^+r$)

Genotype	*Plaques*	*Designation*
$h\ r^+$	42	Parental progeny 76%
h^+r	34	
h^+r^+	12	Recombinants 24%
$h\ r$	12	

Source: Data derived from Hershey and Rotman, 1949.

cycle of the phage. We now know that in wild-type phages, reproduction is inhibited once a particular-sized plaque has been formed. The *r* mutant T2 phages overcome this inhibition, producing larger plaques.

Salvador Luria discovered another bacteriophage mutation, *host range* (*h*). This mutation extends the range of bacterial hosts that the phage can infect. Although wild-type T2 phages can infect *E. coli* B (a unique strain), they normally cannot attach or be adsorbed to the surface of *E. coli* B-2 (a different strain). The *h* mutation, however, provides the basis for adsorption and subsequent infection of *E. coli* B-2. When grown on a mixture of *E. coli* B and B-2, the center of the *h* plaque appears much darker than the h^+ plaque (Figure 9–19).

Table 9.1 lists other types of mutations that have been isolated and studied in the T-even series of bacteriophages (e.g., T2, T4, T6). These mutations are important to the study of genetic phenomena in bacteriophages.

Intergenic Mapping

Genetic recombination in bacteriophages was discovered during **mixed infection experiments** in which two distinct mutant strains were allowed to *simultaneously* infect the same bacterial culture. These studies were designed so that the number of viral particles sufficiently exceeded the number of bacterial cells to ensure simultaneous infection of most cells by both viral strains. Because two loci are involved, recombination is referred to as intergenic.

For example, in one study using the T2/*E. coli* system, the parental viruses were of either the h^+r (wild-type host range, rapid lysis) or hr^+ (extended host range, normal lysis) genotype. If no recombination occurred, these two parental genotypes would be the only expected phage progeny. However, the recombinants h^+r^+ and hr were detected in addition to the parental genotypes (see Figure 9–19). As with eukaryotes, the percentage of recombinant plaques divided by the total number of plaques reflects the relative distance between the genes.

$$\text{recombinational frequency} = h^+r^+ + h\,r/\text{total plaques} \times 100$$

Sample data for the *h* and *r* loci are shown in Table 9.2.

Similar recombinational studies have been conducted with numerous mutant genes in a variety of bacteriophages. Data are analyzed in much the same way as in eukaryotic mapping experiments. Two- and three-point mapping crosses are possible, and the percentage of recombinants in the total number of phage progeny is calculated. This value is proportional to the relative distance between two genes along the DNA molecule constituting the chromosome.

Investigations into phage recombination support a model similar to that of eukaryotic crossing over—a breakage and reunion process between the viral chromosomes. A fairly clear picture of the dynamics of viral recombination is emerging. After the early phase of infection, the chromosomes of the phages begin replication. As this stage progresses, a pool of chromosomes accumulates in the bacterial cytoplasm. If double infection by phages of two genotypes has occurred, then the pool of chromosomes initially consists of the two parental types. Genetic exchange between these two types will occur before, during, and after replication, producing recombinant chromosomes.

In the case of the h^+r and hr^+ example discussed here, recombinant h^+r^+ and hr chromosomes are produced. Each of these chromosomes can undergo replication, with new replicates exchanging with each other and with parental chromosomes. Furthermore, recombination is not restricted to exchanges between two chromosomes—three or more can be involved simultaneously. As phage development progresses, chromosomes are randomly removed from the pool and packed into the phage head, forming mature phage particles. Thus, a variety of parental and recombinant genotypes are represented in progeny phages.

GENETICS, TECHNOLOGY, AND SOCIETY

Eradicating Cholera: Edible Vaccines

Almost lost amid the furor over the cloned sheep, Dolly, was the original purpose of the undertaking: genetically engineering an animal that would produce a valuable pharmaceutical product, eventually leading to a herd of genetically identical drug-producing animals. The goal, in other words, was to be able to turn sheep, cattle, and other animals into living drug factories.

Meanwhile, almost unnoticed by the general public, the genetic engineering of plants to serve a similar role is much closer to reality. When genes are inserted into a plant genome, transgenic plants are created that produce the foreign gene product. One of the most intriguing and potentially life-saving objectives is genetically engineering plants to produce vaccines against human diseases. Immunization would then involve eating foods altered to contain a protein that acts as an antigen and stimulates the production of antibodies to protect against bacterial or viral infection.

Plants have many advantages as vaccine producers. Since plants can be grown in large numbers, plant-produced vaccines should be less expensive than conventional vaccines. Further, the extensive purification and refrigerated transport and storage of vaccines will not be required. This is important in parts of the world where the supply of electricity is unreliable. Finally, since people would simply eat the vaccine-containing food, there would be no need for syringes and needles, or medical staff to give injections.

Leading the effort to develop edible vaccines in plants is Charles J. Arntzen, formerly at Texas A&M University and now president of the Boyce Thompson Institute for Plant Research at Cornell University. Arntzen's research team is focusing on intestinal diseases, especially cholera. Cholera may at first seem an odd target, since it is a disease that has not been a major public health problem in this country for over a century. But cholera remains a leading cause of death of infants and children throughout the Third World, where basic sanitation is lacking and water supplies are often contaminated. For example, in July of 1994, 70,000 cases of cholera were reported among the Rwandans crowded into refugee camps in Goma, Zaire, leading to 12,000 fatalities. A cholera epidemic struck East Africa in the fall of 1997, killing nearly 3000. And after an absence of over 100 years, cholera reappeared in Latin America in 1991, spreading from Peru to Mexico and claiming more than 10,000 lives.

The causative agent of cholera is *Vibrio cholerae*, a curved, rod-shaped bacterium found mostly in rivers and oceans. Most strains of *V. cholerae* are harmless. Only one strain, called O1, is pathogenic. Infection occurs when a person drinks water or eats food contaminated with this strain. Once in the digestive system, the bacteria colonize the small intestine and begin producing proteins called enterotoxins. The cholera enterotoxin binds to the surface of the mucosal cells lining the intestine, triggering the massive secretion of water and dissolved salts from these cells. This results in violent diarrhea, which, if untreated, is followed by severe dehydration, muscle cramps, lethargy, and often death.

The cholera enterotoxin consists of two polypeptides, called the A and B subunits, which individually have no effect. For the toxin to be active, one A subunit must be linked to five B subunits. Since it is the B subunit of this complex that binds to intestinal cells, Arntzen's group decided to use this polypeptide as their antigen, reasoning that antibodies against it would potentially prevent the toxin from binding and render the bacteria harmless.

Arntzen and his team used the B subunit of an *E. coli* enterotoxin, which is similar in structure and immunological properties to the cholera protein. The DNA coding sequence of the B-subunit gene was obtained, to which they attached a promoter that would prompt transcription in all tissues of the plant. They then introduced this hybrid gene into potato plants by means of *Agrobacterium*-mediated transformation. They chose the potato not only because methods for transformation and regeneration of this plant are fairly routine but also so they could assay the effectiveness of the antigen in the edible part of the plant, the tuber. Analysis showed that the engineered plants expressed their new gene and produced the B-subunit.

After feeding mice a few grams of the genetically engineered tubers that had produced the B-subunit, they found that the mice began to produce specific antibodies against it, which were secreted into the small intestine. And, most critically, these mice were later fed purified enterotoxin and were protected from its effects—they did not develop the symptoms of cholera. Clinical trials are now planned to test the efficacy of the potato-produced vaccine in humans.

In the meantime, the Arntzen group is also working toward producing edible vaccines in bananas, which have several advantages over potatoes. First, bananas can be grown almost anywhere throughout the tropical or subtropical developing countries of the world, exactly where they are most needed. And unlike potatoes, bananas are usually eaten raw, avoiding potential inactivation of the antigenic proteins by cooking. Finally, bananas are well-liked by infants and children, making this approach to immunization a feasible one.

Procedures are now being perfected for the transformation of banana cells with genes encoding the cholera B-subunit. However, it will be some time before the engineered bananas are ready to test, since it takes three years to grow a banana crop. If all goes as planned, it may someday be possible to immunize all Third World children against cholera and other intestinal diseases, saving untold thousands of young lives.

References

Arntzen, C. J. 1997. Edible vaccines. *Publ. Health Rep.* 112: 190–97.

Haq, T. A., Mason, H. S., Clements, J. D., and Arntzen, C. J. 1995. Oral immunization with a recombinant bacterial antigen produced in transgenic plants. *Science* 268: 714–16.

Sanchez, J. L., and Taylor, D. N. 1997. Cholera. *Lancet* 349: 1825–30.

Chapter Summary

1. Inherited phenotypic variation in bacteria results from spontaneous mutation.
2. Genetic recombination in bacteria can result from three different modes: conjugation, transformation, and transduction.
3. Conjugation is initiated by a bacterium housing a plasmid called the F factor. If the F factor is in the cytoplasm of a donor cell (F^+), the recipient F^- cell receives a copy of the F factor, converting it to the F^+ status.
4. If the F factor is integrated into the donor cell chromosome (Hfr), recombination is initiated with the recipient F^- cell and genetic information flows unidirectionally to it. Time mapping of the bacterial chromosome is based on the orientation and position of the F factor in the donor chromosome.
5. The products of a group of genes designated as *rec* are directly involved in recombination between the invading DNA and the recipient bacterial chromosome.
6. Plasmids, such as the F factor, are autonomously replicating DNA molecules found in the bacterial cytoplasm. Some plasmids contain unique genes conferring antibiotic resistance, as well as those necessary for their transfer during conjugation.
7. In the phenomenon of transformation, which does not require cell contact, exogenous DNA enters the host chromosome of a recipient bacterial cell. Linkage mapping of closely aligned genes may be performed using this process.
8. Bacteriophages (viruses that infect bacteria) demonstrate a defined life cycle during which they reproduce within the host cell. They are studied using the plaque assay.
9. Bacteriophages can be lytic, where they infect the host cell and reproduce, then lyse the host cell; or they can lysogenize the host cell, where they infect it and integrate their DNA into the host chromosome, but do not reproduce.
10. Transduction is virus-mediated bacterial recombination. When a lysogenized bacterium subsequently reenters the lytic cycle, the new bacteriophages serve as the vehicles for the transfer of host (bacterial) DNA. In the process of generalized transduction, a random part of the bacterial chromosome is transferred.
11. Transduction is also used for bacterial linkage and mapping studies.

Key Terms

adaptation hypothesis, 183
auxotroph, 184
bacteria, 183
bacteriophage, 192
Col plasmid, 191
colicin, 191
competence, 191
conjugation, 184
cotransduction, 197
cotransformation, 192
episome, 195
F^- cell, 186, 189
F^+ cell, 186
F′ cell, 186
F factor, 186
F pilus, 186
fertility factor, 186
filterable agent (FA), 196
fluctuation test, 183
genetic recombination, 184
heteroduplex, 192
high-frequency recombination (Hfr), 187
interrupted mating technique, 187
lag phase, 184
logarithmic (log) phase, 184
lysogenic, 195
lysogeny, 195
merozygote, 190
minimal medium, 183
mixed infection experiments, 198
phage, 192
plaque, 195
plaque assay, 194
plasmid, 191
prophage, 195
prototroph, 184
r-determinants, 191
R plasmids, 191
RecA protein, 190
RecBCD protein, 190
rec genes, 190
resistance transfer factor (RTF), 191
sex pilus, 186
spontaneous mutation, 183
stationary phase, 184
temperate phage, 195
transduction, 185
transformation, 185, 191
virulent phage, 195

INSIGHTS AND SOLUTIONS

1. Time mapping is performed in a cross involving the genes *his*, *leu*, *mal*, and *xyl*. The recipient cells are auxotrophic for all four genes. After 25 minutes, mating is interrupted with the results in recipient cells shown below. Diagram the positions of these genes relative to the origin (*O*) of the F factor and to one another.

90% are xyl^+

80% are mal^+

20% are his^+

none are leu^+

Solution: The *xyl* gene is transferred most frequently, so it is closest to *O* (very close). The *mal* gene is next and reasonably close to *xyl*, followed by the more distant *his* gene. The *leu* gene is far beyond these three, since no recovered recombinants include it. The diagram shows these relative locations along a piece of the circular chromosome.

2. In four Hfr strains of bacteria, all derived from an original F^+ culture grown over several months, a group of hypothetical genes is studied and shown to transfer in the orders shown in the table below. (a) Assuming *B* is the first gene along the chromosome, determine the sequence of all genes shown. (b) One strain creates an apparent dilemma. Which one is it? Explain why the dilemma is only apparent, not real.

Hfr Strain	*Order of Transfer*					
1	*E*	*R*	*I*	*U*	*M*	*B*
2	*U*	*M*	*B*	*A*	*C*	*T*
3	*C*	*T*	*E*	*R*	*I*	*U*
4	*R*	*E*	*T*	*C*	*A*	*B*

Solution: **(a)** The sequence is found by overlapping the genes in each strain.

Strain 2 *U M* ***B A C*** *T*

Strain 3 *C* ***T E R*** *I U*

Strain 1 *E R* ***I*** *U* ***M*** *B*

Starting with *B* in strain 2, the gene sequence is *BACTERIUM*.

(b) Strain 4 creates a dilemma, which is resolved when we realize that the F factor is integrated in the opposite orientation. Thus, the genes enter in the opposite sequence, starting with gene *R*.

$\overrightarrow{RETCAB}$

3. Three strains of bacteria, each bearing a separate mutation, a^-, b^-, or c^- are the sources of donor DNA in a transformation experiment. Recipient cells are wild type for those genes, but express the mutant d^-. (a) Based on the data below and assuming that the location of the *d* gene precedes the *a*, *b*, and *c* genes propose a linkage map for these four genes. (b) If the donor DNA is wild type and the recipient cells are either a^-b^-, a^-c^-, or b^-c^-, in which case would wild-type transformants be expected most frequently?

DNA Donor	*Recipient*	*Transformants*	*Frequency of Transformants*
a^-d^+	a^+d^-	a^+d^+	0.21
b^-d^+	b^+d^-	b^+d^+	0.18
c^-d^+	c^+d^-	c^+d^+	0.63

Solution: **(a)** These data reflect the relative distances between each of the *a*, *b*, and *c* genes and the *d* gene. The *a* and *b* genes are about the same distance from the *d* gene and are thus tightly linked to one another. The *c* gene is more distant. Assuming that the *d* gene precedes the others, the map looks like this:

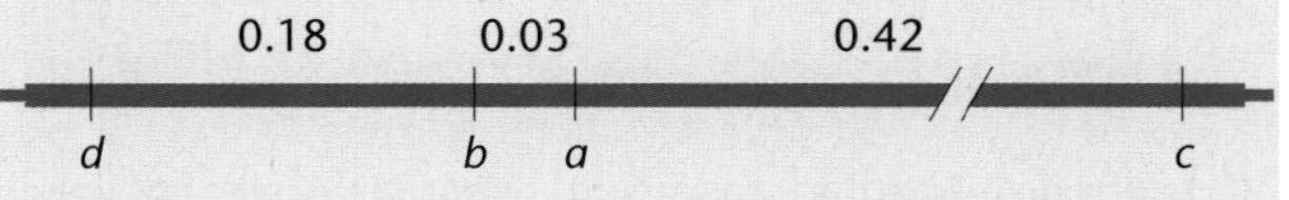

(b) Because the *a* and *b* genes are closely linked, they most likely cotransform in a single event. Thus, recipient cells of a^-b^- are most likely to convert to wild type.

Problems and Discussion Questions

1. Distinguish among the three modes of recombination in bacteria.
2. With respect to F^+ and F^- bacterial matings, (a) How was it established that physical contact was necessary? (b) How was it established that chromosome transfer was unidirectional? (c) What is the genetic basis of a bacterium's being F^+?
3. List all of the differences between $F^+ \times F^-$ and Hfr $\times$ F^- bacterial crosses.
4. List all of the differences between F^+, F^-, Hfr, and F' bacteria.
5. Describe the basis for chromosome mapping in the Hfr $\times$ F^- crosses.
6. Why are the recombinants produced from an Hfr $\times$ F^- cross rarely if ever F^+?
7. Describe the origin of F' bacteria and merozygotes.
8. Describe the mechanism of transformation.
9. In a transformation experiment involving a recipient bacterial strain of genotype a^-b^-, the results below were obtained. What can you conclude about the location of the *a* and *b* genes relative to each other?

	Transformants (%)		
Transforming DNA	a^+b^-	a^-b^+	a^+b^+
a^+b^+	3.1	1.2	0.04
a^+b^- and a^-b^+	2.4	1.4	0.03

10. In a transformation experiment, donor DNA was obtained from a prototroph bacterial strain ($a^+b^+c^+$), and the recipient was a triple auxotroph ($a^-b^-c^-$). What general conclusions can you draw about the linkage relationships among the three genes from the following transformant classes that were recovered?

$a^+b^-c^-$	180
$a^-b^+c^-$	150
$a^+b^+c^-$	210
$a^-b^-c^+$	179
$a^+b^-c^+$	2
$a^-b^+c^+$	1
$a^+b^+c^+$	3

11. The bacteriophage genome consists primarily of genes encoding proteins that make up the head, collar and tail, and tail fibers. When these genes are transcribed following phage infection, how are these proteins synthesized, since the phage genome lacks genes essential to ribosome structure?
12. Describe the temporal sequence of the bacteriophage life cycle.
13. In the plaque assay, what is the precise origin of a single plaque?
14. In the plaque assay, exactly what makes up a single plaque?
15. A plaque assay is perfomed beginning with 1.0 mL of a solution containing bacteriophages. This solution is serially diluted three times by taking 0.1 mL and adding it to 9.9 mL of liquid medium. The final dilution is plated and yields 17 plaques. What is the initial density of bacteriophages in the original 1.0 mL?
16. Describe the difference between the lytic cycle and lysogeny when bacteriophage infection occurs.
17. Define prophage.
18. Explain the observations that led Zinder and Lederberg to conclude that the prototrophs recovered in their transduction experiments were not the result of Hfr-mediated conjugation.
19. Describe generalized transduction and distinguish between abortive and complete transduction.
20. Describe how generalized transduction can be used to map the bacterial chromosome. How does cotransduction play a role in mapping?
21. Two theoretical genetic strains of a virus ($a^-b^-c^-$ and $a^+b^+c^+$) are used to simultaneously infect a culture of host bacteria. Of 10,000 plaques scored, the genotypes in the table below were observed. Determine the genetic map of these three genes on the viral chromosome. Was interference positive or negative?

$a^+b^+c^+$	4100	$a^-b^+c^-$	160
$a^-b^-c^-$	3990	$a^+b^-c^+$	140
$a^+b^-c^-$	740	$a^-b^-c^+$	90
$a^-b^+c^+$	670	$a^+b^+c^-$	110

22. Describe the conditions under which genetic recombination may occur in bacteriophages.
23. If a single bacteriophage infects one *E. coli* cell present in a culture of bacteria and, upon lysis, yields 200 viable viruses, how many phages will exist in a single plaque if three more lytic cycles occur?
24. A phage-infected bacterial culture was subjected to a series of dilutions, and a plaque assay was performed in each case, with the results shown below. What conclusion can be drawn in the case of each dilution?

Dilution Factor	*Assay Results*
10^4	All bacteria lysed
10^5	14 plaques
10^6	0 plaques

25. When the interrupted mating technique was used with five different strains of Hfr bacteria, the order of gene entry during recombination shown below was observed. On the basis of these data, draw a map of the bacterial chromosome. Do the data support the concept of circularity?

Hfr Strain	*Order*				
1	*T*	*C*	*H*	*R*	*O*
2	*H*	*R*	*O*	*M*	*B*
3	*M*	*O*	*R*	*H*	*C*
4	*M*	*B*	*A*	*K*	*T*
5	*C*	*T*	*K*	*A*	*B*

26. In *B. subtilis*, linkage analysis of two mutant genes affecting the synthesis of the two amino acids, tryptophan (trp_2^-) and tyrosine (tyr_1^-) was performed using transformation. (E. Nester, M. Schafer, and J. Lederberg (1963) *Genetics 48:* 529–51.) Examine the data in the table that follows, and draw all possible conclusions regarding linkage.

	Donor DNA	*Recipient Cell*	*Transformants*	*Number of Transformants*
Part A	$trp_2^+ tyr_1^+$	$trp_2^- tyr_1^-$	$trp^+ \ tyr^-$	196
			$trp^- \ tyr^+$	328
			$trp^+ \ tyr^+$	367
Part B	$trp_2^+ tyr_1^-$ and $trp_2^- tyr_1^+$	$trp_2^- tyr_1^-$	$trp^+ \ tyr^-$	190
			$trp^- \ tyr^+$	256
			$trp^+ \ tyr^+$	2

27. What is the role of part B in the experiment shown in Problem 26?
28. An Hfr strain is used to map three genes in an interrupted mating experiment. The cross is $Hfr/a^+b^+c^+rif^s \times F^-/a^-b^-c^-rif^r$ (no map order is implied in the listing of the alleles; *rif* = the antibiotic rifampicin). The a^+ gene is required for biosynthesis of nutrient A, the b^+ gene for nutrient B, and c^+ for nutrient C. The minus alleles are auxotrophs for these nutrients. The cross is initiated at time = 0, and, at various intervals, the mating mixture is plated on three types of medium. Each plate contains minimal medium (MM) *plus* rifampicin *plus* the specific supplements indicated in the table below (results for each time point are shown as number of colonies growing on each plate). (a) What is the purpose of rifampicin in the experiment? (b) Based on these data, determine the approximate location on the chromosome of the *a*, *b*, and *c* genes relative to one another and to the F factor. (c) Can the location of the *rif* gene be determined in this experiment? If not, design an experiment to determine the location of *rif* relative to the F factor and to gene *b*.

Supplements Added to MM	*Time of Interruption (min)* 5	*10*	*15*	*20*
Nutrients A and B	0	0	4	21
Nutrients B and C	0	5	23	40
Nutrients A and C	4	25	60	82

29. In a cotransformation experiment using various combinations of genes, two at a time, the data below were produced. Determine which genes are linked and to whom.

Successful Cotransformation	*Unsuccessful Cotransformation*
a and *d*; *b* and *c*; *b* and *f*	*a* and *b*; *a* and *c*; *a* and *f* *d* and *b*; *d* and *c*; *d* and *f* *a* and *e*; *b* and *e*; *c* and *e* *d* and *e*; *f* and *e*

30. In Problem 29, another gene, *g*, was also studied. It demonstrated positive cotransformation when tested with gene *f*. Predict the results of experiments when it was tested with genes *a*, *b*, *c*, *d*, and *e*.
31. Bacterial conjugation, mediated mainly by conjugative plasmids like F, represents a potential health threat through the sharing of genes for pathogenicity or antibiotic resistance. Given that more than 400 different species of bacteria co-inhabit a healthy human gut and more than 200 co-inhabit human skin, Dionisio and colleagues. investigated the ability of plasmids to undergo between-species conjugal transfer. Data (Dionisio et al. 2002. *Genetics* 162: 1525–32) are presented below involving various species of the enterobacterial genus, *Escherichia*. The data are presented as "log base 10" values, where for example, −2.0 would be equivalent to 10^{-2} as a rate of transfer. Assume that all differences between values are statistically significant. (a) What general conclusion(s) can be drawn from these data? (b) In what species is within-species transfer most likely? In what species pair is between-species transfer most likely? (c) What is the significance of these findings in terms of human health?

	Donor			
Recipient	*E. chrysanthemi*	*E. blattae*	*E. fergusonii*	*E. coli*
E. chrysanthemi	−2.4	−4.7	−5.8	−3.7
E. blattae	−2.0	−3.4	−5.2	−3.4
E. fergusonii	−3.4	−5.0	−5.8	−4.2
E. coli	−1.7	−3.7	−5.3	−3.5

32. A study was conducted in an attempt to determine which functional regions of a particular conjugative transfer gene (*tra1*) are involved in the transfer of plasmid R27 in *Salmonella enterica*. The R27 plasmid is of significant clinical interest because it is capable of encoding multiple-antibiotic resistance to typhoid fever. To identify functional regions responsible for conjugal transfer, an analysis (modified from Lawley et al. 2002. *J. Bacteriol.* 184: 2173–80) was conducted whereby particular regions of the *tra1* gene were mutated and tested for their impact on conjugation. Shown at the bottom of the page is a map of the regions tested and believed to be involved in conjugative transfer of the R27 plasmid. Similar shading indicates related function, and the numbers correspond to each functional region subjected to mutation analysis. Above the map is a table showing the effects of these mutations on R27 conjugation. (a) Given the data, do all functional regions appear to influence conjugative transfer similarly? (b) Which regions appear to have the most impact on conjugation? (c) Which regions appear to have a limited impact on conjugation? (d) What general conclusions might one draw from these data?

Effects of Mutations in Functional Regions of Transfer Region 1 (*tra1*) on R27 Conjugation

R27 Mutation in Region	*Conjugative Transfer*	*Relative Conjugation Frequency (%)*
1	+	100
2	+	100
3	−	0
4	+	100
5	−	0
6	−	0
7	+	12
8	−	0
9	−	0
10	−	0
11	+	13
12	−	0
13	−	0
14	−	0

Selected Readings

Adelberg, E. A. 1960. *Papers on Bacterial Genetics*. Boston: Little, Brown.

Birge, E. A. 1988. *Bacterial and Bacteriophage Genetics—An Introduction*. New York: Springer-Verlag.

Brock, T. 1990. *The Emergence of Bacterial Genetics*. Cold Spring Harbor: Cold Spring Harbor Press.

Broda, P. 1979. *Plasmids*. New York: W. H. Freeman.

Bukhari, A. I., Shapiro, J. A., and Adhya, S. L., (eds.), 1977. *DNA Insertion Elements, Plasmids, and Episomes*. Cold Spring Harbor: Cold Spring Harbor Press.

Cairns, J., Stent, G. S., and Watson, J. D., (eds.), 1966. *Phage and the Origins of Molecular Biology*. Cold Spring Harbor: Cold Spring Harbor Press.

Campbell, A. M. 1976. How viruses insert their DNA into the DNA of the host cell. *Sci. Am.* (Dec.) 235: 102–13.

Fox, M.S. 1966. On the mechanism of integration of transforming deoxyribonucleate. *J. Gen. Physiol.* 49: 183–96.

Hayes, W. 1968. *The Genetics of Bacteria and Their Viruses*. 2nd ed. New York: Wiley.

Hershey, A. D., and Rotman, R. 1949. Genetic recombination between host range and plaque-type mutants of bacteriophage in single cells. *Genetics* 34: 44–71.

Hotchkiss, R. D., and Marmur, J. 1954. Double marker transformations as evidence of linked factors in deoxyribonucleate transforming agents. *Proc. Natl. Acad. Sci.* (*USA*) 40: 55–60.

Jacob, F., and Wollman, E. L. 1961. Viruses and genes. *Sci. Am.* (June) 204: 92–106.

Kruse, H., and Sorum, H. 1994. Transfer of multiple drug resistance plasmids between bacteria of diverse origins in natural microenvironments. *Appl. Environ. Microbiol.* 60: 4015–21.

Lederberg, J. 1986. Forty years of genetic recombination in bacteria: A fortieth anniversary reminiscence. *Nature* 324: 627–28.

Luria, S. E., and Delbruck, M. 1943. Mutations of bacteria from virus sensitivity to virus resistance. *Genetics* 28: 491–511.

Lwoff, A. 1953. Lysogeny. *Bacteriol. Rev.* 17: 269–337.

Miller, J. H. 1992. *A Short Course in Bacterial Genetics*. Cold Spring Harbor: Cold Spring Harbor Press.

Miller, R.V. 1998. Bacterial gene swapping in nature. *Sci. Am.* (Jan.) 278: 66–71.

Morse, M. L., Lederberg, E. M., and Lederberg, J. 1956. Transduction in *Escherichia coli* K12. *Genetics* 41: 141–56.

Novick, R. P. 1980. Plasmids. *Sci. Am.* (Dec.) 243: 102–27.

Smith-Keary, P. F. 1989. *Molecular Genetics of Escherichia coli*. New York: Guilford Press.

Stahl, F. W. 1987. Genetic recombination. *Sci. Am.* (Nov.) 256: 91–101.

Stent, G. S. 1963. *Molecular Biology of Bacterial Viruses*. New York: W. H. Freeman.

————. 1966. *Papers on Bacterial Viruses*. 2nd ed. Boston: Little, Brown.

Visconti, N., and Delbruck, M. 1953. The mechanism of genetic recombination in phage. *Genetics* 38: 5–33.

Wollman, E. L., Jacob, F., and Hayes, W. 1956. Conjugation and genetic recombination in *Escherichia coli* K12. *Cold Spring Harb. Symp. Quant. Biol.* 21: 141–62.

Zinder, N. D. 1958. Transduction in bacteria. *Sci. Am.* (Nov.) 199: 38–46.

CHAPTER 10

CHAPTER CONCEPTS

Bronze sculpture of the Watson-Crick double-helical DNA. *(Richard Megna/Fundamental Photographs)*

DNA Structure and Analysis

Earlier in the text, we discussed the existence of genes on chromosomes that control phenotypic traits and the way in which the chromosomes are transmitted through gametes to future offspring. Logically, genes must contain some sort of information, which, when passed to a new generation, influences the form and characteristics of the offspring; we refer to this as **genetic information**. We might also conclude that this same information in some way directs the many complex processes leading to the adult form.

Until 1944, it was not clear what chemical component of the chromosome makes up genes and constitutes the genetic material. Because chromosomes were known to have both a nucleic acid and a protein component, both were candidates. In 1944, however, there emerged direct experimental evidence that the nucleic acid DNA serves as the informational basis for heredity.

Once the importance of DNA in genetic processes was realized, work intensified with the hope of discerning not only the structural basis of this molecule but also the relationship of its structure to its function. Between 1944 and 1953, many scientists sought information that might answer the most significant and intriguing question in the history of biology: How does DNA serve as the genetic basis for the living process? Researchers believed the answer depended strongly on the chemical structure of the DNA molecule, given the complex but orderly functions ascribed to it.

These efforts were rewarded in 1953 when James Watson and Francis Crick set forth their hypothesis for the double-helical nature of DNA. The assumption that the molecule's functions would be clarified more easily once its general structure was determined proved to be correct. In this chapter, we initially review the evidence that DNA is the genetic material and then discuss the elucidation of its structure.

HOW DO WE KNOW WHAT WE KNOW?

IN THIS CHAPTER, WE WILL FOCUS ON DNA, the molecule that stores genetic information in all living things. We shall define its structure and delve into how we analyze this molecule. As you study this topic, you should try to answer several fundamental questions:

1. How were we able to prove that DNA, and not some other molecule, serves as the genetic material in bacteria and bacteriophages?
2. How do we know that DNA also serves as the genetic material in eukaryotes such as humans?
3. How do we know that the structure of DNA is in the form of a right-handed double-helical molecule?
4. How do we know that complementary base pairs in DNA really exist?
5. How do we know that G pairs with C and that A pairs with T as complementary base pairs are formed?
6. How do we know that repetetive DNA sequences exist in eukaryotes? ■

10.1 The Genetic Material Must Exhibit Four Characteristics

For a molecule to serve as the genetic material, it must possess four major characteristics: **replication**, **storage of information**, **expression of information**, and **variation by mutation**. Replication of the genetic material is one facet of the cell cycle, a fundamental property of all living organisms. Once the genetic material of cells replicates and is doubled in amount, it must then be partitioned equally into daughter cells. During the formation of gametes, the genetic material is also replicated but is partitioned so that each cell gets only one-half of the original amount of genetic material—the process of meiosis. Although the products of mitosis and meiosis differ, these processes are both part of the more general phenomenon of cellular reproduction.

The characteristic of storage can be viewed as a repository of genetic information that may or may not be expressed. It is clear that while most cells contain a complete complement of DNA, at any point in time they express only part of its genetic potential. For example, bacteria "turn on" many genes in response to specific environmental conditions, and turn them off when conditions change. In vertebrates, skin cells may display active melanin genes but never activate their hemoglobin genes; digestive cells activate many genes specific to their function, but do not activate their melanin genes.

Expression of the stored genetic information is the complex process of **information flow** within the cell (Figure 10–1). The initial event is the **transcription** of DNA, in which three main types of RNA molecules are synthesized: messenger RNA (mRNA), ribosomal RNA (rRNA), and transfer RNA (tRNA). Of these, mRNAs are translated into proteins. Each type of mRNA is the product of a specific gene and directs the synthesis of a different protein. **Translation** occurs in conjunction with ribosomes and involves tRNA, which adapts the chemical information in mRNA to the amino acids that make up proteins. Collectively, these processes form the **central dogma of molecular genetics**: "DNA makes RNA, which makes proteins."

The genetic material is also the source of variability among organisms through the process of mutation. If a mutation—a change in the chemical composition of DNA—occurs, the alteration is reflected during transcription and translation, affecting the specific protein. If a mutation is present in gametes, it will be passed to future generations and, with time, become distributed throughout the population. Genetic variation, which also includes rearrangements within and between chromosomes, provides the raw material for the process of evolution.

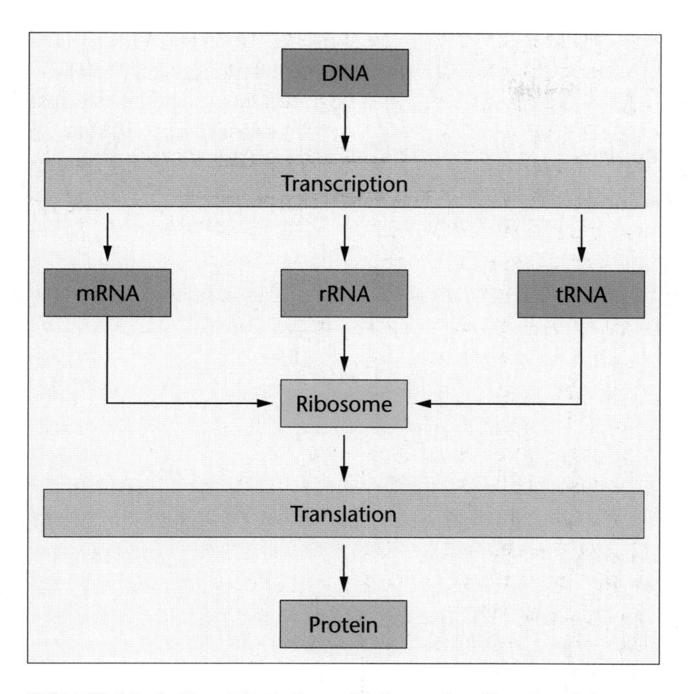

FIGURE 10–1 Simplified view of information flow involving DNA, RNA, and proteins within cells.

10.2 Until 1944, Observations Favored Protein as the Genetic Material

The idea that genetic material is physically transmitted from parent to offspring has been accepted for as long as the concept of inheritance has existed. Beginning in the late nineteenth century, research into the structure of biomolecules progressed considerably, setting the stage for describing the genetic material in chemical terms. Although proteins and nucleic acid were both major candidates for the role of the genetic material, until the 1940s, many geneticists favored proteins. This is not surprising, since a diversity of proteins was known to be abundant in cells, and much more was known about protein chemistry.

DNA was first studied in 1868 by a Swiss chemist, Friedrick Miescher. He isolated cell nuclei and derived an acid substance containing DNA that he called **nuclein**. As investigations progressed, however, DNA, which was shown to be present in chromosomes, seemed to lack the chemical diversity necessary to store extensive genetic information. This conclusion was based largely on Phoebus A. Levene's observations in 1910 that DNA contained approximately equal amounts of four similar molecules called *nucleotides*. Levene postulated incorrectly that identical groups of these four components were repeated over and over, which was the basis of his **tetranucleotide hypothesis** for DNA structure. Attention was thus directed away from DNA, favoring proteins. However, in the 1940s, Erwin Chargaff showed that Levene's proposal was incorrect when he demonstrated that most organisms do not contain precisely equal proportions of the four nucleotides. We shall see later that the structure of DNA accounts for Chargaff's observations.

10.3 Evidence Favoring DNA as the Genetic Material Was First Obtained During the Study of Bacteria and Bacteriophages

The 1944 publication by Oswald Avery, Colin MacLeod, and Maclyn McCarty concerning the chemical nature of a "transforming principle" in bacteria was the initial event that led to the acceptance of DNA as the genetic material. Their work, along with subsequent findings of other research teams, constituted the first direct experimental proof that DNA, and not protein, is the biomolecule responsible for heredity. It marked the beginning of the era of molecular genetics, a period of discovery in biology that made biotechnology feasible and has moved us closer to understanding the basis of life. The impact of the initial findings on future research and thinking paralleled that of the publication of Darwin's theory of evolution and the subsequent rediscovery of Mendel's postulates of transmission genetics. Together, these events constituted the three great revolutions in biology.

Transformation Studies

The research that provided the foundation for Avery, MacLeod, and McCarty's work was initiated in 1927 by Frederick Griffith, a medical officer in the British Ministry of Health. He experimented with several different strains of the bacterium *Diplococcus pneumoniae*.* Some were *virulent strains*, which cause pneumonia in certain vertebrates (notably humans and mice), while others were *avirulent strains*, which do not cause illness.

The difference in virulence depends on the existence of a polysaccharide capsule; virulent strains have this capsule, whereas avirulent strains do not. The nonencapsulated bacteria are readily engulfed and destroyed by phagocytic cells in the animal's circulatory system. Virulent bacteria, which possess the polysaccharide coat, are not easily engulfed; they multiply and cause pneumonia.

The presence or absence of the capsule causes a visible difference between colonies of virulent and avirulent strains. Encapsulated bacteria form **smooth colonies (*S*)** with a shiny surface when grown on an agar culture plate; nonencapsulated strains produce **rough colonies (*R*)** (Figure 10–2). Thus, virulent and avirulent strains are easily distinguished by standard microbiological culture techniques.

Each strain of *Diplococcus* may be one of dozens of different types called *serotypes*. The specificity of the serotype is due to the detailed chemical structure of the polysaccharide constituent of the thick, slimy capsule. Serotypes are identified by immunological techniques and are usually designated by Roman numerals. Griffith used the avirulent type II*R* and the virulent type III*S* in his critical experiments. Table 10.1 summarizes the characteristics of these strains.

*This organism is now designated *Streptococcus pneumonia*.

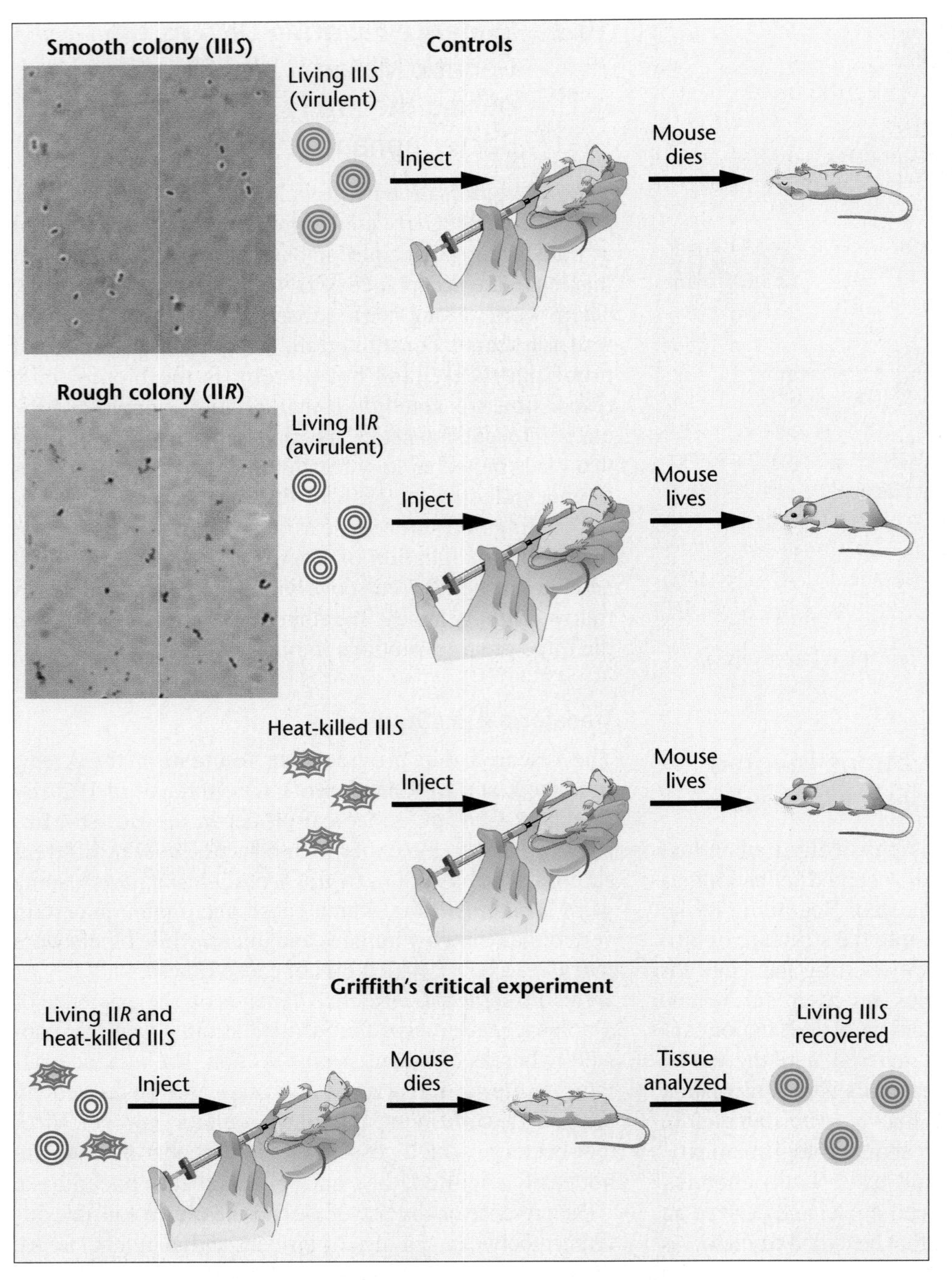

FIGURE 10–2 Griffith's transformation experiment. The photographs show bacterial cells that exhibit capsules (type III*S*) or not (type II*R*). *(Photos: Bruce Iverson/Photomicrography)*

TABLE 10.1 Strains of *Diplococcus pneumoniae* Used by Frederick Griffith in His Original Transformation Experiments

Serotype	*Colony Morphology*	*Capsule*	*Virulence*
II*R*	Rough	Absent	Avirulent
III*S*	Smooth	Present	Virulent

Griffith knew from the work of others that only living virulent cells produced pneumonia in mice. If heat-killed virulent bacteria were injected into mice, no pneumonia resulted, just as living avirulent bacteria failed to produce the disease. Griffith's critical experiment (Figure 10–2) involved injecting mice with living II*R* (avirulent) cells combined with heat-killed III*S* (virulent) cells. Since neither cell type caused death in mice when injected alone, Griffith expected that the double injection would not kill the mice. But, after five days, all of the mice that had received both types of cells were dead. Paradoxically, analysis of their blood revealed large numbers of living type III*S* bacteria.

As far as could be determined, these III*S* bacteria were identical to the III*S* strain from which the heat-killed cell preparation had been made. Control mice, injected only with living avirulent II*R* bacteria, did not develop pneumonia and remained healthy. This ruled out the possibility that the avirulent II*R* cells simply changed (or mutated) to virulent III*S* cells in the absence of the heat-killed III*S* bacteria. Instead, some type of interaction had taken place between living II*R* and heat-killed III*S* cells.

Griffith concluded that the heat-killed III*S* bacteria somehow converted live avirulent II*R* cells into virulent III*S* cells. Calling the phenomenon **transformation**, he suggested that the **transforming principle** might be some part of the polysaccharide capsule or a compound required for capsule synthesis, although the capsule alone did not cause pneumonia. To use Griffith's term, the transforming principle from the dead III*S* cells served as a "pabulum" for the II*R* cells.

Griffith's work led bacteriologists and other physicians to explore the phenomenon of transformation. By 1931, Henry Dawson and his coworkers showed that transformation could occur *in vitro* (in a test tube containing only bacterial cells). That is, injection into mice was not necessary for transformation to occur. By 1933, Lionel J. Alloway had refined the *in vitro* experiments using extracts from *S* cells added to living *R* cells. The soluble filtrate from the heat-killed *S* cells was as effective in inducing transformation as were the intact cells. Alloway and others did not view transformation as a genetic event, but rather as a physiological modification of some sort. Nevertheless, the experimental evidence that a chemical substance was responsible for transformation was quite convincing.

Then, in 1944, after ten years of work, Avery, MacLeod, and McCarty published their results in what is now regarded as a classic paper in the field of molecular genetics. They reported that they had obtained the transforming principle in a highly purified state, and that beyond reasonable doubt it was DNA.

The details of their work are illustrated in Figure 10–3. The researchers began their isolation procedure with large quantities (50–75 L) of liquid cultures of type III*S* virulent cells. The cells were centrifuged, collected, and heat-killed. Following various chemical treatments, a soluble filtrate was derived from these cells, which retained the ability to induce transformation of type II*R* avirulent cells. The soluble filtrate was treated with a protein-digesting enzyme, called a protease, and an RNA-digesting enzyme, called ribonuclease. Such treatment destroyed the activity of any remaining

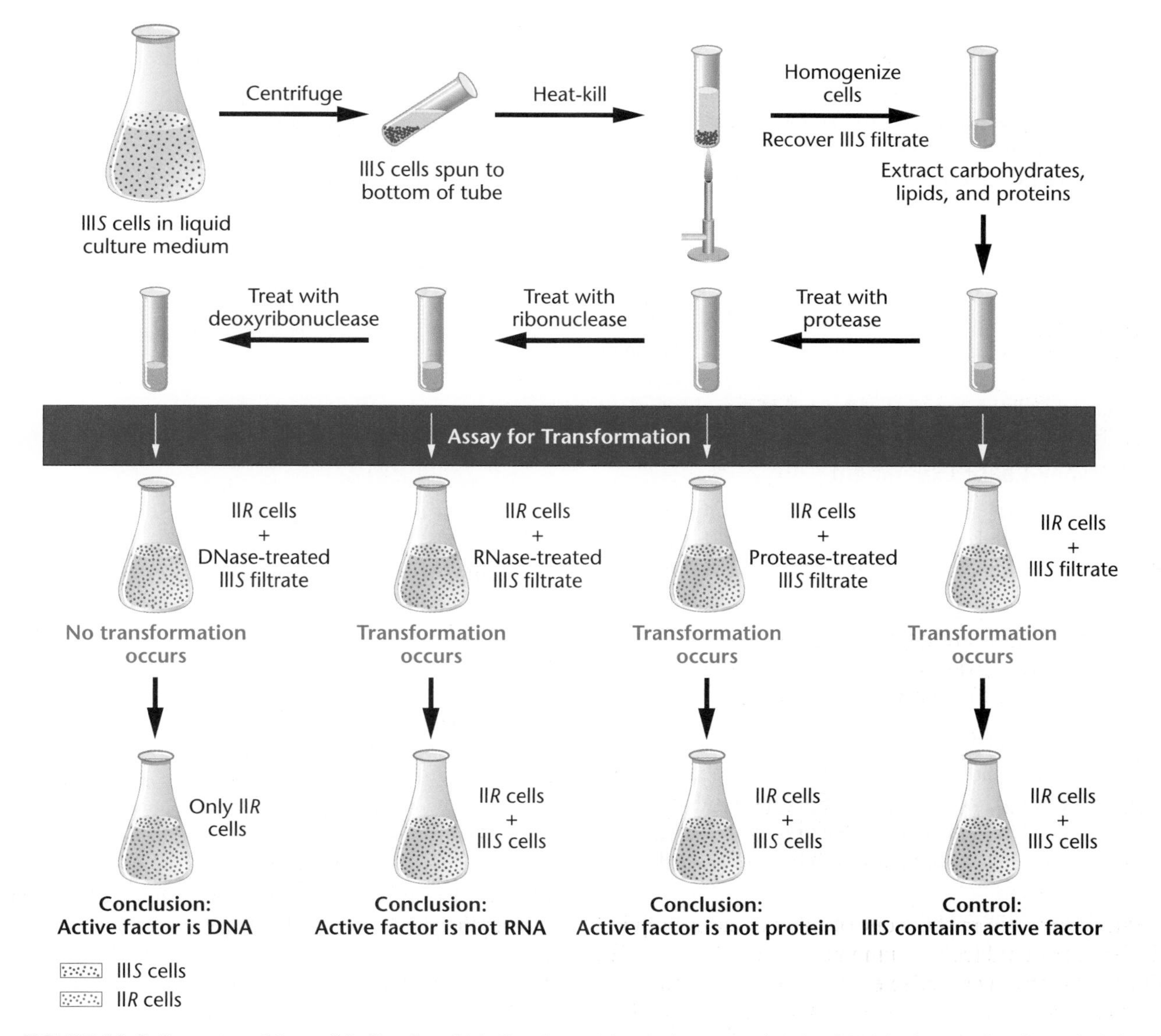

FIGURE 10–3 Summary of Avery, MacLeod, and McCarty's experiment demonstrating that DNA is the transforming principle.

protein and RNA. Nevertheless, transforming activity still remained. They concluded that neither protein nor RNA was responsible for transformation. The final confirmation came with experiments using crude samples of the DNA-digesting enzyme **deoxyribonuclease**, isolated from dog and rabbit sera. Digestion with this enzyme destroyed transforming activity present in the filtrate—thus, Avery and his coworkers were certain that the active transforming principle in these experiments was DNA.

The great amount of work, the confirmation and reconfirmation of the conclusions, and the logic of the experimental design involved in the research of these three scientists are truly impressive. Their conclusion in the 1944 publication, however, was stated very simply: "The evidence presented supports the belief that a nucleic acid of the desoxyribose* type is the fundamental unit of the transforming principle of *Pneumococcus* type III."

They also immediately recognized the genetic and biochemical implications of their work. They suggested that the transforming principle interacts with the II*R* cell and gives rise to a coordinated series of enzymatic reactions that culminates in the synthesis of the type III*S* capsular polysaccharide. They emphasized that, once transformation occurs, the capsular polysaccharide is produced in successive generations. Transformation is therefore heritable, and the process affects the genetic material.

Transformation has now been shown to occur in *Hemophilus influenzae*, *Bacillus subtilis*, *Shigella paradysenteriae*, and *Escherichia coli*, among many other microorganisms. Transformation of numerous genetic traits other than colony morphology has been demonstrated, including those that resist antibiotics or metabolize various nutrients. These observations further strengthened the belief that transformation by DNA is primarily a genetic event, rather than simply a physiological change. We will pursue this idea in the Insights and Solutions section at the end of this chapter.

The Hershey-Chase Experiment

The second major piece of evidence supporting DNA as the genetic material was provided during the study of the bacterium *E. coli* and one of its infecting viruses, bacteriophage T2. Often referred to simply as a *phage*, the virus consists of a protein coat surrounding a core of DNA. Electron micrographs reveal the phage's external structure to be composed of a hexagonal head plus a tail. Figure 10–4 shows the life cycle of a T-even bacteriophage such as T2, as it was known in 1952. Recall that the phage adsorbs to the bacterial cell and some component of the phage enters the bacterial cell. Following infection, the viral information "commandeers" the cellular machinery of the host and undergoes viral reproduction. In a reasonably short time, many new phages are constructed and the bacterial cell is lysed, releasing the progeny viruses.

In 1952, Alfred Hershey and Martha Chase published the results of experiments designed to clarify the events leading

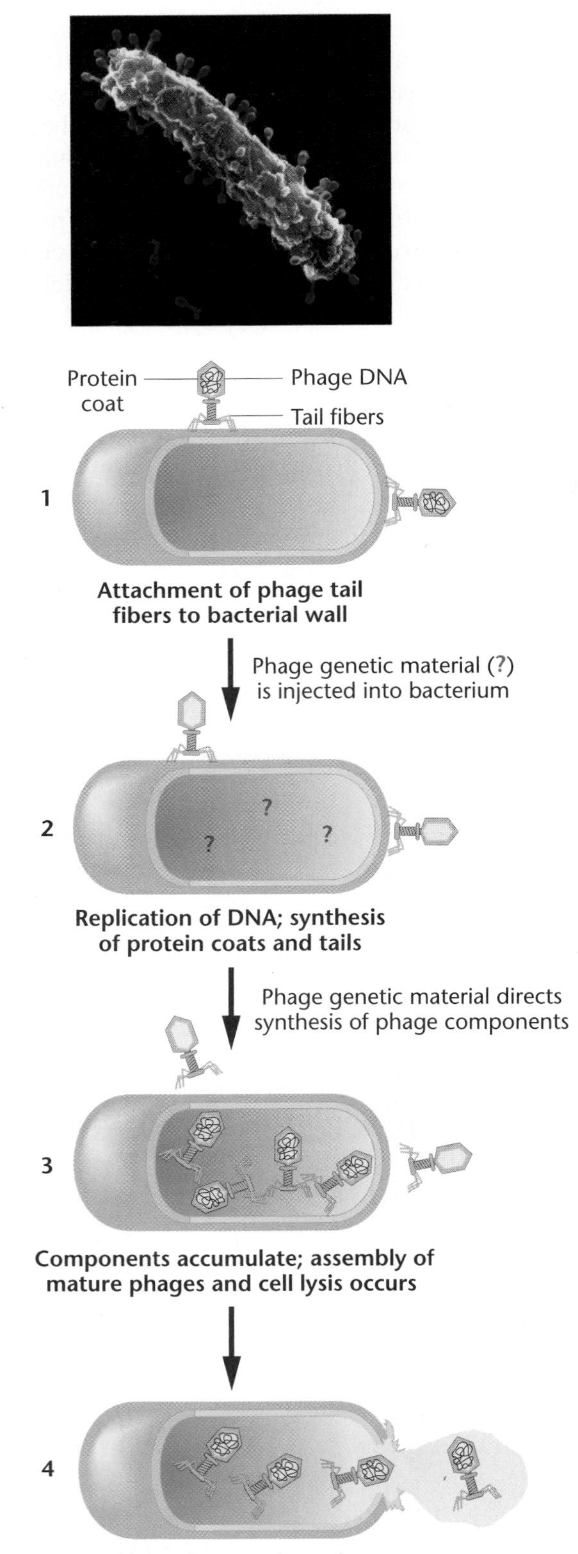

FIGURE 10–4 Life cycle of a T-even bacteriophage. The electron micrograph shows an *E. coli* cell during infection by numerous T2 phages (shown in blue). *(Photo: Oliver Meckes/Max-Planck-Institute-Tubingen/Photo Researchers, Inc.)*

*Desoxyribose is now spelled deoxyribose.

to phage reproduction. Several of the experiments clearly established the independent functions of phage protein and nucleic acid in the reproduction process of the bacterial cell. Hershey and Chase knew from this existing data that:

1. T2 phages consist of approximately 50% protein and 50% DNA.
2. Infection is initiated by adsorption of the phage by its tail fibers to the bacterial cell.
3. The production of new viruses occurs within the bacterial cell.

It appeared that some molecular component of the phage, DNA and/or protein, entered the bacterial cell and directed viral reproduction. Which was it?

Hershey and Chase used radioisotopes to follow the molecular components of phages during infection. Both ^{32}P and ^{35}S, radioactive forms of phosphorus and sulfur, respectively, were used. DNA contains phosphorus but not sulfur, so ^{32}P effectively labels DNA. Because proteins contain sulfur but not phosphorus, ^{35}S labels protein. *This is a key point in the experiment.* If *E. coli* cells are first grown in the presence of *either* ^{32}P *or* ^{35}S and then infected with T2 viruses, the progeny phage will have *either* a labeled DNA core *or* a labeled protein coat, respectively. These radioactive phages can be isolated and used to infect unlabeled bacteria (Figure 10–5).

When labeled phage and unlabeled bacteria were mixed, an *adsorption complex* is formed as the phages attached their tail fibers to the bacterial wall. These complexes were isolated and subjected to a high shear force by placing them in a blender. This force strips off the attached phages, which can then be analyzed separately (Figure 10–5). By tracing the radioisotopes, Hershey and Chase were able to demonstrate that most of the ^{32}P-labeled DNA had transferred into the bacterial cell following adsorption; on the other hand, most all of the ^{35}S-labeled protein remained outside the bacterial cell and was recovered in the phage "ghosts" (empty phage coats) after the blender treatment. After separation, the bacterial cells, which now contained viral DNA, were eventually lysed as new phages were produced. These progeny contained ^{32}P, but not ^{35}S.

Hershey and Chase interpreted these results as indicating that the protein of the phage coat remains outside the host cell and is not involved in the production of new phages. On the other hand, and most important, phage DNA enters the host cell and directs phage reproduction. Hershey and Chase had demonstrated that the genetic material in phage T2 is DNA, not protein.

These experiments, along with those of Avery and his colleagues, provided convincing evidence that DNA is the molecule responsible for heredity. This conclusion has since served as the cornerstone of the field of molecular genetics.

Transfection Experiments

During the eight years following the publication of the Hershey-Chase experiment, additional research with bacterial viruses provided even more solid proof that DNA is the genetic material. In 1957, several reports demonstrated that if *E. coli* is treated with the enzyme lysozyme, the outer wall of the cell can be removed without destroying the bacterium. Enzymatically treated cells are naked, so to speak, and contain only the cell membrane as the outer boundary of the cell—these structures are called **protoplasts** (or **spheroplasts**). John Spizizen and Dean Fraser independently reported that by using protoplasts, they were able to initiate phage multiplication with disrupted T2 particles. That is, provided protoplasts were used, a virus did not have to be intact for infection to occur.

Similar but more refined experiments were reported in 1960 using only DNA purified from bacteriophages. This process of infection by only the viral nucleic acid, called **transfection**, proves conclusively that phage DNA alone contains all the necessary information for producing mature viruses. Thus, the evidence that DNA serves as the genetic material in all organisms was further strengthened, even though all direct evidence had been obtained from bacterial and viral studies.

10.4 Indirect and Direct Evidence Supports the Concept that DNA Is the Genetic Material in Eukaryotes

In 1950, eukaryotic organisms were not amenable to the types of experiments that used bacteria and viruses to demonstrate that DNA is the genetic material. Nevertheless, it was generally assumed that the genetic material would be a universal substance and also serve this role in eukaryotes. Initially, support for this assumption relied on several circumstantial observations that, taken together, indicated that DNA is also the genetic material in eukaryotes. Subsequently, direct evidence established unequivocally the central role of DNA in genetic processes.

Indirect Evidence: Distribution of DNA

The genetic material should be found where it functions—in the nucleus as part of chromosomes. Both DNA and protein fit this criterion. However, protein is also abundant in the cytoplasm, while DNA is not. Both mitochondria and chloroplasts are known to perform genetic functions, and DNA is also present in these organelles. Thus, DNA is found only where primary genetic function is known to occur. Protein, however, is found everywhere in the cell. These observations are consistent with the interpretation favoring DNA over protein as the genetic material.

Because it had been established earlier that chromosomes within the nucleus contain the genetic material, a correlation was expected between the ploidy (n, $2n$, etc.) of cells and the quantity of the molecule that functions as the genetic material. Meaningful comparisons can be made between gametes (sperm and eggs) and somatic or body cells. The latter are recognized as being diploid ($2n$)

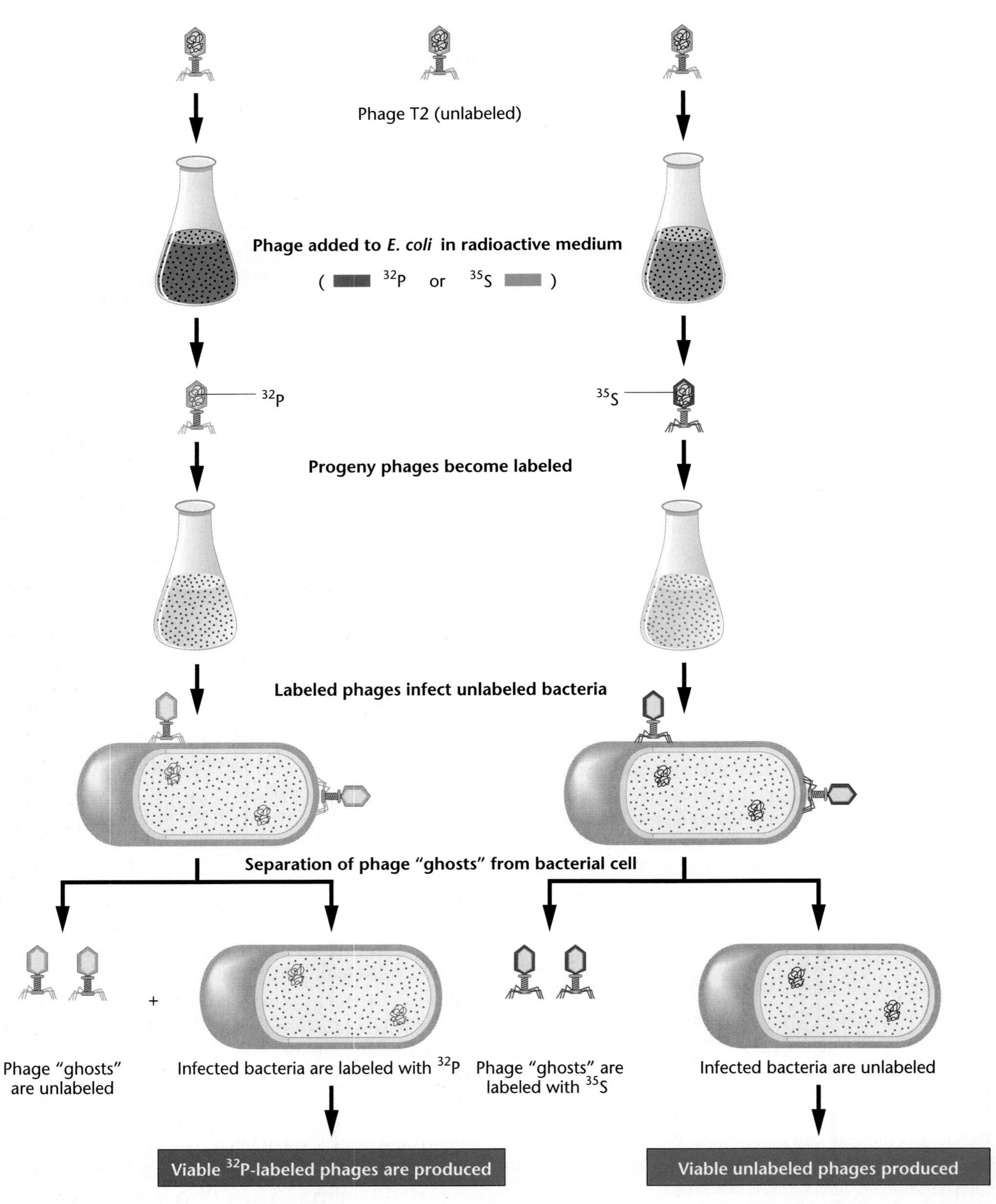

FIGURE 10–5 Summary of the Hershey-Chase experiment demonstrating that DNA, not protein, is responsible for directing the reproduction of phage T2 during the infection of *E. coli*.

and containing twice the number of chromosomes as gametes, which are haploid (n).

Table 10.2 compares the amount of DNA found in haploid sperm and the diploid nucleated precursors of red blood cells from a variety of organisms. The amount of DNA and the number of sets of chromosomes is closely correlated. No consistent correlation can be observed between gametes and diploid cells for proteins, thus again favoring DNA over proteins as the genetic material of eukaryotes.

TABLE 10.2 DNA Content of Haploid Versus Diploid Cells of Various Species (in picograms)

*Organism**	*n*	*2n*
Human	3.25	7.30
Chicken	1.26	2.49
Trout	2.67	5.79
Carp	1.65	3.49
Shad	0.91	1.97

*Sperm (n) and nucleated precursors to red blood cells ($2n$) were used to contrast ploidy levels.

Indirect Evidence: Mutagenesis

Ultraviolet (UV) light is one of many agents capable of inducing mutations in the genetic material. Simple organisms such as yeast and other fungi can be irradiated with various wavelengths of UV light, and the effectiveness of each wavelength can then be measured by the number of mutations it induces. When the data are plotted, an **action spectrum** of UV light as a mutagenic agent is obtained. This action spectrum can then be compared with the **absorption spectrum** of any molecule suspected to be genetic material (Figure 10–6). *The molecule serving as the genetic material is expected to absorb at the wavelengths shown to be mutagenic.*

UV light is most mutagenic at the wavelength (λ) of 260 nanometers (nm), and both DNA and RNA absorb UV light most strongly at 260 nm. On the other hand, protein absorbs most strongly at 280 nm, yet no significant mutagenic effects are observed at this wavelength. This indirect evidence also supports the idea that a nucleic acid is the genetic material and tends to exclude protein.

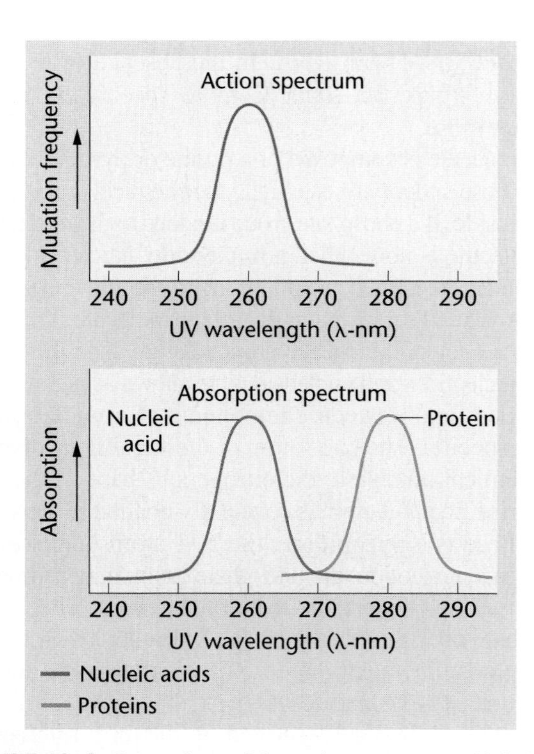

FIGURE 10–6 Comparison of the action spectrum, which determines the most effective mutagenic UV wavelength; and the absorption spectrum, which shows the range of wavelengths where nucleic acids and proteins absorb UV light.

Direct Evidence: Recombinant DNA Studies

Although the circumstantial evidence described above does not constitute direct proof that DNA is the genetic material in eukaryotes, these observations spurred researchers to forge ahead, basing their work on this hypothesis. Today, there is no doubt of the validity of this conclusion. DNA *is* the genetic material in eukaryotes. The strongest evidence is provided by molecular analysis utilizing **recombinant DNA technology**. In this procedure, segments of eukaryotic DNA corresponding to specific genes are isolated and spliced into bacterial DNA. This complex can then be inserted into a bacterial cell, and its genetic expression is monitored. If a eukaryotic gene is introduced, the presence of the corresponding eukaryotic protein product demonstrates directly that this DNA is present and functional in the bacterial cell. This has been shown to be the case in countless instances. For example, the products of the human genes specifying insulin and interferon are produced by bacteria after the human genes that encode these proteins are inserted. As the bacterium divides, the eukaryotic DNA replicates along with the host DNA and is distributed to the daughter cells, which also express the human genes and synthesize the corresponding proteins. The availability of vast amounts of DNA coding for specific genes, available as a result of recombinant DNA research, has led to other direct evidence that DNA serves as the genetic material. Work done in the laboratory of Beatrice Mintz demonstrated that DNA encoding the human β-globin gene, when microinjected into a fertilized mouse egg, is later found to be present and expressed in adult mouse tissue, and it is transmitted to and expressed in that mouse's progeny. These mice are examples of **transgenic animals**. More recent work introduced rat DNA encoding a growth hormone into fertilized mouse eggs. About one-third of the resultant mice grew to twice their normal size, indicating that foreign DNA was present and functional. Subsequent generations of mice inherited this genetic information and also grew to a large size. This clearly demonstrates that DNA meets the requirement of expression of genetic information in eukaryotes. Later we shall discuss exactly how DNA is stored, replicated, expressed, and mutated.

10.5 RNA Serves as the Genetic Material in Some Viruses

Some viruses contain an RNA core rather than a DNA core. In these viruses, it appears that RNA serves as the genetic material—an exception to the general rule that

DNA performs this function. In 1956, it was demonstrated that when purified RNA from **tobacco mosaic virus (TMV)** was spread on tobacco leaves, the characteristic lesions caused by viral infection subsequently appeared on the leaves. Thus, it was concluded that RNA is the genetic material of this virus.

In 1965 and 1966, Norman Pace and Sol Spiegelman demonstrated further that RNA from the phage Qβ can be isolated and replicated *in vitro*. Replication depends on an enzyme, **RNA replicase**, which is isolated from host *E. coli* cells following normal infection. When the RNA replicated *in vitro* is added to *E. coli* protoplasts, infection and viral multiplication (transfection) occur. Thus, RNA synthesized in a test tube serves as the genetic material in these phages by directing the production of all the components necessary for viral replication.

One other group of RNA-containing viruses bears mention. These are the **retroviruses**, which replicate in an unusual way. Their RNA serves as a template for the synthesis of the complementary DNA molecule. The process, **reverse transcription**, occurs under the direction of an RNA-dependent DNA polymerase enzyme called **reverse transcriptase**. This DNA intermediate can be incorporated into the genome of the host cell, and when the host DNA is transcribed, copies of the original retroviral RNA chromosomes are produced. Retroviruses include the human immunodeficiency virus (HIV), which causes AIDS, as well as RNA tumor viruses.

10.6 The Structure of DNA Holds the Key to Understanding Its Function

Having established that DNA is the genetic material in all living organisms (except certain viruses), we turn now to the structure of this nucleic acid. In 1953, James Watson and Francis Crick proposed that the structure of DNA is in the form of a double helix. Their proposal was published in a short paper in the journal *Nature*, reprinted in its entirety (see p. 219). In a sense, this publication was the finish of a highly competitive scientific race to obtain what some consider to be the most significant finding in the history of biology. This race, as recounted in Watson's book *The Double Helix*, demonstrates the human interaction, genius, frailty, and intensity involved in the scientific effort that eventually led to the elucidation of DNA structure.

The data available to Watson and Crick, crucial to the development of their proposal, came primarily from two sources: (1) base composition analysis of hydrolyzed samples of DNA, and (2) X-ray diffraction studies of DNA. The analytical success of Watson and Crick can be attributed to model building that conformed to the existing data. If the correct solution to the structure of DNA is viewed as a puzzle, Watson and Crick, working in the Cavendish Laboratory in Cambridge, England, were the first to fit the pieces together successfully.

Nucleic Acid Chemistry

Before turning to this work, a brief introduction to nucleic acid chemistry is in order. This chemical information was well known to Watson and Crick during their investigation and served as the basis of their model building.

DNA is a nucleic acid, and nucleotides are the building blocks of all nucleic acid molecules. Sometimes called mononucleotides, these structural units have three essential components: a **nitrogenous base**, a **pentose sugar** (a five-carbon sugar), and a **phosphate group**. There are two kinds of nitrogenous bases: the nine-member double-ring **purines** and the six-member single-ring **pyrimidines**.

Two types of purines and three types of pyrimidines are found in nucleic acids. The two purines are **adenine** and **guanine**, abbreviated A and G. The three pyrimidines are **cytosine**, **thymine**, and **uracil** (C, T, and U). The chemical structures of the five bases are shown in Figure 10–7a. Both DNA and RNA contain A, G, and C; but only DNA contains the base T, and only RNA contains the base U. Each nitrogen or carbon atom of the ring structures of purines and pyrimidines is designated by a number. Note that corresponding atoms in the purine and pyrimidine rings are numbered differently.

The pentose sugars found in nucleic acids give them their names. **Ribonucleic acids (RNA)** contain **ribose**, while **deoxyribonucleic acids (DNA)** contain **deoxyribose**. Figure 10–7b shows the chemical structures for these two pentose sugars. Each carbon atom is distinguished by a number with a prime sign ($'$). As you can see in Figure 10–7b, compared with ribose, deoxyribose has a hydrogen atom rather than a hydroxyl group at the C-2$'$ position. The absence of a hydroxyl group at the C-2$'$ position thus distinguishes DNA from RNA. In the absence of the C-2$'$ hydroxyl group, the sugar is more specifically named **2-deoxyribose**.

If a molecule is composed of a purine or pyrimidine base and a ribose or deoxyribose sugar, the chemical unit is called a **nucleoside**. If a phosphate group is added to the nucleoside, the molecule is now called a **nucleotide**. Nucleosides and nucleotides are named according to the specific nitrogenous base (A, G, C, T, or U) that is part of the molecule. The structure of a nucleotide and the nomenclature used in naming DNA nucleotides and nucleosides are shown in Figure 10–8.

The bonding between components of a nucleotide is highly specific. The C-1$'$ atom of the sugar is involved in the chemical linkage to the nitrogenous base. If the base is a purine, the N-9 atom is covalently bonded to the sugar; if the base is a pyrimidine, the N-1 atom bonds to the sugar. In a nucleotide, the phosphate group may be bonded to the C-2$'$, C-3$'$, or C-5$'$ atom of the sugar. The C-5$'$—phosphate configuration is shown in Figure 10–8. It is by far the most prevalent one in biological systems and the one found in DNA and RNA.

Nucleotides are also described by the term **nucleoside monophosphate (NMP)**. The addition of one or two phosphate groups results in **nucleoside diphosphates (NDP)** and **triphosphates (NTP)**, respectively, as shown

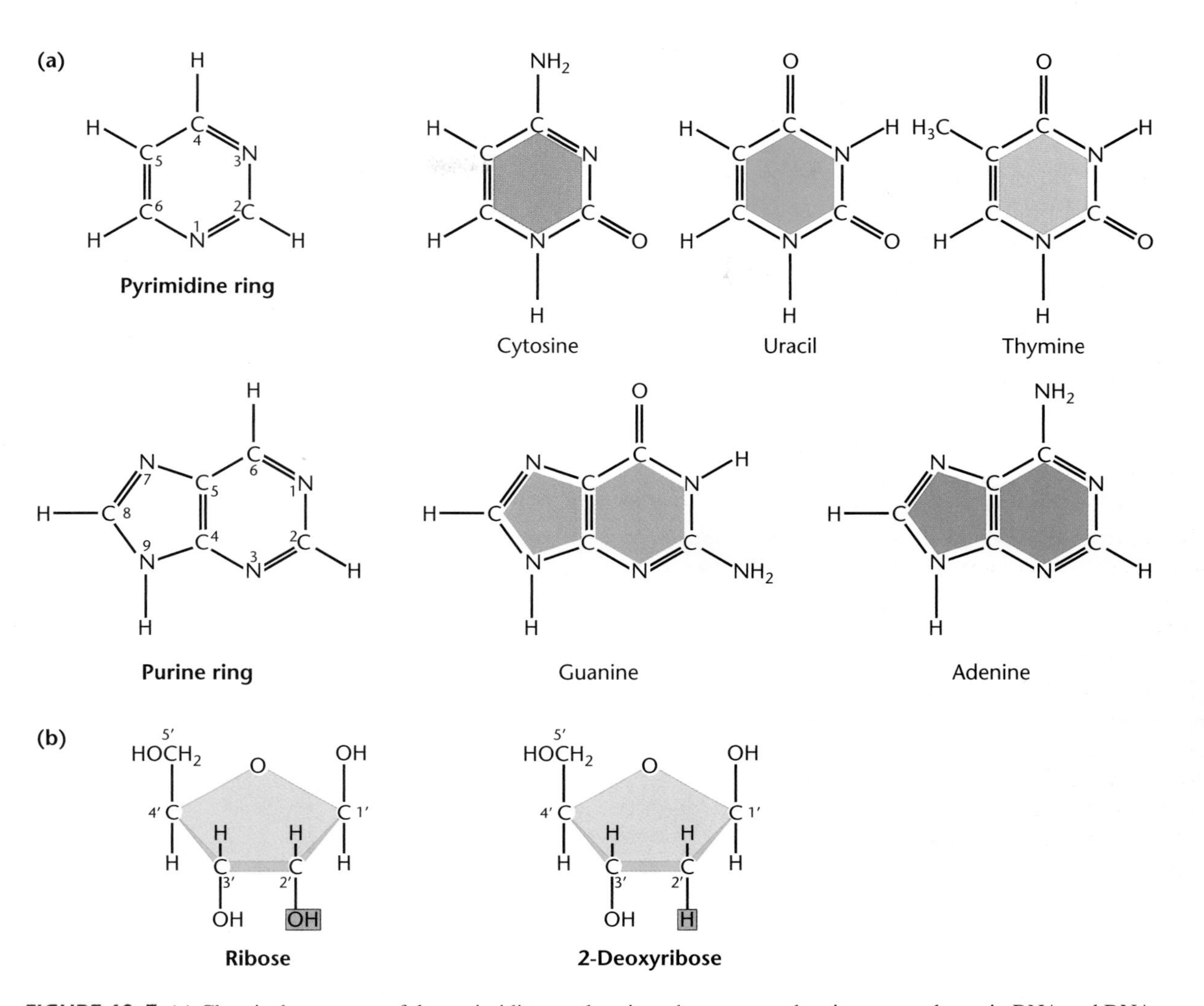

FIGURE 10–7 (a) Chemical structures of the pyrimidines and purines that serve as the nitrogenous bases in RNA and DNA. (b) Chemical ring structures of ribose and 2-deoxyribose, which serve as the pentose sugars in RNA and DNA, respectively.

MEDIA TUTORIAL DNA structure

in Figure 10–9. The triphosphate form is significant because it is the precursor molecule during nucleic acid synthesis within the cell. Additionally, adenosine triphosphate (ATP) and guanosine triphosphate (GTP) are important in cell bioenergetics because of the large amount of energy involved in adding or removing the terminal phosphate group. The hydrolysis of ATP or GTP to ADP or GDP and inorganic phosphate (P_1) is accompanied by the release of a large amount of energy in the cell. When these chemical conversions are coupled to other reactions, the energy produced is used to drive them. As a result, ATP and GTP are involved in many cellular activities.

The linkage between two mononucleotides involves a phosphate group linked to two sugars. A **phosphodiester bond** is formed as phosphoric acid is joined to two alcohols (the hydroxyl groups on the two sugars) by an ester linkage on both sides. Figure 10–10 shows the resultant phosphodiester bond in DNA. Each structure has a C-3′ end and a C-5′ end. Two joined nucleotides form a dinucleotide; three nucleotides, a trinucleotide; and so forth. Short chains consisting of up to 20 nucleotides are called **oligonucleotides**; longer chains are **polynucleotides**.

Long polynucleotide chains account for the large molecular weight of DNA and explain its most important property—storage of vast quantities of genetic information. If each nucleotide position in this long chain can be occupied by any one of four nucleotides, extraordinary variation is possible. For example, a polynucleotide only 1000 nucleotides in length can be arranged 4^{1000} different ways, each one different from all other possible sequences. This potential variation in molecular structure is essential if DNA is to store the vast amounts of chemical information necessary to direct cellular activities.

Base Composition Studies

Between 1949 and 1953, Erwin Chargaff and his colleagues used chromatographic methods to separate the four bases in DNA samples from various organisms. Quantitative methods were then used to determine the amounts of the four nitrogenous bases from each source. Table 10.3(a) lists some of Chargaff's original data. Parts (b) and (c) show more recently derived information that reinforces Chargaff's findings. As we shall see, Chargaff's

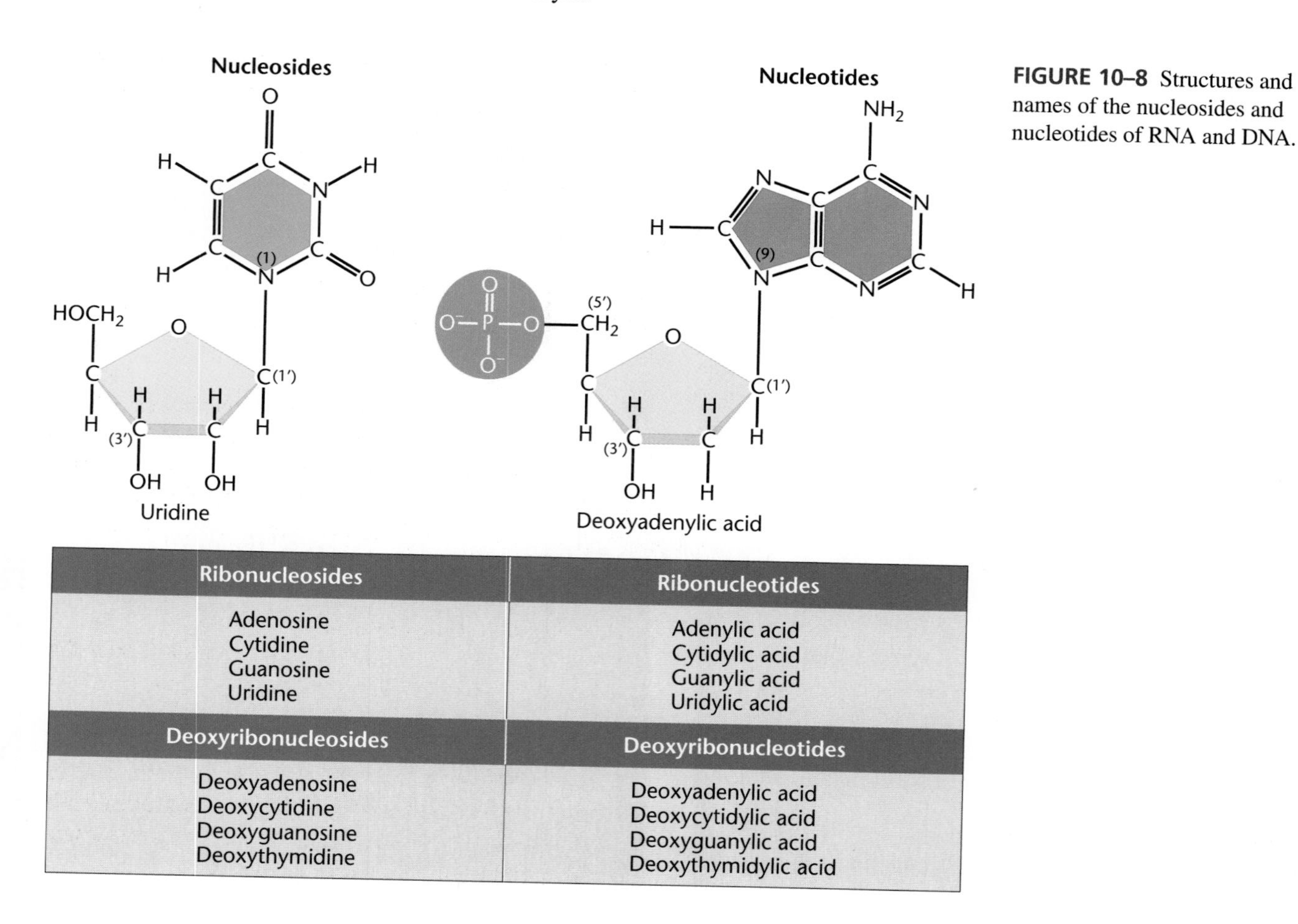

Ribonucleosides	Ribonucleotides
Adenosine	Adenylic acid
Cytidine	Cytidylic acid
Guanosine	Guanylic acid
Uridine	Uridylic acid
Deoxyribonucleosides	**Deoxyribonucleotides**
Deoxyadenosine	Deoxyadenylic acid
Deoxycytidine	Deoxycytidylic acid
Deoxyguanosine	Deoxyguanylic acid
Deoxythymidine	Deoxythymidylic acid

FIGURE 10–8 Structures and names of the nucleosides and nucleotides of RNA and DNA.

data were critical to the successful model of DNA put forward by Watson and Crick. On the basis of these data, the following conclusions may be drawn.

1. The amount of adenine residues is proportional to the amount of thymine residues in DNA (columns 1, 2, and 5). Also, the amount of guanine residues is proportional to the amount of cytosine residues (columns 3, 4, and 6).
2. Based on this proportionality, the sum of the purines (A + G) equals the sum of the pyrimidines (C + T), as shown in column 7.
3. The percentage of (G + C) does not necessarily equal the percentage of (A + T). As you can see, this ratio varies greatly between different organisms, as shown in column 8.

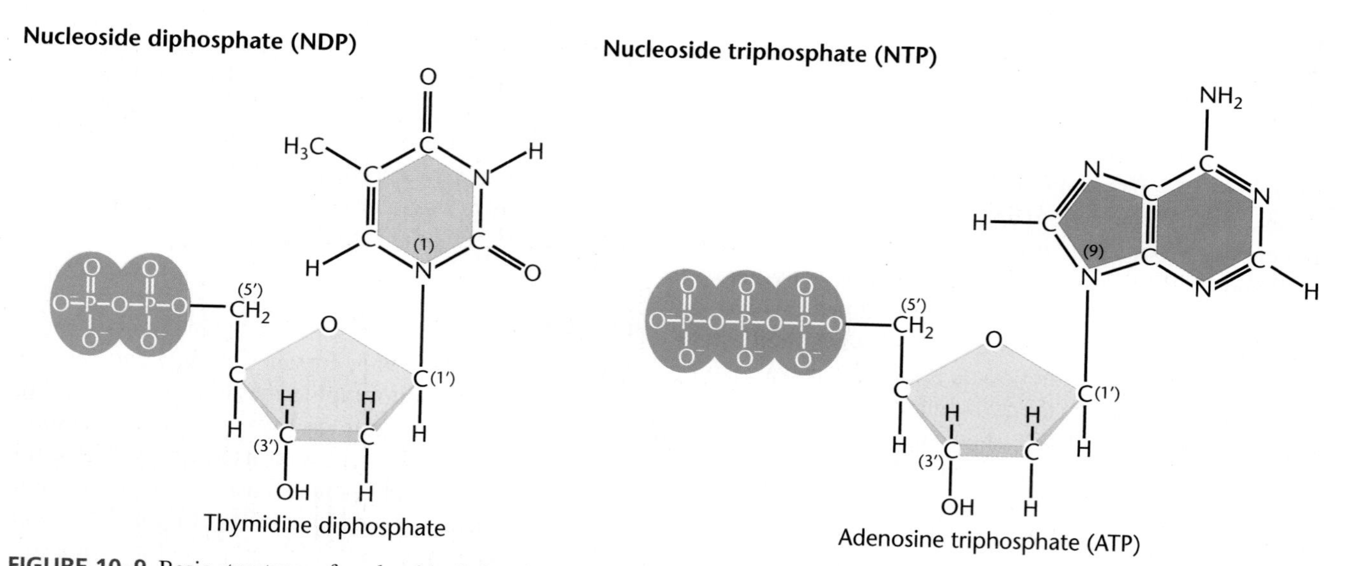

FIGURE 10–9 Basic structures of nucleoside diphosphates and triphosphates. Thymidine diphosphate and adenosine triphosphate are diagrammed here.

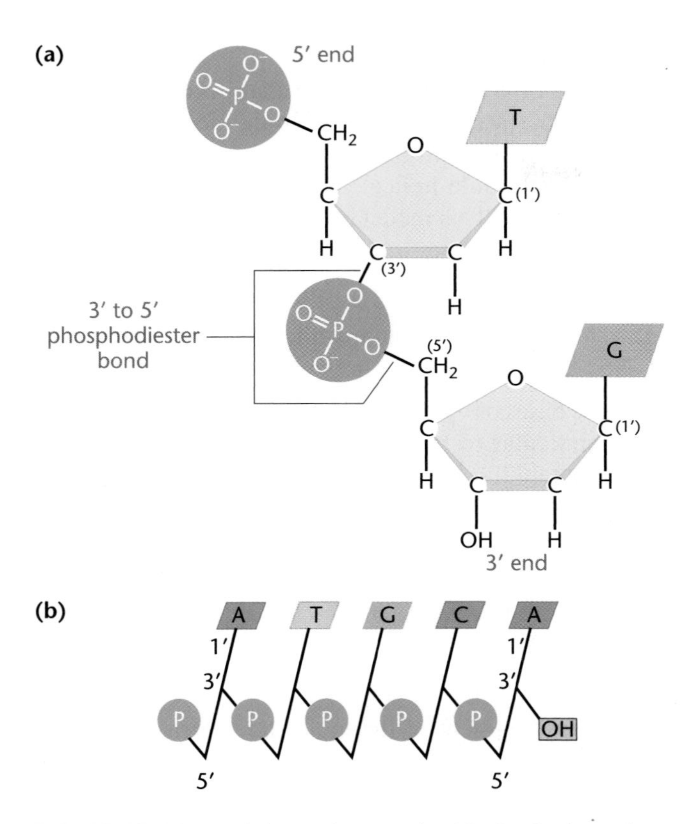

FIGURE 10–10 (a) Linkage of two nucleotides by the formation of a C-3′–C-5′ (3′–5′) phosphodiester bond, producing a dinucleotide. (b) Shorthand notation for a polynucleotide chain.

These conclusions indicate a definite pattern of base composition of DNA molecules. The data provided the initial clue to the DNA puzzle. They also directly refuted Levene's tetranucleotide hypothesis, which stated that all four bases are present in equal amounts.

X-Ray Diffraction Analysis

When fibers of a DNA molecule are subjected to X-ray bombardment, the X-rays scatter according to the molecule's atomic structure. The pattern of scatter can be captured as spots on photographic film and analyzed, particularly for the overall shape of and regularities within the molecule. This process, **X-ray diffraction analysis**, was applied successfully to the study of protein structure by Linus Pauling and other chemists. The technique had been attempted on DNA as early as 1938 by William Astbury. By 1947, he had detected a periodicity of 3.4 angstroms (Å)* within the structure of the molecule, which suggested to him that the bases were stacked like coins on top of one another.

Between 1950 and 1953, Rosalind Franklin, working in the laboratory of Maurice Wilkins, obtained improved X-ray data from more purified samples of DNA (Figure 10–11). Her work confirmed the 3.4 Å periodicity seen by Astbury, and suggested that the structure of DNA was some sort of helix. However, she did not propose a definitive model. Pauling had analyzed the work of Astbury and others and proposed incorrectly that DNA is a triple helix.

*Today, measurement in nanometers (nm) is favored (1 nm = 10 Å).

TABLE 10.3 DNA Base Composition Data

*(a) Chargaff's Data**

	Molar Proportions†			
	1	*2*	*3*	*4*
Source	*A*	*T*	*G*	*C*
Ox thymus	26	25	21	16
Ox spleen	25	24	20	15
Yeast	24	25	14	13
Avian tubercle bacilli	12	11	28	26
Human sperm	29	31	18	18

(c) G + C Content in Several Organisms

Organism	*G + C (%)*
Phage T2	36.0
Drosophila	45.0
Maize	49.1
Euglena	53.5
Neurospora	53.7

(b) Base Compositions of DNAs from Various Sources

	Base Composition				*Base Ratio*		*A + T:G + C Ratio*	
	1	*2*	*3*	*4*	*5*	*6*	*7*	*8*
Source	*A*	*T*	*G*	*C*	*A:T*	*G:C*	*(A + G):(C + T)*	*(A + T):(C + G)*
Human	30.9	29.4	19.9	19.8	1.05	1.00	1.04	1.52
Sea urchin	32.8	32.1	17.7	17.3	1.02	1.02	1.02	1.58
E. coli	24.7	23.6	26.0	25.7	1.04	1.01	1.03	0.93
Sarcina lutea	13.4	12.4	37.1	37.1	1.08	1.00	1.04	0.35
T7 bacteriophage	26.0	26.0	24.0	24.0	1.00	1.00	1.00	1.08

**Source:* Chargaff, 1950.
†Moles of nitrogenous constituent per mole of P (often, the recovery was less than 100%).

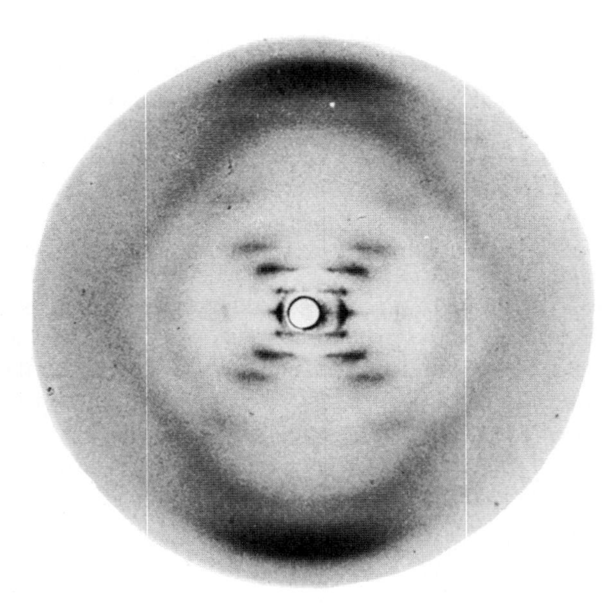

FIGURE 10–11 X-ray diffraction photograph of purified DNA fibers. The strong arcs on the periphery show closely spaced aspects of the molecule, providing an estimate of the periodicity of nitrogenous bases, which are 3.4 Å apart. The inner cross pattern of spots shows the grosser aspect of the molecule, indicating its helical nature. *(M. H. F. Wilkins. Courtesy of Bio-Physics Department, King's College, London, England.)*

The Watson-Crick Model

Watson and Crick published their analysis of DNA structure in 1953 (see p.219). By building models under the constraints of the information just discussed, they proposed the double-helical form of DNA shown in Figure 10–12a. This model has these major features:

1. Two long polynucleotide chains are coiled around a central axis, forming a right-handed double helix.
2. The two chains are antiparallel; that is, their C-5′-to-C-3′ orientations run in opposite directions.
3. The bases of both chains are flat structures, lying perpendicular to the axis; they are "stacked" on one another, 3.4 Å (0.34 nm) apart, and located on the inside of the structure.
4. The nitrogenous bases of opposite chains are *paired* as the result of hydrogen bonds; in DNA, only A=T and G≡C pairs occur.
5. Each complete turn of the helix is 34 Å (3.4 nm) long; thus, 10 bases exist per turn in each chain.
6. In any segment of the molecule, alternating larger **major grooves** and smaller **minor grooves** are apparent along the axis.
7. The double helix measures 20 Å (2.0 nm) in diameter.

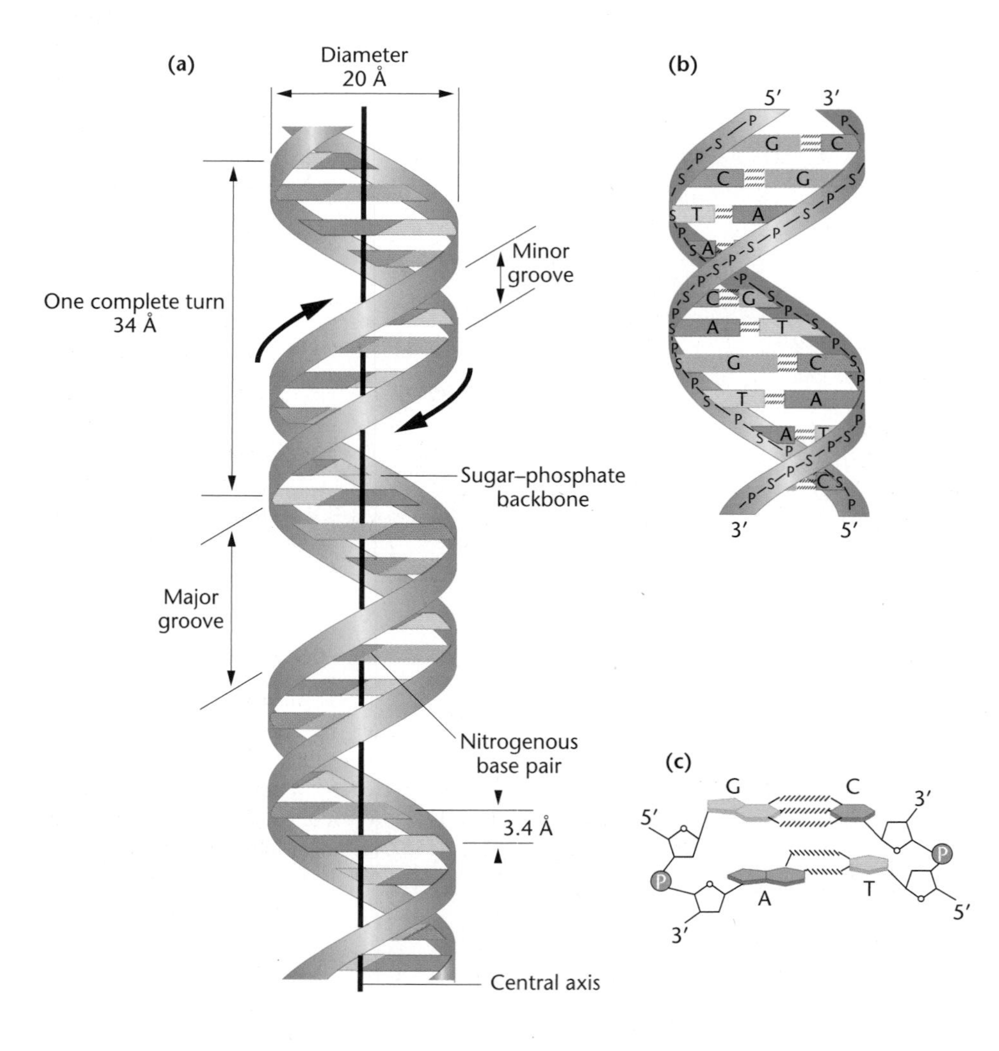

FIGURE 10–12 (a) The DNA double helix as proposed by Watson and Crick. The ribbonlike strands constitute the sugar–phosphate backbones, and the horizontal rungs constitute the nitrogenous base pairs, of which there are 10 per complete turn. The major and minor grooves are shown. The solid vertical bar represents the central axis. (b) A detailed view labeled with the bases, sugars, phosphates, and hydrogen bonds of the helix. (c) A demonstration of the antiparallel nature of the helix and the horizontal stacking of the bases.

Molecular Structure of Nucleic Acids: A Structure for Deoxyribose Nucleic Acid

We wish to suggest a structure for the salt of deoxyribose nucleic acid (D. N. A.). This structure has novel features which are of considerable biological interest. A structure for nucleic acid has already been proposed by Pauling and Corey.[1] They kindly made their manuscript available to us in advance of publication.

Their model consists of three intertwined chains, with the phosphates near the fibre axis, and the bases on the outside. In our opinion, this structure is unsatisfactory for two reasons: (1) We believe that the material which gives the X-ray diagrams is the salt, not the free acid. Without the acidic hydrogen atoms it is not clear what forces would hold the structure together, especially as the negatively charged phosphates near the axis will repel each other. (2) Some of the van der Waals distances appear to be too small.

Another three-chain structure has also been suggested by Fraser (in the press). In his model the phosphates are on the outside and the bases on the inside, linked together by hydrogen bonds. This structure as described is rather ill-defined, and for this reason we shall not comment on it.

We wish to put forward a radically different structure for the salt of deoxyribose nucleic acid. This structure has two helical chains each coiled round the same axis. We have made the usual chemical assumptions, namely, that each chain consists of phosphate diester groups joining β-D-deoxyribofuranose residues with 3′, 5′ linkages. The two chains (but not their bases) are related by a dyad perpendicular to the fibre axis. Both chains follow right-handed helices, but owing to the dyad the sequences of the atoms in the two chains run in opposite directions. Each chain loosely resembles Furberg's[2] model No. 1; that is, the bases are on the inside of the helix and the phosphates on the outside. The configuration of the sugar and the atoms near it is close to Furberg's "standard configuration," the sugar being roughly perpendicular to the attached base. There is a residue on each chain every 3.4 Å in the z-direction. We have assumed an angle of 36° between adjacent residues in the same chain, so that the structure repeats after 10 residues on each chain, that is, after 34 Å. The distance of a phosphate atom from the fibre axis is 10 Å. As the phosphates are on the outside, cations have easy access to them.

The structure is an open one, and its water content is rather high. At lower water content we would expect the bases to tilt so that the structure could become more compact.

The novel feature of the structure is the manner in which the two chains are held together by the purine and pyrimidine bases. The planes of the bases are perpendicular to the fibre axis. They are joined together in pairs, a single base from one chain being hydrogen-bonded to a single base from the other chain, so that the two lie side by side with identical z-co-ordinates. One of the pair must be a purine and the other a pyrimidine for bonding to occur. The hydrogen bonds are made as follows: purine position 1 to pyrimidine position 1; purine position 6 to pyrimidine position 6.

If it is assumed that the bases only occur in the structure in the most plausible tautomeric forms (that is, with the keto rather than the enol configuration) it is found that only specific pairs of bases can bond together. These pairs are: adenine (purine) with thymine (pyrimidine), and guanine (purine) with cytosine (pyrimidine).

In other words, if an adenine forms one member of a pair, on either chain, then on these assumptions the other member must be thymine; similarly for guanine and cytosine. The sequence of bases on a single chain does not appear to be restricted in any way. However, if only specific pairs of bases can be formed, it follows that if the sequence of bases on one chain is given, then the sequence on the other chain is automatically determined.

It has been found experimentally[3, 4] that the ratio of the amounts of adenine to thymine, and the ratio of guanine to cytosine, are always very close to unity for deoxyribose nucleic acid.

It is probably impossible to build this structure with a ribose sugar in place of deoxyribose, as the extra oxygen atom would make too close a van der Waals contact.

The previously published X-ray data[5, 6] on deoxyribose nucleic acid are insufficient for a rigorous test of our structure. So far as we can tell, it is roughly compatible with the experimental data, but it must be regarded as unproved until it has been checked against more exact results. Some of these are given in the following communications. We were not aware of the details of the results presented there when we devised our structure, which rests mainly though not entirely on published experimental data and stereochemical arguments.

It has not escaped our notice that the specific pairing we have postulated immediately suggests a possible copying mechanism for the genetic material. Full details of the structure, including the conditions assumed in building it, together with a set of co-ordinates for the atoms, will be published elsewhere.

We are much indebted to Dr. Jerry Donohue for constant advice and criticism, especially on interatomic distances. We have also been stimulated by a knowledge of the general nature of the unpublished experimental results and ideas of Dr. M. H. F. Wilkins, Dr. R. E. Franklin and their co-workers at King's College, London. One of us (J. D. W.) has been aided by a fellowship from the National Foundation for Infantile Paralysis.

J. D. Watson
F. H. C. Crick
Medical Research Council Unit for the Study of the Molecular Structure of Biological Systems, Cavendish Laboratory, Cambridge, England

[1] Pauling L., and Corey, R. B., *Nature*, 171, 346 (1953); *Proc. U.S. Nat. Acad. Sci.*, 39, 84 (1953).
[2] Furberg, S., *Acta Chem. Scand.*, 6, 634 (1952).
[3] Chargaff, E., for references see Zamenhof, S., Brawerman, G., and Chargaff, E., *Biochim. et Biophys. Acta*, 9, 402 (1952).
[4] Wyatt, G. R., *J. Gen. Physiol.*, 36, 201 (1952).
[5] Astbury, W. T., *Symp. Soc. Exp. Biol. 1, Nucleic Acid*, 66 (Camb. Univ. Press, 1947).
[6] Wilkins, M. H. F., and Randall, J. T., *Biochim. et Biophys. Acta*, 10, 192 (1953).

The nature of *base-pairing* (point 4 above) is the most genetically significant feature of the model. Before we discuss it in detail, several other important features warrant emphasis. First, the antiparallel nature of the two chains is a key part of the double-helix model. While one chain runs in the 5′-to-3′ orientation (what seems right side up to us), the other chain is in the 3′-to-5′ orientation (and thus appears upside down). This is illustrated in Figure 10–12b and c. Given the constraints of the bond angles of the various nucleotide components, the double helix could not be constructed easily if both chains ran parallel to one another.

The key to the model proposed by Watson and Crick is the specificity of base-pairing. Chargaff's data suggested that the amounts of A equaled T and that the amounts of G equaled C. Watson and Crick realized that if A pairs with T and C pairs with G, this would account for these proportions and that such pairing could occur as a result of hydrogen bonding between base pairs (Figure 10–12c), providing the chemical stability necessary to hold the two chains together. Arranged in this way, both major and minor grooves become apparent along the axis. Further, a purine (A or G) opposite a pyrimidine (T or C) on each "rung of the spiral staircase" of the proposed double helix accounts for the 20 Å (2 nm) diameter suggested by X-ray diffraction studies.

The specific A═T and G≡C base-pairing is the basis for **complementarity**. This term describes the chemical affinity provided by hydrogen bonding between the bases. As we shall see, complementarity is very important in DNA replication and gene expression.

It is appropriate to inquire into the nature of a hydrogen bond and to ask whether it is strong enough to stabilize the helix. A **hydrogen bond** is a very weak electrostatic attraction between a covalently bonded hydrogen atom and an atom with an unshared electron pair. The hydrogen atom assumes a partial positive charge, while the unshared electron pair—characteristic of covalently bonded oxygen and nitrogen atoms—assumes a partial negative charge. These opposite charges are responsible for the weak chemical attractions. As oriented in the double helix, adenine forms two hydrogen bonds with thymine, and guanine forms three hydrogen bonds with cytosine. Although two or three individual hydrogen bonds are energetically very weak, 2000-3000 bonds in tandem (typical of two long polynucleotide chains) provide great stability to the helix.

Another stabilizing factor is the arrangement of sugars and bases along the axis. In the Watson-Crick model, the *hydrophobic* ("water-fearing") nitrogenous bases are stacked almost horizontally on the interior of the axis and are thus shielded from water. The *hydrophilic* ("water-loving") sugar–phosphate backbone is on the outside of the axis, where both components can interact with water. These molecular arrangements provide significant chemical stabilization to the helix.

A more recent and accurate analysis of the form of DNA that served as the basis for the Watson-Crick model reveals a minor structural difference. A precise measurement of the number of base pairs (bp) per turn has demonstrated a value of 10.4 bp rather than the 10.0 bp predicted by Watson and Crick. In the classical model, each base pair is rotated 36° around the helical axis relative to the adjacent base pair, whereas the new finding requires a rotation of 34.6°. Thus, there are slightly more than 10 bp per turn.

The Watson-Crick model had an immediate effect on the emerging discipline of molecular biology. Even in their initial 1953 article, the authors noted, "It has not escaped our notice that the specific pairing we have postulated immediately suggests a possible copying mechanism for the genetic material." Two months later, Watson and Crick pursued this idea in a second article in *Nature*, suggesting a specific mode of replication of DNA—the **semiconservative model**. The second article alluded to two new concepts: (1) the storage of genetic information in the sequence of the bases, and (2) the mutations or genetic changes that would result from an alteration of the bases. These ideas have received vast amounts of experimental support since 1953 and are now universally accepted.

Watson and Crick's synthesis of ideas was highly significant with regard to subsequent studies of genetics and biology. The nature of the gene and its role in genetic mechanisms could now be viewed and studied in biochemical terms. Recognition of their work, along with that of Wilkins, led to their receiving the Nobel Prize in Physiology or Medicine in 1962. This was one of many such awards bestowed for their work in the field of genetics.

10.7 Alternative Forms of DNA Exist

Under different conditions of isolation, several conformational forms of DNA have been recognized. At the time Watson and Crick performed their analysis, two forms—**A-DNA** and **B-DNA**—were known. Watson and Crick's analysis was based on X-ray studies of B-DNA performed by Franklin, which is present under aqueous, low-salt conditions and is believed to be the biologically significant conformation.

While DNA studies around 1950 relied on the use of X-ray diffraction, more recent investigations have been performed using **single-crystal X-ray analysis**. The earlier studies achieved limited resolution of about 5 Å, but single crystals diffract X-rays at about 1 Å, near atomic resolution. As a result, every atom is "visible" and much greater structural detail is available during analysis.

With this modern technique, A-DNA, which is prevalent under high-salt or dehydration conditions, has now been scrutinized. In comparison to B-DNA (Figure 10–13), A-DNA is slightly more compact, with 11 bp in each complete turn of the helix, which is 23 Å (2.3 nm) in diameter. It is also a right-handed helix, but the orientation of the bases is somewhat different—they are tilted and displaced laterally in relation to the axis of the helix. As a result, the

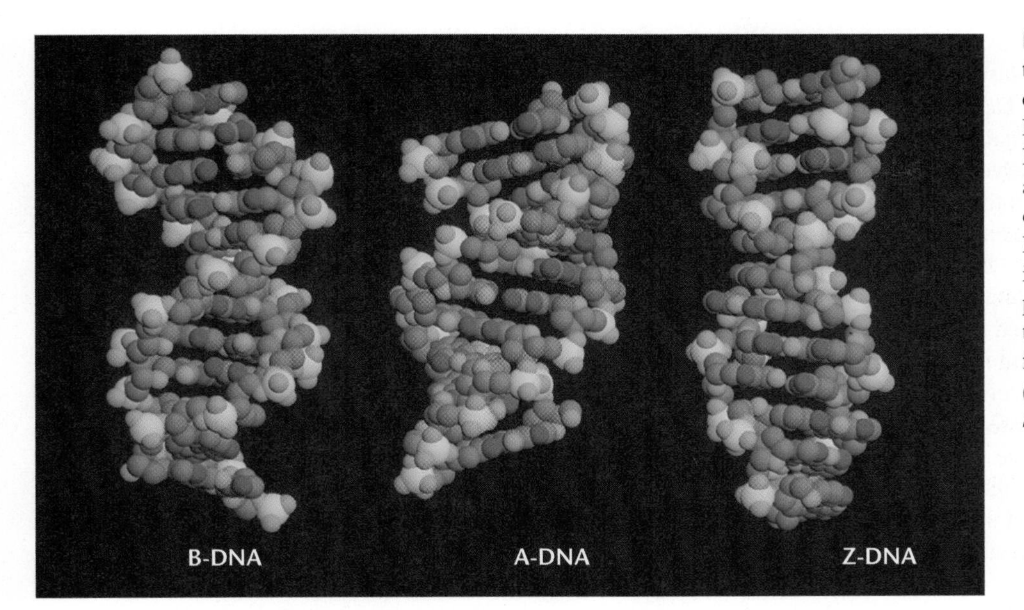

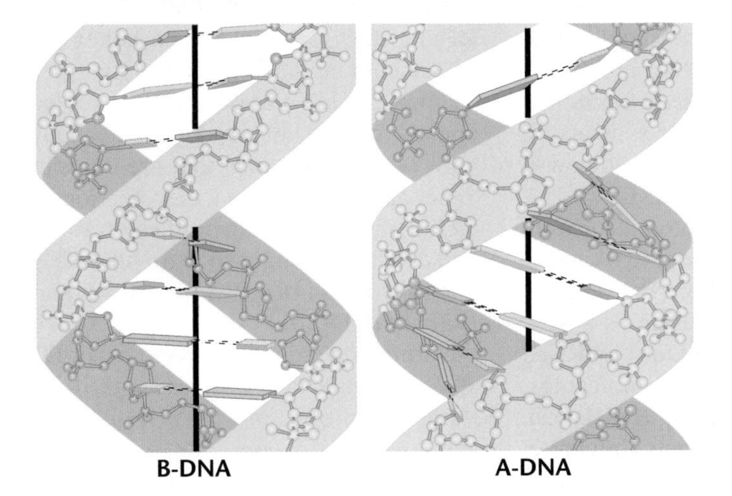

FIGURE 10–13 The top half of the figure shows computer-generated space-filling models of B-DNA, A-DNA, and Z-DNA. Below the photograph is an artist's rendering showing the orientation of the base pairs of B-DNA and A-DNA. Note that in B-DNA the base pairs are perpendicular to the helix, while they are tilted and pulled away from the helix in A-DNA. *(Photo: Ken Edward/Science Source/Photo Researchers, Inc.)*

appearance of the major and minor grooves is modified. It seems doubtful that A-DNA occurs *in vivo* (under physiological conditions).

Other forms of DNA (e.g., C-, D-, E-, and most recently, P-DNA) are now known, but it is **Z-DNA** that has drawn the most attention. Discovered by Andrew Wang, Alexander Rich, and their colleagues in 1979 when they examined a small synthetic DNA fragment containing only G≡C base pairs, Z-DNA takes on the rather remarkable configuration of a left-handed double helix (Figure 10–13). Like A- and B-DNA, Z-DNA consists of two antiparallel chains held together by Watson-Crick base pairs. Beyond these characteristics, Z-DNA is quite different. The left-handed helix is 18 Å (1.8 nm) in diameter, contains 12 bp per turn, and assumes a zigzag conformation (hence its name). The major groove present in B-DNA is nearly eliminated in Z-DNA.

Speculation abounds over the possibility that regions of Z-DNA exist in the chromosomes of living organisms. The unique helical arrangement could provide an important recognition point for the interaction with other molecules. However, it is still not clear whether Z-DNA occurs *in vivo*.

10.8 The Structure of RNA Is Chemically Similar to DNA, but Single-Stranded

The structure of RNA molecules resembles DNA, with several important exceptions. Although RNA also has nucleotides linked with polynucleotide chains, the sugar ribose replaces deoxyribose, and the nitrogenous base uracil replaces thymine. Another important difference is that most RNA is single-stranded, although there are two important exceptions. First, RNA molecules sometimes fold back on themselves to form double-stranded regions of complementary base pairs. Second, some animal viruses that have RNA as their genetic material contain double-stranded helices.

As established earlier (see Figure 10–1), three major classes of cellular RNA molecules function during the expression of genetic information: **ribosomal RNA (rRNA)**, **messenger RNA (mRNA)**, and **transfer RNA (tRNA)**. These molecules all originate as complementary copies of one of the two strands of DNA segments during the process of transcription. That is, their nucleotide sequence is complementary to the deoxyribonucleotide sequence of DNA, which served as the template for their synthesis. Because uracil replaces thymine in RNA, uracil is complementary to adenine during transcription and RNA base-pairing.

Table 10.4 characterizes these major forms of RNA, as found in prokaryotic and eukaryotic cells. Different RNAs are distinguished according to their sedimentation behavior in a centrifugal field and their size, as measured by the number of nucleotides each contains. Sedimentation behavior depends on a molecule's density, mass, and shape, and its measure is called the **Svedberg coefficient** (*S*). While higher *S* values almost always designate molecules of greater molecular weight, the correlation is not direct; that is, a twofold increase in molecular weight does not lead to a twofold increase in *S*. This is because, in addition to a molecule's mass, the size and the shape of the molecule also affect its rate of sedimentation (*S*). As you can see in Table 10.4, a wide variation exists in the size of the three classes of RNA.

Ribosomal RNA is generally the largest of these molecules (reflected in its *S* values) and usually constitutes about 80% of all RNA in the cell. Ribosomal RNAs are important structural components of **ribosomes**, which function as nonspecific workbenches where proteins are synthesized during translation. The various forms of rRNA found in prokaryotes and eukaryotes differ distinctly in size.

Messenger RNA molecules carry genetic information from the DNA of the gene to the ribosome. The mRNA molecules vary considerably in size, which reflects the variation in the size of the protein encoded by the mRNA as well as the gene serving as the template for transcription of mRNA.

Transfer RNA, the smallest class of RNA molecules, carries amino acids to the ribosome during translation. Since more than one tRNA molecule interacts simultaneously with the ribosome, the molecule's smaller size facilitates these interactions.

These RNAs represent the major forms of the molecule involved in genetic expression, but other unique RNAs exist that perform various roles. For example, **small nuclear RNA (snRNA)** participates in processing mRNAs. **Telomerase RNA** is involved in DNA replication at the ends of chromosomes (the telomeres), and **antisense RNA** and **short interfering RNA (siRNA)** are involved in gene regulation. DNA stores genetic information, while RNA most often functions in the expression of that information.

10.9 Many Analytical Techniques Have Been Useful During the Investigation of DNA and RNA

Since 1953, the role of DNA as the genetic material and the role of RNA in transcription and translation have been clarified through detailed analysis of nucleic acids. Let's consider several methods of analysis of these molecules that have been particularly important. Many of them are based on the unique nature of the hydrogen bond that is so integral to the structure of nucleic acids.

Hydrogen bonds impart an interesting and important set of qualities to the chemical behavior of nucleic acids under both laboratory and physiological conditions. For example, if DNA is isolated and subjected to slow heating, the double helix is denatured and unwinds. If a mixture of single strands that are complementary to each other are slowly cooled, they reassociate and re-form the helix. In the laboratory, these transformations can be "tracked" by using a spectrophotometer and monitoring the absorption of UV light (or optical density, OD) at 260 nm (OD_{260}).

During unwinding, the viscosity of DNA decreases and UV absorption increases (called the **hyperchromic shift**). A melting profile, in which OD_{260} is plotted against temperature, is shown for two DNA molecules in Figure 10–14. The midpoint of each curve is called the **melting temperature** (T_m), where 50% of the strands have unwound. The molecule with a higher T_m has a higher percentage of G≡C base pairs than A=T base pairs compared to the molecule with the lower T_m, since G≡C pairs share three hydrogen bonds compared to the two bonds between A=T pairs.

TABLE 10.4 RNA Characterization

RNA Class	*Total RNA* (%)*	*Components (Svedberg Coefficient)*	*Eukaryotic (E) or Prokaryotic (P)*	*Number of Nucleotides*
Ribosomal (rRNA)	80	5*S*	P and E	120
		5.8*S*	E	160
		16*S*	P	1542
		18*S*	E	1874
		23*S*	P	2904
		28*S*	E	4718
Transfer (tRNA)	15	4*S*	P and E	75–90
Messenger (mRNA)	5	varies	P and E	100–10,000

*In *E. coli*.

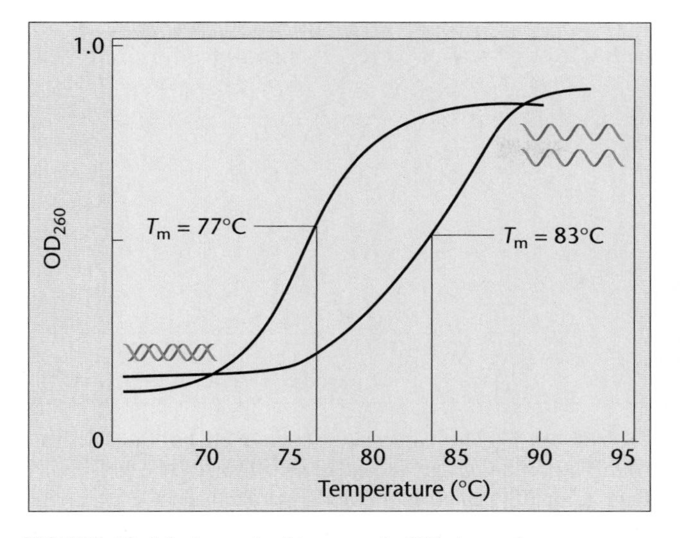

FIGURE 10–14 A graph of increase in UV absorption versus temperature (the hyperchromic effect) for two DNA molecules with different G≡C contents. The molecule with a melting point (T_m) of 83°C has a greater G≡C content than the molecule with T_m = 77°C.

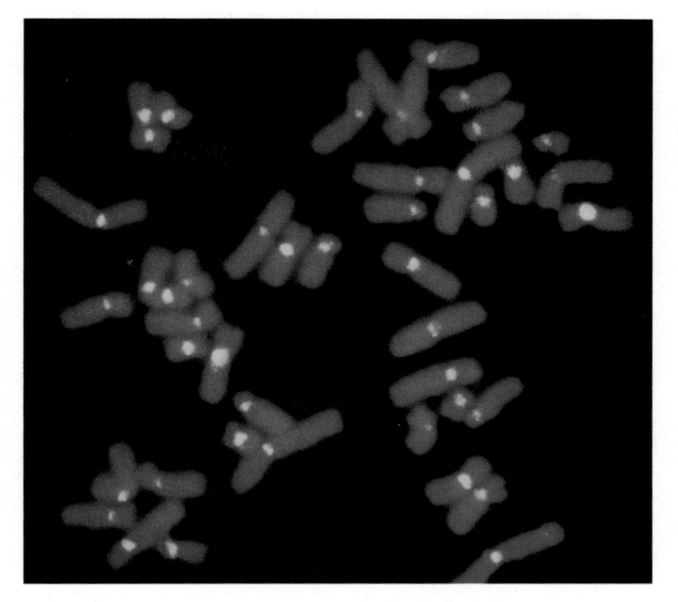

FIGURE 10–15 *In situ* hybridization of human metaphase chromosomes using FISH. The probe, specific to centromeric DNA, produces a yellow fluorescence signal indicating hybridization. The red fluorescence is produced by propidium iodide counterstaining of chromosomal DNA. *(Ventana Medical Systems, Inc.)*

Molecular Hybridization Techniques

The property of denaturation/renaturation of nucleic acids is the basis for one of the most powerful and useful techniques in molecular genetics—**molecular hybridization**. Provided that a reasonable degree of base complementarity exists, under the proper temperature conditions, two nucleic acid strands from different sources will rejoin. As a result, molecular hybridization is possible between DNA strands from different species, and between DNA and RNA strands. For example, an RNA molecule will hybridize with the segment of DNA from which it was transcribed or with a DNA molecule from a different species, as long as its nucleotide sequence is nearly the same.

The technique can even be performed using the DNA in cytological preparations as the "target" for hybrid formation. This process is called ***in situ* molecular hybridization**. Mitotic or interphase cells are first fixed to slides and then subjected to hybridization conditions. Single-stranded DNA or RNA is added, and hybridization is monitored. The nucleic acid that is added may be either radioactive or contain a fluorescent label to allow its detection. In the former case, autoradiography is used.

Figure 10–15 illustrates the use of a fluorescent label. A "probe," a short fragment of DNA that is complementary to DNA in the chromosome's centromere regions, has been hybridized. Fluorescence occurs only in the centromere regions and thus identifies each one along its chromosome. Because fluorescence is used, the technique is known by the acronym **FISH** (**fluorescent *in situ* hybridization**). The use of this technique to identify chromosomal locations housing specific genetic information has been a valuable addition to geneticists' repertoire of experimental techniques.

Reassociation Kinetics and Repetitive DNA

In one extension of molecular hybridization procedures, the *rate of reassociation* of complementary single DNA strands is analyzed. This technique, **reassociation kinetics**, was first refined and studied by Roy Britten and David Kohne.

The DNA used in such studies is first fragmented into small pieces by shearing forces introduced during its isolation. The resultant DNA fragments have an average size of several hundred base pairs. The fragments are then dissociated into single strands by heating (denatured), and when the temperature is lowered, reassociation is monitored. During reassociation, pieces of single-stranded DNA randomly collide. If they are complementary, a stable double strand is formed; if not, they separate and are free to encounter other DNA fragments. The process continues until all possible matches are made.

The results of one such experiment are graphed in Figure 10–16. The percentage of reassociation of DNA fragments is plotted against a logarithmic scale of normalized time, a function referred to as C_0t, where C_0 is the initial concentration of DNA single strands, and t is the time.

A great deal of information can be obtained from studies that compare the reassociation of DNA of different organisms. For example, we can compare the point in the reaction when one-half of the DNA is present as double-stranded fragments. This point is called the **half reaction time** ($C_0t_{1/2}$). Provided that all of the DNA fragments contain unique nucleotide sequences and all are about the same size, $C_0t_{1/2}$ varies directly in relation to the total length of the DNA.

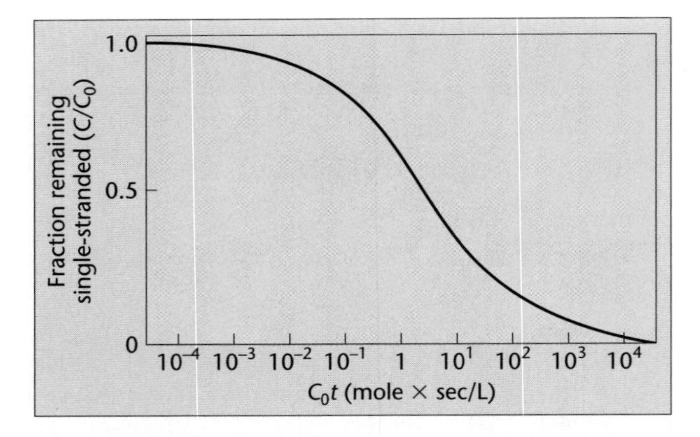

FIGURE 10–16 The ideal time course for reassociation of DNA (C/C_0) when, at time zero, all of the DNA consists of unique fragments of single-stranded complements. Note that the *x*-axis (C_0t) is scaled logarithmically.

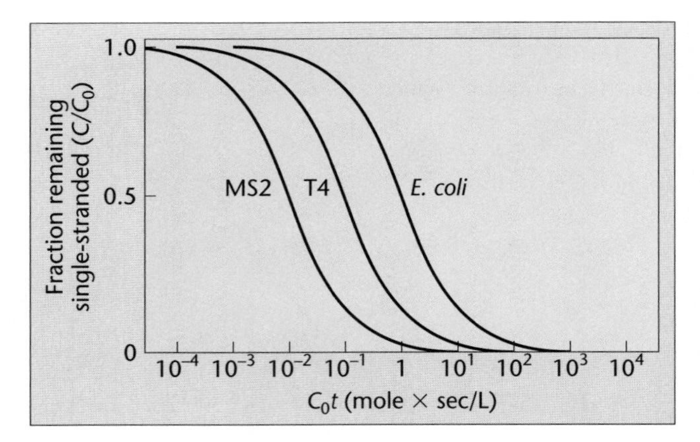

FIGURE 10–17 The reassociation rates (C/C_0) of DNA derived from phage MS2, phage T4, and *E. coli*. The genome of T4 is larger than MS2, and that of *E. coli* is larger than T4.

Figure 10–17 compares DNA from three sources, each with a different genome size. As genome size increases, the curves obtained have a similar shape but are shifted farther and farther to the right, indicative of an extended reassociation time. Reassociation occurs more slowly in larger genomes because it takes longer for initial matches if there are greater numbers of unique DNA fragments. The collisions are random, therefore the more sequences that are present, the greater the number of mismatches before all correct matches are made.

When reassociation kinetics in eukaryotic organisms with much larger genome sizes was first studied, a surprising observation was made. Rather than exhibiting a reduced rate of reassociation, the data revealed that *some* DNA segments reassociated even more rapidly than those derived from *E. coli*. The remaining DNA, as expected because of its greater size and complexity, took longer to reassociate.

For example, Britten and Kohne examined DNA from calf thymus tissue. Based on their observations (Figure 10–18), they correctly hypothesized that the rapidly reassociating fraction might represent repetitive sequences in the calf genome. This interpretation would explain why these segments reassociate so rapidly—multiple copies of the same sequence are much more likely to make matches, thus reassociating more quickly than single copies. On the other hand, the remaining DNA segments consist of unique nucleotide sequences (present only once) in the genome. Because calf thymus DNA has many more unique sequences than *E. coli*, their reassociation takes longer.

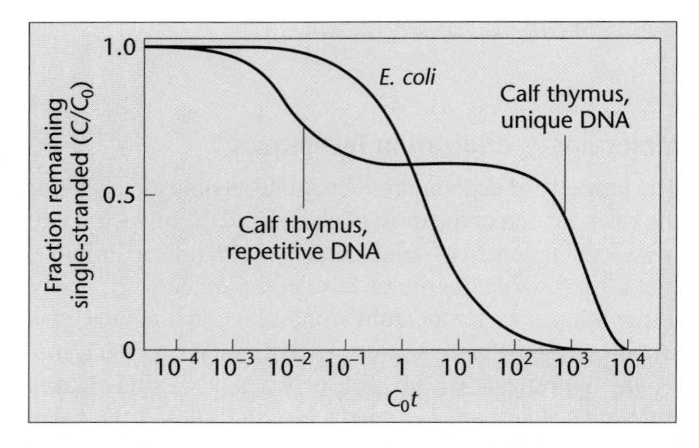

FIGURE 10–18 The C_0t curve of calf thymus DNA compared with *E. coli*. The repetitive fraction of calf DNA reassociates more quickly than that of *E. coli*, while the more complex, unique calf DNA takes longer to reassociate than that of *E. coli*.

Many copies of a sequence in the genome are collectively known as **repetitive DNA**. Repetitive DNA is prevalent in eukaryotic genomes and is key to our understanding of how genetic information is organized in chromosomes. Careful study has shown that various levels of repetition exist. In some cases, short DNA sequences are repeated over a million times. In other cases, longer sequences are repeated only a few times, or intermediate levels of sequence redundancy are present. The discovery of repetitive DNA was one of the first clues that much of the DNA in eukaryotes is not contained in the genes that encode proteins.

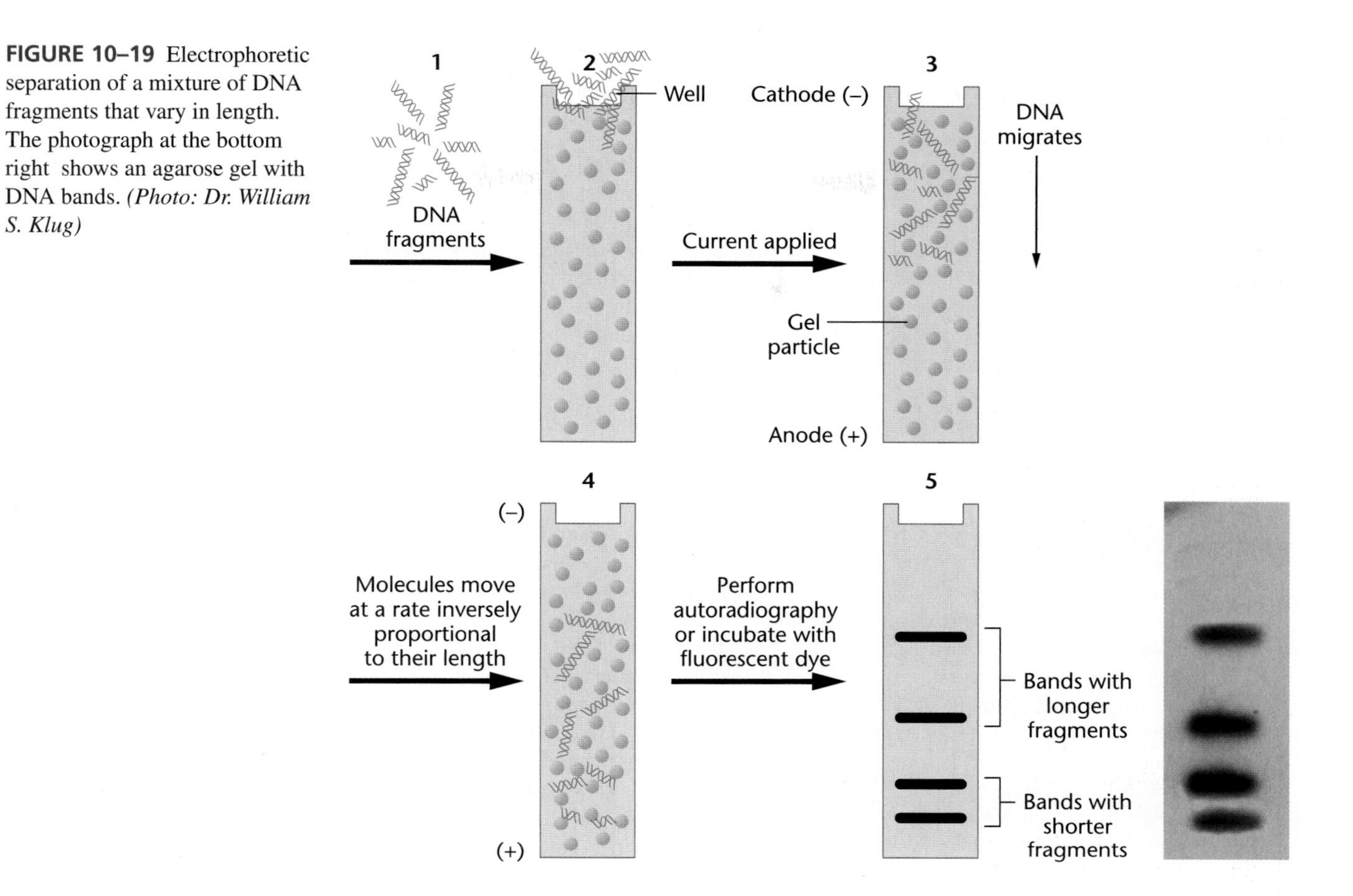

FIGURE 10–19 Electrophoretic separation of a mixture of DNA fragments that vary in length. The photograph at the bottom right shows an agarose gel with DNA bands. *(Photo: Dr. William S. Klug)*

10.10 Nucleic Acids Can Be Separated Using Electrophoresis

The final, essential, technique in the analysis of nucleic acids that we will discuss is **electrophoresis**. This technique separates different-sized fragments of DNA and RNA chains and is invaluable in current research investigations in molecular genetics.

In general, electrophoresis separates the molecules in a mixture by causing them to migrate under the influence of an electric field. A sample is placed on a porous substance (a piece of filter paper or a semisolid gel), which is then placed in a solution that conducts electricity. If two molecules have approximately the same shape and mass, the one with the greater net charge will migrate more rapidly toward the electrode of opposite polarity.

As electrophoretic technology developed from its initial application to protein separation, researchers discovered that using gels of varying pore sizes significantly improved the resolution of this research technique. This advance is particularly useful for mixtures of molecules with a similar charge–mass ratio but of different sizes. For example, two polynucleotide chains of different lengths (e.g., 10 versus 20 nucleotides) are both negatively charged based on the phosphate groups of the nucleotides. Both chains move to the positively charged pole (the anode), but the charge–mass ratio is the same for each chain, and separation due to the electric field is minimal. However, using porous medium such as a **polyacrylamide gel** or an **agarose gel**, which can be prepared with various pore sizes, enables us to separate the two molecules. *Smaller molecules migrate at a faster rate through the gel than larger molecules* (Figure 10–19). The key to separation is based on the matrix (pores) of the gel, which restricts migration of larger molecules more than it restricts smaller molecules. The resolving power is so great that polynucleotides that vary by just one nucleotide in length are separated. Once electrophoresis is complete, bands representing the variously sized molecules are identified either by autoradiography (if a component of the molecule is radioactive) or by the use of a fluorescent dye that binds to nucleic acids.

Electrophoretic separation of nucleic acids is at the heart of a variety of other commonly used research techniques. Of particular note are the various "blotting" techniques (e.g., Southern blots and Northern blots), as well as DNA sequencing methods, which we will discuss in detail later in the text.

GENETICS, TECHNOLOGY, AND SOCIETY

The Twists and Turns of the Helical Revolution

Western civilization is frequently transformed by new scientific ideas that overturn our self-concepts and permanently alter our relationships with each other and the rest of the animate world. For 50 years, we have been in the midst of such a revolution—one as significant as those triggered by Darwin's theory of evolution or the Copernican rejection of Ptolemy's earth-centered universe.

The revolution began in April 1953 with Watson and Crick's discovery of the molecular structure of DNA. Their discovery that the DNA molecule consists of a twisted double helix, held together by weak bonds between specific pairs of bases, suddenly provided elegant solutions to age-old questions about the mechanisms of heredity, mutation, and evolution. Some of the greatest mysteries of life could be explained by the beauty and simplicity of a helix that replicates and shuffles the code of life.

After 1953, the double helix rapidly became the focus of modern science. Using knowledge of DNA's helical structure, molecular biologists quickly devised methods to purify, mutate, cut, and paste DNA in the test tube. They spliced DNA molecules from one organism into those of another, and then introduced these chimeric molecules into bacteria or cells in culture. They read the nucleotide sequences of genes and modified the traits of bacteria, fungi, fruit flies, and mice by removing and mutating their genes, or by introducing genes from other organisms. On the 50th anniversary of Watson and Crick's double-helical DNA model, the Human Genome Project announced the completion of the largest DNA project so far—sequencing the entire human genome.

In a mere 50 years, the helical revolution has touched the lives of millions of people. We can now test for simple genetic diseases, such as Tay-Sachs, cystic fibrosis and sickle-cell anemia. We can manufacture large quantities of medically important proteins, such as insulin and growth hormone, using DNA technologies. DNA forensic tests help convict criminals, exonerate the innocent, and establish paternity. By following a trail of DNA sequences, anthropologists can now trace human origins back in time and place.

The helical revolution has profoundly altered our views of ourselves and our world. Although scientists dismiss the idea that humans are simply the products of their genes, popular culture endows DNA with almost magical powers. Genes are said to explain personality, career choice, criminality, intelligence—even fashion preferences and political attitudes. Advertisements hijack the language of genetics in order to grant inanimate objects a "genealogy" or "genetic advantage." Popular culture speaks of DNA as an immortal force, with the ability to affect morality and fate. The double helix is proclaimed as the essence of life, with the power to shape our future. Simple genetic explanations for our behavior appear to have more resonance for us than explanations involving social influences, economic factors, or free will. The beauty, symmetry, and biological significance of the double helix has insinuated itself into art, movies, advertising, and music. Paintings, sculpture, films such as *Jurassic Park* and *Boys from Brazil*—even video games and perfumes—use the language and imagery of genetics to confer upon the DNA molecule all the power and fears of modern technology.

But what of the future? Can we predict how the double helix and genetics will shape our world over the next 50 years? Although prophecy is certainly a risky business, some scientific developments seem assured. With the completion of the Human Genome Project, we will undoubtedly identify more and more of the genes that control normal and abnormal processes. In turn, this will enhance our ability to diagnose and predict genetic diseases. Over the next 50 years, we can look forward to biotechnologies as complex as gene therapies, prenatal diagnoses, and screening programs for susceptibilities to diseases as complicated as cancer and heart disease. We will continue to expand the applications of genetic engineering to agriculture as we manipulate plant and animal genes for enhanced productivity, disease resistance, and flavor.

The helical revolution will also continue to transform our concepts of ourselves and other creatures. As the human genome is compared to the genomes of other animals, it will become increasingly evident that we are closely related genetically to the rest of the animate world. The nucleotide sequence of the human genome differs only about 1% from that of chimpanzees, and some of our genes are virtually identical to homologous genes in plants, animals, and bacteria. As we realize the extent of our genetic kinship, extending over billions of years in a linear chain from the first life on earth, it is possible that this knowledge will alter our relationships with animals and with each other. When more genes are identified that contribute to phenotypic traits as simple as eye color and as complicated as intelligence or sexual orientation, it is possible that we will define ourselves even more as genetic beings and even less as creatures of free will or as the products of our environment.

Over the next 50 years, we will inevitably be faced with the practical and philosophical consequences of the DNA revolution. Will society harness DNA for everyone's benefit, or will this new genetic knowledge be used as a vehicle for discrimination? At the same time that modern genetics grants us more dominion over life, will it paradoxically increase our feelings of powerlessness? Will our new DNA-centered self-concepts increase our compassion for all life forms, or will it increase our perceived separation from the natural world? We will make our choices, and human history will proceed.

References

Dennis, C., and Campbell, P. 2003. The eternal molecule. (Introduction to a series of feature articles commemorating the 50th anniversary of the discovery of DNA structure). *Nature* 421: 396.

Web Site

A Revolution at 50. [*The New York Times* articles, on the 50th anniversary of the discovery of DNA structure.]
http://www.nytimes.com/indexes/2003/02/25/health/genetics/index.html

Chapter Summary

1. The existence of a genetic material capable of replication, storage, expression, and mutation is deducible from observed patterns of inheritance in organisms. Both proteins and nucleic acids were initially considered as possible candidates for the genetic material.
2. Transformation studies, as well as experiments using bacteria infected with bacteriophages, strongly suggested that DNA is the genetic material for bacteria and most viruses.
3. Initially, only indirect observations supported the hypothesis that DNA controls inheritance in eukaryotes. These included DNA distribution in the cell, quantitative analysis of DNA, and UV-induced mutagenesis. More recent recombinant DNA techniques, as well as experiments with transgenic mice, have provided direct experimental evidence that the eukaryote genetic material is DNA.
4. RNA serves as the genetic material in some viruses, including bacteriophages as well as some plant and animal viruses.
5. By the 1950s, many scientists sought to determine the structure of DNA. These efforts culminated in 1953 with Watson and Crick's proposal. Based on base-pairing information and X-ray diffraction data, they constructed a model. The key features of their model include two antiparallel polynucleotide chains held together in a right-handed double helix by the hydrogen bonds formed between complementary bases. To date, the basic tenets of the double-helix model have held true.
6. RNA varies from DNA by virtue of almost always being single-stranded, containing uracil rather than thymine, and having ribose rather than deoxyribose as its constituent sugar.
7. The structure of DNA lends itself to various forms of analysis, which have in turn led to studies of the functional aspects of the genetic machinery. Absorption of UV light, denaturation–reassociation, and electrophoresis procedures are important tools in the study of nucleic acids. Reassociation kinetics analysis enabled geneticists to postulate the existence of repetitive DNA in eukaryotes, where certain nucleotide sequences are present many times in the genome.

Key Terms

absorption spectrum, 213
action spectrum, 213
adenine (A), 214
A-DNA, 219
agarose gels, 225
antisense RNA, 222
B-DNA, 219
central dogma of molecular genetics, 206
complementarity, 219
cytosine (C), 214
deoxyribonuclease, 211
deoxyribonucleic acid (DNA), 214
deoxyribose, 214
electrophoresis, 225
fluorescent *in situ* hybridization (FISH), 223
genetic information, 206
guanine (G), 214
half reaction time, 223
hydrogen bond, 219
hyperchromic shift, 222
information flow, 206
in situ molecular hybridization, 223
major groove, 218
melting temperature, 222
messenger RNA (mRNA), 220
minor groove, 218
molecular hybridization, 223
nitrogenous base, 214
nuclein, 207
nucleoside, 214
nucleoside diphosphates (NDP), 214
nucleoside monophosphate (NMP), 214
nucleoside triphosphate (NTP), 214
nucleotide, 214
oligonucleotide, 215
pentose sugar, 214
phosphate group, 214
phosphodiester bond, 215
polyacrylamide gels, 225
polynucleotide, 215
protoplast, 211
purine, 214
pyrimidine, 214
reassociation kinetics, 223
recombinant DNA technology, 213
repetitive DNA, 224
replication, 206
retrovirus, 214
reverse transcriptase, 214
reverse transcription, 214
ribonucleic acid (RNA), 214
ribose, 214
ribosomal RNA (rRNA), 220
ribosome, 222
RNA replicase, 214
rough colonies (*R*), 207
semiconservative model of replication, 219
short interfering RNA (siRNA), 222
single-crystal X-ray analysis, 219
small nuclear RNA (snRNA), 222
smooth colonies (*S*), 207
spheroplast, 211
Svedberg coefficient (*S*), 222
telomerase RNA, 222
tetranucleotide hypothesis, 207
thymine (T), 214
tobacco mosaic virus (TMV), 214
transcription, 206
transfection, 211
transfer RNA (tRNA), 220
transformation, 209
transforming principle, 209
transgenic animal, 213
translation, 206
2- deoxiribose, 214
ultraviolet (UV) light, 213
uracil (U), 214
X-ray diffraction analysis, 217
Z-DNA, 220

INSIGHTS AND SOLUTIONS

In contrast to preceding chapters, this chapter does not emphasize genetic problem solving. Instead, it recounts some of the initial experimental analyses that launched the era of molecular genetics. Quite fittingly, then, our Insights and Solutions section shifts its emphasis to experimental rationale and analytical thinking, an approach that will continue throughout the remainder of the text, whenever appropriate.

1. Based strictly on the transformation analysis of Avery, MacLeod, and McCarty, what objection might be made to the conclusion that DNA is the genetic material? What other conclusion might be considered?

Solution: Based solely on their results, we could conclude that DNA is essential for transformation. However, DNA might have been a substance that caused capsular formation by converting nonencapsulated cells *directly* to cells with a capsule. That is, DNA may simply have played a catalytic role in capsular synthesis, leading to cells that display smooth, type III colonies.

2. What observations argue against this objection?

Solution: First, transformed cells pass the trait on to their progeny cells, thus supporting the conclusion that DNA is responsible for heredity, not for the direct production of polysaccharide coats. Second, subsequent transformation studies over the next five years showed that other traits, such as antibiotic resistance, could be transformed. Therefore, the transforming factor has a broad general effect, not one specific to polysaccharide synthesis.

3. If RNA were the universal genetic material, how would this have affected the Avery experiment and the Hershey-Chase experiment?

Solution: In the Avery experiment, rather than ribonuclease (RNase), deoxyribonuclease (DNase) would have eliminated transformation. Had this occurred, Avery and his colleagues would have concluded that RNA was the transforming factor. Hershey and Chase would have obtained identical results, since ^{32}P would also label RNA, but not protein.

4. Sea urchin DNA, which is double-stranded, contains 17.5% of its bases in the form of cytosine (C). What percentages of the other three bases are expected to be present in this DNA?

Solution: The amount of C equals G, so guanine is also present at 17.5%. The remaining bases, A and T, are present in equal amounts and together they represent the remaining bases (100–35). Therefore, A = T = 65/2 = 32.5%.

Problems and Discussion Questions

1. The functions ascribed to the genetic material are replication, expression, storage, and mutation. What does each of these terms mean?
2. Discuss the reasons why proteins were generally favored over DNA as the genetic material before 1940. What was the role of the tetranucleotide hypothesis in this controversy?
3. Contrast the various contributions made to our understanding of transformation by Griffith, Alloway, and Avery.
4. When Avery and his colleagues had obtained what was concluded to be purified DNA from the III*S* virulent cells, they treated the fraction with proteases, ribonuclease, and deoxyribonuclease, followed by the assay for retention or loss of transforming ability. What were the purpose and results of these experiments? What conclusions were drawn?
5. Why were ^{32}P and ^{35}S chosen in the Hershey-Chase experiment? Discuss the rationale and conclusions of this experiment.
6. Does the design of the Hershey-Chase experiment distinguish between DNA and RNA as the molecule serving as the genetic material? Why or why not?
7. Would an experiment similar to that performed by Hershey and Chase work if the basic design were applied to the phenomenon of transformation? Explain why or why not.
8. What observations are consistent with the conclusion that DNA serves as the genetic material in eukaryotes? List and discuss them.
9. What are the exceptions to the general rule that DNA is the genetic material in all organisms? What evidence supports these exceptions?
10. Draw the chemical structure of the three components of a nucleotide, and then link them together. What atoms are removed from the structures when the linkages are formed?
11. How are the carbon and nitrogen atoms of the sugars, purines, and pyrimidines numbered?
12. Adenine may also be named 6-amino purine. How would you name the other four nitrogenous bases, using this alternative system? (O is oxy, and CH_3 is methyl.)
13. Draw the chemical structure of a dinucleotide composed of A and G. Opposite this structure, draw the dinucleotide composed of T and C in an antiparallel (or upside-down) fashion. Form the possible hydrogen bonds.
14. Describe the various characteristics of the Watson-Crick double-helix model for DNA.
15. What evidence did Watson and Crick have at their disposal in 1953? What was their approach in arriving at the structure of DNA?
16. What might Watson and Crick have concluded, had Chargaff's data from a single source indicated the base composition below?

	A	*T*	*G*	*C*
%	29	19	21	31

Why would this conclusion be contradictory to Wilkins and Franklin's data?

17. How do covalent bonds differ from hydrogen bonds? Define base complementarity.
18. List three main differences between DNA and RNA.
19. What are the three types of RNA molecules? How is each related to the concept of information flow?
20. What component of the nucleotide is responsible for the absorption of ultraviolet light? How is this technique important in the analysis of nucleic acids?
21. What is the physical state of DNA after being denatured by heat?
22. What is the hyperchromic effect? How is it measured? What does T_m imply?
23. Why is T_m related to base composition?

24. Compare the curves below, representing reassociation kinetics. What can be said about the DNA represented by each set of data compared with *E. coli?*

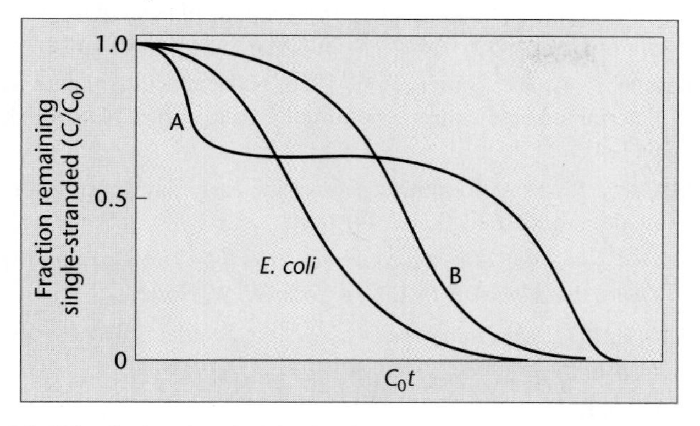

25. What is the chemical basis of molecular hybridization?
26. What did the Watson-Crick model suggest about the replication of DNA?
27. A genetics student was asked to draw the chemical structure of an adenine- and thymine-containing dinucleotide derived from DNA. His answer is shown below. The student made more than six major errors. One of them is circled, numbered 1, and explained. Find five others. Circle them, number them 2–6, and briefly explain each by following the example given.

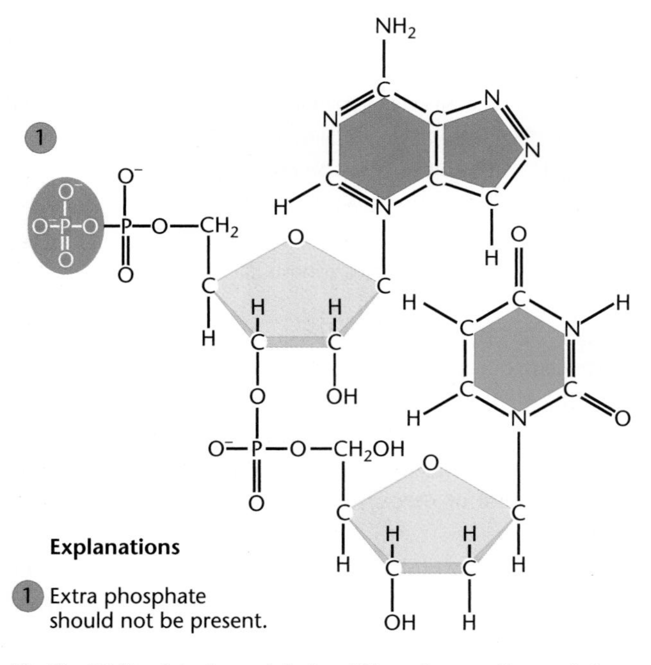

28. The DNA of the bacterial virus T4 produces a $C_0t_{1/2}$ of about 0.5 and contains 10^5 nucleotide pairs in its genome. How many nucleotide pairs are present in the genome of the virus MS2 and the bacterium *E. coli*, whose respective DNAs produce $C_0t_{1/2}$ values of 0.001 and 10.0?
29. A primitive eukaryote was discovered that displayed a unique nucleic acid as its genetic material. Analysis revealed the following observations:
    (i) X-ray diffraction studies display a general pattern similar to DNA, but with somewhat different dimensions and more irregularity.
    (ii) A major hyperchromic shift is evident upon heating and monitoring UV absorption at 260 nm.
    (iii) Base composition analysis reveals four bases in the following proportions:

Adenine	=	8%	Hypoxanthine	=	18%
Guanine	=	37%	Xanthine	=	37%

    (iv) About 75% of the sugars are deoxyribose, while 25% are ribose.

    Attempt to solve the structure of this molecule by postulating a model that is consistent with the foregoing observations.
30. With the information given in this chapter on B- and Z-DNA and the nature of helices, carefully analyze the structures shown here, and draw conclusions about the helical nature of areas (a) and (b). Which is right-handed and which is left-handed?

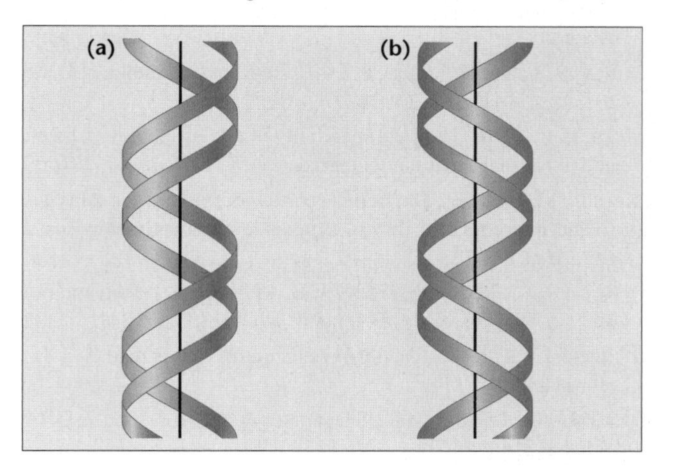

31. One of the most common spontaneous lesions that occurs in DNA under physiological conditions is the hydrolysis of the amino group of cytosine, converting it to uracil. What would be the effect on DNA structure if a uracil group replaced cytosine?
32. In some organisms, cytosine is methylated at carbon 5 of the pyrimidine ring after it is incorporated into DNA. If a 5-methyl cytosine is then hydrolyzed, as described in Problem 31, what base will be generated?
33. *Newsdate: March 1, 2015.* A unique creature has been discovered during exploration of outer space. Recently, its genetic material has been isolated and analyzed, and is similar in some ways to DNA in chemical makeup. It contains in abundance the 4-carbon sugar erythrose and a molar equivalent of phosphate groups. Additionally, it contains six nitrogenous bases: adenine (A), guanine (G), thymine (T), cytosine (C), hypoxanthine (H), and xanthine (X). These bases exist in the following relative proportions:

$$A = T = H \quad \text{and} \quad C = G = X$$

    X-ray diffraction studies have established a regularity in the molecule, and a constant diameter of about 30 Å.

    Together, these data have suggested a model for the structure of this molecule. (a) Propose a general model of this molecule, and briefly describe it. (b) What base-pairing properties must exist for H and for X in the model? (c) Given the constant diameter of 30 Å, do you think *either* (i) both H and X are purines or both pyrimidines, *or* (ii) one is a purine and one is a pyrimidine?
34. You are provided with DNA samples from two newly discovered bacterial viruses. Based on the various analytical techniques discussed in this chapter, construct a research protocol that would be useful in characterizing and contrasting the DNA of both viruses. Indicate the type of information you hope to obtain for each technique included in the protocol.
35. During electrophoresis, DNA molecules can easily be separated according to size because all DNA molecules have the same charge–mass ratio and the same shape (long rod). Would you expect RNA molecules to behave in the same manner as DNA during electrophoresis? Why or why not?

Selected Readings

Adleman, L. M. 1998. Computing with DNA. *Sci. Am.* (Aug.) 279: 54–61.

Avery, O. T., MacLeod, C. M., and McCarty, M. 1944. Studies on the chemical nature of the substance inducing transformation of pneumococcal types. Induction of transformation by a desoxyribonucleic acid fraction isolated from pneumococcus type III. *J. Exp. Med.* 79: 137–58. (Reprinted in Taylor, J. H. 1965. *Selected Papers in Molecular Genetics*. Orlando: Academic Press.)

Britten, R. J., and Kohne, D. E. 1970. Repeated segments of DNA. *Sci. Am.* (Apr.) 222: 24–31.

Chargaff, E. 1950. Chemical specificity of nucleic acids and mechanism for their enzymatic degradation. *Experientia* 6: 201–09.

Dawson, M. H. 1930. The transformation of pneumococcal types: I. The interconvertibility of type-specific *S. pneumococci*. *J. Exp. Med.* 51: 123–47.

DeRobertis, E. M., and Gurdon, J. B. 1979. Gene transplantation and the analysis of development. *Sci. Am.* (Dec.) 241: 74–82.

Dickerson, R. E. 1983. The DNA helix and how it is read. *Sci. Am.* (June) 249: 94–111.

Dickerson, R. E., et al. 1982. The anatomy of A-, B-, and Z-DNA. *Science* 216: 475–85.

Dubos, R. J. 1976. *The Professor, the Institute and DNA: Oswald T. Avery, His Life and Scientific Achievements*. New York: Rockefeller Univ. Press.

Felsenfeld, G. 1985. DNA. *Sci. Am.* (Oct.) 253: 58–78.

Fraenkel-Conrat, H., and Singer, B. 1957. Virus reconstruction: II. Combination of protein and nucleic acid from different strains. *Biochem. Biophys. Acta* 24: 530–48. (Reprinted in Taylor, J. H. 1965. *Selected Papers in Molecular Genetics*. Orlando: Academic Press.)

Franklin, R. E., and Gosling, R. G. 1953. Molecular configuration in sodium thymonucleate. *Nature* 171: 740–41.

Griffith, F. 1928. The significance of pneumococcal types. *J. Hyg.* 27: 113–59.

Guthrie, G. D., and Sinsheimer, R. L. 1960. Infection of protoplasts of *Escherichia coli* by subviral particles. *J. Mol. Biol.* 2: 297–305.

Hershey, A. D., and Chase, M. 1952. Independent functions of viral protein and nucleic acid and in growth of bacteriophage. *J. Gen. Physiol.* 36: 39–56. (Reprinted in Taylor, J. H. 1965. *Selected Papers in Molecular Genetics*. Orlando: Academic Press.)

Judson, H. 1979. *The Eighth Day of Creation: Makers of the Revolution in Biology*. New York: Simon & Schuster.

Levene, P. A., and Simms, H. S. 1926. Nucleic acid structure as determined by electrometric titration data. *J. Biol. Chem.* 70: 327–41.

McCarty, M. 1980. Reminiscences of the early days of transformation. *Annu. Rev. Genet.* 14: 1–16.

————. 1985. *The Transforming Principle: Discovering That Genes Are Made of DNA*. New York: W. W. Norton.

Olby, R. 1974. *The Path to the Double Helix*. Seattle: Univ. of Washington Press.

Palmiter, R. D., and Brinster, R. L. 1985. Transgenic mice. *Cell* 41: 343–45.

Pauling, L., and Corey, R. B. 1953. A proposed structure for the nucleic acids. *Proc. Natl. Acad. Sci. (USA)* 39: 84–97.

Rich, A., Nordheim, A., and Wang, A. H.-J. 1984. The chemistry and biology of left-handed Z-DNA. *Ann. Rev. Biochem.* 53: 791–846.

Spizizen, J. 1957. Infection of protoplasts by disrupted T2 viruses. *Proc. Natl. Acad. Sci. (USA)* 43: 694–701.

Stent, G. S. (ed.) 1981. *The Double Helix: Text, Commentary Review, and Original Papers*. New York: W. W. Norton.

Stewart, T. A., Wagner, E. F., and Mintz, B. 1982. Human β-globin gene sequences injected into mouse eggs, retained in adults, and transmitted to progeny. *Science* 217: 1046–48.

Varmus, H. 1988. Retroviruses. *Science* 240: 1427–35.

Watson, J. D. 1968. *The Double Helix*. New York: Atheneum.

Watson, J. D., and Crick, F. C. 1953a. Molecular structure of nucleic acids. A structure for deoxyribose nucleic acids. *Nature* 171: 737–38.

Watson, J. D., and Crick, F. C. 1953b. Genetic implications of the structure of deoxyribose nucleic acid. *Nature* 171: 964.

Weinberg, R. A. 1985. The molecules of life. *Sci. Am.* (Oct.) 253: 48–57.

Wilkins, M. H. F., Stokes, A. R., and Wilson, H. R. 1953. Molecular structure of desoxypentose nucleic acids. *Nature* 171: 738–40.

Yung, J. 1996. New FISH probes—The end in sight. *Nat. Gen.* 14: 10–12.

CHAPTER 11

CHAPTER CONCEPTS

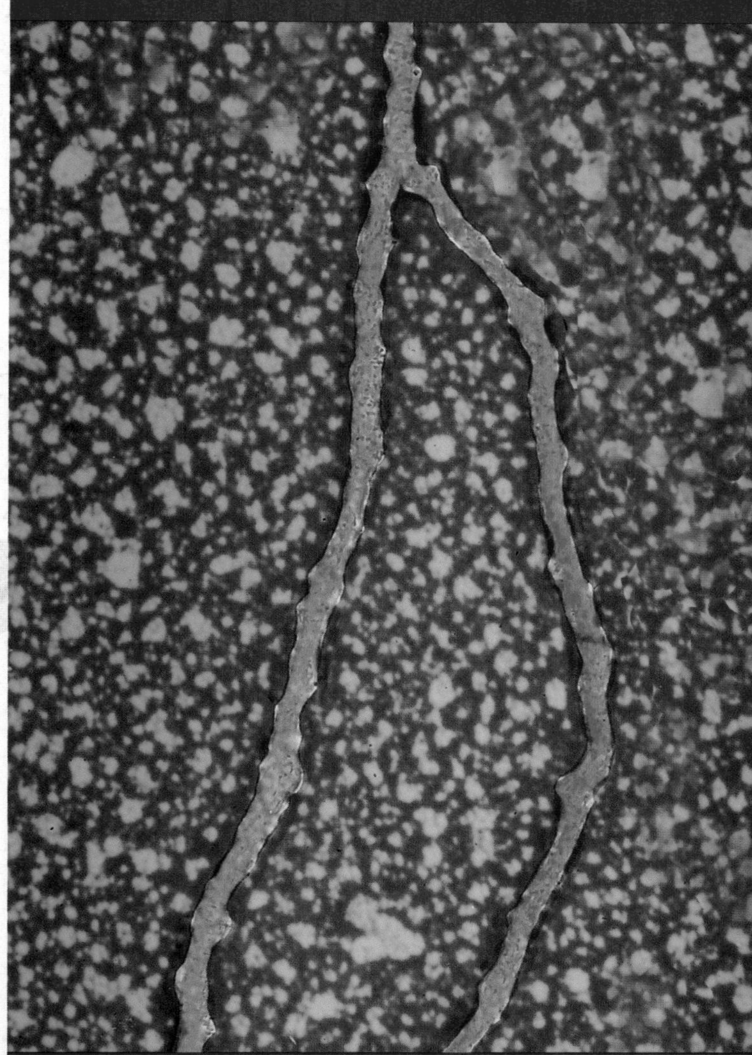

Transmission electron micrograph of human DNA from a HeLa cell, illustrating the replication fork that characterizes DNA replication within a single replicon. *(Dr. Gopal Murti/Science Photo Library/Photo Researchers, Inc.)*

DNA—Replication and Synthesis

After Watson and Crick proposed their model for the structure of DNA, scientists focused attention on how this molecule replicates. Replication is an essential function of the genetic material and must be executed precisely if genetic continuity is to be maintained following cell division. This is an enormous and complex task. Consider for a moment that in the human genome, over 3 billion (3×10^9) base pairs exist within the 23 chromosomes. To duplicate a molecule of this size faithfully requires a mechanism of extreme precision. Even an error rate of only 10^{-6} (one in a million) will still create 3000 errors, obviously an excessive number during each replication cycle. While it is not error-free, an extremely accurate system of DNA replication has evolved in all organisms.

As Watson and Crick wrote in their 1953 paper, the model of the double helix provided their initial insight into how replication could occur. This mode, called semiconservative replication, is strongly supported from numerous studies of viruses, prokaryotes, and eukaryotes.

Once the general mode of replication was clarified, research to determine the precise details of DNA synthesis intensified. What has since been discovered is that numerous enzymes and other proteins are necessary to copy a DNA helix. Because of the complexity of the chemical events during synthesis, this subject remains an extremely active area of research.

In this chapter, we discuss the general mode of replication as well as the specific details of the synthesis of DNA. The research leading to this knowledge is yet another link in our understanding of life processes at the molecular level.

HOW DO WE KNOW WHAT WE KNOW?

IN THIS CHAPTER, WE WILL FOCUS ON HOW DNA is replicated during cell division. We shall elucidate the general mechanism of replication and describe how DNA is synthesized when it is copied. As you study this topic, you should try to answer several fundamental questions:

1. How was the mode of DNA replication determined?
2. How was it demonstrated that DNA synthesis occurs under the direction of DNA polymerase?
3. How do we know the requirements of DNA polymerase in directing DNA synthesis?
4. How do we know that *in vivo* DNA synthesis occurs in the 5′-to-3′ direction?
5. How was it established that DNA polymerase I is *not* the enzyme responsible for *in vivo* DNA synthesis?
6. How do we know that DNA synthesis is discontinuous on one of the two template strands, thus producing Okazaki fragments?
7. How did we learn how the "telomere problem" is solved during DNA synthesis? ■

11.1 DNA Is Reproduced by Semiconservative Replication

It was apparent to Watson and Crick that, because of the arrangement and nature of the nitrogenous bases, each strand of a DNA double helix could serve as a template for the synthesis of its complement (Figure 11–1). They proposed that if the helix were unwound, each nucleotide along the two parent strands would have an affinity for its complementary nucleotide. As we learned in Chapter 10, complementarity is due to the hydrogen bonds that form. If thymidylic acid (T) were present, it would "attract" adenylic acid (A); if guanidylic acid (G) were present, it would attract cytidylic acid (C). The reverse is also true. A would attract T, and C would attract G. If these nucleotides were then linked covalently into polynucleotide chains along both templates, the result would be two new but identical double strands of DNA. Each replicated DNA molecule would consist of one "old" and one "new" strand, hence the reason for the name **semiconservative replication**.

Two other possible modes of replication also rely on the parental strands as a template (Figure 11–2). In **conservative replication**, synthesis of complementary polynucleotide chains occurs as described above. Following synthesis, however, the two newly created strands

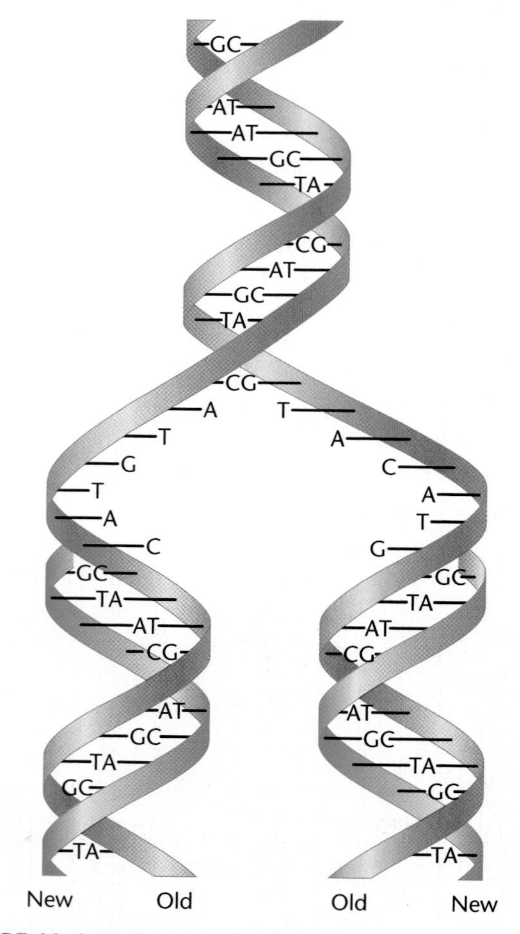

FIGURE 11–1 Generalized model of semiconservative replication of DNA. New synthesis is shown in green.

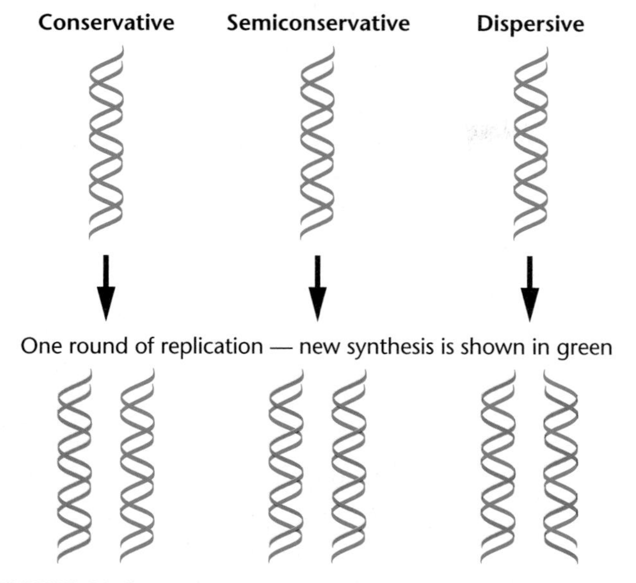

FIGURE 11–2 Results of one round of DNA replication for each of the three possible modes by which replication could be accomplished.

are brought together, and the parental strands reassociate. The original helix is thus "conserved."

In the second mode, called **dispersive replication**, the parental strands are dispersed into two new double helices after replication. Each new strand then consists of both old and new DNA. This mode would involve cleavage of the parental strands during replication. It is the most complex of the three possibilities and is therefore least likely to occur. It could not, however, be ruled out as an experimental model. Figure 11–2 shows the theoretical results of a single round of replication by the three modes.

The Meselson-Stahl Experiment

In 1958, Matthew Meselson and Franklin Stahl published the results of an experiment providing strong evidence that cells use semiconservative replication to produce new DNA molecules. *Escherichia coli* cells were grown for many generations in a medium where $^{15}NH_4Cl$ (ammonium chloride) was the only nitrogen source. A "heavy" isotope of nitrogen, ^{15}N contains one more neutron than the naturally occurring ^{14}N isotope. Unlike radioactive isotopes, ^{15}N is stable and does not decay (i.e., it is not radioactive). After many generations, all nitrogen-containing molecules, including the nitrogenous bases of DNA, contained the heavier isotope in the *E. coli* cells. DNA containing ^{15}N can be distinguished from that containing ^{14}N DNA, by the use of **sedimentation equilibrium centrifugation**, in which centrifugation "forces" samples through a density gradient of a heavy metal salt such as cesium chloride. The denser ^{15}N-DNA reaches equilibrium in the gradient at a point closer to the bottom (where the density is greater) than does ^{14}N-DNA.

In the experiment (Figure 11–3), uniformly labeled ^{15}N cells were transferred to a medium containing only $^{14}NH_4Cl$. Thus, all new syntheses of DNA during replication contained only the lighter isotope of nitrogen. The time of transfer to the new medium was taken as zero ($t = 0$). The *E. coli* cells were allowed to replicate over several generations, with cell samples removed after each replication cycle. DNA was then isolated from each sample and subjected to sedimentation equilibrium centrifugation.

After one generation, the isolated DNA was present only in a single band of intermediate density—the expected result for semiconservative replication. Each replicated

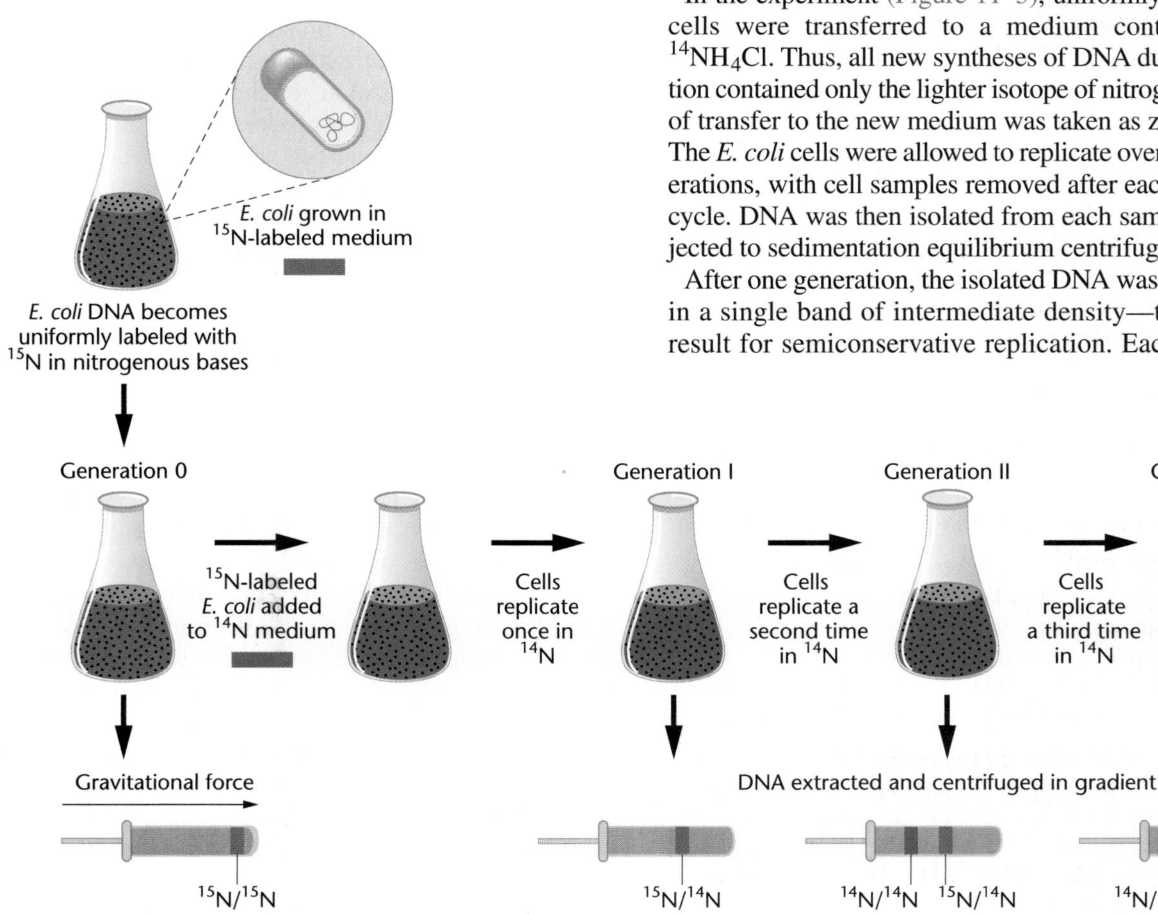

FIGURE 11–3 The Meselson-Stahl experiment.

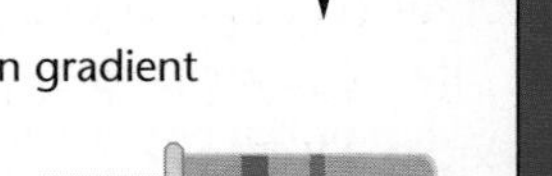

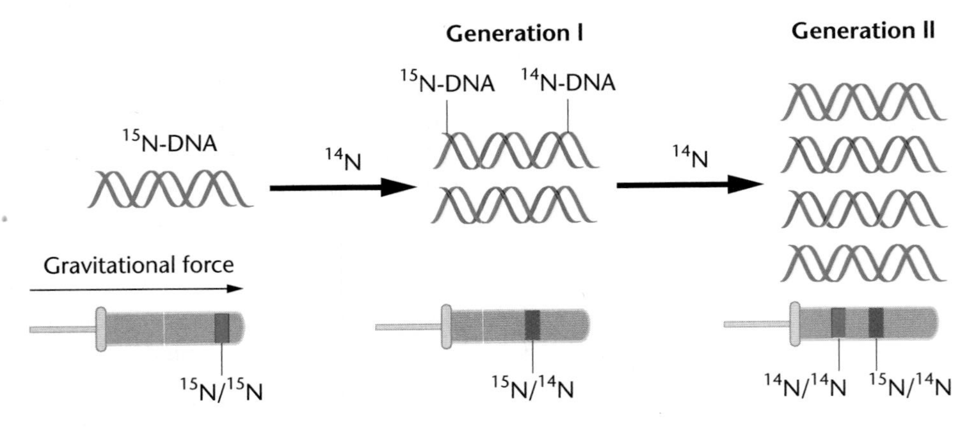

FIGURE 11–4 The expected results of two generations of semiconservative replication in the Meselson-Stahl experiment.

molecule was composed of one new ^{14}N-strand and one old ^{15}N-strand, as seen in Figure 11–4. This result effectively ruled out the conservative replication mode, in which two distinct bands are predicted.

After two cell divisions (generation II), DNA samples showed two density bands: One was intermediate and the other was lighter, corresponding to the ^{14}N position in the gradient. Similar results occurred after a third generation, except that the proportion of the ^{14}N-band increased. If replication were dispersive, all subsequent generations after $t = 0$ would demonstrate DNA of an intermediate density. In each subsequent generation, the ratio of ^{14}N to ^{15}N would increase and the hybrid band would become lighter and lighter, eventually approaching the ^{14}N-band. Since this result was not observed, the dispersive mode was ruled out. Thus, the results of the Meselson-Stahl experiment provided strong support for the semiconservative mode of DNA replication, as postulated by Watson and Crick.

Semiconservative Replication in Eukaryotes

In 1957, the year before the work of Meselson and his colleagues was published, J. Herbert Taylor, Philip Woods, and Walter Hughes presented evidence that semiconservative replication also occurs in eukaryotic organisms. They experimented with root tips of the broad bean *Vicia faba*, which are an excellent source of dividing cells. These researchers examined the chromosomes of these cells following replication of DNA. They monitored the replication process by labeling DNA with ^{3}H-thymidine, a radioactive precursor of DNA, and then performing autoradiography.

Autoradiography is a cytological technique that pinpoints the location of an isotope in a cell. In this procedure, a photographic emulsion is placed over a section of cellular material (root tips in this experiment), and the preparation is stored in a dark place. The slide is then developed, much as photographic film is processed. Because the radioisotope emits energy, the emulsion turns black at the approximate point of emission. The end result is the presence of dark spots or "grains" on the surface of the section, locating the newly synthesized DNA in the cell.

Root tips were grown for approximately one generation in the presence of the radioisotope and then placed in an unlabeled medium, where cell division continued. At the conclusion of each generation, cultures were arrested at metaphase by the addition of colchicine (a chemical derived from the crocus plant, which poisons the mitotic spindle fibers), and the chromosomes were examined by autoradiography. Figure 11–5 shows a single chromosome's replication over two division cycles as well as the distribution of grains. In this experiment, labeled thymidine was found only in association with chromatids that contained newly synthesized DNA.

The results are compatible with the semiconservative mode of replication. After replication I, radioactivity is detected over both sister chromatids. This is expected because each chromatid will contain one new radioactive DNA strand and one old unlabeled strand. After the second replication cycle (in an unlabeled medium), only one of the two new sister chromatids should be radioactive because half of the parent strands are unlabeled. With only the minor exceptions of sister chromatid exchange (see Chapter 8), this result was observed.

Together, the Meselson-Stahl and Taylor-Woods-Hughes experiments, as well as studies with other organisms, soon led to general acceptance of the semiconservative mode of replication. These experiments also strongly supported Watson and Crick's proposal for the double-helix model of DNA.

Origins, Forks, and Units of Replication

Semiconservative replication is the general mode by which DNA is duplicated. To enhance our understanding of this pattern, let's briefly consider some relevant issues. The first issue concerns the **origin of replication**. Where along the chromosome is DNA replication initiated? Is there only a single origin, or does DNA synthesis begin at more than one point? Is a point of origin random, or is it located at a specific region along the chromosome? Second, once replication begins, does it proceed in a single direction or in both directions away from the origin? In other words, is replication **unidirectional** or **bidirectional**?

At the actual point along the chromosome where replication is occurring, the strands of the helix are unwound, creating a **replication fork**. The fork initially appears at the point of origin of synthesis and then moves along the DNA duplex as replication proceeds. The length of DNA that is replicated following one initiation event at a single origin is a unit called the **replicon**. If replication is bidi-

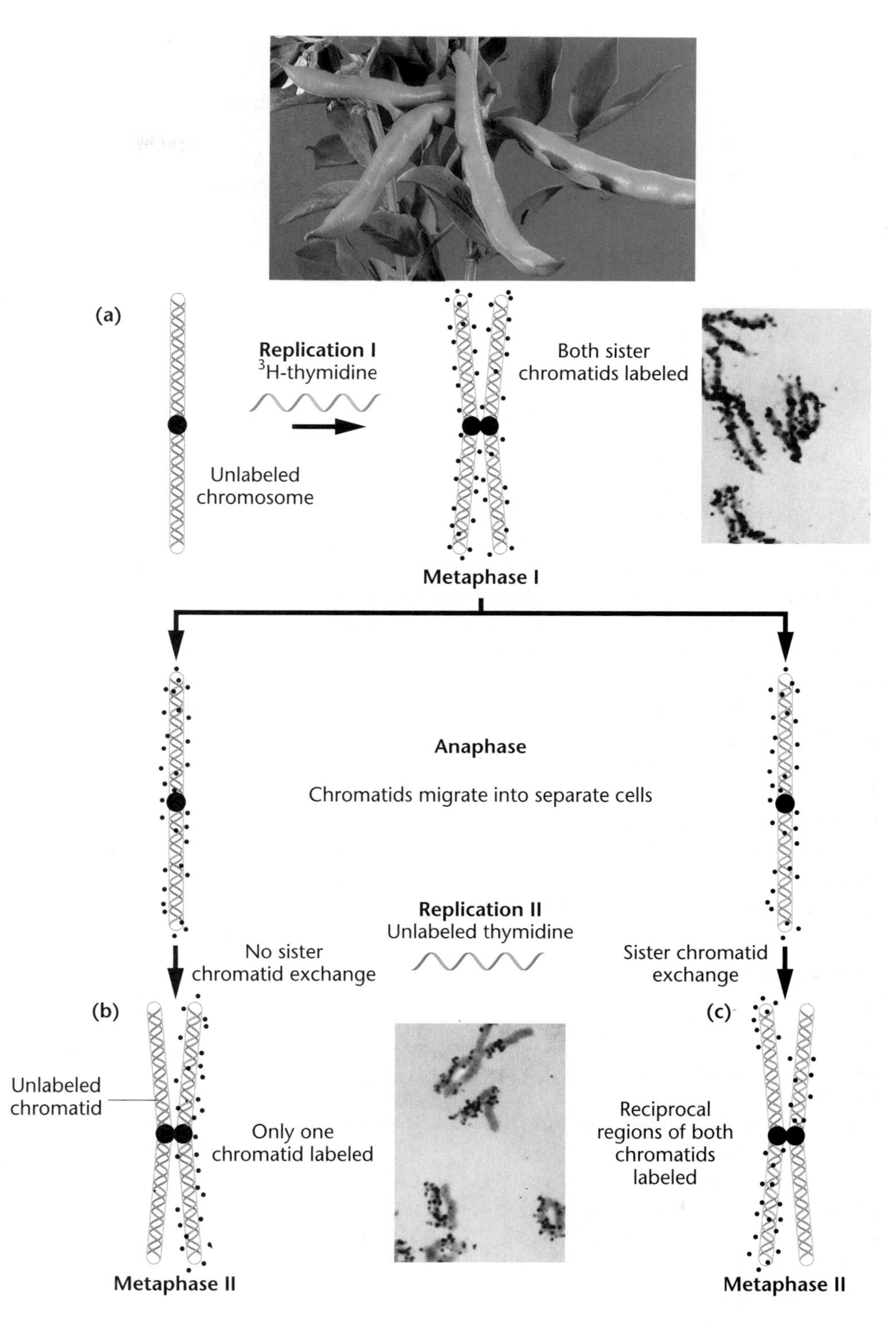

FIGURE 11–5 The Taylor-Woods-Hughes experiment, demonstrating the semiconservative mode of replication of DNA in root tips of *Vicia faba*. A portion of the plant is shown in the top photograph. (a) An unlabeled chromosome proceeds through the cell cycle in the presence of ^{3}H-thymidine. As it enters mitosis, both sister chromatids of the chromosome are labeled, as shown by autoradiography. After a second round of replication (b), this time in the absence of ^{3}H-thymidine, only one chromatid of each chromosome is expected to be surrounded by grains. Except where a reciprocal exchange occurred between sister chromatids (c), the expectation was upheld. The micrographs are of the actual autoradiograms obtained in the experiment. *(Top photo: Walter H. Hodge/Peter Arnold, Inc.; right and bottom photos from,* "Molecular Genetics", Pt. 1, *pp. 74–75, J. H. Taylor, (ed). Copyright © 1963 and renewed 1991, reproduced with permission from Elsevier Science Ltd.)*

rectional, two replication forks will be present, migrating in opposite directions away from the origin.

The evidence for the origin and direction of replication is clear. John Cairns tracked replication in *E. coli*, using both radioisotopes and autoradiography. He demonstrated that replication is initiated in only one region. In *E. coli*, this specific region, called *oriC*, has since been mapped along the chromosome. It has 245 base pairs, though only a small number of them are essential to the initiation of DNA synthesis. In bacteriophages and bacteria, DNA synthesis originates at a single point, so the entire chromosome constitutes one replicon. The presence of only a single origin is characteristic of bacteria, which have only one circular chromosome.

Results put forward by other researchers, again relying on autoradiography, demonstrated that replication is bidirectional, moving away from *oriC* in both directions (Figure 11–6). This creates two replication forks that migrate farther and farther apart as replication proceeds. The forks eventually merge as semiconservative replication of the entire chromosome is completed at a termination region, called *ter*.

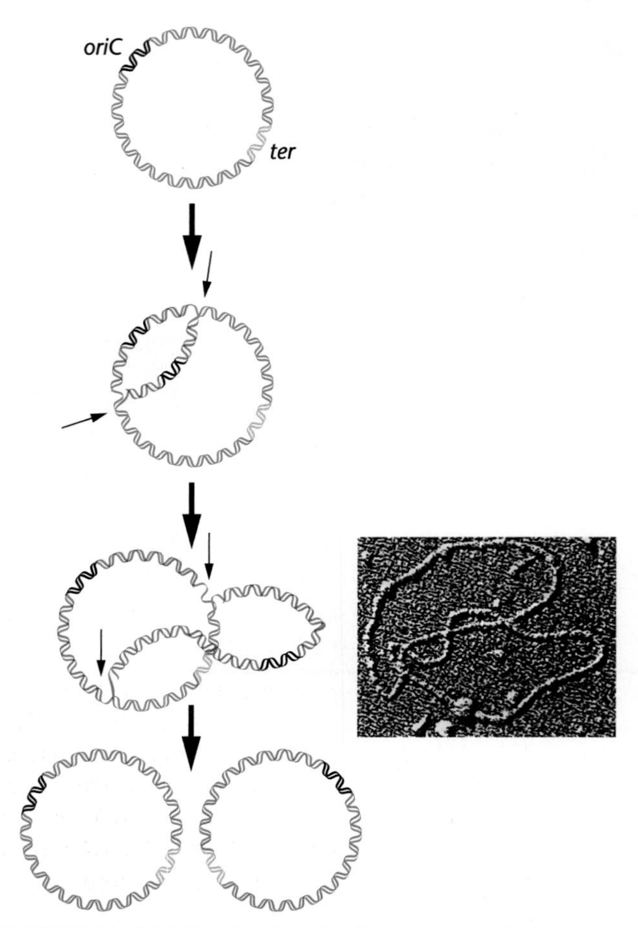

FIGURE 11–6 Bidirectional replication of the *E. coli* chromosome. The thin black arrows identify the advancing replication forks. The electron micrograph is of a bacterial chromosome in the process of replication, comparable to the figure next to it. *(Photo: Sundin and Varshavsky, 1981. Reprinted from* Cell, *25: 659. With permission from Elsevier Science. Courtesy of A. Varshavsky.)*

11.2 DNA Synthesis in Bacteria Involves Three Polymerases, as Well as Other Enzymes

The determination that replication is semiconservative and bidirectional indicates only the *pattern* of DNA duplication and the association of the finished strands with each other once synthesis is complete. A more complex issue is how the actual *synthesis* of long complementary polynucleotide chains occurs. As in most studies of molecular biology, the answer has been found by using microorganisms. Research began about the same time as the Meselson-Stahl work and is still an active area of investigation. What is most apparent in this research is the tremendous chemical complexity of the biological synthesis of DNA.

DNA Polymerase I

Studies of the enzymology of DNA replication were first reported by Arthur Kornberg and his colleagues in 1957. They isolated an enzyme from *E. coli* that directed DNA synthesis in a cell-free (*in vitro*) system. The enzyme is called **DNA polymerase I** since it was the first of several to be isolated. Kornberg determined the major requirements for *in vitro* DNA synthesis under the direction of the enzyme:

- All four deoxyribonucleoside triphosphates (dATP, dCTP, dGTP, dTTP = dNTP)* must be present.
- DNA template must be added.

If any one of the four deoxyribonucleoside triphosphates was omitted from the reaction, no synthesis occurred. If derivatives of precursor molecules other than the nucleoside triphosphate were used (nucleotides or nucleoside diphosphates), synthesis did not occur. If no DNA template was added, synthesis of DNA occurred but was reduced greatly. Template-dependent synthesis directed by Kornberg's enzyme appeared to be exactly the type required for semiconservative replication. The reaction is summarized in Figure 11–7. The enzyme has since been shown to be a single polypeptide containing 928 amino acids.

The way in which each nucleotide is added to the growing chain is a function of the specificity of DNA polymerase I. As shown in Figure 11–8, the precursor dNTP contains the three phosphate groups attached to the 5′-carbon of deoxyribose. As the two terminal phosphates are cleaved during synthesis, the remaining phosphate attached to the 5′-carbon is covalently linked to the 3′-OH

*dNTP designates the deoxyribose forms of the four nucleoside triphosphates; in a similar manner, dNMP refers to the monophosphate forms.

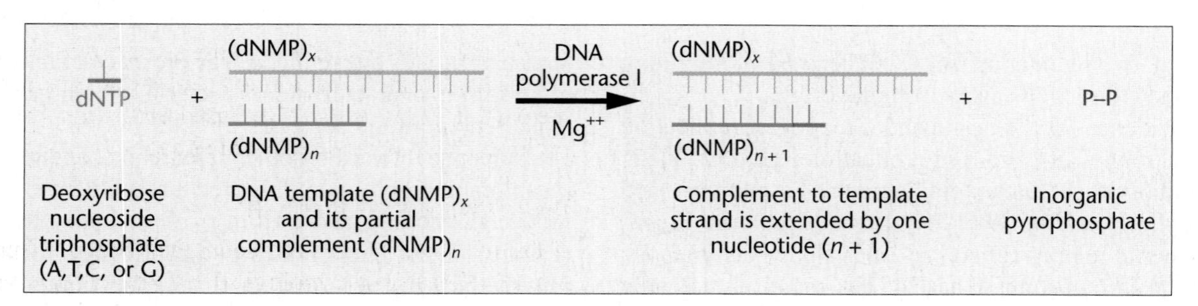

FIGURE 11–7 The chemical reaction catalyzed by DNA polymerase I. During each step, a single nucleotide is added to the growing complement of the DNA template, using a nucleoside triphosphate as the substrate. The release of inorganic pyrophosphate drives the reaction bioenergetically.

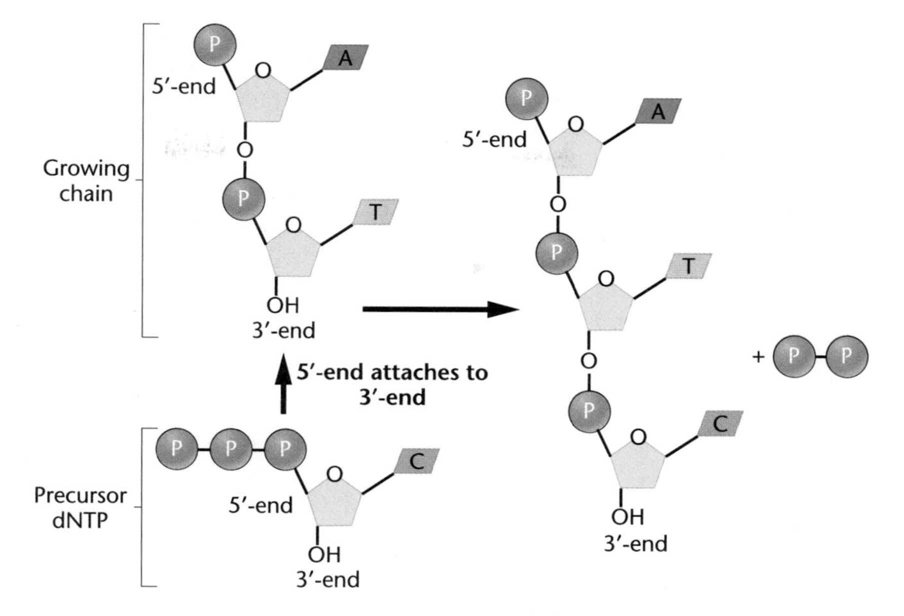

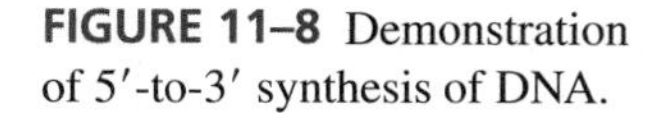
FIGURE 11–8 Demonstration of 5′-to-3′ synthesis of DNA.

group of the deoxyribose to which it is added. Thus, **chain elongation** occurs in the **5′-to-3′ direction** by the addition of one nucleotide at a time to the growing 3′-end. Each step provides a newly exposed 3′-OH group that can participate in the next addition of a nucleotide as DNA synthesis proceeds.

Having shown how DNA was synthesized, Kornberg sought to demonstrate the accuracy, or fidelity, with which the enzyme had replicated the DNA template. In 1957, the nucleotide sequences of the template and the product could not be determined; therefore, he initially relied on several indirect methods.

One of Kornberg's approaches was to compare the nitrogenous base compositions of the DNA template with those of the recovered DNA product. Table 11.1 shows Kornberg's base composition analysis of three DNA templates. These can be compared with the product synthesized in each case. Within experimental error, the base composition of each product agreed with the DNA template used. These data, along with other types of comparisons of template and product, suggested that the templates were replicated faithfully.

TABLE 11.1 Base Composition of the DNA Template and the Product of Replication in Kornberg's Early Work

	Template (T) or	*Percentage*			
Organism	*Product (P)*	*A*	*T*	*G*	*C*
T2	T	32.7	33.0	16.8	17.5
	P	33.2	32.1	17.2	17.5
E. coli	T	25.0	24.3	24.5	26.2
	P	26.1	25.1	24.3	24.5
Calf	T	28.9	26.7	22.8	21.6
	P	28.7	27.7	21.8	21.8

Source: Kornberg (1960).

Synthesis of Biologically Active DNA

Despite Kornberg's extensive work, not all researchers were convinced that DNA polymerase I was the enzyme that replicated DNA within cells (*in vivo*). Their reservations involved observations that the *in vitro* rate of synthesis was much slower than the *in vivo* rate, that the enzyme was much more effective replicating single-stranded DNA than double-stranded DNA, and that the enzyme appeared to *degrade* DNA as well as to synthesize it.

Uncertain of the true cellular function of DNA polymerase I, Kornberg pursued another approach. He reasoned that if the enzyme could be used to synthesize **biologically active DNA** *in vitro*, then DNA polymerase I must be the major catalyzing force for DNA synthesis within the cell. The term *biological activity* means that the DNA synthesized supports metabolic activities and directs reproduction of the organism from which it was originally duplicated.

In 1967, Kornberg, Mehran Goulian, and Robert Sinsheimer showed that the DNA of the small bacteriophage ϕX174 could be completely copied by DNA polymerase I *in vitro* and that the new product could be isolated and used to transfect *E. coli*. This resulted in the production of mature phages under the direction of the synthetic DNA, thus demonstrating biological activity.

This demonstration of biological activity was viewed as a precise assessment of faithful copying. If even a single error had occurred to alter the base sequence of any of the 5386 nucleotides constituting the ϕX174 chromosome, the change might easily have caused a mutation that would prohibit the production of viable phages.

DNA Polymerases II and III

Although DNA synthesized under the direction of polymerase I demonstrated biological activity, a more serious reservation about the enzyme's true biological role was raised in 1969. Peter DeLucia and John Cairns reported the discovery of a mutant strain of *E. coli* that was deficient in

DNA polymerase I activity. The mutation was designated *polA1*. In the absence of the functional enzyme, this mutant strain of *E. coli* still duplicated its DNA and reproduced successfully. Other properties of the mutation led DeLucia and Cairns to conclude that in the absence of DNA polymerase I, these cells were highly deficient in their ability to "repair" DNA. For example, the mutant strain was highly sensitive to UV light and radiation, both of which damage DNA and are therefore mutagenic. Nonmutant bacteria were able to repair a great deal of UV-induced damage. These observations led to two conclusions:

1. At least one other enzyme responsible for replicating DNA *in vivo* is present in *E. coli* cells.
2. DNA polymerase I may serve a secondary function *in vivo*. This function is now believed by Kornberg and others to be critical to the *fidelity* of DNA synthesis, but this enzyme does not actually synthesize the entire complementary strand during replication.

To date, two other unique DNA polymerases have been isolated from cells lacking polymerase I activity and from cells that do contain polymerase I. Table 11.2 shows that the two enzymes, **DNA polymerase II** and **III**, share several characteristics with DNA polymerase I. While none of the three *initiate* DNA synthesis on a template, all three can elongate an existing DNA strand, called a **primer**. As we shall see, RNA is also an adequate primer and is, in fact, used initially.

The DNA polymerase enzymes are all large complex proteins exhibiting a molecular weight in excess of 100,000 daltons (Da). All three possess 3′-to-5′ **exonuclease activity**, which means that they have the potential to polymerize in one direction and then pause and excise nucleotides that were just added. As we will discuss later in this chapter, this activity allows the enzymes to proofread newly synthesized DNA and remove the incorrect nucleotides, which can then be replaced. DNA polymerase I also demonstrates 5′-to-3′ exonuclease activity. Thus, the enzyme can excise nucleotides starting at the end where synthesis begins, and proceeding in the direction of synthesis, can remove the RNA primer.

Why did Kornberg isolate only polymerase I and not polymerase III? Probably because polymerase I is present in greater amounts than is polymerase III, and it is also much more stable.

What are the roles of the three polymerases *in vivo*? Polymerase I is thought to be responsible for removing the primer and for the gap-filling synthesis. These gaps occur naturally as primers are removed. Its exonuclease activity also allows for proofreading during this process. Polymerase II appears to repair DNA damaged by external forces, such as UV light. It is encoded by a gene that may be activated by disruption of DNA synthesis at the replication fork. Polymerase III is the enzyme responsible for the polymerization essential to replication, and its 3′-to-5′ exonuclease activity provides its proofreading function, as described above.

The DNA polymerase III molecule is extremely complex. Its active form, called a **holoenzyme**, consists of two sets (a dimer) of the 10 separate polypeptide subunits shown in Table 11.3 and has a molecular weight in excess of 600,000 Da. The largest subunit, α, has a molecular weight of 140,000 Da and, along with subunits ε and θ, constitutes the *core enzyme* responsible for the polymerization activity of the holoenzyme. The α subunit is responsible for nucleotide polymerization on the template strands, whereas the ε subunit of the core enzyme possesses the 3′-to-5′ exonuclease activity.

A second group of five subunits (γ, δ, δ', χ, and ψ) forms what is called the γ complex, which "loads" the enzyme onto the template at the replication fork. This enzymatic function requires energy and is dependent on the hydrolysis of ATP. The β subunit prevents the core enzyme from falling off the template during polymerization. Finally, the τ subunit holds the two core polymerases together at the replication fork. The holoenzyme and several other proteins at the replication fork form a complex nearly as large as a ribosome known as a **replisome**. We consider the function of DNA polymerase III in more detail later in this chapter.

11.3 Many Complex Issues Must Be Resolved During DNA Replication

We have thus far established that replication is semiconservative and bidirectional along a single replicon in bacteria and many viruses. Also, we know that synthesis is in the 5′-to-3′ mode under the direction of DNA polymerase

TABLE 11.2 Properties of Bacterial DNA Polymerases I, II, and III

Properties	*I*	*II*	*III*
Initiation of chain synthesis	−	−	−
5′-to-3′ Polymerization	+	+	+
3′-to-5′ Exonuclease activity	+	+	+
5′-to-3′ Exonuclease activity	+	−	−
Molecules of polymerase/cell	400	?	15

TABLE 11.3 Subunits of the DNA Polymerase III Holoenzyme

Subunit	*Function*	*Groupings*
α	5′-to-3′ Polymerization	Core enzyme: elongates polynucleotide chain and proofreads
ε	3′-to-5′ Exonuclease	
θ	unknown	
γ, δ, δ', χ, ψ	Loads enzyme on template (serves as clamp loader)	γ complex
β	Sliding clamp structure (processivity factor)	
τ	Dimerizes core complex	

III, creating two replication forks. These move in opposite directions away from the origin of synthesis. But many issues must still be resolved to gain a comprehensive understanding of DNA replication:

1. A mechanism must exist by which the helix is initially unwound (or denatured) and stabilized in an "open" configuration so that synthesis can proceed along both strands.
2. As unwinding and subsequent DNA synthesis proceeds, increased coiling creates tension farther down the helix, which must be reduced.
3. A primer of some sort must be synthesized so that polymerization can commence under the direction of DNA polymerase III. Surprisingly, RNA, not DNA, serves as this primer.
4. Once the RNA primers have been synthesized, DNA polymerase III commences synthesis of the complement of both strands of the parent molecule. Because the strands are antiparallel, continuous synthesis in the direction in which the replication fork moves is possible along only one of the two strands. On the other strand, synthesis is discontinuous in the opposite direction.
5. The RNA primers must be removed prior to completion of replication. The gaps that are temporarily created must be filled with DNA that is complementary to the template at each location.
6. The newly synthesized DNA strand that fills each temporary gap must be ligated to the adjacent strand of DNA.

You should be sure to carefully examine each of the figures in this section to help you understand DNA synthesis.

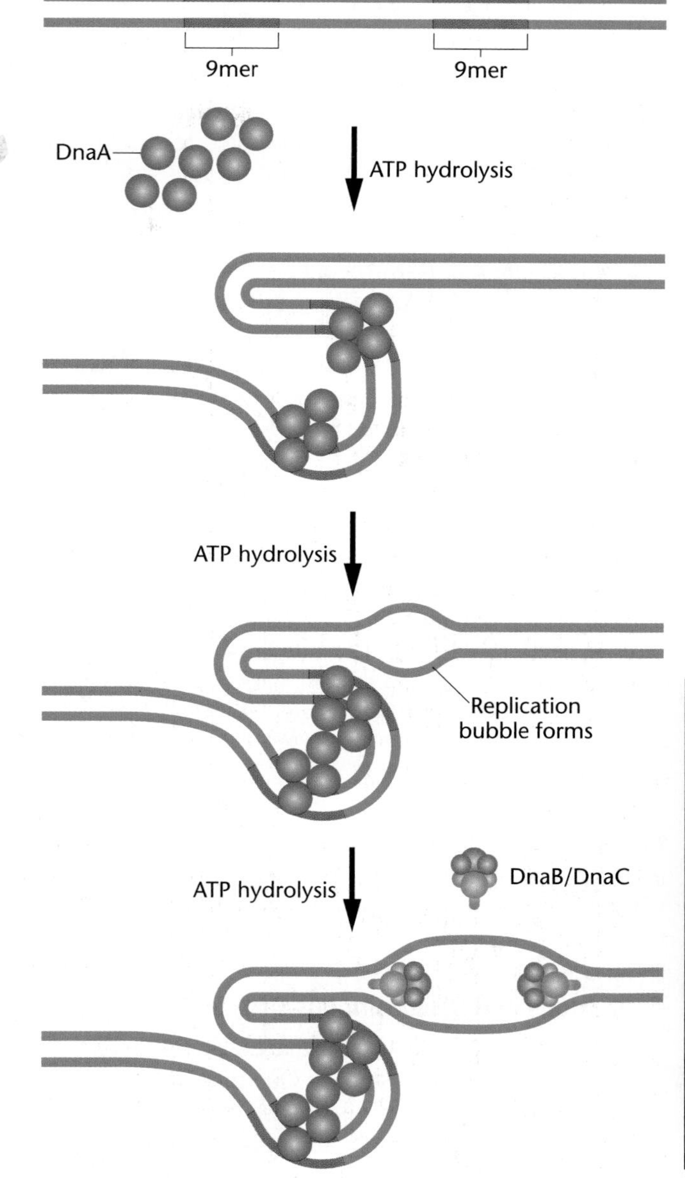

FIGURE 11–9 Helical unwinding of DNA during replication as accomplished by DnaA, DnaB, and DnaC proteins. Initial binding of many monomers of DnaA occurs at DNA sites that contain repeating sequences of 9mers. Not illustrated are 13mers, which are also involved.

MEDIA TUTORIAL DNA replication

Unwinding the DNA Helix

As discussed earlier, DNA synthesis is initiated at a single origin along the circular chromosome of most bacteria and viruses. This region of the *E. coli* chromosome has been particularly well studied. Called *oriC*, it consists of 245 base pairs characterized by repeating sequences of 9 and 13 bases (called **9mers** and **13mers**). One particular protein (called **DnaA protein** because it is encoded by the gene *dnaA*) initiates unwinding of the helix. A number of subunits of the DnaA protein bind to each of several 9mers. This step is essential in facilitating the subsequent binding of **DnaB** and **DnaC proteins** that further open and destabilize the helix (Figure 11–9). These proteins, which require the energy normally supplied by ATP hydrolysis to break hydrogen bonds and denature the double helix, are called **helicases**. Other proteins, called **single-stranded binding proteins (SSBPs)** stabilize the open conformation.

As unwinding proceeds, a coiling tension is created ahead of the replication fork, often producing supercoiling. In circular molecules, supercoiling takes the form of added twists and turns of the DNA, much like the coiling created in a rubber band by stretching it out and then twisting one end. Supercoiling can be relaxed by **DNA gyrase**, a member of a larger group of enzymes referred to as **DNA topoisomerases**. DNA gyrase makes either single- or double-stranded "cuts" and also catalyzes localized movements that "undo" the twists and knots created during supercoiling. The strands are then resealed. These various reactions are driven by the energy released during ATP hydrolysis.

Together, the DNA, the polymerase complex, and associated enzymes make up an array of molecules that participate in DNA synthesis and are part of what we have previously called the replisome.

Initiation of DNA Synthesis

Once a small portion of the helix is unwound, DNA synthesis is initiated. As we have seen, DNA polymerase III requires a primer with a free 3′-end in order to elongate a polynucleotide chain. This fact prompted researchers to investigate how the first nucleotide could be added. Although no free 3′-OH group is initially present, it is now clear that RNA is the primer that initiates DNA synthesis.

A short segment of RNA (about 5–15 nucleotides long), complementary to DNA, is first synthesized on the DNA template. Synthesis of the RNA is directed by a form of RNA polymerase called **primase**, which does not require a free 3′-end to initiate synthesis. It is to this short segment of RNA that DNA polymerase III begins to add 5′-deoxyribonucleotides, initiating DNA synthesis. Figure 11–10 shows how synthesis is initiated on a DNA template. At a later point, the RNA primer must be clipped out and replaced with DNA. This occurs under the direction of DNA polymerase I. RNA priming is a universal phenomenon during the initiation of DNA synthesis that is seen in viruses, bacteria, and all eukaryotic organisms.

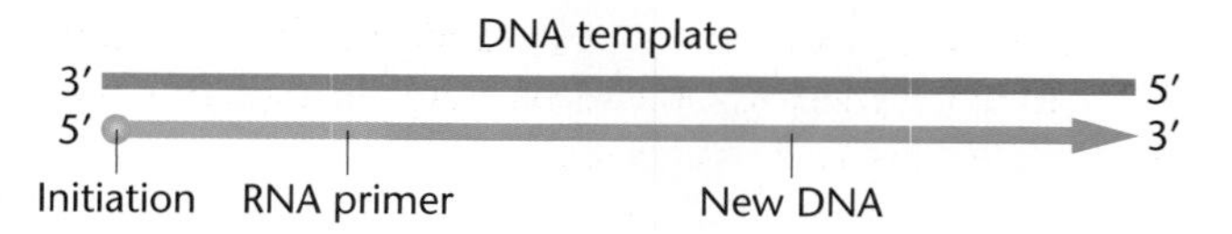

FIGURE 11–10 The initiation of DNA synthesis. A complementary RNA primer is first synthesized, to which DNA is added. All synthesis is in the 5′-to-3′ direction. Eventually, the RNA primer is replaced by DNA under the direction of DNA polymerase I.

Continuous and Discontinuous DNA Synthesis

We must now reconsider the fact that the double helix consists of two **antiparallel** strands. One runs in the 5′-to-3′ direction, while the other runs in the opposite, 3′-to-5′, direction. Because DNA polymerase III synthesizes DNA in only the 5′-to-3′ direction, synthesis along an advancing replication fork simultaneously occurs in one direction on one strand and in the opposite direction on the other. As the strands unwind and the replication fork progresses down the helix (Figure 11–11), only one strand, the **leading DNA strand**, can serve as a template for **continuous DNA synthesis**. As the fork continues along the helix, many points of initiation are necessary on the opposite strand, or **lagging DNA strand**, resulting in **discontinuous DNA synthesis**.

Evidence supporting discontinuous DNA synthesis was first provided by Reiji Okazaki, Tuneko Okazaki, and their colleagues. They discovered that when bacteriophage DNA is replicated in *E. coli*, some of the newly formed DNA that is hydrogen-bonded to the template strand is present as small fragments containing 1000–2000 nucleotides. RNA primers are a part of each fragment. These pieces, called **Okazaki fragments**, convert into longer and longer DNA strands of higher molecular weight as synthesis proceeds.

Discontinuous synthesis of DNA requires enzymes that remove the RNA primer and unite the Okazaki fragments into the continuous lagging strand. As we have noted, DNA polymerase I removes the primer and replaces the missing nucleotides. Joining the fragments appears to be the work of **DNA ligase**, which catalyzes the formation of the phosphodiester bond that closes the gap between the discontinuously synthesized strands. The evidence that DNA ligase performs this function during DNA synthesis is strengthened by observations of a ligase-deficient mutant strain (*lig*) of *E. coli* in which a large number of unjoined Okazaki fragments accumulate.

Concurrent Synthesis on the Leading and Lagging Strands

Given the model just discussed, you might well ask how DNA polymerase III synthesizes DNA on both the leading and lagging strands. Can both strands be replicated simultaneously at the same replication fork, or are the events distinct, involving two separate copies of the enzyme? Evidence suggests that both strands replicate

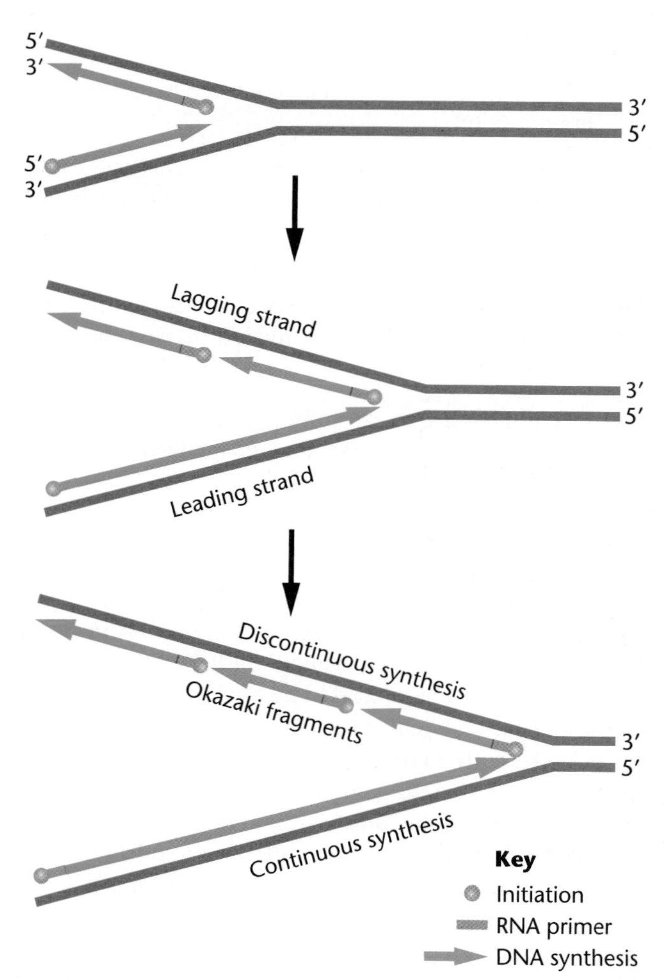

FIGURE 11–11 Opposite polarity of DNA synthesis along the two strands is necessary because the two strands of DNA run antiparallel to one another and DNA polymerase III synthesizes in only one direction (5′-to-3′). On the lagging strand, synthesis must be discontinuous, resulting in the production of Okazaki fragments. On the leading strand, synthesis is continuous. RNA primers initiate synthesis on both strands.

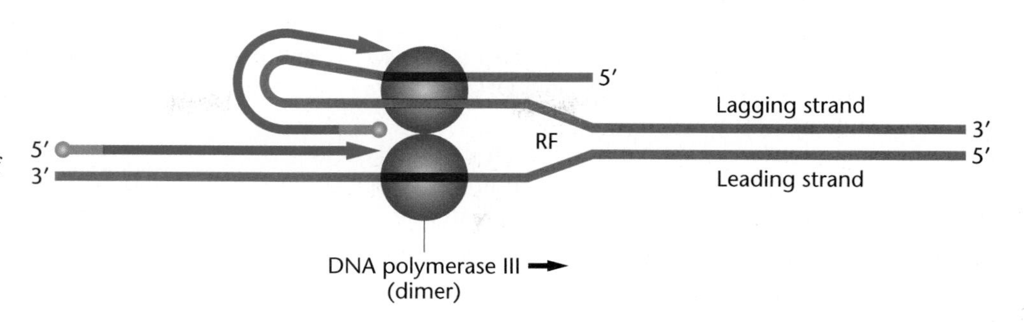

FIGURE 11–12 Illustration of how concurrent DNA synthesis is achieved on both the leading and lagging strands at a single replication fork. The lagging template strand is "looped" over to invert the physical direction of synthesis, but not the biochemical direction. The enzyme functions as a dimer, with each core enzyme achieving synthesis on one or the other strand.

simultaneously. As Figure 11–12 shows, if the lagging strand forms a loop, nucleotide polymerization occurs on both template strands under the direction of a dimer of the enzyme. After the synthesis of 100–200 base pairs, the monomer of the enzyme on the lagging strand encounters a completed Okazaki fragment, at which point it releases the lagging strand. A new loop is then formed with the lagging template strand, and the process repeats. Looping inverts the orientation of the template but not the direction of actual synthesis on the lagging strand, which is always in the 5′-to-3′ direction.

Another important feature of the holoenzyme that facilitates synthesis at the replication fork is a dimer of the β subunit that forms a clamplike structure around the newly formed DNA duplex. The β-subunit clamp prevents the **core enzyme** (the α, ε, and θ subunits responsible for catalysis of nucleotide addition) from falling off the template as polymerization proceeds. Because the entire holoenzyme moves along the parent duplex, advancing the replication fork, the β-subunit dimer is often called a sliding clamp.

Proofreading and Error Correction During DNA Replication

The essence of DNA replication is the synthesis of a new strand that is precisely complementary to the template strand at each nucleotide position. Although the action of DNA polymerases is very accurate, synthesis is not perfect and a noncomplementary nucleotide is sometimes erroneously inserted. To compensate for such inaccuracies, polymerases I and III both possess **3′-to-5′ exonuclease activity**, which enables them to detect and excise a mismatched nucleotide in the 3′-to-5′ direction. Once the mismatched nucleotide is removed, 5′-to-3′ synthesis again proceeds. This process, **exonuclease proofreading**, increases the fidelity of synthesis. In the case of the holoenzyme form of DNA polymerase III, the ε subunit is directly involved in the proofreading step. In strains of *E. coli* where a mutation has rendered the ε subunit nonfunctional, the error (mutation) rate during DNA synthesis is substantially increased.

11.4 A Coherent Model Summarizes DNA Replication

We can now combine the various aspects of DNA replication occurring at a single replication fork into the coherent model shown in Figure 11–13. At the advancing fork, a helicase is unwinding the double helix. Once unwound, single-stranded binding proteins associate with them, preventing the re-formation of the helix. In advance of the replication fork, DNA gyrase diminishes the tension that is created as the helix supercoils. Each half of the dimeric polymerase is a core enzyme bound to one template strand by a β-subunit sliding clamp. Continuous synthesis occurs on the leading strand, while the lagging strand loops over

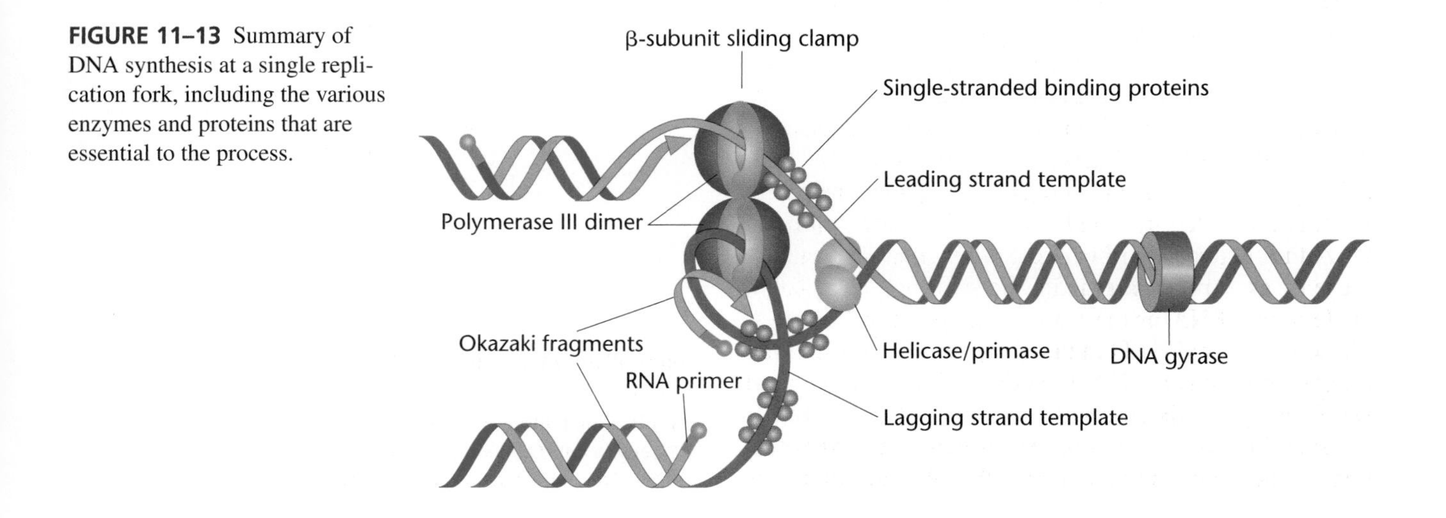

FIGURE 11–13 Summary of DNA synthesis at a single replication fork, including the various enzymes and proteins that are essential to the process.

for simultaneous synthesis to occur on both strands. Not shown in the figure, but essential to replication on the lagging strand, is the action of DNA polymerase I and DNA ligase, which replace the RNA primer with DNA and join the Okazaki fragments, respectively.

The investigation of DNA synthesis is an extremely active area of research, and this model will no doubt be extended in the future. In the meantime, it gives us a summary of DNA synthesis against which we can interpret genetic phenomena.

11.5 Replication Is Controlled by a Variety of Genes

Much of what we know about DNA replication in viruses and bacteria is based on genetic analysis of the process. For example, we have already discussed studies involving the *polA1* mutation, which revealed that DNA polymerase I is not the major enzyme responsible for replication. Many other mutations interrupt or seriously impair some aspect of replication, such as the ligase-deficient and the proofreading-deficient mutations mentioned previously. Genetic analysis frequently uses **conditional mutations**, which are expressed under one condition but not under a different condition. For example, a **temperature-sensitive mutation** may not be expressed at a particular *permissive* temperature; but when mutant cells are grown at a *nonpermissive* (or *restrictive*) temperature, the mutation is expressed. The investigation of temperature-sensitive mutants provides insights into the product and the associated function of the normal, nonmutant gene.

As shown in Table 11.4, a variety of genes in *E. coli* specify the subunits of polymerases I, II, and III along with encoding products involved in specification of the origin of synthesis, helix unwinding and stabilization, initiation and priming, relaxation of supercoiling, repair, and ligation. The discovery of such a large group of genes attests to the complexity of the replication process, even in a relatively simple prokaryote. Given the enormous quantity of DNA that must be unerringly replicated in a very brief time, this level of complexity is not unexpected. As you will see next, the process is even more involved and therefore more difficult to investigate in eukaryotes.

TABLE 11.4 Some of the Various *E. coli* Mutant Genes and Their Products or Role in Replication

Mutant Gene	*Enzyme or Role*
polA	DNA polymerase I
polB	DNA polymerase II
dnaE, N, Q, X, Z	DNA polymerase III subunits
dnaG	Primase
dnaA, I, P	Initiation
dnaB, C	Helicase at *oriC*
oriC	Origin of replication
gyrA, B	Gyrase subunits
lig	Ligase
rep	Helicase
ssb	Single-stranded binding proteins
rpoB	RNA polymerase subunit

11.6 Eukaryotic DNA Synthesis Is Similar to, but More Complex than, Synthesis in Prokaryotes

Research shows that eukaryotic DNA is replicated in a manner similar to that of bacteria. In both systems, double-stranded DNA unwinds at a replication origin, two replication forks form, and bidirectional synthesis of DNA occurs on the leading and lagging strand templates under the direction of DNA polymerase. Eukaryotic polymerases have the same fundamental requirements for DNA synthesis as do bacterial systems: four deoxyribonucleoside triphosphates, a template, and a primer. However, because eukaryotic cells contain much more DNA per cell and because this DNA is complexed with proteins, eukaryotes face many problems not encountered by bacteria. As we might expect, these complications make the process of DNA synthesis much more complex in eukaryotes and more difficult to study. However, a great deal is now known about the process.

Multiple Replication Origins

The most obvious difference between eukaryotic and prokaryotic DNA replication is that eukaryotic chromosomes contain multiple replication origins, in contrast to the single site that is part of the *E. coli* chromosome. The multiple origins are visible under an electron microscope (Figure 11–14). They are essential if the entire genome of a typical eukaryote is to be replicated in a reasonable amount of time. Recall that (1) eukaryotes have much greater amounts of DNA than bacteria (e.g., yeast has four times as much, and *Drosophila* has 100 times as much DNA

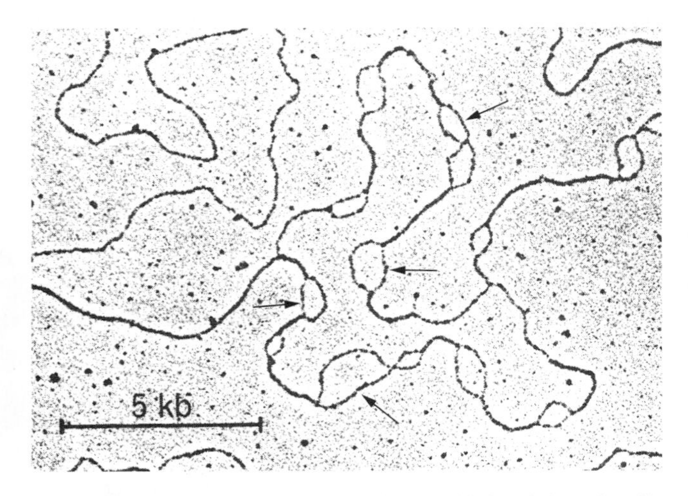

FIGURE 11–14 A demonstration of the multiple origins of replication along a eukaryotic chromosome. Each origin is apparent as a replication bubble along the axis of the chromosome. *(H. J. Kreigstein and D. S. Hogness, 1974.* Proc. Natl. Acad. Sci. (USA) *71: 136. Fig. 2, p. 137.)*

as *E. coli*), and (2) the rate of synthesis by eukaryotic DNA polymerase is much slower—only about 50 nucleotides per second, a rate 20 times less than the comparable bacterial enzyme. Under these conditions, single-origin replication of a typical eukaryotic genome would take up to a month to complete! However, replication is accomplished in as little as three minutes in some eukaryotic organisms.

Many insights concerning the molecular nature of the multiple origins and the initiation of DNA synthesis at these sites are now available. Most information was originally derived from the study of yeast (e.g., *Saccharomyces cerevisiae*), which have between 250 and 400 replicons. Subsequent studies used mammalian cells, which have as many as 25,000 replicons. The origins of replication in yeast have been isolated and are called **autonomously replicating sequences (ARSs)**. They consist of a unit of 11 base pairs, flanked by other short sequences involved in efficient initiation. We know from Chapter 2 that DNA synthesis is restricted to the S phase of the eukaryotic cell cycle. Research has shown that the numerous origins are not all activated at once. Instead, clusters of 20–80 adjacent replicons are activated sequentially throughout the S phase until all DNA is replicated.

How does the polymerase find the ARSs among so much DNA? A mechanism initiated during the G1 phase of the cell cycle exists where all ARSs are initially bound by a group of specific proteins (six in yeast), forming the **origin recognition complex (ORC)**. Mutations in either the ARSs or in any of the genes encoding the proteins of the ORC abolish or reduce initiation of DNA synthesis. Since these recognition complexes are formed in G1 but synthesis is not initiated at these sites until the S phase, there must be still other proteins involved in the actual initiation signal. The most important of these proteins are specific kinases, key enzymes involved in phosphorylation, that are integral parts of cell-cycle control. When bound along with ORC, a prereplication complex is formed that is accessible to DNA polymerase. After these kinases are activated, they complete the initiation complex, directing localized unwinding and triggering DNA synthesis. Activation also inhibits re-formation of the prereplication complexes once DNA synthesis has been completed at each replicon. This is an important mechanism since it distinguishes segments of DNA that have completed replication from unreplicated DNA, thus maintaining orderly and efficient replication. It ensures that replication occurs only once along each stretch of DNA during each cell cycle.

Eukaryotic DNA Polymerases

The most complex aspect of eukaryotic replication is the array of polymerases involved in directing DNA synthesis. Six different forms of the enzyme have been isolated and studied. For the polymerases to access DNA, the topology of the helix must first be modified. As synthesis is triggered at each origin site, the double strands are opened up in an AT-rich region, which allows a helicase enzyme to enter, further unwinding the double-stranded DNA. Before polymerases begin synthesis, histone proteins complexed to the DNA (which form the characteristic **nucleosomes** of chromatin) also must be stripped away or otherwise modified. As DNA synthesis proceeds, histones reassociate with the newly formed duplexes, reestablishing the characteristic nucleosome pattern (Figure 11–15). In eukaryotes, the synthesis of new histone proteins is tightly coupled to DNA synthesis during the S phase of the cell cycle.

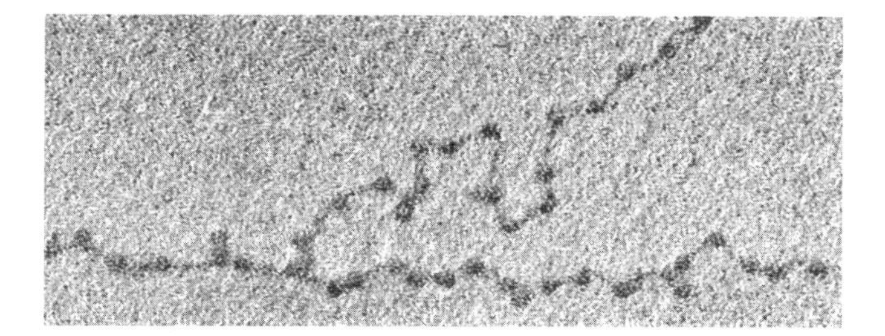

FIGURE 11–15 An electron micrograph of a eukaryotic replicating fork that demonstrates the presence of histone protein-containing nucleosomes on both branches. *(Dr. Harold Weintraub, Howard Hughes Medical Institute, Fred Hutchinson Cancer Center.* Essential Molecular Biology, *2nd ed. Freifelder & Malachinski, Jones & Bartlett, Fig. 7-24, p. 141.)*

Of the six known polymerases, three (Pol α, Pol δ, and Pol ε) are now considered to be essential to nuclear DNA replication in eukaryotic cells. Two others (Pol β and Pol ξ) appear to be involved in DNA repair. The sixth form (Pol γ) is involved in the synthesis of mitochondrial DNA. Presumably, its replication function is limited to that organelle even though it is encoded by a nuclear gene. All but one of the enzyme's six forms (Pol β) consist of multiple subunits. Different subunits perform different functions during replication.

Pol α and Pol δ may be the major forms of the enzyme involved in initiating nuclear DNA synthesis, so we will concentrate our discussion on them. Two of their four subunits function as a primase in the synthesis of RNA primers on both the leading and lagging template strands. Then, another subunit elongates the RNA primer by adding complementary deoxyribonucleotides, constituting the initial phase of DNA synthesis. Pol α is said to possess low **processivity**, a term related to the length of DNA that is synthesized before it dissociates from the template. Thus, after a short DNA sequence is added to the RNA primer, **polymerase switching** occurs, whereby Pol α dissociates from the template and is replaced by Pol δ. This form of the enzyme possesses high processivity as well as 3′-to-5′ exonuclease activity, which provides it with the potential to proofread. It is also capable of a 100-fold increase in the rate of synthesis in comparison to Pol α. Thus, under the direction of Pol δ, elongation and proofreading of the growing DNA strand occurs as synthesis continues. Pol ε, the third essential form, possesses the same general characteristics as Pol δ, but is believed to operate under different cellular conditions. In yeast, mutations that render Pol ε inactive are lethal, attesting to its essential function during replication.

The process just described applies to both leading and lagging strand synthesis. On both strands, RNA primers must be replaced with DNA. On the lagging strand, the Okazaki fragments, about 10 times smaller (100–150 nucleotides) in eukaryotes than in prokaryotes, must be ligated.

To accommodate the increased number of replicons, eukaryotic cells contain many more DNA polymerase molecules than do bacteria. While *E. coli* has about 15 copies of DNA polymerase III per cell, there may be up to 50,000 copies of the α form of DNA polymerase in animal cells. As has been pointed out, the presence of greater numbers of smaller replicons in eukaryotes compared with bacteria compensates for the slower rate of DNA synthesis in eukaryotes. *E. coli* requires 20–40 minutes to replicate its chromosome, while *Drosophila*, with 40 times more DNA, is known to accomplish the same task in only 3 minutes during embryonic cell divisions.

11.7 The Ends of Linear Chromosomes Are Problematic During Replication

A final difference between prokaryotic and eukaryotic DNA synthesis involves the nature of the chromosomes. Unlike the closed, circular DNA of most bacteria and most bacteriophages, eukaryotic chromosomes are linear. During replication, they face a special problem at the ends, or **telomeres**, of these linear molecules. While synthesis proceeds normally to the end of the leading strand, a difficulty arises on the lagging strand as the RNA primer is removed (Figure 11–16). Normally, the newly created gap would be filled by adding a nucleotide to the existing 3′-OH group provided during discontinuous synthesis (to the right of gap b in Figure 11–16). However, there is no strand present to provide the 3′-OH group because this is the end of the chromosome. As a result, each successive round of synthesis theoretically shortens the chromosome by the length of the RNA primer. Because this is such a significant problem, we can suppose, at least for some cells, that a molecular solution would have developed early in evolution and be shared by all eukaryotes, and this is indeed the case.

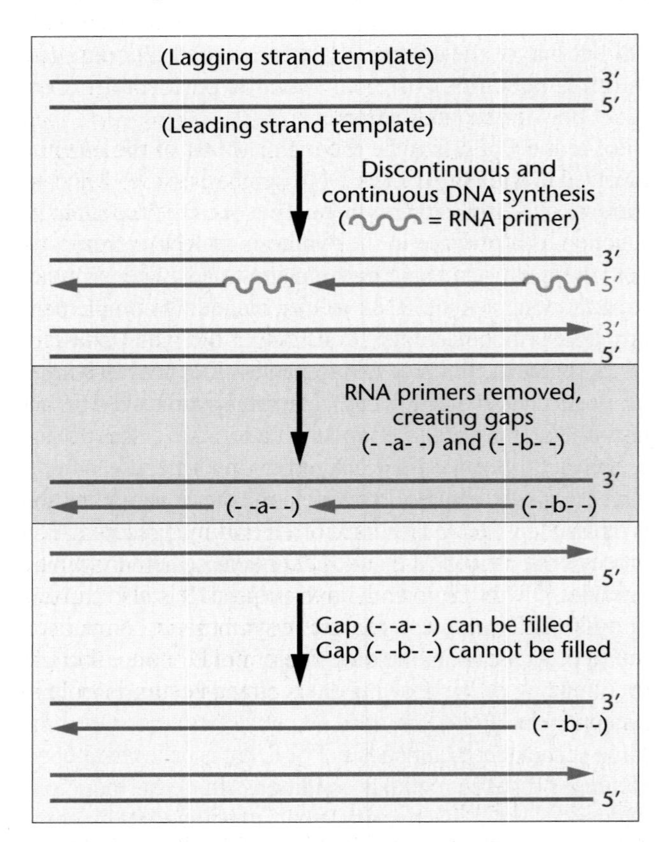

FIGURE 11–16 The difficulty encountered during the replication of the ends of linear chromosomes: A gap (marked - -b- -) cannot be filled following synthesis on the lagging strand.

Discovery of a unique eukaryotic enzyme, **telomerase**, has helped us understand how more complex organisms solve this problem. In the ciliated protozoan *Tetrahymena*, the many telomeres all terminate in the sequence 5′-TTGGGG-3′. Telomerase adds a six-nucleotide repeat (TTGGGG) to the ends of molecules already containing this sequence, thus preventing the telomeric ends from shortening after each replication. As shown in Figure 11–17, telomerase adds several copies of the six-nucleotide repeat to the 3′-end of the lagging strand (using 5′-to-3′ synthesis). These repeats form a "hairpin loop,"

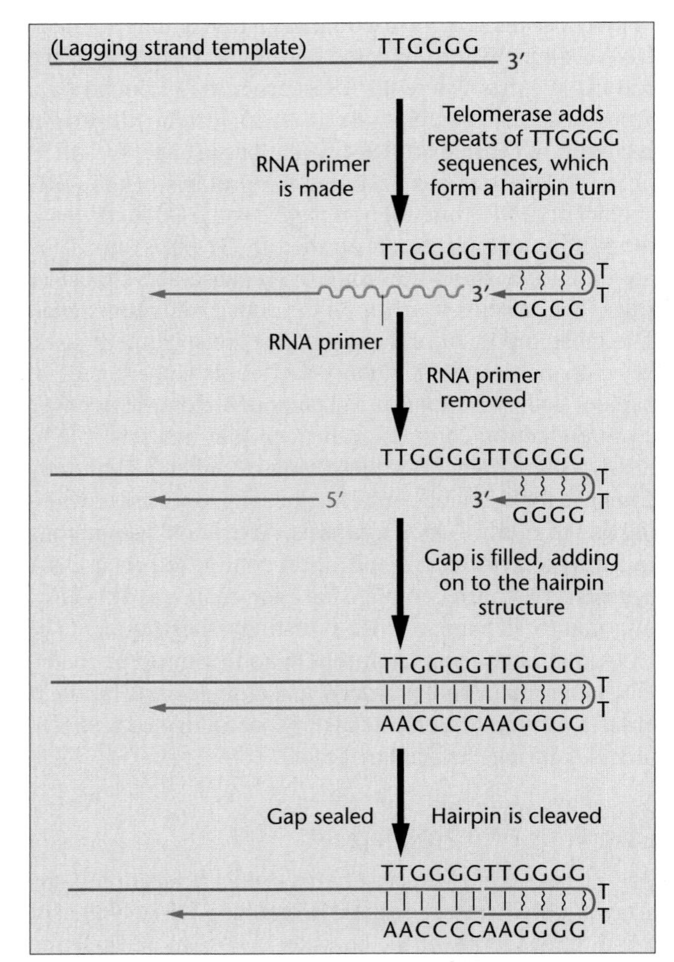

FIGURE 11–17 The predicted solution to the difficulty posed in Figure 11–16. The enzyme telomerase directs synthesis of the TTGGGG sequences, which results in the formation of a hairpin structure. The gap can now be filled, and the hairpin structure is cleaved.

which is stabilized by unorthodox hydrogen bonding between opposite guanine residues (G=G). This creates a free 3′-OH end that serves as a substrate for DNA polymerase I to fill the gap after the RNA primer is removed. The hairpin loop is then cleaved, and the potential loss of DNA in each subsequent replication cycle is averted. This process occurs in other eukaryotes under the direction of a similar enzyme, but the terminal telomeric sequence is specific for the organism.

Further investigation of the *Tetrahymena* telomerase enzyme by Elizabeth Blackburn and Carol Greider has yielded an extraordinary finding. This enzyme adds the same six-nucleotide sequence (TTGGGG) to DNA terminals even if they lack this sequence. Thus, the TTGGGG sequence of the DNA substrate is not the signal for telomerase function.

Blackburn and Greider have now established how the enzyme actually works. The enzyme is unique in that it contains within its molecular structure a short piece of RNA that is essential to its catalytic activity, making it a ribonucleoprotein. The RNA component encodes the sequences used by the enzyme as a template. The RNA contains 159 bases, including the sequence 5′-AACCCC-3′, which is complementary to the sequence whose synthesis it directs. Analogous enzyme functions have been found in other single-celled organisms. The RNA-containing telomerase enzyme behaves in a manner analogous to reverse transcriptase (an enzyme) in that it synthesizes a DNA complement on an RNA template.

Telomeric DNA sequences have been highly conserved throughout evolution, reflecting the critical function of telomeres. In the essay at the end of this chapter, we shall see that telomere shortening has been linked to a molecular mechanism involved in the aging process of cells. In most eukaryotic somatic cells, telomerase is, in fact, not active, and thus with each cell division, the telomeres of each chromosome shorten. After many divisions, the telomere is seriously eroded and the cell loses the capacity for division. Malignant cells, on the other hand, maintain telomerase activity and are immortalized.

11.8 DNA Recombination, Like DNA Replication, Is Directed by Specific Enzymes

We conclude this chapter by returning to a topic we discussed in Chapter 8—**genetic recombination**. There, you learned that the process of crossing over depends on breakage and rejoining of the DNA strands between homologs. Now that we have discussed the chemistry and replication of DNA, it is appropriate to consider how recombination occurs at the molecular level. In general, the following information pertains to genetic exchange between any two homologous, double-stranded DNA molecules, whether they be viral or bacterial chromosomes or eukaryotic homologs undergoing meiosis. Genetic exchange at equivalent positions along two chromosomes with substantial DNA sequence homology is referred to as **general**, or **homologous**, **recombination**.

Several models attempt to explain crossing over, and they all share certain common features. First, all are based on the initial proposals put forth independently by Robin Holliday and Harold L. K. Whitehouse in 1964. They also depend on the complementarity between DNA strands for their precision of exchange. Finally, each model relies on a series of enzymatic processes to accomplish genetic recombination.

One such model is shown in Figure 11–18, where the alleles *A*,*a* and *B*,*b* are included. It begins with two paired DNA duplexes or homologs (part a in the figure), each of which has a single-stranded nick introduced (part b) at an identical position by an endonuclease. The ends of the strands produced by these cuts are then displaced and subsequently pair with their complements on the opposite duplex (part c). A ligase then seals the loose ends (part d), creating hybrid duplexes called **heteroduplex DNA molecules**. The exchange creates a cross-bridged structure. The position of the cross bridge then moves down the chromosome as a result of branch migration (part e). This occurs as a result of a zipperlike action as hydrogen bonds break and then re-form between complementary bases of the displaced strands of each duplex. This migration yields an increased length of heteroduplex DNA on both homologs.

If the duplexes separate (Figure 11–18f) and the bottom portions rotate 180° (part g), an intermediate planar structure called a **chi form** (χ) is created—the characteristic **Holliday structure**. If the two strands on the opposite homologs previously uninvolved in the exchange are now nicked by an endonuclease (part h) and ligation occurs (part i), recombinant duplexes are created. Note that the arrangement of alleles is altered as a result of recombination.

Evidence supporting this model includes electron microscopic visualization of chi-form planar molecules from bacteria where four duplex arms join at a single point of exchange (Figure 11–18g). Additional important evidence comes from the discovery in *E. coli* of the **RecA protein**. This molecule promotes the exchange of reciprocal single-stranded DNA molecules, as occurs in Figure 11–18c. RecA also enhances hydrogen-bond formation during strand displacement, thus initiating heteroduplex formation.

Many other enzymes essential to the nicking and ligation process have also been discovered and investigated. The products of *recB*, *recC*, and *recD* genes are thought to be involved in nicking and unwinding DNA. Numerous mutations that prevent genetic recombination have been found in viruses and bacteria. These mutations are thought to identify genes whose products play an essential role in this process.

Gene Conversion

A modification of the model Figure 11–18 has helped us better understand a unique genetic phenomenon known as **gene conversion**. Initially found in yeast by Carl

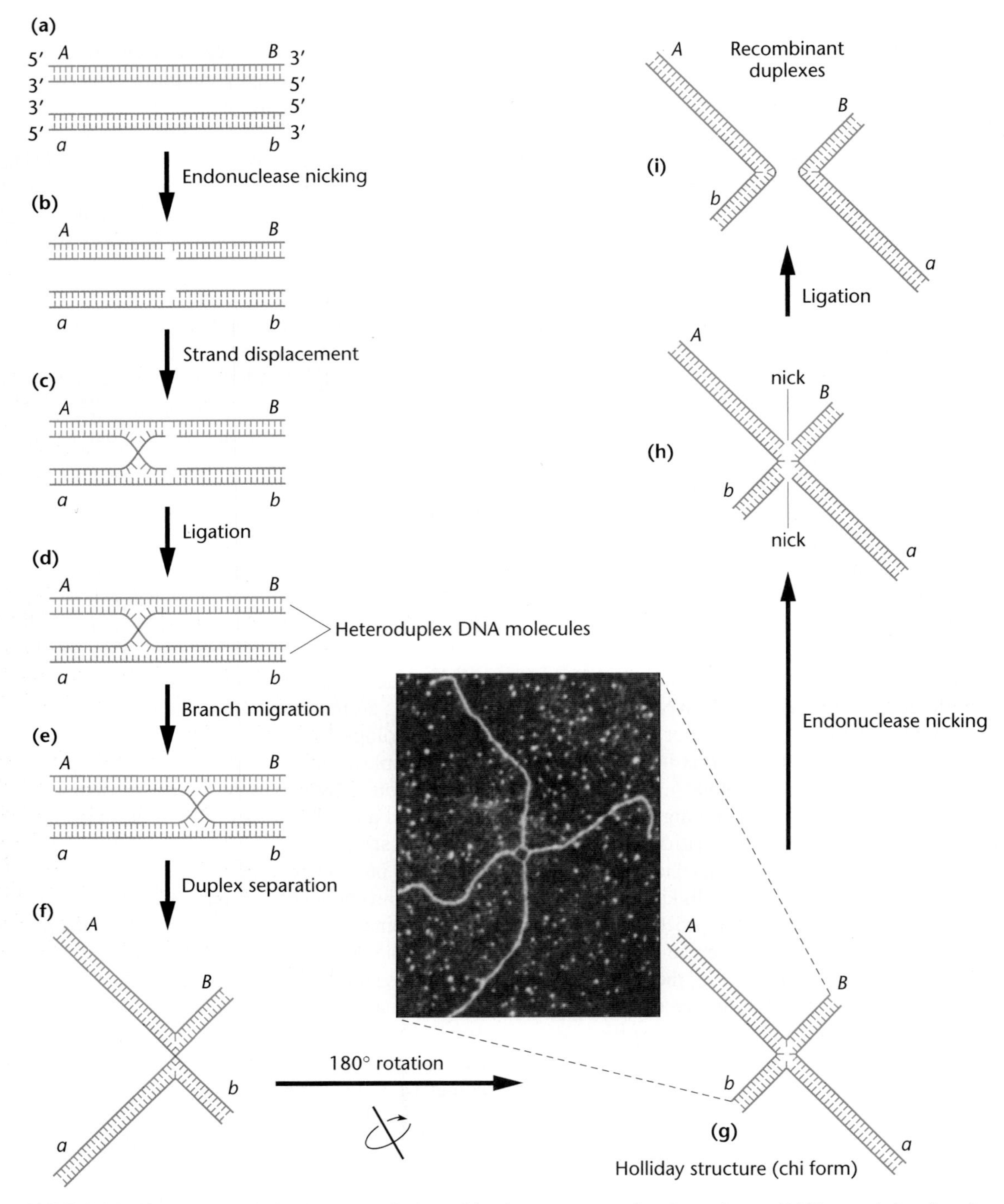

FIGURE 11–18 Model depicting how genetic recombination can occur when heterologous DNA strands break and rejoin. Each stage is described in the text. The electron micrograph shows DNA in a chi-form structure similar to the diagram in (g). The DNA is an extended Holliday structure, derived from the *ColE1* plasmid of *E. coli*. *(Photo: David Dressler, Oxford University, England)*

Lindegren and in *Neurospora* by Mary Mitchell, gene conversion is characterized by a genetic exchange ratio involving two closely linked genes, which is *nonreciprocal*. If we cross two *Neurospora* strains, each bearing a separate mutation ($a+ \times +b$), a *reciprocal* recombination event between the genes yields spore pairs of the ++ and *ab* genotypes. However, a nonreciprocal exchange yields one pair without the other. Working with pyridoxine mutants, Mitchell observed several asci containing spore patterns displaying the ++ genotype, but not the reciprocal product (*ab*). The frequency of these events was higher than the predicted mutation rate, and thus could not be accounted for by that phenomenon, so they were called *gene conversions* because it appeared that one allele had somehow been "converted" into another during an event in which

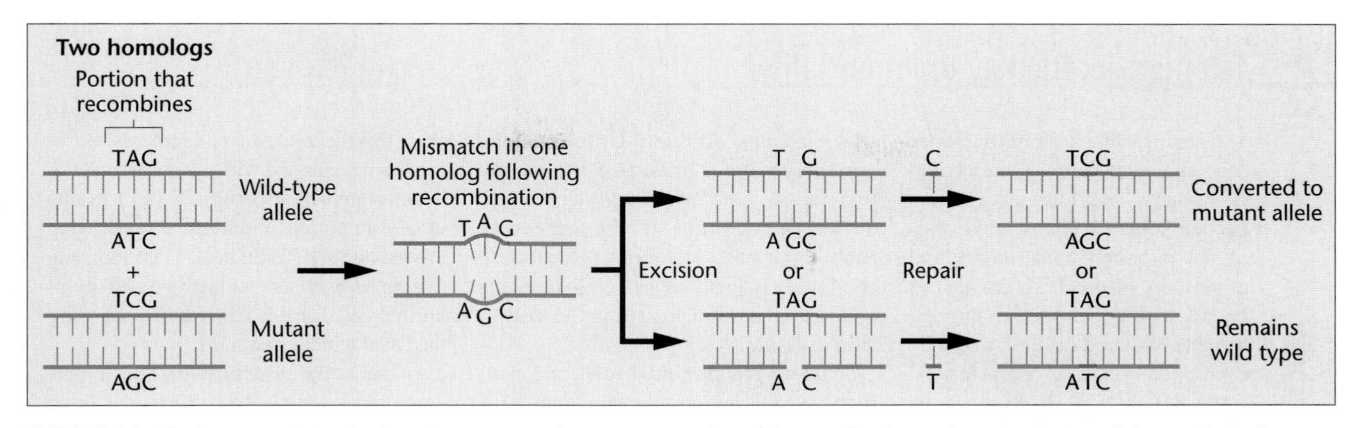

FIGURE 11–19 A proposed mechanism that accounts for gene conversion. A base-pair mismatch occurs in one of the two homologs (bearing the mutant allele) during heteroduplex formation, which accompanies recombination in meiosis. During excision repair, one of the two mismatches is removed and the complement is synthesized. In one case (top), the mutant base pair is preserved. When it is subsequently included in a recombinant spore, the mutant genotype is maintained. In the other case (bottom), the mutant base pair is converted to the wild-type sequence. When included in a recombinant spore, the wild-type genotype is expressed, leading to a nonreciprocal exchange ratio.

genetic exchange also occurred. Similar findings are apparent in the study of other fungi as well.

Gene conversion is now considered to be a consequence of the recombination process. One possible explanation interprets conversion as a mismatch of base pairs during heteroduplex formation, as shown in Figure 11–19. Mismatched regions of hybrid strands are repaired by excising one strand and synthesizing the complement, using the remaining strand as a template. Excision can occur in either strand, yielding two possible corrections. One repairs the mismatched base pair and converts it to restore the original sequence. The other also corrects the mismatch, but does so by copying the altered strand, creating a base-pair substitution. Conversion may have the effect of creating identical alleles on the two homologs that were initially different.

In our example in Figure 11–19, suppose the G≡C pair on one homolog was responsible for the mutant allele, while the A=T pair was part of the wild-type gene sequence on the other homolog. Converting the G≡C pair to A=T changes the mutant allele to wild type, just as Mitchell originally observed.

Gene-conversion events help explain other puzzling genetic phenomena in fungi. For example, when mutant and wild-type alleles of a single gene are crossed, asci should yield equal numbers of mutant and wild-type spores. However, exceptional asci with 3:1 or 1:3 ratios are sometimes observed. These ratios can be explained by gene conversion. This phenomenon has been detected during mitotic events in fungi, as well as in studies of unique compound chromosomes in *Drosophila*.

GENETICS, TECHNOLOGY, AND SOCIETY

Telomerase: The Key to Immortality?

Humans, like all multicellular organisms, grow old and die. As we age, our immune systems become less efficient, wound healing is impaired, and tissues and organs lose resilience. It has always been a mystery why we go through these age-related declines and why each species has a characteristic finite life span. Why do we grow old? Can we reverse this march to mortality? Some recent discoveries suggest that the answers to these questions may lie at the ends of our chromosomes.

The study of human aging begins with an investigation of human cells growing in culture dishes. Like the organisms from which the cells are taken, cells in culture have a finite life span. This "replicative senescence" was noted over 30 years ago when it was reported that normal human fibroblasts lose their ability to grow and divide after about 50 cell divisions. These senescent cells remain metabolically active, but can no longer proliferate—eventually, they die. Although we don't know whether cellular senescence directly causes organismal aging, the evidence is suggestive. For example, cells from young people go through more divisions in culture than cells from older people. Human fetal cells divide 60–80 times before undergoing senescence, whereas cells from older adults divide only 10–20 times. In addition, cells from species with short life spans stop growing after fewer divisions than cells from species with longer life spans. Mouse cells divide 10–15 times in culture, but tortoise cells undergo over 100 divisions. Moreover, cells from patients with genetic premature aging syndromes (such as Werner syndrome) undergo fewer divisions in culture than cells from normal patients.

Another characteristic of aging cells is that their telomeres (the tips of linear chromosomes) become shorter. Telomeres help preserve the structural integrity of chromosomes by protecting their ends from degrading or from fusing to other chromosomes. Telomeres are created and maintained by *telomerase*—a remarkable RNA-containing enzyme that prevents the chromosomes from shrinking into oblivion. Unfortunately, normal somatic cells contain little if any telomerase. As a result, telomere length decreases by about 100 base pairs every time a normal cell divides. Telomere shortening may act as a clock that counts cell divisions and instructs the cell to stop dividing.

Could we gain perpetual youth and vitality by increasing our telomere lengths? A recent study suggests that it may be possible to reverse senescence by artificially increasing the amount of telomerase in our cells. When investigators introduced cloned telomerase genes into normal human cells in culture, telomeres lengthened by thousands of base pairs and the cells continued to divide long past their senescence point. These observations confirm that telomere length acts as a cellular clock. In addition, they suggest that some of the atrophy of tissues that accompanies old age may someday be reversed by activating telomerase genes. However, before we rush out to buy telomerase pills, we must consider a possible consequence of cellular immortality—cancer.

Although normal cells undergo senescence after a specific number of cell divisions, cancer cells do not. It is thought that cancers arise after several genetic mutations accumulate in a cell. These mutations disrupt the normal checks and balances that control cell growth and division. It therefore seems logical that cancer cells would also stop the normal aging clock—if their telomeres became shorter after each cell division, tumor cells would eventually succumb to aging and cease growth. However, if they synthesized telomerase, they would arrest the ticking of the senescence clock and become immortal. In keeping with this idea, over 90% of human tumor cells contain telomerase activity and have stable telomeres. The correlation between uncontrolled tumor-cell growth and the presence of telomerase activity is so strong that telomerase assays are being developed as diagnostic markers for cancer. Although there is currently some debate about whether the presence of telomerase is a prerequisite for or simply a consequence of cell transformation, it is possible that acquiring telomerase activity may be an important step in the development of a cancer cell. In support of this hypothesis, recent studies show that the induction of telomerase activity, when combined with the inactivation of a tumor-suppressor gene ($p16^{INK4a}$), results in cell immortalization—an essential step toward tumor development. Therefore, any attempt to increase telomerase activity in normal cells carries the risk of enhancing the development of tumors.

An attractive possibility is that telomerase may be an ideal target for anticancer drugs. Drugs that inhibit telomerase might destroy cancer cells by allowing their telomeres to shorten, thereby forcing the cells into senescence. Because most normal human cells do not express telomerase, this therapy could be specific for tumor cells and hence less toxic than most current anticancer drugs. Such anti-telomerase anticancer drugs are currently under development by Geron Corporation. Cultured tumor cells that are treated with an anti-telomerase agent lose telomeric sequences and die after about 25 cell divisions.

Before anti-telomerase drugs can be developed and used on humans, several questions must be answered. Is telomerase required by some normal human cells (such as lymphocytes and germ cells)? If so, anti-telomerase drugs may be unacceptably toxic. Could some cancer cells compensate for the loss of telomerase by using other telomere-lengthening mechanisms (such as recombination)? If so, anti-telomerase drugs may be doomed to failure. Even if we inhibit telomerase activity in tumor cells, could they undergo enough divisions before reaching senescence to still damage the host?

Will telomerase allow us to both arrest cancers *and* reverse the descent into old age? Time will tell.

References

Bodnar, A. G. et al. 1998. Extension of life-span by introduction of telomerase into normal human cells. *Science* 279: 349–52.

deLange, T. 1998. Telomeres and senescence: Ending the debate. *Science* 279: 334–35.

Kiyono, T. et al. 1998. Both Rb/$p16^{INK4a}$ inactivation and telomerase activity are required to immortalize human epithelial cells. *Nature* 396: 84–88.

Chapter Summary

1. In theory, three modes of DNA replication are possible: semiconservative, conservative, and dispersive. Though all three rely on base complementarity, semiconservative replication is the most straightforward and was predicted.
2. In 1958, Meselson and Stahl resolved this problem in favor of semiconservative replication in *E. coli*, showing that newly synthesized DNA consists of one old strand and one new strand. Taylor, Woods, and Hughes used root tips of the broad bean to demonstrate semiconservative replication in eukaryotes.
3. During the same period, Kornberg isolated the enzyme DNA polymerase I from *E. coli* and showed that it is capable of directing *in vitro* DNA synthesis, provided that a template and precursor nucleoside triphosphates are supplied.
4. The subsequent discovery of the *polA1* mutant strain of *E. coli*, capable of DNA replication in spite of its lack of polymerase I activity, cast doubt on this enzyme's *in vivo* replicative function. DNA polymerases II and III were then isolated. Polymerase III has been identified as the enzyme responsible for DNA replication *in vivo*.
5. During the process of DNA synthesis, the double helix unwinds, forming a replication fork where synthesis begins. Proteins stabilize the unwound helix and assist in relaxing the coiling tension created ahead of the replication activity.
6. Synthesis is initiated at specific sites along each template strand by the enzyme primase, which results in a short segment of RNA that provides a suitable 3′-end, upon which DNA polymerase III can begin polymerization.
7. Because of the antiparallel nature of the double helix, polymerase III synthesizes DNA continuously on the leading strand in a 5′-to-3′ direction. On the opposite strand, called the lagging strand, synthesis results in short Okazaki fragments that are later joined by DNA ligase.
8. DNA polymerase I removes and replaces the RNA primer with DNA, which is joined to the adjacent polynucleotide by DNA ligase.
9. The isolation of numerous phage and bacterial mutant genes affecting many of the molecules involved in the replication of DNA has helped define the complex genetic control of the entire process.
10. DNA replication in eukaryotes is similar to but more complex than replication in prokaryotes. Multiple replication origins exist, and multiple forms of DNA polymerase direct DNA synthesis.
11. Replication at the telomeres of linear molecules poses a special problem in eukaryotes that can be solved by a unique RNA-containing enzyme called telomerase.
12. Homologous recombination between genetic molecules relies on a series of enzymes that can cut, realign, and reseal DNA strands. The phenomenon of gene conversion may be best explained in terms of mismatched repair synthesis during these exchanges.

Key Terms

antiparallel, 240
autonomously replicating sequences (ARSs), 243
autoradiography, 234
bidirectional replication, 234
biologically active DNA, 237
chain elongation, 237
chi form (χ), 245
conditional mutation, 242
conservative replication, 232
continuous DNA synthesis, 240
core enzyme, 241
discontinuous DNA synthesis, 240
dispersive replication, 233
DnaA protein, 239
DnaB protein, 239
DnaC protein, 239
DNA gyrase, 239
DNA ligase, 240
DNA polymerase I, 236
DNA polymerase II, 238
DNA polymerase III, 238
DNA topoisomerase, 239
exonuclease activity, 238
exonuclease proofreading, 241
5′-to-3′ direction, 237
gene conversion, 246
general recombination, 245
genetic recombination, 245
Holliday structure, 245
helicase, 239
heteroduplex DNA molecules, 245
holoenzyme, 238
homologous recombination, 245
lagging DNA strand, 240
leading DNA strand, 240
9mer, 239
nucleosome, 243
Okazaki fragment, 240
origin of replication, 234
origin recognition complex, 243
polymerase switching, 243
primase, 240
primer, 238
processivity, 243
RecA protein, 245
replication fork, 234
replicon, 234
replisome, 238
sedimentation equilibrium centrifugation, 233
semiconservative replication, 232
single-stranded binding proteins (SSBPs), 239
telomerase, 244
telomere, 244
temperature-sensitive mutation, 242
13mer, 239
3′-to-5′ exonuclease activity, 241
unidirectional replication, 234

INSIGHTS AND SOLUTIONS

1. Predict the theoretical results of conservative and dispersive models of DNA synthesis using the conditions of the Meselson-Stahl experiment. Follow the results through two generations of replication after the cells have been shifted to an ^{14}N-containing medium, using the following sedimentation pattern:

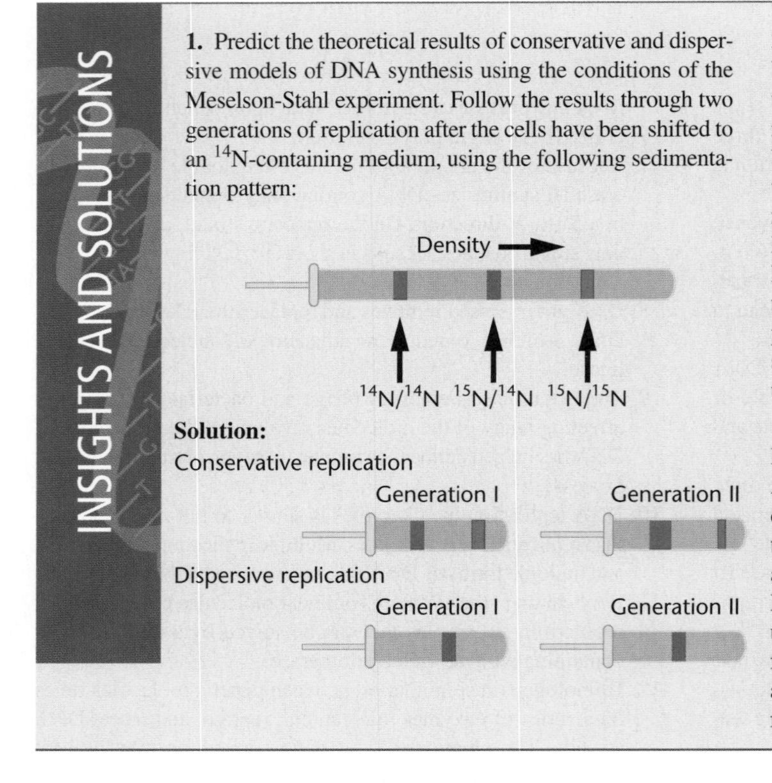

Solution:

Conservative replication

Generation I Generation II

Dispersive replication

Generation I Generation II

2. Mutations in the *dnaA* gene of *E. coli* are lethal and can be studied only after the isolation of conditional, temperature-sensitive mutations. Such mutant strains grow nicely and replicate their DNA at the permissive temperature of 18°C, but they do not grow or replicate their DNA at the restrictive temperature of 37°C. Two observations helped determine the function of the *dnaA* gene product. First, *in vitro* studies using DNA templates that have been unwound do not require the DnaA protein. Second, if intact cells are grown at 18°C and then shifted to 37°C, DNA synthesis continues at this temperature until one round of replication is completed, and DNA synthesis stops. What do these observations suggest about the role of the *dnaA* gene product?

Solution: These observations suggest that *in vivo,* the DnaA protein is essential to the initiation of DNA synthesis. At 18°C (the permissive temperature), the mutation is not expressed and DNA synthesis begins. Following the shift to the restrictive temperature, the DNA synthesis that was already initiated continues, but no new synthesis can begin. Because the DnaA protein is not required to synthesize unwound DNA, this observation suggests that the protein functions during initiation by interacting with the intact helix and somehow facilitating the localized denaturing necessary for synthesis to proceed. In fact, both conclusions are valid.

Problems and Discussion Questions

1. Compare conservative, semiconservative, and dispersive modes of DNA replication.
2. Describe the role of ^{15}N in the Meselson-Stahl experiment.
3. In the Meselson-Stahl experiment, which of the three modes of replication could be ruled out after one round of replication? After two rounds?
4. Predict the results of the experiment by Taylor, Woods, and Hughes if replication were (a) conservative and (b) dispersive.
5. Reconsider Problem 33 in Chapter 10. In the model you proposed, could the molecule be replicated semiconservatively? Why? Would other modes of replication work?
6. What are the requirements for *in vitro* synthesis of DNA under the direction of DNA polymerase I?
7. In Kornberg's initial experiments, it is rumored that he grew *E. coli* in Anheuser-Busch beer vats. (He was working at Washington University in St. Louis.) Why do you think this was helpful to the experiment?
8. How did Kornberg assess the fidelity of DNA by polymerase I in copying DNA template?
9. Which characteristics of DNA polymerase I raised doubts that its *in vivo* function is the synthesis of DNA leading to complete replication?
10. One of Kornberg's tests demonstrating that DNA polymerase I is the enzyme used *in vivo* involved the replication of $\phi X174$ DNA. What is meant by "biologically active" DNA?
11. What was the significance of the *polA1* mutation?
12. Summarize and compare the properties of DNA polymerase I, II, and III.
13. List and describe the function of the 10 subunits constituting DNA polymerase III. Distinguish between the holoenzyme and the core enzyme.
14. Distinguish between (a) unidirectional and bidirectional synthesis, and (b) continuous and discontinuous synthesis of DNA.
15. List the proteins that unwind DNA during *in vivo* DNA synthesis. How do they function?
16. Define and indicate the significance of (a) Okazaki fragments, (b) DNA ligase, and (c) primer RNA during DNA replication.
17. Outline the current model for DNA synthesis.
18. Why is DNA synthesis expected to be more complex in eukaryotes than in bacteria? How is DNA synthesis similar in the two types of organisms?
19. If the analysis of DNA from two different microorganisms demonstrated very similar base compositions, are the DNA sequences of the two organisms also nearly identical?
20. Suppose that *E. coli* synthesizes DNA at a rate of 100,000 nucleotides per minute and takes 40 minutes to replicate its chromosome. (a) How many base pairs are present in the entire *E. coli* chromosomes? (b) What is the physical length of the chromosome in its helical configuration—that is, what is the circumference of the chromosome if it were opened into a circle?
21. Several temperature-sensitive mutant strains of *E. coli* display the characteristics listed below. Predict what enzyme or function is being affected by each mutation.
    (a) Newly synthesized DNA contains many mismatched base pairs.
    (b) Okazaki fragments accumulate, and DNA synthesis is never completed.
    (c) No initiation occurs.
    (d) Synthesis is very slow.
    (e) Supercoiled strands remain after replication, which is never completed.
22. Define gene conversion, and describe how this phenomenon is related to genetic recombination.
23. Many of the gene products involved in DNA synthesis were initially defined by studying mutant *E. coli* strains that could not synthesize DNA. (a) The *dnaE* gene encodes the α subunit of

DNA polymerase III. What effect is expected from a mutation in this gene? How could the mutant strain be maintained? (b) The *dnaQ* gene encodes the ε subunit of DNA polymerase. What effect is expected from a mutation in this gene?

24. In 1994, telomerase activity was discovered in human cancer cell lines. Although telomerase is not active in human somatic tissue, this discovery indicated that humans do contain the genes for telomerase proteins and telomerase RNA. Since inappropriate activation of telomerase can cause cancer, why do you think the genes coding for this enzyme have been maintained in the human genome throughout evolution? Are there any types of human body cells where telomerase activation would be advantageous or even necessary? Explain.
25. The genome of *D. melanogaster* consists of approximately 1.6×10^8 base pairs. DNA synthesis occurs at a rate of 30 base pairs per second. In the early embryo, the entire genome is replicated in five minutes. How many bidirectional origins of synthesis are required to accomplish this feat?
26. Assume a hypothetical organism in which DNA replication is conservative. Design an experiment similar to that of Taylor, Woods, and Hughes that will unequivocally establish this fact. Using the format established in Figure 11–5, draw sister chromatids and illustrate the expected results establishing this mode of replication.
27. DNA polymerases in all organisms add only 5′-nucleotides to the 3′-end of a growing DNA strand, never to the 5′-end. One possible reason for this is the fact that most DNA polymerases have a proofreading function that would not be *energetically* possible if DNA synthesis occurred in the 3′-to-5′ direction. (a) Sketch the reaction that DNA polymerase would have to catalyze if DNA synthesis occurred in the 3′-to-5′ direction. (b) Consider the information in your sketch and speculate as to why proofreading would be problematic.
28. An alien organism was investigated. It displayed characteristics of eukaryotes. When DNA replication was examined, two unique features were apparent: (1) no Okazaki fragments were observed; and (2) there was a telomere problem (i.e., telomeres shortened) but only on one end of the chromosome. Create a model of DNA that is consistent with both of these observations.
29. Assume that the sequence of bases given below is present on one nucleotide chain of a DNA duplex and that the chain has opened up at a replication fork. Synthesis of an RNA primer occurs on this template starting at the base that is underlined. (a) If the RNA primer consists of eight nucleotides, what is its base sequence? (b) In the intact RNA primer, which nucleotide has a free 3′-OH terminus?

    3′.....GGCTACC<u>T</u>GGATTCA.....5′

30. Given the diagram at the top of the next column, assume that the phase G1 chromosome on the left underwent one round of replication in ^{3}H-thymidine and the metaphase chromosome on the right had both chromatids labeled. Which of the replicative models (conservative, dispersive, semiconservative) could be eliminated by this observation?

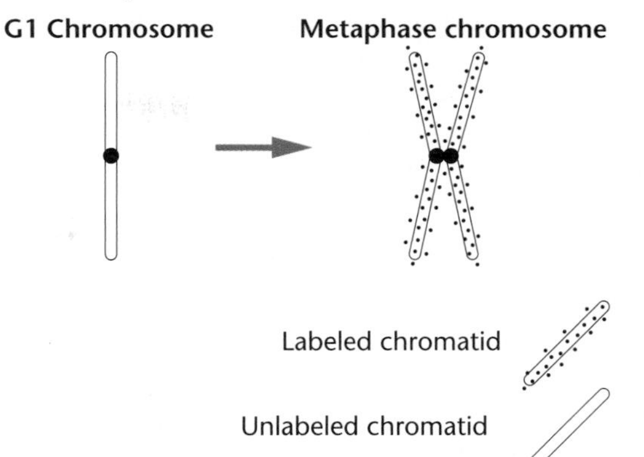

31. Consider the figure of a dinucleotide below. (a) Is it DNA or RNA? (b) Is the arrow closest to the 5′ or the 3′ end? (c) Suppose that the molecule was cleaved with the enzyme spleen diesterase, which breaks the covalent bond connecting the phosphate to C-5′. After cleavage, to which nucleoside is the phosphate now attached, deoxyadenosine or deoxythymidine?

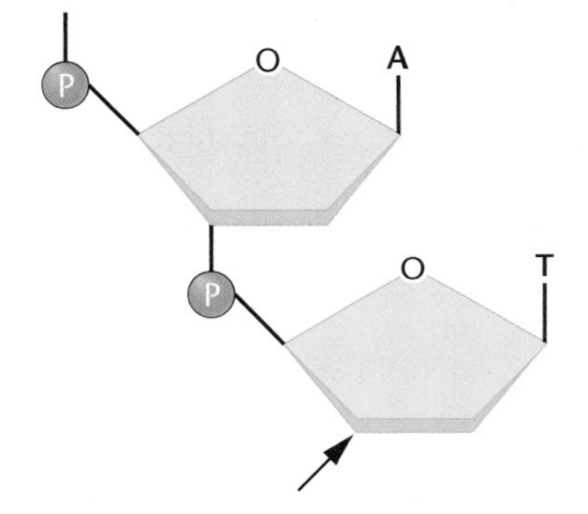

32. DNA is allowed to replicate in moderately radioactive ^{3}H-thymidine for several minutes and is then switched to a highly radioactive medium for several more minutes. Synthesis is stopped and the DNA is subjected to autoradiography and electron microscopy. Interpret as much as you can regarding DNA replication from the drawing of the electron micrograph presented here.

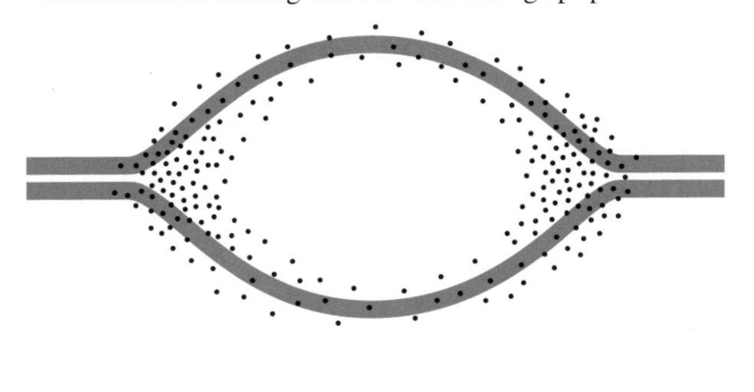

Selected Readings

Blackburn, E. H. 1991. Structure and function of telomeres. *Nature* 350: 569–572.

Bodnar, A. G. et al. 1998. Extension of life-span by introduction of telomerase into normal human cells. *Science* 279: 349–52.

Bramhill, D., and Kornberg, A. 1988. A model for initiation at origins of DNA replication. *Cell* 54: 915–18.

DeLucia, P., and Cairns, J. 1969. Isolation of an *E. coli* strain with a mutation affecting DNA polymerase. *Nature* 224: 1164–66.

Denhardt, D. T., and Faust, E. A. 1985. Eukaryotic DNA replication. *Bioessays* 2: 148–53.

Diffley, J. F. X. 1996. Once and only once upon a time: Specifying and regulating origins of DNA replication in eukaryotic cells. *Genes and Dev.* 10: 2819–30.

Dressler, D., and Potter, H. 1982. Molecular mechanisms in genetic recombination. *Annu. Rev. Biochem.* 51: 727–61.

Gilbert, D. M. 2001. Making sense of eukaryotic DNA replication origins. *Science* 294: 96–100.

Greider, C. W. 1996. Telomeres, telomerase, and cancer. *Sci. Am.* (Feb.) 274: 92–97.

———. 1998. Telomerase activity, cell proliferation, and cancer. *Proc. Natl. Acad. Sci. (USA)* 95: 90–92.

Greider, C. W., and Blackburn, E. H. 1989. A telomeric sequence in the RNA of *Tetrahymena* telomerase required for telomere repeat synthesis. *Nature* 337: 331–36.

Herendeen, D. R., and Kelly, T. J. 1996. DNA polymerase III: Running rings around the fork. *Cell* 84: 5–8.

Hindges, R., and Hÿbscher, U. 1997. DNA polymerase essential for DNA transactions *Biol. Chem.* 378: 345–62.

Holliday, R. 1964. A mechanism for gene conversion in fungi. *Genet. Res.* 5: 282–304.

Holmes, F. L. 2001. *Meselson, Stahl, and Replication of DNA: A History of the "Most Beautiful Experiment in Biology."* New Haven: Yale Univ. Press.

Huberman, J. C. 1987. Eukaryotic DNA replication: A complex picture partially clarified. *Cell* 48: 7–8.

Kornberg, A. 1960. Biological synthesis of DNA. *Science* 131: 1503–8.

———. 1974. *DNA Synthesis*. New York: W. H. Freeman.

———. 1979. Aspects of DNA replication. *Cold Spring Harbor Symp. Quant. Biol.* 43: 1–10.

Kornberg, A., and Baker, T. A. 1992. *DNA Replication*. 2nd ed. New York: W. H. Freeman.

Lehman, I. R. 1974. DNA ligase: Structure, mechanism, and function. *Science* 186: 790–97.

Lindegren, C. C. 1953. Gene conversion in *Saccharomyces*. *J. Genet.* 51: 625–37.

Lodish, H. et al. 1999. *Molecular Cell Biology*. 4th ed. New York: W. H. Freeman.

Meselson, M., and Stahl, F. W. 1958. The replication of DNA in *Escherichia coli*. *Proc. Natl. Acad. Sci. (USA)* 44: 671–82.

Mitchell, M. B. 1955. Aberrant recombination of pyridoxine mutants of *Neurospora*. *Proc. Natl. Acad. Sci. (USA)* 41: 215–20.

Radding, C. M. 1978. Genetic recombination: Strand transfer and mismatch repair. *Annu. Rev. Biochem.* 47: 847–80.

Radman, M., and Wagner, R. 1988. The high fidelity of DNA duplication. *Sci. Am.* (Aug.) 259: 40–46.

Stahl, F. W. 1979. *Genetic Recombination: Thinking About It in Phage and Fungi*. New York: W. H. Freeman.

———. 1987. Genetic recombination. *Sci. Am.* (Feb.) 256: 90–101.

Taylor, J. H., Woods, P. S., and Hughes, W. C. 1957. The organization and duplication of chromosomes revealed by autoradiographic studies using tritium-labeled thymidine. *Proc. Natl. Acad. Sci. (USA)* 48: 122–28.

Thommes, P., and Hubscher, U. 1992. Eukaryotic DNA helicases: Essential enzymes for DNA transactions. *Chromosoma* 101: 467–73.

Wang, J. C. 1982. DNA topoisomerases. *Sci. Am.* (July) 247: 94–108.

Watson, J. D. et al. 1987. *Molecular Biology of the Gene, Vol. 1. General Principles*. 4th ed. Menlo Park: Benjamin/Cummings.

Whitehouse, H. L. K. 1982. *Genetic Recombination: Understanding the Mechanisms*. New York: Wiley.

Zyskind, J. W., and Smith, D. W. 1986. The bacterial origin of replication, *oriC*. *Cell* 46: 489–90.

CHAPTER 12

CHAPTER CONCEPTS

A chromatin fiber viewed using a scanning transmission electron microscope (STEM). *(Science VU/BMRL/Visuals Unlimited)*

Chromosome Structure and DNA Sequence Organization

Once geneticists understood that DNA houses genetic information, it became very important to determine how DNA is organized into genes and how these basic units of genetic function are organized into chromosomes. In short, the major question had to do with how the genetic material was organized as it made up the genome of organisms. There has been much interest in this question because knowledge of the organization of the genetic material and associated molecules is important to the understanding of many other areas of genetics. For example, the way in which the genetic information is stored, expressed, and regulated must be related to the molecular organization of the genetic molecule, DNA. How genomic organization varies in different organisms—from viruses to bacteria to eukaryotes—will undoubtedly provide a better understanding of the evolution of organisms on Earth.

In this chapter, we focus on the various ways DNA is organized into chromosomes. We first survey what we know about chromosomes in viruses and bacteria and in two cellular organelles: mitochondria and chloroplasts. Then, we will examine the large specialized structures called polytene and lampbrush chromosomes. In the second half of the chapter, we discuss how eukaryotic chromosomes are organized. For example, how is DNA complexed with proteins to form chromatin, and how are the chromatin fibers, characteristic of interphase, condensed into chromosome structures visible during mitosis and meiosis? We conclude the chapter by examining aspects of DNA sequence organization characteristic of eukaryotic genomes.

HOW DO WE KNOW WHAT WE KNOW?

IN THIS CHAPTER, WE WILL FOCUS ON chromosome structure and the way DNA is organized within chromosomes. As you study this topic, you should try to answer several fundamental questions:

1. How have we come to know so much about chromosomes?
2. How have we actually seen the organization of DNA within chromosomes, given that DNA has a diameter of only 2 nm?
3. How can we discern the organization of DNA sequence motifs within a molecule of DNA?
4. How do we know that mitochondria and chloroplasts contain DNA?
5. How do we know that puffs are areas of active transcription in polytene chromosomes? ■

12.1 Viral and Bacterial Chromosomes Are Relatively Simple DNA Molecules

The chromosomes of viruses and bacteria are much less complicated than those in eukaryotes. They usually consist of a single nucleic acid molecule, unlike the multiple chromosomes comprising the genome of higher forms. The chromosomes are largely devoid of associated proteins and contain relatively little genetic information. These characteristics have greatly simplified analysis, and we now have a fairly comprehensive view of the structure of viral and bacterial chromosomes.

The chromosomes of viruses consist of a nucleic acid molecule—either DNA or RNA—that can be either single- or double-stranded. They can exist as circular structures (closed loops), or they can take the form of linear molecules. For example, the single-stranded DNA of the **ϕx174 bacteriophage** and the double-stranded DNA of the **polyoma virus** are closed loops housed within the protein coat of the mature viruses. The **bacteriophage lambda (λ)**, on the other hand, possesses a linear double-stranded DNA molecule prior to infection, which closes to form a ring upon its infection of the host cell. Still other viruses, such as the T-even series of bacteriophages, have linear double-stranded chromosomes of DNA that do not form circles inside the bacterial host. Thus, circularity is not an absolute requirement for replication in some viruses.

Viral nucleic acid molecules have been seen with the electron microscope. Figure 12–1 shows a mature bacteriophage λ with its double-stranded DNA molecule in the circular configuration. One constant feature shared by viruses, bacteria, and eukaryotic cells is the ability to package an exceedingly long DNA molecule into a relatively small volume. In λ, the DNA is 17 μm long and must fit into the phage head, which is less than 0.1 μm on any side.

Table 12.1 compares the length of the chromosomes of several viruses to the size of their head structure. In each case, a similar packaging feat must be accomplished. Compare the dimensions given for phage T2 with the micrograph of both the DNA and the viral particle shown in Figure 12–2. Seldom does the space available in the head of a virus exceed the chromosome volume by more than a factor of two. In many cases, almost all of the space is filled, indicating nearly perfect packing. Once packed within the head, the genetic material is functionally inert until it is released into a host cell.

Bacterial chromosomes are also relatively simple in form. They always consist of a double-stranded DNA molecule, compacted into a structure sometimes referred to as the **nucleoid**. *Escherichia coli*, the most extensively studied bacterium, has a large circular chromosome measuring approximately 1200 μm (1.2 mm) in length. When the cell is gently lysed and the chromosome released, it can be visualized under the electron microscope (Figure 12–3).

DNA in bacterial chromosomes is associated with several types of **DNA-binding proteins**. Two, called **HU and H proteins**, are small but abundant in the cell and contain a high percentage of positively charged amino acids that can bond ionically to the negative charges of the phosphate groups in DNA. These proteins are structurally similar to molecules called histones that are associated with eukaryotic DNA. (We will discuss histone organization later in this chapter.) Unlike the tightly packed chromosome present in the head of a virus, the bacterial chro-

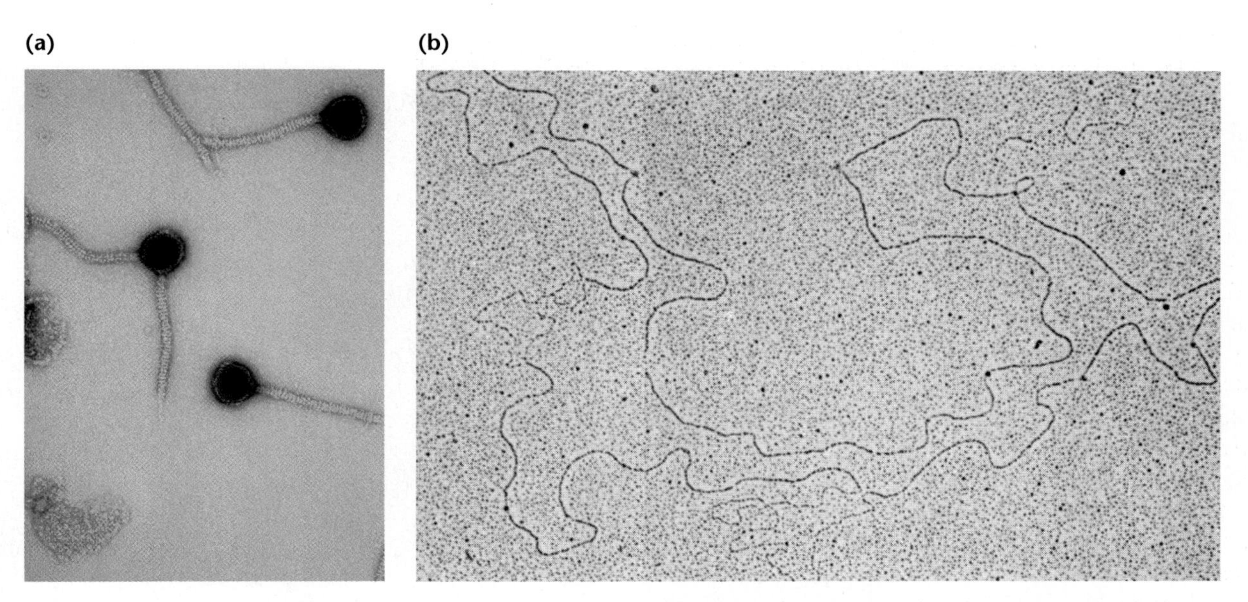

FIGURE 12–1 Electron micrographs of phage λ (left) and the DNA that was isolated from it (right). The chromosome is 17 μm long. Note that the phages are magnified about five times more than the DNA. *(Left: Dr. M. W. Wurtz/Biozentrum, University of Basel/Science Photo Library/Photo Researchers, Inc.; right: Science Source/Photo Researchers, Inc.)*

TABLE 12.1 The Genetic Material of Representative Viruses and Bacteria

		Nucleic Acid			
	Organism	*Type*	*SS or DS**	*Length (μm)*	*Overall Size of Viral Head or Bacteria (μm)*
Viruses	φX174	DNA	SS	2.0	0.025 × 0.025
	Tobacco mosaic virus	RNA	SS	3.3	0.30 × 0.02
	Phage λ	DNA	DS	17.0	0.07 × 0.07
	T2 phage	DNA	DS	52.0	0.07 × 0.10
Bacteria	*Haemophilus influenzae*	DNA	DS	832.0	1.00 × 0.30
	Escherichia coli	DNA	DS	1200.0	2.00 × 0.50

*SS = single-stranded; DS = double-stranded.

mosome is *not* functionally inert. Despite its somewhat compacted condition in the bacterial cell, the chromosome can be readily replicated and transcribed.

12.2 Mitochondria and Chloroplasts Contain DNA Similar to Bacteria and Viruses

Numerous studies demonstrate that both **mitochondria** and **chloroplasts** contain their own DNA and genetic system for expressing this information. This was first suggested by the discovery of mutations in yeast, other fungi, and plants that produce altered phenotypes that can be linked to these organelles. Transmission of such traits was found to occur through the cytoplasm rather than through chromosomes in the nucleus. Because both mitochondria and chloroplasts are inherited through the maternal cytoplasm in most organisms, these observations suggested that the organelles house their own DNA, which, when mutated, may be responsible for the altered phenotypes.

Thus, geneticists set out to look for more direct evidence of DNA in these organelles. Electron microscopists not only documented the presence of DNA in both organelles, they also saw DNA in a form quite unlike that seen in the nucleus of the eukaryotic cells that house these organelles. This DNA looked remarkably similar to that seen in viruses and bacteria. This similarity, along with other observations, led to the idea that mitochondria and chloroplasts arose independently more than a billion years ago from free-living, prokaryote-like organisms that possessed the abililty to undergo aerobic respiration (mitochondria) or photosynthesis (choroplasts). This theory, called the **endosymbiotic hypothesis**, was championed by Lynn Margulis and others. They proposed that the prokaryotes were engulfed by larger primitive eukaryotic cells, which lacked these bioenergetic functions. A symbiotic relationship developed whereby the prokaryotic organisms eventually lost their ability to function independently, while the

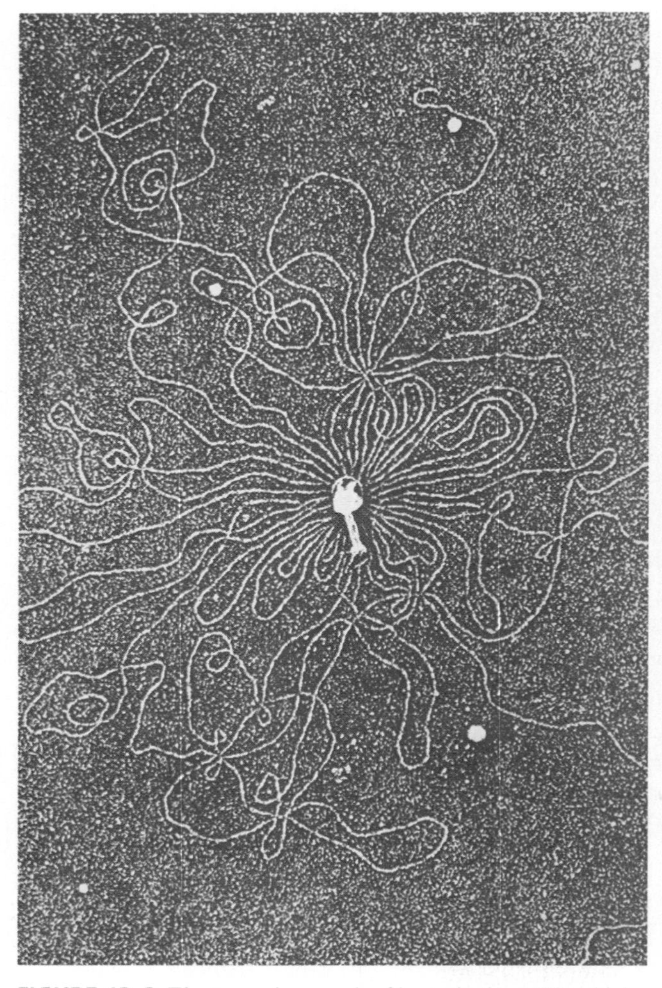

FIGURE 12–2 Electron micrograph of bacteriophage T2, which has had its DNA released by osmotic shock. The chromosome is 52 μm long. *(Science Source/Photo Researchers, Inc.)*

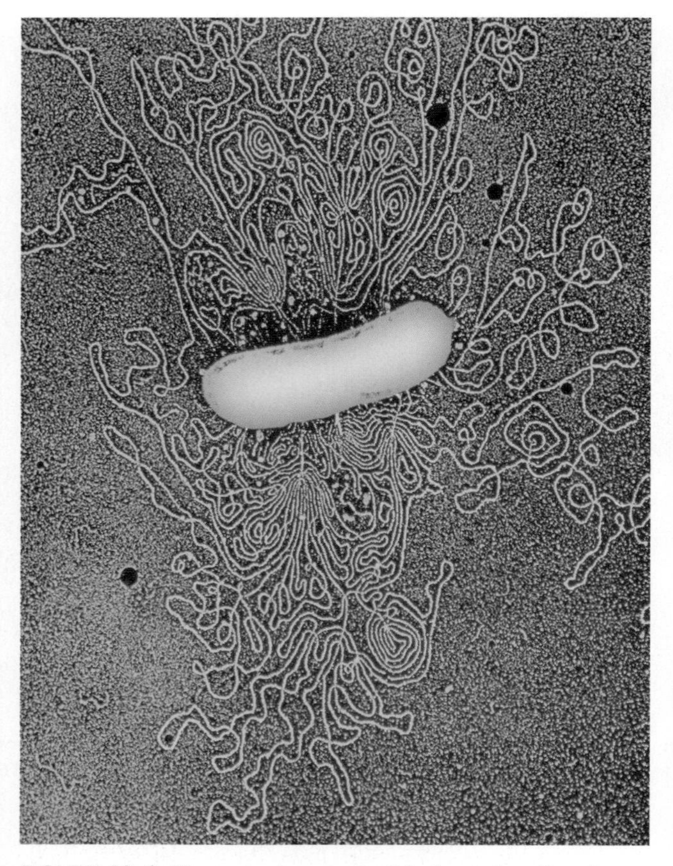

FIGURE 12–3 Electron micrograph of the bacterium *E. coli*, which has had its DNA released by osmotic shock. The chromosome is 1200 μm long. *(Dr. Gopal Murti/Science Photo Library/Photo Researchers, Inc.)*

eukaryotic host cells gained the ability to either respire aerobically or undergo photosynthesis. Although many questions remain unanswered, the basic tenets of this theory are widely accepted.

In the following sections, we shall explore what is known about the DNA found in these cellular organelles and examine what is known about the genes present on these DNA molecules.

Molecular Organization and Gene Products of Mitochondrial DNA

Extensive information is now available about the molecular aspects and gene products of **mitochondrial DNA (mtDNA)**. In most eukaryotes, mtDNA is a double-stranded closed circle (Figure 12–4) that replicates semiconservatively and is free of the chromosomal proteins characteristic of eukaryotic DNA. In size, mtDNA differs greatly among organisms, as demonstrated in Table 12.2. In a variety of animals, including humans, mtDNA consists of about 16,000–18,000 bp (16–18 kb). However, yeast (*Saccharomyces*) contains 75 kb, while up to 367 kb may be present in plant mitochondria such as in *Arabidopsis*. Vertebrates have 5–10

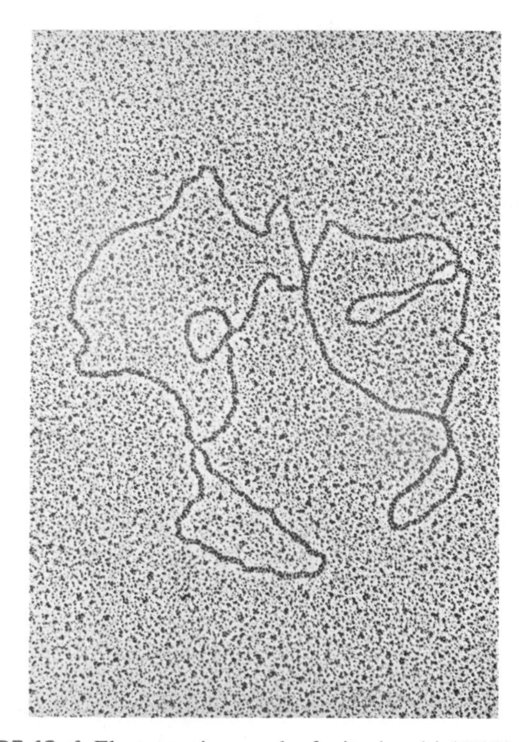

FIGURE 12–4 Electron micrograph of mitochondrial DNA (mtDNA) derived from *Xenopus laevis*. *(Dr. Don W. Fawcett/Kahri/Dawid/Science Source/Photo Researchers, Inc.)*

TABLE 12.2 The Size of mtDNA in Different Organisms

Organism	*Size (kb)*
Human	16.6
Mouse	16.2
Xenopus (frog)	18.4
Drosophila (fruit fly)	18.4
Saccharomyces (yeast)	75.0
Pisum sativum (pea)	110.0
Arabidopsis (mustard plant)	367.0

such DNA molecules per organelle, while plants have 20–40 copies per organelle.

We can say several things about mtDNA. With only rare exceptions, introns, the noncoding regions of genes, do not appear to be present in mitochondrial genes and there are few or no gene repetitions. Expression of mitochondrial genes uses several modifications of the otherwise standard genetic code. Replication of mtDNA is dependent on enzymes encoded by nuclear DNA. In humans, mtDNA encodes 2 ribosomal RNAs (rRNAs), 22 transfer RNAs (tRNAs), as well as numerous polypeptides essential to the cellular respiratory functions of the organelles. In almost every case, the polypeptides are part of multichain proteins, and the other polypeptides of each protein are encoded in the nucleus, synthesized in the cytoplasm, and then transported into the organelle. Thus, the protein-synthesizing apparatus and the molecular components for cellular respiration are jointly derived from nuclear and mitochondrial genes.

As expected then, ribosomes found in the organelle are different from those present in the neighboring cytoplasm. Table 12.3 shows that mitochondrial ribosomes of different species vary considerably in their sedimentation coefficients, ranging from 55*S* to 80*S*. Since many are closer in their coefficient to bacterial counterparts than eukaryotes, this also supports the endosymbiont hypothesis.

The nuclear-coded gene products essential to biological activity in mitochondria include DNA and RNA polymerases,

TABLE 12.3 Sedimentation Coefficients of Mitochondrial Ribosomes

Kingdom	*Examples*	*Sedimentation Coefficient (S)*
Animalia	Vertebrates	55–60
	Insects	60–71
Protista	*Euglena*	71
	Tetrahymena	80
Fungi	*Neurospora*	73–80
	Saccharomyces	72–80
Plantae	Maize	77

initiation and elongation factors essential for translation, ribosomal proteins, aminoacyl tRNA synthetases, and several tRNA species. These imported components are distinct from their cytoplasmic counterparts, even though both sets are coded by nuclear genes. For example, the synthetase enzymes essential for charging mitochondrial tRNA molecules (a process essential to translation) show a distinct affinity for the mitochondrial tRNA species as compared to the cytoplasmic tRNAs. Similar affinity has been shown for the initiation and elongation factors. Furthermore, while bacterial and nuclear RNA polymerases are known to be composed of numerous subunits, the mitochondrial variety consists of only one polypeptide chain. This polymerase is generally susceptible to antibiotics that inhibit bacterial RNA synthesis but not to eukaryotic inhibitors. The contributions of nuclear and mitochondrial gene products are shown in Figure 12–5.

Molecular Organization and Gene Products of Chloroplast DNA

Chloroplasts, like mitochondria, contain an autonomous genetic system distinct from that found in the nucleus and cytoplasm. This system includes DNA as a source of genetic information and a complete protein-synthesizing apparatus. Also similar to mitochondria, the molecular components of

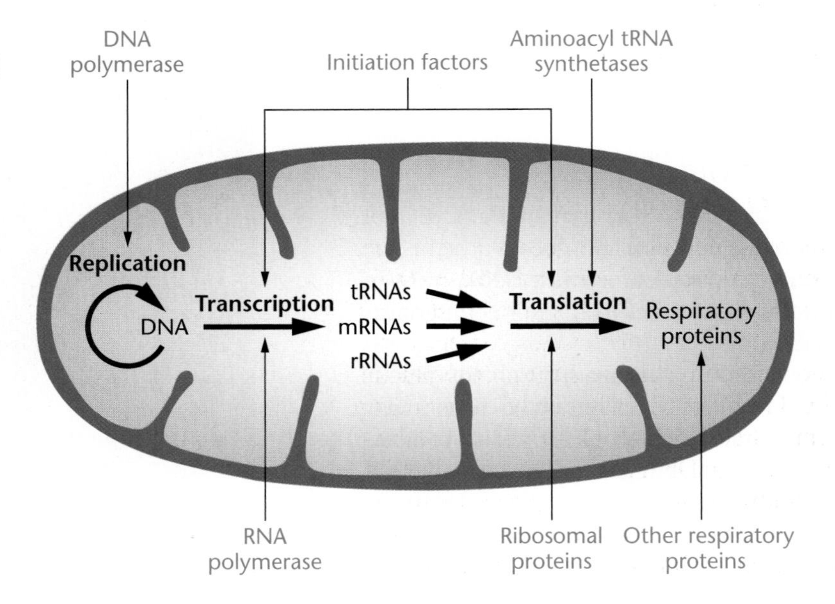

FIGURE 12–5 Gene products that are essential to mitochondrial function. Those shown entering the organelle are derived from the cytoplasm and encoded by the nucleus.

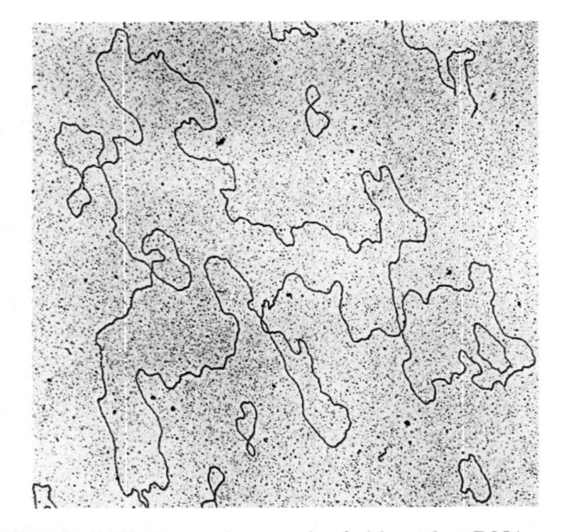

FIGURE 12–6 Electron micrograph of chloroplast DNA obtained from lettuce. *(Dr. Richard D. Kolodnar/Dana-Farber Cancer Institute)*

the chloroplast translation apparatus are jointly derived from both nuclear and organelle genetic information. **Chloroplast DNA (cpDNA)**, shown in Figure 12–6, is much larger than mitochondrial DNA—usually 100–225 kb in length. Nevertheless, it also shares similarities to DNA found in prokaryotic cells. It is circular, double-stranded, replicated semiconservatively, and free of the associated proteins characteristic of eukaryotic DNA. Compared with nuclear DNA from the same organism, it invariably shows a different buoyant density and base composition.

In the green alga *Chlamydomonas*, there are about 75 copies of the chloroplast DNA molecule per organelle, and each copy of DNA is 195,000 bp (195 kb) in length. In higher plants such as the sweet pea, multiple copies of the DNA molecule are present in each organelle, but the molecule is considerably smaller than that in *Chlamydomonas,* at 134 kb. Genetic recombination between the multiple copies of DNA within chloroplasts has been documented in *Chlamydomonas*.

Some chloroplast gene products function during translation. In a variety of higher plants (beans, lettuce, spinach, maize, and oats), two sets of the genes coding for the ribosomal RNAs—5*S*, 16*S*, and 23*S* rRNA—are present. Additionally, chloroplast DNA codes for at least 25 tRNA species and a number of ribosomal proteins specific to the chloroplast ribosomes. These ribosomes have a sedimentation coefficient slightly less than 70*S*, similar to that of bacteria. Even though chloroplast ribosomal proteins are encoded by both nuclear and chloroplast DNA, most, if not all, such proteins are distinct from their counterparts in cytoplasmic ribosomes.

Still other chloroplast genes specific to the photosynthetic function have been identified. Mutations in these genes may inactivate photosynthesis in chloroplasts with this mutation. One of the major photosynthetic enzymes is ribulose-1-5-bisphosphate carboxylase (RuBP). Interestingly, the small subunit of this enzyme is encoded by a nuclear gene, whereas the large subunit is encoded by cpDNA.

The great increase in size in cpDNA compared to mtDNA can be partly explained by an increased number of genes. However, the biggest difference appears to be due to the presence of long noncoding sequences of DNA as well as duplications of many DNA sequences. This observation is indicative of the independent evolution that occurred in chloroplasts and mitochondria following their initial invasion of a primitive eukaryote-like cell.

12.3 Specialized Chromosomes Reveal Variations in the Organization of DNA

We now consider two cases of genetic organization that demonstrate the specialized forms that chromosomes can take. Both types, polytene chromosomes and lampbrush chromosomes, are so large that their organization was discerned using light microscopy long before we understood how mitotic chromosomes form from interphase chromatin. The study of these chromosomes provided many of our initial insights into the arrangement and function of the genetic information.

Polytene Chromosomes

Giant **polytene chromosomes** are found in various tissues (salivary, midgut, rectal, and malpighian excretory tubules) in the larvae of some flies and in several species of protozoans and plants. Such structures were first observed by E. G. Balbiani in 1881. The vast amount of information obtained from studies of these genetic structures provided a model system for subsequent investigations of chromosomes. What is particularly intriguing about polytene chromosomes is that they can be seen in the nuclei of interphase cells.

Each polytene chromosome is 200–600 μm long, and when they are observed under the light microscope, they reveal a linear series of alternating bands and interbands (Figure 12–7). The banding pattern is distinctive for each chromosome in any given species. Individual bands are

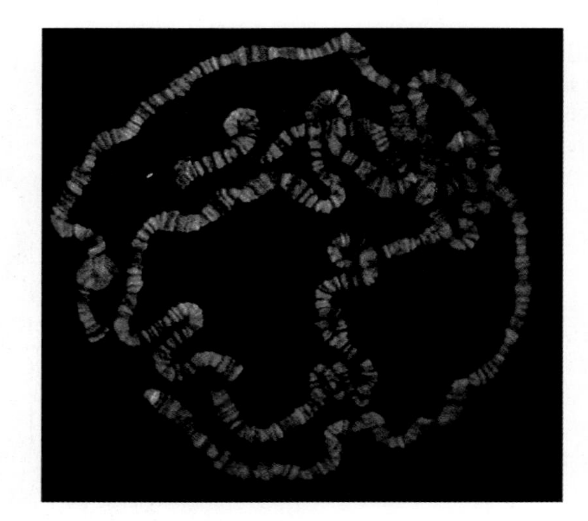

FIGURE 12–7 Polytene chromosomes derived from larval salivary gland cells of *Drosophila*. *(Image courtesy of Brian Harmon and John Sedat University of California San Francisco)*

sometimes called **chromomeres**, a generalized term describing lateral condensations of material along the axis of a chromosome.

Extensive study using electron microscopy and radioactive tracers led to an explanation for the unusual appearance of these chromosomes. First, polytene chromosomes represent paired homologs. This is highly unusual in most organisms since paired homologs are present in somatic cells, where chromosomal material is normally dispersed as chromatin and homologs are not paired. Second, their large size and distinctiveness result from the many DNA strands that compose them. The DNA of these paired homologs undergoes many rounds of replication, *but without strand separation or cytoplasmic division*. As replication proceeds, chromosomes contain 1000–5000 DNA strands that remain in precise parallel alignment with one another. Apparently, the parallel register of so many DNA strands gives rise to the distinctive band pattern along the axis of the chromosome.

The presence of bands on polytene chromosomes was initially interpreted as the visible manifestation of individual genes. The discovery that the strands present in bands undergo localized uncoiling during genetic activity further strengthened this view. Each such uncoiling event results in what is called a **puff** because of its appearance (Figure 12–8). That puffs are visible manifestations of gene activity (transcription that produces RNA) is evidenced by their high rate of incorporation of radioactively labeled RNA precursors, as assayed by autoradiography. Bands that are not extended into puffs incorporate fewer radioactive precursors or none at all.

The study of bands during development in insects, such as *Drosophila* and the midge fly *Chironomus*, reveals differential gene activity. A characteristic pattern of band formation, which is equated with gene activation, is observed as development proceeds. Despite attempts to resolve the issue, it is not yet clear how many genes are contained in each band; however, we do know that a band can contain up to 10^7 bp of DNA, certainly enough DNA to encode 50–100 average-sized genes.

Lampbrush Chromosomes

Another specialized chromosome that has given us insight into chromosomal structure is the **lampbrush chromosome**, so named because its resembles the brushes used to clean kerosene-lamp chimneys in the nineteenth century. Lampbrush chromosomes were first discovered in 1892 in the oocytes of sharks and are now known to be characteristic of most vertebrate oocytes as well as the spermatocytes of some insects. Therefore, they are meiotic chromosomes. Most experimental work has been done with material taken from amphibian oocytes.

These chromosomes are easily isolated from oocytes in the diplotene stage of the first prophase of meiosis, where they are active in directing the metabolic activities of the developing cell. The homologs are seen as synapsed pairs held together by chiasmata. However, instead of condensing, as most meiotic chromosomes do, lampbrush chromosomes often extend to lengths of 500–800 μm. Later in meiosis, they revert to their normal length of 15–20 μm. Based on these observations, lampbrush chromosomes are interpreted as extended, uncoiled versions of the normal meiotic chromosomes.

The two views of lampbrush chromosomes in Figure 12–9 provide significant insights into their morphology. Part (a) shows the meiotic configuration under the light microscope. The linear axis of each structure contains a large number of condensed areas, and as with polytene chromosomes, these are referred to as chromomeres. Emanating from each chromomere is a pair of **lateral loops**, which give the chromosome its distinctive appearance. In part (b), the scanning electron micrograph (SEM) reveals adjacent loops present along one of the two axes of the chromosome. As with bands in polytene chromosomes, much more DNA is present in each loop than is needed to encode a single gene. This SEM provides a clear view of the chromomeres and the chromosomal fibers emanating from them. Each chromosomal loop is thought to be composed of one DNA double helix, while the central axis is composed of two DNA helices. This hypothesis is consistent with the belief that each meiotic chromosome is composed of a pair of sister chromatids. Studies using radioactive RNA precursors have revealed that the loops are active in the synthesis of RNA. The lampbrush loops, in a manner similar to puffs in polytene chromosomes, represent DNA that has been uncoiled from the central chromomere axis during transcription.

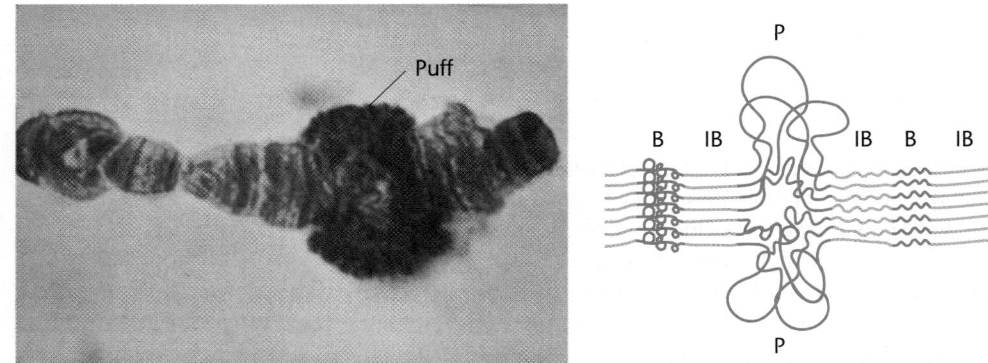

FIGURE 12–8 Photograph of a puff within a polytene chromosome. The diagram depicts the uncoiling of strands within a band (B) region to produce a puff (P) in polytene chromosomes. Interband regions (IB) are also labeled. *(Photo: Science Source/Photo Researchers, Inc.)*

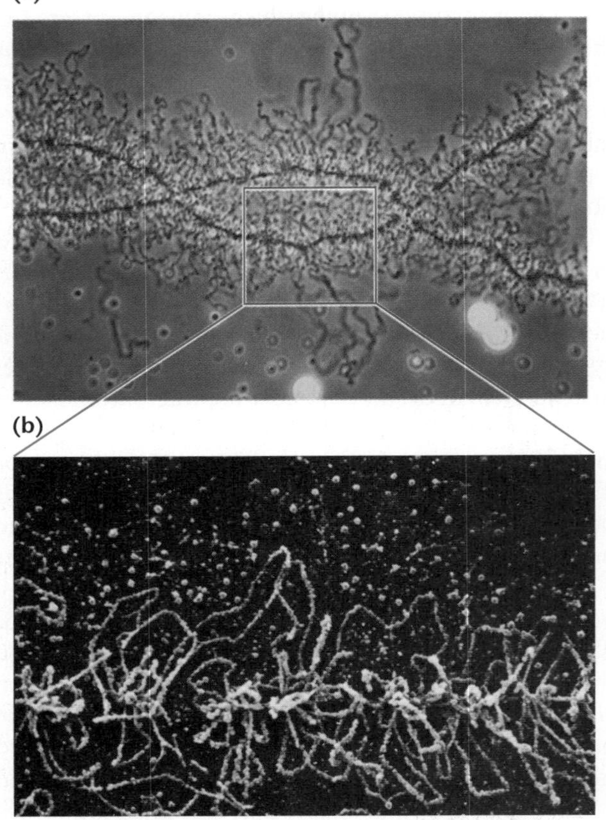

FIGURE 12–9 Lampbrush chromosomes derived from amphibian oocytes. Part (a) is a photomicrograph; part (b) is a scanning electron micrograph. *(Top: Nicole Angelier, Inc.; bottom: Omikron/Photo Researchers)*

12.4 DNA Is Organized into Chromatin in Eukaryotes

We now turn our attention to the way DNA is organized in eukaryotic chromosomes. Our focus will be on conventional eukaryotic cells, in which chromosomes are visible only during mitosis. After chromosome separation and cell division, cells enter the interphase stage of the cell cycle, during which time the components of the chromosome uncoil and are present in the form referred to as **chromatin**. While in interphase, the chromatin is dispersed in the nucleus and the DNA of each chromosome is replicated. As the cell cycle progresses, most cells reenter mitosis, whereupon chromatin coils into visible chromosomes once again. This condensation represents a length contraction of some 10,000 times for each chromatin fiber.

The organization of DNA during the transitions just described is much more intricate and complex than in viruses or bacteria, which never exhibit a process similar to mitosis. This is due to the greater amount of DNA per chromosome, as well as the presence of a large number of proteins associated with eukaryotic DNA. For example, while DNA in the *E. coli* chromosome is 1200 μm long, the DNA in each human chromosome ranges from 19,000–73,000 μm in length. In a single human nucleus, all 46 chromosomes contain sufficient DNA to extend almost 2 meters. This genetic material, along with its associated proteins, is contained within a nucleus that usually measures about 5–10 μm in diameter.

Such intricacy parallels the structural and biochemical diversity of the many types of cells in a multicellular eukaryotic organism. Different cells assume specific functions based on highly specific biochemical activity. While all cells carry a full genetic complement, different cells activate different sets of genes, so a highly ordered regulatory system governing the readout of the information must exist. Such a system must in some way be imposed upon or related to the molecular structure of the genetic material.

Because of the limitation of light microscopy, early studies of the structure of eukaryotic genetic material concentrated on intact chromosomes, preferably large ones, such as the polytene and lampbrush chromosomes. Subsequently, new techniques for biochemical analysis, as well as the examination of relatively intact eukaryotic chromatin and mitotic chromosomes under the electron microscope, have greatly enhanced our understanding of chromosome structure.

Chromatin Structure and Nucleosomes

As we have seen, the genetic material of viruses and bacteria consists of strands of DNA or RNA that are nearly devoid of proteins. In eukaryotic chromatin, a substantial amount of protein is associated with the chromosomal DNA in all phases of the eukaryotic cell cycle. The associated proteins are divided into basic, positively charged **histones** and less positively charged nonhistones. The histones clearly play the most essential structural role of all the proteins associated with DNA. Histones contain large amounts of the positively charged amino acids lysine and arginine, making it possible for them to bond electrostatically to the negatively charged phosphate groups of nucleotides. Recall that a similar interaction has been proposed for several bacterial proteins. The five main types of histones are shown in Table 12.4.

The general model for chromatin structure is based on the assumption that chromatin fibers, composed of DNA and protein, undergo extensive coiling and folding as they are condensed within the cell nucleus. X-ray diffraction studies confirm that histones play an important role in chromatin structure. Chromatin produces regularly spaced

TABLE 12.4 Categories and Properties of Histone Proteins

Histone Type	*Lysine-Arginine Content*	*Molecular Weight (Da)*
H1	Lysine-rich	23,000
H2A	Slightly lysine-rich	14,000
H2B	Slightly lysine-rich	13,800
H3	Arginine-rich	15,300
H4	Arginine-rich	11,300

diffraction rings, suggesting that repeating structural units occur along the chromatin axis. If the histone molecules are chemically removed from chromatin, the regularity of this diffraction pattern is disrupted.

A basic model for chromatin structure was worked out in the mid-1970s. Several observations were particularly relevant to the development of this model:

1. Digestion of chromatin by certain endonucleases, such as micrococcal nuclease, yields DNA fragments that are approximately 200 bp in length or multiples thereof. This demonstrates that enzymatic digestion is not random, for if it were, we would expect a wide range of fragment sizes. Thus, chromatin consists of some type of repeating unit, each of which is protected from enzymatic cleavage, except where any two units are joined. It is the area between units that is attacked and cleaved by the endonuclease.

2. Electron microscopic observations of chromatin reveal that chromatin fibers are composed of linear arrays of spherical particles (Figure 12–10). Discovered by Ada and Donald Olins, the particles occur regularly along the axis of a chromatin strand and resemble beads on a string. These particles, initially referred to as ν-bodies (ν is the Greek letter nu), are now called **nucleosomes**. This conforms nicely to the earlier observation, which suggests the existence of repeating units.

3. Studies of precise interactions of histone molecules and DNA in the nucleosomes constituting chromatin show that histones H2A, H2B, H3, and H4 occur as two types of tetramers, $(H2A)_2 \cdot (H2B)_2$ and $(H3)_2 \cdot (H4)_2$. Roger Kornberg predicted that each repeating nucleosome unit consists of one of each tetramer (creating an octamer) in association with about 200 bp of DNA. Such a structure is consistent with previous observations and provides the basis for a model that explains the interaction of histones and DNA in chromatin.

4. When nuclease digestion time is extended, some of the 200 bp of DNA are removed from the nucleosome, creating a **nucleosome core particle** consisting of 147 bp. The DNA lost in this prolonged digestion is responsible for linking nucleosomes together. This linker DNA is associated with the fifth histone, H1.

5. On the basis of this information, as well as on X-ray and neutron-scattering analyses of crystallized core particles by John T. Finch, Aaron Klug, and others, a detailed model of the nucleosome was put forward in 1984. In this model, the 147-bp DNA core is coiled around an octamer of histones in a left-handed superhelix, which completes about 1.7 turns per nucleosome.

(a)

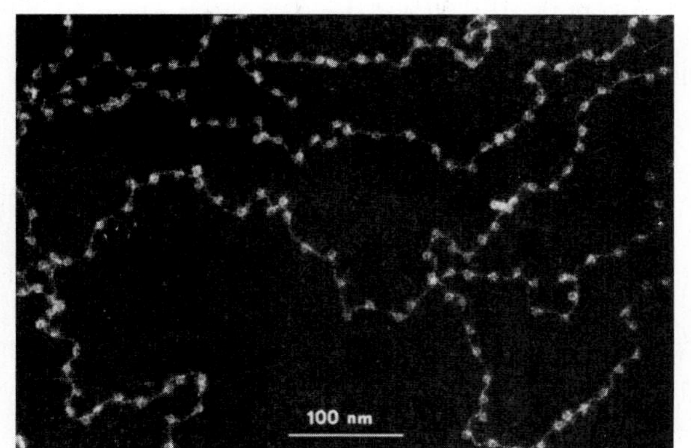

(b)

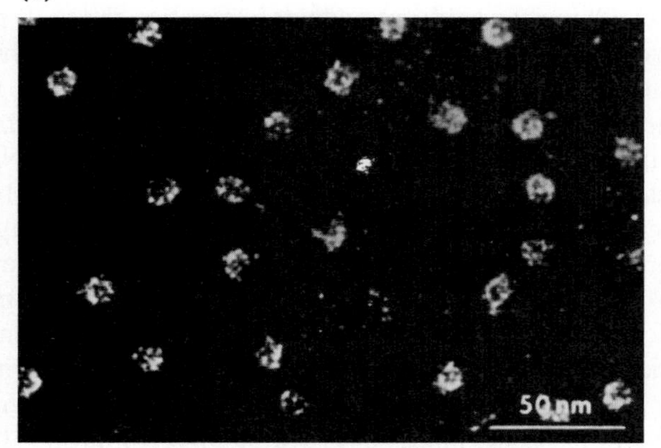

FIGURE 12–10 (a) Dark-field electron micrograph of nucleosomes present in chromatin derived from a chicken erythrocyte nucleus. (b) Dark-field electron micrograph of nucleosomes produced by micrococcal nuclease digestion. *(Olins and Olins, 1978. Figures 1 and 4)*

The extensive investigation of nucleosomes provides the basis for predicting how the chromatin fiber within the nucleus is formed and how it coils up into a mitotic chromosome. The 2-nm DNA molecule is initially coiled into a nucleosome about 11 nm in diameter (Figure 12–11a), consistent with the longer dimension of the ellipsoidal nucleosome. Significantly, the formation of the nucleosome represents the first level of packing, whereby the DNA helix is reduced to about one-third its original length.

In the nucleus, the chromatin fiber seldom, if ever, exists in the extended form described here. Instead, the 11-nm chromatin fiber is further packed into a thicker 30-nm fiber, called a **solenoid** (Figure 12–11b). This larger fiber consists of numerous closely coiled nucleosomes, creating the second level of packing. Solenoids condense the eukaryotic fiber by a factor of five. The exact details of this structure are not completely clear, but 30-nm chromatin fibers are characteristically seen under the electron microscope.

In the transition to the mitotic chromosome, still another level of packing occurs. The 30-nm fiber forms a series of looped domains that condense the structure into the chromatin fiber, which is 300 nm in diameter (Figure 12–11c). The fibers are then coiled into the chromosome arms that constitute a chromatid, which is part of the metaphase chromosome (Figure 12–11d). While we show the chromatid arms to be 700 nm in diameter in this figure, this value undoubtedly varies among different organisms. At a value of 700 nm, a pair of sister chromatids comprising a chromosome measures about 1400 nm.

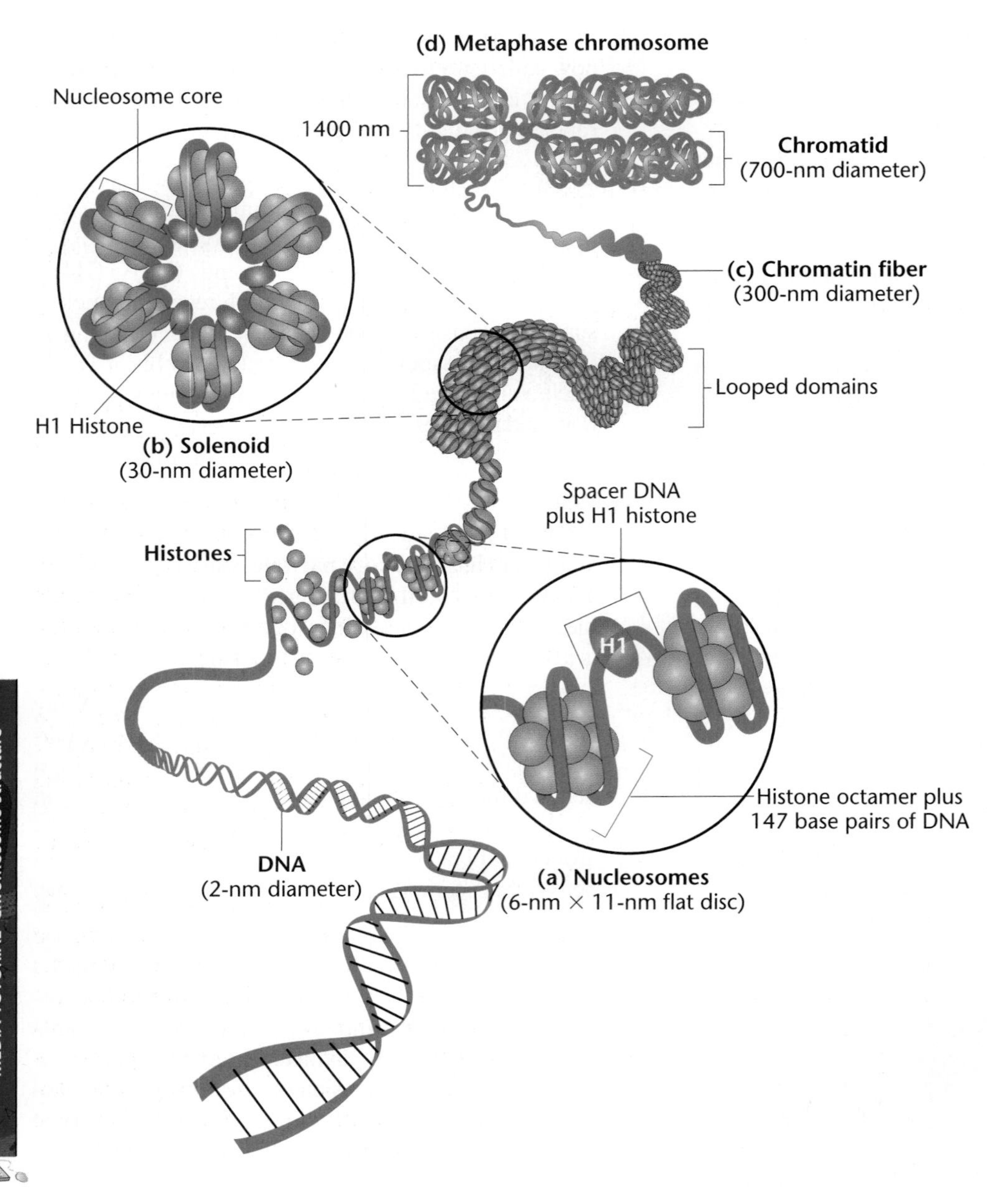

FIGURE 12–11 General model of the association of histones and DNA in the nucleosome, showing how the chromatin fiber can coil into a more condensed structure, ultimately producing a metaphase chromosome.

MEDIA TUTORIAL Chromosome structure

The importance of the organization of DNA into chromatin and chromatin into mitotic chromosomes can be illustrated by considering a human cell that stores its genetic material in a nucleus that is about 5–10 μm in diameter. The haploid genome contains 3.2×10^9 base pairs of DNA distributed among 23 chromosomes. The diploid cell contains twice that amount. At 0.34 nm per base pair, this amounts to an enormous length of DNA (as stated earlier, almost 2 m). One estimate is that about 25×10^6 nucleosomes per nucleus are complexed with the DNA present in a typical human nucleus.

In the overall transition from a fully extended DNA helix to the extremely condensed status of the mitotic chromosome, a packing ratio (the ratio of DNA length to the length of the structure containing it) of about 500:1 must be achieved. In fact, our model accounts for a ratio only one-tenth that. Obviously, the larger fiber can be further bent, coiled, and packed as even greater condensation occurs during the formation of a mitotic chromosome.

Chromatin Remodeling

As with many significant findings in genetics, the study of nucleosomes has answered some important questions, but at the same time, also led us to new ones. For example, in the preceding discussion, we established that histone proteins play an important structural role in packaging DNA into the nucleosomes that make up chromatin. While solving the structural problem of how to organize a huge amount of DNA within the eukaryotic nucleus, a new problem was apparent: the chromatin fiber, when complexed with histones and folded into various levels of compaction, makes the DNA inaccessible to interaction with important nonhistone proteins. The variety of proteins that function in enzymatic and regulatory roles during the processes of replication and gene expression must interact directly with DNA. To accommodate these protein–DNA interactions, chromatin must be induced to change its structure, a process called **chromatin remodeling**. In the case of replication and gene expression, chromatin must relax its compact structure, but be able to reverse the process during periods of inactivity.

Insights into how different states of chromatin structure may be achieved were forthcoming in 1997, when Timothy Richmond and members of his research team were able to significantly improve the level of resolution in X-ray diffraction studies of nucleosome crystals (from 7 Å in the 1984 studies to 2.8 Å in the 1997 studies). One model based on their work is shown in Figure 12–12. At this resolution, most atoms are visible, thus revealing the subtle twists and turns of the superhelix of DNA that encircles the histones. Recall that the double-helical ribbon represents 147 bp of DNA surrounding 4 pairs of histone proteins. This configuration is repeated over and over in the chromatin fiber and is the principal packaging unit of DNA in the eukaryotic nucleus.

The work of Richmond and colleagues, extended to a resolution of 1.9 Å in 2003, has revealed the details of the location of each histone entity within the nucleosome. Of particular interest to chromatin remodeling is that there are unstructured histone tails that are not packed into the folded histone domains within the core of the nucleosome. For example, tails devoid of any secondary structure extending from histones H3 and H2B protrude through the minor groove channels of the DNA helix. The tails of histone H4 appear to make a connection with adjacent nucleosomes. The significance of histone tails is that they provide potential targets for a variety of chemical interactions that may be linked to genetic functions along the chromatin fiber.

Several of these potential chemical interactions are now recognized as important to genetic function. One of the best-studied histone modifications involves **acetylation**, resulting from the action of histone acetyltransferase

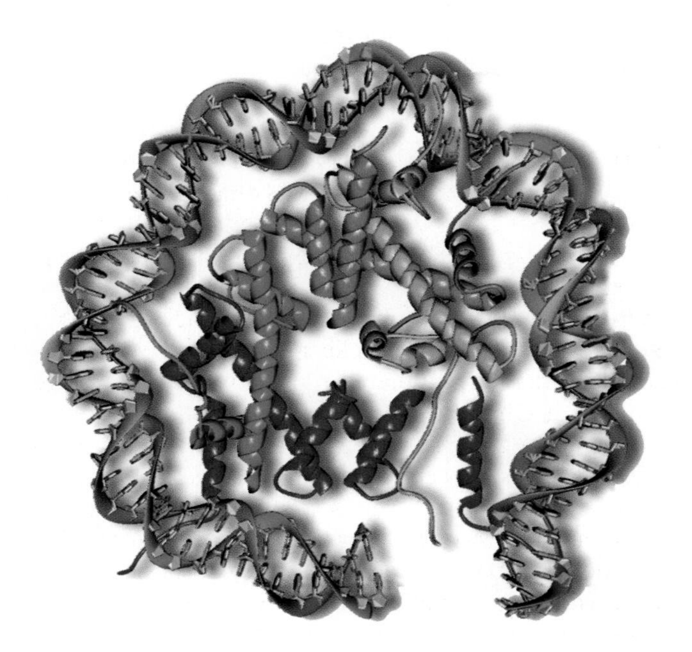

FIGURE 12–12 The nucleosome core particle derived from X-ray crystal analysis at 2.8 Å resolution. The double-helical DNA surrounds four pairs of histones.

(HAT). This enzyme adds an acetyl group to the positively charged amino group present on the side chain of the amino acid lysine, effectively changing the net charge of the protein by neutralizing the positive charge. Lysine is in abundance in histones, and it has been known for some time that acetylation is linked to gene activation. It appears that high levels of acetylation open the chromatin fiber, an effect that occurs in regions of active genes and decreases in inactive regions. A well-known example involves the inactive X chromosome forming a Barr body in mammals, in which histone H4 is known to be greatly underacetylated.

The two other important chemical modifications include the **methylation** and **phosphorylation** of amino acids that are part of histones. These chemical processes result from the action of enzymes called methyltransferases and kinases, respectively. Methyl groups can be added to both arginine and lysine of histones, and this change has been correlated with the inactivation of genes. Phosphate groups can be added to the hydroxyl groups of the amino acids serine and histidine, introducing a negative charge on the protein. During the cell cycle, increased phosphorylation, particularly of histone H3, is known to occur at characteristic times. Such chemical modification is believed to be related to the cycle of chromatin unfolding and condensation that occurs during and after DNA replication. Phosphorylation has also been linked to gene activation during interphase.

Much more work must be done to elucidate the specific involvement of chromatin remodeling during genetic processes—in particular, the way in which modifications are influenced by regulatory molecules within cells. It is clear that the dynamic forms in which chromatin exists are vitally important to the way that all genetic processes directly involving DNA are executed, including replication and transcription and its regulation—processes that are at the heart of genetic function.

Heterochromatin

Evidence that the DNA of each eukaryotic chromosome consists of one continuous double-helical fiber along its entire length might suggest that the whole chromosome is structurally uniform. However, in the early part of the twentieth century, it was observed that some parts of the chromosome remain condensed and stain deeply during interphase, but most parts are uncoiled and do not stain. In 1928, the terms **heterochromatin** and **euchromatin** were coined to describe the parts of chromosomes that remain condensed and those that are uncoiled, respectively.

Subsequent investigation revealed a number of characteristics that distinguish heterochromatin from euchromatin. Heterochromatic areas are genetically inactive because they either lack genes or contain genes that are repressed. Also, heterochromatin replicates later during the S phase of the cell cycle than euchromatin does. The discovery of heterochromatin provided the first clues that parts of eukaryotic chromosomes do not always encode proteins. Instead, some chromosome regions are thought to be involved in maintenance of the chromosome's structural integrity and in other functions, such as chromosome movement during cell division.

Heterochromatin is characteristic of the genetic material of eukaryotes. Early cytological studies showed that areas of the centromeres are composed of heterochromatin. The ends of chromosomes, called *telomeres*, are also heterochromatic. In some cases, whole chromosomes are heterochromatic. Such is the case with the mammalian Y chromosome, much of which is genetically inert. And, as we discussed in Chapter 5, the inactivated X chromosome in mammalian females is condensed into an inert heterochromatic Barr body. In some species, such as mealy bugs, all of the chromosomes in one entire haploid set are heterochromatic.

When certain heterochromatic areas from one chromosome are translocated to a new site on the same or another nonhomologous chromosome, genetically active areas sometimes become genetically inert if they lie adjacent to the translocated heterochromatin. This influence on existing euchromatin is one example of what is more generally referred to as a **position effect**. That is, the position of a gene or group of genes relative to all other genetic material may affect their expression.

12.5 Eukaryotic Genomes Demonstrate Complex Sequence Organization Characterized by Repetitive DNA

Thus far, we have looked at how DNA is organized into chromosomes in bacteriophages, bacteria, and eukaryotes. We now begin an examination of what we know about the organization of DNA sequences within the chromosomes making up an organism's genome, placing our emphasis on eukaryotes. Once we establish the pattern of genome organization, we will focus on how the genes themselves are organized within chromosomes in a later chapter.

We have already established that, in addition to single copies of unique DNA sequences that comprise genes, a great deal of the DNA sequences within chromosomes is repetitive in nature and that various levels of repetition occur within the genome of organisms. Many studies have now provided insights into repetitive DNA, demonstrating various classes of these sequences and their organization within the genome. Figure 12–13 outlines the various categories of repetitive DNA. Some functional genes are present in more than one copy and are therefore repetitive in nature. However, the majority of repetitive sequences are nongenic, and in fact, most serve no known function. We explore three main categories: (1) heterochromatin found associated with centromeres and making up telomeres, (2) tandem repeats of both short and long DNA sequences, and (3) transposable sequences that are interspersed throughout the genome of eukaryotes.

Repetitive DNA and Satellite DNA

The nucleotide composition of the DNA (e.g., the percentage of G≡C versus A═T pairs) of a particular species is reflected in its density, which can be measured with sedimentation equilibrium centrifugation. When eukaryotic DNA is analyzed in this way, the majority is present as a single main band or peak of fairly uniform density. However, one or more additional peaks represent DNA that differs slightly in density. This component, called **satellite DNA**, represents a variable proportion of the total DNA, depending on the species. A profile of main-band and satellite DNA from the mouse is shown in Figure 12–14. By contrast, prokaryotes contain only main-band DNA.

The significance of satellite DNA remained an enigma until the mid-1960s, when Roy Britten and David Kohne developed a technique for measuring the reassociation kinetics of DNA that had previously been dissociated into

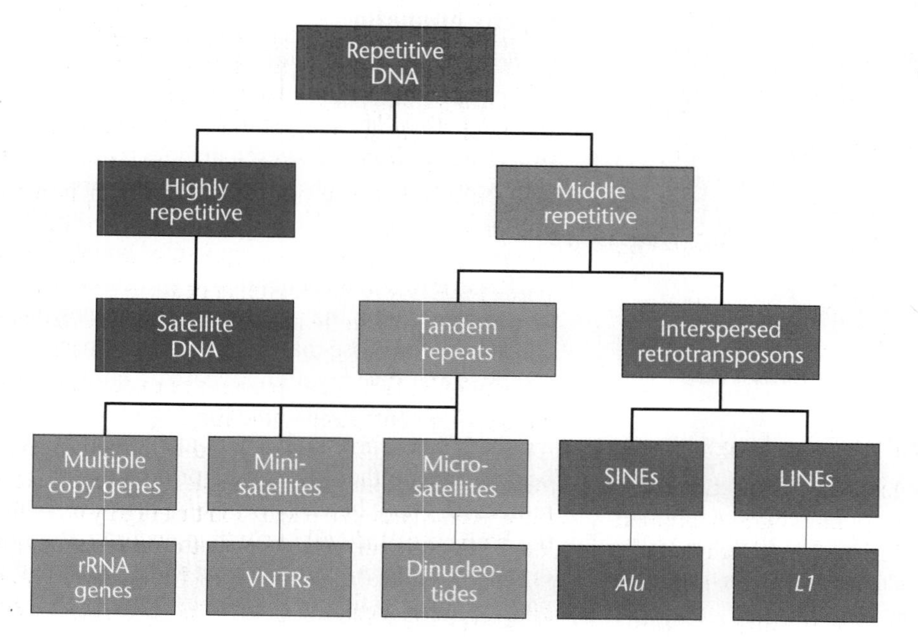

FIGURE 12–13 An overview of the categories of repetitive DNA.

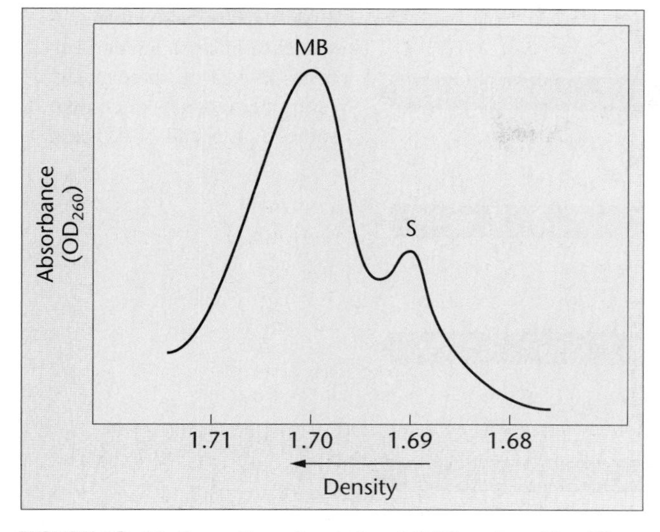

FIGURE 12–14 Separation of main-band (MB) and satellite (S) DNA from the mouse, using ultracentrifugation in a CsCl gradient.

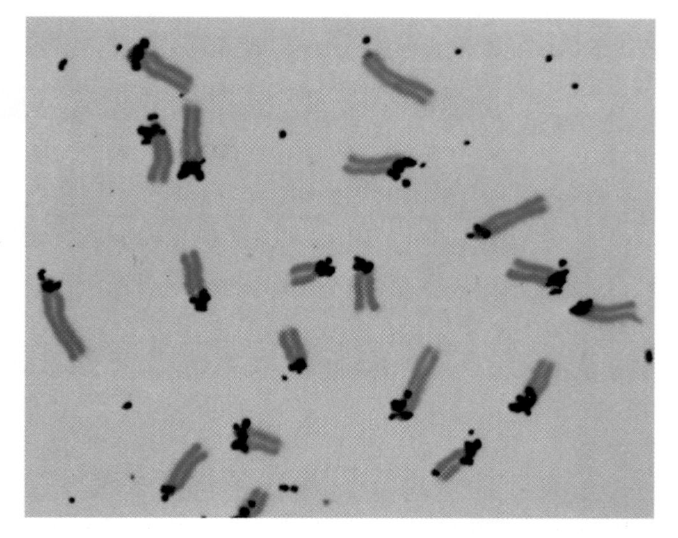

FIGURE 12–15 *In situ* molecular hybridization between RNA transcribed from mouse satellite DNA and mitotic chromosomes. The grains in the autoradiograph localize the chromosome regions (the centromeres) containing satellite DNA sequences. *(Pardue and Gall. 1972. © Keter)*

single strands. They demonstrated that certain portions of DNA reassociated more rapidly than others. They concluded that rapid reassociation was characteristic of multiple DNA fragments composed of identical or nearly identical nucleotide sequences—the basis for the descriptive term **repetitive DNA**.

When satellite DNA is subjected to analysis by reassociation kinetics, it falls into the category of **highly repetitive DNA**, which is known to consist of short sequences repeated a large number of times. Further evidence suggests that these sequences are present as tandem repeats clustered in very specific chromosomal areas known to be heterochromatic—the regions flanking centromeres. This was discovered in 1969 when several researchers, including Mary Lou Pardue and Joe Gall, applied ***in situ* molecular hybridization** to the study of satellite DNA. This technique involves the molecular hybridization between an isolated fraction of radioactively labeled DNA or RNA probes and the DNA contained in the chromosomes of a cytological preparation. Following the hybridization procedure, autoradiography is performed to locate the chromosome areas complementary to the fraction of DNA or RNA.

Pardue and Gall demonstrated that probes made from mouse satellite DNA hybridize with the DNA of centromeric regions of mouse mitotic chromosomes (Figure 12–15). Several conclusions were drawn: Satellite DNA differs from main-band DNA in its molecular composition, as established by buoyant density studies. It is composed of short repetitive sequences. Finally, satellite DNA is found in the heterochromatic centromeric regions of chromosomes.

Centromeric and Telomeric DNA Sequences

Separation of chromatids is essential to the fidelity of chromosome distribution during mitosis and meiosis. Most estimates of infidelity during mitosis are exceedingly low: 1×10^{-5} to 1×10^{-6} or 1 error per 100,000 to 1,000,000 cell divisions. As a result, it has been generally assumed that analysis of the DNA sequence of centromeric regions will provide insights into the rather remarkable features of this chromosomal region. This DNA region is designated the **CEN region**, and its function is now reasonably clear. Its structure within the heterochromatic region binds a platform of proteins forming the centromere, which includes the kinetochore that binds to the spindle fiber during division.

The analysis of the CEN regions of yeast *Saccharomyces cerevisiae* chromosomes provided the basis for a model system first described by John Carbon and Louis Clarke. Each centromere serves an identical function, so it is not surprising that all CENs were found to be remarkably similar in their organization. The CEN region of yeast chromosomes consists of about 225 bp, which can be further divided into three regions (Figure 12–16). The first and third regions (I and III) are relatively short and highly conserved, consisting of only 8 bp and 26 bp, respectively. Region II, which is larger (80–85 bp) and extremely rich in A and T (up to 95%), varies in sequence among different chromosomes.

Mutational analysis suggests that regions I and II are less critical to centromere function than is region III. Mutations in the former regions are often tolerated, but mutations in region III can disrupt centromere function. While the DNA of this region appears to be essential to the eventual binding to the spindle fiber, DNA sequences are not unique to specific chromosomes. They can be exchanged between chromosomes experimentally without altering centromere function.

The amount of DNA associated with the centromeres of multicellular eukaryotes is much more extensive than in yeast. Recall from our discussion that highly repetitive satellite DNA is localized in the centromere regions of mice. Such sequences, absent from yeast but characteristic of most multicellular organisms, vary considerably in size. For

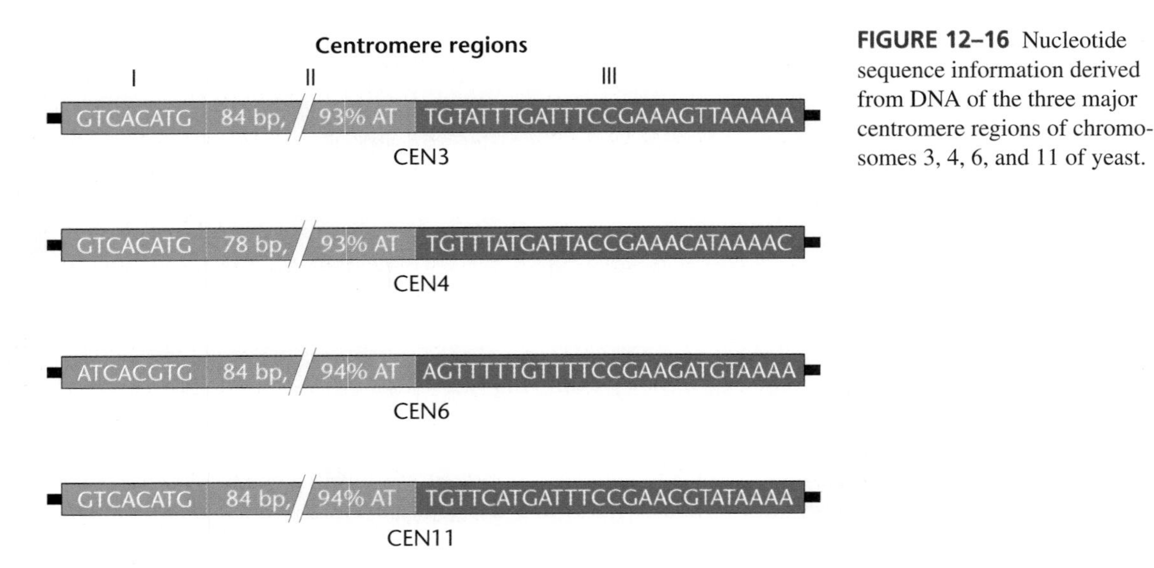

FIGURE 12–16 Nucleotide sequence information derived from DNA of the three major centromere regions of chromosomes 3, 4, 6, and 11 of yeast.

example, the 10-bp sequence AATAACATAG is tandemly repeated many times in the centromeres of all 4 chromosomes of *Drosophila*. In humans, one of the most recognized satellite DNA sequences is the **alphoid family**. Found mainly in the centromere regions, alphoid sequences, each about 170 bp in length, are present in tandem arrays of up to 1 million base pairs. The role of this highly repetitive DNA in centromere function remains unclear.

The other prominent structural component of chromosomes is the **telomere**, found at the ends of linear chromosomes. Telomeres provide stability to the chromosome by rendering chromosome ends generally inert in interactions with other chromosome ends. It is thought that some aspect of the molecular structure of telomeres must be unique compared with most other chromosome regions. As with centromeres, the analysis of telomeres was first approached by investigating the smaller chromosomes of simple eukaryotes, such as protozoans and yeast. The idea that all telomeres of all chromosomes in a given species might share a related nucleotide sequence has now been borne out.

Two types of telomere sequences have been discovered. The first type, called **telomeric DNA sequences**, consists of short tandem repeats. It is this group that contributes to the stability and integrity of the chromosome. In the ciliate *Tetrahymena*, over 50 tandem repeats of the hexanucleotide sequence GGGGTT occur. In humans, the sequence GGGATT is repeated many times. The analysis of telomeric DNA sequences has shown them to be highly conserved throughout evolution, reflecting the critical role they play in maintaining the integrity of chromosomes.

The second type, **telomere-associated sequences**, is also repetitive and is found both adjacent to and within the telomere. These sequences vary among organisms, and their significance remains unknown.

Middle Repetitive Sequences: VNTRs and Dinucleotide Repeats

Still another prominent category of repetitive DNA sheds more light on our understanding of the organization of the eukaryotic genome. In addition to highly repetitive DNA, which constitutes about 5% of the human genome (and 10% of the mouse genome), a second category, recognized by C_0t analysis, **middle** (or *moderately*) **repetitive DNA** is fairly well characterized. Since we are currently learning a great deal about the human genome, we will use our own species to illustrate this category of DNA in genome organization.

Middle repetitive DNA most prominently consists of either tandemly repeated or interspersed sequences. No function has been ascribed to these components of the genome. An example includes those called **variable number tandem repeats (VNTRs)**. The repeating DNA sequence of VNTRs can be 15–100 bp long and is found within and between genes. Many such clusters, often referred to as **minisatellites**, are dispersed throughout the genome.

The number of tandem repeats of each specific sequence at each location varies in individuals, creating localized regions of 1000–5000 bp (1–5 kb) in length. The variation in size (length) of these regions between individuals in humans is the basis of the forensic technique of *DNA fingerprinting*.

Another group of tandemly repeated sequences consists of dinucleotides, or **microsatellites**. Like VNTRs, they are dispersed throughout the genome and vary among individuals in the number of repeats that are present at any site. For example, the most common microsatellite is the dinucleotide $(CA)_n$, where n is the number of repeats—usually between 5 and 50. These clusters are also used forensically, and are useful molecular markers during genome analysis.

Repetitive Transposed Sequences: SINEs and LINEs

Still another category of repetitive DNA consists of sequences that are interspersed throughout the genome, rather than being tandemly repeated. They can be either short or long, and many have the added distinction of being **transposable sequences**, which are mobile and can move to different locations within the genome. A large portion of eukaryotic genomes are composed of these sequences.

For example, **short interspersed elements**, or **SINEs**, are less than 500 bp long and may be present 500,000 times or more in the human genome. The best-known human SINE is a set of closely related sequences called the ***Alu* family** (the name is based on the presence of DNA sequences recognized by the restriction endonuclease *AluI*). Members of this DNA family, which are also found in other mammals, are 200–300 bp long and dispersed rather uniformly throughout the genome, both between and within genes. In humans, this family encompasses more than 5% of the entire genome.

Members of the *Alu* family are sometimes transcribed. The role of this RNA is not certain, but it may relate to their mobility in the genome. In fact, *Alu* sequences are thought to have arisen from an RNA element whose DNA complement was dispersed throughout the genome as a result of the activity of reverse transcriptase (an enzyme that synthesizes DNA on an RNA template).

The group of **long interspersed elements** (**LINEs**) represents still another category of repetitive transposable DNA sequences. The most prominent example in humans is the **L1 family**. Members of this sequence family are about 6400 bp long and are present up to 100,000 times, according to one estimate. Their 5′-end is highly variable, and their role has yet to be defined.

The basis for transposition of L1 elements is now clear. The L1 DNA sequence is first transcribed into an RNA molecule. The RNA then serves as the template for the synthesis of the DNA complement via the enzyme reverse transcriptase. This enzyme is encoded by a portion of the L1 sequence. The new L1 copy then integrates into the DNA of the chromosome at a new site. Because this mechanism of transposition resembles that used by retroviruses, LINEs are referred to as **retrotransposons**.

SINEs and LINEs represent a significant portion of human DNA. Both types of elements share the organizational feature of a mixture of about 70% unique and 30% repeating sequences within the DNA of each entity. Collectively, they make up about 10% of the genome.

Middle Repetitive Multiple Copy Genes

In some cases, middle repetitive DNA includes functional genes tandemly present in multiple copies. For example, many copies exist of the genes encoding ribosomal RNA. *Drosophila* has 120 copies per haploid genome. Single genetic units encode a large precursor molecule that is processed into the 5.8*S*, 18*S*, and 28*S* rRNA components. In humans, multiple copies of this gene are clustered on the p arm of the acrocentric chromosomes 13, 14, 15, 21, and 22. Multiple copies of the genes encoding 5*S* rRNA are transcribed separately from multiple clusters found together on the terminal portion of the p arm of chromosome 1.

12.6 The Vast Majority of a Eukaryotic Genome Does Not Encode Functional Genes

Given the information above involving various forms of repetitive DNA in eukaryotes, we can pose an important question: *What proportion of the eukaryotic genome actually encodes functional genes?* Taken together, the various forms of highly repetitive and moderately repetitive DNA comprise up to 40% of the human genome. Such an observation is not uncommon in eukaryotes. In addition to repetitive DNA, a large amount of single-copy DNA sequences as defined by C_0t analysis appears to be noncoding. A small portion of them are called **pseudogenes**, which represent evolutionary vestiges of duplicated copies of genes that have undergone sufficient mutations to render them untranscrible—in some cases, multiple copies exist.

While the proportion of the genome consisting of repetitive DNA varies among organisms, one feature seems to be shared: Only a very small part of the genome actually codes for proteins. For example, the 20,000–30,000 genes encoding proteins in sea urchin occupy less than 10% of the genome. In *Drosophila*, only 5–10% of the genome is occupied by genes coding for proteins. In humans, it appears that the estimated 30,000 functional genes occupy less than 5% of the genome.

The study of various forms of repetitive DNA has significantly enhanced our understanding of genome organization. In a later chapter, we will explore the organization of genes within chromosomes.

Chapter Summary

1. The organization of the molecular components that form chromosomes is essential to understanding the function of the genetic material. Largely devoid of associated proteins, bacteriophage and bacterial chromosomes contain DNA molecules in a form equivalent to the Watson-Crick model.
2. Mitochondria and chloroplasts contain DNA that encodes products essential to their biological function. This DNA is remarkably similar in form and appearance to some bacterial and bacteriophage DNA, lending support to the endosymbiotic hypothesis, which suggests that these organelles were once free-living prokaryote-like organisms.
3. Polytene and lampbrush chromosomes are examples of specialized structures that have extended our knowledge of genetic organization and function.
4. The eukaryotic chromatin fiber is a nucleoprotein organized into repeating units called nucleosomes. Composed of 147 bp of DNA and an octamer of four types of histones, the nucleosome facilitates the conversion of the extended chromatin fiber characteristic of interphase into the highly condensed chromosome seen in mitosis.
5. The structural heterogeneity of the chromosome axis has been established as a result of both biochemical and cytological investigation. Heterochromatin, prematurely condensed in interphase, is genetically inert. The centromeric and telomeric regions, parts of the Y chromosome, and the Barr body are examples.
6. DNA analysis reveals unique nucleotide sequences in both the centromere and telomere regions of chromosomes, which no doubt impart the heterochromatic characteristic that they share.

7. Eukaryotic genomes demonstrate complex sequence organization characterized by numerous categories of repetitive DNA.
8. Repetitive DNA consists of either tandem repeats clustered in various regions of the genome or single sequences interspersed uniformly throughout the genome. In the former group, the size of each cluster varies among individuals, providing a form of biochemical identity. The latter group of sequences may be short like *Alu* or long such as in L1, and they are transposable elements.
9. The vast majority of a eukaryotic genome does not encode functional genes. In humans, for example, less than 5% of the genome is used to encode the estimated 30,000 genes found in our genome.

Key Terms

ϕX174 bacteriophage, 254
acetylation, 263
alphoid family, 266
Alu family, 266
bacteriophage lambda (λ), 254
CEN region, 265
chloroplast, 255
chloroplast DNA (cpDNA), 258
chromatin, 260
chromatin remodeling, 262
chromomeres, 259
DNA-binding proteins, 254
endosymbiotic hypothesis, 255
euchromatin, 263
heterochromatin, 263
highly repetitive DNA, 265
histone, 260
HU and H proteins, 254
in situ molecular hybribidation, 254
lampbrush chromosome, 259
lateral loops, 259
L1 family, 267
long interspersed elements (LINEs), 267
methylation, 263
microsatellites, 266
middle repetitive DNA, 266
minisatellites, 266
mitochondria, 255
mitochondrial DNA (mtDNA), 256
nucleoid, 254
nucleosome core particle, 261
nucleosome, 261
phosphorylation, 263
polyoma virus, 254
polytene chromosomes, 258
position effect, 264
pseudogenes, 267
puff, 259
repetitive DNA, 265
retrotransposons, 267
satellite DNA, 264
short interspersed elements (SINEs), 266
solenoid, 261
telomere, 266
telomere-associated sequences, 266
telomeric DNA sequences, 266
transposable sequences, 266
variable number tandem repeats (VNTRs), 266

INSIGHTS AND SOLUTIONS

A previously undiscovered single-cell organism was found living at a great depth on the ocean floor. Its nucleus contained only a single linear chromosome with 7×10^6 nucleotide pairs of DNA coalesced with three types of histonelike proteins. Consider the following questions:

1. A short micrococcal nuclease digestion yielded DNA fractions of 700, 1400, and 2100 bp. Predict what these fractions represent. What conclusions can be drawn?

Solution: The chromatin fiber may consist of a variation of nucleosomes containing 700 bp of DNA. The 1400- and 2100-bp fractions, respectively, represent 2 and 3 linked nucleosomes. Enzymatic digestion may have been incomplete, leading to the latter two fractions.

2. The analysis of individual nucleosomes reveals that each unit contained one copy of each protein and that the short linker DNA contained no protein bound to it. If the entire chromosome consists of nucleosomes (discounting any linker DNA), how many are there, and how many total proteins are needed to form them?

Solution: Since the chromosome contains 7×10^6 bp of DNA, the number of nucleosomes, each containing 7×10^2 bp, is equal to

$$7 \times 10^6 / 7 \times 10^2 = 10^4 \text{ nucleosomes}$$

The chromosome contains 10^4 copies of each of the 3 proteins, for a total of 3×10^4 proteins.

3. Further analysis revealed the organism's DNA to be a double helix similar to the Watson-Crick model, but containing 20 bp per complete turn of the right-handed helix. The physical size of the nucleosome was exactly double the volume occupied by that found in all other known eukaryotes, by virtue of increasing the distance along the fiber axis by a factor of two. Compare the degree of compaction of this organism's nucleosome to that found in other eukaryotes.

Solution: The unique organism compacts a length of DNA consisting of 35 complete turns of the helix (700 bp per nucleosome/20 bp per turn) into each nucleosome. The normal eukaryote compacts a length of DNA consisting of 20 complete turns of the helix (200 bp per nucleosome/10 bp per turn) into a nucleosome one-half the volume of that in the unique organism. The degree of compaction is therefore less in the unique organism.

4. No further coiling or compaction of this unique chromosome occurs in the unique organism. Compare this to a eukaryotic chromosome. Do you think an interphase human chromosome 7×10^6 bp in length would be a shorter or longer chromatin fiber?

Solution: The length of the unique chromosome is compacted into 10^4 nucleosomes, each with an axis length twice that of the eukaryotic fiber. The eukaryotic fiber consists of

$$7 \times 10^6 / 2 \times 10^2 = 3.5 \times 10^4 \text{ nucleosomes}$$

which is 3.5 more than the unique organism. However, they are compacted by a factor of five in each solenoid. Therefore, the chromosome of the unique organism is a longer chromatin fiber.

Problems and Discussion Questions

1. Contrast the sizes of the chromosome of bacteriophage λ and T2 with that of *E. coli*. How does this relate to their relative size and complexity?
2. Bacteriophages and bacteria almost always contain their DNA as circular (closed loops) chromosomes. Phage λ is an exception, maintaining its DNA in a linear chromosome within the viral particle. However, as soon as it is injected into a host cell, it circularizes before replication begins. Taking into account the information in Chapter 11, what advantage exists in replicating circular DNA molecules compared to linear molecules?
3. Contrast the appearance of the DNA associated with mitochondria and chloroplasts.
4. Compare the size of DNA and the encoded gene products in mitochondria and chloroplasts.
5. While protein synthesis occurs within mitochondria and chloroplasts, not all necessary gene products are encoded by the organellar DNA. Explain the origin of the genetic machinery within the organelles.
6. Mitochondria and chloroplasts contain ribosomal RNA molecules that are unlike those found in the adjoining cytoplasm. How does this observation relate to the endosymbiotic hypothesis?
7. Describe how giant polytene chromosomes are formed?
8. Salivary gland cells from *Drosophila* are isolated and placed in the presence of radioactive thymidylic acid. Autoradiaography is performed, revealing polytene chromosomes. Predict the distribution of the grains along the chromosomes.
9. What genetic process is occurring in a puff of a polytene chromosome?
10. Describe the structure of LINE sequences. Why are LINEs referred to as retrotransposons?
11. During what genetic process are lampbrush chromosomes present in vertebrates?
12. Why might we predict that the organization of eukaryotic genetic material will be more complex than that of viruses or bacteria?
13. Describe the sequence of research findings that led to the development of the model of chromatin structure.
14. What is the molecular composition and arrangement of the components in the nucleosome?
15. Describe how the transitions that occur as nucleosomes are coiled and folded, ultimately forming a chromatid.
16. Provide a comprehensive definition of heterochromatin, and list as many examples as you can.
17. Mammals contain a diploid genome consisting of at least 10^9 bp. If this amount of DNA is present as chromatin fibers, where each group of 200 bp of DNA is combined with 9 histones into a nucleosome and each group of 6 nucleosomes is combined into a solenoid, achieving a final packing ratio of 50, determine (a) the total number of nucleosomes in all fibers, (b) the total number of histone molecules combined with DNA in the diploid genome, and (c) the combined length of all fibers.
18. Assume that a viral DNA molecule is a 50 μm-long circular strand of a uniform 20 Å diameter. If this molecule is contained in a viral head that is a 0.08 μm-diameter sphere, will the DNA molecule fit into the viral head, assuming complete flexibility of the molecule? Justify your answer mathematically.
19. How many base pairs are in a molecule of phage T2 DNA 52 μm long?
20. If a human nucleus is 10 μm in diameter, and it must hold as much as 2 m of DNA, which is complexed into nucleosomes that are 11 nm in diameter during full extension, what percentage of the volume of the nucleus is occupied by the genetic material?
21. Sun and others (2002. *Proc. Natl. Acad. Sci. (USA)* 99: 8695–8700) studied *Drosophila* in which two normally active genes, w^+ (wild-type allele of the *white*-eye gene) and *hsp26* (a heat-shock gene), were introduced (using a plasmid vector) into euchromatic and heterochromatic chromosomal regions. The relative activity of each gene was assessed, and an approximation of the data obtained is shown below. Considering three characteristics of heterochromatin, which is/are supported by the experimental data?

	Activity (relative percentage)	
Gene	*Euchromatin*	*Heterochromatin*
hsp26	100%	31%
w^+	100%	8%

22. Using molecular methods to "label" chromosomes with fluorescent dyes, Nagele and colleagues (1995. *Science* 270: 1831–35) observed a precise nuclear positioning of chromosomes during an early stage of mitosis (prometaphase) in human fibroblast cells. Below is a sketch modified from this research that describes the relative positions of chromosomes 7, 8, 16, and X. Homologous chromosomes share the same color. Assuming that this pattern is consistent among other cells in humans, what conclusions can be drawn regarding the nuclear positions of the chromosomes during interphase and during the initial phases of mitosis? How could chromosomal localization influence gene function during interphase?

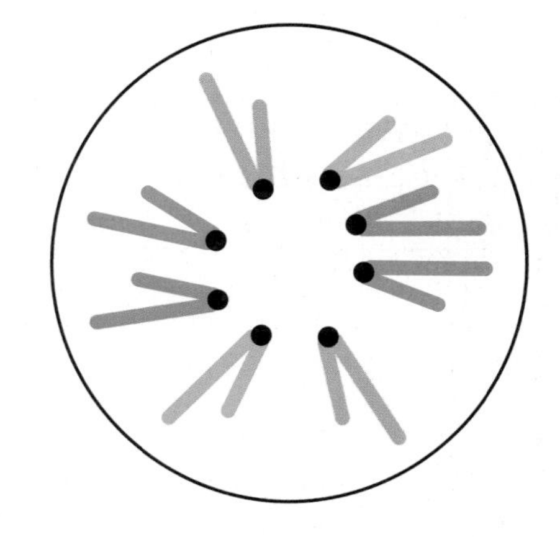

23. While much remains to be learned about the role of nucleosomes and chromatin structure and function, recent research indicates that *in vivo* chemical modification of histones is associated with changes in gene activity. For example, Bernstein and others (2000. *Proc. Natl. Acad. Sci. (USA)* 97: 5340–45) determined that acetylation of H3 and H4 is associated with 21.1% and 13.8% increase in yeast gene activity, respectively, and that yeast heterochromatin is hypomethylated relative to the genome average. Speculate on the significance of these findings in terms of nucleosome–DNA interactions and gene activity.

24. In an article entitled *Nucleosome positioning at the replication fork*, Lucchini and others (2002. *EMBO* 20: 7294–302) state, "both the 'old' randomly segregated nucleosomes as well as the 'new' assembled histone octamers rapidly position themselves (within seconds) on the newly replicated DNA strands. . . ." Given this statement, how would one compare the distribution of nucleosomes and DNA in newly replicated chromatin? How could one experimentally test the distribution of nucleosomes on newly replicated chromosomes?

25. The human genome contains approximately 10^6 copies of an *Alu* sequence, one of the best-studied classes of short interspersed elements (SINEs), per haploid genome. Individual *Alus* share a 282-nucleotide consensus sequence followed by a 3′-adenine-rich tail region (Schmid. 1998. *Nuc. Acids Res.* 26: 4541–50). Given that there are approximately 3×10^9 bp per human haploid genome, about how many base pairs are spaced between each *Alu* sequence?

26. Below is a diagram of the general structure of the bacteriophage λ chromosome. Speculate on the mechanism by which it forms a closed ring upon infection of the host cell.

5′GGGCGGCGACCT—double-stranded region——3′
3′—double-stranded region——CCCGCCGCTGGA5′

27. Tandemly repeated DNA sequences with a repeat sequence of one to six base pairs [e.g., $(GACA)_n$] are called microsatellites and are common in eukaryotes. A particular subset, the trinucleotide repeats, is of great interest because of the role it plays in human neurodegenerative disorders (Huntington disease, myotonic dystrophy, spinal-bulbar muscular atrophy, spinocerebellar ataxia, and fragile X syndrome). Below are data modified from Toth and colleagues (2000. *Gen. Res.* 10: 967–81) regarding the location of microsatellites within and between genes. What general conclusions can be drawn from these data?

Percentage of Microsatellite DNA Sequences Within Genes and Between Genes

Taxonomic Group	*Within Genes*	*Between Genes*
Primates	7.4	92.6
Rodents	33.7	66.3
Arthropods	46.7	53.3
Yeasts	77.0	23.0
Other fungi	66.7	33.3

28. More information from the research effort in Problem 27 produced data regarding the pattern of the length of such repeats within genes. Each value in the following table represents the number of times a microsatellite of a particular sequence length, one to six bases long, is found with genes. For instance, in primates, a dinucleotide sequence (GC, for example) is found 10 times, while a trinucleotide is found 1126 times. In fungi, a repeat motif composed of 6 nucleotides (GACACC, for example) is found 219 times whereas a tetranucleotide repeat (GACA, for example) is found only 2 times. Analyze and interpret these data by indicating what general pattern is apparent regarding the distribution of various microsatellite lengths within genes. Of what significance might this general pattern display?

Distribution of Microsatellites by Unit Length Within Genes

	Length of Repeated Motif (bp)					
Taxonomic Group	*1*	*2*	*3*	*4*	*5*	*6*
Primates	49	10	1126	29	57	244
Rodents	62	70	1557	63	116	620
Arthropods	12	34	1566	0	21	591
Yeasts	36	19	706	7	52	330
Other fungi	9	4	381	2	35	219

29. In spite of the considerable medical and biological significance of repetitive DNA sequences, the factors that determine their genesis and genomic distribution remain uncertain. Misalignment of repetitive DNA strands and DNA polymerase slippage have been described as mechanisms causing variation in repeat number of *existing* repeats. Until recently, there has been little information relating to the *initial* creation of a microsatellite genomic region. Wilder and Holocher (2001. *Mol. Biol. Evol.* 18: 384–92) sequenced DNA surrounding numerous tetranucleotide microsatellite regions in several strains within two species of *Drosophila* and observed the sequences shows below. (a) Identify the microsatellite tetranucleotide motif. Is it a perfect motif? Using Pu to represent a purine and Py to represent a pyrimidine, symbolize the tetranucleotide repeat in a form similar to $(CPuGPy)_n$. (b) What is the sequence of the nonmicrosatellite region? Is it a perfectly conserved region among all the species and strains listed?

Species (strain)	*Base Sequence*
D. nigrodunni-1	5′-TCGATATAGCCATGTCCGTCTGTCCGTCTGT
D. nigrodunni-2	5′-TCGATATAGCCATGTCCGTCTGTCCGTCTGT
D. nigrodunni-3	5′-TCGATATAGCAATGTCCGTCTGTCCGTCTGT
D. dunni-1	5′-TCGATATAGCAATGTCCGTCTGTCCGTCTGT
D. dunni-2	5′-TCGATATAGCCATGTCCGTCTGTCCGTCTGT

30. Regarding the findings and your analysis of data in Problem 29, what significance might there be to having a highly conserved nonmicrosatellite region flanking a specific microsatellite type?

Selected Readings

Angelier, N. et al. 1984. Scanning electron microscopy of amphibian lampbrush chromosomes. *Chromosoma* 89: 243–53.

Beerman, W., and Clever, U. 1964. Chromosome puffs. *Sci. Am.* (Apr.) 210: 50–58.

Callan, H. G. 1986. *Lampbrush Chromosomes*. New York: Springer-Verlag.

Carbon, J. 1984. Yeast centromeres: Structure and function. *Cell* 37: 352–53.

Corneo, G. et al. 1968. Isolation and characterization of mouse and guinea pig satellite DNA. *Biochem.* 7: 4373–79.

DuPraw, E. J. 1970. *DNA and Chromosomes*. New York: Holt, Rinehart & Winston.

Gall, J. G. 1963. Kinetics of deoxyribonuclease on chromosomes. *Nature* 198: 36–38.

———— 1981. Chromosome structure and the C-value paradox. *J. Cell Biol.* 91: 3s–14s.

Green, B. R., and Burton, H. 1970. *Acetabularia* chloroplast DNA: Electron microscopic visualization. *Science* 168: 981–82.

Grivell, L. A. 1983. Mitochondrial DNA. *Sci. Am.* (Mar.) 248: 78–89.

Hewish, D. R., and Burgoyne, L. 1973. Chromatin sub-structure. The digestion of chromatin DNA at regularly spaced sites by a nuclear deoxyribonuclease. *Biochem. Biophys. Res. Comm.* 52: 504–10.

Hill, R. J., and Rudkin, G. T. 1987. Polytene chromosomes: The status of the band–interband question. *BioEss.* 7: 35–40.

Horn, P. J., and Peterson, C. L. 2002. Chromatin higher order folding: Wrapping up transcription. *Science* 297: 1824–27.

Jeffreys, A. J., Wilson, V., and Thein, S. L. 1985. Hypervariable minisatellite regions in human DNA. *Nature* 314: 66–73.

Korenberg, J. R., and Rykowski, M. C. 1988. Human genome organization: *Alu*, LINES, and the molecular organization of metaphase chromosome bands. *Cell* 53: 391–400.

Kornberg, R. D. 1975. Chromatin structure: A repeating unit of histones and DNA. *Science* 184: 868–71.

Kornberg, R. D., and Klug, A. 1981. The nucleosome. *Sci. Am.* (Feb.) 244: 52–64.

Lorch, Y., Zhang, N., and Kornberg, R. D. 1999. Histone octamer transfer by a chromatin-remodeling complex. *Cell* 96: 389–92.

Luger, K. et. al. 1997. Crystal structure of the nucleosome core particle at 2.8 Å resolution. *Nature* 389: 251–56.

Margulis, L. 1970. *Origin of Eukaryotic cells*. New Haven: Yale Univ. Press.

Moyzis, R. K. 1991. The human telomere. *Sci. Am.* (Aug) 265: 48–55.

Olins, A. L., and Olins, D. E. 1974. Spheroid chromatin units (ν bodies). *Science* 183: 330–32.

———— 1978. Nucleosomes: The structural quantum in chromosomes. *Am. Sci.* 66: 704–11.

Richmond, T. J., and Davey, C. A. 2003. The structure of DNA in the nucleosome core. *Nature* 423: 145–50.

Singer, M. F. 1982. SINEs and LINEs: Highly repeated short and long interspersed sequences in mammalian genomes. *Cell* 28: 433–34.

Sullivan, B. A., Blower, M. D., and Karpen, G. H. 2001. Determining centromere identity: Cyclical stories and forking paths. *Nat. Rev. Gen.* 2: 584–96.

Van Holde, K. E. 1989. *Chromatin*. New York: Springer-Verlag.

Verma, R. S. (ed.) 1988. *Heterochromatin: Molecular and Structural Aspects*. Cambridge: Cambridge Univ. Press.

Wolfe, A. 1998. *Chromatin: Structure and Function*. 3rd ed. San Diego: Academic Press.

Zakian, V. A. 1995. Telomeres: Beginning to understand the end. *Science* 270: 1601–06.

CHAPTER 13

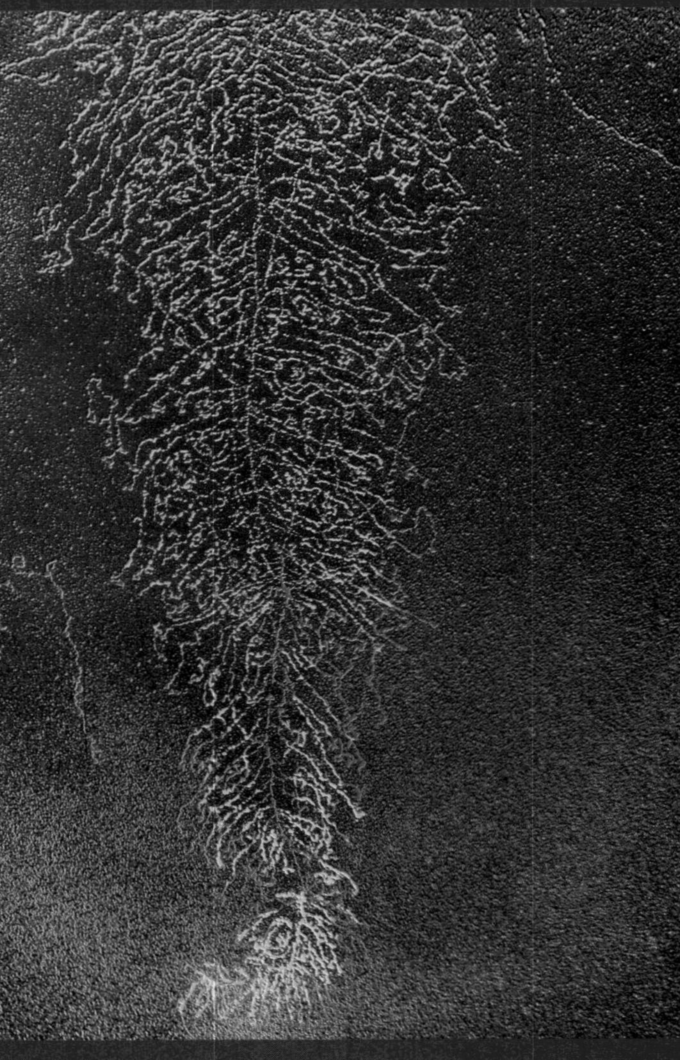

Electron micrograph visualizing the process of transcription. *(Prof. Oscar L. Miller/Science Photo Library/Photo Researchers, Inc.)*

The Genetic Code and Transcription

CHAPTER CONCEPTS

The linear sequence of deoxyribonucleotides making up DNA ultimately dictates the components constituting proteins, the end product of most genes. The central question is how such information stored as a nucleic acid is decoded into a protein. Figure 13–1 gives a simplified overview of how this transfer of information occurs. In the first step in gene expression, information on one of the two strands of DNA (the template strand) is transferred into an RNA complement through transcription. Once synthesized, this RNA acts as a "messenger" molecule bearing the coded information—hence its name, messenger RNA (mRNA). The mRNAs then associate with ribosomes, where decoding into proteins takes place.

In this chapter, we focus on the initial phases of gene expression by addressing two major questions. First, how is genetic information encoded? Second, how does the transfer from DNA to RNA occur, thus defining the process of transcription? As you shall see, ingenious analytical research established that the genetic code is written in units of three letters—ribonucleotides present in mRNA that reflect the stored information in genes. Each triplet code word directs the incorporation of a specific amino acid into a protein as it is synthesized. As we can predict based on our prior discussion of the replication of DNA, transcription is also a complex process dependent on a major polymerase enzyme and a cast of supporting proteins. We will explore what is known about transcription in bacteria and then contrast this prokaryotic model with the differences found in eukaryotes. Together, the information in this and the next chapter provides a comprehensive picture of molecular genetics, which serves as the most basic foundation for the understanding of living organisms. In Chapter 14, we will address how translation occurs and discuss the structure and function of proteins.

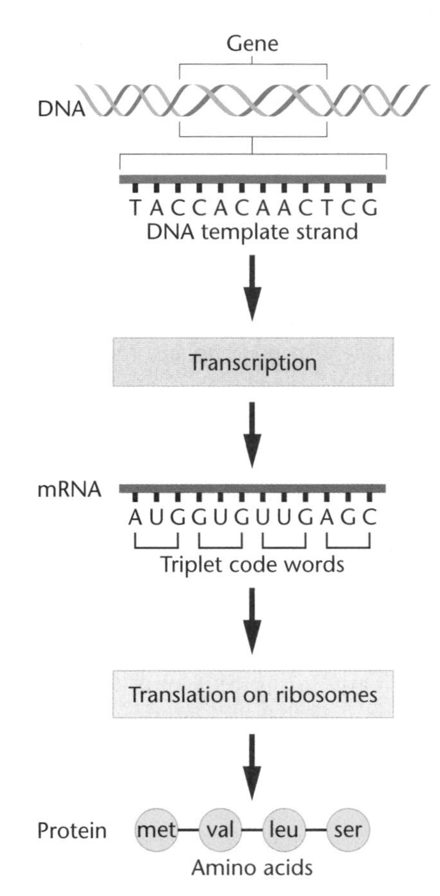

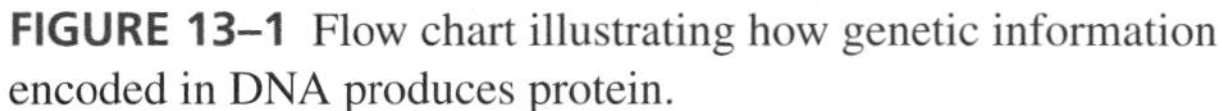
FIGURE 13–1 Flow chart illustrating how genetic information encoded in DNA produces protein.

HOW DO WE KNOW WHAT WE KNOW?

IN THIS CHAPTER, WE WILL FOCUS ON how the genetic information stored in DNA encodes proteins, the end products of most genes. We shall also elucidate the general mechanism by which the genetic code is deciphered in cells, the process of transcription. As you study this topic, you should try to answer several fundamental questions:

1. Why did geneticists believe, even before experimental evidence was obtained, that the genetic code is a triplet?
2. Once the triplet nature of the code seemed certain, how were the 64 possible triplet codes deciphered; that is, how do we know which three-letter code specifies each amino acid?
3. How do we know that expression of the information encoded in DNA involves an RNA intermediate?
4. After the genetic code was predicted experimentally, how did we determine these predictions were correct?
5. How do we know that the initial transcript of a eukaryotic gene contains noncoding sequences that must be removed before accurate translation into proteins can occur? ■

13.1 The Genetic Code Exhibits a Number of Characteristics

Before we consider the various analytical approaches that led to our current understanding of the genetic code, let's summarize the general features that characterize it.

MEDIA TUTORIAL Transcription

1. The genetic code is written in linear form, using the ribonucleotide bases that compose mRNA molecules as "letters." The ribonucleotide sequence is derived from the complementary nucleotide bases in DNA.
2. Each "word" within the mRNA contains three ribonucleotide letters. Each group of three ribonucleotides, called a **codon**, specifies one amino acid; the code is thus a **triplet codon**.
3. The code is **unambiguous**—each triplet specifies only a single amino acid.
4. The code is **degenerate**; that is, a given amino acid can be specified by more than one triplet codon. This is the case for 18 of the 20 amino acids.
5. The code contains "start" and "stop" signals, certain triplets that **initiate** and **terminate** translation.

6. No internal punctuation (such as a comma) is used in the code. Thus, the code is said to be **commaless**. Once translation of mRNA begins, the codons are read one after the other, with no breaks between them.
7. The code is **nonoverlapping**. Once translation commences, any single ribonucleotide at a specific location within the mRNA is part of only one triplet.
8. The code is nearly **universal**. With only minor exceptions, a single coding dictionary is used by almost all viruses, prokaryotes, archaea, and eukaryotes.

13.2 Early Studies Established the Basic Operational Patterns of the Code

In the late 1950s, before it became clear that mRNA is the intermediate that transfers genetic information from DNA to proteins, researchers thought that DNA itself might directly encode proteins during their synthesis. Because ribosomes had already been identified, the initial thinking was that information in DNA was transferred in the nucleus to the RNA of the ribosome, which served as the template for protein synthesis in the cytoplasm. This concept soon became untenable as accumulating evidence indicated that there was an unstable intermediate template. The RNA of ribosomes, on the other hand, was extremely stable. As a result, in 1961 researchers postulated the existence of **messenger RNA (mRNA)**. Once mRNA was discovered, it was clear that even though genetic information is stored in DNA, the code that is translated into proteins resides in RNA. The central question then was how only four letters—the four nucleotides—could specify 20 words—the amino acids.

The Triplet Nature of the Code

In the early 1960s, Sidney Brenner argued on theoretical grounds that the code had to be a triplet since three-letter words represent the minimal use of four letters to specify 20 amino acids. A code of four nucleotides, taken two at a time, for example, provides only 16 unique code words (4^2). A triplet code yields 64 words (4^3)—clearly more than the 20 needed—and is much simpler than a four-letter code which specifies 256 words (4^4).

Experimental evidence supporting the triplet nature of the code was subsequently derived from research by Francis Crick and his colleagues. Using phage T4, they studied **frameshift mutations**, which result from the addition or deletion of one or more nucleotides within a gene and subsequently the mRNA transcribed by it. The gain or loss of letters shifts the *frame of reading* during translation. Crick and his colleagues found that the gain or loss of one or two nucleotides caused a mutation, but when three nucleotides were involved, the frame of reading was reestablished (Figure 13–2). This would not occur if the code was anything other than a triplet. This work also suggested that most triplet codes are not blank, but rather encode amino acids, supporting the concept of a degenerate code.

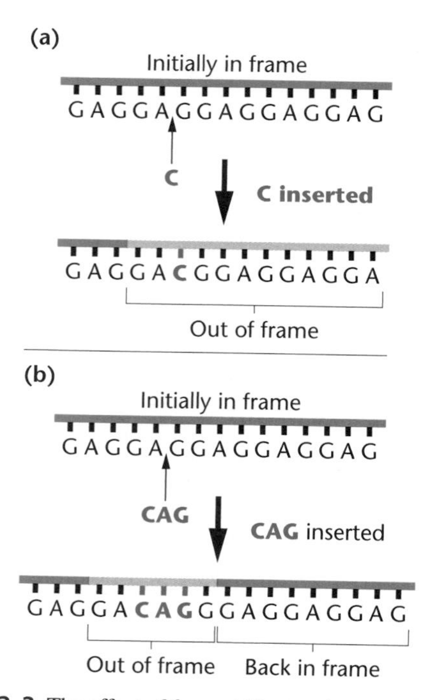

FIGURE 13–2 The effect of frameshift mutations on a DNA sequence with the repeating triplet sequence GAG. (a) The insertion of a single nucleotide shifts all subsequent triplet reading frames. (b) The insertion of three nucleotides changes only two triplets, but the frame of reading is then reestablished to the original sequence.

13.3 Studies by Nirenberg, Matthaei, and Others Deciphered the Code

In 1961, Marshall Nirenberg and J. Heinrich Matthaei deciphered the first specific coding sequences, which served as a cornerstone for the complete analysis of the genetic code. Their success, as well as that of others who made important contributions to breaking the code, was dependent on the use of two experimental tools—an ***in vitro* (cell-free) protein-synthesizing system** and an enzyme, **polynucleotide phosphorylase** which enabled the production of synthetic mRNAs. These mRNAs are templates for polypeptide synthesis in the cell-free system.

Cell-Free Polypeptide Synthesis

In the cell-free system, amino acids are first incorporated into polypeptide chains. This *in vitro* mixture must contain the essential factors for protein synthesis in the cell: ribosomes, tRNAs, amino acids, and other molecules essential to translation. In order to follow (or trace) protein synthesis, one or more of the amino acids must be radioactive. Finally, an mRNA must be added, which serves as the template that will be translated.

In 1961, mRNA had yet to be isolated. However, use of the enzyme polynucleotide phosphorylase allowed the artificial synthesis of RNA templates, which could be added to the cell-free system. This enzyme, isolated from bacteria, catalyzes the reaction shown in Figure 13–3. Discovered in 1955 by Marianne Grunberg-Manago and Severo Ochoa, the enzyme functions metabolically in bacterial cells to

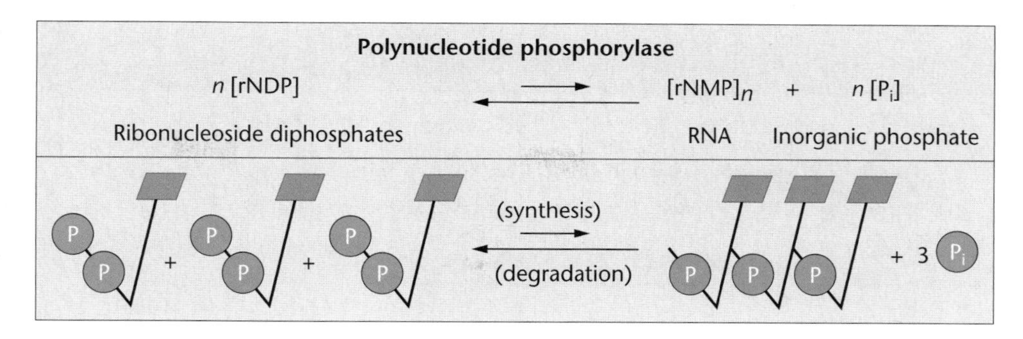

FIGURE 13–3 The reaction catalyzed by the enzyme polynucleotide phosphorylase. Note that the equilibrium of the reaction favors the degradation of RNA but can be "forced" in the direction favoring synthesis.

degrade RNA. However, *in vitro*, with high concentrations of ribonucleoside diphosphates, the reaction can be "forced" in the opposite direction to synthesize RNA, as shown.

In contrast to RNA polymerase, polynucleotide phosphorylase does not require a DNA template. As a result, each addition of a ribonucleotide is random, based on the relative concentration of the four ribonucleoside diphosphates added to the reaction mixtures. The probability of the insertion of a specific ribonucleotide is proportional to the availability of that molecule, relative to other available ribonucleotides. *This point is absolutely critical to understanding the work of Nirenberg and others in the ensuing discussion.*

The cell-free system for protein synthesis and the availability of synthetic mRNAs provided a means of deciphering the ribonucleotide composition of various triplets encoding specific amino acids.

The Use of Homopolymers

In their initial experiments, Nirenberg and Matthaei synthesized **RNA homopolymers**, each with only one type of ribonucleotide. Therefore, the mRNA added to the *in vitro* system was either UUUUUU . . . , AAAAAA . . . , CCCCCC . . . , or GGGGGG . . . . They tested each mRNA and were able to determine which, if any, amino acids were incorporated into newly synthesized proteins. To do this, the researchers labeled 1 of the 20 amino acids added to the *in vitro* system and conducted a series of experiments, each with a different radioactively labeled amino acid.

For example, in experiments using ^{14}C-phenylalanine (Table 13.1), Nirenberg and Matthaei concluded that the message poly U (polyuridylic acid) directed the incorporation of only phenylalanine into the homopolymer polyphenylalanine. Assuming the validity of a triplet code, they determined the first specific codon assignment—UUU

TABLE 13.1 Incorporation of ^{14}C-Phenylalanine into Protein

Artificial mRNA	*Radioactivity (counts/min)*
None	44
Poly U	39,800
Poly A	50
Poly C	38

Source: After Nirenberg and Matthaei (1961).

codes for phenylalanine. Using similar experiments, they quickly found that AAA codes for lysine and CCC codes for proline. Poly G was not an adequate template, probably because the molecule folds back upon itself. Thus, the assignment for GGG had to await other approaches.

Note that the specific triplet codon assignments were possible only because homopolymers were used. This method yields only the composition of triplets, but since three identical letters can have only one possible sequence (e.g., UUU), the actual codons were identified.

Mixed Copolymers

With these techniques in hand, Nirenberg and Matthaei, and Ochoa and coworkers turned to the use of **RNA heteropolymers**. In this type of experiment, two or more different ribonucleoside diphosphates are added in combination to form the artificial message. The researchers reasoned that if they knew the relative proportion of each type of ribonucleoside diphosphate, they could predict the frequency of any particular triplet codon occurring in the synthetic mRNA. If they then added the mRNA to the cell-free system and ascertained the percentage of any particular amino acid present in the new protein, they could analyze the results and predict the *composition* of triplets specifying particular amino acids.

This approach is shown in Figure 13–4. Suppose that A and C are added in a ratio of 1A:5C. The insertion of a ribonucleotide at any position along the RNA molecule during its synthesis is determined by the ratio of A:C. Therefore, there is a $1/6$ chance for an A and a $5/6$ chance for a C to occupy each position. On this basis, we can calculate the frequency of any given triplet appearing in the message.

For AAA, the frequency is $(1/6)^3$, or about 0.4%. For AAC, ACA, and CAA, the frequencies are identical—that is, $(1/6)^2(5/6)$, or about 2.3% for each letter. Together, all three 2A:1C triplets account for 6.9% of the total three-letter sequences. In the same way, each of three 1A:2C triplets accounts for $(1/6)(5/6)^2$, or 11.6% (or a total of 34.8%); CCC is represented by $(5/6)^3$, or 57.9% of the triplets.

By examining the percentages of any given amino acid incorporated into the protein synthesized under the direction of this message, we can propose probable base compositions for each amino acid (Figure 13–4). Since proline appears 69% of the time, we could propose that proline is encoded by CCC (57.9%) and one triplet of 2C:1A (11.6%). Histidine, at 14%, is probably coded by one 2C:1A (11.6%) and one 1C:2A (2.3%). Threonine, at 12%, is likely coded by only one 2C:1A. Asparagine and

Possible compositions	Probability of occurrence of any triplet	Possible triplets	Final %
3A	$(1/6)^3 = 1/216 = 0.4\%$	AAA	0.4
1C:2A	$(5/6)(1/6)^2 = 5/216 = 2.3\%$	AAC ACA CAA	3 x 2.3 = 6.9
2C:1A	$(5/6)^2(1/6) = 25/216 = 11.6\%$	ACC CAC CCA	3 x 11.6 = 34.8
3C	$(5/6)^3 = 125/216 = 57.9\%$	CCC	57.9
			100.0

Chemical synthesis of message

RNA
C C C C C C C C A C C C C C C C A A C C A C C C C C A C C C C C A C C C A A

Translation of message

Percentage of amino acids in protein		Probable base-composition assignments
Lysine	<1	AAA
Glutamine	2	1C:2A
Asparagine	2	1C:2A
Threonine	12	2C:1A
Histidine	14	2C:1A, 1C:2A
Proline	69	CCC, 2C:1A

FIGURE 13–4 Results and interpretation of a mixed copolymer experiment where a ratio of 1A:5C is used (1/6A : 5/6C).

glutamine each appear to be coded by one of the 1C:2A triplets, and lysine appears to be coded by AAA.

Using as many as all four ribonucleotides to construct the mRNA, the researchers conducted many similar experiments. Although determining the *composition* of the triplet code words for all 20 amino acids represented a significant breakthrough, the *specific sequences* of triplets were still unknown—other approaches were needed.

The Triplet Binding Assay

It was not long before more advanced techniques were developed. In 1964, Nirenberg and Philip Leder developed the **triplet binding assay**, which led to specific assignments of triplets. The technique took advantage of the observation that ribosomes, when presented *in vitro* with an RNA sequence as short as three ribonucleotides, will bind to it and form a complex similar to that found *in vivo*. The triplet acts like a codon in mRNA, attracting the complementary sequence within tRNA (Figure 13–5). The triplet sequence in tRNA that is complementary to a codon of mRNA is an **anticodon**.

Although it was not yet feasible to chemically synthesize long stretches of RNA, triplets of known sequence could be synthesized in the laboratory to serve as templates. All that was needed was a method to determine which tRNA–amino acid was bound to the triplet RNA–ribosome complex. The test system Nirenberg and Leder devised was quite simple. The amino acid to be tested was made radioactive, and a charged tRNA was produced. Because

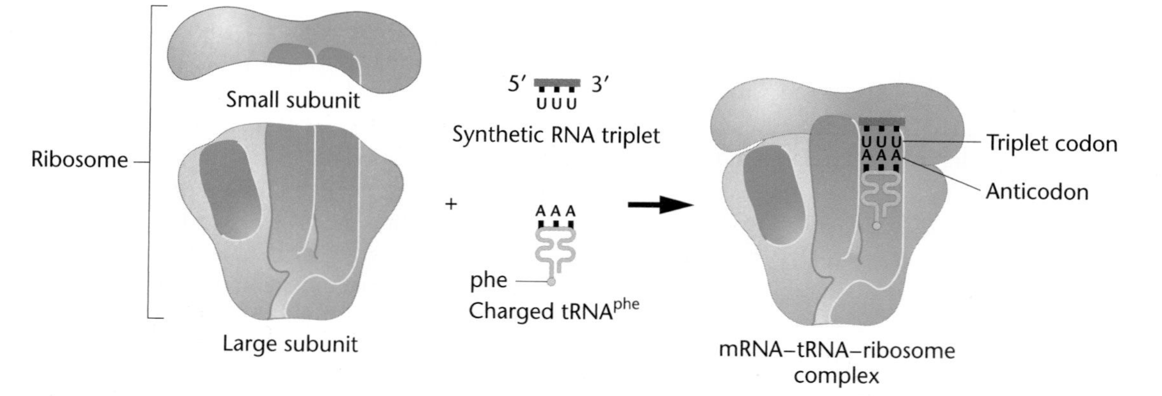

FIGURE 13–5 An example of the triplet binding assay. The UUU triplet acts as a codon, attracting the complementary $tRNA^{phe}$ anticodon AAA.

TABLE 13.2 Amino Acid Assignments to Specific Trinucleotides Derived from the Triplet Binding Assay

Trinucleotides	*Amino Acid*
AAA AAG	Lysine
AUG	Methionine
AUU AUG AUA	Isoleucine
CCG CCA	Proline
CCU CCC	Proline
CUC CUA CUG	Leucine
GAA GAG	Glutamic acid
UCA UCG	Serine
UCU UCC	Serine
UGU UGC	Cysteine
UUA UUG CUU	Leucine
UUU UUC	Phenylalanine

codon compositions were known, researchers could narrow the range of amino acids that should be tested for each specific triplet.

The radioactively charged tRNA, the RNA triplet, and ribosomes are incubated together and then passed through a nitrocellulose filter, which retains the larger ribosomes but not the other smaller components, such as charged tRNA. If radioactivity is not retained on the filter, an incorrect amino acid has been tested. But if radioactivity remains on the filter, it is retained because the charged tRNA has bound to the triplet associated with the ribosome. When this occurs, a specific codon assignment can be made.

Work proceeded in several laboratories, and in many cases clear-cut, unambiguous results were obtained. Table 13.2, for example, shows 26 triplets assigned to 9 amino acids. However, in some cases, the degree of triplet binding was inefficient and assignments were not possible. Eventually, about 50 of the 64 triplets were assigned. These specific assignments of triplets to amino acids led to two major conclusions. First, the genetic code is *degenerate*; that is, one amino acid can be specified by more than one triplet. Second, the code is *unambiguous*. That is, a single triplet specifies only one amino acid. As you shall see later in this chapter, these conclusions have been upheld with only minor exceptions. The triplet binding technique was a major innovation in deciphering the genetic code.

Repeating Copolymers

Yet another innovative technique used to decipher the genetic code was developed in the early 1960s by Gobind Khorana, who chemically synthesized long RNA molecules consisting of short sequences repeated many times. First, he created shorter sequences (e.g., di-, tri-, and tetranucleotides), which were then replicated many times and finally joined enzymatically to form the long polynucleotides. As shown in Figure 13–6, a dinucleotide made in this way is converted to an mRNA with two repeating triplets. A trinucleotide is converted to an mRNA with three potential triplets, depending on the point at which initiation occurs, and a tetranucleotide creates four repeating triplets.

When these synthetic messages were added to a cell-free system, the predicted number of amino acids incorporated was upheld. Several examples are shown in Table 13.3. When the data were combined with those on composition assignment and triplet binding, specific assignments were possible.

One example of specific assignments made from this data demonstrates the value of Khorana's approach. Consider the following three experiments in concert with one another.

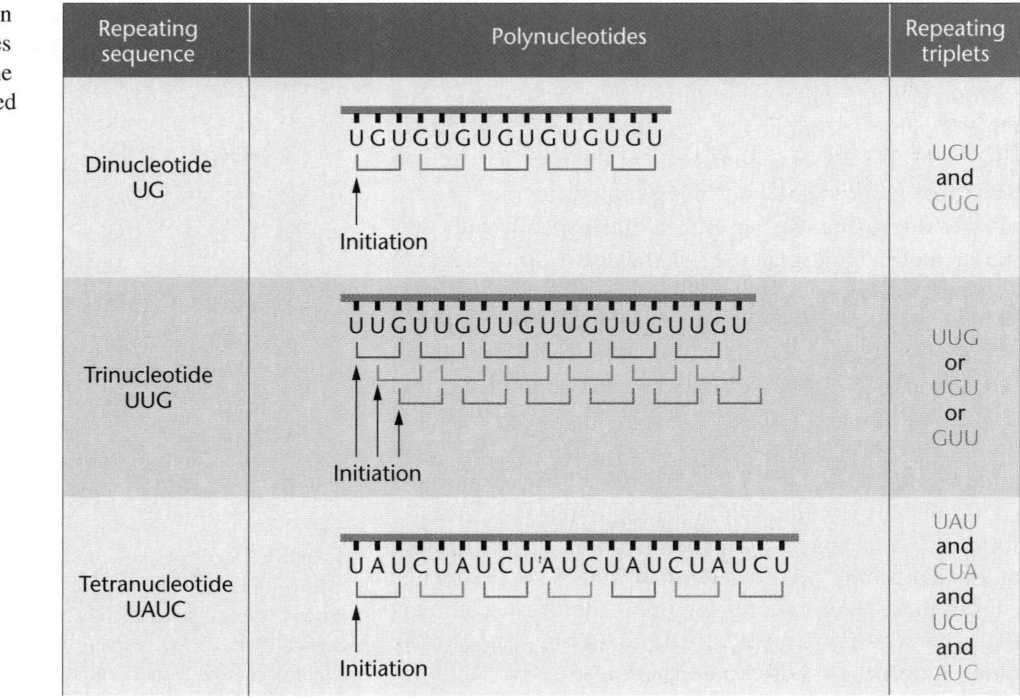

FIGURE 13–6 The conversion of di-, tri-, and tetranucleotides into repeating copolymers. The triplet codons that are produced in each case are shown.

TABLE 13.3 Amino Acids Incorporated Using Repeated Synthetic Copolymers of RNA

Repeating Copolymer	*Codons Produced*	*Amino Acids in Polypeptides*
UG	UGU	Cysteine
	GUG	Valine
AC	ACA	Threonine
	CAC	Histidine
UUC	UUC	Phenylalanine
	UCU	Serine
	CUU	Leucine
AUC	AUC	Isoleucine
	UCA	Serine
	CAU	Histidine
UAUC	UAU	Tyrosine
	CUA	Leucine
	UCU	Serine
	AUC	Isoleucine
GAUA	GAU	—
	AGA	None
	UAG	—
	AUA	—

The repeating trinucleotide sequence UUCUUCUUC... produces three possible triplets: UUC, UCU, and CUU, depending on the initiation point. When placed in a cell-free translation system, the polypeptides containing phenylalanine (phe), serine (ser), and leucine (leu) are produced. On the other hand, the repeating dinucleotide sequence UCUCUCUC... produces the triplets UCU and CUC with the incorporation of leucine and serine into the polypeptide. These results indicate that the triplets UCU and CUC specify leucine and serine, but they don't specify which triplet specifies which amino acid. We can further conclude that *either* the CUU *or* the UUC triplet also encodes leucine or serine, while the other encodes phenylalanine.

To derive more specific information, let's examine the results of using the repeating tetranucleotide sequence UUAC, which produces the triplets UUA, UAC, ACU, and CUU. The CUU triplet is one of the two that interest us. Three amino acids are incorporated: leucine, threonine, and tyrosine. Because CUU must specify only serine or leucine and because, of these two, only leucine appears, we can conclude that CUU specifies leucine.

Once this is established, we can logically determine all other assignments. Of the two triplet pairs remaining (UUC and UCU from the first experiment and UCU and CUC from the second experiment), whichever triplet is common to both must encode serine. This is UCU. By elimination, we find that UUC encodes phenylalanine and CUC encodes leucine. While the logic must be carefully followed, four specific triplets encoding three different amino acids have been assigned from these experiments.

From these interpretations, Khorana reaffirmed triplets that were already deciphered and filled in gaps left from other approaches. For example, the use of two tetranucleotide sequences, GAUA and GUAA, suggested that at least two triplets were *termination codons*. He reached this conclusion because neither of these sequences directed the incorporation of any amino acids into a polypeptide. Since there are no triplets common to both messages, he predicted that each repeating sequence would contain at least one triplet that terminates protein synthesis. Table 13.3 lists the possible triplets for the poly-(GAUA) sequence, of which UAG is a termination codon.

13.4 The Coding Dictionary Reveals the Function of the 64 Triplets

The various techniques used to decipher the genetic code have yielded a dictionary of 61 triplet codon–amino acid assignments. The remaining three triplets are termination signals, not specifying any amino acid.

Degeneracy and the Wobble Hypothesis

A general pattern of triplet codon assignments becomes apparent when we look at the genetic coding dictionary. Figure 13–7 designates the assignments in a particularly illustrative form first suggested by Francis Crick.

Most evident is that the code is degenerate, as the early researchers predicted. That is, almost all amino acids are specified by two, three, or four different codons. Three amino acids (serine, arginine, and leucine) are each

Second position

First position (5′-end)	U	C	A	G	Third position (3′-end)
U	UUU phe	UCU ser	UAU tyr	UGU cys	U
	UUC phe	UCC ser	UAC tyr	UGC cys	C
	UUA leu	UCA ser	UAA Stop	UGA Stop	A
	UUG leu	UCG ser	UAG Stop	UGG trp	G
C	CUU leu	CCU pro	CAU his	CGU arg	U
	CUC leu	CCC pro	CAC his	CGC arg	C
	CUA leu	CCA pro	CAA gln	CGA arg	A
	CUG leu	CCG pro	CAG gln	CGG arg	G
A	AUU ile	ACU thr	AAU asn	AGU ser	U
	AUC ile	ACC thr	AAC asn	AGC ser	C
	AUA ile	ACA thr	AAA lys	AGA arg	A
	AUG met	ACG thr	AAG lys	AGG arg	G
G	GUU val	GCU ala	GAU asp	GGU gly	U
	GUC val	GCC ala	GAC asp	GGC gly	C
	GUA val	GCA ala	GAA glu	GGA gly	A
	GUG val	GCG ala	GAG glu	GGG gly	G

Initiation Termination

FIGURE 13–7 The coding dictionary. AUG encodes methionine, which initiates most polypeptide chains. All other amino acids except tryptophan, which is encoded only by UGG, are represented by two to six triplets. The triplets UAA, UAG, and UGA are termination signals and do not encode any amino acids.

encoded by six different codons. Only tryptophan and methionine are encoded by single codons.

Also evident is the *pattern* of degeneracy. Most often, in a set of codons specifying the same amino acid, the first two letters are the same, with only the third differing. Crick discerned a pattern in the degeneracy at the third position, and in 1966, he postulated the **wobble hypothesis**.

Crick's hypothesis first predicted that the initial two ribonucleotides of triplet codes are more critical than the third in attracting the correct tRNA. He postulated that hydrogen bonding at the third position of the codon–anticodon interaction is less constrained and need not adhere as specifically to the established base-pairing rules. The wobble hypothesis thus proposes a more flexible set of base-pairing rules at the third position of the codon (Table 13.4).

This relaxed base-pairing requirement, or "wobble," allows the anticodon of a single form of tRNA to pair with more than one triplet in mRNA. Consistent with the wobble hypothesis and degeneracy, U at the first position (the 5′-end) of the tRNA anticodon may pair with A or G at the third position (the 3′-end) of the mRNA codon, and G may likewise pair with U or C. Inosine, one of the modified bases found in tRNA, may pair with C, U, or A. Applying these wobble rules, a minimum of about 30 different tRNA species is necessary to accommodate the 61 triplets specifying an amino acid. If nothing more, wobble can be considered a potential economy measure, provided that the fidelity of translation is not compromised. Current estimates are that 30–40 tRNA species are present in bacteria and up to 50 tRNA species exist in animal and plant cells.

Initiation and Termination

Initiation of protein synthesis is a highly specific process. In bacteria (in contrast to the *in vitro* experiments discussed earlier), the initial amino acid inserted into all polypeptide chains is a modified form of methionine—***N*-formylmethionine** (**fmet**). Only one codon, AUG, codes for methionine, and it is sometimes called the **initiator codon**. However, when AUG appears internally in mRNA rather than at an initiating position, unformylated methionine is inserted into the polypeptide chain. Rarely, another triplet, GUG, specifies methionine during initiation, although it is not clear why this happens, since GUG normally encodes valine.

In bacteria, either the formyl group is removed from the initial methionine upon the completion of protein synthesis or the entire formylmethionine residue is removed. In eukaryotes, methionine is also the initial amino acid during polypeptide synthesis. However, it is not formylated.

As mentioned in the preceding section, three other triplets (UAG, UAA, and UGA) serve as **termination codons**, punctuation signals that do not code for any amino acid. They are not recognized by a tRNA molecule, and translation terminates when they are encountered. Mutations that produce any of the three triplets internally in a gene will also result in termination. Consequently, only a partial polypeptide has been synthesized when it is prematurely released from the ribosome. When such a change occurs in the DNA, it is called a **nonsense mutation**.

TABLE 13.4 Codon–Anticodon Base-Pairing Rules

Base at First Position (5′, end) of tRNA	*Base at Third Position (3′, end) of mRNA*
A	U
C	G
G	C or U
U	A or G
I	A, U, or C

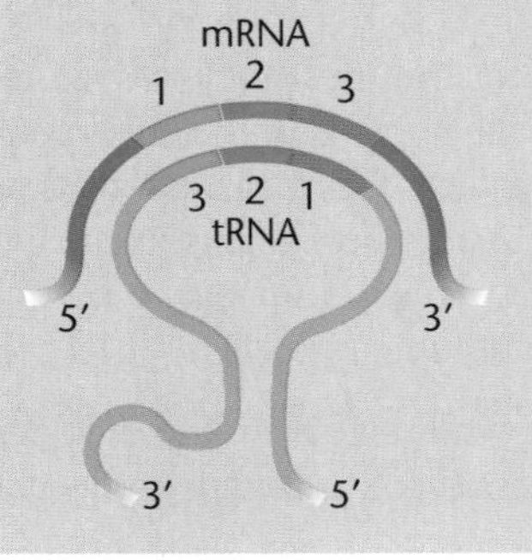

mRNA
1 2 3
3 2 1
tRNA
5′
3′
3′
5′

13.5 The Genetic Code Has Been Confirmed in Studies of Bacteriophage MS2

The various aspects of the genetic code discussed thus far yield a fairly complete picture. The code is triplet in nature, degenerate, unambiguous, and commaless, but it contains punctuation with respect to start and stop signals. These individual principles have been confirmed by a detailed analysis of the RNA-containing **bacteriophage MS2** by Walter Fiers and his coworkers.

MS2 is a bacteriophage that infects *E. coli*. Its nucleic acid (RNA) contains only about 3500 ribonucleotides, making up only 3 genes. These genes specify a coat protein, an RNA-directed replicase, and a maturation protein (the A protein). This simple system of a small genome and few gene products enabled Fiers and his colleagues to sequence the genes and their products. The amino acid sequence of the coat protein was completed in 1970, and the nucleotide sequence of the gene and several nucleotides on each of its ends was reported in 1972.

When the chemical natures of the gene and protein are compared, a *colinear relationship* is evident. That is, based on the coding dictionary, the linear sequence of nucleotides (and thus the sequence of triplet codons) corresponds precisely with the linear sequence of amino acids in the protein. Furthermore, the codon for the first amino acid is AUG, the common initiator codon. The codon for the last amino acid is followed by two consecutive termination codons, UAA and UAG.

By 1976, the other two genes and their protein products were sequenced. The analysis clearly showed that the genetic code in this virus was identical to that established in bacterial systems. Other evidence suggested that the code was also identical in eukaryotes, thus providing confirmation of what seemed to be a universal genetic code.

13.6 The Genetic Code Is Nearly Universal

Between 1960 and 1978, it was generally assumed that the genetic code would be found to be universal, applying equally to viruses, bacteria, archaea, and eukaryotes. Certainly, the nature of mRNA and the translation machinery seemed to be very similar in these organisms. For example, cell-free systems derived from bacteria can translate eukaryotic mRNAs. Poly U stimulates translation of polyphenylalanine in cell-free systems when the components are derived from eukaryotes. Many recent studies involving recombinant DNA technology (see Chapter 17) reveal that eukaryotic genes can be inserted into bacterial cells, which are then transcribed and translated. Within eukaryotes, mRNAs from mice and rabbits have been injected into amphibian eggs and efficiently translated. For the many eukaryotic genes that have been sequenced, notably those for hemoglobin molecules, the amino acid sequence of the encoded proteins adheres to the coding dictionary established from bacterial studies.

However, several 1979 reports on the coding properties of DNA derived from mitochondria of yeast and humans (mtDNA) undermined the principle of the universality of the genetic language. Since then, mtDNA has been examined in many other organisms.

Cloned mtDNA fragments have been sequenced and compared with the amino acid sequences of various mitochondrial proteins, revealing several exceptions to the coding dictionary (Table 13.5). Most surprising is that the codon UGA, normally specifying termination, specifies the insertion of tryptophan during translation in yeast and human mitochondria. In yeast mitochondria, threonine is inserted instead of leucine when CUA is encountered in mRNA. In human mitochondria, AUA, which normally specifies isoleucine, directs the internal insertion of methionine.

In 1985, several other exceptions to the standard coding dictionary were discovered in the bacterium *Mycoplasma capricolum* and in the nuclear genes of the protozoan ciliates *Paramecium*, *Tetrahymena*, and *Stylonychia*. For example, as shown in Table 13.5, one alteration converts the termination codons (UGA) to tryptophan, yet several others convert the normal termination codon (UAA and UAG) to glutamine. These changes are significant because both a prokaryote and several eukaryotes are involved, representing distinct species that have evolved separately over a long period of time.

Note the apparent pattern in several of the altered codon assignments. The change in coding capacity involves only a shift in recognition of the third, or wobble, position. For example, AUA specifies isoleucine in the cytoplasm and methionine in the mitochondrion, but in the cytoplasm, methionine is specified by AUG. Likewise, UGA calls for termination in the cytoplasm, but it specifies tryptophan in the mitochondrion; and in the cytoplasm, tryptophan is specified by UGG. It has been suggested that such changes in codon recognition may represent an evolutionary trend toward reducing the number of tRNAs needed in mitochondria; only 22 tRNA species are encoded in human mitochondria, for example. However, until more examples are found, the differences must be considered to be exceptions to the previously established general coding rules.

TABLE 13.5 Exceptions to the Universal Code

Triplet	*Normal Code Word*	*Altered Code Word*	*Source*
UGA	Termination	Tryptophan	Human and yeast mitochondria; *Mycoplasma*
CUA	Leucine	Threonine	Yeast mitochondria
AUA	Isoleucine	Methionine	Human mitochondria
AGA AGG	Arginine	Termination	Human mitochondria
UAA	Termination	Glutamine	*Paramecium; Tetrahymena; Stylonychia*
UAG	Termination	Glutamine	*Paramecium*

13.7 Transcription Synthesizes RNA on a DNA Template

Even while the genetic code was being studied, it was quite clear that proteins were the end products of many genes. Thus, while some geneticists attempted to elucidate the code, other research efforts focused on the nature of genetic expression. The central question was how DNA, a nucleic acid, could specify a protein composed of amino acids.

The complex multistep process begins with the transfer of genetic information stored in DNA to RNA. The process by which RNA molecules are synthesized on a DNA template is called **transcription**. It results in an mRNA molecule complementary to the gene sequence of one of the double helix's two strands. Each triplet codon in the mRNA is, in turn, complementary to the anticodon region of its corresponding tRNA as the amino acid is correctly inserted into the polypeptide chain during translation. The significance of transcription is enormous, for it is the initial step in the process of information flow within the cell. The idea that RNA is involved as an intermediate molecule in the process of information flow between DNA and protein was suggested by the following observations:

1. DNA is, for the most part, associated with chromosomes in the nucleus of the eukaryotic cell. However, protein synthesis occurs in association with ribosomes located outside the nucleus in the cytoplasm. Therefore, DNA does not appear to participate directly in protein synthesis.
2. RNA is synthesized in the nucleus of eukaryotic cells, where DNA is found, and is chemically similar to DNA.

3. Following its synthesis, most RNA migrates to the cytoplasm, where protein synthesis (translation) occurs.
4. The amount of RNA is generally proportional to the amount of protein in a cell.

Collectively, these observations suggested that genetic information, stored in DNA, is transferred to an RNA intermediate, which directs the synthesis of proteins. As with most new ideas in molecular genetics, the initial supporting experimental evidence was based on studies of bacteria and their phages. It was clearly established that during initial infection, RNA synthesis preceded phage protein synthesis and that the RNA is complementary to phage DNA.

The results of these experiments agree with the concept of a messenger RNA (mRNA) being made on a DNA template and then directing the synthesis of specific proteins in association with ribosomes. This concept was formally proposed by François Jacob and Jacques Monod in 1961 as part of a model for gene regulation in bacteria. Since then, mRNA has been isolated and studied thoroughly. There is no longer any question about its role in genetic processes.

13.8 RNA Polymerase Directs RNA Synthesis

To prove that RNA can be synthesized on a DNA template, it was necessary to demonstrate that there is an enzyme capable of directing this synthesis. By 1959, several investigators, including Samuel Weiss, had independently isolated such a molecule from rat liver. Called **RNA polymerase**, it has the same general substrate requirements as does DNA polymerase, the major exception being that the substrate nucleotides contain the ribose rather than the deoxyribose form of the sugar. Unlike DNA polymerase, no primer is required to initiate synthesis; the initial base remains as a nucleoside triphosphate (NTP). The overall reaction summarizing the synthesis of RNA on a DNA template can be expressed as

$$n(\text{NTP}) \xrightarrow[\text{enzyme}]{\text{DNA}} (\text{NMP})_n + n(\text{PP}_i)$$

As this equation reveals, nucleoside triphosphates (NTPs) are substrates for the enzyme, which catalyzes the polymerization of nucleoside monophosphates (NMPs), or nucleotides, into a polynucleotide chain $(\text{NMP})_n$. Nucleotides are linked during synthesis by 5′-to-3′ phosphodiester bonds (see Figure 10–10). The energy created by cleaving the triphosphate precursor into the monophosphate form drives the reaction, and inorganic pyrophosphates (PP_i) are produced.

A second equation summarizes the sequential addition of each ribonucleotide as the process of transcription progresses:

$$(\text{NMP})_n + \text{NTP} \xrightarrow[\text{enzyme}]{\text{DNA}} (\text{NMP})_{n+1} + \text{PP}_i$$

Here each transcription step involves the addition of one ribonucleotide (NMP) to the growing polyribonucleotide chain $(\text{NMP})_n$, using a nucleoside triphosphate (NTP) as the precursor.

RNA polymerase from *E. coli* has been extensively characterized and shown to consist of subunits designated α, β, β', and σ. The active form of the enzyme, the **holoenzyme**, contains the subunits α_2, β, β', σ and has a molecular weight of almost 500,000 Da. Of these subunits, it is the β and β' polypeptides that provide the catalytic basis and active site for transcription. As we shall see, another subunit called the **σ (sigma) factor** plays a regulatory function in the initiation of RNA transcription.

While there is but a single form of the enzyme in *E. coli*, there are several different σ factors, creating variations of the polymerase holoenzyme. On the other hand, eukaryotes display three distinct forms of RNA polymerase, each consisting of a greater number of polypeptide subunits than in bacteria.

Promoters, Template Binding, and the σ Subunit

Transcription results in the synthesis of a single-stranded RNA molecule complementary to a region along only one strand of the DNA double helix. For simplicity, let's call the transcribed DNA strand the **template strand** and its complement the **partner strand**.

The initial step is **template binding** (Figure 13–8). In bacteria, the site of this initial binding is established when the RNA polymerase σ subunit recognizes specific DNA sequences called **promoters**. These regions are located in the 5′-region upstream from the point of initial transcription of a gene. It is believed that the enzyme "explores" a length of DNA until it recognizes the promoter region and binds to about 60 nucleotide pairs of the helix, 40 of which are upstream from the point of initial transcription. Once this occurs, the helix is denatured or unwound locally, making the DNA template accessible to the action of the enzyme.

The importance of promoter sequences cannot be overemphasized. They govern the efficiency of the initiation of transcription. In bacteria, both strong promoters and weak promoters have been discovered. These promoters cause a variation in time of initiation from once every 1–2 seconds to only once every 10–20 minutes. Mutations in promoter sequences may severely reduce the initiation of gene expression. Because the interaction of promoters with RNA polymerase governs transcription, the nature of the binding between them is at the heart of discussions concerning genetic regulation, the subject of Chapter 16. While we will pursue more detailed information involving promoter-enzyme interactions, we must address three points here.

The first point is the concept of **consensus sequences** of DNA. These sequences are similar (homologous) in different genes of the same organism or in one or more genes of related organisms. Their conservation throughout evolution attests to the critical nature of their role in biological processes. Two such sequences have been found in bacterial promoters. One, TATAAT, is located 10 nucleotides upstream from the site of initial transcription (the **−10 region**, or **Pribnow box**). The other, TTGACA, is located 35 nucleotides upstream

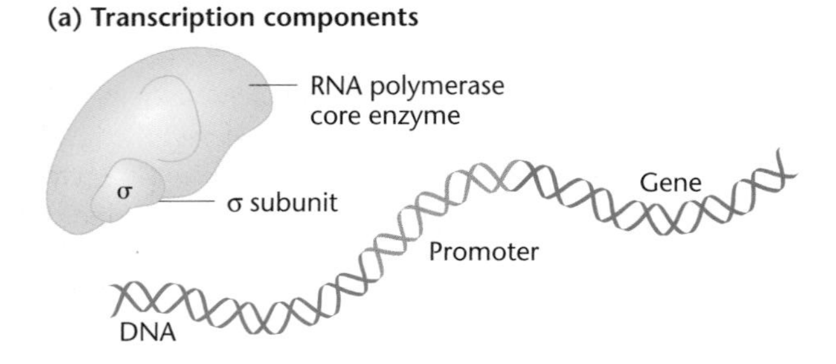

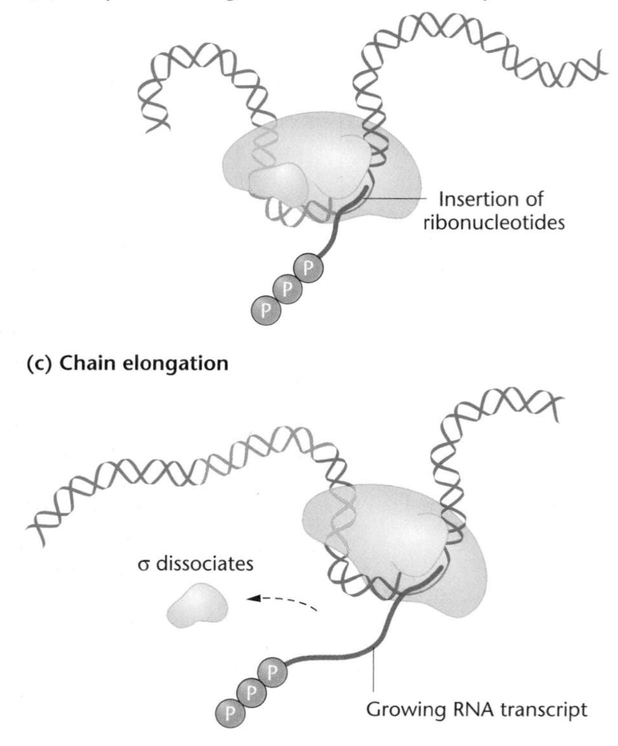

FIGURE 13–8 The early stages of transcription in prokaryotes, showing (a) the components of the process; (b) template binding at the −10 site involving the σ subunit of RNA polymerase and subsequent initiation of RNA synthesis; and (c) chain elongation, after the σ subunit has dissociated from the transcription complex and the enzyme moves along the DNA template.

(the **−35 region**). Mutations in either region diminish transcription, often severely. In most eukaryotic genes studied, a consensus sequence comparable to that in the −10 region has been recognized and is called the TATA box.

The second point is that the degree of RNA polymerase binding to different promoters varies greatly, which causes the variable gene expression mentioned earlier. Currently, this is attributed to sequence variation in the promoters.

A final general point involves the σ subunit in bacteria. The major form is designated as σ^{70}, based on its molecular weight of 70 kilodaltons (kDa). The promoters of most bacterial genes recognize this form; however, several alternative forms of RNA polymerase in *E. coli* have unique σ subunits associated with them (e.g., σ^{28}, σ^{32}, σ^{38}, and σ^{54}). Each recognize different promoter sequences and provide specificity to the initiation of transcription.

Initiation, Elongation, and Termination of RNA Synthesis

Once it has recognized and bound to the promoter (Figure 13–8b), RNA polymerase catalyzes **initiation**, the insertion of the first 5′-ribonucleoside triphosphate, which is complementary to the first nucleotide at the start site of the DNA template strand (Figure 13–8b). As we noted earlier, no primer is required. Subsequent ribonucleotide complements are inserted and linked by phosphodiester bonds as RNA polymerization proceeds. This process of **chain elongation** (Figure 13–8c), continues in the 5′-to-3′ direction, creating a temporary DNA/RNA duplex whose chains run antiparallel to one another.

After a few ribonucleotides have been added to the growing RNA chain, the σ subunit dissociates from the holoenzyme and elongation proceeds under the direction of the core enzyme. In *E. coli,* this process proceeds at the rate of about 50 nucleotides/second at 37°C.

Eventually, the enzyme traverses the entire gene until it encounters a specific nucleotide sequence that acts as a termination signal. The termination sequences, about 40 base pairs in length, are extremely important in prokaryotes because of the close proximity of the end of one gene and the upstream sequences of the adjacent gene. In some cases, the termination of synthesis is dependent on the **termination factor, ρ (rho)**—a large hexameric protein that physically interacts with the growing RNA transcript. At the point of termination, the transcribed RNA molecule is released from the DNA template and the core polymerase enzyme dissociates. The synthesized RNA molecule is precisely complementary to a DNA sequence representing the template strand of a gene. Wherever an A, T, C, or G residue existed, a corresponding U, A, G, or C residue, respectively, is incorporated into the RNA molecule. These RNA molecules ultimately provide the information leading to the synthesis of all proteins present in the cell.

It is important to note that in bacteria, groups of genes whose products are related are often clustered along the chromosome. In many such cases, they are contiguous and all but the last gene lack the encoded signals for termination. The result is that during transcription, a large mRNA is produced that encodes more than one protein. Since genes in bacteria are sometimes called cistrons, the RNA is called a **polycistronic mRNA**. The products of genes transcribed in this fashion are usually all needed at the same time, so this is an efficient way to transcribe and subsequently translate the needed genetic information. In eukaryotes, **monocistronic mRNAs** are the rule.

13.9 Transcription in Eukaryotes Differs from Prokaryotic Transcription in Several Ways

Much of our knowledge of transcription was gained through studies of prokaryotes. The general aspects of the mechanics of these processes are similar in eukaryotes, although there are several notable differences. First we will summarize some of the major differences and then expand on some of them in the following sections.

1. Transcription in eukaryotes occurs within the nucleus under the direction of three separate forms of RNA polymerase. In eukaryotes the RNA transcript is not free to associate with ribosomes prior to the completion of transcription, as it is in prokaryotes. For the mRNA to be translated, it must move out of the nucleus into the cytoplasm.
2. Initiation and regulation of transcription involve a more extensive interaction between upstream DNA sequences and protein factors involved in stimulating and initiating transcription. In addition to promoters, other control units called enhancers may be located in the 5′ regulatory region upstream from the initiation point, but they have also been found within the gene or even in the 3′ downstream region beyond the coding sequence.
3. Maturation of eukaryotic mRNA from the primary RNA transcript involves many complex stages referred to generally as "processing." An initial processing step involves the addition of a 5′-cap and a 3′-tail to most transcripts destined to become mRNAs. Other extensive modifications occur in the internal nucleotide sequence of eukaryotic RNA transcripts that eventually serve as mRNAs. The initial (or primary) transcripts are most often much larger than those that are eventually translated. Sometimes called **pre-mRNAs**, they are part of a group of molecules found only in the nucleus—collectively known as **heterogeneous nuclear RNA (hnRNA)**. Such RNA molecules are of variable but large size (up to 10^7 Da) and are complexed with proteins, forming **heterogeneous nuclear ribonucleoprotein particles (hnRNPs)**. Only about 25% of hnRNA molecules are converted to mRNA. In those that are converted, substantial amounts of the ribonucleotide sequence are excised, and the remaining segments are spliced back together prior to translation. This phenomenon has given rise to the concepts of split genes and **splicing** in eukaryotes.

In the remainder of this chapter, we will elaborate on these differences. Several are related to the regulation of transcription in eukaryotes.

Initiation of Transcription in Eukaryotes

The recognition of certain highly specific DNA regions by RNA polymerase is the basis of orderly genetic function in all cells. Eukaryotic RNA polymerase exists in three unique forms, each of which transcribes different types of genes, as indicated in Table 13.6. Each enzyme is larger and more complex than the single prokaryotic polymerase, consisting of 2 large subunits and 10–15 smaller subunits. In regard to the initial template binding step and promoter regions, most is known about polymerase II, which transcribes all mRNAs in eukaryotes.

At least three **cis-acting elements** of a eukaryotic gene aid the efficient initiation of transcription by polymerase II. Recall from our discussion of the cis–trans test in Chapter 4 that the term *cis* is drawn from organic chemistry nomenclature, meaning "next to" or on the same side as, in contrast to being "across from" or *trans*, to other functional groups. In molecular genetics, then, cis-elements are adjacent parts of the same DNA molecule.

One cis-acting element found within the promoter region is called the **Goldberg-Hogness (TATA) box**, located about 35 nucleotide pairs upstream (−35) from the start point of transcription. The consensus sequence is a heptanucleotide consisting solely of A and T residues (TATAAAA). The sequence and function are analogous to that found in the −10 promoter region of prokaryotic genes. Because this region is common to most eukaryotic genes, the TATA box is thought to be nonspecific and is responsible only for fixing the site of transcription initiation by facilitating denaturation of the helix. Such a conclusion is supported by the fact that A═T base pairs are less stable than G≡C pairs.

A second common cis-acting element is the **CAAT box**, located upstream within the promoter of many genes at about 80 nucleotides from the start of transcription (−80) It contains the consensus sequence GGCCAATCT. Still other upstream regulatory regions have been found, and most genes contain one or more of them. They influence the efficiency of the promoter, along with the TATA and CAAT box. The location of each element is based on studies of deletions of particular regions of the promoter, each of which reduces the efficiency of transcription.

DNA regions called **enhancers** represent yet another cis-acting element. Although their locations can vary, enhancers are often found even farther upstream than the regions we have already mentioned or even downstream or within the gene. Thus, they can modulate transcription from a distance. Although they may not participate directly in template binding, they are essential to highly efficient initiation of transcription.

Complementing the cis-acting regulatory sequences are various **trans-acting factors** that facilitate template binding and, therefore, the initiation of transcription. These proteins are referred to as **transcription factors**. They are essential because RNA polymerase II cannot bind directly to eukaryotic promoter sites and initiate transcription without them. The transcription factors involved with human RNA polymerase II binding have been well characterized and are designated TFIIA, TFIIB, and so on. One of these, TFIID, binds directly to the TATA-box sequence and is sometimes called the **TATA-binding protein (TBP)**. TFIID consists of about 10 polypeptide subunits. Once initial binding to DNA occurs, at least seven other transcription factors bind sequentially to TFIID, forming an extensive pre-initiation complex, which is then bound by RNA polymerase II.

TABLE 13.6 RNA Polymerases in Eukaryotes

Form	*Product*	*Location*
I	rRNA	Nucleolus
II	mRNA, snRNA	Nucleoplasm
III	5*S* rRNA tRNA	Nucleoplasm

Transcription factors with similar activity have been discovered in a variety of eukaryotes, including *Drosophila* and yeast. These factors appear to supplant the role of the σ factor in the prokaryotic enzyme and play an important role in eukaryotic gene regulation.

Recent Discoveries Concerning RNA Polymerase Function

Most recently, interest in the process of transcription has focused on details of the structure and function of RNA polymerase. The ability to crystallize large nucleic acid–protein structures and perform X-ray diffraction analysis at a resolution below 5 Å has led to some remarkable observations. In particular, the work of Roger Kornberg and colleagues, studying the enzyme isolated from yeast at a resolution of 2.8 Å, has been informative. Note that achieving such a low resolution allowed him to see each amino acid of every protein in the complex.

Kornberg's discovery provided a highly detailed account of the most critical processes of transcription. RNA polymerase II in yeast contains two large subunits and ten smaller ones, forming a huge three-dimensional complex with a molecular weight of about 500 kDa. The promoter region of the DNA duplex that is to be transcribed enters a positively charged cleft formed between the two large subunits of the enzyme. The subunits form a structure resembling a pair of jaws. Prior to association with DNA, the cleft is open; once associated with DNA, the cleft partially closes, securing the duplex during the initiation of transcription. The critical region of the enzyme involved in the transition is about 50 kDa in size and is called the *clamp*.

Once secured by the clamp, a small duplex region of DNA separates at the *active center*, a position within the enzyme where complementary RNA synthesis is initiated on the DNA template strand. However, the entire complex is unstable, and transcription most often terminates following the incorporation of only a few ribonucleotides. It is not clear why, but *abortive transcription* is repeated a number of times before a stable DNA–RNA hybrid that includes a transcript of 11 ribonucleotides is formed. Once this occurs, abortive transcription is overcome, a stable complex is achieved, and elongation of the RNA transcript proceeds in earnest. Transcription at this point is said to have achieved a level of *highly processive RNA polymerization*.

As transcription proceeds, the enzyme moves along the DNA, and at any given time, about 40 base pairs of DNA and 18 residues of the growing RNA chain are part of the enzyme complex. The RNA that was synthesized earliest runs through a groove in the enzyme and exits under the *lid*, a structure at the top and back of the enzyme. Another area, called the *pore*, has been found at the bottom of the enzyme and is the point through which RNA precursors gain entry into the complex.

Eventually, as transcription proceeds, the portion of the DNA that signals termination is encountered and the complex once again becomes unstable, much as it was during the state of abortive transcription. The clamp opens, and both DNA and RNA are released from the enzyme as transcription is terminated. This completes a cycle that constitutes the paradigm of transcription: An unstable complex is formed during the initiation of transcription, stability is established once elongation creates a duplex of sufficient size, elongation proceeds, and then instability again characterizes termination of transcription.

These findings extend our knowledge of transcription considerably. It is significant that of the 10 subunits that are part of Kornberg's yeast model, all are very similar to their counterparts in the human enzyme. Nine of the 10 have also been conserved in the other forms of eukaryotic RNA polymerase (I and III). Based on the preceding description, try to mentally visualize the process of transcription from the time DNA associates with the enzyme until the transcript is released from the large molecular complex. If you have developed a clear set of images of this process in your mind, you no doubt have acquired an understanding of transcription in eukaryotes, which is more complex than in prokaryotes.

Heterogeneous Nuclear RNA and Its Processing: Caps and Tails

The genetic code is written in the ribonucleotide sequence of mRNA, which originated in the template strand of DNA, where complementary sequences of deoxyribonucleotides exist. In bacteria, the relationship between DNA and RNA appears to be quite direct. The DNA base sequence is transcribed into an mRNA sequence, which is then translated into an amino acid sequence according to the genetic code. By contrast, complex processing of mRNA in eukaryotes occurs before it is transported to the cytoplasm to participate in translation.

By 1970, accumulating evidence showed that eukaryotic mRNA is transcribed initially as a precursor molecule much larger than that which is translated. This notion was based on the observation by James Darnell and his coworkers of heterogeneous nuclear RNA (hnRNA) in mammalian nuclei that contained nucleotide sequences common to the smaller mRNA molecules present in the cytoplasm. They proposed that the initial transcript of a gene results in a large RNA molecule that must first be processed in the nucleus before it appears in the cytoplasm as a mature mRNA molecule. The various processing steps, are summarized in Figure 13–9.

The initial **posttranscriptional modification** of eukaryotic RNA transcripts destined to become mRNAs involves the 5′-end of these molecules, where a **7-methylguanosine (7mG) cap** is added (Figure 13–9, step 2). The cap, which is added even before the initial transcript is complete, appears to be important to subsequent processing within the nucleus, perhaps by protecting the 5′-end of the molecule from nuclease attack. Subsequently, this cap may be involved in the transport of mature mRNAs across the nuclear membrane into the cytoplasm. The cap is fairly complex and distinguished by a unique 5′-to-5′ bonding between the cap and the initial ribonucleotide of the RNA. Some eukaryotes also contain a methyl group (CH_3) on the 2′-carbon of the ribose sugars of the first two ribonucleotides of the RNA.

Further insights into the processing of RNA transcripts during the maturation of mRNA came from the discovery that both hnRNAs and mRNAs have a stretch of as many as 250

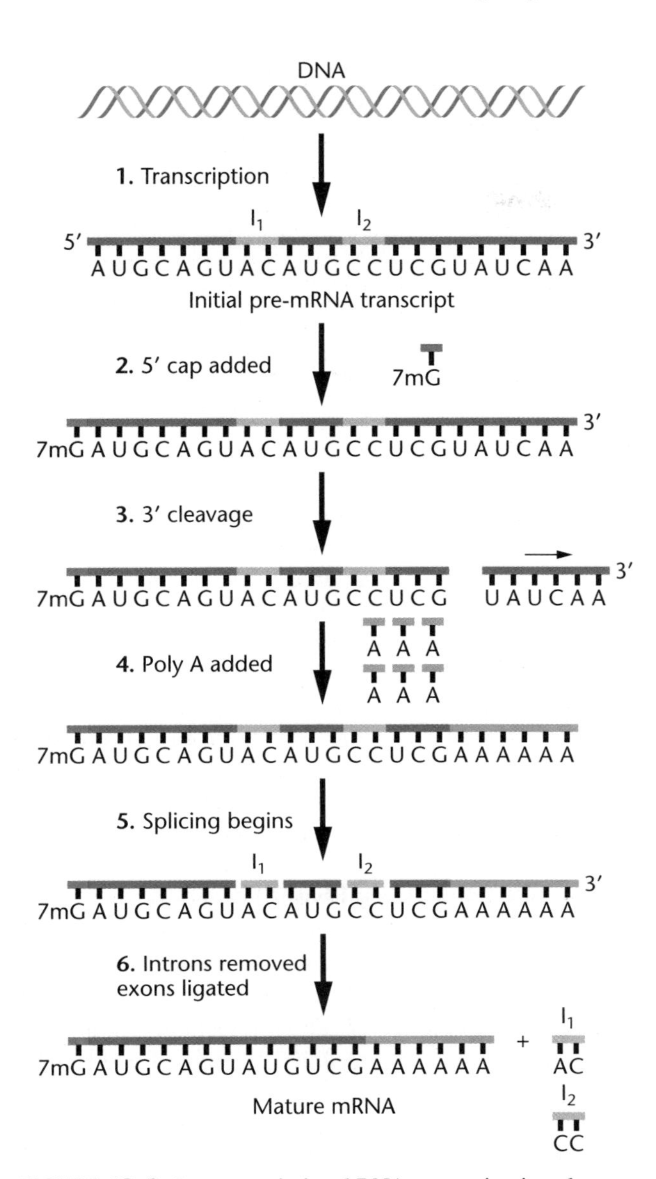

FIGURE 13–9 Posttranscriptional RNA processing in eukaryotes. Heterogeneous nuclear RNA (pre-mRNA) is converted to mRNA, which contains a 5′-cap and a 3′-poly-A tail. The introns are then spliced out.

adenylic acid residues at their 3′-end. Such **poly-A sequences** are added after the 5′-7mG cap has been added. First, the 3′-end of the initial transcript is cleaved enzymatically at a point 10–35 ribonucleotides from a highly conserved AAUAAA sequence (step 3). Then polyadenylation occurs by the sequential addition of a poly-A sequence (step 4). Poly A has been found at the 3′-end of almost all mRNAs studied in a variety of eukaryotic organisms. The exceptions seem to be the products of histone genes.

While the AAUAAA sequence is not found on all eukaryotic transcripts, mutations in that sequence of those transcripts that do have it cannot add the poly-A tail. In the absence of this tail, the RNA transcripts are rapidly degraded. Therefore, both the 5′-cap and the 3′-poly-A tail are critical if an RNA transcript is to be further processed and transported to the cytoplasm.

13.10 The Coding Regions of Eukaryotic Genes Are Interrupted by Intervening Sequences

One of the most exciting breakthroughs in the history of molecular genetics occurred in 1977 when Susan Berget, Philip Sharp, and Richard Roberts presented direct evidence that the genes of animal viruses contain *internal* nucleotide sequences that are not expressed in the amino acid sequence of the proteins they encode. These internal DNA sequences are present in initial RNA transcripts, but they are removed before the mature mRNA is translated (Figure 13–9, steps 5 and 6). These nucleotide segments are called **intervening sequences** (I_1 and I_2 in Figure 13–9), and the genes that contain them are **split genes**. DNA sequences that are not represented in the final mRNA product are also called **introns** (*int* for intervening), and those retained and expressed are called **exons** (*ex* for expressed). Splicing involves the removal of the ribonucleotide sequences present in introns as a result of an excision process, and the rejoining of exons.

Similar discoveries were soon made in a variety of eukaryotic genes. Two approaches have been most fruitful. The first involves the molecular hybridization of purified, functionally mature mRNAs with DNA containing the genes specifying that mRNA. Hybridization between nucleic acids that are not perfectly complementary results in **heteroduplexes**, in which introns present in the DNA but absent in the mRNA loop out and remain unpaired. Such structures can be visualized with the electron microscope, as shown in Figure 13–10. The chicken ovalbumin shown in the figure is a heteroduplex with seven loops (A–G), representing seven introns whose sequences are present in DNA but not in the final mRNA.

The second approach provides more specific information. It involves a comparison of nucleotide sequences of DNA with those of mRNA and the correlation with amino acid sequences. Such an approach allows the precise identification of all intervening sequences.

Thus far, most eukaryotic genes have been shown to contain introns. One of the first to be identified was the **β-globin gene** in mice and rabbits, studied independently by Philip Leder and Richard Flavell. Both the mouse and rabbit genes contain two introns of approximately the same size and same location within the gene. The rabbit gene is diagrammed in Figure 13–11, revealing the intron locations. Similar introns have been found in the β-globin gene in all mammals examined.

The **ovalbumin gene** of chickens has been extensively characterized by Bert O'Malley in the United States and Pierre Chambon in France. As shown in Figure 13–11, the gene contains seven introns. In fact, the majority of the gene's DNA sequence are composed of introns and are thus "silent." The initial RNA transcript is nearly three times the length of the mature mRNA. Compare the ovalbumin gene in Figures 13–10 and 13–11. Can you match the unpaired loops in Figure 13–10 with the sequence of introns specified in Figure 13–11?

The list of genes containing intervening sequences is long. In fact, few eukaryotic genes seem to lack introns. An

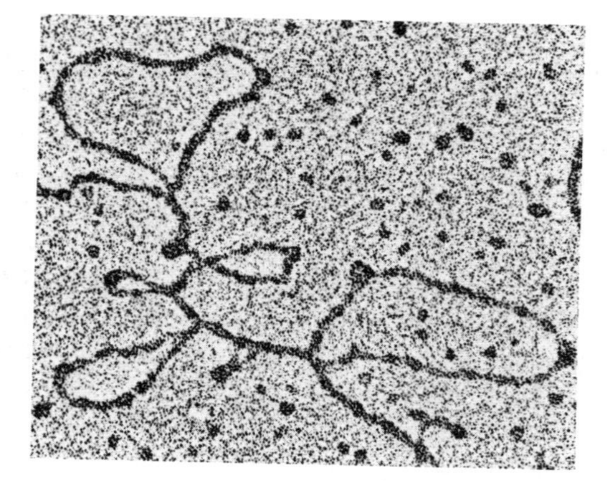

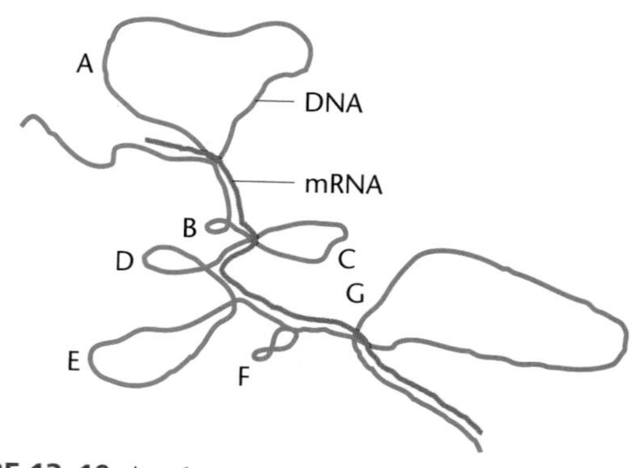

FIGURE 13–10 An electron micrograph and interpretive drawing of the hybrid molecule (heteroduplex) formed between the template DNA strand of the chicken ovalbumin gene and the mature ovalbumin mRNA. Seven DNA introns, labeled A–G, produce unpaired loops. *(Courtesy of Bert W. O'Malley, M.D)*

extreme example of the number of introns in a single gene is found in the gene coding for one subunit of collagen, the major connective tissue protein in vertebrates. The *pro-α-2(1) collagen* gene contains 50 introns. The precision of cutting and splicing that occurs must be extraordinary if errors are not to be introduced into the mature mRNA. What is equally noteworthy is a comparison of the size of genes with the size of the final mRNA once the introns are removed. As shown in Table 13.7, about 15% of the collagen gene consists of exons that finally appear in mRNA.

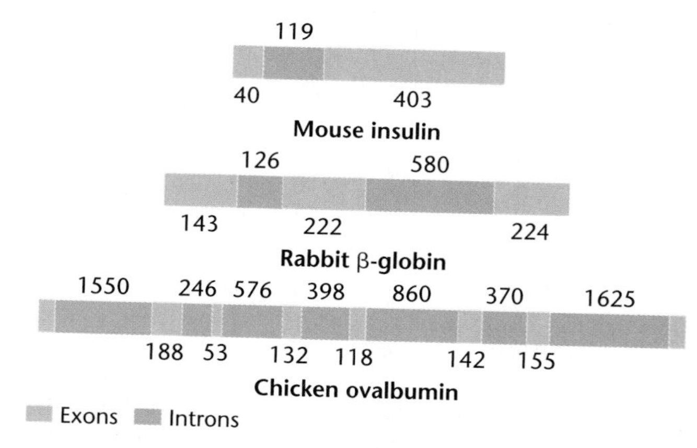

FIGURE 13–11 Intervening sequences in various eukaryotic genes. The numbers indicate the number of nucleotides present in various intron and exon regions.

TABLE 13.7 Comparing Human Gene Size, mRNA Size, and the Number of Introns

Gene	*Gene Size (kb)*	*mRNA Size (kb)*	*Number of Introns*
Insulin	1.7	0.4	2
Collagen [*pro-α-2(1)*]	38.0	5.0	50
Albumin	25.0	2.1	14
Phenylalanine hydroxylase	90.0	2.4	12
Dystrophin	2000.0	17.0	50

For other proteins, an even more extreme picture emerges. Only about 8% of the albumin gene remains to be translated, and in the largest human gene known, dystrophin (the missing protein product in Duchenne muscular dystrophy), less than 1% of the gene sequence is retained in the mRNA. Several other human genes are also included in Table 13.7.

While the vast majority of eukaryotic genes examined thus far contain introns, there are several exceptions. Notably, the genes coding for histones and interferon do not appear to contain introns. It is not clear why or how the genes encoding these molecules have been maintained throughout evolution without acquiring the extraneous information characteristic of almost all other genes.

Splicing Mechanisms: Autocatalytic RNAs

The discovery of split genes led to intensive attempts to elucidate the mechanism by which introns of RNA are excised and exons are spliced back together, and a great deal of progress has been made. Interestingly, it appears that somewhat different mechanisms exist for different types of RNA, as well as for RNAs produced in mitochondria and chloroplasts.

Introns can be categorized into several groups based on their splicing mechanisms. Group I, represented by introns that are part of the primary transcript of rRNAs, requires no additional components for intron excision; the intron itself is the source of the enzymatic activity necessary for its own removal. This amazing discovery, which contradicted expectations, was made in 1982 by Thomas Cech and his colleagues during a study of the ciliate protozoan *Tetrahymena*. Reflecting their autocatalytic properties, RNAs that are capable of splicing themselves are sometimes called **ribozymes**.

The **self-excision process** is shown in Figure 13–12. Chemically, two nucleophilic reactions (called transesterification reactions) take place. The first is an interaction between guanosine, which acts as a cofactor in the reaction, and the primary transcript (Figure 13–12a). The 3′-OH group of guanosine is transferred to the nucleotide adjacent to the 5′-end of the intron. The second reaction involves the interaction of the newly acquired 3′-OH group on the left-hand exon and the phosphate on the 3′-end of the right intron (Figure 13–12b). The intron is spliced out, and the two exon regions are ligated, leading to the mature RNA (Figure 13–12c).

Self-excision of group I introns, as described, is now known to apply to pre-rRNAs from other protozoans. Self-excision also seems to govern the removal of introns present in the primary mRNA and tRNA transcripts produced in the mitochondria and chloroplasts. These are referred to as group II introns. Like group I molecules, splicing involves two autocatalytic reactions leading to the excision of introns. However, guanosine is not involved as a cofactor in group II.

Splicing Mechanisms: The Spliceosome

Introns are a major component of nuclear-derived pre-mRNA transcripts. Compared with the other RNAs we have discussed, introns in nuclear-derived mRNA can be much larger—up to 20,000 nucleotides—and they are more plentiful. Their removal appears to require a much more complex mechanism, which has been more difficult to define.

Nevertheless, many clues are now emerging and the model in Figure 13–13 illustrates the removal of one intron. The nucleotide sequences near the perimeters of this type of intron are often similar. Many begin at the 5′-end with a GU dinucleotide sequence and terminate at the 3′-end with an AG dinucleotide sequence. These, as well as other consensus sequences shared by introns, attract specific molecules that form a molecular complex, a **spliceosome**, essential to splicing. Spliceosomes have been identified in extracts of yeast as well as mammalian cells, and are very large, being 40*S* in yeast and 60*S* in mammals. Perhaps the most essential component of spliceosomes is the unique set of **small nuclear RNAs (snRNAs)**. These RNAs are usually 100–200 nucleotides or less and are often complexed with proteins to form **small nuclear ribonucleoproteins (snRNPs**, or **snurps)**. These are found only in the nucleus. Because they are rich in uridine residues, the snRNAs have been arbitrarily designated U1, U2, . . . , U6.

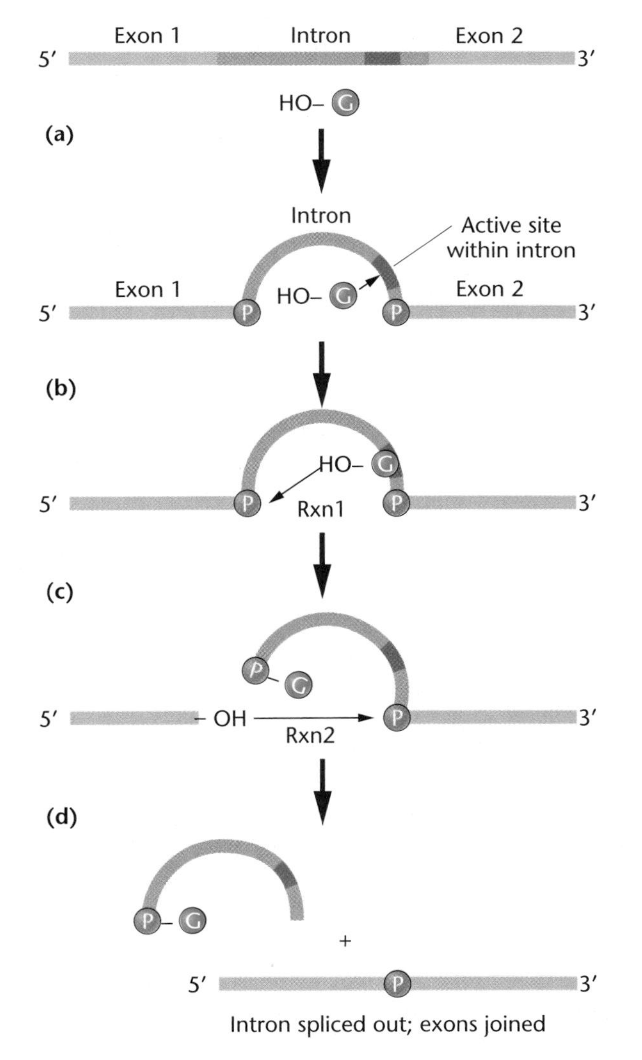

FIGURE 13–12 Splicing mechanism of pre-rRNA involving group I introns that are removed from the initial transcript. The process is one of self-excision involving two transesterification reactions.

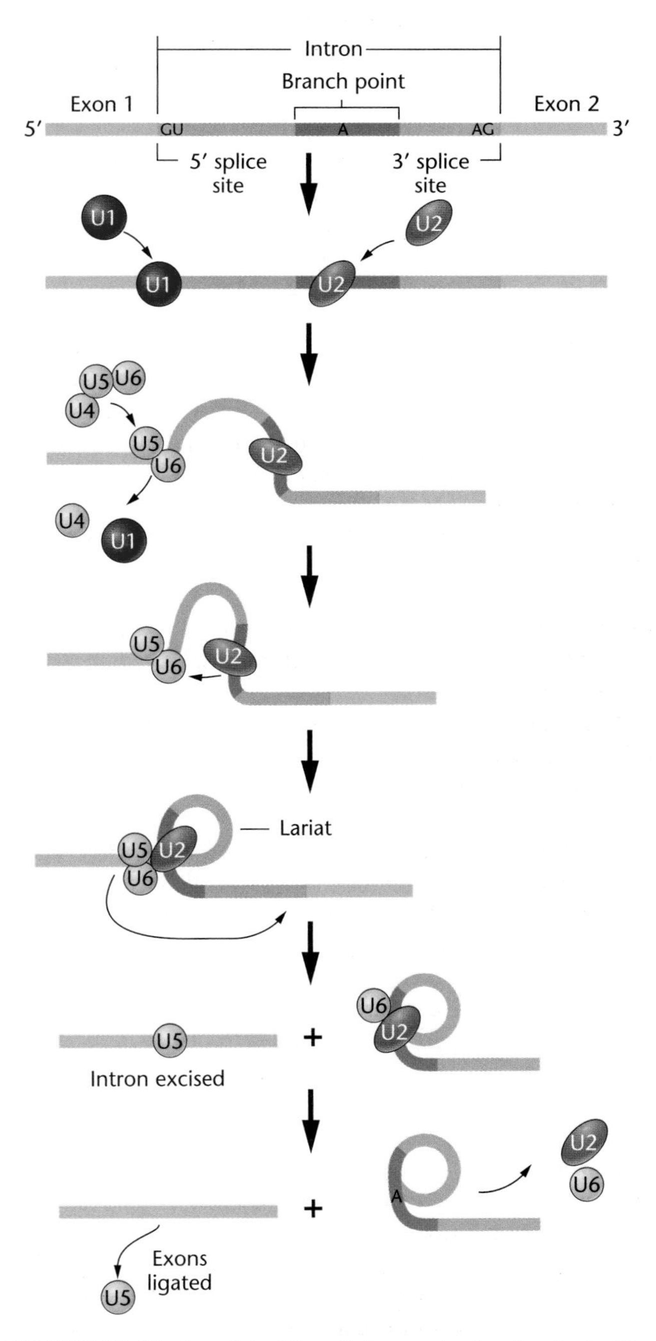

FIGURE 13–13 A model of the splicing mechanism involved with the removal of an intron from a pre-mRNA. Excision is dependent on various snRNAs (U1, U2, . . . , U6) that combine with proteins to form snurps, which are part of the spliceosome. The lariat structure in the intermediate stage is characteristic of this mechanism.

The snRNA of U1 bears a nucleotide sequence that is homologous to the 5′-end of the intron. Base pairing resulting from this homology promotes binding that represents the initial step in formation of the spliceosome. After the other snurps (U2, U4, U5, and U6) are added, splicing commences. As with group I splicing, two transesterification reactions are involved. The first involves the interaction of the 3′-OH group from an adenine (A) residue present within the **branch point** region of the intron. The A residue attacks the 5′-splice site, cutting the RNA chain. In a subsequent step involving several other snurps, an intermediate structure is formed and the second reaction ensues, linking the cut 5′-end of the intron to the A. This results in the formation of the characteristic loop structure called a *lariat*, which contains the excised intron. The exons are then ligated, and the snurps are released.

The processing involved in splicing represents a potential regulatory step during gene expression. For example, several cases are known where introns present in pre-mRNAs *derived from the same gene* are spliced *in more than one way*, thereby yielding different collections of exons in the mature mRNA. This process of **alternative splicing** yields a group of mRNAs that, upon translation, results in a series of related proteins called **isoforms**. A growing number of examples have been found in organisms ranging from viruses to *Drosophila* to humans. Alternative splicing of pre-mRNAs provides the basis for producing related proteins from a single gene, thus increasing the number of gene products that can be derived from an organism's genome.

13.11 RNA Editing Modifies the Final Transcript

In the late 1980s, still another quite unexpected form of posttranscriptional RNA processing was discovered. In this form, called **RNA editing**, the nucleotide sequence of a pre-mRNA is actually changed prior to translation. As a result, the ribonucleotide sequence of the mature RNA differs from the sequence encoded in the exons of the DNA from which the RNA was transcribed.

Other variations of RNA editing exist, but there are two main types—**substitution editing**, in which the identities of individual nucleotide bases are altered; and **insertion/deletion editing**, where nucleotides are added to or subtracted from the total number of bases. Substitution editing is used in some nuclear-derived eukaryotic RNAs and is prevalent in mitochondrial and chloroplast RNAs transcribed in plants. *Physarum polycephalum*, a slime mold, uses both substitution and insertion/deletion editing for its mitochondrial mRNAs.

Trypanosoma, a parasite that causes African sleeping sickness, and its relatives use extensive insertion/deletion editing in mitochondrial RNAs. The number of uridines added to an individual transcript can make up more than 60% of the coding sequence, usually forming the initiation codon and placing the rest of the sequence into the proper reading frame. Insertion/deletion editing in trypanosomes is directed by **gRNA (guide RNA) templates**, which are transcribed from the mitochondrial genome. These small RNAs are complementary to the edited region of the final, edited mRNAs. They base-pair with the pre-edited mRNAs to direct the editing machinery to make the correct changes.

The best-studied examples of substitutional editing occur in mammalian nuclear-encoded mRNA transcripts. Apolipoprotein B (apo B) exists in both a long and a short form, although a single gene encodes both proteins. In human intestinal cells, apo B mRNA is edited by a single C-to-U change, which converts a CAA glutamine codon into a UAA stop codon and terminates the polypeptide at approximately half its genomically encoded length. The editing is performed by a complex of proteins that bind to a "mooring sequence" on the mRNA transcript just downstream of the editing site. In a second system, the synthesis of subunits constituting the glutamate receptor channels (GluR) in mammalian brain tissue is also affected by RNA editing. In this case, adenosine (A) to inosine (I) editing occurs in pre-mRNAs prior to their translation, where I is read as guanosine (G) during translation. A family of three ADAR (adenosine deaminase acting on RNA) enzymes is believed to be responsible for editing various sites within the glutamate channel subunits. The double-stranded RNAs required for editing by ADARs are provided by intron/exon pairing of the GluR mRNA transcripts. The editing changes alter the physiological parameters (solute permeability and desensitization response time) of the receptors containing the subunits.

The importance of RNA editing resulting from the action of ADARs is most apparent when we examine situations in which these enzymes have lost their functional capacity as a result of mutation. In several investigations, the loss of function was shown to have a lethal impact in mice. In one study, embryos heterozygous for a defective *ADAR1* gene die during embryonic development as a result of a defective hematopoietic system. In another study, mice with two defective copies of *ADAR2* progress through development normally, but are prone to epileptic seizures and die while still in the weaning stage. Their tissues contain the unedited version of one of the GluR products. The defect leading to death is believed to be in the brain. Heterozygotes for the mutation are normal.

Findings such as these in mammals have established that RNA editing provides still another important mechanism of posttranscriptional modification and that this process is not restricted to small or asexually reproducing genomes such as mitochondria. Several new examples of RNA editing have been found each year since its discovery, and this trend is likely to continue. Further, the process has important implications for the regulation of genetic expression.

GENETICS, TECHNOLOGY, AND SOCIETY

Antisense Oligonucleotides: Attacking the Messenger

Standard chemotherapies for diseases such as cancer and AIDS are often accompanied by toxic side effects. Conventional therapeutic drugs target both normal and diseased cells, with diseased or infected cells being only slightly more susceptible than the patient's normal cells. Scientists have long wished for a magic bullet that could seek out and destroy the virus or cancer cell, leaving normal cells alive and healthy. Over the last decade, one particularly promising candidate for magic-bullet status has emerged—the *antisense oligonucleotide*.

Antisense therapies have arisen through an understanding of the molecular biology of gene expression. Gene expression is a two-step process. First, a single-stranded messenger RNA (mRNA) is copied from one strand of the duplex DNA molecule. Second, the mRNA is transported to the cytoplasm and complexed with ribosomes, and its genetic information is translated into the amino acid sequence of a polypeptide.

Normally, a gene is transcribed into RNA from only one strand of the DNA duplex. The resulting RNA is known as sense RNA. However, it is sometimes possible for the other DNA strand to be copied into RNA. The RNA produced by the transcription of the "wrong" strand of DNA is called antisense RNA. As with complementary strands of DNA molecules, complementary strands of RNA can form double-stranded molecules.

The formation of duplex structures between a sense and an antisense RNA may affect the sense RNA in at least two ways. Binding antisense RNA to sense RNA may physically block ribosome binding or elongation, hence inhibiting translation. Or binding antisense RNA to sense RNA may trigger the degradation of the sense RNA because double-stranded RNA molecules are attacked by intracellular ribonucleases. In both cases, gene expression is blocked.

What makes the antisense approach so exciting is its potential specificity. Scientists can design antisense RNA (or DNA) molecules of known nucleotide sequence and can then synthesize large amounts of these antisense nucleic acids *in vitro*. Usually, oligonucleotides up to 20 nucleotides are used. It is theoretically possible to treat cells with these synthetic antisense oligonucleotides, have to the oligonucleotides enter the cell and bind precise target mRNAs, and to turn off the synthesis of one specific protein. If that protein is necessary for virus reproduction or cancer cell growth (but is not necessary in normal cells), the antisense oligonucleotide should have only therapeutic effects.

In the past few years, laboratory tests of antisense drugs have been so promising that several clinical trials are in progress. In addition, one antisense drug (Vitravene©, from ISIS Pharmaceuticals and CIBA Vision Corp.) has been approved for sale in the United States. Vitravene is an antisense oligonucleotide used to treat cytomegalovirus-induced retinitis in AIDS patients. Cytomegalovirus (CMV) is a common virus that infects most people. Although it causes few problems in people with normal immune systems, it can cause serious symptoms in people with conditions that impair the immune system (such as AIDS). Up to 40% of AIDS patients develop retinitis or blindness as a result of CMV infections of the eye. Clinical trials indicate that the introduction of antisense CMV oligonucleotides into the eyes of AIDS patients with CMV-induced retinitis significantly delays the disease progression compared to untreated groups. It is not known how the oligonucleotides operate, but laboratory studies suggest that antisense oligonucleotides may trigger the destruction of CMV mRNA and interfere with the absorption of the virus to the surface of host cells.

Antisense oligonucleotides may also act as anti-inflammatory drugs. Clinical trials are in progress to test antisense compounds in the treatment of asthma, rheumatoid arthritis, psoriasis, ulcerative colitis, and transplant rejection. One interesting antisense oligonucleotide (now in clinical trials conducted by ISIS and Boehringer Ingelheim) binds to the mRNA that encodes ICAM-1, a cell-surface glycoprotein that helps activate immune system and inflammatory cells. It is often overexpressed in body tissues that exhibit extreme inflammatory responses. Researchers hope that antisense oligonucleotides might temper inflammatory responses by reducing the expression of ICAM-1. The results of a recent clinical trial suggest that antisense ICAM-1 may be an effective treatment for Crohn disease, a form of inflammatory bowel disease that affects about 200,000 people in the United States. It is a debilitating, chronic disease, and treatments such as steroids and immunosuppressive drugs are often ineffective and have toxic side effects. In one clinical trial, almost half of the Crohn's patients went into remission after treatment with antisense ICAM-1, whereas none of the placebo group did so. In addition, one-third of the treated patients were well enough to stop steroid treatments by the end of the trial period. In this trial, the antisense ICAM-1 drugs appeared to be well-tolerated and safe.

Some interesting clinical trials are in progress to test antisense oligonucleotides as treatments for some types of cancer. These oligonucleotides are designed to reduce the synthesis of proteins that are either overexpressed in cancer cells or are present as mutant forms. These drugs will be evaluated as treatments for tumors of the ovary, prostate, breast, brain, colon, and lung. Other antisense drugs in the pipeline are designed to attack the hepatitis B, hepatitis C, human papilloma, and AIDS viruses.

If antisense therapeutics pass the scrutiny of these scientific and clinical trials, we may indeed have acquired a molecular magic bullet to use in the battle against a variety of diseases.

References

Edgington, S. M. 1997. Antisense '97: A roundtable on the state of the industry. *Nature Biotech.* 15: 519–24.

Nyce, J. W., and Metzger, W. J. 1997. DNA antisense therapy for asthma in an animal model. *Nature* 385: 721–25.

Roush, W. 1997. Antisense aims for a renaissance. *Science* 276: 1192–93.

Chapter Summary

1. The genetic code stored in DNA is copied to RNA, where it is used to direct the synthesis of polypeptide chains. It is degenerate, unambiguous, nonoverlapping, and commaless.
2. The complete coding dictionary, determined by using various experimental approaches, reveals that of the 64 possible codons, 61 encode the 20 amino acids found in proteins, while 3 triplets terminate translation. One of the 61 codons is the initiation codon and specifies methionine.
3. The observed pattern of degeneracy often involves only the third letter of a triplet series. It led Francis Crick to propose the wobble hypothesis.
4. Confirmation for the coding dictionary, including codons for initiation and termination, was obtained by comparing the complete nucleotide sequence of phage MS2 with the amino acid sequence of the corresponding proteins. Other findings support the belief that, with minor exceptions, the code is universal for all organisms.
5. Transcription, the initial step in gene expression, describes the synthesis of a strand of RNA complementary to a DNA template, under the direction of RNA polymerase.
6. The processes of transcription, like DNA replication, can be subdivided into the stages of initiation, elongation, and termination. Also like DNA replication, the process relies on base-pairing affinities between complementary nucleotides.
7. Initiation of transcription is dependent upon an upstream 5′-region of DNA called the promoter that represents the initial binding site for RNA polymerase. Promoters contain specific DNA sequences such as the TATA box essential to polymerase binding.
8. Transcription is more complex in eukaryotes than in prokaryotes. The primary transcript is a pre-mRNA that must be modified in various ways before it can be efficiently translated. Processing, which produces a mature mRNA, includes the addition of a 7mG cap and a poly-A tail; and the removal, through splicing, of intervening sequences, or introns. RNA editing of pre-mRNA prior to its translation also occurs in some systems.

Key Terms

alternative splicing, 288
anticodon, 275
bacteriophage MS2, 279
β-globin gene, 285
CAAT box, 283
chain elongation, 282
cis-acting elements, 283
codon, 273
consensus sequence, 282
degenerate code, 273
enhancer, 283
exon, 285
frameshift mutation, 274
guide RNA (gRNA) template, 288
Goldberg-Hogness (TATA) box, 283
heteroduplex, 285
heterogeneous nuclear RNA (hnRNA), 283
heterogeneous ribonucleoprotein (hnRNP), 283
holoenzyme, 281
initiation (of transcription), 282
initiator codon, 279
insertion/deletion editing, 288
intervening sequence, 285
intron, 285
in vitro (cell-free) protein-synthesizing system, 274
isoform, 288
messenger RNA (mRNA), 274
7-methylguanosine (7mG) cap, 285
monocistronic mRNA, 283
N-formylmethionine (fmet), 279
nonsense mutation, 279
ovalbumin gene, 286
partner strand, 281
poly-A sequence, 285
polycistronic mRNA, 283
polynucleotide phosphorylase, 274
posttranscriptional modification, 285
pre-mRNA, 283
Pribnow box, 282
promoter, 281
ribozyme, 287
RNA editing, 288
RNA heteropolymer, 275
RNA homopolymer, 275
RNA polymerase, 281
self-excision process, 287
σ (sigma) factor, 281
small nuclear ribonucleoprotein (snRNP, or snurp), 288
small nuclear RNA (snRNA), 288
spliceosome, 288
splicing, 283
split gene, 285
substitution editing, 288
TATA-binding protein (TBP), 284
TATA box, 283
template binding, 281
template strand, 281
-10 region, 282
termination codon, 279
termination factor, ρ (rho), 282
-35 region, 282
trans-acting factor, 284
transcription, xx
transcription factor, 284
triplet binding assay, 275
triplet codon, 273
universality (of the code), 274
wobble hypothesis, 278

INSIGHTS AND SOLUTIONS

1. Calculate how many triplet codons would be possible had evolution seized on six bases (three complementary base pairs) rather than four bases within the structure of DNA. Would 6 bases accommodate a 2-letter code, assuming 20 amino acids and start and stop codons?

Solution: Six things taken three at a time will produce $(6)^3$ or 216 triplet codes. If the code was a doublet, there would be $(6)^2$ or 36 2-letter codes, more than enough to accommodate 20 amino acids and start and stop signals.

2. In a heteropolymer experiment using 1/2C : 1/4A : 1/4G, how many different triplets will occur in the synthetic RNA molecule? How frequently will the most frequent triplet occur?

Solution: There will be $(3)^3$, or 27, triplets produced. The most frequent will be CCC, present $(1/2)^3$, or 1/8, of the time.

3. In a regular copolymer experiment, where UUAC is repeated over and over, how many different triplets will occur in the synthetic RNA, and how many amino acids will occur in the polypeptide when this RNA is translated? (Consult Figure 13–7.)

Solution: The synthetic RNA will repeat four triplets—UUA, UAC, ACU, and CUU—over and over. Because both UUA and CUU encode leucine, while ACU and UAC encode threonine and tyrosine, respectively, the polypeptides synthesized under the directions of this RNA would contain three amino acids in the repeating sequence leu-leu-thr-tyr.

4. Actinomycin D inhibits DNA-dependent RNA synthesis. This antibiotic is added to a bacterial culture where a specific protein is being monitored. Compared to a control culture, where no antibiotic is added, translation of the protein declines over a period of 20 minutes, until no further protein is made. Explain these results.

Solution: The mRNA, which is the basis for the translation of the protein, has a lifetime of about 20 minutes. When actinomycin D is added, transcription is inhibited and no new mRNAs are made. Those already present support the translation of the protein for up to 20 minutes.

Problems and Discussion Questions

1. Early proposals regarding the genetic code considered the possibility that DNA served directly as the template for polypeptide synthesis (Gamow, 1954, in Selected Readings). In eukaryotes, what difficulties would such a system pose? What observations and theoretical considerations argue against such a proposal?
2. In studies of frameshift mutations, Crick, Barnett, Brenner, and Watts-Tobin found that either three nucleotide insertions or deletions restored the correct reading frame. (a) Assuming the code is a triplet, what effect would the addition or loss of six nucleotides have on the reading frame? (b) If the code were a sextuplet (consisting of six nucleotides), would the reading frame be restored by the addition or loss of three, six, or nine nucleotides?
3. In a mixed copolymer experiment using polynucleotide phosphorylase, 3/4G : 1/4C was added to form the synthetic message.Using the resulting amino acid composition of the ensuing protein shown below, (a) indicate the percentage of time each possible triplet will occur in the message, and (b) determine one consistent base-composition assignment for the amino acids present. (c) Considering the wobble hypothesis, predict as many specific triplet assignments as possible.

Glycine	36/64	(56%)
Alanine	12/64	(19%)
Arginine	12/64	(19%)
Proline	4/64	(6%)

4. When repeating copolymers are used to form synthetic mRNAs, dinucleotides produce a single type of polypeptide that contains only two different amino acids. On the other hand, a trinucleotide sequence produces three different polypeptides, each consisting of only a single amino acid. Why? What will be produced when a repeating tetranucleotide is used?
5. The mRNA formed from the repeating tetranucleotide UUAC incorporates only three amino acids, but the use of UAUC incorporates four amino acids. Why?
6. In studies using repeating copolymers, AC . . . incorporates threonine and histidine, and CAACAA . . . incorporates glutamine, asparagine, and threonine. What triplet code can definitely be assigned to threonine?
7. In a coding experiment using repeating copolymers (as shown in Table 13.3), the data below were obtained. AGG is known to code for arginine. Taking into account the wobble hypothesis, assign each of the four remaining different triplet codes to its correct amino acid.

Copolymer	*Codons Produced*	*Amino Acids in Polypeptide*
AG	AGA, GAG	arg, glu
AAG	AGA, AAG, GAA	lys, arg, glu

8. In the triplet binding assay technique, radioactivity remains on the filter when the amino acid corresponding to the experimental triplet is labeled. Explain the basis of this technique.
9. When the amino acid sequences of insulin isolated from different organisms were determined, some differences were noted. For example, alanine was substituted for threonine, serine was substituted for glycine, and valine was substituted for isoleucine at corresponding positions in the protein. List the single-base changes that could occur in triplets to produce these amino acid changes.
10. In studies of the amino acid sequence of wild-type and mutant forms of tryptophan synthetase in *E. coli*, the following changes have been observed:

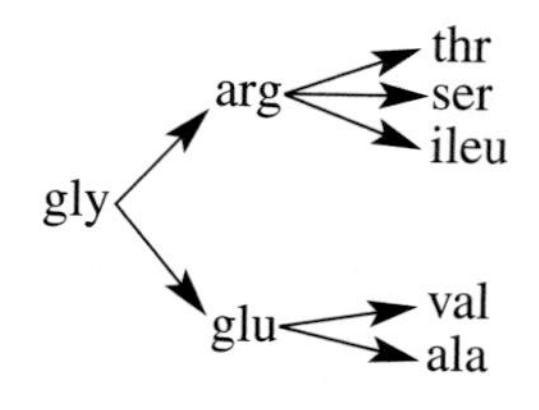

Determine a set of triplet codes in which only a single nucleotide change produces each amino acid change.

11. Why doesn't polynucleotide phosphorylase (Ochoa's enzyme) synthesize RNA *in vivo*?
12. Refer to Table 13.1. Can you hypothesize why a mixture of (Poly U) + (Poly A) would not stimulate incorporation of ^{14}C-phenylalanine into protein?
13. Predict the amino acid sequence produced during translation of the short theoretical mRNA sequences below. (Note that the second sequence was formed from the first by a deletion of only one nucleotide.) What type of mutation gave rise to sequence 2?

Sequence 1: AUGCCGGAUUAUAGUUGA
Sequence 2: AUGCCGGAUUAAGUUGA

14. A short RNA molecule was isolated that demonstrated a hyperchromic shift indicating secondary structure. Its sequence was determined to be

AGGCGCCGACUCUACU

(a) Propose a two-dimensional model for this molecule. (b)What DNA sequence would give rise to this RNA molecule through transcription? (c) If the molecule were a tRNA fragment containing a CGA anticodon, what would the corresponding codon be? (d) If the molecule were an internal part of a message, what amino acid sequence would result from it following translation? (Refer to the code chart in Figure 13–7.)

15. A glycine residue exists at position 210 of the tryptophan synthetase enzyme of wild-type *E. coli*. If the codon specifying glycine is GGA, how many single-base substitutions will result in an amino acid substitution at position 210, and what are they? How many will result if the wild-type codon is GGU?
16. Shown here is a theoretical viral mRNA sequence:

5′-AUGCAUACCUAUGAGACCCUUGGA-3′

(a) Assuming that it could arise from overlapping genes, how many different polypeptide sequences can be produced? Using the chart in Figure 13–7, what are the sequences? (b) A base substitution mutation that altered the sequence in part (a) eliminated the synthesis of all but one polypeptide. The altered sequence is shown below. Use Figure 13–7 to determine why it was altered.

5′-AUGCAUACCUAUGUGACCCUUGGA-3′

17. Most proteins have more leucine than histidine residues, but more histidine than tryptophan residues. Correlate the number of codons for these three amino acids with this information.
18. Define the process of transcription. Where does this process fit into the central dogma of molecular genetics?
19. What was the initial evidence for the existence of mRNA?
20. Describe the structure of RNA polymerase in bacteria. What is the core enzyme? What is the role of the σ subunit?
21. In a written paragraph, describe the abbreviated chemical reactions that summarize RNA polymerase-directed transcription.
22. Messenger RNA molecules are very difficult to isolate from prokaryotes because they are quickly degraded. Can you suggest a reason why this occurs? Eukaryotic mRNAs are more stable and exist longer in the cell than do prokaryotic mRNAs. Is this an advantage or disadvantage for a pancreatic cell making large quantities of insulin?
23. Deoxyribonucleotide sequences derived from a template strand of DNA are shown below. (a) Determine the mRNA sequence that would be derived from transcription of each DNA sequence. (b) Using Figure 13–7, determine the amino acid sequence that is encoded by each mRNA. (c) For sequence 1, what is the sequence of the partner DNA strand?

Sequence 1: 5′-CTTTTTTGCCAT-3′
Sequence 2: 5′-ACATCAATAACT-3′
Sequence 3: 5′-TACAAGGGTTCT-3′

24. In a mixed copolymer experiment, messengers were created with either 4/5C : 1/5A or 4/5A : 1/5C. These messages yielded proteins with the following amino acid compositions shown in the table below. Using these data, predict the most specific coding composition for each amino acid.

4/5C:1/5A		4/5A:1/5C	
Proline	63.0%	Proline	3.5%
Histidine	13.0%	Histidine	3.0%
Threonine	16.0%	Threonine	16.6%
Glutamine	3.0%	Glutamine	13.0%
Asparagine	3.0%	Asparagine	13.0%
Lysine	0.5%	Lysine	50.0%
	98.5%		99.1%

25. Shown below are the amino acid sequences of the wild-type and three mutant forms of a short protein. (a) Using Figure 13–7, predict the type of mutation that created each altered protein. (b) Determine the specific ribonucleotide change that led to the synthesis of each mutant protein. (c) The wild-type RNA consists of nine triplets. What is the role of the ninth triplet? (d) For the first eight wild-type triplets, which, if any, can you determine specifically from an analysis of the mutant proteins? In each case, explain why or why not. (e) Another mutation (mutant 4) is isolated. Its amino acid sequence is unchanged, but mutant cells produce abnormally low amounts of the wild-type proteins. As specifically as you can, predict where this mutation exists in the gene.

Wild type: met-trp-tyr-arg-gly-ser-pro-thr
Mutant 1: met-trp
Mutant 2: met-trp-his-arg-gly-ser-pro-thr
Mutant 3: met-cys-ile-val-val-val-gln-his

26. The genetic code is degenerate. Amino acids are encoded by either 1, 2, 3, 4, or 6 triplet codons. (See Figure 13–7.) An interesting question is whether the frequency of triplet codes is in any way correlated with the frequency that amino acids appear in proteins? That is, is the genetic code optimized for its intended use? Some approximations of the frequency of appearance of nine amino acids in proteins in *E. coli* are shown in the table on the next page. (a) Determine how many triplets encode each amino acid in the table. (b) Devise a way to graphically express the data from part (a) and the table. (c) Analyze your data to determine what, if any, correlations can be drawn between the relative frequency of amino acids making up proteins with the number of triplets for each. Write a paragraph that states your specific and general conclusions. (d) How would you proceed with your analysis if you wanted to pursue this problem further?

Amino Acid	*Percentage of Appearance in Proteins*
met	2
cys	2
gln	5
pro	5
arg	5
ile	6
glu	7
ala	8
leu	10

27. As described in Chapter 12, Alu elements proliferate in the human genome by a process called retrotransposition, in which the *Alu* DNA sequence is transcribed into RNA, copied into double-stranded DNA, and then inserted back into the genome at a site distant from that of its "parent" *Alu* gene. This seems to have been an extremely efficient process, since *Alu* genes have proliferated to about 10^6 copies in the human genome. This efficiency is largely due to the fact that *Alus*, like many small structural RNAs, carry their promoter sequences *within the transcribed region of the gene*, rather than 5′ to the transcription start site. If *Alus* carried promoters upstream to the transcription site, similar to those of protein-coding genes, what would happen once they were retrotransposed? Would a retrotransposed *Alu* gene be able to proliferate? Explain.
28. DNA and RNA base compositions were analyzed from a hypothetical bacterial species with the results shown in the table below. (a) On the basis of these data, what can you conclude about the DNA and RNA of the organism? (b) Are the data consistent with the Watson-Crick model of DNA? (c) Is the RNA single-stranded or double-stranded? (d) If we assume that the entire length of DNA has been transcribed, do the data suggest that RNA has been derived from the transcription of one or both DNA strands. Can we determine this from these data?

Ratio	*DNA*	*RNA*
(A + G)/(T + C)	1.0	
(A + T)/(C + G)	1.2	
(A + G)/(U + C)		1.3
(A + U)/(C + G)		1.2

29. M. Klemke and others (2001. EMBO J. 20: 3849–60.) discovered an interesting coding phenomenon in an exon within a neurologic hormone receptor gene in mammals, by which two protein entities (XLαs and ALEX) appear to be produced from the same exon. Below is the DNA sequence of the exon's 5′-end derived from a rat. The lowercase letters represent the initial coding portion for the XLαs protein, and the uppercase letters indicate the portion where the ALEX entity initiates. (For simplicity, and to correspond with the RNA coding dictionary, it is customary to represent the noncoding, partner strand of the DNA segment.)

5′-gtcccaaccatgcccaccgatcttccgcctgcttctgaagATGCGGGCCCAG

(a) Convert the noncoding DNA sequence to the coding RNA sequence. (b) Locate the initiator codon within the XLαs segment. (c) Locate the initiator codon within the ALEX segment. Are the two initiator codons in frame? (d) Provide the amino acid sequence for each coding sequence. In the region of overlap, are the two amino acid sequences the same? (e) Are there any evolutionary advantages to having the same DNA sequence code for two protein products? Are there any disadvantages?
30. The concept of consensus sequences of DNA was introduced in this chapter as those sequences that are similar (homologous) in different genes of the same organism or in genes of different organisms. Examples were the Pribnow box and the −35 region in prokaryotes and the TATA-box region in eukaryotes. Work by Novitsky and colleagues (2002. Virology. *J. Virol.* 76: 5435–51) indicates that among 73 isolates of HIV-Type 1C (a major contributor to the AIDS epidemic), a GGGNNNNNCC consensus sequence exists (where N equals any nitrogenous base) in the promoter–enhancer region of the NF-κB transcription factor, a cis-acting motif which is critical in initiating HIV transcription in human macrophages. The authors contend that finding this and other conserved sequences may be of value in designing an AIDS vaccine. What advantages would finding these consensus sequences confer? What disadvantages?
31. Theoretically, antisense oligodeoxynucleotides (relatively short, 7–20 nucleotides, single-stranded DNAs) can selectively block disease-causing genes. They do so by base-pairing with complementary regions of the RNA transcripts and inhibiting their function. The RNA/DNA hybrid is recognized by intracellular RNase H and the RNA portion of the DNA/RNA hybrid is degraded. Cancer genes have often been chosen as potential targets for antisense drugs. Data from Cho and coworkers (2001. *Proc. Natl. Acad. Sci.* (USA) 98: 9819–23.) (see the table below) established that in response to a single population of antisense oligodeoxynucleotides that target a specific kinase mRNA in a prostate cancer cell line, there was as much as a 20-fold (plus and minus) response differential of unrelated gene expression (as measured by mRNA production). (a) Diagram your concept of the intracellular action of an antisense oligodeoxynucleotide. (b) Diagram your concept of the action of RNase H. (c) What might cause nonrelated gene expression to be decreased, as in the case of collagen and catalase? (d) What might cause nonrelated gene expression to be increased, as in the case of myosin and G protein? (e) Given the data presented here, what drawbacks might you expect in the use of antisense therapy for genetic diseases?

Gene Assayed	*Change in Expression*
Myosin light chain	+14.4X
G protein receptor	+3.7X
Collagen	−4.0X
Catalase	−6.7X

32. Recent observations indicate that alternative splicing is a common mechanism for eukaryotes to expand their repertoire of gene functions. Studies by Xu and colleagues (2002. *Nuc. Acids Res.* 30: 3754–66.) indicate that approximately 50% of human genes use alternative splicing, and approximately 15% of disease-causing mutations involve aberrant alternative splicing. Different tissues show remarkably different frequencies of alternative splicing, with the brain accounting for approximately 18% of such events. (a) What does alternative splicing mean, and what is an isoform? (b) What evolutionary strategy does alternative splicing offer, and why might some tissues engage in more alternative splicing than others?

Selected Readings

Barralle, F. E. 1983. The functional significance of leader and trailer sequences in eukaryotic mRNAs. *Int. Rev. Cytol.* 81: 71–106.

Barrell, B. G., Air, G., and Hutchinson, C. 1976. Overlapping genes in bacteriophage $\phi X174$. *Nature* 264: 34–40.

Barrell, B. G., Banker, A. T., and Drouin, J. 1979. A different genetic code in human mitochondria. *Nature* 282: 189–94.

Bass, B. L. (ed.) 2000. *RNA Editing*. Oxford: Oxford Univ. Press.

Birnstiel, M., Busslinger, M., and Strub, K. 1985. Transcription termination and 3′ processing: The end is in site. *Cell* 41: 349–59.

Bonitz, S. G. et al. 1980. Codon recognition rules in yeast mitochondria. *Proc. Natl. Acad. Sci. (USA)* 77: 3167–70.

Breitbart, R. E., and Nadal-Ginard, B. 1987. Developmentally induced, muscle-specific *trans* factors control the differential splicing of alternative and constitutive troponin T-exons. *Cell* 49: 793–803.

Brenner, S. 1989. *Molecular Biology: A Selection of Papers*. Orlando: Academic Press.

Brenner, S., Jacob, F., and Meselson, M. 1961. An unstable intermediate carrying information from genes to ribosomes for protein synthesis. *Nature* 190: 575–80.

Cattaneo, R. 1991. Different types of messenger RNA editing. *Annu. Rev. Genet.* 25: 71–88.

Cech, T. R. 1986. RNA as an enzyme. *Sci. Am.* (Nov.) 255(5): 64–75.

————. 1987. The chemistry of self-splicing RNA and RNA enzymes. *Science* 236: 1532–39.

Cech, T. R., Zaug, A. J., and Grabowski, P. J. 1981. *In vitro* splicing of the ribosomal RNA precursor of *Tetrahymena*. Involvement of a guanosine nucleotide in the excision of the intervening sequence. *Cell* 27: 487–96.

Chambon, P. 1975. Eucaryotic nuclear RNA polymerases. *Annu. Rev. Biochem.* 44: 613–38.

————. 1981. Split genes. *Sci. Am.* (May) 244: 60–71.

Cold Spring Harbor Laboratory. 1966. The genetic code. *Cold Spring Harbor Symp. Quant. Biol.* Vol. 31.

Cramer, P., Bushnell, D. A., and Kornberg, R. D. 2001. Structural basis of transcription: RNA polymerase II at 2.8 Å resolution. *Science* 292: 1863–76.

Crick, F. H. C. 1962. The genetic code. *Sci. Am.* (Oct.) 207: 66–77.

————. 1966a. The genetic code: III. *Sci. Am.* (Oct.) 215: 55–63.

————. 1966b. Codon-anticodon pairing: The wobble hypothesis. *J. Mol. Biol.* 19: 548–55.

Crick, F. H. C., Barnett, L., Brenner, S., and Watts-Tobin, R. J. 1961. General nature of the genetic code for proteins. *Nature* 192: 1227–32.

Darnell, J. E. 1983. The processing of RNA. *Sci. Am.* (Oct.) 249: 90–100.

Dickerson, R. E. 1983. The DNA helix and how it is read. *Sci. Am.* (Dec.) 249: 94–111.

Dugaiczk, A. et al. 1978. The natural ovalbumin gene contains seven intervening sequences. *Nature* 274: 328–33.

Fiers, W. et al. 1976. Complete nucleotide sequence of bacteriophage MS2 RNA: Primary and secondary structure of the replicase gene. *Nature* 260: 500–07.

Gamow, G. 1954. Possible relation between DNA and protein structures. *Nature* 173: 318.

Guthrie, C., and Patterson, B. 1988. Spliceosomal snRNAs. *Annu. Rev. Genet.* 22: 387–419.

Hodges, P., and Scott, J. 1992. Apolipoprotein B mRNA editing: A new tier for the control of gene expression. *Trends Biochem. Sci.* 17: 77–81.

Horton, H. R. et al. 2002. *Principles of Biochemistry*, 3rd ed. Upper Saddle River: Prentice-Hall.

Humphrey, T., and Proudfoot, N. J. 1988. A beginning to the biochemistry of polyadenylation. *Trends Genet.* 4: 243–45.

Judson, H. F. 1979. *The Eighth Day of Creation*. New York: Simon & Schuster.

Jukes, T. H. 1963. The genetic code. *Am. Sci.* 51: 227–45.

Kable, M. L. et al. 1996. RNA editing: A mechanism for gRNA-specified uridylate insertion into precursor mRNA. *Science* 273: 1189–95.

Keegan, L. P., Gallo, A., and O'Connell, M. A. 2000. Survival is impossible without an editor. *Science* 290: 1707–08.

Khorana, H. G. 1967. Polynucleotide synthesis and the genetic code. *Harvey Lec.* 62: 79–105.

Lodish, H. et al. 2000. *Molecular Cell Biology*. New York: W. H. Freeman.

Maniatis, T., and Reed, R. 1987. The role of small nuclear ribonucleoprotein particles in pre-mRNA splicing. *Nature* 325: 673–78.

Nikolov, D. B., and Burley, S. K. 1997. RNA polymerase II transcription initiation: A structural view. *Proc. Natl. Acad. Sci. (USA)* 94: 15–22.

Nilsen, T. W. 1994. RNA–RNA interactions in the spliceosome: Unraveling the ties that bind. *Cell* 78: 1–4.

Nirenberg, M. W. 1963. The genetic code: II. *Sci. Am.* (Mar.) 190: 80–94.

Nirenberg, M. W., and Leder, P. 1964. RNA code words and protein synthesis. *Science* 145: 1399–1407.

Nirenberg, M. W., and Matthaei, H. 1961. The dependence of cell-free protein synthesis in *E. coli* upon naturally occurring or synthetic polyribosomes. *Proc. Natl. Acad. Sci. (USA)* 47: 1588–1602.

O'Malley, B. et al. 1979. A comparison of the sequence organization of the chicken ovalbumin and ovomucoid genes. In *Eucaryotic Gene Regulation*, (ed.) R. Axel et al. pp. 281–99. Orlando,: Academic Press.

Padgett, R. A. et al. 1986. Splicing of messenger RNA precursors. *Annu. Rev. Biochem.* 55: 1119–50.

Proudfoot, N. J., Furger, A., and Dye, M. J. 2002. Integrating mRNA processing with transcription. *Cell* 108: 501–13.

Reed, R., and Maniatis, T. 1985. Intron sequences involved in lariat formation during pre-mRNA splicing. *Cell* 41: 95–105.

Sharp, P. A. 1987. Splicing of messenger RNA precursors. *Science* 235: 766–71.

————. 1994. Nobel Lecture: Split genes and RNA splicing. *Cell* 77: 805–15.

Sharp, P. A., and Eisenberg, D. 1987. The evolution of catalytic function. *Science* 238: 729–30.

Steitz, J. A. 1988. Snurps. *Sci. Am.* (June) 258: 56–63.

Volkin, E., Astrachan, L., and Countryman, J. L. 1958. Metabolism of RNA phosphorus in *E. coli* infected with bacteriophage T7. *Virology* 6: 545–55.

Watson, J. D. 1963. Involvement of RNA in the synthesis of proteins. *Science* 140: 17–26.

Woychik, N. A., and Hampsey, M. 2002. The RNA polymerase II machinery: Structure illuminates function. *Cell* 108: 453–63.

CHAPTER 14

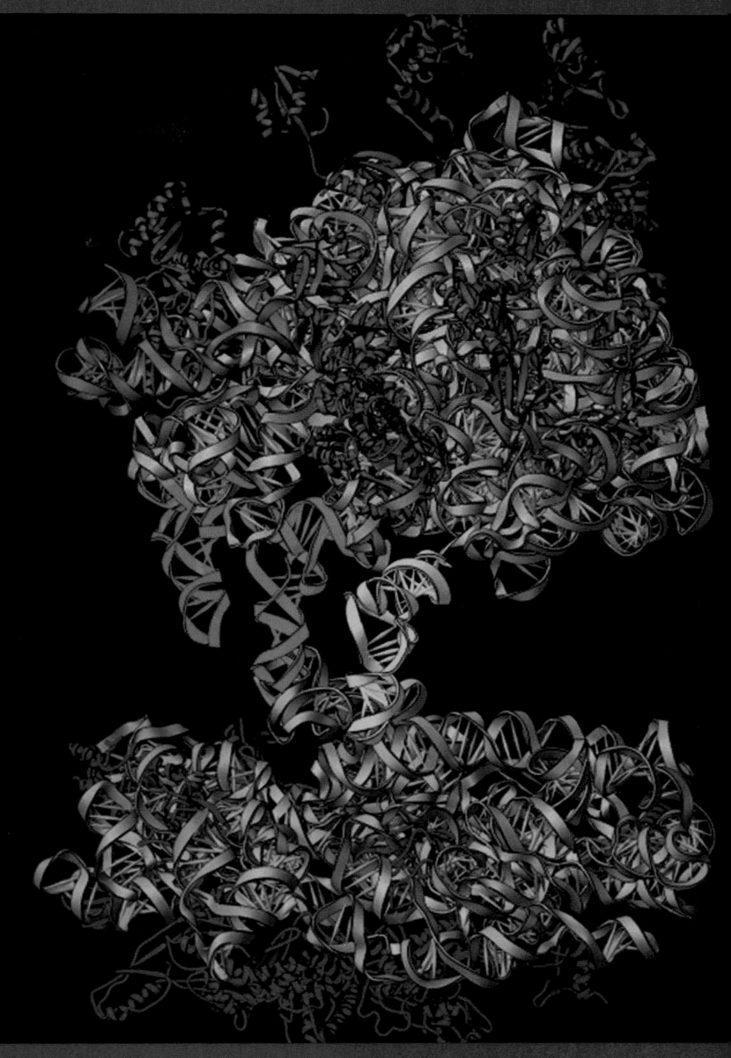

Crystal structure of a *Thermus thermophilus* 70S ribosome containing three bound transfer RNAs. *(Image provided by Dr. Albion Baucon (baucon@biology.ucsc.edu). Copyright American Association for the Advancement of Science. Reprinted from the front cover of Science, Vol.292 May 4 2001.)*

Translation and Proteins

CHAPTER CONCEPTS

We have already learned that a genetic code exists that stores information in the form of triplet nucleotides in DNA and that this information is initially expressed through the process of transcription into a messenger RNA that is complementary to one strand of the DNA helix. However, the final product of gene expression, in almost all instances, is a polypeptide chain consisting of a linear series of amino acids whose sequence has been prescribed by the genetic code. In this chapter, we will examine how the information present in mRNA is processed to create polypeptides, which then fold into protein molecules. We will also review the evidence that confirms that proteins are the end products of gene expression and briefly discuss the various levels of protein structure, diversity, and function. This information extends our understanding of gene expression and provides an important foundation for interpreting how the mutations that arise in DNA can result in the diverse phenotypic effects observed in organisms.

HOW DO WE KNOW WHAT WE KNOW?

IN THIS CHAPTER, WE WILL FOCUS ON HOW GENETIC information, transferred from DNA to mRNA, is expressed through the production of proteins, the end product of most gene expression. As you study this topic, you should try to answer several fundamental questions:

1. How do we know that the process of translation occurs in association with ribosomes?
2. How have the mechanics of translation been deciphered?
3. How do we know that proteins are the end products of genetic expression?
4. How did we come to understand that a single gene ultimately encodes a single polypeptide chain?
5. How have we confirmed that the triplet codes present in mRNA direct amino acid insertions—one after the other—into a polypeptide chain during translation?
6. How do we know that the structure of a protein is intimately related to the function of that protein? ■

14.1 Translation of mRNA Depends on Ribosomes and Transfer RNAs

Translation of mRNA is the biological polymerization of amino acids into polypeptide chains. This process, alluded to in our discussion of the genetic code in Chapter 13, occurs only in association with ribosomes, which serve as nonspecific workbenches. The central question in translation is how triplet ribonucleotides of mRNA direct specific amino acids into their correct position in the polypeptide. This question was answered once **transfer RNA (tRNA)** was discovered. This class of molecules adapts specific triplet codons in mRNA to their correct amino acids. The *adaptor hypothesis* for the role of tRNA was postulated by Francis Crick in 1957.

In association with a ribosome, mRNA presents a triplet codon that calls for a specific amino acid. A specific tRNA molecule contains within its nucleotide sequence three consecutive ribonucleotides complementary to the codon, the **anticodon**, which can base-pair with the codon. Another region of this tRNA is covalently bonded to its corresponding amino acid.

Inside the ribosome, hydrogen bonding of tRNAs to mRNA holds the amino acids in proximity so that a peptide bond can be formed. This process occurs over and over as mRNA runs through the ribosome and amino acids are polymerized into a polypeptide. Before we discuss the actual process of translation, let's first consider the structures of the ribosome and tRNA.

Ribosomal Structure

Because of its essential role in the expression of genetic information, the **ribosome** has been extensively analyzed. One bacterial cell contains about 10,000 ribosomes, and a eukaryotic cell contains many times more. Electron microscopy reveals that the bacterial ribosome is about 25 μm at its largest diameter and consists of two subunits, one large and one small. Both subunits consist of one or more molecules of rRNA and an array of **ribosomal proteins**. When the two subunits are associated with each other in a single ribosome, the structure is sometimes called a **monosome**.

The specific differences between prokaryotic and eukaryotic ribosomes are summarized in Figure 14–1. The subunit and rRNA components are most easily isolated and characterized on the basis of their sedimentation behavior in sucrose gradients (their rate of migration, or Svedberg coefficient *S*, which reflects their density, mass, and shape). In prokaryotes, the monosome is a 70*S* particle; and in eukaryotes, it is approximately 80*S*. Sedimentation coefficients, which reflect the variable rate of migration of different-sized particles and molecules, are not additive. For example, the prokaryotic 70*S* monosome consists of a 50*S* and a 30*S* subunit, and the eukaryotic 80*S* monosome consists of a 60*S* and a 40*S* subunit.

The larger subunit in prokaryotes consists of a 23*S* rRNA molecule, a 5*S* rRNA molecule, and 31 ribosomal proteins. In the eukaryotic equivalent, a 28*S* rRNA molecule is accompanied by a 5.8*S* and 5*S* rRNA molecule and 49 proteins. The smaller prokaryotic subunits consist of a 16*S* rRNA component and 21 proteins. In the eukaryotic equivalent, an 18*S* rRNA component and 33 proteins are found. The approximate molecular weights (MW) and the numbers of nucleotides of these components are also shown in Figure 14–1.

Regarding the components of the ribosome, it is now clear that the RNA molecules perform the all-important catalytic functions associated with translation. The many proteins, whose functions were long a mystery, are thought to promote the binding of the various molecules involved in translation and in general, to fine-tune the process.

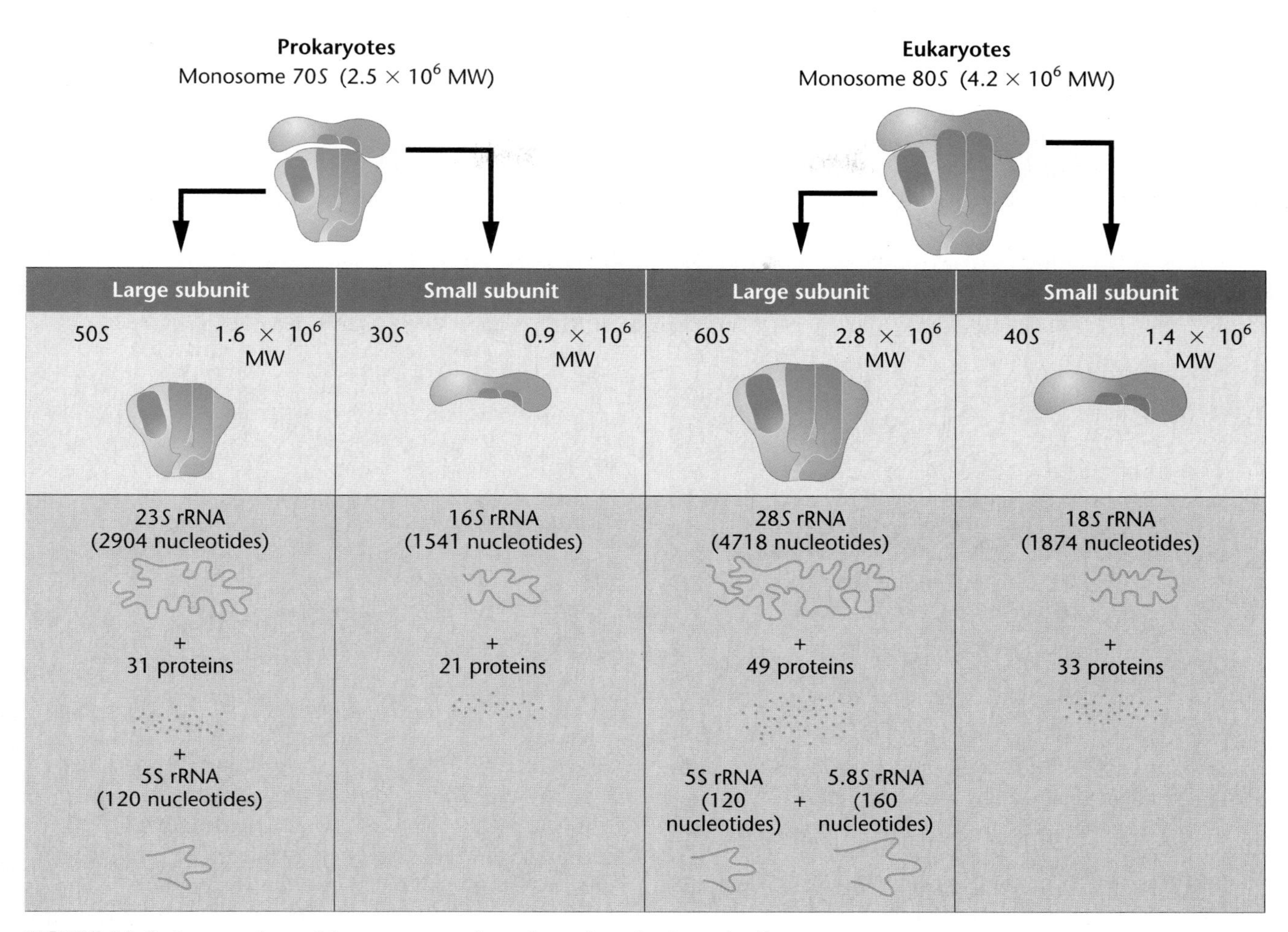

Prokaryotes
Monosome 70S (2.5 × 10^6 MW)
Eukaryotes
Monosome 80S (4.2 × 10^6 MW)
Large subunit
Small subunit
Large subunit
Small subunit
50S 1.6 × 10^6 MW
30S 0.9 × 10^6 MW
60S 2.8 × 10^6 MW
40S 1.4 × 10^6 MW
23S rRNA
(2904 nucleotides)
16S rRNA
(1541 nucleotides)
28S rRNA
(4718 nucleotides)
18S rRNA
(1874 nucleotides)
+
31 proteins
21 proteins
49 proteins
33 proteins
5S rRNA
(120 nucleotides)
5S rRNA
(120 nucleotides)
5.8S rRNA
(160 nucleotides)

FIGURE 14–1 A comparison of the components in prokaryotic and eukaryotic ribosomes.

This conclusion is based on the observation that some of the catalytic functions in ribosomes still occur in experiments involving "ribosomal protein-depleted" ribosomes.

Molecular hybridization studies have established the degree of redundancy of the genes coding for the rRNA components. The *E. coli* genome contains seven copies of a single sequence that encodes all three components—23*S*, 16*S*, and 5*S*. The initial transcript of these genes produces a 30*S* RNA molecule that is enzymatically cleaved into these smaller components. Coupling of the genetic information encoding these three rRNA components ensures that after multiple transcription events, equal quantities of all three will be present as ribosomes are assembled.

In eukaryotes, many more copies of a sequence encoding the 28*S* and 18*S* components are present. In *Drosophila*, approximately 120 copies per haploid genome are each transcribed into a molecule of about 34*S*. This molecule is then processed into the 28*S*, 18*S*, and 5.8*S* rRNA species. These species are homologous to the three rRNA components of *E. coli*. In *Xenopus laevis*, over 500 copies of the 34*S* component are present per haploid genome. In mammalian cells, the initial transcript is even larger at 45*S*. The rRNA genes, called **rDNA**, are part of the moderately repetitive DNA fraction and are present in clusters at various chromosomal sites.

Each cluster in eukaryotes consists of **tandem repeats**, with each unit separated by a noncoding **spacer DNA** sequence. In humans, these gene clusters have been localized near the ends of chromosomes 13, 14, 15, 21, and 22. The unique 5*S* rRNA component of eukaryotes is not part of this larger transcript. Instead, genes coding for this ribosomal component are distinct and located separately. In humans, a gene cluster encoding them has been located on chromosome 1.

Despite the detailed knowledge available about the structure and genetic origin of the ribosomal components, a complete understanding of the function of these components has thus far eluded geneticists. This is not surprising; the ribosome is perhaps the most intricate of all cellular structures. For example, the bacterial monosome has a combined molecular weight of 2.5 million Da.

tRNA Structure

Because of their small size and stability in the cell, transfer RNAs (tRNAs) have been investigated extensively and are the best-characterized RNA molecules. They are composed of only 75–90 nucleotides, displaying a nearly identical structure in bacteria and eukaryotes. In both types of organisms, tRNAs are transcribed as larger precursors, which are cleaved into mature 4*S* tRNA molecules. In *E. coli*, for

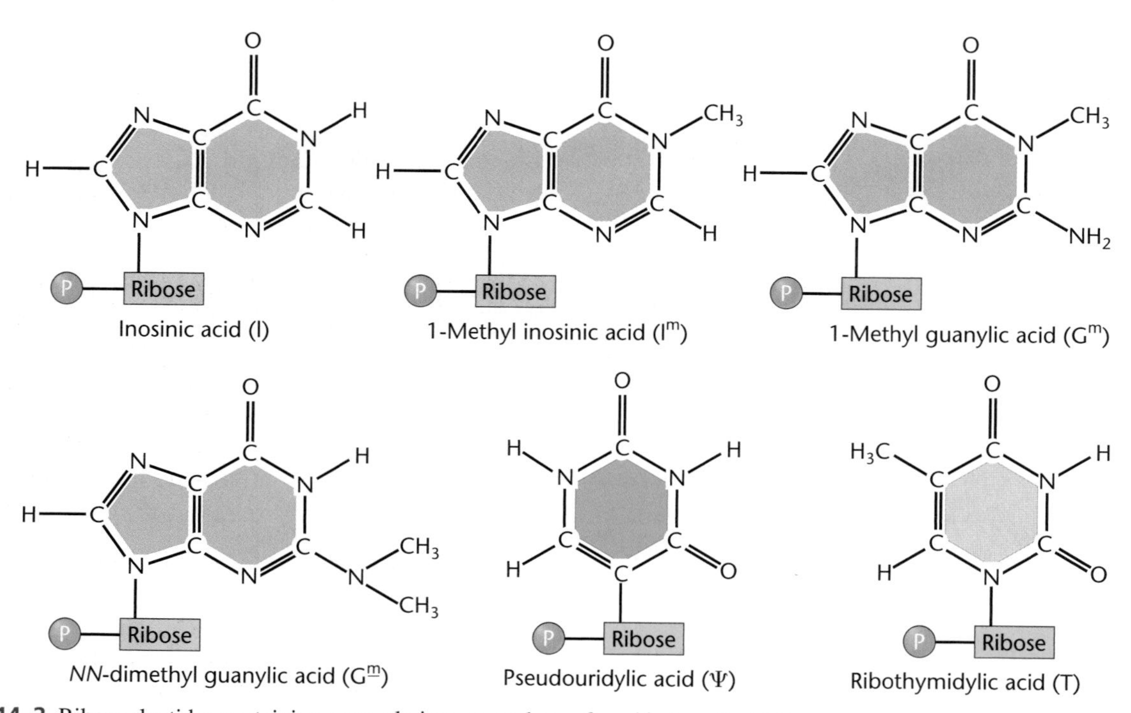

FIGURE 14–2 Ribonucleotides containing unusual nitrogenous bases found in transfer RNA.

example, tRNA^{tyr} (the superscript identifies the specific tRNA and the cognate amino acid that binds to it) is composed of 77 nucleotides, yet its precursor contains 126 nucleotides.

In 1965, Robert Holley and his colleagues reported the complete sequence of tRNA^{ala} isolated from yeast. They found that a number of nucleotides are unique to tRNA. As shown in Figure 14–2, each nucleotide is a modification of one of the four nitrogenous bases normally present in RNA (G, C, A, and U). These include inosinic acid, which contains the purine hypoxanthine, ribothymidylic acid and, pseudouridine, among others. These modified structures, variously referred to as *unusual*, *rare*, or *odd bases*, are created *after* transcription, illustrating the more general concept of **posttranscriptional modification**. In this case, the unmodified base is inserted during transcription of tRNA, and subsequently, enzymatic reactions catalyze the chemical modifications to the base.

MEDIA TUTORIAL Transfer RNAs

Holley's sequence analysis led him to propose the two-dimensional **cloverleaf model of tRNA**. It was known that tRNA demonstrates a secondary structure due to base-pairing. Holley discovered that he could arrange the linear model in such a way that several stretches of base-pairing would result. This arrangement created a series of paired stems and unpaired loops resembling the shape of a cloverleaf. Loops consistently contained modified bases that did not generally form base pairs. Holley's model is shown in Figure 14–3.

The triplets GCU, GCC, and GCA specify alanine; therefore, Holley looked for an anticodon sequence complementary to one of these codons in his tRNA^{ala} molecule. He found it in the form of CGI (the 3′-to-5′ direction) in one loop of the cloverleaf. The nitrogenous base I (inosinic acid) can form hydrogen bonds with U, C, or A, the third members of the triplets. Thus, the **anticodon loop** was established.

Studies of other tRNA species reveal many constant features. At the 3′-end, all tRNAs contain the sequence (. . . pCpCpA-3′). This is the end of the molecule where the amino acid is covalently joined to the terminal adenosine residue. All tRNAs contain the nucleotide (5′-Gp . . .) at the other end of the molecule. In addition, the lengths of various stems and loops are very similar. Each tRNA that has been examined also contains an anticodon complementary to the known amino acid codon for which it is specific, and all anticodon loops are present in the same position of the cloverleaf.

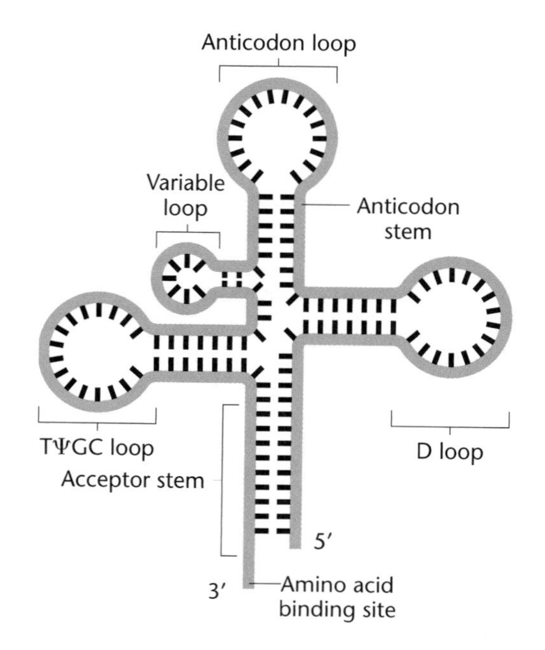

FIGURE 14–3 Holley's two-dimensional cloverleaf model of transfer RNA. Black pegs represent nitrogenous bases.

Because the cloverleaf model was predicted strictly on the basis of nucleotide sequence, there was great interest in the X-ray crystallographic examination of tRNA, which reveals a three-dimensional structure. By 1974, Alexander Rich and his colleagues in the United States, and J. Roberts, B. Clark, Aaron Klug, and their colleagues in England had succeeded in crystallizing tRNA and performing X-ray crystallography at a resolution of 3 Å. At this resolution, the pattern formed by individual nucleotides is discernible.

As a result of these studies, a complete three-dimensional model of tRNA is shown in Figure 14–4. At one end of the molecule is the anticodon loop and stem, and at the other end is the 3′-acceptor region where the amino acid is bound. Geneticists speculate that the shapes of the intervening loops may be recognized by the specific enzymes responsible for adding the amino acid to tRNA—a subject to which we now turn our attention.

Charging tRNA

Before translation can proceed, the tRNA molecules must be chemically linked to their respective amino acids. This activation process, called **charging**, occurs under the direction of enzymes called **aminoacyl tRNA synthetases**. There are 20 different amino acids, so there must be at least 20 different tRNA molecules and as many different enzymes. In theory, because there are 61 triplet codes, there could be the same number of specific tRNAs and enzymes. However, because of the ability of the third member of a triplet code to "wobble," it is now thought that there are at least 32 different tRNAs. It is also believed that there are only 20 synthetases, one for each amino acid, regardless of the greater number of corresponding tRNAs.

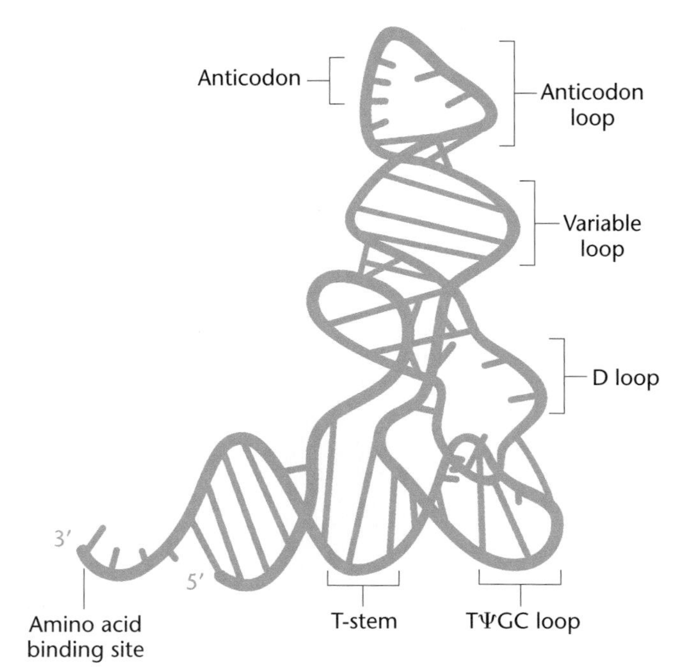

FIGURE 14–4 A three-dimensional model of transfer RNA.

The charging process is outlined in Figure 14–5. In the initial step, the amino acid is converted to an activated form, reacting with ATP to create an **aminoacyladenylic acid**. A covalent linkage is formed between the 5′-phosphate group of ATP and the carboxyl end of the amino acid. This molecule remains associated with the synthetase enzyme, forming a complex that then reacts with a specific tRNA molecule. During this next step, the amino acid is transferred to the appropriate tRNA and bonded covalently to the adenine residue at the 3′-end. The charged tRNA may now participate directly in protein synthesis. Aminoacyl tRNA synthetases are highly specific enzymes because they recognize only one amino acid and the subset of corresponding tRNAs called **isoaccepting tRNAs**. This is crucial if fidelity of translation is to be maintained.

14.2 Translation of mRNA Can Be Divided into Three Steps

In a way similar to transcription, the process of translation can be best described by breaking it into discrete phases. We will consider three phases, each with its own illustration, but keep in mind that translation is a dynamic, continuous process. You should correlate the following

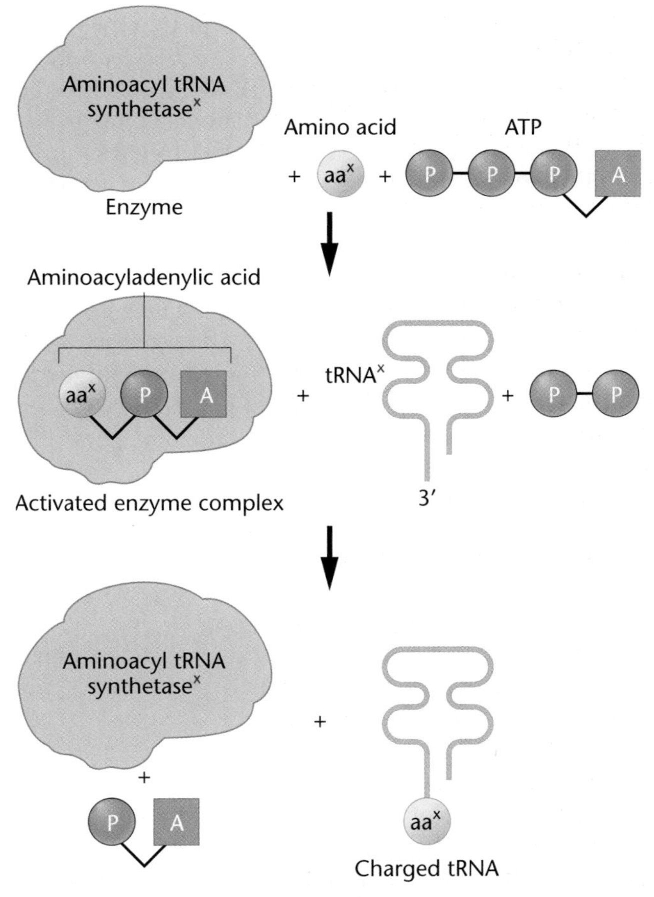

FIGURE 14–5 Steps involved in charging tRNA. The superscript x denotes that only the corresponding specific tRNA and specific aminoacyl tRNA synthetase enzyme are involved in the charging process for each amino acid.

TABLE 14.1 Various Protein Factors Involved During Translation in *E. coli*

Process	*Factor*	*Role*
Initiation of translation	IF1	Stabilizes 30*S* subunit
	IF2	Binds fmet-tRNA to 30*S*-mRNA complex; binds to GTP and stimulates hydrolysis
	IF3	Binds 30*S* subunit to mRNA
Elongation of polypeptide	EF-Tu	Binds GTP; brings aminoacyl-tRNA to the A site of ribosome
	EF-Ts	Generates active EF-Tu
	EF-G	Stimulates translocation; GTP-dependent
Termination of translation and release of polypeptide	RF1	Catalyzes release of the polypeptide chain from tRNA and dissociation of the translocation complex; specific for UAA and UAG termination codons
	RF2	Behaves like RF1; specific for UGA and UAA codons
	RF3	Stimulates RF1 and RF2

discussion with the step-by-step characterization in the figures. Many of the protein factors and their roles in translation are summarized in Table 14.1.

Initiation

Initiation of translation is depicted in Figure 14–6. Recall that the ribosome serves as a nonspecific workbench for the translation process. Most ribosomes, when they are not involved in translation, are dissociated into their large and small subunits. Initiation of translation in *E. coli* involves the small ribosome subunit, an mRNA molecule, a specific charged tRNA, GTP, Mg^{++}, and at least three proteinaceous **initiation factors (IFs)** that enhance the binding affinity of the various translational components. In prokaryotes, the initiation codon of mRNA (AUG) calls for the modified amino acid **formylmethionine (fmet)**.

The small ribosomal subunit binds to several initiation factors, and this complex then binds to mRNA (step 1). In bacteria, this binding involves a sequence of up to six ribonucleotides (AGGAGG, not shown), which *precedes* the initial AUG start codon of mRNA. This sequence (containing only purines and called the **Shine-Dalgarno sequence**) base-pairs with a region of the 16*S* rRNA of the small ribosomal subunit, facilitating initiation.

Another initiation protein then enhances the binding of charged fmet-tRNA to the small subunit in response to the AUG triplet (step 2). This step sets the reading frame so that all subsequent groups of three ribonucleotides are translated accurately. This aggregate represents the **initiation complex**, which then combines with the large ribosomal subunit. At

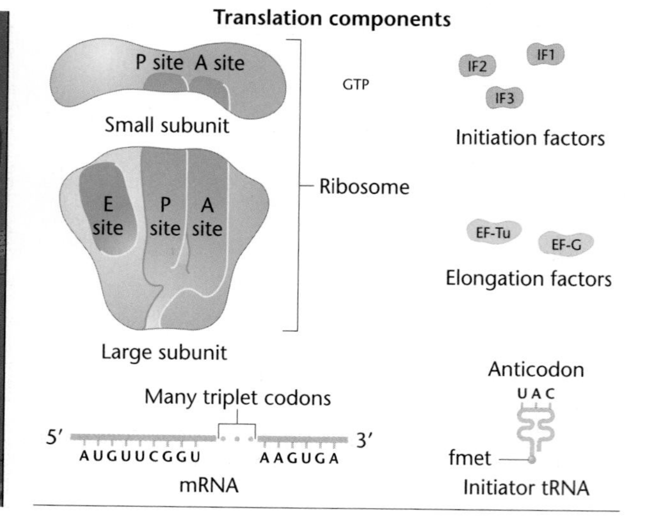

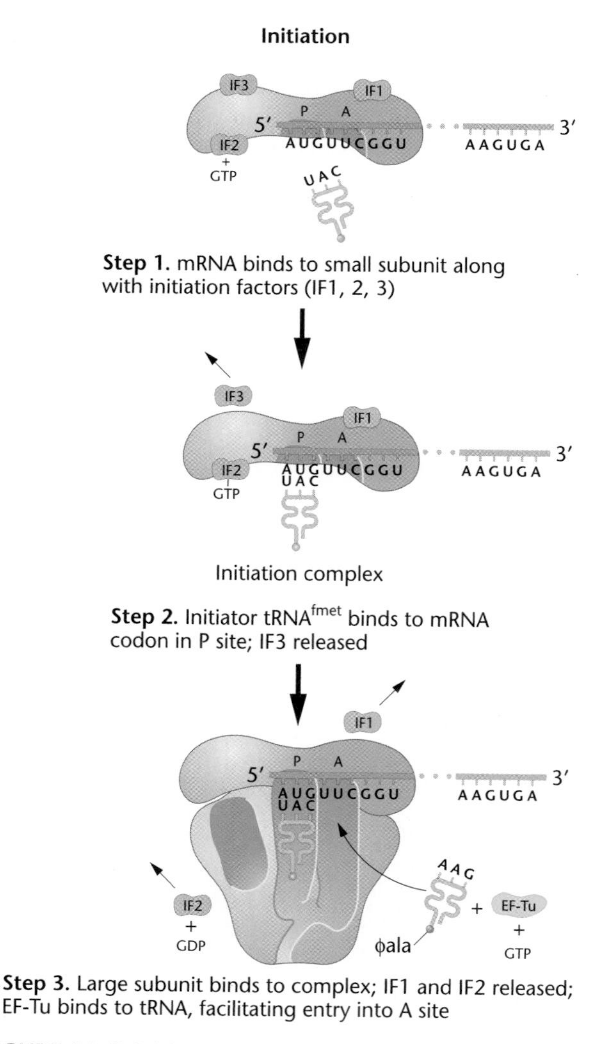

FIGURE 14–6 Initiation of translation. The components are depicted at the left of the figure.

this point, a molecule of GTP is hydrolyzed, providing the required energy, and the initiation factors are released (step 3).

Elongation

The second phase of translation, elongation, is depicted in Figure 14–7. Once both subunits of the ribosome are assembled with the mRNA, binding sites for two charged tRNA molecules are formed. These are the **P (peptidyl) site** and the **A (aminoacyl) site**. The charged initiator tRNA binds to the P site, provided that the AUG triplet of mRNA is in the corresponding position of the small subunit.

Increasing the growing polypeptide chain by one amino acid is called **elongation**. The sequence of the second triplet in mRNA dictates which charged tRNA molecule will become positioned at the A site (step 1). Once it is present, an enzyme within the large subunit of the ribosome, called **peptidyl transferase**, catalyzes the formation of the peptide bond, which links the two amino acids (step 2). At the same time, the covalent bond between the amino acid and the tRNA occupying the P site is hydrolyzed (broken) producing a dipeptide, which is attached to the 3′-end of tRNA still residing in the A site.

Before elongation can be repeated, the tRNA attached to the P site, which is now uncharged, must be released from the large subunit. The uncharged tRNA moves transiently through a third site on the ribosome, called the **E (exit) site**. The entire *mRNA–tRNA–*aa_2*–*aa_1 *complex* then shifts in the direction of the P site by a distance of three nucleotides (step 3). This event requires several protein **elongation factors (EFs)** as well as the energy derived from hydrolysis of GTP. The result is that the third triplet of mRNA is now in a position to accept another specific charged tRNA into the A site (step 4). One simple way to distinguish the two sites is to remember that, *following the shift*, the P site (P for peptide) contains a tRNA attached to a peptide chain, whereas the A site (A for amino acid) contains a tRNA with an amino acid attached.

The sequence of elongation is repeated over and over (steps 5 and 6). An additional amino acid is added to the growing polypeptide chain each time the mRNA advances through the ribosome. Once a polypeptide chain of reasonable size is assembled (about 30 amino acids), it begins to emerge from the base of the large subunit, as illustrated in step 6. A tunnel exists within the large subunit, from which the elongating polypeptide emerges.

As we have seen, the role of the small subunit during elongation is one of "decoding" the triplets present in mRNA, while that of the large subunit is peptide bond synthesis. The efficiency of the process is remarkably high. The observed error rate is only about 10^{-4}—an incorrect amino acid will occur only once in every 20 polypeptides of an average length

Elongation

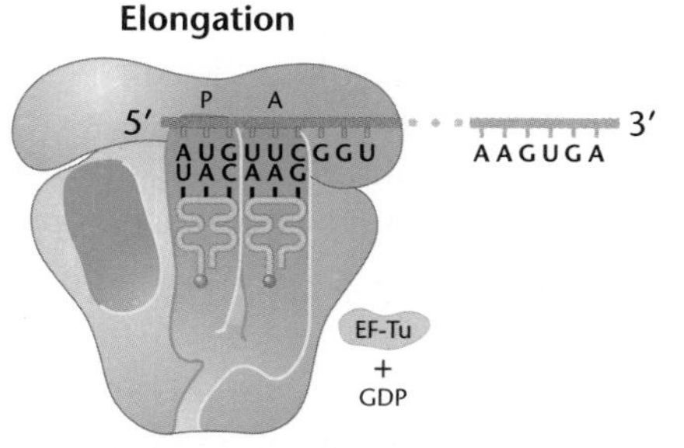

Step 1. Second charged tRNA has entered A site, facilitated by EF-Tu; first elongation step commences

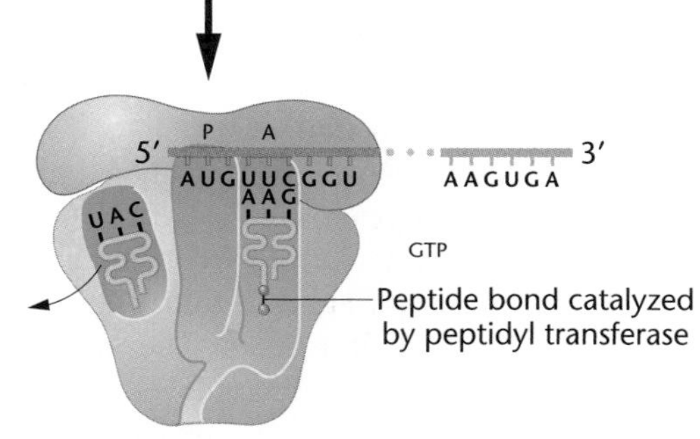

Step 2. Dipeptide bond forms; uncharged tRNA moves to E-site and then out of ribosome

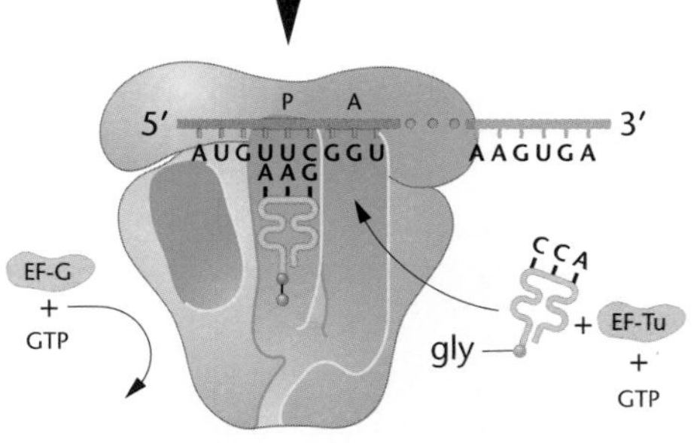

Step 3. mRNA has shifted by three bases; EF-G facilitates the translocation step; first elongation step completed

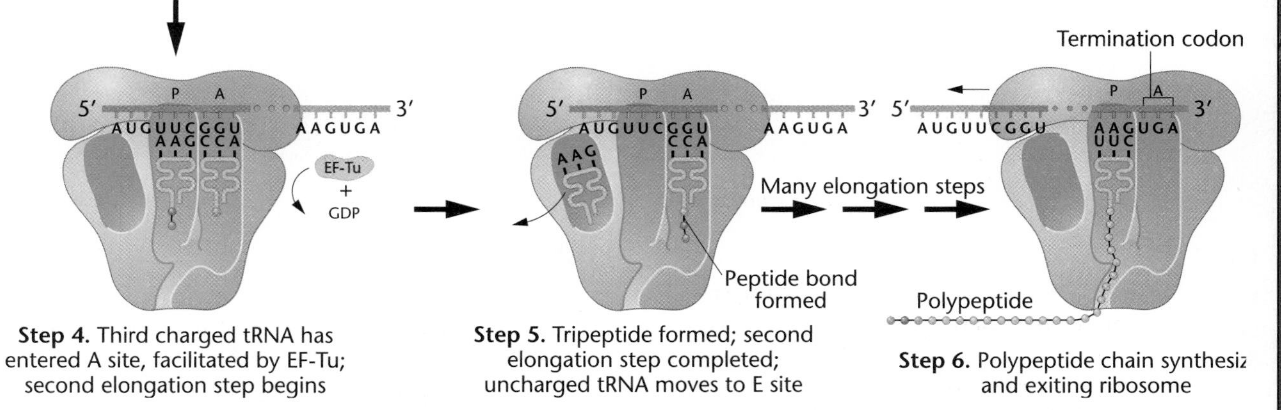

Step 4. Third charged tRNA has entered A site, facilitated by EF-Tu; second elongation step begins

Step 5. Tripeptide formed; second elongation step completed; uncharged tRNA moves to E site

Step 6. Polypeptide chain synthesiz and exiting ribosome

FIGURE 14–7 Elongation of the growing polypeptide chain during translation.

MEDIA TUTORIAL Translation of mRNA

of 500 amino acids. In *E. coli*, elongation occurs at a rate of about 15 amino acids per second at 37°C.

Termination

Termination, the third phase of translation, is depicted in Figure 14–8. Termination of protein synthesis is signaled by one or more of three triplet codes in the A site:

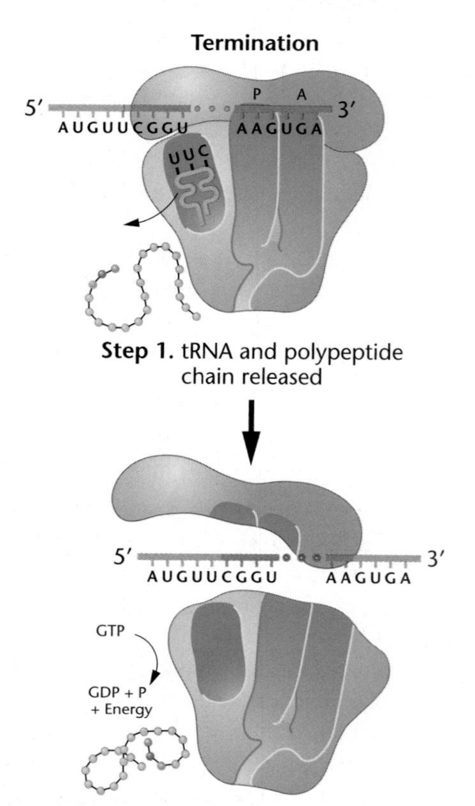

FIGURE 14–8 Termination of the process of translation.

MEDIA TUTORIAL Translation of mRNA

UAG, UAA, or UGA. These codons do not specify an amino acid, nor do they call for a tRNA in the A site. They are called **stop codons**, **termination codons**, or **nonsense codons**. The finished polypeptide is therefore still attached to the terminal tRNA at the P site, and the A site is empty. The termination codon signals the action of **GTP-dependent release factors**, which cleave the polypeptide chain from the terminal tRNA, releasing it from the translation complex (step 1). Once cleavage occurs, the tRNA is released from the ribosome, which then dissociates into its subunits (step 2). If a termination codon appears in the middle of an mRNA molecule as a result of mutation, cleavage occurs, and the polypeptide chain is prematurely terminated.

Polyribosomes

As elongation proceeds and the initial portion of mRNA has passed through the ribosome, this mRNA is free to associate with another small subunit to form a second initiation complex. This process can be repeated several times with a single mRNA and results in what are called **polyribosomes** or just **polysomes**.

Polyribosomes can be isolated and analyzed following a gentle lysis of cells. The photos in Figure 14–9 show these complexes as seen under an electron microscope. In Figure 14–9(a), you can see the thin lines of mRNA between the individual ribosomes. The micrograph in Figure 14–9(b) is even more remarkable, for it shows the polypeptide chains emerging from the ribosomes during translation. The formation of polysome complexes represents an efficient use of the components available for protein synthesis during a particular unit of time. Using the analogy of a song recorded on a tape and a tape recorder, in polysome complexes one tape (mRNA) would be

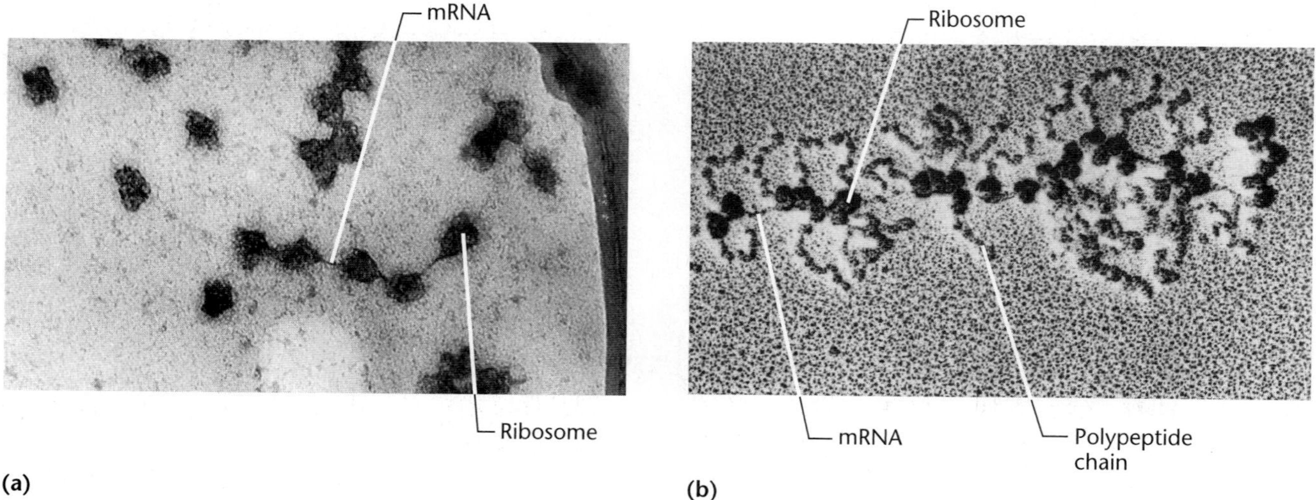

FIGURE 14–9 Polyribosomes as seen under the electron microscope. Those in (a) were derived from rabbit reticulocytes engaged in the translation of hemoglobin mRNA. The polyribosomes in (b) were taken from giant salivary gland cells of the midgefly, *Chironomus thummi*. Note that the nascent polypeptide chain is apparent as it emerges from each ribosome. Its length increases as translation proceeds from left (5′) to right (3′) along the mRNA. *(Left photo: "The Structure and Function of Polyribosomes." Alexander Rich, Jonathan R. Warner and Howard M. Goodman, 1963. Reproduced by permission of the Cold Spring Harbor Laboratory Press. Cold Spring Harbor Symp. Quant. Biol. 28 (1963) fig. 4c (top), p. 273, © 1964. Right photo: E. V. Kiseleva)*

played simultaneously by several recorders (the ribosomes), but at any given moment, each song (the polypeptide being synthesized in each ribosome) would be at a different point in the lyrics.

14.3 Crystallographic Analysis Has Revealed Many Details About the Functional Prokaryotic Ribosome

Our knowledge of the process of translation and the structure of the ribosome, as described in the previous sections, is based primarily on biochemical and genetic observations, in addition to visualization of ribosomes under the electron microscope. Because of the tremendous size and complexity of the functional ribosome during active translation, obtaining the crystals needed to perform X-ray diffraction studies was extremely difficult. Nevertheless, great strides have been made in the past several years. First, the individual ribosomal subunits were crystallized and examined in several laboratories, most prominently that of V. Ramakrishnan. Then, in 2001, the crystal structure of the intact 70*S* ribosome, complete with associated mRNA and tRNAs, was examined by Harry Noller and his colleagues—in essence, the entire translational complex was seen at the atomic level. Both Ramakrishnan and Noller derived the ribosomes from the bacterium *Thermus thermophilus*.

Many noteworthy observations have been made from these investigations. One of the models based on Noller's findings is shown as the opening photograph of this chapter. For example, the sizes and shapes of the subunits, measured at atomic dimensions, are in agreement with earlier estimates based on high-resolution electron microscopy. Further, the shape of the ribosome changes during different functional states, attesting to the dynamic nature of the process of translation. A great deal has also been learned about the prominence and location of the RNA components of the subunits. For example, about one-third of the 16*S* RNA is responsible for producing a flat projection within the smaller 30*S* subunit referred to as the "platform," which modulates movement of the mRNA–tRNA complex during translocation.

More information supports the concept that RNA is the major "player" in the ribosome during translation. The interface between the two subunits, considered to be the location in the ribosome where polymerization of amino acids occurs, is composed almost exclusively of RNA. In contrast, the numerous ribosomal proteins are found mostly on the periphery of the ribosome. These observations confirm what has been predicted on genetic grounds—the catalytic steps that join amino acids during translation occur under the direction of RNA, not proteins.

Another interesting finding involves the actual location of the three sites predicted to house tRNAs during translation. All three sites (A, P, and E) have been identified, and in each case, the RNA of the ribosome makes direct contact with the various loops and domains of the tRNA molecule. This observation points to the importance of the different regions of tRNA and helps us understand why the specific three-dimensional conformation of all tRNA molecules has been preserved throughout evolution.

A final observation takes us back almost 50 years, to when Francis Crick proposed the wobble hypothesis. The Ramakrishnan research group has identified the precise location along the 16*S* RNA of the 30*S* subunit involved in the decoding step between mRNA and tRNA. At this location, two particular nucleotides of the 16*S* RNA actually flip out and probe the codon–anticodon region and are also believed to check the accuracy of base-pairing during this interaction. According to the wobble hypothesis, the stringency of this step is high for the first two base pairs, but less stringent for the third (or wobble) base pair.

These landmark studies provide us with a much better picture of the dynamic changes that must occur within the ribosome during translation. However, numerous questions still remain about ribosome structure and function. In particular, the role of the many ribosomal proteins has yet to be clarified. Nevertheless, the models that are emerging based on the work of Noller, Ramakrishnan, and their many colleagues provide us with a much better understanding of the mechanism of translation.

14.4 Translation Is More Complex in Eukaryotes

The general features of the model we just discussed were initially derived from investigations of the translation process in bacteria. As we have seen, one main difference between translation in prokaryotes and eukaryotes is that in the latter, translation occurs on larger ribosomes whose rRNA and protein components are more complex than those of prokaryotes (see Figure 14–1).

Several other differences are also important. Eukaryotic mRNAs are much longer-lived than their prokaryotic counterparts. Most exist for hours rather than minutes prior to degradation by nucleases in the cell; thus they remain available much longer to orchestrate protein synthesis.

Several aspects involving the initiation of translation are also different in eukaryotes. First, as we discussed in Chapter 13, the 5′-end of mRNA is capped with a 7-methylguanosine (7mG) residue at maturation. The presence of the 7mG cap, absent in prokaryotes, is essential to efficient translation, since RNAs that lack the cap are translated poorly. In addition, most eukaryotic mRNAs contain a short recognition sequence that surrounds the initiating AUG codon—5′-ACCAUGG. Named after Marilyn Kozak, who discovered it, this Kozak sequence appears to function during initiation in the same way that the Shine-Dalgarno sequence functions in prokaryotic mRNA. Both greatly facilitate the initial binding of mRNA to the small subunit of the ribosome.

Another difference is that the amino acid formylmethionine is not required to initiate eukaryotic translation. However, as in prokaryotes, the AUG triplet, which encodes methionine, is essential to the formation of the translational complex, and a unique transfer RNA ($tRNA_i^{met}$) is used during initiation.

Protein factors similar to those in prokaryotes guide the initiation, elongation, and termination of translation in eukaryotes. Many of these eukaryotic factors are clearly homologous to their counterparts in prokaryotes. However, a greater number of factors are usually required during each step, and some are more complex than in prokaryotes.

Finally, recall that in eukaryotes a large proportion of the ribosomes are found in association with the membranes that make up the endoplasmic reticulum (forming the rough ER). Such membranes are absent from the cytoplasm of prokaryotic cells. This association in eukaryotes facilitates the secretion of newly synthesized proteins from the ribosomes directly into the channels of the endoplasmic reticulum. Recent studies using cryo-electron microscopy have established how this occurs. A **tunnel** in the large subunit of ribosomes begins near the point where the two subunits interface and exits near the back of the large subunit. The location of the tunnel within the large subunit is the basis for the belief that it provides the conduit for the movement of the newly synthesized polypeptide chain out of the ribosome. In studies in yeast, newly synthesized polypeptides enter the ER through a membrane channel formed by a specific protein, Sec61. This channel is perfectly aligned with the exit point of the ribosomal tunnel. In prokaryotes, the polypeptides are released by the ribosome directly into the cytoplasm.

14.5 The Initial Insight that Proteins Are Important in Heredity Was Provided by the Study of Inborn Errors of Metabolism

Let's consider how we know that proteins are the end products of genetic expression. The first insight into the role of proteins in genetic processes was provided by observations made by Sir Archibald Garrod and William Bateson early in the twentieth century. Garrod was born into an English family of medical scientists. His father was a physician with a strong interest in the chemical basis of rheumatoid arthritis, and his eldest brother was a leading zoologist in London. It is not surprising, then, that as a practicing physician, Garrod became interested in several human disorders that seemed to be inherited. Although he also studied albinism and cystinuria, we shall describe his investigation of the disorder **alkaptonuria**. Individuals afflicted with this disorder cannot metabolize the alkapton 2,5-dihydroxyphenylacetic acid, also known as homogentisic acid. As a result, an important metabolic pathway (Figure 14–10) is blocked. Homogentisic acid accumulates in cells and tissues and is excreted in the urine. The molecule's oxidation products are black and easily detectable in the diapers of newborns. The products tend to accumulate in cartilaginous areas, causing the ears and nose to darken. The deposition of homogentisic acid in joints leads to a benign arthritic condition. This rare disease is not serious, but it persists throughout an individual's life.

Garrod studied alkaptonuria by either increasing dietary protein or adding the amino acids phenylalanine or tyrosine to the diet; both of these amino acids are chemically related to homogentisic acid. Under these conditions, homogentisic

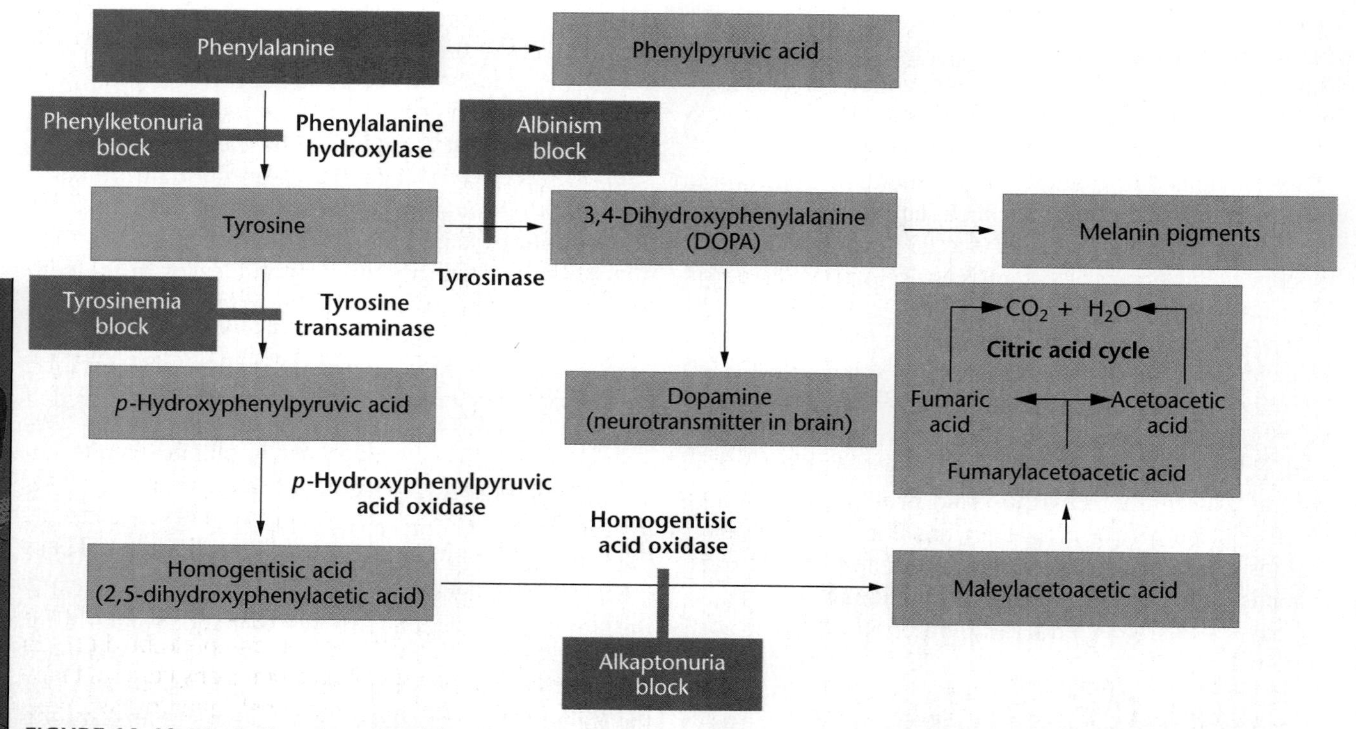

FIGURE 14–10 Metabolic pathway involving phenylalanine and tyrosine. Various metabolic blocks resulting from mutations lead to the disorders phenylketonuria, alkaptonuria, albinism, and tyrosinemia.

acid levels increased in the urine of alkaptonurics but not in unaffected individuals. Garrod concluded that normal individuals are able to break down, or catabolize, this alkapton, but afflicted individuals are not. By studying the disorder's pattern of inheritance, Garrod further concluded that alkaptonuria is inherited as a simple recessive trait.

On the basis of these conclusions, Garrod hypothesized that hereditary information controls chemical reactions in the body and that the inherited disorders he studied are the result of alternative modes of metabolism. While the terms *genes* and *enzymes* were not familiar during Garrod's time, he used the corresponding concepts *unit factors* and *ferments*. Garrod published his initial observations in 1902.

Only a few geneticists, including Bateson, were familiar with or referred to Garrod's work. Garrod's ideas fit nicely with Bateson's belief that inherited conditions are caused by the lack of some critical substance. In 1909, Bateson published *Mendel's Principles of Heredity*, in which he linked Garrod's ferments with heredity. However, for almost 30 years, most geneticists failed to see the relationship between genes and enzymes. Garrod and Bateson, like Mendel, were ahead of their time.

Phenylketonuria

The inherited human metabolic disorder, **phenylketonuria (PKU)**, results when another reaction in the pathway shown in Figure 14–10 is blocked. Described first in 1934, this disorder can result in mental retardation and is transmitted as an autosomal recessive disease. Afflicted individuals are unable to convert the amino acid phenylalanine to the amino acid tyrosine. These molecules differ by only a single hydroxyl group (OH) that is present in tyrosine, but absent in phenylalanine. The reaction is catalyzed by the enzyme **phenylalanine hydroxylase**, which is inactive in affected individuals and active at about a 30% level in heterozygotes. The enzyme functions in the liver. The normal blood level of phenylalanine is about 1 mg/100 mL; phenylketonurics show levels as high as 50 mg/100 mL.

As phenylalanine accumulates, it can be converted to phenylpyruvic acid and subsequently to other derivatives. These are less efficiently resorbed by the kidney and tend to spill into the urine more quickly than phenylalanine. Both phenylalanine and its derivatives subsequently enter the cerebrospinal fluid, resulting in elevated levels in the brain. The presence of these substances during early development is thought to cause mental retardation.

Screening newborns for PKU is routine throughout the United States, preventing retardation through early detection. Phenylketonuria occurs in approximately one in 11,000 births. When the condition is detected in the analysis of an infant's blood, a strict dietary regimen is instituted. A low-phenylalanine diet can reduce by-products such as phenylpyruvic acid, and the abnormalities characterizing the disease can be diminished.

Our knowledge of inherited metabolic disorders such as alkaptonuria and phenylketonuria has caused a revolution in medical thinking and practice. Human diseases, once believed to be solely attributable to the action of invading microorganisms, viruses, or parasites, clearly can have a genetic basis. We now know that thousands of medical conditions are caused by errors in metabolism resulting from mutant genes. These human biochemical disorders include all classes of organic biomolecules.

14.6 Studies of *Neurospora* Led to the One-Gene:One-Enzyme Hypothesis

In two separate investigations beginning in 1933, George Beadle provided the first convincing experimental evidence that genes are directly responsible for the synthesis of enzymes. The first investigation, conducted in collaboration with Boris Ephrussi, involved *Drosophila* eye pigments. Together, they confirmed that mutant genes that alter the eye color of fruit flies could be linked to biochemical errors that, in all likelihood, involved the loss of enzyme function. Encouraged by these findings, Beadle then joined with Edward Tatum to investigate nutritional mutations in the pink bread mold *Neurospora crassa*. This investigation led to the **one-gene:one-enzyme hypothesis**.

Beadle and Tatum: *Neurospora* Mutants

In the early 1940s, Beadle and Tatum chose to work with *Neurospora* because much was known about its biochemistry and because mutations could be induced and isolated with relative ease. By inducing mutations, they produced strains that had genetic blocks of reactions essential to the growth of the organism.

Beadle and Tatum knew that this mold could manufacture nearly everything necessary for normal development. For example, using rudimentary carbon and nitrogen sources, this organism can synthesize 9 water-soluble vitamins, 20 amino acids, numerous carotenoid pigments, all essential purines, and pyrimidines. Beadle and Tatum irradiated asexual conidia (spores) with X-rays to increase the frequency of mutations and allowed them to be grown on "complete" medium containing all the necessary growth factors (e.g., vitamins and amino acids). Under such growth conditions, a mutant strain unable to grow on minimal medium was able to grow by virtue of supplements present in the enriched complete medium. All the cultures were then transferred to minimal medium. If growth occurred on the minimal medium, the organisms were able to synthesize all the necessary growth factors themselves, and the researchers concluded that the culture did not contain a mutation. If no growth occurred, then they concluded that the culture contained a nutritional mutation, and the only task remaining was to determine its type. These results are shown in Figure 14–11(a).

Many thousands of individual spores from this procedure were isolated and grown on complete medium. In subsequent tests on minimal medium, many cultures failed to grow, indicating that a nutritional mutation had been induced. To identify the mutant type, the mutant strains were then tested on a series of different minimal media

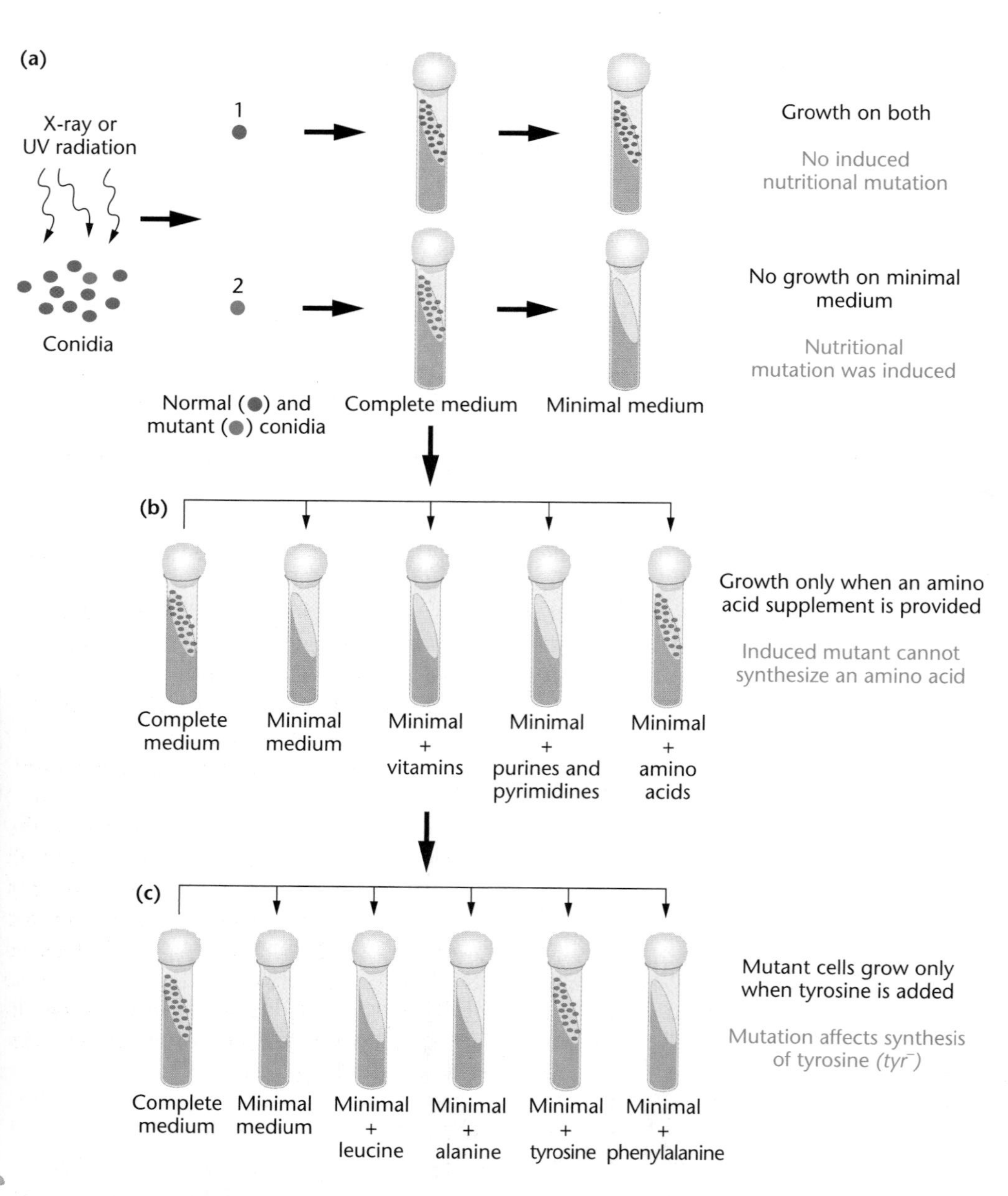

FIGURE 14–11 Induction, isolation, and characterization of a nutritional auxotrophic mutation in *Neurospora*. (a) Most conidia are not affected, but one conidium (shown in red) contains the mutation. In (b) and (c), the precise nature of the mutation is established and found to involve the biosynthesis of tyrosine.

(Figure 14–11b), each containing groups of supplements, and subsequently on media containing single vitamins, purines, pyrimidines, and amino acids (Figure 14–11c) until one specific supplement that permitted growth was found. Beadle and Tatum reasoned that the supplement that restored growth would be the molecule that the mutant strain could not synthesize.

The first mutant strain they isolated required vitamin B_6 (pyridoxine) in the medium, and the second required vitamin B_1 (thiamine). Using the same procedure, Beadle and Tatum eventually isolated and studied hundreds of mutants deficient in the ability to synthesize other vitamins, amino acids, or other substances.

The findings derived from testing over 80,000 spores convinced Beadle and Tatum that genetics and biochemistry have much in common. It seemed likely that each nutritional mutation caused the loss of the enzymatic activity that facilitated an essential reaction in wild-type organisms. It also appeared that a mutation could be found for nearly any enzymatically controlled reaction. Beadle and Tatum had thus provided sound experimental evidence for the hypothesis that *one gene specifies one enzyme*, an idea alluded to over 30 years earlier by Garrod and Bateson. With modifications, this concept was to become another major principle of genetics.

Genes and Enzymes: Analysis of Biochemical Pathways

The one-gene:one-enzyme concept and its attendant methods have been used over the years to work out many details of metabolism in *Neurospora*, *E. coli*, and a number of other microorganisms. One of the first metabolic pathways to be investigated in detail was that leading to the synthesis of the amino acid arginine in *Neurospora*. By studying seven mutant strains, each requiring arginine for growth (arg^-), Adrian Srb and Norman Horowitz ascertained a partial biochemical pathway that leads to the synthesis of this molecule. Their work demonstrates how genetic analysis can be used to establish biochemical information.

Srb and Horowitz tested each mutant strain's ability to grow if either citrulline or ornithine, two compounds with close chemical similarity to arginine, was used to supplement a minimal medium. If either compound was able to substitute for arginine, they reasoned that it must be involved in the biosynthetic pathway of arginine. They found that both molecules could be substituted in one or more strains.

Of the seven mutant strains, four of them (*arg4–7*) grew if supplied with either citrulline, ornithine, or arginine. Two of them (*arg2* and *arg3*) grew if supplied with citrulline or arginine. One strain (*arg1*) would grow only if arginine was supplied—neither citrulline nor ornithine could substitute for it. From these experimental observations, the following pathway and metabolic blocks for each mutation were deduced:

$$\text{Precursor} \xrightarrow[\text{Enzyme A}]{arg4\text{–}7} \text{Ornithine} \xrightarrow[\text{Enzyme B}]{arg2 \text{ and } arg3} \text{Citrulline} \xrightarrow[\text{Enzyme C}]{arg1} \text{Arginine}$$

The logic supporting these conclusions is as follows: If mutants *arg4* through *arg7* can grow regardless of which of the three molecules is supplied as a supplement to minimal medium, the mutations preventing growth must cause a metabolic block that occurs *prior* to the involvement of ornithine, citrulline, or arginine in the pathway. When any one of these three molecules is added, its presence bypasses the block. As a result, both citrulline and ornithine appear to be involved in the biosynthesis of arginine. However, the sequence of their participation in the pathway cannot be determined on the basis of these data.

On the other hand, the *arg2* and *arg3* mutations grow if supplied with citrulline, but not if they are supplied with only ornithine. Therefore, ornithine must be synthesized in the pathway *prior* to the block. Its presence will not overcome the block. Citrulline, however, does overcome the block, so it must be synthesized beyond the point of blockage. Therefore, the conversion of ornithine to citrulline represents the correct sequence in the pathway.

Finally, we can conclude that *arg1* represents a mutation preventing the conversion of citrulline to arginine. Neither ornithine nor citrulline can overcome the metabolic block because both participate earlier in the pathway.

Together, these reasons support the sequence of biosynthesis outlined here. Since Srb and Horowitz's work in 1944, the detailed pathway has been worked out and the enzymes controlling each step characterized; an abbreviated metabolic pathway is shown in Figure 14–12.

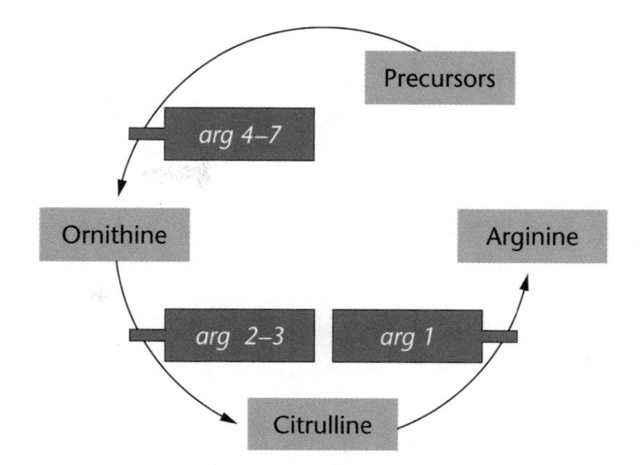

FIGURE 14–12 Abbreviated pathway resulting in the biosynthesis of arginine in *Neurospora*.

14.7 Studies of Human Hemoglobin Established that One Gene Encodes One Polypeptide

The concept of the one-gene:one-enzyme hypothesis that developed in the early 1940s was not immediately accepted by all geneticists. This is not surprising because it was not yet clear how mutant enzymes could cause variation in many phenotypic traits. For example, *Drosophila* mutants demonstrate altered eye size, wing shape, wing-vein pattern, and so on. Plants exhibit mutant varieties of seed texture, height, and fruit size. How an inactive mutant enzyme could result in such phenotypes puzzled many geneticists.

Two factors soon modified the one-gene:one-enzyme hypothesis. First, while nearly all enzymes are proteins, not all proteins are enzymes. As the study of biochemical genetics progressed, it became clear that all proteins are specified by the information stored in genes, leading to the more accurate phraseology, **one-gene:one-protein hypothesis**. Second, proteins often show a substructure consisting of two or more polypeptide chains. This is the basis of the quaternary protein structure, which we will discuss later in this chapter. Because each distinct polypeptide chain is encoded by a separate gene, a more accurate statement of Beadle and Tatum's basic tenet is **one-gene:one-polypeptide chain hypothesis**. These modifications of the original hypothesis became apparent during the analysis of hemoglobin structure in individuals afflicted with sickle-cell anemia.

Sickle-cell Anemia

The first direct evidence that genes specify proteins other than enzymes came from work on mutant hemoglobin molecules found in humans afflicted with the disorder **sickle-cell anemia**. Affected individuals have erythrocytes that, under low oxygen tension, become elongated and curved because of the polymerization of hemoglobin. The sickle shape of these erythrocytes is in contrast to the biconcave disc shape characteristic in unaffected individuals (Figure 14–13). Those with the disease suffer attacks when red blood cells aggregate in the venous side of capillary systems, where oxygen tension is very low. As a result, a variety of tissues are deprived of oxygen and suffer severe damage. When this occurs, an individual is said to experience a sickle-cell crisis. If left untreated, a crisis can be fatal. The kidneys, muscles, joints, brain, gastrointestinal tract, and lungs can be affected.

In addition to suffering crises, these individuals are anemic because their erythrocytes are destroyed more rapidly than are normal red blood cells. Compensatory physiological mechanisms include increased red blood cell production by bone marrow, along with accentuated heart action. These mechanisms lead to abnormal bone size and shape, as well as dilation of the heart.

FIGURE 14–13 A comparison of erythrocytes from (a) healthy individuals, and (b) those with sickle-cell anemia. *(Left photo: Dennis Kunkel/Phototake NYC; Right photo: Francis Leroy/Biocosmos/Science Photo Library/Photo Researchers, Inc.)*

In 1949, James Neel and E. A. Beet demonstrated that the disease is inherited as a Mendelian trait. Pedigree analysis revealed three genotypes and phenotypes controlled by a single pair of alleles, Hb^A and Hb^S. Unaffected and affected individuals result from the homozygous genotypes Hb^AHb^A and Hb^SHb^S, respectively. The red blood cells of heterozygotes, who exhibit the **sickle-cell trait** but not the disease, undergo much less sickling because over half of their hemoglobin is normal. Although they are largely unaffected, heterozygotes are "carriers" of the defective gene, which is transmitted on average to 50% of their offspring.

In that same year, Linus Pauling and his coworkers provided the first insight into the molecular basis of the disease. They showed that hemoglobins isolated from diseased and normal individuals differ in their rates of electrophoretic migration. In this technique, charged molecules migrate in an electric field. If the net charge of two molecules is different, their rates of migration will be different. On this basis, Pauling and his colleagues concluded that a chemical difference exists between normal **(HbA)** and sickle-cell **(HbS)** hemoglobin.

Figure 14–14(a) shows the migration pattern of hemoglobin derived from individuals of all three possible genotypes when it was subjected to **starch gel electrophoresis**. The gel provides the supporting medium for the molecules during migration. In this experiment, samples were placed at a point of origin between a cathode (−) and an anode (+), and an electric field was applied. The migration pattern revealed that all of the molecules moved toward the anode, indicating a net negative charge. However, HbA migrated farther than HbS, suggesting that its net negative charge was greater. The electrophoretic pattern of hemoglobin derived from carriers revealed the presence of both HbA and HbS and confirmed their heterozygous genotype.

Pauling's findings suggested two possibilities. It was known that hemoglobin consists of four nonproteinaceous, iron-containing *heme groups* and a *globin portion* that contains four polypeptide chains. The alteration in net charge in HbS had to be due, theoretically, to a chemical change in one of these components.

Work carried out between 1954 and 1957 by Vernon Ingram resolved this question. He demonstrated that the chemical change occurs in the primary structure of the globin portion of the hemoglobin molecule. Using the **protein fingerprinting technique** shown in Figure 14–14a, Ingram showed that HbS differs in amino acid composition compared to HbA. Human adult hemoglobin contains 2 identical α chains of 141 amino acids, and 2 identical β chains of 146 amino acids in its quaternary structure.

The fingerprinting technique involves enzymatic digestion of the protein into peptide fragments. The mixture is then placed on absorbent paper and exposed to an electric field, where migration occurs according to net charge. The paper is then turned at a right angle to its first exposure and placed in a solvent, where chromatographic action causes the migration of the peptides in the second direction. The end result is a two-dimensional separation of the peptide fragments into a distinctive pattern of spots or a "fingerprint." Ingram's work revealed that HbS and HbA differed by only a single peptide fragment (Figure 14–14b). Further analysis revealed just a single amino acid change: Valine was substituted for glutamic acid at the sixth position of the β chain, thus accounting for the peptide difference (Figure 14–14c).

The significance of this discovery has been multifaceted. It clearly establishes that a single gene provides the genetic information for a single polypeptide chain. Studies of HbS also demonstrate that a mutation can affect the phenotype by directing a single amino acid substitution. Also, by providing the explanation for sickle-cell anemia, the concept of **inherited molecular disease** was firmly established. Finally, this work has led to a thorough study of human hemoglobins, which has provided valuable genetic insights.

In the United States, sickle-cell anemia is found almost exclusively in the African-American population. It affects about 1 in every 625 African-American infants. Currently, about 50,000–75,000 individuals are afflicted. In 1 of about every 145 African-American married couples, both partners are heterozygous carriers. In these cases, each of their children has a 25% chance of having the disease.

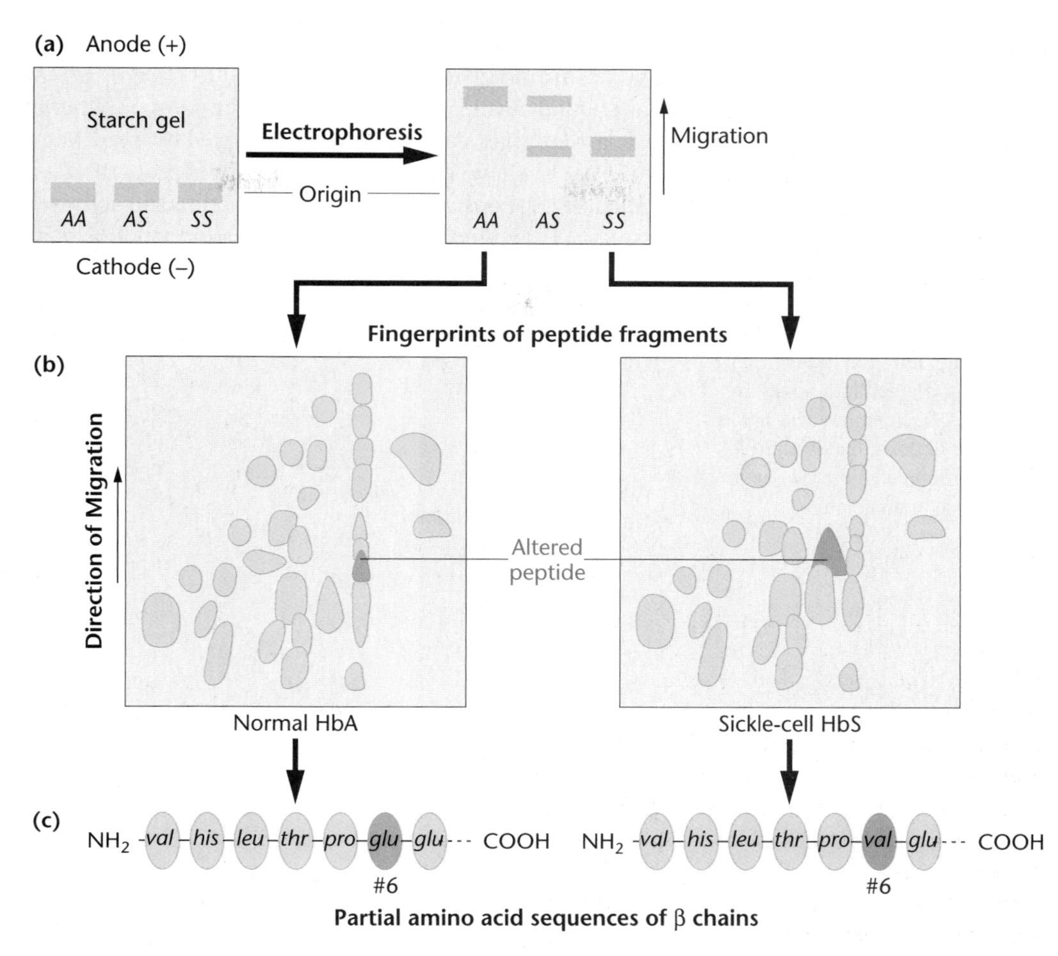

FIGURE 14–14 Investigation of hemoglobin derived from $Hb^A Hb^A$ and $Hb^S Hb^S$ individuals by using electrophoresis, protein fingerprinting, and amino acid analysis. Hemoglobin from individuals with sickle-cell anemia $(Hb^S Hb^S)$ (a) migrates differently in an electrophoretic field, (b) shows an altered peptide in fingerprint analysis, and (c) shows an altered amino acid, valine, at the sixth position in the β chain. During electrophoresis, heterozygotes $(Hb^A Hb^S)$ reveal both forms of hemoglobin.

14.8 The Nucleotide Sequence of a Gene and the Amino Acid Sequence of the Corresponding Protein Exhibit Colinearity

Once it was established that genes specify the synthesis of polypeptide chains, the next logical question was how the genetic information contained in a gene's nucleotide sequence can be transferred to the amino acid sequence of a polypeptide chain. It seemed most likely that a **colinear relationship** would exist between the two molecules. That is, the order of nucleotides in the DNA of a gene would correlate directly with the order of amino acids in the corresponding polypeptide.

The initial experimental evidence in support of this concept was derived from Charles Yanofsky's studies of the *trpA* gene that encodes the A subunit of the enzyme **tryptophan synthetase** in *E. coli*. Yanofsky isolated many independent mutants that had lost the activity of the enzyme, mapped them, and established their location with respect to one another within the gene. He then determined where the amino acid substitution occurred in each mutant protein. When the two sets of data were compared, the colinear relationship was apparent. The location of each mutation in the *trpA* gene correlates with the position of the altered amino acid in the A polypeptide of tryptophan synthetase.

14.9 Protein Structure Is the Basis of Biological Diversity

Having established that the genetic information is stored in DNA and influences cellular activities through the proteins it encodes, we turn now to a brief discussion of protein structure. How can these molecules play such a critical role in determining the complexity of cellular activities? As we shall see, the fundamental aspects of the structure of proteins provide the basis for incredible complexity and diversity. At the outset, we should differentiate between **polypeptides** and **proteins**. Both are molecules composed of amino acids. They differ, however, in their state of assembly and functional capacity. Polypeptides are the precursors of proteins. As it is assembled on the ribosome during translation, the molecule is called a *polypeptide*. When released from the ribosome following translation, a polypeptide folds up and assumes a higher order of structure. When this occurs, a three-dimensional conformation emerges. In many cases, several polypeptides interact to produce this conformation. When the final conformation is achieved, the molecule is now fully functional and is appropriately called a *protein*. It is its three-dimensional conformation that is essential to the function of the molecule.

The polypeptide chains of proteins, like nucleic acids, are linear nonbranched polymers. There are 20 amino acids that serve as the subunits (the building blocks) of proteins. Each amino acid has a **carboxyl group**, an **amino group**, and an **R (radical) group** (a side chain) bound covalently to a **central carbon (C) atom**. The R group gives each amino acid its chemical identity. Figure 14–15 shows the 20 R groups, which exhibit a variety of configurations and can be divided into 4 main classes: (1) *nonpolar* (hydrophobic), (2) *polar* (hydrophilic), (3) *positively charged*, and (4) *negatively charged*. Because polypeptides are often long polymers and because each position may be occupied by any 1 of the 20 amino acids with their unique chemical properties, enormous variation in chemical conformation and activity is possible. For example, if an average polypeptide is composed of 200 amino acids

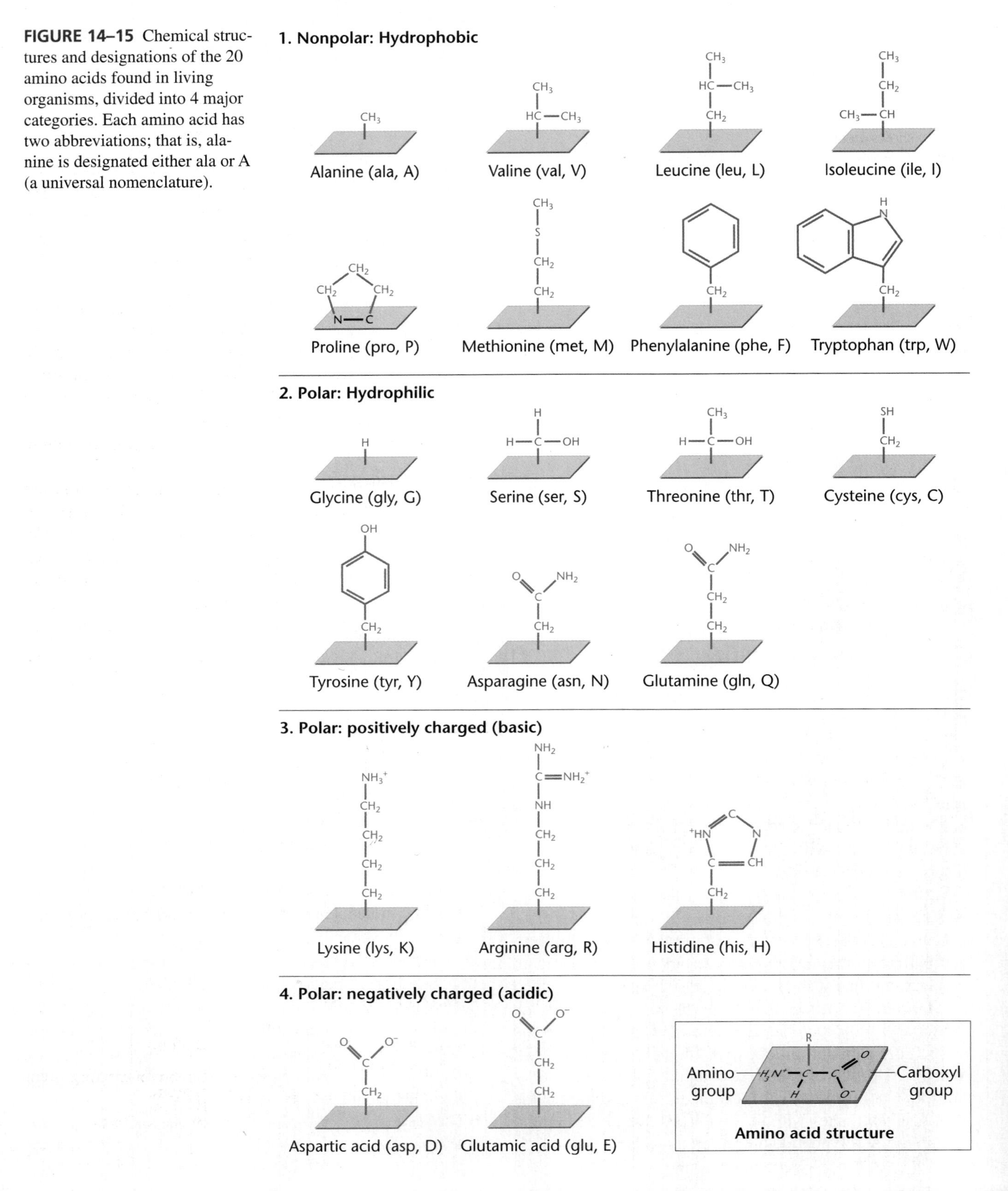

FIGURE 14–15 Chemical structures and designations of the 20 amino acids found in living organisms, divided into 4 major categories. Each amino acid has two abbreviations; that is, alanine is designated either ala or A (a universal nomenclature).

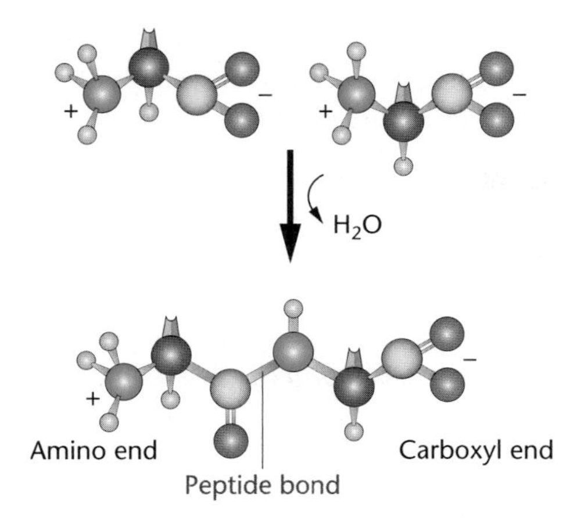

FIGURE 14–16 Peptide bond formation between two amino acids, resulting from a dehydration reaction.

(molecular weight of about 20,000 Da), 20^{200} different molecules, each with a unique sequence, can be created using the 20 different building blocks.

Around 1900, German chemist Emil Fischer determined the manner in which the amino acids are bonded together. He showed that the amino group of one amino acid reacts with the carboxyl group of another amino acid during a dehydration reaction, releasing a molecule of H_2O. The resulting covalent bond is a peptide bond (Figure 14–16). Two amino acids linked together constitute a dipeptide, three a tripeptide, and so on. Once 10 or more amino acids are linked by peptide bonds, the chain is referred to as a polypeptide. Generally, no matter how long a polypeptide is, it will contain a free amino group at one end (the N-terminus) and a free carboxyl group at the other end (the C-terminus).

Four levels of protein structure are recognized: primary, secondary, tertiary, and quaternary. The sequence of amino acids in the linear backbone of the polypeptide constitutes its **primary structure**. It is specified by the sequence of deoxyribonucleotides in DNA via an mRNA intermediate. The primary structure of a polypeptide helps determine the specific characteristics of the higher orders of organization as a protein is formed.

The **secondary structure** refers to a regular or repeating configuration in space assumed by amino acids closely aligned in the polypeptide chain. In 1951, Linus Pauling and Robert Corey predicted, on theoretical grounds, an ***α* helix** as one type of secondary structure. The *α*-helix model (Figure 14–17a) has since been confirmed by X-ray crystallographic studies. The helix is composed of a spiral chain of amino acids stabilized by hydrogen bonds; it is rodlike and has the greatest possible theoretical stability.

The side chains (the R groups) of amino acids extend outward from the helix, and each amino acid residue occupies a vertical distance of 1.5 Å in the helix. There are 3.6 residues per turn. While left-handed helices are theoretically possible, all proteins seen with an *α* helix are right-handed.

Also in 1951, Pauling and Corey proposed a second structure, the ***β*-pleated sheet**. In this model, a single polypeptide chain folds back on itself or several chains run in either parallel or antiparallel fashion next to one another. Each such structure is stabilized by hydrogen bonds formed between atoms on adjacent chains (Figure 14–17b). A zigzagging plane is formed in space with adjacent amino acids 3.5 Å apart.

As a general rule, most proteins demonstrate a mixture of *α*-helix and *β*-pleated-sheet structures. Globular proteins, most of which are round in shape and water soluble, usually contain a core of *β*-pleated-sheet structure as well as many areas with *α*-helical structures. The more structurally rigid proteins, many of which are water insoluble, rely on more extensive *β*-pleated-sheet regions for their

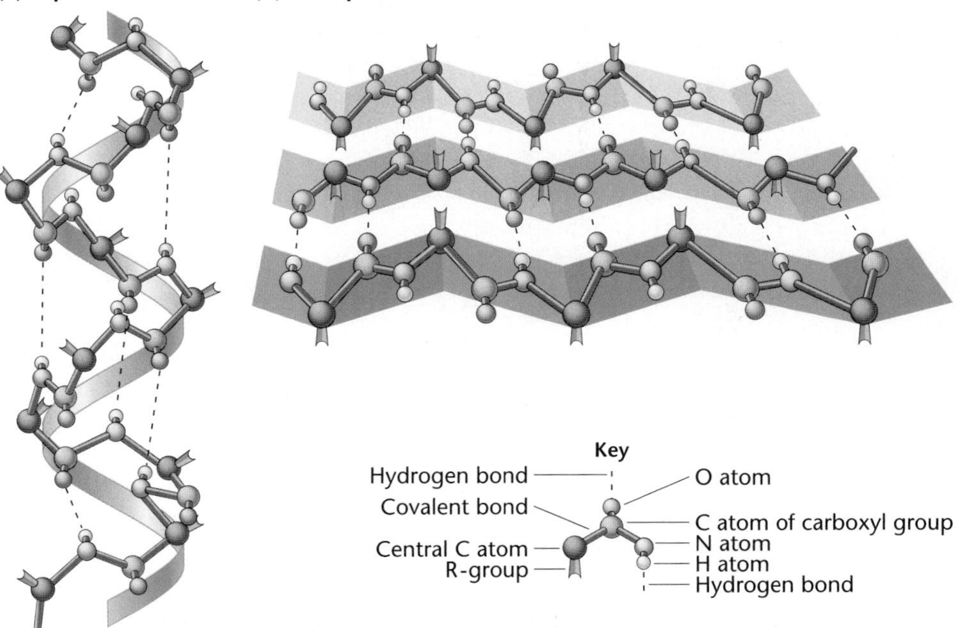

FIGURE 14–17 (a) The right-handed *α* helix, which represents one form of secondary structure of a polypeptide chain. (b) The *β*-pleated sheet, an alternative form of secondary structure of polypeptide chains. To maintain clarity, not all atoms are shown.

rigidity. For example, **fibroin**, the protein made by the silk moth, depends extensively on this form of secondary structure.

The secondary structure describes the arrangement of amino acids within certain areas of a polypeptide chain, but **tertiary protein structure** defines the three-dimensional conformation of the entire chain in space. Each protein twists and turns and loops around itself in a very particular fashion, characteristic of the specific protein. A model of the three-dimensional tertiary structure of the respiratory pigment myoglobin is shown in Figure 14–18. Three aspects of this level of structure are most important in stabilizing the molecule and in determining its conformation.

1. Covalent disulfide bonds form between closely aligned cysteine residues to make the unique amino acid cystine.
2. Nearly all of the polar hydrophilic R groups are located on the surface, where they can interact with water.
3. The nonpolar hydrophobic R groups are usually located on the inside of the molecule, where they interact with one another, avoiding interaction with water.

It is important to emphasize that the three-dimensional conformation achieved by any protein is a product of the *primary structure* of the polypeptide. Thus, the genetic code need only specify the sequence of amino acids to encode information that leads ultimately to the final assembly of proteins. The three stabilizing factors depend on the location of each amino acid relative to all others in the chain. As the polypeptide is folded, the most thermodynamically stable conformation possible results. This level of organization is essential because the specific function of any protein is directly related to its tertiary structure.

The **quaternary level** of organization applies only to proteins composed of more than one polypeptide chain and indicates the conformation of the various chains in relation to one another. This type of protein is *oligomeric*, and each chain is a *protomer* or, less formally, a subunit. Individual protomers have conformations that fit together with other subunits in a specific complementary fashion. Hemoglobin, an oligomeric protein consisting of four polypeptide chains, has been studied in great detail—its **quaternary protein structure** is shown in Figure 14–19. Most enzymes, including DNA and RNA polymerase, demonstrate quaternary structure.

14.10 Posttranslational Modification Alters the Final Protein Product

Polypeptide chains, like RNA transcripts, are often modified after they have been synthesized. This additional processing is broadly described as **posttranslational modification**. Although many of these alterations are detailed biochemical transformations beyond the scope of this discussion, you should be aware that they occur and that they are critical to the functional capability of the final protein product. Several examples of posttranslational modification are listed below.

- **The N-terminus and C-terminus amino acids are usually removed or modified.** For example, the initial N-terminal formylmethionine residue in bacterial polypeptides is usually removed enzymatically. Often, the amino group of the initial methionine residue is removed, and the amino group of the N-terminal residue is chemically modified (acetylated) in eukaryotic polypeptide chains.

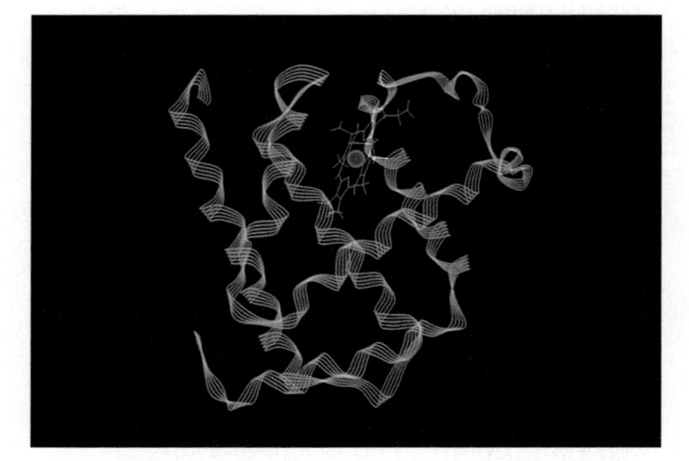

FIGURE 14–18 The tertiary level of protein structure in a respiratory pigment, myoglobin. The bound oxygen atom is shown in red. *(Horton, et al.* Principles of Biochemistry, *3rd ed. © 2002. Reprinted by permission of Prentice-Hall, Inc., Upper Saddle River, NJ)*

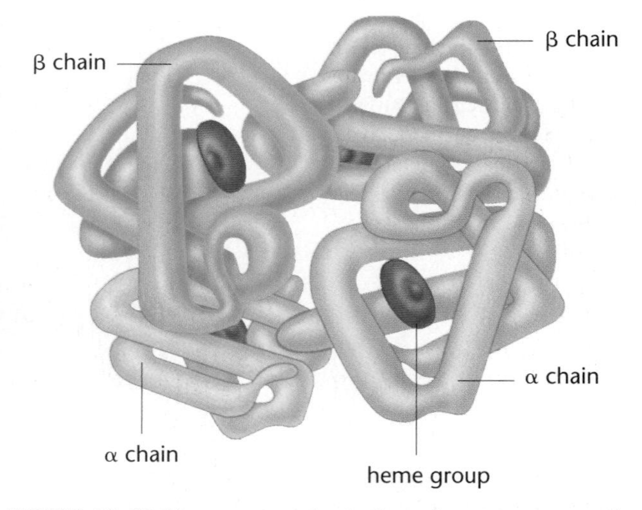

FIGURE 14–19 The quaternary level of protein structure as seen in hemoglobin. Four chains (two α and two β) interact with four heme groups to form the functional molecule.

- **Individual amino acid residues are sometimes modified.** For example, phosphates may be added to the hydroxyl groups of certain amino acids, such as tyrosine. Modifications such as these create negatively charged residues that bond ionically with other molecules. The process of phosphorylation is extremely important in regulating many cellular activities and results from the action of enzymes called **kinases**. In other proteins, methyl groups may be added enzymatically.
- **Carbohydrate side chains are sometimes attached.** These are added covalently, producing **glycoproteins**, an important category of molecules that includes many antigenic determinants, such as those specifying the antigens in the ABO blood-type system in humans.
- **Polypeptide chains may be trimmed.** For example, insulin is first translated into a longer molecule that is enzymatically trimmed to its final form of 51 amino acids.
- **Signal sequences are removed.** At the N-terminus of some proteins, a sequence of up to 30 amino acids plays an important role in directing the protein to the location in the cell where it functions. This is called a **signal sequence**, and it determines the final destination of a protein within the cell in a process called **protein targeting**. For example, proteins whose fate involves secretion or that are to become part of the plasma membrane are dependent on specific sequences for their initial transport into the lumen of the endoplasmic reticulum. While the signal sequence of various proteins with a common destination might differ in their primary amino acid sequence, they do share many chemical properties. For example, proteins destined for secretion contain a string of up to 15 hydrophobic amino acids preceded by a positively charged amino acid at the N-terminus of the signal sequence. Once the polypeptides are transported, but prior to achieving functional status as proteins, the signal sequence is enzymatically removed from these polypeptides.
- **Polypeptide chains are often complexed with metals.** The tertiary and quaternary levels of protein structure often include and are dependent on metal atoms. The function of the protein is thus dependent on the molecular complex that includes both polypeptide chains and metal atoms. Hemoglobin, containing four iron atoms along with four polypeptide chains, is a good example.

These types of posttranslational modifications are important in achieving the functional status specific to any given protein. Because the final three-dimensional structure of the molecule is intimately related to its specific function, how polypeptide chains ultimately fold into their final conformations is also an important topic. For many years, it was thought that protein folding was a spontaneous process whereby the molecule achieved maximum thermodynamic stability based largely on the combined chemical properties inherent in the amino acid sequence of the polypeptide chain(s) composing the protein. However, numerous studies have shown that for many proteins, folding is dependent upon members of a family of still other, ubiquitous proteins called **chaperones**. These proteins (sometimes called *molecular chaperones* or *chaperonins*) facilitate the folding of other proteins. While the mechanism by which chaperones function is not yet clear, like enzymes, they do not become part of the final product. Initially discovered in *Drosophila*, where they are called **heat-shock proteins (HSP)**, chaperones have been discovered in a variety of organisms, including bacteria, animals, and plants.

14.11 Protein Function Is Directly Related to the Structure of the Molecule

The essence of life on Earth rests at the level of diverse cellular function. One can argue that DNA and RNA simply serve as vehicles to store and express genetic information. However, proteins are at the heart of cellular function. And it is the capability of cells to assume diverse structures and functions that distinguishes most eukaryotes from less evolutionarily advanced organisms such as bacteria. Therefore, an introductory understanding of protein function is critical to a complete view of genetic processes.

Proteins are the most abundant macromolecules found in cells. As the end products of genes, they play many diverse roles. For example, the respiratory pigments **hemoglobin** and **myoglobin** transport oxygen, which is essential for cellular metabolism. **Collagen** and **keratin** are structural proteins associated with the skin, connective tissue, and hair of organisms. **Actin** and **myosin** are contractile proteins, found in abundance in muscle tissue. Still other examples are the **immunoglobulins**, which function in the immune system of vertebrates; **transport proteins**, involved in movement of molecules across membranes; some of the **hormones** and their **receptors**, which regulate various types of chemical activity; and **histones**, which bind to DNA in eukaryotic organisms.

The largest group of proteins with a related function are the **enzymes**. Since we have referred to these molecules throughout this chapter, it may be useful to extend our discussion and include a more detailed description of their biological role.

Enzymes specialize in catalyzing chemical reactions within living cells. They increase the rate at which a chemical reaction reaches equilibrium but do not alter the end

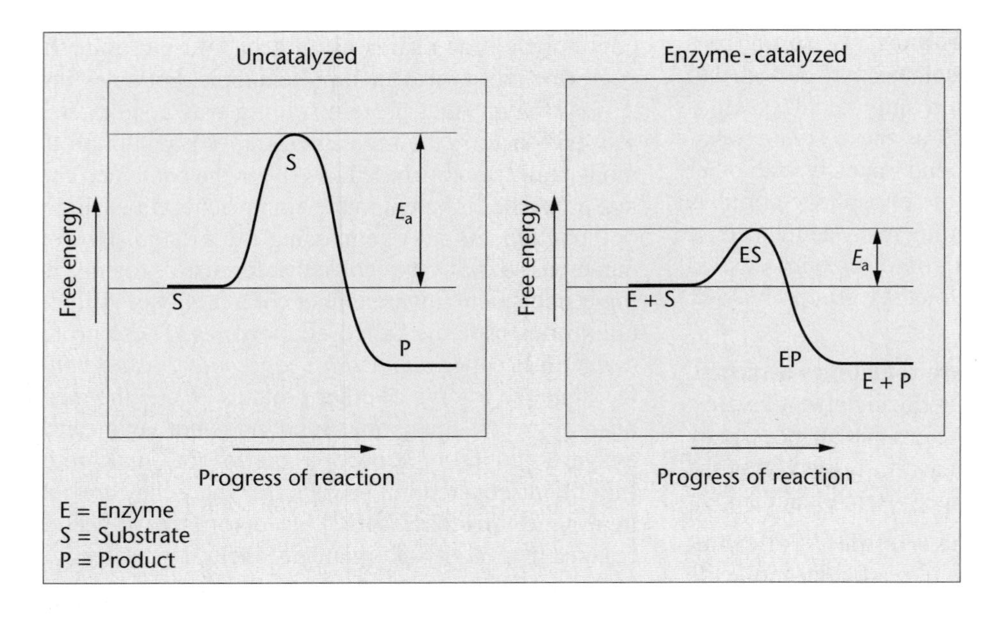

FIGURE 14–20 Energy requirements of an uncatalyzed versus an enzymatically catalyzed chemical reaction. The energy of activation (E_a) necessary to initiate the reaction is substantially lower as a result of catalysis.

point of the chemical equilibrium. Their remarkable, highly specific catalytic properties largely determine the metabolic capacity of any cell type. The specific functions of many enzymes involved in the genetic and cellular processes of cells are described throughout this text.

Biological catalysis is a process whereby the **energy of activation (E_a)** for a given reaction is lowered (Figure 14–20). The energy of activation is the increased kinetic energy state that molecules usually must reach before they react with one another. This state can be attained as a result of elevated temperatures, but enzymes allow biological reactions to occur at lower physiological temperatures. In this way, enzymes make life as we know it possible.

The catalytic properties and specificity of an enzyme are determined by the chemical configuration of the molecule's **active site**. This site is associated with a crevice, a cleft, or a pit on the surface of the enzyme, which binds the reactants, or substrates, facilitating their interaction. Enzymatically catalyzed reactions control metabolic activities in the cell. Each reaction is either catabolic or anabolic. **Catabolism** is the degradation of large molecules into smaller, simpler ones with the release of chemical energy. **Anabolism** is the synthetic phase of metabolism and yields the various components that make up nucleic acids, proteins, lipids, and carbohydrates.

14.12 Proteins Consist of Functional Domains

We conclude our discussion of proteins by briefly discussing the important finding that regions made up of specific amino acid sequences are associated with specific functions in protein molecules. Such sequences, usually between 50 and 300 amino acids, constitute **protein domains** and are represented by modular portions of the protein that fold into stable, unique conformations independently of the rest of the molecule. Different domains impart different functional capabilities. Some proteins contain only a single domain, while others contain two or more.

The significance of domains rests at the tertiary structure level of proteins. Each such modular unit can be a mixture of secondary structures, including both α helices and β-pleated sheets. The unique conformation that is assumed in a single domain imparts a specific function to the protein. For example, a domain may serve as the catalytic basis of an enzyme or it may impart the capability to bind to a specific ligand as part of a membrane or another molecule. Thus, in the study of proteins, you will hear of *catalytic domains*, *DNA-binding domains*, and so on. A protein must be envisioned as being composed of a series of structural and functional modules. Obviously, the presence of multiple domains in a single protein increases the versatility of each molecule and adds to its functional complexity.

Exon Shuffling

An interesting proposal to explain the genetic origin of protein domains was put forward by Walter Gilbert in 1977. Gilbert suggested that the functional regions of genes in higher organisms consist of collections of exons originally present in ancestral genes that were brought together through recombination during the course of evolution. Referring to this process as **exon shuffling**, Gilbert proposed that exons, like protein domains, are also modular, and that during evolution, exons may have been reshuffled between genes in eukaryotes with the result that different genes share similar domains.

Since 1977, a serious research effort has been directed toward the analysis of gene structure. In 1985, more direct evidence in favor of Gilbert's proposal of exon modules was presented. For example, the human gene encoding the membrane receptor for low-density lipoproteins (LDL) was isolated and sequenced. The

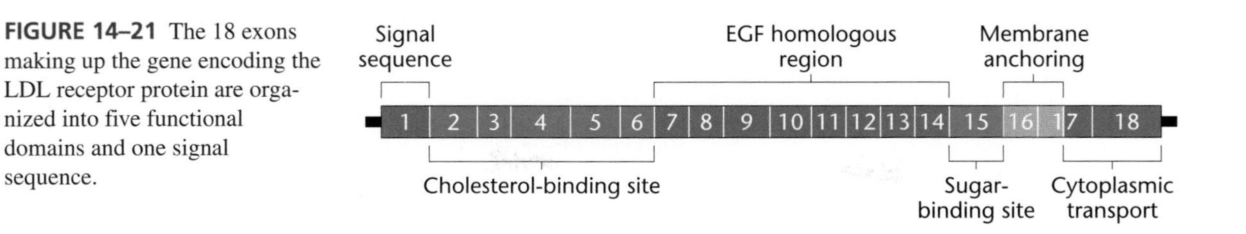

FIGURE 14–21 The 18 exons making up the gene encoding the LDL receptor protein are organized into five functional domains and one signal sequence.

LDL receptor protein is essential to the transport of plasma cholesterol into the cell. It mediates endocytosis and seems to have numerous functional domains. These include the capability to bind specifically to the LDL substrates and to interact with other proteins at different levels of the membrane during transport across it. In addition, this receptor protein is modified posttranslationally by the addition of a carbohydrate; a domain must exist that links to this carbohydrate.

Detailed analysis of the gene encoding this protein supports the concept of exon modules and their shuffling during evolution. The gene is quite large—45,000 nucleotides—and contains 18 exons. These represent only slightly fewer than 2600 nucleotides. These exons are related to the functional domains of the protein *and* appear to have been recruited from other genes during evolution. Figure 14–21 shows these relationships. The first exon encodes a signal sequence that is removed from the protein before the LDL receptor becomes part of the membrane. The next five exons represent the domain specifying the binding site for cholesterol. This domain is made up of a sequence of 40 amino acids repeated 7 times. The next domain consists of a sequence of 400 amino acids bearing a striking homology to the peptide-hormone epidermal growth factor (EGF) in mice. This region is encoded by 8 exons and contains 3 repetitive sequences of 40 amino acids. A similar sequence is also found in three blood-clotting proteins. The fifteenth exon specifies the domain for the posttranslational addition of the carbohydrate, while the next two specify regions of the protein that are part of the membrane, anchoring the receptor to specific sites on the cell surface.

These observations concerning the LDL exons are fairly compelling support for the theory of exon shuffling during evolution. Certainly, there is no disagreement that protein domains are responsible for specific molecular interactions.

GENETICS, TECHNOLOGY, AND SOCIETY

Mad Cows and Heresies: The Prion Story

In March 1996, the British government announced that a new brain disease had killed 10 young Britons, and that the victims might have caught the disease by eating infected beef. Bovine spongiform encephalopathy (BSE), popularly known as "mad cow disease," slowly destroys brain cells and is always fatal. Recent studies confirm that BSE and the human disease, new variant Creutzfeldt-Jakob disease (nvCJD), are so similar at the molecular and pathological levels that they are very likely to be the same disease. Mad cow disease triggered political turmoil in Europe, a worldwide ban on British beef, and the near-collapse of the $8.9-billion British beef industry. The European Union demanded the slaughter and incineration of 4.7 million British cattle, a campaign that cost the government over $12 billion to compensate farmers, import milk, and buy stock for new herds. Most European countries, as well as the Middle East, Japan, and Canada have discovered BSE in at least one cow. Although most nvCJD cases have occurred in Britain, cases have also appeared in France, Italy, Ireland, South America, Canada, and the United States Over 125 people have now died of nvCJD, and epidemiologists estimate that between 10,000 and 150,000 cases may appear in the next two decades.

BSE, nvCJD, and Creutzfeldt-Jakob disease (CJD) are all members of a group of neurological diseases known as spongiform encephalopathies that affect animals (BSE) and humans (CJD and nvCJD). In this group of diseases, the infected brain tissue eventually resembles a sponge (hence, spongiform) and is riddled with proteinaceous deposits. Victims of the disease lose motor function, become demented, and ultimately die. CJD and nvCJD differ symptomatically in that victims of nvCJD display psychological symptoms such as depression or anxiety prior to developing neurological symptoms like shaking or paralysis. Also, patients with nvCJD are usually young (16–40 years), whereas CJD normally affects people over 55. A number of CJD cases arise spontaneously and randomly, at a rate of one per million per year worldwide, but CJD can also be inherited as an autosomal dominant condition.

CJD can be transmitted through corneal or nervous tissue grafts or by injection of a growth hormone derived from human pituitary glands. Kuru, a CJD-like disease of the Fore people of New Guinea, was transmitted from person to person through ritualistic cannibalism. Transmissible spongiform encephalopathies in animals include scrapie (sheep and goats), chronic wasting disease (deer and elk), and BSE. Like Kuru, BSE is passed from animal to animal by ingestion of diseased animal remains, particularly neural tissue. The epidemic of BSE in Britain occurred because diseased cows and sheep were processed and fed to cattle as a protein supplement. In 1998, the British government banned the use of cows and sheep as feed for other cows and sheep, and the epidemic has subsided. The European Union has now banned the use of feeds containing animal products for all livestock. In contrast, the United States and Canada still allow nonruminant animals to consume feed containing ruminants, and ruminant animals to consume feed containing nonruminants, as well as certain ruminant by-products including blood, gelatin, and fat. These regulations may change after the discovery of a BSE case in Canada in 2003. In addition, recent studies suggest that BSE may be transmitted via blood or tallow.

For many years, the spongiform encephalo-pathies defied the best efforts of scientists to analyze them. The diseases are difficult to study because they require injection of infected brain material into the brains of experimental animals and the diseases, take months or years to develop. In addition, the infectious agent is apparently not a virus or bacterium, and infected animals do not develop antibodies against these mysterious agents. There are no treatments for the diseases, and the only way to make a firm diagnosis is to examine brain tissue after death. The infectious material is unaffected by radiation or nucleases that damage nucleic acids; however, it is destroyed by some reagents that hydrolyze or modify proteins. In the early 1980s, American scientist Stanley Prusiner purified the infectious agent and concluded that it consists of only protein. He proposed that scrapie is spread by an infectious protein particle that he called a **prion**. His hypothesis was dismissed by most scientists, as the idea of an infectious agent with no DNA or RNA as genetic material was heretical. However, Prusiner and others presented evidence supporting the prion hypothesis, and the notion that the disease can be transmitted by an infectious particle that contains no genetic material has gained acceptance.

If prions are composed of protein only, how do they cause disease? The answer may be as strange as the disease itself. The protein that makes up a prion (PrP) is a version of a normal protein that is synthesized in neurons and found in the brains of all adult animals. The difference between normal PrP and prion PrP lies in their secondary protein structures. Normal, noninfectious PrP folds into α helices, whereas infectious prion PrP folds into β-pleated sheets. When a normal PrP molecule contacts a prion PrP molecule, the normal protein is somehow unfolded and refolded into the abnormal PrP conformation. Once the normal PrP molecule has been transformed into an abnormal PrP molecule, it spreads its lethal conformation to neighboring normal PrP molecules, and the process takes off in a chain reaction. Normal PrP is a soluble protein that is easily destroyed by heat or enzymes that digest proteins. However, abnormal infectious PrP is insoluble in detergents, resists both heat and protease digestion, and is nearly indestructible. Hence, spongiform encephalopathies can be considered diseases of secondary protein structure.

Many urgent questions need to be addressed. How extensive is BSE contamination of the world's food supply? How many humans are infected with nvCJD but don't show symptoms? Can humans or animals act as asymptomatic carriers of prion diseases? Can prions exist in other parts of the body besides the brain and spinal cord, and if so, can nvCJD be spread through blood transfusions, from mother to fetus, or by sterilized surgical instruments? Can we develop diagnostic tests and therapies for BSE and nvCJD? Are we near the end of the BSE story, or is it just beginning?

References

Balter, M. 2000. Tracking the human fallout from "mad cow disease." *Science* 289: 1452–54.

Belay, E. D. 1999. Transmissible spongiform encephalopathies in humans. *Annu. Rev. Microbiol.* 53: 283–314.

Spencer, C. A. 2004. *Mad Cows and Cannibals: A Guide to the Transmissible Spongiform Encephalopathies*. Upper Saddle River: Pearson/Prentice-Hall.

Web Sites

[Online compilation of news reports about mad cow disease.]
http://organicconsumers.org/madcow.htm

Chapter Summary

1. Translation describes the synthesis of polypeptide chains, under the direction of mRNA and in association with ribosomes. This process ultimately converts the information stored in the genetic code of the DNA that makes up a gene into the corresponding sequence of amino acids making up the polypeptide.
2. Translation is a complex energy-requiring process that also depends on charged tRNA molecules and numerous protein factors. Transfer RNA (tRNA) serves as the adaptor molecule between an mRNA triplet and the appropriate amino acid.
3. The processes of translation, like transcription, can be subdivided into the stages of initiation, elongation, and termination. Translation relies on base-pairing affinities between complementary nucleotides and is more complex in eukaryotes than in prokaryotes.
4. The first insight that proteins are the end products of gene expression was provided by the study of inherited metabolic disorders in humans early in the twentieth century by Garrod. Inborn errors of metabolism leading to cystinuria, albinism, and alkaptonuria were the basis of his studies.
5. The investigation of nutritional requirements in *Neurospora* by Beadle and colleagues made it clear that mutations cause the loss of enzyme activity. Their work led to the one-gene:one-enzyme hypothesis.
6. The one-gene:one-enzyme hypothesis was later revised as the one-gene:one polypeptide chain hypothesis. Pauling and Ingram's investigations of hemoglobins from patients with sickle-cell anemia led to the discovery that one gene directs the synthesis of only one polypeptide chain.
7. The proposal suggesting that a gene's nucleotide sequence specifies in a colinear manner the sequence of amino acids in a polypeptide chain was confirmed by experiments involving mutations in the tryptophan synthetase gene in *E. coli*.
8. Proteins, the end products of gene expression, demonstrate four levels of structural organization that together provide the chemical basis for their three-dimensional conformation, which is the basis of the molecule's function.
9. Of the myriad functions performed by proteins, the most influential role is assumed by enzymes. These highly specific, cellular catalysts play a central role in the production of all classes of molecules in living systems.
10. Proteins consist of one or more functional domains, which are shared by different molecules. The origin of these domains may be the result of exon shuffling during evolution.

Key Terms

actin, 313
active site, 314
alkaptonuria, 304
amino group, 310
A (aminoacyl) site, 301
aminoacyl tRNA synthetases, 299
aminoacyladenylic acid, 299
anabolism, 314
anticodon, 296
anticodon loop, 298
β-pleated sheet, 311
biological catalysis, 314
carboxyl group, 310
catabolism, 314
central carbon (C) atom, 310
chaperone, 313
charging (of tRNA), 299
cloverleaf model of tRNA, 298
colinear relationship, 309
collagen, 313
E (exit) site, 301
elongation, 301
elongation factors, 301
energy of activation (Ea), 314
enzyme, 313
exon shuffling, 314
fibroin, 312
formylmethionine (fmet), 300
glycoprotein, 313
GTP-dependent release factor, 302
HbA, 308
HbS, 308
heat-shock proteins (HSPs), 313
hemoglobin, 313
histone, 313
immunoglobulin, 313
inherited molecular disease, 308
initiation, 300
initiation complex, 300
initiation factors, 300
isoaccepting tRNA, 299
keratin, 313
kinase, 313
LDL receptor protein, 315
monosome, 296
myoglobin, 313
myosin, 313
nonsense codon, 302
one-gene:one-enzyme hypothesis, 305
one-gene:one-polypeptide chain hypothesis, 307
one-gene:one-protein hypothesis, 307
peptidyl transferase, 301
phenylalanine hydroxylase, 305
phenylketonuria (PKU), 305
polypeptide, 309
polyribosome, 302
polysome, 302
posttranscriptional modification, 298
posttranslational modification, 312
P (peptidyl) site, 301
primary structure, 311
prion, 316
protein, 309
protein domain, 314
protein fingerprinting technique, 308
protein targeting, 313
quaternary protein structure, 312
rDNA, 297
receptors, 313
ribosomal proteins, 296
ribosome, 296
R (radical) group, 310
secondary structure, 311
Shine-Dalgarno sequence, 300
sickle-cell anemia, 307
sickle-cell trait, 308
signal sequence, 313
spacer DNA, 297
starch gel electrophoresis, 308
stop codon, 302
tandem repeats, 297
termination, 302
termination codon, 302
tertiary protein structure, 312
transfer RNA (tRNA), 296
translation, 296
transport protein, 313
tryptophan synthetase, 309
tunnel, 304

INSIGHTS AND SOLUTIONS

1. The growth responses in the chart below were obtained using four mutant strains of *Neurospora* and the related compounds A, B, C, and D. None of the mutations grow on minimal medium. Draw all possible conclusions from this data.

	Growth Product			
Mutation	**A**	**B**	**C**	**D**
1	–	–	–	–
2	+	+	–	+
3	+	+	–	–
4	–	+	–	–

Solution: Nothing can be concluded about mutation *1* except that it lacks some essential growth factor, perhaps even unrelated to the biochemical pathway represented by mutations *2*, *3*, and *4*. Nor can anything be concluded about compound C. If it is involved in the pathway, it is a product that was synthesized prior to compounds A, B, and D.

We now analyze these three compounds and the control of their synthesis by the enzymes encoded by mutations *2*, *3*, and *4*. Because product B allows growth in all three cases, it may be considered the "end product"—it bypasses the block in all three instances. Using similar reasoning, product A precedes B in the pathway, since it bypasses the block in two of the three steps, and product D precedes B; yielding a partial solution:

$$C(?) \qquad D \longrightarrow A \longrightarrow B$$

Now let's determine which mutations control which steps. Since mutation *2* can be alleviated by products D, B, and A, it must control a step prior to all three products, perhaps the direct conversion to D (although we cannot be certain). Mutation *3* is alleviated by B and A, so its effect must precede them in the pathway. Thus, we assign it as controlling the conversion of D to A. Likewise, we can assign mutation *4* to the conversion of A to B, leading to a more complete solution:

$$C(?) \xrightarrow{2(?)} D \xrightarrow{3} A \xrightarrow{4} B$$

Problems and Discussion Questions

1. List and describe the role of all of the molecular constituents present in a functional polyribosome.
2. Contrast the roles of tRNA and mRNA during translation, and list all enzymes that participate in the transcription and translation process.
3. Francis Crick proposed the adaptor hypothesis for the function of tRNA. Why did he choose that description?
4. During translation, what molecule bears the anticodon? The codon?
5. The α chain of eukaryotic hemoglobin is composed of 141 amino acids. What is the minimum number of nucleotides in an mRNA coding for this polypeptide chain? Assuming that each nucleotide is 0.34 nm long in the mRNA, how many triplet codes can simultaneously occupy space in a ribosome that is 20 nm in diameter?
6. Summarize the steps involved in charging tRNAs with their appropriate amino acids.
7. Each transfer RNA requires at least four specific recognition sites that must be inherent in its tertiary protein structure in order for it to carry out its role. What are these sites?
8. Discuss the potential difficulties involved in designing a diet to alleviate the symptoms of phenylketonuria.
9. Phenylketonurics cannot convert phenylalanine to tyrosine. Why don't these individuals exhibit a deficiency of tyrosine?
10. Phenylketonurics are often more lightly pigmented than are normal individuals. Can you suggest a reason why this is so?
11. The synthesis of flower pigments is known to be dependent on enzymatically controlled biosynthetic pathways. Postulate the role of mutant genes and their products in producing the observed phenotypes in the crosses shown.

(a)	P_1:	white strain A × white strain B
	F_1:	all purple
	F_2:	9/16 purple : 7/16 white
(b)	P_1:	white × pink
	F_1:	all purple
	F_2:	9/16 purple : 3/16 pink : 4/16 white

12. A series of mutations in the bacterium *Salmonella typhimurium* results in the requirement of either tryptophan or some related molecule in order for growth to occur. From the data shown here, suggest a biosynthetic pathway for tryptophan.

	Growth Supplement				
Mutation	***Minimal Medium***	***Anthranilic Acid***	***Indole Glycerol Phosphate***	***Indole***	***Tryptophan***
trp-8	–	+	+	+	+
trp-2	–	–	+	+	+
trp-3	–	–	–	+	+
trp-1	–	–	–	–	+

13. The study of biochemical mutants in organisms such as *Neurospora* has demonstrated that some pathways are branched. The data below illustrate the branched nature of the pathway resulting in the synthesis of thiamine. Why don't the data support a linear pathway? Can you postulate a pathway for the synthesis of thiamine in *Neurospora?*

	Growth Supplement			
Mutation	***Minimal Medium***	***Pyrimidine***	***Thiazole***	***Thiamine***
thi-1	–	–	+	+
thi-2	–	+	–	+
thi-3	–	–	–	+

14. Explain why the one-gene:one-enzyme hypothesis is no longer considered to be totally accurate.
15. Why is an alteration of electrophoretic mobility interpreted as a change in the primary structure of the protein under study?

16. Hemoglobin is a tetramer consisting of two α and two β chains. How does this information relate to each of the four levels of protein structure?
17. Using sickle-cell anemia as a basis, describe what is meant by a genetic or inherited molecular disease. What are the similarities and dissimilarities between this type of a disorder and a disease caused by an invading microorganism?
18. Contrast the contributions Pauling and Ingram made to our understanding of the genetic basis for sickle-cell anemia.
19. Hemoglobins from two individuals are compared by starch gel electrophoresis and with protein fingerprinting. Electrophoresis reveals no difference in migration, but fingerprinting shows an amino acid difference. How is this possible?
20. Describe what colinearity means. What is the significance of this concept in the study of genetics?
21. Certain mutations called *amber* in bacteria and viruses result in premature termination of polypeptide chains during translation. Many *amber* mutations have been detected at different points along the gene that codes for a head protein in phage T4. How might this system be further investigated to demonstrate and support the concept of colinearity?
22. In your opinion, which of the four levels of protein organization is the most critical to a protein's function? Defend your choice.
23. List as many different categories of protein functions as you can. Wherever possible, give an example of each category.
24. How does an enzyme function? Why are enzymes essential for living organisms?
25. Does Fiers', as discussed work with phage MS2, discussed in Chapter 13, constitute more direct evidence in support of colinearity than Yanofsky's work with the *trpA* locus in *E. coli*, as discussed in this chapter? Explain.
26. Shown below are several amino acid substitutions in the α and β chains of human hemoglobin. Use the genetic code table in Figure 13–7 to determine how many of them can occur as a result of a single nucleotide change.

Hb Type	*Normal Amino acid*	*Substituted Amino Acid*
HbJ Toronto	ala	asp (α-5)
HbJ Oxford	gly	asp (α-15)
Hb Mexico	gln	glu (α-54)
Hb Bethesda	tyr	his (β-145)
Hb Sydney	val	ala (β-67)
HbM Saskatoon	his	tyr (β-63)

27. Early detection and adherence to a strict dietary regime has relieved much of the mental retardation that once occurred in people afflicted with phenylketonuria (PKU). Now, affected individuals often lead normal lives and have families. For various reasons, some individuals adhere less rigorously to their diet as they get older. Predict the effect of such dietary neglect on the newborns of mothers with PKU.
28. In 1962, F. Chapeville and others (1962. *Proc. Natl. Acad. Sci. (USA)* 48: 1086–93.) reported an experiment in which they isolated radioactive ^{14}C-cysteinyl-tRNA^{cys} (charged tRNA^{cys} + cysteine). They then removed the sulfur group from the cysteine, creating alanyl-tRNA^{cys} (charged tRNA^{cys} + alanine). When alanyl-tRNA^{cys} was added to a synthetic mRNA calling for cysteine, but not alanine, a polypeptide chain was synthesized containing alanine. What can you conclude from this experiment?
29. Three independently assorting genes are known to control the biochemical pathway here that provides the basis for flower color in a hypothetical plant:

$$\text{colorless} \xrightarrow{A-} \text{yellow} \xrightarrow{B-} \text{green} \xrightarrow{C-} \text{speckled}$$

Homozygous recessive mutations, which interrupt each step, are known. Determine the phenotypic results in the F_1 and F_2 generations resulting from the P_1 crosses involving true-breeding plants given below.

(a) speckled	(*AABBCC*) × (*AAbbCC*)	yellow
(b) yellow	(*AAbbCC*) × (*AABBcc*)	green
(c) colorless	(*aaBBCC*) × (*AABBcc*)	green

30. How would the results in cross (a) of Problem 29 vary if genes A and B were linked with no crossing over between them? How would the results of cross (a) vary if genes A and B were linked and 20 map units apart?
31. HbS results from the amino acid change of glutamic acid to valine at the number 6 position in the β chain of human hemoglobin. HbC is the result of a change at the same position in the β chain, but lysine replaces glutamic acid. Return to the genetic code table in Figure 13–7, and determine whether single nucleotide changes can account for these mutations. Then turn to Figure 14–15, and examine the R groups in the amino acids glutamic acid, valine, and lysine. (a) Describe the chemical differences between the three amino acids, and predict how the changes might alter the structure of the molecule and lead to altered hemoglobin function. (b) HbS results in anemia and resistance to malaria, whereas in those with HbA, the parasite *Plasmodium falciparum* invades red blood cells and causes the disease. Predict whether those with HbC are likely to be anemic and whether they would be resistant to malaria.
32. Deep in a previously unexplored South American rain forest, a species of plant was discovered with true-breeding varieties whose flowers were either pink, rose, orange, or purple. A very astute plant geneticist made a single cross, carried to the F_2 generation, as shown below. Based solely on these data, he was able to propose both a mode of inheritance for flower pigmentation, and a biochemical pathway for the synthesis of these pigments. Carefully study the data. Create your own hypothesis to explain the mode of inheritance, and then propose a biochemical pathway consistent with your hypothesis. What other crosses would enable you to test the hypothesis?

P_1:	purple × pink
F_1:	all purple
F_2:	27/64 purple
	16/64 pink
	12/64 rose
	9/64 orange

33. The emergence of antibiotic-resistant strains of *Enterococci* and transfer of resistant genes to other bacterial pathogens have highlighted the need for new generations of antibiotics to combat serious infections. To grasp the range of potential sites for the action of existing antibiotics, sketch the components of the translation machinery (e.g., see step 3 of Figure 14–6), and using a series of numbered pointers, indicate the specific location for the action of the antibiotics shown in the following table.

Antibiotic	*Action*
1. Streptomycin	Binds to 30*S* ribosomal subunit.
2. Chloramphenicol	Inhibits peptidyl transferase of 70*S* ribosome
3. Tetracycline	Inhibits binding of charged tRNA to ribosome
4. Erythromycin	Binds to free 50*S* particle and prevents formation of 70*S* ribosome
5. Kasugamycin	Inhibits binding of $tRNA^{fmet}$
6. Thiostrepton	Prevents translocation by inhibiting EF-G

34. Development of antibiotic resistance by pathogenic bacteria represents a major health concern. One potential new antibiotic is evernimicin, which was isolated from *Micromonospora carbonaceae*. Evernimicin is an oligosaccharide antibiotic with activity against a broad range of gram-positive pathogenic bacteria. To determine the mode of action of this drug, Adrian and others (2000. *Antimicrob. Ag. and Chemo.* 44: 3101–06.) analyzed 23*S* ribosomal DNA mutants that showed reduced sensitivity to evernimicin. They and others discovered two classes of mutants that conferred resistance: 23*S* rRNA nucleotides 2475–2483 and ribosomal protein L16. This suggests that these two ribosomal components are structurally and functionally linked. It turns out that the tRNA anticodon stem-loop appears to bind to the A site of the ribosome at rRNA bases 2465–2485. This finding conforms to the proposed function of L16, which appears to be involved in attracting the aminoacyl stem of the tRNA to the ribosome at its A site. Using your sketch of the translation machinery from Problem 33 along with this information, designate where the proposed antibacterial action of evernimicin is likely to occur.
35. The flow of genetic information from DNA to protein is mediated by messenger RNA. If you introduce short DNA strands (called antisense oligonucleotides) which are complementary to mRNAs, hydrogen bonding may occur and "label" the DNA/RNA hybrid for ribonuclease-H degradation of the RNA. Lloyd and others (2001. *Nuc. Acids Res.* 29: 3664–73.) compared the effect of different length antisense oligonucleotides upon ribonuclease-H–mediated degradation of tumor necrosis factor (*TNFα*) mRNA. TNFα exhibits antitumor and proinflammatory activities. The graph below indicates the efficacy of various-sized antisense oligonucleotides in causing ribonuclease-H cleavage. (a) Describe how antisense oligonucleotides interrupt the flow of genetic information in a cell. (b) What general conclusion is apparent in the graph? (c) What factors other than oligonucleotide length are likely to influence antisense efficacy *in vivo*?

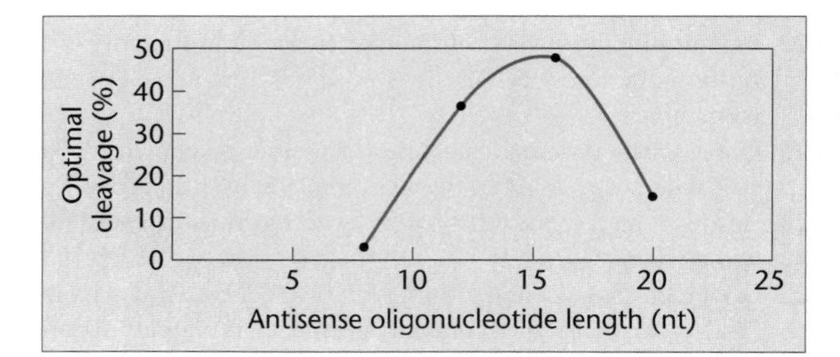

Selected Readings

Anfinsen, C. B. 1973. Principles that govern the folding of protein chains. *Science* 181: 223–30.

Bartholome, K. 1979. Genetics and biochemistry of phenylketonuria—Present state. *Hum. Genet.* 51: 241–45.

Bateson, W. 1909. *Mendel's Principles of Heredity*. Cambridge: Cambridge Univ. Press.

Beadle, G. W. 1945. Genetics and metabolism in *Neurospora*. *Physiol. Rev.* 25: 643.

Beadle, G. W., and Tatum, E. L. 1941. Genetic control of biochemical reactions in *Neurospora*. *Proc. Natl. Acad. Sci. (USA)* 27: 499–506.

Beet, E. A. 1949. The genetics of the sickle-cell trait in a Bantu tribe. *Ann. Eugenics* 14: 279–84.

Boyer, P. D. (ed.) 1974. *The Enzymes*, Vol 10. 3rd ed. Orlando: Academic Press.

Bray, D. 1995. Protein molecules as computational elements in living cells. *Nature* 376: 307–12.

Chapeville, F. et al. 1962. On the role of soluble ribonucleic acid in coding for amino acids. *Proc. Natl. Acad. Sci. (USA)* 48: 1086–93.

Cigan, A. M., Feng, L., and Donahue, T. F. 1988. $tRNA^{met}$ functions in directing the scanning ribosome to the start site of translation. *Science* 242: 93–98.

Dahlberg, A. E. 1989. The functional role of ribosomal RNA in protein synthesis. *Cell* 57: 525–29.

Dickerson, R. E., and Geis, I. 1983. *Hemoglobin: Structure, Function, Evolution, and Pathology*. Menlo Park: Benjamin/Cummings.

Doolittle, R. F. 1985. Proteins. *Sci. Am.* (Oct.) 253: 88–99.

Ezzell, C. 1994. Evolutions: Molecular chaperones and protein folding. *J. NIH Res.* 6: 103.

Fisher, J., and Arnold, J. 1999. *Instant Notes in Chemistry for Biologists*. New York: Springer-Verlag.

Frank, J. 1998. How the ribosome works. *Amer. Scient.* 86: 428–39.

Garrod, A. E. 1902. The incidence of alkaptonuria: A study in chemical individuality. *Lancet* 2: 1616–20.

————. 1909. *Inborn Errors of Metabolism*. London: Oxford Univ. Press. (Reprinted 1963, Oxford Univ. Press, London.)

Garrod, S. C. 1989. Family influences on A. E. Garrod's thinking. *J. Inher. Metab. Dis.* 12: 2–8.

Horton, H. R. et al. 2002. *Principles of Biochemistry*. 3rd ed. Upper Saddle River: Prentice-Hall.

Ingram, V. M. 1957. Gene mutations in human hemoglobin: The chemical difference between normal and sickle cell hemoglobin. *Nature* 180: 326–28.

Koshland, D. E. 1973. Protein shape and control. *Sci. Am.* (Oct.) 229: 52–64.

LaDu, B. N., Zannoni, V. G., Laster, L., and Seegmiller, J. E. 1958. The nature of the defect in tyrosine metabolism in alkaptonuria. *J. Biol. Chem.* 230: 251.

Lake, J. A. 1981. The ribosome. *Sci. Am.* (Aug.) 245: 84–97.

Maniatis, T. et al. 1980. The molecular genetics of human hemoglobins. *Annu. Rev. Genet.* 14: 145–78.

Moore, P. B. 1988. The ribosome returns. *Nature* 331: 223–27.

Murayama, M. 1966. Molecular mechanism of red cell sickling. *Science* 153: 145–49.

Neel, J. V. 1949. The inheritance of sickle-cell anemia. *Science* 110: 64–66.

Nirenberg, M. W., and Leder, P. 1964. RNA code words and protein synthesis. *Science* 145: 1399–1407.

Noller, H. F. 1973. Assembly of bacterial ribosomes. *Science* 179: 864–73.

Nomura, M. 1984. The control of ribosome synthesis. *Sci. Am.* (Jan.) 250: 102–14.

Pauling, L., Itano, H. A., Singer, S. J., and Wells, I. C. 1949. Sickle-cell anemia: A molecular disease. *Science* 110: 543–48.

Porce, B. T., and Garrett, R. A. 1999. Ribosomal mechanics, antibiotics, and GTP hydrolysis. *Cell* 97: 423–26.

Ramakrishnan, V. 2002. Ribosome structure and the mechanism of translation. *Cell* 108: 557–72.

Rich, A., and Houkim, S. 1978. The three-dimensional structure of transfer RNA. *Sci. Am.* (Jan.) 238: 52–62.

Rich, A., Warner, J. R., and Goodman, H. M. 1963. The structure and function of polyribosomes. *Cold Spring Harbor Symp. Quant. Biol.* 28: 269–85.

Richards, F. M. 1991. The protein folding problem. *Sci. Am.* (Jan.) 264: 54–63.

Rould, M. A. et al. 1989. Structure of *E. coli* glutaminyl-tRNA synthetase complexed with $tRNA^{gln}$ and ATP at 2.8 Å resolution. *Science* 246: 1135–42.

Saks, M. E., Sampson, J. R., and Abelson, J. N. 1994. The transfer RNA identity problem: A search for rules. *Science* 263: 191–97.

Scott-Moncrieff, R. 1936. A biochemical survey of some Mendelian factors for flower colour. *J. Genet.* 32: 117–70.

Scriver, C. R., and Clow, C. L. 1980. Phenylketonuria and other phenylalanine hydroxylation mutants in man. *Annu. Rev. Genet.* 14: 179–202.

Sharp, P. A., and Eisenberg, D. 1987. The evolution of catalytic function. *Science* 238: 729–30.

Srb, A. M., and Horowitz, N. H. 1944. The ornithine cycle in *Neurospora* and its genetic control. *J. Biol. Chem.* 154: 129–39.

Uchino, T. et al., 1995. Molecular basis of phenotypic variation in patients with argininemia. *Hum. Genet.* 96: 255–60.

Warner, J., and Rich, A. 1964. The number of soluble RNA molecules on reticulocyte polyribosomes. *Proc. Natl. Acad. Sci. (USA)* 51: 1134–41.

Wimberly, B.T. et al. 2000. Structure of the 30*S* ribosomal subunit. *Nature* 407: 327–33.

Yanofsky, C., Drapeau, G., Guest, J., and Carlton, B. 1967. The complete amino acid sequence of the tryptophan synthetase A protein and its colinear relationship with the genetic map of the *A* gene. *Proc. Natl. Acad. Sci. (USA)* 57: 296–98.

Yusupov, M. M. et al. 2001. Crystal structure of the ribosome at 5.5 Å resolution. *Science* 292: 883–96.

Ziegler, I. 1961. Genetic aspects of ommochrome and pterin pigments. *Adv. Genet.* 10: 349–403.

Zubay, G. L., and Marmur, J. (eds.), 1973. *Papers in Biochemical Genetics*. 2nd ed. New York: Holt, Rinehart & Winston.

CHAPTER 15

Mutant erythrocytes derived from an individual with sickle-cell anemia. *(Francis Leroy/Biocosmos/Science Photo Library/Photo Researchers, Inc.)*

Gene Mutation, DNA Repair, and Transposable Elements

CHAPTER CONCEPTS

We have previously defined the four characteristics or functions ascribed to the genetic information as replication, storage, expression, and variation by mutation. In a sense, mutation is a failure to store the genetic information faithfully. If a change occurs in the stored information, it may be reflected in the expression of that information and will be propagated following replication. Historically, the term *mutation* includes both chromosomal changes and changes within single genes. Changes in chromosomes are collectively referred to as **chromosomal mutations** or **aberrations**. In this chapter, we are concerned with **gene mutations**. A change may be a simple substitution of one nucleotide or may involve the insertion or deletion of one or more nucleotides within the normal sequence of DNA.

Mutations form the basis for genetic studies. The resulting phenotypic variability enables geneticists to identify and study the genes that control the traits that have been modified. Without the phenotypic variability that mutations provide, genetic analysis would be impossible. For example, if all pea plants displayed a uniform phenotype, Mendel would have had no basis for his experimentation. Because of the importance of mutations, great attention has been given to their origin, induction, and classification.

Certain organisms lend themselves to induction of mutations that can be detected easily and studied throughout reasonably short life cycles. Viruses, bacteria, fungi, fruit flies, other invertebrates, certain plants, and mice fit these criteria. Thus, these organisms have been widely used to study mutation and mutagenesis, and through other studies they have also contributed to more general aspects of genetic knowledge.

Once we have discussed mutation, we will turn our attention to two related topics—DNA repair and transposable genetic elements. These topics are logical extensions of our consideration of gene mutation. Repair processes serve to counteract mutation. Transposable genetic elements often disrupt the normal structure of the gene and therefore create mutations.

HOW DO WE KNOW WHAT WE KNOW?

IN THIS CHAPTER, WE WILL FOCUS ON one of the major sources of genetic variation: gene mutations. We shall discuss how mutations arise and how we study them. We will also focus on the repair mechanisms that have evolved to counteract the DNA damage that lead to altered phenotypes. As you study this topic, you should try to answer several fundamental questions:

1. How do we determine that variations among organisms are due to mutations within genes?
2. How do we know what the rate of either spontaneous or induced mutation is in an organism such as *Drosophila*?
3. How do we distinguish between various types of mutations in human genes?
4. How do we know if a chemical compound is mutagenic?
5. How do we know that mechanisms exist to repair DNA damage?
6. How did we learn that mobile elements exist within an organism's genome? ■

15.1 Mutations May Be Classified in Various Ways

Mutations are classifed by various schemes. These schemes are not mutually exclusive, but instead depend simply on which aspects of mutation are being investigated or discussed. In this section, we describe several distinctions that are used to classify mutations.

Spontaneous Versus Induced Mutations

All mutations are described as either *spontaneous* or *induced*. Although these two categories overlap to some degree, **spontaneous mutations** are those that just happen in nature. No specific agents are associated with their occurrence, and they are generally assumed to be random changes in the nucleotide sequences of genes. Most are linked to normal chemical processes in the organism that alter the structure of the nitrogenous bases that are part of the existing genes. The majority of spontaneous mutations are thought to occur during the enzymatic process of DNA replication, an idea that we will discuss later in this chapter. Once an error is present in the genetic code, it may be reflected in the amino acid composition of the specified protein. If the changed amino acid is present in a part of the molecule critical to the structure or biochemical activity, a functional alteration can result.

It is generally agreed that any natural phenomenon that heightens chemical reactivity in cells will induce mutations. For example, radiation from cosmic and mineral sources and ultraviolet radiation from the sun are energy sources that most organisms are exposed to and, as such, may be factors that cause spontaneous mutations.

In contrast to spontaneous events, those that result from the influence of *any* artificial factor are considered to be **induced mutations**. The earliest demonstration of an artificially induced mutation occurred in 1927, when Hermann J. Muller reported that X-rays could cause mutations in *Drosophila*. In 1928, Lewis J. Stadler reported that X-rays had the same effect on barley. In addition to various forms of radiation, a wide spectrum of chemical agents is also known to be mutagenic, as we shall see later in this chapter.

Gametic Versus Somatic Mutations

When we consider the effects of mutation in eukaryotic organisms, it is important to distinguish whether the change occurs in somatic cells or in gametes. Mutations arising in somatic cells are not transmitted to future generations, and mutations occurring in somatic cells that create recessive autosomal alleles are rarely of any consequence to the organism. The expression of most such mutations is likely to be masked by the dominant allele. Somatic mutations will have a greater impact if they are dominant or if they are X-linked, since such mutations are most likely to be immediately expressed. Similarly, the impact will be more noticeable if such somatic mutations occur early in development, when undifferentiated cells give rise to several differentiated tissues or organs. Mutations occurring in adult tissues are often masked by the thousands upon thousands of nonmutant cells performing the normal function.

Mutations in gametes or gamete-forming tissues are part of the germline and are of greater concern because they are transmitted to offspring. **Dominant mutations** may be expressed phenotypically in the first generation. **X-linked recessive mutations** arising in the gametes of a heterogametic female may be expressed in hemizygous male offspring. This will occur provided that the male offspring receives the affected X chromosome. Because of heterozygosity, the occurrence of an **autosomal recessive mutation** in the gametes of either males or females (even one resulting in a lethal allele) may be unnoticed for many generations until the resultant allele has become widespread in the population. The new allele will become evident only when a chance mating brings two copies of it together in the homozygous condition.

Other Categories of Mutation

Various types of mutations are classified on the basis of their effect on the organism. Note that a single mutation may fall into more than one category. The most easily observed mutations are those affecting a **morphological trait**. For example, all of Mendel's pea characters and many genetic variations encountered in the study of *Drosophila* fit this designation. They cause obvious changes in morphology.

A second broad category of mutations includes those that exhibit **nutritional** or **biochemical variations** in phenotype. In bacteria and fungi, the inability to synthesize a particular amino acid or vitamin is an example of a typical nutritional mutation. In humans, sickle-cell anemia and hemophilia are examples of biochemical mutations. While such mutations in these organisms are not visible and do not always affect specific morphological characters, they can have a more general effect on the well-being and survival of the affected individual.

A third category consists of mutations that affect behavior patterns of an organism. For example, mating behavior or circadian rhythms of animals may be altered. The primary effect of a **behavior mutation** is often difficult to discern. For example, the mating behavior of a fruit fly can be impaired if it cannot beat its wings. However, the defect may be in (1) the flight muscles, (2) the nerves leading to them, or (3) the brain, where the nerve impulses that initiate wing movements originate. The study of behavior and the genetic factors influencing it has benefited immensely from investigations of behavior mutations.

Another type of mutation can affect the regulation of genes. A regulatory gene can produce a product that controls the transcription of another gene. In other instances, a region of DNA either close to or far from a gene can modulate its activity. In either case, **regulatory mutations** can disrupt normal regulatory processes and permanently activate or inactivate a gene. Our knowledge of genetic regulation depends on the study of mutations that disrupt this process.

Still another group consists of **lethal mutations**. Nutritional and biochemical mutations can also fall into this category. A mutant bacterium that cannot synthesize a specific amino acid will be unable to grow and divide if plated on a medium lacking that amino acid. Various human biochemical disorders, such as Tay-Sachs disease and Huntington disease, are lethal at different points in the life cycle.

Finally, any of these categories can exist as **conditional mutations**. Even though a mutation is present in the genome of an organism, it may not be evident under certain conditions. Among the best examples are **temperature-sensitive mutations**, found in a variety of organisms. At certain "permissive" temperatures, a mutant gene product functions normally, only to lose its functional capability at a different, "restrictive" temperature. When shifted to a restrictive temperature, the impact of the mutation becomes apparent, even lethal, and is then amenable to investigation. The study of conditional mutations has been extremely important in experimental genetics, particularly in understanding the function of genes essential to the viability of organisms.

15.2 Genetic Techniques, Cell Cultures, and Pedigree Analysis Are All Used to Detect Mutations

Before geneticists can study the mutational process directly or obtain mutant organisms for genetic investigations, they must be able to detect mutations. The ease and efficiency of detecting mutations in a particular organism has generally determined the organism's usefulness in genetic studies. In this section, we use several examples to show how mutations are detected.

Detection in Bacteria and Fungi

Detection of mutations is most efficient in haploid microorganisms such as bacteria and fungi. Detection depends on a selection system by which mutant cells can be isolated easily from nonmutant cells. The general principles are similar in bacteria and fungi. To illustrate, we describe how nutritional mutations in the fungus *Neurospora crassa* are detected.

Neurospora is a pink mold that normally grows on bread, but it can also be cultured in the laboratory. This eukaryotic mold is haploid in the vegetative phase of its life cycle. Thus, mutations can be detected without the complications generated by heterozygosity in diploid organisms. Wild-type *Neurospora* grows on a **minimal culture medium** of glucose, a few inorganic acids and salts, a nitrogen source such as ammonium nitrate, and the vitamin biotin. Induced nutritional mutants will not grow on minimal medium, but will grow on a supplemented or **complete medium** that contains numerous amino acids, vitamins, nucleic acid derivatives, and so forth. Microorganisms that are nutritional wild types (requiring only minimal medium) are **prototrophs**, while those mutants that require a specific supplement to the minimal medium are **auxotrophs**.

Nutritional mutants can be detected and isolated by their failure to grow on minimal medium and their ability to grow on complete medium, because the mutant cells cannot synthesize some essential compound that is absent in minimal medium but present in complete medium. Once a nutritional mutant is detected and isolated, the missing compound is determined by attempts to grow the mutant strain in a series of tubes, each containing minimal medium supplemented with a single compound. The auxotrophic mutation can be specified in this way.

This detection technique was used by Beadle and Tatum to support their one-gene : one-enzyme hypothesis (Chapter 14), which we discussed in the context of isolation of mutant *Neurospora* strains (see Figure 14–11). Similar techniques have been useful in the analysis of microorganisms, which have been particularly critical during the study of molecular genetics.

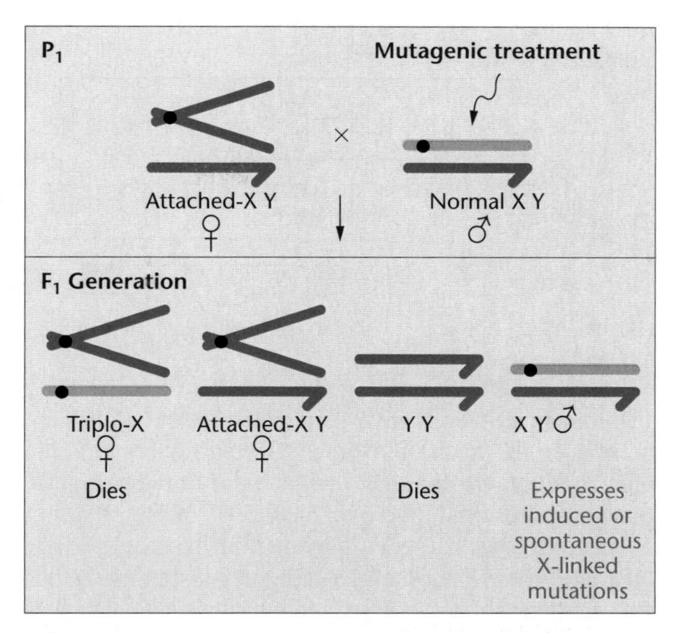

FIGURE 15–1 The attached-X method for detection of induced morphological mutations in *Drosophila*.

Detection in *Drosophila*

Muller, in his studies demonstrating that X-rays are mutagenic, developed a number of detection systems in *Drosophila melanogaster*. These systems are used to estimate both spontaneous and induced rates of X-linked and autosomal recessive lethal mutations. Let's consider the **attached-X procedure**, which assesses X-linked mutations.

Attached-X females have two X chromosomes attached to a single centromere and one Y chromosome, in addition to the normal diploid complement of autosomes. When attached-X females are mated to males with normal sex chromosomes (XY), four types of progeny result: triplo-X females that die, viable attached-X females, YY males that also die, and viable XY males. Figure 15–1 shows how P_1 males that have been treated with a mutagenic agent produce F_1 male offspring that express any spontaneous or induced X-linked recessive mutation. The same approach works equally well whether or not a mutagenic agent has been used. Spontaneous mutations are expressed in the first generation.

In detection techniques devised for recessive autosomal lethals in *Drosophila*, dominant marker mutations are followed through a series of three generations. Although these techniques are more cumbersome to perform, they are fairly efficient.

Detection in Plants

Genetic variation in plants is extensive. Mendel's peas were the basis for the fundamental postulates of transmission genetics, and subsequent studies of plants have enhanced our understanding of gene interaction, polygenic inheritance, linkage, sex determination, chromosome rearrangements, and polyploidy. Many variations are detected simply by visual observation, but techniques also exist for detecting biochemical mutations in plants.

One technique involves the analysis of a plant's biochemical composition. For example, the isolation of proteins from maize endosperm, hydrolysis of the proteins, and determination of the amino acid composition have revealed that the *opaque*-2 mutant strain contains significantly more lysine than do other, nonmutant lines. Since maize protein is usually low in lysine, this mutation significantly improves the nutritional value of the plant. Once this mutant was discovered, plant geneticists and other specialists analyzed the amino acid compositions of various strains of other grain crops, including rice, wheat, barley, and millet. The results of such analyses are useful in combating malnutrition diseases resulting from inadequate protein or the lack of essential amino acids in the diet.

Other detection techniques involve the tissue culture of plant-cell lines in defined culture medium. The plant cells are treated like microorganisms, and their resistance to herbicides or disease toxins can be determined by adding these compounds to the culture medium. The techniques associated with conditional lethal mutants can be used on plant cells in tissue culture and then applied to the genetics of higher plants, providing a means of detection that would not be possible in an intact plant.

Detection in Humans

Humans are obviously not suitable experimental organisms. Therefore, the techniques that have been developed for detecting mutations in organisms such as *Drosophila* are not useful to human geneticists. To determine the genetic basis for any human characteristic or disorder, geneticists analyze a pedigree that traces the family history back as many generations as possible. If a trait is shown to be inherited, it is then possible to predict whether the mutant allele is behaving as a dominant or a recessive and whether it is X-linked or autosomal.

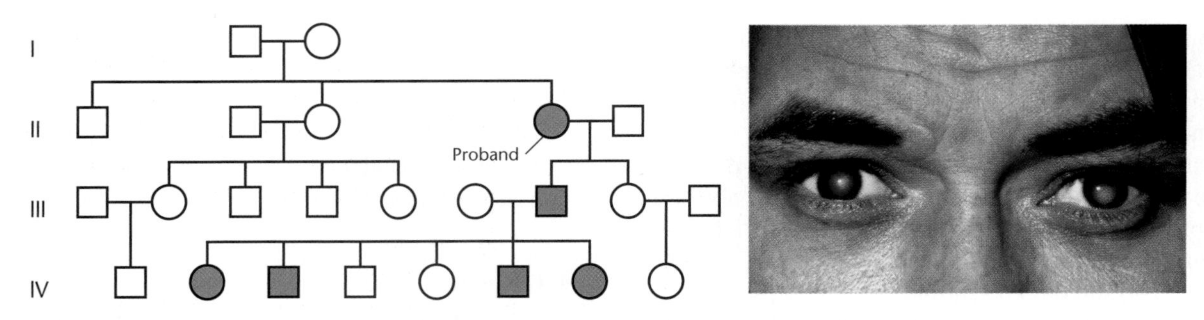

FIGURE 15–2 A hypothetical pedigree of inherited cataract of the eye in humans. *(Photo: Sue Ford/Science Photo Library/Photo Researchers, Inc.)*

Dominant mutations are the simplest to detect. If they are present on the X chromosome, affected fathers pass the phenotypic trait to all their daughters. If dominant mutations are autosomal, approximately 50% of the offspring of an affected heterozygous individual are expected to show the trait. Figure 15–2 shows a pedigree illustrating the initial occurrence of an autosomal dominant allele for cataracts of the eye. The parents in generation I were unaffected, but one of three offspring (generation II) developed cataracts. This female, the proband, produced two children, of which the male child was affected. Of his six offspring, four were affected (generation IV). These observations are consistent with, but do not prove, an autosomal dominant mode of inheritance. However, the high percentage of affected offspring in generation IV favors this conclusion. Also, the unaffected daughter in this generation argues against X-linkage because she received her X chromosome from her affected father. This conclusion is sound, provided that the mutant allele is completely penetrant, that is, the mutant phenotype is always expressed.

X-linked recessive mutations can also be detected by pedigree analysis. The most famous case of an X-linked mutation in humans is that of **hemophilia**, which was found in the descendants of Queen Victoria. The recessive mutation for hemophilia has occurred many times in human populations, but the political consequences of the mutation that occurred in the royal family were sweeping. Inspection of the pedigree in (Figure 15–3) leaves little doubt that Victoria was heterozygous (*Hh*) for the trait. Her father was not affected, and there is no reason to believe that her mother

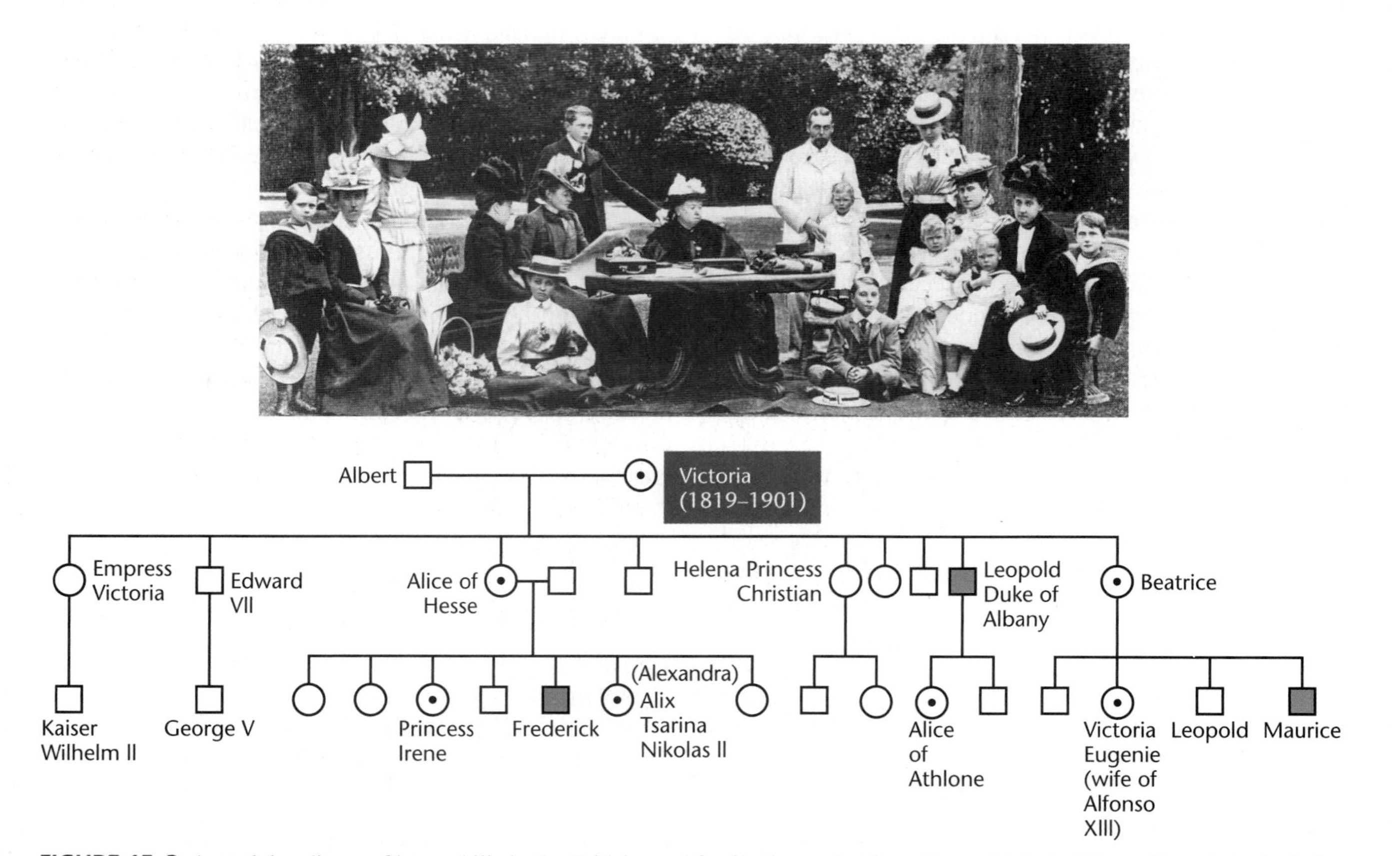

FIGURE 15–3 A partial pedigree of hemophilia in the British royal family descended from Queen Victoria. The pedigree is typical of the transmission of X-linked recessive traits. Circles with a dot in them indicate presumed female carriers heterozygous for the trait. The photograph shows Queen Victoria (seated at the table, center) and some of her immediate family. *(Photo: Mary Evans Picture Library/Photo Researchers, Inc.)*

was a carrier, as Victoria was. Robert Massie's *Nicholas and Alexandra* and Robert and Suzanne Massie's *Journey* (see this chapter's Selected Readings) provide fascinating reading on the topic of hemophilia.

It is also possible to detect autosomal recessive alleles. Because this type of mutation is "hidden" when it is heterozygous, it is not unusual for the trait to appear only intermittently through a number of generations. A mating between an affected individual and a homozygous normal individual will produce unaffected heterozygous carrier children. Matings between two carriers will produce, on average, one-fourth affected offspring.

In addition to pedigree analysis, human cells are now routinely cultured *in vitro*. This procedure allows the detection of many more mutations than any other form of analysis. Analysis of enzyme activity, protein migration in electrophoretic fields, and direct sequencing of DNA and proteins are among the techniques that have demonstrated wide genetic variation between individuals in human populations.

15.3 The Spontaneous Mutation Rate Varies Greatly Among Organisms

The types of detection systems we have been discussing allow geneticists to estimate mutation rates. However we can ascertain induced mutations only when the induced rate clearly exceeds the spontaneous mutation rate for the organism under study. Determining the rate of spontaneous mutation provides the baseline for measuring the rate of experimentally induced mutation.

Examination of the spontaneous rate in a variety of organisms reveals many interesting points. First, the rate is exceedingly low for all organisms studied. Second, the rate is seen to vary considerably in different organisms. Third, even within the same species, the spontaneous mutation rate varies from gene to gene.

Viral and bacterial genes undergo spontaneous gene mutation, on average, about once in 100 million (10^{-8}) cell divisions. *Neurospora* exhibits a similar rate, but maize, *Drosophila*, and humans demonstrate a rate several orders of magnitude higher. The genes studied in these groups average between 1/1,000,000 and 1/100,000 (10^{-6} to 10^{-5}) mutations per gamete formed. Mouse genes are still another order of magnitude higher in their spontaneous mutation rate, 1/100,000 to 1/10,000 (10^{-5} to 10^{-4}) It is not clear why such a large variation occurs in mutation rate. The variation might reflect the relative efficiency of enzyme systems whose function is to repair errors created during replication. We will discuss repair systems later in this chapter.

15.4 Mutations Occur in Many Forms and Arise in Different Ways

Even though we are aware that the gene is a reasonably complex genetic unit, particularly in eukaryotes, we will use a *simplified* definition of a gene in our description of the molecular basis of mutation. Let's consider a gene as a linear sequence of nucleotide pairs representing stored chemical information. Because the genetic code is a triplet, each sequence of three nucleotides specifies a single amino acid in the corresponding polypeptide. Any change that disrupts these sequences or the coded information provides sufficient basis for a mutation. The least complex change is the substitution of a single nucleotide. In Figure 15–4, such a change is compared with our written language, using three-letter words to be consistent with the genetic code. A change of one letter alters the meaning of the sentence: "THE CAT SAW THE DOG." It becomes "THE CAT SAW THE HOG" or "THE BAT SAW THE DOG," creating *missense*. These sentences are analogies to what are most appropriately referred to as **base substitutions**, or **point mutations**. The mutation has changed the sense of the information into various forms of missense.

We also use two more formal terms to describe nucleotide substitutions. If a purine replaces a purine or a pyrimidine replaces a pyrimidine, a **transition** has occurred. If a purine and a pyrimidine are interchanged, a **transversion** has occurred.

A second type of change that can occur in the nucleotide sequence is the insertion or deletion of a single nucleotide at any point along the gene. As illustrated in Figure 15–4, the remainder of the three-letter (code) words become garbled, creating much more extensive missense. These examples are called **frameshift mutations** because the frame of reading has been altered.

The analogy in Figure 15–4 demonstrates that insertions and deletions have the potential to change all subsequent

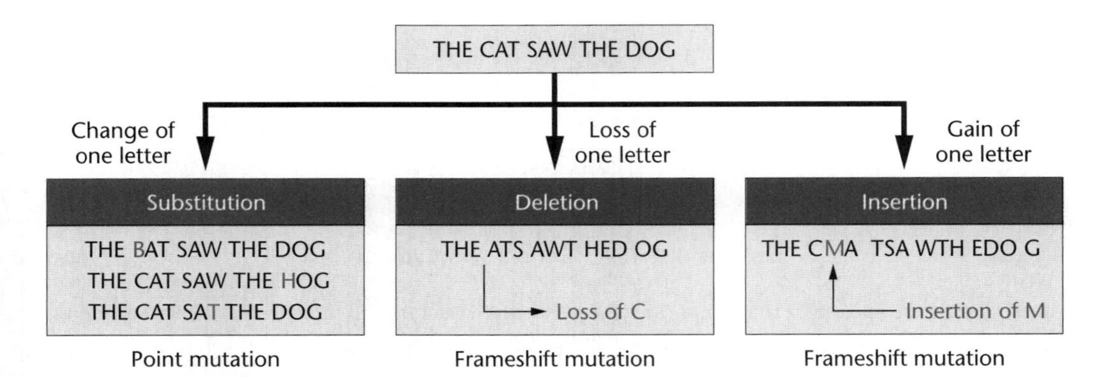

FIGURE 15–4 Analogy of the impact of the substitution, deletion, and insertion of one letter in a sentence composed of three-letter words demonstrating point and frameshift mutations.

MEDIA TUTORIAL Mutations at the DNA level

triplet codes in a gene. It is probable that when this occurs, of the 64 possible triplets, one of many altered triplets will be either UAA, UAG, or UGA. These are termination codons. When one is encountered during translation, polypeptide synthesis is terminated. Obviously, the results of frameshift mutations can be very severe!

Tautomeric Shifts

In 1953, immediately after they proposed a molecular structure of DNA, Watson and Crick published a paper in which they discussed the genetic implications of this structure. They recognized that the purines and pyrimidines found in DNA could exist in *tautomeric forms*; that is, each can exist in alternative chemical forms, differing by only a single proton shift in the molecule. Watson and Crick suggested that **tautomeric shifts** could result in base-pair changes or mutations.

The most stable tautomers of the nitrogenous bases result in the standard hydrogen bonds that are the basis of the double-helical model of DNA. The less-frequently occurring tautomers are capable of hydrogen bonding with noncomplementary bases. However, the pairing is always between a pyrimidine and a purine. Figure 15–5 compares the normal base-pairing relationships with the rare unorthodox pairings. The biologically important unstable tautomers involve keto–enol pairs for thymine and guanine, and amino–imino pairs for cytosine and adenine.

The effect leading to mutation occurs during DNA replication when a rare tautomer in the template strand matches with a noncomplementary base. In the next round of replication, the "mismatched" members of the base pair are separated, and each specifies its normal complementary base. The end result is a transition mutation (see Figure 15–6).

Base Analogs

Mutagenic chemicals called **base analogs** are molecules that can substitute for purines or pyrimidines during nucleic acid biosynthesis. The halogenated derivative of uracil in the number 5 position of the pyrimidine ring **5-bromouracil (5-BU)*** is a good example. Figure 15–7 compares the structure of this thymine analog with the structure of thymine. The presence of the bromine atom in place of the methyl group increases the probability that a tautomeric shift will occur. If 5-BU is incorporated into DNA in place of thymine and a tautomeric shift to the enol form occurs,

*If 5-BU is chemically linked to deoxyribose, the nucleoside analog bromodeoxyuridine (BUdR) is formed.

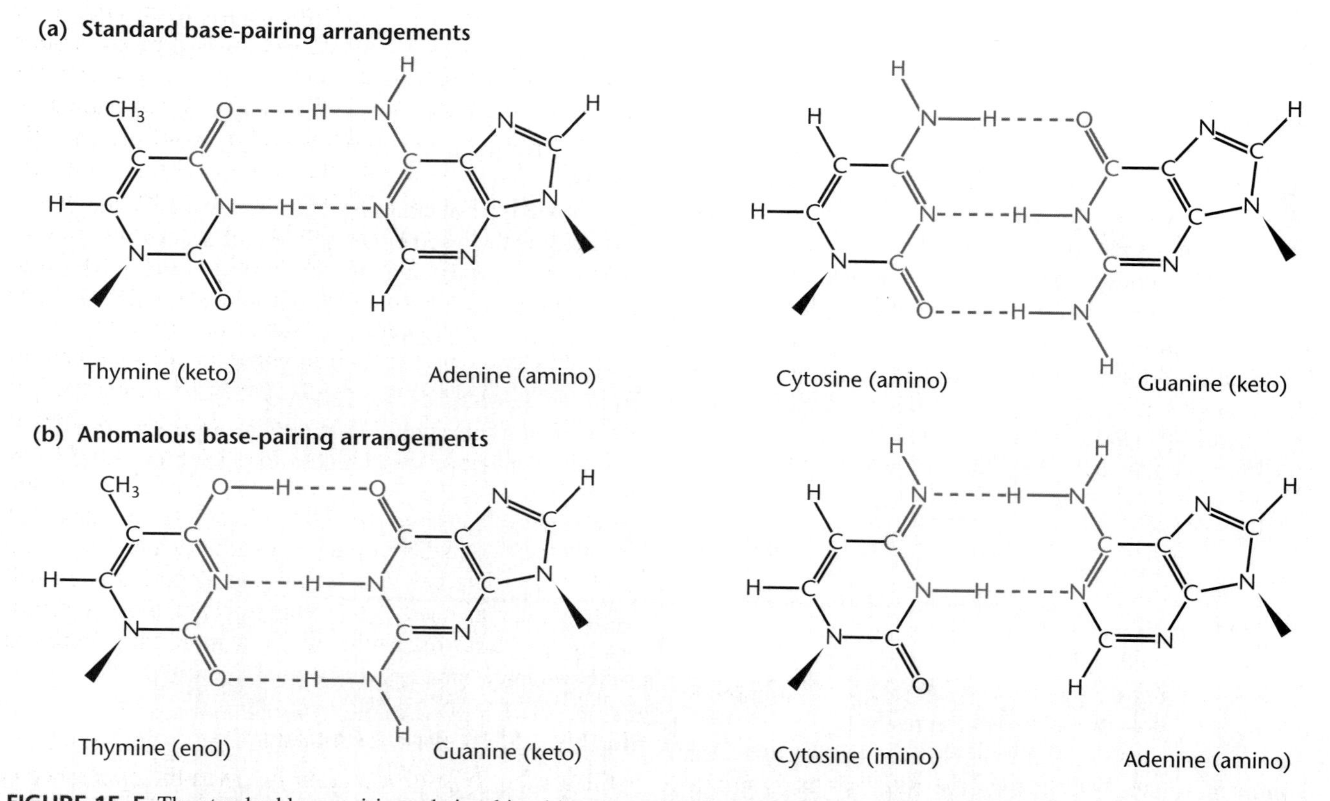

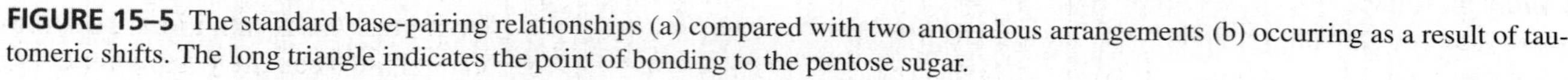
FIGURE 15–5 The standard base-pairing relationships (a) compared with two anomalous arrangements (b) occurring as a result of tautomeric shifts. The long triangle indicates the point of bonding to the pentose sugar.

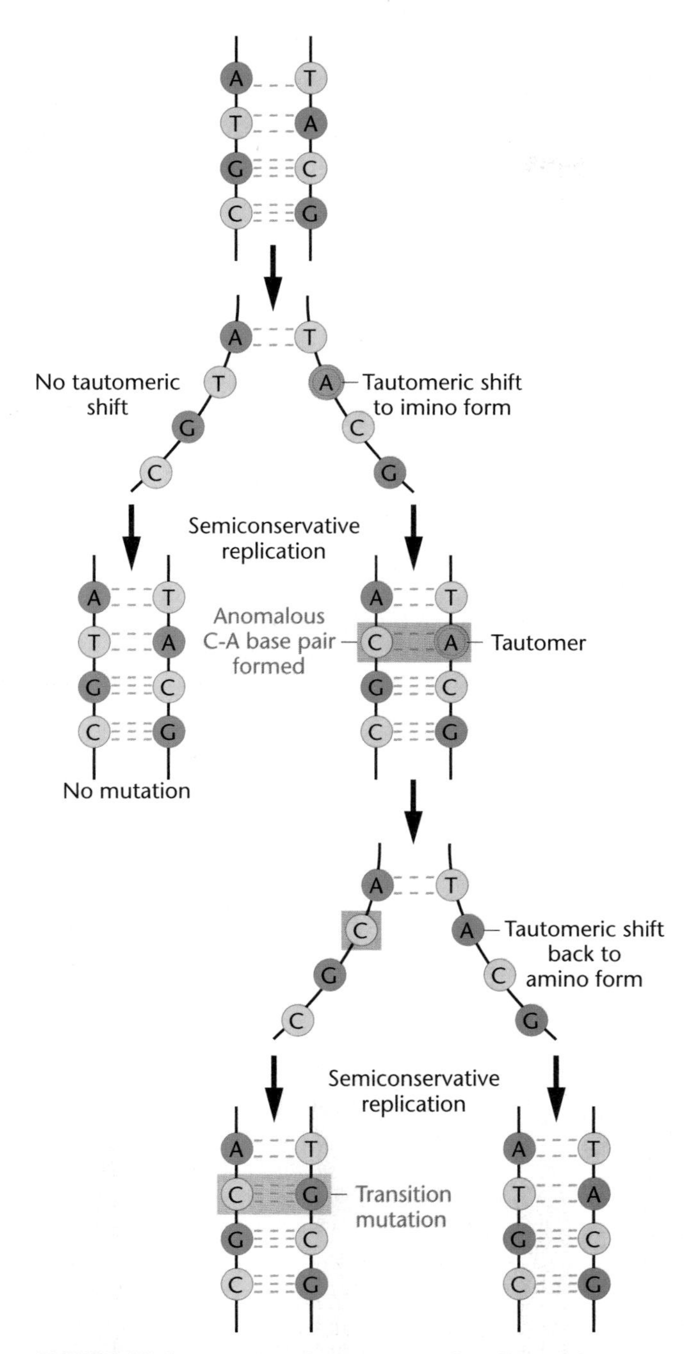

FIGURE 15–6 Formation of a T=A to a C≡G transition mutation as a result of a tautomeric shift in adenine.

5-BU base-pairs with guanine. After one round of replication, an A=T to G≡C transition results.

There are other base analogs that are mutagenic. One, **2-amino purine (2-AP)**, can serve successfully as an analog of adenine. In addition to its base-pairing affinity with thymine, 2-AP can also base-pair with cytosine. As such, transitions from A=T to G≡C can result following replication.

Because of the specificity by which base analogs such as 2-AP induce transition mutations, base analogs can also be used to induce reversion to the wild-type nucleotide sequence. This alteration is called **reverse mutation**. The process also occurs spontaneously, but at a much lower rate.

Alkylating Agents

The sulfur-containing mustard gases were one of the first groups of chemical mutagens discovered. This discovery was made in studies involving chemical warfare during World War I. Mustard gases are **alkylating agents**; that is, they donate an alkyl group such as a methyl group (CH_3) or an ethyl group ($CH_3—CH_2$) to amino or keto groups in nucleotides. For example, **ethylmethane sulfonate (EMS)**, often used as an experimental mutagen, alkylates the keto group in the number 6 position of guanine and in the number 4 position of thymine (Figure 15–8). As with base analogs, base-pairing affinities are altered and transition mutations result. In the case of **6-ethylguanine**, this molecule acts like a base analog of adenine, causing it to pair with thymine.

Acridine Dyes and Frameshift Mutations

Other chemical mutagens cause frameshift mutations, as we saw in Figure 15–4. These result from adding or removing one or more base pairs in the polynucleotide sequence of the gene. Inductions of frameshift mutations have been studied in detail with a group of aromatic molecules known as **acridine dyes**. Proflavin, the most widely studied acridine mutagen, and acridine orange are examples. Acridine dyes are of about the same dimension as a nitrogenous base pair and are known to intercalate or wedge between purines and pyrimidines of intact DNA. Intercalation of acridine dyes induces contortions in the DNA helix, which causes deletions and insertions.

One model suggests that the resultant frameshift mutations are generated at gaps produced in the DNA during replication, repair, or recombination. During these events, there is the possibility of slippage and improper base-pairing of one strand with the other. This model suggests that intercalation of the acridine into an improperly base-paired region can extend the existence of these slippage structures. If so, the probability increases that the mispaired configuration will exist when synthesis and rejoining occurs, thereby resulting in an addition or deletion of one or more bases from one of the strands.

Apurinic Sites and Deamination

Still another type of mutation involves the spontaneous loss of one of the nitrogenous bases in an intact double-helical DNA molecule. Most frequently, such an event

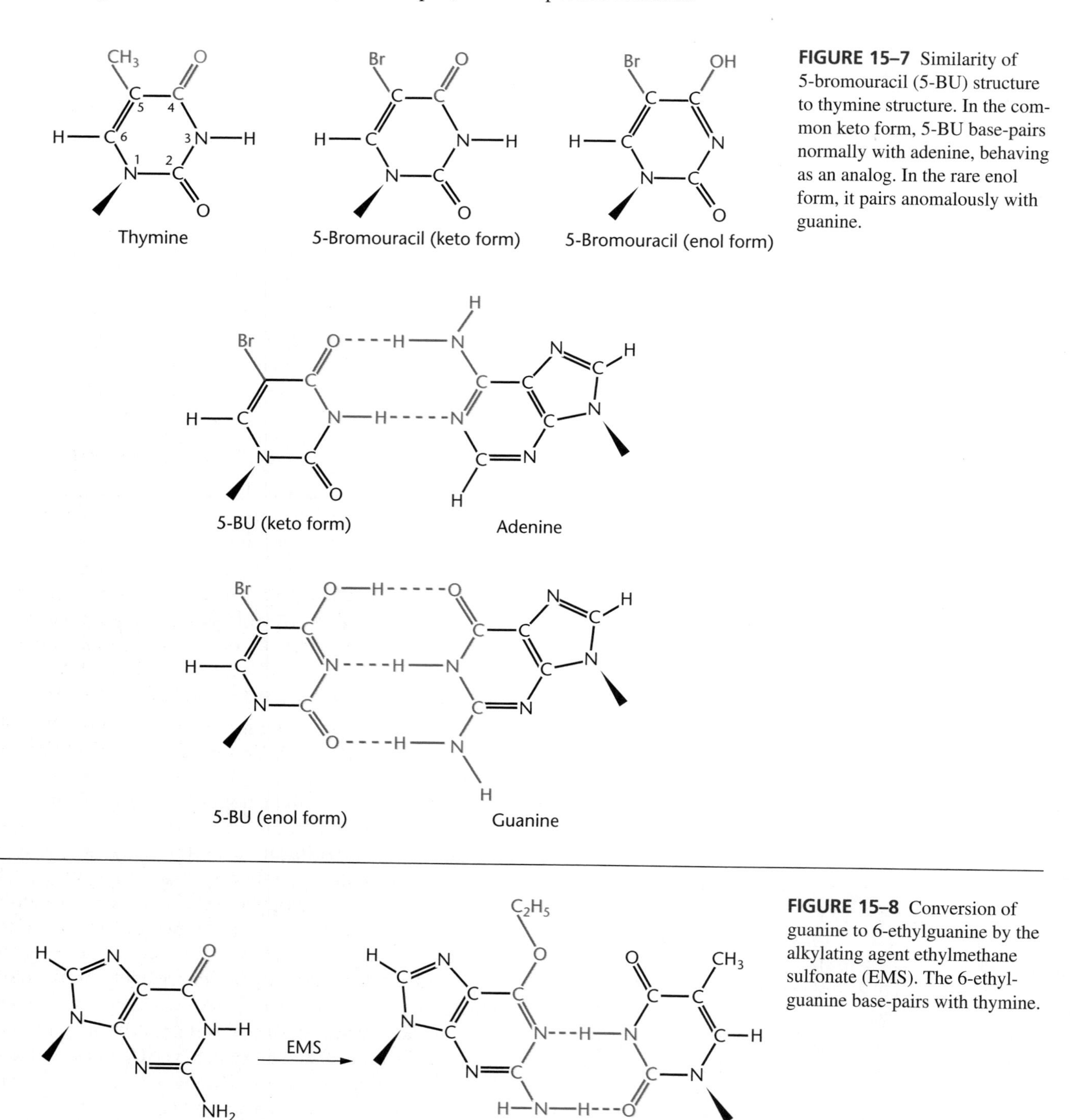

FIGURE 15–7 Similarity of 5-bromouracil (5-BU) structure to thymine structure. In the common keto form, 5-BU base-pairs normally with adenine, behaving as an analog. In the rare enol form, it pairs anomalously with guanine.

FIGURE 15–8 Conversion of guanine to 6-ethylguanine by the alkylating agent ethylmethane sulfonate (EMS). The 6-ethylguanine base-pairs with thymine.

involves either guanine or adenine. These sites, called **apurinic (AP) sites,*** are created by the "breaking" of the glycosidic bond linking the 1′-C of deoxyribose and the number 9 position of the purine ring. Geneticists estimate that thousands of such spontaneous lesions are formed daily in the DNA of mammalian cells in culture.

The absence of a nitrogenous base at an AP site will alter the genetic code if the strand involved is transcribed and translated. If replication occurs, the AP site is an inadequate template and can cause replication to stall. If a nucleotide is inserted, it is frequently incorrect, causing still another mutation. Fortunately, as we shall soon see, cells have repair systems that often counteract and correct this type of lesion.

Another type of chemical change is known to cause some mutations. In the process of **deamination**, an amino group is converted to a keto group in cytosine and adenine. In these two cases, cytosine is converted to uracil and adenine is changed to hypoxanthine. The major effect of these changes is to alter the base-pairing specificities of these two molecules during DNA replication. For example, cytosine normally pairs with guanine. Following its conversion to uracil, which pairs with adenine, the original

*Apyrimidinic sites, where a pyrimidine has been lost, occur as well, and are also referred to as AP sites.

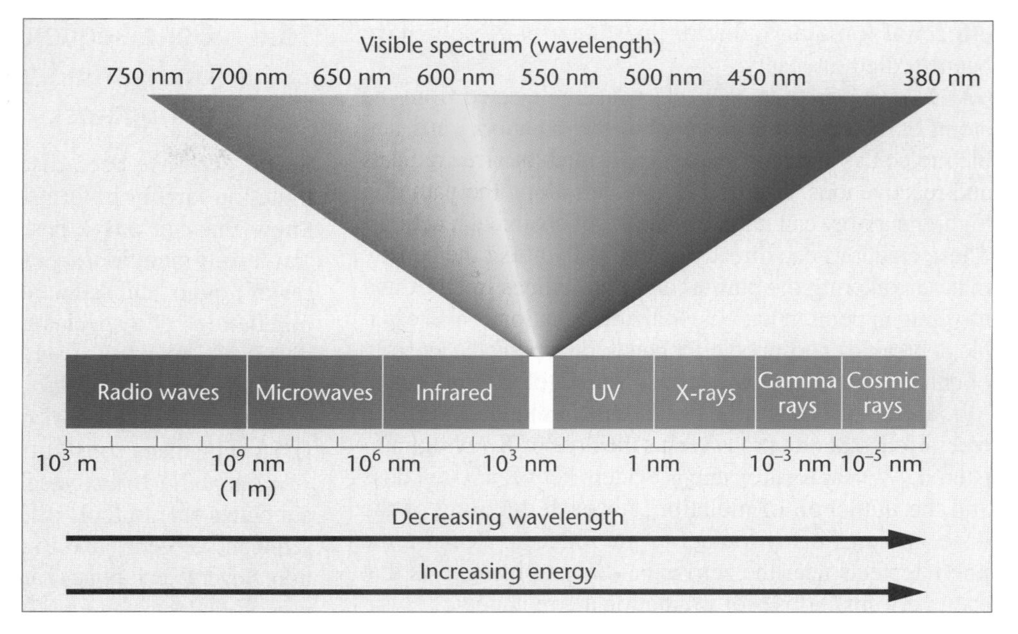

FIGURE 15–9 The components of the electromagnetic spectrum and their associated wavelengths.

G≡C pair is converted to an A=U pair and then, following an additional replication, converts to an A=T pair. When adenine is deaminated, an original A=T pair is converted to a G≡C pair because hypoxanthine pairs naturally with cytosine. Nitrous acid is a known mutagen capable of inducing deamination of bases in DNA.

15.5 Ultraviolet and Ionizing Radiation Are Mutagenic

All energy on Earth consists of a series of electromagnetic components of varying wavelength (Figure 15–9). Referred to as the **electromagnetic spectrum**, the energy associated with the various components varies inversely with wavelength. Everything longer than, and including visible light, is benign when it interacts with most organic molecules. However, everything of shorter wavelength than visible light, which is inherently more energetic, has a disruptive impact on organic molecules, such as those comprising living tissue. For example, we know that purines and pyrimidines absorb **ultraviolet (UV) radiation** most intensely at a wavelength of about 260 nm, a property that has been useful in the detection and analysis of nucleic acids. In 1934, as a result of studies involving *Drosophila* eggs, it was discovered that UV radiation is mutagenic, and by 1960, several studies concerning the *in vitro* effect on the components of nucleic acids had been completed. The major effect of UV radiation is to induce the formation of **pyrimidine dimers**, particularly between two thymine residues (Figure 15–10). While cytosine–cytosine and thymine–cytosine dimers can also be formed, they are less prevalent. The dimers distort the DNA conformation and inhibit normal replication. As a result, errors can be introduced into the base sequence of DNA during replication, and when extensive, these are responsible, at least in part, for the killing effects of UV radiation on microorganisms.

As we shall see later in this chapter, several mechanisms have evolved to correct UV-induced lesions. We illustrate the essential nature of such "repair" processes in humans by discussing the inherited disorder xeroderma pigmentosum, where mutation has inactivated one UV-repair mechanism. Such individuals have been described as "children of the night," since they cannot risk exposure to the ultraviolet component of sunlight without the risk of epidermal malignancy.

Ionizing Radiation

Within the electromagnetic spectrum, energy varies inversely with wavelength. As shown in Figure 15–9, X-rays, gamma rays, and cosmic rays have even shorter wavelengths than does UV radiation and are therefore more energetic. As a result, they are strong enough to penetrate deeply into tissues, causing ionization of the molecules encountered along the way. Hermann J. Muller and Lewis J. Stadler established in the 1920s that these sources of **ionizing radiation** are mutagenic. Since that time, the

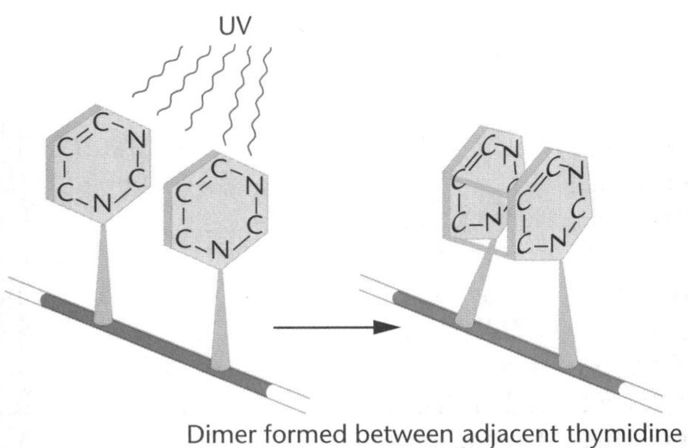

FIGURE 15–10 Induction of a thymine dimer by UV radiation, leading to distortion of the DNA. The atoms of the pyrimidine ring illustrate the formation of the covalent cross-links.

MEDIA TUTORIAL DNA repair: photoreactivation and excision

effects of ionizing radiation, particularly X-rays, have been studied intensely.

As X-rays penetrate cells, electrons are ejected from the atoms of molecules encountered by the radiation. Thus, stable molecules and atoms are transformed into free radicals and reactive ions. The trail of ions left along the path of a high-energy ray can initiate a variety of chemical reactions. These reactions can directly or indirectly affect the genetic material, altering the purines and pyrimidines in DNA and resulting in point mutations. Ionizing radiation is also capable of breaking phosphodiester bonds, disrupting the integrity of chromosomes and producing a variety of aberrations.

Figure 15–11 graphs the percentage of induced X-linked recessive lethal mutations versus the dose of X-rays administered. A linear relationship is evident between X-ray dose and the induction of mutation; for each doubling of the dose, twice as many mutations are induced. Because the line intersects near the zero axis, this graph suggests that even very small doses of irradiation are mutagenic.

These observations can be interpreted in the form of the **target theory**, first proposed in 1924 by J. A. Crowther and F. Dessauer. The theory proposes that there are one or more sites, or targets, within cells and that a single event of irradiation at one site will bring about a damaging effect, or mutation. In effect, the target theory suggests that the X-rays interact directly with the genetic material.

An observation concerning irradiation effects during the cell cycle is of particular interest. Cells that have entered mitosis are much more susceptible to radiation effects than cells in G1, S, or G2. This is because condensed chromosomes present a more susceptible target than dispersed chromatin. X-rays can break chromosomes, resulting in terminal or intercalary deletions, translocations, and general chromosome fragmentation. This property is one of the reasons why radiation is used to treat human malignancy. Tumorous cells are undergoing division more often than their nonmalignant counterparts, so they are more susceptible to the damaging effect of radiation.

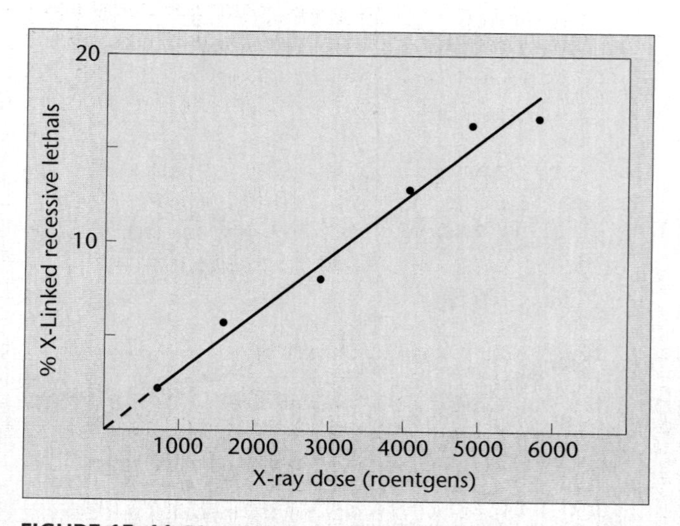

FIGURE 15–11 Plot of the percentage of X-linked recessive mutations induced by increasing doses of X-rays. If extrapolated, the graph intersects the zero axis as shown by the dashed line.

15.6 Gene Sequencing Has Enhanced Understanding of Mutations in Humans

So far, we have been discussing the molecular basis of mutation largely in terms of nucleic acid chemistry. We know that various types of nucleotide conversions or frameshift mutations occur, primarily because our analyses of amino acid sequences of many proteins within the populations of a species show substantial diversity. This diversity, which has arisen during evolution, is a reflection of changes in the triplet codons following substitution, insertion, or deletion of one or more nucleotides in the DNA sequences constituting genes.

As our ability to analyze DNA more directly increases, we are better able to look specifically at the actual nucleotide sequence of genes and to gain greater insights into mutation. Several techniques capable of accurate, rapid sequencing of DNA have greatly extended our knowledge of molecular genetics. In this section, we examine the results of a number of studies that have investigated the actual gene sequence of various mutations that affect humans.

ABO Blood Types

The ABO system is based on a series of antigenic determinants found on erythrocytes and other cells, particularly epithelial types. Three alleles of a single gene exist, the product of which modifies the H substance. The modification involves glycosyltransferase activity, converting the H substance to either the A or B antigen, as a result of the product of the I^A or I^B allele, respectively, or failing to modify the H substance as a result of the I^O allele.

The responsible gene has been sequenced in 14 cases of varying ABO status. When DNAs from I^A and I^B alleles were compared, four consistent nucleotide substitutions were found. It is assumed that the resulting changes in the amino acid sequence of the glycosyltransferase gene product lead to the different modifications of the H substance.

The I^O allele situation is unique and intriguing. Individuals homozygous for this allele have type O blood, lack glycosyltransferase activity, and fail to modify the H substance. Analysis of the DNA of this allele shows one consistent change that is unique compared with the sequence of the other alleles, the deletion of a single nucleotide early in the coding sequence, causing a frameshift mutation. A complete messenger RNA is transcribed, but at translation, the frame of reading shifts at the point of deletion and continues out of frame for about 100 nucleotides before a stop codon is encountered. At this point, the polypeptide chain terminates prematurely, resulting in a nonfunctional product.

These findings provide a direct molecular explanation of the ABO allele system and the basis for the biosynthesis of the corresponding antigens. The molecular basis for the antigenic phenotypes is clearly the result of structural alterations, or mutations, of the nucleotide sequence of the gene encoding the glycosyltransferase enzyme.

Muscular Dystrophy

Muscular dystrophy is characterized by severe progressive muscular degeneration, or myopathy, resulting in the death of the affected individuals by early adulthood. Because the condition is recessive and X-linked and because affected males die before they can reproduce, females are rarely affected by the disorder. The incidence of 1/3500 live male births makes muscular dystrophy one of the most common life-shortening hereditary disorders known. Two related forms exist. **Duchenne muscular dystrophy (DMD)** is more severe and more common than the allelic form, **Becker muscular dystrophy (BMD)**.

The region containing the gene, which consists of over 2 million bp, has been analyzed extensively. In normal (unaffected) individuals, transcription results in a messenger RNA containing about 14,000 bases (14 kb) that is translated into the protein *dystrophin*, consisting of 3685 amino acids. This protein can be detected in most cases of the less severe BMD, but is rarely found in DMD. This has led to the hypothesis that most mutations causing BMD do not alter the reading frame, but that most DMD mutations change the reading frame early in the gene, resulting in the premature termination of dystrophin translation. This hypothesis is consistent with the observed differences in severity of these two forms of the disorder.

In an extensive analysis of the DNA of 194 patients (160 DMD and 34 BMD), J. T. Den Dunnen and associates found that 128 of these mutations (65%) consisted of substantial deletions or insertions. Of 115 deletions, 17 occurred in BMD cases, and of 13 insertions, 1 occurred in BMD, with the remainder in DMD patients. In most cases, the results were consistent with the reading-frame hypothesis: With few exceptions, DMD mutations changed the frame of reading, whereas BMD mutations usually did not alter the reading frame.

Perhaps the most noteworthy of Den Dunnen's findings is the high percentage of deleterious mutations that represent the deletion or insertion of nucleotides within the gene. This observation reflects the fact that a mutation caused by a random single-nucleotide substitution within a gene is more likely to be tolerated without the devastating effect of muscular dystrophy than the addition or loss of numerous nucleotides that can alter the frame of reading. There are three reasons for this.

1. A nucleotide substitution may not change the encoded amino acid, since the code is degenerate.
2. If an amino acid substitution does result, the change may not be present at a location within the protein that is critical to its function.
3. Even if the altered amino acid is present at a critical region, it may still have little or no effect on the function of the protein. For example, an amino acid might be changed to another with nearly identical chemical properties or to one with very similar recognition properties, such as shape.

As a result, single-base substitutions sometimes have little or no effect on protein function or they may simply reduce the efficiency but not eliminate the functional capacity of the gene product. As more mutant genes are analyzed directly, our picture of mutation will become increasingly clear.

Trinucleotide Repeats in Fragile X Syndrome, Myotonic Dystrophy, and Huntington Disease

Beginning about 1990, the molecular analysis of the DNA representing the genes responsible for a number of inherited human disorders provided a remarkable set of observations. Mutant genes were sometimes characterized by an expansion of a simple **trinucleotide repeat sequence**, usually from fewer than 15 copies in normal individuals to a large number in affected individuals. For example, a different trinucleotide DNA sequence that is repeated many times is present in each of the three genes thought to be responsible for the X-linked fragile X syndrome, and the autosomal disorders myotonic dystrophy, Huntington disease, and spinobulbar muscular atrophy (Kennedy disease). While repeated sequences are also present in the nonmutant (normal) allele of each gene, the mutations are characterized by a significant variable increase in the number of times the trinucleotide is repeated. Table 15.1 summarizes the various diseases that are discussed next, with particular emphasis on the size of the repeats in normal and diseased individuals.

We have previously discussed the variable onset of the expression of various phenotypes. In several cases, a correlation has been found between the number of repeats and the age of manifestation of mutant phenotypes. The greater the number of repeats, the earlier disease onset occurs. Further, in affected individuals, the number of repeats may increase in each subsequent generation. This general phenomenon, **genetic anticipation**, reflects a unique form of mutation related to an instability of the specific regions in each of the three different genes.

TABLE 15.1 Summary of Trinucleotide-Repeat Disorders

	Trinucleotide Repeat	*Number (normal)*	*Number (affected)*	*Inheritance Pattern*
Huntington disease	CAG	10–35	36–120	Autosomal dominant
Myotonic dystrophy	CTG	5–35	> 230	Autosomal dominant
Fragile X syndrome	CGG	6–54	> 230	X-linked dominant
Spinobulbar muscular atrophy	CAG	10–35	35–60	X-linked dominant

We begin with a short discussion of **fragile X syndrome**. The responsible gene, *FMR-1*, may have several hundred to several thousand copies of the trinucleotide sequence CGG. Individuals with up to 54 copies are normal and do not display the mental retardation associated with the syndrome. Individuals with 54–230 copies are considered carriers. Although they are normal, their offspring may contain even more copies and express the syndrome.

Myotonic dystrophy, or **DM** (its original name was dystrophia myotonica), is the most common form of adult muscular dystrophy. Caused by a dominant mutant gene located on the long arm of chromosome 19, the disorder is not as severe as DMD and it is highly variable both in symptoms and age of onset. Mild myotonia—atrophy and weakness—of the musculature of the face and extremities is most common. Cataracts, reduced cognitive ability, and cutaneous and intestinal tumors are also part of the syndrome.

The affected gene is believed to be *MDPK*, which encodes a serine–threonine protein kinase, MDPK. This protein is the product of 15 exons of the gene, the last of which encodes the 3′-untranslated RNA sequence of the mRNA. It is this sequence that houses the multiple copies of the trinucleotide, reflected in the gene as the DNA sequence CTG. Individuals with 5–37 copies are normal, and the number of copies is stable from generation to generation. Above this number of copies, symptoms range from mild to severe, with onset occurring anywhere between birth and age 60. Both the severity and onset are directly correlated with the size of the repeated sequence. Minimally affected patients are known to contain up to 150 repeats, while severely affected patients have up to 1500 copies of the the CTG triplet. Genetic anticipation is exhibited in the offspring.

More recently, the gene responsible for **Huntington disease** has been found to demonstrate a similar mutational pattern. Behaving as an autosomal dominant and located on chromosome 4, the gene contains the trinucleotide CAG repeat 10 to 35 times in normal individuals. The sequence exists in significantly increased numbers (up to 120) in diseased individuals, and much earlier onset occurs when the number of copies present is closer to the upper range. Interestingly, in still another disorder, **spinobulbar muscular atrophy (Kennedy disease)**, the involved gene (different from that in Huntington disease) also contains repeated copies of the CAG triplet. However, only 35–60 copies of the sequence cause individuals to be affected in Kennedy disease.

The role of repeated sequences in normal and mutant genes remains a mystery. Their locations within the gene vary in each case. In Huntington and Kennedy diseases, the repeat lies within the coding portion of the gene. In Huntington disease, this causes the mutant *huntingtin* protein to contain an excess of glutamine residues. Such is not the case in the other two disorders, however. In the gene responsible for fragile X syndrome, the repeat is upstream (the 5′-end) in an area that is most often involved in regulating gene expression. In the case of myotonic dystrophy, the repeat is downstream (the 3′-end).

The mechanism by which the repeated sequence expands from generation to generation is also of great interest. How such an instability during DNA replication affects only specific areas of certain genes is currently an important research topic. Current thinking is that expansion can result from both errors during replication as well as in the mechanisms responsible for repairing damaged DNA. Whatever the cause may be, this general instability seems to be more prevalent in humans than in many other organisms.

15.7 The Ames Test Is Used to Assess the Mutagenicity of Compounds

There is great concern about the possible mutagenic properties of any chemical that enters the human body, whether through the skin, digestive system, or respiratory tract. For example, the residual materials of air and water pollution, food preservatives and additives, artificial sweeteners, herbicides, pesticides, and pharmaceutical products have been scrutinized. As we have seen, mutagenicity can be tested in various organisms, including *Drosophila*, mice, and cultured mammalian cells. The most common test, which involves bacteria, was devised by Bruce Ames.

The **Ames test** (Figure 15–12) uses any of four strains of the bacterium *Salmonella typhimurium* that were selected for sensitivity to and specificity for mutagenesis. One strain is used to detect base-pair substitutions, and the other three detect various frameshift mutations. Each mutant strain is unable to synthesize histidine and therefore requires histidine for growth (his^-) The assay measures the frequency of reverse mutation, which yields wild-type (his^+) bacteria. Greater sensitivity to mutagens occurs because these strains bear other mutations that eliminate both the DNA excision-repair system (discussed later in this chapter) and the lipopolysaccharide barrier that coats and protects the surface of the bacteria.

It is very interesting to note that many substances entering the human body are relatively innocuous until activated metabolically, usually in the liver, to a more chemically reactive product. Thus, the Ames test includes a step in which the test compound is incubated *in vitro* in the presence of a mammalian liver extract. Or test compounds can be injected into a mouse, where they are modified by liver enzymes and then recovered.

In the initial use of Ames testing in the 1970s, a large number of known carcinogens were examined and over 80% were shown to be strong mutagens. This is not surprising. The transformation of cells to the malignant state undoubtedly occurs as a result of some alteration of DNA. Although a positive response as a mutagen does not prove that a compound is carcinogenic, the Ames test is useful as a preliminary screening device. It is used extensively in conjunction with the industrial and pharmaceutical development of chemical compounds.

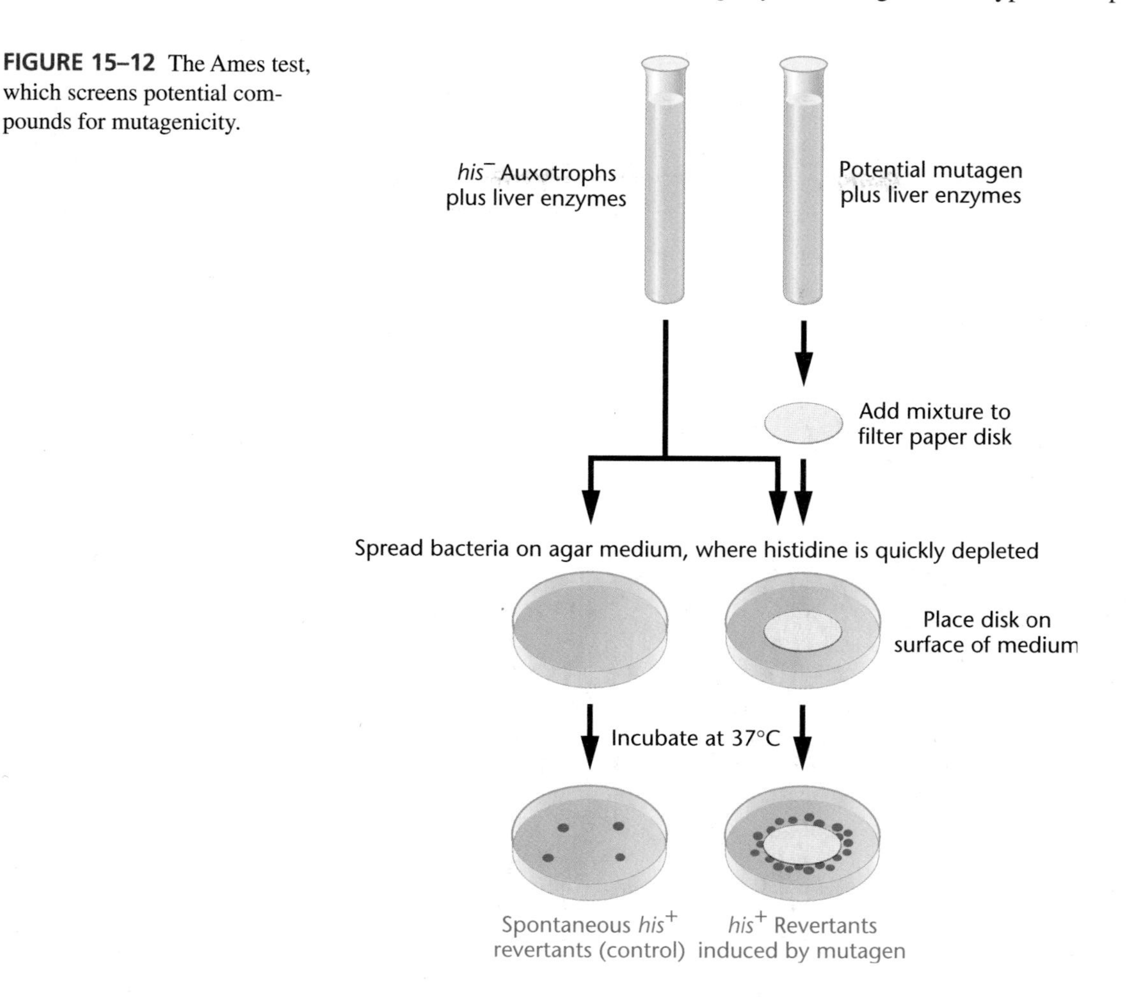

his⁻ Auxotrophs plus liver enzymes
Potential mutagen plus liver enzymes
Add mixture to filter paper disk
Spread bacteria on agar medium, where histidine is quickly depleted
Place disk on surface of medium
Incubate at 37°C
Spontaneous *his*+ revertants (control)
his+ Revertants induced by mutagen

FIGURE 15–12 The Ames test, which screens potential compounds for mutagenicity.

MEDIA TUTORIAL Chemical mutagenesis: Ames test

15.8 Organisms Can Counteract DNA Damage by Activating Several Types of Repair Systems

In previous sections of this chapter, we established that replicating and nonreplicating DNA molecules are vulnerable to various forms of errors, which induce lesions that lead to gene mutations. Living systems have evolved a variety of very elaborate repair systems that are able to counteract many of the forms of DNA damage that lead to mutation. As you shall see, such repair systems are essential to the maintenance of the genetic integrity of organisms, and as such, to the survival of organisms on the earth.

Photoreactivation Repair: Reversal of UV Damage in Prokaryotes

Recall that UV light is mutagenic as a result of the creation of pyrimidine dimers (see Figure 15–10). The study of the mutagenicity of UV radiation paved the way for the discovery of many forms of natural repair of DNA damage. The first relevant discovery concerning UV repair in bacteria was made in 1949 when Albert Kelner observed the phenomenon of **photoreactivation repair**. He showed that the UV-induced damage to *E. coli* DNA could be partially reversed if the cells were exposed briefly to light in the blue range of the visible spectrum following irradiation. The photoreactivation repair process was subsequently shown to be temperature-dependent, suggesting that the light-induced mechanisms involve an enzymatically controlled chemical reaction. Thus, visible light appears to induce the repair process of the DNA damaged by UV radiation.

Further studies of photoreactivation have revealed that the process is dependent on the activity of a protein called the **photoreactivation enzyme (PRE)**. This molecule can be isolated from extracts of *E. coli* cells. The enzyme's mode of action is to cleave the bonds between thymine dimers, thus reversing the effect of UV radiation on DNA (Figure 15–13). Although the enzyme will associate with a dimer in the dark, it must absorb a photon of light to cleave the dimer. In spite of the potential for diminishing UV-induced mutations, photoreactivation repair is not essential in *E. coli* since a null mutation in the gene coding for the PRE is not lethal. In addition, the enzyme cannot be detected in humans and other eukaryotes, who must rely on other repair mechanisms to reverse the effects of UV radiation.

Excision Repair in Prokaryotes and Eukaryotes

Investigations in the early 1960s suggested that in addition to PRE, a *light-independent* repair system exists in prokaryotes such as *E. coli* that repairs damage to DNA caused by both exogenous and endogenous agents. The basic mechanism of this type of repair, referred to as

MEDIA TUTORIAL DNA repair: photoreactivation and excision

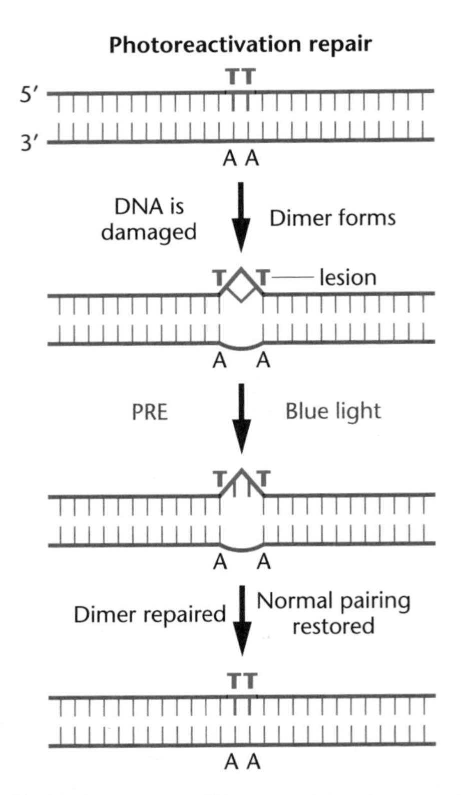

FIGURE 15–13 Damaged DNA repaired by photoreactivation repair. The bond creating the thymine dimer is cleaved by the photoreactivation enzyme (PRE), which must be activated by blue light in the visible spectrum.

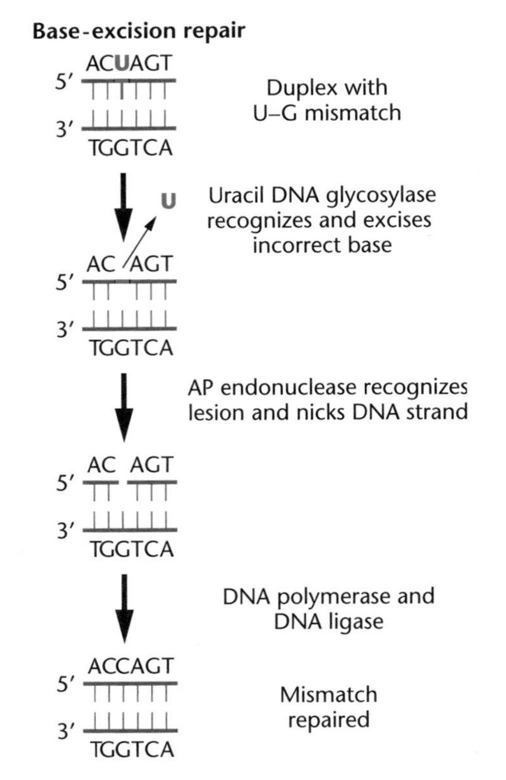

FIGURE 15–14 Base-excision repair (BER) accomplished by uracil DNA glycosylase, AP endonuclease, DNA polymerase, and DNA ligase. Uracil is recognized as a noncomplementary base, excised, and replaced with the complementary base (C).

excision repair, has been conserved throughout evolution, occurring in all prokaryotic and eukaryotic organisms. The process consists of three basic steps:

1. The distortion or error is recognized and enzymatically clipped out by a nuclease. This excision may affect only a single base or a single nucleotide, or it may include several nucleotides adjacent to the error as well, leaving a gap in the helix.

2. DNA polymerase I fills this gap by inserting deoxyribonucleotides complementary to those on the intact strand. The enzyme adds these bases to the 3′-OH end of the clipped DNA.

3. The joining enzyme DNA ligase seals the final "nick" that remains at the 3′-OH end of the last base inserted, closing the gap.

DNA polymerase I, the enzyme discovered by Arthur Kornberg, was once assumed to be the universal DNA-replication enzyme. However, it was the discovery of the *polA1* mutation that demonstrated the role of this enzyme in repairing UV-induced lesions in *E. coli*. Cells carrying the *polA1* mutation lack functional polymerase I, replicate their DNA normally, and are unusually sensitive to UV radiation. Apparently, such cells are unable to fill the gap created by the excision of the thymine dimers.

There are two types of excision repair: *base-excision repair* and *nucleotide-excision repair*. **Base-excision repair (BER)** corrects damage to nitrogenous bases created by spontaneous hydrolysis or by agents that chemically alter them. As shown in Figure 15–14, the first step of the BER pathway in *E. coli* involves recognition of the chemically altered base by **DNA glycosylases**, which are specific to different types of DNA damage. For example, the enzyme uracil-DNA glycosylase recognizes the unusual presence of uracil when it is part of a nucleotide in DNA. The enzyme first cuts the glycosidic bond between the base and the sugar, creating an apyrimidinic (AP) site. Such a sugar with a missing base is then recognized by an enzyme called **AP endonuclease**. The endonuclease makes a cut in the sugar backbone at the AP site. This creates a distortion in the DNA helix that is recognized by the excision-repair system, which is then activated, leading ultimately to the correction of the error.

Although much has been learned about glycosylases in *E. coli*, very little is known about DNA glycosylases in eukaryotes. In addition, unlike the nucleotide excision pathway we are going to discuss next, there is no human or animal disease model available that has a defective BER pathway, so it is difficult to assess the importance of BER in eukaryotes.

While base-excision repair recognizes and replaces modified bases in DNA, the **nucleotide-excision repair (NER)** pathway repairs "bulky" lesions in DNA that alter or distort the double helix, such as the UV-induced pyrimidine dimers discussed previously. The NER pathway, diagrammed in Figure 15–15, was first discovered in *E. coli* by Paul Howard-Flanders and coworkers who were able

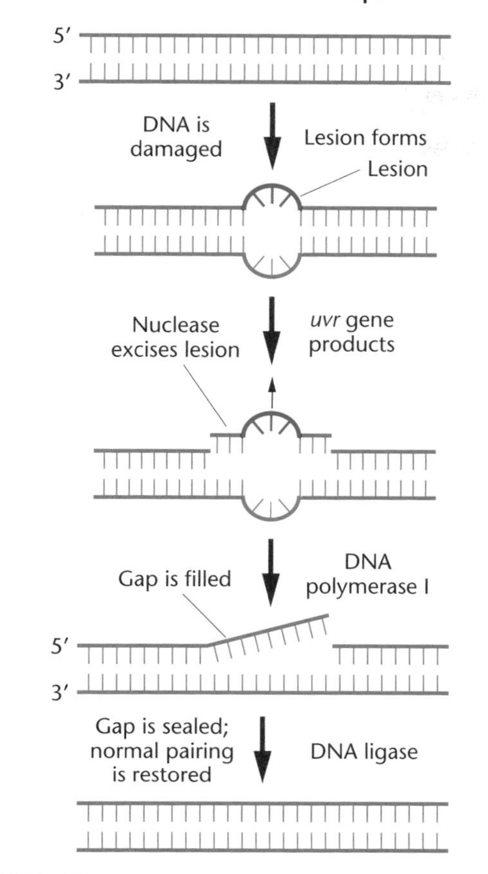

FIGURE 15–15 Nucleotide-excision repair (NER) of a UV-induced thymine dimer. In actuality, 13 bases are excised in prokaryotes and 28 bases are excised in eukaryotes during repair.

to isolate several independent mutants that demonstrated sensitivity to UV radiation. One group of genes was designated *uvr* (ultraviolet repair) and included the *uvrA*, *uvrB*, and *uvrC* mutations. In the NER pathway, the *uvr* gene products recognize and clip out the lesions in the DNA. Usually, a very specific number of nucleotides are clipped out around both ends of the lesion. In *E. coli*, there are usually a total of 13 nucleotides removed, which include the lesion. The repair is then completed by DNA polymerase I and DNA ligase, in a manner similar to BER. The undamaged strand opposite the lesion is used as a template for the replication, resulting in repair.

Xeroderma Pigmentosum and Nucleotide-Excision Repair

The mechanism of nucleotide-excision repair (NER), which has been elucidated in eukaryotes, is much more complicated than prokaryotic NER because it involves many more proteins. Much of what is known about this repair system in humans has resulted from detailed studies of individuals with **xeroderma pigmentosum (XP)**, a rare genetic disorder that predisposes individuals to severe skin abnormalities. These individuals have lost their ability to undergo NER; as a result, individuals suffering from XP who are exposed to the UV radiation present in sunlight exhibit reactions that range from initial freckling and skin ulceration to the subsequent development of skin cancer. Figure 15–16 shows two XP individuals, one of whom had early diagnosis of XP and was protected from sunlight throughout her life.

The condition is very severe and can be lethal, although early detection and protection from sunlight can arrest it. Because sunlight contains UV radiation, a causal relationship had been predicted between XP and the production of thymine dimers. Thus, the ability to repair UV-induced lesions was investigated in human fibroblast cultures derived from XP and normal individuals. (Fibroblasts are undifferentiated connective tissue cells.) The results suggested that the XP phenotype may be caused by more than one mutant gene.

In 1968, James Cleaver showed that cells from XP patients were deficient in **unscheduled DNA synthesis**—DNA synthesis other than that occurring during chromosome replication—which is elicited in normal cells by UV radiation. This type of synthesis was thought to represent the activity of the excision-repair system, and the suggestion was that XP cells are deficient in excision repair.

The link between xeroderma pigmentosum and inadequate excision repair has been strengthened by the use of **somatic cell hybridization** studies. Cultured fibroblast cells from any two unrelated XP patients are induced to fuse, forming a heterokaryon where the two nuclei

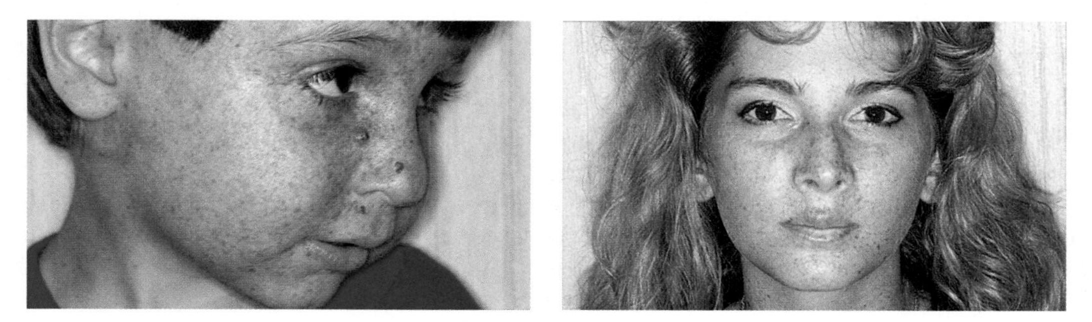

FIGURE 15–16 Two individuals with xeroderma pigmentosum. The 4-year-old boy on the left shows marked skin lesions induced by sunlight. Mottled redness (erythema) and irregular pigment changes in response to cellular injury are apparent. Two nodular cancers are present on his nose. The 18-year-old girl on the right has been carefully protected from sunlight since her diagnosis of xeroderma pigmentosum in infancy. Several cancers have been removed and she has worked as a successful model. *(W. Clark Lambert, M.D., Ph.D., University of Medicine & Dentistry of New Jersey)*

share a common cytoplasm. After fusion, excision repair, as assayed by unscheduled DNA synthesis, may be reestablished in the heterokaryon. When this occurs, the two variants are said to demonstrate **complementation**. Alone, neither cell type demonstrates excision repair, but together (in a heterokaryon) the process does occur. In genetic terms, this is strong evidence that in the two patients from whom the cells were derived, different affected genes led to the disease. Complementation occurs because the heterokaryon has at least one normal copy of each gene.

Based on many studies, patients have been divided into seven complementation groups, suggesting that at least seven different genes may be involved in excision repair. Any two strains in different complementation groups will complement one another, but if two strains contain separate mutations that are in the same complementation group (gene), no complementation occurs.

These seven human genes, and their protein products, have now been identified. The genes are found in disparate regions of the genome, and a homologous gene for each has been identified in yeast. Approximately 20% of XP patients do not fall into any of the seven groups. They manifest similar symptoms, but their fibroblasts do not demonstrate defective excision repair. There is some evidence that instead, their fibroblasts are less efficient in normal DNA replication.

As a result of the study of the defective genes in xeroderma pigmentosum, a great deal is known about how NER counteracts DNA damage in normal cells. The first step in humans is the recognition of the damaged DNA by a specific protein called XPA (*X*erodema *P*igmentosum gene *A*). When XPA binds to the damaged DNA helix, it causes other proteins in the pathway to be recruited to the site. The transcription factor TFIIH is involved in creating the repair complex, which excises a 28-nucleotide-long fragment from the helix, including the lesion. Recall that NER in bacteria is also quite specific in the number of nucleotides excised, although less so than in eukaryotes.

Proofreading and Mismatch Repair

MEDIA TUTORIAL DNA repair: proofreading, mismatch, and SOS

One of the most common types of error in DNA is the one made during replication when an incorrect (noncomplementary) nucleotide is inserted by DNA polymerase. The enzyme in bacteria (DNA polymerase III) is known to make such an error approximately once every 100,000 insertions, leading to an error rate of 10^{-5}. Fortunately, the enzyme polices its own synthesis by **proofreading** each step, catching 99% of those errors. During polymerization, when an incorrect nucleotide is inserted, the enzyme complex has the potential to recognize the error and reverse its direction and behave as an exonuclease, cutting out the incorrect nucleotide and then replacing it. This improves the efficiency of replication 100-fold, creating only $1/10^7$ mismatches immediately following DNA replication, for a final error rate of 10^{-7}.

To cope with errors that remain after proofreading, **mismatch repair** may be activated. This mechanism was proposed over 20 years ago by Robin Holliday, and the molecular basis of this process is now well-established. As in other DNA lesions, the alteration or mismatch must be detected, the incorrect nucleotide(s) removed, and replacement with the correct nucleotide(s) must occur. But a special problem exists with correction of a mismatch. How does the repair system recognize which strand is correct (the template) and which strand contains the mismatched base (the newly synthesized strand)? How the repair system discriminates and recognizes the "new" nucleotide puzzled geneticists for decades. Statistically, if the mismatch was recognized but no discrimination occurred and excision was random, the strand bearing the correct base would be clipped out 50% of the time. The concept of strand discrimination by a repair enzyme is thus a critical step.

This process has been elucidated at least in some bacteria, including *E. coli*, and is based on **DNA methylation**. These bacteria contain an enzyme, adenine methylase, that recognizes the DNA sequence

$$5' \ldots \text{GATC} \ldots 3'$$

$$3' \ldots \text{CTAG} \ldots 5'$$

as a substrate. Upon recognition, a methyl group is added to each of the adenine residues. This modification is stable throughout the cell cycle.

Following a further round of replication, the newly synthesized strands remain temporarily unmethylated. It is at this point that the repair enzyme recognizes the mismatch and preferentially binds to the unmethylated strand. A nick is made by an endonuclease protein, either 5′ or 3′, to the mismatch on the unmethylated strand. The nicked DNA strand is then unwound, degraded, and replaced until the mismatch is reached and excised. The proteins involved in this process are slightly different depending on which side of the mismatch the nick was made.

A series of *E. coli* gene products, MutH, MutL, MutS, and MutU are involved in the discrimination step. Mutations in each gene result in strains deficient in mismatch repair. While the mechanism described here is based on studies of *E. coli*, similar mechanisms involving homologous proteins are known to exist in yeast and mammals.

Post-Replication Repair and the SOS Repair System

Still other types of repair have been discovered, illustrating the diversity of mechanisms that have evolved to overcome DNA damage. One system, **post-replication repair**, was discovered in an excision-defective strain of *E. coli* and first proposed by Miroslav Radman. This system responds *after* damaged DNA has escaped repair and failed to be completely replicated, hence its name. As shown in Figure 15–17, when DNA bearing a lesion of some sort (such as a pyrimidine dimer) is being replicated, DNA polymerase at first stalls at the lesion and then skips

Post-replication repair

Step 1. Lesion present in DNA unwound prior to replica

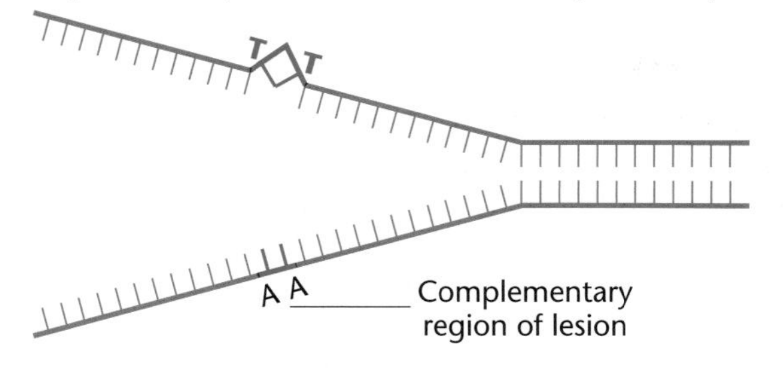

Step 2. Replication skips over lesion and continues

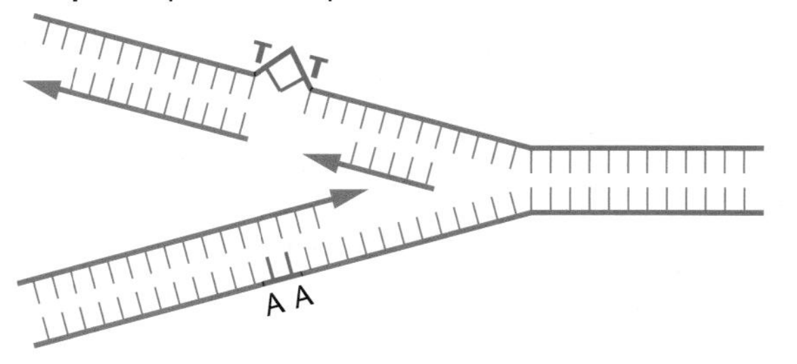

Step 3. Undamaged complementary region of parental strand is recombined

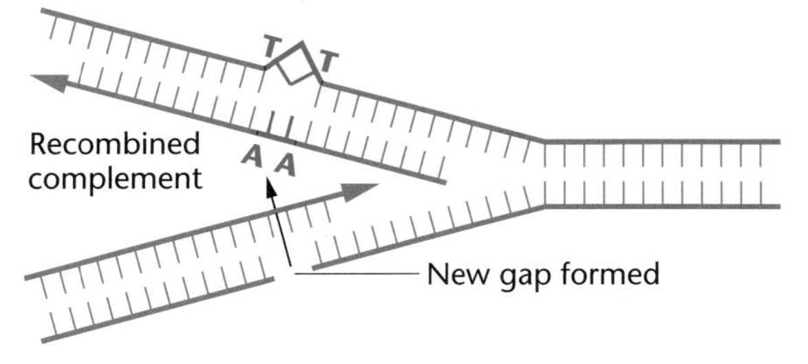

Step 4. New gap is filled by DNA polymerase and DNA

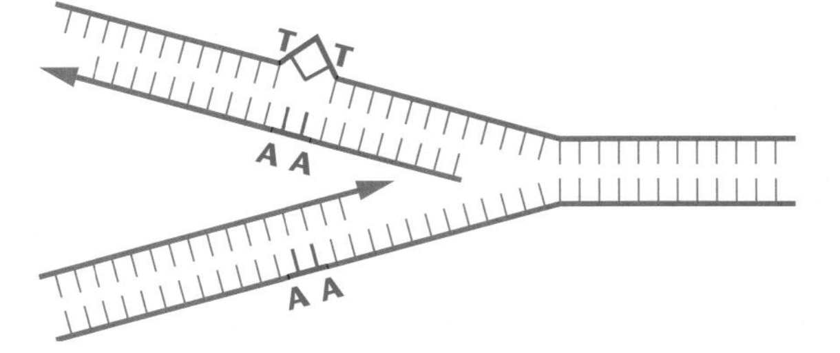

FIGURE 15–17 Post-replication repair that occurs if DNA replication has skipped over a lesion, such as a thymine dimer. Through the process of recombination, the correct complementary sequence is recruited from the parental strand and inserted into the gap opposite the lesion. The new gap created is filled by DNA polymerase and DNA ligase.

over it, leaving a gap along the newly synthesized strand. To counteract this, the RecA protein directs a recombinational exchange process whereby this gap is filled as a result of the insertion of a segment initially present on the undamaged complementary strand. This event creates a gap on the "donor" strand, which can then be filled by repair synthesis as replication proceeds. Because a recombinational event is involved, post-replication is also referred to as *homologous recombination repair*, a more general category.

Another repair pathway in *E. coli* is the **SOS repair system**, which responds to damaged DNA in a different way. Instead of skipping over a lesion during DNA synthesis, as in post-replication repair, this system creates conditions that allow DNA polymerase to continue replication across the lesion. While no gap is created, the fidelity of replication is compromised, and the system is described as being *error-prone*. Phil Hanawalt and Paul Howard-Flanders, among others, have established that as many as 20 different genes (including *lexA*, *recA*, and *uvr*) may be induced by DNA damage, so that their products become involved in the SOS response.

Double-Strand Break Repair in Mammals

Thus far, we have discussed only repair pathways that deal with damage *within* a particular strand of DNA. We conclude our discussion of DNA repair by considering what happens when both strands of the DNA helix are cleaved, say, as a result of exposure to ionizing radiation. This causes what are called double-strand breaks. This damage occurs in bacteria as well, but we will focus our discussion on eukaryotic cells. Fortunately, a specialized process, the DNA **double-strand break repair** (**DSB repair**) pathway, is activated and is responsible for ultimately reannealing the two DNA segments. Recently, interest has grown in DNA DSB repair because defects in this pathway are associated with X-ray hypersensitivity and immune deficiency. Such defects may also underlie familial disposition to breast and ovarian cancer.

Similar to post-replication repair, one pathway involved in double-strand break repair is referred to as **homologous recombinational repair**—damaged DNA is actually recombined and replaced with homologous undamaged DNA. This is necessary because when both strands are broken, there is no undamaged template strand available to use as the source of the complementary DNA sequence during repair. Therefore, the genetic information present in the homologous region of either a sister or nonsister homolog is "recruited" to replace the damaged double-stranded break. The undamaged homologous region is then recombined into the damaged DNA molecule. The process usually occurs during the late S/G2 phase of the cell cycle, after DNA replication, thus allowing recruitment of the undamaged copy of the "sister" chromosome to be utilized. In yeast, the system depends upon a complex of at least five proteins, called the RAD52 complex. As a result of homologous recombinational repair, the complex is able to restore the integrity of the broken double helix.

A second pathway, called **nonhomologous recombinational repair** or **end joining**, achieves the same type of repair. However, as the name implies, the mechanism does not recruit a homologous region of DNA during repair. This system is activated in G1, prior to DNA replication, and two different protein complexes are involved.

15.9 Site-Directed Mutagenesis Allows Researchers to Investigate Specific Genes

A useful experimental technique, **site-directed mutagenesis**, has enabled researchers to introduce a designed mutation at a prescribed site within a gene of interest. The technique relies on the availability of a cloned gene and utilizes a number of manipulations involving recombinant DNA technology (see Chapter 17).

The goal of this technique is to alter one or more specific nucleotides within a gene to change a specific triplet codon. Upon transcription and translation, this change causes the insertion of a "mutant" amino acid into the protein encoded by the original gene. Designed mutations are particularly useful in studying the effects of genetic change on protein function. Various approaches can be used to accomplish the goal. Most are variations on the theme we are about to describe (Figure 15–18).

The initial step is to determine the nucleotide sequence of the gene being studied (Step 1). This can be accomplished by DNA sequencing techniques, or it can be predicted if the amino acid sequence of the protein is known by utilizing our knowledge of the genetic code.

The next step is to isolate one of the two complementary strands from the DNA of known sequence and decide which nucleotides are to be changed (Step 2). Then, a synthetic oligonucleotide (a small piece of DNA) complementary to that region at all points except in the triplet sequence or sequences that are to be altered is chemically synthesized (Step 3). This sequence includes the triplet encoding the amino acid that will be changed in the protein.

In Step 4, the synthetic oligonucleotide is hybridized with the original parent strand, forming a partial duplex because of the complementarity along most of its length. If DNA polymerase and DNA ligase are then added to this hybrid complex, the short sequence is extended so that a duplex of the entire gene is formed; the two strands are now perfectly complementary except at the point of alteration.

If this DNA is replicated (Step 5), two types of duplexes form: One is like the original, unaltered gene and the other contains the newly designed sequence. Recombinant DNA technology makes it possible not only to complete the above manipulations but also to allow the altered gene to be expressed so that large amounts of the desired protein are available for study. Or, as we shall see in the next section, the altered gene can be introduced into an organism and studied.

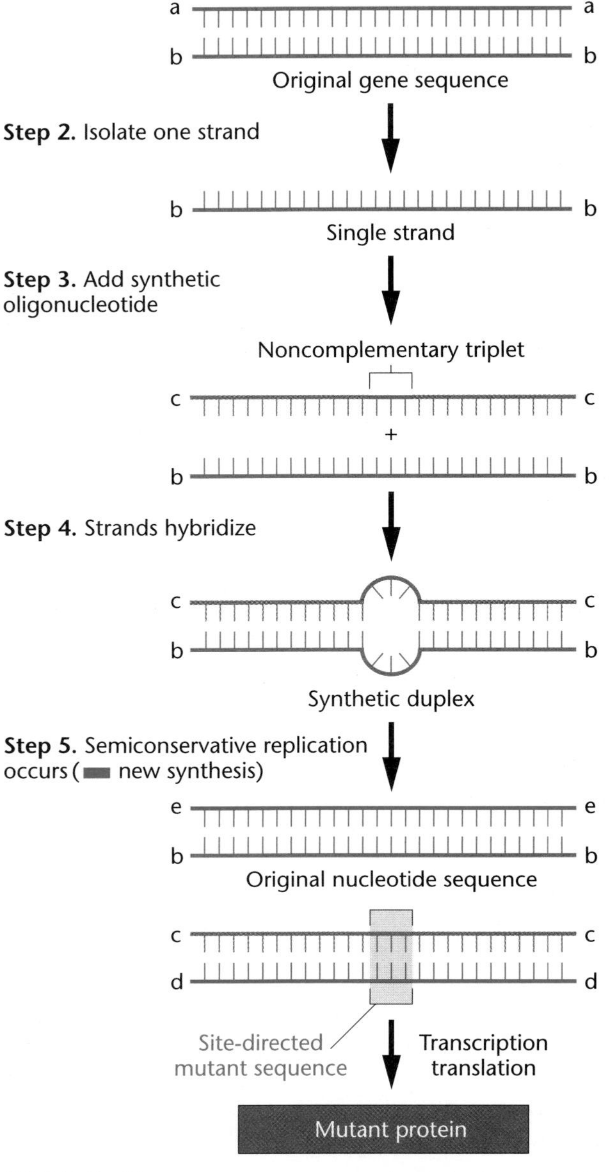

FIGURE 15–18 Site-directed mutagenesis. A single strand of DNA from a gene of interest is initially isolated. This is hybridized with a synthetic oligonucleotide containing a triplet altered to encode an amino acid of choice. After semiconservative replication, a different complementary base pair is present in one of the new duplexes. Upon transcription and translation, a mutant protein that was "designed" in the laboratory will be produced.

Knockout Genes and Transgenes

One of the most useful applications of site-directed mutagenesis is to alter a gene sufficiently so as to render it nonfunctional. (Recall that such a "loss-of-function" condition is referred to as a **null allele**.) Either deletions within the gene are created, or a disruptive sequence is inserted. Techniques have been developed where altered genes can be inserted into the germline of an organism, allowing for the assessment of the gene's function. When insertion into the genome involves *replacement* of the comparable gene of the organism from which it originated, the process is a **gene knockout**. The organism now contains a *knockout mutation*, and the genetic alteration creates a **knockout organism**.

For example, a knockout mouse is a member of a true-breeding strain that lacks function of the gene that has been replaced with a null allele. Knockout mice serve as models for studying some human genetic disorders. In such cases, the mouse gene comparable to the gene that causes the human disorder is isolated, subjected to site-directed mutagenesis, and used to replace the normal mouse gene. Cystic fibrosis and Duchenne muscular dystrophy have been investigated in this way. In mice, the application of gene knockout techniques has been particularly fruitful in studies involving the genetic control of early development and behavior.

When a gene is inserted into an organism *in addition* to its normal copies, it is called a **transgene**, and the organism is called a **transgenic organism** (e.g., a transgenic plant). In such cases, the gene may have undergone site-directed mutagenesis or it may be a foreign gene isolated from another organism. This technology has been used more extensively than gene knockout since it is easier to accomplish because specific replacement is not required.

The study of transgenic organisms is performed routinely in plants, *Drosophila*, mice, and a variety of other organisms. Of particular note is the potential provided in agricultural studies, as well as gene therapy techniques in our own species.

15.10 Transposable Genetic Elements Move Within the Genome and May Disrupt Genetic Function

We conclude this chapter by discussing the phenomenon of **transposable genetic elements**, sometimes referred to as **transposons** or **Tn elements**, genetic units that can move or be transposed within the genome. It is appropriate to discuss Tn elements here because the movement of genetic units from one place in the genome to another often disrupts genetic function and results in phenotypic variation. As such, the impact of the transposition of genetic units often fits into a broad definition of mutation.

Barbara McClintock first studied transposons in maize almost 50 years ago. However, even in the 1950s and 1960s the idea that genetic information was *not* fixed within the genome of an organism was slow to find acceptance. Such a notion was quite alien to the classical interpretation of genes on chromosomes. It was not until other transposable genetic elements were discovered—and their molecular basis revealed—that the phenomenon was found to be nearly universal.

Insertion Sequences

Although the presence of transposable elements in maize had been predicted earlier, the first observation at the molecular level did not come until the early 1970s. A number of independent researchers, including Peter Starlinger and James Shapiro, visualized a unique class of mutations affecting different genes in various bacterial strains. For example, the expression of a cluster of related genes involving galactose metabolism in *E. coli* was repressed as a result of one such mutation. The phenotypic effect was heritable, but was found not to be caused by a base-pair change characteristic of conventional gene mutations. Instead, it was shown that a short, specific DNA segment had been inserted into the bacterial chromosome at the beginning of the galactose gene cluster. When this segment was excised spontaneously from the bacterial chromosome, wild-type function was restored. Such a segment came to be called an **insertion sequence** (**IS**).

It was subsequently revealed that several other distinct DNA segments could behave in a similar fashion, by inserting themselves into the chromosome and affecting gene function. These DNA segments are relatively short and do not exceed 2000 bp (2 kb). The first insertion sequence to be characterized in *E. coli*, IS1, is about 800 bp long; IS2, IS3, IS4, and IS5 are about 1250–1400 bp in length.

Analysis of the DNA sequences of most IS units reveals a feature important to their mobility. The nucleotide sequences contain **inverted terminal repeats** (**ITRs**) of one another; that is, the final sequence at the 5′-end of one strand is the same as the final sequence at the 5′-end of the other strand, except that the sequences run in opposite directions (Figure 15–19). Although the figure shows the terminal-repeating unit to consist of only a few nucleotides, many more are actually involved. For example, in *E. coli* the IS1 termini contain about 20 nucleotide pairs, IS2 and IS3 about 40 pairs, and IS4 about 18 pairs. It seems likely that these terminal sequences are an integral part of the mechanism of the insertion of IS units into DNA. The fact that insertion of IS units is more likely to occur at certain DNA regions than at others suggests that IS terminals can recognize certain target sequences in the DNA during the process of insertion.

Careful investigation has revealed IS units in the wild-type *E. coli* chromosome as well as in the plasmids of bacterial DNA. Thus, their presence does not always result in mutation. In the *E. coli* chromosome, five or more copies of IS1, IS2, and IS3 are present—the exact number of copies varies, depending on the strain examined.

Bacterial Transposons

In addition to their potential mutational effects, IS units play an even more significant role in the formation and movement of the larger *transposon Tn elements*. Transposons in bacteria consist of IS units that contain within their internal DNA sequence genes whose functions are unrelated to the insertion process. Like IS units, Tn elements are mobile in

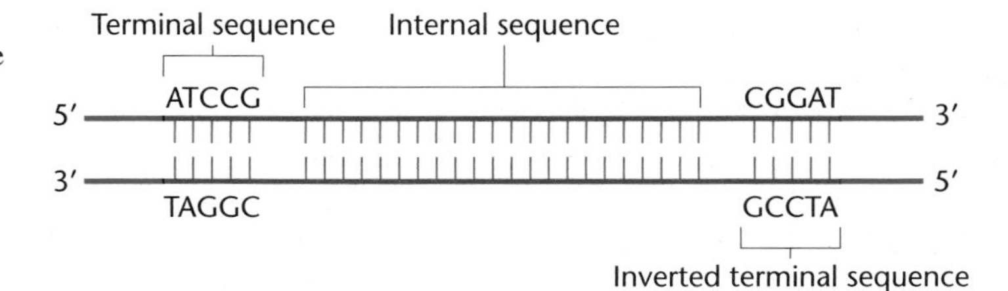

FIGURE 15–19 An insertion sequence (IS), shown in red. The terminal sequences are perfect inverted repeats of one another.

both bacterial and viral chromosomes and in plasmids. The Tn elements provide a mechanism for movement of genetic information from place to place both within and between organisms. Transposons were first discovered to move between DNA molecules as a result of observations of antibiotic-resistant bacteria. In the mid-1960s, Susumu Mitsuhashi first suggested that the genes responsible for resistance to several antibiotics were mobile and could move between bacterial plasmids and chromosomes.

Transposons have become the focus of increased interest, particularly because they have been found in organisms other than bacteria. Bacteriophages that demonstrate the ability to insert their genetic material into the host chromosome behave in a similar fashion. The bacteriophage mu, with over 35,000 nucleotides, can insert its DNA at various places in the *E. coli* chromosome. Like IS units, if insertion occurs within a gene, mutant behavior at that locus results. Transposons have also been discovered in higher organisms, including yeast, corn, *Drosophila*, and humans.

The *Ac–Ds* System in Maize

With our knowledge of insertion sequences and transposons in bacteria, it is no surprise that mobile genetic units also exist in eukaryotes. They are, in fact, more widespread, and like bacteria, they often have the effect of altering the expression of genetic information.

About 20 years before the discovery of transposons in prokaryotic organisms, Barbara McClintock analyzed the genetic behavior of two mutations, *Dissociation* (*Ds*) and *Activator* (*Ac*), in corn plants (maize). She examined the phenotypes of maize kernels that resulted from genes expressed in either the endosperm or aleurone layers (Figure 15–20) and correlated her observations with a cytological examination of the maize chromosomes. Initially, McClintock determined that *Ds* is located on chromosome 9. If *Ac* is also present in the genome, *Ds* induces breakage at a point on the chromosome adjacent to its own location. If breakage occurs in somatic cells during their development, progeny cells often lose part of chromosome 9, causing a variety of phenotypic effects.

Subsequent analysis suggested to McClintock that both the *Ds* and *Ac* genes are sometimes transposed to different chromosomal locations. While *Ds* moves only if *Ac* is also present, *Ac* is capable of autonomous movement. The location at which *Ds* comes to reside determines its genetic effect. That is, it might cause chromosome breakage or it might inhibit gene expression. In cells where gene expression is inhibited, *Ds* might move yet again, thus releasing this inhibition. In these cases, the *Ds* element is believed to insert into a gene and subsequently to depart from it, causing changes in gene expression.

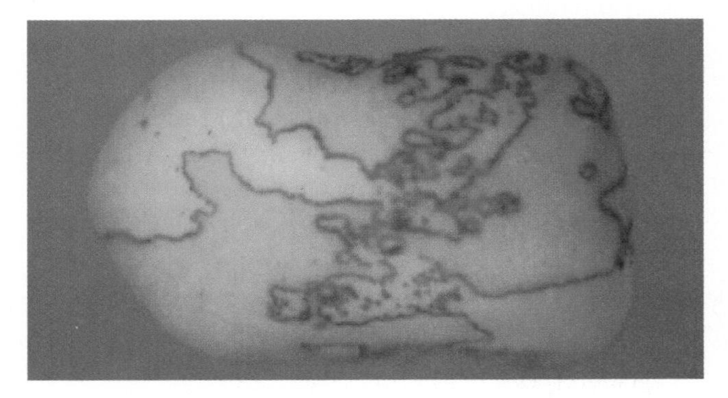

FIGURE 15–20 Maize kernel showing spots of colored aleurone produced by genetic transposition involving the *Ac–Ds* system. *(Dr. Nina Federoff/Department of Embryology, Carnegie Institute of Washington)*

Figure 15–21 shows the sort of movements and effects of the *Ds* and *Ac* elements just described. In McClintock's original observation, when the *Ds* element jumped out of chromosome 9, this excision event restored normal gene function. In maize cells, pigment synthesis was restored. McClintock concluded that the *Ds* and *Ac* genes are *transposable controlling elements*.

It was not until many years later that anything comparable to transposable controlling elements in maize was recognized in other organisms. When bacterial insertion sequences and transposons were discovered, many parallels were evident. Transposons and insertion sequences were seen to move into and out of chromosomes, to insert at different positions, and to affect gene expression at the point of insertion.

Several *Ac* and *Ds* elements have been isolated and carefully analyzed, and the relationship between the two elements has been clarified (Figure 15–22). One *Ac* element sequence is 4563 bp long and strikingly similar to one of the known bacterial transposons. This sequence contains two 11-bp imperfect ITRs, two **open reading frames (ORFs)**, and three noncoding regions. Open reading frames contain initiation and termination sequences and are considered to encode gene products.

The first *Ds* element studied (*Ds-a*) is nearly identical in structure to *Ac* except for a 194-bp segment that has been deleted from the largest open reading frame. There is some evidence that this gene encodes a **transposase enzyme**, essential to transposition of both the *Ac* and *Ds* elements. The deletion of part of this gene in the *Ds-a* element explains its dependence on the *Ac* element for transposition. Several other *Ds* elements have also been sequenced, and each reveals an even larger deletion in the same region. In each case, however, the terminal repeats are retained and seem to be essential for transposition, provided that a functional transposase enzyme is supplied by the gene in the *Ac* element.

Although the validity of Barbara McClintock's proposed mobile elements was questioned by other researchers, molecular analysis has since verified her conclusions. Barbara McClintock was awarded the Nobel Prize in Physiology or Medicine in 1983 for her work.

Copia and P Elements in *Drosophila*

As noted above, transposable genetic elements have been discovered in other eukaryotic organisms, such as yeast, *Drosophila*, and primates (including humans). In 1975, David Hogness and his colleagues, David Finnigan, Gerald Rubin, and Michael Young, identified a class of genes in *Drosophila melanogaster* that they designated as ***copia* elements**. These

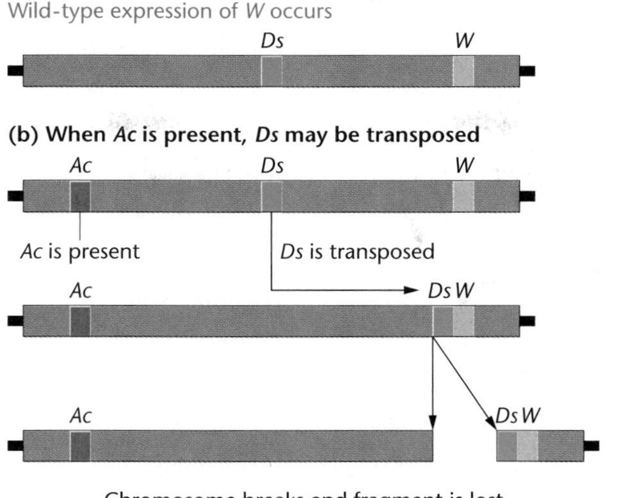
(a) In absence of Ac, Ds is not transposable
Wild-type expression of W occurs
Ds
W
(b) When Ac is present, Ds may be transposed
Ac
Ds
W
Ac is present
Ds is transposed
Ac
Ds W
Ac
Ds W

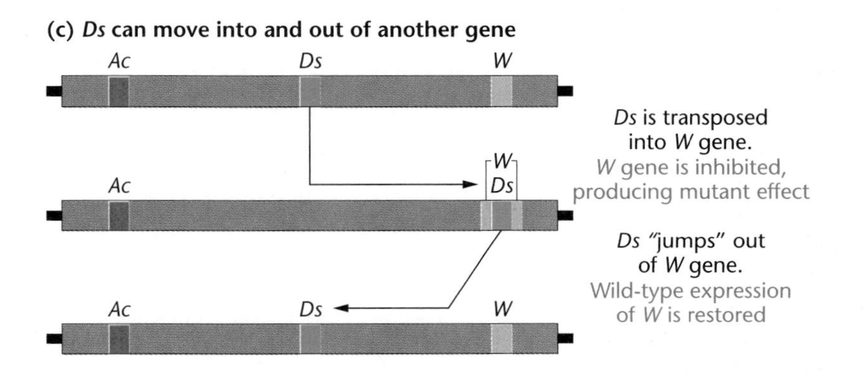
Chromosome breaks and fragment is lost
Expression of W ceases, producing mutant effect
(c) Ds can move into and out of another gene
Ac
Ds
W
Ds is transposed
into W gene.
W gene is inhibited,
producing mutant effect
W
Ds
Ac
Ds "jumps" out
of W gene.
Wild-type expression
of W is restored
Ac
Ds
W

FIGURE 15–21 Consequences of the influence of the *Ac* element on the *Ds* element. (b) In the presence of *Ac*, *Ds* is transposed to a region adjacent to a theoretical gene *W*. Subsequent chromosomal breakage is induced, the *W*-bearing segment is lost, and no gene expression occurs. (c) *Ds* is transposed to a region within the *W* gene, causing immediate mutant expression. *Ds* may also jump out of the *W* gene, and *W* gene activity and its wild-type expression are restored.

genes transcribe copious amounts of RNA (hence their name). *Copia* elements are present up to 30 times in the genome of cells and are nearly identical in nucleotide sequence. Mapping studies show that they are transposable to different chromosomal locations and are dispersed throughout the genome.

Copia genes appear to be only 1 of approximately 30 families of transposable elements in *Drosophila*, each of which is present 20–50 times in the genome. Together, these families constitute about 5% of the *Drosophila* genome and over 50% of the middle repetitive DNA of this organism. One estimate projects that 50% of all visible mutations in *Drosophila* are the result of the insertion of transposons into otherwise wild-type genes!

Despite the variability in DNA sequence between the members of different families, they share a common structural

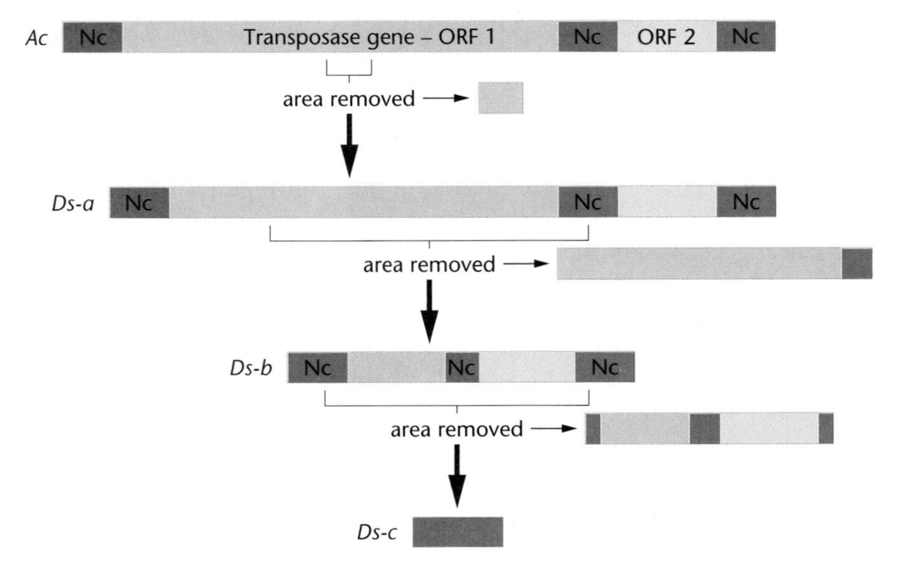
Ac
Nc
Transposase gene – ORF 1
Nc
ORF 2
Nc
area removed
Ds-a
Nc
Nc
Nc
area removed
Ds-b
Nc
Nc
Nc
area removed
Ds-c

FIGURE 15–22 A comparison of the structure of an *Ac* element with three *Ds* elements, all of which have been isolated and sequenced. The imperfect inverted repeats are at the ends of the *Ac* element. The transposase gene is in an open reading frame (ORF 1). No function has yet been assigned to ORF 2. Noncoding regions are designated Nc. As this scheme shows, *Ds-a* appears to be an *Ac* element containing a small deletion in the gene encoding the transposase enzyme.

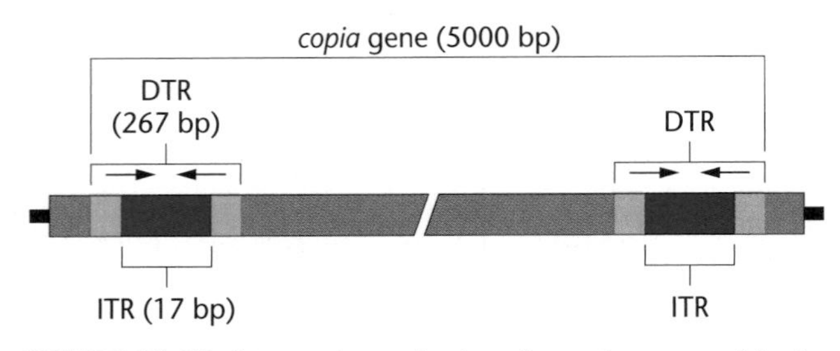

FIGURE 15–23 Structural organization of a *copia* transposable element in *Drosophila melanogaster*, showing the terminal repeats.

organization thought to be related to the insertion and excision processes of transposition. Each *copia* gene consists of approximately 5000–8000 bp of DNA, including a long family-specific **direct terminal repeat (DTR)** sequence of 267 bp at each end. Within each repeat is a short ITR of 17 bp. These features are shown in Figure 15–23. The DTR sequences are found in other transposons in other organisms but are not universal. However, the shorter ITR sequences are considered universal.

Insertion of *copia*, as with other transposons, appears to be dependent on ITR sequences and also appears to occur at specific target sites in the genome. In general, eukaryotic transposons are strikingly similar to one another and share many features with those in bacteria.

Still another interesting category of transposable elements in *Drosophila* is the family of **P elements**. These were discovered while studying the phenomenon of **hybrid dysgenesis**, a condition that causes sterility, elevated mutation rate, and chromosome rearrangement in the offspring of crosses between certain strains of fruit flies.

Hybrid dysgenesis is due to high rates of P-element transposition in the germline, where these mobile DNA elements insert themselves into or near genes, thereby causing mutations. P elements are 2.9 kb long, with 31-bp ITRs. The elements encode at least two proteins, one of which is the transposase enzyme that is required for transposition. The enzyme is expressed only in the germline, accounting for the tissue specificity of P-element transposition.

P elements are useful as vectors to introduce transgenes into *Drosophila*. A common method of generating P-element-induced transformants is to inject *Drosophila* embryos with a mixture of 2 P elements. One P element lacks the transposase gene and cannot move by itself, while the other P element contains the transposase gene but cannot insert itself into the genome. The two injected P elements are taken up by the embryo's cells, transposase will be expressed in the germline of the fly, and the P element inserts itself; into the fly's DNA. The next generation of flies will contain stable copies of P-element DNA, including the transgene, inserted at random locations. Researchers then screen the flies for mutant phenotypes of interest, isolate the fly's DNA, and clone the gene that was mutated. This technology has been used to identify genes involved in *Drosophila* development, behavior, and regulation of gene expression.

Mutations can arise from several kinds of insertional events. If a P element inserts itself into the coding region of a gene, it can destroy the normal gene product. If it is inserted into the promoter region of a gene, it can affect the level of expression of the gene. Insertions into introns can affect splicing or cause premature termination of transcription.

Transposable Genetic Elements in Humans

The final class of transposable elements that we will discuss is represented by the ***Alu* family** of **short interspersed elements (SINEs)**, which are characteristic of moderately repetitive DNA in mammals. Human DNA contains some 500,000 copies of this 200–300-bp sequence in the genome. Originally detected using reassociation kinetic analysis (see Chapter 10), these transposable elements were so named because they contain specific nucleotide sequences that are cleaved by a restriction endonuclease called *Alu*I.

The *Alu* family is a significant set of elements in the genome—they have been found in the DNA of all the primates and rodents studied. Within the *Alu* elements in mammals, a specific 40-bp DNA sequence has been conserved. Also, *Alu* sequences are represented in some nuclear transcripts.

The *Alu* elements are considered to be transposable based on several lines of evidence. The most important is that their 200–300-bp sequence is flanked on either side by direct repeat sequences consisting of 7–20 bp, paralleling bacterial insertion sequences. These flanking regions are related to the insertion process during transposition. Second, clustered regions of *Alu* sequences vary in the DNA of both normal and diseased individuals and in different tissues of the same individual. In addition, the sequences have been found extrachromosomally.

The potential mobility and mutagenic effects of these and other elements have far-reaching implications, as can be seen in a recent example of a transposon "caught in the act." The case involves a male child with hemophilia. One cause of hemophilia is a defect in blood-clotting factor VIII, the product of an X-linked gene. Haig Kazazian and his colleagues found a transposable element much longer than *Alu* sequences inserted at two points within the gene. Also present elsewhere in the genome, this sequence was shown to be a **long interspersed element (LINE)**. Both LINEs and SINEs were discussed in detail in Chapter 12.

Researchers were very interested in determining if one of the mother's X chromosomes contained this specific LINE. If so, the unaffected mother would be heterozygous and had passed the LINE-containing chromosome to her son. The startling finding was that the LINE sequence is *not* present on either of her X chromosomes, but *was* detected on chromosome 22 of both parents. This suggests that this mobile element may have moved from one chromosome to another in the gamete-forming cells of the mother, prior to being transmitted into the son's germline.

Many questions remain concerning this and other transposable elements. What is their origin? Were they once some sort of retrovirus? Exactly how do they move, and what has been their role during evolution? These and other questions will intrigue researchers for many years to come.

GENETICS, TECHNOLOGY, AND SOCIETY

Chernobyl's Legacy

On April 26, 1986, Reactor 4 of the Chernobyl Nuclear Power Station exploded and ejected massive amounts of radioactive material into the surrounding countryside and throughout the Northern Hemisphere. The accident killed 31 emergency workers and caused acute radiation sickness in over 200 others. In the nine days following the initial explosion, as the reactor's temperature approached meltdown, radioactive fission products including radioactive iodine, xenon, strontium, and cesium, were released into the atmosphere. Fallout traveled through central Europe, reaching Finland and Sweden three days after the initial explosion and the United Kingdom, and North America within a week. Millions of people were exposed to measurable amounts of radioactivity. People living within 30 km (about 14 miles) of Chernobyl were exposed to high levels of radioactivity prior to their evacuation 36 hours after the accident. Equally high were the exposures of some of the 600,000 military and civilian workers who were sent to Chernobyl to decontaminate the area and encase the shattered reactor in a sarcophagus.

The Chernobyl incident was the world's largest accidental release of radioactive material. The question that remains is whether Chernobyl's radioactive pollution directly threatens the long-term health of millions of people.

More is known about the effects of ionizing radiation on human health than any other toxic agent (except perhaps cigarette smoke). Ionizing radiation (such as X-rays and gamma rays) is a subset of radiation that possesses sufficient energy to eject electrons from atoms. Ionizing radiation can damage any cellular component, alter nucleotides, and induce double-strand breaks in DNA. These DNA lesions can lead to mutations or chromosomal translocations.

High doses of ionizing radiation increase the risk of developing certain cancers. The survivors of the atomic-bomb blasts of Hiroshima and Nagasaki showed an increased incidence of leukemia within two years of the bombing. Breast cancers increased 10 years after exposure, as did cancers of the lung, thyroid, colon, ovary, stomach, and nervous system. Since ionizing radiation induces DNA damage, the offspring of the survivors were expected to show increases in birth defects, yet these increases were not detected.

The problem in extrapolating from Hiroshima to Chernobyl is the difference in dose. In atomic-bomb survivors, the cancer rate increased among people exposed to at least 200 mSv (mSv = millisievert, a unit dose of absorbed radiation). It is estimated that people living in the most contaminated areas close to Chernobyl may have received about 50 mSv, and some cleanup workers were exposed to approximately 250 mSv. Outside the Chernobyl area, radiation doses were estimated at 0.4–0.9 mSv in Germany and Finland, 0.01 mSv in the United Kingdom, and 0.0006 mSv in the United States.

To put these numbers in perspective, the average dose for medical diagnostic procedures (such as chest and dental X-rays) is 0.39 mSv per year. A person's radiation exposure from natural background sources (cosmic rays, rocks, and radon gas) is about 2–3 mSv per year. Smokers expose themselves to an average dose of about 2.8 mSv per year from the intake of naturally occurring radioactive materials in tobacco smoke.

It appears that mutation rates among plants and animals exposed to Chernobyl's radioactive waste may be 2- to 10-fold higher than normal. Also, Chernobyl cleanup workers exhibit a 25% higher mutation frequency at the *HPRT* locus. Mutation rates at minisatellites among children born in polluted areas are 2-fold higher than those from controls in the United Kingdom. But do these increased mutation rates translate into health effects?

Estimates of 26 leukemia cases higher than the 25–30 that would occur spontaneously were projected in the 115,000 people evacuated from Chernobyl. It was also estimated that up to 17,000 additional cancers might occur among Europeans, above the 123 million that would occur normally. At present, however, there have been no detectable increases in leukemias or solid tumors in contaminated areas of the former Soviet Union, Finland, or Sweden, or in the 600,000 Chernobyl cleanup workers.

Despite the lack of detectable increases in leukemias and solid tumors, one type of cancer has increased in the Chernobyl area. The rate of childhood thyroid cancer has reached over 100 cases per million children per year, whereas normal rates are expected to be between 0.5 and 3 cases per million children per year. Although epidemiologists debate whether these increases are entirely due to radiation or to increased reporting, the scale of the increase and the fact that radioactive isotopes of iodine made up a significant portion of the Chernobyl fallout make the link between thyroid cancer and Chernobyl's fallout a plausible one.

The most profound immediate effects of Chernobyl have been psychological. Chernobyl cleanup workers have demonstrated a 50% increase in suicides and a detectable increase in smoking- and alcohol-related disease. This mirrors other studies showing that 45% of people living within 300 km of Chernobyl believe that they have a radiation-induced illness. Health effects such as depression, sleep disturbance, hypertension, and altered perception have been documented. Posttraumatic stress may be a greater threat to health than the actual radiation exposure from the accident. People feel that they live in constant danger and are simply awaiting the results of a cancer "lottery." Even if cancer rates and genetic defects do not increase dramatically, the indirect health effects from the Chernobyl Nuclear Power Station explosion have been and continue to be immense.

References

Anspaugh, L. R., Catlin, R. J., and Goldman, M. 1988. The global impact of the Chernobyl reactor accident. *Science* 242: 1513–19.

Ginzberg, H. M. 1993. The psychological consequences of the Chernobyl accident—Findings from the International Atomic Energy Agency study. *Publ. Health Rep.* 108: 184–92.

Jacob, P. et al. 2000. Thyroid cancer risk in Belarus after the Chernobyl accident: Comparison with external exposures. *Rad. and Env. Biophys.* 39: 25–31.

Rahu, M. et al. 1997. The Estonian study of Chernobyl cleanup workers: II. Incidence of cancer and mortality. *Radiation Res.* 147: 653–57.

Williams, D. 1994. Chernobyl, eight years on. *Nature* 371: 556.

Website

UNSCEAR focuses on Chernobyl accident in General Assembly Report. June 6, 2000. [Press release of the United Nations Information Service.]
www.un.org/ha/chernobyl/unsceare.htm

Chapter Summary

1. The phenomenon of mutation not only provides the basis for most of the inherent variation present in living organisms but also serves as the working tool of the geneticist in studying and understanding the nature of genetic processes.
2. Mutations are distinguished by the tissues that are affected. Somatic mutations may affect the individual but are not heritable. Mutations arising in gametes may produce new alleles that can be passed on to offspring and can enter the gene pool.
3. Another classification of mutations relies on their effect. Morphological mutations, for example, may be detected visibly. Other types include biochemical, lethal, conditional, and regulatory mutations, groups that are not mutually exclusive.
4. Organisms in which mutations can be easily induced and detected are most often used in genetic studies. Viruses, bacteria, fungi, *Drosophila*, certain plants, and mice are frequently used because of these properties, as well as the fact that they have short life cycles.
5. Spontaneous mutations may arise naturally as a result of rare chemical rearrangements of atoms, or tautomeric shifts, and as the result of errors occurring during DNA replication. Although spontaneous mutations are very rare, their rates of occurrence may be increased experimentally by a variety of mutagenic agents.
6. Mutagenic agents, such as base analogs as well as alkylating and deaminating agents, cause chemical changes in nucleotides that alter their base-pairing affinities. As a result, base substitution mutations arise after DNA replication. Mutagenicity of chemicals can be assessed using the Ames test.
7. Frameshift mutations, induced specifically by acridine dyes, arise when the addition or deletion of one or more nucleotides (but not multiples of three) occurs.
8. Direct analysis of DNA from individuals with specific ABO blood types and muscular dystrophy has been informative. Complete loss of function, as in the case in blood type O and the Duchenne form of muscular dystrophy, occurs when deletions or duplications of nucleotides have shifted the reading frame during translation.
9. Another form of mutation, discovered in several human disorders, includes unstable trinucleotide repeats that increase in size during DNA replication and are inherited in this form through successive generations. Increasing numbers of repeats often correlate with severity and early onset of disease.
10. Ultraviolet light and high-energy radiation from gamma, cosmic, or X-ray sources are also potent mutagenic agents. UV light induces the formation of pyrimidine dimers in DNA, while high-energy radiation causes the ionization of molecules in its path and is more penetrating than UV.
11. Various forms of repair of DNA lesions, such as pyrimidine dimers, have been discovered, including photoreactivation, excision repair, mismatch repair, proofreading, and various forms of recombinational repair, including double-stranded break repair. The significance of repair mechanisms is apparent when, in humans, excision repair is abolished because of mutation. The result is the severe human disorder xeroderma pigmentosum.
12. Site-directed mutagenesis is a technique that allows researchers to create specific alterations in the nucleotide sequence of the DNA of genes.
13. The insertion sequences in bacteria and other transposable genetic elements in eukaryotes are mobile genetic units characterizing the genomes of all organisms. They have a profound effect on genetic expression, thus serving as a distinct category of mutagenic agent.

Key Terms

acridine dye, 329
alkylating agent, 329
Alu family, 344
Ames test, 334
2-amino purine (2-AP), 329
AP endonuclease, 336
apurinic site (AP site), 330
attached-X procedure, 325
auxotroph, 325
base analog, 329
base-excision repair (BER), 336
base substitution, 327
Becker muscular dystrophy (BMD), 333
behavior mutation, 324
5-bromouracil (5-BU), 329
chromosomal aberration, 323
complementation, 338
complete medium, 324
conditional mutation, 324
copia elements, 342
deamination, 330
direct terminal repeat (DTR), 344
DNA glycosylase, 336
DNA methylation, 338
double-strand break repair (DSB repair), 339
Duchenne muscular dystrophy (DMD), 333
electromagnetic spectrum, 331
end joining, 339
6-ethylguanine, 329
ethylmethane sulfonate (EMS), 329
excision repair, 336
fragile X syndrome, 334
frameshift mutation, 327
gene knockout, 340
gene mutation, 323
genetic anticipation, 333
hemophilia, 326
homologous recombinational repair, 339
Huntington disease, 334
hybrid dysgenesis, 344
induced mutation, 323
insertion sequence (IS), 341
inverted terminal repeat (ITR), 341
ionizing radiation, 331
knockout organism, 340
lethal mutation, 324
long interspersed element (LINE), 344
minimal culture medium, 324
mismatch repair, 338
myotonic dystrophy (DM), 334
nucleotide-excision repair (NER), 336
nonhomologons recombinational repair, 339
nutritional variation, 324
null allele, 340
open reading frame (ORF), 342
P element, 344
photoreactivation enzyme (PRE), 335

photoreactivation repair, 335
point mutation, 327
post-replication repair, 338
proofreading, 338
prototroph, 325
pyrimidine dimer, 331
regulatory mutation, 324
reverse mutation, 329
short interspersed element (SINE), 344
site-directed mutagenesis, 340
somatic cell hybridization, 337
SOS repair system, 339
spinobulbar muscular atrophy (Kennedy disease), 334
spontaneous mutation, 323
target theory, 332
tautomeric shift, 328
temperature-sensitive mutation, 324
transgene, 341
transgenic organism, 341
transition, 327
transposable genetic element, 341
transposase enzyme, 342
transposon, 341
Tn element, 341
transversion, 327
trinucleotide repeat sequence, 333
ultraviolet (UV) radiation, 331
unscheduled DNA synthesis, 337
xeroderma pigmentosum (XP), 337

INSIGHTS AND SOLUTIONS

1. How could you isolate a mutant strain of bacterial cells that is resistant to penicillin, an antibiotic that inhibits cell wall synthesis?

Solution: Grow a culture of bacterial cells in liquid medium, and plate the cells on agar medium to which penicillin has been added; only penicillin-resistant cells will reproduce and form colonies. Each colony will, in all likelihood, represent a cloned group of cells with the identical mutation. Then, isolate members of each colony. To enhance the chance of such a mutation's, arising, you might want to add a mutagen to the liquid culture.

2. The base analog 2-amino purine (2-AP) substitutes for adenine during DNA replication, but it may base-pair with cytosine. The base analog 5-bromouracil (5-BU) substitutes for thymine, but it may base-pair with guanine. Follow the double-stranded trinucleotide sequence shown here through three rounds of replication, assuming that in the first round, both analogs are present and become incorporated wherever possible. In the second and third round of replication, they are removed. What final sequences occur? The solution appears in the right-hand column.

3. A rare dominant mutation expressed at birth was studied in humans. Records showed that 6 cases were discovered in 40,000 live births. Family histories revealed that in 2 cases, the mutation was already present in one of the parents. Calculate the spontaneous mutation rate for this mutation.

Solution: Only four cases represent a new mutation. Because each live birth represents 2 gametes, the sample size is from 80,000 meiotic events. The rate is equal to

$$\frac{4}{80,000} = \frac{1}{20,000} = 5 \times 10^{-5}$$

2. Solution:

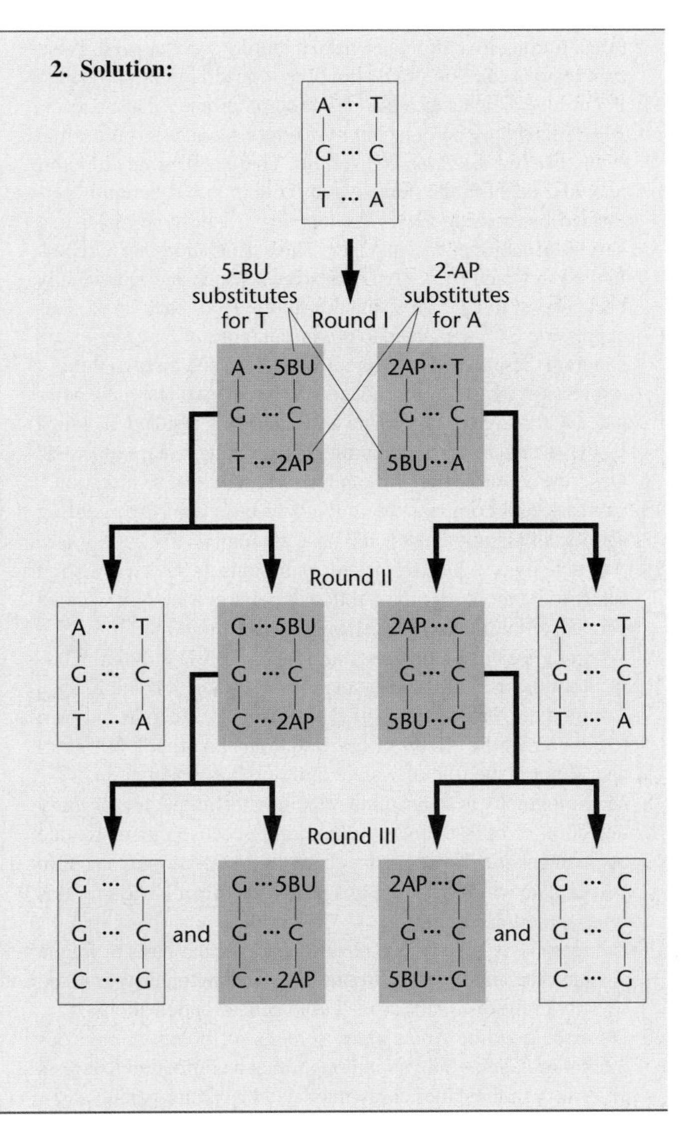

Problems and Discussion Questions

1. What is the difference between a chromosomal mutation and a gene mutation?
2. Discuss the importance of mutations to the successful study of genetics.
3. Describe the technique for the detection of nutritional mutants in *Neurospora*.
4. Most mutations are thought to be deleterious. Why, then, is it reasonable to state that mutations are essential to the evolutionary process?
5. Why is a random mutation more likely to be deleterious than beneficial?
6. Most mutations in a diploid organism are recessive. Why?
7. What is meant by a conditional lethal mutation?
8. Contrast the concerns about mutation in somatic and gametic tissue.
9. In the attached-X technique for detection of mutations in *Drosophila*, (a) explain the rationale behind detection. (b) What type of mutation may be detected?
10. In an experiment using the attached-X system, a student irradiated wild-type males and subsequently scored 500 F_1 cultures, finding that all males in each culture were normal. What conclusions can you draw about the induced mutation rate?
11. In *Drosophila*, induced mutations on chromosome 2 that are recessive lethals may be detected by using a second chromosomal stock, *Curly Lobe/Plum* (*Cy L /Pm*). These alleles are all dominant and lethal in the homozygous condition. Detection is performed by crossing *Cy L /Pm* females to wild-type males that have been subjected to a mutagen. Three generations are required. In the F_1 of each cross, *Cy L* males are selected and individually backcrossed to *Cy L /Pm* females. In the F_2 of each cross, flies expressing *Cy L* are mated to produce a series of F_3 generations. Diagram these crosses, and predict how an F_3 culture would vary if a recessive lethal were induced in the original test male compared with the case wherein no lethal mutation resulted. In which F_3 flies would a recessive morphological mutation be expressed?
12. Describe a tautomeric shift and how it may lead to mutation.
13. Contrast and compare the mutagenic effects of deaminating agents, alkylating agents, and base analogs.
14. Acridine dyes induce frameshift mutations. Is such a mutation likely to be more detrimental than a point mutation in which a single pyrimidine or purine has been substituted?
15. Why are X-rays a more potent mutagen than UV radiation?
16. Contrast the induction of mutations by UV radiation and X-rays.
17. Contrast the various types of DNA repair mechanisms known to counteract the effects of UV radiation and other DNA damage. What is the role of visible light in photoreactivation?
18. Mammography is an accurate screening technique for the early detection of breast cancer in humans. Because this technique uses diagnostic X-rays, it has been highly controversial. Can you explain why? What reasons justify the use of X-rays for this type of medical diagnosis?
19. Compose a short essay that relates the molecular basis of fragile X syndrome, myotonic dystrophy, and Huntington disease to the severity of the disorders, as well as to genetic anticipation.
20. Describe how the Ames assay screens for potential environmental mutagens. Why is it thought that a compound that tests positively in the Ames assay may also be carcinogenic?
21. Describe the general approach used in site-directed mutagenesis.
22. What genetic defect results in xeroderma pigmentosum (XP) in humans? How does this defect relate to the phenotype associated with this disorder?
23. Differentiate between point mutations that are transitions and those that are transversions. Using the DNA bases (A, T, C, and G), list the four types of transitions and the eight types of transversions.
24. In a bacterial culture in which all cells were unable to synthesize leucine (leu^-), a potent mutagen was added and the cells were allowed to undergo one round of replication. At that point, samples were taken, a series of dilutions was made, and the cells were plated on either minimal medium or leucine-enhanced minimal medium. The first culture condition (minimal medium) allowed the detection of mutations from leu^- to leu^+, while the second culture condition (minimal medium plus leucine) allowed the determination of the total cells, since all bacteria can grow. From the results in the table below, (a) determine the frequency of mutant cells. (b) What is the rate of mutation at the locus involved with leucine biosynthesis?

Culture Condition	*Dilution*	*Colonies*
Minimal medium	10^{-1}	18
Minimal medium + leucine	10^{-7}	6

25. Contrast the various transposable genetic elements in bacteria, maize, *Drosophila*, and humans. What properties do they share?
26. *Ty*, a transposable genetic element in yeast, has been found to contain an open reading frame (ORF) that encodes the enzyme reverse transcriptase. This enzyme synthesizes DNA from an RNA template. Speculate on the role of this gene product in the transposition of *Ty* within the yeast genome.
27. Hypothetical findings from studies of heterokaryons formed from seven human xeroderma pigmentosum cell strains are presented below. Complementing groups are marked +; noncomplementing groups are marked −. These data represent the occurrence of unscheduled DNA synthesis in the fused heterokaryon when neither of the strains alone showed synthesis. (a) What does unscheduled DNA synthesis represent? (b) Which strains fall into the same complementation groups? (c) How many different groups are revealed based on these limited data? (d) How do we interpret the presence of these complementation groups?

	XP1	*XP2*	*XP3*	*XP4*	*XP5*	*XP6*	*XP7*
XP1	−						
XP2	−	−					
XP3	−	−	−				
XP4	+	+	+	−			
XP5	+	+	+	+	−		
XP6	+	+	+	+	−	−	
XP7	+	+	+	+	−	−	−

28. Imagine yourself as one of the team of geneticists who launched the study of the genetic effects of high-energy radiation on the surviving Japanese population immediately following the atomic-bomb attacks at Hiroshima and Nagasaki in 1945. Demonstrate your insights into both chromosomal and gene mutation by outlining comprehensive short-term and long-term studies that address this topic. Be sure to include strategies for considering the effects on both somatic and germline tissues.
29. Cystic fibrosis (CF) is a severe autosomal recessive disorder in humans that results from a chloride ion–channel defect in epithelial cells. Over 500 sequence alterations have been identified in the 24 exons of the responsible gene *CFTR* (cystic fibrosis transmembrane regulator), including dozens of different missense mutations and frameshift mutations, as well as numerous splicing defects. Although all affected CF individuals demonstrate chronic obstructive lung disease, there is variation in pancreatic enzyme insufficiency (PI). Speculate which types of observed mutations are likely to give rise to less severe symptoms of CF, including only minor PI. A number of the 500 sequence alterations within the exon regions of the *CFTR* gene do not give rise to cystic fibrosis. Using your accumulated knowledge of the genetic code, gene expression, protein function, and mutation, explain to a freshman biology major how this might be.
30. Electrophilic oxidants are known to create the highly deleterious lesion in DNA named 7,8-dihydro-8-oxoguanine (oxoG). (Bruner, S. D. et al. 2000. Nature 403: 859–62.) Normally, guanine base-pairs with cytosine, but oxoG base-pairs with either cytosine or adenine. (a) What are the sources of reactive oxidants within cells that cause this type of lesion? (b) Drawing on your knowledge of nucleotide chemistry, draw the molecular structure of oxoG, and, below it, draw guanine. Opposite guanine, sketch cytosine, including the hydrogen bonds that allow these two molecules to base-pair. Does the structure of oxoG, in contrast to guanine, provide any hint as to why it base pairs with adenine? (c) Assume that an unrepaired oxoG lesion is present in the helix of DNA opposite cytosine. Predict the type of mutation that will occur after several rounds of replication. (d) Drawing on your knowledge of DNA repair mechanisms, consider which cellular approaches might work to counteract an oxoG lesion and determine which of them is likely to be most effective.
31. Among the Betazoids in the world of *Star Trek*®, the ability to read minds is under the control of a *mindreader* (*mr*) gene. Most Betazoids can read minds, but rare recessive mutations in the *mr* gene result in two alternative phenotypes: delayed receivers and insensitives. Delayed receivers have some mind-reading ability but perform the task much more slowly than normal Betazoids; insensitives cannot read minds at all. Betazoid genes do not have introns, so the gene contains only coding DNA. It is 3332 nucleotides in length, arranged in a *four*-letter genetic code.

    The table at the top of the next column shows data from unrelated *mr* mutations. For each mutation, provide a plausible explanation for why it gives rise to its associated phenotype and not to the other phenotype. For example, hypothesize why the *mr-1* nonsense mutation in codon 829 gives rise to the milder delayed-receiver phenotype rather than the more severe insensitive phenotype. (More than one explanation is possible, so be creative within plausible bounds!)

Description of Mutation	***Phenotype***
mr-1 Nonsense mutation in codon 829	Delayed receiver
mr-2 Missense mutation in codon 52	Delayed receiver
mr-3 Deletion of nucleotides 83–150	Delayed receiver
mr-4 Missense mutation in codon 192	Insensitive
mr-5 Deletion of nucleotides 83–93	Insensitive

32. Approximately 5% of live-born human offspring will have a genetic disorder, and of these, 20% are due to new germline mutations. Below are data derived from a study by Walter and colleagues (1998. *Proc. Natl. Acad. Sci. (USA)* 95: 10,015–19.) of the spontaneous mutation frequency for an introduced gene in mouse spermatids at various life stages. (a) What conclusions can be drawn from the data? (b) Given the nature of spermatogenesis, what factors might contribute to the results?

Age (months)	***No. Animals Tested***	***Mutation Frequency***
2	2	0.4×10^{-5}
15	6	0.5×10^{-5}
28	7	4.0×10^{-5}*

*indicates a statistical difference

33. Skin cancer carries a lifetime risk nearly equal to all other cancers combined. Below is a graph (modified from Kraemer. 1997. *Proc. Natl. Acad. Sci. (USA)* 94:11–14.) depicting the age of onset of skin cancers in patients with or without xeroderma pigmentosum (XP), where cumulative percentage is plotted against age. The non-XP curve is based on 29,757 cancers surveyed by the National Cancer Institute, and the curve representing those with XP is based on 63 skin cancers from the Xeroderm Pigmentosum Registry. (a) Provide an overview of the information contained in the graph. (b) Explain why individuals with XP show such an early age of onset.

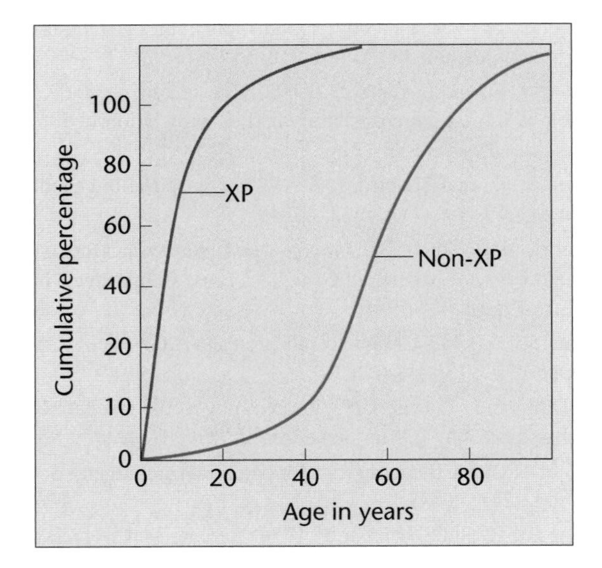

34. Flower color in morning glories is primarily dependent on a complex flavonoid biosynthetic pathway leading to the production of anthocyanin. Flower colors appear to be significantly influenced by multiple genes involving epistasis and typical dominant/recessive patterns. Flowers may be pigmented in shades of red/magenta or variegated with pigmented patches randomly or uniformly distributed. The table below outlines the known causes of variations in morning glory flower pigmentation. (Clegg and Durbin. 2000. *Proc. Natl. Acad. Sci. (USA)* 97: 7016–23.) If these mutations represent virtually all the known types of mutation causing variation in morning glory pigmentation, what general conclusions can you reach from the data? Defend your conclusions.

Species	*Phenotype*	*Cause*
Ipomoea purpurea	Pink sectors on white corolla	Transposon in intron
	Few purple sectors on white corolla	Transposon in intron
	Stable white	Two transposons in intron
	Stable white	Genomic rearrangement
	Pink corolla	Transposon in exon
Ipomoea nil	Magenta corolla	Point mutation
	Round spots on unpigmented corolla	Transposon insertion
	Colored flecks and sectors	Transposon insertion
	Double corolla	Transposon insertion
	White corolla	Transposon insertion

Selected Readings

Ames, B. N., McCann, J., and Yamasaki, E. 1975. Method for detecting carcinogens and mutagens with the *Salmonella*/mammalian microsome mutagenicity test. *Mut. Res.* 31: 347–64.

Auerbach, C. 1978. Forty years of mutation research: A pilgrim's progress. *Heredity* 40: 177–87.

Bates, G., and Lehrach, H. 1994. Trinucleotide repeat expansions and human genetic disease. *BioEssays* 16: 277–83.

Beadle, G. W., and Tatum, E. L. 1945. *Neurospora* II. Methods of producing and detecting mutations concerned with nutritional requirements. *Am. J. Bot.* 32: 678–86.

Berg, D., and Howe, M. (eds.), 1989. *Mobile DNA*. Washington, DC: American Soc. of Microbiology.

Carter, P. 1986. Site-directed mutagenesis. *Biochem. J.* 237: 1–7.

Cleaver, J. E. 1968. Defective repair replication of DNA in xeroderma pigmentosum. *Nature* 218: 652–56.

Cleaver, J. E., and Karentz, D. 1986. DNA repair in man: Regulation by a multiple gene family and its association with human disease. *Bioessays* 6: 122–27.

Cohen, S. N., and Shapiro, J. A. 1980. Transposable genetic elements. *Sci. Am.* (Feb.) 242: 40–49.

Comfort, N. C. 2001. *The Tangled Field: Barbara: McClintock's Search for the Patterns of Genetic Control*. Cambridge: Harvard Univ. Press.

Deering, R. A. 1962. Ultraviolet radiation and nucleic acids. *Sci. Am.* (Dec.) 207: 135–44.

Den Dunnen, J. T. et al. 1989. Topography of the Duchenne muscular dystrophy (DMD) gene. *Am. J. Hum. Genet.* 45: 835–47.

Devoret, R. 1979. Bacterial tests for potential carcinogens. *Sci. Am.* (Aug.) 241: 40–49.

Doring, H. P., and Starlinger, P. 1984. Barbara McClintock's controlling elements: Now at the DNA level. *Cell* 39: 253–59.

Drake, J. W. 1991. A constant rate of spontaneous mutation in DNA-based microbes, *Proc. Natl. Acad. Sci. (USA)* 88: 7160–64.

Drake, J. W. et al. 1975. Environmental mutagenic hazards. *Science* 187: 505–14.

Drake, J. W., Glickman, B. W., and Ripley, L. S. 1983. Updating the theory of mutation. *Am. Sci.* 71: 621–30.

Eyre-Walker, A., and Keightley, P. D. 1999. High genomic deleterious mutation rates in hominids. *Nature* 397: 344–47.

Federoff, N. V. 1984. Transposable genetic elements in maize. *Sci. Am.* (June) 250: 85–98.

Friedberg, E. C. 2003. DNA damage and repair. *Nature* 421: 436–40.

Friedberg, E. C., Walker, G. C., and Siede, W. 1995. *DNA Repair and Mutagenesis*. Washington, DC: ASM Press.

Hanawalt, P. C., and Haynes, R. H. 1967. The repair of DNA. *Sci. Am.* (Feb.) 216: 36–43.

Haseltine, W. A. 1983. Ultraviolet light repair and mutagenesis revisited. *Cell* 33: 13–17.

Hendrickson, E. A. 1997. Cell-cycle regulation of mammalian DNA double-strand break repair. *Am. J. Hum. Genet.* 61: 795–800.

Howard-Flanders, P. 1981. Inducible repair of DNA. *Sci. Am.* (Nov.) 245: 72–80.

Jiricny, J. 1998. Eukaryotic mismatch repair: An update. *Mutation Res.* 409: 107–21.

Kelner, A. 1951. Revival by light. *Sci. Am.* (May) 184: 22–25.

Knudson, A. G. 1979. Our load of mutations and its burden of disease. *Am. J. Hum. Genet.* 31: 401–13.

Kraemer, F. H. et al. 1975. Genetic heterogeneity in xeroderma pigmentosum: Complementation groups and their relationship to DNA repair rates. *Proc. Natl. Acad. Sci. (USA)* 72: 59–63.

Little, J. W., and Mount, D. W. 1982. The SOS regulatory system of *E. coli*. *Cell* 29: 11–22.

Massie, R. 1967. *Nicholas and Alexandra*. New York: Atheneum.

Massie, R. and Massie, S. 1975. *Journey*. New York: Knopf.

McCann, J., Choi, E., Yamasaki, E., and Ames, B. 1975. Detection of carcinogens as mutagens in the *Salmonella*/microsome test: Assay of 300 chemicals. *Proc. Natl. Acad. Sci. (USA)* 72: 5135–39.

McClintock, B. 1956. Controlling elements and the gene. *Cold Spring Harbor Symp. Quant. Biol.* 21: 197–216.

Macdonald, M. E. et al. 1993. A novel gene containing a trinucleotide repeat that is expanded and unstable in Huntington's disease chromosome. *Cell* 72: 971–80.

McKusick, V. A. 1965. The royal hemophilia. *Sci. Am.* (Aug.) 213: 88–95.

Miki, Y. 1998. Retrotransposal integration of mobile genetic elements in human disease. *J. Hum. Genet.* 43: 77–84.

Muller, H. J. 1927. Artificial transmutation of the gene. *Science* 66: 84–87.

———. 1955. Radiation and human mutation. *Sci. Am.* (Nov.) 193: 58–68.

Nickolofff, J.A., and Hoekstra, M. F. (eds.), 2001. *DNA Damage and Repair. Vol. III. Advances from Phage to Humans*. Totowa: Humana Press.

O'Hare, K. 1985. The mechanism and control of P element transposition in *Drosophila*. *Trends Genet.* 1: 250–54.

Osanna, N., Peterson, K. R., and Mount, D. W. 1986. Genetics of DNA repair in bacteria. *Trends Genet.* 2: 55–58.

Radman, M., and Wagner, R. 1988. The high fidelity of DNA duplication. *Sci. Am.* (Aug.) 259(2): 40–6.

Sherratt, D. J. (ed.), 1995. *Mobile Genetic Elements*. New York: Oxford Univ. Press.

Shortle, D., DiMario, D., and Nathans, D. 1981. Directed mutagenesis. *Annu. Rev. Genet.* 15: 265–94.

Sigurbjornsson, B. 1971. Induced mutations in plants. *Sci. Am.* (Jan.) 224: 86–95.

Spradling, A. C., Stern, D. M., Kiss, I., Roote, J., Laverty, T., and Rubin, G. M. 1995. Gene disruptions using P transposable elements: An integral component of the *Drosophila* Genome Project. *Proc. Natl. Acad. Sci. (USA)* 92: 10,824–30.

Stadler, L. J. 1928. Mutations in barley induced by X-rays and radium. *Science* 66: 84–7.

Sutherland, B. M. 1981. Photoreactivation. *Bioscience* 31: 439–44.

Tomlin, N. V., and Aprelikova, O. N. 1989. Uracil DNA glycosylases and DNA uracil repair. *Int. Rev. Cytol.* 114: 81–124.

Vogel, F. 1992. Risk calculations for hereditary effects of ionizing radiation in humans. *Hum. Genet.* 89: 127–46.

Wells, R. D. 1994. Molecular basis of genetic instability of triplet repeats. *J. Biol. Chem.* 271: 2875–78.

Wills, C. 1970. Genetic load. *Sci. Am.* (Mar.) 222: 98–107.

Yamamoto, F. et al. 1990. Molecular genetic basis of the histo-blood group ABO system. *Nature* 345: 229–33.

CHAPTER 16

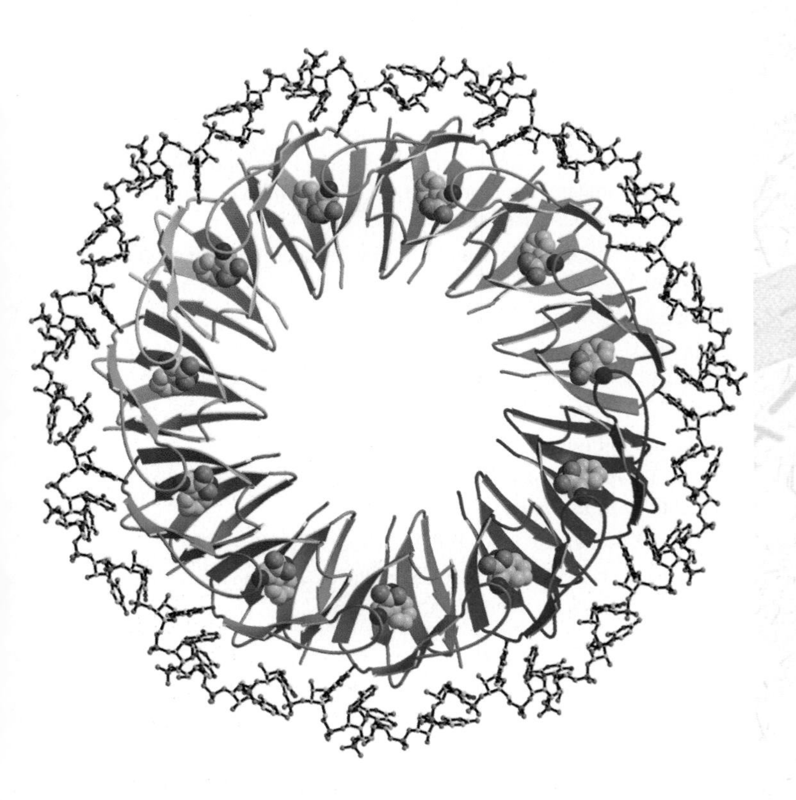

Model of the trp RNA-binding attenuation protein (TRAP). *Nature Vol 401 Sept 16, 1999 (Cover). Article: "Structure of the trp RNA-binding attenuation protein, TRAP, bound to RNA" by Alfred A. Antson, Eleanor J. Dodson, Guy Dodson, Richard B. Greaves, Xiao-ping Chen & Paul Gollnick.*

Regulation of Gene Expression

CHAPTER CONCEPTS

16.1 Prokaryotes Exhibit Efficient Genetic Mechanisms to Respond to Environmental Conditions

16.2 Lactose Metabolism in *E. coli* Is Regulated by an Inducible System
- Structural Genes
- The Discovery of Regulatory Mutations
- The Operon Model: Negative Control
- Genetic Proof of the Operon Model
- Isolation of the *lac* Repressor
- The CAP Protein: Positive Control of the *lac* Operon

16.3 Crystal Structure Analysis of Repressor Complexes Has Confirmed the Operon Model

16.4 Tryptophan Metabolism in *E. coli* Is Controlled by a Repressible Gene System
- Genetic Evidence for the *trp* Operon
- Attenuation

16.5 Eukaryotic Gene Regulation Differs from Regulation in Prokaryotes

16.6 Regulatory Elements and Transcription Factors Control the Expression of Eukaryotic Genes
- Promoters
- Enhancers
- Transcription Factors
- Structural Motifs of Transcription Factors
- Assembly of the Transcription Complex
- Chromatin Remodeling, DNA Methylation, and Gene Expression

16.7 Steroid Hormones Regulate Some Genes

16.8 Posttranscriptional Events Also Regulate Gene Expression
- Alternative Splicing of mRNA

In previous chapters, we established how DNA is organized into genes, how genes store genetic information, and how this information is expressed. Let's now consider one of the most fundamental issues in molecular genetics: *How is genetic expression regulated*? A functional bacterial cell contains thousands of proteins present at widely different concentrations, yet each polypeptide is encoded by a single gene. Synthesis of bacterial gene products changes dramatically in response to environmental conditions. Bacterial cells normally synthesize the enzymes to metabolize lactose only when it is present in the environment and other, more readily metabolized carbon sources are not. In the absence of lactose, the enzymes needed to metabolize this sugar are also absent. These observations lend support to the idea that gene expression is precisely regulated.

In eukaryotic organisms, cells of the pancreas do not make retinal pigment and retinal cells do not make insulin. In multicellular eukaryotes, genetic regulation is at the heart of cellular specialization, a general process called *differentiation*. Differential gene expression serves as the basis for phenotypic differentiation at the cellular and tissue levels in both plants and animals.

In this chapter, we will discuss examples of gene regulation in prokaryotes (specifically bacteria) and eukaryotes. Pivotal to this discussion is our knowledge that all cells in an organism contain a complete set of genes characteristic of that species. Regulation is *not* accomplished by eliminating unused genetic information; instead, mechanisms have evolved to control the expression of genes. Some of these mechanisms, particularly in bacterial systems and their phages, have been extensively characterized.

HOW DO WE KNOW WHAT WE KNOW?

IN THIS CHAPTER, WE WILL FOCUS ON how the expression of genetic information stored in DNA is regulated. We shall elucidate a number of mechanisms by which the process of transcription is either activated or repressed. As you study this topic, you should try to answer several fundamental questions:

1. How do we know that transcription of bacterial genes can be induced by external stimuli?
2. How were we able to determine how induction occurs in bacteria and how the repressed state is maintained?
3. Ultimately, many bacterial systems rely on a "repressor" molecule that regulates transcription. How do we know that repressor molecules exist?
4. Regulation of eukaryotic genes is known to involve the binding of various proteins to "promoters" and "enhancers." How do we know that promoters and enhancers exist in eukaryotic genes?
5. How do we know that posttranscriptional regulation in eukaryotes occurs? ■

16.1 Prokaryotes Exhibit Efficient Genetic Mechanisms to Respond to Environmental Conditions

Regulation of gene expression has been studied extensively in prokaryotes, particularly in *E. coli*. Highly efficient mechanisms have evolved that turn genes on and off, depending on the cell's metabolic needs in particular environments. Detailed analysis of proteins in *E. coli* has shown that for the more than 4000 polypeptide chains encoded by the genome, a vast range of concentration of gene products exists. Some proteins may be present in as few as 5–10 molecules per cell, whereas others, such as ribosomal proteins and the many proteins involved in the glycolytic pathway, are present in as many as 100,000 copies per cell.

That microorganisms regulate the synthesis of gene products can be illustrated by considering the utilization of lactose (a galactose-glucose–containing disaccharide) as a carbon source. When lactose is present in the culture medium, many bacteria and yeast produce enzymes specific to lactose metabolism, but when it is absent, the enzymes are not manufactured. These organisms thus "adapt" to their environment, producing certain enzymes only when specific chemical substrates are present. Such enzymes are said to be **inducible enzymes**, reflecting the role of the substrate, which is the **inducer** of enzyme production. In contrast, other enzymes that are produced continuously, regardless of the chemical makeup of the environment, are *constitutive* enzymes.

Studies have also revealed cases where the presence of a specific molecule causes inhibition of genetic expression. This is often the case for molecules that are the end products of biosynthetic pathways. Amino acids can be synthesized by bacterial cells, but if the amino acids are present in the growth medium, they can be taken up and used. In such cases, it is inefficient for the cell to produce the enzymes necessary for the synthesis of those amino acids, and transcription of mRNA for the appropriate biosynthetic enzymes is repressed. This is an example of a **repressible system** of gene regulation.

Regulation, whether it is inducible or repressible, may be under either **negative** or **positive control**. Under negative control, genetic expression occurs *unless it is shut off by some form of a regulator molecule*. In contrast, under positive control, transcription occurs *only if a regulator molecule directly stimulates RNA production*. In theory, either type of control can govern inducible or repressible systems. Our discussion in the ensuing sections of this chapter will clarify these contrasting systems of regulation.

16.2 Lactose Metabolism in *E. coli* Is Regulated by an Inducible System

Beginning in the 1950s, and continuing through the next decade, Jacques Monod, Joshua Lederberg, François Jacob, and Andre L'woff amassed genetic and biochemical evidence involving lactose metabolism. These researchers and others provided insights into how gene activity is repressed when lactose is absent but induced

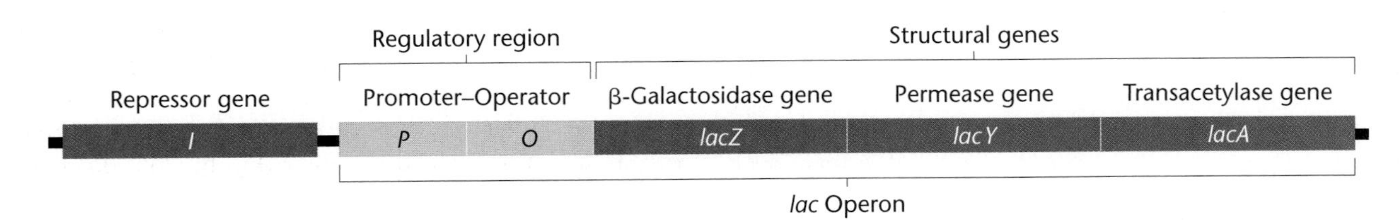

FIGURE 16–1 A simplified overview of the genes and regulatory units involved in the control of lactose metabolism (not to scale).

when it is available. In the presence of lactose, the concentration of the enzymes responsible for lactose metabolism increases rapidly from a few molecules to thousands per cell. The enzymes responsible for lactose metabolism are thus *inducible*, and lactose serves as the *inducer*.

In prokaryotes, genes that code for enzymes with related functions (e. g., genes involved with lactose metabolism) tend to be organized in clusters, and they are often under the coordinated genetic control of a single regulatory unit. The site of this **cis-acting element** is almost always linked upstream to the gene cluster it controls. Interactions at the site of this element involve binding molecules called **trans-acting elements** that control transcription of the gene cluster. Actions at the regulatory site determine whether the genes are expressed and thus whether the corresponding enzymes or other protein products are present. Binding a trans-acting element at a cis-acting site can regulate the gene cluster with either negative or positive control. In this section, we discuss how bacterial gene clusters are coordinately regulated.

Paramount to understanding how gene expression is controlled in this system was the discovery of a regulatory gene and a regulatory site that are part of the gene cluster. Neither of these elements encodes enzymes necessary for lactose metabolism. That is the function of the three genes in the cluster. As shown in Figure 16–1, the three structural genes and the adjacent regulatory site constitute the **lactose**, or **lac, operon**. Together, the entire gene cluster functions in an integrated fashion to provide a rapid response to the presence or absence of lactose.

Structural Genes

There are three structural genes in the lac operon. The *lacZ* gene encodes **β-galactosidase**, an enzyme that converts the disaccharide lactose to the monosaccharides glucose and galactose (Figure 16–2). This conversion is essential if lactose is to serve as the primary energy source in glycolysis. The second gene, *lacY*, specifies the primary structure of **permease**, an enzyme that facilitates the entry of lactose into the bacterial cell. The third gene, *lacA*, codes for the enzyme **transacetylase**. While its physiological role is not completely clear, some researchers believe that it helps remove toxic by-products of lactose metabolism from the cell.

To study the genes coding for these three enzymes, researchers isolated numerous mutations to eliminate the function of one enzyme. Such lac^- mutants were first isolated and studied by Joshua Lederberg. Mutant cells that fail to produce active β-galactosidase $(lacZ^-)$ or permease $(lacY^-)$ cannot use lactose as an energy source. Mutations were also found in the transacetylase gene $(lacA^-)$. Mapping studies by Lederberg established that all three genes are closely linked or contiguous to one another in the order *Z–Y–A* (see Figure 16–1).

Knowledge of their close linkage led to the discovery that all three genes are transcribed as a single unit, resulting in a single mRNA representing the DNA sequence of all three. This is called a **polycistronic mRNA** (Figure 16–3). Thus, the regulation of the *lac* genes is coordinated because a single message is the basis for translation of all three gene products.

The Discovery of Regulatory Mutations

How does lactose activate structural genes and induce the synthesis of the related enzymes? A partial answer comes from the discovery and study of **gratuitous inducers**, chemical analogs of lactose such as the sulfur analog **isopropylthiogalactoside (IPTG)**, shown in (Figure 16–4). Gratuitous inducers behave like natural inducers, but they do not serve as substrates for the enzymes that are subsequently synthesized. Their discovery provides strong evidence that the primary induction event does not depend on the interaction between the inducer and the enzyme.

What, then, is the role of lactose in induction? The answer to this question required the study of another class

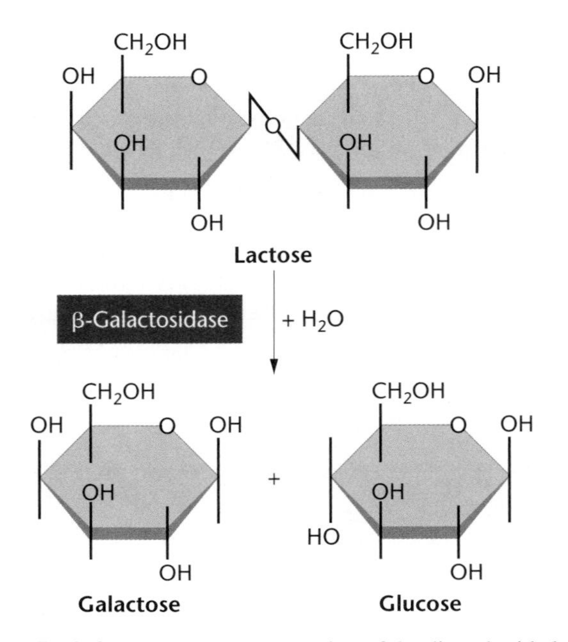

FIGURE 16–2 The catabolic conversion of the disaccharide lactose into its monosaccharide units, galactose and glucose.

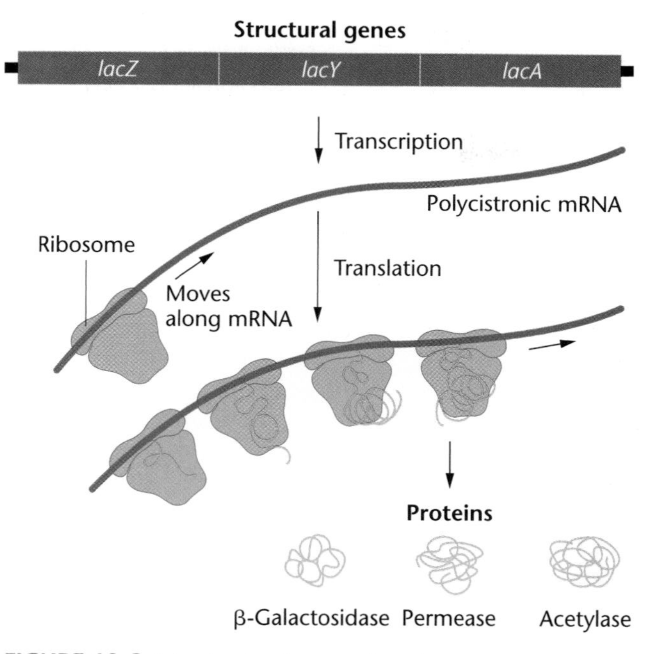

FIGURE 16–3 The structural genes of the *lac* operon are transcribed into a single polycistronic mRNA, which is translated simultaneously by several ribosomes into the three enzymes encoded by the operon.

of mutations called **constitutive mutants**. In this type of mutant, the enzymes are produced regardless of the presence or absence of lactose. Maps of the first type of constitutive mutation, *lacI*$^-$, showed that it is located at a site on the DNA close to, but distinct from, the structural genes. The *lacI* gene is called a **repressor gene**. We shall soon see why this name is appropriate. A second set of constitutive mutations producing identical effects was found in a region immediately adjacent to the structural genes. This class of mutations, designated *lacO*^{C}, identifies the **operator region** of the operon. Because inducibility has been eliminated in both types of constitutive mutation (the enzymes are continually produced), clearly regulation has been disrupted by genetic changes.

The Operon Model: Negative Control

Around 1960, Jacob and Monod proposed the **operon model**, a scheme involving negative control whereby a group of genes is regulated and expressed together as a unit. As we saw in Figure 16–1, the *lac* operon they proposed consists of the *lacZ*, *lacY*, and *lacA* structural genes

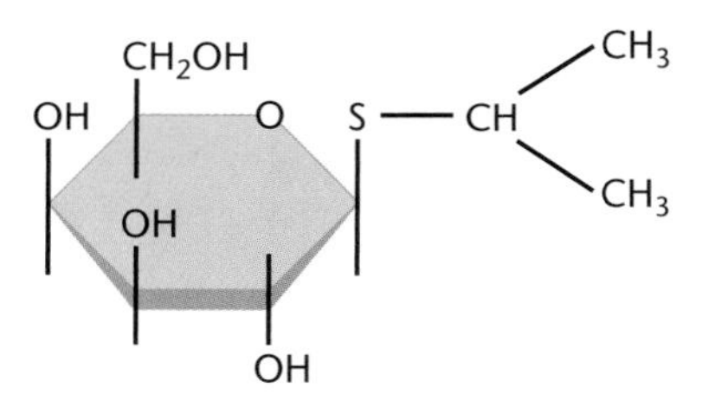

FIGURE 16–4 The gratuitous inducer isopropylthiogalactoside (IPTG).

as well as the adjacent sequences of DNA referred to as the regulatory region, including the operator. They argued that the *lacI* gene regulates the transcription of the structural genes by producing a **repressor molecule** and that it is an **allosteric repressor** meaning that the molecule reversibly interacts with another molecule, causing both a conformational change in three-dimensional shape and a change in chemical activity. Figure 16–5 illustrates the components of the *lac* operon as well as the action of the *lac* repressor in the presence and absence of lactose.

Jacob and Monod suggested that the repressor normally interacts with the DNA sequence of the operator region. When it does so, it inhibits the action of RNA polymerase, effectively repressing the transcription of the structural genes (Figure 16–5b). However, when lactose is present, it binds to the repressor and causes an allosteric conformational change. This change alters the binding site of the repressor, rendering it incapable of interacting with operator DNA (Figure 16–5c). In the absence of the repressor–operator interaction, RNA polymerase transcribes the structural genes and the enzymes necessary for lactose metabolism are produced. Since transcription occurs only when the repressor fails to bind to the operator region, negative control is exerted.

The operon model uses potential molecular interactions to explain the efficient regulation of the structural genes. In the absence of lactose, the enzymes encoded by the genes are not needed and their transcription is repressed. When lactose is present, it indirectly induces the activation of the genes by binding with the repressor.* If all lactose is metabolized, none is available to bind to the repressor, which is again free to bind to operator DNA and repress transcription.

Both the *lacI*$^-$ and *lacO*^{C} constitutive mutations interfere with these molecular interactions, allowing continuous transcription of the structural genes. In the case of the *lacI*$^-$ mutant, seen in (Figure 16–6a), the repressor protein is altered and cannot bind to the operator region, so the structural genes are always turned on. In the case of the *lacO*^{C} mutant (Figure 16–6b), the nucleotide sequence of the operator DNA is altered and will not bind with a normal repressor molecule. The result is the same: Structural genes are always transcribed.

Genetic Proof of the Operon Model

The operon model is a good one because it leads to three major predictions that can be tested to determine its validity. The major predictions to be tested are that (1) the *lacI* gene produces a diffusible cellular product, (2) the *lacO* region is involved in regulation but does not produce a product, and (3) the *lacO* region must be adjacent to the structural genes in order to regulate transcription.

The construction of partially diploid bacteria enables us to assess these assumptions, particularly those that predict

*Technically, allolactose, an isomer of lactose, is the inducer. Allolactose is produced during the initial step in the metabolism of lactose by β-galactosidase.

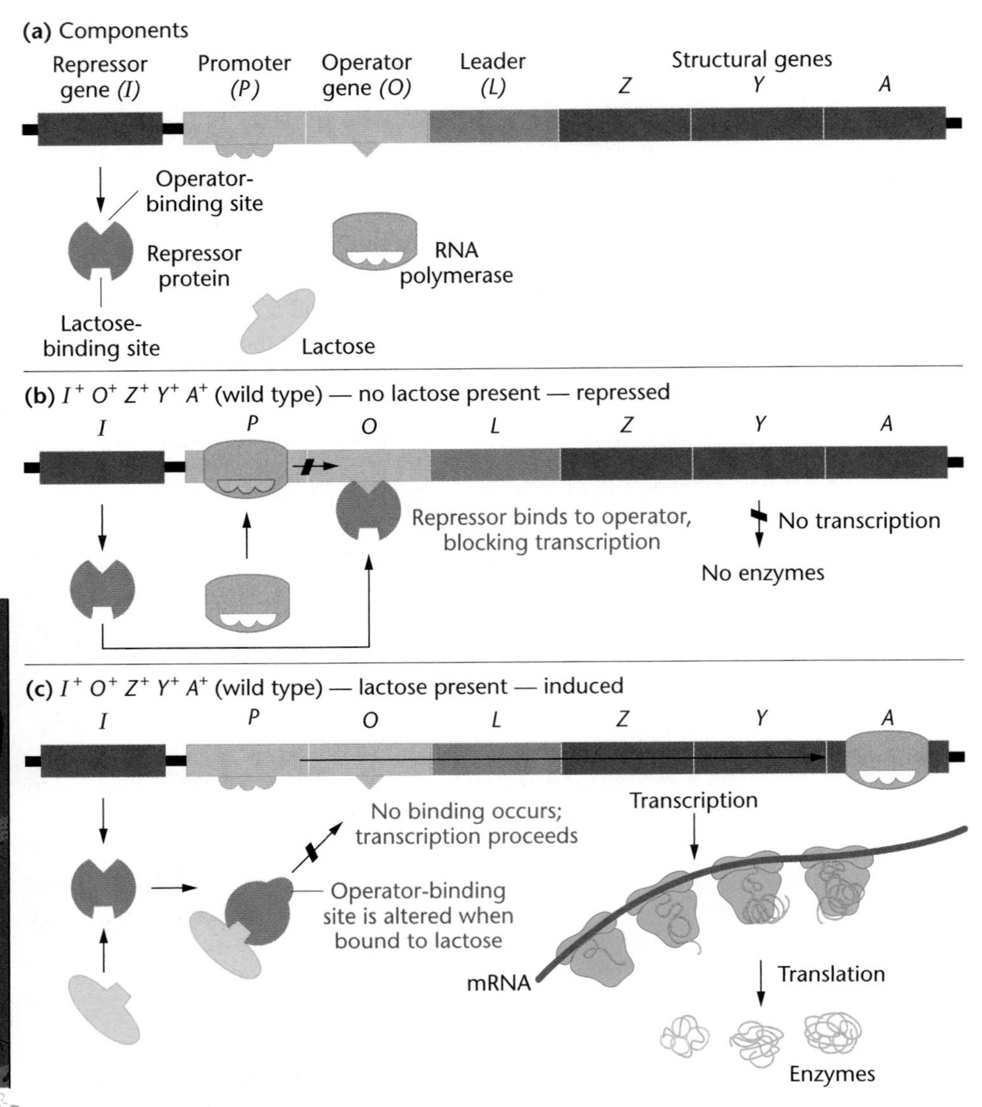

FIGURE 16–5 The components of the wild-type *lac* operon and the response in the absence and the presence of lactose.

MEDIA TUTORIAL Regulation of gene expression: prokaryotes

trans-acting regulatory elements. As you learned in our discussion of bacterial genetics (Chapter 9), a bacterium whose F plasmid contains chromosomal genes is designated F′. When an F^- cell acquires such a plasmid, it will contain its own chromosome plus one or more additional genes present in the plasmid, creating a host cell called a **merozygote**, which is diploid for those genes. The use of this plasmid makes it possible, for example, to introduce an I^+ gene into a host cell whose genotype is I^- or to introduce an O^+ gene into a host cell of genotype O^C. The Jacob-Monod operon model predicts how regulation should be affected in such cells. Adding an I^+ gene to an I^- cell should restore inducibility because a normal repressor, which is a *trans*-acting factor, would again be produced. Adding an O^+ gene to an O^C cell should have no effect on constitutive enzyme production since regulation depends on an O^+ gene immediately adjacent to the structural genes; that is, O^+ is a *cis-acting regulator*.

Results of these experiments are shown in Table 16.1, where Z represents the structural genes. (The inserted genes are listed after the designation F′.) In both cases described just above, the Jacob-Monod model is upheld

TABLE 16.1 A Comparison of Gene Activity (+ or −) in the Presence or Absence of Lactose for Various *E. coli* Genotypes

	Presence of β-Galactosidase Activity	
Genotype	*Lactose Present*	*Lactose Absent*
$I^+O^+Z^+$	+	−
A. $I^+O^+Z^-$	−	−
$I^-O^+Z^+$	+	+
$I^+O^CZ^+$	+	+
B. $I^-O^+Z^+$/F′ I^+	+	−
$I^+O^CZ^+$/F′ O^+	+	+
C. $I^+O^+Z^+$/F′ I^-	+	−
$I^+O^+Z^+$/F′ O^C	+	−
D. $I^SO^+Z^+$	−	−
$I^SO^+Z^+$/F′ I^+	−	−

Note: In parts B, C, and D, most genotypes are partially diploid, containing an F factor plus attached genes (F′).

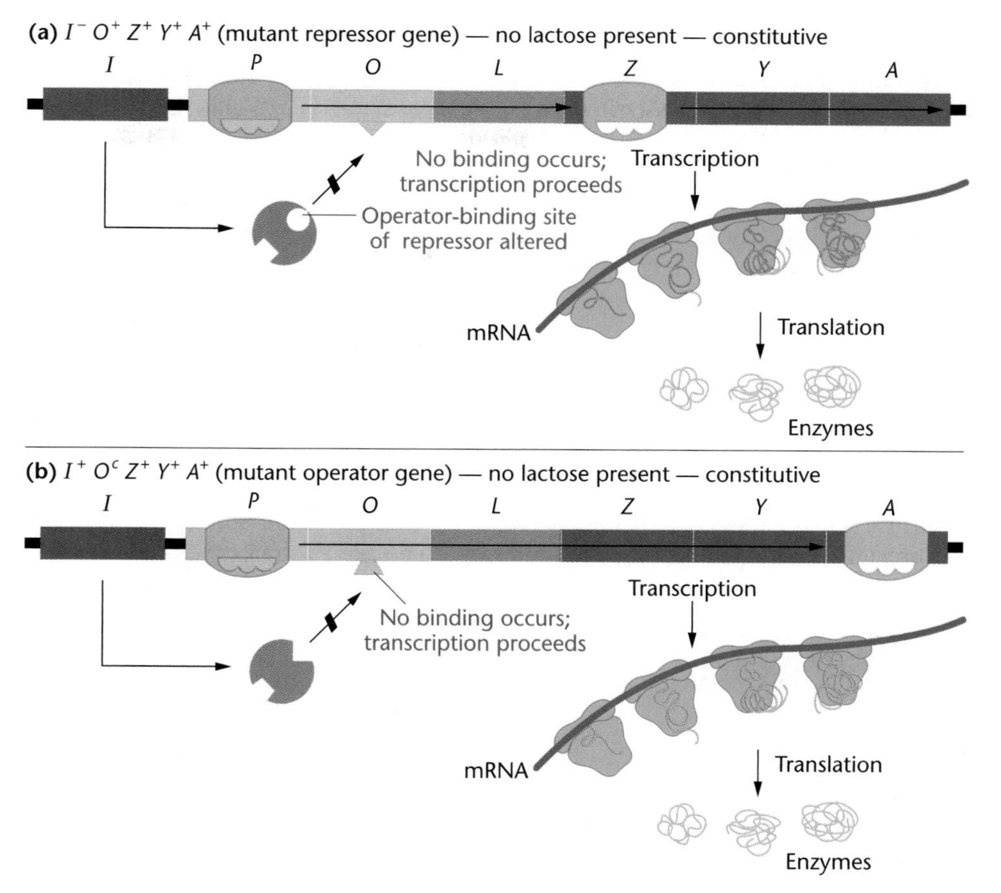

FIGURE 16–6 The response of the *lac* operon in the absence of lactose when a cell bears either the I^- or the O^C mutation.

(part B of Table 16.1). Part C shows the reverse experiments, where either an I^- gene or an O^C region is added to cells of normal inducible genotypes. As the model predicts, inducibility is maintained in these partial diploids.

Another prediction of the operon model is that certain mutations in the *I* gene should have the opposite effect of I^-. That is, instead of being constitutive by failing to interact with the operator, mutant repressor molecules should be produced that cannot interact with the inducer, lactose. As a result, the repressor would always bind to the operator sequence, and the structural genes would be permanently repressed (Figure 16–7). If this were the case, the presence of an additional I^+ gene would have little or no effect on repression.

In fact, an I^S mutation was discovered, where the operon is "superrepressed," as shown in part D of Table 16.1. An additional I^+ gene does not effectively relieve repression of gene activity. These observations again support the operon model for gene regulation.

Isolation of the *lac* Repressor

Although the Jacob-Monod operon theory succeeded in explaining many aspects of genetic regulation in prokaryotes, the nature of the repressor molecule was not known when their landmark paper was published in 1961. While they had assumed that the allosteric repressor was a protein, RNA was also a candidate because activity of the molecule required the ability to bind to DNA. Despite many attempts to isolate and characterize the hypothetical repressor molecule, no direct chemical evidence was immediately forthcoming. A single *E. coli* cell contains no more than about 10 copies of the *lac* repressor; direct chemical identification of

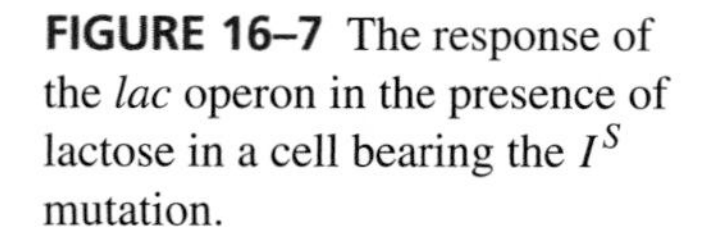
FIGURE 16–7 The response of the *lac* operon in the presence of lactose in a cell bearing the I^S mutation.

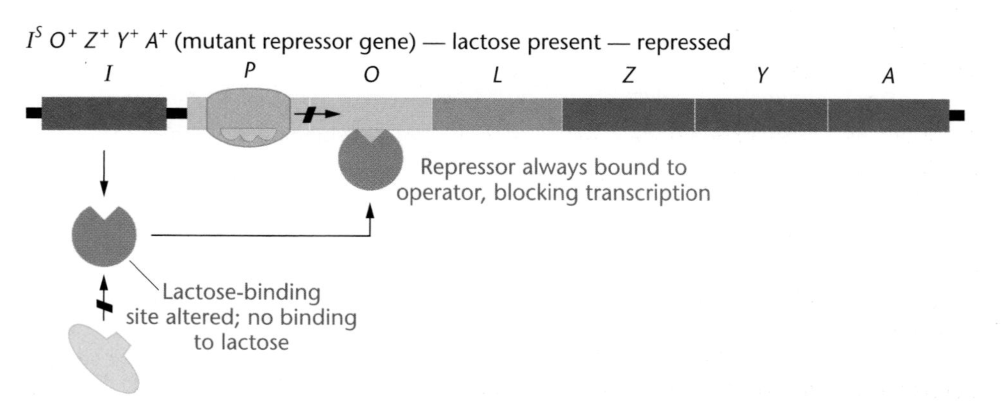

10 molecules in a population of millions of proteins and RNAs in a single cell was a tremendous challenge.

In 1966, Walter Gilbert and Benno Müller-Hill reported the isolation of the *lac* repressor in partially purified form. To achieve the isolation, they used a *regulator quantity* (I^q) mutant strain that contains about 10 times as much repressor as do wild-type *E. coli* cells. Also instrumental in their success were the use of the gratuitous inducer, IPTG, which binds to the repressor, and the technique of **equilibrium dialysis**. In this technique, extracts of I^q cells are placed in a dialysis bag and allowed to attain equilibrium with an external solution of radioactive IPTG, which is small enough to diffuse freely in and out of the bag. At equilibrium, the concentration of IPTG is higher inside the bag than in the external solution, indicating that an IPTG-binding complex is present in the cell extract and that this material is too large to diffuse across the wall and out of the bag.

Ultimately the IPTG-binding complex was purified and shown to have various characteristics of a protein. In contrast, extracts of I^- constitutive cells having no *lac* repressor activity did not exhibit IPTG-binding activity, strongly suggesting that the isolated protein was the repressor molecule.

The CAP Protein: Positive Control of the *lac* Operon

As we discussed at the beginning of this section, the role of β-galactosidase is to cleave lactose into its components glucose and galactose. Then, for galactose to be used by the cell, it is converted to glucose. What if the cell finds itself in an environment that contains an ample amount of lactose *and* glucose? It is not energetically efficient for a cell to be induced by lactose to make β-galactosidase, since what it really needs—glucose—is already present. As we shall see, still another molecular component, the **catabolite-activating protein (CAP)**, effectively represses the expression of the *lac* operon when glucose is present. This inhibition, called **catabolite repression**, reflects the greater simplicity with which glucose may be metabolized in comparison to lactose. The cell "prefers" glucose, and if it is available, the *lac* operon is not activated, even when lactose is present.

To understand CAP and its role in regulation, let's backtrack for a moment. When the *lac* repressor is bound to the inducer, the *lac* operon is activated and RNA polymerase transcribes the structural genes. As we learned previously, transcription is initiated as a result of the binding that occurs between RNA polymerase and the nucleotide sequence of the **promoter region**, found upstream (5′) from the initial coding sequences. Within the *lac* operon, the promoter region is found between the *I* gene and the operator region (*O*, see Figure 16–1). Careful examination has revealed that polymerase binding is never very efficient unless CAP is also present to facilitate the process.

The mechanism is summarized in Figure 16–8. In the absence of glucose and under inducible conditions, CAP

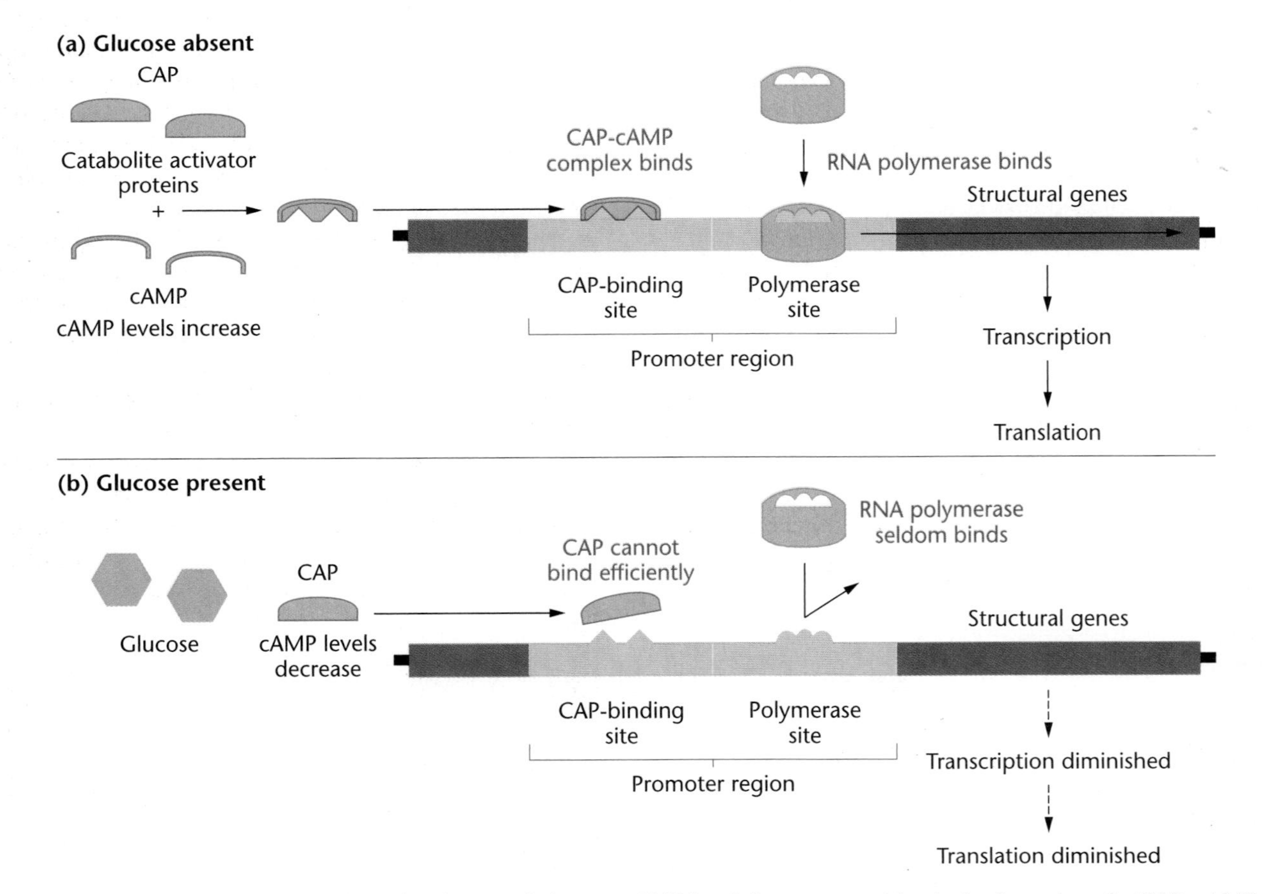

FIGURE 16–8 Catabolite repression. (a) In the absence of glucose, cAMP levels increase, resulting in the formation of a CAP–cAMP complex, which binds to the CAP site of the promoter, stimulating transcription. (b) In the presence of glucose, cAMP levels decrease, CAP–cAMP complexes are not formed, and transcription is not stimulated.

exerts *positive control* by binding to the CAP site, facilitating RNA polymerase binding at the promoter, and thus transcription. Therefore, for maximal transcription, the repressor must be bound by lactose (so as not to repress operon expression) and CAP must be bound to the CAP-binding site.

This leads to the central question about CAP: What role does glucose play in inhibiting CAP binding when it is present? The answer involves still another molecule, **cyclic adenosine monophosphate (cAMP)**, on which CAP binding is dependent. In order to bind to the promoter, CAP must be linked to cAMP. The level of cAMP is itself dependent on an enzyme **adenyl cyclase**, which catalyzes the conversion of ATP to cAMP.*

The role of glucose in catabolite repression is now clear. It inhibits the activity of adenyl cyclase, causing a decline in the level of cAMP in the cell. Under this condition, CAP cannot form the CAP–cAMP complex essential to the positive control of transcription.

Recently, CAP and CAP–cAMP have been examined using X-ray crystallography, revealing important information. CAP is a dimer that inserts into adjacent regions of a specific nucleotide sequence of the DNA making up the promoter. When the CAP–cAMP complex is bound to DNA, it bends the DNA and causes it to assume a new conformation. We shall learn more about this in the next section.

Binding studies performed in solution further clarify the mechanism of gene activation. Alone, neither CAP–cAMP nor RNA polymerase has a strong affinity to bind to *lac* promoter DNA. Nor does either molecule have a strong affinity to bind to the other. However, when both molecules are together in the presence of the *lac* promoter DNA, a tightly bound complex is formed. This is an example of what, in biochemical terms, is called *cooperative binding*. In the case of CAP–cAMP and the *lac* operon, the phenomenon illustrates the high degree of specificity that is involved in the genetic regulation of just one small group of genes.

Regulation of the *lac* operon by catabolite repression results in efficient energy use because the presence of glucose overrides the need for the metabolism of lactose, even if lactose is available to the cell. Catabolite repression involving CAP has also been observed for other inducible operons, including those controlling the metabolism of galactose and arabinose.

*Because of its involvement with cAMP, CAP is also called cyclic AMP receptor protein (CRP), and the gene encoding the protein is named *crp*. Because the protein was first named CAP, we will adhere to the initial nomenclature.

16.3 Crystal Structure Analysis of Repressor Complexes Has Confirmed the Operon Model

We now have a thorough understanding of the biochemical nature of the regulatory region of the *lac* operon, identifying the precise locations of its various components relative to one another (Figure 16–9). In 1996, Mitchell Lewis, Ponzy Lu, and their colleagues succeeded in determining the crystal structure of the *lac* repressor, as well as the structure of the repressor that is bound to the inducer and to operator DNA. As a result, previous information based on genetic and biochemical data has been complemented with the missing structural interpretation, and a nearly complete picture of the regulation of the operon has emerged.

The repressor, as the gene product of the *I* gene, is a monomer consisting of 360 amino acids. Within the monomer, the region of inducer binding has been identified (Figure 16–10a). The functional repressor contains four monomer subunits, creating a homotetramer. The tetramer can be cleaved with a protease under controlled conditions and then yields five fragments. Four of them are derived from the N-terminals of the tetramer, and they bind to operator DNA. The fifth fragment is the remaining core of the tetramer, derived from the COOH-terminals; it binds to lactose and gratuitous inducers such as IPTG. Analysis has revealed that, at any single time, each tetramer can bind to two symmetrical operator DNA helices (Figure 16–10b).

The operator DNA that was previously defined by mutational studies ($lacO^C$) and confirmed by DNA-sequencing analysis is located just downstream from the start of the *lacZ* gene, but upstream from the beginning of the actual coding sequence. The crystallographic studies show that the actual region of repressor binding of this primary operator (O_1) consists of 21 base pairs. Two other auxiliary operator regions have been identified (shown in Figure 16–9). One, O_2, is 401 base pairs downstream from the primary operator within the *lacZ* gene. The other, O_3, is 93 base pairs upstream from O_1, just above the CAP site. *In vivo*, all three operators must be bound for maximum repression.

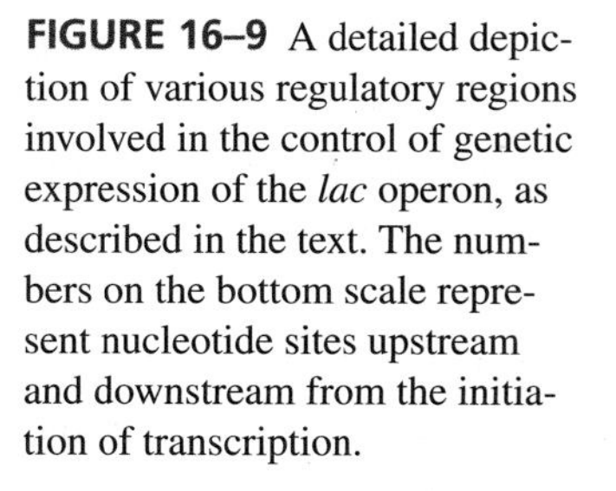

FIGURE 16–9 A detailed depiction of various regulatory regions involved in the control of genetic expression of the *lac* operon, as described in the text. The numbers on the bottom scale represent nucleotide sites upstream and downstream from the initiation of transcription.

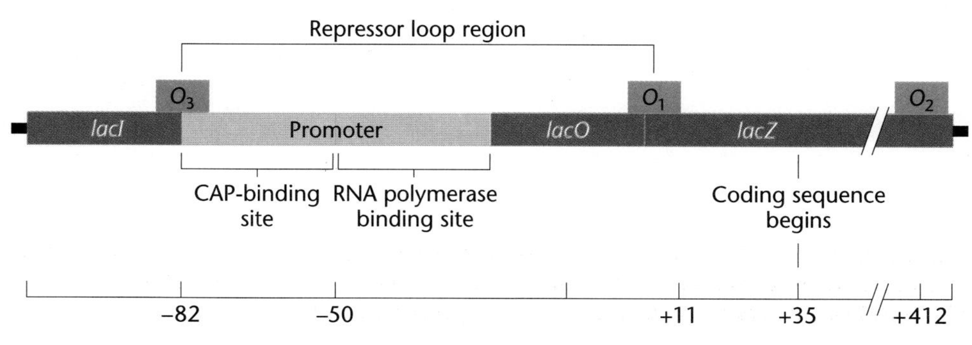

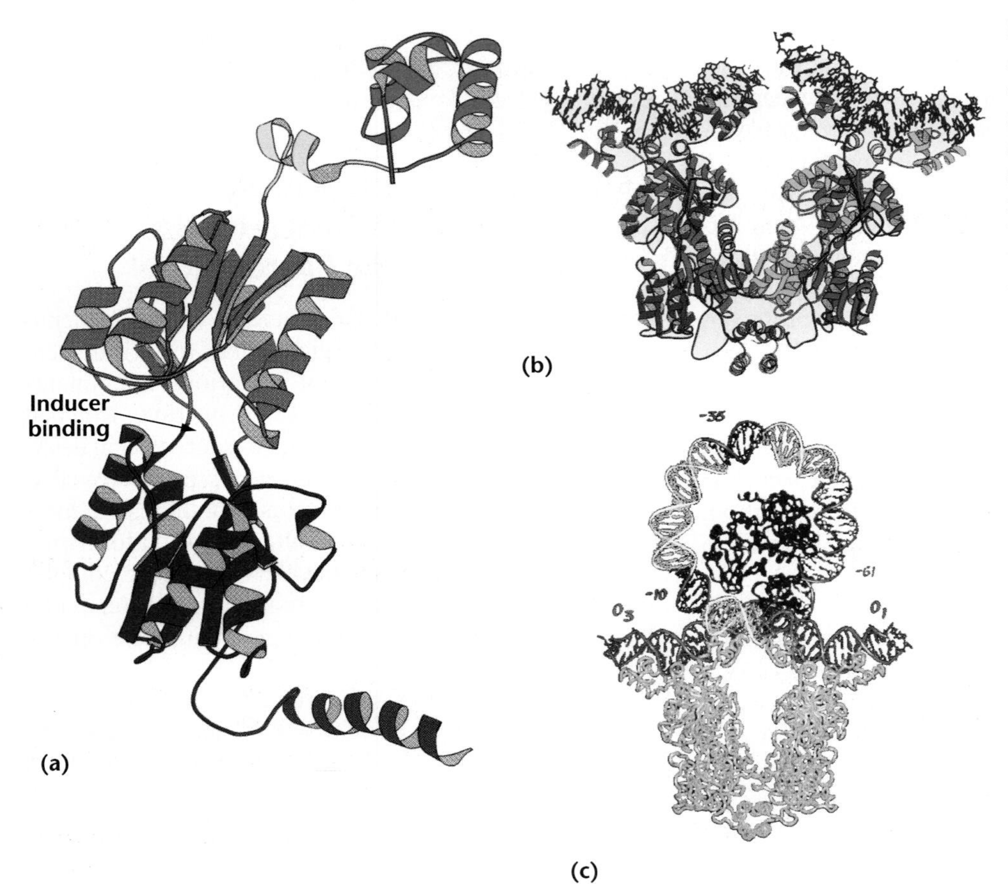

FIGURE 16–10 Models of the *lac* repressor and its binding to operator sites with DNA, as generated from crystal structure analysis. (a) The repressor monomer; the arrow points to the inducer-binding site. The DNA-binding region is shown in red. (b) The repressor dimer bound to two 21–base-pair segments of operator DNA (shown in blue). (c) The repressor and CAP (shown in dark blue) bound to the *lac* DNA. Binding to operator regions O_1 and O_3 creates a 93–base-pair repression loop of promoter DNA. *Science, Lewis, et al 271 pg. 1247-1254/Johnson Research Foundation.*

Repressor binding at two operator sites distorts the conformation of DNA, causing it to bend away from the repressor. When a model is created involving dual binding of operators O_1 and O_3, the 93 base pairs of DNA that intervene must jut out, forming what is called a **repression loop** (Figure 16–10c). This model positions the promoter region that binds RNA polymerase on the inside of the loop, which prevents access during repression. In addition, the repression loop positions the CAP-binding site to facilitate CAP interaction with RNA polymerase upon subsequent induction. DNA looping during repression is similar to transitions that are predicted to occur in eukaryotic systems.

Studies have also defined the three-dimensional conformational changes that accompany the allosteric transitions that occur during the interactions with the inducer molecules. Taken together, the crystallographic studies bring us to a new level of understanding of the regulatory process occurring within the *lac* operon, confirming the findings and predictions of Jacob and Monod in their model set forth over 40 years ago and based strictly on genetic grounds.

16.4 Tryptophan Metabolism in *E. coli* Is Controlled by a Repressible Gene System

Although the process of induction had been known for some time, it was not until 1963 that Monod and colleagues discovered a repressible operon. Wild-type *E. coli* are capable of producing the enzymes necessary for the biosynthesis of amino acids as well as other essential macromolecules. Focusing his studies on the amino acid tryptophan and the enzyme **tryptophan synthetase**, Monod discovered that if tryptophan is present in sufficient quantity in the culture medium, the enzymes necessary for its synthesis are not produced. Energetically, repression of the genes involved in the production of these enzymes is highly economical for the cell when ample tryptophan is present.

Further investigation showed that a series of enzymes encoded by five contiguous genes on the *E. coli* chromosome is involved in tryptophan synthesis. These genes are part of an operon, and in the presence of tryptophan, they are all coordinately repressed and none of the enzymes is produced. Because of the great similarity between this repression and the induction of enzymes for lactose metabolism, Jacob and Monod proposed a model of gene regulation analogous to the *lac* system (Figure 16–11).

To account for repression, they suggested the presence of a *normally inactive repressor* that alone cannot interact with the operator region of the operon. However, the repressor is an allosteric molecule that can bind to tryptophan. When this amino acid is present, the resultant complex of repressor and tryptophan attains a new conformation that binds to the operator, repressing transcription. Thus, when tryptophan, the end product of this anabolic pathway, is present, the system is repressed and enzymes are not made. Because the regulatory complex inhibits transcription of the operon,

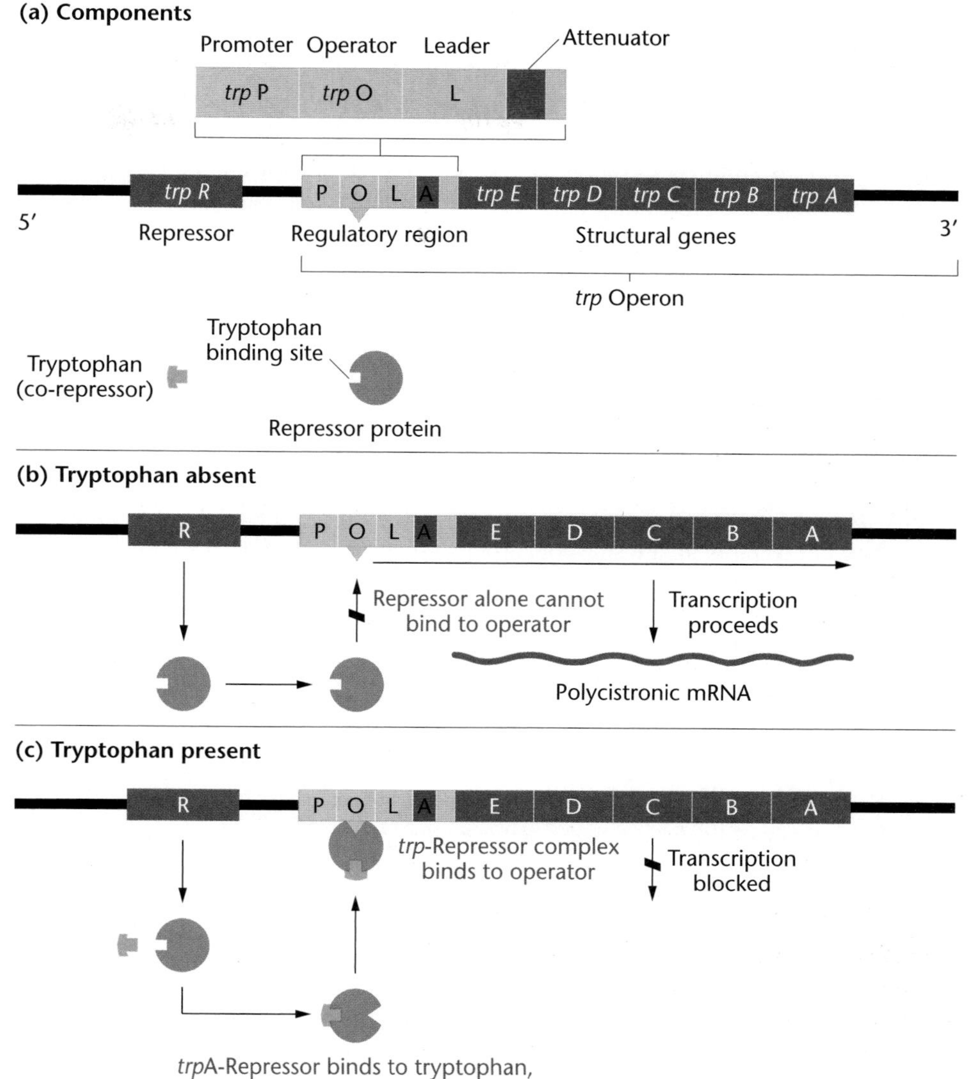

FIGURE 16–11 (a) The components involved in the regulation of the tryptophan operon. Regulatory conditions are depicted involving either activation or repression of the structural genes. (b) In the absence of tryptophan, an inactive repressor is made that cannot bind to the operator (O), thus allowing transcription to proceed. (c) In the presence of tryptophan, it binds to the repressor, causing an allosteric transition to occur. This complex binds to the operator region, leading to repression of the operon.

this repressible system is under *negative control*. And, as tryptophan participates in repression, it is referred to as a **co-repressor** in this regulatory scheme.

Genetic Evidence for the *trp* Operon

Support for the concept of a repressible operon was soon forthcoming, based primarily on the isolation of two distinct categories of constitutive mutations. The first class, $trpR^-$, maps at some distance from the structural genes. This locus represents the gene coding for the repressor. Presumably, the mutation either inhibits the interaction of the repressor with tryptophan or inhibits repressor formation entirely. Whichever the case, no repression ever occurs in cells with the $trpR^-$ mutation. As expected, if the *trpR* gene encodes a repressor molecule, the presence of an additional $trpR^+$ gene restores repressibility.

The second constitutive mutant is analogous to that of the operator of the lactose operon because it maps immediately adjacent to the structural genes. Furthermore, the addition of a wild-type operator gene into mutant cells (as a trans-acting element) does not restore enzyme repression. This is predictable if the mutant operator can no longer interact with the repressor–tryptophan complex.

The entire *trp* operon has now been well defined, as shown in Figure 16–11. Five contiguous structural genes (*trpE*, *-D*, *-C*, *-B*, and *-A*) are transcribed as a polycistronic message that directs translation of the enzymes that catalyze the biosynthesis of tryptophan. As in the *lac* operon, a promoter region (*trpP*) represents the binding site for RNA polymerase, and an operator region (*trpO*) binds the repressor. In the absence of binding, transcription is initiated within the overlapping *trpP–trpO* region and proceeds along a **leader sequence** 162 nucleotides prior to the first structural gene (*trpE*). Within this leader sequence, still another regulatory site has been found, called an attenuator. As you shall see, this regulatory unit is an integral part of the control mechanism of this operon.

Attenuation

Charles Yanofsky, his coworker Kevin Bertrand, and their colleagues observed that in *E. coli*, even when tryptophan is present and the *trp* operon is repressed, initiation of transcription of the leader sequence of the operon usually still occurs. Thus, while the activated repressor binds to the operator region, it does not strongly inhibit the *initial expression* of the operon, suggesting that there must be a subsequent mechanism by which tryptophan inhibits enzyme synthesis. Yanofsky discovered that in the presence of high concentrations of tryptophan, transcription of the leader sequence of the operon proceeds, but transcription of mRNA is usually terminated prematurely. This process is called **attenuation**, indicative of the effect of diminishing genetic expression of the operon. However, when tryptophan is absent or present in very low concentrations, the repressor is inactive and does not bind to the promoter. Transcription is initiated, but *not* subsequently terminated; instead it proceeds through the leader sequence and into the structural genes. As a result, a polycistronic mRNA is transcribed and the enzymes essential to the biosynthesis of tryptophan are subsequently translated.

Identification of the site involved in attenuation was made possible by the isolation of deletion mutations (which abolish attenuation), in the region of the leader sequence. Thus, the site is referred to as the **attenuator**. An explanation of how attenuation occurs and how it is overcome, put forward by Yanofsky and colleagues, involves folding of the RNA transcribed from the leader sequence. During attenuation (when tryptophan is abundant), the structure of the RNA transcribed *from the leader sequence* mimics that found at the end of mRNA molecules, forming a "hairpin loop," or *antiterminator hairpin*. This configuration leads to the premature termination of transcription, reducing the amount of mRNA produced.

The question, of course, is how the absence (or a low concentration) of tryptophan allows attenuation to be bypassed. A key point in Yanofsky's model is that the leader transcript must be translated for the antiterminator hairpin to form. He discovered that the leader transcript includes two triplets (UGG) that encode tryptophan preceded upstream by an initial AUG sequence that prompts the initiation of translation by ribosomes. When adequate tryptophan is present, charged $tRNA^{trp}$ is also present. As a result, translation proceeds past these triplets and a *terminator hairpin* is formed. If cells are starved of tryptophan, charged $tRNA^{trp}$ is unavailable. The ribosome then "stalls" during the translation of triplets as charged $tRNA^{trp}$ is called for, but is unavailable because of the lack of tryptophan. This event induces the formation of the antiterminator hairpin within the transcript. As a result, attenuation is overcome and transcription proceeds, leading to expression of the entire set of structural genes.

Yanofsky and colleagues have more recently investigated attenuation in another bacterial species, *Bacillus subtilis* and have made an interesting discovery. This species also utilizes attenuation and hairpin structures to regulate its *trp* operon. In contrast to *E. coli*, however, *B. subtilis* relies on attenuation as the sole mechanism for regulation, lacking a mechanism that represses transcription entirely in this operon.

However, the molecular signals that cause attenuation in *B. subtilis* do not invoke the process of translation, as does *E. coli*, to induce the hairpin that terminates transcription. Instead, a specific protein, ***trp* RNA-binding attenuation protein (TRAP)**, isolated in the 1990s by Charles Yanofsky and coworkers, either binds or does not bind to the attenuator leader sequence, thereby inducing the alternative terminator or antiterminator configurations, respectively. When bound, an RNA "belt" is formed around TRAP, preventing the antiterminator hairpin from forming. This results in the formation of the terminator configuration, thus leading to premature termination of transcription and, consequently, attenuation of expression of the operon. A molecular model of TRAP is illustrated in the opening photograph of this chapter.

The phenomenon of attenuation appears to be a mechanism common to other bacterial operons that encode the enzymes essential to the biosynthesis of amino acids. In addition to tryptophan, operons involved in threonine, histidine, leucine, and phenylalanine display attenuators in their leader sequences.

16.5 Eukaryotic Gene Regulation Differs from Regulation in Prokaryotes

The 1960 discovery of the operon in *E. coli* by Jacob and Monod was the first step in unraveling the mechanisms that regulate gene expression. While some of the principles of regulation found in the *lac* operon are also exhibited in eukaryotes, a more complex system of gene regulation has evolved in eukaryotes.

As we pointed out in the introduction of this chapter, in multicellular eukaryotes, differential regulation of gene expression is at the heart of cellular differentiation and function. The central question is how an organism expresses a subset of genes in one cell type and a different subset of genes in another cell type. At the cellular level, this regulation is *not* accomplished by eliminating unused genetic information; instead, mechanisms have evolved to activate specific portions of the genome and to repress the expression of other genes. The activation and repression of selected loci is a delicate balancing act for an organism; the expression of a gene at the wrong time, in the wrong cell type, or in abnormal amounts can lead to a deleterious phenotype or death, even when the gene itself is normal.

There are numerous reasons for gene regulation's greater complexity in eukaryotes:

1. Eukaryotic cells contain a much greater amount of genetic information than do prokaryotic cells, and this DNA is complexed with histones and other proteins to form chromatin. The structure of chromatin, either open (decondensed) and available to be transcribed or closed (condensed) and unavailable, is an important factor during gene regulation.

2. Genetic information in eukaryotes is carried on many chromosomes (rather than on just one), and these chromosomes are enclosed within a nucleus bound by a double membrane.

3. In eukaryotes, transcription is spatially and temporally separated from translation—transcription occurs in the nucleus and translation occurs later in the cytoplasm. Thus, mRNA must be transported out of the nucleus in order to serve as the basis of protein synthesis.
4. The transcripts of eukaryotic genes are processed and reduced in size before transport to the cytoplasm.
5. Eukaryotic mRNA has a much longer half-life ($t_{1/2}$) than does prokaryotic mRNA. If transcription is turned off in prokaryotes, the mRNA decays within minutes and translation ceases.
6. Most eukaryotes are multicellular with differentiated cell types, which typically use different sets of genes to make different proteins even though each cell contains a complete set of genes.

As a result of these factors, eukaryotic gene expression potentially can be controlled in many different ways. As shown in Figure 16–12, these include regulation during: (1) transcription (as in prokaryotes); (2) splicing and processing of the pre-mRNA, called posttranscription control; (3) transport to the cytoplasm; (4) stability of mRNA; (5) translation (e. g., selecting which mRNAs are translated); and (6) posttranslational modification of the protein product.

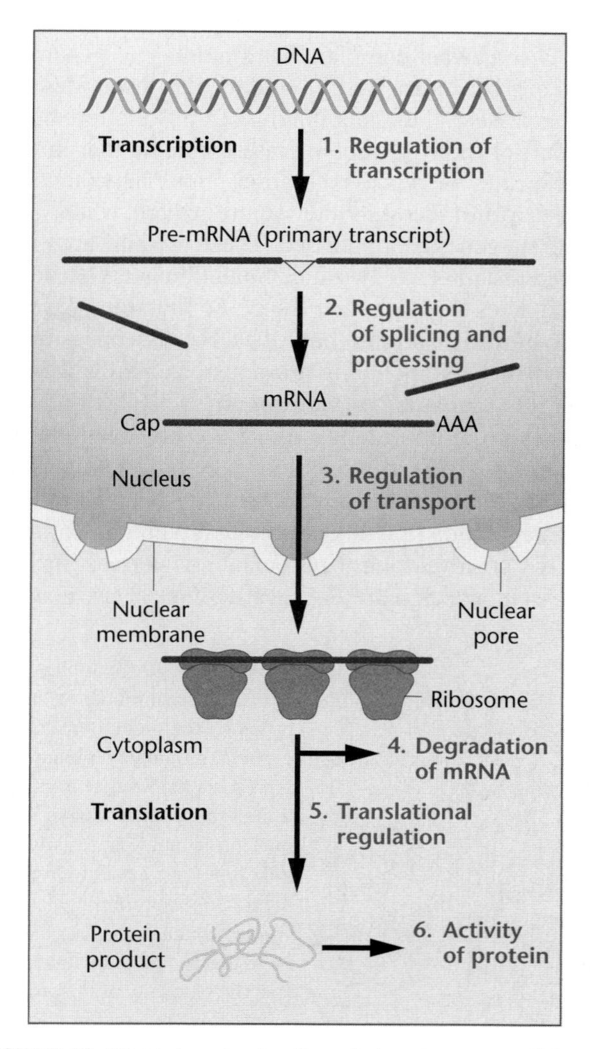

FIGURE 16–12 Various levels of regulation that are possible during the expression of the genetic material in eukaryotes.

16.6 Regulatory Elements and Transcription Factors Control the Expression of Eukaryotic Genes

The internal structure of eukaryotic genes includes several forms of regulatory elements controlling transcription. These are divided into two main types: promoters and enhancers. They interact with a wide variety of proteins called transcription factors that serve to modulate their activity. In the following sections, we will demonstrate how regulatory elements and transcription factors function during eukaryotic genetic expression.

Promoters

Promoters, which have counterparts in bacteria, consist of cis-acting nucleotide sequences that serve as the recognition point for RNA polymerase binding. Therefore, they represent the region necessary to *initiate* transcription and are located immediately adjacent to the genes they regulate. Eukaryotic promoters recognized by RNA polymerase II (which transcribes genes into mRNA) consist of short modular DNA sequences usually located within 100 bp upstream (in the 5′-direction) from the point of inititiation (Figure 16–13). The core region of the promoter contains the **TATA box** sequence, located about 25–30 bp upstream from the initial point of transcription (upstream regions are designated with a minus sign). It consists of an 8-bp consensus sequence, a sequence conserved in most or all genes studied, composed only of T=A pairs, often flanked on either side by G≡C-rich regions. The critical importance of the TATA box is demonstrated by mutational studies (Figure 16–14). Mutations within the nucleotide sequence of this element severely reduce transcription, while those in adjacent bases have little or no effect on gene expression. Such a reduction is considered to be the result of losing the ability to bind to trans-acting transcription factors, which are responsible for stimulating transcription.

Farther upstream from the core region of the promoter are several proximal elements that are also critical to the initiation of transcription. One of these is the **CAAT box**. Its consensus sequence is CAAT or CCAAT, and it frequently appears in the region −70 to −80 bp from the start site of transcription. Like the TATA box, mutations in the sequence of the CAAT box severely diminish transcription (Figure 16–14). Mutations on either side of this element have little or no effect. Another proximal element is the **GC box**, which has the consensus sequence GGGCGG and is usually found around position −110 Often present in multiple copies, its importance has also been documented by mutational analysis.

MEDIA TUTORIAL Regulation of gene expression: eukaryotes

Transcription start site (+1)

GC box GGGCGG | CAAT box GGCCAATC | TATA box TATAAA

−110 −70 −30

FIGURE 16–13 The promoter at the 5′-end of eukaryotic genes consists of several modular elements, including the TATA box (−30), the CAAT box (−70), and the GC box (about −110).

In different genes, there is significant variation in the number and orientation of promoter elements and the distance between them. In addition to RNA polymerase II, eukaryotic genes are transcribed by other polymerases, classified as type I (yielding rRNAs) and type III (yielding tRNAs, 5*S* rRNA, and several other small cellular RNAs). The promoters for both types of polymerase have a different sequence and bind different transcription factors.

Enhancers

Besides the effect of the promoter regions, transcription of most, if not all, eukaryotic genes is affected by additional cis-acting DNA sequences called **enhancers**. Like promoters, these regions interact with trans-acting regulatory proteins. The impact is to greatly increase the efficiency of initiation, thus increasing the overall rate of transcription. Enhancers can be distinguished from promoters by several characteristics.

1. The position of the enhancer is not fixed; it can be found upstream, downstream, or within the gene it regulates.
2. Often, enhancers operate from quite a distance from their target gene—as much as 50 kb.
3. The orientation of an enhancer can be inverted without significant effect on its action.
4. If an enhancer is moved to another location in the genome or if an unrelated gene is placed near an enhancer, transcription of the adjacent gene is positively regulated.

Several examples serve to illustrate the varied nature of enhancer location. In the immunoglobulin heavy-chain genes, an enhancer is located *within* the gene it regulates, in an intron between two coding regions. Downstream enhancers are found in the human β-globin gene; and in chickens, an enhancer is located between the β-globin and the ϵ-globin genes.

The most intriguing question about enhancers is how they exert control over transcription when they are located at some distance from either promoters or the transcriptional start site. As we shall see, enhancers are bound by transcription factors that can alter the configuration of chromatin by bending or looping the DNA. This is thought to bring distant enhancers and promoters into close proximity in order to form activated transcription complexes. In the new configuration, transcription is stimulated above a basal level, increasing the rate of RNA synthesis.

Transcription Factors

Overall, the picture of transcriptional regulation in eukaryotes is somewhat complex, but a number of generalizations can be drawn. Many proteins have been identified that are essential to the initiation of transcription, but are not part of the RNA polymerase molecule that initiates and executes the process of transcription. These are called **transcription factors**—they control where, when, and at what rate genes are expressed. These proteins are modular and usually have two functional domains (clusters of amino acids that carry out a specific function). One, the **DNA-binding domain**, binds to DNA sequences in regulatory regions, including promoters and enhancers; the other, the **trans-activating domain**, activates transcription via protein–protein interaction. For example, this domain of a transcription factor may bind to other transcription factors at the promoter or to RNA polymerase.

The regulation of transcription by protein factors is primarily a positive-control system, although transcriptional repressors, which represent negative control, are now also

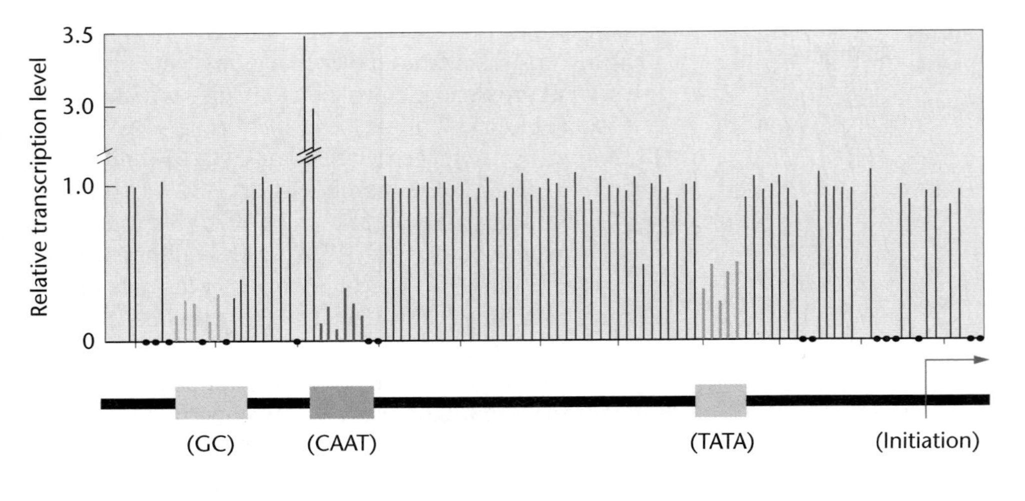

FIGURE 16–14 Summary of the effects of point mutations in the promoter region on transcription of the β-globin gene. Each line represents the level of transcription produced by a single nucleotide mutation (relative to wild type) in a separate experiment. Dots represent nucleotides where no mutation was obtained. Note that mutations within the specific elements of the promoter have the greatest effect on the level of transcription.

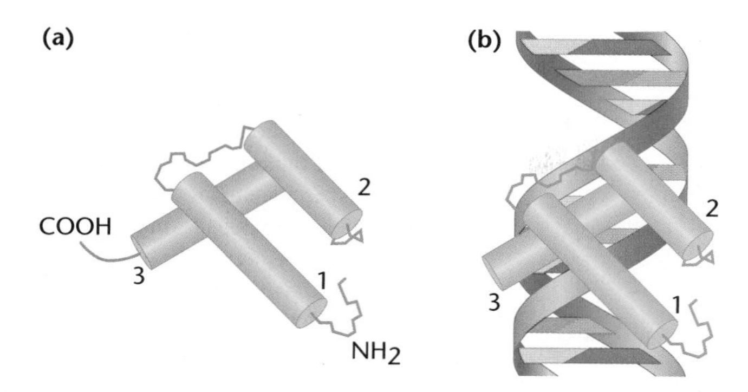

FIGURE 16–15 Illustration of an HTH motif where (a) three planes of the α helix of the protein are established. (b) These domains bind in the grooves of the DNA molecule.

being recognized as important components of gene regulation. Before we provide specific examples of such factors and discuss how they regulate gene expression, let's explore what is known about *how* they bind to nucleic acids—a critical feature in genetic function.

Structural Motifs of Transcription Factors

The DNA-binding domains of eukaryotic transcription factors take on several forms. They have distinctive three-dimensional structural patterns or motifs. There are three major types of these structural motifs: helix–turn–helix, zinc finger, and basic leucine zippers. This classification is not exhaustive, and other new groups will undoubtedly be established as new factors are characterized.

The first DNA-binding domain to be discovered was the **helix–turn–helix (HTH) motif**. In prokaryotes, HTH motifs have been identified in the *lac* repressor, the *trp* repressor, and other proteins (Figure 16–15); and studies indicate that the HTH motif is present in many eukaryotic DNA-binding proteins. This motif is characterized by its geometric conformation rather than a distinctive amino acid sequence. Two adjacent α helices separated by a "turn" of several amino acids enables the protein to bind to DNA and provides the motif's name. Unlike several of the other DNA-binding patterns, the HTH motif cannot fold or function alone, but is always part of a larger DNA-binding domain.

The potential for forming helix–turn–helix geometry has been recognized in distinct regions of a large number of eukaryotic genes known to regulate developmental processes. The **homeobox**, a stretch of 180 bp, is present almost universally in eukaryotic organisms. It specifies a **homeodomain** sequence of 60 amino acids that can form a helix–turn–helix structure. Of the 60 amino acids, many are basic (arginine and lysine), and a conserved sequence is found among these genes. We will discuss homeobox-containing genes in Chapter 20 because of their significance in developmental processes.

Zinc fingers are one of the major structural families of eukaryotic transcription factors, and they are involved in many aspects of gene regulation. Originally discovered in a transcription factor in the frog, *Xenopus laevis*, this structural motif has been identified in proto-oncogenes that regulate cell growth in *Drosophila* and in proteins whose synthesis is induced by growth factors and differentiation signals. There are several types of zinc-finger proteins, each with a distinctive structural pattern.

One zinc-finger protein contains clusters of two cysteine and two histidine residues at repeating intervals (Figure 16–16). The interspersed cysteine and histidine residues covalently bind zinc atoms, folding the amino acids into loops (the zinc fingers). Each finger consists of approximately 23 amino acids, with a loop of 12–14 amino acids between the cysteine and histidine residues and a linker between the loops consisting of 7 or 8 amino acids. The amino acids in the loop interact with and bind to specific DNA sequences. Studies have shown that zinc fingers bind in the major groove of the DNA helix and wrap at least partway around the DNA. Within the major groove, the zinc finger makes contact with a set of DNA bases and may form hydrogen bonds with the bases, especially in G-rich strands. The number of fingers in a zinc-finger transcription factor varies from 2 to 13; the length of the DNA-binding sequence also varies in size.

A third type of domain is represented by the **basic leucine zipper (bZIP)**, first seen as a stretch of 35 amino acids in a nuclear protein in rat liver. In the bZIP, four leucine residues are spaced seven amino acids apart and flanked by basic amino acids. The leucine-rich regions form a helix with leucine residues protruding at every other turn. When two of these molecules dimerize (Figure 16–17), the leucine residues are "zipped" together. The dimer contains two α-helical regions adjacent to the zipper, which bind to phosphate residues and specific bases in DNA, thus making the dimer look like a pair of scissors.

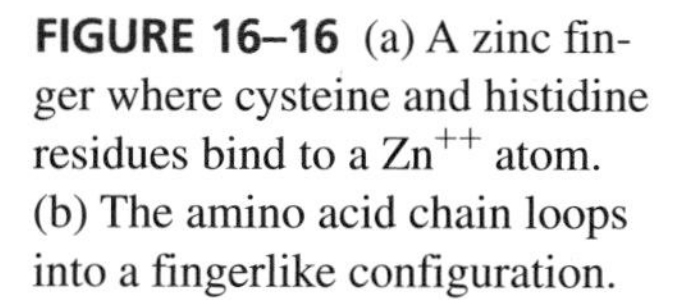

FIGURE 16–16 (a) A zinc finger where cysteine and histidine residues bind to a Zn^{++} atom. (b) The amino acid chain loops into a fingerlike configuration.

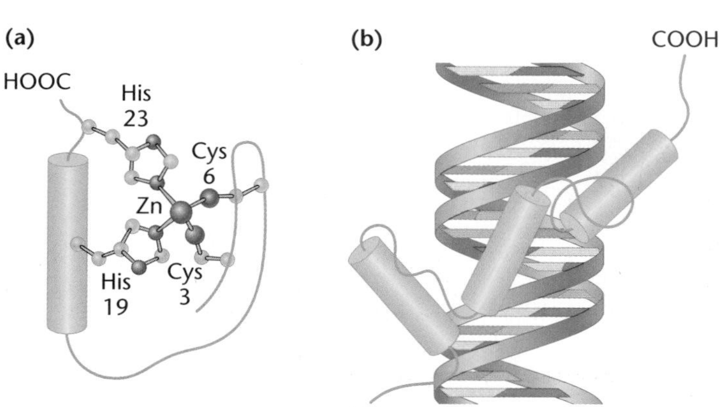

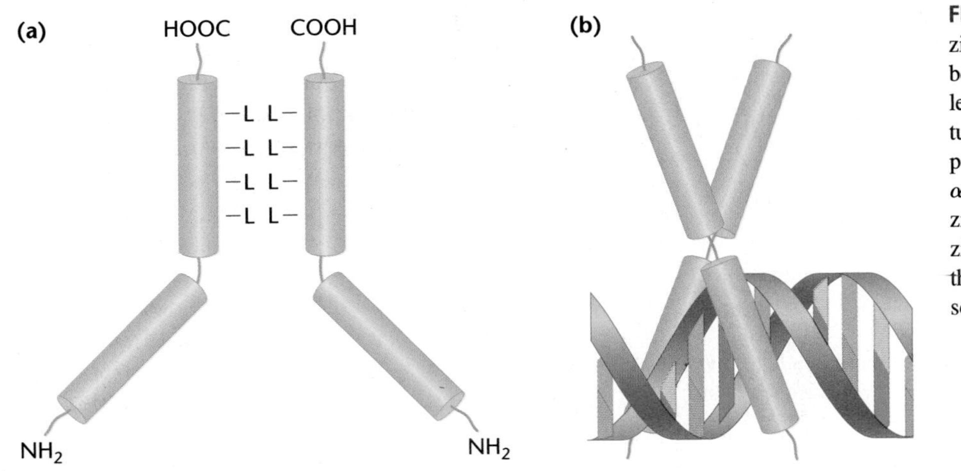

FIGURE 16–17 (a) A leucine zipper. Dimer formation occurs because of the presence of leucine residues at every other turn of the α helix in facing polypeptide chains. (b) When the α-helical regions form a leucine zipper, the regions beyond the zipper form a Y-shaped region that grips the DNA in a scissorslike configuration.

In addition to domains that bind DNA, transcription factors contain domains that activate transcription. These regions can occupy from 30 to 100 amino acids and are distinct from the DNA-binding domains. Such stretches of amino acids interact with other transcription factors (such as those that bind to the TATA sequence) or directly with the RNA polymerase.

Assembly of the Transcription Complex

We can now examine specifically how eukaryotic gene regulation occurs. A number of insights into the transcriptional apparatus of type II genes in eukaryotes (those transcribed by RNA polymerase II) are now available, providing the basis for a model for the initiation of mRNA synthesis (Figure 16–18). Central to the model are the var-

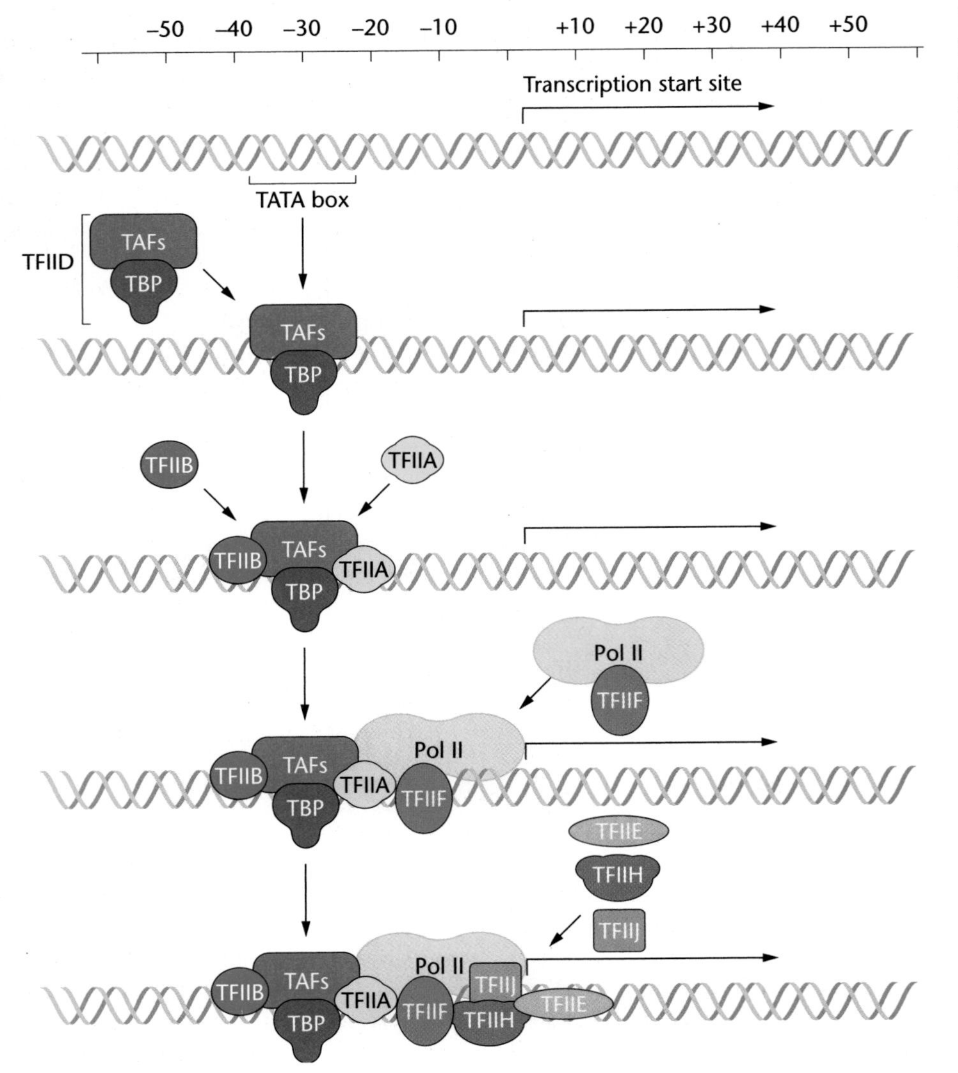

FIGURE 16–18 Assembly of the transcription complex in eukaryotes. The initial step involves binding TBP to the TATA box, which bends the DNA. A combination of transcription factors (TFIIA, TFIIB, etc.) and RNA polymerase II are added in stepwise fashion prior to the initiation of transcription.

ious trans-acting protein factors that facilitate template binding by RNA polymerase II. These transcription factors are well-known and designated TFIIA, TFIIB, and so on. One of them, TFIID, is a multi-subunit protein complex that includes a protein that binds directly to the TATA box of the promoter. Thus, it is called the **TATA-binding protein (TBP)**. TFIID includes at least 11 **TBP-associated factors (TAFs)**. In all genes studied, binding of the TBP at the TATA box of the promoter is essential to subsequent RNA polymerase binding at the start site. About 20 bp of DNA are involved in binding the TBP, and the other subunits then bind to the growing complex.

Other transcriptional factors are then assembled at the promoter region, in a specific order. TFIID (which includes TBP) responds to contact with activator proteins by undergoing conformational changes that expedite the binding of other transcription factors and RNA polymerase. As shown in Figure 16–18, TFIIB and TFIIA bind first, followed by TFIIF, which is complexed with RNA polymerase. After addition of TFIIE and TFIIH, the initiation complex is complete, and transcription is initiated at the start site just downstream from the TATA box. TFIIH is a large multi-subunit complex that demonstrates helicase activity, facilitating the separation of the DNA helix at the start site. As the transcription proceeds, TBF remains bound to the TATA sequence of the promoter, but the other transcription factors dissociate from the complex.

At this point, transcription of the DNA downstream will ensue at a low *basal* level. The final stage involves the achievement of the *induced* state, where transcription is stimulated at a much higher level. This state involves other areas of the promoter region, enhancers, and still other transcription factors, which together influence the rate at which RNA polymerase initiates transcription. Multiple transcription factors bind cooperatively to enhancers and then to the transcription complex.

This general model is believed to constitute the mechanism for eukaryotic gene regulation. As you will see, there are still other aspects of the genetic machinery that also come into play.

Chromatin Remodeling, DNA Methylation, and Gene Expression

We know that DNA in eukaryotes is complexed with proteins at interphase in the nucleus and is present in the form of **chromatin**. You already learned that chromatin is characterized by the presence of repeating structures called **nucleosomes**, each of which consists of an octamer of 4 different histone proteins and about 200 bp of DNA. When complexed to histone proteins in this way, eukaryotic DNA is relatively inaccessible to transcription factors, and thus the formation of an active transcription complex cannot occur. Therefore, the normal structure of chromatin seems to be sufficient to significantly repress gene activity. As a result, activation of genes requires **chromatin remodeling**, whereby the conformation of chromatin is altered in a way that the proteins of the nucleosome are released from the DNA, allowing it to become accessible to transcription factors and RNA polymerase.

There are several ways in which this might occur. One involves specialized proteins that actually disrupt the nucleosome particles and remove the histones, thus freeing up the DNA. For example, several specific proteins in yeast, called **SWI** and **SNF proteins**, are part of a larger complex that functions to disrupt chromatin structure by displacing or removing the proteins making up the nucleosome, thereby facilitating transcription. Similar complexes have been found in other eukaryotes. In another mechanism, **acetylation** of histone proteins by specific enzymes lessens the attraction of histones to DNA, resulting in a remodeling of chromatin structure. Conversely, deacetylation involves specific enzymes that reverse this process and is thought to repress gene activity.

Another type of change in chromatin that plays a role in gene regulation involves adding or removing methyl groups to the bases in DNA. The DNA of most eukaryotic organisms is modified after replication by the enzyme-mediated addition of methyl groups to bases and sugars. **DNA methylation** most often involves cytosine. Approximately 5% of the cytosine residues are methylated in the genome of any given eukaryotic species.

The ability of base methylation to alter gene expression is known from studies on the *lac* operon in *E. coli*. Methylation of DNA in the operator region, even at a single cytosine residue, can cause a marked change in the affinity of the repressor for the operator. Methylation of cytosine occurs at the number 5 carbon of the pyrimidine ring, causing the methyl group to protrude into the major groove of the DNA helix, where it can alter the binding of proteins to the DNA. This happens most often when the cytosine residue is part of a CG doublet in DNA, where the cytosines on both strands are affected:

5′-mCpG-3′
3′-GpCm-5′

Consistent with this information, analysis of the methylation of a given gene in different tissues shows that, in general, if a gene is expressed, its promoter region is not methylated or has a low level of methylation. Further, in mammalian females, the X chromosome that forms a Barr body is almost totally inactive in gene expression and has been shown to have a significantly higher level of methylation than its counterpart X chromosome, which is transcriptionally active. Within the chromosome condensed into the Barr body, small regions remain active and have much lower levels of methylation than is seen in adjacent, inactive regions.

The absence of methyl groups in DNA is related to increased gene expression, but methylation cannot be regarded as a universal mechanism for gene regulation because methylation is not a phenomenon characterizing all eukaryotes. In *Drosophila*, for example, little or no methylation of DNA has been observed. Thus, methylation may represent only one of several ways in which gene expression can be regulated by genomic changes.

16.7 Steroid Hormones Regulate Some Genes

Our final consideration of eukaryotic gene regulation (at the level of transcription) involves a short discussion of how **steroid hormones** affect their target cells. Our knowledge of this system provides a nice recap of the various aspects of eukaryotic regulation discussed thus far. Steroid hormones are used to regulate growth and development and to maintain homeostasis. The major sex hormones, vitamin D, and the homeostatic adrenal hormones that regulate glucose metabolism and mineral utilization are all steroids.

The general scheme of steroid hormone action is illustrated in Figure 16–19. Hormones enter the cell by passing through the plasma membrane and binding to a specific **hormone receptor protein** in the cytoplasm. The receptor–hormone complex is translocated to the nucleus and activates transcription of one or more specific genes.

Hormone receptor proteins and the DNA sequences to which they bind have been studied intensively for more than a decade. All of the receptors analyzed have three functional domains: a variable N-terminus domain, unique to each receptor; a short, highly conserved central domain that binds to DNA; and a C-terminus domain that binds to the hormone. The DNA-binding central domain contains two zinc fingers. The zinc fingers in hormone receptors bind to specific DNA sequences known as **hormone-responsive elements (HREs)**.

HREs share some characteristics with promoters and enhancers. They are composed of short consensus sequences that are related but not always identical. HREs are often located several hundred bases upstream from the transcription start site and may be present in multiple copies. They are also often present in promoter or enhancer sequences.

While binding the receptor to the HRE is necessary for activating a specific gene, it may not be sufficient on its own to cause activation. Binding the receptor to the HRE may serve simply to facilitate the interaction of other transcription factors by altering the chromatin structure in the HRE and adjacent regions. Such an alteration may make other binding sites, including the promoter, available to bind transcription factors and RNA polymerase II, resulting in the initiation of transcription.

In some ways, then, steroid hormone regulation of gene transcription in eukaryotes is similar to the positive-control systems found in some of the operons in prokaryotes. An external effector binds to a cytoplasmic receptor, changes the configuration of the receptor, and the receptor moves to the DNA, where it acts as a transcription factor. In other ways, however, the regulation of gene expression is quite different in that the DNA-binding sequence is often at a great distance from the regulated gene and alterations in chromatin structure mediate the action of the receptor.

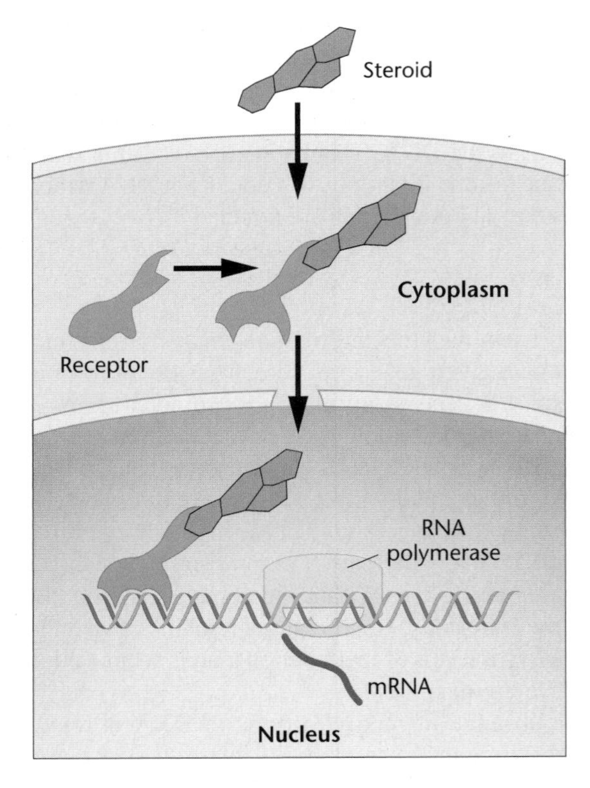

FIGURE 16–19 The effect of a steroid hormone on gene expresssion. The hormone enters the cell and binds to a specific receptor, leading to a conformational change in a DNA binding domain of the receptor. The complex enters the nucleus and behaves as a transcription factor, stimulating a target gene to initiate transcription.

16.8 Posttranscriptional Events Also Regulate Gene Expression

As we have seen, the regulation of genetic expression occurs at many points along the pathway from DNA to protein. Although transcriptional control is perhaps the most obvious and widely used mode of regulation in eukaryotes, **posttranscriptional regulation** also occurs in many organisms. Eukaryotic nuclear RNA transcripts are modified prior to translation, noncoding introns are removed, the remaining exons are precisely spliced together, and the mRNA is modified by the addition of a cap at the 5′-end and a poly-A tail at the 3′-end. The message is then complexed with proteins and exported to the cytoplasm. Each of these processing steps offers several opportunities for regulation. We shall examine one mechanism that is especially important in eukaryotes: alternative splicing of a single mRNA transcript to give multiple mRNAs.

Alternative Splicing of mRNA

Alternative splicing can generate different forms of a protein, so that expression of one gene can give rise to a family of related proteins. Figure 16–20 illustrates an example where the polypeptide products derived from a single type of pre-mRNA are distinct from one another. Here, an initial bovine pre-mRNA transcript is processed into one or two **preprotachykinin mRNAs (PPT mRNAs)**.

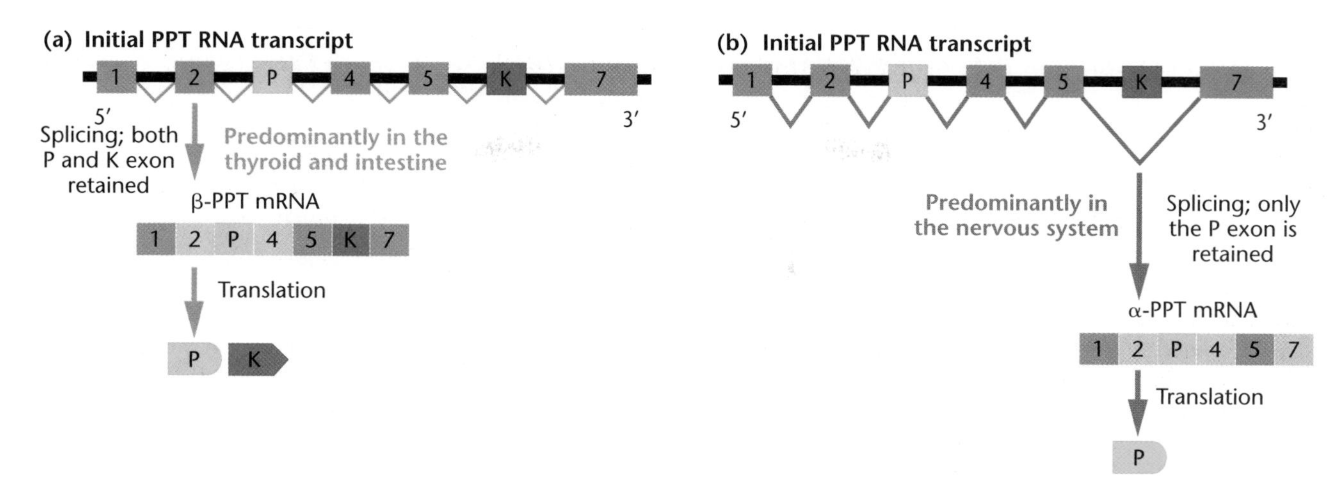

FIGURE 16–20 Alternative splicing of the initial RNA transcript of PPT mRNA. Exons are either numbered or designated by the letters P or K. (a) The inclusion of P and K exons leads to β-PPT mRNA, which, upon translation, yields both the P and K tachykinin neuropeptides. (b) When the K exon is excluded, α-PPT mRNA is produced and only the P neuropeptide is synthesized.

This precursor mRNA molecule potentially includes the genetic information specifying **P** and **K neuropeptides**. These two peptides, which are members of the family of sensory neurotransmitters referred to as **tachykinins**, are believed to play different physiological roles. While the P neuropeptide is largely restricted to tissues of the nervous system, the K neuropeptide is found more predominantly in the intestine and thyroid.

The RNA sequences for both neuropeptides are derived from the same gene. However, the processing of the initial PPT RNA transcript can occur in two ways. In Figure 16–20a, processing that includes both the P and K exons yields β-PPT mRNA, which upon translation results in the synthesis of both the P and K neuropeptides. Conversely, exclusion of the K exon during processing results in the α-PPT mRNA, which upon translation yields neuropeptide P but not K (Figure 16–20b). The analysis of the relative levels of the two types of RNA demonstrates striking differences between tissues. In nervous-system tissues, α-PPT mRNA predominates by as much as a factor of three, while β-PPT mRNA is the predominant type in the thyroid and intestine. This should not be surprising, since it parallels the location of the P and K neuropeptides.

Given the existence of alternative splicing, how many different polypeptides can be derived from the same pre-mRNA? Work on α-tropomyosin, an accessory protein that regulates muscle contraction, has provided a partial answer to this question. In rats, the α-tropomyosin gene contains a total of 14 exons, 6 of which make up 3 pairs that are alternatively spliced. Only one member of each pair ends up in the finished mRNA, never both. Alternative splicing of this pre-mRNA results in 10 different forms of α-tropomyosin, many of which are tissue-specific. Another gene, for the muscle form of troponin T, a protein that regulates the calcium needed for contraction, produces 64 known forms of the protein from a single pre-mRNA by alternative splicing.

GENETICS, TECHNOLOGY, AND SOCIETY

Why Is There No Effective AIDS Vaccine?

AIDS is the deadliest plague of the last century, surpassing the great influenza epidemic of 1918. It causes unimaginable human suffering and threatens to destabilize vast areas of the world—economically, socially, and politically. Since the epidemic began in the 1970s, it has progressed into every country and continent of the world. Over 6 million people are infected each year. At present, 36 million people are infected with HIV worldwide and over 100 million will be infected by the end of the decade. Approximately 90% of AIDS cases occur in developing countries, particularly in sub-Saharan Africa and Southeast Asia. One tenth of these are children, and one-half are 15–24-year-olds. According to a recent United Nations report, one-third of the work force in Zimbabwe will die from AIDS by the year 2005. Since the beginning of the epidemic, 13 million children have been orphaned following the death of their HIV-infected parents. It is estimated that there will be over 40 million AIDS orphans by the end of the decade. Since the beginning of the epidemic, over 21 million people have died. Despite preventive measures, AIDS continues to increase in North America. Approximately 45,000 people are infected each year. At present, about 1 million people in North America are living with AIDS.

Ironically, we know how to prevent the spread of HIV infection: modification of sexual practices, needle exchanges, screening of the blood supply, treatment of infected pregnant women. But these measures have only slowed, not stopped, HIV spread. The powerful drug combinations now used to treat AIDS are extremely costly, preventing their use in parts of the world that need treatments the most. In addition, the virus is developing drug resistance.

The need for a vaccine to stop this global pandemic is obvious. If medical science can eradicate smallpox and control polio and tuberculosis, then why can we not control AIDS? It is not from lack of trying. Since 1987, more than 60 vaccine clinical trials were conducted worldwide. These trials show that candidate HIV vaccines are safe and result in production of antibodies that recognize the protein used as the vaccine. But no vaccine to date has been able to effectively neutralize a real virus infection.

Historically, the best viral vaccines have been prepared from live-attenuated viruses or whole viruses that have been killed. However, in the case of HIV, there are obvious safety concerns; hence, no human clinical trials have been performed with either live-attenuated or whole-killed viruses.

To date, most candidate HIV vaccines use the HIV envelope protein—called gp120 or gp160—as the immunogen. This protein is present in the external coating of the virus and is the viral surface protein that binds to the human cells that HIV infects. The gp120/160 vaccines take many forms: small synthetic peptides, recombinant proteins synthesized within organisms such as vaccinia virus or bacteria, or even naked DNA encoding gp160 protein. So far, tests of these vaccines have yielded disappointing results. Although the gp120/160 vaccines elicit antibody responses to the protein in the vaccine, they are not effective against real viruses isolated from the wild.

Why is it so difficult to develop an HIV vaccine? The reasons are a combination of biological and social factors. Perhaps the greatest impediment to developing an effective AIDS vaccine is the genetic plasticity of the virus. The reverse transcriptase of HIV-1 is extremely error-prone, introducing one or more mutations into the HIV genome during each round of replication. In addition, the highest rate of mutation occurs in the env gene, which encodes the gp120/160 protein. As a result, gp120/160 is continually mutating, both within each infected individual and throughout the world. A vaccine that triggers production of antibodies against one type of gp120/160 protein will be less effective against subtypes in other parts of the world or against HIV mutants that arise spontaneously within each person. To further complicate the story, gp120/160 proteins are heavily glycosylated, with sugar molecules protecting the envelope proteins from detection by circulating antibodies.

HIV's stealthy life cycle also interferes with vaccine effectiveness. Although HIV is transmitted as free virus in body fluids, it rapidly becomes sequestered within host cells. After infection, HIV's genetic material integrates into the host cell's genome, remaining quiescent for long periods of time. During this latent period, the virus is undetectable by the host's immune system. The final punch in HIV's biological assault is its choice of host cell. By infecting and destroying the host's helper T cells, the virus disables the immune system and ensures its ultimate escape from immune clearance.

HIV vaccine development has been handicapped by the lack of a relevant animal model system in which to test candidate vaccines. The only nonhuman animal that can be infected with HIV is the chimpanzee; however, serious economic and ethical considerations affect the use of these higher primates as experimental animals. Scientists also test HIV vaccines in monkeys using an HIV-like virus called simian immunodeficiency virus (SIV) or a recombinant virus of SIV and HIV. However, SIV may be too far removed genetically from HIV and may not be a relevant model for HIV vaccine action. Many scientists argue that the only true experimental animal for testing HIV vaccines is the human and that extensive clinical trials of multiple vaccine types should be conducted as quickly as possible.

But this is not as simple as it sounds. Phase III clinical trials (efficacy trials) must be conducted on thousands of volunteers who are not presently infected with HIV, but are at risk of infection. These volunteers must be followed over many years, and such trials require tens of millions of dollars to conduct. Volunteers may ultimately test positive for HIV as a result of vaccine injection. This has raised fears that they may suffer discrimination in obtaining employment or insurance. In industrialized countries, people with the highest risk of HIV infection are the hardest to recruit to vaccine trials. They include injection drug users, high-risk male homosexuals, and those with other sexually transmitted diseases. These populations are difficult to identify and do not trust government or academia. In some developing countries, the incidence of HIV infection is extremely high, providing a larger population from which to recruit for vaccine trials. However, the lack of trained investigators and medical infrastructure make trials difficult to conduct in these countries. In addition, most candidate vaccines have been developed by Western researchers and are based on strains of HIV found predominantly in industrialized countries. Understandably, few developing countries want to participate in trials when there is no certainty that they will benefit from or have access to any vaccine that emerges from these trials. The scientific uncertainties and difficulties in recruiting for clinical trials has led to a vicious circle. Pharmaceutical companies are reluctant to invest millions of dollars in research and clinical trials when it is uncertain whether effective vaccines can be devised against HIV. However, without extensive clinical trials of many types of candidate vaccines, it is unlikely that we will ever know whether an AIDS vaccine will be effective. The type of global, high-risk research and clinical testing required to develop effective AIDS vaccines will involve extensive private and public sector cooperation and high levels of investment from industrialized nations. Given the horrific human and social consequences of the global AIDS pandemic, there appears to be little choice.

References

Vastag, B. 2001. HIV vaccine efforts inch forward. *J.A.M.A.* 286: 1826–28.

Johnston, M. I., and Flores, J. 2001. Progress in HIV vaccine development. *Curr. Opin. Pharmac.* 1: 504–10.

Nabel, G. J. 2001. Challenges and opportunities for development of an AIDS vaccine. *Nature* 410: 1002–07.

Web Site

HIV vaccine development status report, NIH.
http://www.niaid.nih.gov/aidsvacine/_whsummarystatus.htm

Chapter Summary

1. A system of genetic regulation must exist if the complete genome is not to be continuously active in transcription throughout the life of every cell of all species. Highly refined mechanisms have evolved that regulate transcription, maintaining genetic efficiency.
2. Both inducible and repressible operons, illustrated by the *lac* and *trp* gene complexes, respectively, have been documented and studied in *E. coli*. The complexes involve genes of a regulatory nature in addition to structural genes that code for the enzymes of the system. Both operons are under the negative control of a repressor molecule.
3. The catabolite-activating protein (CAP) facilitates the binding of RNA polymerase to the promoter in the *lac* and other operons. Catabolite repression, a mechanism that represses the operon when glucose is present, has evolved presumably because glucose can be used more efficiently than lactose.
4. Transcription in eukaryotes is controlled by regulatory DNA sequences known as promoters and enhancers. Regulatory sequences, including the TATA, GC, and CAAT boxes, are elements of promoters found near the transcriptional starting point. Enhancer elements, which appear to control the degree of transcription, can be located before, after, or within the gene expressed.
5. Transcription factors are proteins that bind to DNA recognition sequences within the promoters and enhancers and activate transcription through protein–protein interaction. Such factors display various types of motifs that are related to their potential binding to DNA.
6. Alteration of chromatin conformation is one way in which gene expression can be regulated. The degree of methylation of cytosine residues in DNA is believed to alter chromatin, with greater methylation ultimately reducing transcription.
7. Gene expression in eukaryotes can be regulated by steroid hormones originating outside the cell. The signals are transduced by proteins that shuttle between the cytoplasm and nucleus, regulating patterns of transcription.
8. Several types of posttranscriptional control of gene expression are possible in eukaryotes. One such mechanism is alternative splicing of a single class of pre-mRNAs to generate multiple mRNA species.

Key Terms

acetylation, 367
adenyl cyclase, 359
allosteric repressor, 355
alternative splicing, 368
attenuation, 362
attenuator, 362
β-galactosidase, 354
basic leucine zipper (bZIP), 365
CAAT box, 363
catabolite-activating protein (CAP), 358
catabolite repression, 358
chromatin, 367
chromatin remodeling, 367
cis-acting element, 354
constitutive mutant, 355
co-repressor, 361
cyclic adenosine monophosphate (cAMP), 359
DNA-binding domain, 364
DNA methylation, 367
enhancer, 364
equilibrium dialysis, 358
GC box, 363
gratuitous inducer, 354
helix–turn–helix (HTH) motif, 365
homeobox, 365
homeodomain, 365
hormone-receptor protein, 368
hormone-responsive element (HRE), 368
inducer, 353
inducible enzyme, 353
isopropylthiogalactoside (IPTG), 354
K neuropeptide, 369
lac operon, 354
lactose operon, 354
leader sequence, 361
merozygote, 355
negative control, 353
nucleosome, 367
operator region, 355
operon model, 355
P neuropeptide, 369
permease, 354
polycistronic mRNA, 354
positive control, 353
posttranscriptional regulation, 368
preprotachykinin mRNAs (PPT mRNA), 368
promoter region, 358

promoter, 363
repressible system, 353
repression loop, 360
repressor gene, 355
repressor molecule, 355
SNF protein, 367
SWI protein, 367
steroid hormone, 368
tachykinin, 369
TATA-binding protein (TBP), 367
TATA box, 363
TBP-associated factor (TAF), 367
transacetylase, 354
trans-acting element, 354
trans-activating domain, 364
transcription factor, 364
trp RNA-binding attenuation protein (TRAP), 362
tryptophan synthetase, 360
zinc finger, 365

INSIGHTS AND SOLUTIONS

1. A theoretical operon (*theo*) in *E. coli* contains several structural genes encoding enzymes that are involved sequentially in the biosynthesis of an amino acid. Unlike the *lac* operon, where the repressor region is separate from the operon, the gene encoding the regulator molecule is contained within the *theo* operon. When the end product (the amino acid) is present, it combines with the regulator molecule. This complex then binds to the operator region, repressing the operon. In the absence of the amino acid, the regulatory molecule fails to bind to the operator and transcription proceeds.

Characterize this operon; then consider the effect of (a) a mutation in the operator region, (b) a mutation in the promoter region, and (c) a mutation in the regulator gene. In each case, consider also the situation in which the wild-type gene is present along with the mutant gene in partially diploid cells (F′). Specify if the operon will be active or inactive in transcription, assuming that the mutation affects the regulation of the *theo* operon, and compare each response to the equivalent situation in the *lac* operon.

Solution: The operon is under negative control and is repressible. The regulatory molecule, when bound to the amino acid, binds to the operator region and inhibits gene expression.

(a) As in the *lac* operon, a mutation in the *theo* operator region inhibits binding with the repressor complex, and transcription occurs constitutively. The presence of an F′ plasmid bearing the wild-type allele would have no effect.

(b) A mutation in the *theo* promoter region may well inhibit binding to RNA polymerase and therefore inhibit transcription. This would also happen in the *lac* operon. A wild-type promoter present in an F′ plasmid would have no effect.

(c) A mutation in the *theo* regulator gene, as in the *lac* system, may inhibit either its binding to the repressor or its binding to the operator region. In both cases, transcription will be constitutive because the *theo* system is repressible. Both cases result in the failure of the regulator to bind to the operator, allowing transcription to proceed. In the *lac* system, failure to bind the corepressor lactose would permanently repress the system. The addition of a wild-type allele would restore repressibility, provided that this gene was transcribed constitutively.

Problems and Discussion Questions

1. Contrast the need for the enzymes involved in the metabolism of lactose and tryptophan in bacteria (a) in the presence of lactose and tryptophan, and (b) in their absence.
2. Contrast positive and negative control systems.
3. Contrast the role of the repressor in an inducible system and in a repressible system.
4. For the following *lac* genotypes, predict whether the structural genes of the operon are constitutive, permanently repressed, or inducible in the presence of lactose.

Genotype	*Constitutive*	*Repressed*	*Inducible*
$I^+O^+Z^+$			X
$I^-O^+Z^+$			
$I^+O^CZ^+$			
$I^-O^+Z^+$/F′ I^+			
$I^+O^CZ^+$/F′ O^+			
$I^SO^+Z^+$			
$I^SO^+Z^+$/F′ I^+			

5. Predict whether functional enzymes are made, nonfunctional enzymes are made, or no enzymes are made for the genotypes and condition in the table below.

Genotype	*Condition*	*Functional Enzyme*	*Nonfunctional Enzyme*	*No Enzym*
$I^+O^+Z^+$	No lactose			X
$I^+O^CZ^+$	Lactose			
$I^-O^+Z^-$	No lactose			
$I^-O^+Z^-$	Lactose			
$I^-O^+Z^+$/F′ I^+	No lactose			
$I^+O^CZ^+$/F′ O^+	Lactose			
$I^+O^+Z^-$/F′ $I^+O^+Z^+$	Lactose			
$I^-O^+Z^-$/F′ $I^+O^+Z^+$	No lactose			
$I^SO^+Z^+$/F′ O^+	No lactose			
$I^+O^CZ^+$/F′ O^+Z^+	Lactose			

6. In a theoretical bacterial operon, regions *a*, *b*, *c*, and *d* represent the repressor gene, the promoter sequence, the operator region, and the structural gene, but not necessarily in that order. This operon regulates the metabolism of a theoretical molecule (tm). From the data given below, first decide if the operon is inducible or repressible. Then assign *a*, *b*, *c*, and *d* to the four parts of the operon.

Genotype	*tm Present*	*tm Absent*
$a^+b^+c^+d^+$	AE	NE
$a^-b^+c^+d^+$	AE	AE
$a^+b^-c^+d^+$	NE	NE
$a^+b^+c^-d^+$	IE	NE
$a^+b^+c^+d^-$	AE	AE
$a^-b^+c^+d^+/F'\ a^+b^+c^+d^+$	AE	AE
$a^+b^-c^+d^+/F'\ a^+b^+c^+d^+$	AE	NE
$a^+b^+c^-d^+/F'\ a^+b^+c^+d^+$	AE + IE	NE
$a^+b^+c^+d^-/F'\ a^+b^+c^+d^+$	AE	NE

AE = active enzyme, IE = inactive enzyme, NE = no enzyme.

7. Predict the level of genetic activity of the *lac* operon as well as the status of the *lac* repressor molecule and the CAP protein under the cellular conditions listed in the accompanying table.

	Lactose	*Glucose*
(a)	−	−
(b)	+	−
(c)	−	+
(d)	+	+

8. Even though the *lacZ*, *lacY*, and *lacA* structural genes are transcribed as a single polycistronic mRNA, each gene contains the appropriate initiation and termination signals essential for translation. Predict what will happen when a cell growing in the presence of lactose contains a deletion of one nucleotide (a) early in the *lacZ* gene and (b) early in the *lacA* gene.
9. Predict the effect on the inducibility of the *lac* operon of a mutation that disrupts the function of (a) the *crp* gene, which encodes the CAP protein, and (b) the CAP-binding site within the promoter.
10. Describe how the *lac* repressor was isolated.
11. Describe the evidence that the *lac* repressor, once isolated, indeed serves as a repressor molecule within the operon scheme.
12. A bacterial operon is responsible for the production of the biosynthetic enzymes needed to make the theoretical amino acid tisophane (tis). The operon is regulated by a separate gene product, *R*. Deletion of the *R* gene causes the loss of enzyme synthesis. In the wild-type condition, when tis is present, no enzymes are made. In the absence of tis, the enzymes are made. Mutations in the operator gene (O^-) result in repression regardless of the presence of tis. (a) Is the operon under positive or negative control? Propose a model for (b) repression of the genes in the presence of tis in wild-type cells, and (c) the O^- mutations.
13. The SOS DNA repair genes in *E. coli* are negatively regulated by the *lexA* gene product, called the LexA repressor. When a cell sustains extensive damage to its DNA, the LexA repressor is inactivated by the *recA* gene product, and transcription of the SOS genes is increased dramatically. One of the SOS genes is the *uvrA* gene. You are studying the function of the *uvrA* gene product in DNA repair. You isolate a mutant strain that shows constitutive expression of the UvrA protein. You name this mutant strain $uvrA^C$. The simple diagram below shows the *lexA* and *uvrA* operons. (a) Describe two different mutations that would result in a *uvrA* constitutive phenotype. Indicate the actual genotypes involved. (b) Outline a series of genetic experiments, using partial diploid strains that would allow you to determine which of the two possible mutations you have isolated.

$P^{Lex\ A}$ $O^{Lex\ A}$ Lex A P^{uvrA} O^{uvrA} uvrA

14. A fellow student considers the issues in Problem 13 and argues that there is a more straightforward, nongenetic experiment that could differentiate between the two types of mutations. While the experiment requires no fancy genetics techniques, you must be able to easily assay the products of the other SOS genes. Propose such an experiment.
15. (a) Why is gene regulation assumed to be more complex in a multicellular eukaryote than in a prokaryote? (b) Why is the study of this phenomenon in eukaryotes more difficult?
16. List and define the potential levels of gene regulation in eukaryotes.
17. Distinguish between promoters and enhancers.
18. Describe the role of transcription factors in the regulation of gene expression.
19. Is the binding of a transcription factor to its DNA recognition sequence necessary and sufficient for initiation of transcription of a regulated gene? What else plays a role in this process?
20. How are genes regulated by steroid hormones?
21. Contrast the modification of chromatin structure and posttranscriptional modes of gene regulation in eukaryotes.
22. Contrast and compare gene regulation in prokaryotes and eukaryotes.
23. A marine bacterium is isolated and shown to contain an inducible operon whose genetic products metabolize oil when it is encountered in the environment. Investigation demonstrates that the operon is under positive control and that there is a *reg* gene whose product interacts with an operator region (*O*) to regulate the structural genes designated *sg*. To understand how the operon functions, a constitutive mutant strain and several partial diploid strains are isolated and tested with the results shown below. Draw all possible conclusions about the mutation as well as the nature of regulation of the operon. Is the constitutive mutation in the trans-acting *reg* element or in the cis-acting *O* element?

Host Chromosome	*F′ Factor*	*Phenotype*
Wild type	None	Inducible
Wild type	*reg* Gene from mutant strain	Inducible
Wild type	Operon from mutant strain	Constitutive
Mutant strain	*reg* Gene from wild type	Constitutive

24. You are interested in studying transcription factors and have developed an *in vitro* transcription system by using a defined segment of DNA that is transcribed under the control of a eukaryotic promoter. The transcription of this DNA occurs when you add purified RNA polymerase II, TFIID (the TATA-binding factor), and TFIIB and TFIIE (which bind to RNA polymerase). You perform a series of experiments that compare the efficiency of transcription in the "defined system" with the efficiency of transcription in a crude nuclear extract. You test the two systems with your template DNA and with various deletion templates that you have generated. The results of your study are shown below. (a) Why is there no transcription from the −11 deletion template? (b) How do the results for the nuclear extract and the defined system differ from the *undeleted* template? How would you interpret these results? (c) Compare the results with the nuclear extract and the purified system for the various deleted templates. How would you interpret the results with the deleted templates? Be very specific about what you can conclude from these data.

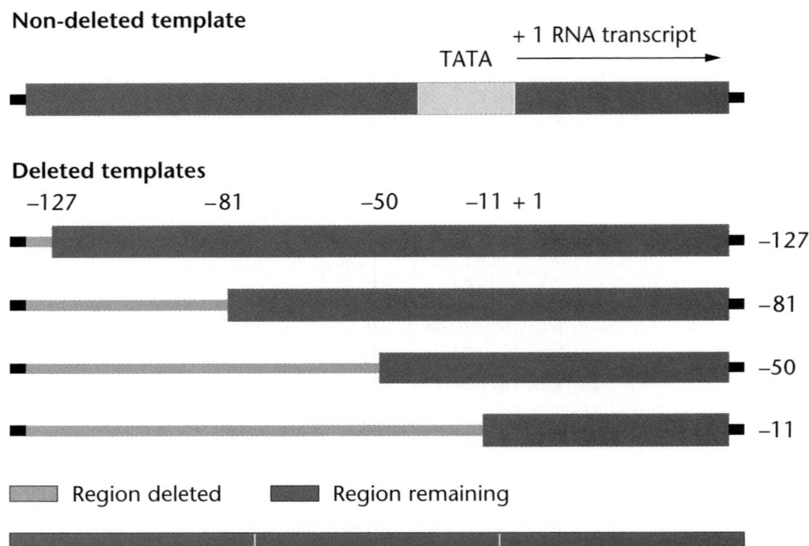

DNA added	Nuclear extract	Purified system
Undeleted	++++	+
–127 deletion	++++	+
–81 deletion	++++	+
–50 deletion	+	+
–11 deletion	o	o

Key: + Low-efficiency transcription
++++ High-efficiency transcription
o No transcription

25. While it is customary to consider transcriptional regulation in eukaryotes as resulting from the positive or negative influence of different factors binding to DNA, a more complex picture is emerging. For instance, Ducret and others (1999. *Mol. and Cell. Biol.* 19: 7076–87) described the action of a transcriptional repressor (Net) that is regulated by nuclear export. Under neutral conditions, Net inhibits transcription of target genes; however, when phosphorylated, Net stimulates transcription of target genes. When stress conditions exist in a cell (ultraviolet light or heat shock), Net is excluded from the nucleus, and target genes are transcribed. Devise a model that includes diagrams that provide a consistent explanation of these three conditions.

26. Given that alternative splicing can lead to different populations of RNAs from a single primary transcript, there has been considerable appeal to regard selective nuclear transport of RNAs as a form of genetic regulation in eukaryotes. The discovery of what appears to be an array of nuclear fibrous elements adds to that appeal. Pederson (2000. *Molecular Biology of the Cell.* 11: 799–805) reviewed literature from the late 1990s, which reported on various labeling experiments designed to determine whether gene regulation in eukaryotes is related to nuclear sorting of pre-mRNAs. The vast majority of the results indicated that nuclear poly-A RNA moves by diffusion. What influence would these results have on models that relate gene regulation to nuclear RNA transport?

27. Noncoding DNA sequences in eukaryotes are viewed more by what is absent from the DNA (promoters, exons, genes, terminators) than by what possible functions might exist. Nevertheless, it is possible that noncoding DNA may be involved in cell-specific genetic regulation by generating particular chromatin folding patterns that open or conceal certain genetic regions for transcription. What is an inherent weakness in this suggestion, and how does this weakness apply to other suggested models for regulation in eukaryotes?

28. Compartmentalization of eukaryotic cellular contents theoretically provides an opportunity for gene regulation. One focus for determining regulation is the import and export of proteins and nucleic acids through nuclear pores. Molecular cargo appears to be transported primarily by two classes of proteins, importin and exportin, as well as other proteins. (Macara, I. 2001. *Microbiol. and Mol. Biol. Rev.* 65: 570–94.) Suggest a role for such transport proteins in eukaryotic gene regulation.
29. DNA supercoiling occurs when coiling tension is generated ahead of the replication fork where it is relieved by DNA gyrase. Supercoiling may also be involved in genetic regulation. Liu and colleagues (2001. *Pro. Natl. Acad. Sci. (USA)* 98: 14,883–88) discovered that transcriptional enhancers operating over a long distance (2500 base pairs) are dependent on DNA supercoiling while enhancers operating over shorter distances (110 base pairs) are not so dependent. Using a diagram, suggest a way in which supercoiling may positively influence enhancer activity over long distances.
30. DNA methylation is commonly associated with reduction of transcription. Irvine and colleagues (2002. *Mol. and Cell. Biol.* 22: 6689–96) studied the impact of the location of DNA methylation relative to gene activity in human cells. The data below describe the relative expression of a reporter gene (luciferase) with variable DNA methylation outside and within the transcription region. What general conclusions can be drawn from this data?

DNA Segment	*Patch size of Methylation (kb)*	*Number of Methylated CpGs*	*Relative Luciferase Expression*
Outside transcription unit	0.0	0	490X
(0–7.6 kb away)	2.0	100	290X
	3.1	102	250X
	12.1	593	2X
Inside transcription unit	0.0	0	490X
	1.9	108	80X
	2.4	134	5X
	12.1	593	2X

31. Attenuation of the *trp* operon was viewed as a relatively inefficient way to achieve genetic regulation when it was first discovered in the 1970s. Since then, however, attenuation has been found to be a relatively common, effective regulatory strategy. There are also examples in eukaryotes with similarities to attenuation in prokaryotic systems. Assuming that attenuation is a relatively inefficient way to achieve genetic regulation, what might explain its widespread use?

Selected Readings

Aso, T., Shilatifard, A., Conaway, J. W., and Conaway, R. C. 1996. Transcription syndromes and the role of RNA polymerase II general transcription factors in human disease. *J. Clin. Investig.* 97: 1561–69.

Beato, M. 1989. Gene regulation by steroid hormones. *Cell* 56: 335–44.

Beckwith, J. R., and Zipser, D. (eds.), 1970. *The Lactose Operon.* Cold Spring Harbor: Cold Spring Harbor Laboratory Press.

Berget, S. M. 1995. Exon recognition in vertebrate splicing. *J. Biol. Chem.* 270: 2411–14.

Bertrand, K. et al. 1975. New features of the regulation of the tryptophan operon. *Science* 189: 22–26.

Black, D. L. 2000. Protein diversity from alternative splicing: A challenge for bioinformatics and post-genomic biology. *Cell* 103: 367–70.

Blumenthal, T. 1995. *Trans*-splicing and polycistronic transcription in *Caenorhabditis elegans*. *Trends Genet.* 11: 132–36.

Buratowski, S. 1995. Mechanisms of gene activation. *Science* 270: 1773–74.

Busch, S. J., and Sassone-Corsi, P. 1990. Dimers, leucine zippers and DNA-binding domains. *Trends Genet.* 6: 36–40.

Cremer, T., and Cremer, C. 2001. Chromosome territories, nuclear architecture and gene regulation in mammalian cells. *Nat. Rev. Gen.* 2: 292–301.

Dillon, N., and Sabbattini, P. 2000. Functional gene expression domains: Defining the functional unit of eukaryotic gene regulation. *Bioess.* 22: 657–65.

Drapkin, R., Merino, A., and Reinberg, D. 1993. Regulation of RNA polymerase II transcription. *Curr. Opin. Cell Biol.* 5: 469–76.

Dynan, W. S. 1988. Modularity in promoters and enhancers. *Cell* 58: 1–4.

Edelson, E. 1990. Transcription factors: Governors for the genetic engine. *Mosaic* 21: 2–9.

Gilbert, W., and Müller-Hill, B. 1966. Isolation of the *lac* repressor. *Proc. Natl. Acad. Sci. (USA)* 56: 1891–98.

———. 1967. The *lac* operator in DNA. *Proc. Natl. Acad. Sci. (USA)* 58: 2415–21.

Gilbert, W., and Ptashne, M. 1970. Genetic repressors, *Sci. Am.* (June) 222: 36–44.

Hayes, J. J., and Wolffe, A. P. 1992. The interaction of transcription factors with nucleosomal DNA. *BioEss.* 14: 597–603.

Jacob, F., and Monod, J. 1961. Genetic regulatory mechanisms in the synthesis of proteins. *J. Mol. Biol.* 3: 318–56.

Jacobsen, A., and Peltz, S. 1996. Interrelationships of the pathway of mRNA decay and translation in eukaryotic cells. *Ann. Rev. Biochem.* 65: 693–739.

Jacobson, R. H., and Tjian, R. 1996. Transcription factor IIA: A structure with multiple functions. *Science* 272: 830–36.

Kass, S. U., Pruss, D., and Wolffe, A. P. 1997. How does DNA methylation repress transcription? *Trends in Genet.* 13: 444–49.

Lee, T. I., and Young, R. A. 2000. Transcription of eukaryotic protein-coding genes. *Ann. Rev. Genet.* 34: 77–137.

Lewis, M. et al. 1996. Crystal structure of the lactose operon repressor and its complexes with DNA and inducer. *Science* 271: 1247–54.

Littlewood, T. D., and Evans, G. I. 1994. Transcription factors 2: Helix-loop-helix. *Protein Profile* 1: 639–709.

Lopez, A. J. 1995. Developmental role of transcription factor isoforms generated by alternate splicing. *Dev. Bio.* 172: 396–411.

Lohr, D. 1997. Nucleosome transactions on the promoters of the yeast *GAL* and *PHO* genes. *J. Biol. Chem.* 272: 26,795–98.

Lugar, K. et. al. 1997. Crystal structure of the nucleosome core particle at 2.8 Å resolution. *Nature* 389: 251–56.

Maniatis, T., Goodbourn, S., and Fischer, J. A. 1987. Regulation of inducible and tissue-specific expression. *Science* 236: 1237–45.

Maniatis, T., and Ptashne, M. 1976. A DNA operator–repressor system. *Sci. Am.* (Jan.) 234: 64–76.

McCarthy, J. E., and Brimacombe, R. 1994. Prokaryotic translation: The interactive pathway leading to initiation. *Trends Genet.* 10: 402–7.

Meehan, R. et. al. 1992. Transcriptional repression by methylation of CpG. *J. Cell Sci.* (Suppl.) 16: 9–14.

Miller, J. H., and Reznikoff, W. S. 1978. *The Operon.* Cold Spring Harbor: Cold Spring Harbor Laboratory Press.

Mitchell, P. J., and Tjian, R. 1989. Transcriptional regulation in mammalian cells by sequence-specific DNA binding proteins. *Science* 245: 371–78.

Müller-Hill, B. 1996. *The* lac *Operon: A Short History of a Genetic Paradigm.* Hawthorne: Walter de Gruyter.

Pieler, T., and Bellefroid, E. 1994. Perspectives on zinc finger protein function and evolution—an update. *Mol. Biol. Rep.* 20: 1–8.

Ptashne, M., and Gann, A. 2002. *Genes and Signals.* Cold Spring Harbor: Cold Spring Harbor Laboratory Press.

Ptashne, M., Johnson, A. D., and Pabo, C. O. 1982. A genetic switch in a bacterial virus. *Sci. Am.* (Nov.) 247: 128–40.

Stringer, K. F., Ingles, C. J., and Greenblatt, J. 1990. Direct and selective binding of an acidic transcriptional activation domain to the TATA-box factor TFIID. *Nature* 345: 783–86.

Struhl, K. 1993. Yeast transcription factors. *Curr. Opin. Cell Biol.* 5: 513–20.

Tate, P., and Bird, A. 1993. Effects of DNA methylation on DNA-binding proteins and gene expression. *Curr. Opin. Genet. Dev.* 3: 226–31.

Tjian, R. 1995. Molecular machines that control genes. *Sci. Am.* 272: 54–61.

Valbuzzi, V., and Yanofsky, C. 2001. Inhibition of the *B. subtilis* regulatory protein TRAP by the TRAP-inhibitory protein. *Science* 293: 2057–61.

Workman, J. L., and Kingston, R. E. 1998. Alteration of nucleosome structure as a mechanism of transcriptional control. *Ann. Rev. Biochem.* 67: 545–79.

Yankofsky, C. 1981. Attenuation in the control of expression of bacterial operons. *Nature* 289: 751–58.

Yanofsky, C., and Kolter, R. 1982. Attenuation in amino acid biosynthetic operons. *Annu. Rev. Genet.* 16: 113–134.

CHAPTER 17

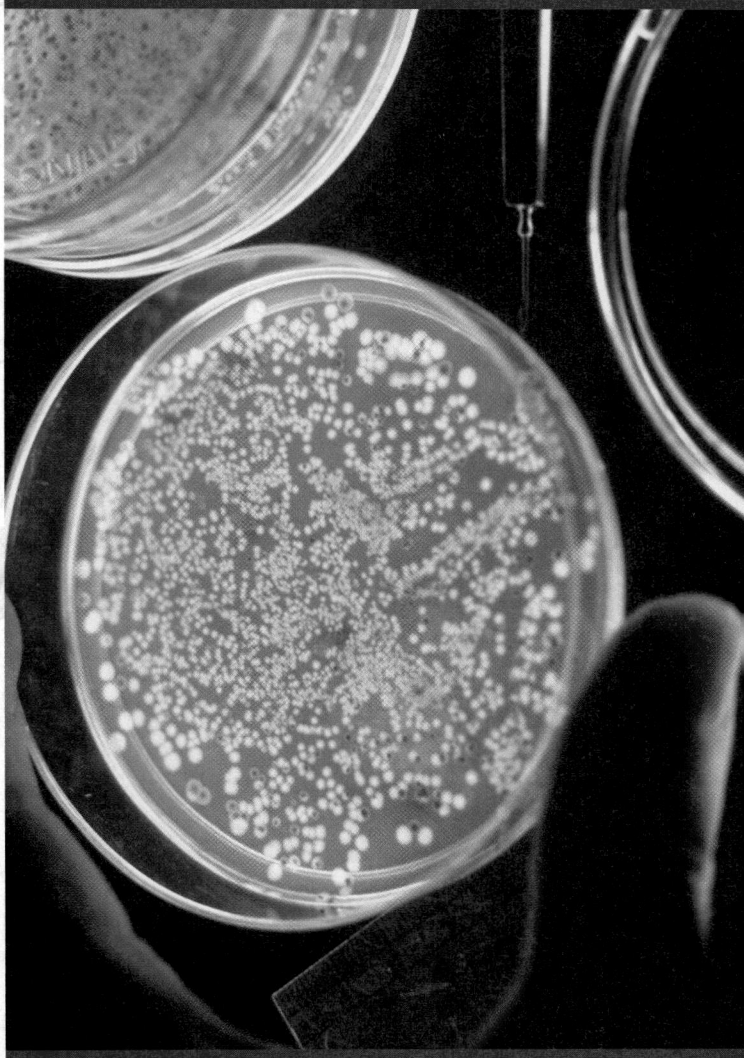

A Petri dish showing the growth of host cells after uptake of recombinant plasmids. *(Michael Gabridge/Visuals Unlimited)*

Recombinant DNA Technology

CHAPTER CONCEPTS

Techniques to create, replicate, and analyze recombinant DNA molecules were developed in the mid to late 1970s as a way for researchers to isolate and study specific DNA sequences. Recombinant DNA technology makes it possible to identify and isolate a single gene from the thousands or tens of thousands present in a genome and to produce large quantities of this gene in the form of cloned DNA molecules. In addition, the cloned gene can be transferred into cells that will synthesize its encoded gene product, which can then be recovered and purified for use in research, medicine, or industry.

Clones are identical organisms, cells, or molecules descended from a single ancestor. Cloning a gene produces many identical copies that can be used for numerous purposes, including research into its structure and organization or the commercial production of its encoded protein. In this chapter, we review the basic methods of recombinant DNA technology used to isolate, replicate, and analyze genes. In Chapter 19, we will discuss some applications of this technology to research, medicine, the legal system, agriculture, and industry.

HOW DO WE KNOW WHAT WE KNOW?

IN THIS CHAPTER, WE WILL FOCUS ON the essential techniques of recombinant DNA technology, including the methods used to clone DNA molecules and the ways in which clones of interest can be identified and analyzed. As you study this topic, you should try to answer several fundamental questions:

1. How do we know that restriction enzymes recognize palindromic sequences running in opposite directions on complementary strands of a DNA molecule?
2. How do we know that DNA fragments have been successfully incorporated into vectors such as plasmids?
3. How do we know that PCR requires both a primer and a heat-resistant polymerase?
4. How do we know that a cloned DNA fragment contains a specific gene? ■

17.1 An Overview of Recombinant DNA Technology

The term **recombinant DNA** refers to a combination of DNA molecules that are not found together in nature. Although genetic processes such as crossing over produce recombined DNA molecules, the term recombinant DNA is generally reserved for molecules produced by joining DNA obtained from different biological sources.

Recombinant DNA technology uses methods derived from nucleic acid biochemistry, coupled with genetic techniques originally developed for the study of bacteria and viruses. This technology is a powerful tool for isolating potentially unlimited quantities of a gene. Although several methods are available, the basic procedure involves the following steps:

1. DNA to be cloned is purified from cells or tissues.
2. Proteins called **restriction enzymes** are used to generate specific DNA fragments. These enzymes recognize and cut DNA molecules at specific nucleotide sequences.
3. The fragments produced by restriction enzymes are joined to other DNA molecules that serve as **vectors**, or carrier molecules. A vector joined to a DNA fragment is a **recombinant DNA molecule**.
4. The recombinant DNA molecule is transferred to a host cell. Within the host cell, the recombinant molecule replicates, producing dozens of identical copies, or clones, of the recombinant molecule.
5. As host cells replicate, the recombinant DNA molecules within them are passed on to all their progeny, creating a population of host cells, each of which carries copies of the cloned DNA sequence.
6. The cloned DNA can be recovered from host cells, purified, and analyzed.
7. The cloned DNA can then be transcribed, its mRNA translated, and the encoded gene product isolated and used for research or sold commercially.

17.2 Recombinant DNA Molecules Are Constructed Using Several Components

Recombinant DNA and gene-cloning technology provides scientists with methods for isolating large quantities of specific genes or other DNA sequences; this has facilitated studies of gene organization, structure, and expression. It has also given rise to the multibillion dollar biotechnology industry.

Restriction Enzymes

The cornerstone of recombinant DNA technology is a class of enzymes called restriction enzymes. These enzymes, isolated from bacteria, restrict or prevent viral infection in bacterial cells by degrading the invading viral DNA. Restriction enzymes are endonucleases that recognize a specific nucleotide sequence and cut both strands of the DNA within that sequence. The 1978 Nobel Prize in Physiology or Medicine was awarded to Werner Arber, Hamilton Smith, and Daniel Nathans for their work on restriction enzymes. To date, over 200 restriction enzymes have been identified. Their usefulness in cloning derives from their ability to reproducibly cut DNA into fragments.

One of the first restriction enzymes to be identified was isolated from *Escherichia coli* and is designated *Eco*RI (pronounced echo-r-one). Its nucleotide recognition sequence and cleavage pattern are shown in Figure 17–1.

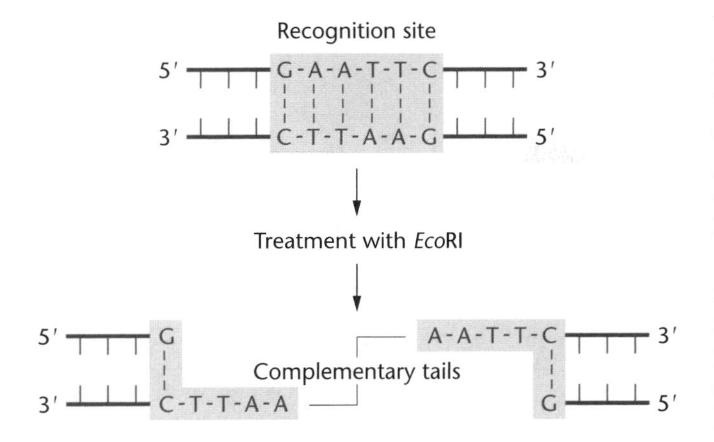

FIGURE 17–1 The restriction enzyme *Eco*RI recognizes and binds to the palindromic nucleotide sequence GAATTC. Cleavage of the DNA at this site produces complementary single-stranded tails. These single-stranded tails can anneal with single-stranded tails from other DNA fragments to form recombinant DNA molecules.

DNA fragments produced by *Eco*RI digestion have overhanging single-stranded tails ("sticky ends") that can form hydrogen bonds with complementary single-stranded tails on other DNA fragments. If they are mixed under the proper conditions, DNA fragments from two sources form recombinant molecules by hydrogen bonding of their sticky ends. The fragments can be covalently linked to form recombinant DNA molecules by the enzyme **DNA ligase** (Figure 17–2). Some restriction enzymes and their recognition sequences are shown in Figure 17–3.

Vectors

Fragments of DNA produced by restriction enzyme digestion cannot directly enter bacterial cells for cloning. However, when a DNA fragment is joined to a vector, it can gain entry to a host cell, where it can be replicated or cloned into many copies. Vectors are, in essence, carrier DNA molecules.

To serve as a vector, a DNA molecule must be able to independently replicate itself and the DNA segment it carries. It should also contain several cleavage sites that are present only once in the vector. One of these sites is cleaved with a restriction enzyme and used to insert a DNA fragment from another source that has been cut with the same enzyme. A vector should carry a selectable marker gene (usually antibiotic resistance genes or genes for enzymes absent from the host cell), to distinguish host cells that carry vectors from host cells that do not contain vectors. Finally, it should be easy to recover from the host cell. Many different vectors are used, including those derived from **plasmids** and **bacteriophages**.

Genetically modified plasmids were the first vectors developed and are still widely used for cloning. Plasmids used as vectors are derived from naturally occurring, extrachromosomal, double-stranded DNA molecules that replicate autonomously within bacterial cells (Figure 17–4). Although only a single plasmid may enter a host cell, once inside, some plasmids can increase their copy number so that several hundred copies are present. When used as vectors, such plasmids allow more copies of cloned DNA to be produced and are genetically engineered to contain a number of convenient restriction

FIGURE 17–2 DNA from different sources is cleaved with *Eco*RI and mixed to allow annealing to form recombinant molecules. The enzyme DNA ligase then chemically bonds these annealed fragments into an intact recombinant DNA molecule.

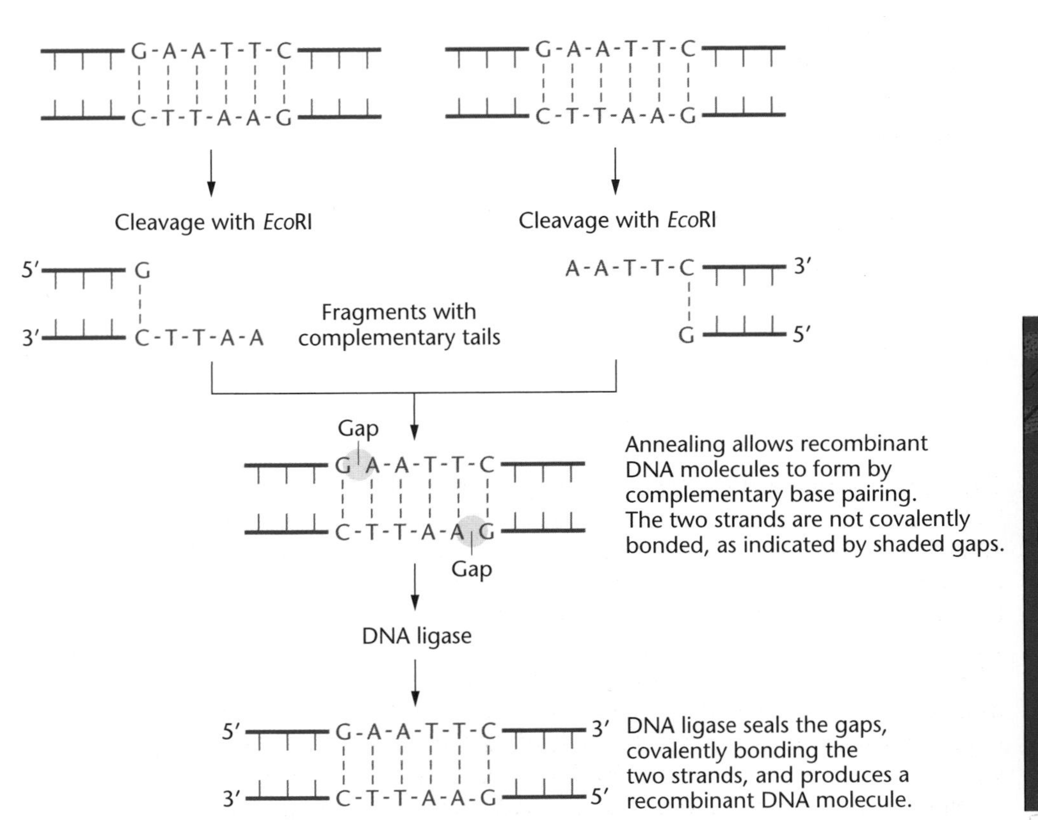

MEDIA TUTORIAL Recombinant DNA technology: restriction enwymes

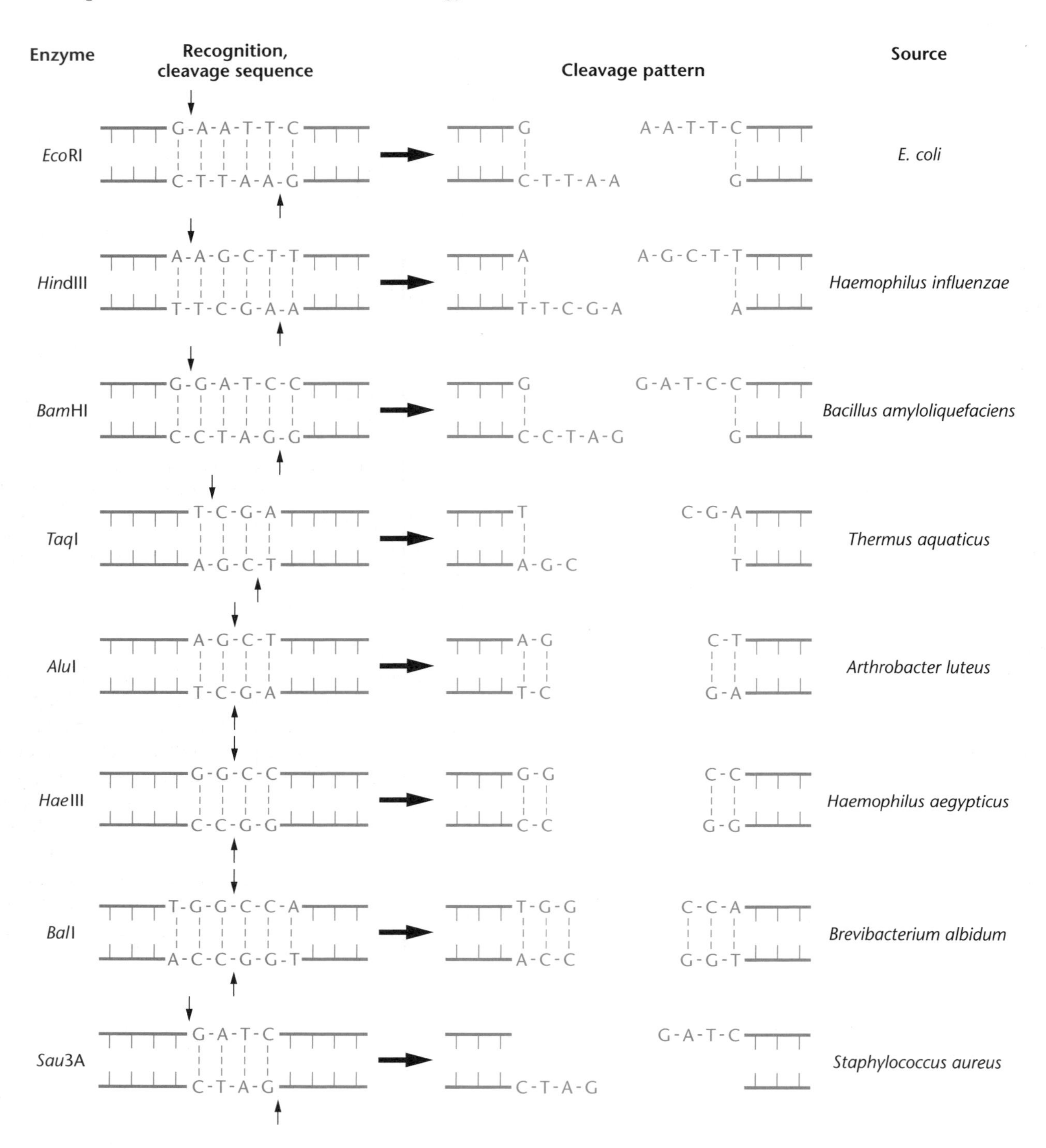

FIGURE 17–3 Some common restriction enzymes, with their recognition sequences, cutting sites, cleavage patterns, and sources.

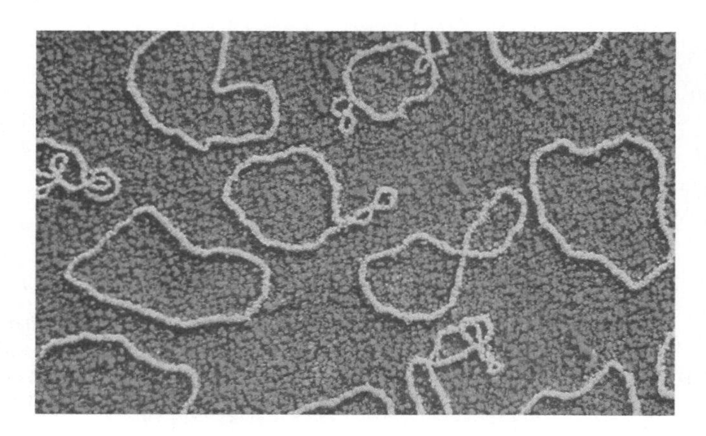

FIGURE 17–4 A color-enhanced electron micrograph of circular plasmid molecules isolated from *E. coli*. Genetically engineered plasmids are used as vectors for cloning DNA. *(K.G. Murti/Visuals Unlimited)*

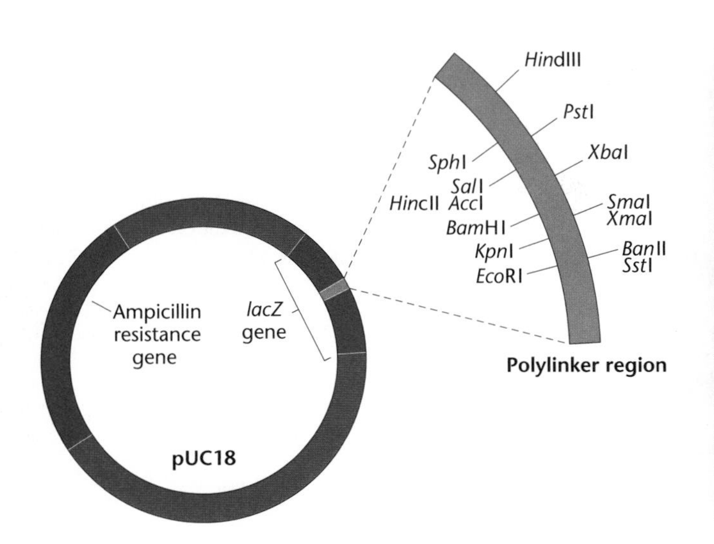

FIGURE 17–5 A diagram of the plasmid pUC18 showing the polylinker region, located within a *lacZ* gene. DNA inserted into the polylinker region disrupts the *lacZ* gene, resulting in white colonies that allow direct identification of bacterial colonies carrying cloned DNA inserts.

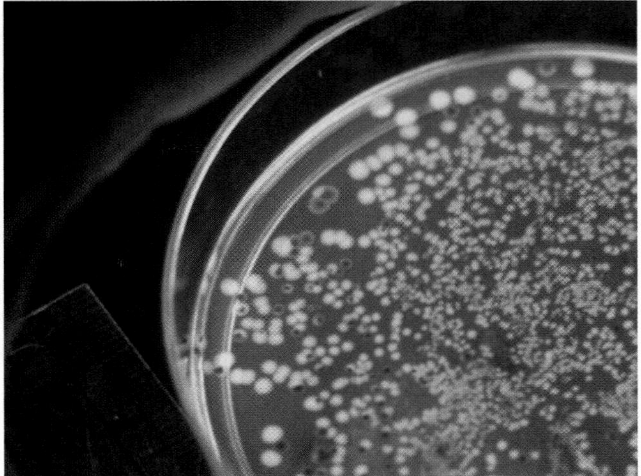

FIGURE 17–6 A Petri dish showing the growth of bacterial cells after uptake of recombinant plasmids. The medium on the plate contains a compound called Xgal. DNA inserts into the pUC18 vector disrupt the gene responsible for the formation of blue colonies. Cells in the blue colonies do not carry any cloned DNA inserts, whereas the white colonies contain vectors carrying DNA inserts. *(Michael Gabridge/Visuals Unlimited)*

sites and marker genes that reveal the presence of plasmids in host cells.

Many genetically engineered plasmid vectors are now available with features that make it easy to identify host cells carrying a plasmid with an inserted DNA fragment. One such plasmid is pUC18 (Figure 17–5), which has several useful features as a vector.

- It is small (2686 bp), so it can carry relatively large DNA inserts.
- In a host cell, it replicates up to 500 copies per cell, thus producing many copies of inserted DNA fragments.
- A large number of restriction enzyme sites have been engineered into pUC18, conveniently clustered in one region called a **polylinker site**.
- It allows recombinant plasmids to be easily identified. For example, pUC18 carries a fragment of the bacterial *lacZ* gene, and the polylinker is inserted into this fragment. Expression of *lacZ* causes bacterial host cells carrying pUC18 to produce blue colonies when grown on medium containing a compound known as Xgal. If a DNA fragment is inserted into the polylinker site, the *lacZ* gene is inactivated and a bacterial cell carrying pUC18 with an inserted DNA fragment forms white colonies, making them easy to identify (Figure 17–6).

Plasmid vectors generally carry up to 10 kb of inserted DNA, but for many experiments, larger pieces of DNA are necessary. For these purposes, genetically modified strains of **phage λ** are used as vectors (Figure 17–7). All of the genes in phage λ have been identified, mapped, and sequenced. The central third of its chromosome can be replaced with foreign DNA without affecting the phage's ability to infect cells and form plaques (Figure 17–8).

To clone DNA using this vector, the phage DNA is cut with a restriction enzyme such as *Eco*RI, resulting in a left arm, a right arm, and the dispensable central region. The arms are isolated and mixed with DNA from another source that also has been cut with *Eco*RI. Ligation with DNA ligase follows, producing recombinant molecules. Phage λ vectors containing inserted DNA are packaged into phage heads *in vitro* and then introduced into bacterial host cells, where they reproduce to form particles of infective phage, each of which carries a DNA insert. As they reproduce, the phages form plaques from which the cloned DNA can be recovered.

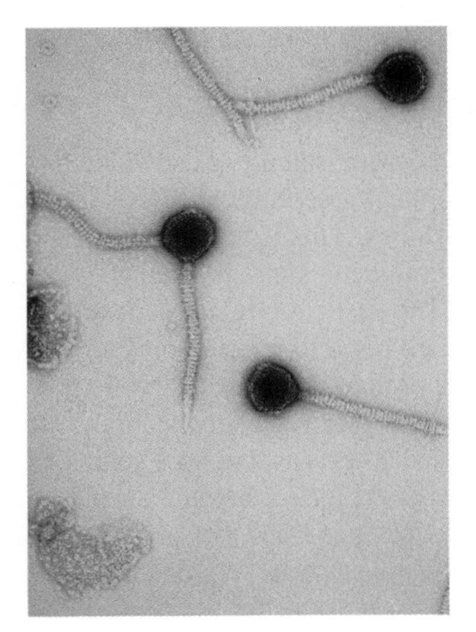

FIGURE 17–7 A colorized electron micrograph of several phage λ, widely used as a vector in recombinant DNA work. *(Dr. M. Wurtz/Biozentrum, University of Basel/Science Photo Library/Photo Researchers, Inc.)*

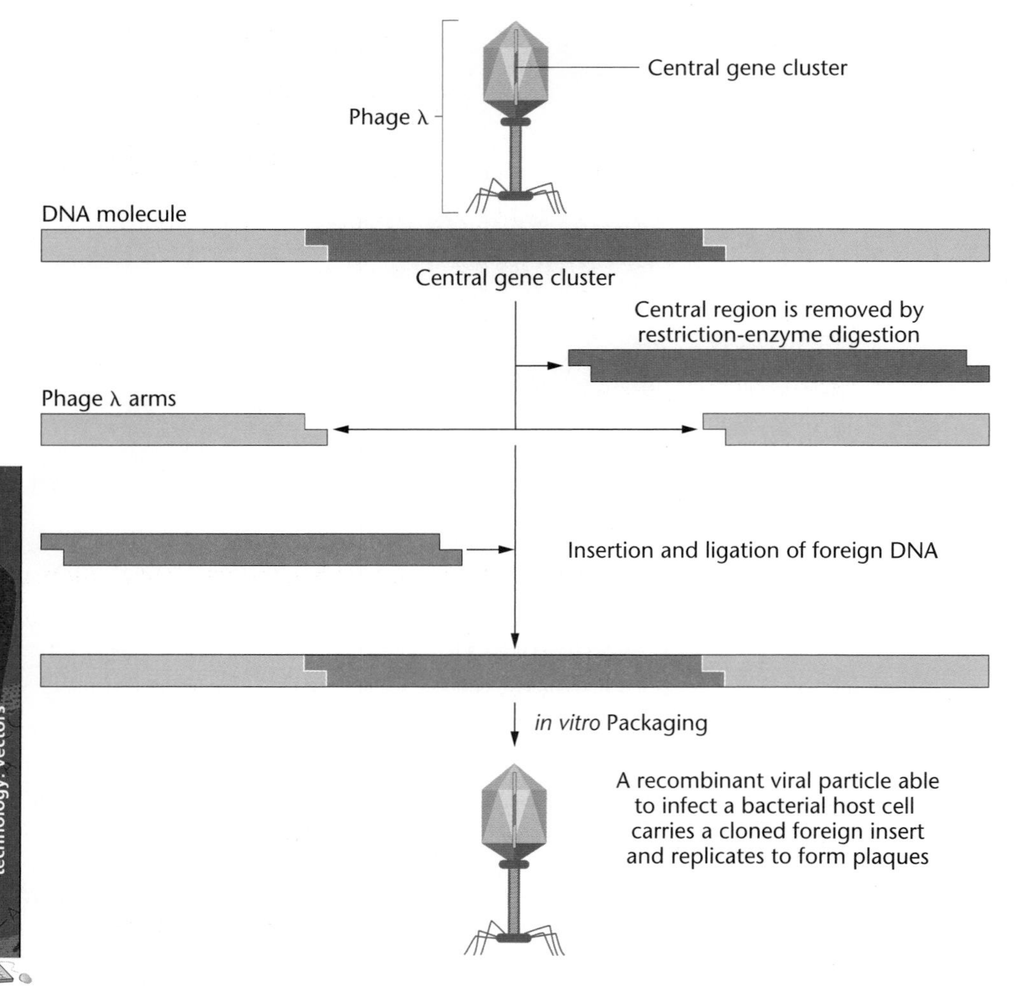

FIGURE 17–8 Phage λ as a vector. DNA is extracted from the phage, the central gene cluster is removed, and the DNA to be cloned is ligated into the arms of the phage λ chromosome. The recombinant chromosome is then packaged into phage proteins to form a recombinant virus.

MEDIA TUTORIAL Recombinant DNA technology: vectors

Phage vectors can carry inserts of about 20 kb, more than twice as long as DNA inserts in plasmid vectors. This is an important advantage when cloning entire genomes. In addition, some phage vectors accept only inserts of a minimum size; they do not carry relatively useless inserts of a few dozen to a few hundred nucleotides in length.

17.3 Cloning in Prokaryotic Host Cells

As discussed earlier, scientists use biotechnology to construct *and* replicate recombinant DNA molecules to produce many clones. This is accomplished by transferring recombinant molecules into host cells, where replication takes place. The first cloning method, cell-based cloning, has been widely used to make copies of DNA fragments.

A variety of prokaryotic and eukaryotic cells can serve as hosts for recombinant vector replication. One of the most commonly used hosts is a laboratory strain of the bacterium *E. coli* known as K12. *E. coli* strains such as K12 are genetically well characterized, and they are hosts for a wide range of vectors. Several steps are required to create recombinant DNA molecules and transfer them to an *E. coli* host cell (Figure 17–9). First, the DNA to be cloned is isolated and treated with a restriction enzyme to create fragments ending in a specific sequence. The fragments are then ligated to plasmid molecules that have been cut with the same restriction enzyme, creating a recombinant vector. The recombinant vector is transferred into an *E. coli* host cell, where the recombinant plasmid replicates to form dozens of copies. The final step is to plate the bacteria, allow them to form colonies, and analyze them to identify those that have taken up the recombinant plasmids. Because the cells in each colony are derived from a single ancestral cell, all the cells in the colony and the plasmids they contain are genetically identical clones. In a similar manner, phages containing foreign DNA are used to infect *E. coli* host cells, and each resulting plaque represents a cloned descendant of a single ancestral bacteriophage.

17.4 Cloning in Eukaryotic Host Cells

The yeast *Saccharomyces cerevisiae* is extensively used as a host cell for the cloning and expression of eukaryotic genes. While yeast is a eukaryotic organism, it can be grown and manipulated in much the same way as bacterial cells. Yeast also has been intensively studied, providing a large catalog of mutations and a highly developed genetic map. In addition, the entire yeast genome has been sequenced, and most genes in the organism have been identified. To study the function of some eukaryotic proteins, it is necessary to use a host cell that can posttranslationally modify the protein to convert it into a functional form—bacterial host cells

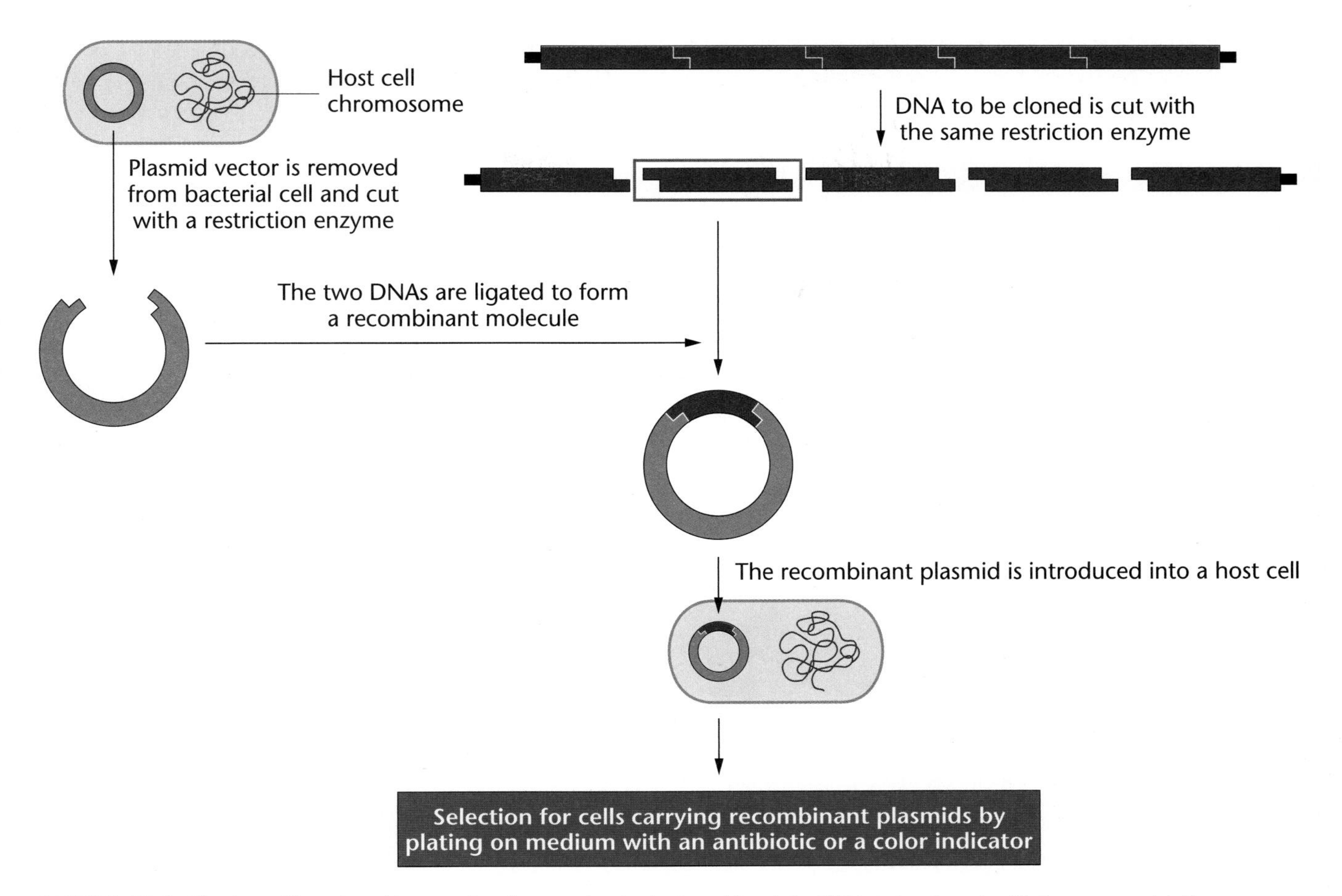

FIGURE 17–9 Cloning with a plasmid vector involves cutting both plasmid and the DNA to be cloned with the same restriction enzyme. The DNA to be cloned is spliced into the vector and transferred to a bacterial host for replication.

cannot carry out these modifications. Finally, yeast has been used for centuries in the baking and brewing industries and is considered to be a safe organism for producing proteins for vaccines and therapeutic agents. Table 17.1 lists some of the products of cloning in yeast.

Several types of yeast cloning vectors have been developed, one of which is the **yeast artificial chromosome (YAC)** (Figure 17–10). Like natural chromosomes, a YAC has telomeres at each end, an origin of replication (which initiates DNA synthesis), and a centromere. These components are joined to selectable marker genes (*TRP1* and *URA3*) and a cluster of restriction sites for DNA inserts. Yeast chromosomes range from 230 kb to over 1900 kb, making it possible to clone DNA inserts from 100 to 1000 kb in YACs. The ability to clone large pieces of DNA into these vectors makes them an important tool in genome projects, including the Human Genome Project.

TABLE 17.1 Recombinant Proteins Synthesized in Yeast Host Cells

Hepatitis B virus surface protein
Malaria parasite protein
Insulin
Epidermal growth factor
Platelet-derived growth factor
α_1-Antitrypsin
Clotting factor XIIIA

Although cloning in yeast vectors and host cells is currently the most advanced eukaryotic system used, other systems, including human artificial chromosome vectors using mammalian cells as hosts, are being developed.

17.5 The Polymerase Chain Reaction Permits Cloning Without Host Cells

Developed in the early 1970s, recombinant DNA techniques revolutionized the way geneticists and molecular biologists conducted research and gave birth to the booming biotechnology industry. However, these techniques are often labor-intensive and time-consuming. In 1986, another technique, called the **polymerase chain reaction (PCR)**, was developed. This advance again revolutionized recombinant DNA methodology and further accelerated the pace of biological research. The significance of this method is underscored by the awarding of the 1993 Nobel Prize in Chemistry to Kary Mullis for developing the PCR technique.

PCR is a rapid method of DNA cloning that has extended the power of recombinant DNA research and eliminated the need for host cells in DNA cloning. Although cell-based

MEDIA TUTORIAL Polymerase chain reaction

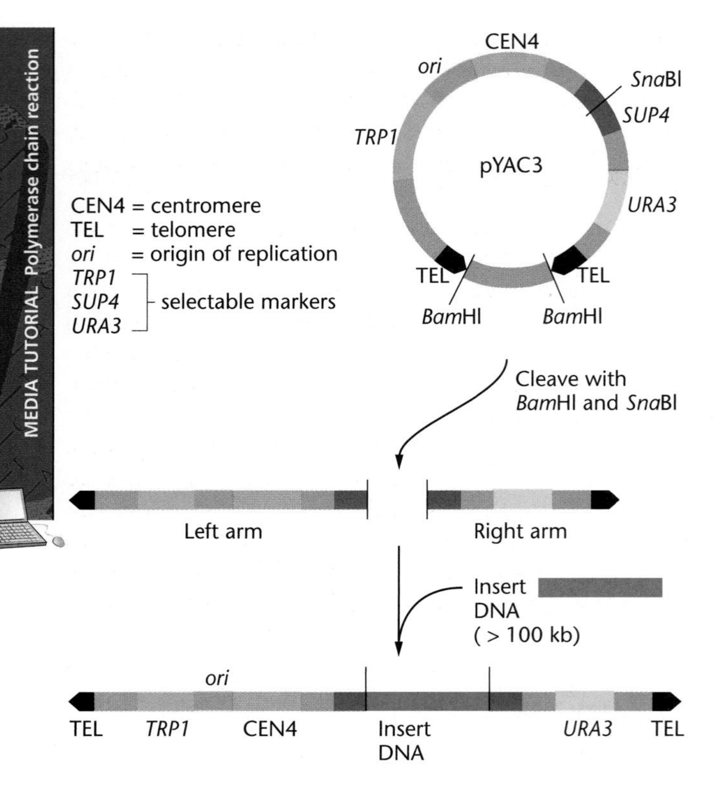

FIGURE 17–10 The yeast artificial chromosome pYAC3 contains telomere sequences (TEL), a centromere (CEN4) derived from yeast chromosome 4, and an origin of replication (*ori*). These elements give the cloning vector the properties of a chromosome. *TRP1*, and *URA3* are yeast genes that are selectable markers for the left and right arms of the chromosome. Within the *SUP4* gene is a restriction site for the enzyme *Sna*B1. Two *Bam*H1 restriction sites flank a spacer segment. Cleavage with *Sna*B1 and *Bam*H1 breaks the artificial chromosome into two arms. The DNA to be cloned is treated with *Sna*B1 producing a collection of fragments. The arms and fragments are ligated together, and the artificial chromosome is inserted into yeast host cells. Because yeast chromosomes are large, the artificial chromosome accepts inserts in the million base-pair range.

cloning is still widely used, PCR is the method of choice in many applications, including molecular biology, human genetics, evolution, development, conservation, and forensics.

PCR generates many copies of a specific DNA sequence through a series of *in vitro* reactions and can amplify target DNA sequences present in infinitesimally small quantities in a population of other DNA molecules. For PCR, some information about the nucleotide sequence of the target DNA is required. This information is used to synthesize two complementary oligonucleotide primers (one for each end of the target DNA sequence) that are added to a single-stranded DNA sample. In the reaction, the primers hybridize to nucleotides that flank the sequence that is to be amplified. A heat-stable form of DNA polymerase synthesizes a second strand of the target DNA (Figure 17–11).

The PCR reaction involves three basic steps, and the amount of amplified DNA produced is theoretically limited only by the number of times these steps are repeated.

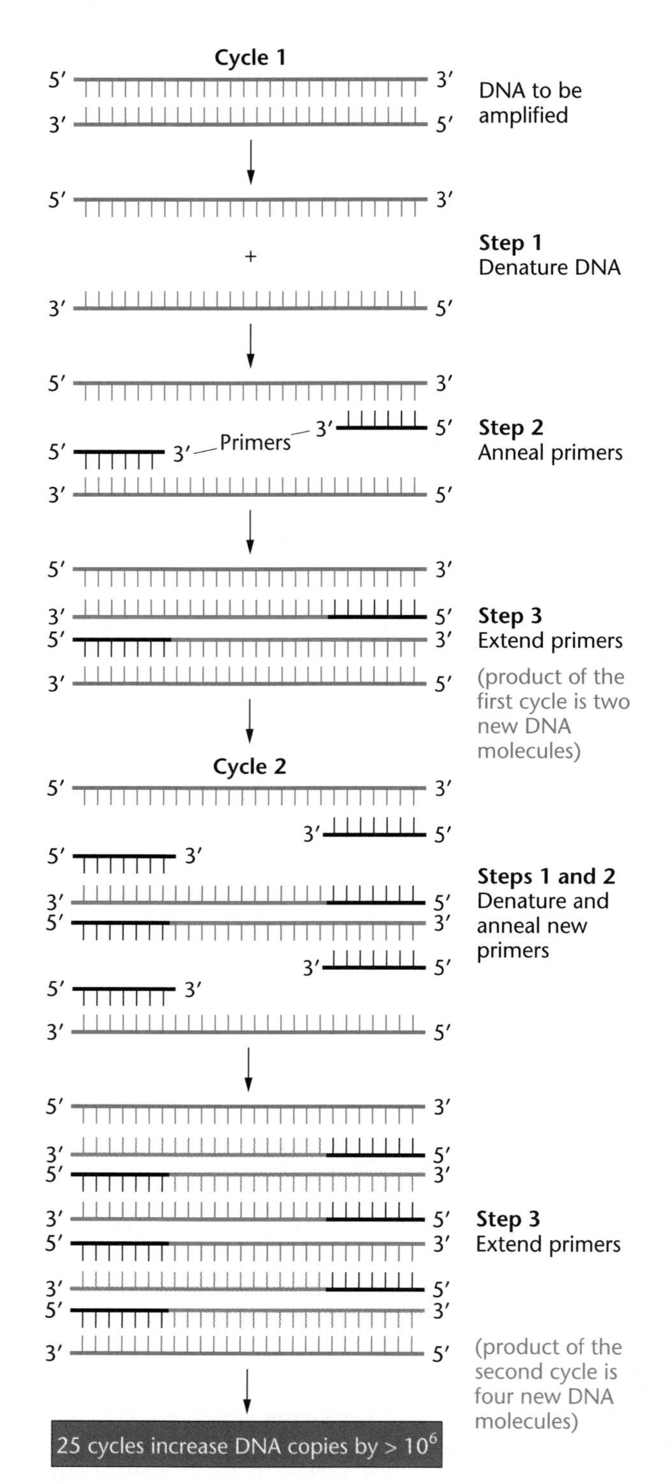

FIGURE 17–11 In the polymerase chain reaction (PCR), the target DNA is denatured into single strands; each strand is then annealed to a short, complementary primer. DNA polymerase and nucleotides extend the primers in the 3′ direction, using the single-stranded DNA as a template. The result is a newly synthesized double-stranded DNA molecule with the primers incorporated into it. Repeated cycles of PCR can amplify the original DNA sequence by more than a millionfold.

1. The DNA to be amplified is *denatured* into single strands. This DNA need not be pure and can come from many sources including genomic DNA, mummified remains, fossils, or forensic samples such as dried blood or semen, single hairs or dried samples from medical records. Heating to 90–95°C denatures

the double-stranded DNA, which dissociates into single strands (usually in about 5 minutes).

2. The temperature of the reaction is lowered to between 50°C and 70°C, and at this *annealing* temperature, the primers bind to the denatured DNA. These primers are synthetic oligonucleotides (15–30 nucleotides long) that bind specifically to sequences flanking the target segment. The primers will serve as starting points for synthesizing new DNA strands complementary to the target DNA.
3. A heat-stable form of DNA polymerase (*Taq* polymerase)* is added to the reaction mixture. DNA synthesis is carried out at temperatures between 70°C and 75°C. The *Taq* polymerase *extends* the primers by adding nucleotides in the 5′-to-3′ direction, making a double-stranded copy of the target DNA.

Each set of three steps—**denaturation** of the double-stranded product, **annealing** of primers, and **extension** by polymerase—is a cycle. PCR is a chain reaction because the number of new DNA strands is doubled in each cycle, and the new strands, along with the old strands, serve as templates in the next cycle. Each cycle, which takes about 5 minutes, can be repeated, and in less than 3 hours, 25–30 cycles result in an over 1 millionfold increase in the amount of DNA (see Figure 17–11). Machines called *thermocyclers* have automated this process. They can be programmed to carry out a predetermined number of cycles, yielding large amounts of the target DNA, which can be used in cloning, sequencing, clinical diagnosis, and genetic screening.

PCR-based DNA cloning has several advantages over cell-based cloning. PCR is rapid and can be carried out in a few hours, rather than the days required for cell-based cloning. In addition, the design of PCR primers uses computer software, and the commercial synthesis of the oligonucleotides is also fast and economical. If desired, the products of PCR can be cloned into plasmid vectors for further use.

PCR is also very sensitive and amplifies target DNA from vanishingly small DNA samples, including the DNA in a single cell. This feature of PCR is invaluable in several areas, including genetic testing, forensics, and molecular paleontology. DNA samples that have been partly degraded, mixed with other materials, or embedded in a medium (such as amber) can be used when conventional cloning would be difficult or impossible.

Although PCR is a valuable technique, it does have limitations: some information about the nucleotide sequence of the target DNA must be known, and even minor contamination of the sample with DNA from other sources can cause problems. For example, cells shed from the skin of a laboratory worker can contaminate samples gathered from a crime scene or taken from fossils, making it difficult to obtain accurate results. PCR reactions must always be performed with carefully designed and appropriate controls.

* *Taq* polymerase is an enzyme from a bacterium, *Thermus aquaticus*, which lives in hot springs

PCR DNA cloning is now one of the most widely used techniques in genetics and molecular biology. PCR and its variations have many other applications. It quickly identifies restriction-site variants and variations in tandemly repeated DNA sequences that can serve as genetic markers for gene mapping studies. Using gene-specific primers, PCR screens for mutations in genetic disorders, allowing the location and nature of the mutation to be determined quickly. Primers can be designed to distinguish target sequences that differ by only a single nucleotide, making it possible to develop allele-specific probes for genetic testing. Random primers allow the indiscriminate amplification of DNA, which is particularly advantageous when studying samples from single cells, fossils, or from crime scenes where a single hair or even a saliva-moistened postage stamp is the source of the DNA. Researchers also explore uncharacterized DNA regions adjacent to known regions and even sequence DNA with PCR. The technique has been used to enforce the worldwide ban on the sale of certain whale products and to settle arguments about the pedigree background of purebred dogs. In short, PCR is one of the most versatile techniques in modern genetics.

17.6 Libraries Are Collections of Cloned Sequences

Each cloned DNA segment is relatively small and may represent only a single gene or a portion of a gene, thus many separate clones are needed to explore even a small fraction of an organism's genome. A set of DNA clones derived from a single individual is a cloned *library*. These libraries can represent an entire genome, a single chromosome, or a set of genes that are expressed in a single cell type.

Genomic Libraries

Ideally, a **genomic library** contains at least one copy of all the sequences in an organism's genome. Genomic libraries are constructed using host cell cloning methods, since PCR-cloned DNA fragments are relatively small. In the process, DNA is extracted from cells or tissues, then cut with restriction enzymes and the fragments ligated into vectors. Since some vectors (such as plasmids) carry only a few thousand base pairs of inserted DNA, selecting the vector so that it contains a genome in the smallest number of clones is an important consideration in preparing a genomic library.

How big does a genomic library have to be to have a 95 or 99% chance of containing all the sequences in a genome? The number of clones required to carry all sequences in a genome depends on several factors, including the average size of the cloned inserts, the size of the genome to be cloned, and the level of probability desired. The number of clones in a library can be calculated as

$$N = \frac{\ln(1 - P)}{\ln(1 - f)}$$

where N is the number of required clones, P is the probability of recovering a given sequence, and f is the fraction of the genome in each clone.

MEDIA TUTORIAL Cloning libraries

Suppose we want to prepare a library of the human genome large enough to have a 99% chance of containing all the sequences in the genome. The choice of vector is a primary consideration in making this library. If we construct the library using a plasmid vector with an average insert size of 5 kb, then over 2.4 million clones would be required for a 99% probability of recovering any given sequence from the genome. Because of its size, this library would be difficult to screen efficiently. If a phage vector with an average insert size of 17 kb is used as a vector, then around 800,000 clones would be required for a 99% probability of finding any given human sequence. While it is much smaller than a plasmid library, screening a phage library of this size would still be a labor-intensive chore. However, if the library was constructed with a YAC vector with an average insert size of 1 Mb, then the library would contain only about 14,000 YACs, making it relatively easy to screen. Vectors with large cloning capacities such as YACs are an essential part of the Human Genome Project.

cDNA Libraries

A library of the genes expressed in a specific cell type at a specific time can be constructed by taking advantage of the fact that almost all eukaryotic mRNA molecules contain a poly-A tail at their 3′-ends. This mRNA can be isolated and used to synthesize **complementary DNA (cDNA)** molecules that are subsequently cloned to form a **cDNA library**. After the mRNA containing a poly-A tail is isolated, a poly-dT primer is added to pair with the poly-A residues (Figure 17–12). The poly-dT primer is the starting point for synthesizing a complementary DNA strand with **reverse transcriptase**. The result is an RNA–DNA double-stranded molecule. The RNA strand is removed, and the remaining single-stranded DNA becomes the template for DNA synthesis with the enzyme **DNA polymerase I**. The 3′-end of the single-stranded cDNA often loops back upon itself to form a hairpin loop. This loop serves as a primer for synthesizing the second strand by DNA polymerase I. The product is a double-stranded DNA molecule with the strands joined at one end. This loop can then be opened with the enzyme S_1 nuclease, producing a double-stranded DNA molecule (the complementary DNA), that can be cloned into a plasmid or phage vector.

PCR methods also generate cDNA from the 3′- or 5′-end of mRNA molecules. This strategy, called RACE (*r*apid *a*mplification of *c*DNA *e*nds), is dependent upon knowing a short nucleotide sequence contained in the coding region of the mRNA to be amplified. RACE identifies mRNAs that may be present in only one or two copies per cell, and the ends of cDNA molecules generated by RACE are used to search for cloned genes in genomic libraries.

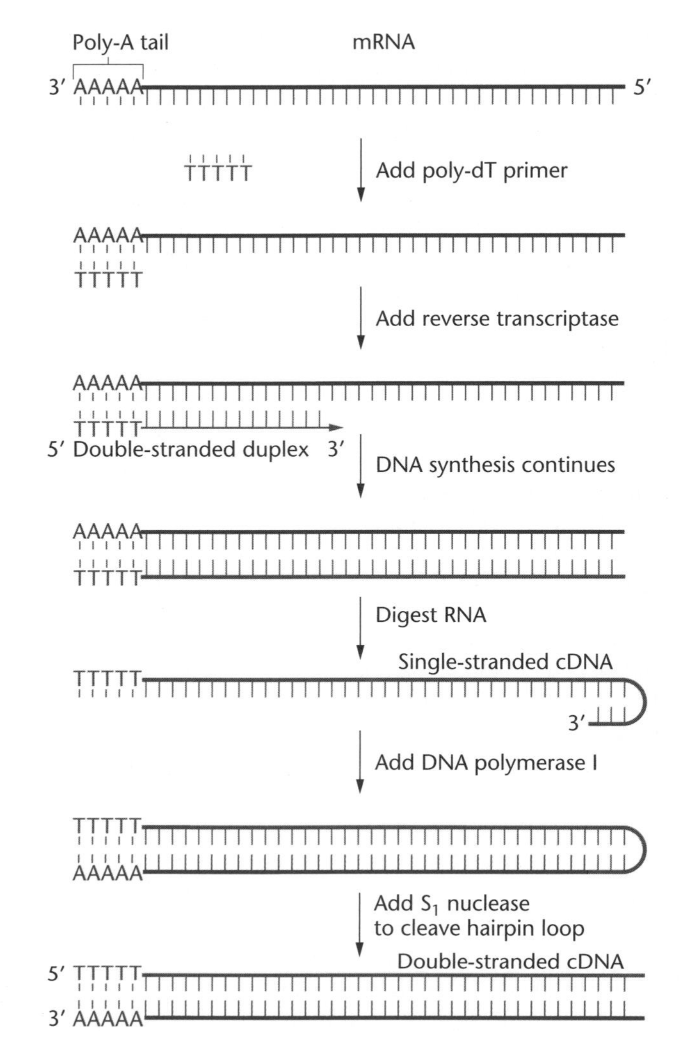

FIGURE 17–12 Reverse transcriptase uses mRNA as a template to synthesize a complementary DNA strand (cDNA) and forms an mRNA–cDNA double-stranded duplex. The mRNA is digested, and the 3′-end of the cDNA often folds back to form a hairpin loop. After the second DNA strand is synthesized, S_1 nuclease opens the hairpin loop. The result is a double-stranded cDNA molecule that can be cloned into a suitable vector or used as a probe for library screening.

17.7 Specific Clones Can Be Recovered from a Library

A genomic library can contain up to several hundred thousand clones. To find a specific gene, we need to identify and isolate only the clone or clones containing that gene. We also must be able to determine whether a clone contains all or only part of the gene. Several methods allow us to sort through a library to recover clones of interest; the choice of method often depends on the circumstances and available information about the gene being sought.

Probes Identify Specific Clones

Many screening procedures employ probes to select specific clones from a library. A **probe** is any piece of DNA or RNA that has been labeled and is complementary to some part of a cloned sequence present in the library. When used in a hybridization reaction, the probe identifies complementary nucleic acid sequences present in one or more clones. Probes can be labeled with radioactivity, or they can cause chemical or color reactions to indicate the location of a specific clone.

Probes are derived from a variety of sources—even related genes isolated from other species can be used if enough of the DNA sequence has been conserved. For example, extrachromosomal copies of the ribosomal genes of the clawed frog *Xenopus laevis* can be isolated by centrifugation and cloned into plasmid vectors. Because ribosomal gene sequences have been highly conserved during evolution, clones carrying the human ribosomal genes can be recovered from a human genomic library using cloned fragments of *Xenopus* ribosomal DNA as probes.

If the gene to be selected from a genomic library is expressed in certain cell types, a cDNA probe can be used. This technique is particularly helpful when purified or enriched mRNA for a gene product can be obtained. For example, β-globin mRNA is present in high concentrations in certain stages of red blood cell development. The mRNA purified from these cells can be copied by reverse transcriptase into a cDNA molecule or amplified by PCR RACE for use as a probe. In fact, a cDNA probe was originally used to recover the structural gene for human β-globin from a cloned genomic library.

Screening a Library

To screen a *plasmid library*, clones from the library are grown on nutrient agar plates, where they form hundreds or thousands of colonies (Figure 17–13). A replica of the

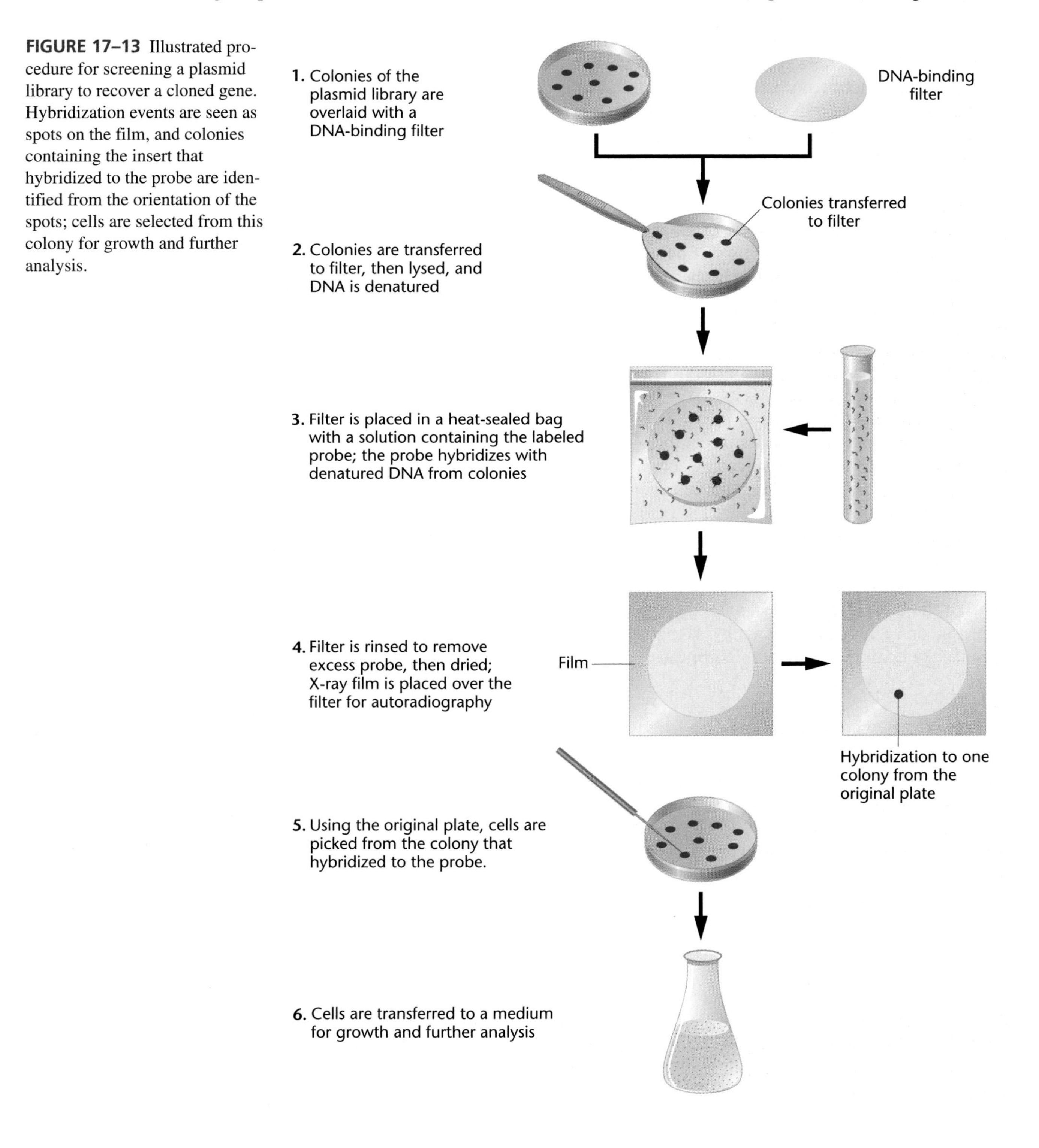
1. Colonies of the plasmid library are overlaid with a DNA-binding filter
DNA-binding filter
Colonies transferred to filter
2. Colonies are transferred to filter, then lysed, and DNA is denatured
3. Filter is placed in a heat-sealed bag with a solution containing the labeled probe; the probe hybridizes with denatured DNA from colonies
4. Filter is rinsed to remove excess probe, then dried; X-ray film is placed over the filter for autoradiography
Film
Hybridization to one colony from the original plate
5. Using the original plate, cells are picked from the colony that hybridized to the probe.
6. Cells are transferred to a medium for growth and further analysis

FIGURE 17–13 Illustrated procedure for screening a plasmid library to recover a cloned gene. Hybridization events are seen as spots on the film, and colonies containing the insert that hybridized to the probe are identified from the orientation of the spots; cells are selected from this colony for growth and further analysis.

colonies is made by gently pressing a nylon or nitrocellulose filter onto the plate's surface; this transfers the pattern of bacterial colonies from the plate to the filter. The filter is then passed through various solutions to lyse the bacteria, denature the double-stranded DNA into single strands, and bind these strands to the filter.

The DNA on the filter is screened by incubation with a nucleic acid probe. If a radioactive DNA probe is used, it is denatured to form single strands and then added to a solution that contains the filter. If the nucleotide sequence of any of the DNA on the filter is complementary to the probe, a double-stranded DNA–DNA hybrid molecule will form between the probe and the cloned DNA. After incubation, unbound and/or excess probe molecules are washed away, and the filter is assayed to detect the hybrid molecules. If a radioactive probe has been used, the filter is overlaid with a piece of X-ray film. Radioactive decay in the probe molecules hybridized to DNA on the filter will expose the film, producing dark spots on it; these spots represent colonies on the plate containing the cloned gene of interest (Figure 17–13). The corresponding colony is identified and recovered from the original nutrient plate, and the cloned DNA it contains can be used in further experiments. With nonradioactive probes, a chemical reaction emits photons of light (chemiluminescence) to expose the photographic film and reveal the location of colonies carrying the gene of interest.

To screen a *phage library*, a slightly different method, called **plaque hybridization**, is used. A solution of phage carrying DNA inserts is spread over a lawn of bacteria growing on a plate. The phages infect the bacterial cells and form plaques as they replicate. Each plaque, which appears as a clear spot on the plate, represents the progeny of a single phage and is a clone. The plaques are transferred to a nylon or nitrocellulose filter, and the phage DNA on the filter is denatured into single strands and screened with a labeled probe. Phage plaques are much smaller than plasmid colonies, and many plaques can be screened on a single filter, making this method more efficient for screening large genomic libraries.

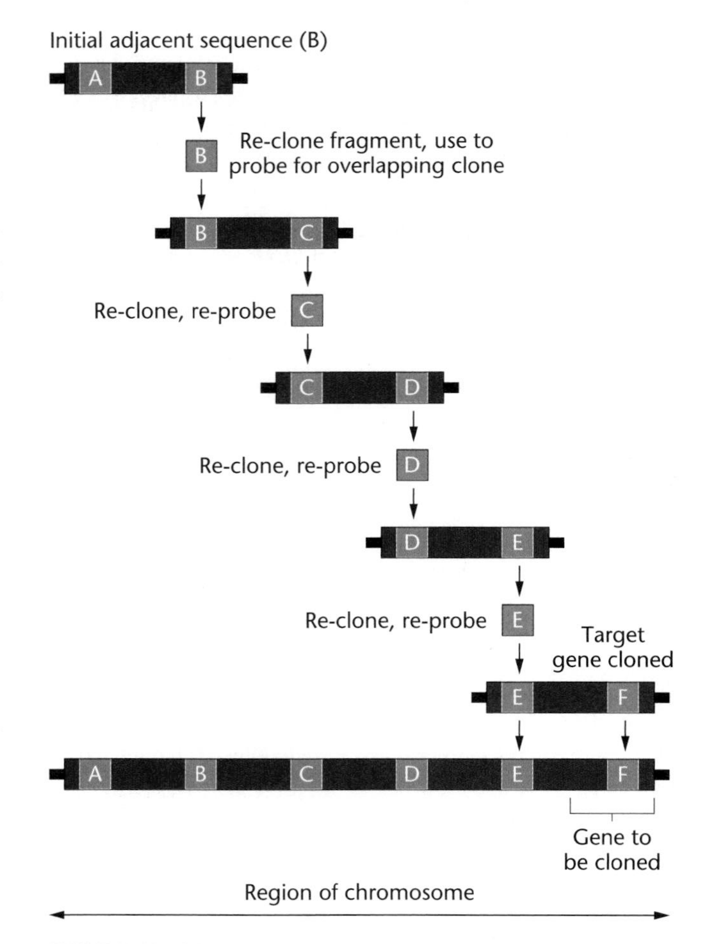

FIGURE 17–14 In chromosome walking, re-cloned fragment B is used to probe for an overlapping clone C. Re-cloning C and probing for D in the genomic library is repeated until the gene in question, F, has been reached.

Chromosome Walking

In some cases, when the approximate location of a gene is known, it is possible to clone the gene by first cloning nearby sequences. Often these nearby sequences are identified by linkage analysis and serve as starting points for **chromosome walking**. In a chromosome walk, the end piece of a cloned DNA fragment is re-cloned and used as a probe to recover another set of overlapping clones from the genomic library (Figure 17–14). These clones are analyzed to determine their degree of overlap. A subfragment from one end of the overlapping cloned DNA is used to recover another set of overlapping clones, and the analysis is repeated. In this way, it is possible to "walk" along the chromosome, clone by clone.

Genes recovered in a chromosome walk can be identified by nucleotide sequencing of the recovered clones and searching for an **open reading frame** (**ORF**). An ORF is a stretch of nucleotides that begins with a start codon followed by amino acid-encoding codons, and ends with one or more stop codons. Although it is laborious and time-consuming, chromosome walking has identified the genes involved in several human genetic disorders, including those for cystic fibrosis and Duchenne muscular dystrophy.

Chromosome walking has several limitations in complex eukaryotic genomes, including the human genome. If a probe contains a repetitive sequence, such as an *Alu* sequence, it will hybridize to other clones in the library containing that sequence. Most of these clones will *not* be adjacent to the clone from which the probe was derived, and the chromosome walk will be terminated. In some cases, a technique called **chromosome jumping** skips over the region containing the repetitive sequences and continues the walk.

17.8 Cloned Sequences Can Be Characterized in Several Ways

The recovery and identification of genes and other specific DNA sequences by cloning or by PCR is a powerful tool for analyzing genomic structure and function. In fact, much of the Human Genome Project is based on such techniques. In the following sections, we consider some of these methods, which are used to answer questions about the organization and function of cloned sequences.

Restriction Mapping

One of the first steps in characterizing a DNA clone is the construction of a **restriction map**. A restriction map establishes the number, order, and distance between sites of restriction enzyme cleavage along a cloned segment of DNA. Restriction maps for different cloned DNAs are usually different enough to serve as an identity tag for that clone. Recall that restriction map units are expressed in base pairs or, for longer lengths, kilobase pairs. Restriction maps provide information about the length of a cloned insert and the location of restriction enzyme sites within the DNA. These data can be used to re-clone fragments of a gene or compare its internal organization with that of other cloned sequences.

Figure 17–15 shows the construction of a restriction map from a cloned DNA segment. For this map, let's begin with a cloned DNA segment 7.0 kb in length. Three samples of the cloned DNA are digested with restriction enzymes—one with *Hin*dIII, one with *Sal*I, and one with both *Hin*dIII and *Sal*I. The fragments are separated by gel electrophoresis (Chapters 10 and 14), stained with ethidium bromide, and photographed. Their size is measured by comparing them to a set of molecular weight standards run in adjacent lanes. The map is then constructed by analyzing the fragments. When the DNA is cut with *Hin*dIII, two fragments (0.8 and 6.2 kb) are produced; this confirms

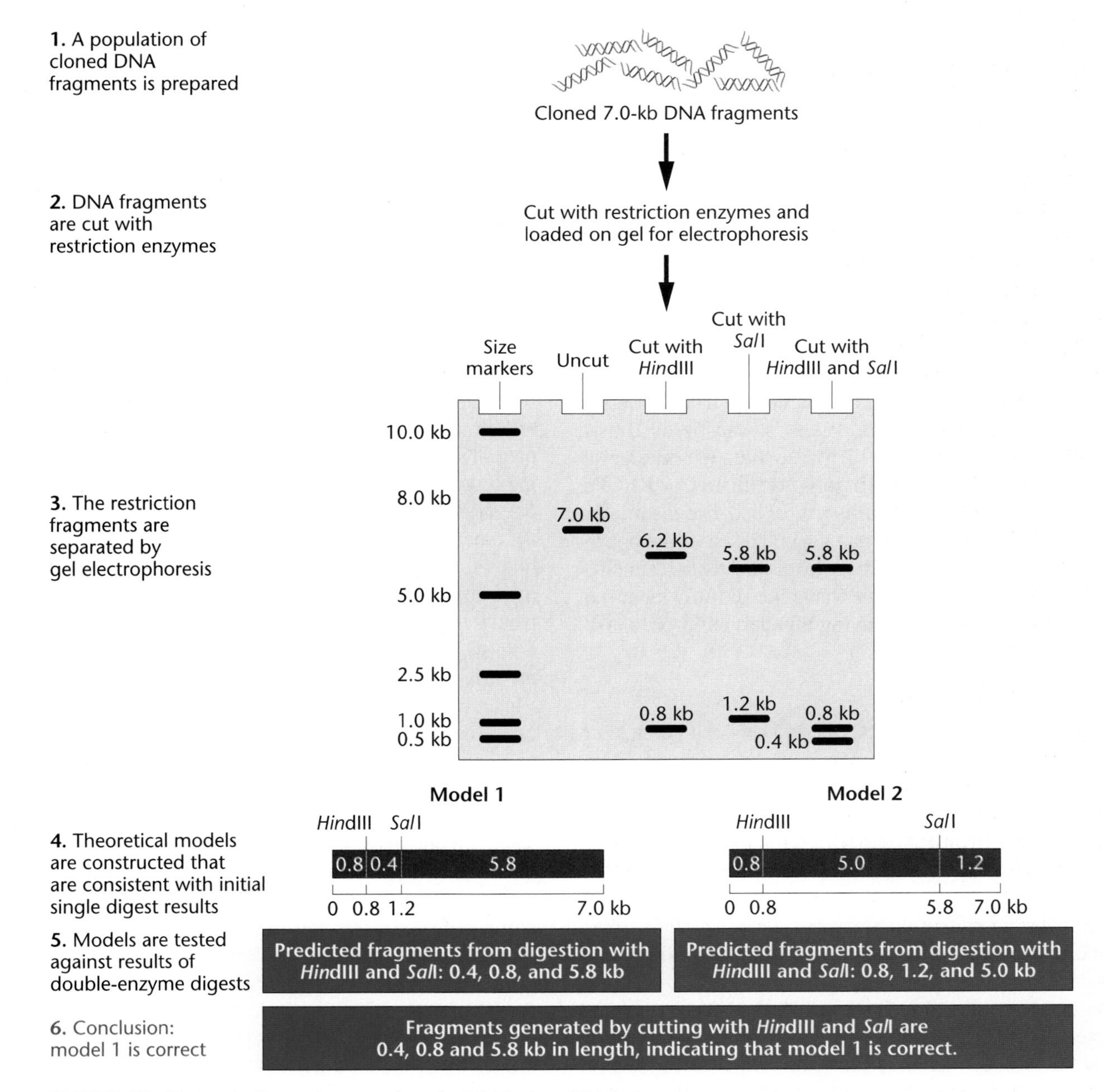

FIGURE 17–15 In this illustration, samples of a 7.0-kb cloned DNA fragment are used to construct a restriction map. Models are constructed to predict the fragment sizes that will be generated by cutting with restriction enzymes. Comparing the predicted fragments with those observed on the gel indicates the correct restriction map.

MEDIA TUTORIAL Restriction mapping

that the cloned insert is 7.0 kb in length and that there is only one cleavage site for this enzyme (located 0.8 kb from one end). When the DNA is cut with *Sal*I, two fragments (1.2 and 5.8 kb) result, indicating that this enzyme also has one cutting site, located 1.2 kb from one end of the inserted DNA segment.

These results show that each enzyme has one restriction site, but the relationship between the two sites is unknown. From the information available, two different maps are possible. In Figure 17–15, model 1, indicates that the *Hin*dIII site is located 0.8 kb from one end and the *Sal*I site is 1.2 kb from the same end. The alternative map, model 2, locates the *Hin*dIII site 0.8 kb from one end and the *Sal*I site 1.2 kb from the other end.

The correct model is selected by analyzing the results from the sample digested with both *Hin*dIII and *Sal*I. Model 1 predicts that digestion with both enzymes will generate three fragments: 0.4, 0.8, and 5.8 kb in length; model 2 predicts that these fragments will be 0.8, 5.0, and 1.2 kb in length. The actual fragment pattern observed on the gel medium indicates that model 1 is correct (see Figure 17–15).

Restriction maps are an important way to characterize a cloned DNA segment and can be constructed in the absence of information about the coding capacity or function of the mapped DNA. In conjunction with other techniques, restriction mapping can define the boundaries of a gene, dissect the molecular organization of a gene and its flanking regions, and locate mutational sites within genes.

Restriction sites play an important role in mapping genes to specific human chromosomes and to defined regions of individual chromosomes. In addition, if a restriction site maps close to a mutant gene, it can be used as a marker in a diagnostic test. These sites are known as **restriction fragment length polymorphisms**, or **RFLPs**. RFLPs have proven especially useful because the mutant genes underlying many human genetic diseases are poorly characterized at the molecular level and restriction sites closely linked to mutant genes have successfully identified heterozygotes at risk for having affected children.

Nucleic Acid Blotting

Many of the techniques described in this chapter rely on hybridization between complementary nucleic acid (DNA or RNA) molecules. One of the most widely used methods for detecting such hybrids is called **Southern blotting** (after Edward Southern, who devised it). This technique has two components: the separation of DNA fragments by gel electrophoresis and after transfer to a filter, hybridization of the fragments using labeled probes. Gel electrophoresis characterizes the number of fragments produced by restriction digestion and their molecular weights. Hybridization characterizes the DNA sequences present in the fragments. The DNA to be characterized by Southern blot hybridization can come from several sources, including clones selected from a library and genomic DNA. Our discussion will use examples from both sources to show how Southern blots are used to characterize the number, size, organization, and sequence content of DNA fragments.

To make a Southern blot, DNA is cut into fragments with one or more restriction enzymes and the fragments are separated by gel electrophoresis (Figure 17–16). The DNA in the gel is usually stained and photographed or scanned to reveal the number and molecular weights of the restriction fragments (Figure 17–16). The DNA in the gel is then denatured to form single-stranded fragments by treating the gel with an alkaline solution. The gel is then overlaid with a membrane of DNA-binding material, usually nitrocellulose or a nylon derivative. To transfer the fragments, the membrane and gel are placed on a wick (often a sponge) in contact with a buffer solution. The buffer flows through the wick, gel, and membrane by capillary action (see Figure 17–16). As the buffer solution flows, the DNA fragments move out of the gel and become immobilized on the membrane.

The DNA fragments on the membrane are hybridized with a labeled, single-stranded DNA probe. Only DNA fragments complementary to the probe's nucleotide sequence will form double-stranded hybrids. Excess probe is washed away, and the hybridized fragments are visualized on a piece of film (Figures 17-16 and 17–17). The Southern blot method can be used to identify which clones in a library contain a given DNA sequence (such as ribosomal DNA, a globin gene, etc.), and to characterize the size of the fragments, thus deriving a restriction map of the cloned DNA. Southern blots can also be used to determine whether a clone contains all or only part of a gene and to ascertain the overall size and sequence organization of a gene or DNA sequence of interest.

In Figure 17–17, researchers had cut genomic DNA from two strains of *E. coli* with several restriction enzymes. The pattern of fragments with each restriction enzyme is shown in Figure 17–17a. The number and sizes of fragments from one strain are shown in lanes a, c, and e; and fragments from the other strain are shown in lanes b, d, and f. The Southern blot of this gel is shown in Figure 17–17b. The probe hybridized to only one band in each lane, and the size of the band is the same in the two strains. These results indicate that both strains contain the DNA sequence of interest and that the restriction pattern is very similar for the two strains.

In addition to characterizing cloned DNAs, Southern blots can be used to map restriction sites within and near a gene and to identify DNA fragments carrying all or parts of a single gene in a mixture of fragments. Southern blots also detect rearrangements, deletions, and duplications in genes associated with human genetic disorders and cancers.

To determine whether a cloned gene is transcriptionally active in a given cell or tissue type, a related blotting technique probes for the presence of mRNA that is complementary to a cloned gene. This technique first extracts mRNA from a specific cell or tissue type and separates the RNA molecules by gel electrophoresis. The resulting pattern of RNA bands is transferred to a sheet of membrane, as in the Southern blot. The membrane is then hybridized to a single-stranded DNA probe, derived from the cloned gene. If RNA complementary to the DNA probe is present, it is detected as a band on the film.

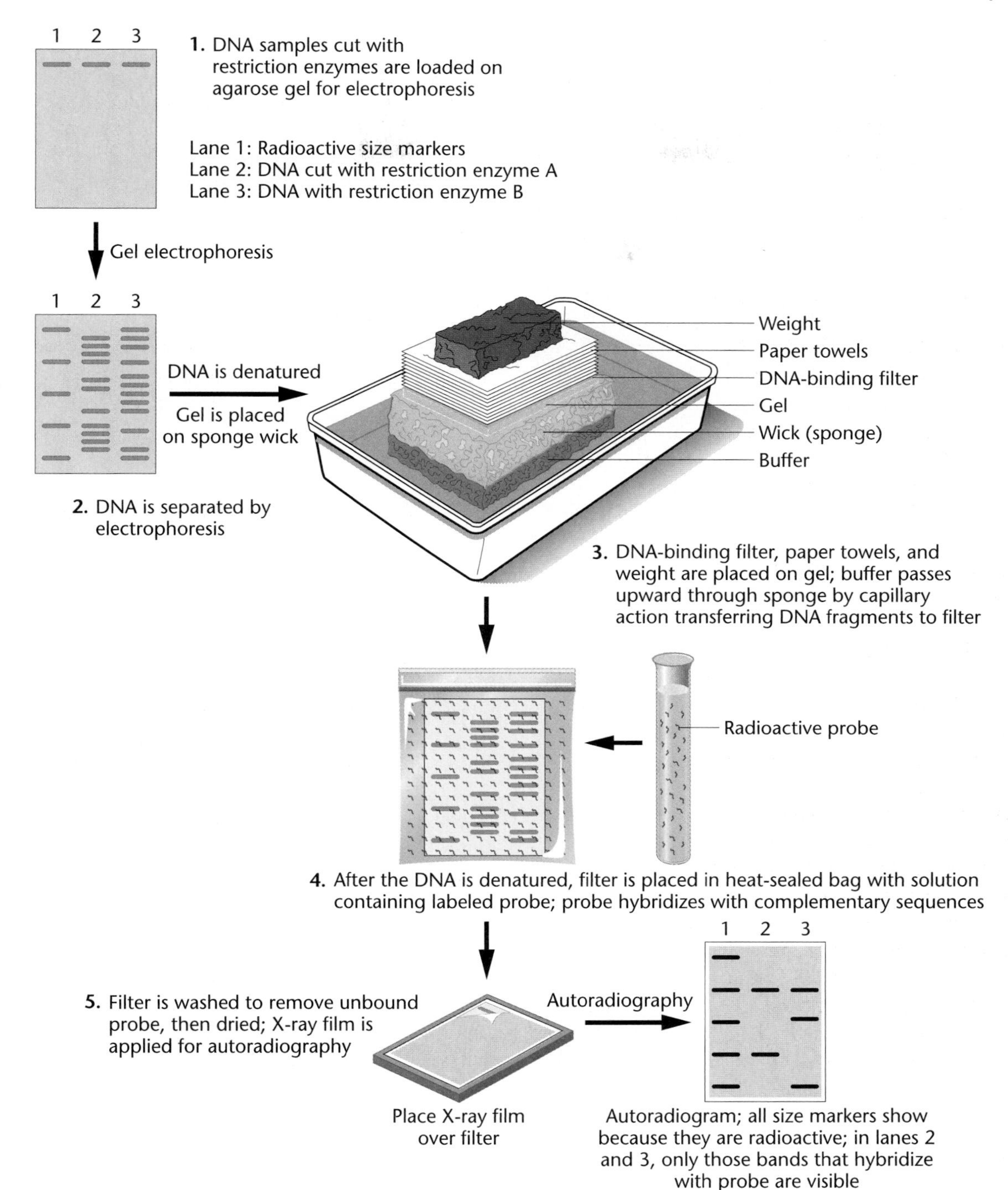
1 2 3
1. DNA samples cut with restriction enzymes are loaded on agarose gel for electrophoresis
Lane 1: Radioactive size markers
Lane 2: DNA cut with restriction enzyme A
Lane 3: DNA with restriction enzyme B
Gel electrophoresis
1 2 3
DNA is denatured
Gel is placed on sponge wick
2. DNA is separated by electrophoresis
Weight
Paper towels
DNA-binding filter
Gel
Wick (sponge)
Buffer
3. DNA-binding filter, paper towels, and weight are placed on gel; buffer passes upward through sponge by capillary action transferring DNA fragments to filter
Radioactive probe
4. After the DNA is denatured, filter is placed in heat-sealed bag with solution containing labeled probe; probe hybridizes with complementary sequences
1 2 3
5. Filter is washed to remove unbound probe, then dried; X-ray film is applied for autoradiography
Autoradiography
Place X-ray film over filter
Autoradiogram; all size markers show because they are radioactive; in lanes 2 and 3, only those bands that hybridize with probe are visible

FIGURE 17–16 In the Southern blotting technique, samples of the DNA to be probed are cut with restriction enzymes and the fragments separated by gel electrophoresis. The pattern of fragments is visualized and photographed under ultraviolet illumination by staining the gel with ethidium bromide. The gel is then placed on a sponge wick in contact with a buffer solution and covered with a DNA-binding filter. Layers of paper towels or blotting paper are placed on top of the filter and held in place with a weight. Capillary action draws the buffer through the gel, transferring the pattern of DNA fragments from the gel to the filter. The DNA fragments on the filter are then denatured into single strands and hybridized with a labeled DNA probe. The filter is washed to remove excess probe and overlaid with a piece of X-ray film for autoradiography. The hybridized fragments show as bands on the X-ray film.

Because the original procedure (DNA bound to a filter) is known as a Southern blot, this procedure (RNA bound to a filter) is called a **northern blot**. (Following this somewhat perverse logic, another procedure involving proteins bound to a filter is known as a **western blot**.)

Northern blots provide information about the expression of specific genes and are used to study patterns of gene expression in embryonic and adult tissues. Northern blots also detect alternatively spliced mRNAs and multiple types of transcripts derived from a single gene and are used to derive other information about transcribed mRNAs. If marker RNAs of known size are run in an adjacent lane, the size of a gene's mRNA can be measured. In addition, the amount of transcribed RNA present in the cell or tissue being studied is related to the density of the RNA band on the film. Measuring band density gives the relative

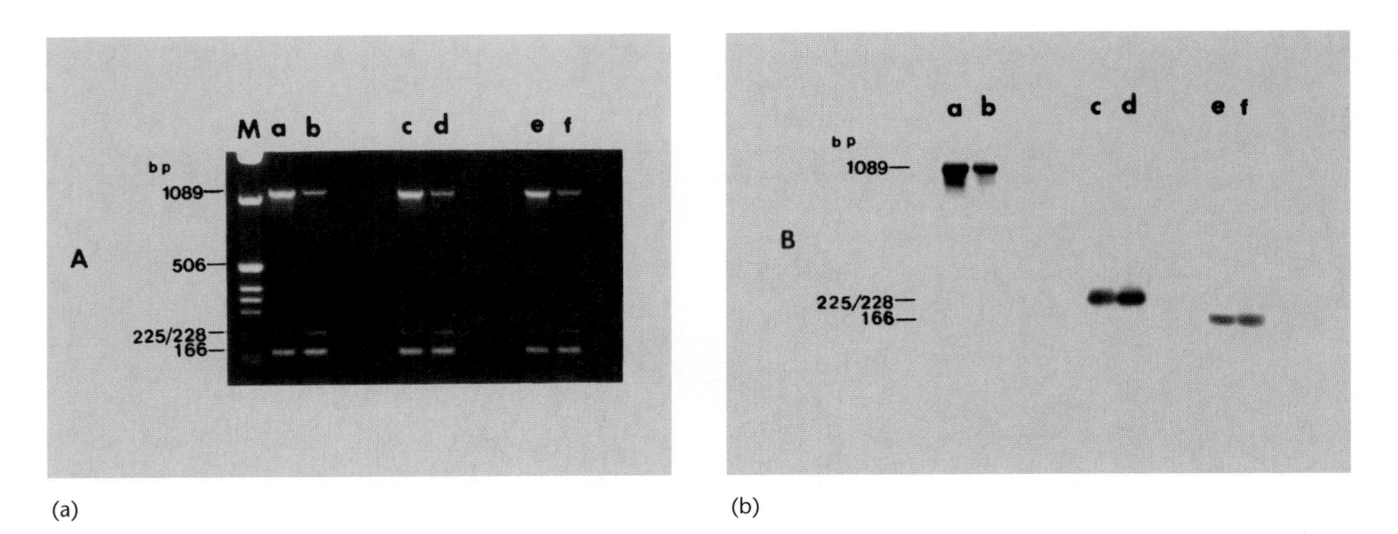

FIGURE 17–17 (a) Agarose gel stained with ethidium bromide to show DNA fragments. Lane M contains the size markers. Lanes a, c, and e contain DNA from one bacterial strain of *E. coli*, and lanes b, d, and f contain DNA from another strain. (b) A Southern blot prepared from the gel in part (a). Only those bands containing DNA sequences complementary to the probe show hybridization. The size markers were not radioactive and are not visible on the Southern blot. *Reprinted with permission from* Journal of Food Protection, *vol 59, No.6, 1996, p. 573. Copyright held by the International Association for Food Protection, Des Moines, Iowa, U.S.A. Courtesy of Dr. Pina M. Fratamico.*

transcriptional activity. Thus, northern blots characterize and quantify the transcriptional activity of genes in different cells, tissues, and organisms.

17.9 DNA Sequencing: The Ultimate Way to Characterize a Clone

In a sense, a cloned DNA sequence is completely characterized when its nucleotide sequence is known. The ability to sequence cloned DNA has greatly enhanced our understanding of gene structure, gene function, and the mechanisms of regulation.

The most common method of **DNA sequencing** is based on **chain termination**. In this procedure, a single-stranded DNA molecule whose sequence is to be determined is used as a template for synthesizing a series of complementary strands. Each of these strands randomly terminates at a different, specific nucleotide (Figure 17–18). In the first step of this reaction, a short primer is annealed to the single-stranded template, and the primer-bound DNA is distributed into four

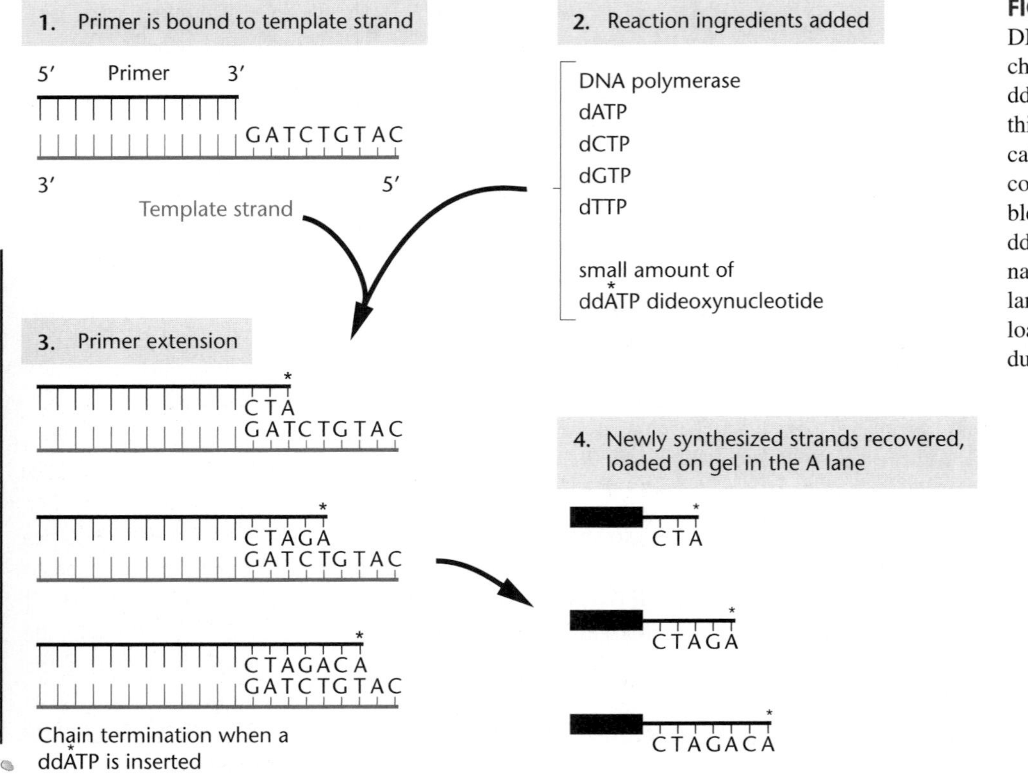

FIGURE 17–18 Illustration of DNA sequencing using the chain-termination method. Here, ddATP and the A inserted from this didexoynucleotide are indicated with an asterisk. Over the course of the reaction, all possible termination sites will have a ddNTP inserted. Chains terminating in A are loaded in the A lane, those ending in C are loaded in the C lane, and so forth during electrophoresis.

tubes. In the second step, this primer is elongated in the 5′-to-3′ direction by adding the enzyme DNA polymerase and the four deoxyribonucleotide triphosphates (dATP, dCTP, dGTP, and dTTP). In addition, each tube contains a small amount of one of the four base-specific analogs, called **dideoxynucleotides** (e. g., ddATP). As DNA synthesis takes place, the DNA polymerase occasionally inserts a dideoxynucleotide (instead of a deoxynucleotide) into a growing DNA strand. Since the analog lacks a 3′-hydroxyl group, it cannot form a 3′ bond to add another nucleotide to the strand and DNA synthesis terminates. For example, if ddATP is present, termination takes place at sites opposite thymidine in the template strand (Figure 17–18). As the reaction proceeds, the tubes accumulate a series of DNA molecules that differ in length at their 3′-ends. The fragments from the four reaction tubes (one for each dideoxynucleotide) are separated in adjacent lanes by gel electrophoresis. Electrophoresis separates DNA fragments in each lane that differ in size by a single nucleotide. The result is a series of bands forming a ladderlike pattern (Figure 17–19). The nucleotide sequence of the DNA can be read directly from bottom to top, corresponding to the 5′-to-3′ sequence of the DNA strand complementary to the template.

Large-scale genome sequencing projects use automated machines that sequence several hundred thousand nucleotides per day. In this procedure, the four dideoxynucleotide analogs are labeled with a fluorescent dye (Figure 17–20) so that chains terminating in adenosine are labeled with one color, those ending in cytosine with another color, and so forth. All four labeled dideoxynucleotides are added to a single tube, and after synthesis with DNA polymerase, the reaction products are loaded into one lane on a gel. Each band fluoresces a different color. A detector in the sequencing machine reads the color of each band and determines whether it represents an A, T, C, or G. The data are stored and analyzed using the appropriate software or printed for reading (Figure 17–21).

DNA sequencing projects have now been completed on the genomes of an ever-increasing number of organisms. Sequencing provides information about the number, nature, and organization of genes in a genome and elucidates the mutational events that alter both genes and gene products.

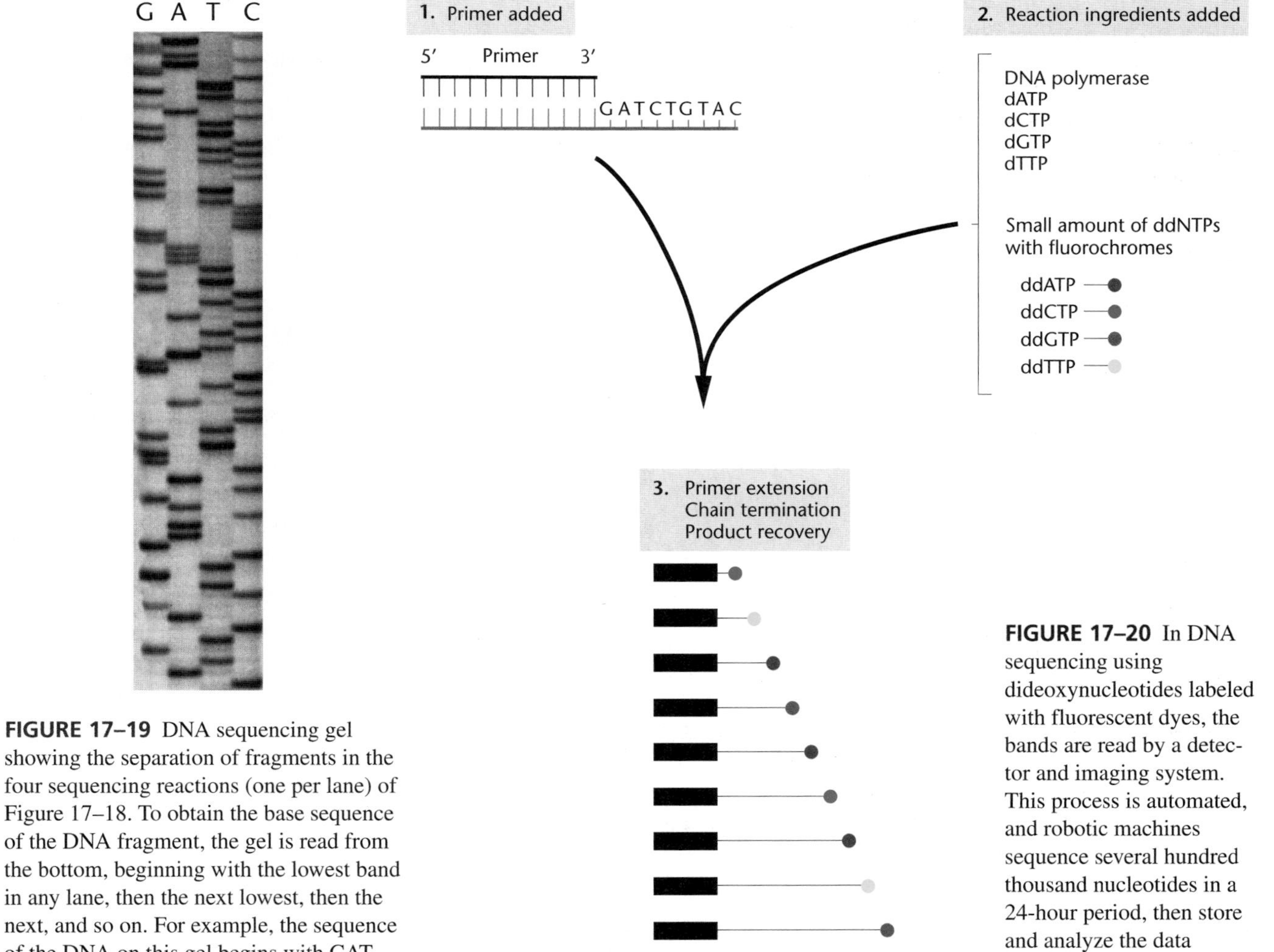

FIGURE 17–19 DNA sequencing gel showing the separation of fragments in the four sequencing reactions (one per lane) of Figure 17–18. To obtain the base sequence of the DNA fragment, the gel is read from the bottom, beginning with the lowest band in any lane, then the next lowest, then the next, and so on. For example, the sequence of the DNA on this gel begins with CATGTCAG. *(Dr. Suzanne McCutcheon)*

FIGURE 17–20 In DNA sequencing using dideoxynucleotides labeled with fluorescent dyes, the bands are read by a detector and imaging system. This process is automated, and robotic machines sequence several hundred thousand nucleotides in a 24-hour period, then store and analyze the data automatically.

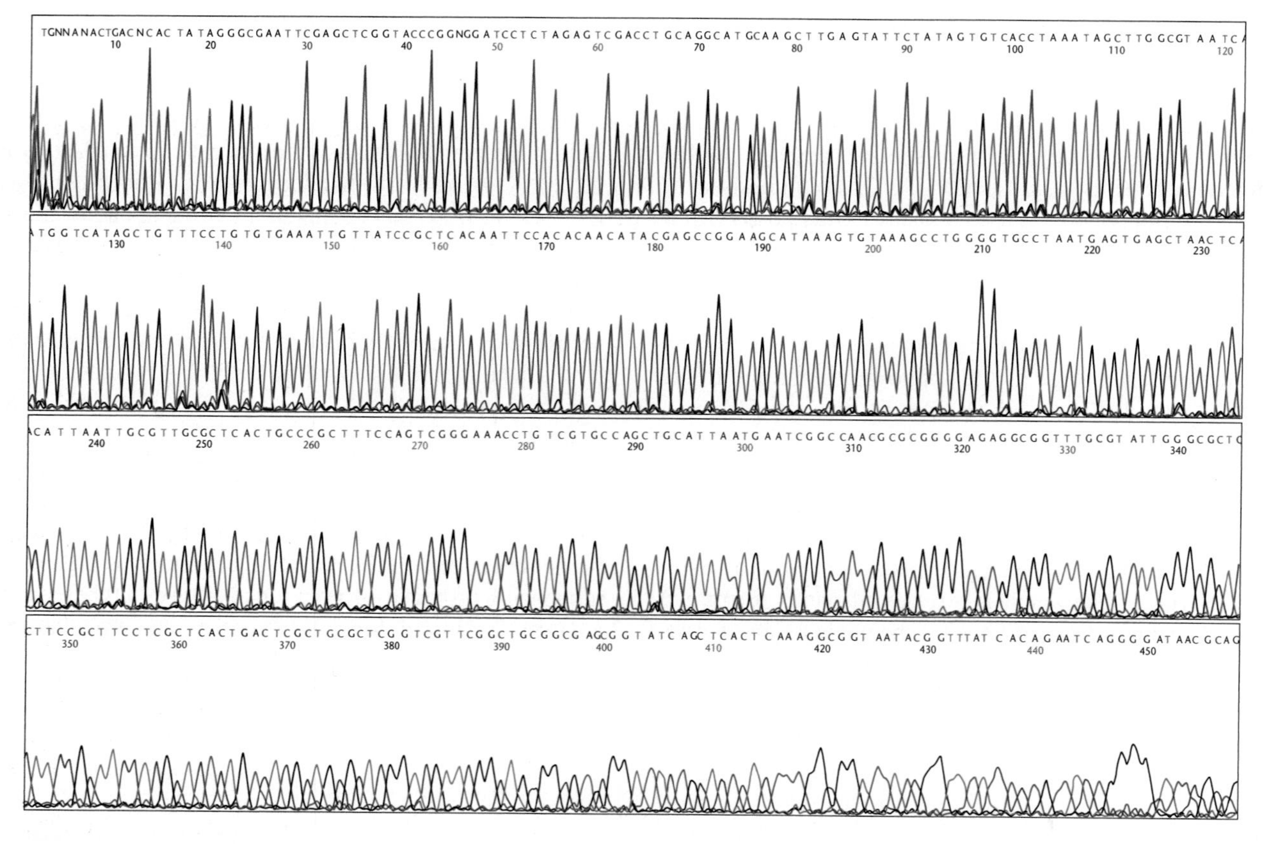

FIGURE 17–21 Automated DNA sequencing using fluorescent dyes, one for each base. Each peak represents the correct nucleotide in the sequence. Numbers indicate length of the sequence. The separated bases are read in order along the axis from left to right.

GENETICS, TECHNOLOGY, AND SOCIETY

Beyond Dolly: The Cloning of Humans

The death of Dolly the sheep, in February 2003, marked a poignant end to the beginning of the human cloning debate.

Six years earlier, her birth took the world by surprise. Before Dolly, the idea that an animal could be cloned from the cells of an adult animal was science fiction—something from *Brave New World* or *The Boys from Brazil*. For decades, scientists believed that it would be impossible to clone mammals, as DNA from adult cells could not be reprogrammed to code for the development of a new, complete organism. But then Dolly appeared.

Dolly, who was cloned from a frozen udder cell of a long-dead sheep, was brought into being by a group of embryologists led by Ian Wilmut and Keith Campbell at the Roslin Institute in Scotland. Their goal was to clone transgenic farm animals that secrete pharmaceutical products, such as blood-clotting factors or insulin, into their milk. In this way, herds of identical animals might be used as bioreactors to synthesize large quantities of medically important proteins.

The cloning method that Wilmut and Campbell used to create Dolly—a procedure called *nuclear transfer*—was first suggested by embryologist Hans Spemann in 1938. The method is simply to replace the nucleus of an egg with the nucleus from an adult cell, thereby creating a hybrid zygote. In theory, the genetic information in the donor nucleus should direct all further embryonic development, and the new organism should be a genetic replica of the adult that donated the nucleus. Although the procedure sounds simple in theory, it proved to be extremely difficult in practice because adult nuclei express only a small subset of the genes required for embryonic development. Wilmut and Campbell overcame this limitation by reprogramming the adult nuclei before transferring them into recipient eggs. They did this by starving the donor cells so they became quiescent. In addition, they passed an electric current through the recipient egg. For unknown reasons, these procedures turned on the silent genes within the differentiated cell nucleus. To create Dolly, over 200 udder-cell nuclei were transferred into eggs. Of these nuclear transfers, only 29 developed into embryos; and 13 of them were implanted into surrogate mother ewes. One pregnancy resulted, which culminated in the birth of Dolly. Although Dolly was the first mammal to be cloned from adult cells, the method has since been used to clone mice, pigs, cattle, rabbits, goats, an ox, a mule, a horse, and a cat.

Dolly's birth not only shattered scientists' views of mammalian cloning but it triggered an avalanche of controversy. The idea that humans might be cloned was denounced as immoral, repugnant, and ethically wrong. Within days of the announcement of Dolly's birth, bills were introduced into the United States Congress to prohibit research into human cloning, and worldwide bans were called for. Frightening scenarios were proposed—rich and powerful people cloning themselves for reasons of vanity, people with serious illnesses cloning replicas to act as organ donors, and legions of human clones suffering loss of autonomy, individuality, and kinship ties.

Is it really possible to clone humans? And if so, *should* we create human clones? The answer to the first question is simple: The same technology used to create Dolly could be used to clone a human. Most of the technical procedures are already used for human *in vitro* fertilizations, and it seems likely that adult human nuclei could be reprogrammed similarly to the adult sheep nuclei that created Dolly. However, the cloning process is extremely inefficient. Nuclear transfers into hundreds of human eggs would be required to yield one full-term pregnancy. As yet, there have been no verified successful attempts to clone a human, despite a few high-profile claims by several doctors and a religious sect.

The answer to the second question is not as simple. To begin with, it is necessary to understand the differences between two kinds of cloning—reproductive and therapeutic. Reproductive cloning results in the creation of an animal, such as Dolly. In contrast, therapeutic cloning creates early-stage embryos for the purpose of harvesting stem cells—cells that have the potential to treat various diseases. As legislators and the public tend to equate these two forms of cloning, research into human therapeutic cloning has been affected by all-encompassing bans that target reproductive cloning.

Although most scientists believe that research into therapeutic cloning should continue, they conclude that we should refrain from human reproductive cloning for both scientific and ethical reasons. The most serious concerns are that many cloned animals appear to suffer developmental defects. About 12% of cloned mice and 38% of cloned goats show congenital abnormalities such as oversized internal organs, and respiratory, circulatory, and immunological defects. The ethical arguments against human reproductive cloning involve concerns about potential abuses of the technology and threats to the dignity of human procreation. Some people worry that human clones might be discriminated against or cannibalized for spare parts and even question whether they would have the same personhood as nonclones. In contrast, proponents of human cloning suggest that the technology should be considered merely another fertility treatment, like *in vitro* fertilization—enabling infertile couples to have children that are genetically related to them or who do not carry a genetic defect of one parent.

As the debate over human cloning continues, Dolly will take her place among the important players in the history of modern science. Her remains have been preserved and are now on display in the National Museum of Scotland in Edinburgh. She is survived by several of her six lambs and the sadness of those who knew her and cared for her.

References

Jaenisch, R., and I. Wilmut, 2001. Don't clone humans! *Science* 291: 2552.

Chapter Summary

1. The cornerstone of recombinant DNA technology is a class of enzymes called restriction enzymes, which cut DNA at specific recognition sites. The fragments thus produced are joined with DNA vectors to form recombinant DNA molecules.
2. Vectors replicate autonomously in host cells and facilitate the manipulation of the newly created recombinant DNA molecules. Vectors are constructed from many sources, including bacterial plasmids and phages.
3. Recombinant DNA molecules are transferred into a host, and cloned copies are produced during host-cell replication. A variety of host cells may be used for replication, including bacteria, yeast, and mammalian cells. Cloned copies of foreign DNA sequences are recovered, purified, and analyzed.
4. The polymerase chain reaction (PCR) is a method for amplifying a specific DNA sequence that is present in a collection of DNA sequences, such as genomic DNA. The PCR method allows DNA to be cloned without host cells and is a rapid, sensitive method with wide-ranging applications.
5. Once cloned, DNA sequences are analyzed through a variety of methods, including restriction mapping and DNA sequencing. Other methods such as Southern blotting use hybridization to identify genes and flanking regulatory regions within the cloned sequences.

Key Terms

annealing, 385
bacteriophage, 379
cDNA library, 386
chain termination, 392
chromosome jumping, 388
chromosome walking, 388
clone, 378
complementary DNA (cDNA), 386
denaturation, 385
dideoxynucleotide, 393
DNA ligase, 379
DNA polymerase I, 386
DNA sequencing, 392
extension, 385
genomic library, 385
northern blot, 391
open reading frame (ORF), 388
phage λ, 381
plaque hybridization, 388
plasmid, 379
polylinker site, 381
polymerase chain reaction (PCR), 383
probe, 386
recombinant DNA, 378
recombinant DNA molecule, 378
restriction enzyme, 378
restriction fragment length polymorphism (RFLP), 390
restriction map, 389
reverse transcriptase, 386
Southern blotting, 390
vector, 378
western blot, 391
yeast artificial chromosome (YAC), 383

INSIGHTS AND SOLUTIONS

The recognition site for the restriction enzyme *Sau*3A is GATC (see Figure 17–3); the recognition site for the enzyme *Bam*HI is GGATCC, where the four internal bases are identical to the *Sau*3A site. Therefore, the single-stranded ends produced by the two enzymes are identical. Suppose you have a cloning vector that contains a *Bam*HI site and foreign DNA that you have cut with *Sau*3A. (a) Can this DNA be ligated into the *Bam*HI site of the vector, and if so, why? (b) Can the DNA segment cloned into this site be cut from the vector with *Sau*3A? With *Bam*HI? What potential problems do you see with the use of *Bam*HI?

Solution: (a) DNA cut with *Sau*3A can be ligated into the vector's *Bam*HI site because the single-stranded ends generated by the two enzymes are identical.

(b) The DNA can be cut from the vector with *Sau*3A because the recognition sequence for this enzyme (GATC) is maintained on each side of the insert. Recovering the cloned insert with *Bam*HI is more problematic. In the ligated vector, the conserved sequences are GGATC (left) and GATCC (right). The correct base will *follow* the conserved sequence (to produce GGATCC on the left) only about 25% of the time, and the correct base will *precede* the conserved sequence (and produce GGATCC on the right) about 25% of the time as well. Thus, *Bam*HI will be able to cut the insert from the vector ($0.25 \times 0.25 = 0.0625$), or only about 6% of the time.

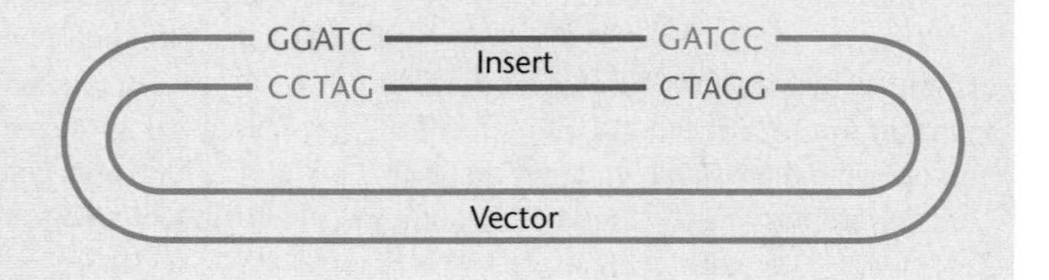

Problems and Discussion Questions

1. What roles do restriction enzymes, vectors, and host cells play in recombinant DNA studies?
2. Why is poly-dT an effective primer for reverse transcriptase?
3. The human insulin gene contains a number of introns. In spite of the fact that bacterial cells do not excise introns from mRNA, explain how a gene like this can be cloned into a bacterial cell and produce insulin.
4. Restriction enzymes recognize palindromic sequences in intact DNA molecules and cleave the double-stranded helix at these sites. Inasmuch as the bases are internal in a DNA double helix, how is this recognition accomplished?
5. Although the potential benefits of cloning in higher plants are obvious, the development of this field has lagged behind cloning in bacteria, yeast, and mammalian cells. Can you think of a reason for this?
6. Using DNA sequencing on a cloned DNA segment, you recover the nucleotide sequence shown below. Does this segment contain a palindromic recognition site for a restriction enzyme? What is the double-stranded sequence of the palindrome? What enzyme would cut at this site? (Consult Figure 17–3 for a list of restriction enzyme recognition sites.)

   CAGTATCCTAGGCAT
7. Restriction enzyme sites are palindromic; that is, they read the same in the 5′-to-3′ direction on each strand of DNA. What is the advantage of having restriction sites organized in this way?
8. List the advantages and disadvantages of using plasmids and YACs as cloning vectors.
9. Listed below are the cleavage sites for some restriction enzymes that recognize sequences of four bases and six bases. Assuming random distribution and equal amounts of each nucleotide in the DNA to be cut, on average, how far apart are each of these restriction sites?

*Taq*I	TCGA	*Hin*dIII	AAGCTT
*Alu*I	AGCT	*Bal*I	TGGCCA
*Hae*III	GGCC	*Bam*HI	GGATCC

10. Some restriction enzymes have recognition sites that are specific and unambiguous, while others have sites that are ambiguous, such that any purine, pyrimidine, or nucleotide can occupy a given position in the cutting site. In the examples below, *Not*I has an unambiguous cutting site, *Hin*fI has an ambiguous cutting site (N = any nucleotide), and *Xho*II also has an ambiguous cutting site (Pu = any purine, and Py = any purine). Assuming random distribution and equal amounts of each nucleotide in the DNA to be cut, on average, how far apart are each of these restriction cutting sites?

*Not*I	GCGGCCGC
*Hin*fI	GANTC
*Xho*II	PuGATCPy

11. What are the advantages of using a restriction enzyme with relatively few cutting sites? When would you use such enzymes?
12. An ampicillin-resistant, tetracycline-resistant plasmid, pBR322, is cleaved with *Pst*I, which cleaves within the ampicillin gene. The cut plasmid is ligated with *Pst*I-digested *Drosophila* DNA to prepare a genomic library, and the mixture is used to transform *E. coli* K12. (a) Which antibiotic should be added to the medium to select cells that have incorporated a plasmid? (b) What growth pattern should be selected to obtain plasmids containing *Drosophila* inserts? (c) How can you explain the presence of colonies that are resistant to both antibiotics?

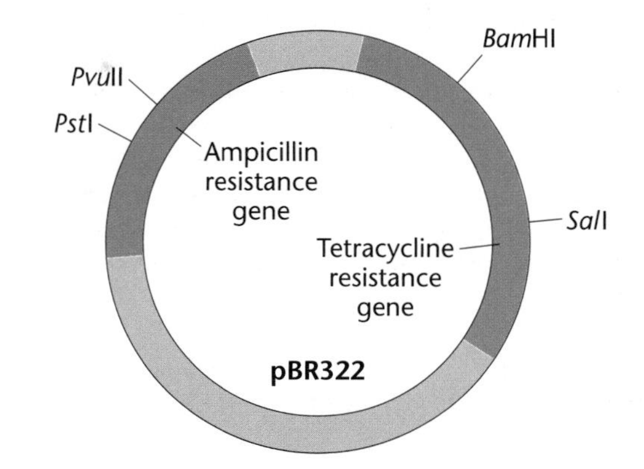

13. Plasmids isolated from the clones in Problem 12 are found to have an average length of 5 kb. Given that the *Drosophila* genome is 1.5×10^5 kb long, how many clones would be necessary to give a 99% probability that this library contains all genomic sequences?
14. In a control experiment, a plasmid containing a *Hin*dIII site within a kanamycin resistance gene is cut with *Hin*dIII, re-ligated, and used to transform *E. coli* K12 cells. Kanamycin-resistant colonies are selected, and plasmid DNA from these colonies is subjected to electrophoresis. Most of the colonies contain plasmids that produce single bands that migrate at the same rate as the original intact plasmid. A few colonies, however, produce two bands, one of original size and one that migrates much higher in the gel. Diagram the origin of this slow band as a product of ligation.
15. You have just created the world's first genomic library from the African okapi, a relative of the giraffe. No genes from this genome have been previously isolated or described. You wish to isolate the gene encoding the oxygen-transporting protein β-globin from the okapi library. This gene has been isolated from humans, and its nucleotide sequence and amino acid sequence are available in databases. Using the information available about the human, β-globin gene, what two strategies can you use to isolate this gene from the okapi library?
16. When making cDNA, the single-stranded DNA produced by reverse transcriptase can be made double stranded by treatment with DNA polymerase I. However, no primer is required with the DNA polymerase. Why is this?
17. What should you consider in deciding which vector to use in constructing a genomic library of eukaryotic DNA?
18. You are given a cDNA library of human genes prepared in a bacterial plasmid vector. You are also given the cloned yeast gene that encodes EF-1a, a protein that is highly conserved in protein sequence among eukaryotes. Outline how you would use these resources to identify the human cDNA clone encoding EF-1a.
19. Once you have isolated the human cDNA clone for EF-1a in Problem 18, you sequence the clone and find that it is 1384

nucleotide pairs long. Using this cDNA clone as a probe, you isolate the DNA encoding EF-1a from a human genomic library. The genomic clone is sequenced and found to be 5282 nucleotides long. What accounts for the difference in length observed between the cDNA clone and the genomic clone?

20. You have recovered a cloned DNA segment and determine that the insert is 1300 bp in length. To characterize this cloned segment, you isolate the insert and decide to construct a restriction map. Using enzyme I and enzyme II, followed by gel electrophoresis, you determine the number and size of the fragments produced by enzymes I and II alone and in combination shown in the table below. Construct a restriction map from these data, showing the positions of the restriction sites relative to one another and the distance between them in units of base pairs.

Enzymes	*Restriction Fragment Sizes (bp)*
I	350, 950
II	200, 1100
I and II	150, 200, 950

21. To create a cDNA library, cDNA can be cloned into vectors. In analyzing cDNA clones, it is often difficult to find clones that are full length—that is, extending to the 5′-end of the mRNA. Why is this so?

22. Although the capture and trading of great apes has been banned in 112 countries since 1973, it is estimated that about 1000 chimpanzees are removed annually from Africa and smuggled into Europe, the United States, and Japan. This illegal trade is often disguised by simulating births in captivity. Until recently, genetic identity tests to uncover these illegal activities were not used because of the lack of highly polymorphic markers and the difficulties of obtaining chimpanzee blood samples. Recently, a study was reported in which DNA samples were extracted from freshly plucked chimpanzee hair roots and used as templates for PCR. The primers used in these studies flank highly polymorphic sites in human DNA that result from variable numbers of tandem nucleotide repeats. Several offspring and their putative parents were tested to determine whether the offspring are "legitimate" or the product of illegal trading. The data are shown in the Southern blot below.

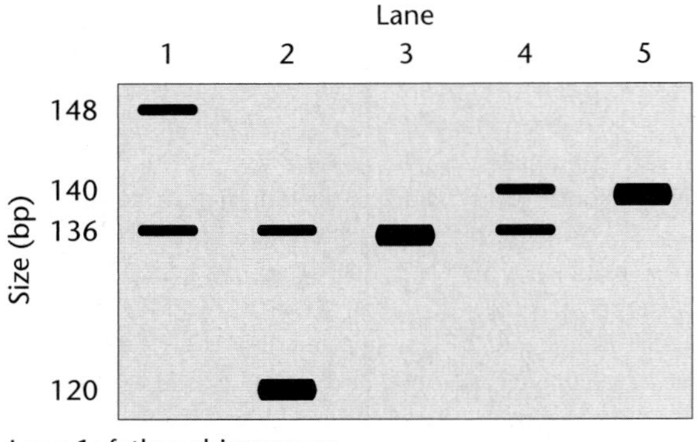

Examine the data carefully, and choose the best conclusion.
(a) None of the offspring are legitimate.
(b) Offspring B and C are not the products of these parents and were probably purchased on the illegal market. The data are consistent with offspring A being legitimate.
(c) Offspring A and B are products of the parents shown, but C is not and was therefore probably purchased on the illegal market.
(d) There are not enough data to draw any conclusions. Additional polymorphic sites should be examined.
(e) No conclusion can be drawn because "human" primers were used.

23. You have obtained a clone of a human gene A that is linked (within 200 kb) to a gene that causes breast cancer. Choose and correctly order 6 of the items from the following list to outline how you could use chromosome walking to obtain clones of all genes within 200 kb of gene A.
(a) Partially digest DNA with restriction enzyme *Bam*HI to obtain overlapping fragments of about 20 kb.
(b) Completely digest DNA with restriction enzyme *Eco*RI.
(c) Isolate human genomic DNA.
(d) Isolate *E. coli* genomic DNA.
(e) Probe northern blot with A.
(f) Repeat steps a–e until the desired number of clones is obtained.
(g) Insert fragment mixture into phage λ to make a phage library.
(h) Screen library with gene probe A, and isolate a hybridizing clone.
(i) Re-clone a small fragment from the end of this clone, use it to rescreen the library, and isolate a hybridizing clone.

24. The accompanying partial restriction map shows a recombinant plasmid, pBIO220, formed by cloning a piece of *Drosophila* DNA (striped box), including the gene *rosy* into the vector pBR322, which also contains the penicillin resistance gene, *pen*. The vector part of the plasmid contains only the two E sites shown, and no A or B sites. The gel shows several restriction digests of pBIO220.

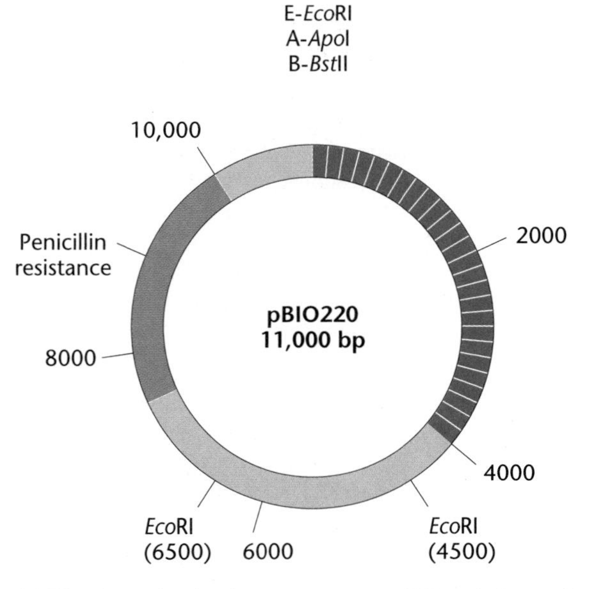

(a) Use the stained gel pattern on page 399 to deduce where restriction sites are located in the cloned fragment. (b) A PCR-amplified copy of the entire 2000-bp *rosy* gene was used to probe a Southern blot of the same gel. Use the Southern-blot results to deduce the locations of *rosy* in the cloned fragment. Redraw the map showing the location of the *rosy* gene.

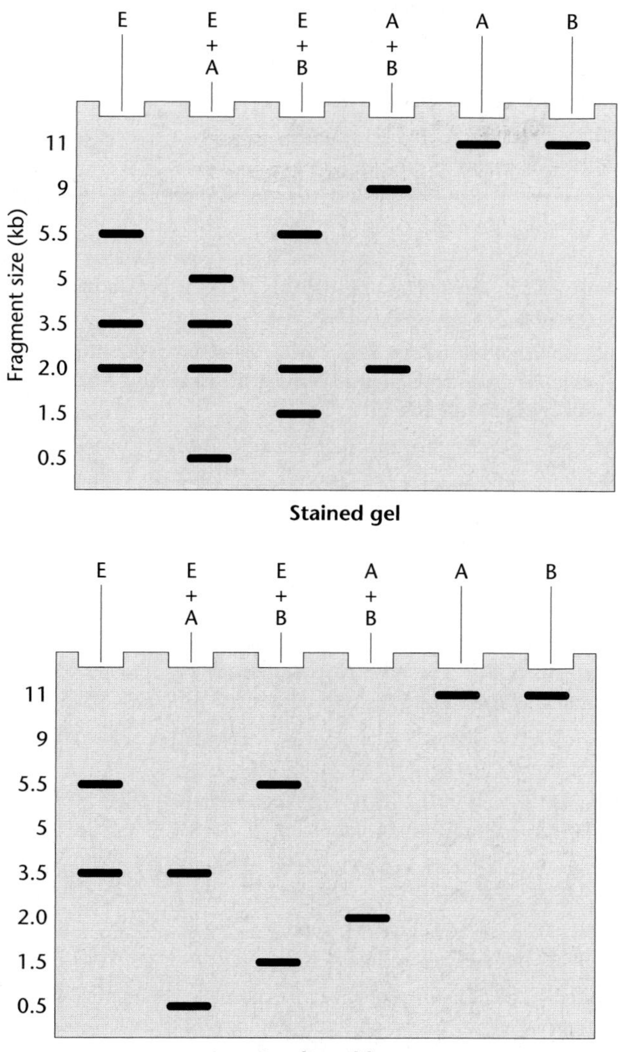

25. One of the six restriction maps shown here is consistent with the pattern of bands shown in the accompanying figure in the gel after digestion with several restriction endonucleases. The enzymes that were used are shown.

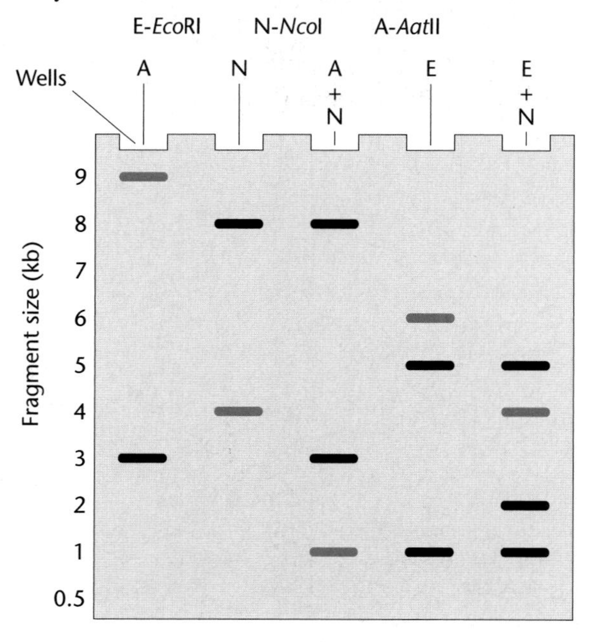

(a) From your analysis of the pattern of bands on the gel, select the correct map and explain your reasoning. (b) In a Southern blot prepared from this gel, the highlighted bands (pink) hybridized with the gene *pep*. Where is the *pep* gene located?

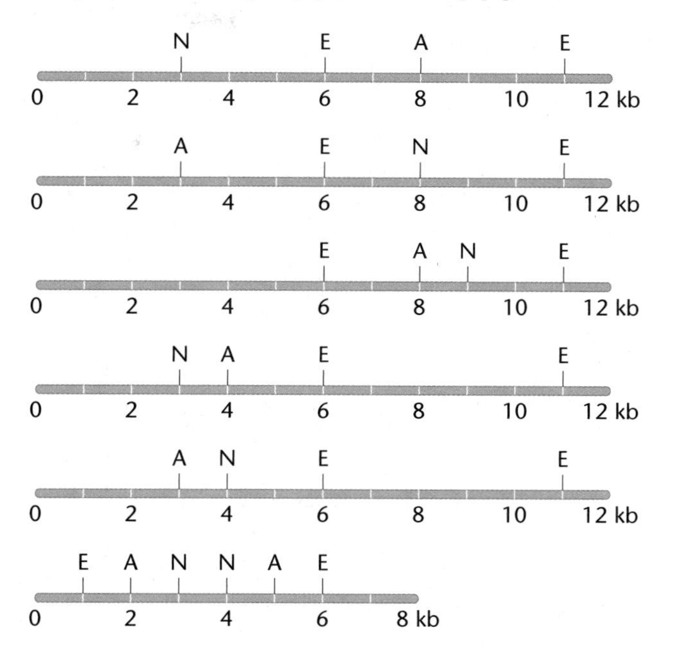

26. Briefly describe the problem that a stretch of repeated sequences would cause in a chromosome walk, and name the procedure used to overcome this problem.
27. List the steps involved in screening a genomic library. What must be known before starting such a procedure? What are the potential problems with such a procedure, and how can they be overcome or minimized?
28. To estimate the number of cleavage sites in a particular piece of DNA with a known size, you can apply the formula $N/4^n$, where N is the number of base pairs in the target DNA, and n is the number of bases in the recognition sequence of the restriction enzyme. If the recognition sequence for *Bam*HI is GGATCC and the phage λ DNA contains approximately 48,500 bp, how many restriction sites would you expect?
29. In a typical PCR reaction, what phenomena are occurring at temperature ranges (a) 90–95°C, (b) 50–70°C, and (c) 70–75°C?
30. A widely used method for calculating the annealing temperature for a primer used in PCR is 5 degrees below the T_m(°C), which is computed by the equation $81.5 + 0.41(\%GC) - (675/N)$, where %GC is the percentage of GC nucleotides in the oligonucleotide, and N is the length of the oligonucleotide. Notice from the formula that both the GC content *and* the length of the oligonucleotide are variables. Assuming you have the oligonucleotide shown below as a primer, compute the annealing temperature for PCR. What is the relationship between T_m(°C) and %GC? Why? (*Note:* In reality, this computation provides only a starting point for empirical determination of the most useful annealing temperature.)

5′-TTGAAAATATTTCCCATTGCC-3′

31. We usually think of enzymes as being most active at around 37°C, yet in PCR the DNA polymerase is subjected to multiple exposures of relatively high temperatures and seems to function appropriately at 70–75°C. What is special about the DNA polymerizing enzymes typically used in PCR?

32. How are dideoxynucleotides (ddNTPs) used in the chain-temination method of DNA sequencing?
33. Assume you have conducted a standard DNA sequencing using the chain-temination method. You performed all the steps correctly and electrophoresed the resulting DNA fragments correctly, but when you looked at the sequencing gel, many of the bands were duplicated (in terms of length) in other lanes. What might have happened?

Selected Readings

Alton, E. W., and D. M. Geddes, 1995. Gene therapy for cystic fibrosis: A clinical perspective. *Gene Ther.* 2: 88–95.

Antonarakis, S. 1989. Diagnosis of genetic disorders at the DNA level. *N. Engl. J. Med.* 320: 153–63.

Barker, D. et al. 1987. Gene for von Recklinghausen neurofibromatosis is in the pericentromeric region of chromosome 17. *Science* 236: 1001–162.

Burger, S. L., and Kimmel, A. R. 1987. *Guide to Molecular Cloning Techniques. Methods in Enzymology*. Vol. 152. San Diego: Academic Press.

Carter, P. J., and Samulski P. J., 2000. Adeno-associated viral vectors as gene delivery vehicles. *Int. J. Mol. Med.* 6: 17–27.

Colosimo, A., Goncz, K. K., Holmes, A. R., et al. 2000. Transfer and expression of foreign genes in mammalian cells. *BioTech.* 29: 314–31.

Dube, I. D., and Cournoyer, D., 1995. Gene therapy: Here to stay. *Can. Med. Assoc. J.* 152: 1605–13.

Eisensmith, R. C., and Woo, S. 1995. Molecular genetics of phenylketonuria: From molecular anthropology to gene therapy. *Adv. Genet.* 32: 199–271.

Guyer, M. S., and Collins, F. S. 1993. The Human Genome Project and the future of medicine. *Am. J. Dis. Child.* 147: 1145–52.

Knorr, D., and Sinskey, A. J. 1985. Biotechnology in food production and processing. *Science* 229: 1224–29.

McKusick, V. A. 1988. The new genetics and clinical medicine. *Hosp. Pract.* 23: 177–91.

Mullis, K. B. 1990. The unusual origin of the polymerase chain reaction. *Sci. Am.* (Apr.) 262: 56–65.

Old, R. W., and Primrose, S. B. 1994. *Principles of Genetic Manipulation: An Introduction to Genetic Engineering*. 5th ed. Palo Alto: Blackwell Scientific.

Oste, C. 1988. Polymerase chain reaction. *BioTech.* 6: 162–67.

Reiss, J., and Cooper, D. N. 1990. Application of the polymerase chain reaction to the diagnosis of human genetic disease. *Hum. Genet.* 85: 1–8.

Sambrook, Fitch, J., E. F. and Maniatis, T. 1989. *Molecular Cloning: A Laboratory Manual*. 2nd ed. Cold Spring Harbor: Cold Spring Harbor Press.

Southern, E. 1975. Detection of specific sequences among DNA fragments separated by gel electrophoresis. *J. Mol. Biol.* 98: 503–07.

Tal, J. 2000. Adeno-associated virus-based vectors in gene therapy. *J. Biomed. Sci.* 7: 279–91.

Thomas, T. L., and Hall, T. C. 1985. Gene transfer and expression in plants: Implications and potential. *BioEss.* 3: 149–53.

Torrey, J. G. 1985. The development of plant biotechnology. *Am. Sci.* 73: 354–63.

Welsh, J., and McClelland, M. 1990. Fingerprinting genomes using PCR with arbitrary primers. *Nucl. Acids Res.* 18: 7213–18.

White, R. 1985. DNA sequence polymorphisms revitalize linkage approaches in human genetics. *Trends Genet.* 1: 177–80.

Williams, J. G. K. et al. 1990. DNA polymorphisms amplified by arbitrary primers are useful as genetic markers. *Nucl. Acids Res.* 18: 6531–35.

CHAPTER 18

Arabidopsis thaliana, one of the model organisms used in genetic studies whose genome has been sequenced. *(Dr. Jeremy Burgess/Science Photo Library/Photo Researchers, Inc.)*

Genomics, Bioinformatics, and Proteomics

CHAPTER CONCEPTS

18.1 Genomics: Sequencing Helps Identify and Map All Genes in a Genome

18.2 Bioinformatics Provides Tools for Analyzing Genomic Information
- Compiling the Sequence
- Annotating the Sequence

18.3 Functional Genomics Classifies Genes and Identifies Their Functions
- Functional Genomics of a Bacterial Genome
- Strategies for Functional Assignment of Unknown Genes

18.4 Prokaryotic Genomes Have Some Unexpected Features
- Size Range of Eubacterial Genomes
- Genomes of Eubacteria
- Genomes of Archaea

18.5 Eukaryotic Genomes Have a Mosaic of Organizational Patterns
- General Features of the Eukaryotic Genome
- The *C. elegans* Genome
- Genomes of Higher Plants
- The Human Genome: The Human Genome Project
- Chromosomal Organization of Human Genes

18.6 Genomics Provides Insight into Genome Evolution
- The Minimum Genome for Living Cells
- Origin of the Eukaryotic Genome
- Genomie Duplications in Eukaryote Evolution
- Gene Duplications

18.7 Comparative Genomics: Multigene Families Diversify Gene Function
- The Globin Gene Family
- The Immunoglobulin Gene Family

18.8 Proteomics Identifies and Analyzes the Proteins in a Cell
- Proteomics Technology
- The Bacterial Proteome Changes with Alterations in the Environment
- Proteome Analysis of an Organelle: The Nucleolus

Geneticists began to identify and map genes in experimental organisms in the early years of the twentieth century. Although a wide range of organisms were originally studied, emphasis has gradually been placed on a few organisms, such as *Drosophila*, maize, mice, bacteria, and yeast. For these studies, geneticists developed two main approaches. The first approach identified spontaneous mutations in these organisms or developed mutant strains created by using chemical or physical agents. Once a set of mutations was available, the second approach, the generation of genetic maps by linkage analysis was started. In some organisms, such as *Drosophila*, physical maps of genes on chromosomes were also created. These approaches, developed 70–90 years ago, were efficient and are still widely used in genetics. The limitation of these methods is that at least one mutation for each gene is required to identify all genes in a genome, and obtaining mutations can be a time-consuming endeavor. In addition, mutations often have a lethal phenotype or no clear phenotype, making it difficult or impossible to map the mutated gene.

Beginning in the mid-1980s, geneticists moved away from the classical approaches of mutagenesis and mapping by linkage studies and began using recombinant DNA technology as a third approach to genetic analysis. In this approach, a collection of clones representing all the genes in a genome is established. The clones are pieced together into overlapping sets and assembled into genetic and physical maps that encompass the entire genome. In the final step, the clones are sequenced, with all genes in the genome identified by their nucleotide sequence. Collectively, these methods are called **genomics**. A number of genomic methods were discussed in Chapter 17, and more will be described here. Once obtained, genomic sequence data must be stored and analyzed. This is a formidable task: The human genome sequence consists of more than 3 billion nucleotides. **Bioinformatics** is an emerging field concerned with the development and application of computer hardware and software for acquisition, storage, analysis, and visualization of biological information such as genome sequence data. Public on-line databases for the storage and analysis of genome information are now essential tools for geneticists. **Proteomics** is the study of proteins encoded by a genome, including a list of which genes are expressed, their time of expression, the type and extent of any posttranslational modification of the gene product, the function of the encoded protein, and its location in various cellular compartments.

In many cases, gathering and analyzing genome information is a large-scale, labor-intensive endeavor, often requiring the coordinated effort of many laboratories. For this reason, geneticists have formed collaborations, called genome projects. One of the largest and best-known efforts is the **Human Genome Project (HGP)**. The Human Genome Project is a coordinated international effort to determine the sequence of the 3.2 billion base pairs in the haploid human genome and identify all the genes in the genome. In the United States, a project to map and sequence the human genome was proposed in 1986, and in 1988, the National Institutes of Health (NIH) and the United States Department of Energy created a joint committee to develop a plan for the project. The HGP got under way in 1990. Other countries, notably France, Britain, and Japan, began similar projects, all of which are now coordinated by an international organization, the Human Genome Organization (HUGO). Under the umbrella of the HGP, there are genome projects for a number of organisms, including bacteria (*Escherichia coli*), yeast (*Saccharomyces cerevisiae*), nematode (*Caenorhabditis elegans*), fruit fly (*Drosophila melanogaster*), and mouse (*Mus musculus*). The time-line for these projects is shown in Figure 18–1. The Human Genome Project and related programs are discussed further in Chapter 19.

In this chapter, we outline the technology of genome sequence analysis and review the significant findings from a variety of genome projects using prokaryotic and eukaryotic organisms. This review includes insights into the organization of genes in chromosomes, genome evolution, findings from comparative genomics, and the minimum number of genes necessary for life, as well as the evolution of gene families. We also discuss the emerging field of proteomics and examine how geneticists are studying patterns of gene expression as well as the applications of this knowledge in various fields.

HOW DO WE KNOW WHAT WE KNOW?

IN THIS CHAPTER, WE WILL FOCUS ON the methods used in sequencing genomes, analysis of the sequence, and assignment of gene function. We also examine what is known about prokaryotic and eukaryotic genomes. As you study this topic, you should try to answer several fundamental questions:

1. How do we know that a sequenced genome is accurate?
2. How do we know when all the genes in a sequenced genome have been identified?
3. How do we know the function of proteins encoded in a genome?
4. How do we know the minimum gene set required for life?
5. How do we know which cellular proteins are found in the nucleolus? ■

18.1 Genomics: Sequencing Helps Identify and Map All Genes in a Genome

Geneticists use two different methods for sequencing genomes. The **clone-by-clone method** was the first to be developed (Figure 18–2a), and it begins with the construction of genomic libraries covering all the DNA of an organism. Next, the clones are assembled into genetic and physical maps encompassing the entire genome. The nucleotide sequence is determined clone by clone until the entire genome is sequenced. The clone-by-clone method

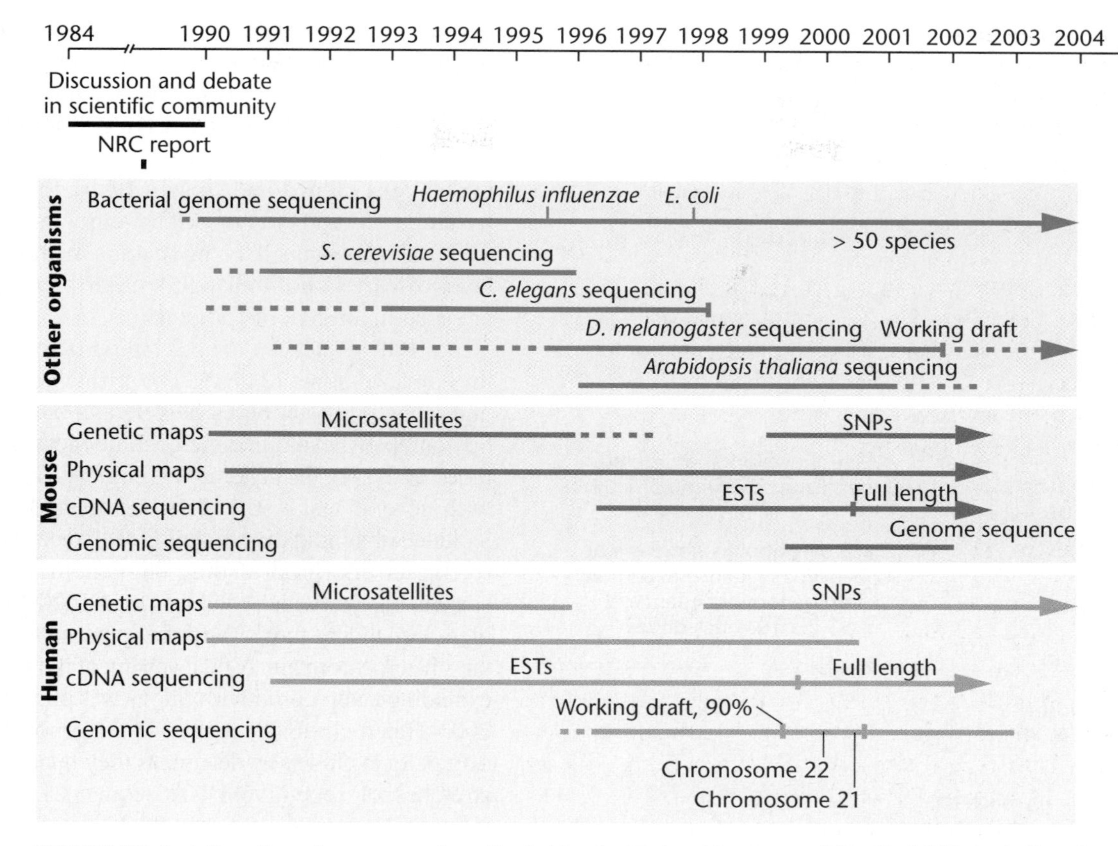

FIGURE 18–1 A time-line of genome projects funded by the National Institutes of Health (NIH), including the Human Genome Project. Key: NRC, National Research Council; SNPs, single nucleotide polymorphisms; ESTs, expressed sequence tags. *(Redrawn from International Human Genome Sequencing Consortium. 2001. Initial sequencing and analysis of the human genome.* Nature *409: 860–921, Figure 1, p. 862.)*

was chosen for the publicly funded Human Genome Project sponsored by the NIH and the Department of Energy.

In the **shotgun cloning method** (Figure 18–2b), genomic libraries are prepared and randomly selected clones are sequenced until all or nearly all segments of the genome have been sequenced repeatedly. Computers with assembler software organize the nucleotide sequence information into a genome sequence. This method, developed by Craig Venter and his colleagues at The Institute for Genome Research (TIGR), was used to sequence the genome of *Haemophilus influenzae* in 1995. This was the first organism to have its genome completely sequenced. After refining the method and using it to sequence the genomes of other prokaryotes, the shotgun method was used to sequence eukaryotic genomes, including *Drosophila* and humans. In a later section of this chapter, we will discuss some of the properties of the human genome that were discovered with these sequencing methods.

18.2 Bioinformatics Provides Tools for Analyzing Genomic Information

Traditionally, geneticists use laboratory notebooks to record and analyze the results of their experiments. Technology has now generated methods that sequence entire genomes or measure expression of all genes in a genome. These technologies produce data on a scale that was unimaginable only a few years ago. The ability to generate large amounts of experimental data has fundamentally changed the way results from genetic experiments are stored and analyzed. Instead of lab notebooks, geneticists and other biologists may now use large-scale public and private databases to store, analyze, and visualize their experimental results.

As mentioned earlier, bioinformatics unites computer science with biology. One of the basic challenges in bioinformatics is managing the flood of information from genome projects. In the next sections, we will discuss the analysis of data from genome projects, including compiling genome sequences, identifying genes, and assigning functions to identified genes. Access to genomic databases as well as some databases explaining more about bioinformatics is provided at this book's Web site.

Compiling the Sequence

To ensure that the nucleotide sequence of a genome is complete and error-free, the genome is sequenced more than once. For example, using the shotgun method on the genome of the bacterium *Pseudomonas aeruginosa*, researchers sequenced the 6.3 million nucleotides 7 times to ensure that the sequence was accurate. Even with this level of redundancy, the assembler software recognized 1604 regions that required further clarification. These regions were reanalyzed and resequenced to improve accuracy. Finally, the shotgun method's sequence was compared with the sequence of two widely separated genomic regions obtained by conventional cloning. The sequence of the 81,843 nucleotides cloned and sequenced by the clone-by-clone method was in perfect agreement with the sequence obtained by the shotgun method. This level of care is not unusual; similar precautions are used in all genome projects.

The HGP sequenced the 3.2 billion base pairs of the human genome 12 times. The privately run shotgun-cloning project based at Celera, a biotechnology company, examined the genome more than 35 times. A draft of the genome was completed in 2001, with several tasks still to be completed. These included obtaining the remaining sequence and correcting errors (proofreading the genome), filling sequence gaps (which total less than 150 Mb), and then sequencing the 15% of the genome that contains heterochromatin. A final version of the genome, excluding heterochromatic regions was published in 2003. Heterochromatic regions of the genome were originally excluded by design, as they contain long stretches of repetitive DNA sequences initially thought to contain no genes. However, while sequencing the *Drosophila* genome, researchers discovered that heterochromatic regions contain a small number of genes (about 50 in *Drosophila*). As a result, heterochromatic regions of the human genome must now be sequenced to ensure that all genes are identified. Once the human genome or any other genome is sequenced, compiled, and proofread, the next stage—**annotation**—begins.

Annotating the Sequence

After a genome has been sequenced, organized, and checked for accuracy, the next task is to find all the genes that encode products (proteins and RNA). This is the first step in annotation—a process that identifies genes, their regulatory sequences, and their function(s). Annotation also identifies nonprotein coding genes (including ribosomal RNA, transfer RNA, and small nuclear RNAs) and finds and characterizes the mobile genetic elements and repetitive-sequence families that may be present in the genome.

Locating protein-coding genes is done by inspecting the sequence, either by using computer software

(a) Clone-by-clone method

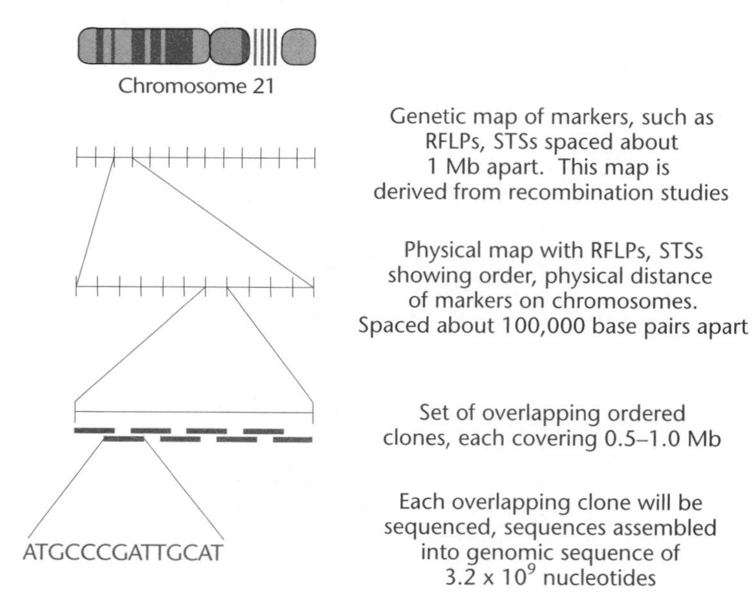
Chromosome 21
Genetic map of markers, such as RFLPs, STSs spaced about 1 Mb apart. This map is derived from recombination studies
Physical map with RFLPs, STSs showing order, physical distance of markers on chromosomes. Spaced about 100,000 base pairs apart
Set of overlapping ordered clones, each covering 0.5–1.0 Mb
Each overlapping clone will be sequenced, sequences assembled into genomic sequence of 3.2 x 10⁹ nucleotides
ATGCCCGATTGCAT

(b) Shotgun method

Isolate bacterial chromosomes
Fragment by sonication
Sonicated fragments are cloned into a vector
Prepare clone library

Clones are selected at random from library
Clones are sequenced

Computer with compiler software
Assembler software used to assemble sequence

FIGURE 18–2 (a) In the clone-by-clone method, clones from a genomic library are organized into genetic and physical maps of each chromosome. After the clones are arranged into physical maps, they are broken into smaller, overlapping clones that cover each chromosome. Each smaller clone is sequenced, and the genomic sequence is assembled by stringing together the nucleotide sequence of the clones. (b) In the shotgun method, a genomic library is constructed from fragments of genomic DNA. Clones are selected from the library at random and sequenced. The sequence is assembled by looking for sequence overlaps between clones from different libraries. This is done with a computer, using assembler software designed for genomic analysis.

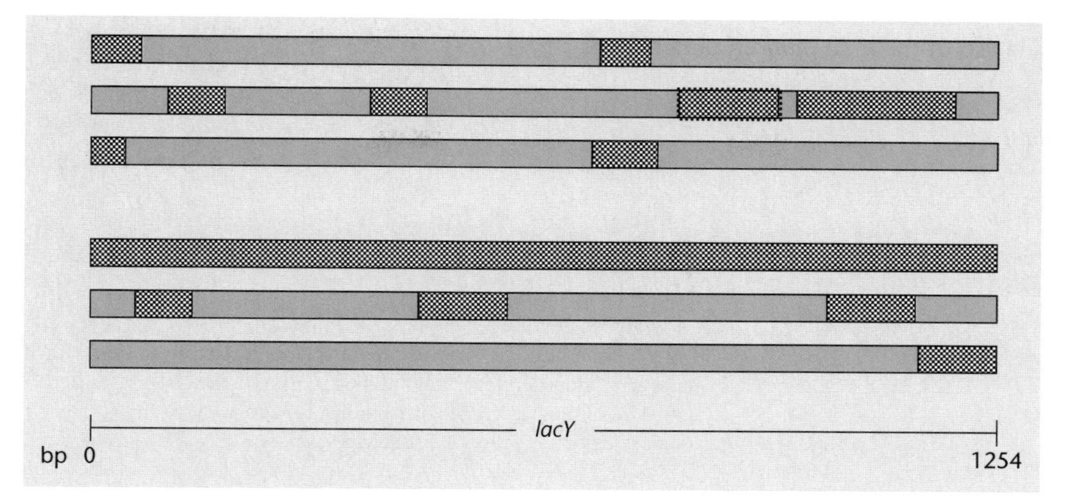

FIGURE 18–3 ORF analysis of the *lacY* gene of *E. coli*. ORFs for each of the six possible reading frames, three on each DNA strand, are shown as dark, stippled regions. Only ORFs longer than 50 codons are marked. Because most genes are much longer than 50 codons, ORF length is one indication of an actual gene. In this case, only one ORF covers the entire sequence and shows the location and orientation of the *lacY* gene. *(On-line: http://www.ncbi.nlm.gov/gorf/)*

or by eye. Genes have several identifying features. Protein-coding genes are composed of **open reading frames (ORFs)**, a series of nucleotides that specify an amino acid sequence. ORFs begin with an initiation sequence (usually ATG) and end with a termination sequence (TAA, TAG, or TGA). Scanning a DNA sequence for ORFs beginning with an ATG and followed by a termination codon is one strategy for finding genes. Scanning for ORFs, usually by computer, is an effective method for annotating bacterial genomes (Figure 18–3). However, the task is complicated by the fact that any DNA sequence has six different reading frames with possible ORFs: three on one strand and three in the opposite direction on the complementary strand.

The organization of genes in eukaryotic genomes (including the human genome) makes direct searching for ORFs more difficult. First, many eukaryotic genes have a series of exons (coding regions) interrupted by introns (noncoding regions). As a result, these genes are not organized as continuous ORFs. Scanning software can interpret each exon as a separate gene because termination codons are often found in introns. Second, genes in humans and other eukaryotes are often widely spaced, increasing the chances of finding false ORFs. For example, over 70% of the human genome is composed of DNA located between genes, called intergenic spacer DNA.

Newer versions of ORF-scanning software designed for eukaryotic genomes are more efficient. These programs search for features such as codon bias, the nucleotide sequences characteristic of intron–exon junctions, upstream regulatory sequences such as the TATA box, 3′ AATAAA poly-A signals, and in some genomes, CpG islands. Codon bias is the selective use of one or two codons to encode amino acids that can be encoded by a number of different codons. For example, alanine can be encoded by GCA, GCT, GCC, and GCG. If the codons were used randomly, each would be used about 25% of the time. Yet in the human genome, GCC is used 41% of the time, and GCG only 11% of the time. Codon bias is present in exons but should not be present in introns or intergenic spacers. **CpG islands** are DNA regions containing many copies of this nucleotide doublet and are found in gene-rich regions of the mammalian genome, often adjacent to genes. However, searching for codon bias and other features of eukaryotic genes such as CpG islands is not foolproof, and scanning by eye is a common backup technique.

18.3 Functional Genomics Classifies Genes and Identifies Their Functions

After a genomic sequence is annotated, the next task is to assign functions to all the genes in the sequence. Some of the genes have functions assigned by the classic method of mutagenesis and linkage mapping, but many other genes have no clear-cut function assigned. One approach to assigning functions to these genes is the use of homology searches. This analysis has several components:

- Search databases such as GenBank to find similar genes isolated from other organisms.
- Compare the sequence of an ORF with that of a well-characterized gene from another organism.
- Scan the ORF for functional motifs, regions of DNA that encode protein domains such as ion channels, DNA-binding regions, or secretion/export signals.

Let's examine how this strategy is used in the analysis of a prokaryotic genome.

Functional Genomics of a Bacterial Genome

Table 18.1 shows the results of an ORF analysis in the *P. aeruginosa* genome. In this analysis, as in all functional genomic analyses, the ORFs are classified on a scale

TABLE 18.1 Analysis of the *P. aeruginosa* Genome

General Features			
Genome size (bp)	6,264,403		
G + C content	66.6%		
Coding regions	89.4%		
Coding Sequences			
Confidence level	***ORFs***	***(%)***	***Definition***
1	372	6.7	*P. aeruginosa* genes with demonstrated function
2	1059	19.0	Strong homologs of genes with demonstrated function from other organisms
3	1590	28.5	Genes with proposed function based on motif searches or limited homology
4a	769	13.8	Homologs of reported genes of unknown function
4b	1780	32.0	No homology to any reported sequences
Total	5570	100	

Source: Stover, et al. 2000. Complete genome sequence of *Pseudomonas aeruginosa* PA01, an opportunistic pathogen. *Nature* 406: 959–64. Table 1, p. 961.

according to confidence levels. Level 1 genes are those whose function is already known from conventional genetic analysis or is otherwise known to have a demonstrated function. Level 2 genes have a strong homology to genes with known functions in other organisms. Level 3 genes have a proposed function based on encoded motifs or limited homology to genes in other organisms. Level 4 genes have no known or proposed function. In a functional analysis of the *P. aeruginosa* genome (Table 18.2), about half the genes have no known or proposed function.

TABLE 18.2 Functional Classes of Predicted Genes in *P. aeruginosa*

	ORFs	
Functional Class	***Number***	***%***
Adaptation, protection (e.g., cold shock proteins)	60	1.1
Amino acid biosynthesis and metabolism	150	2.7
Antibiotic resistance and susceptibility	19	0.3
Biosynthesis of cofactors, prosthetic groups, and carriers	119	2.1
Carbon compound catabolism	130	2.3
Cell division	26	0.5
Cell wall	83	1.5
Central intermediary metabolism	64	1.1
Chaperones and heat shock proteins	52	0.9
Chemotaxis	43	0.8
DNA replication, recombination, modification, and repair	81	1.5
Energy metabolism	166	3.0
Fatty acid and phospholipid metabolism	56	1.0
Membrane proteins	7	0.1
Motility and attachment	65	1.2
Nucleotide biosynthesis and metabolism	60	1.1
Protein secretion/export apparatus	83	1.5
Putative enzymes	409	7.3
Related to phage, transposon, or plasmid	38	0.7
Secreted factors (toxins, enzymes, alginate)	58	1.0
Transcriptional regulators	403	7.2
Transcription, RNA processing, and degradation	45	0.8
Translation, posttranslational modification, and degradation	149	2.7
Transport of small molecules	555	10.0
Two-component regulatory systems	118	2.1
Hypothetical	1774	31.8
Unknown (conserved hypothetical)	757	13.6
Total	5570	100

Source: Stover, et al. 2000. Complete genome sequence of *Pseudomonas aeruginosa* PA01, an opportunistic pathogen. *Nature* 406: 959–64.

This situation is not unique to *P. aeruginosa*; in almost all organisms whose genomes have been sequenced to date, about half of their genes have no known function.

The goal in the analysis of the *P. aeruginosa* genome, as in all genomic analyses, is to move all of the genes up to Level 1 or 2 and to understand how these genes and their products interact in the biology of the organism. The levels used in functional genomics do not reflect the importance of a gene to an organism's survival or reproduction but are assignments based on our knowledge of a gene's function.

Strategies for Functional Assignments of Unknown Genes

Yeast (*S. cerevisiae*) is one of the model eukaryotic organisms in genetic analysis. Despite decades of study using classical genetic analysis, only about one-third of the 6200 genes revealed by genomic sequencing had been fully characterized (assigned as Level 1 genes). In the years following publication of the yeast genome, the number of characterized genes has risen to over 3700, and an additional 500 or so genes have been assigned as Level 2 or 3 genes based on homology. This leaves about 1900–2000 genes that encode proteins of no known function.

Because an array of genetic, biochemical, and molecular techniques are well-established for studying yeast, it is instructive to see how these are being applied in functional analyses of the yeast genome. The use of these techniques has changed—they are now being integrated and applied to carry out functional analysis on a genomic scale, instead of studying just one gene or protein at a time. It is now possible to use these techniques to analyze hundreds or thousands of genes in a single experiment. These experiments are analyzing gene expression, protein–protein interactions, post–translational modifications, as well as other approaches summarized in Figure 18–4. The goal of these studies is to assign each gene to one or more of the several hundred core processes that occur in a living cell (DNA replication, transcription, translation, transport, biosynthesis, etc.).

As an example, let's examine one large-scale method for studying DNA-protein interactions in yeast, specifically, the mapping of transcription factor binding sites (Figure 18–5). To do this, growing yeast cells are treated with formaldehyde, then if transcription factors are bound to their DNA target sequences, the formaldehyde will cross-link them in place. The DNA is extracted from the yeast cells and sheared into small fragments. A transcription factor of interest with its bound DNA is immunoprecipitated by using an antibody against that transcription factor. The attached DNA is extracted, amplified by PCR, and labeled with a fluorescent probe. The labeled DNA is used as a target in a DNA microarray containing yeast genes. The intensity of the fluorescent probe at each spot is a measure of its binding activity. In one application of

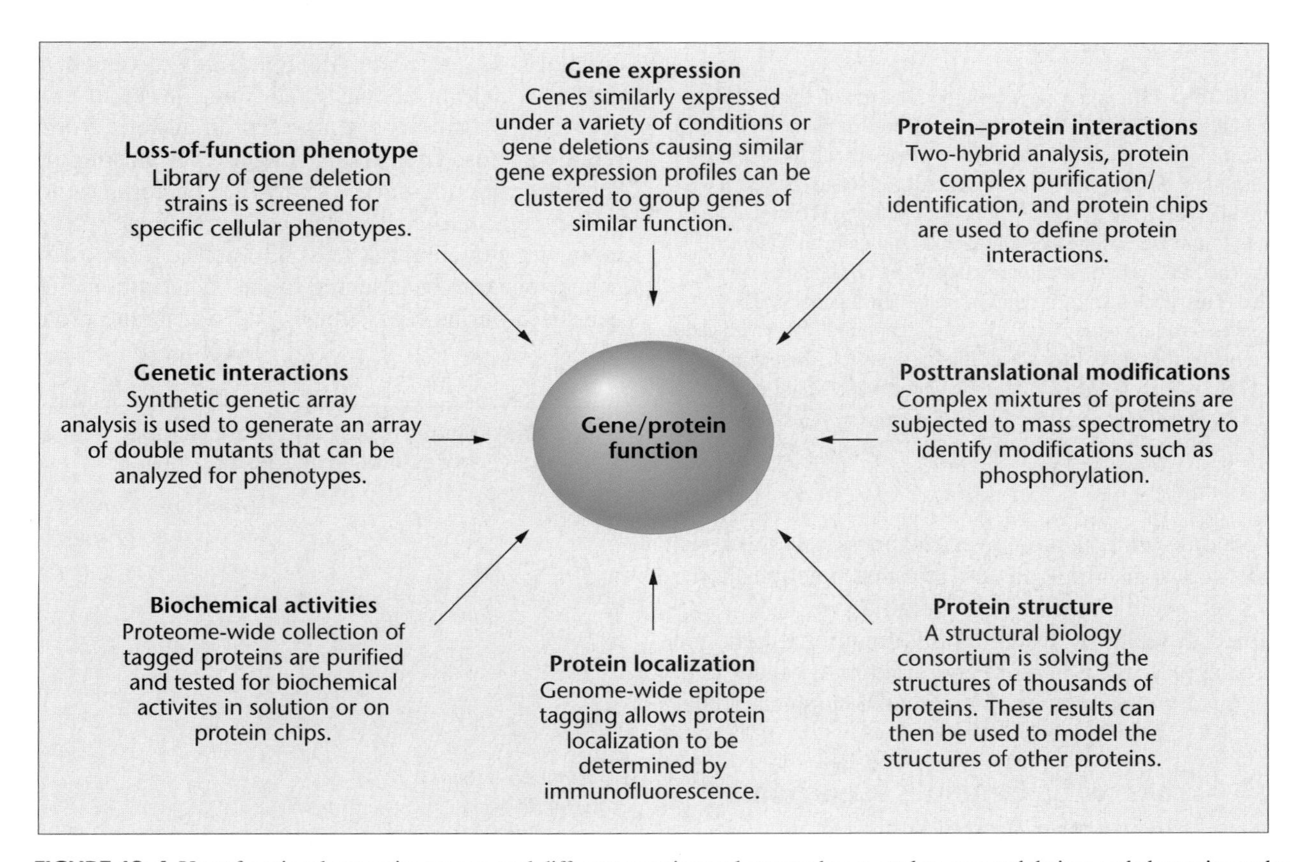

FIGURE 18–4 Yeast functional genomics uses several different experimental approaches to study genes and their encoded proteins and to assign functions to all genes in the yeast genome. Most of these methods are scaled up to analyze hundreds or thousands of genes or proteins in a single experiment, producing large amounts of data for analysis. *(Reprinted from* Current Opinion in Cell Biology, *Vol. 15, Martin AC and Drubin DG, Impact of genome-wide functional analysis on cell biology, pp. 6-13, Copyright (2003), with permission from Elsevier.)*

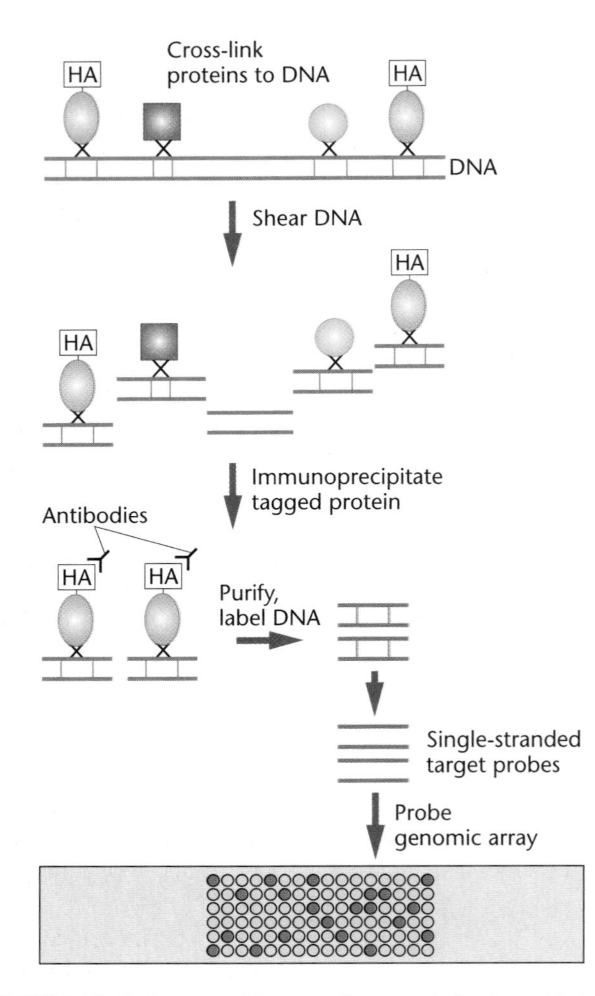

FIGURE 18–5 Genome-wide screen for transcription factor binding sites. In the top line, DNA-binding proteins, including one tagged with a protein fragment (HA), are captured by cross-linking them to their chromosomal binding sites. The DNA is extracted and sheared. The sheared DNA is treated with an antibody against the HA protein fragment to immunoprecipitate the DNA-binding protein and its bound DNA. The DNA is purified, amplified by PCR, and fluorescently labeled. This labeled DNA is hybridized to known genomic DNA sequences on a microarray. Hybridization signals indicate which genes are bound by the probe. In this way, binding sites for a transcription factor can be analyzed across the entire genome in a single experiment. *(Reprinted by permission from Nature Reviews Genetics 2(4):302–312, copyright 2001, Macmillan Magazines Ltd.)*

this genome-wide method, over 200 previously unknown targets of a transcriptional activator that functions at the G1/S cell cycle interface were identified.

Although we have discussed this functional genomic approach in yeast, it is also being used in the analysis of other eukaryotic genomes, including the human genome.

18.4 Prokaryotic Genomes Have Some Unexpected Features

The sequencing of almost one hundred prokaryotic and eukaryotic genomes has been completed. The results indicate that there are significant differences in genome organization between prokaryotes and eukaryotes. Therefore, we will consider the organization of these genomes separately. Since most prokaryotes have small genomes amenable to shotgun cloning and sequencing, most of the completed projects have focused on two types of prokaryotes, Eubacteria and Archaea (Table 18.3), and there are more than 200 additional projects under way. The prokaryotic genomes already sequenced include several that cause human diseases, such as cholera, tuberculosis, and leprosy.

Size Range of Eubacterial Genomes

Based on genome project results, we can make a number of generalizations concerning the size and organization of the genomes of Eubacteria. Traditionally, the bacterial genome has been thought of as relatively small (less than 5 Mb) and contained within a single circular DNA molecule. The flood of genomic information now available has challenged this viewpoint. Although most prokaryotic genomes are small, genome sizes vary widely. In fact, there is some overlap between larger bacterial genomes (30 Mb in *Bacillus megaterium*) and smaller eukaryotic genomes (12.1 Mb in yeast).

Additionally, while most prokaryotic genomes sequenced to date are circular DNA molecules, an increasing number of genomes composed of linear DNA molecules are being identified, including *Borrelia burgdorferi*, the organism that causes Lyme disease. Perhaps more important, new findings about plasmids are redefining our view of the bacterial genome as a single, circular DNA molecule. Most plasmids carry nonessential genes and can be transferred from one cell to another. The same plasmid is often present in bacteria from different species. These facts suggest that plasmid genes should not be included as part of a bacterial genome. However, *B. burgdorferi* carries approximately 17 plasmids, which contain at least 430 genes. Some of these genes appear to be essential to the bacterium, including genes for purine biosynthesis and membrane proteins.

TABLE 18.3 Genome Size and Gene Number in Selected Prokaryotes

	Genome Size (Mb)	*Number of Genes*
Archaea		
Archaeoglobus fulgidis	2.17	2493
Methanococcus jannaschii	1.66	1738
Thermoplasma acidophilum	1.56	1509
Eubacteria		
E. coli	4.64	4397
Bacillus subtilis	4.21	4212
H. influenzae	1.83	1791
Aquifex aeolicus	1.55	1552
Rickettsia prowazekii	1.11	834
Mycoplasma pneumoniae	0.82	710
Mycoplasma genitalium	0.58	503

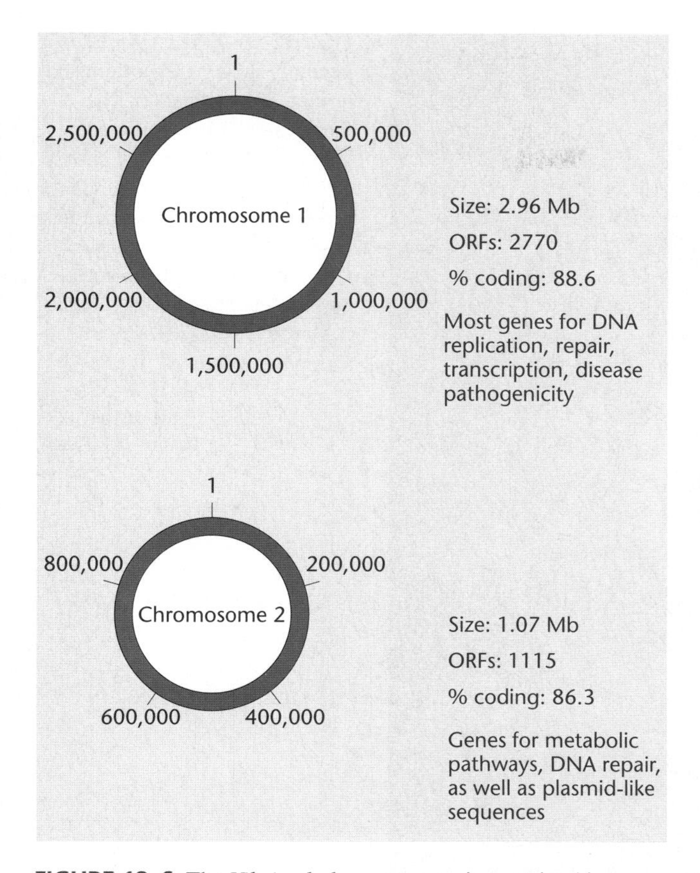

FIGURE 18–6 The *Vibrio cholerae* genome is contained in two chromosomes. The larger chromosome (chromosome 1) contains most of the genes for essential cellular functions and infectivity. Most of the genes on chromosome 2 (52% of 115) are of unknown function. The bias in gene content and the presence of plasmid-like sequences on chromosome 2 suggest that this chromosome was a megaplasmid captured by an ancestral *Vibrio* species. *(Redrawn from Heidelberg, J. F. et al. 2000. DNA sequence of both chromosomes of the cholera pathogen,* Vibrio cholerae. Nature *406: 477–83, Figure 2, p. 478.)*

Recently, sequencing the genome of *Vibrio cholerae*, the organism responsible for cholera, revealed the presence of two circular chromosomes (Figure 18–6). Chromosome 1 (2.96 Mb) contains 2770 ORFs, and chromosome 2 (1.07 Mb) contains 1115 ORFs, some of which encode essential genes, such as ribosomal proteins. Based on its nucleotide sequence, chromosome 2 is apparently derived from a plasmid captured by an ancestral species, or it could represent a chromosome with many inserted plasmid genes. Two chromosomes are also present in other species of *Vibrio*. The existence of prokaryotic genomes containing multiple chromosomes, where at least one chromosome is derived from a plasmid, raises several important questions. For example, when should a plasmid be considered a chromosome, and what regulatory mechanisms control gene expression and metabolism in a multichromosome prokaryotic genome? Answers to these questions and others will redefine some ideas about prokaryotic genomes and the nature of plasmids and may provide clues about the evolution of multichromosome eukaryotic genomes.

Genomes of Eubacteria

We can make two generalizations about the organization of protein-coding genes in bacterial chromosomes. First, gene density is very high, averaging about one gene per kilobase of DNA-*E. coli* has a 4.6 Mb genome containing 4288 protein–coding genes, a density very close to one gene per kilobase (Figure 18–7); *Mycoplasma genitalium* has a small genome (0.58 Mb), with 503 genes, and its gene density is also close to one gene per kilobase pair. This close packing of genes in prokaryotic genomes results in a very high proportion of the DNA (approximately 85–90%) serving as coding DNA. In *M. genitalium*, there are only about 110–125 bp of DNA between genes. Typically, less than 1% of bacterial DNA is noncoding DNA, usually in the form of transposable elements that can move from one place to another in the genome. Introns are extremely rare in bacterial genomes. A second generalization we can make is that bacterial genomes are characterized by the presence of operons. In *E. coli*, 27% of the predicted transcription units are contained in operons (almost 600 operons).

In other bacterial genomes, the organization of genes into transcriptional units is challenging our ideas about the nature of operons. In *Aquifex aeolicus*, most genes are parts of polycistronic transcription units we would usually call operons. However, our working definition that operons contain genes that are part of a single biochemical pathway does not hold true in this organism. In *A. aeolicus* (Figure 18–8), one operon contains six genes: two for DNA recombination, one for lipid synthesis, one for nucleic acid synthesis, one for protein synthesis, and one that encodes a protein for cell motility. Other operons in this species also contain genes with several different functions. This finding, combined with similar results from other genome projects, provides new insights into the nature of operons and their role in biochemical processes in bacterial cells and may redefine our concept of the operon.

The organization of a small segment of the *E. coli* genome is shown in Figure 18–9. The lines above the genes show how those genes are organized into transcription units with promoters that locate the start of transcription and the direction of transcription for each unit. Some of these units are organized as operons. Note that there are three cases of overlapping genes, where one gene is nested inside another on the same DNA strand (shown in green). Overlapping genes are common in viruses (where space is at a premium), but are infrequent in bacteria and most other organisms.

Genomes of Archaea

Archaea (formerly known as Archaebacteria) is one of the three major domains of living organisms. The other two are the Eubacteria (true bacteria, without a membrane-bound nucleus, such as *E. coli* and *Bacillus*) and Eukarya (containing a membrane-bound nucleus). Archaea, like Eubacteria, are prokaryotes; that is, they have no nucleus. They have only recently been recognized as a separate domain of life.

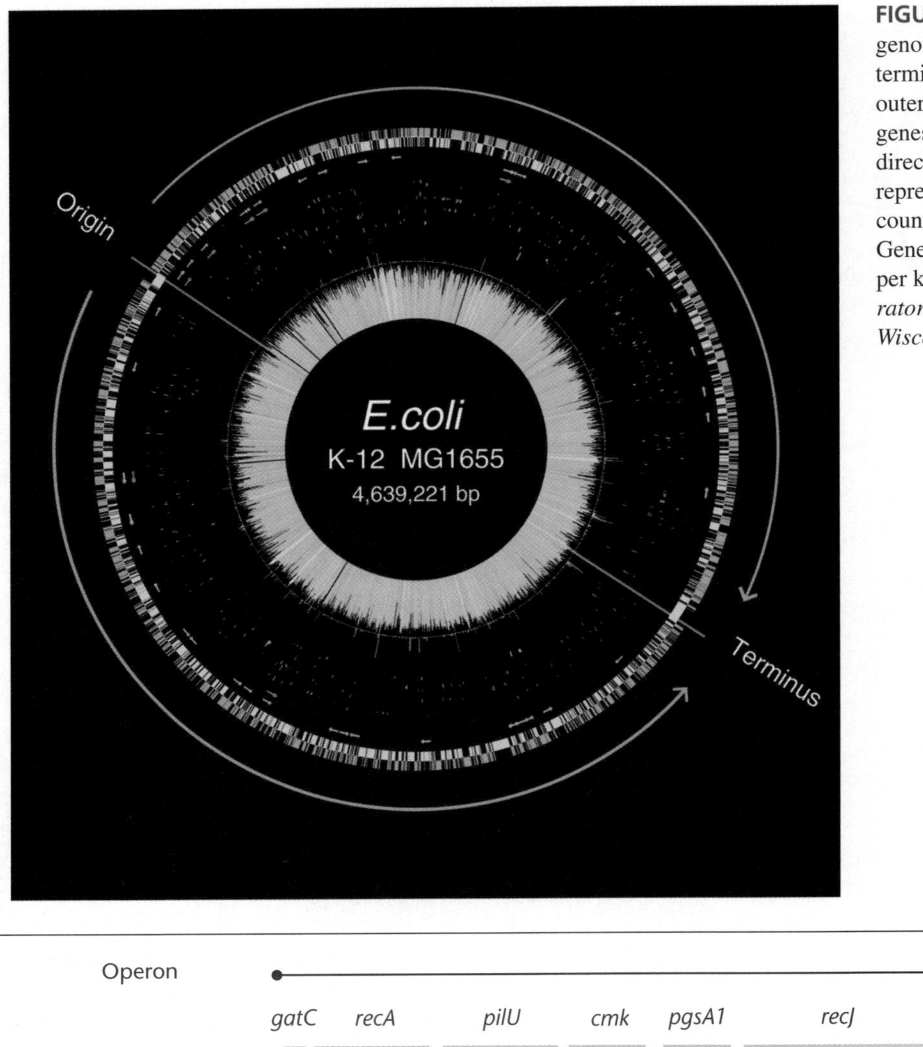

FIGURE 18–7 The *E. coli* genome, showing the origin and terminus of replication. The outer circle of bars represents genes transcribed in a clockwise direction, and the inner circle represents genes transcribed in a counterclockwise direction. Gene density is about one gene per kb. *(Dr. Fred Blattner, Laboratory of Genetics, University of Wisconsin, Madison, WI)*

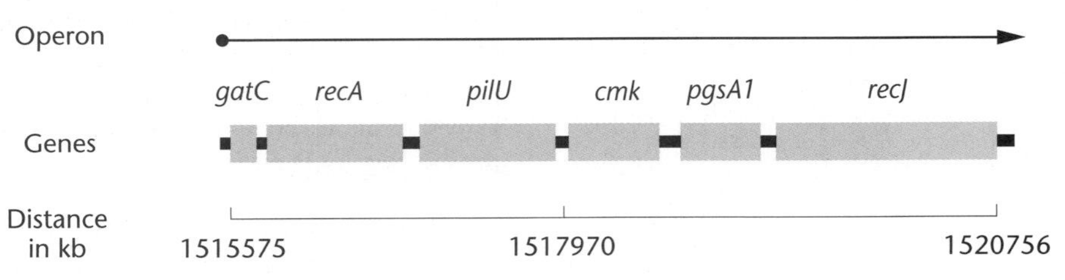

FIGURE 18–8 The operon from the *A. aeolicus* genome contains genes for protein synthesis (*gatC*), for DNA recombination (*recA* and *recJ*), for a motility protein (*pilU*), for nucleotide biosynthesis (*cmk*), and for lipid biosynthesis (*pgsA1*). This organization challenges the conventional idea that genes in an operon encode products that control a common biochemical pathway. *(On-line: http://www.ncbi.nlm. nih.gov/Entrez; and from Deckert, G. et al. 1998. The complete genome of the hyperthermophilic bacterium* Aquifex aeolicus. Nature *392: 353–58, Figure 1, p. 358.)*

Typically, Archaea are extremophiles; they live in extreme environments with very high temperature, high salt content, high pressure, or extreme pH. Even though Archaea are structurally similar to Eubacteria, it was recognized early on that in many aspects of metabolism, they more closely resemble eukaryotes. With the completion of several genome projects from the Archaea, we can now examine this relationship more closely. The archaeon, *Methanococcus jannaschii* (see Table 18.3) has a circular, double-stranded DNA genome of 1.66 Mb with 1738 protein-coding genes. This organism was originally isolated from a high-temperature deep-sea vent 2600 meters below sea level. It has an optimum growth temperature of 85°C, and can survive at 94°C (water boils at 100°C). The *M. jannaschii* genome contains three chromosomes: a large circular chromosome of 166 Mb and two small circular chromosomes of 58.4 and 16.5 kb. Most of the genes in this organism (58%) do not match any other known genes, but the majority of genes involved in energy production, cell division, and general metabolism closely resemble those of Eubacteria. Gene organization also resembles Eubacteria; they are densely packed, have operons, and do not have introns.

On the other hand, Archaea genes also have significant similarities to eukaryotic genes. The genes involved in

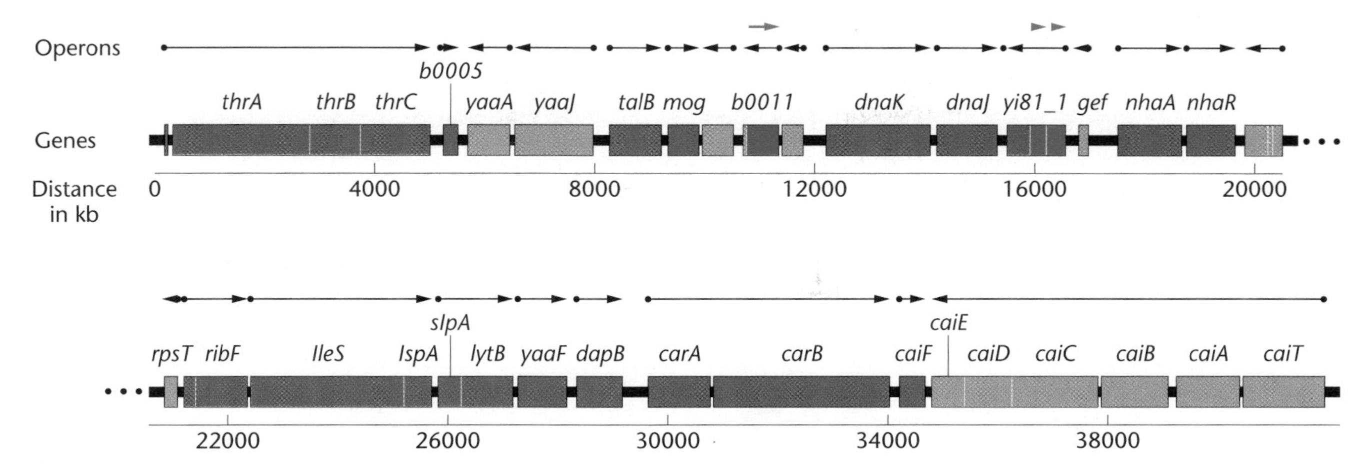

FIGURE 18–9 A portion of the *E. coli* chromosome showing genes and operons. A dot indicates the promoter for each gene or operon. Arrows and color indicate the direction of transcription: dark purple genes are transcribed left to right; light purple are transcribed right to left. Overlapping genes are shown in green. *(On-line: http://www.ncbi.nlm.nih.gov/Entrez)*

RNA synthesis, protein synthesis, and DNA synthesis more closely resemble those in eukaryotes. Most surprising is the presence of histone chromosomal proteins and evidence that chromosomal DNA is organized into chromatin. Although they have no introns in their protein-coding genes, Archaea do have introns in their tRNA genes, as do eukaryotes.

18.5 Eukaryotic Genomes Have a Mosaic of Organizational Patterns

Eukaryotic nuclear genomes are usually divided into a series of linear DNA molecules, each contained in an individual chromosome. In addition, eukaryotes have a second genome, the mitochondrial genome, present as a circular DNA molecule. Plants and other photosynthetic organisms also have a third genome, a circular DNA molecule carried in the chloroplast. Our focus in this chapter is the nuclear genome. Although the basic features of the eukaryotic genome are similar in different species, genome size is highly variable in eukaryotes (Table 18.4). Genome sizes range from about 10 Mb in fungi to over 100,000 Mb in some flowering plants (a 10,000 fold range), the number of chromosomes per genome ranges from two into the hundreds (about a 100 fold range) but the number of genes varies much less than either genome size or chromosome number.

TABLE 18.4 Genome Size and Gene Number in Selected Eukaryotes

Organism	*Genome Size (Mb)*	*Number of Genes*
S. cerevisiae (yeast)	12.1	6548
Plasmodium falciparum (malaria)	30	≈6500
C. elegans (nematode)	97	>20,000
Arabidopsis thalania (mustard plant)	120	≈20,000
D. melanogaster (fruit fly)	170	≈16,000
Oryza sativa (rice)	415	≈20,000
Zea mays (maize)	2500	≈20,000
Homo sapiens (human)	3300	≈35,000
Hordeum vulgare (barley)	5300	≈20,000

General Features of the Eukaryotic Genome

Compared with prokaryotes, eukaryotes have a relatively low gene density as shown in Figure 18–10. In general, more complex eukaryotes have less compact genomes with lower levels of gene density. If we compare regions of different eukaryotic genomes, several features become apparent:

- **Gene density** The 50-kb region of the yeast genome from chromosome III in Figure 18–10(b) contains over 20 genes, while the 50-kb segment of the human genome from chromosome 16 in Figure 18–10(c) contains only 6 genes.
- **Introns** None of the yeast genes shown in Figure 18–10(b) contain introns, whereas most human genes contain introns. In fact, the entire yeast genome has only 239 introns, while just a single gene in the human genome can contain over 100 introns.
- **Repetitive sequences** The presence of introns and the existence of repetitive sequences are two major reasons for the wide range of genome sizes in eukaryotes. In some plants, such as maize, repetitive sequences are the dominant feature of the genome. The maize genome has about 2500 Mb of DNA, over two-thirds of which is composed of repetitive DNA, resulting in very low gene density in the genome of this species (Figure 18–10d).

The *C. elegans* Genome

Although eukaryotic genomes share many features, specific genomic features vary widely. The genome of the nematode *C. elegans* contains 97 Mb of DNA organized into six chromosomes, each with about 20,000 genes—19,099 of which are protein-coding genes. The gene density is much lower than in yeast: —3 times more genes and 8 times more DNA, with an average density of about

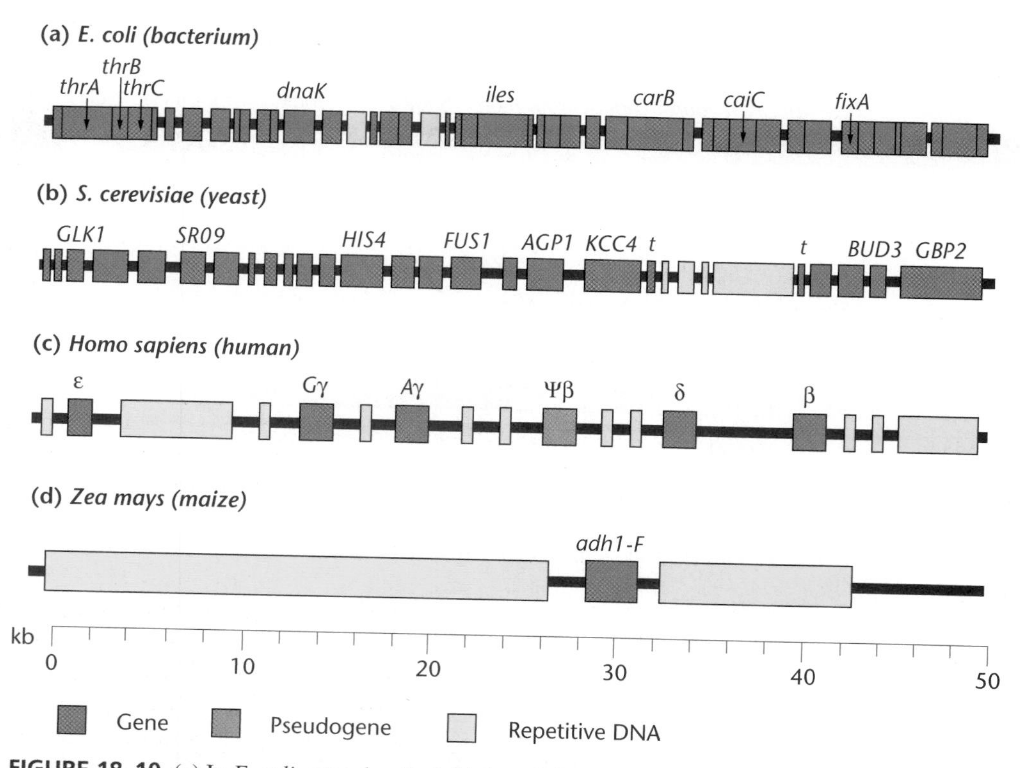

FIGURE 18–10 (a) In *E. coli*, gene density is high, and there are very few repetitive sequences. (b) A 50-kb region from chromosome III of yeast contains over 20 genes and little repetitive DNA. (c) A 50-kb region from human chromosome 11 contains 6 genes and stretches of repetitive DNA. (d) 50 kb of the maize genome surrounding the *Adh* locus. This gene is surrounded by long stretches of repetitive DNA. In eukaryotes (b–d), gene density is lower, and portions of the genome are occupied by repetitive DNA sequences. *(Parts a and b, on-line: http://www.ncbi.nlm.nih.gov/Entrez. Part c, redrawn from Margot, J. B. et al. 1989. Complete nucleotide sequence of the rabbit β-like globin gene cluster. Analysis of intergenic sequences and comparison with the human β-like gene cluster.* J. Mol. Biol. *205: 15–40, Figure 8, p. 37. Part d, redrawn from SanMiguel, P. et al. 1996. Nested retrotransposons in the intergenic region of the maize genome.* Science *274: 756–68 Figure 1, and Figure 3 p. 756, p.766.)*

1 gene per 5 kb. About half of the *C. elegans* genome is composed of intergenic spacer DNA, and a significant fraction of the genome is highly repetitive simple-sequence DNA (sequences such as ATAT repeated tens of thousands of times).

The organization of *C. elegans* genes differs dramatically from most eukaryotes. Approximately 25% of the genes in *C. elegans* genes are organized into polycistronic transcription units, or operons, like those in bacteria (Figure 18–11). In addition, compared with yeast, introns are far more prevalent in *C. elegans* genes, whose average gene has five introns. In fact, 26% of the *C. elegans* genome is introns; in addition, many of its genes are found within the introns of other genes (Figure 18–12).

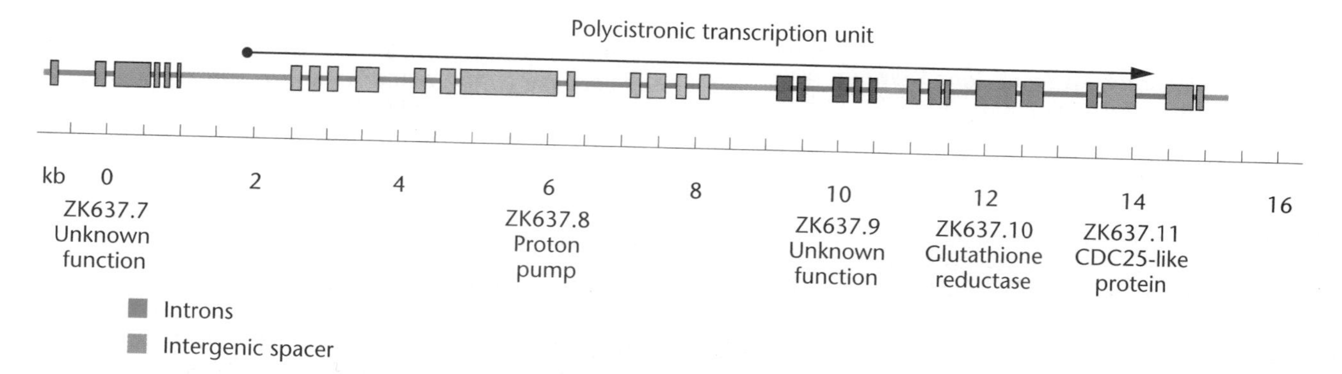

FIGURE 18–11 A region on chromosome III of *C. elegans* showing the location and organization of five genes with introns for each gene indicated by a different color. The central genes, ZK637.8, ZK637.9, and ZK637.10 are part of an operon and are transcribed as a set to form a polycistronic mRNA. Operons are common in bacterial genomes but rare in most eukaryotic genomes. In *C. elegans*, however, about 25% of all genes are part of operons.

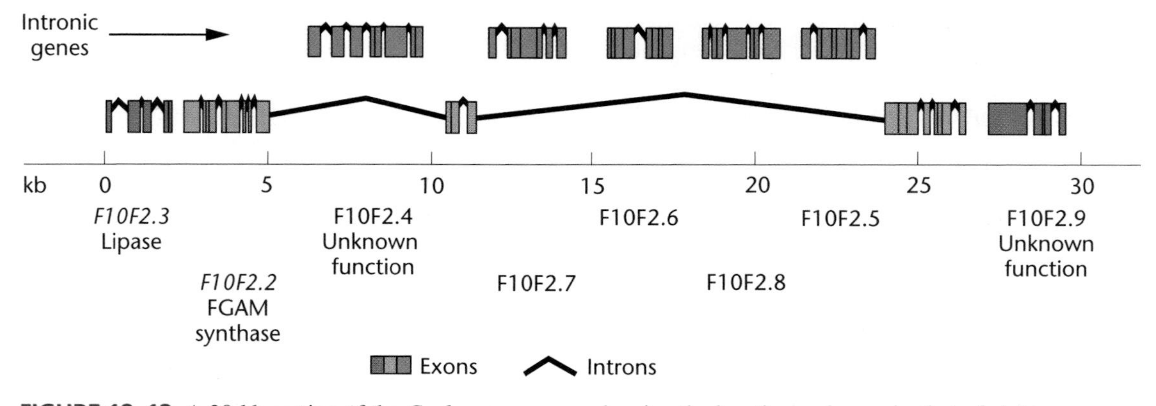

FIGURE 18–12 A 30-kb section of the *C. elegans* genome showing the location and organization of eight genes. The gene *F10F2.3* encodes a lipase, spans just over 25 kb, and contains two introns. Five other genes (FI0F2.4, F10F2.7, F10F2.6, F10F2.8, and FI0F2.5) are located in the introns of the lipase gene.

Genomes of Higher Plants

To geneticists, the small flowering plant *Arabidopsis thaliana* is the *Drosophila* of the plant world. Its very small genome, 120 Mb distributed in 5 chromosomes, is comparable in gene number and size to those of *C. elegans* and *Drosophila*. It contains an estimated 20,000 genes with a gene density of 1 gene per 5 kb, which also resembles that of *C. elegans* and *Drosophila*. At least half of its genes are closely related to genes found in bacteria and humans. Because of its compact genome, *Arabidopsis* is a model organism for studying other plants with larger genomes, such as the economically important grasses, rice, maize, and barley.

Analysis of the nucleotide sequence of chromosomes 2 and 4 of *Arabidopsis* reveals a dynamic genome. Both chromosomes contain many tandem gene duplications (239 duplications on chromosome 2, involving 539 genes) as well as larger duplications involving 4 blocks of DNA sequences, spanning 2.5 Mb. There is some interchromosomal gene duplication as well; genes on chromosome 4 are also present on chromosome 5. Genomic and gene duplications have played a large role in evolution that will be discussed in a later section.

Other plants, such as maize and barley, have genomes more than an order of magnitude larger than *Arabidopsis*, but have about the same number of genes (see Table 18.4). Genes in these large-genome plants are clustered in stretches of DNA separated by long stretches of intergenic DNA (Figure 18–13). Collectively, the gene clusters (known as the gene space) occupy only 12–24% of the

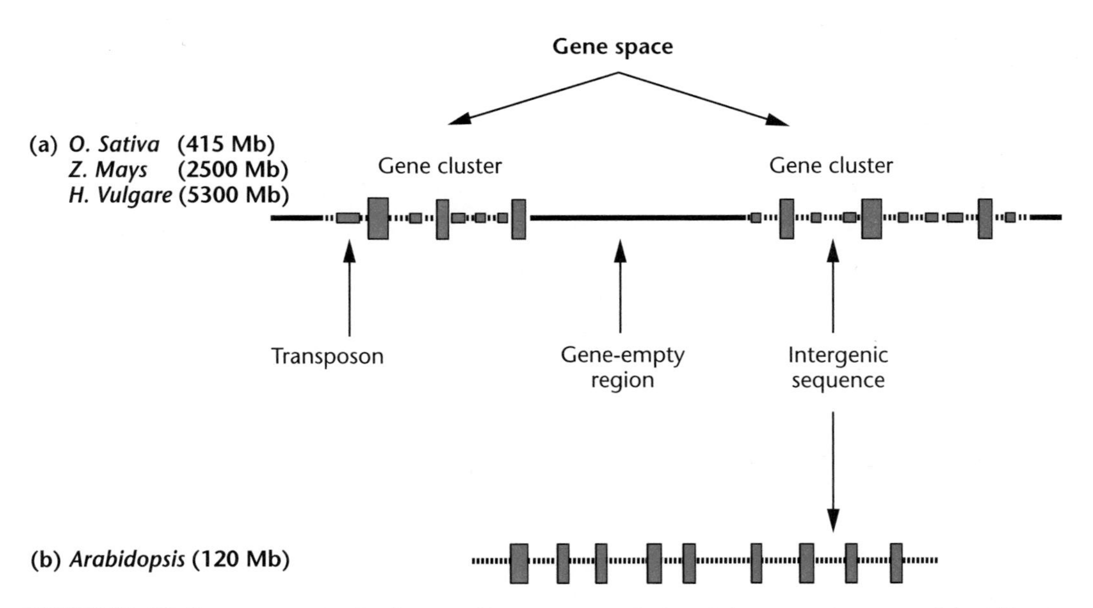

FIGURE 18–13 Genome organization in several large-genome plants and the compact genome of *Arabidopsis*. (a) The vertical boxes are genes located in clusters, separated by long, gene-empty spaces of repetitive DNA sequences. Within the gene clusters, the intergenic spaces contain many transposons (small horizontal boxes). (b) In the *Arabidopsis* genome, gene-empty regions have been lost, and transposable elements have been lost or reduced. The result is a much smaller genome with genes at a much higher density throughout the genome. *(Barakat et al. 1998.* Proc. Nat. Acad. Sci. (USA) *95: 10044–49. Figure 4, p. 10048.)*

genome. In maize, the intergenic DNA is composed mainly of transposons. Because plants closely related to *Arabidopsis* have larger genomes but similar numbers of genes, the small genome of *Arabidopsis* is thought to derive from a genomic contraction, in which almost all of the gene-empty regions disappeared, along with most of the transposon sequences. In addition, there may have been a reduction in the size and number of introns, accounting for the compact genome size in this plant.

The Human Genome: The Human Genome Project

In June 2000, the public and private genome projects jointly announced the completion of a draft sequence of the human genome; and in February 2001, they each published an analysis covering about 96% of the euchromatic region of the genome. The remaining work of completing the sequence by filling in the gaps was completed in 2003, when the finished sequence was published. Attention is now directed at analyzing and interpreting the vast amount of data gathered in this project. The human genome is about 25 times larger than the genome of any organism that has been sequenced to date, and it is the first vertebrate genome to be completed. Here, we present a summary of the major features of the human genome.

Although the human genome contains over 3 billion nucleotides, the DNA sequences that encode proteins make up only about 5% of the genome. At least 50% of the genome is derived from transposable elements, such as LINE and *Alu* sequences. Genes are distributed over 24 chromosomes, with clusters of gene-rich regions separated by gene-poor regions (referred to as gene deserts). Gene deserts correlate with the G bands seen in karyotypes. Chromosome 19 has the highest gene density, and chromosomes 13 and the Y chromosome have the lowest gene density.

Not all of the genes in the genome have been identified, but it appears that humans have about 30,000 genes, a number much lower than the previous estimate of 50,000 to 100,000. The average gene size, including introns and exons, is about 27 kb. Human genes tend to be larger and to contain more and larger introns than the genes and introns in invertebrate genomes such as *Drosophila*. The largest human gene known so far is the gene encoding dystrophin. This gene, associated in mutant form with muscular dystrophy, is 2.5 Mb in length and is larger than many bacterial chromosomes. Most of the transcription unit in this gene is composed of introns. It is not uncommon to find 30, 40, or even 50 introns in some human genes.

The number of introns in human genes ranges from 0 (histone genes) to 234 (titin, a gene that encodes a muscle protein). An unexpected finding is that hundreds of genes have been transferred directly from bacteria into vertebrate genomes, including our own, by a mechanism that remains unknown. Functions have been assigned to about 60% of the identified genes (Figure 18–14), and annotation of the rest is an ongoing part of the project.

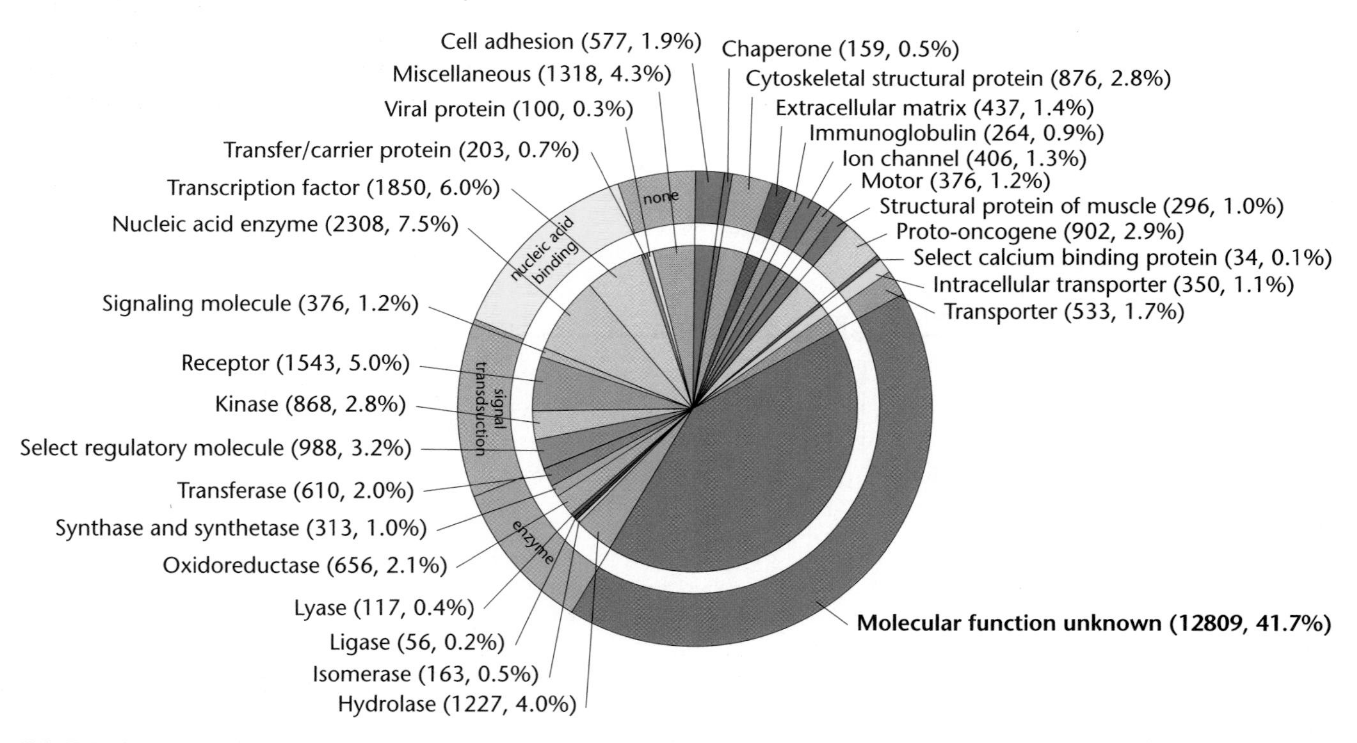

FIGURE 18–14 A preliminary list of assigned functions for 26,588 genes in the human genome based on similarity to proteins of known function. Among the most common genes are those involved in nucleic acid metabolism (7.5% of all genes identified), transcription factors (6.0%), receptors (5%), protein kinases (2.8%) and cytoskeletal structural proteins (2.8%). A total of 12,809 predicted proteins (41%) have unknown functions, reflecting the work that is still needed to fully decipher our genome. *(Venter et al. 2001. The sequence of the human genome.* Science *291: 1304–51. Figure 15, p. 1335.)*

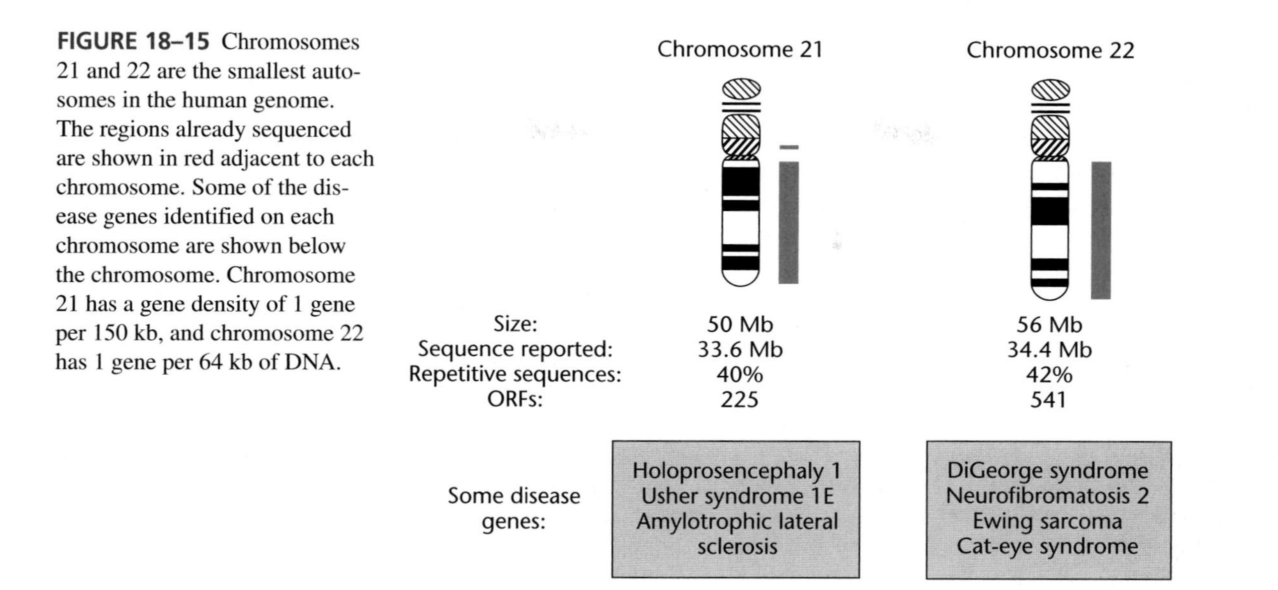

FIGURE 18–15 Chromosomes 21 and 22 are the smallest autosomes in the human genome. The regions already sequenced are shown in red adjacent to each chromosome. Some of the disease genes identified on each chromosome are shown below the chromosome. Chromosome 21 has a gene density of 1 gene per 150 kb, and chromosome 22 has 1 gene per 64 kb of DNA.

Chromosomal Organization of Human Genes

Sequencing of the long arms of human chromosomes 21 and 22 reveals some interesting features about the genetic landscape of these two chromosomes (Figure 18–15). One of the most intriguing findings is the difference in gene density on these two chromosomes. The long arms of these 2 chromosomes are similar in size (chromosome 21 is about 33.65 Mb, chromosome 22 is about 34.65 Mb), but chromosome 21 has only 225 genes (about 1 gene per 150 kb of DNA), while chromosome 22 has 541 genes (about 1 gene per 64 kb of DNA).

Closer examination of gene distribution reveals that the genes on these chromosomes are not evenly spaced. The upper region of the long arm of chromosome 21 (corresponding to the large G band) is gene-poor and averages 1 gene per 304 kb of DNA (Figure 18–16). The lower (telomeric) region of the long arm has a much higher gene density, with a gene every 95 kb of DNA. In addition, several regions of chromosome 21 are almost empty of genes; 1 region spanning 7 Mb contains only 1 gene, and 3 other regions of 1 Mb each contain no genes. Together, these gene-poor regions add up to 10 Mb, or about 1/3 the length of the long arm. Chromosome 22 has a 2.5-Mb region near the telomere and 2 smaller regions (about 1 Mb each) located elsewhere that lack genes.

Chromosomes 21 and 22 both contain duplicated regions. On chromosome 21, a 220-kb region is duplicated near both ends of the long arm, and another 10-kb region is duplicated near the centromere. On chromosome 22, a 60-kb segment

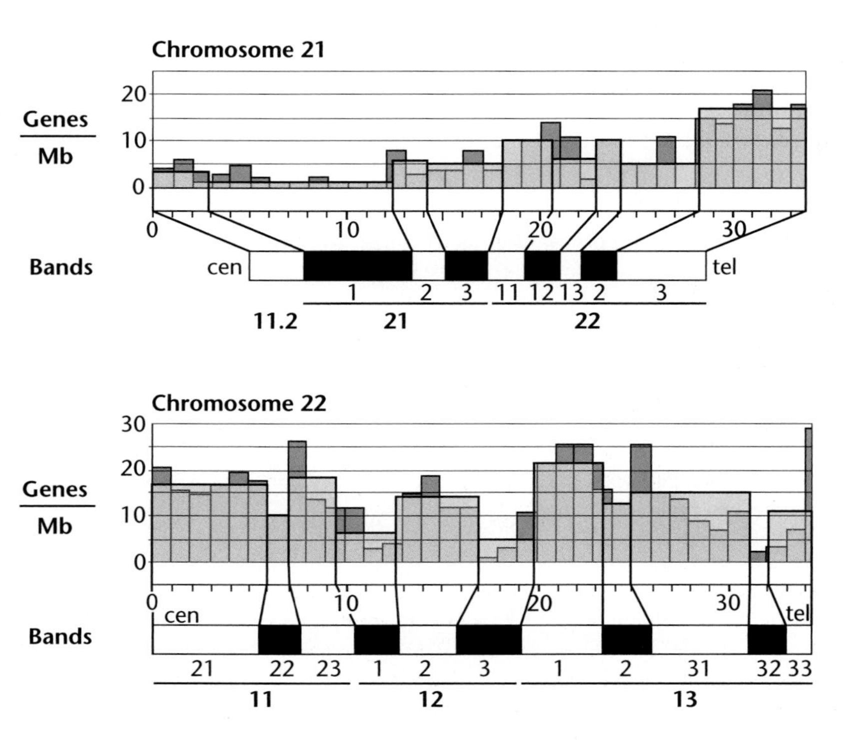

FIGURE 18–16 Gene density in the long arms of human chromosomes 21 and 22 (genes per Mb of DNA). In chromosome 21, the region in the dark staining band is relatively gene-poor, and the region near the distal tip is gene-rich. On chromosome 22, there is a small region near the chromosome tip that is gene-poor and two other regions, closer to the centromere that are relatively gene poor. Band numbers are shown under the chromosome bands. *(Redrawn from Saccone, S. et al. 2001. Genes, isochores and bands in human chromosomes 21 and 22.* Chromosome Res. *9: 533–39, Figure 1, p. 536.)*

is duplicated. A study of chromosome breakpoints associated with translocations and deletions on chromosome 21 indicates that breakpoints cluster within and near duplicated regions, suggesting that these regions may mediate events involved in chromosomal rearrangement. Future analysis of the duplicated regions and their associated repetitive sequences may provide a molecular explanation for the events in chromosome breakage.

18.6 Genomics Provides Insight into Genome Evolution

Fossil records indicates that cells similar to bacteria were present on our planet about 3.5 billion years ago. We must assume that the genomes of these organisms or of organisms that appeared shortly thereafter were composed of a double-stranded DNA molecule. The first eukaryotic fossils (resembling single-celled algae) date from about 1.4 billion years ago. As both prokaryotes and eukaryotes underwent changes in size, shape, and complexity, their genomes also changed dynamically. These changes were driven by genetic mechanisms that include mutation, recombination, transposition, gene transfer, as well as gene deletion and duplication. By examining and comparing genomes of organisms that exist today, we can gain insights into how these mechanisms have shaped genomes.

The Minimum Genome for Living Cells

What is the minimum number of genes necessary to support life? We cannot fully answer this question yet because we don't know the functions of each gene in the genomes that have been sequenced. However, using genes with known functions, we can speculate on the minimum number of genes required to maintain life. To do this, we can compare sequence information from two of the smallest bacterial genomes, *M. genitalium* and *M. pneumoniae*. These two closely-related organisms are among the simplest self-replicating prokaryotes known. *M. genitalium* has a genome of 0.58 Mb, while that of *M. pneumoniae* is 0.82 Mb. They are members of a group of bacteria that lack a cell wall and invade other organisms, often causing disease in a wide range of hosts, including insects, plants, and humans (genital and respiratory infections).

The *M. genitalium* genome has 480 protein-coding genes. The genome of *M. pneumoniae* has those 480 genes plus an additional 197 genes, for a total of 677 protein-coding genes. In contrast, *E. coli* has a 4.6-Mb genome with 4288 ORFs, and *H. influenzae* has a 1.8-Mb genome with 1727 ORFs. Table 18.5 summarizes the functions of some genes in these bacteria.

It is fascinating to compare the genes required for a given function among species. The greatest difference among *H. influenzae*, *M. genitalium*, and *E. coli* involves a marked difference in biosynthetic capability. For example, *H. influenzae* has 68 genes involved in amino acid biosynthesis, yet *M. genitalium* has only 1, and *E. coli* has 131 genes for this function. Despite having nearly four times as many total genes, the physiological capabilities of *H. influenzae* are very similar to *M. genitalium*, although the latter must rely on numerous metabolic products from its host.

These comparisons combined with experimental results enable us to speculate on the minimal number of genes necessary for organisms to exist as independent self-reproducing organisms. Using transposon mutagenesis, Clair Fraser, Clyde Hutchinson, and colleagues mutated genes in *M. genitalium* and *M. pneumoniae* one at a time, then tested each resulting mutant strain for viability. Although they did not mutate every gene in these genomes, they concluded that free-living organisms require a minimum of 250–350 genes.

The *M. genitalium* genome is the smallest one known for a free-living organism. Smaller genomes have been found in symbiotic bacteria; one of these (*Buchnera*, a symbiont of aphids) has a genome of only 450 kb. So the question of the minimum number of genes required to support life has to be followed by another question: What is the minimum number of genes for free-living organisms or symbionts? Better answers to these questions will

TABLE 18.5 Functional Classes of Genes in Three Bacterial Species

Functional Class	*E. coli*	*M. genitalium*	*H. influenzae*
Protein-coding genes	4288	470	1,727
DNA replication, repair	115	32	87
Transcription	55	12	27
Translation	182	101	141
Regulatory proteins	178	7	64
Amino acid biosynthesis	131	1	68
Nucleic acid biosynthesis	58	19	53
Lipid metabolism	48	6	25
Energy metabolism	243	31	112
Uptake, transport proteins	427	34	123

Sources: Blattner, F. et al. 1997. The complete genome sequence of *Escherichia coli* K-12. *Science* 277: 1453–62. Table 4, p. 1458; Fraser, C.M. et al. 1995. The minimal gene complement of *Mycoplasma genitalium*. *Science* 270: 397–403. Table 2, p. 400.

come only when more prokaryotic genomes have been sequenced and when functions have been assigned to each of the genes in these genomes.

Origin of the Eukaryotic Genome

Eukaryotes are traditionally distinguished from prokaryotes by several morphological features: a membrane-bound nucleus, cytoplasmic membrane systems such as the endoplasmic reticulum, and a cytoskeleton. As genome projects provide more data, comparisons between prokaryotic and eukaryotic genomes are providing clues to the origins of the eukaryotic genome and how eukaryotes arose from prokaryotes. Several lines of evidence, including amino acid analysis of proteins, gene sequences, and metabolic pathways, indicate that the eukaryotic genome is actually a mosaic that has received major contributions from both the Archaea and the Eubacteria. For example, the eukaryotic nuclear genome has few, if any, operons, and its genes contain introns, both of which are features of the Archaea.

The eukaryotic mitochondrial genome strongly resembles that of alpha proteobacteria. To explain these observations, it has been proposed that eukaryotes arose as a consequence of a symbiotic association between an anaerobic archaebacterial host and an alpha proteobacterium (like *Rickettsia prowazekii*), which evolved into the mitochondrion. The single evolutionary origin of all eukaryotes suggests that this event occurred successfully only once in the history of the earth. This topic is discussed in detail in Chapter 23.

Genome Duplications in Eukaryote Evolution

The events just outlined may account for the origin of the eukaryotic genome but not for the size and complexity that distinguish eukaryotic genomes from those of prokaryotes. Gene duplication has played an important role in the evolution of eukaryotic genomes. Most attention has focused on duplications of single genes, but Susumo Ohno proposed that whole genome duplications are an important evolutionary mechanism. Analysis of genome sequence data supports the idea that much of the difference in gene number separating prokaryotes from eukaryotes resulted from genome expansions. We now recognize that a major expansion in eukaryote genomic size resulted from a genome duplication event that accompanied the appearance of vertebrates in the fossil record. However, genome duplications have occurred at other times throughout eukaryotic evolution.

An analysis of the yeast genome shows traces of an ancient expansion by genomic duplication. Kenneth Wolfe and Denis Shields searched through the yeast genome and compared the sequence of each yeast gene with every other yeast gene. Unexpectedly, they uncovered 55 duplicated regions containing 376 genes, which cover 50% of the genome. Of the 55 regions, 50 are in the same relative position on different chromosomes (Figure 18–17). For example, chromosomes XI and XIII contain a duplicated block of genes (block 43) in which gene order and orientation with respect to the centromere has been conserved. Other chromosomes contain internal duplications, such as block 53, which is located on either side of the centromere on chromosome XII. Evolutionary analysis indicates that this genome duplication event took place about 100 million years ago. That there is extensive duplication in the yeast genome was unexpected because it was widely believed that the yeast genome was close to the minimum size for eukaryotes.

Analysis of the human genome also shows evidence of an ancient large-scale duplication, followed by rearrangements and gene loss in some duplicated regions. These duplications range from small segments covering only a few genes up to large stretches that cover almost an entire chromosome. The larger duplications date to the origin of vertebrates, about 500 million years ago. Together, there are 1077 blocks of duplicated regions in the human genome, containing just over 10,000 genes (about one-third of the genome). One such duplication between chromosomes 18 and 20 is shown in Figure 18–18.

Gene Duplications

The almost uninterrupted flow of sequence data from genome projects is providing evidence that multigene families are present in many, if not all, genomes. In addition to genome-wide duplications, small blocks of genes and single genes can be duplicated by several mechanisms, including:

- **Unequal crossing over**. A recombination event between members of a homologous pair of chromosomes in which a DNA segment is duplicated in one of the recombination products.
- **Replication errors**. During replication of a template molecule, slippage can cause the insertion of a short segment into the newly synthesized strand.

Once they have been generated, members of multigene families may remain linked on a single chromosome or they may disperse to other parts of the genome. Several mechanisms drive this process, including inversion, translocation, and transposition by mobile elements.

Molecular phylogenetics has traced the ancestry and relationships among members of gene families. One of the best-studied examples of divergence is the globin gene superfamily (Figure 18–19). In this family, duplication of an ancestral gene encoding an oxygen transport protein occurred some 800 million years ago. This split produced two sister genes, one of which evolved into the modern-day myoglobin gene. (Myoglobin is an oxygen-carrying protein found in muscle.) The other gene became the ancestral globin gene. About 500 million years ago, the ancestral globin gene duplicated to form prototypes of the α-globin and β-globin gene subfamilies. The **α-globin** and **β-globin genes** encode the proteins found in hemoglobin, the oxygen-carrying molecule in red blood cells. Additional duplications within these genes occurred within the last 200 million

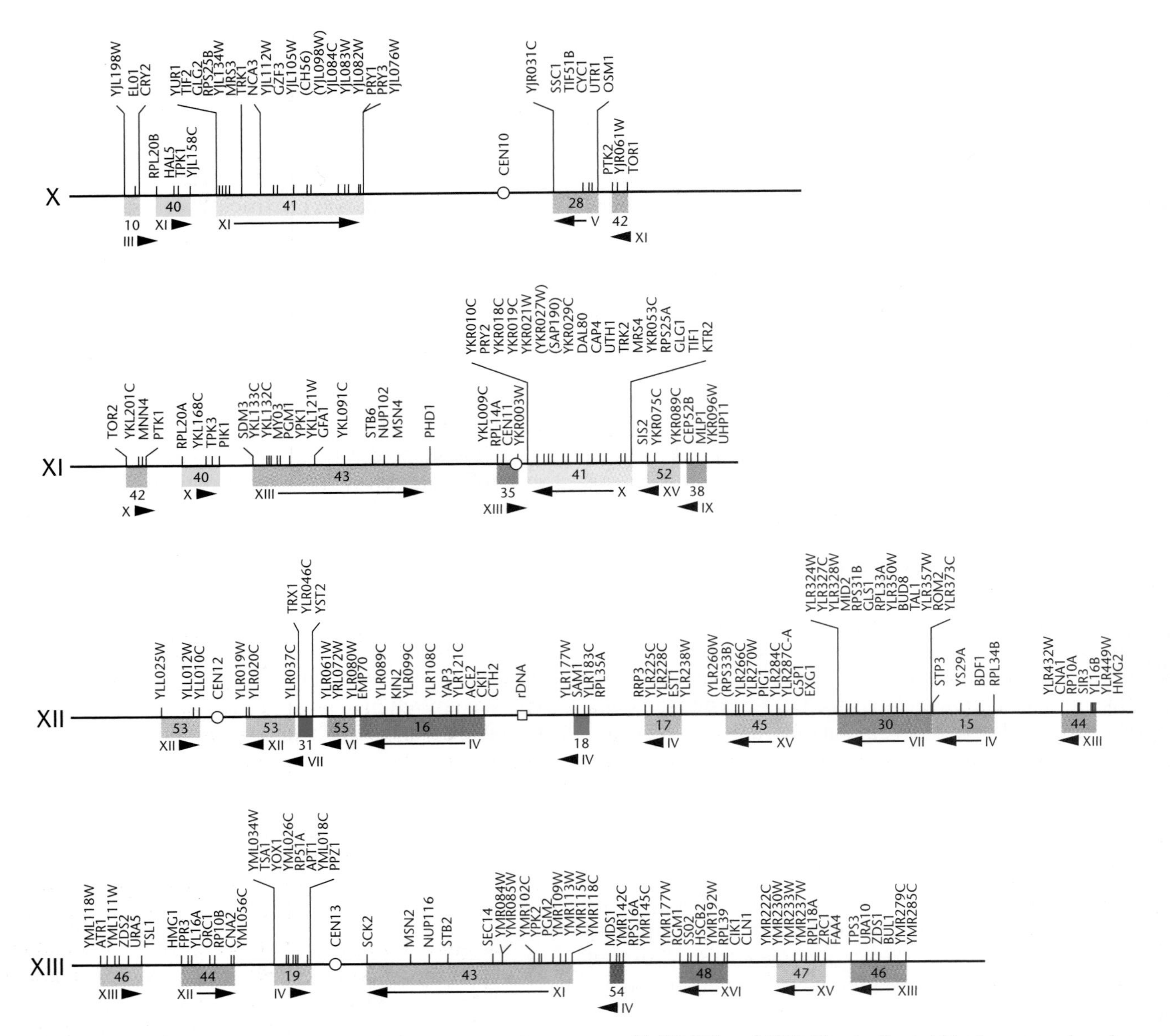

FIGURE 18–17 Location of duplicated regions on yeast chromosomes X, XI, XII, and XIII. The duplicated blocks are numbered with Arabic numerals. The Roman numerals below the blocks are chromosome numbers and give the location of the duplicate block. Arrows show the relative orientation of the blocks. Genome-wide orientation relative to the centromere is conserved, except for five blocks. *(Wolfe and Shields. 1997. Molecular evidence for an ancient duplication of the entire yeast genome,* Nature *387: 708–13. Figure 2, p. 710.)*

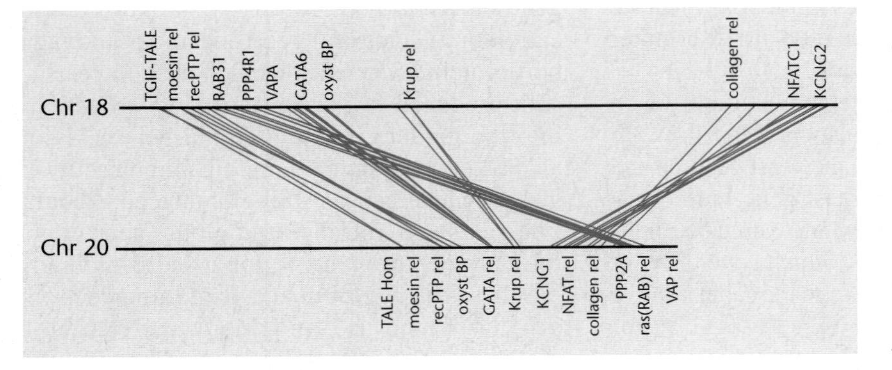

FIGURE 18–18 A segmental duplication of genes on human chromosomes 18 and 20. This duplication involves a total of 64 genes. For clarity, only 12 duplicated pairs are shown here. In the human genome, there are 1077 duplicated blocks of genes, containing a total of 10,310 genes. *(Venter et al. 2001. The sequence of the human genome.* Science *291: 1304–51. Figure 13, p. 1332.)*

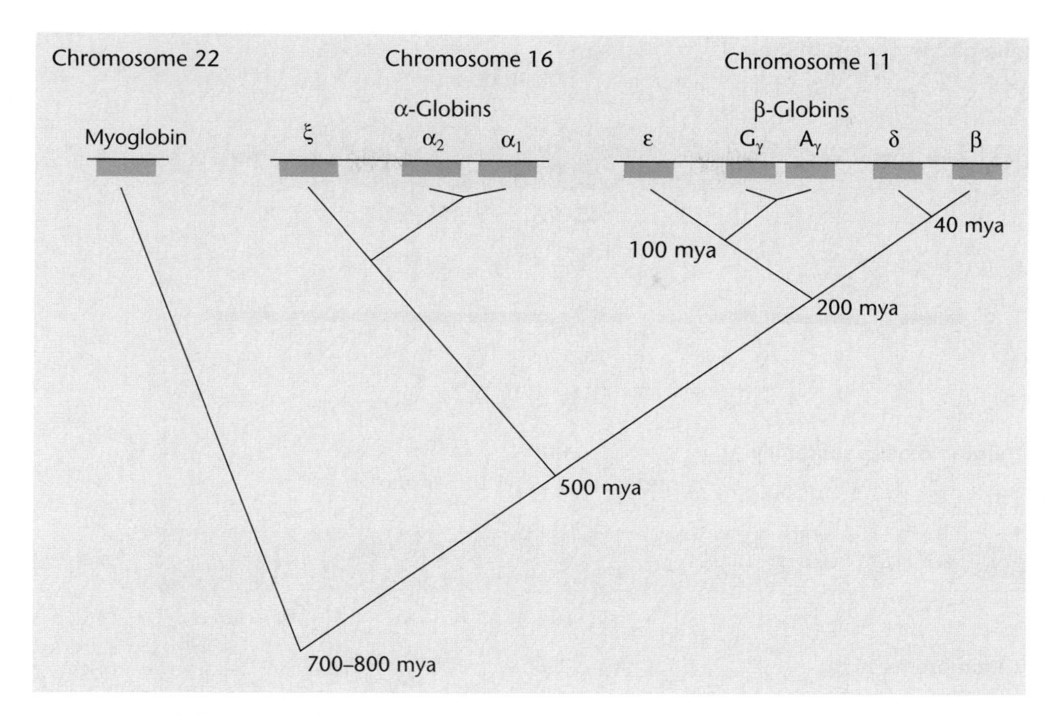

FIGURE 18–19 In this diagram of the evolutionary history of the globin gene superfamily, a duplication event in an ancestral gene gave rise to 2 lineages.about 700–800 million years ago (mya). One line led to the myoglobin gene, which is located on chromosome 22 in humans; the other underwent a second duplication event about 500 mya, giving rise to the ancestors of the α-globin and β-globin gene subfamilies. Duplications about 200 mya produced the α- and β-globin gene subfamilies. In humans, the α-globin genes are located on chromosome 16 and the β-globin genes are on chromosome 11.

years. Subsequent events dispersed members of this superfamily, and each gene now resides on a separate chromosome.

Similar patterns of evolution are observed in other gene families, including the trypsin–chymotrypsin family of proteases, the homeotic selector genes of animals, and the rhodopsin family of visual pigments.

18.7 Comparative Genomics: Multigene Families Diversify Gene Function

Members of multigene families share DNA-sequence homology and descend from a single ancestral gene, and their gene products frequently have similar functions. Members of multigene families are often but not always found together in a single location along a chromosome. To see how analysis of multigene families provides insight into eukaryotic genome organization and evolution, we first examine cases where groups of genes encode very similar but not identical polypeptide chains that become part of proteins with closely related functions. Multiple proteins that arise from single-gene duplications are known as **paralogs**. The globin gene family responsible for encoding the various polypeptides in hemoglobin molecules exemplifies a paralogous multigene family that arose by duplication and dispersal to different chromosomal sites.

The Globin Gene Family

The human α-globin and β-globin genes are two of the most intensively studied regions of the human genome. There is a cluster of 3 α-globin genes on the short arm of chromosome 16, and another cluster with 5 β-globin genes on the short arm of chromosome 11. Members of both subfamilies share nucleotide sequence similarity, but members of the same subfamily have the greatest amount of sequence similarity.

Hemoglobin is a tetramer, containing two α- and two β-polypeptides. Each polypeptide incorporates a heme group that reversibly binds oxygen. Within each subfamily, genes are coordinately turned on and off during embryonic, fetal, and adult stages of development. For both the α- and β-globin gene subfamilies, this expression occurs in the same order in which the genes are arranged on the chromosome.

The α-globin gene subfamily (Figure 18–20a) spans more than 30 kb and contains 3 genes: the ζ (zeta) gene, expressed only in the early embryonic stage; and 2 copies of the α gene, expressed during the fetal (α_1) and adult stages (α_2). In addition, two **pseudogenes** ($\psi\zeta$ and $\psi\alpha_1$) are present in the cluster. Pseudogenes are designated by the prefix ψ (psi), followed by the symbol of the gene they most resemble. Thus, the designation $\psi\alpha_1$ indicates a pseudogene of the adult α_1 gene. Pseudogenes are nonfunctional versions of genes that resemble other gene sequences, but contain significant nucleotide substitutions, deletions, and duplications that prevent their expression.

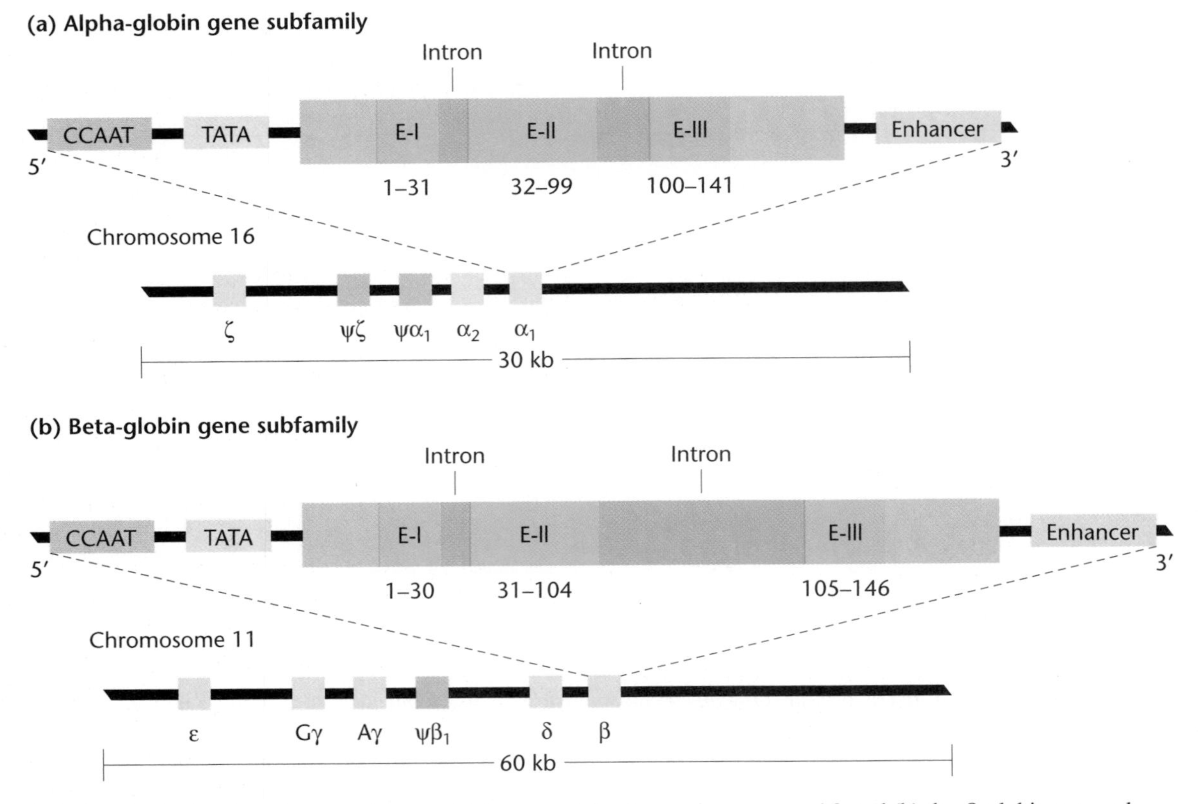

FIGURE 18–20 (a) Organization of the α-globin gene subfamily on chromosome 16, and (b) the β-globin gene subfamily on chromosome 11. Also shown is the internal organization of the α_1 gene and the β gene. Each gene contains three exons (E-I, E-II, E-III) and two introns. The numbers below the exons indicate the amino acids in the gene product encoded by each exon.

The organization of the α-globin subfamily members and the location of their introns and exons reveal several features. First, as is common in eukaryotes, the DNA encoding the three functional α genes occupies only a small portion of the region containing the subfamily. Most of the DNA in this region is intergenic spacer. Second, each functional gene in this subfamily contains two introns at precisely the same positions. Third, the nucleotide sequences within corresponding exons are nearly identical in the ζ and α genes. Both of these genes encode polypeptide chains of 141 amino acids. However, their intron sequences are highly divergent, even though they are about the same size. Note that much of the nucleotide sequence of each gene is contained in these noncoding introns.

The human β-globin gene cluster is longer than the α-globin cluster and contains five genes spaced over 60 kb of DNA (Figure 18–20b). As with the α-globin gene subfamily, the order of genes on the chromosome parallels their order of expression during development. Of the five genes, three are expressed prior to birth. The ε gene is expressed only during embryogenesis, while the two nearly identical γ genes (G_γ and A_γ) are expressed only during fetal development. The polypeptide products of the two γ genes differ only by a single amino acid. The two remaining genes, δ and β, are expressed after birth. Finally, a single pseudogene $\psi\beta_1$ is present within the subfamily. All 5 functional genes encode proteins with 146 amino acids and have 2 similarly sized introns at exactly the same positions. The second intron in the β-globin genes is significantly larger than its counterpart in the functional α-globin genes. These similarities reflect the evolutionary history of each subfamily and the events such as gene duplication, nucleotide substitution, and chromosome translocations that produced the present-day globin superfamily.

The Immunoglobulin Gene Family

The immune system is an effective barrier against successful invasion by potentially harmful foreign substances. These invaders are recognized as nonself and are subsequently destroyed by the immune system. A highly specialized case of localized genomic alterations is a hallmark of maturation and function in the immune system. From a set of less than 300 genes, alterations—including recombination events, deletions, and mismatching of rejoined DNA segments in immature cells of the immune system—create a vast potential for immune response.

One part of the immune system produces antibodies against foreign substances. Anything that causes antibody production is termed an **antigen**. Many different molecules can act as antigens, including proteins, polysaccharides, and nucleic acids. Usually, a distinctive structural

TABLE 18.6 Classes and Components of Immunoglobulins

Ig Class	*Light Chain*	*Heavy Chain*	*Tetramers*
IgM	κ or λ	μ	$\kappa_2\mu_2$, $\lambda_2\mu_2$
IgD	κ or λ	δ	$\kappa_2\delta_2$, $\lambda_2\delta_2$
IgG	κ or λ	γ	$\kappa_2\gamma_2$, $\lambda_2\gamma_2$
IgE	κ or λ	ε	$\kappa_2\varepsilon_2$, $\lambda_2\varepsilon_2$
IgA	κ or λ	α	$\kappa_2\alpha_2$, $\lambda_2\alpha_2$

feature of an antigen, called an **epitope**, stimulates antibody production. Antigens can be free molecules, or they can be part of the surface of a cell, microorganism, or virus. **Antibodies** are proteins produced and secreted by the B cells of the immune system. Among vertebrates, each individual can produce millions of different antibodies, each responding to a different antigen. In the next sections, we examine the molecular basis of antibody diversity.

Antibodies are produced by plasma B cells, a type of lymphocyte (white blood cell). Five classes of antibodies or **immunoglobulins (Ig)** are recognized: IgM, IgD, IgG, IgE, and IgA (Table 18.6). Secretion of IgM antibodies is the first response of plasma cells to an antigen, and they are part of the early stages of the immune response. IgD antibodies are found on the surface of B cells and may regulate their action, but little is known about these antibodies. The IgG class represents about 80% of the antibodies found in the blood and is the most intensively characterized group of antibodies. Immunoglobulins of the IgA class (found in breast milk) can be secreted across plasma membranes, and they help resist infections of the respiratory and digestive tracts. IgE antibodies fight parasitic infections and are also associated with allergic responses.

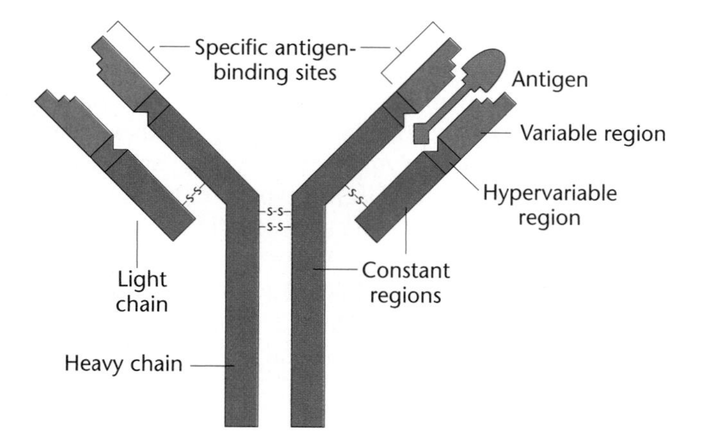

FIGURE 18–21 A typical antibody (IgG) molecule is Y-shaped and contains four polypeptide chains. The longer arms are H chains, and the shorter arms are L chains. The chains are joined by disulfide bonds. Each chain contains a variable region and a constant region. The variable and hypervariable regions of a pair of L and H chains form a combining site that interacts with a specific antigen.

A typical antibody molecule (Figure 18–21) contains two different polypeptide chains, each present in two copies and held together by disulfide bonds. Each larger, or **heavy (H) chain**, molecule contains approximately 440 amino acids. The sequence of the first 110 amino acids at the N-terminus differs among heavy chains and is known as a variable region (V_H). The remaining C-terminal amino acids are the same in all H chains and make up the constant region (C_H). In humans, genes on the long arm of chromosome 14 encode the H chains.

Each **light (L) chain** contains 220 amino acids, with the first 110 amino acids making up the variable region (V_L). The remaining amino acids at the C-terminus make up the constant region (C_L). Two types of L chains exist: κ chains, encoded by genes on human chromosome 2, and λ chains, encoded by genes on chromosome 22. Together, the variable and hypervariable regions of the heavy and light chains form the **antibody-combining site**. Each combining site has a unique structural conformation that allows it to bind to a specific antigen, like a key in a lock.

In response to an antigen, the immune system can generate cells that synthesize a new, unique antibody. Because billions of different antibodies can be made, it is impossible for separate genes to encode each of them; there is simply not enough DNA in the human genome to encode all of these gene products.

The key to understanding how tens of millions of antibodies can be produced lies in understanding immunoglobulin gene structure. In humans, the κ-chain gene (Figure 18–22) has several components: the leader-variable (L-V) region, the joining (J) region, and a constant (C) region. There are 70–100 DNA segments in the L-V region, each with a different nucleotide sequence and a promoter, but no enhancer. The J region contains six different segments, and the C region contains only one C segment. An enhancer, but no promoter, is located between the last J segment and the C segment. This feature ensures that segments cannot be transcribed individually.

During B-cell maturation, one of the 100 L-V regions (L_2–V_2 in Figure 18–22) and its promoter are randomly joined via a recombination event to one of the six J regions (J_3 in the figure), and to the C region (and its enhancer). The recombination event forms a functional L-chain gene (the second line in Figure 18–22). All DNA between the selected L-V segment and the J_3 segment is excised and destroyed. This type of recombination differs from that seen in meiosis and involves different enzymes. The joining event between the V and J regions is imprecise and can occur over a span of about 6 bp. In addition, during this recombination event, the ends of the DNA are subjected to a **break-nibble-add mechanism**, in which a few bases are randomly removed and a few bases randomly added. The newly created gene contains three exons (L, V-J, and C) and two introns, one between L and V-J, another between J_3 and C. The unselected segments, J_4–J_6, become part of the second intron. The L-chain gene is transcribed, processed, and translated to form κ-chain proteins that become part of

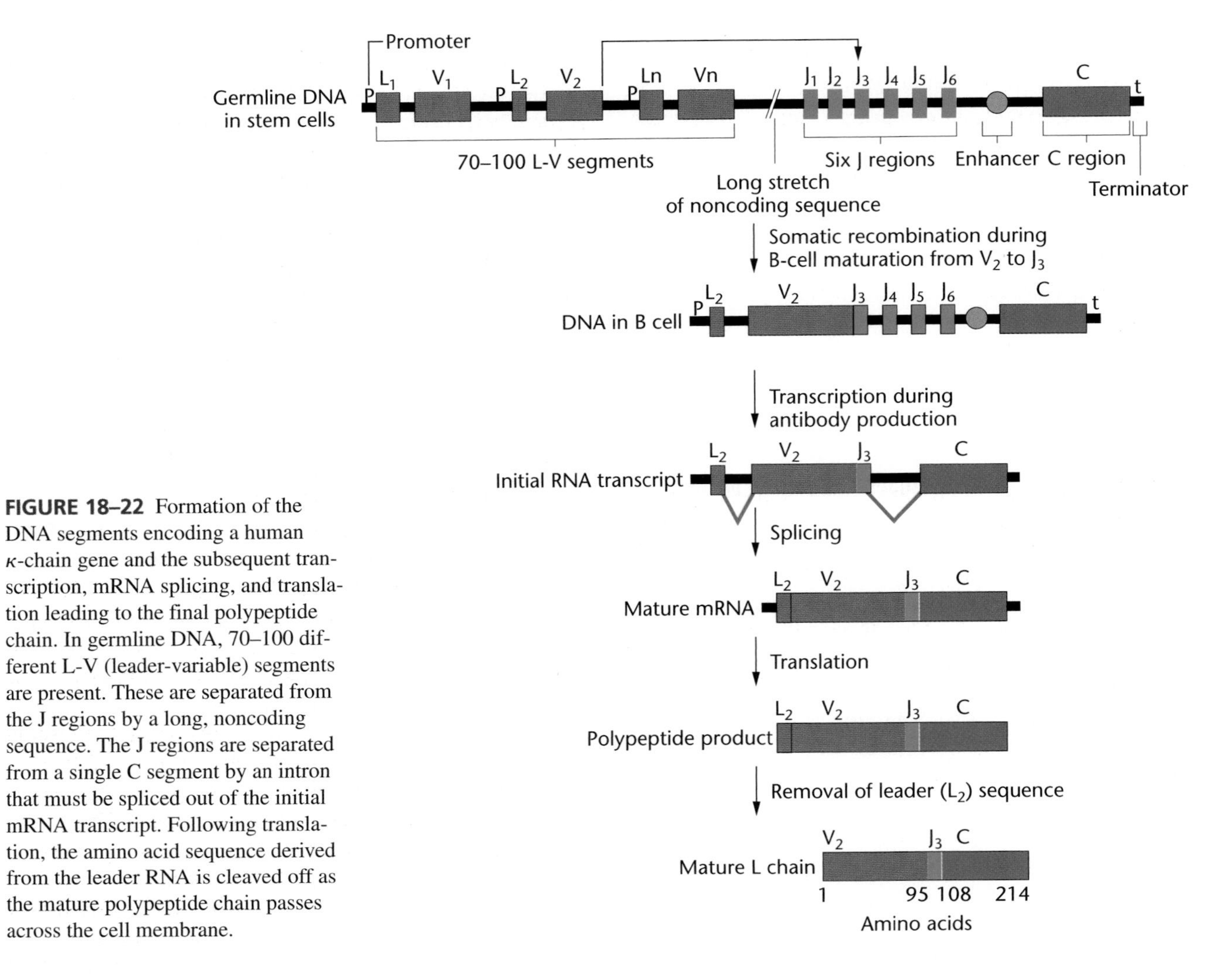

FIGURE 18–22 Formation of the DNA segments encoding a human κ-chain gene and the subsequent transcription, mRNA splicing, and translation leading to the final polypeptide chain. In germline DNA, 70–100 different L-V (leader-variable) segments are present. These are separated from the J regions by a long, noncoding sequence. The J regions are separated from a single C segment by an intron that must be spliced out of the initial mRNA transcript. Following translation, the amino acid sequence derived from the leader RNA is cleaved off as the mature polypeptide chain passes across the cell membrane.

an antibody molecule. The rearranged gene is stable and is passed on to all progeny of the B cell in which this event occurs.

Antibody diversity in κ-chain production results from combining any of the 100 segments in the L-V regions with any of the 6 segments in the J regions, and generates about 600 different κ-chain genes. Since each joining reaction can occur over several base pairs, the number of possible genes increases to several thousand. Thus, additional diversity comes from the break-nibble-add mechanism. The organization of the λ-chain genes is somewhat different from that of the κ-chain genes, but the recombination events generate a similar number of different λ-chain genes.

The heavy-chain genes extend over a large span of DNA and include four types of regions: L-V, D, J, and C. During B-cell maturation, random recombination of H-chain components generates a large number of H-chain genes. If we assume there are 25 D segments, combining any of these with any of the 300 V segments and 6 J segments generates 45,000 H-chain genes. With imprecise joining and the break-nibble-add mechanism, the total number of different genes climbs markedly. The potential for overall antibody diversity is estimated by multiplying the combination of all H-chain genes with all L-chain genes, resulting in hundreds of millions of possible antibody genes from a few hundred coding sequences.

To summarize, antibody diversity is due to four features of the immunoglobulin gene system: (1) multiple numbers of variable segments, (2) multiple numbers of diversity and joining segments, (3) multiple splice locations with break-nibble-add joining, and (4) multiple combining of L chains with H chains. As antibody-forming B cells mature, DNA recombination rearranges these genes so that each mature B lymphocyte encodes, synthesizes, and secretes only one specific type of antibody. Each mature B cell can make one type of light chain (κ or λ) and one type of heavy chain. When an antigen is present, it stimulates the B cell and populations of differentiated cells are produced, all of which synthesize one type of antibody that interacts with the antigen.

18.8 Proteomics Identifies and Analyzes the Proteins in a Cell

Proteome is a relatively new term, coined to define the complete set of proteins encoded by a genome. In a narrower sense, it also describes the set of proteins expressed in a cell at a given time. **Proteomics**, the study of the proteome, uses high-throughput technologies to achieve a full description of the molecular and biochemical events in a cell. As genome projects provide information about the number and kinds of genes present in prokaryotic and eukaryotic genomes, the question of protein function is becoming a central issue in biology. As stated earlier, in most of the genomes sequenced to date, newly discovered genes have no known function and others have only presumed functions assigned by analogy with known genes made by sequence comparisons from databases, not experimental evidence from laboratory experiments. For example, in *E. coli* and *S. cerevisiae*, more than half of the genes encoded by their genomes have no known function. In the human genome, about 41% of the genes are of unknown function.

Understanding gene function involves more than just identifying the gene products. Once made, many gene products are modified by cleavage of end groups (such as signal sequences, propeptides, or initiator methionine residues), by the addition of chemical groups (e.g., methyl, acetyl, phosphoryl), or by linkage to sugars and lipids. Proteins are internally and externally cross-linked and in some rare and interesting cases, even processed by removing internal amino acid sequences (called **inteins**). Over a hundred mechanisms of posttranslational modification are known in addition to the high level of diversity produced by the alternate splicing of mRNA. Thus, the human genome, which may have only about 30,000 protein-coding genes, may produce over 350,000 different gene products. Analysis of protein function is complicated by the fact that many proteins work via protein–protein interactions or as part of a large molecular complex. Even enzymes are often localized to specific regions of the cell by these interactions. The goal of proteomics is to provide information about a protein's function, structure, posttranslational modifications, protein–protein interactions, cellular localization, variants, and relationships (shared domains, evolutionary history) to other proteins—for every protein encoded in a genome.

Proteomics Technology

The basic techniques in proteomics involve separating and identifying proteins isolated from cells. The most commonly used combination of techniques involves two-dimensional gel electrophoresis (2DGE) and mass spectrometry (MS). In 2DGE (Figure 18–23), proteins extracted from a cell are loaded onto a polyacrylamide gel and the proteins are separated according to their electrical charge. When completed, the gel is rotated 90°; and during another round of electrophoresis, the proteins are separated in a second dimension according to their molecular weight. When the gels are stained, proteins are revealed as spots; typical gels show 200–10,000 spots (Figure 18–24). To identify individual proteins, spots are cut from the gel and then digested with the enzyme trypsin to produce a characteristic set of peptide fragments. The fragments are analyzed by MS. A mass spectrometer ionizes a gaseous peptide sample and analyzes the mass-to-charge ratio of the sample as it passes by a detector. In peptide mass fingerprinting, the mass of the peptides analyzed is compared with mass data in protein databases to identify the protein. High-performance instruments can identify hundreds of proteins per day, and banks of spectrometers can process thousands of samples in a single day. New instruments with faster sample processing times and increased sensitivity and accuracy are under development and will allow proteomics to have a significant impact on many areas of biology.

The Bacterial Proteome Changes with Alterations in the Environment

As outlined earlier, *M genitalium*, with a genome of 480 genes, represents one of the simplest, independently living organisms known. Valerie Wasinger and her colleagues used proteomics to provide a snapshot of which genes are expressed under two different growth conditions in *M. genitalium*: exponential growth and the stationary phase following rapid growth.

Using 2 DGE, Wasinger's group identified 427 protein spots in exponentially growing cells. Of these, 201 were analyzed and identified by peptide digestion, mass spectrometry, and comparison to known proteins. The analysis uncovered 158 known proteins (33% of the proteome) and 17 unknown proteins. The remaining spots included fragments derived from larger proteins, different forms of the same protein (isoforms), and posttranslationally modified products. The identified proteins included enzymes involved in energy metabolism, DNA replication, transcription, translation, and transport of materials across the cell membrane.

During the transition from exponential growth to the stationary phase, there was a 42% reduction in the number of proteins synthesized. In addition, some new proteins appeared, and other proteins underwent dramatic changes in abundance. These changes are apparently a consequence of nutrient depletion, increased acidity of the growth medium, and other adaptations to environmental changes.

Wasinger's analysis helps establish the minimum number of expressed genes required for independent existence and the changes in gene expression that accompany the transition to the stationary phase. This study also points up some of the limitations of current proteomics technology. Only the most abundantly expressed proteins can be detected with 2DGE. In this study, most of the proteome was unexpressed or undetected because the proteins were present in very low abundance (too low to be detected on gels) or were not solubilized and recovered by the extraction methods used in Wasinger's study. In spite of these limitations, proteomic analysis provides a wide range of information that cannot be obtained by genome sequencing.

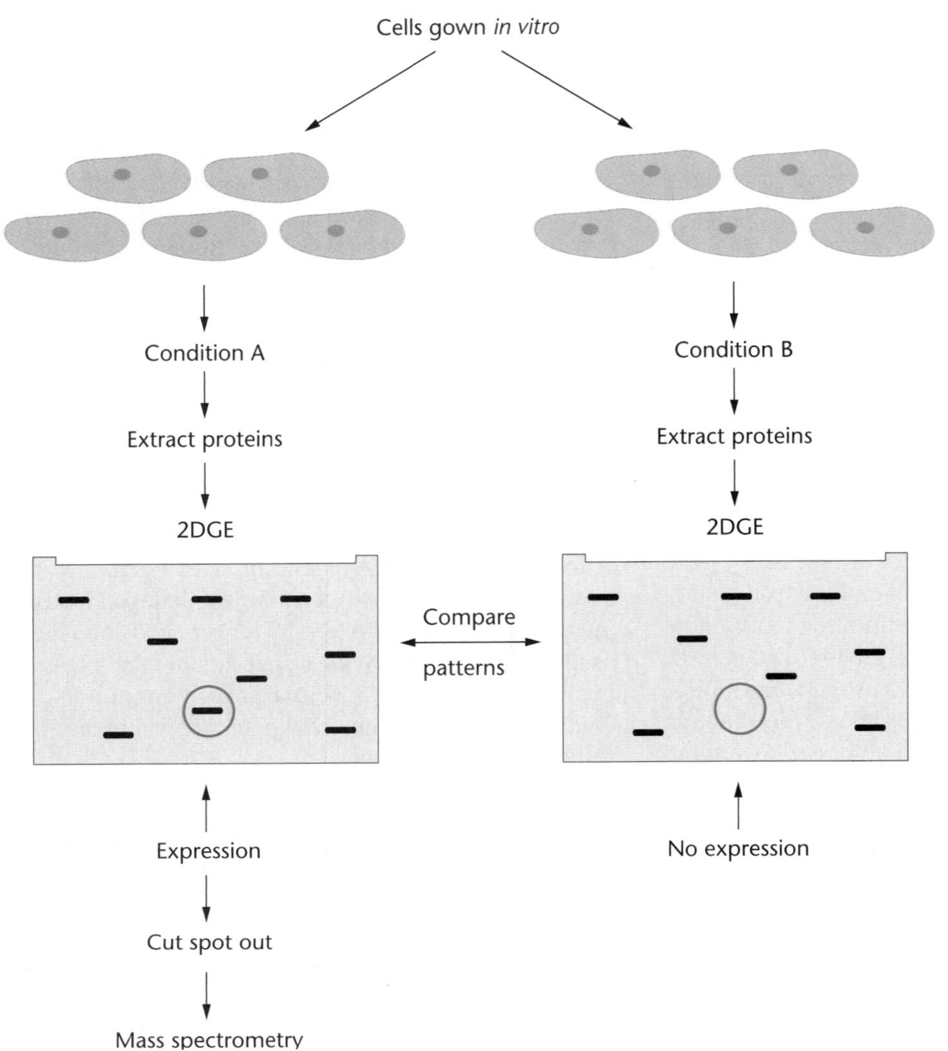

FIGURE 18–23 In a typical proteomic analysis, cells are exposed to two different conditions (such as growth conditions, drugs, or hormones). After treatment, proteins are extracted and separated by 2DGE. The pattern of spots is then compared for evidence of differential gene expression. Spots of interest are cut out from the gel, digested into peptide fragments, and analyzed by mass spectrometry to identify the protein in the spot.

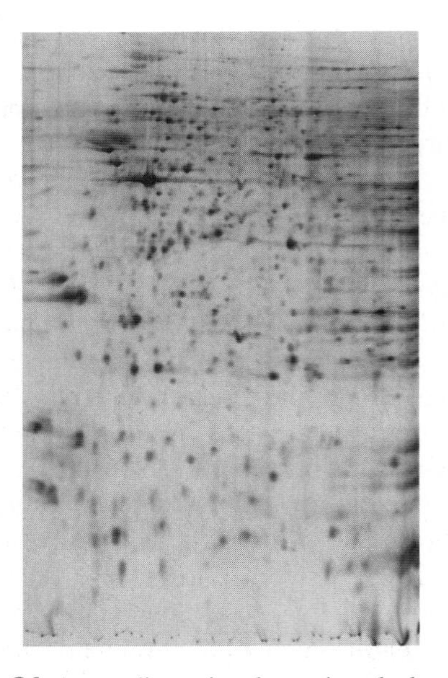

FIGURE 18–24 A two-dimensional protein gel, showing the separated proteins as spots. Several thousand proteins can be displayed on such gels. *(Dr. Carl Merril/Laboratory of Biochemical Genetics/National Institute of Mental Health, NIH)*

Proteome Analysis of an Organelle: The Nucleolus

Proteome analysis faces several inherent problems because the proteome is a very dynamic system. Cellular proteins have a vast range of concentrations (more than a millionfold) and hundreds of posttranslational and alternative splicing forms, many of which may be difficult to separate from one another. In addition, there are cell-cycle specific variations in abundance and many protein–protein interactions to deal with. Another problem is technical; for several reasons, some of which were outlined earlier, proteins displayed by 2DGE usually represent only a fraction of those present in a cell.

Isolation of a portion of the cell's proteome is one way to reduce the impact of these limitations. This approach reduces the complexity of the sample and improves the separation and quantitative analysis of the sample's peptide fragments. One way to do this is by isolating subcellular organelles and components. Subproteome analysis has been used to successfully study the proteomes of nuclear pores, cell surfaces, the nucleolus, and other cellular components.

The nucleolus is usually the largest and most prominent organelle in the eukaryotic nucleus and is a dynamic structure. The nucleolus forms early in the G1 phase of the

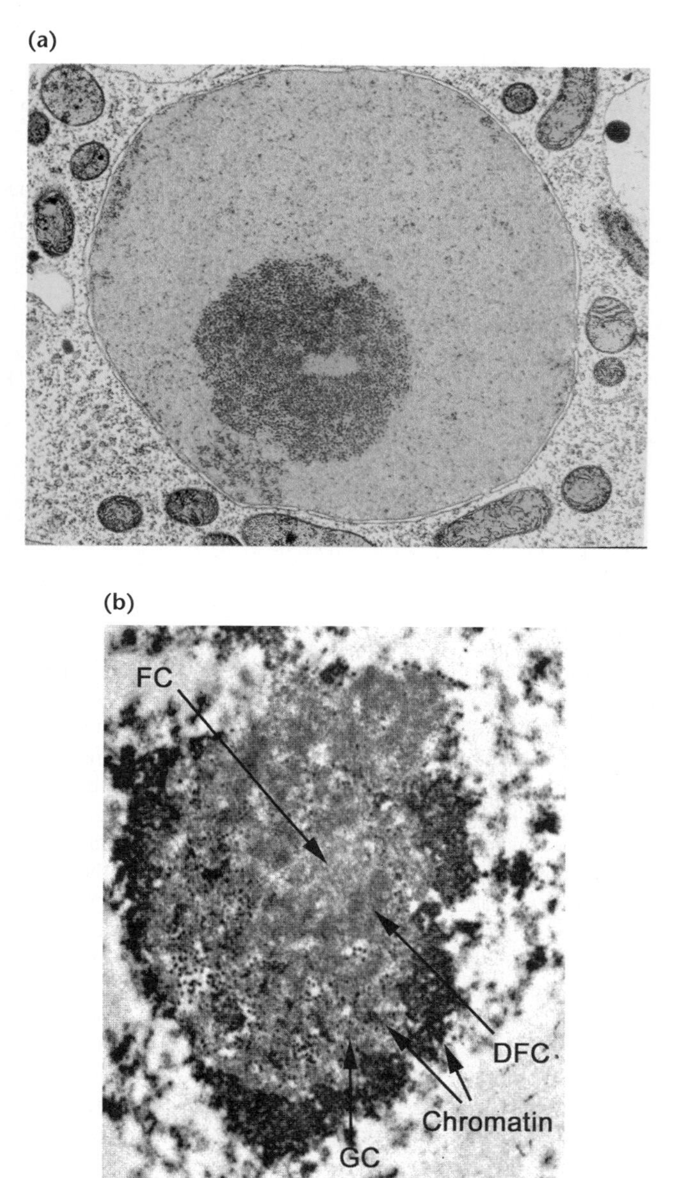

FIGURE 18–25 (a) A transmission electron micrograph showing nucleoli in the nucleus of a eukaryotic cell. (b) A transmission electron micrograph of the nucleolus showing the nucleolar compartments. Key: FC, fibrillar center; DFC, dense fibrillar compartment; and GC, granular compartment. *(BioPhoto/Photo Researchers. Inc. (a) Dr. David Pulak/Cellnucleus.com (b)).*

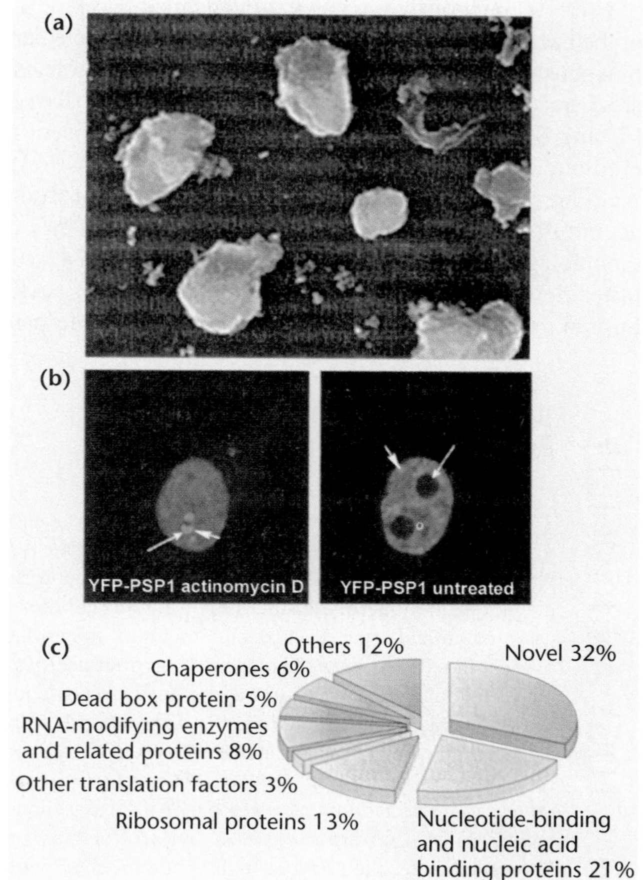

FIGURE 18–26 (a) Scanning electron micrograph showing purified human nucleoli. Proteins were solubilized, fractionated, and analyzed by mass spectrometry. (b) At right, localization of one newly identified nucleolar protein (PSP1), within nuclear structures called paraspeckles. At left, inhibition of transcription by treating the cells with actinomycin D causes the nucleolus to reorganize and causes the formation of cap structures containing PSP1. (c) Functional assignments of the more than 400 proteins identified in this proteomic analysis. Dead box proteins are enzymes that participate in folding RNA, including spliceosome formation. Note that 32% of the proteins identified were previously unknown. *(From Aebersold, R., and Mann, M. 2003. Mass-spectrometry based proteomics.* Nature *422: 198–207. Figure 4, p. 203, http://www.nature.com).*

cell cycle at the chromosomal sites of the rDNA genes (Figure 18–25a), and is disassembled just before mitosis. Three distinct subcompartments of the nucleolus have been described: the fibrillar center (FC), the dense fibrillar component (DFC), and the granular center (GC) (see Figure 18–25b). Traditionally, the nucleolus has been regarded solely as a ribosome factory, in which the ribosomal RNAs are transcribed in the FC, processed in the DFC and along with 5*S* RNA and ribosomal proteins, assembled into ribosomal subunits in the GC.

Recent studies have suggested that the nucleolus may have other important functions, including assembly of spliceosomes, telomerase, and other small nuclear RNA-protein complexes. An analysis of the nucleolar proteome would help confirm these new functions and define others.

To study the nucleolar proteome, Jens Andersen and his colleagues isolated nucleoli from HeLa cells, a human cancer cell line, and analyzed the sample purity using several methods, including electron microscopy (Figure 18–26a). Nucleolar proteins were separated using 2DGE. A variety of methods were used to prepare the sample prior to MS, thus enabling the investigators to identify more than 400 nucleolar proteins (Figure 18–26c).

This study identified many proteins already known to be associated with the nucleolus, as well as many nuclear proteins not known to exist in the nucleolus. The role of the nucleolus in assembly of spliceosomes and other RNA-protein complexes was confirmed. More important, discovery of over 100 previously unknown proteins making up 32% of the nucleolar proteome indicates that the list of nucleolar functions will probably grow much larger as these proteins are characterized and as functional roles are assigned to them.

Using fluorescent tagging of one of the newly discovered nucleolar proteins (Paraspeckle Protein 1, PSP 1), Andersen's group was able to show that this protein accumulates in a previously unknown compartment of the nucleus, called paraspeckles, along with at least two other newly identified proteins (Figure 18–26b). All human cells examined to date contain 10–20 nuclear paraspeckles in the space between chromosomes, and these are associated with RNA splicing components. The studies show that PSP 1 and the other two proteins move between the nucleolus and the nucleus in a transcription-dependent fashion. When transcription is inhibited, these three proteins move to the nucleolus and localize to a cap structure at the periphery of the nucleolus (Figure 18–25b). This study emphasizes that we have much to learn about the dynamic interactions between the nucleus and nucleolus and that the nucleolus is much more than a ribosome factory.

GENETICS, TECHNOLOGY, AND SOCIETY

Footprints of a Killer

For millennia, anthrax has afflicted humans. It is likely that anthrax was one of the deadly Egyptian plagues at the time of Moses, and cases of anthrax have been documented since Roman times. Throughout history, human anthrax cases were mostly sporadic and resulted from contact with animal products contaminated with spores of *Bacillus anthracis*—a bacterium that preferentially infects grazing herbivores. Spores can enter the body through cuts in the skin, by inhalation, or by ingestion. Although skin anthrax is usually curable, inhalation anthrax and ingestion anthrax are often fatal.

The hostile use of infectious biological agents such as anthrax reaches back into antiquity. The Romans deliberately contaminated food and water supplies with animal carcasses during the Carthaginian Wars in the fifth century B.C. During the Middle Ages, bodies of plague-infected people and animals were catapulted into cities under siege and were used to contaminate the enemy's drinking wells. The use of infectious biological projectiles continued into the twentieth century during the South African Boer wars and the Russian Revolution.

During World War I, Germany infected livestock with *B. anthracis*, with the aim of exporting infected meat to the Allies. During World War II, Japan killed up to 10,000 prisoners by infecting them with biological and chemical agents, including anthrax. The Japanese also sprayed anthrax and other biological agents over 11 Chinese cities. From 1940 to 1942, approximately 700 Chinese people (along with 1700 Japanese soldiers) died in these attacks.

The Allies also developed extensive biological weapons programs during this time. Britain manufactured 5 million cattle-food cakes that contained anthrax and conducted anthrax explosives testing on Gruinard Island, lying off the coast of Scotland. This small island remained contaminated with anthrax spores and was uninhabitable for over four decades. In 1986, Gruinard was finally decontaminated with 280 tons of formaldehyde and 2000 tons of seawater.

The United States conducted biological weapons programs beginning in the early 1940s and had manufactured large quantities of anthrax by 1946, including 5000 bombs filled with *B. anthracis*. Between 1949 and 1968, cities in the United States were secretly used as experimental sites to test aerosolization and dispersal of biological agents. Bacteria of size similar to *B. anthracis* (such as *Serratia marcescens* and *B. globi*) were sprayed over San Francisco and into the ventilator shafts of the New York City subway system. These tests showed that small amounts of bacteria could be rapidly and easily dispersed over urban areas.

In 1969, President Nixon terminated the offensive biological weapons program, and stocks of biological weapons were destroyed between 1971 and 1973. However, small quantities of anthrax and other pathogens were retained to develop therapies to treat biological weapons attacks. In addition, the U. S. Army Medical Research Institute of Infectious Diseases (USAMRIID) was established to conduct research into biological weapons defenses. USAMRIID programs were conducted in collaboration with numerous universities and research institutes.

Although the United States has halted its offensive biological weapons program, over 15 countries currently have or are developing biological weapons. These include Russia, Israel, Egypt, China, Iran, Iraq, Libya, Syria, India, and North Korea. In the 1980s, the Soviet Union was capable of manufacturing thousands of tons of weapons-grade anthrax annually in 20- to 50-ton reactors. In 1979, anthrax spores were accidentally released from a Soviet biological weapons facility in Sverdlovsk. The anthrax cloud dispersed in a 3- mile radius from the site of release, killing at least 70 people. In the 1990s, it was found that Iraq had manufactured 8000 liters of anthrax spores—enough to kill every human on earth. Less is known about nongovernment sources of anthrax weapons. The Japanese terrorist group, Aum Shinrikyo, conducted research on anthrax and is thought to have made unsuccessful attempts to use anthrax as a weapon. Also, the Egyptian Islamic Jihad apparently obtained anthrax from an East Asian country.

In addition to its use as a biological weapon, along with the widespread knowledge of how to turn it into a lethal weapon, *B. anthracis* itself is a formidable foe. Anthrax is relatively easy to produce in large quantities, the spores are extremely stable, and easy to transport, conceal, and disperse. One gram of spores contains 100 million lethal doses.

Against this backdrop emerged the post-September 11 anthrax attacks– and the scientific efforts to track the source of the attacks. Between October 2001 and January 2002, 23 anthrax cases were reported to the U. S. Centers for Disease Control and Prevention. Eleven of these were inhalation anthrax cases, and 12 were skin anthrax cases. There were 5 fatalities, and approximately 32,000 people were given prophylactic antibiotic therapy. Anthrax spores were discovered in post offices, government buildings, and media centers in Florida, Washington, D.C., New Jersey, and New York. The spores appeared to come from a handful of letters that passed through high-speed mail-sorting equipment, contaminating postal workers as well as people in the offices to which the letters were addressed. But who sent the letters, and where did they get the weapons-grade bacterial spores?

Few physical clues emerged from the anthrax-laced envelopes. Although the handwriting and threats were similar in all of the letters, investigators found no fingerprints, hair, or fibers on them. The envelopes were sealed with tape, so there was no saliva with which to perform DNA tests to identify the sender.

The anthrax bacterium itself may yield the clearest evidence with which to apprehend the killer or killers. Using PCR to identify restriction fragment length polymorphisms in bacterial DNA, scientists discovered that the same strain—the Ames strain—was present in all the letters. This strain was first isolated in Ames, Iowa, and has been maintained by the U. S. Army and many research labs since 1980. However, identifying the specific lab from which the anthrax was obtained may be a more challenging task. Most organisms undergo random genetic variation over generations, and these variations can distinguish isolates grown in one location from those grown in another—unfortunately, *B. anthracis* is one of the most genetically homogeneous organisms known. Using genetic fingerprinting of 15 different DNA fragment markers, scientists have been unable to distinguish the Ames strain used in the attacks from other Ames strains. At present, scientists are sequencing the genomes of the Ames strain used in the attacks and comparing the sequence to that of other Ames strains maintained in research labs and government facilities. If the source of the anthrax strain used in the post-September 11 attacks can be identified by DNA sequence comparison, it could lead investigators directly to the perpetrators of these incidents.

References

Enserink, M. 2001. Taking anthrax's genetic fingerprints. *Science* 294: 1810–12.

Enserink, M. 2002. TIGR begins assault on the anthrax genome. *Science* 295: 1442–43.

Lesho, E., Dorsey, D., and Bunner, D. 1998. Feces, dead horses, and fleas: Evolution of the hostile use of biological agents. *Western J. of Med.* 168: 512.

Chapter Summary

1. Once obtained, genome sequences are analyzed in several steps to ensure that the sequence is accurate, to identify all encoded genes, and to classify known genes into functional categories. They are then deposited into searchable databases.
2. Bacterial genomes have very high gene density, averaging one gene per kilobase pair of DNA. Typically, as much as 90% of the chromosome encodes genes. Many genes are organized into polycistronic transcription units that do not contain introns. Archaea are a prokaryotic domain. In some ways, their chromosome and gene organization resembles eukaryotic genes and genomes.
3. Eukaryotic genomes are organized into two or more chromosomes, each containing a linear double-stranded DNA molecule. The gene density is much lower than that in bacteria. Genes typically are not organized into operons, rather, each is a separate transcription unit. However, the nematode *C. elegans* has many genes organized into operons. Eukaryotic genes are often interrupted with introns.
4. Complex multicellular eukaryotes differ from the less complex yeast in a number of ways. These eukaryotes have more genes and much more DNA. This results in gene densities falling to 1 gene per 5 kb or even 1 gene per 10–20 kb or more. A higher proportion of eukaryotic genes have introns, the number of introns per gene increases, and the size of introns increases as complexity increases from yeast to *C. elegans* to humans. Some plants, such as *Arabidopsis*, have a gene structure and organization that is indistinguishable from animals. Other plants, such as maize, have a different organization, with vast blocks of transposable elements separating islands of genes.
5. In the human genome, large differences exist in gene density on different chromosomes, with gene-rich regions alternating with gene-poor regions. Duplicated segments are a common feature of the chromosomes sequenced to date.
6. Many eukaryotic genes have undergone duplication followed by sequence divergence, leading to multigene families. The globin and immunoglobulin gene clusters are prime examples of this phenomenon.
7. Proteomics is used to study the expression of genes in bacterial cells under different growth conditions and is providing insight into the gene sets cells use during growth. These techniques have also been used to study the architecture of subcellular elements, such as the nucleolus.

Key Terms

annotation, 404
antibody, 412
antibody-combining site, 421
antigen, 420
bioinformatics 402
break-nibble-add mechanism, 421
clone-by-clone method, 402
CpG island, 405
epitope, 421
gene density, 411
genomics, 402
α-globin gene, 417
β-globin gene, 417
heavy chain (H), 421
Human Genome Project (HGP), 402
immunoglobulin (Ig), 421
intein, 423
intron, 411
light (L) chain, 421
open reading frames (ORFs), 405
paralog, 419
proteome, 422
proteomics, 402
pseudogene, 419
repetitive sequence, 411
replication error, 417
shotgun cloning method, 403
unequal crossing over, 417

INSIGHTS AND SOLUTIONS

An antibody molecule contains two identical H chains and two identical L chains, resulting in antibody specificity. Recall that there are five classes of H-chain genes and two classes of L-chain genes. Because of the high degree of variability in the genes that encode the H and L chains, it is possible (even likely) that an antibody-producing cell contains two different alleles of the H-chain gene and two different alleles of the L-chain gene. Yet the antibodies produced by the plasma cell contain only a single type of H chain and a single type of L chain. How can you account for this, based on the number of classes of H- and L-chain genes and the possibility of heterozygosity?

Solution: Although an antibody-producing cell contains different classes of H- and L-chain genes and although it is entirely possible and even likely that a given antibody-producing cell will contain different alleles for the H-chain gene and for the L-chain gene, a phenomenon known as allelic exclusion allows the expression of only one H-chain allele and one L-chain allele in any given plasma cell at a given time. That is not to say that the same antibody is produced over the life span of the antibody-producing cell, however. At first, many plasma cells produce antibodies of the IgM class, yet at later times, they may produce antibodies of a different class (e. g., IgG). Even in switching between different H-chain genes, allelic exclusion is maintained, with only one allele of an H-chain gene (or L-chain gene) being expressed at a given time.

Problems and Discussion Questions

1. Used in gene annotation, gene prediction programs allow researchers to identify likely coding regions in DNA sequences. Annotation is complicated when genes are complex, contain multiple initiation sites, or contain numerous exons. For example, Pavy and others (1999. *Bioinformatics* 15: 887–99) determined that even for the most well-studied organisms, such programs can predict correct exon boundaries only about 80% of the time. Given this percentage, what is the likelihood of determining the correct exon boundaries in a gene with five exons?
2. Recent genome-sequencing efforts have provided considerable insight to the molecular nature of living systems, However, it has become increasingly apparent that to fully comprehend the genome, re-annotation, and manual verification, more and more advanced techniques will be needed. For instance, Haas and colleagues (2002. *Genome Biology3(6)@genomebiology.com/2002/3/6/RESEARCH/0029*) recently re-annotated the *Arabidopsis* genome and found 240 new genes, 92 of which are homologous to known proteins. In addition, they identified a new class of exons, called micro-exons, which vary in length from 3 to 25 base pairs. At the beginning of the DNA sequence project, it was often stated that to "know the sequence of DNA is to know the blueprint of life." In what way would you qualify this statement?
3. In a recent draft annotation and overview of the human genome sequence, Wright and others (2001. *Genome Biology 2(7)@genomebiology.com/2001/2/7/RESEARCH/0025*) presented a graph similar to the one shown below. The graph provides the approximate number of embryo-specific genes for each chromosome. Review earlier information in the text on human chromosomal aneuploids, and correlate that information with the graph. Does this graph provide insight as to why some aneuploids occur and others do not?

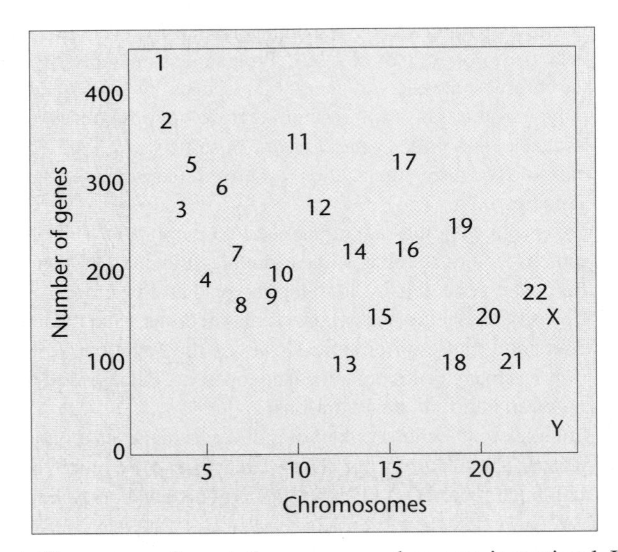

4. The process of annotating a sequenced genome is continual. In March 2000, the first annotated sequence of the *Drosophila* genome was released, which predicted 13,601 protein-coding genes within the euchromatic region of the genome. Shown on p. 429 are selected data from Release 2 (October, 2000) and Release 3. (Modified from Misra et al. 2002. *Genome Biology 3(12)@genomebiology.com/2002/3/12/RESEARCH/0083*). (a) Assuming a uniform distribution for Release 3, approximately how many base pairs of DNA are between protein-coding genes in *Drosophila*? (b) On average, approximately how many exons are there per gene for Release 3? (c) Approximately how many introns are there per gene? (d) What appears to be the most significant difference between Release 2 and Release 3? (e) What are alternative transcripts?

Criteria	*Release 2*	*Release 3*
Total length of euchromatin	116.2 Mb	116.8 Mb
Total protein-coding genes	13,474	13,379
Protein-coding exons	50,667	54,934
Introns	48,381	48,257
Genes with alternative transcripts	689	2729

5. The data given in Problem 4 indicate that the more closely researchers examine genome sequences, the more complex the interpretations of those data will become. Misra and colleague's data (2002) found that nested and overlapping genes are common in *Drosophila*. They determined that approximately 7.5% of all Release 3 genes were included within the introns of other genes and the majority are transcribed from the opposite strand of the including gene. In addition, they found that about 15% of the annotated genes involve the overlap of mRNAs on opposite strands. What impact will this information have on genome annotation, and what clinical significance might it have?
6. Recent sequencing of the heterochromatic regions (repeat-rich sequences concentrated in centric and telomeric areas) of the *Drosophila* genome indicates that within 20.7 Mb, there are 297 protein-coding genes (Bergman et al. 2002. *Genome Biology 3(12)@genomebiology.com/2002/3/12/RESEARCH/0086*). Given that the euchromatic regions of the genome contain 13,379 protein-coding genes in 116.8 Mb, what general conclusion is apparent?
7. In April 2003, scientists announced that the Human Genome Project had finished its mission and that the human genome was completely known. One of the leaders of the project was quoted as saying, "We have before us the instruction set that carries each of us from the one-cell egg through adulthood to the grave." However, from the beginning the HGP excluded the heterochromatic regions at the tips of chromosomes and the regions surrounding the centromeres. The human genome is about 3000 Mb, and heterochromatic regions comprise about 15% of this total. If gene density in human heterochromatin is the same as it is in *Drosophila*, how many genes remain to be discovered in humans? Would you be comfortable stating that all human genes have been identified?
8. One of the main problems in annotation is deciding how long an ORF must be before it is accepted as a gene. Shown below are three different ORF scans of the *E. coli* genome region containing the *lacY* gene. Hatched regions indicate ORFs. The scans have been set to accept ORFs of 50, 100, and 300 nucleotides as genes. How many putative genes are detected in each scan? The longest ORF covers 1254 bp; the next longest covers 234 bp; and the shortest covers 54 bp. How can you decide how many genes are actually in this region? In this type of ORF scan, is it more likely that the number of genes in the genome will be overestimated or underestimated? Why?

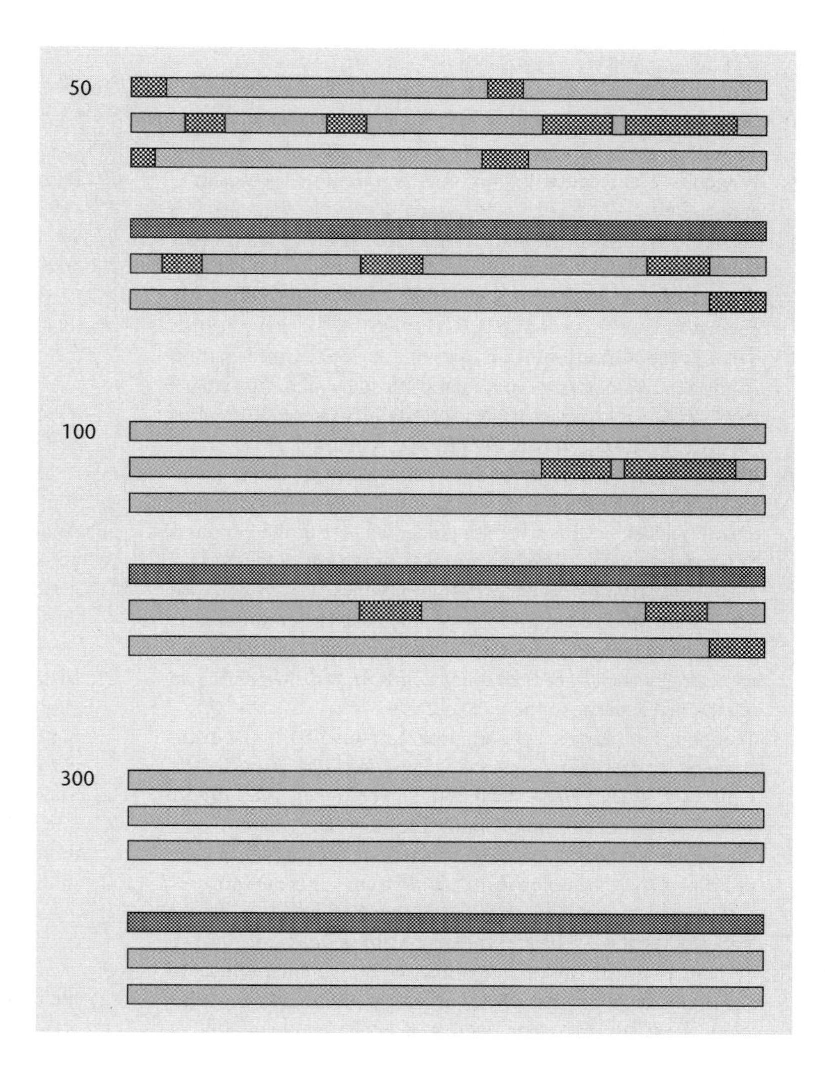

9. To deal with the problems of correctly annotating microbial genomes, Marie Skovgaard and her colleagues (2001. *Trends Genet.* 17: 425–28) compared the annotated number of genes in bacterial genomes derived from sequence analysis to the number of known proteins in each organism as reported in a protein database. The results of their study are summarized in the graph below. The errors range from a few percent for *M. genitalium* to almost 100% for *Aeropyrum pernix*. The general trend shown in the graph is that the error rate increases as the GC content of the genome increases. What explanation might account for this? What precautions should be taken in annotating the genomes of bacteria with high GC content?

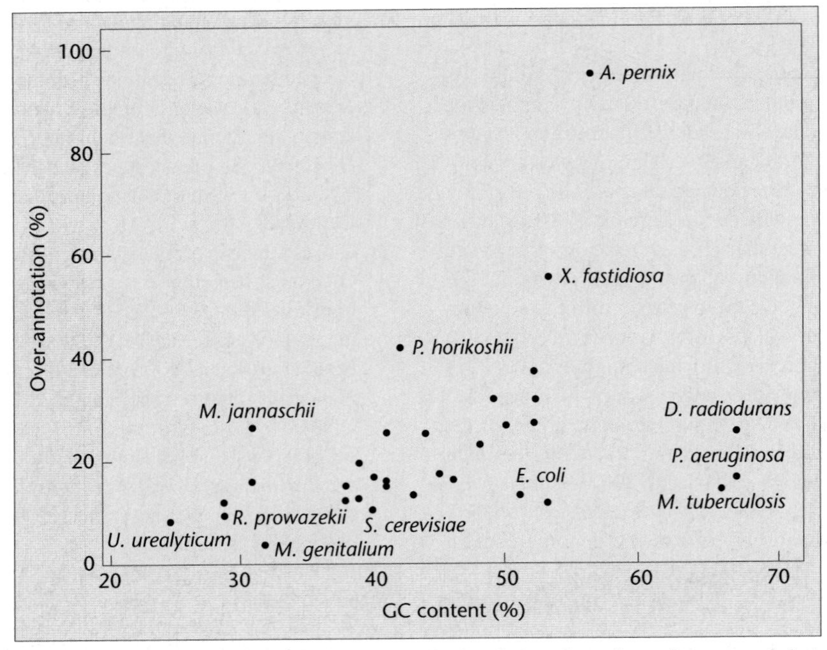

(*Reprinted from* Trends in Genetics, *Vol. 17(8), Skovgaard M et al., On the total number of genes and their length distribution in complete microbial genomes, pp. 425-429, Copyright (2001), with permission from Elsevier.*)

10. What are CpG islands, and how are they used in analysis of genomes?
11. What is functional genomics? How does it differ from comparative genomics?
12. What features do Archaea genomes share with eukaryotic genomes?
13. What is the definition of an operon? How does information from the *A. aeolicus* genome alter our ideas about operons? How do findings about transcription units in the *C. elegans* genome fit into the classic definition of operons?
14. Plasmids can be transferred between species of bacteria and most carry nonessential genes. For these and other reasons, plasmid genes have not been included as part of the genomes of bacterial species. The bacterium *B. burgdorferi* contains 17 plasmids carrying 430 genes, some of which are essential for life. Should plasmids carrying essential genes be considered as part of an organism's genome? What about other plasmids that do not carry such genes? In other words, how do we define an organism's genome in these cases?
15. Compare and contrast the chemical nature, size, and form assumed by the genetic material of Eubacteria and yeast. Do the same with Archaea and yeast. Which group has more differences, and which has more similarities to yeast?
16. Why might we predict that the organization of eukaryotic genetic material is more complex than that of viruses or bacteria?
17. Compare the gene organization of bacterial genes to that of eukaryotic genes. What are the major differences?
18. *C. elegans* is a eukaryotic organism with a genome of 97 Mb and about 20,000 genes. What organizational features of this genome are unusual when compared to the genomes of other eukaryotes, such as yeast and *Drosophila*?
19. Based on the completion of the genome sequence of *Arabidopsis* in 2000, Simillion and others (2002. *Proc. Nat. Acad. Sci. (USA)* 99: 13627–32) have estimated that the *Arabidopsis* genome has undergone three rounds of genome duplication or polyploidization events in the last 100 million years. Presently, of the approximately 25,000 genes in *Arabidopsis*, these researchers estimate that about 80% are in duplicated regions of the genome. Examination of Table 18.4 reveals that *Arabidopsis*, rice, maize, and barley have approximately the same number of genes, but that *Arabidopsis* has much less DNA. What type of genomic organization appears to account for the vast differences in DNA content with similar gene numbers in these species?
20. Annotation of the human genome sequence reveals that our genome contains 30,000–33,000 genes. Proteomic analysis indicates that human cells are capable of synthesizing more than 300,000 different proteins. How can this discrepancy be reconciled?
21. List some of the following general features of the human genome: size, how much codes for proteins, how much is composed of repetitive sequences, where genes are distributed on chromosomes, and how many genes it contains.
22. The discovery that *M. genitalium* has a genome of 0.58 Mb and only 470 protein-coding genes has sparked interest in determining the minimum number of genes needed for a living cell. In the search for organisms with smaller and smaller genomes, a new species of Archaea, *Nanoarchaeum equitans*, was discovered in a high-temperature vent on the ocean floor (Huber, et al., 2002. *Nature* 417: 63–67). This prokaryote has one of the smallest cell sizes ever discovered, and its genome is only about 0.5 Mb. However, organisms such as *M. genitalium*, *N. equitans*, and other microbes with very small genomes are

either parasites or symbionts. How does this affect the search for a minimum genome? Should the definition of the minimum genome size for a living cell be redefined?

23. In the search for the smallest bacterial genome and the minimum number of genes necessary for life, attention has turned to species of *Buchnera*, which live as intracellular symbionts in aphid intestinal cells. As symbionts, they need not maintain the genes necessary for infection and for evasion of the host's immune system as do parasites or pathogens and may have smaller genomes. The genome of one species of *Buchnera*, designated APS, has been sequenced. It has 564 genes in a circular chromosome of 640 kb (Shigenobou, et al. 2000. *Nature* 407: 81–86). To determine whether *Buchnera* genome size is conserved across different groups of aphids, Gil, and colleagues (2002. *Proc. Nat. Acad. Sci. (USA)* 99: 4454–58) physically mapped the genome sizes of nine *Buchnera* genomes that were isolated from five aphid families. The genomes were sized by digestion with restriction enzymes, and separation of the resulting fragments was done by gel electrophoresis. The data for some *Buchnera* species are given in the following table. Although there are some discrepancies in sizes, the sum of the fragments correspond to the size of the chromosome that appeared on the gel without restriction digestion. From your analysis of the data, is genome reduction in *Buchnera* still occurring? How do the genome sizes obtained for these species compare with the genome of *M. genitalium*? The APS species of Buchnera contains 564 coding genes in a 641 kb genome. How many genes should be present in species CCE? How does this compare to the number of genes in *M. genitalium*? Are there other ways to determine the minimum genome needed for life without searching for other bacterial species with small genomes?

	Size of DNA Fragments Produced by Restriction Enzymes, (kb)			
Buchnera	***Apa*I**	***Rsr*II**	***ApaI* + *Rsr*II**	***Total DNA Length kb***
APS	286, 226, 73, 52, 3.4	277, 264, 99	240, 104, 99, 73, 52, 45, 24, 3.4	640±
THS	545	545	320, 234	544±
CCE	440	450	405, 46	448±
CCU	265, 135, 50, 25	475	200, 136, 64, 50, 30	476±

24. How are gene duplications generated, and how do they contribute to genome evolution?
25. In addition to comparisons of nucleotide sequences for determining phylogenetic relationships among organisms, studies of gene order have become an informative source for genome studies. In addition to providing evidence of evolutionary relationships, gene-order data have been used to predict gene function and functional interactions of proteins. Tamames (2001) determined that, in prokaryotes, loss of gene order occurs when phylogenetic distance increases, but contrary to expectation, significant conservation is maintained in distant groups. What factors might you expect to contribute to the conservation of gene order among distantly related species?
26. Genomic sequencing has opened doors to numerous studies that help us understand the evolutionary forces that shape the genetic makeup of organisms. Using databases containing the sequences of 27 genomes, Kreil and Ouzounis (2001) examined the relationship between GC content and global amino acid composition. They found that it is possible to identify thermophilic species on the basis of their amino acid compositions alone, which suggests that evolution in a hot environment selects for a certain whole-organism amino acid composition. In what way might evolution in extreme environments influence genome and amino acid composition? How might evolution in extreme environments influence the interpretation of genome sequence data?
27. What are pseudogenes, and how are the produced?
28. The β-globin gene family consists of 60 kb of DNA, yet only 5% of the DNA encodes β-globin gene products. Account for as much of the remaining 95% of the DNA as you can.
29. What do V_L, C_H, IgG, J, and D represent in immunoglobulin structure?
30. If germline DNA contains 10 V, 30 D, 50 J, and 3 C segments, how many unique DNA sequences can be formed by recombination?
31. If there are 5 V, 10 D, and 20 J regions available to form a heavy-chain gene and 10 V and 100 J regions available to form a L-chain gene, how many unique antibodies can be formed?
32. Annotation of the proteome attempts to relate each protein to function in time and space. Traditionally, protein annotation depended on an amino acid sequence comparison between a query protein and a protein with known function. If two proteins shared a considerable portion of their sequence, the query would inherit the function of the annotated protein. Below is a representation of the "look-the-same" method of protein annotation involving a query sequence and three different human proteins (modified from Rigoutsos, et al. 2002. *Nucl. Acids Res.* 30: 3901–16). Note that the query sequence aligns to common domains within the three other proteins. What argument might you present to suggest that the function of the query is not related to the function of the other three proteins?

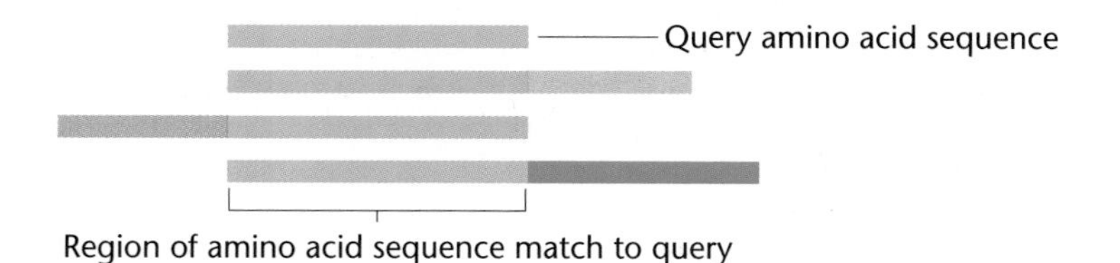

Selected Readings

Aebersold, R., and Mann, M. 2003. Mass spectrometry-based proteomics. *Nature* 422: 198–207.

Andersen, J. et al. 2002. Directed proteomic analysis of the human nucleolus. *Curr. Biol.* 12: 1–11.

Baltimore, D. 2001. Our genome. *Nature* 409: 814–16.

Blackstock, W. P. et al. 1999. Proteomics: Quantitative and physical mapping of cellular proteins. *Trends Biotechnol.* 17: 121–27.

Blattner, F. R. et al. 1997. The complete genome sequence of *Escherichia coli* K-12. *Science* 277: 1453–74.

Brown, P. O., and Botstein, D. 1999. Exploring the world of the genome with DNA microarrays. *Nature Genet.* (Suppl.) 21: 33–7.

Bult, C. J. et al. 1996. Complete genome sequence of the methanogenic Archaeon, *Methanococcus jannaschii. Science* 273: 1058–72.

Fraser, C. M. et al. 1995. The minimal gene complement of *Mycoplasma genitalium. Science* 270: 397–403.

Freeman, W. M. et al. 2000. Fundamentals of DNA hybridization arrays for gene expression analysis. *BioTech.* 29: 1042–55.

Gellert, M. 1996. A new view of V-D-J recombination. *Genes to Cells* 1: 269–75.

Gil, R. et al. 2002. Extreme genome reduction in *Buchnera* spp.: Toward the minimal genome needed for symbiotic life. *Proc. Nat. Acad. Sci. (USA)* 99: 4454–58.

Goffeau, A. et al. 1996. Life with 6000 genes. *Science* 274: 546–67.

Golub, T. et al. 2000. Molecular classification of cancer: Class discovery and class prediction by gene expression monitoring. *Science* 286: 531–37.

Hamedeh, H., and Mshari, C. 2000. Gene chips and functional genomics. *Am. Scientist* 88: 508–15.

Hanash, S. 2003. Disease proteomics. *Nature* 422: 226–32.

International Human Genome Sequencing Consortium. 2001. Initial sequencing and analysis of the human genome. *Nature* 409: 860–921.

Iyer, V. R. et al. 1999. The transcriptional program in the response of human fibroblasts to serum. *Science* 283: 83–87.

Kumar, A., and Snyder, M. 2001. Emerging technologies in yeast genomics. *Nature Rev. Genet.* 2: 302–12.

Martin, A., and Drubin, D. 2003. Impact of genome-wide functional analyses on cell biology research. *Curr. Opin. Cell Biol.* 15: 6–13.

Meinke, D. W. et al. 1998. *Arabidopsis thaliana*: A model plant for genome analysis. *Science* 282: 662–82.

Mewes, H. W. et al. 1997. Overview of the yeast genome. *Nature* 387: 5–105.

Myers, E. W. et al. 2000. A whole-genome assembly of *Drosophila. Science* 287: 2196–2204

San Miguel, P. et al. 1998. The paleontology of intergene retrotransposons of maize. *Nature Genet.* 20: 43–47.

Stover, C. K., et al. 2000. Complete genome sequence of *Pseudomonas aeruginosa* PAO1, an opportunistic pathogen. *Nature* 406: 959–64.

The *C. elegans* Sequencing Consortium. 1998. Genome sequence of the nematode *C. elegans*: A platform for investigating biology. *Science* 282: 2012–18.

Tyers, M., and Mann, M. 2003. From genomics to proteomics. *Nature* 422: 193–97.

Venter, J. C. et al. 1998. Shotgun sequencing of the human genome. *Science* 280: 1540–42.

———. 2001. The sequence of the human genome. *Science* 291: 1304–51.

Wolfe, K. H., and Shields, D. C. 1997. Molecular evidence for an ancient duplication of the entire yeast genome. *Nature* 387: 708–13.

CHAPTER 19

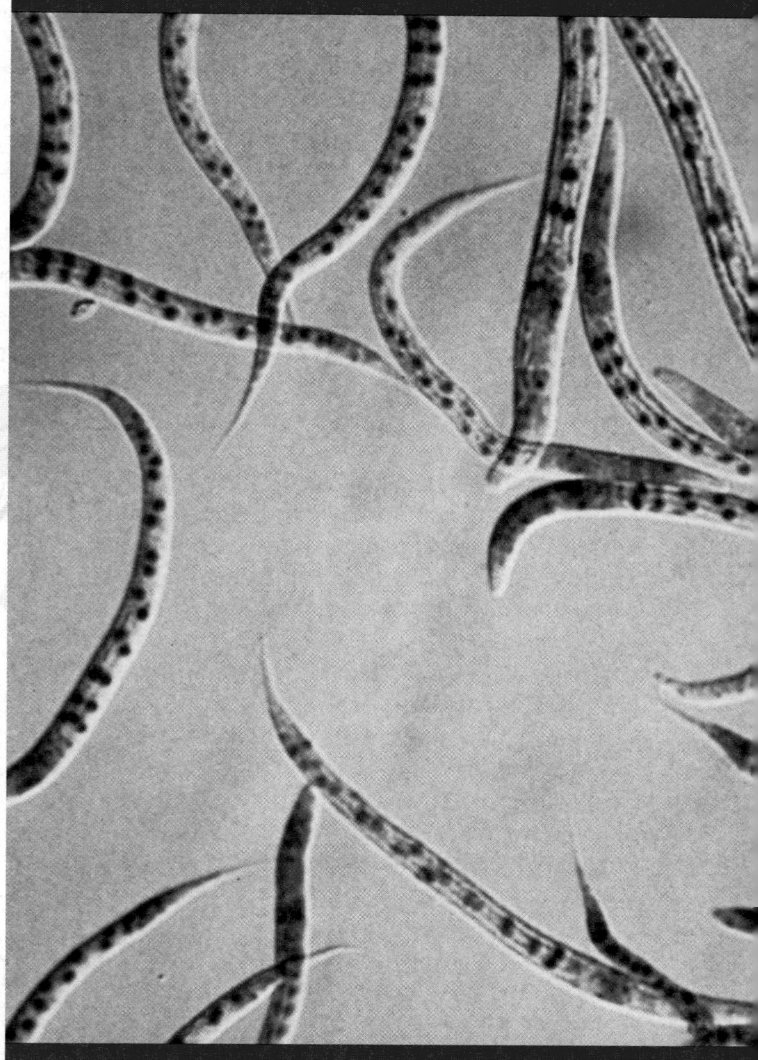

Genetically engineered *Caenorhabditis elegans* (roundworms) that have turned blue in response to environmental stress. *(Dr Eve G. Stringham and Dr. Peter M. Candido, Journal of experimental Zoology 266: 227-233 (1993). Publication of Wiley-Liss Inc. University of British Columbia.)*

Applications and Ethics of Biotechnology

CHAPTER CONCEPTS

In 1971, a paper published by Hamilton Smith, Daniel Nathans, and Walter Arber marked the beginning of the recombinant DNA era. The paper described the isolation of an enzyme from a bacterial strain and the use of the enzyme to cleave viral DNA at specific sites. It contained the first published photograph of DNA cut with a restriction enzyme. From this modest beginning, recombinant DNA technology has revolutionized almost all fields of experimental biology and has spread far beyond the research lab to become the foundation of a major industry, biotechnology. In the intervening years, biotechnology has become part of everyday life; it is used to produce medicines, industrial chemicals, and even milk and other foods found in the supermarket.

In Chapter 17, we described the methods used to create and analyze recombinant DNA molecules, and in Chapter 18 we examined the use of this technology in genome analysis. This chapter describes how these tools are being used in commercial and scientific applications in the biotechnology industry and the ethical problems posed by its use. We will consider a cross section of applications that illustrate the power of biotechnology in generating new plants and animals, producing food, diagnosing and treating diseases, and in forensics and mapping human genes. Biotechnology is also being used to produce therapeutic proteins in genetically altered plants and animals. We begin by explaining how biotechnology has changed basic methods of food production and generated economic, social, and environmental controversies. Next, we examine the impact of recombinant DNA on medicine, from the prenatal analysis of genotypes, genome scanning, and the treatment of genetic disorders by gene therapy. In addition, we will consider the impact of biotechnology on forensics. Finally, we will discuss some of the ways biotechnology is being used in basic research, using human gene mapping as an example.

HOW DO WE KNOW WHAT WE KNOW?

IN THIS CHAPTER, WE WILL FOCUS ON the applications of recombinant DNA technology to agriculture, medicine, and other areas. As you study this topic, you should try to answer several fundamental questions:

1. How do we know that genetically modified organisms are safe?
2. How do we know that people can be vaccinated against disease by eating certain genetically modified foods?
3. How do we know that gene therapy delivers a cloned normal gene to target cells?
4. How do we know that an individual can be identified with certainty by DNA fingerprinting? ■

19.1 Biotechnology Has Revolutionized Agriculture

Although recombinant DNA techniques were originally developed to facilitate basic research on gene organization and regulation of expression, gene transfer methods are also the foundation of the biotechnology industry. Biotechnology is used to manufacture a wide range of products, including hormones, clotting factors, herbicide-resistant plants, enzymes for food production, and vaccines. In the last decade, this industry has grown into a multibillion-dollar segment of the economy. In this section, we will review some of the current uses of biotechnology in agriculture and the issues raised by this application of recombinant DNA technology.

Transgenic Crops and Herbicide Resistance

Damage from weed infestation destroys about 10% of crops worldwide. To combat this problem, more than 100 different herbicides are used, at an annual cost of more than 10 billion dollars. Some of these herbicides also kill crop plants, while others remain in the environment or are carried by water runoff to contaminate local water supplies and cause abiotic "dead zones" in the ocean at river mouths. Creating herbicide-resistant plants is one way to improve crop yields, while at the same time reducing the environmental impact of herbicides. To do this, vectors carrying genes for herbicide resistance are transferred to crop plants.

We will examine how resistance to one herbicide was transferred to crops including soybeans and maize. The herbicide **glyphosate** is effective at very low concentrations, is not toxic to humans, and is rapidly degraded by soil microorganisms. Glyphosate works by inhibiting the action of a chloroplast enzyme called EPSP synthase. This enzyme is important in amino acid biosynthesis in both bacteria and plants. Without the ability to synthesize vital amino acids, plants wither and die.

The development of glyphosate-resistant crop plants began with the isolation and cloning of an EPSP synthase gene from a glyphosate-resistant strain of *E. coli*. This gene was cloned into a vector adjacent to plant virus promoter sequences and upstream from plant transcription-termination sequences. The recombinant vector was transferred into the bacterium *Agrobacterium tumifaciens* (Figure 19–1). Plasmid-carrying bacteria were, in turn, used to infect cells in discs cut from plant leaves. Calluses formed by interaction with *A. tumifaciens* were then selected for their ability to grow on glyphosate. Transgenic plants generated from glyphosate-resistant calluses were grown and sprayed with glyphosate at concentrations four times higher than that needed to kill wild-type plants. The transgenic plants that overproduced the EPSP synthase grew and developed, while the control plants withered and died. Glyphosate-resistant corn and soybeans developed in this way are now on the market in the United States and other countries. Since their introduction in 1996, genetically engineered crops modified to tolerate herbicides

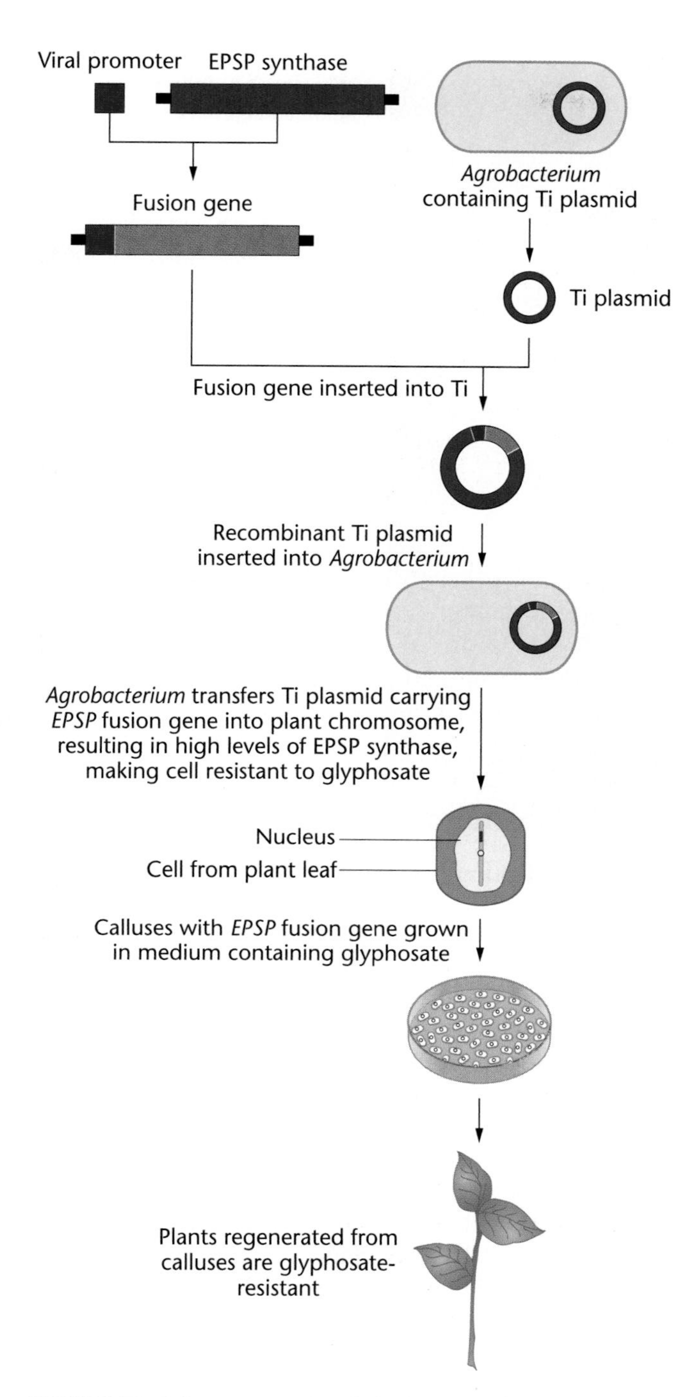

FIGURE 19–1 In gene transfer of glyphosate resistance, the *EPSP* gene is fused to a promoter from cauliflower mosaic virus. This fusion gene is then transferred to a Ti plasmid vector, and the recombinant vector is inserted into an *Agrobacterium* host. *Agrobacterium* infection of cultured plant cells transfers the *EPSP* fusion gene into a plant-cell chromosome. Cells that acquire the gene are able to synthesize large quantities of EPSP synthase, making them resistant to the herbicide glyphosate. Resistant cells are selected by growth in herbicide-containing medium. Plants regenerated from these cells are herbicide-resistant.

and resist insect pests have been planted in several countries (Figure 19–2). In the United States, the three major genetically engineered crops are corn, soybeans, and cotton. In 2000, 26% of the corn, 68% of the soybeans, and 69% of the cotton planted in the United States was genetically modified.

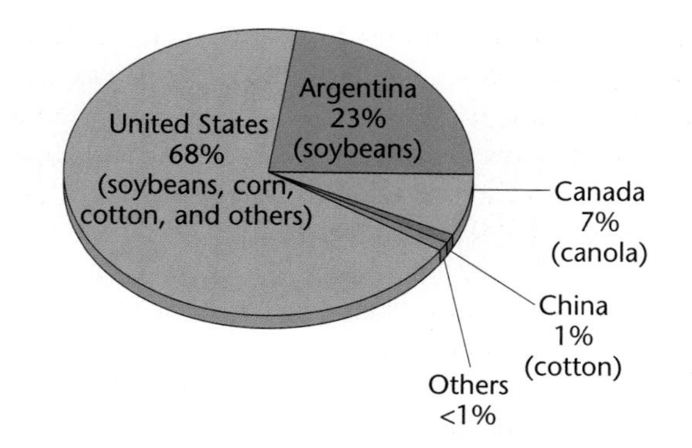

FIGURE 19–2 A pie chart showing the percentage of global land area planted in genetically modified crop plants in 2000. The United States accounts for two-thirds of the world production of genetically modified crops. *(From PEW Initiative on* Food and Biotechnology *[on-line] http://pewagbiotech.org/resources/factsheets/display.php3?FactsheetID=1)*

Developing countries have also engineered genetically modified crops that have been widely adopted by farmers. In China, for example, over 30% of the cotton crop planted in 2000 was made up of strains genetically modified to resist insect pests.

Nutritional Enhancement of Crop Plants

In the last 50–100 years, genetic improvement of crop plants through the traditional methods of artificial selection and genetic crosses has resulted in dramatic improvements in productivity and nutritional enhancement. For example, corn yields have increased 4-fold over the last 60 years, and more than half of this increase is due to genetic improvement by artificial selection. Broccoli contains gluconsinolates, compounds thought to have a role in cancer prevention by activation of the anti-cancer marker enzyme quinolase. New hybrid varieties produced by conventional crosses have a 100-fold increase in enzyme levels.

Gene transfer by recombinant DNA techniques offers a new way to enhance the nutritional value of plants. Many crop plants are deficient in some of the nutrients required in the human diet, and biotechnology is being used to produce crop plants with these dietary requirements. One major example of this is the production of "golden rice" with enhanced levels of β-carotene, a precursor to vitamin A (Figure 19–3). In this case, three genes encoding enzymes in the biosynthetic pathway leading to carotenoid synthesis were transferred to the rice genome using methods of recombinant DNA technology. Two of these genes came from the daffodil, and one from a bacterium. Other work is directed at enhancing the levels of key fatty acids, antioxidants, and other vitamins and minerals already present in crop plants. These efforts are directed at addressing the dietary lack of nutrients affecting more than 40% of the world's population. More important, the idea that crops can be grown for health as well as food is changing our view of agriculture. As discussed later, this idea extends to the production of plant-based human proteins, antibodies, and vaccines.

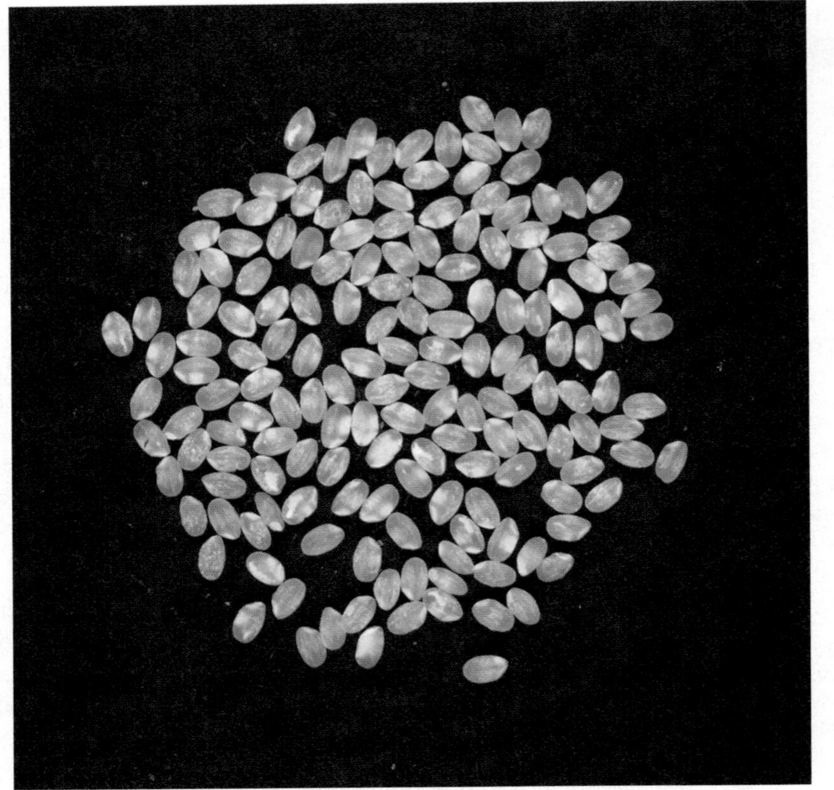

FIGURE 19–3 A photograph of golden rice, a strain genetically modified to produce β-carotene, a precursor to vitamin A. *(Reprinted with permission from "New Genes Boost Rice Nutrients" by I. Potrykus and P. Beyer, Science, August 13, 1999, Vol. 285, pp 994.)*

Concerns About Genetically Modified Organisms

Most genetically modified food products contain an introduced gene that encodes a protein that confers a desired trait (herbicide resistance, insect resistance, etc.). Much of the concern over genetically modified plants centers on issues of safety and environmental consequences. Are genetically modified plants containing a new protein safe to eat? In general, if the proteins are not toxic, allergenic and do not have any other negative physiological effects, they are not considered to be a significant hazard to health. In the case of herbicide-resistant EPSP-containing food plants, the protein is readily degraded in digestive fluids, is nontoxic to mice at doses thousands of times higher than any potential human exposure, and has no amino acid sequence similarity to known protein toxins and allergens. Regulatory oversight by appropriate government agencies using standardized methods for the evaluation of proteins in genetically modified foods is being developed in the United States and in Europe.

What about the environmental risks associated with genetically modified plants? Environmental risks include gene transfer by crossbreeding with wild plants, toxicity, and invasiveness of the modified plant, resulting in loss of natural species (loss of biodiversity). In fact, these problems are the same problems facing the use of conventional crop plants, and there is no current scientific evidence that genetically modified crops are inherently different from nonmodified crops. However, as gene transfer technology becomes more sophisticated and multiple traits are transferred, novel traits or combinations of traits may be generated and these crops may require specific management procedures.

Plants and animals were domesticated 8000—10,000 years ago, and we have been genetically modifying these organisms by selective breeding ever since, producing the diversity of domesticated plants and animals we have today. Biotechnology has changed the rate at which new plants and animals can be developed and, by enabling gene transfer between species, has altered the types of changes that can be made. Unlike selective-breeding programs, however, which have introduced thousands of genetically altered strains, biotechnology has generated concerns about the release of genetically modified organisms into the environment and about the safety of eating such products. If biotechnology is to achieve a new green revolution, these concerns need to be addressed through prudent research and education of the public.

19.2 Pharmaceutical Products Are Synthesized in Genetically Altered Organisms

The first human gene product manufactured by using recombinant DNA and licensed for therapeutic use was human insulin, which became available in 1982. Insulin is a protein hormone that regulates glucose metabolism. Individuals who cannot produce insulin have diabetes, a disease that, in its more severe form, affects more than 2 million individuals in the United States.

Clusters of cells embedded in the pancreas synthesize a precursor peptide known as preproinsulin. As this polypeptide is secreted from the cell, amino acids are cleaved from the end and the middle of the chain. This process produces the mature insulin molecule, which consists of two polypeptide chains (*A* and *B* chains), joined by disulfide bonds. Insulin circulating in the blood regulates the uptake of glucose. Diabetics are unable to produce sufficient insulin and must take insulin as a medication. Before the development of recombinant DNA technology, insulin was extracted from pancreatic glands that were recovered from cows and pigs being slaughtered for meat.

Insulin Production in Bacteria

Although synthetic human insulin is now produced by another process, a look at the original method is instructive, as it shows both the promise and the difficulty of applying recombinant DNA technology. A functional insulin molecule contains two polypeptide chains, *A* and *B*. The *A* subunit has 21 amino acids, and the *B* subunit has 30. In the original bacterial process, synthetic genes for the *A* and *B* subunits were constructed by oligonucleotide synthesis (63 nucleotides for the *A* polypeptide and 90 nucleotides for the *B* polypeptide). Each synthetic oligonucleotide was inserted into a vector adjacent to a gene encoding the bacterial form of the enzyme, β-galactosidase. When transferred to a bacterial host, the *β-gal* gene and the adjacent synthetic oligonucleotide were transcribed and translated as a unit. The product, a **fusion polypeptide**, consisted of the amino acid sequence for β-galactosidase attached to the amino acid sequence for one of the insulin subunits (Figure 19–4). The fusion proteins were purified from bacterial extracts and treated with cyanogen bromide to cleave the fusion protein from the β-galactosidase. Each insulin subunit was produced separately by this process. When they were mixed, the two subunits spontaneously united, forming an intact, active insulin molecule. The purified insulin was then packaged for use by diabetics.

Several genetically engineered proteins for therapeutic use have been produced by similar methods (Table 19.1). In most cases, cloning a human gene into a plasmid and inserting the recombinant vector into a bacterial host produces the proteins. After ensuring that the transferred gene is expressed, large quantities of the transformed bacteria are produced, and the human protein is recovered and purified.

Transgenic Animal Hosts and Pharmaceutical Products

Bacterial hosts were used to produce the first generation of therapeutic proteins, even though there are some disadvantages in using prokaryotic hosts to synthesize eukaryotic proteins. For example, bacterial cells are unable

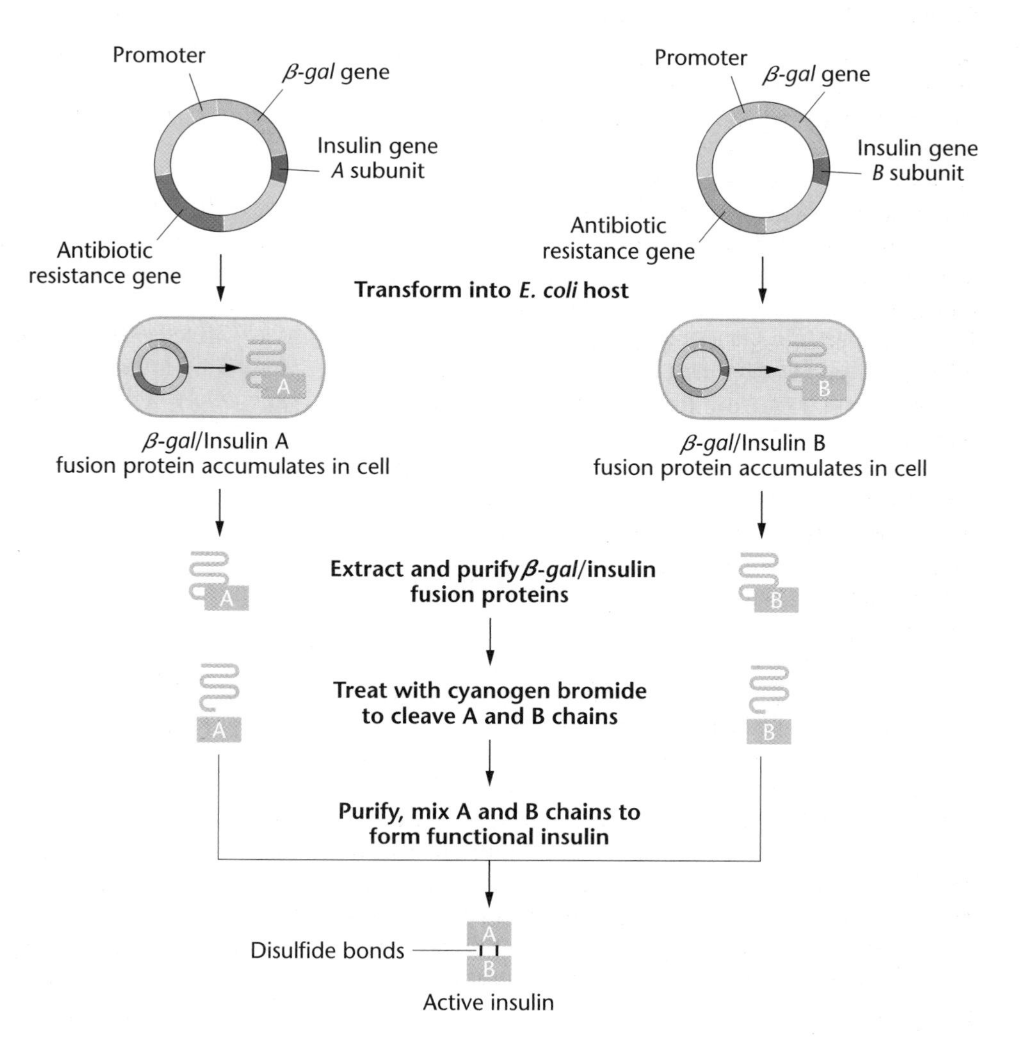

FIGURE 19–4 To synthesize recombinant human insulin, synthetic oligonucleotides encoding the insulin A and B chains were inserted at the tail end of a cloned *E. coli* *β-gal* gene. The recombinant plasmids were transferred to *E. coli* hosts, where the β-gal/insulin fusion protein was synthesized and accumulated in the host cells. Fusion proteins were then extracted from the host cells and purified. Insulin chains were released from the β-galactosidase by treatment with cyanogen bromide. The insulin subunits were purified and mixed to produce a functional insulin molecule.

TABLE 19.1 Genetically Engineered Pharmaceutical Products Now Available or in Clinical Trials

Gene Product	*Condition Treated*
Atrial natriuretic factor	Heart failure, hypertension
Epidermal growth factor	Burns, skin transplants
Erythropoietin	Anemia
Factor VIII	Hemophilia
Gamma interferon	Cancer
Granulocyte colony-stimulating factor	Cancer
Hepatitis B vaccine	Hepatitis
Human growth hormone	Dwarfism
Insulin	Diabetes
Interleukin-2	Cancer
Superoxide dismutase	Transplants
Tissue plasminogen activator	Heart attack

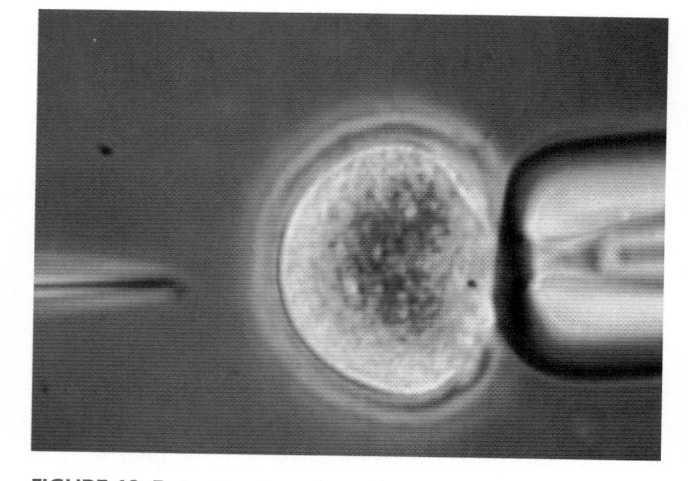

FIGURE 19–5 A micropipette is used to transfer cloned genes into the nucleus of a mammalian zygote. The injected zygote will then be transferred to the uterus of a surrogate mother for development. *(Hank Morgan/Photo Researchers, Inc.)*

to process and modify many eukaryotic proteins. Thus, they cannot add the sugars and phosphate groups that are often needed for full biological activity. In addition, eukaryotic proteins produced in prokaryotic cells often don't fold into the proper three-dimensional configuration and, as a result, are inactive. To overcome these difficulties and increase yields, second-generation methods use eukaryotic hosts. Rather than being produced by host cells grown in tissue culture, human proteins such as α_1-antitrypsin are produced in the milk of livestock.

A deficiency of the enzyme α_1-antitrypsin is associated with the heritable form of emphysema, a progressive and fatal respiratory disorder common among people of European ancestry. To produce α_1-antitrypsin, the human gene was cloned into a vector at a site adjacent to a sheep promoter sequence that regulates the expression of milk-associated proteins. Genes placed next to this promoter are expressed only in mammary tissue. This fusion gene was microinjected into sheep zygotes fertilized *in vitro* (Figure 19–5). The fertilized zygotes were transferred to the uterus of a surrogate mother. The resulting **transgenic** sheep developed normally and after mating produced milk that contained high concentrations of functional human α_1-antitrypsin. This human protein is present in concentrations of up to 35 grams per liter of milk, and can be easily extracted and purified. A small herd of lactating sheep can easily provide an adequate supply of this protein. Herds of other transgenic animals acting as biofactories are becoming part of the pharmaceutical industry. In fact, the famous sheep, Dolly, was cloned to facilitate the establishment of a flock of sheep that would consistently produce high levels of human proteins.

Human proteins produced in transgenic animals undergo clinical testing as a first step in the therapeutic use of recombinant human proteins. A recombinant human enzyme, α-glucosidase, produced in rabbit milk, was clinically tested in children with Pompe disease. This progressive and fatal metabolic disorder is caused by a lack of α-glucosidase production and is inherited as an autosomal recessive condition. The recombinant enzyme was given weekly, and there were no significant side effects. All of the children showed normal enzyme activity in the tissues analyzed, and all showed improvements in their symptoms. If large-scale trials are successful, recombinant α-glucosidase from transgenic animals will become the preferred method of treatment for this disorder.

Transgenic Plants and Edible Vaccines

One of the most beneficial applications of biotechnology may be in the production of vaccines. Vaccines stimulate the immune system to produce antibodies against a disease-causing organism and thereby confer immunity against the disease. Two types of vaccines are commonly used—**inactivated vaccines**, prepared from killed samples of the infectious virus or bacteria; and **attenuated vaccines**, which are live viruses or bacteria that can no longer reproduce, but can cause a usually mild form of the disease.

Biotechnology is being used to produce a new type of vaccine called a **subunit vaccine**. These vaccines consist of one or more surface proteins of the virus or bacterium. The protein acts as an antigen to stimulate the immune system into making antibodies against the virus or bacterium. One of the first subunit vaccines was for hepatitis B, a virus that causes liver damage and cancer. The gene for the hepatitis B surface protein was cloned into a yeast-expression vector and produced by using yeast as a host. The protein is extracted and purified from the host cells and then packaged for use as a vaccine.

Subunit vaccines produced by biotechnology provide a source of pure vaccine manufactured under controlled conditions and are widely used. However, vaccination programs in developing countries are faced with serious problems of cost, transportation, and storage. Vaccines must be kept refrigerated, and sterile conditions must be maintained

FIGURE 19–6 Transgenic tobacco plants carrying an antigenic subunit of hepatitis B virus were generated. Leaves from transgenic plants were treated with antibodies against hepatitis B antigen, showing that the plants produced the antigen. The central leaf in the photograph is a leaf from a normal tobacco plant and is unstained. *(Courtesy of Charles J. Arntzen, Florence Ely Nelson Distinguished Professor and Founding Director, Arizona Biomedical Institute at Arizona State)*

during injection. In the rural areas of many countries, refrigeration and facilities for sterilization of instruments are not available. To overcome these problems, biotechnology is used to develop inexpensive vaccines synthesized in edible food plants. Such vaccines are inexpensive to produce, do not require refrigeration, and do not have to be given under sterile conditions by trained medical personnel.

As a model system, the gene encoding antigenic subunit of hepatitis B vaccine has been transferred to the tobacco plant and expressed in its leaves (Figure 19–6). For use as a source of vaccine, the gene would be inserted into food plants such as grains or vegetables. In clinical trials with a vaccine against the bacteria causing diarrhea, genetically engineered potatoes that carry recombinant bacterial antigens (Figure 19–7) were successfully used to vaccinate human volunteers who ate small quantities (50–100 g) of the potatoes. In another trial, transgenic spinach expressing rabies virus antigens was fed to volunteers. Eight of 14 volunteers showed significant increases in rabies-specific antibodies. Tests using genetically engineered bananas are currently under way. These proof-of-principle trials establish that edible vaccines are feasible. The success of these tests means that genetically engineered edible plants will soon be available to vaccinate infants, children, and adults against many infectious diseases.

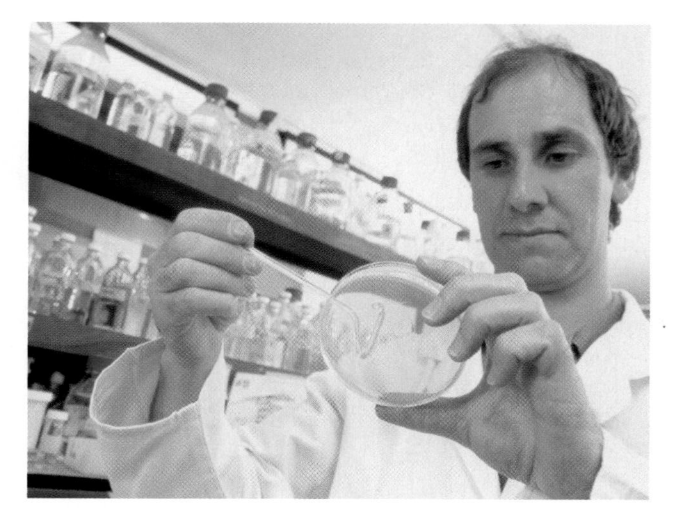

FIGURE 19–7 A researcher prepares bacterial cultures carrying a cloned vaccine. *(David Parker/Photo Researchers, Inc.)*

19.3 Biotechnology Is Used to Diagnose and Screen Genetic Disorders

Many genetic disorders can be prenatally diagnosed using techniques of recombinant DNA technology. The most widely used methods of obtaining samples for prenatal diagnosis are **amniocentesis** and **chorionic villus sampling (CVS)**. In amniocentesis, a needle is used to withdraw amniotic fluid (Figure 19–8), and the fluid and the cells it contains are analyzed for chromosomal or genetic disorders. In CVS, a catheter is inserted into the uterus, and a small tissue sample of the fetal chorion is retrieved. Cytogenetic, biochemical, and recombinant DNA-based testing is then performed on the tissue.

When coupled with these sample-recovery methods, biotechnology has proved to be a highly sensitive and accurate tool for the prenatal detection of genetic disorders. The fetal genotype is examined directly, rather than relying on the few tests available for normal or mutant gene products; thus cloned DNA sequences have expanded the range of prenatal testing. This capability is particularly important because frequently the gene product cannot be detected before birth, even when tests are available. For example, defects in the adult β-globin protein cannot be detected prenatally because the β-globin gene is not expressed until a few days after birth.

Prenatal Diagnosis of Sickle-cell Anemia

Sickle-cell anemia is an autosomal recessive condition common in people with family origins in areas of west Africa, the Mediterranean basin, and parts of the Middle East and India. Sickle-cell anemia is caused by a single amino acid substitution in the β-globin gene. This change is brought about by a single-nucleotide substitution that, coincidentally, eliminates a cutting site for the restriction enzymes *Mst*II and *Cvn*I. As a result, the mutation alters the pattern of restriction fragments seen on Southern blots. These differences in restriction cutting sites can be used

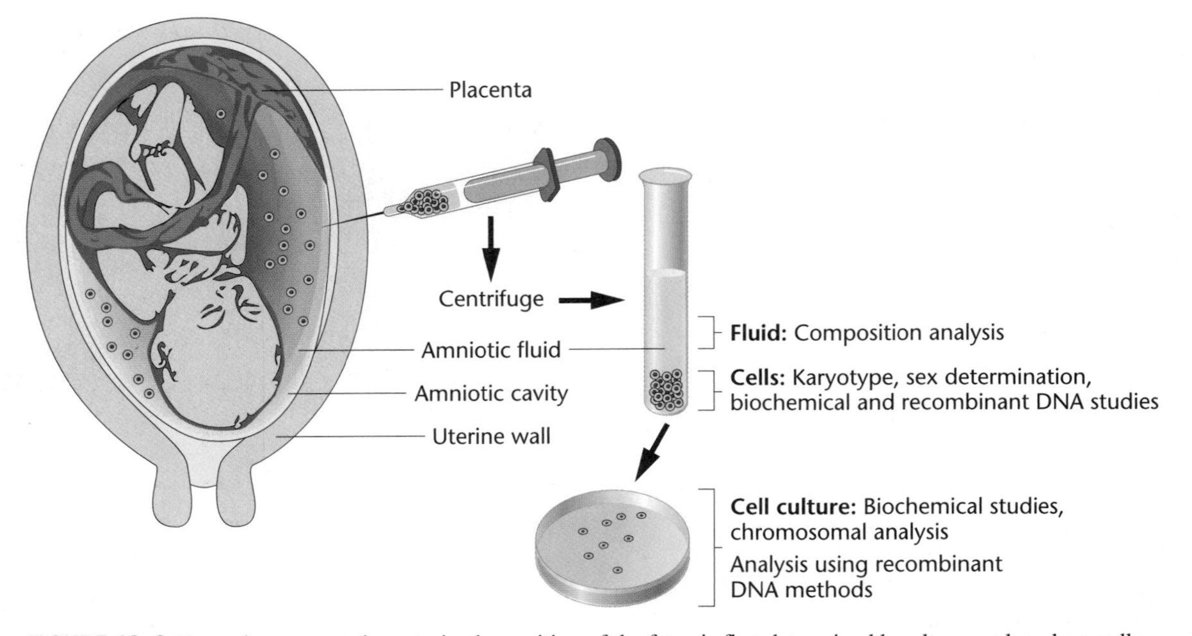

FIGURE 19–8 To perform an amniocentesis, the position of the fetus is first determined by ultrasound, and a needle is then inserted through the abdominal and uterine walls to recover fluid and fetal cells for cytogenetic or biochemical analysis.

to prenatally diagnose sickle-cell anemia and also to determine the genotypes of parents and other family members who may be heterozygous carriers of this condition.

For prenatal diagnosis, fetal cells are obtained by amniocentesis or CVS; for carrier analysis of family members, a blood sample is collected. DNA is extracted from the sample and digested with *Mst*II. This enzyme cuts three times in the region of the normal β-globin gene, producing two small DNA fragments. In the mutant allele, the middle *Mst*II site has been destroyed by the mutation, and one large restriction fragment is produced by digestion with *Mst*II (Figure 19–9). The restriction-digested DNA fragments are separated by gel electrophoresis, transferred to a nylon membrane, and visualized by Southern blot hybridization.

In Figure 19–9, the parents (I-1 and I-2) are both heterozygous carriers of the mutation. Digestion with *Mst*II of the DNA from each parent produces a large band (the mutant allele) and two smaller bands (the normal allele). The parents' first child (II-1) is homozygous normal because she has only the two smaller bands. The second child (II-2) has sickle-cell anemia; he has only one large band and is homozygous for the mutant allele. The fetus (II-3) has a large band and two small bands and is therefore heterozygous for sickle-cell anemia. He or she will be unaffected, but will be a carrier.

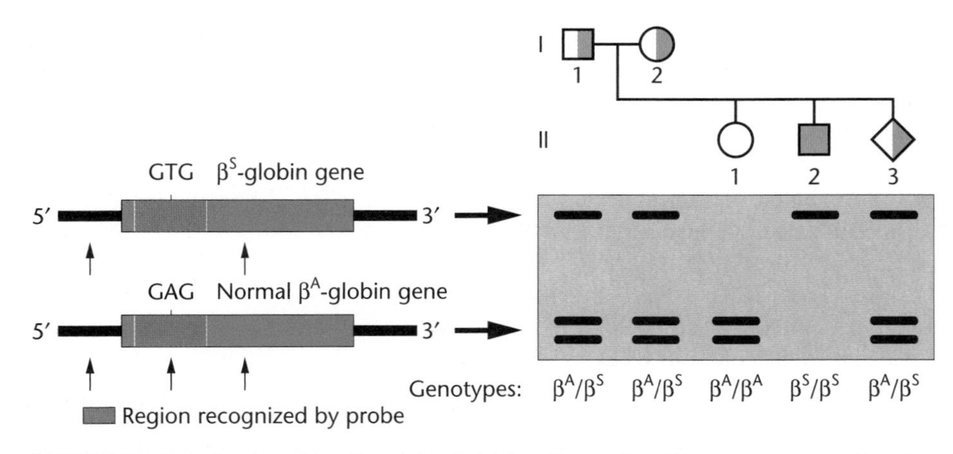

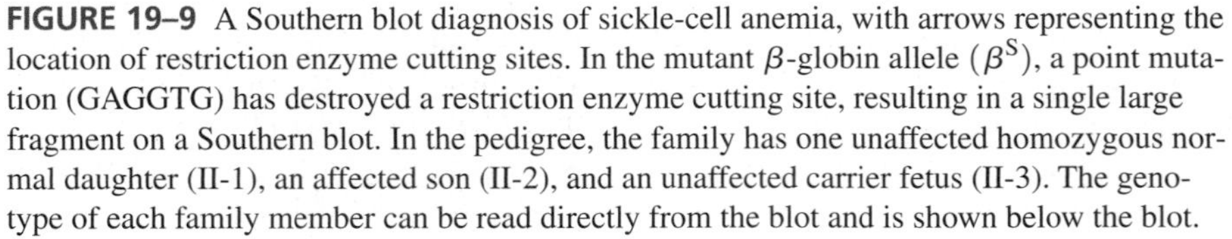

FIGURE 19–9 A Southern blot diagnosis of sickle-cell anemia, with arrows representing the location of restriction enzyme cutting sites. In the mutant β-globin allele (β^S), a point mutation (GAGGTG) has destroyed a restriction enzyme cutting site, resulting in a single large fragment on a Southern blot. In the pedigree, the family has one unaffected homozygous normal daughter (II-1), an affected son (II-2), and an unaffected carrier fetus (II-3). The genotype of each family member can be read directly from the blot and is shown below the blot.

Only about 5–10% of all point mutations can be detected by restriction enzyme analysis. However, if a mutant gene has been well characterized and the mutated region has been sequenced, synthetic oligonucleotides can be used as probes to detect mutant alleles.

Single-Nucleotide Polymorphisms and Genetic Screening

Synthetic probes known as **allele-specific oligonucleotides (ASO)** can identify alleles that differ by as little as a single nucleotide. In contrast to restriction enzyme analysis, which is limited to cases for which a mutation changes a restriction site, ASOs detect single-nucleotide changes of all types, including those that do not affect restriction enzyme cutting sites. As a result, this method offers increased resolution and wider application. Under proper conditions, an ASO will hybridize only with its complementary sequence and not with other sequences, which might vary by as little as a single nucleotide. A method using ASOs and PCR analysis is now available to screen for many disorders, including sickle-cell anemia.

In this procedure, DNA from white blood cells is extracted and denatured into single strands. This DNA is used to amplify a region of the β-globin gene by PCR. A small amount of the amplified DNA is spotted onto filters, and each filter is hybridized to an ASO (Figure 19–10). After visualization, the genotype can be read directly from the filters. With an ASO used for the normal sequence (Figure 19–10a), the homozygous normal (*AA*) genotype produces a dark spot (two copies of the normal allele), and the heterozygous genotype (*AS*) produces a light spot (one copy of the normal allele). The homozygous recessive sickle-cell genotype will not bind the probe, so no spot will be visible; with a probe used for the mutant allele (Figure 19–10b), the pattern is reversed. This rapid, inexpensive, and highly accurate technique is used to diagnose a wide range of genetic disorders caused by point mutations.

In cases where the nucleotide sequence of the normal allele is known and where the molecular nature of the mutant allele has been identified, ASOs are directly synthesized from normal and mutant copies of the gene to screen for heterozygous carriers of the genetic disorder. For example, in people with cystic fibrosis (CF), a deletion called *Δ508* is found in 70% of all mutant copies of the gene. CF is an autosomal recessive disorder associated with a defect in a protein called the **cystic fibrosis transmembrane conductance regulator (CFTR)**, which regulates chloride ion transport across the plasma membrane. To detect heterozygous carriers of the *Δ508* mutation, allele-specific oligonucleotides are made by PCR from cloned samples of the normal allele and the mutant allele. DNA extracted from white blood cells of the individuals to be tested is spotted on a nylon filter and hybridized to each ASO (Figure 19–11). In affected individuals, only the ASO made from the mutant allele hybridizes; in heterozygotes, both ASOs hybridize; and in normal homozygotes, only the ASO from the normal allele hybridizes.

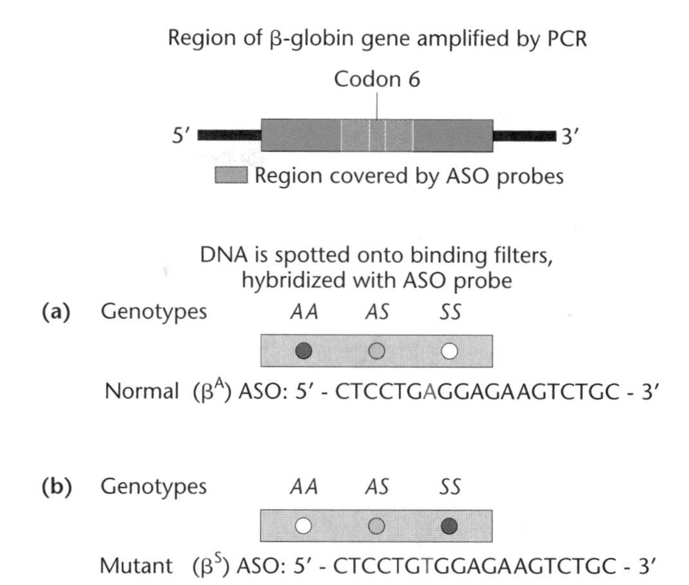

FIGURE 19–10 To determine genotype with allele-specific oligonucleotides (ASOs), the β-globin gene is amplified by PCR, using DNA extracted from blood cells. The amplified DNA is denatured and spotted onto strips of DNA-binding filters. Each strip is hybridized to a specific ASO and visualized on X-ray film after hybridization and exposure. If all three genotypes are hybridized to an ASO from the normal β-globin allele, the pattern in (a) will be observed: *AA*-homozygous individuals have normal hemoglobin that has two copies of the normal β-globin gene and will show heavy hybridization; *AS*-heterozygous individuals carry one normal β-globin allele and one mutant allele and will show weaker hybridization; *SS*-homozygous sickle-cell individuals carry no normal copy of the β-globin gene and will show no hybridization to the ASO probe for the normal β-globin allele. (b) The same genotypes hybridized to the probe for the sickle-cell β-globin allele will show the reverse pattern: no hybridization by the *AA* genotype, weak hybridization by the heterozygote (*AS*), and strong hybridization by the homozygous sickle-cell genotype (*SS*).

CF affects approximately 1 in 2000 individuals of northern European descent, and screening for CF can be used in these populations to detect carriers and counsel people about their genetic status with respect to CF. However, not all of the known mutations for this gene (over 500 mutations have been identified) can be screened, so a negative result does not eliminate someone as a heterozygous carrier—and it is likely that more CF mutations remain to be identified. Consequently, CF screening is not widepread, but will no doubt become commonplace when tests can cover 98–99% of all possible CF mutations.

DNA Microarrays and Genetic Screening

The use of allele-specific nucleotides has been coupled with the technology of the semiconductor industry to produce **DNA microarrays** (also called DNA chips) that can be used to test hundreds or thousands of genes in a single assay. The microarrays are made of glass divided into fields (small squares); each field can be as small as half the width of a

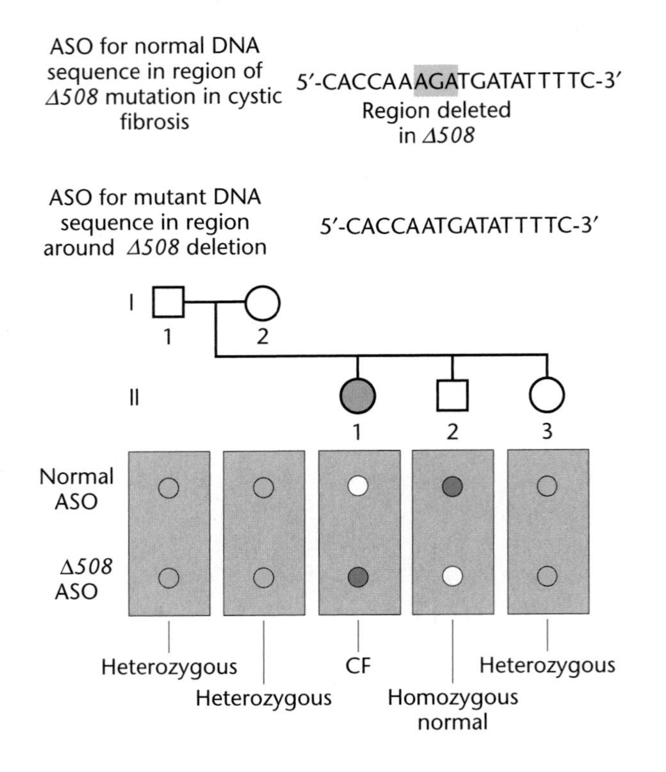

FIGURE 19–11 ASOs for the region spanning the most common mutation in CF, (the *Δ508* allele), are prepared from cloned copies of the normal allele and a *Δ508* allele and then spotted on a DNA-binding membrane. In screening, the CF alleles carried by an individual are amplified by PCR labeled and hybridized to the membrane. The genotype of each family member can then be read directly from the filter. DNA from the parents (I-1 and I-2) hybridizes to both ASOs, indicating that they each carry one normal allele and one mutant allele and are therefore heterozygous. DNA from II-1 hybridizes only to the *Δ508* ASO, indicating that this family member is homozygous for the mutation and has cystic fibrosis. DNA from II-2 hybridizes only to the ASO from the normal CF allele, indicating that this individual carries two normal alleles. DNA from II-3 shows two hybridization spots and is thus heterozygous.

human hair. A field contains copies of a specific synthetic DNA probe about 20 nucleotides in length that is attached to the glass (Figure 19–12). Arrays can be prepared with different nucleotide combination as well as different alleles of specific genes. Along a row of fields, the sequence of the probe differs by one nucleotide from field to field. Thus, a set of four fields (one for each nucleotide) is necessary to test the nucleotide content of a given position in a DNA molecule. The current generation of DNA chips can hold between 280,000 and 560,000 fields, but chips with several million fields are in development.

For this type of genetic testing, DNA is extracted from cells and amplified by PCR. The PCR products are tagged with a fluorescent dye, denatured into single strands, and pumped into the microarray. Fragments with a nucleotide sequence that exactly matches the probe sequence, bind, and those with a sequence that doesn't match are washed off. The microarray is then scanned by a laser, and the fields where hybridization occurs fluoresce (Figure 19–13). Software linked to the microarray analyzes the pattern of hybridization, and the data can be presented in several forms.

DNA microarrays are already used to scan for mutations in the *p53* gene, which is mutated in 60% of all cancers, and to screen for mutations in the *BRCA1* gene, which predispose women to breast cancer. In addition to testing for mutations in single genes, DNA chips can be made to screen thousands of different genes simultaneously. These chips are being used to analyze gene expression patterns of cells during development in cancer and in other applications.

Using a microarray carrying 18,000 genes involved in normal and abnormal white blood cell development, Ash Allzadeh and colleagues analyzed gene expression in a form of non-Hodgkin's lymphoma, a cancer of the white blood cells. About 40% of non-Hodgkin's patients have an aggressive form of cancer called diffuse large B-cell lymphoma (DLBCL). About 40% of these patients respond well to therapy and have extended survival, while the rest do not respond to therapy and die from the disease. Based on the analysis of almost 1.8 million measurements of gene expression patterns of 18,000 genes found in tumors and normal cells, researchers identified two distinct types of cells in DLCBL patients, with almost inverse patterns of expression (Figure 19–14). One type, called GC B-like, had an expression pattern similar to those of B cells in lymph glands, where the B cells are produced. The sec-

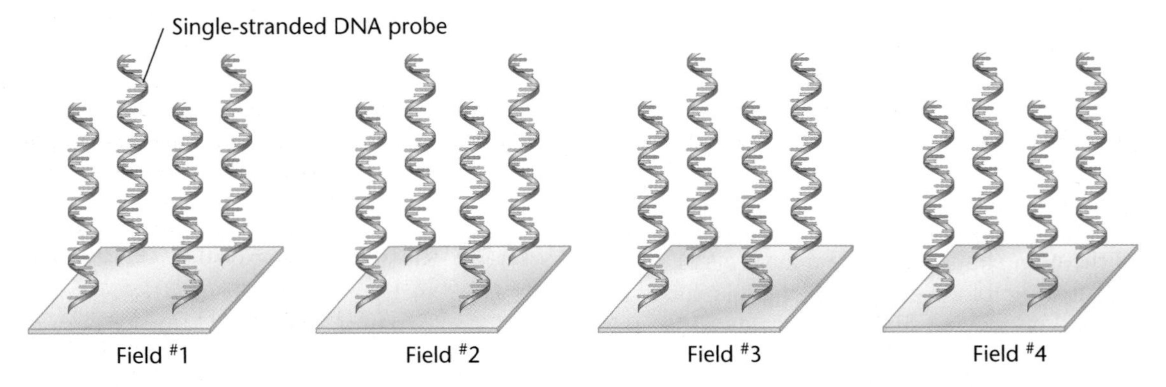

FIGURE 19–12 In making a DNA microarray, short, 15–30 base-pair, single-stranded DNA molecules of known sequence are attached to a glass substrate. Each cluster of identical molecules occupies an area known as a field on the microarray. Each field is about half the width of a human hair.

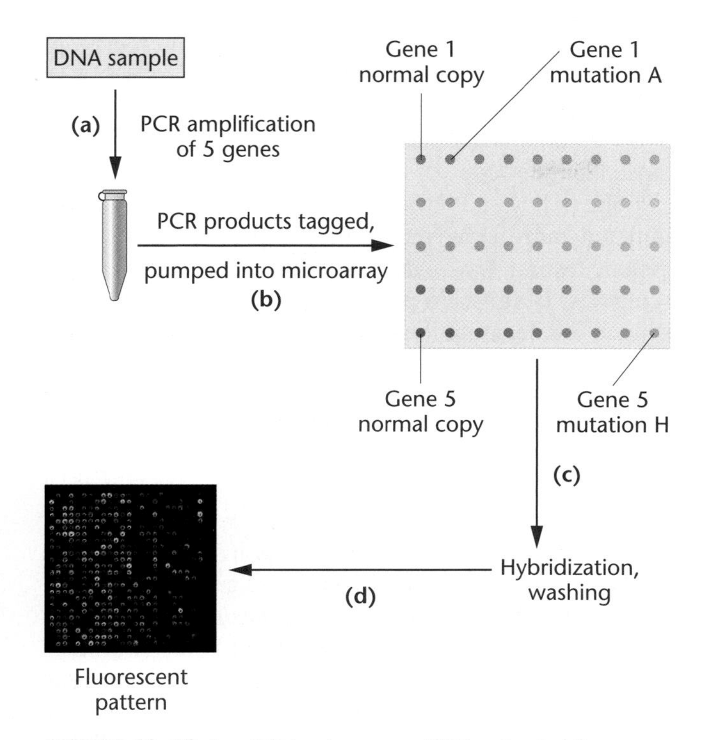

FIGURE 19–13 In a DNA microarray, DNA extracted from a blood sample is amplified by PCR. (a) In this example, primers for five genes are used. (b) The microarray contains single-stranded probes for the normal allele of each of the genes (column 1) and eight mutant alleles for each of the five genes (1 row = 1 gene). (c) The single-stranded PCR products are tagged with fluorescent probes and pumped into the microarray. (d) The resulting hybridization is revealed by the pattern and color of the spots on the microarray. *(Photo Courtesy of Cancer Genetics Branch/National Human Genome Research Institute/NIH)*

ond type, called activated B-like cells, had an expression pattern similar to that seen in B cells within the circulatory system. A plot of patient survival against gene expression patterns revealed that patients with the activated B-like pattern had much lower survival rates (Figure 19–15). This analysis showed that DLCBL is actually two different diseases with different outcomes. The differences in gene expression patterns are being studied to assist in developing drugs specific for each of these cancer types. Similar analyses using microchip arrays are being performed on many types of cancer and may revolutionize the diagnosis and treatment of this disease.

DNA chips have been programmed with up to 30,000 genes in order to study gene expression during embryonic development in the mouse. Chips that carry all the genes in the human genome (estimated to be 30,000--35,000) have also been developed. This technology, called genome scanning, makes it possible to analyze someone's DNA for dozens or hundreds of disease alleles, including those that predispose the person to heart attacks, diabetes, Alzheimer disease, and other genetically defined disease subtypes. Genome scanning is expected to become widely available in the next few years. It will then be possible to scan an individual's genome and define risks for specific diseases years or decades before such conditions appear.

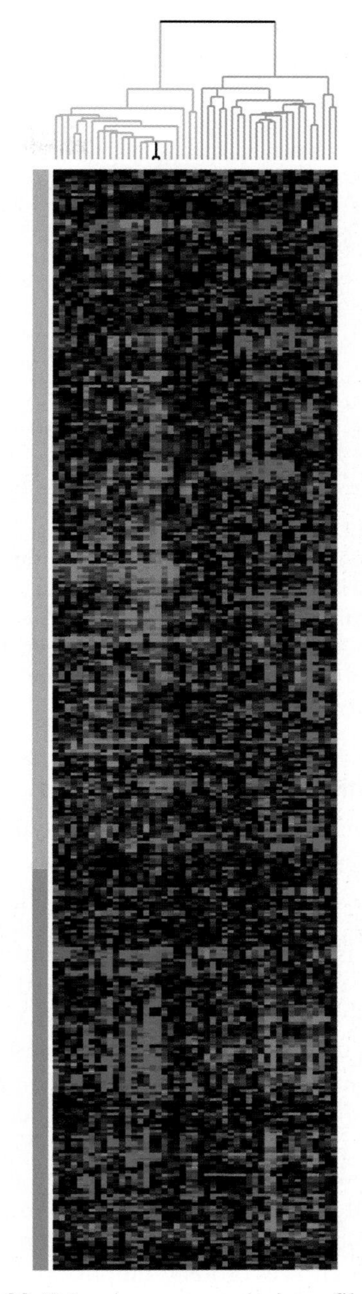

FIGURE 19–14 This microarray analysis profiled 18,000 genes expressed in normal and cancerous lymphocytes. The analysis was repeated to cover 1.8 million individual assays of gene expression. The malignant cells fall into two clusters. The orange cluster contains cells with expression profiles for GC B-like DLBCL cells. The blue cluster contains cells with expression profiles for activated B-like DLBCL cells. The cells within each cluster (shown across the top of the figure) are grouped by how closely their expression profiles resemble each other. The closer together they are, the more closely the profiles resemble each other. The colors represent ratios of relative gene expression compared to normal control cells. Red represents expression greater than the mean level in controls, green represents expression lower than the mean level in controls, and the color intensity represents the magnitude of the difference from the mean. *(From Allzadeh, A. et al. 2000. Distinct types of diffuse large B-cell lymphoma identified by gene expression profiling.* Nature *403: 503–11. Figure 3c, p. 507.)*

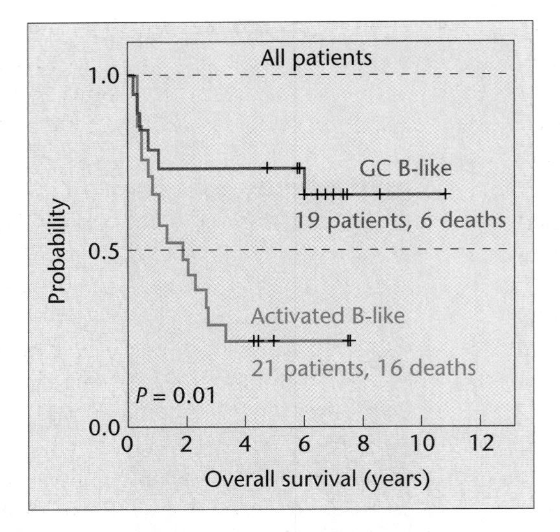

FIGURE 19–15 This graph sorted patients by gene expression profile and survival probability. Those with activated B-like profiles have a much higher rate of death (16 in 21) than those with GC B-like profiles (6 in 19). *(From Allzedah, A. et al. 2000. Distinct types of diffuse large B-cell lymphoma identified by gene expression profiling.* Nature *403: 503–11. Figure 5a, p. 509, http://www.nature.com).*

Genetic Testing and Ethical Dilemmas

We have considered examples of two types of genetic testing: prenatal diagnosis and screening for heterozygous carriers of recessive disorders. With current technology, genetic testing can also be used to predict someone's risk of disease, thereby identifying people who are presently healthy but at high risk of contracting a genetic disease in the future. DNA microarrays are capable of testing for 50–100 diseases at a time, including many that may not develop for years. These advances will affect our health, reproductive patterns, and medical care in fundamental ways. The use of this technology also raises legal, social, and ethical issues that will not be easy to resolve. For example, what should people know before deciding to have a genetic test? How can we protect the information that is revealed by a genetic test? How can we define and prevent genetic discrimination? We know that heterozygotes for sickle-cell anemia are more resistant to malaria than people who are homozygous for the normal allele; this type of protection may be true of other genetic disorders as well. How do we keep the beneficial aspects of mutations as we strive to eliminate their destructive aspects? Some mutations have horrific consequences; others gave rise to our very existence as humans. We know a great deal, yet we know very little. Thoughtful and wide-ranging public debate is essential as we explore the use of biotechnology.

Many of the potential risks and benefits of genetic testing are still unknown. We can test for many genetic diseases for which there are no effective treatments to cure or mitigate the clinical consequences. Should we test people for these disorders? With present technology, a negative result does not necessarily rule out future development of a disease; nor does a positive result always mean that an individual will get the disease. How can we effectively communicate the results of testing and the risks to those being tested?

Public policy and laws on genetic testing are being formulated more slowly than the technology and use of genetic testing. Lawmakers and other groups made up of scientists, health-care professionals, ethicists, and consumers are debating these issues and formulating policy options.

19.4 Genetic Disorders Can Be Treated by Gene Therapy

For decades, *gene products* such as insulin have been made for the therapeutic treatment of genetic disorders. Methods for transferring specific *genes* into mammalian cells, originally developed as research tools, are now being used to treat genetic disorders, a process known as **gene therapy**. In theory, gene therapy transfers a normal allele into a somatic cell that carries one or more mutant alleles. Expression of the normal allele results in a functional gene product whose action produces a normal phenotype. Delivery of these structural genes and their regulatory sequences is accomplished by using a vector or gene transfer system.

In the first generation of gene-therapy trials, the most common method of gene transfer used genetically modified retroviruses as vectors. The first retroviral vectors were based on a mouse virus called Moloney murine leukemia virus (MLV) (Figure 19–16). To create this vector, a cluster of three genes was removed from the

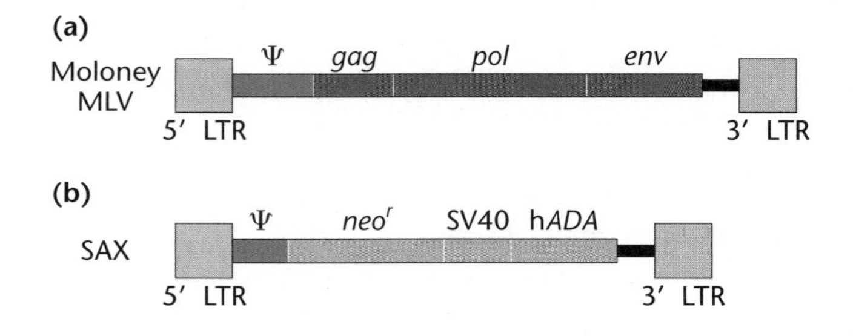

FIGURE 19–16 The native Moloney MLV genome contains a ψ sequence required for encapsulation, as well as genes that encode viral coat proteins (*gag*), an RNA-dependent DNA polymerase (*pol*), and surface glycoproteins (*env*). At each end, the genome is flanked by long terminal repeat (LTR) sequences that control transcription and integration into the host genome. The SAX vector retains the LTR and ψ sequences and includes a bacterial neomycin resistance (neo^r) gene that can be used as a selective marker. As shown, the vector carries a cloned human adenosine deaminase (h*ADA*) gene, which is fused to an SV40 early region promoter–enhancer. The SAX construct is typical of retroviral vectors that are used in human gene therapy.

virus, allowing a cloned human gene to be inserted. After being packaged into a viral protein coat, the recombinant vector can infect cells but cannot replicate itself because of the missing viral genes. Once inside the cell, the recombinant virus with the inserted human gene moves to the nucleus of a cell and integrates into a chromosome, where it becomes part of the genome. In initial attempts at gene therapy, several heritable disorders, including **severe combined immunodeficiency (SCID)**, **familial hypercholesterolemia**, and **cystic fibrosis**, were treated. Let's examine the first attempt to use gene therapy to treat a young girl with SCID and then review the trials that are currently under way using a new generation of viral vectors.

Gene Therapy for Severe Combined Immunodeficiency (SCID)

Gene therapy began in 1990 with the treatment of a young girl named Ashanti DeSilva (Figure 19–17), who has a genetic disorder called severe combined immunodeficiency (SCID). Affected individuals have no functional immune system and usually die from what would normally be minor infections. An autosomal form of SCID is caused by a mutation in the gene encoding the enzyme **adenosine deaminase (ADA)**. Ashanti's gene therapy began with the isolation of white blood cells, or T cells (Figure 19–18). These cells, which are part of the immune system, were mixed with a retroviral vector carrying an inserted copy of the normal *ADA* gene. The virus infected the T cells, and a normal copy of the *ADA* gene was inserted into the genome of some of the T cells. The T cells were first grown in the laboratory and assayed to ensure that the transferred *ADA* gene was expressed. The final step in gene therapy was to inject a billion or so of the altered T cells into her bloodstream. Some of the treated T cells migrated to Ashanti's bone marrow and began dividing. She has maintained normal ADA protein levels in 25–30% of her T cells, and now leads a normal life.

FIGURE 19–17 Ashanti DeSilva, the first person to be treated by gene therapy *(Courtesy of Van de Silva).*

Unfortunately, a second child treated a short time later produced the normal gene in only 0.1–1% of her white blood cells after treatment—a level not high enough to be effective. In later trials, attempts were made to transfer the *ADA* gene into the bone marrow cells that form T cells, but these attempts were mostly unsuccessful. Although gene therapy was originally developed as a treatment for single-gene (monogenic) inherited diseases, this technique was rapidly adapted for the treatment of acquired diseases such as cancer, neurodegenerative diseases, cardiovascular disease, and infectious diseases, such as HIV. As a result, most gene therapy treatments and trials involve these disorders (Figure 19–19a). In fact, gene therapy is used to treat cancer more often than any other condition.

In addition to retroviral vectors, other viruses and other methods are being used to transfer genes into human cells (Figure 19—19b). These methods include the use of viral vectors, chemically assisted transfer of genes across cell membranes, and fusion of cells with artificial vesicles that contain cloned DNA sequences.

Problems and Failures in Gene Therapy

Over a 10-year period, from 1990 to 1999, more than 4000 people underwent gene transfer for a variety of genetic disorders. These trials often failed and thus led to a loss of confidence in gene therapy. Hopes for gene therapy plummeted even further in September 1999 when a teenager, Jesse Gelsinger, died during gene therapy. His death was triggered by a massive inflammatory response to the vector, a modified adenovirus. (Adenoviruses cause colds and respiratory infections.)

In 2000, a group of French researchers reported the first large-scale success in gene therapy. Using a retroviral vector to treat a fatal X-linked form of SCID, three patients developed immune systems after they were treated with a retroviral vector carrying a normal gene. This success was repeated in subsequent treatments of several other patients, but two of these patients later developed leukemia-like disorders. An analysis of cancerous cells from the two patients showed that the retroviral vector had inserted near or into an oncogene called *LMO2*. This insertion activated the oncogene, causing uncontrolled white blood cell proliferation and development of the leukemia-like disorder.

The Future of Gene Therapy: New Vectors and Target-Cell Strategies

Most problems with gene therapy have been traced to the vectors. These first-generation vectors, such as MLV and adenovirus, have several drawbacks. First, integration of the retroviral genome (including the cloned human gene) into the host cell's genome occurs only if the host cells

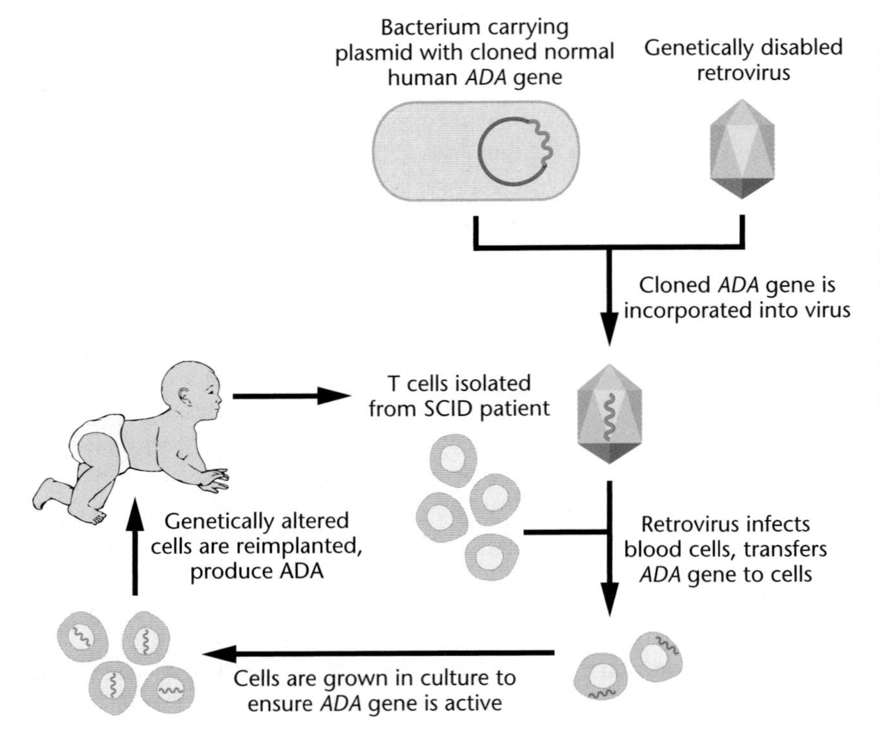

FIGURE 19–18 To treat SCID using gene therapy, a cloned human *ADA* gene is transferred into a viral vector, which is then used to infect white blood cells removed from the patient. The transferred *ADA* gene is incorporated into a chromosome and becomes active. After growth to enhance their numbers, the cells are reimplanted into the patient, where they produce ADA, allowing the development of an immune response.

are replicating their DNA. Under *in vivo* conditions, there is little DNA synthesis in many of the highly differentiated cell types that are suitable target tissues. Second, most of these viruses eventually elicit an immune response in the host, as happened in Jesse Gelsinger's case. Third, insertion of viral genomes into the host chromosome can inactivate or mutate an indispensable gene, the plight of the two French patients. Fourth, retroviruses have a low cloning capacity and cannot carry inserted sequences much larger than 8 kb—many human genes, even discounting introns, exceed this size. Finally, there is the possibility of producing an infectious virus if a recombination event takes place between the vector and any retroviral genomes already present in the host cell.

The disappointing and deadly results from gene therapy trials that used first-generation vectors led to a widespread crisis of confidence in gene therapy in the medical and other scientific communities. To overcome these prob-

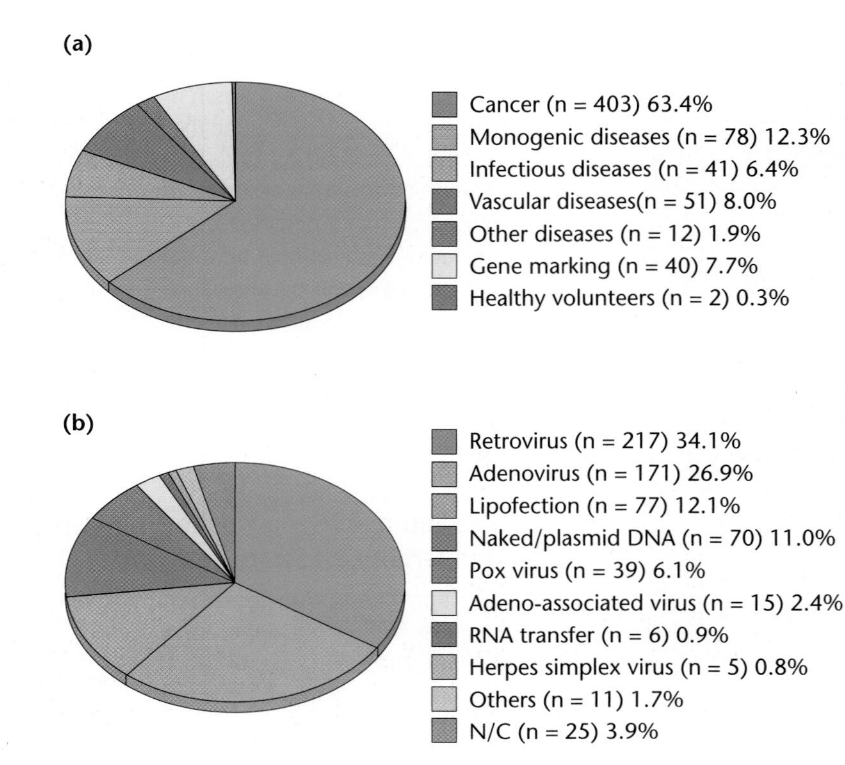

FIGURE 19–19 (a) Pie chart summarizing, by disease, over 630 gene therapy trials under way worldwide. Most trials involve cancer treatment. (b) Gene therapy trials ranked by the vectors used. Retroviruses are the most widely used vectors, accounting for 34% of the total. Newer vectors, such as adeno-associated virus account for only a small percentage of vectors. N/C= not classified *(From Gene Therapy; Clinical Trials; Charts, and Statistics,* Journal of Gene Medicine *website. 2003. © John Wiley & Sons Limited. Reproduced with permission.)*

TABLE 19.2 New Vectors for Gene Therapy

Vector	*Cell Targets*	*Cloning Capacity*	*Advantages*	*Disadvantages*
Adenovirus	Lung, respiratory tract	7.5 kb	Efficient transfection	Strong immune response
Adeno-associated virus	Fibroblasts, T cells, others	4.5 kb	Transfects many cell types	Small insert size
Retroviruses	Proliferating cells	8 kb	Prolonged expression	Low transfection efficiency
Lentiviruses	Stem cells, proliferating cells	8 kb	Efficient transfection	Related to HIV

lems, new viral vectors and strategies for targeting cells are being developed. The properties of current viral vectors and those under development are summarized in Table 19.2. In general, these vectors fall into two categories: those that integrate into the host-cell genome (retroviruses and others) and those that remain in the nucleus but do not integrate into the chromosomes (adeno-associated virus and others). Researchers hope that the use of new vectors will circumvent several of the problems encountered with earlier vectors and will also have new design features to allow regulation of insertion sites and the levels of gene product.

In one application of new strategies for gene therapy, scientists reported the successful production of the blood hormone erythropoeitin in rhesus monkeys and mice whose muscle cells had been injected with a viral vector containing the erythropoeitin gene. These animals produced the hormone only when they also received the antibiotic rapamycin. These results are particularly encouraging because the stimulated levels of the hormone are quite high, the presence of the foreign DNA has not triggered an immune response, and the gene can be repeatedly activated.

This trial used one of the most promising new viral vectors—adeno-associated virus (AAV). This virus can enter nonreplicating cells, does not integrate into the host-cell genome, and is a very small virus, so it does not typically elicit an immune response. In this case, the vector incorporated a rapamycin-responsive promoter that is used to switch on the erythropoietin gene. The gene is expressed only when cells are exposed to rapamycin. This therapy is being considered for patients who have low blood counts, such as dialysis patients who currently receive regular injections of the erythropoietin hormone.

19.5 Gene Therapy Raises Many Ethical Concerns

Gene therapy raises many ethical concerns, and many therapies are still sources of intense debate. As a result, all gene therapy trials currently under way or in the planning stages are restricted to the use of somatic cells as targets for gene transfer. This form of gene therapy is called **somatic gene therapy**; only one individual is affected, and the therapy is done with the permission and informed consent of the patient. The ethical guidelines for gene therapy, as it is currently performed, have been revised and strengthened in the wake of Jesse Gelsinger's death. This sort of therapy can be initiated only after careful review by several levels of administrators, and the trials are monitored to protect the interests of the patient.

Two other forms of gene therapy have *not* been approved, primarily because of the unresolved ethical issues surrounding them. The first is called **germline therapy**, whereby germ cells (the cells that give rise to the gametes, i.e., sperms and eggs) or mature gametes are used as targets for gene transfer. In this approach, the transferred gene would be incorporated into all of the individual's cells that are produced from the genetically altered gamete, including his or her own germ cells. This means that individuals in future generations will also be affected, without their consent. Is this procedure ethical? Do we have the right to make this decision for future generations? Thus far, the concerns have outweighed the potential benefits, and such research is prohibited.

The second unapproved form of gene therapy—which raises an even greater ethical dilemma—is termed **enhancement gene therapy**, whereby human potential might be enhanced for some desired trait. This use of gene therapy is extremely controversial and is strongly opposed by many people. Should genetic technology be used to enhance human potential? For example, should it be permissible to use gene therapy to increase height, enhance athletic ability, or extend intellectual potential? Presently, the consensus is that enhancement therapy, like germline therapy, is an unacceptable use of gene therapy. However, there is an ongoing debate and the issues are still unresolved. For example, the U. S. Food and Drug Administration has permitted the use of growth hormone produced through recombinant DNA technology as a growth enhancer, in addition to its current medical use for the treatment of growth-associated genetic disorders. Critics charge that the use of a gene product for enhancement will lead to the use of transferred genes for the same purpose. The outcome of these debates may affect not only the fate of individuals but that of our species as well.

19.6 Ethical Issues Are an Outgrowth of the Human Genome Project

Geneticists now use recombinant DNA technology to identify genes, diagnose genetic disorders, screen populations for heterozygous carriers, and treat disorders by gene therapy. Knowledge gained by sequencing the human genome will greatly advance our understanding of human genetic, and will have a great impact on biomedical research and health care. However, applications of the knowledge gained from the project raise ethical, social, and legal issues that must be identified, debated, and resolved. Resolutions often take the form of laws or public policy. The ethical debate surrounding gene therapy discussed in the previous section represents a subset of the broader ethical issues raised by knowledge gained as an outcome of the Human Genome Project.

The Ethical, Legal, and Social Implications (ELSI) Program

When the **Human Genome Project** (**HGP**) was first discussed, scientists and the general public raised concerns about how genome information would be used and how the interests of both individuals and society can be protected. To address these concerns, the **Ethical, Legal, and Social Implications** (**ELSI**) **Program** was established as an adjunct to the Human Genome Project. The ELSI program considers a number of issues, including the impact of genetic information on individuals, the privacy and confidentiality of genetic information, implications for medical practice, genetic counseling, and reproductive decision making. Through research grants, workshops, and public forums, ELSI is formulating policy options to address these issues.

ELSI focuses on four areas: (1) privacy and fairness in the use and interpretation of genetic information, (2) ways of transferring genetic knowledge from the research laboratory to clinical practice, (3) ways to ensure that participants in genetic research know and understand the potential risks and benefits of their participation and give informed consent, and (4) public and professional education. Hopefully, as the HGP moves from generating information about the genetic basis of disease to improving treatments, promoting prevention and developing cures, these and other ethical issues will be identified and an international consensus will be developed on appropriate policies and laws.

19.7 Mapping Human Genes with Recombinant DNA Technology

When the first human biochemical genetic disorders were mapped, the phenotype was usually associated with a mutant gene product that could be identified in affected individuals. By using the mutant protein as a marker, pedigree analysis was used to establish the pattern of inheritance and in a few cases, to establish the chromosomal locus of the gene. However, in the majority of human genetic disorders, the function of the normal gene product is unknown, and without a marker, conventional methods of mapping cannot be used. Although the genetic and molecular basis for a handful of diseases was known before the recombinant DNA revolution, real progress in mapping these genes has come only in the last two decades. Now, with our increasingly detailed knowledge of the genome, we can map a gene without information about its product.

MEDIA TUTORIAL Restriction fragment lenght polymorphisms

RFLPs as Genetic Markers

Variations in nucleotide sequences occur throughout the human genome (mostly in noncoding regions) with a frequency of about 1 in 200 nucleotides. These nucleotide changes occur at specific sites and are created by substitutions, deletions, or insertions of one or more nucleotides. Such variations can sometimes create or destroy restriction enzyme cutting sites. If a restriction enzyme site created by nucleotide variation is present on one chromosome but absent on its homolog, the two chromosomes can be distinguished from one another by their pattern of restriction fragments on a Southern blot (Figure 19–20). The region of

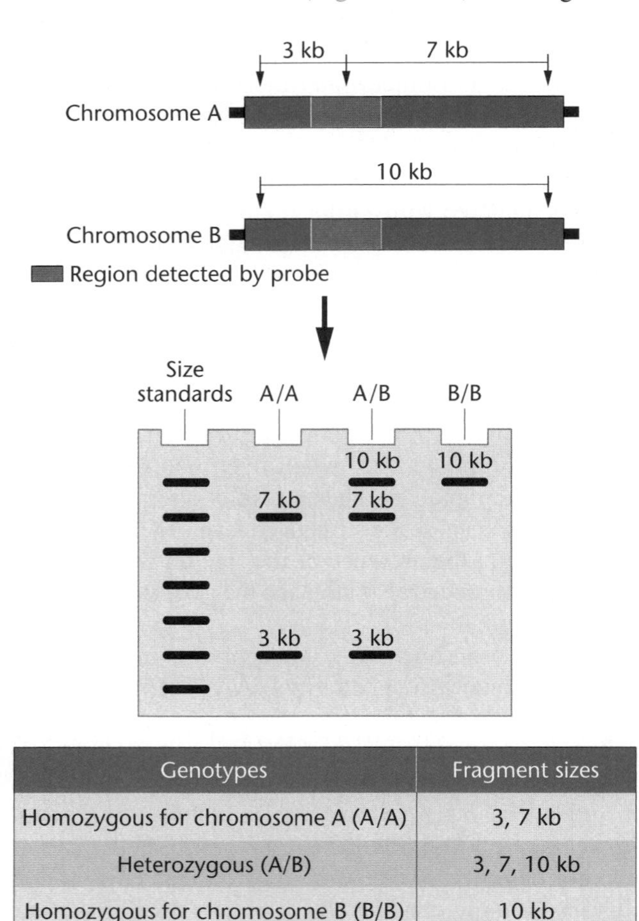

Genotypes	Fragment sizes
Homozygous for chromosome A (A/A)	3, 7 kb
Heterozygous (A/B)	3, 7, 10 kb
Homozygous for chromosome B (B/B)	10 kb

FIGURE 19–20 The alleles on chromosomes A and B represent DNA segments from homologous chromosomes. The region that hybridizes to a probe is shown in green; arrows indicate the location of restriction enzyme cutting sites that define the alleles. On chromosome A, three cutting sites generate fragments of 7 kb and 3 kb. On chromosome B, only two cutting sites are present, generating a single fragment that is 10 kb in length. The absence of the cutting site in chromosome B could be the result of a single base mutation within the enzyme recognition or cutting site. Because these differences in restriction cutting sites are inherited in a codominant fashion, there are three possible genotypes: *AA*, *AB*, and *BB*. The allele combination carried by any individual can be detected by restriction digestion of genomic DNA (obtained, for example, from a blood sample or skin fibroblasts), followed by gel electrophoresis, transfer to a DNA binding filter, and hybridization to the appropriate probe. The fragment patterns for the three possible genotypes are shown as they would appear on a Southern blot.

chromosome A shown in the figure contains three *Bam*HI sites; its homolog, chromosome B, contains two such sites. When chromosome A is cut with *Bam*HI, 3-kb and 7-kb fragments are generated, whereas only a single 10-kb fragment is generated when chromosome B is cut. These fragments can be visualized on a Southern blot by using a probe from this chromosomal region.

Variations in DNA fragment length generated by restriction enzymes are called **restriction fragment-length polymorphisms (RFLPs)**. RFLPs are quite common; thousands have been identified in the human genome, and most have been assigned to specific chromosome regions. These variations are inherited as codominant alleles and can be used as markers to follow the inheritance of genetic disorders from generation to generation in an affected family.

RFLPs can be used to map the chromosomal locus of a genetic disorder if the RFLP cosegregates with the genetic disorder in a multigenerational family (three generations or more). Selecting which RFLP to use in the family being studied is a matter of trial and error. Many different RFLPs may have to be tested to find a set for which most individuals of the family are heterozygous. If family members are heterozygous for an RFLP assigned to a specific chromosome, each member of the chromosome pair being tested can be identified as it passes from generation to generation. This provides a way to establish linkage between a specific chromosome (identified by the RFLP marker) and the phenotype of the disease.

Linkage Analysis Using RFLPs

To map a genetic disorder, the inheritance of a given RFLP and the disorder are traced through a family (Figure 19–21). If loci for the RFLP and the disorder are near each other on the same chromosome, they will show linkage and be inherited together. However, most RFLPs will not show linkage to the disease phenotype because RFLP and the disease loci are on different chromosomes or there is frequent recombination between the RFLP and the disease loci. These data however, are not useless— they help to identify chromosomes and chromosome regions that do *not* carry the disease locus.

For analysis of RFLP linkage with a genetic disorder, researchers use methods of probability and statistics. The probabilities are ultimately expressed as the **logarithm of the odds**, and the method of analysis is referred to as the

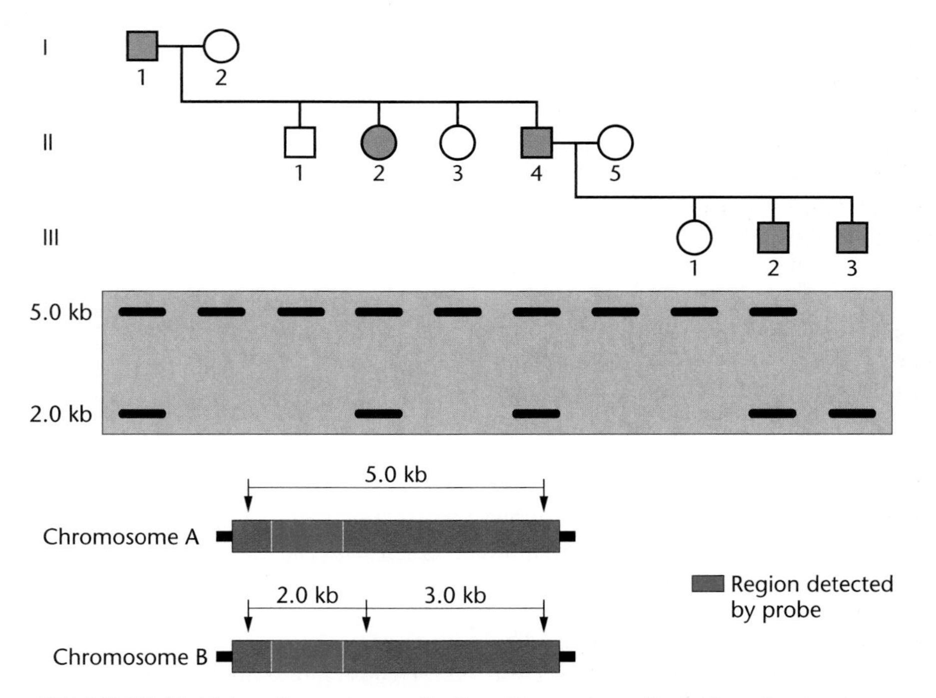

FIGURE 19–21 This pedigree shows a family with members affected by a dominant trait (filled symbols). Family members also carry two alleles of an RFLP locus assigned to a specific chromosome. A 5.0-kb allele (allele *A*) is present on one homolog; the *B* allele on the other homolog consists of two fragments (2.0 kb and 3.0 kb). The probe used in the Southern blot detects the 5-kb *A* allele and the 2.0-kb portion of the *B* allele. The RFLP pattern for each family member is shown below the appropriate pedigree symbol. Individual III-1, who is unaffected, probably received an *A* allele from her father and a *B* allele from her mother. Individual III-2 is affected and probably received an *A* allele from his mother and a *B* allele from his father. The youngest son (III-3), who is affected, received a *B* allele from each parent. The pedigree and Southern blot suggest that the mutant allele for the dominant trait and the RFLP *B* allele are on the same homolog and are therefore linked. Assigning a mutant allele to a chromosome by RFLP analysis is the first step in mapping a gene.

lod score method. A lod score of 3 means that it is 1000 times more likely that the gene and the RFLP marker are linked than not linked; a lod score of 4 means that the odds are 10,000 times greater in favor of linkage. Lod scores of 3-4 are usually taken as evidence that the gene and the marker are linked.

Many genes have been mapped in humans by using techniques such as RFLP analysis, and by compiling studies from many families, the location of many markers can be determined and genetic maps for human chromosomes can be constructed (Figure 19–22). The unit for linkage is the **centimorgan (cM)**, named after the geneticist T. H. Morgan. One centimorgan is equal to a recombination frequency of 1% between two loci.

Positional Cloning: The Gene for Neurofibromatosis

The search for the chromosomal locus for the gene causing **type 1 neurofibromatosis (NF1)** is an example of RFLP analysis in gene mapping. NF1 is inherited as an autosomal dominant condition affecting 1 in 3,000 individuals, and is associated with nervous system defects, including benign tumors and a high incidence of learning disorders.

The *NF1* gene was mapped by RFLP analysis in several steps. First, a group of cooperating research laboratories around the world compared the inheritance of NF1 with dozens of RFLP markers in families with NF1. Each RFLP marker represented a specific human chromosome or chromosome region. The first round of analysis did not identify the chromosome carrying the *NF1* gene, but did produce an **exclusion map**, indicating which RFLPs were not linked to the disease and therefore which chromosomes do *not* carry the NF1 locus. This analysis also produced some evidence of linkage and pointed to chromosomes 5, 10, and 17 as sites where the *NF1* gene might reside. The second step focused attention on these three chromosomes and produced conclusive evidence that the disorder is closely linked to an RFLP near the centromere of chromosome 17 (Figure 19–23). In a third step, more than 30 RFLP markers from this region were used to analyze 13,000 individuals from NF1 families, and the gene was mapped to region 17q11.2. Finally, the locus for NF1 was identified by using a collection of genomic clones that spanned a small region of 17q11.2. The gene's identity was confirmed when mutant versions of the gene were found in individuals with NF1.

Once the gene was identified, the amino acid sequence of the gene product was reconstructed from the DNA sequence. Databases with protein-sequence information were scanned to identify similar proteins in other organisms. Researchers determined that the NF1 gene product is similar to proteins that play a role in signal transduction. Further analysis confirmed that the NF1 protein, **neurofibromin**, is involved in the transduction of intracellular signals and the down-regulation of a gene that controls cell growth. Mutation in the *NF1* gene leads to a loss of control of cell growth, causing the production of the small tumors that are characteristic of this disorder.

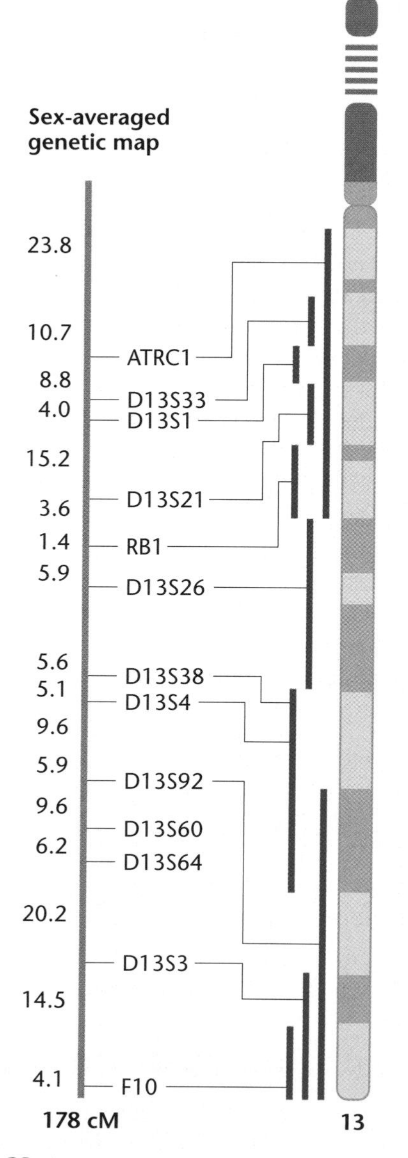

FIGURE 19–22 A genetic and a physical map of human chromosome 13, showing the location of markers. The genetic map for females is 203 cM, and that for males is 158 cM, reflecting the difference in recombination frequencies between females and males. When the two maps are averaged together, the result is the sex-averaged map of 178 cM shown on the left. The location of markers on the physical map is indicated by the blue lines adjacent to the chromosome.

Using recombinant DNA is a departure from previous methods of mapping, which worked from an identified gene product to the gene locus. The mapping, cloning, and sequencing of the NF1 gene and the identification of the gene product took a little over three years, and researchers had begun this project with no direct knowledge of the nature of the gene product or the mutational events that produce the NF1 phenotype. The mapping and cloning process of this gene is an example of **positional cloning**, where the gene can be mapped, isolated, and cloned with no prior knowledge of the gene product. Through use of this strategy, an ever-increasing number of human genes are being mapped and isolated.

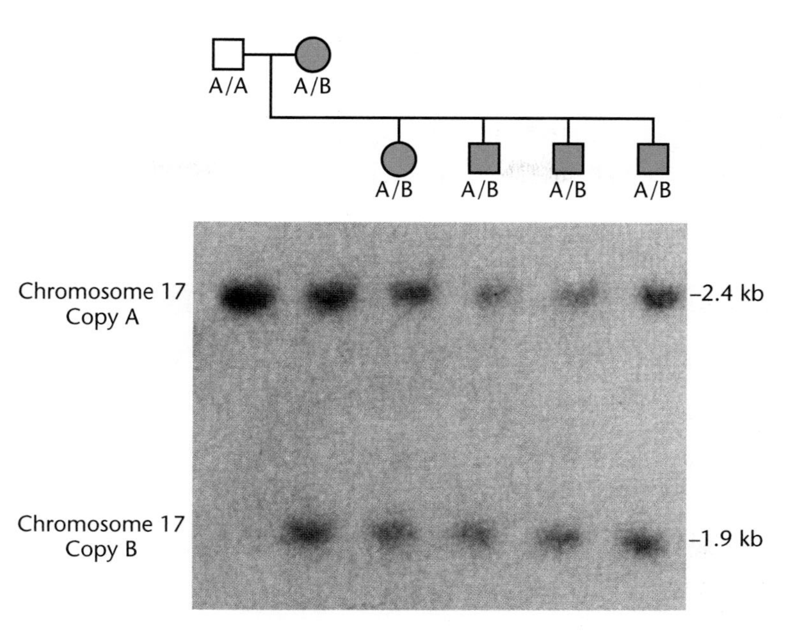

FIGURE 19–23 Segregation of a 1.9-kb RFLP allele with type 1 neurofibromatosis (NF1) in each of four affected offspring and their mother. This RFLP is detected by probe pA10-41, which is a DNA segment from a region near the centromere of human chromosome 17. On the basis of this result and also results from tests using other probes, the locus for NF1 was assigned to chromosome 17. *(From Barker, D. et al. 1987. Gene for von Recklinghausen neurofibromatosis is in the pericentromic region of chromosome 17. Science 236: 1100-1102. Fig 1 (c) 1987 by the American Association for the Advancement of Science.)*

19.8 DNA Fingerprints Can Identify Individuals

As discussed in the previous section, the presence or absence of restriction sites in the human genome can be used as genetic markers. Another type of nucleotide sequence variation, discovered in the mid-1980s, depends on variation in the length of repetitive DNA sequence clusters. These polymorphisms in human DNA serve as the basis for **DNA fingerprinting**. DNA fingerprinting is now used for a wide variety of applications, from identifying paternity to forensics to conservation biology.

Minisatellites (VNTRs) and Microsatellites (STRs)

One of the most useful forms of RFLPs arises from variations in the number of tandemly repeated DNA sequences present between two restriction enzyme sites. These sequences are **minisatellites**, or clusters of 10–100 nucleotides. For example, the sequence

5′-GACTGCCTGCTAAGATGACTGCCTGCTAA
GATGACTGCCTGCTAAGATGACTGCCTGCT
AAGATGACTGCCTGCTAAGATGACTGCCTGC
TAAGATGACTGCCTGCTAAGATGACTGCCTGC
TAAGATGACTGCCTGCTAAGAT-3′

is composed of 9 tandem repeats of the 16 base-pair sequence GACTGCCTGCTAAGAT. Clusters of such sequences are widely dispersed in the human genome, and the number of repeats at each locus ranges from 2 to more than 100. These loci, known as **variable-number tandem repeats (VNTRs)** were introduced in Chapter 12 as examples of middle repetitive DNA. The number of repeats at a given locus is variable, and each variation constitutes a VNTR allele. Many loci have dozens of alleles each; as a result, heterozygosity is common.

A pattern of bands is produced when DNA is cut with restriction enzymes and visualized by Southern blotting using VNTR sequences as probes. This pattern is the DNA fingerprint (Figure 19–24) and is the equivalent of actual fingerprints because the pattern of bands is always the same for a given individual, no matter what tissue is used as the source of the DNA, but the pattern varies from individual to individual, as do real fingerprints. In fact, there is so much variation in the band pattern from individual to individual that, theoretically, each person's pattern is unique. This technique was developed during the 1980s and was first used to solve the murders of two schoolgirls in a high-profile case in Great Britain.

An important limitation of DNA fingerprint analysis is that it requires a relatively large sample of DNA (10,000 cells or about 50 μg)—more than is usually found at a typical crime scene—and the DNA must be relatively intact (non degraded). Thus, DNA fingerprinting has been more useful in the area of paternity testing than in forensics. In paternity testing, blood drawn from the child, mother, and alleged father provide an unlimited source of fresh, intact cells for DNA extraction and analysis.

Forensic scientists have recently developed a different set of markers that are analyzed by PCR (rather than RFLP analysis), allowing trace evidence samples to be typed. These markers, short tandem repeats (STRs), are very similar to VNTRs, but the repeated motif is shorter— between two and nine base pairs. Thirteen tetrameric (four base-pair repeat) STRs have been developed into a marker panel (called the CODIS panel) that is currently used by the FBI and other law enforcement agencies to do DNA typing of crime suspects (Figure 19–25) and create a database of the DNA profiles of convicted felons. It is also used routinely in forensic casework to generate DNA profiles from trace samples (e.g., single hairs, saliva left on a cigarette butt or toothbrush) and from samples that are old and/or degraded (e.g., a skull found in a field, ancient Egyptian mummies). As a result, STRs have replaced VNTRs in most forensic laboratories. In addition, STR typing is less expensive, less labor-intensive, and much faster than

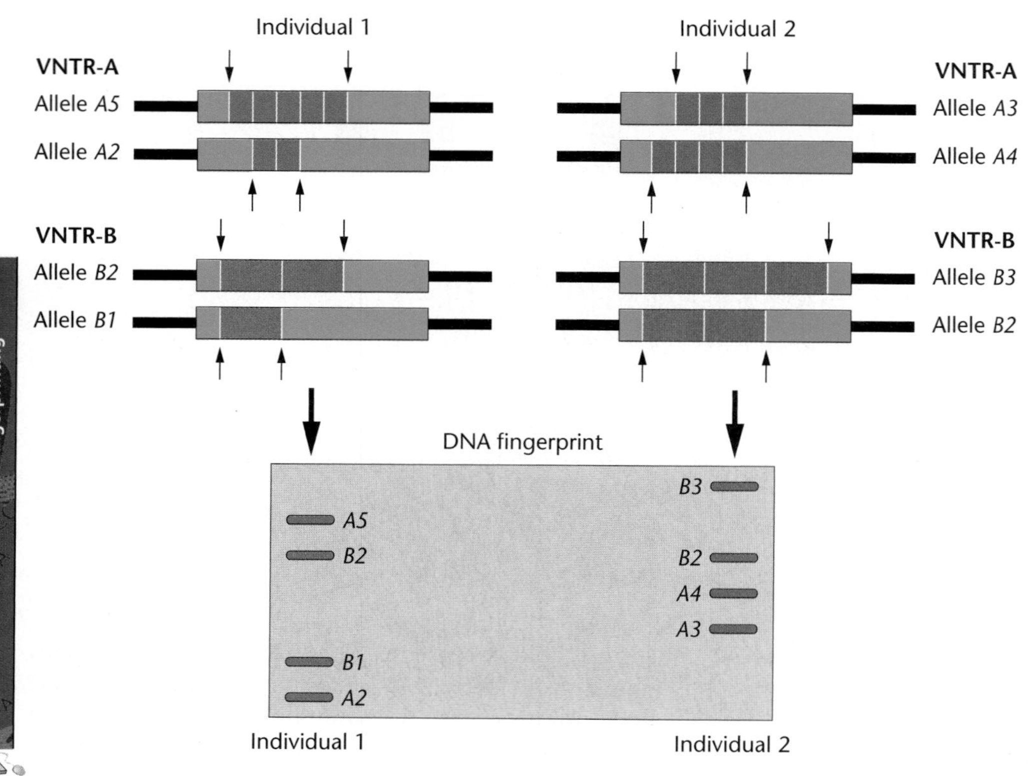

FIGURE 19–24 VNTR alleles at two loci (*A* and *B*) are shown for each individual. (Arrows mark restriction enzyme cutting sites flanking the VNTRs.) Restriction enzyme digestion produces a series of fragments that can be detected as bands on a Southern blot (bottom). Differences in the number of repeats at each locus differ, so the overall pattern of bands is distinct for each individual, even though one band is shared (the *B2* allele band). The pattern is a DNA fingerprint.

MEDIA TUTORIAL DNA fingerprinting

RFLP analysis, so it is rapidly replacing VNTR typing in most paternity laboratories as well.

The results of STR analysis are analyzed and interpreted using statistics, probability, and population genetics. The population frequency of each STR allele in the standard set has been measured in many groups of people across the United States. Using this information, the probability of having any combination of these alleles can be calculated. For example, if an allele of locus 1 is carried by 1 in 333 individuals, and an allele of locus 2 is carried by 1 in 83 individuals, the probability that someone will carry both alleles is equal to the product of their individual frequencies, or 1 in 28,000. This overall probability may not be especially convincing, but if an allele of a third locus is carried by 1 in 100 individuals and an allele of a fourth locus is carried by 1 in 25 individuals and included in the calculations, the combined frequency becomes 1 in 70 million. That is, only about 4 individuals in the U.S. population will carry this combination of alleles. When the genotype of the alleles of all thirteen CODIS STR loci (26 alleles in all) are generated to produce a complete STR profile, the chance of anyone having the same combination is 1 in 100 trillion. Since the planet's population is only about 6 billion, it is easy to see why STR analysis is often referred to as human DNA identification testing.

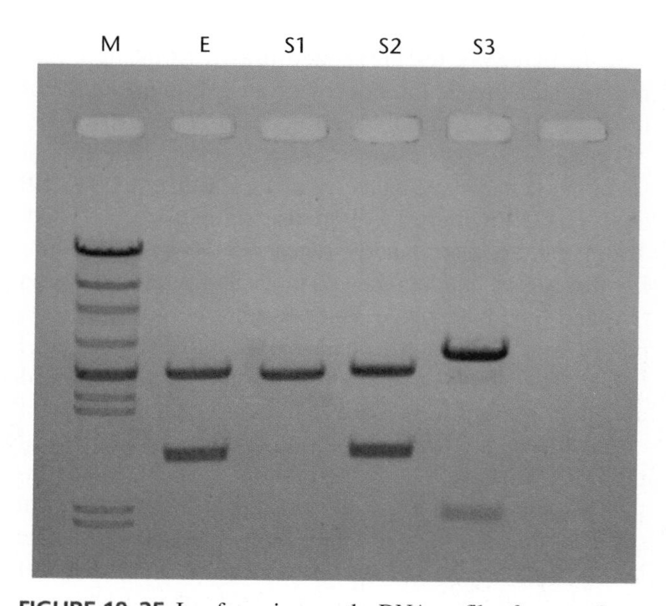

FIGURE 19–25 In a forensic case, the DNA profile of suspect 2 (S2) matches the DNA of the blood sample obtained as evidence. M=markers, E=evidence, S1=suspect 1, S2=suspect 2, S3=suspect 3. *(Edvotek-The Biotechnology Education Company)*

Forensic Applications of DNA Fingerprints

In a criminal case, if a suspect's DNA fingerprints do not match those of the evidence, that individual can be excluded as the criminal (exclusions occur in about 30% of all cases). When a match *is* made between DNA obtained at a crime scene and the DNA of a suspect, there are two possible interpretations: The DNA fingerprint came from the suspect, or it is from someone else with the same pattern of bands. The DNA fingerprint evidence does not prove the suspect is guilty; it is just one piece of evidence that must be considered along with the other facts in the case. However, with the advent of STR typing, more forensic scientists (particularly in the United States) are willing to offer their expert opinion and testify that a DNA profile could have come only from one person and that it provides a direct link between the crime scene sample and a suspect.

GENETICS, TECHNOLOGY, AND SOCIETY

Gene Therapy—Two Steps Forward or Two Steps Back?

In September 1999, 18-year-old Jesse Gelsinger received his first dose of gene therapy. Large numbers of adenovirus vectors bearing the ornithine transcarbamylase (*OTC*) gene were injected into his hepatic artery. The virus vectors were expected to lodge in his liver, enter liver cells and trigger the production of OTC protein. In turn, the OTC protein might correct his genetic defect and perhaps cure him of his liver disease. However, within hours, a massive immune reaction surged through Jesse's body. He developed a high fever, his lungs filled with fluid, multiple organs shut down, and he died four days later of acute respiratory failure. In the aftermath of the tragedy numerous government and scientific inquiries were initiated. Investigators learned that clinical trial scientists had not reported many other adverse reactions to gene therapy, and that some of the scientists were affiliated with private companies that could benefit financially from the trials. They found that serious side effects in animal studies were not explained to patients during informed-consent discussions and some clinical trials were proceeding too quickly in the face of data that suggested caution. The U. S. Food and Drug Administration (FDA) scrutinized gene therapy trials across the country, halted a number of trials and shut down several gene therapy programs. Other research groups voluntarily suspended their gene therapy studies.

Jesse's death dealt a severe blow to the struggling field of gene therapy—a blow from which it was still reeling when a second tragedy hit.

In April 2000, a French group announced that gene therapy for X-linked severe combined immunodeficiency (X-SCID) had succeeded. It was the first unequivocal success in the gene therapy field. In this study, the young patients' bone marrow cells were removed, treated with a retrovirus bearing the γc transmembrane protein gene, and transplanted back to the patients. Nine of 11 patients were cured of their immune deficiency and were able to lead normal lives. Published reports of the study were greeted with enthusiasm by the gene therapy community. But elation turned to despair in 2003, when it became clear that two of the nine children who had been cured of X-SCID had developed leukemia as a direct result of their therapy. The FDA immediately halted 27 similar gene therapy clinical trials and, once again, gene therapy underwent a profound reassessment.

Up until the apparent success of the French X-SCID clinical trials, gene therapy had suffered not only from the scandals and scrutiny that emerged from Jesse Gelsinger's death but also from the skepticism of scientists and the general public about the feasibility of this much-promoted therapeutic technique.

Since the first clinical trial for gene therapy in 1990, almost 500 gene therapy clinical trials involving over 4000 patients have been initiated. These trials aimed to cure cancers, inherited diseases such as hemophilia and cystic fibrosis, and infectious diseases such as AIDS. Despite high expectations and intense publicity, therapeutic benefits have been unclear at best, and more frequently, absent.

The most significant positive outcome of gene therapy came from the first clinical trial in 1990. Ashanti DeSilva, who had received retroviral-transduced T cells for severe combined immunodeficiency (SCID), now leads a normal life, but the reasons for her success are not completely resolved. She had been given a new drug treatment that replaced her missing ADA enzyme prior to and after gene therapy. Hence, it is still not known how much of her cure is due to gene therapy and how much is due to drug treatment.

To date (September, 2003), no human gene-therapy product has been approved for sale. Critics of gene therapy continue to criticize research groups for undue haste, conflicts of interest, sloppy clinical trial management, and for promising much but delivering little. In the mid-1990s, a National Institutes of Health review committee concluded that significant problems remain in all basic aspects of gene therapy, including those aspects that led to Jesse Gelsinger's death and the X-SCID leukemias—adverse immune reactions to viral vectors and the side effects of retroviral vector integration into the host genome.

The question remains whether gene therapy can ever recover from these setbacks and fulfill its promise as a cure for genetic diseases. At present, about 200 gene therapy clinical trials are under way in the United States—most are for cancer, and are in phase I trials (which examine safety and dosage, but not efficacy). Tighter restrictions on clinical trial protocols are designed to correct some of the procedural problems that emerged from the Gelsinger case. In addition, basic science is proceeding with development of safe effective vectors, such as those that insert vector sequences into specific regions of the genome (reducing the possibility of cancer or vector-gene silencing) and those bearing receptors on their surfaces that allow them to infect specific cell types. However, the steps leading to successful gene therapy are many. There is a need to optimize tissue-specific expression of therapeutic genes, to efficiently transduce blood cells in culture, to predict and control immune reactions to vectors, and to develop better animal models in which to test gene therapies prior to clinical trials.

Many scientists feel that we should continue gene therapy research and clinical trials despite the setbacks. However, they have a more sober view of its progress. Clinical trials for any new therapy are potentially dangerous, and often animal studies will not accurately reflect the reaction of individual humans to a new drug or procedure. Inevitably, more adverse reactions to gene therapy will emerge in the clinical trials as methods become more effective. Those in the field now believe that the road ahead will be longer and more difficult than first imagined, but not impossible. Perhaps we should view gene therapy as we have antibiotics, organ transplants, and manned space travel. There will be setbacks and even tragedies, but step by small step, we will move toward a technology that could—someday—provide cures for many severe genetic diseases.

References

Thomas, C. E., Ehrhardt, A., and Kay, M. A. 2003. Progress and problems with the use of viral vectors for gene therapy. *Nat. Rev. Gen.* 4: 346–58.

Web Site

Thompson, L. Sept./Oct. 2000. *Human gene therapy: Harsh lessons, high hopes [on-line]. FDA Consumer Magazine.* ***http://www.fda.gov/fdac/features/2000/500_gene.html***

Chapter Summary

1. The biotechnology industry uses recombinant DNA methods to improve crop plants by transferring herbicide resistance, insect resistance and enhancing the nutritional value of plants. Biotechnology is also used to produce human gene products in a variety of hosts, ranging from bacteria to farm animals. In the near future, it will be possible to use food plants to vaccinate people against infectious disease.
2. Recombinant DNA technology offers a new approach to genetic analysis. Instead of relying on the isolation and mapping of mutant genes, large cloned segments of the genome are manipulated to make genetic and physical maps that use molecular markers rather than phenotypes visible at the level of the organism.
3. Recombinant DNA methods are used in the prenatal diagnosis of human genetic disorders, allowing direct examination of the genotype, whereas previous methods relied on gene expression and the identification of the gene product. It can also identify carriers of genetic disorders, the basis for proposals to screen the population for a number of genetic disorders, including sickle-cell anemia and cystic fibrosis.
4. The availability of cloned human genes has led to their use in gene therapy. In somatic gene therapy, a cloned normal copy of a gene is transferred into a vector, the vector then transfers the gene to a target tissue that takes up and expresses the cloned copy of the gene, thereby altering the mutant phenotype. The development of new and more effective vector systems means that gene therapy probably will become a standard method for treatment of genetic disorders.
4. Gene therapy and the Human Genome Project have raised many ethical issues that are yet to be resolved. The development and application of biotechnology in medicine have moved faster than a consensus about how to use and interpret this technology. Research and education are needed to make informed decisions.
6. Cloned DNA is finding a wide range of applications, including gene mapping and the identification and isolation of the genes responsible for genetic disorders. Positional cloning, which is based on recombinant DNA technology, allows a gene to be mapped and identified with no knowledge of the nature or function of the gene product.
7. Recombinant DNA techniques that detect allelic variants of variable tandem nucleotide repeats (DNA fingerprints) have found applications in forensics, paternity testing, and a wide range of other fields, including archaeology, conservation biology, and public health.

Key Terms

adenosine deaminase (ADA), 445
allele-specific oligonucleotide (ASO), 441
amniocentesis, 439
attenuated vaccines, 438
centimorgan (cM), 450
chorionic villus sampling (CVS), 439
cystic fibrosis, 445
cystic fibrosis transmembrane conductance regulator (CFTR), 441
DNA fingerprinting, 451
DNA microarrays, 441
enhancement gene therapy, 447
Ethical, Legal, and Social Implications (ELSI) Program, 449
exclusion map, 450
familial hypercholesterolemia, 445
fusion polypeptide, 437
gene therapy, 444
germline therapy, 447
glyphosate, 434
Human Genome Project (HGP), 448
inactivated vaccines, 438
lod score method, 450
logarithm of the odds (lod), 449
minisatellite, 451
neurofibromin, 450
positional cloning, 450
restriction fragment-length polymorphism (RFLP), 449
severe combined immunodeficiency (SCID), 445
somatic gene therapy, 447
subunit vaccine, 438
transgenic, 438
type 1 neurofibromatosis (NF1), 450
variable-number tandem repeat (VNTR), 451

INSIGHTS AND SOLUTIONS

1. Probes for DNA fingerprinting can be derived from a single locus or multiple loci. Two multiple-loci probes have been widely employed in both criminal and civil cases and derive from minisatellite loci on chromosome 1 (1cen-q24) and chromosome 7 (7q31.3). These probes, which are used because they produce a highly individual fingerprint, have determined paternity in thousands of cases over the last few years. The results of DNA fingerprinting of a mother (M), putative father (F), and child (C), using the aforementioned probes, are shown in the accompanying figure. The child has 6 maternal bands, 11 paternal bands, and 5 bands shared between the mother and the alleged father. Can you conclude that this man is the father of the child based on this fingerprint?

M C F

Maternal band
Paternal band
Band from either or both parents
Unassigned band
Band common to both parents
M Mother
C Child
F Alleged father

Solution: All bands present in the child can be assigned as coming from either the mother or the father. In other words, all the bands in the child's DNA fingerprint that are not maternal are present in the father. Since the father and the child share 11 bands and the child has no unassigned bands, paternity can be assigned with confidence. In fact, the chance that this man is not the father is on the order of 10^{-13}.

2. The DNA fingerprints of a mother, a child, and the alleged father in a second case are shown in the figure below. The child has 8 maternal bands, 15 paternal bands, 6 bands that are common to both the mother and the alleged father, and 1 band that is not present in either the mother or the alleged father. What are the possible explanations for the presence of the last band? Based on your analysis of the band pattern, which explanation is most likely?

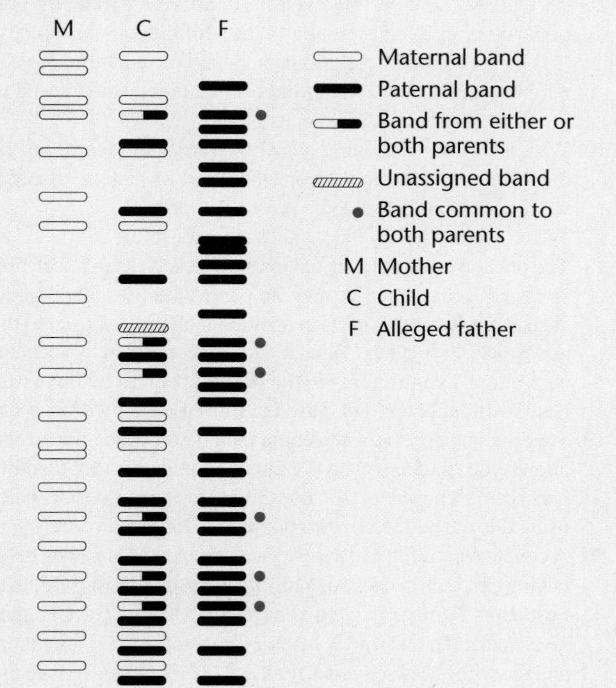

Solution: In this case, one band in the child cannot be assigned to either parent. Two possible explanations are that the child is mutant for one band or that the man tested is not the father. To estimate the probability of paternity, the mean number of resolved bands (n) is determined and the mean probability (x) that a band in individual A matches that in a second, unrelated individual B is calculated. In this case, because the child and the father share 15 bands in common, the probability that the man tested is not the father is very low (probably 10^{-7} or lower). As a result, the most likely explanation is that the child is mutant for a single band. In fact, in 1419 cases of genuine paternity resolved by the minisatellite probes on chromosomes 1 and 7, single-mutant bands in the children were recorded in 399 cases, accounting for 28% of all cases.

3. Infection by HIV-1 (human immunodeficiency virus) is responsible for the destruction of cells in the immune system and results in the symptoms of AIDS (acquired immunodeficiency syndrome). HIV infects and kills cells of the immune system that carry a cell-surface receptor known as CD4. An HIV surface protein known as gp120 binds to the CD4 receptor and allows the virus to enter into the cell. The gene encoding the CD4 protein has been cloned. How might this clone be used along with recombinant DNA techniques to combat HIV infection?

Solution: Several methods that use the CD4 gene are being explored to combat HIV infection. First, because infection depends on an interaction between the viral gp120 protein and the CD4 protein, the cloned CD4 gene has been modified to produce a soluble form of the protein (sCD4). The idea is that HIV can be prevented from infecting cells if the gp120 protein of the virus is bound up with the soluble form of the CD4 protein and thus is unable to bind to CD4 proteins on the surface of immune system cells. Studies in cell culture systems indicate that the presence of sCD4 effectively prevents HIV infection of tissue culture cells. However, studies in HIV-positive humans have been somewhat disappointing, mainly because the strains of HIV used in the laboratory are different from those found in infected individuals. Since HIV-infected cells carry the viral gp120 protein on their surface, the CD4 gene has been fused with genes encoding bacterial toxins to kill them. The resulting fusion protein contains CD4 regions that bind to gp120 and toxin regions that should kill the infected cell. In tissue culture experiments, cells infected with HIV are killed by the fusion protein, whereas uninfected cells survive. Researchers hope that the targeted delivery of drugs and toxins can be used in therapeutic applications to treat HIV infection.

Problems and Discussion Questions

1. In attempting to vaccinate people against diseases by having them eat antigens, the antigen (such as the cholera toxin) must reach the cells of the small intestine. What are some potential problems of this method? Why don't absorbed food molecules stimulate the immune system and make you allergic to the food you eat?
2. One of the main safety issues associated with genetically modified crops is the potential for allergenicity caused by introduction of an allergen or caused by changing the level of expression of a host allergen. Since common allergenic proteins often have identical stretches of a few (6 – 7) amino acids in common, Kleter and Peijnenburg (2002. *BMC Struct. Biol.* 2: 8.) developed a method for screening transgenic crops to evaluate potential allergenic properties. How do you think they accomplished this?
3. There are now more than 1000 cloned farm animals in the United States. Within the next two years, milk from cloned cows and their offspring (born naturally) may be available in supermarkets. These animals are clones, but have not been transgenically modified, and they are no different than are identical twins. Should milk from such animals and their naturally born offspring be labeled as coming from cloned cows or their descendants? Why?
4. One of the major causes of sickness, death, and economic loss in the cattle industry is from *Mannheimia haemolytica*, which causes bovine pasteurellosis (shipping fever). Noninvasive means of delivering a vaccine using transgenic plants expressing immunogens would reduce labor costs and trauma to livestock. An early step toward developing an edible vaccine is to determine whether an injected version of an antigen (usually a derivative of the pathogen) is capable of stimulating the development of antibodies

in a test organism. The table below assesses the ability of a transgenic portion of a toxin (Lkt) of *M. haemolytica* cloned into white clover to stimulate development of specific antibodies in rabbits. (Table modified from Lee et al. 2001. *Infect. and Immunity* 69: 5786–93.) (a) What general conclusion can you draw from the data? (b) With regards to development of a usable edible vaccine, what work remains to be done?

Immunogen injected	*Antibody Production In Serum*
Lkt50*—saline extract	+
Lkt50—column extract	+
Mock injection	−
Pre-injection	−

*Lkt50 is a smaller derivative of Lkt that lacks all hydrophobic regions. + indicates at least 50% neutralization of toxicity of Lkt; − indicates no neutralization activity.

5. Outline the steps involved in transferring glyphosate resistance to a crop plant. Do you envision that this trait can escape from the crop plant and make weeds glyphosate-resistant? Why or why not?
6. Now that enhancement therapy using one gene product has been approved, is this justification for using enhancement gene therapy?
7. Recombinant adenoviruses have been used in a number of preclinical studies to determine the efficacy of gene therapy for rheumatoid arthritis and osteoarthritis. Genes can be delivered by injection to the tissues that need them, and according to Evans, and others (2001. *Arthritis Res.* 3: 142–46.), approximately 20% of all human gene therapy trials have used adenoviruses for gene delivery. The death of a patient in 1999 after infusion of adenoviral vectors has caused concern. As you consider the use of viral vectors as therapy-delivery vehicles for human pathologies, what factors seem of paramount concern?
8. Define somatic gene therapy, germline therapy, and enhancement gene therapy. Which of these is currently in use?
9. *Transductional targeting* is a preferred route for the delivery of therapeutics for human diseases. It involves the development of tissue-specific interactions between the viral vector and a specific tissue. A genetic approach used by Ponnazhagan, and colleagues (2002. *J. Virol.* 76: 12,900–07.) involves engineering the capsid of an adeno-associated virus (type 2) vector to target specific human cell types. Nongenetic approaches are also possible. Speculate on problems associated with the genetic approach of capsid alteration and problems that might be associated with nongenetic approaches to transductional targeting.
10. The development of safe vectors for human gene therapy has been a goal since 1990. Of the variety of problems associated with viral-based vectors, many such viruses (i.e., SV40) have transformation properties thought to be mediated by binding and inactivating gene products such as p53, retinoblastoma protein (pRB) and others. Cooper et al. (1997. *Proc. Nat. Acad. Sci. (USA)* 94: 6450–55.) developed SV40-based vectors that are deficient in binding the p53, pRB, and other proteins. Why would you specifically want to avoid inactivating p53, pRB, and related proteins?
11. Gene therapy for human genetic disorders involves transferring a copy of the normal human gene into a vector and using the vector to transfer the cloned human gene into target tissues. Presumably, the gene enters the target tissue and becomes active, and the gene product relieves the symptoms. (a) Why are disorders such as muscular dystrophy difficult to treat by gene therapy? (b) What are the potential problems by using retroviruses as vectors? (c) Should gene therapy involve germ tissue instead of somatic tissue? What are some of the potential ethical problems associated with the former approach?
12. Sequencing the human genome and the development of microarray technology promises to improve our understanding of normal and abnormal cell behavior. In an article on the applications of microarray technology in breast cancer research (2001. *Breast Can. Res.* 3: 158–175.), Cooper stated that "data from such studies could revolutionize cancer diagnosis." (a) What is microarray technology? (b) In what way might this technology revolutionize our understanding of cancer?
13. What are the advantages of using STRs instead of VNTRs for DNA identification?
14. In producing physical maps of markers and cloned sequences, what advantage does *Drosophila* offer that other organisms, including humans, do not?
15. (a) Outline the steps involved in identifying a gene by positional cloning. What steps may cause difficulty in this process? (b) Once a region on a chromosome has been identified as containing a given gene, what kind of mutations would speed the process of identifying the locus?
16. Why is positional cloning a useful strategy in identifying and mapping a mutant gene? For what kind of genetic disorders would positional cloning be most appropriate?
17. What is an exclusion map, and why is it useful?
18. The phenotype of many behavioral traits, such as manic depression and schizophrenia, may be controlled by several genes, each at a different locus. Can positional cloning be used to map and isolate such genes? What if a trait is controlled by six genes, each equally contributing to the phenotype in an additive way? Can positional cloning be used in this case? Why or why not?
19. Suppose you develop a screening method for cystic fibrosis that allows you to identify the predominant mutation *Δ508* and the next six most prevalent mutations. What must you consider before using this method to screen a population for this disorder?
20. A couple with European ancestry seeks genetic counseling before having children, because of a history of cystic fibrosis (CF) in the husband's family. They each have ASO testing for CF, and it reveals that the husband is heterozygous for the *Δ508* mutation, and the wife is heterozygous for the *R117* mutation. You are their genetic counselor. In meeting with you, they are convinced they are not at risk for having an affected child because they each carry different mutations and cannot have a child who is homozygous for either mutation. What would you say to them?
21. Dominant mutations can be categorized according to whether they increase or decrease the overall activity of a gene or gene product. Although a loss-of-function mutation (a mutation that inactivates the gene product) is usually recessive, for some genes, one dose of the gene product is not sufficient to produce a normal phenotype. In this case, a loss-of-function mutation in the gene will be dominant, and the gene is said to be *haploinsufficient*. A second category of dominant mutations is gain-of-function mutations, which result in increased activity or expression of the gene or gene product. The phenotype of such a mutation results from too much gene product. The gene therapy technique currently used in clinical trials involves the "addition" to somatic cells of a normal copy of a gene. In other words, a normal copy of the gene is inserted into the genome of the mutant somatic cell, but the mutated copy of the gene is not removed or replaced. Will this strategy work for either of the two aforementioned types of dominant mutation?
22. Why are most recombinant human proteins produced in animal or plant hosts instead of bacterial host cells?

23. The DNA sequence surrounding the site of the sickle-cell mutation in the β-globin for normal and mutant genes is shown below.

5′-GACTCCTGAGGAGAAGT-3′
3′-CTGAGGACTCCTCTTCA-5′
Normal DNA

5′-GACTCCTGTGGAGAAGT-3′
3′-CTGAGGACACCTCTTCA-5′
Sickle-cell DNA

Each type of DNA is denatured into single strands and applied to a filter. The paper containing the two spots is hybridized to an ASO of the sequence

5′-GACTCCTGAGGAGAAGT-3′

Which spot, if either, will hybridize to this probe?

24. One form of hemophilia, an X-linked disorder of blood clotting, is caused by a mutation in clotting factor VIII. Many single-nucleotide mutations of this gene have been described, making the detection of mutant genes by Southern blots inefficient. There is, however, an RFLP for the enzyme *Hin*dIII contained in an intron of the factor VIII gene that can often be used in screening, as shown in the figure below.

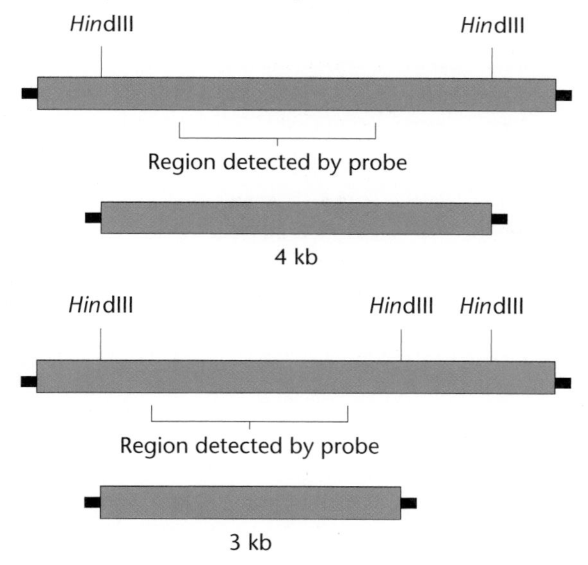

A female whose brother has hemophilia has a 50% risk of being a carrier of this disorder. To test her status, DNA is obtained from her white blood cells and those of family members, cut with *Hin*dIII, and the fragments are probed and visualized by Southern blotting. Using the results below, determine whether either of the females in generation II is a carrier for hemophilia.

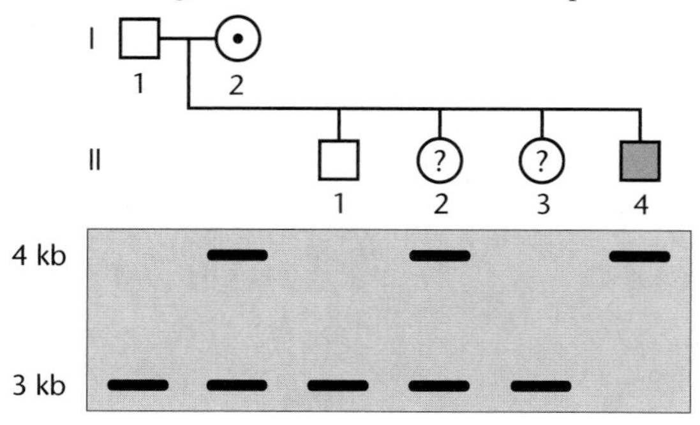

25. The human insulin gene contains introns; since bacterial cells will not excise introns from mRNA, how can a gene like this be cloned into a bacterial cell and produce insulin?

26. In mice transfected with the rabbit β-globin gene, the rabbit gene is active in several tissues, including the spleen, brain, and kidney. In addition, some mice suffer from thalassemia (a form of anemia) caused by an imbalance in the coordinate production of α- and β-globins. Which problems associated with gene therapy are illustrated by these findings?

27. Genome scanning to detect susceptibility to diseases is already in use. As this technology becomes used more widely, medical records will contain the results of such testing. Who should have access to this information? Should employers, potential employers, or insurance companies be allowed to have this information? Would you favor or oppose having the government establish and maintain a central database containing the results of genome scanning on members of the population?

28. What limits the use of differences in restriction enzyme sites as a way of detecting point mutations in human genes?

29. You are asked to assist with a prenatal genetic test for a couple, each of whom is found to be a carrier for a deletion in the β-globin gene that produces β-thalassemia when homozygous. They already have one child who is unaffected and is not a carrier. The woman is pregnant, and the couple wants to know the status of the fetus. You receive DNA samples obtained from the fetus by amniocentesis, and from the rest of the family by extraction from white blood cells. Using a probe for the deletion, you obtain the blot shown below. Is the fetus affected? What is its genotype for the β-globin gene?

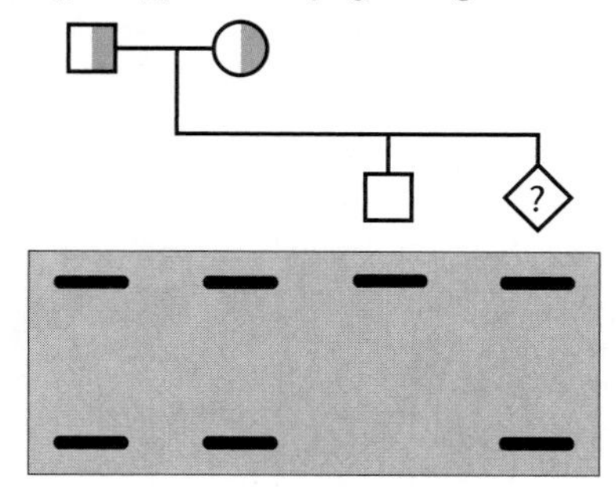

30. Shown below is a pedigree tracking the inheritance of a rare disease. (a) Which mode or modes of inheritance are excluded by or consistent with this pedigree?

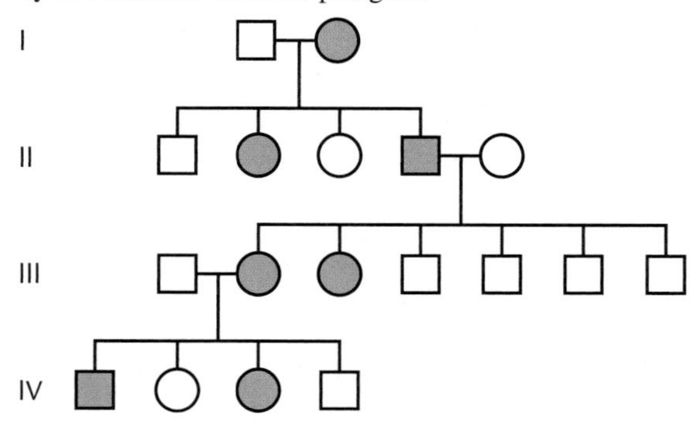

DNA samples from generations II and III above are obtained and subjected to RFLP linkage analysis. One RFLP is found on chromosome 10 (identified by probe *A*), and the other is found on chromosome 21 (identified by probe *B*). The results of the analysis are shown on the next page. Assume that additional data were gathered on this family and that they were consistent with the data shown and statistically significant. (b) On which chromosome is the disease gene located? (c) Individual

III-1 is married to a normal man whose RFLP genotype is *A1A2* and *B1B1*. What kind of prenatal diagnostic test can be done to determine whether the child they are expecting will be normal? Be sure to describe what you can conclude regarding the result, and indicate the accuracy of the test. (d) Individual III-4 marries a woman of genotype *B1B1*. They have a child who is *B1B1*. The father assumes that the child is illegitimate, but the mother, who has taken a genetics course, argues that the child could be the result of an event that occurred in the father's germline and cites two possibilities. What are they?

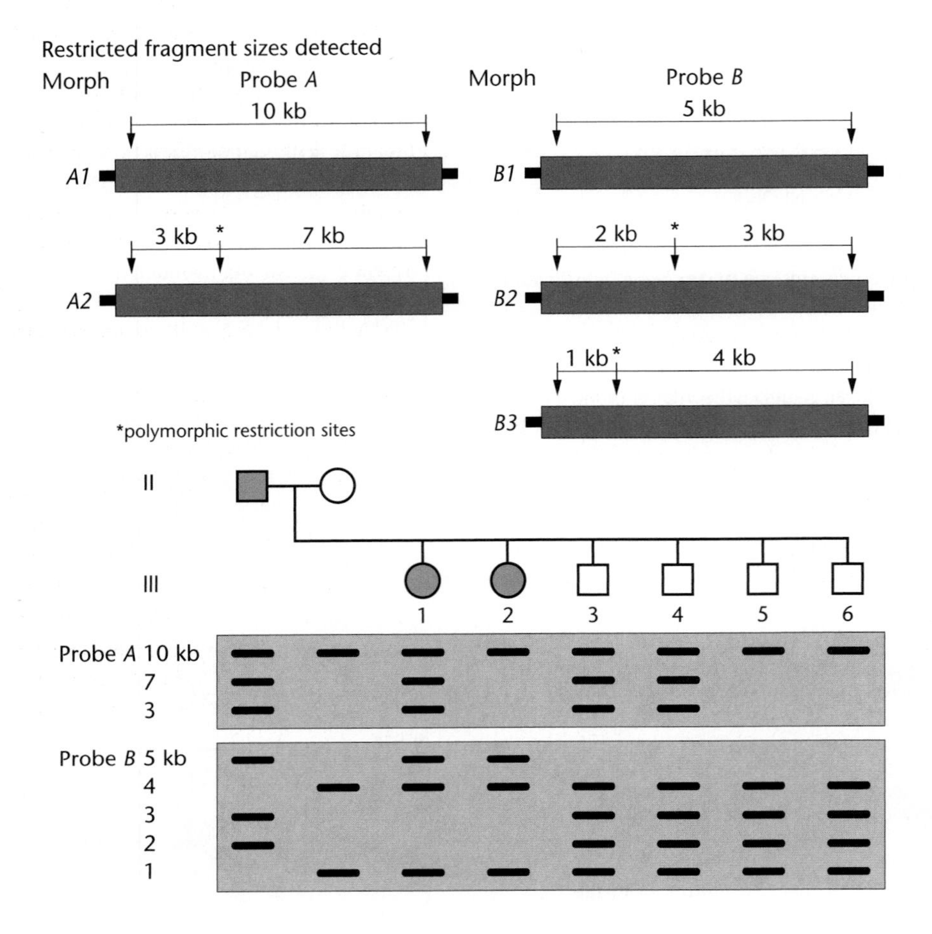

Selected Readings

Anderson, W. F. 2000. The best of times, the worst of times. *Science* 288: 627–28.

Atherton, K. T. 2002. Safety assessment of genetically modified crops. *Toxicol.* 181/182: 421–26.

Barker, D. et al. 1987. Gene for von Recklinghausen neurofibromatosis is in the pericentromeric region of chromosome 17. *Science* 236: 1100–02.

Beaudet, A. L. 1999. Making genomic medicine a reality. *Am. J. Hum. Genet.* 64: 1–13.

Bleck, O., McGrath, J. A., and South, A. P. 2001. Searching for candidate genes in the new millennium. *Clin. Exp. Dermatol.* 26: 279–83.

Cavanna-Calvo, M. et al. 2000. Gene therapy of severe combined immunodeficiency (SCID)-XI disease. *Science* 288: 669–72.

Chang, J. C., and Kan, Y. W. 1981. Antenatal diagnosis of sickle-cell anemia by direct analysis of the sickle mutation. *Lancet* 2: 1127–29.

Dale, P. J., Clarke, B., and Fontes, E. M. G. 2002. Potential for the environmental impact of transgenic crops. *Nat. Biotechnol.* 20: 567–74.

Daniell, H., Streatfield, S. J., and Wycoff K., 2001. Medical molecular farming: Production of antibodies, biopharmaceuticals and edible vaccines in plants. *Trends Plant Sci.* 6: 219–26.

Danna, K., and Nathans, D. 1971. Specific cleavage of simian virus 40 DNA by restriction endonuclease of *Hemophilus influenzae*. *Proc. Natl. Acad. Sci. (USA)* 68: 2913–17.

Engler, O. B. et al. 2001. Peptide vaccines against hepatitis B virus: From animal model to human studies. *Mol. Immunol.* 38: 457–65.

Estruch, J. J. 1997. Transgenic plants: An emerging approach to pest control. *Nature Biotech.* 15: 137–41.

Kmiec, E. B. 1999. Gene therapy. *Amer. Scient.* 87: 240–47.

Mason, H., Lam, D., and Arntzen, C. 1992. Expression of hepatitis B surface antigen in transgenic plants. *Proc. Natl. Acad. Sci. (USA)* 89: 11,745–49.

Moldoveanu, Z. et al. 1993. Oral immunization with influenza virus in biodegradable microspheres. *J. Infect. Dis.* 167: 84–90.

Phillips, M. I. 2001. Gene therapy for hypertension: The preclinical data. *Hypertension* 38: 543–48.

Riley, J. H. et al. 2000. The use of single nucleotide polymorphisms in the isolation of common disease genes. *Pharmacogen.* 1: 39–47.

Thomas, C. E., Ehrhardt, A., and Kay, M. A. 2003. Progress and problems with the use of viral vectors for gene therapy. *Nat. Rev. Genet.* 4: 346–58.

Tucker, G. 2003. Nutritional enhancement of plants. *Curr. Opin. Biotechnol.* 14: 221–25.

Velander, W. H., Lubon, H., and Drohan, W. N. 1997. Transgenic livestock as drug factories. *Sci. Amer.* (Jan) 276: 70–75.

Villa-Komaroff, L. et al. 1978. A bacterial clone synthesizing proinsulin. *Proc. Natl. Acad. Sci. (USA)* 75: 3727–31.

Walmsley, A. M., and Artzen, C. J. 2003. Plant cell factories and mucosal vaccines. *Curr. Opin. Biotechnol.* 14: 145–50.

Wisniewski, J-P., Frangne, N., Massonneau, A., and Dumas, C. 2002. Between myth and reality: Genetically modified maize, an example of a sizeable scientific controversy. *Biochimie* 84: 1095–103.

CHAPTER 20

This unusual four-winged *Drosophila* has developed an extra set of wings as a result of a homeotic mutation, (*Courtesy of E. B. Lewis, California Institute of Technology Archives.*)

Genes and Development

CHAPTER CONCEPTS

In multicellular plants and animals, a fertilized egg initiates a cycle of mitotic divisions and developmental events that ultimately give rise to an adult member of the species from which the gametes were derived. Thousands, millions, or even billions of specialized cells are generated and organized into a cohesive and coordinated unit that we perceive as a living organism. Developmental biologists study the processes that govern transitions from one stage of an organism's life cycle to another. Analysis of developmental processes draws on many different biological disciplines—such as molecular, cellular, and organismal biology—it also uses systems biology to help explain how interconnected networks of genes control biological processes over developmental time and ultimately transform the zygote into an adult organism.

Over the last several decades, genetic analysis coupled with recombinant DNA technology has identified, mapped, cloned, and sequenced genes that regulate developmental processes. We are now organizing these genes into complex networks and systems to explain how the action and interaction of these genes control development in a wide range of organisms.

In this chapter, we will emphasize the role of gene action in regulating development. Genetics is making tremendous strides in analyzing developmental processes because genetic information is required for both the molecular and cellular functions mediating developmental events and for determining the phenotype of the newly formed organism. We will examine the molecular events in several developmental processes: the establishment of the anterior–posterior axis of the body, the progressive limitation of developmental potential, and the role of master genes in specifying adult structures.

HOW DO WE KNOW WHAT WE KNOW?

IN THIS CHAPTER, WE WILL FOCUS ON both the large-scale as well as intercellular events that take place during embryogenesis and the development of adult structures. As you study this topic, you should try to answer several fundamental questions:

1. How do we know how many genes control development in an organism like *Drosophila*?
2. How do we know that molecular gradients in the egg control development?
3. How do we know that a genetic program specifying a body part can be changed?
4. How do we know that chemical signals between cells affect developmental events? ■

20.1 Basic Concepts in Developmental Genetics

In higher eukaryotes, development begins with the formation of a **zygote**, the cell generated by the fusion of a sperm and an oocyte. The cytoplasm of the oocyte is heterogeneous and nonuniform in distribution, a situation known as cell polarity. Following fertilization by the sperm and the early rounds of cell divisions, the nuclei of progeny cells find themselves in different environments as the oocyte's maternal cytoplasm is distributed into the new cells. Evidence suggests that the cytoplasm in these cells exerts influence on the genetic material, causing differential transcription at specific times in development. Therefore, **cytoplasmic localization** of particular cytoplasmic components by individual embryonic cells plays a major role in development.

Gene products synthesized soon after zygote formation further alter the cytoplasm of each cell, producing yet another cellular environment that in turn activates other gene sets, leading to a cascade of gene expression. In addition, as cells increase in number, they influence one another through **cell–cell interaction**. Thus, the environment acting on embryonic nuclei includes molecular components from the maternal genome, gene products transcribed after fertilization, as well as signals from other cells.

Although in the early stages of embryogenesis most cells show no evidence of structural or functional specialization, the combination of localized cytoplasmic components and position in the developing embryo appears to determine their ultimate form. It is as if their fate has been programmed prior to the actual events leading to specialization.

As cells respond to their continually changing external and internal environments, they embark on a pathway of developmental events leading to the formation of an adult cell. The most important stage in this pathway is **determination**, when in response to many external and internal cues, the cell's specific developmental fate becomes fixed. We now know that determination precedes the actual events of **differentiation**, the process by which the cell achieves its final form and function.

In the following sections, we examine how patterns of gene expression control determination, differentiation, and cell–cell interaction during development. Let's begin by examining the role of differential gene expression in establishing the anterior–posterior axis in *Drosophila*.

20.2 Maternal and Zygotic Genes Interact to Establish the Body Axis in *Drosophila*

Why certain cells turn specific genes on or off at specific stages of development is a central question in developmental biology. At present, there is no simple answer to this question. However, information derived from the study of model organisms including plants, marine invertebrates, insects, amphibians, and mammals has provided a general picture of what happens and how it happens. The genetic and molecular analysis of embryonic development in *Drosophila* highlights the key role played by molecular components placed in the oocyte cytoplasm during oogenesis in controlling gene expression.

Overview of *Drosophila* Development

Beginning with the fertilized egg, *Drosophila* passes through five stages of preadult development: the embryo, three larval stages, and the pupal stage. An adult fly emerges from the pupal case about 10 days after fertilization (Figure 20–1). Externally, the *Drosophila* egg has a number of structures that delineate its anterior, posterior, dorsal, and ventral regions (Figure 20–2). The anterior end of the egg contains the micropyle, a conical structure specialized for sperm entry, while the posterior end is rounded and marked by a series of aeropyles (openings that allow gas exchange during development). The dorsal side of the egg is flattened and slightly concave, and carries two appendages; the ventral side of the egg is convex. Internally, the cytoplasm is organized into a series of maternally derived molecular gradients. These gradients play a key role in establishing the developmental fates of nuclei that migrate into specific regions of the embryo.

Immediately after fertilization, the zygote nucleus undergoes a series of divisions(Figure 20–3a, b). After 9 rounds of division, the zygote cytoplasm contains approximately 512 nuclei within a single cell. The nuclei migrate to the egg's outer surface, or cortex, where further divisions take place (Figure 20–3c, d), forming the **nuclear syncytial blastoderm** (a syncytium is any cell with more than one nucleus). Arranged around the periphery of the egg, the nuclei are surrounded by cytoplasm that contains localized gradients of maternally derived transcripts and proteins. A program of gene expression is initiated in these blastoderm nuclei under the control of these cytoplasmic components, leading to different developmental programs in different cells.

The formation of the germ cells at the posterior pole of the embryo demonstrates the regulatory role of these localized cytoplasmic components (Figure 20–3d, e). Experiments have shown that any nuclei placed in the cytoplasm at the posterior of the egg (the pole plasm) form cells that will differentiate into germ cells. Hence the cytoplasm of the posterior pole contains maternal components that direct nuclei to form germ cells. The transcriptional programs triggered in the rest of the migrating nuclei form the embryo's anterior–posterior and dorsal–ventral axes of symmetry.

Following formation of the syncytial blastoderm, plasma membranes organize around individual nuclei, creating the **cellular blastoderm** (Figure 20–3e). Later, the embryo organizes into a series of body segments, defined by the expression patterns of various embryonic genes (Figure 20–4). These then give rise to the adult segments, though they are slightly shifted in position. Within the segments, cells first become determined to form either the anterior or posterior portion (called a **compartment**) of the segment. In later stages, the cell's developmental options become further restricted, so that cells in each compartment eventually become part of a single structure in the adult body.

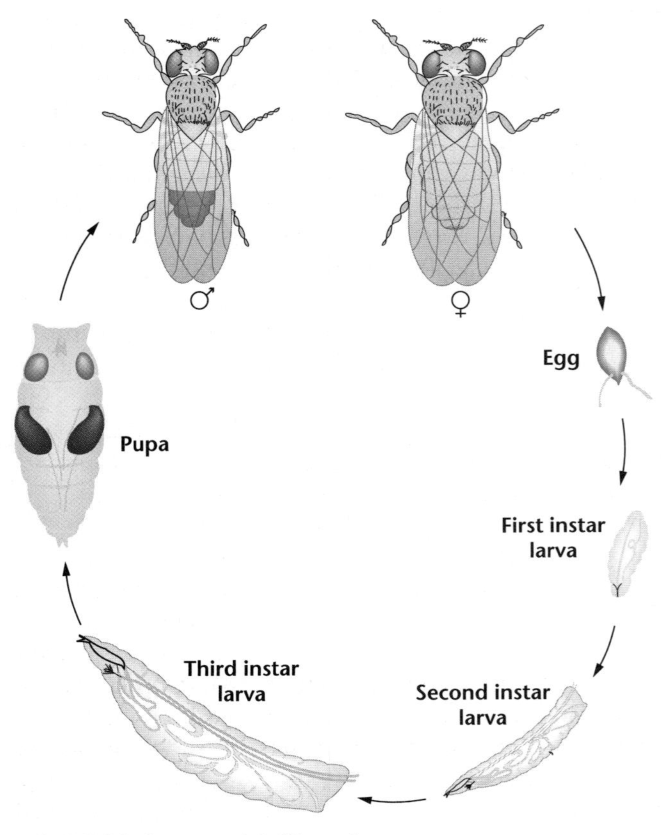

FIGURE 20–1 *Drosophila* life cycle.

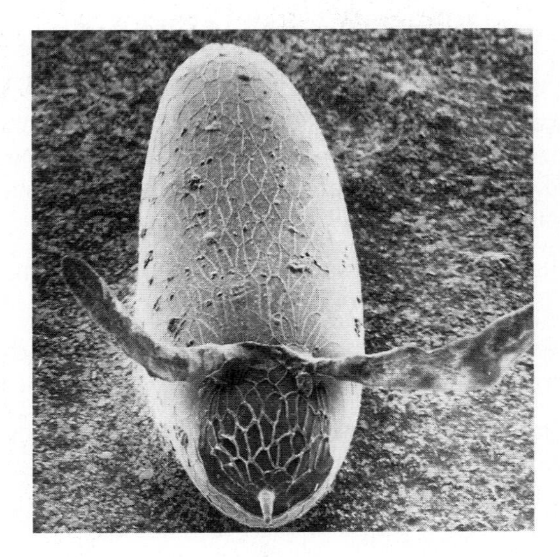

FIGURE 20–2 Scanning electron micrograph of a newly oviposited *Drosophila* egg. The micropyle is the small white projection at the anterior tip (the bottom). The dorsal surface with the two appendages faces up *(Photo: Dr. William S. Klug. The College of New Jersey).*

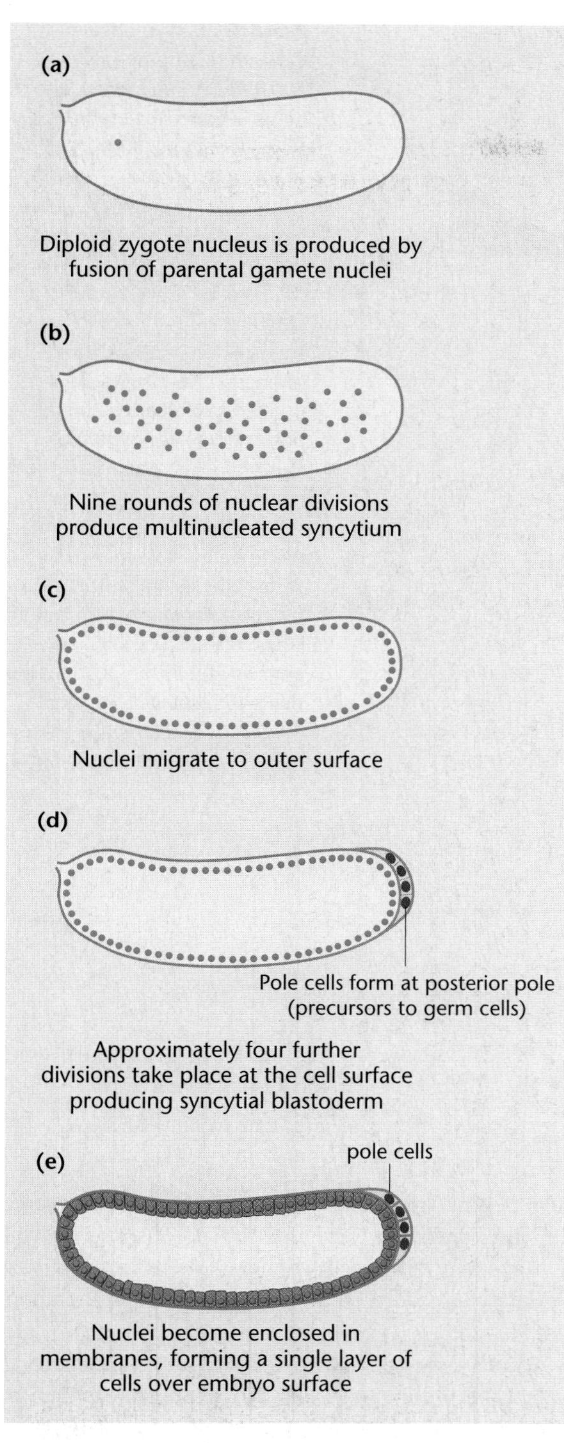

FIGURE 20–3 Early stages of embryonic development in *Drosophila*. (a) Fertilized egg with zygotic nucleus, about 30 minutes after fertilization. (b) Nuclear divisions occur about every 10 minutes, producing a multinucleate cell, the syncytial blastoderm. (c) After approximately 9 divisions (512 nuclei), the nuclei migrate to the outer surface or cortex of the egg. (d) At the surface, 4 additional rounds of nuclear division occur. A small cluster of cells, the pole cells, form at the posterior pole about 2.5 hours after fertilization. These cells will form the germ cells of the adult. (e) About 3 hours after fertilization, the nuclei become enclosed in plasma membranes and form a single layer of cells over the embryo surface, creating the cellular blastoderm.

The adult body plan of *Drosophila* derives its overall organization from the larval body plan and is composed of head, thoracic, and abdominal segments. Much of the larval body is broken down during the pupal stage, and many adult body structures form from small clusters of cells, or **imaginal discs**, formed during larval stages. There are 12 bilaterally paired discs—the eye-antennal discs, leg discs, wing discs, and so forth—and one genital disc. Figure 20–5 shows which imaginal discs will give rise to which adult structures.

Genetic Analysis of Embryogenesis

Our knowledge of anatomical development in *Drosophila* is useful, but genetic analysis has provided a wealth of information about events in embryogenesis. Genes that control embryonic development are of two types: maternal-effect genes and zygotic genes. Products of **maternal-effect genes** (mRNA and/or proteins) are deposited in the developing egg during oogenesis. These products are often distributed in a gradient or concentrated in specific regions of the egg cytoplasm. Female flies carrying deleterious mutations in maternal-effect genes should be sterile since none of the embryos of females homozygous for a recessive mutation would receive wild-type gene products from their mother and would therefore develop abnormally. In *Drosophila*, these maternal-effect genes encode transcription factors, receptors, and proteins that regulate translation. During embryonic development, these gene products activate or repress the expression of zygotic genes in a temporal and spatial sequence.

Zygotic genes are expressed in the developing embryo and are therefore transcribed after fertilization. In this class, the phenotype of genes with deleterious mutations shows embryonic lethality. In a cross between two flies heterozygous for a recessive zygotic mutation, one-fourth of the embryos (the homozygotes) fail to develop normally. In *Drosophila*, many zygotic genes are transcribed in specific regions of the embryo that depend on the distribution of maternal-effect proteins.

Much of what we know about the zygotic genes that regulate development comes from a mutagenic analysis performed by Christiane Nüsslein-Volhard and Eric Wieschaus, who systematically screened for mutations that affect embryonic development. They examined thousands of dead F_2 offspring of mutagenized flies for recessive embryonic lethal mutations with defects in external structures. The parents were thus identified as carriers of these mutations, which they grouped into three classes: gap genes, pair-rule genes, and segment polarity genes. In a paper published in 1980, these two scientists proposed a model in which embryonic development is initiated by gradients of maternal-effect gene products. The positional information laid down by these molecular gradients along the anterior–posterior axis of the embryo is interpreted by two sets of zygotic genes: (1) those identified in their screen—gap, pair-rule, and segment polarity genes—collectively called segmentation genes, which divide the embryo into a series

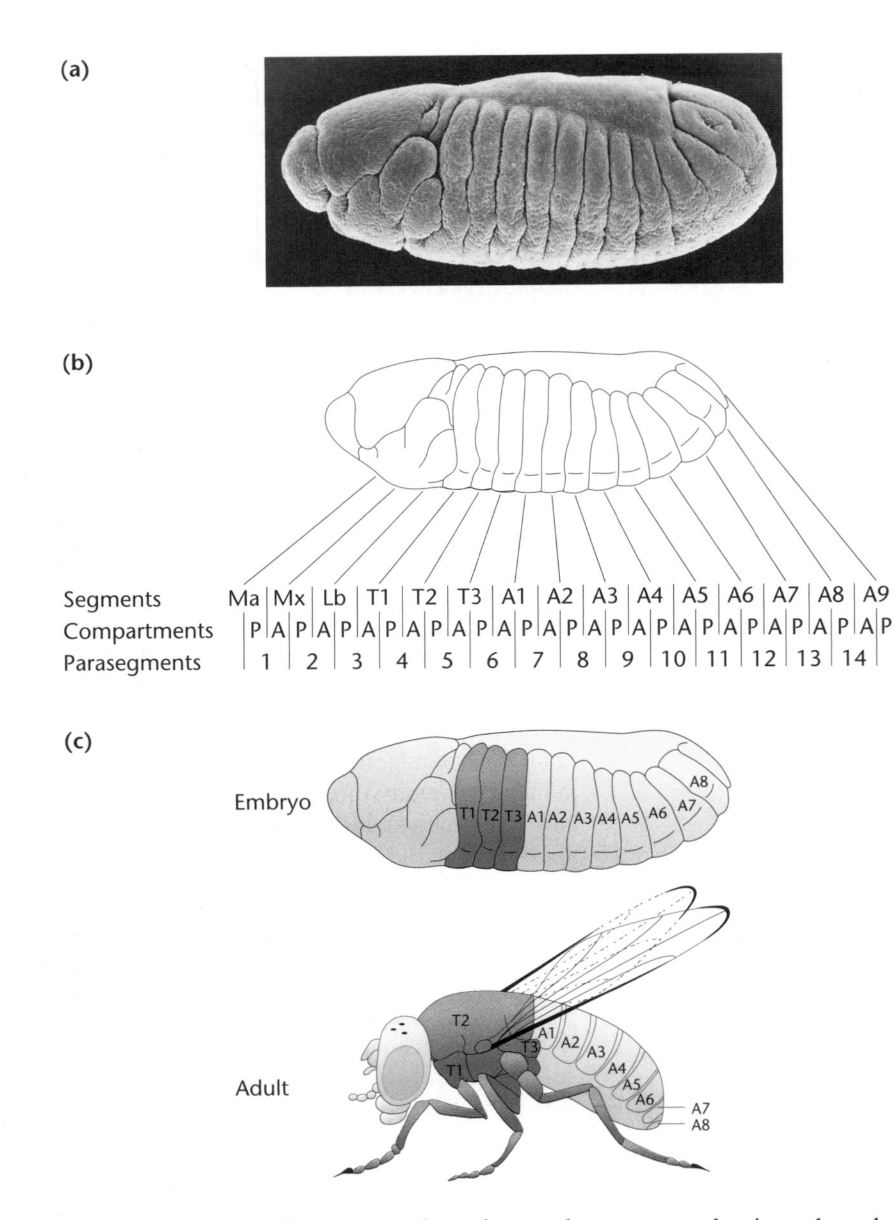

FIGURE 20–4 (a) Scanning electron micrograph of a *Drosophila* embryo at about 10 hours after fertilization. By this stage, the segmentation pattern of the body is clearly established. (b) The segments, compartments, and parasegments of the *Drosophila* embryo. The Ma, Mx, and Lb segments form head structures; T1–T3 are thoracic segments; A1–A9 are abdominal segments. Each segment is divided into anterior (A) and posterior (P) compartments. Parasegments represent an early pattern specification that is later refined into the segmental plan of the body. Note that all the parasegments are shifted forward by one compartment to form the segments. (c) The segmented embryo and the adult structures that will form each segment. *(Photo: F. R. Turner/Department of Biology/Indiana University)*

of stripes or segments and define the number, size, and polarity of each segment, and (2) homeotic genes, which specify the identity or fate of each segment (Figure 20–6).

The model developed by Nüsslein-Volhard and Wieschaus is shown in Figure 20–7. The developmental process begins when the maternal-effect gene products responsible for forming the anterior–posterior axis, which were placed in the egg during oogenesis, are activated immediately after fertilization (Figure 20–7a). Their activity restricts cells to form either anterior or posterior structures. Gene products from this gradient activate the transcription of the gap genes that divide the embryo into a limited number of broad regions (Figure 20–7b). Gap proteins are transcription factors that activate pair-rule genes whose products divide the embryo into regions about two segments wide (Figure 20–7c). The combined action of all gap genes defines segment borders. The pair-rule genes in turn activate the segment polarity genes that divide the segments into anterior and posterior compartments (Figure 20–7d). The collective action of the maternal genes that form the anterior–posterior axis and the segmentation genes defines the fields of action for the homeotic selector (*Hox*) genes (Figure 20–7e).

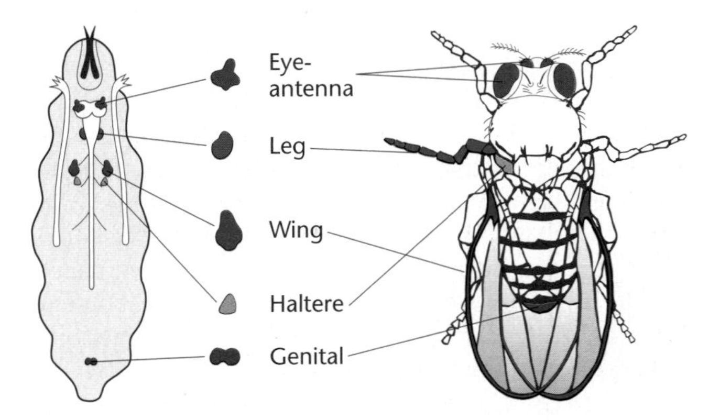

FIGURE 20–5 Imaginal discs of the *Drosophila* larva and the adult structures that derive from them.

In a second screening, Wieschaus and Trudi Schupbach screened thousands of flies for maternal-effect mutations that affected the external structures of the embryo. They

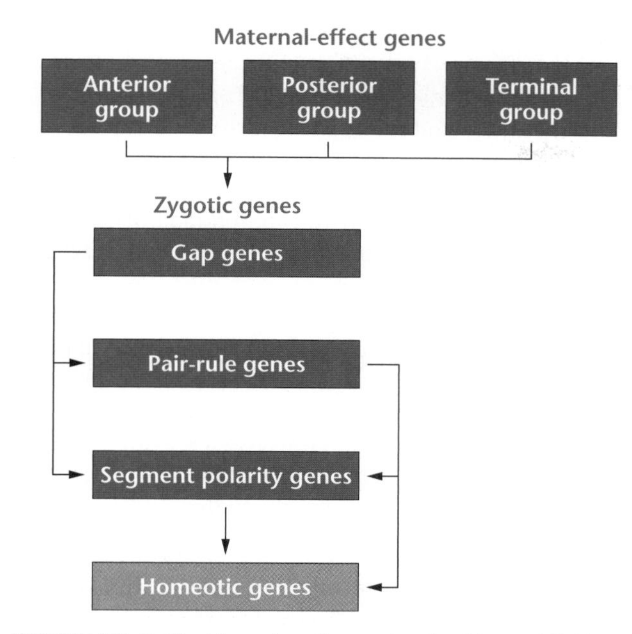

FIGURE 20–6 The hierarchy of genes involved in establishing the segmented body plan in *Drosophila*. Gene products from the maternal genes regulate the expression of the gap, pair-rule, and segment polarity groups of zygotic genes that in turn control expression of the *Hox* genes.

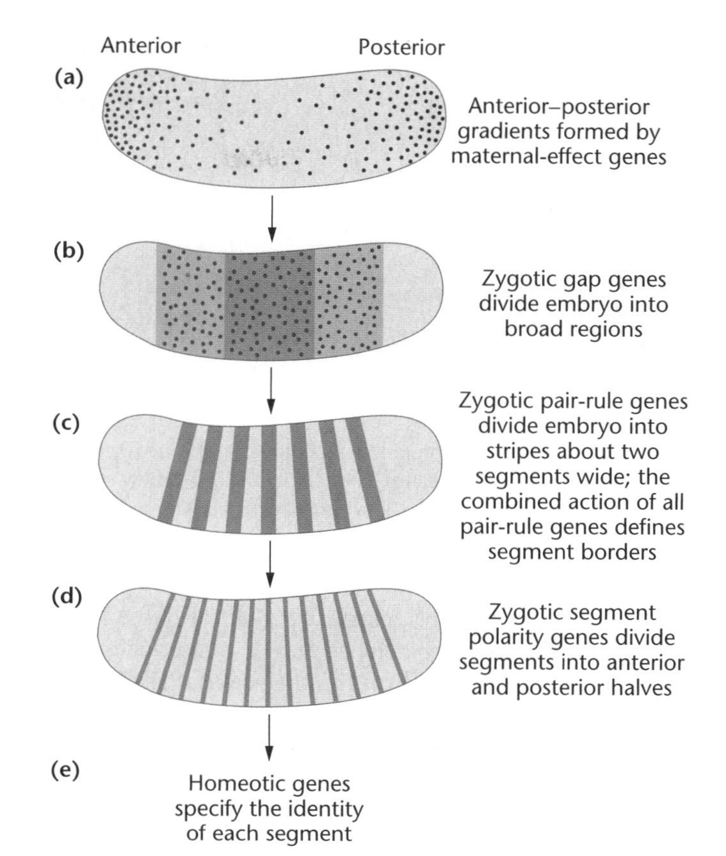

FIGURE 20–7 Progressive restriction of cell fate during development in *Drosophila*. (a) Gradients of maternal proteins are established along the anterior–posterior axis of the embryo. (b–d) The three groups of segmentation genes progressively define the body segments. (e) Individual segments are given identity by the homeotic selector genes.

estimated that about 40 maternal-effect genes and 50–60 zygotic genes regulate embryonic development. This means that only about 100 genes control normal embryogenesis, a surprisingly low number given the complexity of the structures formed from the fertilized egg. For their work on the genetic control of development in *Drosophila*, Nüsslein-Volhard and Wieschaus, along with E. B. Lewis—the geneticist who initially identified and studied many of these genes in the 1970s—were awarded the 1995 Nobel Prize for Physiology or Medicine.

Zygotic Genes and Segment Formation

Zygotic genes are activated or repressed in a positional gradient by the maternal-effect gene products. The expression of three subsets of these genes divides the embryo into a series of segments along the anterior–posterior axis. The **segmentation genes** are normally transcribed in the developing embryo, and their mutations have embryonic lethal phenotypes.

Over 20 segmentation loci have been identified (Table 20.1). They are classified on the basis of their mutant phenotypes: Gap genes delete a group of adjacent segments; pair-rule genes affect every other segment and eliminate a specific part of each affected segment; segment polarity genes cause defects in homologous portions of each segment.

- **Gap genes.** The transcription of gap genes is activated or inactivated by gene products previously expressed along the anterior–posterior axis and by other genes of the maternal gradient systems. When mutated, these genes produce large gaps in the embryo's segmentation pattern. *Hunchback* mutants lose head and thorax structures, *Krüppel* mutants lose thoracic and abdominal structures, and *knirps* mutants lose most abdominal structures. Gap gene transcription divides the embryo into a series of broad regions (the head, thorax, and abdomen). Within these regions, different combinations of gene expression eventually specify both the type of segment that forms and the proper order of segments in

TABLE 20.1 Segmentation Genes in *Drosophila*

Gap Genes	*Pair-Rule Genes*	*Segment Polarity Genes*
Krüppel	*hairy*	*engrailed*
knirps	*even-skipped*	*wingless*
hunchback	*runt*	*cubitis interruptus*^{D}
giant	*fushi-tarazu*	*hedgehog*
tailless	*odd-paired*	*fused*
huckebein	*odd-skipped*	*armadillo*
	sloppy-paired	*patched*
		gooseberry
		paired
		naked
		disheveled

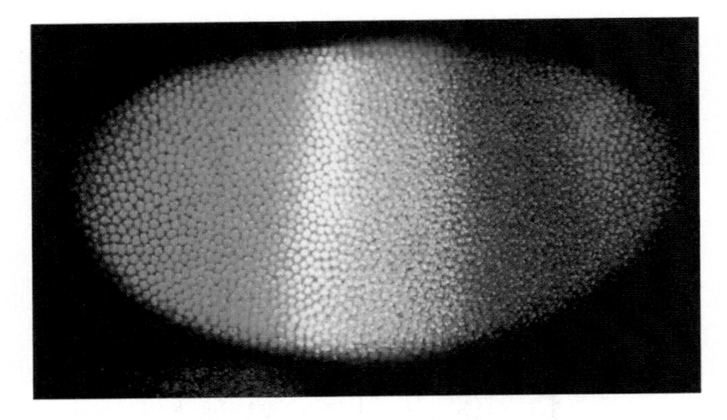

FIGURE 20–8 Expression of several gap genes in a *Drosophila* embryo. The hunchback protein is shown in orange, the Krüppel protein in green, and the Knirps protein in red. The yellow stripe is created when cells contain both hunchback and Krüppel proteins. *(Jim Langeland, Stephen Paddock, and Sean Carroll, University of Wisconsin at Madison.)*

the body of the larva, pupa, and adult. To date, all gap genes that have been cloned encode transcription factors with zinc-finger DNA-binding motifs. The expression of gap genes correlates roughly with their mutant phenotypes: *hunchback* at the anterior, *Krüppel* in the middle, and *knirps* at the posterior (Figure 20–8). Gap genes also control the transcription of pair-rule genes.

- **Pair-rule genes.** Pair-rule genes divide the broad regions established by gap genes into sections about one segment wide. Mutations in the pair-rule genes eliminate segment-size sections at every other segment. The pair-rule genes are expressed in narrow bands or stripes of nuclei that extend around the circumference of the embryo. Expression of this gene set first establishes the boundaries of segments and then establishes the developmental fate of the cells within each segment by controlling the segment polarity genes. At least eight pair-rule genes act to divide the embryo into a series of stripes. However, the boundaries of these stripes overlap, meaning that cells in the stripes express different combinations of pair-rule genes in an overlapping fashion (Figure 20–9). Many of the pair-rule genes encode transcription factors containing helix–turn–helix homeodomains. The transcription of pair-rule genes is mediated by the action of gap gene products, but their pattern is resolved into highly delineated stripes by interaction among the gene products of the pair-rule genes themselves (Figure 20–10).
- **Segment polarity genes.** Expression of segment polarity genes is controlled by transcription factors encoded by the pair-rule genes. Each segment polarity gene becomes active in a single band of cells within each segment that extends around the embryo

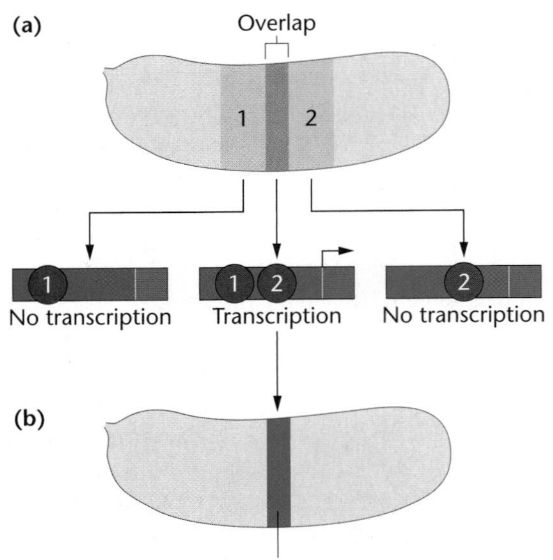

FIGURE 20–9 New patterns of gene expression can be generated by overlapping regions containing gene products. (a) Transcription factors 1 and 2 are present in an overlapping region of expression. If both transcription factors must bind sites in a promoter to trigger the expression of a target gene, the gene will be active only in cells containing both factors (most likely in the zone of overlap). (b) The expression of the target gene in the restricted region of the embryo.

(a)

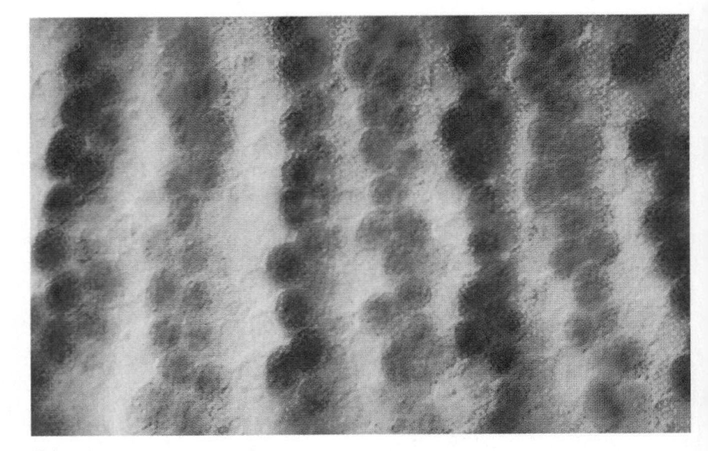

(b)

FIGURE 20–10 Stripe pattern of pair-rule gene expression in the *Drosophila* embryo. This embryo is stained to show expression patterns of the genes *even-skipped* and *fushi-tarazu (ftz)*. (a) Low-power view (b) and high-power view of the same embryo. *(Lawrence, P. A., and Johnson P., 1989. Pattern formation in the* Drosophila *embryo: allocation of cells to parasegments by* even-skipped *and* fushi-tarazu. Development *105: 761–67.)*

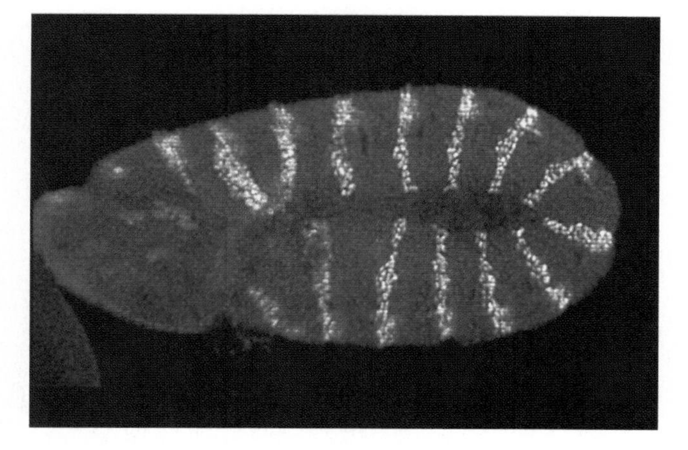

FIGURE 20–11 The 14 stripes of expression of the segment polarity gene *engrailed* in a *Drosophila* embryo. *(Jim Langeland, Stephen Paddock, and Sean Carroll University of Wisconsin at Madison)*

(Figure 20–11). These expression patterns divide the embryo into 14 segments, and the gene products control cellular identity within each segment. Some segment polarity genes, including *engrailed*, encode transcription factors. Rather than activating transcription, however, the engrailed protein competitively inhibits activation by other homeodomain proteins, resulting in the establishment of a segment border.

20.3 Homeotic Genes Control Pattern Formation Along the Anterior–Posterior Body Axis

As segments are formed by the expression of the segmentation genes, another class of genes, the **homeotic genes** are activated as targets of the zygotic genes. Expression of homeotic genes determines the pattern of adult structures to be formed by each body segment. In *Drosophila*, this includes the antennae, mouthparts, legs, wings, thorax, and abdomen. Mutants of these genes are called **homeotic mutants** (from the Greek word for "same") because the identity of one segment is transformed into that of a neighboring segment. For example, the wild-type allele of *Antennapedia* (*Antp*) specifies the formation of a leg on the second segment of the thorax. Dominant gain-of-function *Antp* mutations cause this gene to be expressed in the head as well, and mutant flies have legs on their head in place of antennae (Figure 20–12).

The *Drosophila* genome contains two clusters of homeotic genes on chromosome 3 (Table 20.2). One cluster, the ***Antennapedia* complex *(ANT-C)*** contains five genes required for specifying structures in the head and first two thoracic segments (Figure 20–13a). The second cluster, the ***bithorax* complex *(BX-C)*** contains three genes required for specifying structures formed by the posterior portion of the second thoracic segment, the entire third thoracic segment, and the abdominal segments (Figure 20–13b).

These gene clusters have two properties in common: First, each gene listed in Table 20.2 encodes a transcription factor that includes a DNA-binding domain encoded by a 180-bp sequence known as a **homeobox**. The homeobox encodes a sequence of 60 amino acids known as a **homeodomain**. Second, expression of the genes in

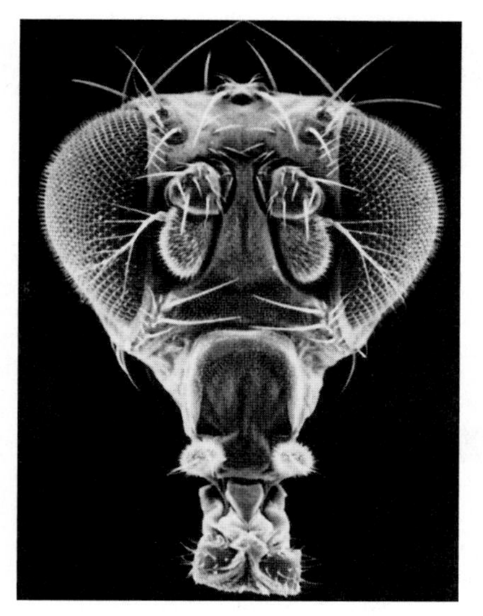

(a)

(b)

FIGURE 20–12 *Antennapedia* (*Antp*) mutation in *Drosophila*. (a) Head from wild-type *Drosophila* showing the structure of the antenna and other head parts. (b) Head from an *Antp* mutant, showing the replacement of normal antenna structures with legs, caused by activation of the *Antp* gene in the head region. *((a) (b) Reproduced by permission from T. Kaufmann at al.* Advances in Genetics *27: 309-362, 1990. Image courtesy of F. Rudolph Turner, Indiana University.)*

TABLE 20.2 **Homeotic Selector Genes of *Drosophila***

ANT-C	*BX-C*
Labial	*Ultrabithorax*
Antennapedia	*Abdominal A*
Sex comb reduced	*Abdominal B*
Deformed	
Proboscipedia	

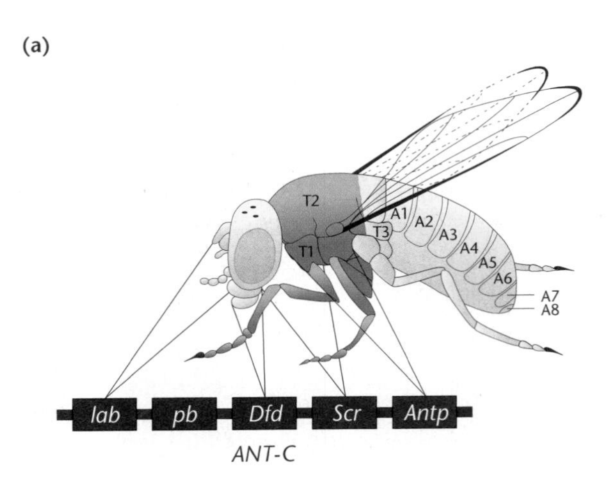

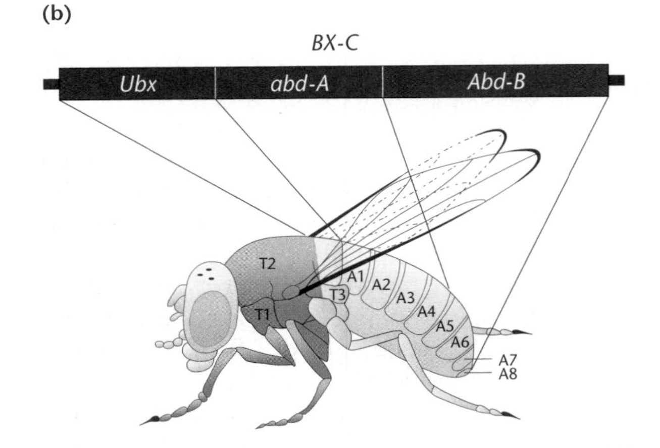

FIGURE 20–13 (a) Genes of the *Antennapedia* complex (*ANT-C*) and the adult structures they specify. The *labial* (*lab*) and *Deformed* (*Dfd*) genes control the formation of head segments. The *Sex comb reduced* (*Scr*) and *Antennapedia* (*Antp*) genes specify the identity of the first two thoracic segments. The remaining gene in the complex, *Proboscipedia* (*Pb*), may not act during embryogenesis but may be required to maintain the differentiated state in adults. In mutants, the labial palps are transformed into legs. (b) Genes of the *bithorax* complex (*BX-C*) and the adult structures they specify. *Ultrabithorax* (*Ubx*) controls structures in the posterior compartment of T2 and structures in T3. The two other genes, *abdominal A* (*abd-A*) and *Abdominal B* (*Abd-B*), specify the segmental identities of the eight abdominal segments (A1–A8).

Table 20.2 is colinear. Genes at the 3′-end of a cluster are expressed at the anterior of the embryo, those in the middle of the cluster are expressed in the middle of the embryo, and genes at the 5′-end of the cluster are expressed in the embryo's posterior region (Figure 20–14). Genes with these properties are called *Hox* genes.

***Hox* gene clusters** are found in the genomes of most eukaryotes with segmented body plans, including *Xenopus*, chickens, mice, and humans (Figure 20–15). Homeodomains from all of the animals examined to date are very similar in amino acid sequence and encode a protein associated with the transcriptional regulation of a specific gene set. This suggests that the segmented body plan of animals may have evolved only once.

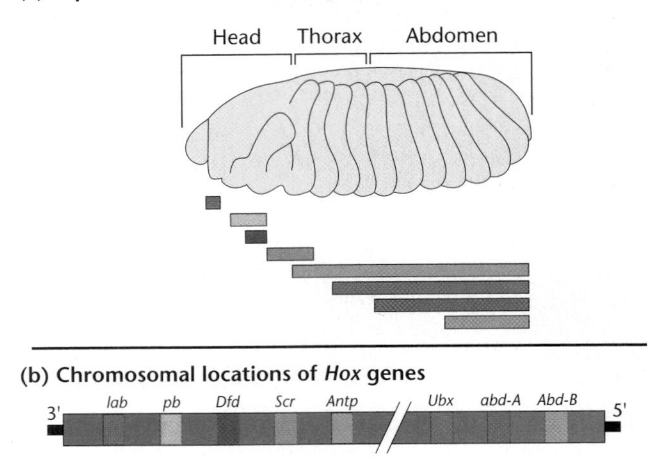

FIGURE 20–14 Colinear relationship between the spatial pattern of expression and chromosomal locations of homeotic genes in *Drosophila*. (a) *Drosophila* embryo and the domains of homeotic gene expression in the embryonic epidermis and central nervous system. (b) Chromosomal location of homeotic genes. Note that the order of genes on the chromosome correlates with the chronological pattern of their expression.

Appendage Formation in *Drosophila*

In addition to the homeobox genes in the *Hox* gene clusters, there is a large and diverse family of other homeobox genes in eukaryotic genomes. In *Drosophila*, one of these, *Distal-less* (*Dll*), plays an important role in the development of the appendages, including the antennae, mouthparts, legs, and wings. This gene, which maps to chromosome 2 and encodes a homeobox transcription factor, is the earliest known gene expressed in appendage formation. Mutations in *Dll* produce a wide range of phenotypes, including transformation of the antennae into legs and the formation of shortened legs that are missing distal structures.

In the antennae, expression of the wild-type *Dll* allele has two roles: It directs cells to form an antenna instead of a leg and controls the proximal (closest to the body) to distal (farthest from the body) specification of antennal structures.

The antenna is the ear and nose of the fly and has several components, which include the arista and three antennal segments (Figure 20–16a). The arista vibrates in response to sound waves; these vibrations initiate signals that are transferred to the first antennal segment and through the antennal nerve (and second antennal segment) to the brain. The third antennal segment is covered with olfactory receptors. These receptors are stimulated by odors, which generate signals that are transferred to the brain through the first two antennal segments and the nerves connected to the brain.

Expression of *Dll* in the late third larval instar and early pupal stage (Figure 20–16b and c) is restricted to the second and third antennal segments, and the arista. At this stage of development, five genes activated by *Dll* are expressed, mostly in a segment-specific pattern (Figure 20–16c). Inter-

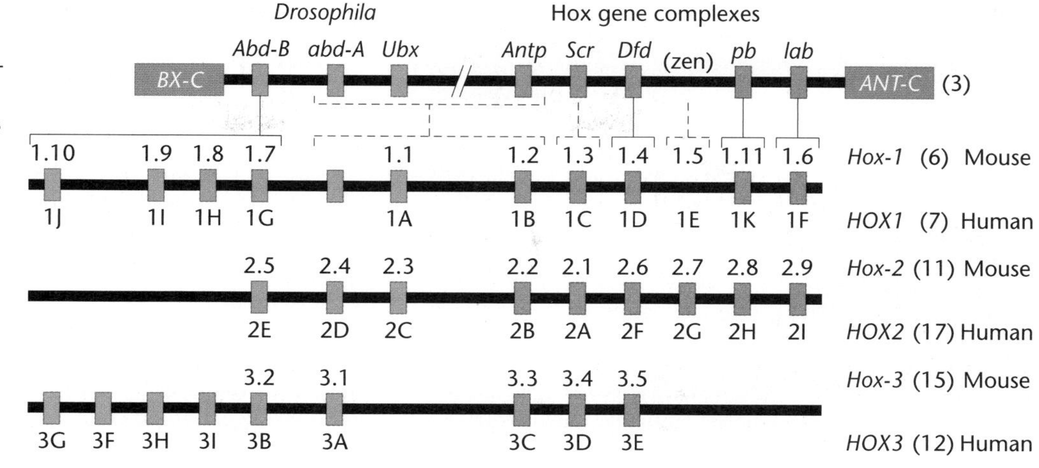

FIGURE 20–15 The *Hox* genes of mammals (human and mouse) and their organizational alignment with the *BX-C* and *ANT-C* complexes of *Drosophila*. Numbers in parentheses represent chromosome numbers.

estingly, four of these genes (*spalt*, *spineless*, *bric a brac*, and *aristaless*) encode transcription factors. The fifth, *dachshund*, encodes a protein of unknown function that is confined to the nucleus. It seems likely that the genes activated by *Dll* expression initiate a cascade of segment-specific gene expression to specify and differentiate each of the antennal structures, but the details of how this is accomplished are still unknown.

Two *Hox* genes, *Ultrabithorax* (*Ubx*) and *abdominal A* (*abd-A*), repress the expression of *Dll* in abdominal segments. This partly explains why flies do not form appendages on their abdominal segments.

To summarize, genes that control development in *Drosophila* act in a temporally and spatially ordered cascade, beginning with genes that establish the anterior–posterior and dorsal–ventral axes of the egg and early embryo. Gradients of maternal mRNAs and proteins along the anterior–posterior axis activate the gap genes, which subdivide the embryo into broad bands. Gap genes in turn activate the pair-rule genes, which divide the embryo into segments. The final group of segmentation genes, the segment polarity genes, divides each segment into anterior and posterior regions arranged linearly along the anterior–posterior axis. These segments are then given identity by homeotic genes. Therefore, this progressive restriction of the developmental potential of the *Drosophila* embryo's cells (all of which occur during the first one-third of embryogenesis) involves a cascade of gene action, with regulatory proteins acting at both the transcriptional and translational levels.

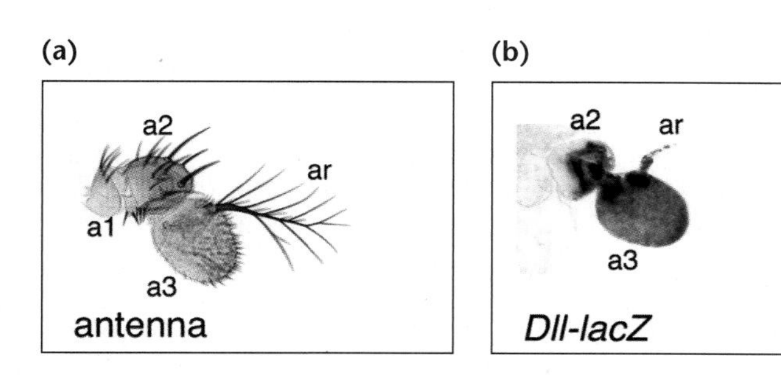

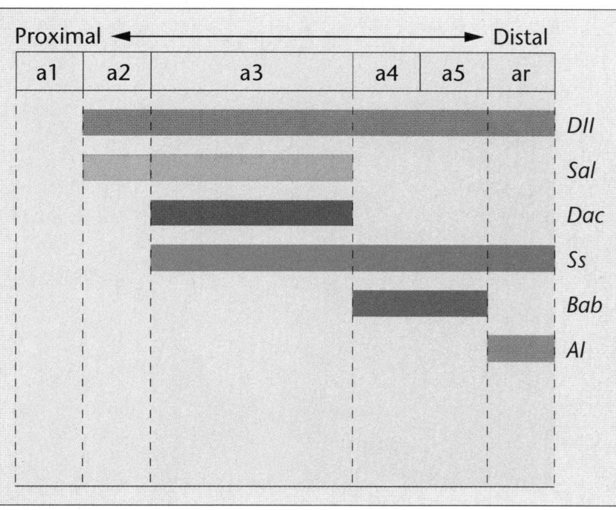

FIGURE 20–16 Action of the *Dll* gene in *Drosophila* produces (a) the wild-type antenna, divided into the arista (ar) and three antennal segments (a1, a2, a3). (b) *Dll* expression (shown in blue) in a late pupal antenna. Expression at this stage is limited to the arista and a3. (c) Activation of *Dll* target genes in antennal segments during the third larval instar. At this stage, *Dll* is expressed in all segments except a1. Each of these target genes is a transcription factor, which in turn activates a cascade of other genes. *(Parts (a) and (b) from Panganiban, G., and J. L. R. Rubenstein, 2002. Developmental functions of the* Distal-less/Dlx *homeobox genes.* Development *129: 4371–86. Figure 1, p. 4372. Part (c)* Distal-less *function during* Drosophila *appendage and sense organ development, Panganiban, G, Copyright © 2000. Reprinted by permission of Wiley-Liss, Inc., a subsidiary of John Wiley & Sons, Inc.)*

Flower Development in *Arabidopsis*

Flower development in *Arabidopsis thaliana* (Figure 20–17), a small plant in the mustard family, has been used to study pattern formation in flowering plants. A cluster of undifferentiated cells called the floral meristem gives rise to flowers (Figure 20–18). Each flower consists of four organs—sepals, petals, stamens, and carpels—that develop from four concentric whorls of cells within the meristem (Figure 20–19); each organ develops from a different whorl. Three classes of floral homeotic genes control

FIGURE 20–17 The flowering plant, *Arabidopsis thaliana*, used as a model organism in plant genetics. *(Elliot M. Meyerowitz/ California Institute of Technology, Div. of Biology)*

the formation of these organs. These genes are activated in an overlapping pattern in the whorls to specify identity of the floral organs. In the outermost whorl, expression of Class A genes gives rise to sepals. In the second whorl, expression of Class A *and* Class B genes gives rise to petals. In the third whorl, expression of Class B *and* Class C genes gives rise to stamens, and in the innermost whorl, expression of Class C genes gives rise to carpels (Figure 20–20a). The genes in each class are listed in Table 20.3.

As in *Drosophila*, mutations in homeotic genes cause normal organs to form in abnormal locations. For example, instead of the normal sepal, petal, stamen, carpel arrangement, a Class A mutation *APETALA2* (*AP2*) (Figure 20–20b), has a carpel, stamen, stamen, carpel arrangement. Mutants of the Class B gene *PISTILLATA* (*PI*) have sepal, sepal, carpel, carpel flowers (Figure 20–20c). In *AGAMOUS* (*AG*) mutants, which lack activity of a Class C gene, the order of organs is sepal, petal, petal, sepal (Figure 20–20d). If mutation eliminates activity of both A- and B-class genes, then the whorls will have only Class C activity (which spread to the first and second whorls because class A is absent), and only carpels will form.

The homeotic ABC genes of *Arabidopsis* are transcription factors expressed in overlapping patterns, just like the homeotic genes in *Drosophila*. In *Arabidopsis*, however, the floral homeotic genes are not homeobox genes, they belong to a different family of transcription factors called the **MADS-box proteins**. Each member of the MADS box contains a common sequence of 58 amino acids that are unrelated to the homeobox sequence. Thus, plants and animals independently evolved the process of

FIGURE 20–18 (a) Parts of the *Arabidopsis* flower. The floral organs are arranged concentrically. The sepals form the outermost ring, followed by petals and stamens, with carpels on the inside. (b) View of the flower from above. *([On-line]* www.salk.edu/LABS/pbio-w/gallery/wt_flower.jpg) *(Max-Planck-Institut fur Entwicklungsbiologie)*

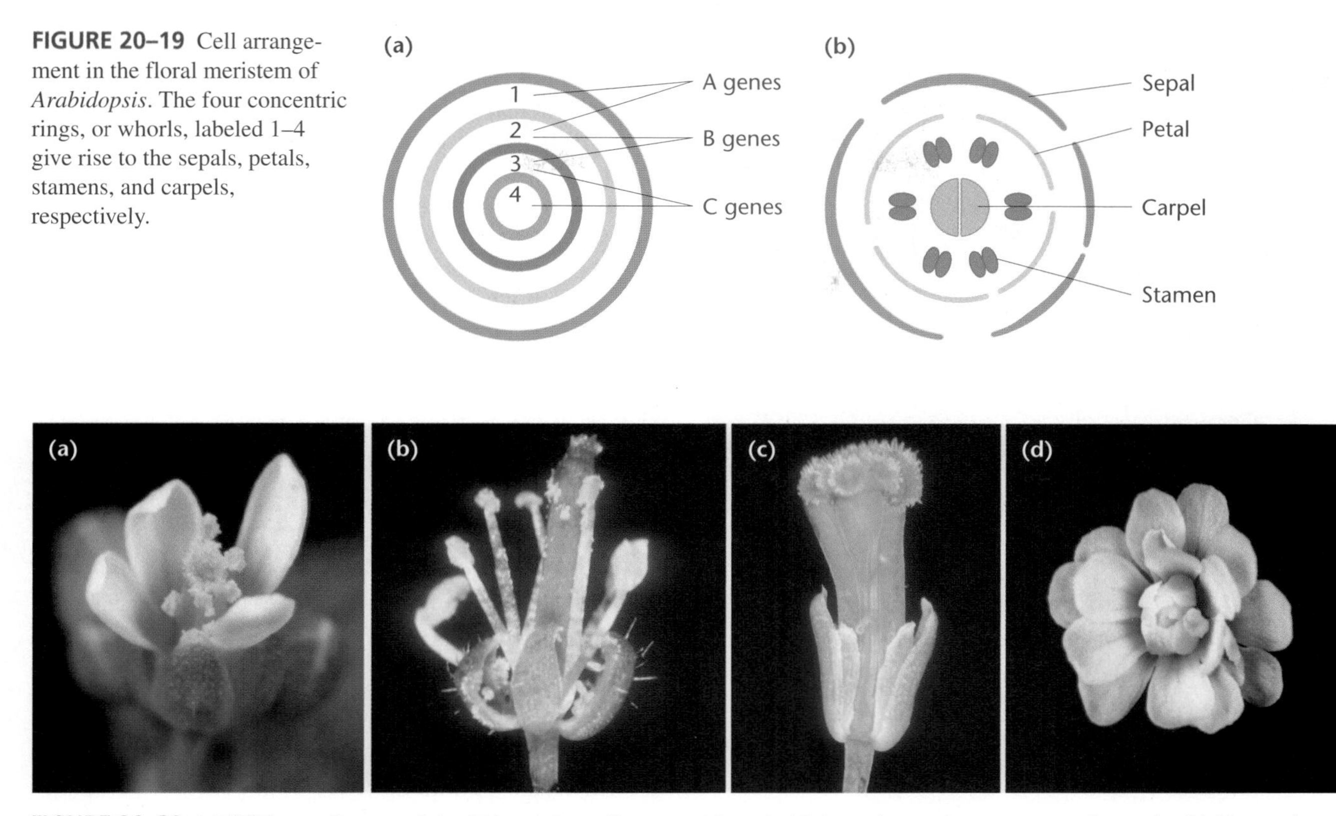

FIGURE 20–19 Cell arrangement in the floral meristem of *Arabidopsis*. The four concentric rings, or whorls, labeled 1–4 give rise to the sepals, petals, stamens, and carpels, respectively.

FIGURE 20–20 (a) Wild-type flowers of *Arabidopsis* have (from outside to inside) sepals, petals, stamens, and carpels. (b) Homeotic *APETALA2* mutant flower, with carpels, stamens, stamens, and carpels. (c) *PISTILLATA* mutants have sepals, sepals, carpels, and carpels. (d) *AGAMOUS* mutants have petals and sepals at places where stamens and carpels should form. *(Dr. Jose Luis Riechmann, Division of Biology California Institute of Technology from Science 2002. 295, pp. 1482-85)*

pattern formation by similar mechanisms starting with different families of transcription factors.

Both *Drosophila* and *Arabidopsis* use members of the *Polycomb* gene family to maintain correct spatial expression of their respective homeotic gene sets. In *Drosophila*, *Polycomb* genes induce changes in the chromatin that inactivate *Hox* gene expression. In *Arabidopsis*, expression of the floral homeotic genes is controlled by *CURLY LEAF* (*CLF*), a gene with significant homology to members of the *Drosophila Polycomb* gene family. *CLF* works by altering the chromatin structure to regulate expression of *AG* and other floral genes. These parallel functions indicate that the mechanisms of chromatin regulation are conserved between plants and animals.

TABLE 20.3 Homeotic Selector Genes in *Arabidopsis*

Class A	*Class B*	*Class C*
*APETALA1 (AP1)**	*APETALA3 (AP3)*	*AGAMOUS (AG)*
APETALA2 (AP2)	*PISTILLATA (PI)*	

*By convention, we use capital letters for wild-type genes in *Arabidopsis*.

20.4 Cell–Cell Interactions Can Control Developmental Fate

During development in multicellular organisms, cells influence the transcriptional patterns and developmental fate of other cells; these interactions involve the generation and reception of signal molecules. Cell–cell interaction is an important process in the development of most eukaryotic organisms, including *Drosophila,* as well as vertebrates such as *Xenopus*, mice, and humans. We will examine cell–cell interaction in one well-studied organism, *Caenorhabditis elegans*.

Overview of *C. elegans* Development

The nematode *C. elegans* is widely used to study the genetic control of development. This organism has two advantages for such studies: Its genetics is well known, and adults have a small number of cells that follow a developmental program that does not change from individual to individual. The adult is about 1 mm long and matures from a fertilized egg in about 2 days (Figure 20–21).

The life cycle consists of an embryonic stage that lasts for about 16 hours, 4 larval stages (L1–L4), and the adult

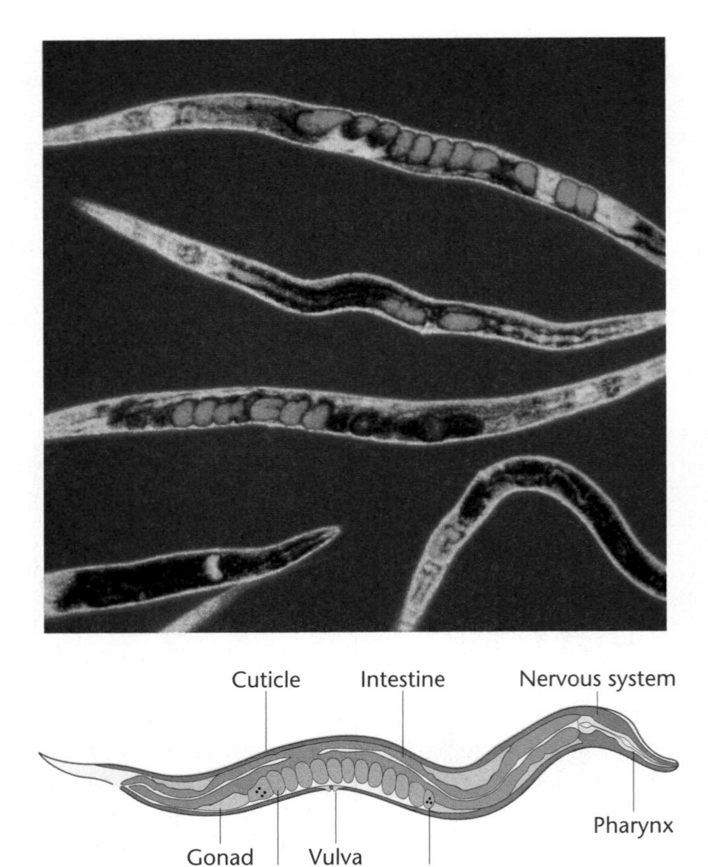

FIGURE 20–21 An adult *C. elegans* nematode, about 1 mm in length, consists of 959 somatic cells and has been used to study many aspects of the genetic control of development. *(James King-Holmes Photo Researcher, Inc.)*

stage. Adults are of two sexes—XX self-fertilizing hermaphrodites that can make both eggs and sperm, and XO males.

The adult hermaphrodite consists of 959 somatic cells and about 2000 germ cells. The exact cell lineage from fertilized egg to adult has been mapped (Figure 20–22) and does not vary between individuals. Knowing the lineage of each cell, we can easily follow the events that result either from mutational alterations in cell fate or from killing cells with laser microbeams or ultraviolet irradiation.

In *C. elegans* hermaphrodites, the fate of cells in the developing reproductive system is determined by cell–cell interactions and gives us insight into how gene expression and cell–cell interactions work together to specify developmental pathways.

Cell–Cell Interactions in Vulva Formation

Adult hermaphrodites lay eggs through a structure called the **vulva**, an opening located about midbody (see Figure 20–21). The vulva is formed from ventral epidermal cells, and the ventral epidermis is formed from 12 cells, P1.p–P12.p. Six of these (P3.p–P8.p) can participate in vulva formation. Normally, only the central three cells (P5.p–P7.p) form the vulva. This process involves several rounds of cell–cell interactions: signaling to establish the anchor cell, inductive signaling from the anchor cell to the P5.p–P7.p ventral cells, and lateral signaling among the vulval precursor cells. To establish the anchor cell in the embryo, two neighboring cells, Z1.ppp and Z4.aaa, interact. As a result of this interaction, one becomes the anchor cell and the other a precursor to the uterus. The *lin-12* gene encodes a cell-surface receptor protein and controls the determination of which cell becomes the anchor cell and which becomes the uterine precursor. In recessive *lin-12(0)* mutants (loss-of-function mutants), both cells become anchor cells. The dominant mutant *lin-12(d)* (a gain-of-function mutation) causes both cells to become uterine precursors. Thus, expression of *lin-12* causes the uterine pathway to be selected since in the absence of the LIN-12 protein both cells become anchor cells. As

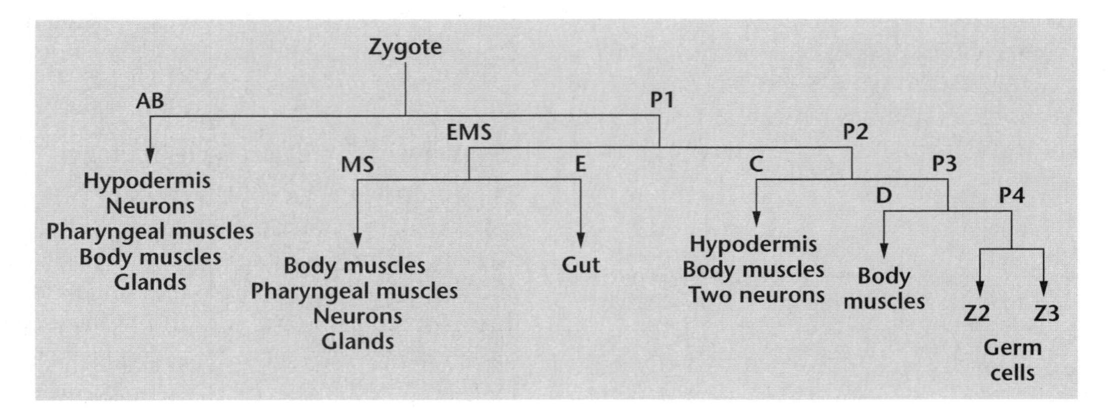

FIGURE 20–22 A truncated lineage chart for *C. elegans* that shows early cell divisions and the tissues and organs formed. Each vertical line represents a cell division, and horizontal lines connect the two cells produced. For example, the first cell division creates two new cells from the zygote, AB and P1. The cells in this chart refer to those present in the first-stage larva L1. During subsequent larval stages, further cell divisions will produce the 959 somatic cells of the adult hermaphrodite worm.

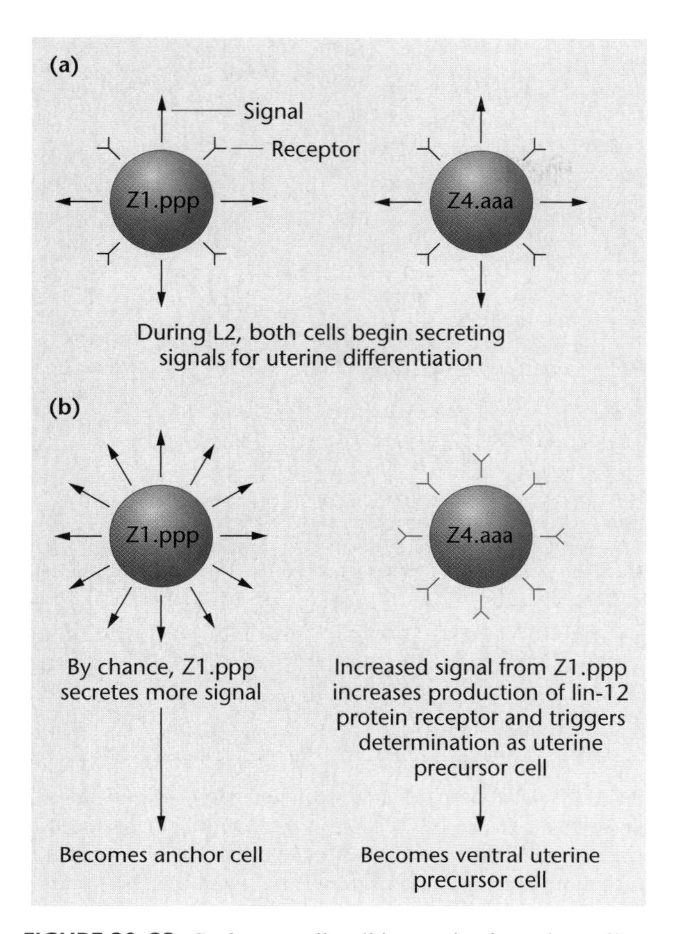

FIGURE 20–23 *C. elegans* cell–cell interaction in anchor cell determination. (a) During L2, two neighboring cells secrete chemical signals that induce uterine differentiation. (b) By chance, cell Z1.ppp secretes more of these signals, causing cell Z4.aaa to increase production of the receptor. The action of increased signals causes Z4.aaa to become the ventral uterine precursor cell and allows Z1.ppp to become the anchor cell.

shown in Figure 20–23, both cells normally synthesize and secrete a chemical signal for uterine differentiation and also synthesize the LIN-12 protein, which is a receptor for the signal. By chance, the cell that secretes more of the signal causes its neighbor to increase transcription of *lin-12*, increasing production of the receptor. The cell with more LIN-12 receptors becomes the uterine precursor, and the remaining cell becomes the anchor cell. The critical factor in this first round of cell–cell interaction and determination is the LIN-12 gene product.

A second round of cell–cell interactions involves the anchor cell and six precursor cells (P3.p–P8.p) located just below the anchor cell. The fate of each of these cells is specified by its position relative to the anchor cell. Figure 20–24 shows the determination pathway described in the following paragraphs. Sometime in larval stage 3, the *lin-3* gene is expressed in the anchor cell. The gene product is a signal protein related to vertebrate epidermal growth factor (EGF). All six precursor cells express a cell surface receptor to the LIN-3 protein, encoded by *let-23*, a gene similar to the vertebrate EGF receptor. Binding the LIN-3 protein to the LET-23 receptor triggers an intracellular cascade of events that determines whether the precursor cells will form the primary vulval precursor cell or secondary vulval cells. Recessive loss-of-function mutations in *let-23* cause the precursors to form nonvulval structures. In other words, in *let-23* mutants, precursor cells act as though they have not received a signal from the anchor cell, and no vulva is formed.

The signals that are transmitted from anchor cells to precursor cells also involve the *let-60* gene. Recessive mutations in *let-60* cause precursor cells to develop as though they have not received a signal from the anchor cell. Dominant *let-60* alleles have the opposite phenotype, causing all precursor cells to respond, thus forming multiple vulvas. The *let-60* gene is the *C. elegans* homolog of *ras*, a human proto-oncogene. The *let-60* dominant gain-of-function mutant that causes multiple vulva formation has a Gly-Glu mutation at amino acid 13, the same mutation that converts *c-ras* into an oncogene.

Normally, the cell closest to the anchor cell (P6.p) receives the strongest signal initiated by binding of LIN-3 to LET-23. This signal activates expression of the *Vulvaless* (*Vul*) gene (named for its mutant phenotype) in P6.p, and the cell adopts the primary pathway—it divides three times to produce vulva cells. The two neighboring cells (P5.p and P7.p) receive a lower amount of signal and initiate a secondary pathway—these cells divide asymmetrically to form additional vulva cells.

To reinforce these developmental pathways, a third level of cell-cell interaction is used. The P6.p cell (the primary vulval cell) activates the *lin-12* gene in the two neighboring cells (P5.p and P7.p). This signal prevents P5.p and P7.p from adopting the division pattern of the primary cell. In other words, cells in which both *Vul* and *lin-12* are active cannot become primary vulval cells. The three remaining Pn.p cells (P3.p, P4.p, and P8.p) receive no signal from the anchor cell. In these cells, the *Multivulva* (*Muv*) gene is expressed, *Muv* represses *Vul*, and the three cells develop as skin cells.

Thus, three levels of cell–cell interaction are used in the developmental pathway leading to vulva formation in *C. elegans*. First, two neighboring cells interact to establish the identity of the anchor cell. Second, the anchor cell interacts with three vulval precursor cells to establish the primary vulval precursor cell (usually P6.p) and two secondary cells (usually P5.p and P7.p). Third, the primary vulval cell interacts with the secondary cells to suppress their ability to adopt the pathway of the primary cell. Each interaction is accompanied by the secretion of molecular signals and by the reception and processing of these signals by neighboring cells. This theme of cell–cell interaction acting in a spatial and temporal cascade to specify the developmental fates of individual cells is repeated over and over in organisms from prokaryotes to higher vertebrates.

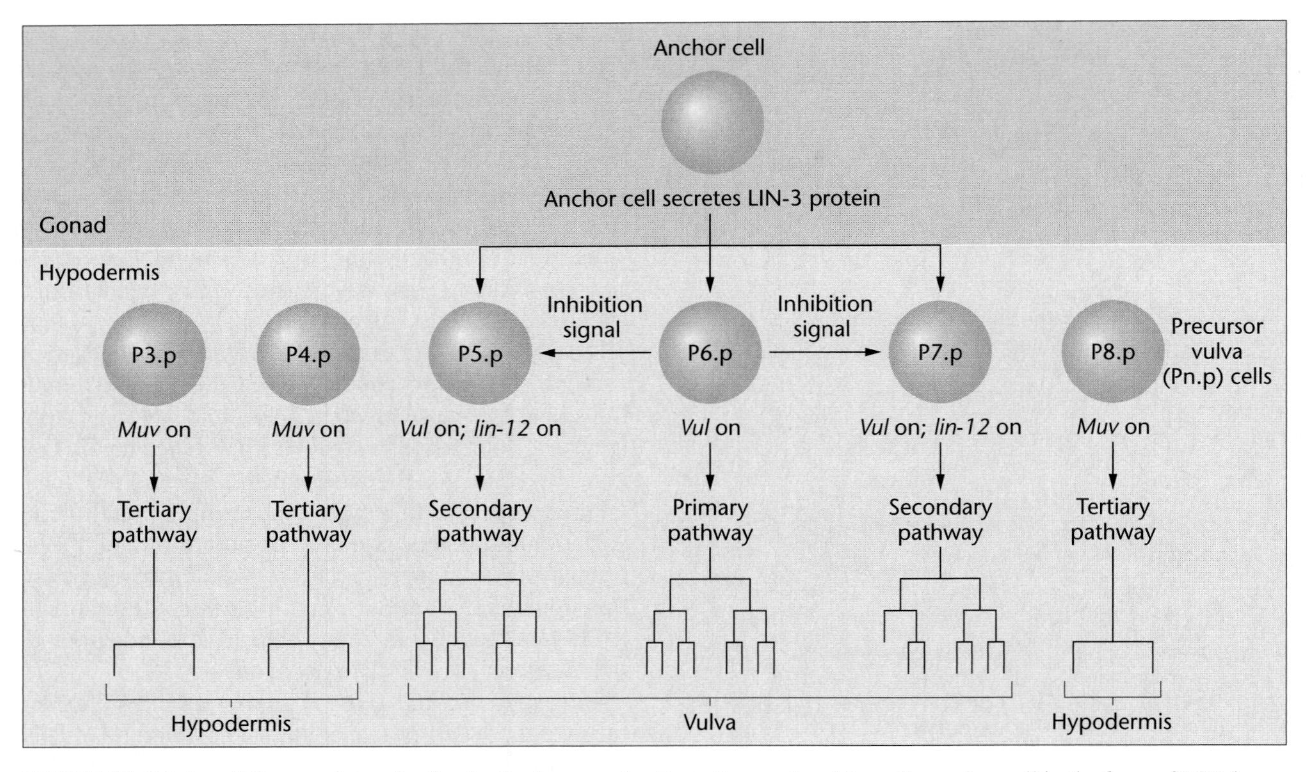

FIGURE 20–24 In cell-lineage determination in *C. elegans* vulva formation, a signal from the anchor cell in the form of LIN-3 protein is received by three precursor vulval cells (Pn.p cells). The cells closest to the anchor cell become primary vulval precursor cells, and adjacent cells become secondary precursor cells. Primary cells secrete a signal that activates the *lin-12* gene in secondary cells, preventing them from becoming primary cells. Flanking Pn.p cells, which receive no signal from the anchor cell *Muv* gene, become hypodermis cells instead of vulval cells.

GENETICS, TECHNOLOGY, AND SOCIETY

Stem Cell Wars

Stem cell research is at the center of a battle fought by scientists, politicians, advocacy groups, religious leaders, and ethicists. Proponents fight for the right to carry out stem cell research, claiming that it is revolutionary, if not miraculous—with the potential to cure diabetes, Parkinson's disease and spinal cord injuries, and also improve the quality of life for millions. Critics lobby for an end to stem cell research, warning that it will propel us down the slippery slope toward disregard for human life. Although stem cell research is the focus of presidential proclamations, highly publicized media campaigns and legislative bans, few of us understand it sufficiently to evaluate its pros and cons. What is stem cell research, and why does it spark such intense controversy?

Stems cells are primitive cells that replicate indefinitely and have the unique capacity to differentiate into cells with specialized functions, such as those found in the heart, brain, liver, and muscle. Stem cells are the origin of all the cells that make up the approximately 200 distinct types of tissues in our bodies. In contrast to stem cells, mature, fully differentiated cells do not replicate or undergo transformations into different cell types. Some types of stem cells are defined as *totipotent*, meaning that they have the ability to differentiate into any mature cell type in the body. Other types of stem cells are pluripotent, and are able to differentiate into only one of several mature cell types.

In the last few years, several research teams have isolated and cultured human pluripotent stem cells. These cells remain undifferentiated and grow indefinitely in culture dishes. When treated with growth factors or hormones, these pluripotent stem cells differentiate into cells that have characteristics of neural, bone, kidney, liver, heart or pancreatic cells.

The fact that pluripotent stem cells grow prolifically in culture and differentiate into more specialized cells has created great excitement. Some foresee a day when stem cells may be a cornucopia from which to harvest unlimited numbers of specialized cells to replace cells in damaged, and diseased tissues. Hence, stem cells could be used to treat Parkinson's disease, type 1 diabetes, chronic heart disease, kidney and liver failure, Alzheimer disease, Duchenne muscular dystrophy, and spinal cord injuries. Some predict that stem cells will be genetically modified to eliminate transplant rejection or to deliver specific gene products, thereby correcting genetic defects or treating cancers. The excitement about stem cell therapies has been fueled by reports of dramatically successful experiments in animals. For example, mice with spinal cord injuries regained their mobility, and bowel and bladder control after they were injected with human stem cells. Both proponents and critics of stem cell research agree that stem cell therapies could be revolutionary. Why, then, should stem cell research be so contentious?

The answer to that question lies in the source of pluripotent stem cells. To date, all pluripotent stem cell lines have been derived from five-day embryonic blastocysts. Blastocysts at this stage consist of about 200 cells, most of which will develop into placental and supporting tissues for the early embryo. The inner cell mass of the blastocyst consists of about 30–40 pluripotent stem cells that develop into all tissues of the embryo. *In vitro* fertilization clinics grow fertilized eggs to the five-day blastocyst stage prior to uterine transfer. Embryonic stem (ES) cell lines are created by dissecting out the inner cell mass of five-day blastocysts and growing the undifferentiated cells in culture dishes. All human ES cell lines have been derived from unused five-day blastocysts that were discarded by *in vitro* fertilization clinics.

The fact that early embryos are destroyed in the process of establishing human ES cell lines disturbs people who believe that preimplantation embryos are persons with rights; however, it does not disturb people who believe that these embryos are too primitive to have an inherent moral status. Both sides in the debate put forth lengthy arguments revolving around the fundamental question of what constitutes a human being.

Critics of ES cell research argue that we may be able to benefit from stem cell therapies without resorting to the use of ES cells. This argument is based on recent reports about the plasticity of adult stem cells. Adult stem cells are undifferentiated cells that are present in differentiated tissues such as blood and brain. They divide within the differentiated tissue and differentiate into mature cells that make up the tissue in which they are found. Adult stem cells have been found in bone marrow, blood, the retina, the brain, skeletal muscle, the liver, skin, and the pancreas. The best-known adult stem cells are hematopoietic stem cells (HSCs) which are found in bone marrow, peripheral blood, and umbilical cords. HSCs differentiate into mature blood cell types such as red blood cells, lymphocytes, and macrophages. HSCs have been used clinically for many years, as transplant material to reconstitute the immune systems of patients undergoing treatment for cancer and autoimmune diseases. Interestingly, recent studies suggest that adult stem cells may have the capacity to differentiate into other cell types.

Although these reports are intriguing, it is still too early to know whether adult stem cells will hold the same pluripotent promise as ES cells. Adult stem cells are rare, are difficult to identify and isolate, and grow poorly, if at all, in culture. However, if these obstacles can be overcome, adult stem cells may provide an ethical alternative to ES cell and calm the raging debate. On the other hand, new philosophical dilemmas could be created. If adult stem cells are found to exhibit the same pluripotency as ES cells, they could have the same potential to create a human embryo—dragging critics and proponents of stem cell research back into the same moral quagmire. At the present time, it is impossible to predict whether either adult or embryonic stem cells will be as miraculous as predicted by scientists and the popular press. But if stem cell research progresses at its current rapid pace, we won't have long to wait.

References

Robertson, J. A. 2001. Human embryonic stem cell research: ethical and legal issues. *Nat. Rev. Gen.* 2: 74–78.

Freed, C. R. 2002. Will embryonic stem cells be a useful source of dopamine neurons for transplant into patients with Parkinson's disease? *Proc. Natl. Acad. Sci. (USA)* 99: 1755–57.

Web Sites

National Institutes of Health. 2001. "Stem Cells: Scientific Progress and Future Research Directions."
http://www.nih.gov/news/stemcell/_scireport.htm

National Institutes of Health. 2000. "Stem Cells: A Primer."
http://www.nih.gov/news/stemcell/_primer.htm

Chapter Summary

1. The role of genetic information during development and differentiation is a major research topic in biology and has been studied extensively. Geneticists are isolating developmental mutations and identifying the genes that control developmental processes.
2. Determination is the regulatory event whereby cell fate becomes fixed during early development. Determination precedes the actual differentiation or specialization of distinctive cell types.
3. During embryogenesis, the internal environment of the cell appears to affect specific gene activity. The regulation of early events is mediated by the maternal cytoplasm, which then influences zygotic gene expression. As development proceeds, both the cell's internal and external environments become further altered by the presence of early gene products and by communication with other cells.
4. In *Drosophila*, both genetic and molecular studies confirm that the egg contains information that specifies the body plan of the larva and adult, and that interactions of embryonic nuclei with the maternal cytoplasm initiate transcriptional programs characteristic of specific developmental pathways.
5. Extensive genetic analysis of embryonic development in *Drosophila* has identified maternal-effect genes that lay down the anterior–posterior axis of the embryo. In addition, these maternal-effect genes activate sets of zygotic segmentation genes, initiating a cascade of gene regulation that ends with homeotic genes determining segment identity.
6. The invariant lineage of all cells in *C. elegans* allows developmental biologists to study the cell–cell signaling required for organ formation.

Key Terms

Antennapedia complex (*ANT-C*), 467
bithorax complex (*BX-C*), 467
cell–cell interaction, 461
cellular blastoderm, 462
compartment, 462
cytoplasmic localization, 461
determination, 461
differentiation, 461
gap gene, 465
homeobox, 467
homeodomain, 467
homeotic gene, 467
homeotic mutant, 467
Hox gene clusters, 468
imaginal disc, 463
MADS-box proteins, 471
maternal-effect genes, 463
pair-rule gene, 466
segment polarity gene, 466
segmentation gene, 465
syncytial blastoderm, 462
vulva, 472
zygote, 461

INSIGHTS AND SOLUTIONS

The timing of differential gene action during development is the key to normal developmental programs. If a gene has been cloned, the time of action and range of cell types in which the gene is active can be determined using molecular techniques. However, when a cloned gene or its transcripts are not available, genetic techniques can establish a comparative order of gene action among two or more genes. By extending this analysis to include several genes, a relative order of gene action can be established. This order serves as a starting point for investigations at the molecular level by providing developmental time scales for gene action. The following example shows how a time scale for the action of two genes is constructed.

In *Drosophila*, the autosomal recessive gene *lozenge-clawless* (*lzcl*) produces multiple abnormalities of the female genitals, eyes, and tarsal regions of the legs. Homozygous females have abnormal genitals, no sperm storage organs, no ovarian glands, and they are sterile. Homozygous mutant males, on the other hand, have normal genitals and are fully fertile. A second autosomal recessive gene, *transformer* (*tra*), converts XX females into phenotypic males (recall that in *Drosophila*, XX flies are female, XY flies are male). Thus, XX flies homozygous for both *lzcl* and *tra* are phenotypically male with normal genitals. What do these results say about the normal sequence of action of these two genes? Which one acts first during development?

Solution: These flies are genetically female since they have two X chromosomes. Females that are homozygous for the *lzcl* gene are expected to have abnormal genitalia. In this case, the action of the *tra* gene that changes the female phenotype into the male phenotype (with the development of male genitalia) must take place before the action of the *lzcl* gene since the XX flies have a normal male phenotype.

Problems and Discussion Questions

1. Carefully distinguish between the terms differentiation and determination. Which phenomenon occurs initially during development?
2. Nuclei from almost any source may be injected into *Xenopus* oocytes. Studies have shown that these nuclei remain active in transcription and translation. How can such an experimental system be useful in developmental genetic studies?
3. The homunculus doctrine postulated that miniature adult entities are contained within the egg and merely unfold and grow to give rise to a mature organism. What sorts of isolated evidence presented in this chapter might have led to this doctrine? Why is the epigenetic theory held as correct today?
4. (a) What are the imaginal disks of *Drosophila*? (b) When do they form, how many are there, and what structures do they form in the adult?
5. Distinguish between the syncytial blastoderm stage and the cellular blastoderm stage in *Drosophila* embryogenesis.
6. (a) What are maternal-effect genes? (b) When are gene products from these genes made, and where are they located? (c) What aspects of development do maternal-effect genes control? (d) What is the phenotype of maternal-effect mutations?
7. Suppose you initiate a screen for maternal-effect mutations in *Drosophila* affecting external structures of the embryo and your screen identifies over 100 mutations that affect external structures. Weischaus and Schupbach estimated from their screening that there are about 40 maternal-effect genes. How do you reconcile these different results?
8. (a) What are zygotic genes, and when are their gene products made? (b) What is the phenotype associated with zygotic gene mutations? Does the maternal genotype contain zygotic genes?
9. List the main classes of zygotic genes. What is the function of each class of these genes?
10. Experiments have shown that any nuclei placed in the polar cytoplasm at the posterior pole of the *Drosophila* egg will differentiate into germ cells. If polar cytoplasm is transplanted into the anterior end of the egg just after fertilization, what will happen to nuclei that migrate into this cytoplasm at the anterior pole?
11. In the sea urchin, early development up to gastrulation may occur even in the presence of actinomycin D, which inhibits RNA synthesis. However, if actinomycin D is present early in development but removed at the end of blastula formation, gastrulation does not proceed. In fact, if actinomycin D is present only between the sixth and eleventh hours of development, gastrulation (normally occurring at the fifteenth hour) is arrested. What conclusions can be drawn concerning the role of gene transcription between hours six and 15?
12. How can you determine whether a particular gene is being transcribed in different cell types?
13. You observe that a particular gene is being transcribed during development, how can you tell whether the expression of this gene is under transcriptional or translational control?
14. Define what is meant by a homeotic mutant. If it were possible to introduce one of the homeotic genes from *Drosophila* into an *Arabidopsis* embryo that was homozygous for a homeotic flowering gene, would you expect any of the *Drosophila* genes to negate (rescue) the *Arabidopsis* mutant phenotype? Why or why not?
15. What are *Hox* genes? What properties do they have in common? Are all homeotic genes *Hox* genes?
16. The homeotic mutation *Antennapedia* causes mutant *Drosophila* to have legs in place of antennae and is a dominant gain-of-function mutation. What are the properties of such mutations? How does the *Antennapedia* gene change antennae into legs?
17. The *Drosophila* homeotic mutation, *spineless aristapedia* (ss^a), results in the formation of a miniature tarsal structure (normally part of the leg) on the end of the antenna. From your knowledge of imaginal discs, what insight is provided by ss^a concerning the role of genes during determination?
18. Embryogenesis and oncogenesis (generation of cancer) share a number of features including cell proliferation, apoptosis (cell death), cell migration and invasion, formation of new blood vessels, and differential gene activity. Embryonic cells are relatively undifferentiated and cancer cells appear to be undifferentiated or dedifferentiated. Homeotic gene expression directs early development, and mutant expression leads to loss of the differentiated state or an alternative cell identity. Lewis (2000. *Breast Can. Res.* 2: 158–69.) suggested that breast cancer may be caused by the altered expression of homeotic genes. When he examined 11 such genes in cancers, 8 were underexpressed while 3 were overexpressed compared with controls. Given what you know about homeotic genes, what is the likelihood that they are involved in oncogenesis?
19. In *Drosophila*, both *fushi tarazu* (*ftz*) and *engrailed* encode homeobox transcription factors and are capable of eliciting the expression of other genes. Both genes work at about the same time during development and in the same region to specify cell fate in body segments. To discover if *ftz* regulates the expression of *engrailed*, if *engrailed* regulates *ftz;* or if both are regulated by another gene, you perform a mutant analysis. In ftz^- embryos (*ftz*/*ftz*), engrailed protein is absent; in $engrailed^-$ embryos (*eng*/*eng*), *ftz* expression is normal. What does this tell you about the regulation of these two genes—does the *engrailed* gene regulate *ftz*, or does the *ftz* gene regulate *engrailed*?
20. Early development depends on the temporal and spatial interplay between maternally supplied material and mRNA and the onset of zygotic gene expression. Maternally encoded mRNAs must be produced, positioned, and degraded (Surdej and Jacobs-Lorena, 1998. *Mol. Cell Biol.* 18: 2892–2900.). For example, transcription of the *bicoid* gene that determines anterior–posterior polarity in *Drosophila* is maternal. The mRNA is synthesized in the ovary by nurse cells and then transported to the oocyte, where it localizes to the anterior ends of oocytes. After egg deposition, *bicoid* mRNA is translated and unstable bicoid protein forms a decreasing concentration gradient from the anterior end of the embryo, where *gap* genes along the anterior half of the embryo are activated. At the start of gastrulation, *bicoid* mRNA has been degraded. Consider two models to explain the degradation of *bicoid* mRNA: (1) Degradation may result from signals within the mRNA (intrinsic model), or (2) degradation may result from the mRNA's position within the egg (extrinsic model). Experimentally, how could one distinguish between these two models?
21. Formation of germ cells in *Drosophila* and many other embryos is dependent on their position in the embryo and their exposure to localized cytoplasmic determinants. Nuclei exposed to cytoplasm in the posterior end of *Drosophila* eggs (the pole plasm) form cells that develop into germ cells under the direction of maternally derived components. Amikura et al. (2001.

Proc. Nat. Acad. Sci. (USA) 98: 9133–38.) consistently found mitochondria-type ribosomes outside mitochondria in the germ plasma of *Drosophila* embryos and postulated that they are intimately related to germ-cell specification. If you were studying this phenomenon, what would you want to know about the activity of these ribosomes?

22. One of the most interesting aspects of early development is the remodeling of the cell cycle from rapid cell divisions, apparently lacking G1 and G2 phases, to slower cell cycles with measurable G1 and G2 phases and checkpoints. During this remodeling, maternal mRNAs that specify cyclins are deadenylated and zygotic genes are activated to produce cyclins. Audic et al. (2001. *Mol. and Cell. Biol.* 21: 1662–71.) suggest that deadenylation requires transcription of zygotic genes. Present a diagram that captures the significant features of these findings.
23. In studying gene action during development, it is desirable to be able to position genes in a hierarchy or pathway of action to establish which genes are primary and in what order genes act. There are several ways of doing this. One is to make double mutants and study the outcome. The gene *fushi-tarazu* (*ftz*) is expressed in early embryos at the seven-stripe stage. All of the genes involved in forming the anterior–posterior pattern affect the expression of this gene, as do the *gap* genes. However, expression of segment-polarity genes is affected by *ftz*. What is the location of *ftz* in this hierarchy?
24. A number of genes that control expression of *Hox* genes in *Drosophila* have been identified. One of these homozygous mutants is *extra sex combs*, where some of the head and all of the thorax and abdominal segments develop as the last abdominal segment. In other words, all affected segments develop as posterior segments. What does this phenotype tell you about which set of *Hox* genes is controlled by the *extra sex combs* gene?
25. The *apterous* gene in *Drosophila* encodes a protein required for wing patterning and growth. It is also known to function in nerve development, fertility, and viability. When human and mouse genes whose protein products closely resemble *apterous* were used to generate transgenic *Drosophila* (Rincon-Limas et al. 1999. *Proc. Nat. Acad. Sci. (USA)* 96: 2165–70.), the apterous mutant phenotype was *rescued*. In addition, the whole-body expression patterns in the transgenic *Drosophila* were similar to normal *apterous*. (a) What is meant by the term *rescued* in this context? (b) What do these results indicate about the molecular nature of development?
26. In *Arabidopsis*, flower development is controlled by sets of homeotic genes. How many classes of these genes are there, and what structures are formed by their individual and combined expression?
27. The floral homeotic genes of *Arabidopsis* are MADS-box proteins, while in *Drosophila*, they are *Hox* genes, belonging to the homeobox gene family. In both *Arabidopsis* and *Drosophila*, members of the *Polycomb* gene family control expression of these divergent homeotic genes. How do *Polycomb* genes control expression of two very different sets of homeotic genes?
28. Vulval development in *C. elegans* is initiated when two neighboring cells (Z1.ppp and Z4.aaa) interact with each other by cell–cell signaling. This signaling involves two components: a membrane-bound signal molecule and a membrane-bound receptor. Initially the cells are developmentally equivalent and produce low levels of both signal and receptor. By chance, the cell that produces more signal causes its neighbor to produce more receptor. The cell producing more signal adopts one developmental fate (anchor cell); the cell producing more receptor adopts another fate (uterine precursor). This form of cell–cell interaction is called the Notch/Delta signaling system and is widely used in metazoan organisms in blood cell development, neurogenesis, retinal development, and other pathways of differentiation. Although it is a widely used signaling mechanism, this pathway works only in adjacent cells. Why is this so, and what are the advantages and disadvantages of such a system?
29. The identification and characterization of genes that control sex determination has been another focus of investigators working with *C. elegans*. As with *Drosophila*, sex in this organism is determined by the ratio of X chromosomes to sets of autosomes. A diploid wild-type male has one X chromosome, and a diploid wild-type hermaphrodite has two X chromosomes. Many different mutations have been identified that affect sex determination. Loss-of-function mutations in a gene called *her-1* cause an XO nematode to develop into a hermaphrodite and have no effect on XX development. (That is, XX nematodes are normal hermaphrodites.) In contrast, loss-of-function mutations in a gene called *tra-1* cause an XX nematode to develop into a male. Deduce the roles of these genes in wild-type sex determination from this information.
30. Based on the information in Problem 29 and the analysis of the phenotypes of single- and double-mutant strains, a model for sex determination in *C. elegans* has been generated. This model proposes that the *her-1* gene controls sex determination by establishing the level of activity of the *tra-1* gene, which in turn, controls the expression of genes involved in generating the various sexually dimorphic tissues. Given this information, (a) does the *her-1* gene product have a negative or a positive effect on the activity of the *tra-1* gene? (b) What would be the phenotype of a *tra-1*, *her-1* double mutant?

Selected Readings

Akam, M. 1998. *Hox* genes: From master genes to micromanagers. *Curr. Biol.* 8: R676–78.

Beachy, P., Helfand, S., and Hogness, D. 1985. Segmental distribution of bithorax complex proteins during *Drosophila* development. *Nature* 313: 545–51.

Brenner, S. 1974. The genetics of *Caenorhabditis elegans*. *Genetics* 77: 71–94.

DeRobertis, E., Oliver, G., and Wright, C. 1990. Homeobox genes and the vertebrate body plan. *Sci. Am.* (July) 262: 46–52.

Duboule, D., and Morata, G. 1994. Colinearity and functional hierarchy among genes of the homeotic complexes. *Trends Genet.* 10: 358–64.

Fay, D. S., and Han, W. 2000. The synthetic multivulval genes of *C. elegans*: Functional redundancy, Ras-antagonism, and cell fate determination. *Genesis* 26: 279–84.

Ferrier, D., and Holland, W. 2001. Ancient origin of the Hox gene cluster. *Nat. Rev. Genet.* 2: 33–38.

Grant, K. et al. 2000. *sem-4* promotes vulval cell-fate determination in *Caenorhabditis elegans* through regulation of *lin-39 Hox*. *Dev. Biol.* 2244: 496–506.

Gurdon, J. 1968. Transplanted nuclei and cell differentiation. *Sci. Am.* (Dec.) 219: 24–35.

Honma, T., and Goto, K. 2000. The *Arabidopsis* floral homeotic gene PISTILLA is regulated by discrete cis-elements responsive to induction and maintenance signals. *Development* 127: 2021–30.

Jenik, P. D., and Irish, V. F. 2000. Regulation of cell proliferation patterns by homeotic genes during *Arabidopsis* floral development. *Development* 127: 1267–76.

Köhler, C., and Grossniklaus, U. 2002. Epigenetic inheritance of expression states in plant development: The role of *Polycomb* group proteins. *Curr. Opin. Cell Biol.* 14: 773–79.

Koornneef, M. et al. 1998. Genetic control of flowering time in *Arabidopsis*. *Ann. Rev. Plant Mol. Biol.* 49: 345–70.

Lawrence, P. A., and Morata, G. 1994. Homeobox genes: Their function in *Drosophila* segmentation and pattern formation. *Cell* 78: 181–89.

Manseau, L., and Schupbach, T. 1989. The egg came first, of course! Anterior–posterior pattern formation in *Drosophila* embryogenesis and oogenesis. *Trends Genet.* 5: 400–05.

Meyerowitz, E. 1994. The genetics of flower development. *Sci. Am.* (Nov) 271: 56–65.

Meyerowitz, E. 2002. Plants compared to animals: The broadest comparative study of development. *Science* 295: 1482–85.

McGinnis, W., and Krumlauf, R. 1992. Homeobox genes and axial patterning. *Cell* 68: 283–302.

Panganiban, G. 2000. *Distal-less* function during *Drosophila* appendage and sense organ development. *Develop. Dynam.* 218: 554–62.

Panganiban, G., and Rubenstein, J. L. R. 2002. Developmental functions of the *Distal-less*/Dlx homeobox genes. *Develop.* 129: 4371–86.

Parcy, F. et al. 1998. A genetic framework for floral patterning. *Nature* 395: 561–66.

Sommer, R. J. 2001. As good as they get: Cells in nematode vulva development and evolution. *Curr. Opin. Cell Biol.* 13: 715–20.

CHAPTER 21

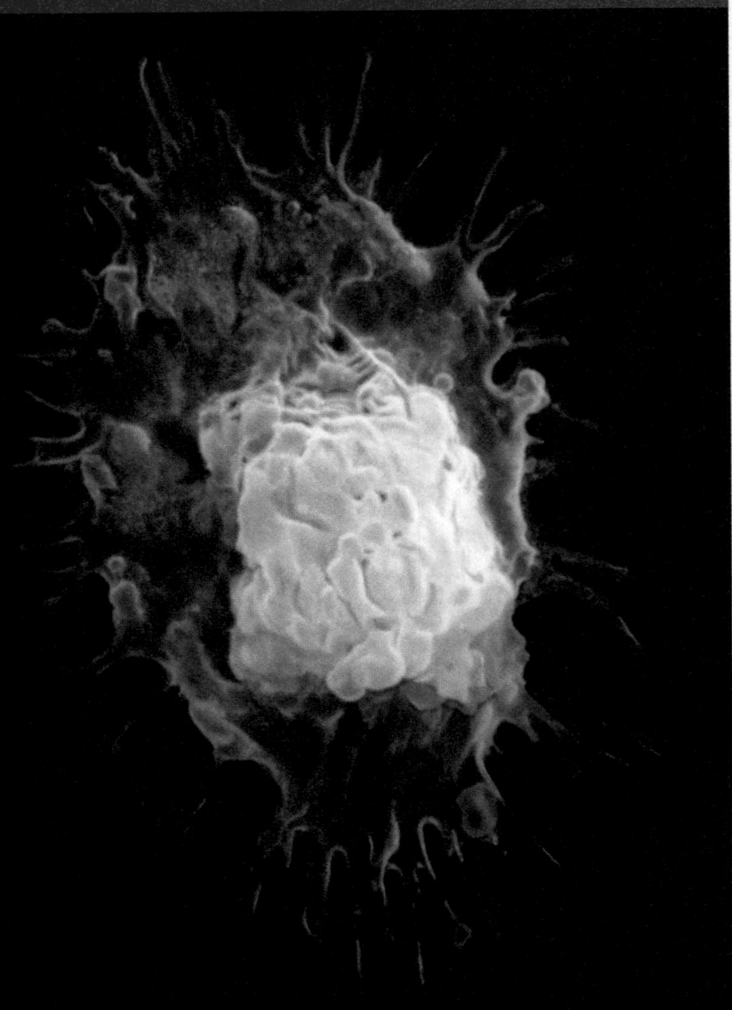

A human breast cancer cell. *(AMC/Albany Medical College/Custom Medical Stock Photo, Inc.)*

The Genetic Basis of Cancer

CHAPTER CONCEPTS

Cancer is a complex group of diseases affecting a wide range of cells and tissues. It is also a serious health problem. In the United States, the lifetime risk of developing cancer is one in two for males and one in three for females (Table 21.1). Mutations that alter the expression of certain genes or their gene products are now regarded as a common feature of all cancers. In most cancers, these mutations occur in somatic cells and are not passed on to future generations. However, in about 1% of cancers, germline mutations are transmitted to offspring and cause susceptibility to cancer. Studies of these mutations provide insights into the origins of cancer.

Genomic alterations associated with cancer can involve small-scale changes, such as single-nucleotide substitutions; or large-scale events, such as chromosome rearrangements, chromosome gain or loss, or even the integration of viral genomes into chromosomal sites. Large-scale genomic alterations are a common feature of cancer; the majority of human tumors are characterized by visible chromosomal changes (Figure 21–1).

It has been known for over 200 years that some types of cancer run in families. An analysis of familiar cancers has led to the identification of a class of genes called **cancer susceptibility genes** that increase the risk of cancer. Variant alleles of these susceptibility genes have an important role in sporadic cancers as well as familial forms of cancer. The likelihood that an individual will ultimately develop cancer depends upon the particular mutant allele, mutations in other genes, and environmental factors. These variables may influence the age of onset and the severity of the disease.

Cancer cells have two properties in common: (1) uncontrolled growth and (2) the ability to **metastasize** (spread) from their original site to other locations in the body. Cell division is the result of cells traversing the cell cycle; in cancer cells, control over the cell cycle is lost, and cells proliferate in an uncontrolled fashion. Investigations into genetic control of the cell cycle are providing insights into the origins of cancer.

Metastasis of cancer cells is controlled by gene products that become localized on the cell surface, controlling how these cells interact with the extracellular matrix and with other cells through cell-surface molecules. Loss of cell–cell contact allows cancer cells to leave the site of tumor formation, spread throughout the body, and invade other tissues. The study of metastasis is less well developed than that of cell-cycle regulation, but research is beginning to provide insights into the secondary events in tumor progression.

We will consider the relationship between genes and cancer, with emphasis on the relationship between the cell cycle and genetic disorders. We will also examine how mutations, chromosomal changes, and environmental agents play a role in the development of cancer.

HOW DO WE KNOW WHAT WE KNOW?

IN THIS CHAPTER, WE WILL EXAMINE the cell cycle, the regulatory checkpoints in the cycle, and the genes that control them along with a discussion of how cancer is related to the cell cycle and the role of the environment in cancer. As you study this topic, you should try to answer several fundamental questions:

1. How do we know that specific proteins in the cell control progress through the cell cycle?
2. How do we know that genes that control the cell cycle are also involved in cancer?
3. How do we know that cancer results from the accumulation of specific mutations in a single cell?
4. How do we know that environmental factors such as viruses play a role in the development of cancer? ■

21.1 Cancer as a Disorder of the Cell Cycle

The cell cycle represents the sequence of events that occurs between mitotic divisions in a eukaryotic cell. Because this cycle is closely related to the genetics of cancer, we will first discuss events in the cell cycle, and the genes that regulate progression through the cycle.

TABLE 21.1 Cancer Probabilities in the United States

Cancer Site		*Birth to 39*	*40–59*	*60–79*	*Birth To Death*
All sites	Male	1 in 62	1 in 12	1 in 3	1 in 2
	Female	1 in 52	1 in 11	1 in 4	1 in 3
Breast	Female	1 in 235	1 in 25	1 in 15	1 in 8
Prostate	Male	<1 in 10,000	1 in 53	1 in 7	1 in 6
Colon-rectum	Male	1 in 1500	1 in 124	1 in 29	1 in 18
	Female	1 in 1900	1 in 149	1 in 33	1 in 18
Lung-bronchus	Male	1 in 2500	1 in 78	1 in 16	1 in 12
	Female	1 in 2900	1 in 106	1 in 25	1 in 18

Source: American Cancer Society.

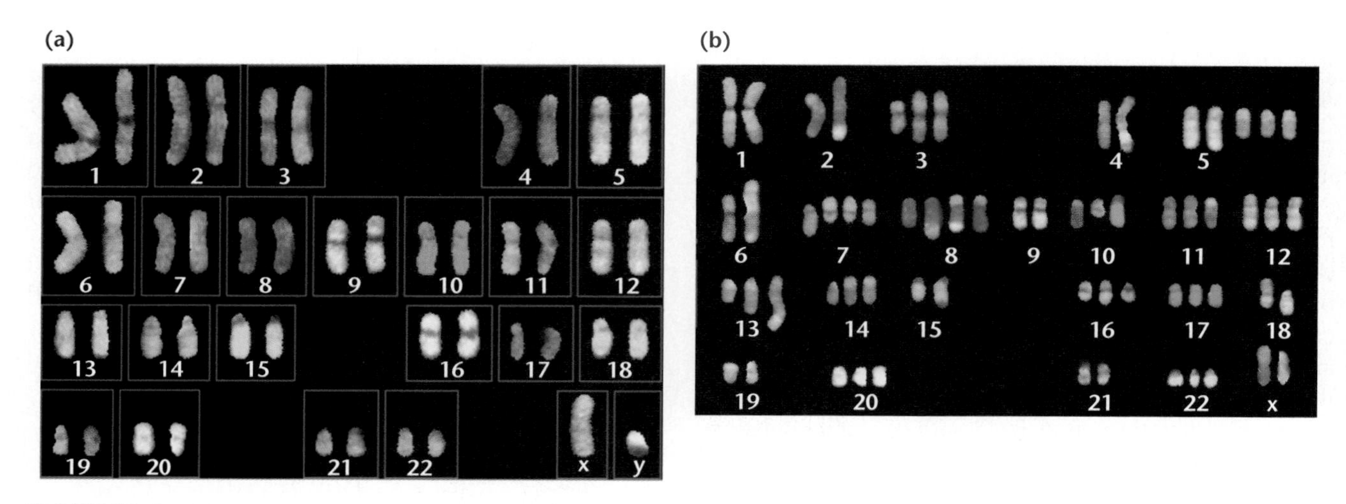

FIGURE 21–1 (a) Spectral karyotype of a normal cell. (b) Karyotype of a cancer cell shows the translocations, deletions, and aneuploidy that are characteristic of cancer cells. *Courtesy of Hesed M. Padilla-Nash, Antonio Fargiano, and Thomas Ried. Section of Cancer Genomics, Genetics Branch, Center for Research, National Cancer Institute, National Institutes of Health, Bethesda, MD.*

The cell cycle progresses from a period of chromosomal DNA replication (S phase) to the segregation of chromosomes into two nuclei during mitosis (M phase). Interspersed between these phases are two gaps, G1 and G2. Together, G1, S, and G2 make up the interphase (Figure 21–2). G1 begins after mitosis; the synthesis of many cytoplasmic elements including ribosomes, enzymes, and membrane-derived organelles occurs at this time. In S, DNA replication produces a duplicate copy of each chromosome. Then the second period of growth and synthesis, G2, occurs as a prelude to mitosis.

Regulatory steps in the cell cycle control when and where normal cells divide and also control when and where such cells stop dividing. Most cells do not cycle continuously, but withdraw from the cycle and enter a nondividing state called G0. These cells can be stimulated to reenter the cycle in response to external signals. These signals, in the form of molecules that bind to cell-surface receptors, are transferred from the plasma membrane through the cytoplasm into the nucleus by signal transduction molecules. These signals initiate a program of gene expression that moves the cell back into the cycle. Highly differentiated cells, such as nerve cells, usually do not divide and remain permanently in G0. These observations of normal cells suggest that the cell cycle is tightly regulated and is dependent on a cell's life history and its differentiated state. Before discussing the cell cycle in cancer, we will examine what we currently know about the genetic regulation of the cell cycle in normal cells.

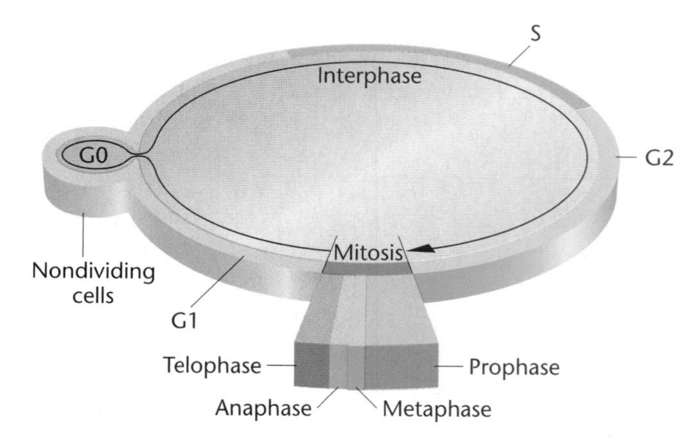

FIGURE 21–2 The cell cycle is controlled at several checkpoints, including one at the G2–mitosis transition and another in late G1 before entry into S phase. These checkpoints involve interactions between transitory proteins (cyclins) and kinases that add phosphate groups to proteins. Phosphorylation of target proteins triggers a cascade of events, allowing progress through the cell cycle.

Control of the Cell Cycle

Much of the basic research on the control of the cell cycle has been conducted by two groups: geneticists working with yeasts, especially *Saccharomyces cerevisiae* and *Schizosaccharomyces pombe*, and developmental biologists studying the newly fertilized eggs of organisms such as frogs, sea urchins, and newts. Both groups have succeeded in identifying and characterizing genes involved in the cell cycle, and their work is now converging and overlapping with important areas of cancer biology, particularly studies on growth factors and the genes that suppress or promote tumor formation.

Cell-Cycle Checkpoints

The cell cycle is regulated at two main checkpoints: the G1/S transition and the G2/mitosis transition (see Figure 21–2). At both points, a decision is made to proceed or to halt progression through the cell cycle. This decision is controlled through the interaction of two classes of proteins. One class of enzymes, the **protein kinases**, selectively phosphorylate target proteins when activated. Although a large number of different protein kinases exist in the cell, only a few are involved in cell-cycle regulation; these are **cyclin-dependent kinases** (**CDKs**). The second class of proteins is **cyclins**, which control progression

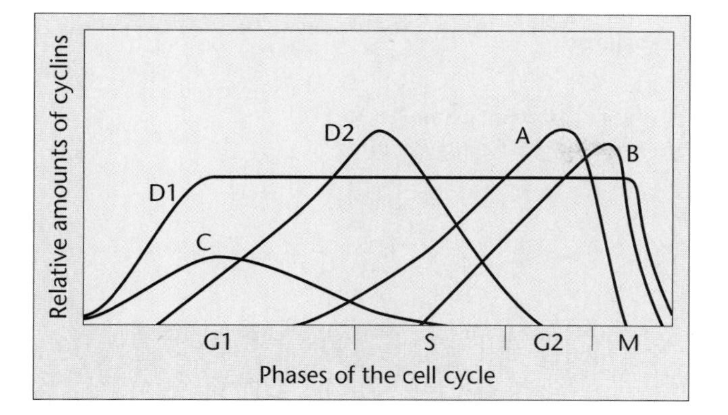

FIGURE 21–3 Relative expression times and amounts of cyclins during the cell cycle. D1 accumulates early in G1 and is expressed at a constant level through most of the cycle. Cyclin C accumulates in G1, reaches a peak, and declines by mid S phase. Cyclin D2 begins accumulating in the last half of G1, reaches a peak just after the beginning of S, and then declines by early G2. Cyclin A appears in late G1, accumulates through S, reaches a peak in G2, and is degraded rapidly as M phase begins. Cyclin B appears in mid S phase, peaks at the G2/M transition, and is rapidly degraded.

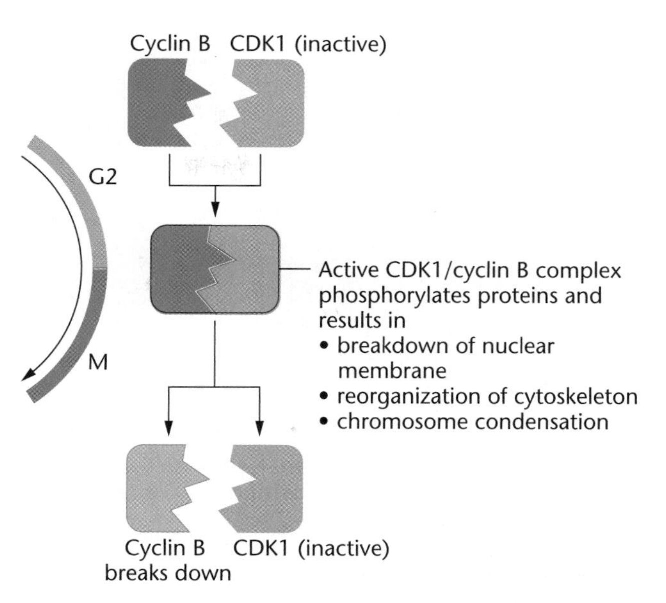

FIGURE 21–4 Transition from G2 to M is controlled by CDK1 and cyclin B. These molecules interact to form a complex that adds phosphate groups to cellular components that break down the nuclear membrane, reorganize the cytoskeleton, and initiate chromosome condensation (histone H1).

through the cell cycle. First identified in the embryos of developing invertebrates, cyclins are continuously synthesized but are periodically degraded in a pattern synchronized with stages of the cell cycle (Figure 21–3). Altogether, almost two dozen different cyclins have been identified, and a growing number of cyclin-dependent kinases are being described, indicating that multiple checkpoints exist in the cell cycle or that these kinases and cyclins have multiple functions.

When kinases bind with cyclins, they become a regulatory molecule that controls the cell's movement through the cycle. At the G1/S control point, a kinase known as CDK4 binds to cyclin D1 and activates the transcription of a set of genes required for S phase. The onset of mitosis (M phase) in most eukaryotic cells is controlled by a kinase called CDK1, which has been biochemically characterized in maturing amphibian eggs and genetically identified in yeast as the product of the *cdc2* gene.

Several events mark the passage from G2 into mitosis (M), including the condensation of chromatin into chromosomes, breakdown of the nuclear membrane, and reorganization of the cytoskeleton. Major events in this transition are regulated by the formation of a CDK1/cyclin B complex. When bound to cyclin B, CDK1 phosphorylates a number of cytoplasmic proteins, which in turn bring about nuclear membrane breakdown and rearrangement of the cytoskeleton. CDK1/cyclin B also phosphorylates histone H1, which may play a role in chromatin condensation (Figure 21–4). Cyclin B may also specify cellular localization of target molecules. Although several experiments suggest that cyclin A is also involved in the progression from G2 to M, its functions are not clearly understood.

21.2 Genes that Control the Cell Cycle Are Involved in Cancer

Mutations that disrupt any step in cell-cycle regulation are candidates for the study of cancer-causing genes. For example, mutations in genes that encode the kinases and cyclins or their target genes are candidates for cancer-causing genes. Evidence is accumulating that the G1 checkpoint is aberrant in many forms of cancer, and mutant G1 cyclins and kinases and the proteins that regulate their activity are good candidates for cancer-causing genes. Studies on predisposition to other cancers have led researchers to conclude that the number of mutations required in a single cell to initiate the development of cancer ranges from 2 to perhaps as many as 20 (Table 21.2).

A number of genes have been identified that when mutated confer a predisposition to specific cancers (Table 21.3). It is clear that the two main properties of cancer—uncontrolled cell division and the ability to spread or metastasize—are the result of genetic alterations. In general, cell division is regulated by genes that normally function to *suppress* cell division and by genes that normally function to *promote* cell division. The first group of regulatory genes is called **tumor-suppressor genes**. Products of these genes halt passage through the cell cycle and prevent mitotic division. For cell division to take place, these genes (and/or their gene products) must be inactive or absent. If tumor-suppressor genes become permanently inactivated or deleted through mutation, control over cell division is lost and the mutant cell begins to proliferate in an uncontrolled fashion.

Genes that normally promote cell division are **proto-oncogenes**. These genes can be "on" or "off"; when they

TABLE 21.2 Number of Mutations Associated with Some Cancers

Cancer	*Chromosome Sites*	*Minimum Number of Mutations Required*
Retinoblastoma	13q	2
Wilms tumor	11p	2
Colon cancer	5p, 12p, 17p, 18q	4–5
Small-cell lung cancer	3p, 11p, 13q, 17p	10–15

TABLE 21.3 Inherited Predispositions to Cancer

Tumor Predisposition Syndromes	*Chromosome*
Early-onset familial breast cancer	17q
Familial adenomatous polyposis	5q
Familial melanoma	9p
Gorlin syndrome	9q
Hereditary nonpolyposis colon cancer	2p
Li-Fraumeni syndrome	17p
Multiple endocrine neoplasia, type 1	11q
Multiple endocrine neoplasia, type 2	22q
Neurofibromatous, type 1	17q
Neurofibromatous, type 2	22q
Retinoblastoma	13q
Von Hippel-Lindau syndrome	3p
Wilms tumor	11p

are on, their products promote cell division. To halt cell division, these genes and/or their gene products must be inactivated. If these genes or their products become permanently switched on, uncontrolled cell division occurs, leading to tumor formation. Mutant forms of proto-oncogenes are permanently switched on and are called **oncogenes**.

21.3 Tumor-Suppressor Genes Repress Cell Division

Many studies document families with high frequencies of certain types of cancers, such as breast, colon, or kidney cancers. Genetic studies indicate that in some of these cases, mutations in single genes predispose cells to become cancerous (see Table 21.3). Let's explore in some detail the inheritance of a predisposition to **retinoblastoma (RB)**, a cancer of the retinal cells of the eye. Then we will discuss how mutations in the *p53* gene play an important role in many forms of cancer.

Retinoblastoma

Retinoblastoma occurs with a frequency of 1 in 14,000–20,000 individuals and most often appears between the ages of 1 and 3 years. Two forms of retinoblastoma are known. In the familial form (about 40% of all cases), individuals inherit one mutant *RB1* allele and are far more susceptible to developing retinoblastoma than those with two normal wild-type *RB1* alleles. Thus, predisposition to retinoblastoma is inherited as an autosomal dominant trait, although development of the cancer itself is actually a recessive trait, as we will explain below. About 85% of those who inherit a mutant *RB1* allele will develop retinal tumors, usually in both eyes. In addition, individuals carrying a mutant *RB1* allele are predisposed to developing other forms of cancer, such as osteosarcoma (a bone cancer) even if they do not develop retinoblastoma.

The second form of retinoblastoma (the remaining 60% of cases) is not associated with a family history of the disease, and tumors develop spontaneously. This sporadic form is characterized by the appearance of tumors in only one eye, and onset occurs at a somewhat later age than in the familial form.

By studying the two types of retinoblastoma, Alfred Knudson and his colleagues developed a model that requires the presence of two mutated copies of the *RB1* gene in the same retinal cell for tumor formation; in other words, tumor development is a recessive trait. In the familial form, one mutant *RB1* allele is inherited and carried by all cells of the body, including cells of the retina (Figure 21–5a). If the second *RB1* allele mutates in any retinal cell, retinoblastoma results. Therefore, individuals carrying an inherited mutation of the *RB1* gene are predisposed to developing retinoblastoma because only one additional mutation is required to cause tumor formation. This does not happen in all cases—about 15% of those inheriting a mutant *RB1* allele do not develop cancer. In these cases, the wild-type *RB1* allele does not mutate in any retinal cells.

In nonfamilial (sporadic) cases (Figure 21–5b), both of the normal *RB1* alleles in a retinal cell must undergo mutation for a tumor to develop. As might be expected, the chance that both copies of the *RB1* gene in a single cell will undergo mutation is a far less frequent event, and tumor formation occurs at a later age. As predicted by Knudson's model, such sporadic forms of retinoblastoma are more likely to occur in a single eye.

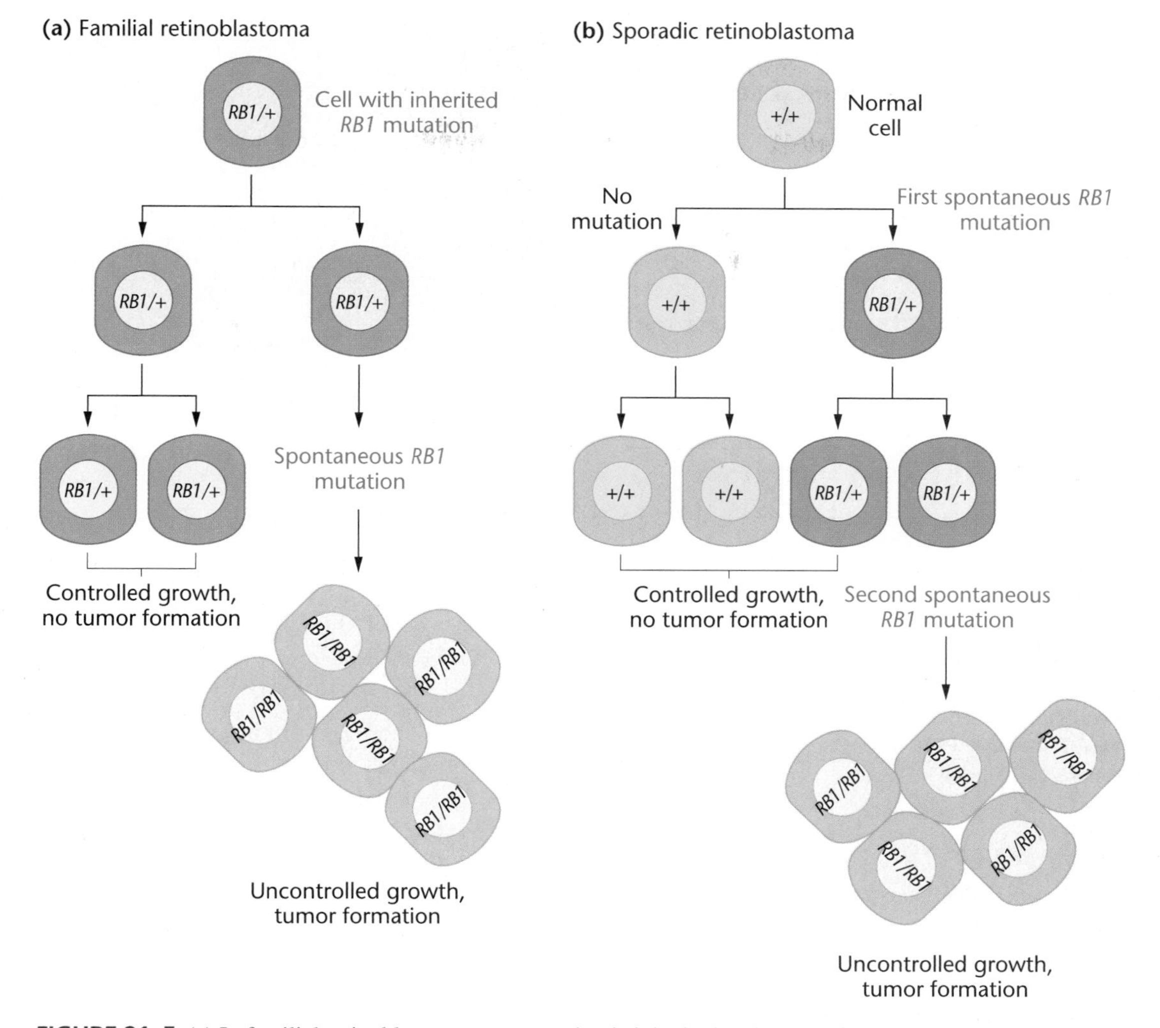

FIGURE 21–5 (a) In familial retinoblastoma, one mutation is inherited and present in all cells. A second mutation at the retinoblastoma locus in any retinal cell results in uncontrolled cell growth and tumor formation. (b) In spontaneous retinoblastoma, two mutations in the retinoblastoma gene in a single cell are acquired sequentially, causing uncontrolled cell growth and division, which results in tumor formation.

The *Retinoblastoma* Gene and the Cell Cycle

The retinoblastoma gene (*RB1*) located on chromosome 13 encodes a **pRB protein**. This protein is found in the nuclei of retinal cells and all cell types examined so far and is present at all stages of the cell cycle. It is part of a regulatory pathway that regulates cell-cycle progression, and pRB acts as a molecular switch controlling the passage of cells from G1 into S phase. Cells can progress through the G1/S transition only when pRB is inactivated by phosphorylation. The activity of pRB is regulated by cyclin-dependent kinases (CDK). When CDK4 binds to cyclin D1, the kinase phosphorylates pRB. In the S, G2, and M phases, pRB is phosphorylated and inactive, but in the G0 and G1 phases it is not phosphorylated and is active. In its active form, pRB binds to members of the E2F family of transcription factors. Recall that transcription factors are proteins that bind to the promoter region of genes and regulate transcription. E2F transcription factors control the expression of the approximately 30 genes required to move the cell from G1 into S phase. When the active form of pRB binds to E2F, transcription is blocked and the cell remains in G1 (Figure 21–6). Late in G1, the CDK4/cyclin D1 complex phosphorylates pRB. The addition of phosphate groups causes pRB to release E2F, which then initiates transcription of genes that move the cell from G1 into S phase.

In normal retinal cells, pRB is active and prevents passage into S phase by interacting with E2F, effectively stopping the cell cycle and cell division. In retinoblastoma cells, both copies of the *RB1* allele are defective, pRB is inactive or absent, and progression through the cell cycle is not regulated. As a result, E2F continuously activates the genes required for passage through the G1/S checkpoint. This important checkpoint is thus overridden, resulting in uncontrolled cell growth and tumor formation.

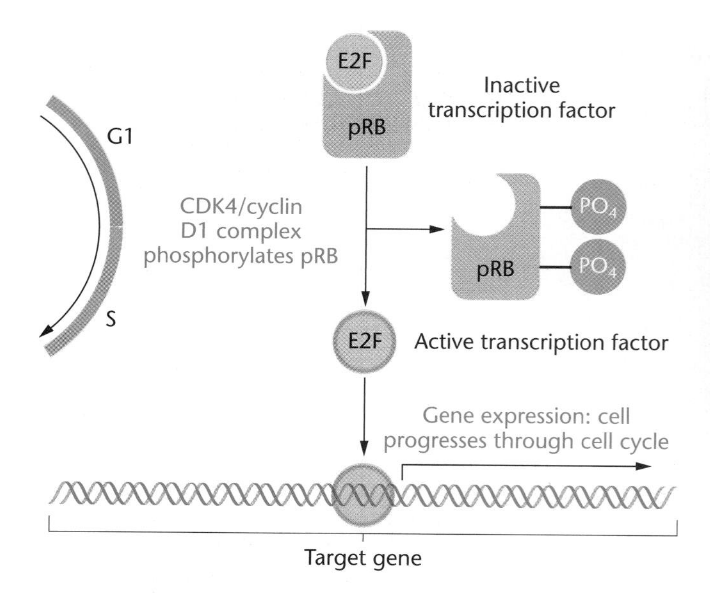

FIGURE 21–6 In the nucleus during G1, pRB interacts with and inactivates transcription factor E2F. As the cell moves from G1 to S, a CDK4/cyclin D1 complex forms and adds phosphate groups to pRB. As pRB becomes phosphorylated, E2F is released and becomes transcriptionally active, allowing the cell to pass through S phase. Phosphorylation of pRB is transitory; as cyclin is degraded, phosphorylation declines.

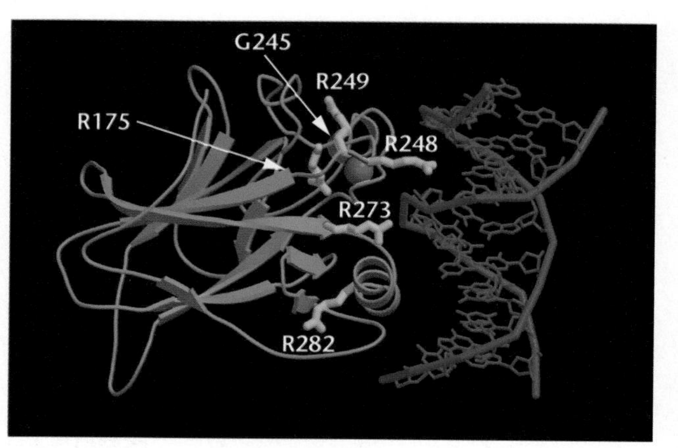

FIGURE 21–7 A computer-generated image of the p53 protein (left), showing it complexed with DNA (right), with the six most frequently mutated regions of the protein shown. R is the abbreviation for arginine, G is the abbreviation for glycine, and the numbers represent the amino acid positions of the mutations. *(Reproduced with permission from Cho, Y., Gorina, S., Jeffrey, P.D., and Pavletich, N.P. Science 265: pp 364-355, Fig. 6b p. 352 Copyright 1994. American Association for the advancement of Science.)*

Guardian of the Genome: *p53*

The ***p53* gene** encodes a nuclear protein that acts as a transcription factor (Figure 21–7). Mutations of *p53* are found in a wide range of cancers, including breast, lung, bladder, and colon cancers. It is estimated that 50–60% of all cancers are associated with mutations in the *p53* gene, which suggests that *p53* controls one or more key events in cell division, and that it is not involved in a cell- or tissue-specific form of regulation. Inherited mutations in the *p53* gene are associated with the Li-Fraumeni syndrome, an autosomal dominant condition with a predisposition to cancer in several tissues.

Normally, the p53 protein is continuously synthesized but rapidly degraded and therefore is present in cells at a low level in an inactive form. Several types of signals cause a shutdown of p53 degradation, leading to a rapid increase in the nuclear concentration of p53. Chemical damage to DNA, double-stranded breaks in DNA induced by ionizing radiation, or the presence of DNA-repair intermediates generated by exposure to ultraviolet light all shut down p53 degradation. Activation of the p53 protein generates several possible responses, including (1) DNA repair; (2) cell-cycle arrest; and (3) **apoptosis**, a genetically programmed pathway of cell death. Each of these tasks is accomplished by the activation of specific target genes by the action of p53 as a transcription factor. The expression of more than 60 primary target genes is mediated by the action of p53.

Cell-cycle arrest by p53 can occur at several phases, including G1. To arrest the cell cycle in G1, p53 stimulates transcription of a gene encoding a CDK inhibitor called p21. The p21 protein targets a number of CDK/cyclin complexes, including the CDK4/cyclin D1 complex we discussed earlier. Inhibition of CDK4/cyclin D1 activity keeps pRB in its active configuration, repressing transcription of the genes needed to move the cell from G1 into S phase. Cells lacking functional p53 are unable to arrest in G1, and they move immediately from G1 into S. These cells do not repair any DNA damage that may be present, and as a result, they have a high rate of mutation. Thus the *p53* gene is often referred to as the "guardian of the genome." The activated p53 protein is a tetramer of p53 subunits, thus mutation of one *p53* allele usually abolishes all p53 activity since almost all tetramers will contain at least one defective subunit. This means that mutations in the *p53* gene act as dominant negative mutations, and that individuals heterozygous for a *p53* mutation will develop cancer with a frequency of 90–95%. The central role of the *p53* gene in controlling the cell cycle emphasizes the relationships between cancer and the cell cycle and between genes that regulate cell growth and cancer.

Breast Cancer Genes

Mutations in *BRCA1*, a gene that maps to the long arm of chromosome 17, are associated with a predisposition to breast cancer; this predisposition is inherited as an autosomal dominant trait. In about 85% of women carrying one mutant *BRCA1* allele, a mutation in the second *BRCA1* allele occurs—these women will develop breast cancer and have an increased risk of ovarian cancer.

A second breast cancer gene, *BRCA2*, located on the long arm of chromosome 13, is also inherited as an autosomal dominant predisposition to breast cancer but is not associated with an increased risk of ovarian cancer. These two genes account for a large majority of breast cancer cases associated with a genetic predisposition (about 10%

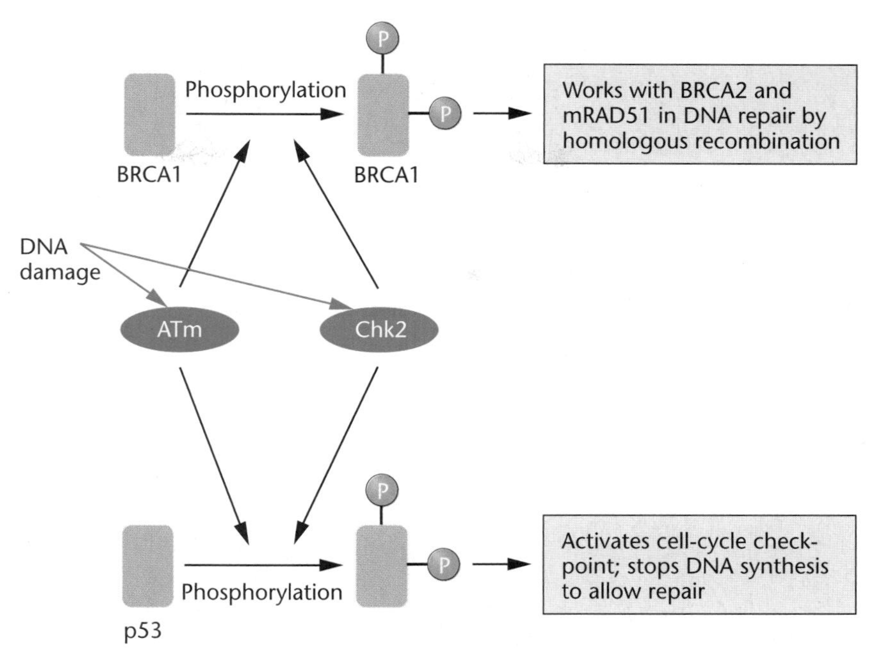

FIGURE 21–8 DNA damage activates the kinase ATm, which in turn activates the kinase Chk2. Once activated, these kinases stimulate the phosphorylation of the nuclear proteins p53 and BRCA1. Phosphorylation stabilizes p53, leading to increased levels of the protein. As the concentration of p53 increases, it activates a cell-cycle control point, thus halting DNA replication. The phosphorylated BRCA1 protein interacts with a number of other proteins, including BRCA2 and mRAD51 to bring about the repair of double-stranded DNA breaks by homologous recombination.

of all cases of breast cancer). However, homozygous mutant forms of these genes play no role in sporadic cases of breast cancer, which make up 90% of all cases.

Both the *BRCA1* and *BRCA2* genes encode large proteins found in the nucleus and are expressed in many tissues. Expression of these genes is highest during S phase. Evidence suggests that both genes have similar functions and are involved in DNA repair and a number of other processes, including recombination, regulation of the cell cycle, and transcription. Some direct evidence for the role of the BRCA1 protein in DNA repair is available. In cells exposed to ionizing radiation (which produces double-stranded DNA breaks), BRCA1 is activated by phosphorylation. A number of different kinases can carry out this phosphorylation.

Recent evidence suggests that two pathways can lead to BRCA1 phosphorylation in response to double-stranded DNA breaks (Figure 21–8). In this model, DNA damage activates a kinase called ATM and a kinase called Chk2. In turn, these kinases phosphorylate BRCA1 and p53. The activated p53 protein arrests replication during S phase to allow DNA repair to take place. The activated BRCA1 protein participates in DNA repair with the BRCA2 protein, a protein called mRAD51, and additional nuclear proteins.

It has been postulated that mutations in *BRCA1 or BRCA2* generate genomic instability through defects in DNA repair. This instability may increase mutations in other genes, giving rise to breast cancer. In sporadic cases there is no association with *BRCA1* and *BRCA2* mutations. These cases involve other, as yet undiscovered genes that may be related to those that act downstream of *BRCA1* and *BRCA1*.

21.4 Proto-oncogenes Promote or Maintain Cell Division

The other category of genes involved in cell-cycle regulation is the **proto-oncogenes**. When expressed, these genes activate or maintain cell division. The protein products of proto-oncogenes can be found in the plasma membrane, cytoplasm, and nucleus (Table 21.4). In spite of their wide-ranging locations, which suggest varying functions, all proto-oncogene proteins characterized to date alter gene expression either directly or indirectly. To stop cell division, expression of these genes and the activity of their gene products must be altered. **Oncogenes** are mutant proto-oncogenes that induce or maintain the uncontrolled cell divisions associated with cancer. Oncogenes are gain-of-function mutations; that is, the mutation results in overexpression, aberrant

TABLE 21.4 Cellular Location of Protooncogene and Oncogene Proteins

Gene	*Location of protooncogene Protein*	*Location of oncogene Protein*
src	Membranes	Membranes
ras	Membranes	Membranes
myc	Nucleus	Nucleus
fps	Cytoplasm	Cytoplasm and membranes
abl	Nucleus	Cytoplasm
erbB	Plasma membrane	Plasma membrane and Golgi

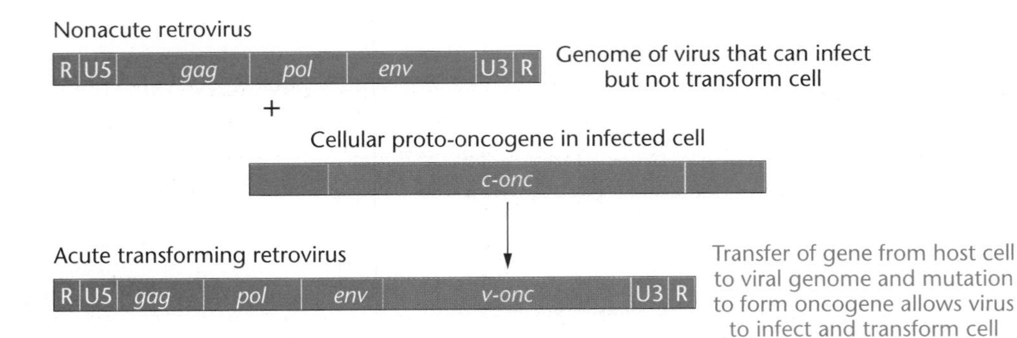

FIGURE 21–9 A retrovirus has acquired a copy of a gene from the host genome, converting it from a *c-onc* into an oncogene that gives the virus the ability to transform a specific type of host cell into a cancerous cell.

expression, or expression of an altered gene product. Unlike most tumor-suppressor genes, where mutations in both alleles of a gene are necessary to promote the development of cancer, only one of the two copies of a proto-oncogene must mutate to induce malignancy, resulting in a dominant cancer phenotype.

Rous Sarcoma Virus and Oncogenes

In 1910, Francis Peyton Rous was the first to infer the existence of specific genes that transform normal cells into cancer cells. Rous studied a connective tissue tumor in chickens, known as a **sarcoma**. He injected cell-free extracts from these tumors into healthy chickens and induced the formation of sarcomas. He postulated the existence of an agent that transmitted the disease, which decades later was shown by other investigators to be a virus, now known as the **Rous sarcoma virus (RSV)**. Rous received the Nobel Prize in Physiology or Medicine in 1966 for his pioneering work in establishing the relationship between viruses and cancer.

RSV infects cells and reproduces within them. Once inside a cell, the single-stranded RNA genome of RSV is transcribed by reverse transcriptase, converting the RNA genome into a single-stranded DNA molecule. This single-stranded DNA is used as a template to synthesize the complementary strand, creating a double-stranded DNA molecule that integrates into the genome of the infected cell, forming a **provirus**. At a later time, the DNA is transcribed into RNA, which is translated into viral proteins. Packaging the viral RNA molecules into these proteins forms new RSV particles. Because the replication cycle of viruses like RSV "reverses" the flow of genetic information, they are called **retroviruses**.

The tumor-forming ability in RSV results from a single gene, the *src* gene, present in the viral genome. This gene, which is responsible for inducing tumor formation in chicken cells, is an oncogene. Retroviruses that carry oncogenes are known as **acute transforming viruses**. Other retroviruses that do not carry oncogenes but induce the activity of cellular genes that bring about tumor formation are known as **nonacute viruses**.

Origin of Viral Oncogenes

Oncogenes (*onc*) carried by acute transforming viruses were not originally viral genes but were acquired from the host's genome during infection, when a cellular proto-oncogene was transferred to the viral genome (Figure 21–9). The proto-oncogene may be mutated during the exchange or be inappropriately expressed as a result of being placed in the viral genome, forming an oncogene. Oncogenes carried by retroviruses are called *v-onc*; the normal cellular version of the gene—the proto-oncogene—is abbreviated *c-onc*. Retroviruses that carry a *v-onc* can infect and transform a host cell into a tumor cell. In the case of RSV, the oncogene captured from the chicken genome, *v-src*, caused infected chicken cells to form cancerous sarcomas. More than 20 oncogenes have been identified in retroviral genomes; overall, several dozen oncogenes are known, and some of them are listed in Table 21.5. It is important to note that although oncogenes were first identified in retroviruses, not all oncogenes are mutant versions of cellular genes carried by retroviruses. Some oncogenes arise by spontaneous mutation in cells—without the involvement of viruses. We must, therefore, consider a broad range of mechanisms in the formation of oncogenes.

TABLE 21.5 Representative Viral Oncogenes

Oncogene	*Origin*	*Species*	*Cellular function*
abl	Abelson murine leukemia virus	Mouse	Tyrosine kinase
erbB	Avian erythroblastosis virus	Chicken	EGF receptor
fos	FBJ osteosarcoma virus	Mouse	Transcription factor
myc	Avian myelocytomatosis virus	Chicken	Transcription factor
N-ras	Neuroblastoma, leukemia	Human	GTP-binding protein
sis	Simian sarcoma virus	Monkey	Platelet-derived growth factor
src	Rous sarcoma virus	Chicken	Tyrosine kinase signal protein

Formation of Oncogenes

At least three mechanisms can explain how proto-oncogenes are converted into oncogenes: **point mutations**, **translocations**, and **overexpression** (see Table 21.6). Some of these events are mediated by viruses, and others by intracellular events that occur in the absence of retroviruses.

Mutations in proteins encoded by *ras* genes demonstrate how a point mutation converts a proto-oncogene into an oncogene. The ***ras* gene family** encodes signal transduction proteins that play a major role in regulating cell growth and division. Ras proteins act as molecular switches that transmit signals from the extracellular environment to the cytoplasm. More than 30% of all human cancers carry a mutant *ras* oncogene. Ras proteins are embedded in the plasma membrane and cycle between an inactive (switched-off) state and an active (switched-on) state. Activated Ras proteins interact with cytoplasmic proteins and initiate a cascade of events that transduces a signal from the external environment through the cytoplasm to the nucleus. In the nucleus, the signal activates the transcription of the genes that start cell division.

TABLE 21.6 Conversion of the Proto-oncogene *c-onc* to Oncogenes

Mechanism	*Oncogene*
Point mutation	*ras*
Translocation	*abl*
Overexpression of gene product	
New promoter by viral insertion	*mos, myb*
New enhancer by viral insertion	*myc*
Amplification of proto-oncogene	*myc*

Comparing the amino acid sequences of Ras proteins from a number of different human carcinomas (tumors of epithelial tissue) reveals that *ras* mutations have single amino acid substitutions at either position 12 or 61 (Figure 21–10) in the 189 amino acid Ras protein. Each of these amino acid changes can be created by a single nucleotide substitution in the *ras* gene. Mutant Ras proteins cannot cycle from the active to the inactive form and are locked into the on position, thus continually signaling for cell division (Figure 21–11). This mutation is one of the first steps in transforming a normal cell to a malignant one.

The creation of an oncogene by translocation is illustrated by the chromosomal events that result in chronic myelogenous leukemia (CML). In this case, described in

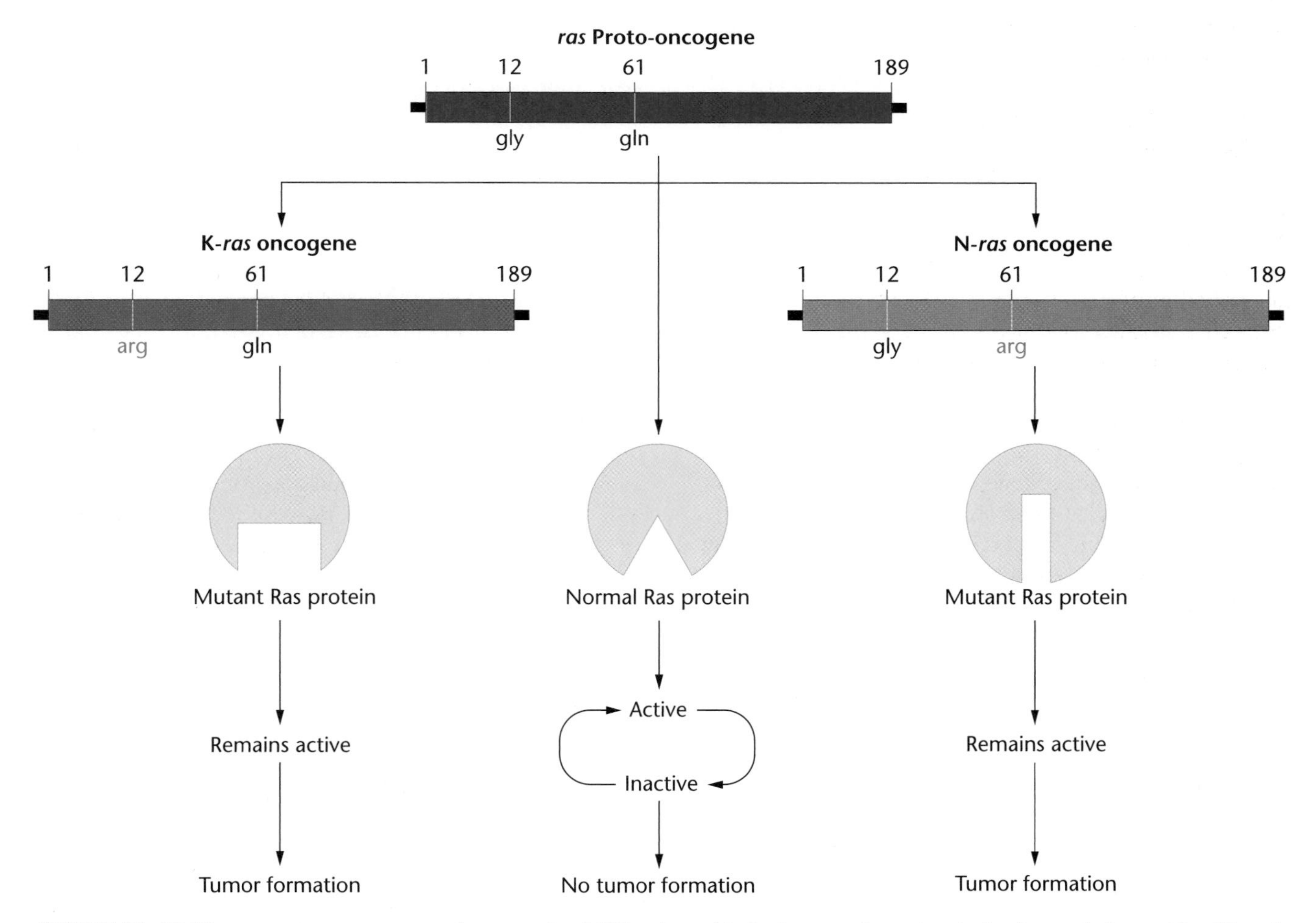

FIGURE 21–10 The *ras* proto-oncogene encodes a protein of 189 amino acids. In the normal protein, glycine is encoded at position 12, and glutamine at position 61. Analysis of *ras* oncogene proteins from several tumors shows a single amino acid substitution at one of these positions, which converts a proto-oncogene into a tumor-promoting oncogene. K-*ras* and N-*ras* are mutant alleles of the *ras* proto-oncogene.

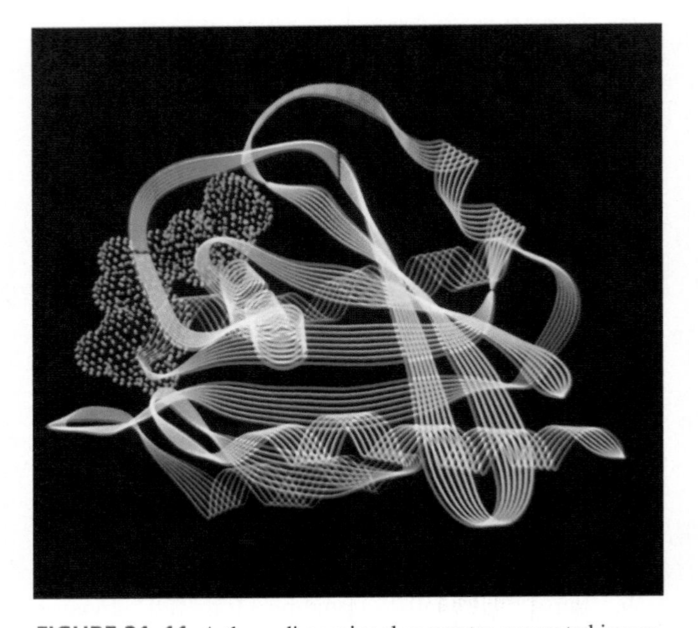

FIGURE 21–11 A three-dimensional computer-generated image of Ras proteins in two different conformations. Normal Ras proteins act as molecular switches controlling cell growth and differentiation. The switch is on (conformation shown in blue) when GTP binds to the protein, and off (conformation shown in yellow) when the GTP is hydrolyzed to GDP. Switching the protein between states alters the conformation of the protein in the two regions (blue and yellow). Oncogenic mutations of *ras* are stuck in the on state, and continuously signal for cell growth. *(Sung-Hou Kim, University of California, Dept. of Chemistry and Lawrence Berkeley National Laboratory, Berkeley, California)*

detail in a later section, the translocation results in the formation of a hybrid gene whose expression causes tumor formation.

Conversion of a proto-oncogene into an oncogene by overexpression occurs by at least three different mechanisms: (1) The proto-oncogene may acquire a new promoter, which causes the level of transcription to increase or causes a silent locus to activate. This is the case in avian leukosis, where strong retroviral promoters are inserted adjacent to a proto-oncogene, causing an increase in both mRNA production and the amount of the gene product. (2) Overexpression involves the acquisition of new upstream regulatory sequences, including enhancers. (3) Amplification of the proto-oncogene results in overexpression. In human tumors, members of the *myc* family of oncogenes are frequently amplified, for instance, the *c-myc* proto-oncogene is amplified up to several hundred copies in some human tumors.

21.5 Colon Cancer as a Genetic Model of Cancer

Cancer is a multistep process that results from mutations in specific genes. Studies of tumors such as retinoblastoma have established that in some cases, only a few steps are required to transform a normal cell into a malignant one. However, most cancers develop in several steps, with intermediate levels of genetic and cellular transformation.

Colorectal cancer has been studied to obtain detailed information about the nature and order of the genetic and cellular events that result in cancer. This cancer has been selected as a model of the multistep process of carcinogenesis for several reasons. First, the malignant form of colorectal tumors develops from preexisting benign tumors. Second, in the development of colon cancer, several distinctive precancerous stages occur, and these stages can be isolated and studied. Third, both hereditary and sporadic forms of colorectal cancer exist. As a result, colorectal cancer is a useful model for studying the interaction of genetic and environmental factors in tumor formation.

Two forms of genetic predisposition to colon cancer are known: (1) The autosomal dominant trait, **familial adenomatous polyposis (FAP)**; and (2) The genetically complex trait, **hereditary nonpolyposis colorectal cancer (HNPCC)**. FAP is associated with about 1% of all cases, and HNPCC is responsible for 3–4% of all colon cancers. The remaining 95% are sporadic. Researchers have developed genetic models of both forms of colon cancer, which provide insight into the events that cause normal cells to become cancerous.

FAP and Colon Cancer

According to the model, sporadic cases of FAP-associated colon cancer begin with a mutation in the *APC* gene on the long arm of chromosome 5. This mutation takes place in a normal epithelial cell lining the colon. The presence of a heterozygous *APC* mutation causes the epithelial cell to partially escape cell-cycle control, and the cell divides to form a small cluster of cells called a **polyp**, or adenoma (Figure 21–12). In FAP, individuals inherit one mutant copy of *APC* and carry this mutation in all cells of the body, including all cells of the colon. These heterozygous individuals form hundreds or thousands of polyps in the colon. In both sporadic cases and FAP, each polyp is a clone of cells, all of which carry an *APC* mutation, and the development of polyps is a dominant trait. It is *not* necessary for the second copy of the *APC* gene to mutate for polyp formation. However, in the majority of cases, the second *APC* allele becomes mutant in a later

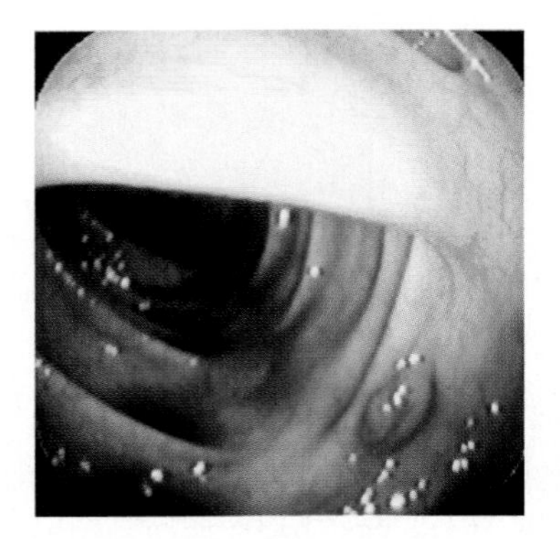

FIGURE 21–12 Polyps in the colon. *(Albert Paglialunga/Phototake N.Y.C.)*

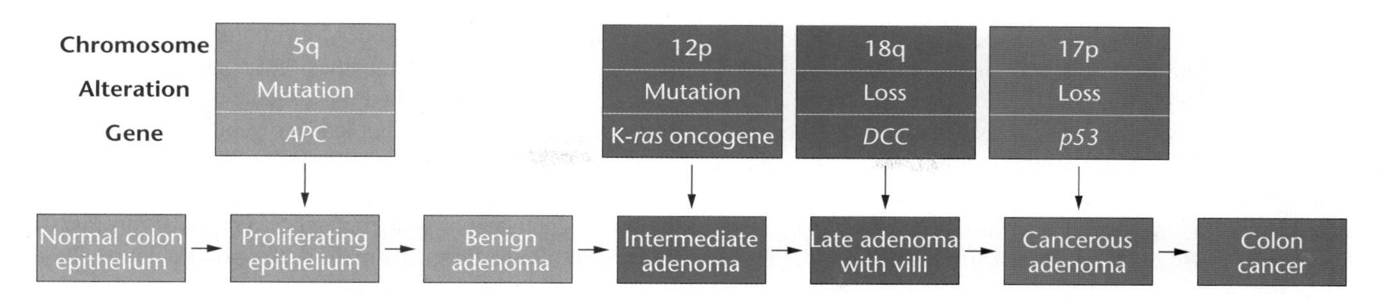

FIGURE 21–13 A model for the multistep production of colon cancer. The first step is the loss or inactivation of one allele of the *APC* gene on chromosome 5. In familial cases, one mutant *APC* allele is inherited. Subsequent mutations involving genes on chromosomes 12, 17, and 18 in cells of the benign adenomas can lead to a malignant transformation that results in colon cancer. Although the mutations in chromosomes 12, 17, and 18 usually occur at a later stage than those involving chromosome 5, the sum of the changes is more important than the order in which they occur.

stage of cancer development. The relative order of mutations in the development of colon cancer is shown in Figure 21–13.

Subsequent mutations of genes within the polyp cause intermediate stages of tumor formation. Dominant-acting mutations in one allele of the *ras* proto-oncogene on chromosome 12 in polyp cells carrying an *APC* mutation cause the polyp to grow larger, forming an intermediate adenoma. To progress further, a polyp cell carrying an *APC* mutation and a *ras* mutation must then acquire mutations in a gene on chromosome 18, called *DCC* (*d*eleted in *c*olon *c*ancer). Mutations in both *DCC* alleles result in the formation of late-stage adenomas with a number of fingerlike outgrowths (villi). Finally, a mutation involving the *p53* gene on 17p causes the transition to a cancerous cell. As discussed earlier, mutations in the *p53* gene are pivotal to the development of a number of cancers, including lung, brain, and breast cancers, as well as colon cancer. Metastasis occurs after the formation of colon cancer, and it involves an unknown number of additional mutations.

Genomic Instability and HNPCC

HNPCC is more common than FAP-associated colon cancer. Families with HNPCC are defined as those in which at least 3 relatives in 2 generations have been diagnosed with colon cancer, with 1 relative being diagnosed at less than 50 years of age. A pedigree of HNPCC is shown in Figure 21–14. HNPCC develops after only a small number of polyps have formed, rather than the hundreds or thousands seen with FAP. At least eight genes associated with HNPCC have been mapped to multiple loci. Mutations at four of these loci generate a cascade of mutations in short, tandemly repeated microsatellite sequences (Chapter 12) located throughout the genome.

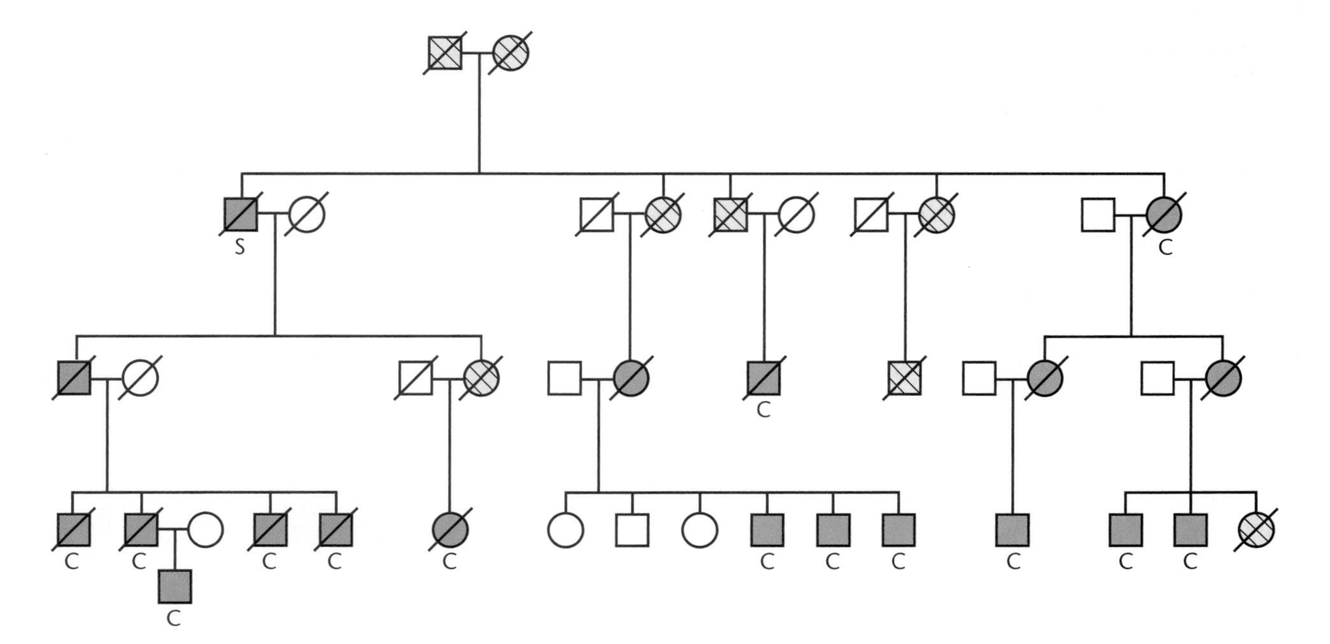

FIGURE 21–14 Pedigree of a family with HNPCC. Filled symbols indicate family members with colon cancer; diagonal stripes mean that diagnosis is uncertain; crosshatching indicates other tumors associated with HNPCC. Symbols with slashes indicate deceased individuals. Colon cancer: C; stomach cancer: S. *(Reprinted with permission from Aaltonon et al. Clues to the pathogenesis of familial colorectal cancer.* Science *260:812-819. Copyright 1993 AAAS.)*

The gene on chromosome 2 associated with HNPCC is a DNA-repair gene called *MSH2*. Inactivation of this gene causes a rapid accumulation of mutations in other parts of the genome and the subsequent development of colorectal cancer. A second DNA-repair gene, *MLH1*, has been mapped to chromosome 3, and two other DNA-repair genes related to *MLH1* and *MSH2* have also been identified. These genes map to 14q22 (*MLH3*) and 2p16 (*MSH6*). All of these genes belong to a family called *mismatch repair* genes (*MMR* genes). Mutations in any of the *MMR* genes promote genome-wide genetic instability in microsatellites, which accelerates the rate of genome-wide mutations, with colon cancer as just one outcome. It may turn out that mutations in any one of these genes is enough to cause HNPCC.

HNPCC is similar to FAP-associated colon cancer, but there are several important differences. FAP begins with a mutation in *APC*, and hundreds or thousands of polyps form. Each polyp progresses slowly toward cancer by accumulating mutations in other genes. However, because of the presence of so many polyps, there is a high probability that at least one polyp will accumulate all of the mutations necessary to cause colon cancer. HNPCC begins with a mutation in the DNA-repair genes in an epithelial cell of the colon, leading to genome-wide mutations, one of which includes a mutation in *APC*. Mutation in *APC* is a relatively rare event, so only a few polyps will form. Even though the polyps are relatively few in number, the continued high mutation rate ensures that at least one polyp will acquire the mutations necessary to become cancerous. Therefore, FAP-associated colon cancer is a disease of rapid tumor initiation, with a slow progression to cancer, while HNPCC is a disease with slow initiation of tumors but rapid progression to cancer. Both forms of colon cancer are mediated by mutations in the *APC* gene.

21.6 The Pathway to Cancer Leads Through Gatekeeper and Caretaker Genes

The two pathways to colon cancer in FAP and HNPCC offer an insight into the nature of genes that control cancer predisposition. The differences in these pathways have generated the idea that mutations in two classes of genes, **gatekeeper genes** and **caretaker genes**, cause predisposition to cancer. The *APC* gene is a gatekeeper; it normally inhibits cell growth. In general, tumor-suppressor genes are gatekeepers. In different cell types, only a few genes serve as gatekeepers; and if both copies of a gatekeeper mutate, a specific cancer such as retinoblastoma or colorectal cancer develops. Individuals predisposed to a specific form of cancer inherit one mutant copy of such a gatekeeper and need only one additional mutation to initiate tumor formation. In spontaneous cancers, both copies of the gatekeeper gene must mutate for cancer to develop.

Caretaker genes have a different role; they maintain the integrity of the genome. DNA-repair genes like *MSH2* and *MLH1* are caretaker genes. These genes normally function to repair DNA damage caused by environmental agents such as ultraviolet light or mismatches that occur during DNA replication. Mutation of a caretaker gene does not directly promote tumor formation but leads to genetic instability of repetitive sequences (microsatellites) that increases the mutation rate of all of the genes in a cell, including the gatekeeper genes. If a gatekeeper gene mutates as a result of this instability and allows tumor formation, the process is accelerated by the high rate of mutation in the cell.

21.7 Chromosomal Translocations Are a Hallmark of Leukemia

Alterations in chromosome structure and/or number are associated with many forms of cancer, but in most cases the relationship between changes in chromosome number or structure and the development of cancer is not understood. For example, individuals with Down syndrome carry an extra copy of chromosome 21 and experience a 20-fold increased risk of leukemia as compared to the general population. How an extra chromosome results in an increased risk of leukemia is not known. Chromosome translocations are common in many forms of cancer, but their relationship to cancer development is unclear. In other cases, the relationship between a chromosome aberration and the development and/or maintenance of the cancerous state is known. Such a connection is most clearly seen in leukemias (Table 21.7).

One of the best-studied examples is the translocation between chromosomes 9 and 22 that is associated with **chronic myelogenous leukemia** (**CML**) (Figure 21–15). Originally, this translocation was described as an abnormal chromosome 21 and called the **Philadelphia chromosome** (because it was discovered in that city). Later, Janet Rowley showed that the Philadelphia chromosome actually is the product of a translocation between chromosomes 9 and 22. Thus, CML may originate from a single cell bearing this translocated chromosome.

By examining a large number of Philadelphia chromosomes, the exact location of the breakpoints on

TABLE 21.7 Specific Chromosome Aberrations and Cancer

Cancer	*Chromosome Alteration*
Acute lymphocytic leukemia	t(4;11)
Acute myelogenous leukemia	t(8;21)
Acute promyelocytic leukemia	t(15;17)
Chronic myelogenous leukemia	t(9;22)
Prostate cancer	del(10q)
Retinoblastoma	del(13q)
Synovial sarcoma	t(X;18)
Testicular cancer	inv(12p)
Wilms tumor	del(11p)

Key: t = translocation; del = deletion; inv = inversion.

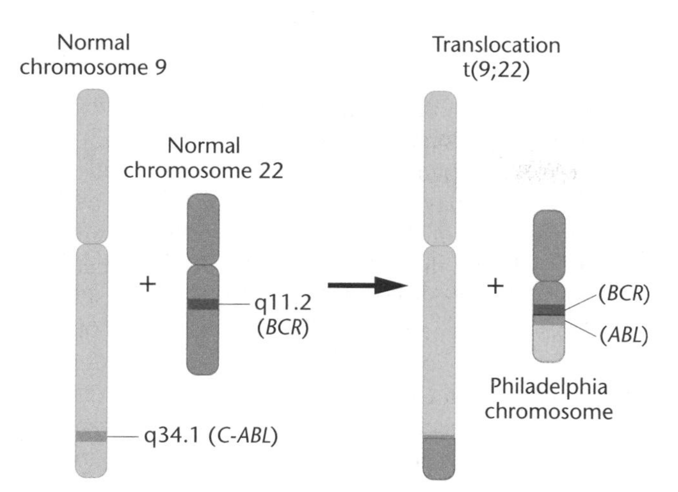

FIGURE 21–15 A reciprocal translocation involving the long arms of chromosomes 9 and 22 results in the production of a characteristic chromosome, the Philadelphia chromosome, which is associated with chronic myelogenous leukemia (CML). The t(9;22) translocation results in the fusion of the *C-ABL* oncogene on chromosome 9 with the *BCR* gene on chromosome 22. The fusion protein is a powerful hybrid molecule that allows cells to escape control of the cell cycle, resulting in leukemia.

chromosomes 9 and 22 were established. Genetic mapping studies using recombinant DNA techniques established that the *C-ABL* proto-oncogene maps to the breakpoint region on chromosome 9 and that the *BCR* gene maps near the breakpoint on chromosome 22. The c-Abl protein is a kinase and the normal Bcr protein activates a phosphorylation reaction (catalyzed by a kinase). In the translocation event, all or most of the *C-ABL* gene is translocated to a region within the *BCR* gene, generating a hybrid *BCR/ABL* oncogene. This oncogene produced by the fusion of two normal genes is transcriptionally active and expresses a hybrid 200-kDa protein product. This protein has been implicated in the generation of CML.

In **Burkitt lymphoma**, both cytogenetic and molecular approaches have been applied to study the outcome of translocations involving chromosome 8 [these include t(8;14), t(8;22), and t(2;8)]. The breakpoint on chromosome 8 in all of these translocations is the same, and the *c-myc* proto-oncogene has been mapped to this locus. The loci at the breakpoints on the other chromosomes involved in this series of translocations all have immunoglobulin genes located at the breakpoints. The movement of the *c-myc* gene to a position near these immunoglobulin genes leads to overexpression of the *c-myc* gene, resulting in the transformation of the lymphoid cells.

Other leukemias and lymphomas exhibit characteristic translocation breakpoints, some of which have been isolated and characterized at the molecular level. In these cases, as in CML, the translocation results in an oncogene produced by fusion of two normal genes and formation of a hybrid protein that causes the cell to undergo a malignant transformation. As researchers identify hybrid gene products at translocation sites, therapeutic strategies can be developed to inactivate the abnormal gene product, which is found only in the cancerous cells. In the case of CML, a drug called Gleevec ™ has been designed. The first step in this process was determining the structure of the hybrid BCR/ABL gene product. Once this was known, it was possible to develop an effective drug that binds to and inactivates the hybrid protein, which is produced only in CML-associated cancerous white blood cells. The success of this treatment has generated interest in developing drugs for other diseases associated with hybrid genes.

21.8 Environmental Factors Contribute to Cancer

The relationship between environmental agents and the genesis of cancer is often elusive, and in the early stages of an investigation, this relationship is based on indirect evidence. Such studies often begin with an epidemiological survey, that compares the cancer death rates among different geographic populations. When differences are found in death rates for a type of cancer within an age group or a cluster of related occupations (such as chemical workers), further research is necessary to identify one or more environmental factors that may correlate with these cancer deaths. These correlations are not conclusive, but they identify factors that may directly relate to the development of cancer. Finally, extensive laboratory investigations may establish the mechanism by which an environmental agent generates cancer. In some cases, the relationship between cancer and the environment is more straightforward.

Hepatitis B and Cancer

Epidemiological surveys show that individuals who develop a form of liver cancer known as **hepatocellular carcinoma** (**HCC**) have previously been infected with the **hepatitis B virus** (**HBV**) (Figure 21–16). In fact, the

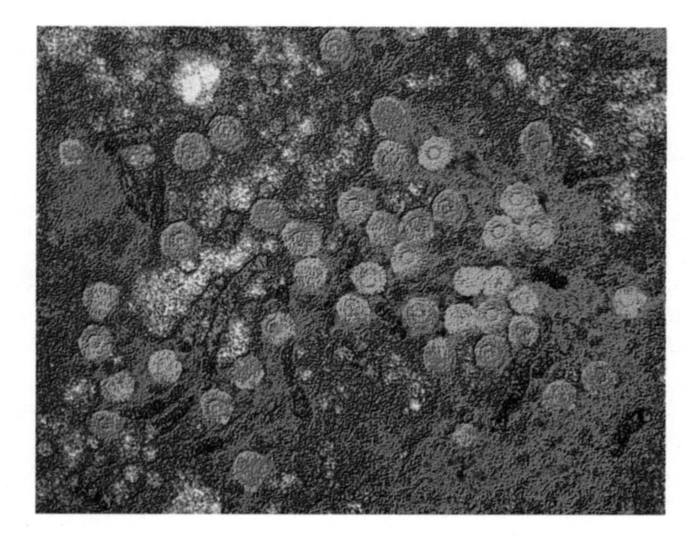

FIGURE 21–16 A false-color transmission electron micrograph of a hepatitis B virus. Infection with this virus often results in cancer of the liver. *(Oliver Meckes & Nicole Ottawa/Photoresearchers, Inc.)*

risk of cancer increases by a factor of 100 for those carrying HBV. Aside from the risk of cancer, HBV infection is a public health risk affecting 350 million people worldwide; it is most prevalent in Asia and tropical regions of Africa. There are two types of infection with HBV, chronic or acute, which produce a wide range of responses, from a chronic, self-limiting infection with few symptoms to active hepatitis and cirrhosis to fatal liver disorders and hepatocellular carcinoma. In the last few years, efforts have centered on understanding the mechanism by which HBV replicates and its role in causing liver cancer. Worldwide, viral infections of all kinds are thought to be responsible for about 15% of all cancers.

The genome of HBV is a mostly double-stranded DNA molecule of 3200 nucleotides. In a chronic infection, the HBV DNA moves to the nucleus, where it inserts into a chromosome. The viral DNA is transcribed into an RNA molecule and packaged into a viral capsid. The capsid containing the RNA pregenome moves to the cytoplasm, where it is reverse-transcribed into a DNA strand, which in turn is made double-stranded. The copied DNA genome is repackaged into a new capsid for release from the cell or returns to the nucleus for another round of replication.

The key to the role of HBV in carcinogenesis apparently resides in its integration into the genome. In the nucleus, the HBV genome can insert itself at many different sites into human chromosomes and cause insertional mutagenesis (Figure 21–17). Insertion often results in cytogenetic alterations of the host genome that include translocations, deletions, or amplifications of adjacent regions. Recall that many forms of cancer are associated with chromosomal rearrangements of these types, and such rearrangements may trigger the development of a cancerous transformation. Insertion of the HBV genome into critical genes can disrupt the cell cycle and normal control of proliferation. Similarly, HBV integration can activate proto-oncogene loci, and altered expression of such loci is clearly implicated in the development of malignant cell growth. In some cases, HBV infection kills liver cells and stimulates regeneration by cell division. Aberrant regulation of regeneration may also lead to cancer. Another proposal is that a viral protein may stimulate cell division and cause the cell to escape cell-cycle control, thus leading to cancer. Whatever the cause, there is a strong link between the HBV virus as an environmental agent and the development of cancer in chronically infected individuals.

Environmental Agents

Cancer is induced through the interaction between an individual's genotype and environmental agents. Many surveys of cancer death rates (Table 21.8) point to the role of genotype/environmental factors that contribute to the development of cancer. The differential expression of a

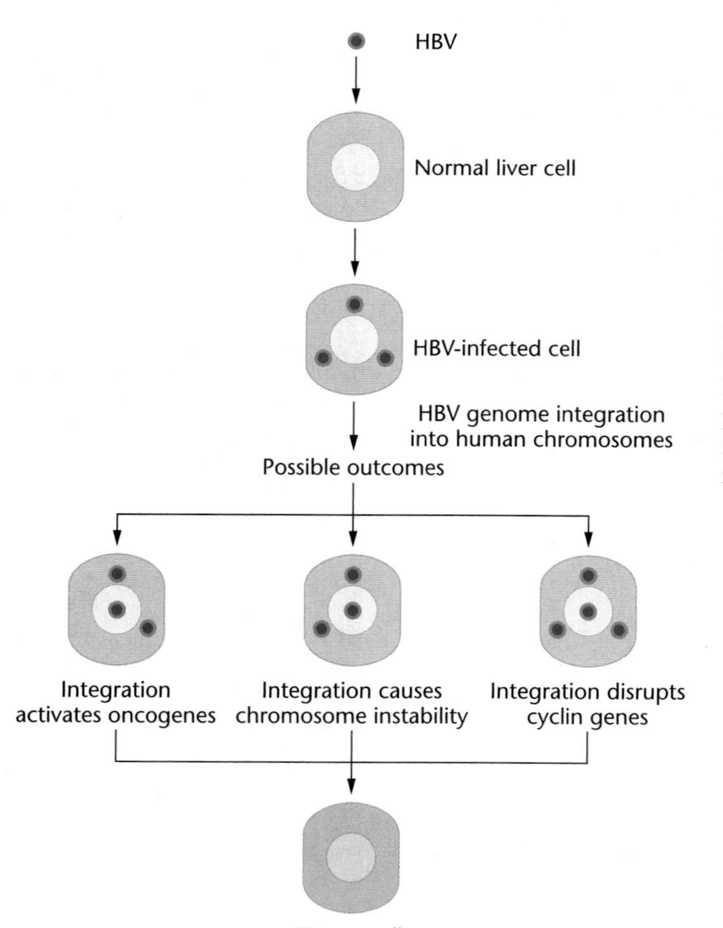

FIGURE 21–17 Some possible outcomes following cellular infection with hepatitis B virus (HBV). Following the integration of the viral chromosome into the human genome, oncogenes can be activated and result in tumor formation. Integration of the HBV genome can also cause chromosome instability, including the production of translocations and deletions. Insertions can also activate proto-oncogenes, causing abnormal regulation of the cell cycle, resulting in cancer formation.

TABLE 21.8 Epidemiology of Cancer in Various Countries

Country	*Death Rate per 100,000*	
	Male	*Female*
Australia	212	125
Canada	214	136
Dominican Republic	54	48
Egypt	39	18
England	248	156
Greece	188	103
Israel	174	145
Japan	190	109
Nicaragua	22	35
Portugal	180	108
Singapore	249	130
United States	216	137
Venezuela	135	128

specific genotype in various environments is often neglected in addressing the role of environment in cancer, but researchers estimate that at least 50% of all cancers are environmentally induced. Environmental agents responsible for cancer include background levels of radiation, long-term occupational exposure to physical and chemical agents, exposure to sunlight, and personal behavior such as diet and tobacco use. To show how environmental factors are identified, let's consider a recent study on the risk of colorectal cancer.

Epidemiological surveys reveal that this cancer occurs with a much higher frequency in North America and Western Europe than in Asia, Africa, or other parts of the world. When people migrate from low-risk areas to high-risk areas, their risk of colon cancer increases to match that of residents in the high-risk areas. This finding suggests that environmental factors, including diet, may play a role in the incidence of colon cancer. In a long-range study called the Nurses Health Study, the dietary habits of more than 88,000 registered nurses across the United States have been monitored since 1980. In that population, 150 cases of colon cancer were observed through 1986. Detailed analysis indicated that nurses who consumed daily meals of pork, beef, or lamb had a 2.5-fold higher risk of colon cancer than those who consumed such meals less than once a month. A more detailed analysis of diets strongly suggests that animal fat in the diet is the environmental risk factor for colon cancer. At the same time, a negative correlation exists between eating skinless chicken meat and the incidence of colon cancer. Laboratory studies have been undertaken to identify the mechanisms by which fat brings about a cancerous transformation of the intestinal epithelium.

Research on the role of external factors as a cause of cancer indicates that neither the environment in general nor pollution in particular is responsible for a large fraction of cancer cases. Rather, it is diet, tobacco, drugs, and other agents (such as excessive medical and dental X-rays or sun exposure) that play significant roles in cancer development. It is estimated that these agents cause 50% of all cancer cases; thus, personal choices figure prominently in the cancer equation.

A number of environmental agents associated with cancer damage DNA. X-rays can damage DNA directly or indirectly, potentially impacting tumor-suppressor genes or proto-oncogenes or causing chromosome breaks that lead to translocations or deletions. Cigarette smoking is highly correlated with cancer incidence, particularly lung cancer. A specific tumor-suppressor gene, *FHIT*, is mutated in a high proportion of smoking-associated lung cancers. Education and judicious changes in lifestyles could prevent a high percentage of all human cancers.

GENETICS, TECHNOLOGY, AND SOCIETY

The Double-Edged Sword of Genetic Testing: The Case of Breast Cancer

These are exhilarating times for genetics and biotechnology. Close on the heels of the completion of the Human Genome Project has come a rush of optimism about future applications of genetics. Scientists and the media predict that gene technologies will soon diagnose and cure diseases as diverse as diabetes, asthma, heart disease, and Parkinson disease.

The prospect of using genetics to prevent and cure a whole range of diseases is exciting. However, in our enthusiasm, we often forget that these new technologies have significant limitations and profound ethical concerns. The story of genetic testing for breast cancer illustrates how we must temper our high expectations with respect for uncertainty.

Breast cancer is the most common cancer among women and the second leading cause of all cancer deaths (after lung cancer). Each year, more than 190,000 new cases are diagnosed in the United States. Breast cancer is not limited to women; about 1400 men are also diagnosed with the disease each year. A woman's lifetime risk of developing breast cancer is about 12%, and the risk increases with age.

Approximately 5–10% of breast cancers are familial, defined by the appearance of several cases of breast or ovarian cancer among near blood relatives and the early onset of these diseases. In 1994, two genes were identified that show linkage to familial breast cancers. Germ-line mutations in these genes (*BRCA1* and *BRCA2*) are associated with the majority of familial breast cancers. The molecular functions of *BRCA1* and *BRCA2* are still uncertain, although they appear to be involved in repairing damaged DNA. Mutations in these genes act as autosomal dominants with variable penetrance. Women who bear mutations in *BRCA1* or *BRCA2* have a 36–85% lifetime risk of developing breast cancer and a 16–60% risk of developing ovarian cancer. Men with germline mutations in *BRCA2* have a 6% lifetime breast cancer risk—a 100-fold increase over the general male population.

BRCA1 and *BRCA2* genetic tests detect any of the over 2000 different mutations that are known to occur within the coding regions of these genes, but the tests have limitations. They do not detect mutations in regulatory regions outside the coding region—mutations that could cause aberrant expression of these genes. Also, little is known about how any particular mutation manifests itself in terms of cancer risk, and the effects of each mutation may be modified by environmental factors and by interactions with other susceptibility genes.

Many patients at risk for familial breast cancer opt to undergo genetic testing. These patients feel that test results will help them to prevent breast or ovarian cancers, will guide them in childbearing decisions, and allow them to inform family members at risk. But none of these benefits is clear-cut.

A woman whose *BRCA* test results are negative may be relieved and feel that she is not subject to familial breast cancer. However, her risk of developing breast cancer is still 12% (the population risk), and she should continue to monitor for the disease. Also, a negative *BRCA* genetic test does not eliminate the possibility that she bears an inherited mutation in another gene that increases breast cancer risk or that *BRCA1* or *BRCA2* gene mutations exist in regions of the genes that are inaccessible to current genetic tests.

A woman whose test results are positive faces difficult choices. Her treatment options are poor, consisting of close monitoring, prophylactic mastectomy or oophorectomy (removal of breasts and ovaries respectively), and taking drugs such as tamoxifen. Prophylactic surgery reduces her risks, but does not eliminate them, as cancers can still occur in tissues that remain after surgery. Drugs such as tamoxifen reduce her risks, but have serious side effects. Genetic tests not only affect the patient but also affect the patient's entire family. People often experience fear, anxiety, and guilt on learning that they are carriers of a genetic disease. Studies show that people who refuse genetic test results often suffer from even more anxiety than those who opt to learn the results. Confidentiality is also a major concern. Patients fear that their genetic test results may be leaked to insurance companies or employers, jeopardizing their prospects for jobs or affordable health and life insurance. One study shows that a quarter of eligible patients refuse *BRCA* gene testing because of concerns about cost, confidentiality, and potential discrimination.

Genetic testing is such a new development that the health system has lagged behind the science. Because genetic testing has both psychological and medical ambiguities, genetic counseling is imperative for patients and their families. However, there are insufficient numbers of genetic counselors with experience in genetic testing, and even in the most qualified hands, issues are complex and difficult. Physicians often have limited knowledge of human clinical genetics and feel inadequate to advise their patients. The federal government and the insurance industries have yet to develop comprehensive policies concerning genetic tests and genetic information. Given the unclear interpretation of *BRCA* genetic tests, the relatively ineffective treatment options, and the potential for psychological and societal side effects, it is not surprising that only about 60% of familial breast cancer patients and their families decide to take the genetic tests.

The unanswered questions about *BRCA1* and *BRCA2* genetic testing are many and important. What cancer risks are associated with which mutations? Should all people have access to *BRCA* tests, or only those at high risk? How can we ensure that the high costs of genetic tests and counseling do not limit this new technology to only a portion of the population? Our struggle with these issues is just beginning, as we develop genetic tests for more and more diseases over the next few decades.

References

Surbone, A. 2001. Ethical implications of genetic testing for breast cancer susceptibility. *Crit. Rev. in Onc./Hem.* 40: 149–57.

Web Sites

Genetic Testing for BRCA1 and BRCA2: It's Your Choice [on-line]. National Institutes of Health. ***http://cis.nci.nih.gov/fact/3_62.htm***

Chapter Summary

1. Cancer is a genetic disorder at the cellular level that can result from the mutation of a subset of genes or from alterations in the timing and amount of gene expression. Although some forms of cancer show familial patterns of inheritance, few show clear evidence for Mendelian inheritance.
2. Cancer results from the uncontrolled proliferation of cells and from the ability of such cells to metastasize and form secondary tumors. Mutant forms of genes involved in regulating the cell cycle are obvious candidates for cancer-causing genes. The cell cycle is regulated at several checkpoints. Two gene products, cyclins and kinases, are involved in regulating some checkpoints. The actions of these genes and the genes they control are important in regulating cell division, and links between these genes and the process of tumor formation are under intense scrutiny.
3. Although most cancers do not show clear-cut patterns of Mendelian inheritance, certain mutant genes predispose individuals to cancer. Studies of retinoblastoma provide insight into the action of mutations that result in the development of tumors and confirm that cell-cycle regulation and cancer are linked.
4. Tumor-suppressor genes normally act to suppress cell division. When these genes mutate or alter expression, control over cell division is lost. Mutations in both alleles of a tumor-suppressor gene are usually necessary to promote the development of cancer.
5. Oncogenes normally function to initiate or maintain cell division; these genes must be mutated or inactivated to halt cell division. If these genes escape control and become permanently switched on, cell division occurs in an uncontrolled fashion. In contrast to tumor-suppressor genes, mutation or inappropriate expression of only one copy of an oncogene can promote the development of cancer.
6. In most cases of cancer, a series of mutations is necessary to cause the malignant state. Colon cancer is a useful model for demonstrating the multistep nature of cancer.
7. The cells of most tumors have visible chromosomal alterations, and the study of these aberrations provides insight into the steps involved in the development of cancer. This relationship between cancer and chromosome alterations has been best studied in leukemias, where the formation of hybrid genes or substitution of regulatory sequences is associated with the transformation of normal cells into malignant tumors.
8. The environment appears to be an important factor in cancer induction. Occupational exposure to physical or chemical substances and viruses, and also diet and other personal choices are important factors in the development of cancer.

Key Terms

acute transforming virus, 488
apoptosis, 486
Burkitt lymphoma, 493
caretaker gene, 492
cancer susceptibility genes, 481
chronic myelogenous leukemia (CML), 492
cyclin, 482
cyclin-dependent kinase (CDK), 482
familial adenomatous polyposis (FAP), 490
gatekeeper gene, 492
hepatitis B virus (HBV), 493
hepatocellular carcinoma (HCC), 493
hereditary nonpolyposis colorectal cancer (HNPCC), 490
metastasize, 481
nonacute virus, 488
oncogene, 484, 487
overexpression, 489
p53 gene, 486
Philadelphia chromosome, 492
point mutation, 489
polyp, 490
pRB protein, 485
protein kinase, 482
proto-oncogene, 483, 487
provirus, 488
ras gene family, 489
retinoblastoma (RB), 484
retrovirus, 488
Rous sarcoma virus (RSV), 488
sarcoma, 488
translocation, 489
tumor-suppressor gene, 483

INSIGHTS AND SOLUTIONS

In disorders such as retinoblastoma, a mutation in one allele of the *RB1* gene can be inherited from the germline, causing an autosomal dominant predisposition to the development of eye tumors. To develop tumors, a somatic mutation in the second copy of the *RB1* gene is necessary, indicating that the mutation itself acts as a recessive trait. Given that the first mutation can be inherited, in what ways can a second mutational event occur?

Solution: In considering how this second mutation arises, we must look at several levels of mutational events, including changes in nucleotide sequence and events that involve whole chromosomes or chromosome parts. Retinoblastoma results when both copies of the *RB1* locus are lost or inactivated. With this in mind, you must first list the phenomena that can result in a mutational loss or the inactivation of a gene.

One way the second *RB1* mutation can occur is by a nucleotide alteration that converts the remaining normal *RB1* allele to a mutant form. This alteration can occur through a nucleotide substitution or by a frameshift mutation caused by the insertion or deletion of nucleotides during replication. A second mechanism involves the loss of the chromosome carrying the normal allele. This event would take place during mitosis, resulting in chromosome 13 monosomy, leaving the mutant copy of the gene as the only *RB1* allele. This mechanism does not necessarily involve loss of the entire chromosome; deletion of the long arm (*RB1* is on 13q) or an interstitial deletion involving the *RB* locus and some surrounding material would have the same result. Alternatively, a chromosome aberration involving loss of the normal copy of the *RB1* gene might be followed by duplication of the chromosome carrying the mutant allele. Two copies of chromosome 13 would be restored to the cell, but the normal *RB1* allele would not be present. Finally, a recombination event followed by chromosome segregation could produce a homozygous combination of mutant *RB1* alleles.

More can be discovered about the mechanisms involved in RB by analyzing tumors with a combination of cytogenetic and molecular techniques (such as RFLP analysis and hybridizations to look for deletions). This would indicate which mechanisms are actually found in tumors and to what extent they are involved in generating the second mutation. Although such analysis is still in the preliminary stages, all of the mechanisms that have been proposed have been found in tumors, indicating that a variety of spontaneous events brings about the second mutation that triggers retinoblastoma.

Problems and Discussion Questions

1. As a genetic counselor, you are asked to assess the risk for a couple with a family history of retinoblastoma who are thinking about having children. Both the husband and wife are phenotypically normal, but the husband has a sister with familial retinoblastoma in both eyes. What is the probability that this couple will have a child with retinoblastoma? Are there any tests that you could recommend to help in this assessment?
2. What events occur in each phase of the cell cycle? Which phase is most variable in length?
3. Where are the major regulatory points in the cell cycle?
4. List the functions of kinases and cyclins, and describe how they interact to cause cells to move through the cell cycle.
5. (a) How does pRb function to keep cells at the G1 checkpoint? (b) How do cells get past the G1 checkpoint to move into S phase?
6. What is the difference between saying that cancer is inherited and saying that predisposition to cancer is inherited?
7. Define tumor-suppressor genes. Why is a mutation in a single copy of a tumor-suppressor gene expected to behave as a recessive gene?
8. In the Rous sarcoma virus (RSV) genome, the host-cell proto-oncogene is converted into an oncogene. List some of the ways in which this conversion can occur.
9. Part of the Ras protein is embedded in the plasma membrane, and part extends into the cytoplasm. How does the Ras protein transmit a signal from outside the cell into the cytoplasm? What happens in cases where the *ras* gene is mutated ?
10. If a cell suffers damage to its DNA while in the S phase, how can this damage be repaired before the cell enters mitosis?
11. Review the differences between transcriptional activity and functions of the tumor-suppressor genes associated with retinoblastoma and breast cancer. Do they have any properties in common? Which properties are different?
12. How can a mutation in one allele of the *p53* gene cause loss of function in almost all p53 protein molecules?
13. Distinguish between oncogenes and proto-oncogenes. In what ways can proto-oncogenes be converted to oncogenes?
14. Of the two classes of genes associated with cancer, tumor-suppressor genes and oncogenes, mutations in which group can be considered gain-of-function mutations, and which are loss-of-function mutations? Explain.
15. How do translocations such as the Philadelphia chromosome lead to oncogenesis?
16. What step in the development of cancer do FAP-associated colon cancer and HNPCC have in common?
17. Compare the mechanisms by which mutations in gatekeeper genes and caretaker genes result in colorectal cancer.
18. A study by Bose, and colleagues (1998. *Blood* 92: 3362–67.) and a previous study by Biernaux, and others 1996 [*Bone Marrow Transplant* 17: (Suppl. 3) S45–S47.] showed that *BCR/ABL* fusion gene transcripts can be detected in 25–30% of healthy adults who do not develop chronic myelogenous leukemia (CML). Explain how these individuals can carry a fusion gene that is transcriptionally active yet do not develop CML.
19. In CML, leukemic blood cells can be distinguished from other cells of the body by the presence of a functional BCR/ABL hybrid protein. Explain how this provides an opportunity to develop a therapeutic approach to a treatment for CML.
20. Cytogenetic and molecular studies indicate that reciprocal translocations between the long arms of chromosomes 9 and 22 are very common events. What molecular and cytological factors may underlie this observation?
21. Infection with hepatitis B virus (HBV) is associated with outcomes ranging from a self-limiting infection to fatal liver disease and liver cancer. Explain how infection with HBV can have such divergent outcomes.
22. Given that up to 50% of all cancers are environmentally induced and most of these are caused by lifestyle choices such as smoking, sun exposure, and diet, what percentage of the money spent on cancer research do you think should be devoted to research and education on preventing cancer rather than on finding a cure?
23. Those who inherit a mutant allele of the *RB1* gene are at risk for a bone cancer called osteosarcoma. You suspect that in these cases, osteosarcoma is caused by a mutation in the second *RB1* allele and have cultured the osteosarcoma cells and clones carrying the *RB1* gene. A colleague sends you a research paper revealing that cancer-prone mice develop malignant tumors when injected with osteosarcoma cells, and you obtain mice from that strain. Using these resources, what experiments would you perform to determine (a) whether osteosarcoma cells carry two *RB1* mutations, (b) whether osteosarcoma cells produce any pRB protein, and (c) if the addition of a normal *RB1* gene will change the cancer-causing potential of osteosarcoma cells.
24. The compound benzo[*a*]pyrene is found in cigarette smoke. This compound chemically modifies guanine bases in DNA. Such abnormal bases are typically removed by an enzyme that hydrolyzes the base, leaving an apurinic site. If such a site is left unrepaired, an adenine is preferentially inserted *across from* the apurinic site. In a study of lung cancer patients (1991. *Nature* 350: 377–78.), tumor cells from 15 out of 25 patients had a G—T transversion in the *p53* gene, which has a known role in cancer formation. You are testifying as an expert witness in a court

case where the widow of a man who was a lifelong smoker and died of lung cancer is suing a company for manufacturing the tobacco products that killed her husband. What would you tell the jury? (Please, no personal expositions on lawyers or the legal system!)

25. Table 21.9 (1994. *Science* 266: 66–71.) summarizes some of the data that have been collected on *BRCA1* mutations in families with a high incidence of both early-onset breast and ovarian cancer. Table 21.10 (ibid.) shows neutral polymorphisms found in control families (with no increased frequency of breast and ovarian cancer). (a) Note the coding effect of the mutation found in kindred group 2082 in Table 21.9; this results from a single base-pair substitution. Draw the normal double-stranded DNA sequence for this codon (with the 5′-and 3′-ends labeled), and show the sequence of events that generated this mutation, assuming that it resulted from an uncorrected mismatch event during DNA replication. (b) Examine the types of mutations that are listed in Table 21.9, and determine if the *BRCA1* gene is likely to be a tumor-suppressor gene or an oncogene. (c) Although the mutations in Table 21.9 are clearly deleterious and cause breast cancer in women at very young ages, each of the kindred groups had at least one woman who carried the mutation but lived until age 80 without developing cancer. Name at least two different mechanisms (or variables) that could underlie variation in the expression of a mutant phenotype, and propose an explanation for the incomplete penetrance of this mutation. How do these mechanisms or variables relate to this explanation?
26. Examine Table 21.10. (a) What is meant by a neutral polymorphism? (b) What is the significance of this table in the context of examining a family or population for *BRCA1* mutations that predispose an individual to cancer? (c) Is the PM2 polymorphism likely to result in a neutral missense mutation or a silent mutation? (d) Answer part (c) for the PM3 polymorphism.
27. Describe the difference between an acute transforming virus and a nonacute virus.
28. What steps are often employed in establishing a relationship between an environmental factor and cancer?
29. In cancer studies, one often sees the term *variable penetrance*. What is meant by this term, and how might it be applied to cancer incidence data?
30. Assume that a young woman in a suspected breast cancer family takes the *BRCA1* and *BRCA2* genetic tests and receives negative results. In other words, she does not test positive for the mutant alleles of *BRCA1* or *BRCA2*. Can she consider herself free of risk for breast cancer?

TABLE 21.9 Predisposing Mutations in *BRCA1*

		Mutation		
Kindred	*Codon*	*Nucleotide Change*	*Coding Effect*	*Frequency in Control Chromosomes*
1901	24	−11 bp	Frameshift or splice	0/180
2082	1313		Gln → Stop	0/170
1910	1756	Extra C	Frameshift	0/162
2099	1775	T → G	Met → Arg	0/120
2035	NA*	?	Loss of transcript	NA*

*NA indicates not applicable, as the regulatory mutation is inferred, and the position has not been identified.

TABLE 21.10 Neutral Polymorphisms in *BRCA1*

			*Frequency in Control Chromosomes**			
Name	*Codon Location*	*Base in Codon*†	*A*	*C*	*G*	*T*
PM1	317	2	152	0	10	0
PM6	878	2	0	55	0	100
PM7	1190	2	109	0	53	0
PM2	1443	3	0	115	0	58
PM3	1619	1	116	0	52	0

*The number of chromosomes with a particular base at the indicated polymorphic site (A, C, G, or T) is shown.
†Position 1, 2, or 3 of the codon.

Selected Readings

Ames, B., Magraw, R., and Gold, L. 1990. Ranking possible cancer hazards. *Science* 236: 71–80.

Ames, B., Profet, M. and Gold, L. 1990. Dietary pesticides (99.99% all natural). *Proc. Natl. Acad. Sci. (USA)* 87: 7777–81.

Anwar, S. et al. 2000. Hereditary non-polyposis colorectal cancer: An updated review. *Eur. J. Surg. Oncol.* 26: 635–45.

Barnes, D. J., and Melo, J. V. 2002. Cytogenetic and molecular aspects of chronic myeloid leukemia. *Acta Haematol.* 108: 180–202.

Barth, A., and Nelson, W. J. 2002. What can humans learn from flies about adenomatous polyposis coli? *BioEss.* 24: 771–74.

Beijersbergen, R. L., and Bernards, R. 1996. Cell cycle regulation by the retinoblastoma family of growth inhibitory proteins. *Biochim. Biophys. Acta* 1287: 103–20.

Bieche, I., and Lidereau, R. 1995. Genetic alterations in breast cancer. *Genes Chromosomes Cancer* 14: 227–51.

Bohlander, S. K. 2000. Fusion genes in leukemia: An emerging network. *Cytogenet. Cell Genet.* 91: 52–56.

Cavenee, W. K., and White, R. L. 1995. The genetic basis of cancer. *Sci. Am.* (Mar.) 272: 72–79.

Compagni, A., and Christofori, G. 2000. Recent advances in research on multistage tumorigenesis. *Brit. J. Cancer* 83: 1–5.

Cornelis, J. F. et al. 1998. Metastasis. *Am. Scient.* 86: 130–41.

Elledge, S. J. 1996. Cell cycle checkpoints: Preventing an identity crisis. *Science* 274: 1664–72.

Elkorn, S.V., and Reed, S. I. 2000. Regulation of G_1 cyclin-dependent kinases in the mammalian cell cycle. *Curr. Opin. Cell Biol.* 12: 676–84.

Evan, G., and Littlewood, T. 1998. A matter of life and cell death. *Science* 281: 1317–22.

Fearon, E. R. 1997. Human cancer syndromes: Clues to the origin and nature of cancer. *Science* 278: 1043–50.

Kastan, M. B. 2001. Checking two steps. *Nature* 410: 766–67.

Kinzler, K. W., and Vogelstein, B. 1997. Gatekeepers and caretakers. *Nature* 386: 761–63.

Lengauer, C., Kinzler, K. W., and Vogelstein, B. 1997. Genetic instability in colorectal cancer. *Nature* 386: 623–27.

Nurse, P. 1997. Checkpoint pathways come of age. *Cell* 91: 865–67.

Pines, L. 1996. Cyclin from sea urchin to HeLas: Making the human cell cycle. *Biochem. Soc. Trans.* 24: 15–33.

Scully, R., and Puget, N. 2002. *BRCA1* and *BRCA2* in hereditary breast cancer. *Biochimie* 84: 95–102.

Sherr, C. J. 1996. Cancer cell cycles. *Science* 274: 1672–77.

Shibata, D., and Aaltonen, L. A. 2001. Genetic predisposition and somatic diversification in tumor development and progression. *Adv. Cancer Res.* 80: 83–114.

Shito, K. et al. 2000. Pathogenesis of non-familial colorectal carcinomas with high microsatellite instability. *J. Clin. Pathol.* 53: 841–45.

Smith, M. L., and Fornace, A. J., Jr. 1996. The two faces of tumor suppressor *p53*. *Am. J. Pathol.* 148: 1019–22.

Smits, V. A. J., and Medema, R. H. 2001. Checking out the G_2/M transition. *Biochim. Biophys. Acta* 1519: 1–12.

Soussi, T. 2000. The *p53* tumor suppressor gene: From molecular biology to clinical investigation. *Ann. N.Y. Acad. Sci.* 910: 121–37.

Stahl, A. et al. 1994. The genetics of retinoblastoma. *Ann. Genet.* 37: 172–78.

Walworth, N. C. 2000. Cell cycle checkpoint kinases: Checking in on the cell cycle. *Curr. Opin. Cell Biol.* 12: 697–704.

Zang, H., Tombline, G., and Weber, B. L. 1998. *BRCA1*, *BRCA2*, and DNA damage response: Collision or collusion? *Cell* 92: 433–36.

CHAPTER 22

CHAPTER CONCEPTS

These lady-bird beetles, from the Chiricahua Mountains in Arizona, show considerable phenotypic variation. *(Edward S. Ross/California Academy of Sciences)*

Population Genetics

Alfred Russel Wallace and Charles Darwin were the first to identify natural selection as the mechanism of adaptive evolution in the mid-nineteenth century, based on their observations of populations: (1) Phenotypic variations exist among individuals within populations; (2) these differences are passed from parents to offspring; (3) more offspring are born than will survive and reproduce; and (4) some variants are more successful at surviving and reproducing than others. In populations where all four factors operate, the relative abundance of the population's different phenotypes may change across generations. In other words, the population evolves.

Although Wallace and Darwin described how organisms evolve by natural selection, there was no accurate model describing the mechanisms responsible for the origin of variation and its inheritance. Gregor Mendel had published his work on the inheritance of traits in 1866, but it received little notice at the time. When Mendel's work was rediscovered in 1900, biologists immediately recognized its importance and began a 30-year effort to reconcile Mendel's concept of genes and alleles with the theory of evolution. In a key insight, biologists realized that changes in the relative frequency of different phenotypic traits in a population are tied to changes in the relative frequency of the alleles that influence the traits. The discipline within evolutionary biology that studies changes in allele frequencies is known as **population genetics**.

Early in the twentieth century, a number of workers, including Gudney Yule, William Castle, Godfrey Hardy, and Wilhelm Weinberg, formulated the basic principles of population genetics. For many years, theorists focused on mathematical models that would describe the genetic structure of populations and later prove useful in practical applications, such as animal and plant breeding. Prominent among the theoreticians who developed these models were Sewall Wright, Ronald Fisher, and J. B. S. Haldane. Following the work of the theorists, fieldworkers and experimentalists tested these models. (We will discuss some of these field studies, such as those done by Dobzhansky and colleagues on *Drosophila* chromosome inversions, in Chapter 23.) Their experiments examined allele frequencies and the forces that alter the frequencies, including selection, mutation, migration, and random genetic drift. In this chapter, we consider some general aspects of population genetics and also discuss other areas of genetics that relate to evolution.

HOW DO WE KNOW WHAT WE KNOW?

IN THIS CHAPTER, WE WILL FOCUS ON the use of the Hardy-Weinberg law to determine allele and genotype frequencies in populations and explore how factors such as mutation, migration, drift, and natural selection alter a population's gene pool. As you study this topic, you should try to answer several fundamental questions:

1. How do we know if the assumptions of the Hardy-Weinberg law are operating in a population?
2. How do we know when allele frequencies and genotype frequencies can be predicted for future generations?
3. How do we know how natural selection operates to alter allele frequencies in a population?
4. How do we know whether inbreeding has an effect on the fitness of individuals in a population? ■

22.1 Populations and Gene Pools

Members of a species often range over a wide geographic area, clustered into populations. A **population** is a group of individuals from the same species that lives in the same geographic area. If we consider a single genetic locus in this population, we may find that individuals within the population have different genotypes. To study population genetics, we compute frequencies at which various alleles and genotypes occur and how these frequencies change from one generation to the next.

When we consider generational changes in alleles and genotypes, we examine gene pools. A **gene pool** consists of all gametes made by all the breeding members of a population in a single generation. These gametes will combine to form the zygotes that become the next generation. The eggs and sperm that constitute the gene pool are haploid and thus contain only a single allele for each genetic locus. When we consider a single locus, we may find that different gametes carry different alleles. The proportion of gametes in a gene pool that carry a particular allele represents the frequency with which this allele occurs in the population.

Populations are dynamic; they expand and contract through changes in birth and death rates, migration, or contact with other populations. The dynamic nature of populations has important consequences and can, over time, lead to changes in the population's gene pool.

22.2 Calculating Allele Frequencies

In analyzing the genetic structure of a population, most geneticists first measure the frequencies at which alleles occur at a particular locus. To do this, the genotypes of a large number of individuals in the population must be determined. In some cases, researchers can infer genotypes directly from phenotypes. In other cases, proteins or DNA sequences are analyzed to determine genotypes. To understand how allele frequencies are calculated in a population, we will consider an example that involves HIV infection rates.

Most people in a population are very susceptible to HIV infection, but in some populations, a small number of individuals are highly resistant to infection even after repeated exposure. Researchers quickly focused on these individuals to determine if genetic factors were involved in resistance to HIV-1, the virus responsible for most AIDS cases worldwide. In 1996, Rong Liu and colleagues discovered exposed

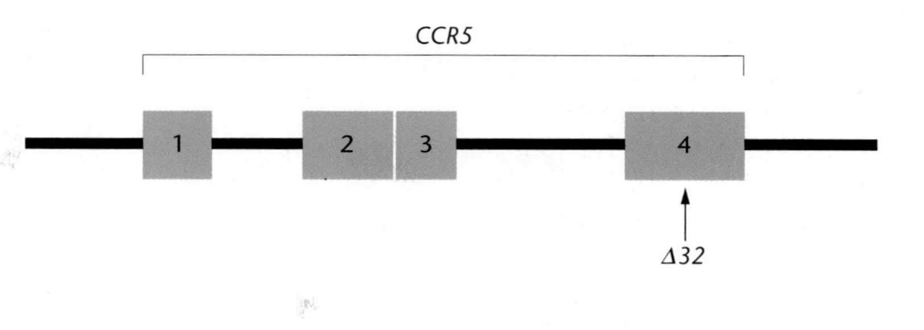

FIGURE 22–1 Organization of the *CCR5* gene in region 3p21.3. The gene contains 4 exons and 2 introns. There is no intron between exons 2 and 3. The arrow shows the location of the 32-bp deletion in exon 4 that confers resistance to HIV-1 infection. *(Redrawn from Klitz, W. et al. 2001. Evolution of the CCR5-Δ32 mutation based on haplotype variation in Jewish and northern European population samples.* Human Immunol. *62: 530–38. Figure 1, p. 532.)*

but uninfected individuals who carried mutant alleles of a gene called *CC-CKR-5* (or *CCR5*). This gene, located on chromosome 3 (Figure 22–1), encodes a protein called the C–C chemokine receptor-5, often abbreviated CCR5. Chemokines are signaling molecules associated with the immune system. When chemokines bind to CCR5 receptors on the surface of white blood cells, the cells respond by moving into inflamed tissues to fight infection.

The CCR5 protein is also a receptor for strains of HIV-1. To gain entry into white blood cells, an HIV surface protein called Env (short for envelope protein) binds to the CD4 protein on the surface of the host cell. Binding to CD4 causes Env to change shape and bind to CCR5, which in turn initiates the fusion of the viral protein coat with the host-cell membrane. Merging the viral envelope with the cell membrane moves the viral core into the cytoplasm, thus infecting the cell.

The mutant allele of the *CCR5* gene discovered by Liu and colleagues carries a 32-bp deletion in one of its exons (Figure 22–2). As a result, the protein encoded by the mutant allele is shortened and so is not inserted into the plasma membrane. Because there is no CCR5 protein present on the cell surface, HIV-1 cannot enter these cells. The gene's normal allele is called *CCR51* (or *1*), and its allele with the 32-bp deletion is called *CCR5-Δ32* (or *Δ32*). The two uninfected individuals described by Liu both had the genotype *Δ32/Δ32*. As a result, they had no CCR5 receptors on the surface of their cells and were resistant to infection by strains of HIV-1 that use CCR5 as a co-receptor.

Curiously, researchers have not discovered any adverse effects associated with this genotype. Heterozygotes with genotype *1/Δ32* are susceptible to HIV-1 infection, but evidence suggests that they progress more slowly to full-blown AIDS. Table 22.1 summarizes the genotypes possible at the *CCR5* locus and the phenotypes associated with each.

The discovery of the *CCR5-Δ32* allele and the fact that it provides some protection against AIDS generates two important questions: Which human populations carry the *Δ32* allele, and how common is it? To address these questions, several teams of researchers surveyed a large number of people from several populations. Genotypes were determined by direct analysis of their DNA to detect the presence or absence of the 32-bp deletion (Figure 22–3). Let's look at one sample population of 100 French individuals from a survey in Brittany.

In this population, 79 individuals have genotype *1/1*, 20 have genotype *1/Δ32*, and 1 has genotype *Δ32/Δ32*. This sample has 158 *1* alleles (each of the 79 *1/1* individuals carries two alleles) plus 20 *1* alleles carried

TABLE 22.1 ***CCR5* Genotypes and Phenotypes**

Genotype	*Phenotype*
1/1	Susceptible to sexually transmitted strains of HIV-1
1/Δ32	Susceptible, but may slowly progress to AIDS
Δ32/Δ32	Resistant to most sexually transmitted strains of HIV-1

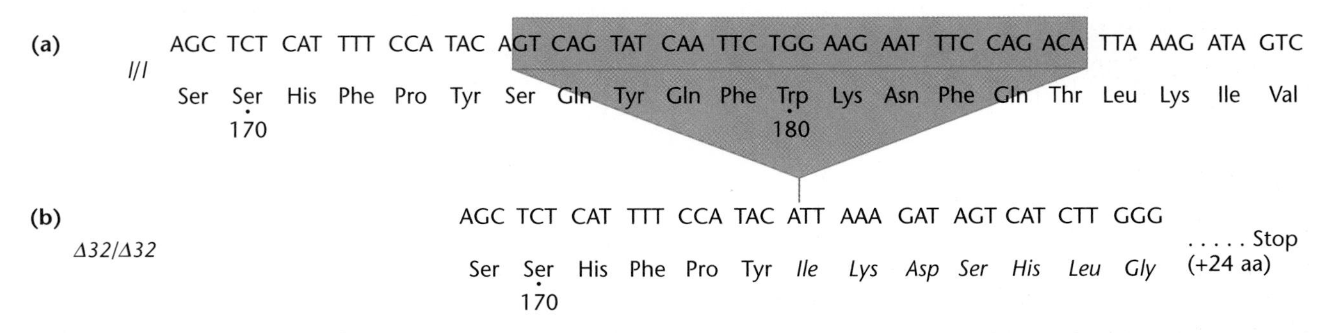

FIGURE 22–2 (a) Nucleotide sequence and amino acid sequence of a region in exon 4 of the normal (*I/I*) allele of the *CCR5* gene. (b) The deletion in the *Δ32* allele, associated with HIV-1 resistance. The first seven amino acids encoded out of frame by this 32-bp deletion are shown in italics. The amino acid number for serine (170) is shown for orientation. The truncated protein of the *Δ32* allele is not inserted into the plasma membrane and cannot therefore act as a co-receptor for HIV-1. *(Reprinted from* Cell, *Vol. 86, Liu et al., Homozygous defect in HIV-1 coreceptor accounts for resistance of some multiply-exposed individuals to HIV-1 infection, pp 367-377, Copyright (1996), with permission from Elsevier.)*

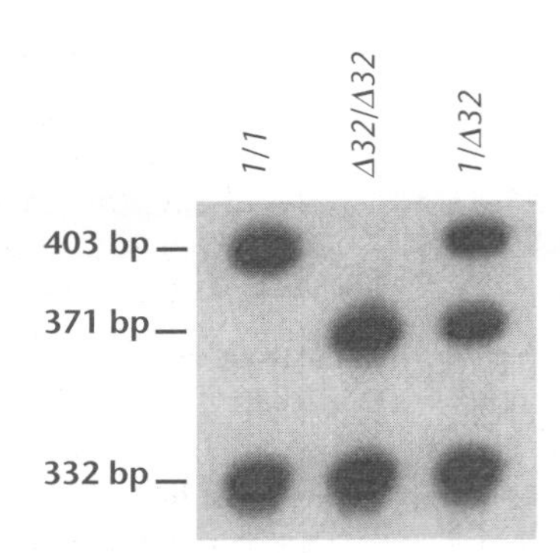

FIGURE 22–3 Allelic variation in the *CCR5* gene. Michel Samson and colleagues used PCR to amplify a part of the *CCR5* gene containing the site of the 32-bp deletion, cut the resulting DNA fragments with a restriction enzyme, and ran the fragments on an electrophoresis gel. Each lane reveals the genotype of a single individual. The *1* allele produces a 332-bp fragment and a 403-bp fragment; the *Δ32* allele produces a 332-bp fragment and a 371-bp fragment. Heterozygotes produce three bands. *(From Michel Samson, Frederick Libert, et al. 1996. Resistance to HIV-1 infection in Caucasian individuals bearing mutant alleles of the CCR-5 chemokine receptor gene. Reprinted with permission from* Nature *[Vol. 382, 22 August. 1996, Fig. 3, p. 725.] © 1996 Macmillan Magazines, Limited.)*

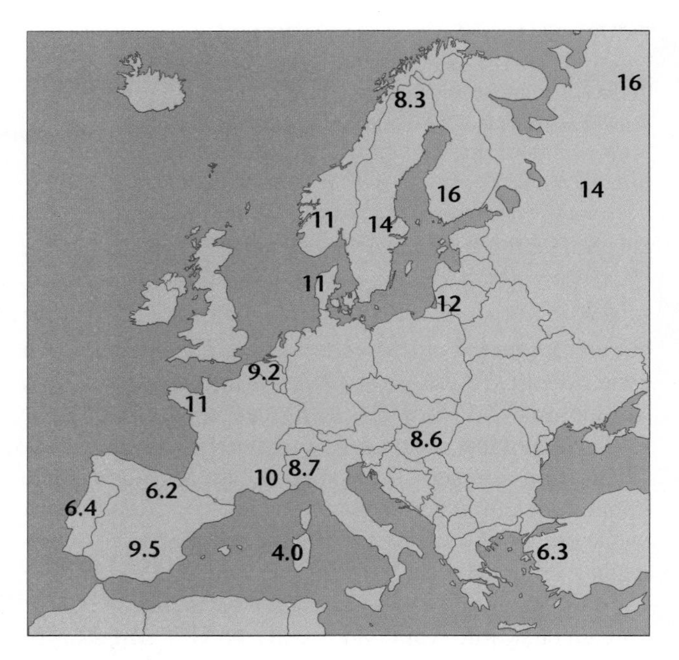

FIGURE 22–4 The frequency (percentage) of the *CCR5-Δ32* allele in 18 European populations. *(From F. Leibert et al., 1998. The* CCR5 *mutation conferring protection against HIV-1 in Caucasian populations has a single and recent origin in northeastern Europe.* Hum. Molec. Genet. *7: 399–406, by permission of Oxford Univ. Press.)*

by the *1*/Δ*32* individuals, for a total of 178. The frequency of the *CCR51* allele in the sample population is thus 178/200 = 0.89 = 89%. The *CCR5*-Δ*32* allele count shows 20 carried by the *1*/Δ*32* individuals, plus 2 carried by the Δ*32*/Δ*32* individual, for a total of 22. The frequency of the *CCR5*-Δ*32* allele is thus 22/200 = 0.11 = 11%. Note that 0.89 + 0.11 = 1.00 = 100%, which confirms that we have accounted for all the alleles of this gene in the gene pool. Table 22.2 shows two methods for computing the frequencies of the *1* and Δ*32* alleles in the Brittany sample.

Figure 22–4 shows the frequency of the *CCR5*-Δ*32* allele in the 18 European populations surveyed. Populations in Northern Europe around the Baltic Sea have the highest frequencies of the Δ*32* allele. There is a sharp gradient in allele frequency from north to south, and populations in Sardinia and Greece have very low frequencies. In populations without European ancestry, the Δ*32* allele is essentially absent. The global distribution of the Δ*32* allele presents an evolutionary puzzle that we'll return to later in the chapter.

TABLE 22.2 Methods of Determining Allele Frequencies from Data on Genotypes

A. From Numbers of Alleles

Genotype:	*1/1*	*1/Δ32*	*Δ32/Δ32*	*Total*
Number of individuals:	79	20	1	100
Number of *1* alleles:	158	20	0	178
Number of *Δ32/Δ32* alleles:	0	20	2	22
Total number of alleles:	158	40	2	200

Frequency of *CCR51* in sample: 178/200 = 0.89 = 89%
Frequency of *CCR5-Δ32* in sample: 22/200 = 0.11 = 11%

B. From Genotype Frequencies

Genotype:	*1/1*	*1/Δ32*	*Δ32/Δ32*	*Total*
Number of individuals:	79	20	1	100
Genotype frequency:	79/100 = 0.79	20/100 = 0.20	1/100 = 0.01	1.00

Frequency of *CCR51* in sample: 0.79 + (0.5)0.20 = 0.89 = 89%
Frequency of *CCR5-Δ32* in sample: (0.5)0.20 + 0.01 = 0.11 = 11%

22.3 The Hardy-Weinberg Law

The large variation in frequency of the *CCR5*-$\Delta 32$ allele among European populations raises some questions. For example, can we expect the allele to increase in populations in which it is currently rare? Population genetics explores such questions by using a mathematical model developed independently by the British mathematician Godfrey H. Hardy and the German physician Wilhelm Weinberg. This model, called the **Hardy-Weinberg law**, shows what happens to alleles and genotypes in an "ideal" population—one that is free of many of the complications that affect real populations—and uses a set of simple assumptions.

1. Individuals of all genotypes have equal rates of survival and equal reproductive success; that is, there is no selection.
2. No new alleles are created or converted from one allele into another by mutation.
3. Individuals do not migrate into or out of the population.
4. The population is infinitely large, which in practical terms means that the population is large enough that sampling errors and other random effects are negligible.
5. Individuals in the population mate randomly.

The Hardy-Weinberg law demonstrates that an ideal population has two properties:

1. The frequency of alleles does not change from generation to generation.
2. After one generation of random mating, offspring genotype frequencies can be predicted from the parent allele frequencies and would be expected to remain constant from that point.

What makes the Hardy-Weinberg law useful is its assumptions. By specifying the assumptions under which the population *cannot* evolve, the Hardy-Weinberg law identifies the real-world forces that cause allele and genotype frequencies to change. In other words, by holding certain conditions constant, the Hardy-Weinberg law isolates the forces of evolution and allows them to be quantified.

Demonstration of the Hardy-Weinberg Law

To demonstrate how the Hardy-Weinberg law works, we will begin with a specific case and then consider the general case. In both examples, we focus on a single locus with two alleles, *A* and *a*.

Imagine a population in which the frequency in both eggs and sperm of allele *A* is 0.7 and allele *a* is 0.3. Note that $0.7 + 0.3 = 1.0$, indicating that all the alleles for the gene present in the gene pool are accounted for. We assume, per Hardy-Weinberg requirements, that individuals mate randomly, which you can visualize as placing all of the gametes in the gene pool into a barrel and stirring them. We then randomly draw eggs and sperm from the barrel and pair them to make zygotes. What genotype frequencies does this give us? For any one zygote, the probability or chance that the egg will contain *A* is 0.7 and the probability that the sperm will contain *A* is 0.7. The probability that *both* egg and sperm will contain *A* is the product of their independent probabilities, or $0.7 \times 0.7 = 0.49$. In other words, we predict that genotype *AA* will occur 49% of the time. The probability that a zygote will be formed from an egg carrying *A* and a sperm carrying *a* (forming an *Aa* heterozygote) is therefore $0.7 \times 0.3 = 0.21$. The probability that a zygote will be formed from an egg carrying *a* and a sperm carrying *A* (forming an *aA* heterozygote) is $0.7 \times 0.3 = 0.21$. Thus, the frequency of all heterozygotes (*Aa* or *aA*) is $0.21 + 0.21 = 0.42 = 42\%$. Finally, the probability that a zygote will be formed from an egg carrying *a* and a sperm carrying *a* is again the product of their independent probabilities, or $0.3 \times 0.3 = 0.09$, meaning that the predicted frequency of genotype *aa* is 9%. To check our calculations, we note that $0.49 + 0.42 + 0.09 = 1.0$, which confirms that we have accounted for all of the zygotes. These calculations are summarized in Figure 22–5.

We started with the frequency of a particular allele in a specific gene pool and calculated the probability that certain genotypes would be produced from this pool. When the zygotes develop into adults and reproduce, what will be the frequency distribution of alleles in the new gene pool? Recall that under the Hardy-Weinberg law, we must assume that all genotypes have equal rates of survival and reproduction. This means that in the next generation, all adults of the current generation contribute equally to the new gene pool. The *AA* individuals constitute 49% of the population, and we can predict that the gametes they produce will also constitute 49% of the gene pool. These gametes all carry allele *A*. Likewise, *Aa* individuals constitute 42% of the population, so we predict that their gametes will constitute 42% of the new gene pool. Half (0.5) of these gametes will carry allele *A*.

MEDIA TUTORIAL Population Genetics

		Sperm	
		fr(*A*) = 0.7	fr(*a*) = 0.3
Eggs	fr(*A*) = 0.7	fr(*AA*) = 0.7×0.7 = 0.49	fr(*Aa*) = 0.7×0.3 = 0.21
	fr(*a*) = 0.3	fr(*aA*) = 0.3×0.7 = 0.21	fr(*aa*) = 0.3×0.3 = 0.09

FIGURE 22–5 Calculating genotype frequencies (fr) from allele frequencies. Gametes represent withdrawals from the gene pool to form the genotypes of the next generation. In this population, the frequency of the *A* allele is 0.7 and the frequency of the *a* allele is 0.3. The frequencies of the genotypes in the next generation are calculated as 0.49 for *AA*, 0.42 for *Aa*, and 0.09 for *aa*. Under the Hardy-Weinberg law, the frequencies of *A* and *a* remain constant from generation to generation.

Thus, the frequency of allele A in the gene pool is $0.49 + (0.5)0.42 = 0.7$. The other half of the gametes produced by Aa individuals will carry allele a. The aa individuals constitute 9% of the population, so their gametes will constitute 9% of the new gene pool. These gametes all carry allele a. Thus, we can predict that allele a in the new gene pool is $(0.5)0.42 + 0.09 = 0.3$. As a check on our calculation, note that $0.7 + 0.3 = 1.0$ accounts for all of the gametes in the gene pool of the new generation. Notice that *the allele frequencies in this gene pool are the same as those in the preceding generation.*

In other words, we have arrived where we began—with a gene pool in which the frequency of allele A is 0.7 and the frequency of allele a is 0.3. These calculations demonstrate the Hardy-Weinberg law. Allele frequencies in our population do not change from one generation to the next, and after just one generation of random mating, the genotype frequencies can be predicted from the allele frequencies. Therefore, this population does not evolve with respect to the locus we have examined.

In considering the general case, we use variables instead of numerical values for the allele frequencies. Imagine a gene pool in which the frequency of allele A is p, and the frequency of allele a is q, such that $p + q = 1.0$ If we randomly draw an egg and a sperm from the gene pool and pair them to make a zygote, the probability that both egg and sperm will carry allele A is $(p \times p)$. Thus, the frequency of genotype AA among the zygotes is p^2. The probabilities that the egg carries A and the sperm carries a is $(p \times q)$ and that the egg carries a and the sperm carries A is $(q \times p)$. Thus, the frequency of genotype Aa among the zygotes is $2pq$. Finally, the probability that the egg and sperm will both carry a is $(q \times q)$, making the frequency of genotype aa among the zygotes q^2. Hence, the distribution of genotypes among the zygotes is

$$p^2 + 2pq + q^2 = 1$$

The calculations are shown in Figure 22–6.

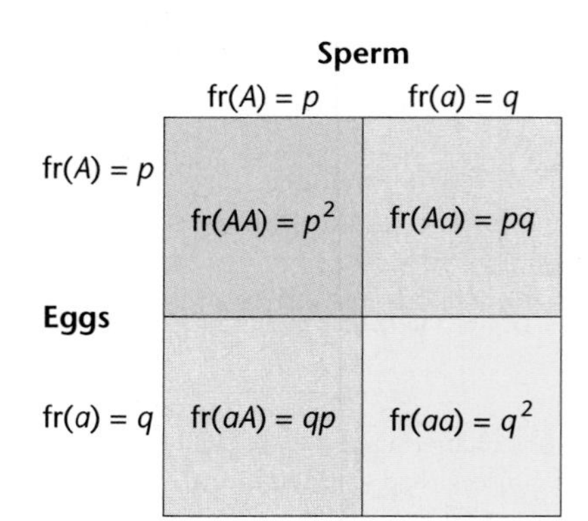

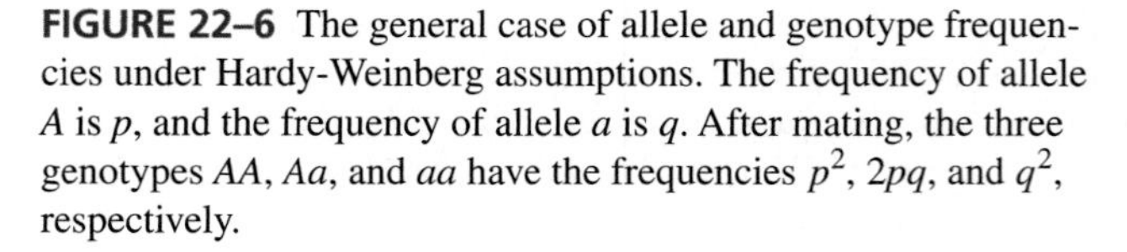
FIGURE 22–6 The general case of allele and genotype frequencies under Hardy-Weinberg assumptions. The frequency of allele A is p, and the frequency of allele a is q. After mating, the three genotypes AA, Aa, and aa have the frequencies p^2, $2pq$, and q^2, respectively.

For our general case, we can ask what the allele frequencies in the new gene pool will be when these zygotes develop into adults and reproduce. All gametes from AA individuals carry allele A, as do half of the gametes from Aa individuals. Thus, we predict that the frequency of allele A in the new gene pool will be

$$p^2 + \frac{1}{2}(2pq) = p^2 + pq$$

Recalling that $p + q = 1.0$, we can therefore substitute $(1 - p)$ for q in the previous equation, which gives

$$p^2 + p(1 - p) = p^2 + p - p^2 = p$$

Likewise, half of the gametes from Aa individuals carry allele a, as do all of the gametes from aa individuals. Thus, we predict that the frequency of allele a in the new gene pool will be

$$\frac{1}{2}(2pq) + q^2 = pq + q^2$$

We substitute $(1 - q)$ for p in this equation and get

$$(1 - q)q + q^2 = q - q^2 + q^2 = q$$

Once again, we have arrived where we began, with a gene pool in which the predicted proportion of allele A is p, and that of allele a is q. The Hardy-Weinberg law holds true for any frequencies of A and a, as long as the frequencies add to 1.0 and the five assumptions are invoked. A population in which the allele frequencies remain constant from generation to generation and in which the genotype frequencies can be predicted from the allele frequencies is said to be in a state of **Hardy-Weinberg equilibrium** for that locus.

Consequences of the Law

The Hardy-Weinberg law has several important consequences. First, it shows that dominant traits do not necessarily increase from one generation to the next. Second, it demonstrates that **genetic variability** can be maintained in a population since, once established in an ideal population, allele frequencies remain unchanged. Third, if we invoke the Hardy-Weinberg assumption of random mating, then knowing the frequency of just one genotype enables us to calculate the frequencies of all other genotypes when there are only two alleles at the locus in question. This relationship is particularly useful in human genetics because it allows us to calculate the frequency of heterozygous carriers for recessive genetic disorders even when all we know is the frequency of affected individuals.

We began this discussion by asking whether we can expect the *CCR5-Δ32* allele to increase in populations in which it is currently rare. From what we now know about the Hardy-Weinberg law, we can say for the general case that if all five assumptions are used, then the frequency of the Δ32 allele will not change.

The general case demonstrates the most important role of the Hardy-Weinberg law: It is the foundation upon which population genetics is built. By showing that there are loci that do not evolve, we can use the Hardy-Weinberg law to identify forces that cause populations to evolve. To do this, we examine those loci that do show changes in allele frequency over time. As we shall see in the second half of this chapter, when the assumptions of the Hardy-Weinberg law are invalid—because of a variety of real-world factors we will discuss there—the allele frequencies in a population may change from one generation to the next. For instance, nonrandom mating does not by itself alter *allele* frequencies, but by altering *genotype* frequencies, it indirectly affects the course of evolution. The Hardy-Weinberg law tells geneticists where to look to find the causes of evolution in populations.

Testing for Equilibrium

One way we establish whether one or more of the Hardy-Weinberg assumptions will hold in a given population is by determining whether the population's genotypes are in equilibrium. To do this, we first determine the frequencies of the genotypes—either directly from the phenotypes (if heterozygotes are recognizable) or by analyzing proteins or DNA sequences. We then calculate the allele frequencies from the genotype frequencies, as we did earlier. Finally, we use the allele frequencies in the parental generation to predict the offspring's genotype frequencies. According to the Hardy-Weinberg law, the genotype frequencies are predicted to fit the $p^2 + 2pq + q^2 = 1.0$ relationship. If they do not, then one or more of the assumptions are invalid for the population in question. We will use the *CCR5* genotypes of a population in Britain to demonstrate the Hardy-Weinberg law. The population includes 283 individuals, of which 223 have genotype *1/1*, 57 have genotype *1/Δ32*, and 3 have genotype *Δ32/Δ32*, These numbers represent genotype frequencies of $223/283 = 0.788$, $57/283 = 0.201$, and $3/283 = 0.011$, respectively. From the genotype frequencies, we compute the *CCR51* allele frequency as 0.89 and the frequency of the *CCR5-Δ32* allele as 0.11. From these allele frequencies, we can use the Hardy-Weinberg law to determine whether this population is in equilibrium. The allele frequencies predict the genotype frequencies as follows:

Expected frequency of genotype *1/1*

$$= p^2 = (0.89)^2 = 0.792$$

Expected frequency of genotype *1/Δ32*

$$= 2pq = 2(0.89)(0.11) = 0.196$$

Expected frequency of genotype *Δ32/Δ32*

$$= q^2 = (0.11)^2 = 0.012$$

These expected frequencies are nearly identical to the observed frequencies, and our test of this population has failed to provide evidence that Hardy-Weinberg assumptions are being violated. This conclusion is confirmed by a chi-square analysis. The χ^2 value in this case is tiny: 0.00023. To be statistically significant at even the most generous accepted level, $p = 0.05$, the χ^2 value would have to be 3.84. (In a test for Hardy-Weinberg equilibrium, the degrees of freedom are given by $k - 1 - m$, where k is the number of genotypes and m is the number of independent allele frequencies estimated from the data. Here, $k = 3$ and $m = 1$ since calculating only one allele frequency allows us to determine the other by subtraction. Thus, we have $3 - 1 - 1 = 1$ degree of freedom.)

On the other hand, if the Hardy-Weinberg test had demonstrated that the population is *not* in equilibrium, it would indicate that one or more assumptions were not being met. To illustrate this, imagine two hypothetical populations: one living on East Island, the other living on West Island. The East Island population all have genotype *1/1*, so the *CCR51* allele constitutes 100% of the population. All of the West Island population have the genotype *Δ32/Δ32* so the frequency of the *CCR5-Δ32* allele is 100%. Both island populations are in Hardy-Weinberg equilibrium.

Now imagine that 500 people from each island move to the previously uninhabited Central Island. The allele and genotype frequencies for the source populations and the new population on Central Island appear in Table 22.3. Both the *1* allele and the *Δ32* allele occur at frequencies of 0.5 in the new population. Given these allele frequencies, the Hardy-Weinberg law predicts that the genotype frequencies will be $0.25 + 0.50 + 0.25 = 1.00$. These expected frequencies obviously differ from the observed frequencies; there are no heterozygotes at all on Central Island.

TABLE 22.3 Frequencies of *CCR51* and *CCR5-Δ32* Alleles in a Hypothetical Population Composed of 500 *1/1* Individuals and 500 *Δ32/Δ32* Individuals

	Allele Frequencies		*Genotype Frequencies*		
Population	*CCR51*	*CCR5-Δ32*	*1/1*	*1/Δ32*	*Δ32/Δ32*
Source Populations					
East Island	1.0	0.0	1.0	0.0	0.0
West Island	0.0	1.0	0.0	0.0	1.0
New Population on Central Island					
Observed	$p = 0.5$	$q = 0.5$	0.5	0.0	0.5
Expected			$p^2 = 0.25$	$2pq = 0.50$	$q^2 = 0.25$

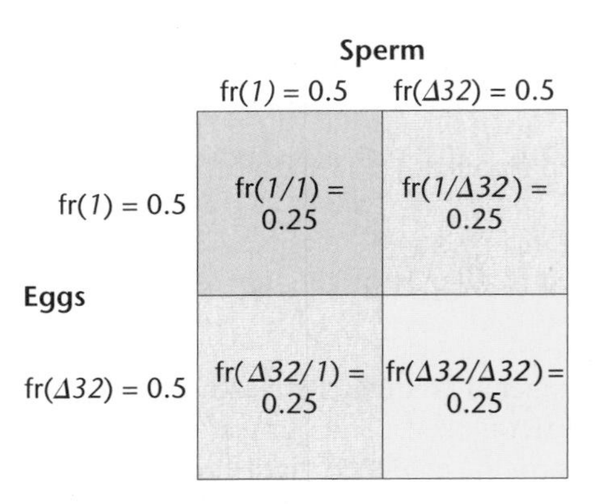

FIGURE 22–7 If mating is random in a population that also meets the other assumptions of the Hardy-Weinberg law, equilibrium will be reached in one generation. Here the genotype frequencies (fr) after one generation of random mating are 0.25, 0.5, and 0.25; compare these with the genotype frequencies shown in Table 22–5.

Therefore, the population is not in Hardy-Weinberg equilibrium (a chi-square analysis would confirm this).

Two assumptions have been violated in the new population. First, everyone living on Central Island is a migrant. Second, Central Island's population is not the product of random mating. Everyone on Central Island moved there and has either two parents from East Island or two parents from West Island. However, it would take only one generation of random mating on Central Island to bring the offspring to the expected allele frequencies, as shown in Figure 22–7.

Are there human populations that are not in Hardy-Weinberg equilibrium? Therese Markow and colleagues studied 122 members of the Havasupai, a population of Native Americans in Arizona. They determined the genotype of each individual at two loci in the major histocompatibility complex (MHC). These genes, *HLA-A* and *HLA-B*, encode proteins involved in the immune system's discrimination between self and nonself. Individuals heterozygous at MHC loci have immune systems that recognize a greater diversity of foreign invaders and thus may be better able to fight disease. Markow and colleagues observed significantly more individuals heterozygous at both loci and significantly fewer homozygous individuals than expected under the Hardy-Weinberg law. Violation of either or both of two Hardy-Weinberg assumptions could explain the excess of heterozygotes among the Havasupai. First, Havasupai fetuses, children, and adults who are heterozygous for *HLA-A* and *HLA-B* may have higher rates of survival than homozygous individuals (i.e., selection is occurring). The other possibility is that rather than choosing their mates randomly, Havasupai people may somehow prefer mates whose MHC genotypes differ from their own (nonrandom mating).

22.4 Extensions of the Hardy-Weinberg Law

Some genes have more than two alleles, and in such cases we can find several alleles of a single locus in a population. The ABO blood group in humans is one such example. The locus I (isoagglutinin) has three alleles I^A, I^B, and I^O, yielding six possible genotypic combinations (I^AI^A, I^BI^B, I^OI^O, I^AI^B, I^AI^O, I^BI^O). In this case, I^A and I^B are codominant alleles, and both of them are dominant to I^O. The result is that homozygous I^AI^A and heterozygous I^AI^O individuals are phenotypically identical, as are I^BI^B and I^BI^O individuals, so we can distinguish only four phenotypic combinations (blood types A, B, AB, and O).

We can calculate both the genotype and allele frequencies for a situation involving three alleles by adding another variable to the Hardy-Weinberg equation. Let p, q, and r represent the frequencies of alleles A, B, and O, respectively. Note that because there are three alleles, our equation becomes

$$p + q + r = 1.0$$

Under the Hardy–Weinberg assumptions, the frequencies of the genotypes are given by

$$(p + q + r)^2 = p^2 + q^2 + r^2 + 2pq + 2pr + 2qr = 1.0$$

If we know the frequencies of blood types for a population, we can then estimate frequencies for the three alleles of the ABO system. For example, in one population that was sampled, the following blood type frequencies were observed: $A = 0.53$, $B = 0.13$, $O = 0.26$. Because the I^O allele is recessive, the population's frequency of type O blood equals the proportion of the recessive genotype r^2. Thus,

$$r^2 = 0.26$$
$$r = \sqrt{0.26}$$
$$r = 0.51$$

Using r, we can then estimate the allele frequencies for the I^A and I^B alleles. The I^A allele is present in two genotypes, I^AI^A and I^AI^O alleles. The frequency of the I^AI^A genotype is represented by p^2, and the I^AI^O genotype by $2pr$. Therefore, the combined frequency of type A blood and type O blood is given by

$$p^2 + 2pr + r^2 = 0.53 + 0.26$$

If we factor the left side of the equation and take the sum of the terms on the right, we get

$$(p + r)^2 = 0.79$$
$$p + r = \sqrt{0.79}$$
$$p = 0.89 - r$$
$$p = 0.89 - 0.51 = 0.38$$

Having calculated p and r (the frequencies of allele I^A and allele I^O), we can now estimate the frequency for the I^B allele:

$$p + q + r = 1.0$$
$$q = 1.0 - p - r$$

TABLE 22.4 Calculating Genotype Frequencies for Multiple Alleles (Allele I^A = 0.38, Allele I^B = 0.11, and Allele I^O = 0.51)

Genotype	Genotype Frequency	Phenotype	Phenotype Frequency
I^AI^A	$p^2 = (0.38)^2 = 0.14$	A	0.53
I^AI^O	$2pr = 2(0.38)(0.51) = 0.39$		
I^BI^B	$q^2 = (0.11)^2 = 0.01$	B	0.12
I^BI^O	$2qr = 2(0.11)(0.51) = 0.11$		
I^AI^B	$2pq = 2(0.38)(0.11) = 0.084$	AB	0.08
I^OI^O	$r^2 = (0.51)^2 = 0.26$	O	0.26

$$= 1.0 - 0.38 - 0.51$$
$$= 0.11$$

The phenotypic and genotypic frequencies for this population are summarized in Table 22.4.

22.5 Using the Hardy-Weinberg Law: Calculating Heterozygote Frequency

In one important application, the Hardy-Weinberg law enables us to estimate the frequency of heterozygotes in a population. The phenotype of a recessive trait is usually distinctive, and the frequency of the homozygous recessive genotype can be determined by counting such individuals in a sample of the population. With this information and the Hardy-Weinberg law, we can then calculate the allele and genotype frequencies.

Cystic fibrosis, an autosomal recessive trait, has an incidence of about $1/2500 = 0.0004$ in people of northern European ancestry. Individuals with cystic fibrosis are easily distinguished from the population at large by symptoms such as salty sweat, excess amounts of thick mucus in the lungs, and susceptibility to bacterial infections. Because this is a recessive trait, individuals with cystic fibrosis must be homozygous. The frequency of recessive homozygotes in a population is represented by q^2, provided that mating has been random in the previous generation. The frequency of the recessive allele, q, in the population therefore can be calculated as

$$q = \sqrt{q^2} = \sqrt{0.0004} = 0.02$$

Since $p + q = 1.0$, then the frequency of the dominant allele, p, is

$$p = 1.0 - q = 1.0 - 0.02 = 0.98$$

In the Hardy-Weinberg equation, the frequency of heterozygotes is $2pq$ and

$$2pq = 2(0.98)(0.02)$$
$$= 0.04(4\%, \text{ or } 1 \text{ in } 25)$$

Thus, heterozygotes for cystic fibrosis are rather common in the population (4%), even though the incidence of homozygous recessives is only 1/2500, or 0.04%.

In general, the frequencies of all three genotypes can be estimated once the frequency of either allele is known and Hardy-Weinberg assumptions are invoked. The relationship between genotype and allele frequency is shown in Figure 22–8. It is important to note from this graph that heterozygotes increase rapidly in a population as the values of p and q move away from 0 or 1.0. This observation confirms our conclusion that when a recessive trait such as cystic fibrosis is rare, the majority of those carrying the allele are heterozygotes. Notice also that in populations in which the frequencies of p and q are between 0.33 and 0.67, heterozygotes occur at higher frequency than either homozygote.

22.6 Natural Selection Alters Allele Frequencies

We have stated that the Hardy-Weinberg law establishes an ideal population that allows us to estimate allele and genotype frequencies at a given locus in populations in which the assumptions of equal reproductive success, absence of mutation, lack of migration, large population size, and random mating hold. Obviously, it is difficult to find natural populations in which all of these assumptions hold for all loci. In nature, populations are dynamic, and changes in size and gene pool are common. The Hardy-Weinberg law allows us to identify and investigate

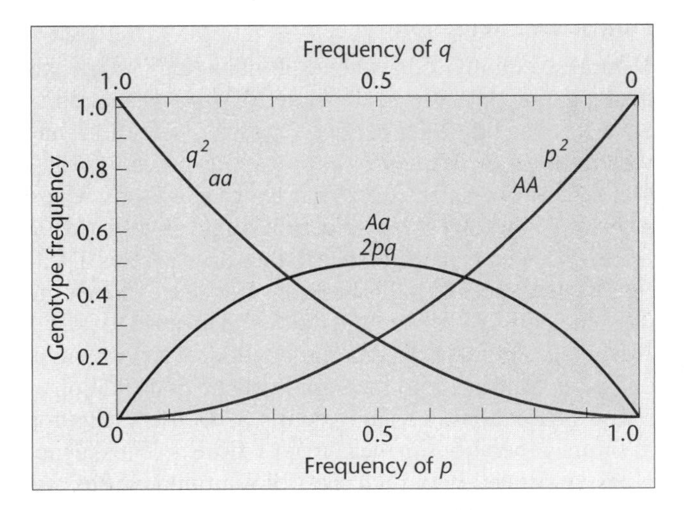

FIGURE 22–8 The relationship between genotype and allele frequencies derived from the Hardy-Weinberg equation.

populations that vary from equilibrium. In the following sections, we will discuss factors that prevent populations from reaching Hardy-Weinberg equilibrium or that drive populations toward a different equilibrium and also the relative contribution of these factors to evolutionary change.

Natural Selection

The first assumption of the Hardy-Weinberg law is that individuals of all genotypes have equal rates of survival and equal reproductive success. If this assumption does not hold, allele frequencies may change from one generation to the next. To see why, let's imagine a population of 100 individuals in which the frequency of allele *A* is 0.5 and that of allele *a* is 0.5. Assuming the previous generation mated randomly, we find that the genotype frequencies in the present generation are $(0.5)^2 = 0.25$ for *AA*, $2(0.5)(0.5) = 0.5$ for *Aa*, and $(0.5)^2 = 0.25$ for *aa*. Our population contains 100 individuals, so we have 25 *AA* individuals, 50 *Aa* individuals, and 25 *aa* individuals. Now suppose that individuals with different genotypes have different rates of survival: All 25 *AA* individuals survive to reproduce, 90% (45) of the *Aa* individuals survive to reproduce, and 80% (20) of the *aa* individuals survive to reproduce. When the survivors reproduce, each contributes two gametes to the new gene pool, giving us $2(25) + 2(45) + 2(20) = 180$ gametes. What are the frequencies of the two alleles in the surviving population? We have 50 *A* gametes from *AA* individuals plus 45 *A* gametes from *Aa* individuals, so the frequency of allele *A* is $(50 + 45)/180 = 0.53$. We have 45 *a* gametes from *Aa* individuals plus 40 *a* gametes from *aa* individuals, so the frequency of allele *a* is $(45 + 40)/180 = 0.47$. You can see that allele *A* has increased, while allele *a* has declined.

A difference among individuals in rate of survival and reproduction is called **natural selection**. Natural selection is an important force shifting allele frequencies within large populations and, along with genetic drift, is one of the most important factors in evolutionary change.

Fitness and Selection

Natural selection occurs whenever individuals with a particular genotype enjoy an advantage in survival and reproduction over other genotypes. However, selection may vary from very weak to very strong. In the previous example, selection was strong. What about situations where selection is weak? Such selection might involve just a fraction of a percent difference in the survival rates of different genotypes. Will this be enough to cause changes in the gene pools of future generations? The answer is yes: Advantages in survival and reproduction, even very small ones, can translate into increased genetic contribution to future generations. An individual's genetic contribution to future generations is measured by **fitness**. Thus, genotypes associated with high rates of reproductive success are said to have high fitness, whereas genotypes associated with low reproductive success are said to have low fitness.

Hardy-Weinberg analysis allows us to examine fitness. By convention, population geneticists use the letter w to represent fitness. Thus, w_{AA} represents the relative fitness of genotype *AA*, w_{Aa} the relative fitness of genotype *Aa*, and w_{aa} the relative fitness of genotype *aa*. For example, assigning the values $w_{AA} = 1.0$, $w_{Aa} = 0.9$, and $w_{aa} = 0.8$ could mean that all *AA* individuals survive, 90% of the *Aa* individuals survive, and 80% of the *aa* individuals survive, as in the previous example.

Let's consider selection against deleterious alleles. Fitness values $w_{AA} = 1.0$, $w_{Aa} = 1.0$, and $w_{aa} = 0$ describe a situation in which allele *a* is a lethal recessive. Homozygous recessive individuals die without leaving offspring, so the frequency of allele *a* will decline. The decline in frequency of allele *a* is described by the equation

$$q_g = \frac{q_0}{1 + gq_0}$$

where q_g is the frequency of allele *a* in generation g, q_0 is the starting frequency of allele *a* (i.e., the frequency of *a* in generation zero), and g is the number of generations that have passed.

Figure 22–9 shows what happens to a lethal recessive allele with an initial frequency of 0.5. At first, because of the high percentage of *aa* genotypes, the frequency of allele *a* declines

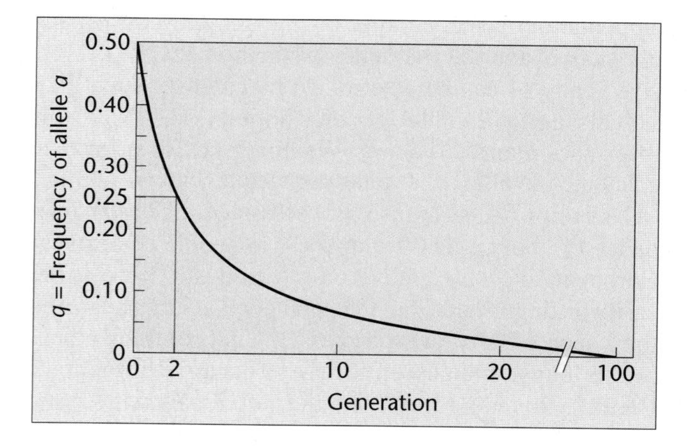

Generation	p	q	p^2	$2pq$	q^2
0	0.50	0.50	0.25	0.50	0.25
1	0.67	0.33	0.44	0.44	0.12
2	0.75	0.25	0.56	0.38	0.06
3	0.80	0.20	0.64	0.32	0.04
4	0.83	0.17	0.69	0.28	0.03
5	0.86	0.14	0.73	0.25	0.02
6	0.88	0.12	0.77	0.21	0.01
10	0.91	0.09	0.84	0.15	0.01
20	0.95	0.05	0.91	0.09	< 0.01
40	0.98	0.02	0.95	0.05	< 0.01
70	0.99	0.01	0.98	0.02	< 0.01
100	0.99	0.01	0.98	0.02	< 0.01

FIGURE 22–9 The change in the frequency of a lethal recessive allele, *a*, is halved in two generations and halved again by the sixth generation. Subsequent reductions occur slowly because the majority of *a* alleles are carried by heterozygotes.

rapidly—the frequency of a (q) is halved in only two generations. By the sixth generation, the frequency is halved again. By now, however, the majority of a alleles are carried by heterozygotes. Because a is recessive, these heterozygotes are not selected against. This means that as time continues to pass, the frequency of allele a declines more slowly. As long as heterozygotes continue to mate, it is difficult for selection to completely eliminate a recessive allele from a population.

Of course, a deleterious allele need not be recessive; many other scenarios are possible. A deleterious allele may be codominant, so that heterozygotes have intermediate fitness; or an allele may be deleterious in the homozygous state but beneficial in the heterozygous state, like the allele for sickle-cell anemia. Examples in which a deleterious allele is codominant but the intensity of selection varies from strong to weak are shown in Figure 22–10. In each case, the frequency of the deleterious allele, a, starts at 0.99 and declines over time. However, the rate of decline depends heavily on the strength of selection. When only 90% of the heterozygotes and 80% of the aa homozygotes survive (red curve), the frequency of allele a drops from 0.99 to less than 0.01 in 85 generations. However, when 99.8% of the heterozygotes and 99.6% of the aa homozygotes survive (blue curve), it takes 1000 generations for the frequency of allele a to drop from 0.99 to 0.93. Two important conclusions can be drawn from this graph. First, given thousands of generations, even weak selection can cause substantial changes in allele frequencies; because evolution generally occurs over a large number of generations, selection is a powerful force of evolution. Second, for selection to produce rapid changes in allele frequencies—changes measurable within a human life span—the differences in fitness among genotypes must be large or the generation time must be short.

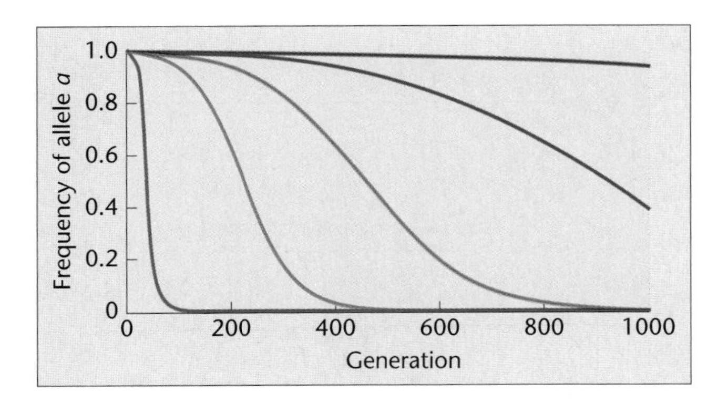

Selection Against Allele a					
	Strong ←				→ Weak
	—	—	—	—	—
w_{AA}	1.0	1.0	1.0	1.0	1.0
w_{Aa}	0.90	0.98	0.99	0.995	0.998
w_{aa}	0.80	0.96	0.98	0.99	0.996

FIGURE 22–10 The rate at which a deleterious allele is removed from a population depends heavily on the strength of selection, as shown in this graph and table.

The manner in which selection affects allele frequencies allows us to make some inferences about the *CCR5-Δ32* allele that we discussed earlier. Because individuals with genotype *Δ32/Δ32* are resistant to most sexually transmitted strains of HIV-1 while individuals with genotypes *1/1* and *1/Δ32* are susceptible, we might expect AIDS to act as a selective force causing the frequency of the *Δ32* allele to increase over time. Indeed it probably will, but the increase in frequency is likely to be slow in human terms.

Let's do a rough calculation in which we assume that we have a population where the current frequency of the *Δ32* allele is 0.10. Under Hardy-Weinberg assumptions, the genotype frequencies in this population are 0.81 for *1/1*, 0.18 for *1/Δ32*, and 0.01 for *Δ32/Δ32*. Let's also assume that 1% of the *1/1* and *1/Δ32* individuals in this population will contract HIV and die of AIDS before reproducing. Based on our assumptions, we can assign fitness levels to the genotypes as follows: $w_{1/1} = 0.99$, $w_{1/\Delta 32} = 0.99$; $w_{\Delta 32/\Delta 32} = 1.0$. Given the assigned fitness, we can predict that the frequency of the *CCR5-Δ32* allele in the next generation will be 0.100091. In fact, it will take about 100 generations (about 2000 years) for the frequency of the *Δ32* allele to reach just 0.11 (Figure 22–11). In other words, the frequency of the *Δ32* allele will probably not change much over the next few generations in most of the populations that currently harbor it. A population genetics perspective sheds light on the *CCR5-Δ32* story in other ways as well. Two research groups have analyzed genetic variation at marker loci closely linked to the *CCR5* gene. Both groups concluded that most, if not all, present-day copies of the *Δ32* allele are descended from a single ancestral copy that appeared in northeastern Europe, a few thousand years ago at most. In fact, one group estimates that the common ancestor of all *Δ32* alleles existed just 700 years ago. How could a new allele rise from a frequency of virtually zero to as high as 20% in roughly 30 generations?

There must have been strong selection in favor of the *Δ32* allele, most likely in the form of an infectious disease. The agent of selection could not have been HIV-1, because

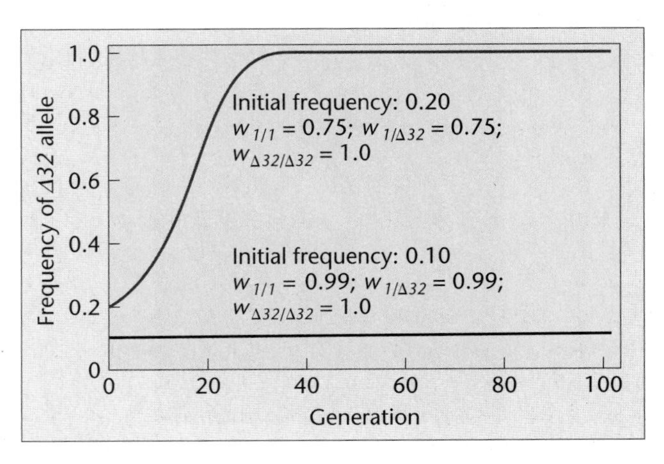

FIGURE 22–11 This graph reveals the rate at which the frequency of the *CCR5-Δ32* allele changes in hypothetical populations with different initial frequencies and different fitnesses.

HIV-1 moved from chimpanzees to humans too recently. Because selection occurred about 700 years ago, J. C. Stephens suggests that the agent of selection was bubonic plague. During the Black Death of 1346–1352, between a quarter and a third of all Europeans died from bubonic plague, which is caused by the bacterium *Yersinia pestis*. This bacterium manufactures a protein that kills certain types of white blood cells, and Stephens hypothesizes that the process through which the bacterial protein kills white cells involves the *CCR5* gene product. If this hypothesis is correct, some mechanism made individuals homozygous for the $\Delta 32$ allele more likely to survive plague epidemics.

In another study, William Klitz and his colleagues concluded that the $\Delta 32$ allele may have arisen as early as the eighth century. They studied the *CCR5* locus in Jewish and European populations and suggested that smallpox may have been the selective agent responsible for the rapid increase in the frequency of this allele. Both HIV and variola (the virus responsible for smallpox) use the CCR5 receptor for infecting cells, and with a fatality rate of 25%, waves of smallpox epidemics would have been a powerful selective agent.

Selection in Natural Populations

Geneticists have done a great deal of research on the effect of natural selection on allele frequencies in both laboratory and natural populations. Among the most detailed studies of natural populations are those that involve insects exposed to pesticides. Christine Chevillon and her colleagues, for example, studied the effect of the insecticide chlorpyrifos on allele frequencies in populations of the house mosquito, *Culex pipiens* (Figure 22–12). Chlorpyrifos kills mosquitoes by interfering with the function of the enzyme acetylcholinesterase (*ACE*), which under normal circumstances breaks down the neurotransmitter acetylcholine. An allele of *ACE* called Ace^R encodes a slightly altered version of *ACE* that is immune to interference by chlorpyrifos. Chevillon measured the frequency of the Ace^R allele in nine populations. In 4 locations chlorpyrifos had been used to control mosquitoes for 22 years; in the other 5 locations, chlorpyrifos had never been used. Chevillon's group predicted that the frequency of Ace^R would be higher in the exposed populations. The researchers also predicted that the allele frequencies for enzymes unrelated to the physiological effects of chlorpyrifos would show no such pattern. Among the control enzymes they studied was aspartate amino transferase 1. The results appear in Figure 22–13. As the researchers predicted, the frequency of the Ace^R allele was significantly higher in the exposed populations. Also as predicted, the frequencies of the most common alleles of the control enzyme gene showed no such trends. The expla-

FIGURE 22–12 Pesticides that are used to control insects act as selective agents, changing allele frequencies of resistance genes in the populations exposed to the pesticide. *(Larsh K. Bristol/Visuals Unlimited, Inc.)*

House mosquito, *Culex pipiens.*

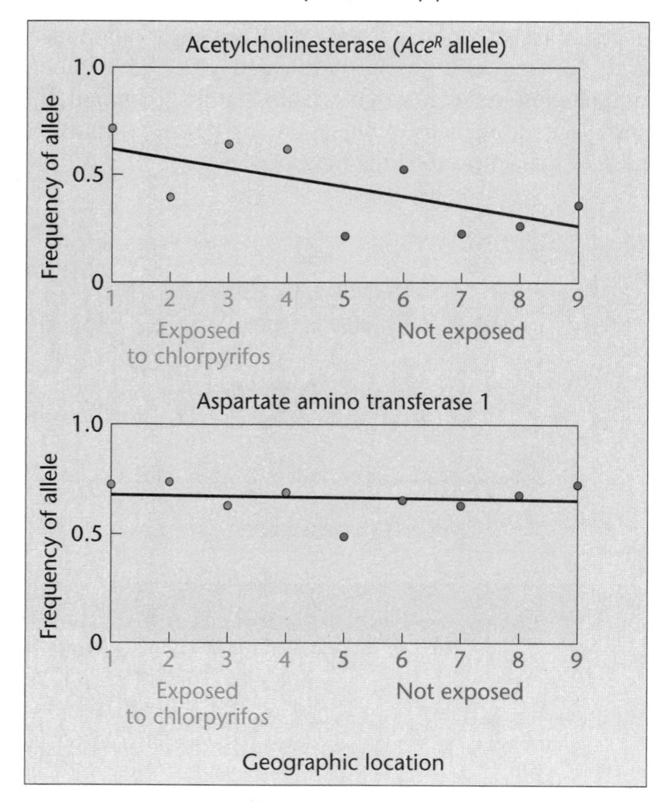

FIGURE 22–13 In natural populations, (top graph) the frequency of the Ace^R allele, which confers resistance to the insecticide chlorpyrifos, is higher in house mosquito populations exposed to chlorpyrifos. (bottom graph) The frequency of an allele for an enzyme unrelated to chlorpyrifos metabolism (aspartate amino transferase 1) shows no such pattern. *(Photo: Hans Pfletschinger/Peter Arnold, Inc.)*

nation is that during the 22 years of exposure to the pesticide, mosquitoes had higher rates of survival if they carried the Ace^R allele. In other words, the Ace^R allele had been favored by natural selection.

Natural Selection and Quantitative Traits

Most phenotypic traits are controlled not by alleles at a single locus but by the combined influence of the individual's genotype at many different loci and the environment. Because selection is a consequence of the organism's genotypic/phenotypic combination, polygenic or quantitative traits (those controlled by a number of genes that might be susceptible to environmental influences) also respond to selection. Such quantitative traits, including adult body height and weight in humans, often demonstrate a continuously varying distribution resembling a bell-shaped curve. Selection for such traits can be classified as (1) directional, (2) stabilizing, or (3) disruptive.

In **directional selection** (important to plant and animal breeders), desirable traits, often representing phenotypic extremes, are selected. If the trait is polygenic, the most extreme phenotypes that the genotype can express will appear in the population only after prolonged selection. An example of directional selection is the long-running experiment at the State Agricultural Laboratory in Illinois selecting for high and low oil content in corn kernels. At the start of the experiment in 1896, a population of 163 corn ears was surveyed. The 24 corn ears that were highest in oil content were used to produce the next generation. In each succeeding generation, the ears with the highest oil content were used to breed. In this line of upward directional selection, the oil content has been raised after 50 generations from about 4% to just over 16%, and there is no sign that a plateau has been reached.

A second population was founded using 12 ears with the lowest oil content, and downward directional selection from this founding population has lowered the oil content from 4% to less than 1% in the same 50-generation period. In this example, the high and low graphs have diverged to the point where the distribution of oil content in one line does not overlap with that of the other line (Figure 22–14).

In the upward directional selection population, the alleles for high oil content increase in frequency and replace the alleles for lower oil content. At some point, all individuals in the line will have the genotype for the highest oil content, and all of them will express the most extreme phenotype for high oil content. In the downward directional selection population, the opposite situation will occur, eventually producing a population with a genotype and phenotype for the lowest oil content. When the two populations fail to respond to further selection, no genetic variance for oil content will remain and each will have a homozygous genotype for oil content.

In nature, directional selection can occur when one of the phenotypic extremes becomes selected for or against, usually as a result of changes in the environment. A carefully documented example comes from research by Peter and Rosemary Grant and their colleagues, who used medium ground finches (*Geospiza fortis*) living on the island of Daphne Major in the Galapagos Islands. The beak size of these birds varied enormously. In 1976, for example, some birds in the population had beaks less than 7 mm deep, while others had beaks more than 12 mm deep. Beak size is heritable, which means that large-beaked parents tend to have large-beaked offspring and small-beaked parents tend to have small-beaked offspring. In 1977, a severe drought on Daphne Major killed about 80% of the finches. Big-beaked birds survived at higher rates than small-beaked birds because when food became scarce, the big-beaked birds were able to eat a greater variety of seeds. When the drought ended in 1978 and the survivors paired off and bred, the offspring inherited their parents' big beaks. Between 1976 and 1978, the beak depth of the average finch in the Daphne Major population increased by just over 0.5 mm, thus the average beak size was shifted toward one phenotypic extreme.

Stabilizing selection, in contrast, tends to favor intermediate types—both extreme phenotypes are selected against. One of the clearest demonstrations of stabilizing selection is provided by the data of Mary Karn and Sheldon Penrose on human birthweight and survival for 13,730 children born over an 11-year period. Figure 22–15 shows the distribution

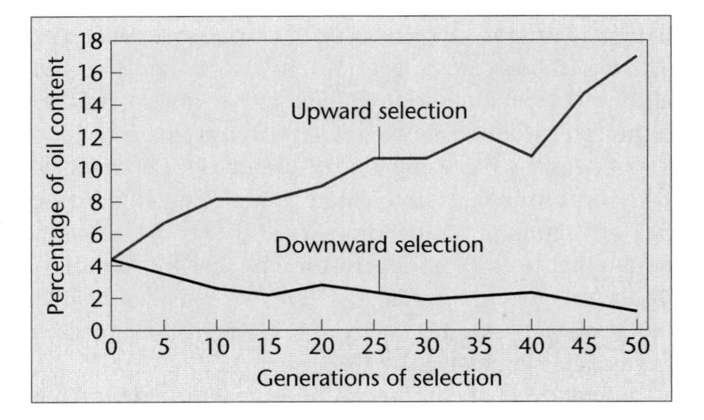

FIGURE 22–14 This graph, which encompasses selection experiments performed since 1896, illustrates directional selection to alter the oil content of corn ears.

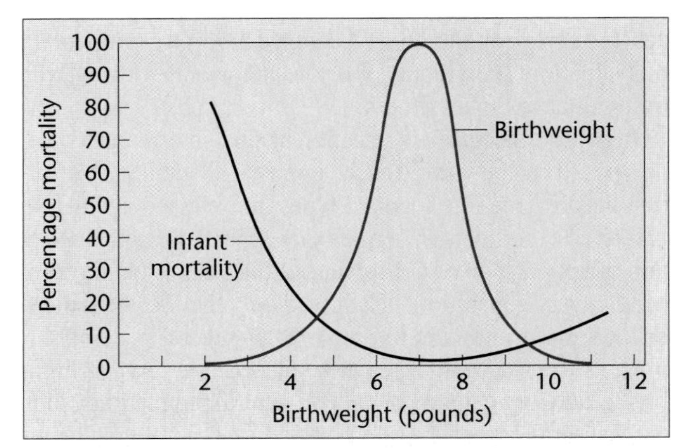

FIGURE 22–15 Stabilizing selection is illustrated by the relationship between birthweight and mortality in humans.

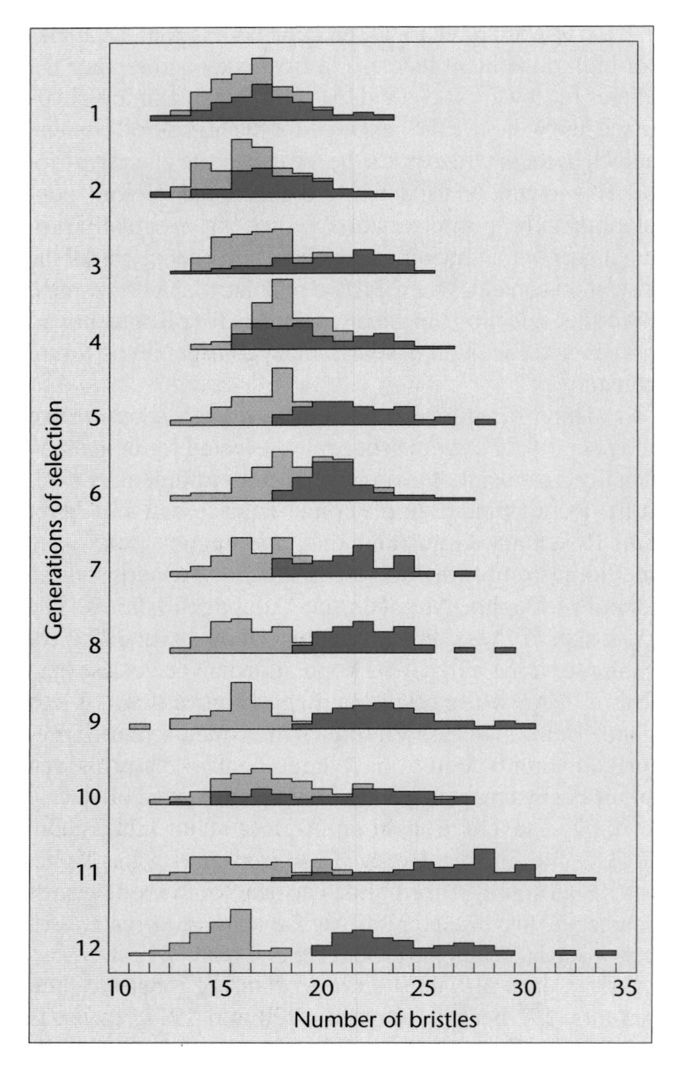

FIGURE 22–16 In an experiment to demonstrate the effect of disruptive selection on bristle numbers in *Drosophila*, individuals with the highest and lowest bristle numbers were selected; the population showed a nonoverlapping divergence within only 12 generations.

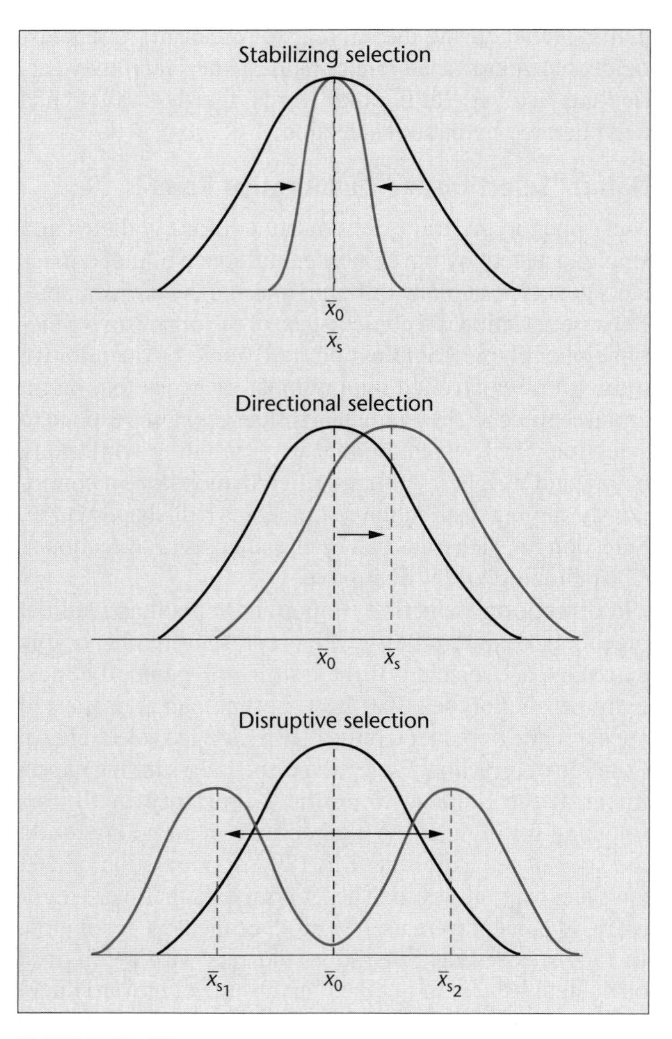

FIGURE 22–17 The impact of stabilizing, directional, and disruptive selection. In each case, the mean of an original population (black) and the mean of the population following selection (red) are shown.

of birthweight and the percentage of mortality at four weeks of age. Infant mortality increases on either side of the optimal birthweight of 7.5 pounds and quite dramatically so at low birthweights. At the genetic level, stabilizing selection acts to keep a population well adapted to its environment. In this situation, individuals closer to the average for a given trait will have higher fitness.

Disruptive selection is selection against intermediate phenotypes. It can be viewed as the opposite of stabilizing selection because the intermediate types are selected against. In one set of experiments, John Thoday applied disruptive selection to a population of *Drosophila* on the basis of bristle number. In every generation, he allowed only the flies with high- or low-bristle numbers to breed. After several generations, most of the flies could be easily placed in a low- or high-bristle category (Figure 22–16). In natural populations, such a situation might exist for a population in a heterogeneous environment. See Figure 22–17 for a graphical summary of the types and effects of selection.

22.7 Mutation

Within a population, the gene pool of each generation is reshuffled to produce new genotypes in the offspring. Because the number of possible genotypic combinations is so large, population members alive at any given point in time represent only a fraction of all possible genotypes. The enormous genetic reserve present in the gene pool allows Mendelian assortment and recombination to continuously produce new genotypic combinations. But assortment and recombination do not produce new alleles. Only **mutation** can act to create new alleles. It is important to keep in mind that mutational events occur at random—that is, without regard for any possible benefit or disadvantage to the organism. Here, we consider whether mutation is, by itself, a significant factor in causing allele frequencies to change.

To determine whether mutation is a significant force in changing allele frequencies, we must measure the rate at which mutations are produced. As most mutations are recessive, it is difficult to directly observe mutation rates in diploid organisms. Therefore, we employ indirect methods that use probability and statistics or large-scale screening programs. For certain dominant mutations, however, a direct method of measurement can be used. To ensure accuracy, three conditions must be met:

1. The allele must produce a distinctive phenotype that can be distinguished from similar phenotypes produced by recessive alleles.
2. The trait must be fully expressed or completely penetrant so that mutant individuals can be identified.
3. An identical phenotype must never be producible by nongenetic agents such as drugs or chemicals.

Mutation rates can be stated as the number of new mutant alleles per a given number of gametes. Suppose that for a gene that undergoes mutation to a dominant allele, 2 out of 100,000 births exhibit a mutant phenotype; the parents of these 2 offspring are phenotypically normal. Because the zygotes that produced these births each carried 2 copies of the gene, we have actually surveyed 200,000 copies of the gene (or 200,000 gametes). If we assume that the affected births are each heterozygous, we have uncovered 2 mutant alleles out of 200,000. Thus, the mutation rate is $2/200{,}000$ or $1/100{,}000$ (1×10^{-5}).

In humans, a dominant form of dwarfism known as **achondroplasia** fulfills the three requirements for measuring mutation rates. Individuals with this skeletal disorder have an enlarged skull and short arms and legs and can be diagnosed at birth by X-ray examination. In a survey of almost 250,000 births, the mutation rate (μ) for achondroplasia has been calculated as

$$\mu = 1.4 \times 10^{-5} \pm 0.5 \times 10^{-5}$$

Note that even in a survey of so many births, the estimate of the mutation rate has a large margin of error $(\pm 0.5 \times 10^{-5})$. For our purposes, we will assume that 1.4×10^{-5} is the rate of mutation for the achondroplasia gene. Knowing the rate of mutation, we can then estimate the extent to which mutation can cause allele frequencies to change from one generation to the next. We represent the normal allele as *d* and the allele for achondroplasia as *D*.

Imagine a population of 500,000 individuals in which everyone has genotype *dd*. The initial frequency of *d* is 1.0, and the initial frequency of *D* is 0. If each individual contributes 2 gametes to the gene pool, the gene pool will contain 1,000,000 gametes, all carrying allele *d*. While the gametes are in the gene pool, 1.4 of every 100,000 *d* alleles mutates into a *D* allele. The frequency of allele *d* is now $(1{,}000{,}000 - 14)/1{,}000{,}000 = 0.999986$, and the frequency of *D* is $14/1{,}000{,}000 = 0.000014$, or 1.4×10^{-5}. From these numbers, it will clearly be a long time before mutation can cause any appreciable change in the allele frequencies in this population.

More generally, if we have two alleles, *A* with frequency *p* and *a* with frequency *q*, and if *m* represents the rate of mutations converting *A* into *a*, then the frequencies of the alleles in the next generation are given by

$$p_{g+1} = p_g - \mu p_g \quad \text{and} \quad q_{g+1} = q_g + \mu p_g$$

where p_{g+1} and q_{g+1} represent the allele frequencies in the next generation and p_g and q_g represent the allele frequencies in the present generation.

Figure 22–18 shows the replacement rate (change over time) in allele *A* for a population in which the initial frequency of *A* is 1.0 and the rate of mutation (μ) converting *A* into *a* is 1.0×10^{-5}. At this mutation rate, it will take about 70,000 generations to reduce the frequency of *A* to 0.5. Even if the rate of mutation increases through exposure to higher levels of radioactivity or chemical mutagens, the impact of mutation on allele frequencies will be extremely weak. Mutation provides the raw material for evolution and is the ultimate source of genetic variability, but *by itself* mutation plays a relatively insignificant role in changing allele frequencies. Instead, the fate of alleles is more likely to be determined by natural selection (discussed previously) and genetic drift (discussed later).

An evolutionary perspective on mutation can lead to medical discoveries. The autosomal recessive disease cystic fibrosis, which we discussed previously in relation to calculating heterozygote frequencies, is caused by a loss-of-function mutation in the gene for a cell-surface protein called the cystic fibrosis transmembrane conductance regulator (CFTR). The frequency for the mutant alleles causing cystic fibrosis is about 2% in European populations. Until recently, most individuals with two mutant alleles died before reproducing, meaning that selection against homozygous recessive individuals was rather strong. This creates a puzzle: In the face of selection against them, what has maintained the mutant alleles at an overall frequency of 2%?

One hypothesis, the **mutation-selection balance hypothesis**, posits that mutation is constantly creating new alleles to replace the ones that are eliminated by selection. However, for this scenario to work, the rate of mutations creating new alleles would have to be high, on the order of 5×10^{-4}, to counteract the effect of selection. Many evolutionary geneticists prefer an alternative explanation, the **heterozygote superiority hypothesis**. According to this hypothesis,

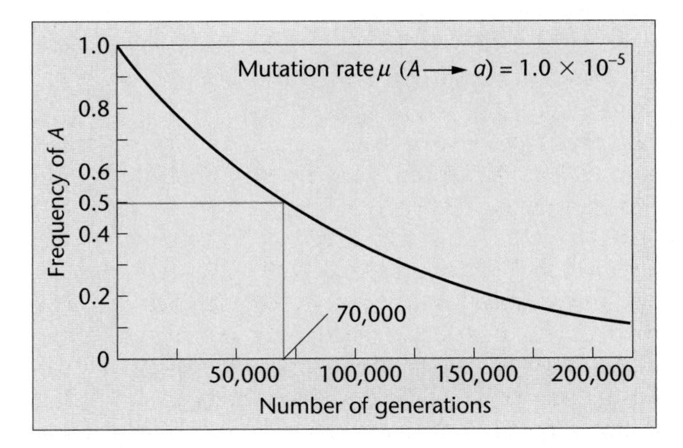

FIGURE 22–18 Replacement rate of an allele by mutation alone, assuming an average mutation rate of 1.0×10^{-5}.

selection against homozygous mutant individuals is counterbalanced by selection in favor of heterozygotes. The most popular agent of selection in favor of heterozygotes is resistance to an as-yet-unidentified disease.

Recent work suggests that cystic fibrosis heterozygotes may have enhanced resistance to typhoid fever. Typhoid fever is caused by the bacterium *Salmonella typhi*, which infiltrates cells of the intestinal lining. In laboratory studies, mouse intestinal cells that were heterozygous for *CFTR*-Δ508 (the analog of the most common cystic fibrosis mutation in humans) were infected by 86% fewer bacteria than did cells homozygous for the wild-type allele. Whether humans that are heterozygous for *CFTR*-Δ508 also enjoy resistance to typhoid fever remains to be established. If they do, cystic fibrosis will join sickle-cell anemia as an example of heterozygote superiority.

22.8 Migration

Most species are divided into populations that to some extent are separated geographically. Various evolutionary forces—including natural selection—can establish different allele frequencies in such populations. **Migration** occurs when individuals move between populations. Imagine a species in which a single locus has two alleles, *A* and *a*. There are two populations of this species, one on a mainland and one on an island. The frequency of *A* on the mainland is represented by p_m, and the frequency of *A* on the island is p_i. Under the influence of migration from the mainland to the island, the frequency of *A* in the next generation on the island (p_{i+1}) is given by

$$p_{i+1} = (1 - m)p_i + mp_m$$

where *m* represents migrants from the mainland to the island. Under these conditions, the frequency of *A* in the next generation on the island (p_{i+1}) will be affected by migration.

For example, assume that $p_i = 0.4$ and $p_m = 0.6$ and that 10% of the parents of the next generation are migrants from the mainland, so that $m = 0.1$. In the next generation, the frequency of allele *A* on the island will be

$$\begin{aligned} p_{i+1} &= [(1 - 0.1) \times 0.4] + (0.1 \times 0.6) \\ &= 0.36 + 0.06 \\ &= 0.42 \end{aligned}$$

Migration from the mainland has changed the frequency of *A* on the island from 0.40 to 0.42 in a single generation. If either *m* is large or p_m is very different from p_i, then a large change in the frequency of *A* can occur in a single generation. If migration is the *only* force acting to change the allele frequency on the island, then an equilibrium will be attained only when $p_i = p_m$.

These calculations reveal that the change in allele frequency attributable to migration is proportional to the differences in allele frequency between the donor and recipient populations and to the rate of migration. Since *m* can have a wide range of values, the effect of migration can substantially alter allele frequencies in populations, as shown for the *B* allele of the ABO blood group in Figure 22–19. Although migration can be difficult to quantify, it can often be estimated.

Migration can also be regarded as the flow of genes between populations that were once geographically isolated.

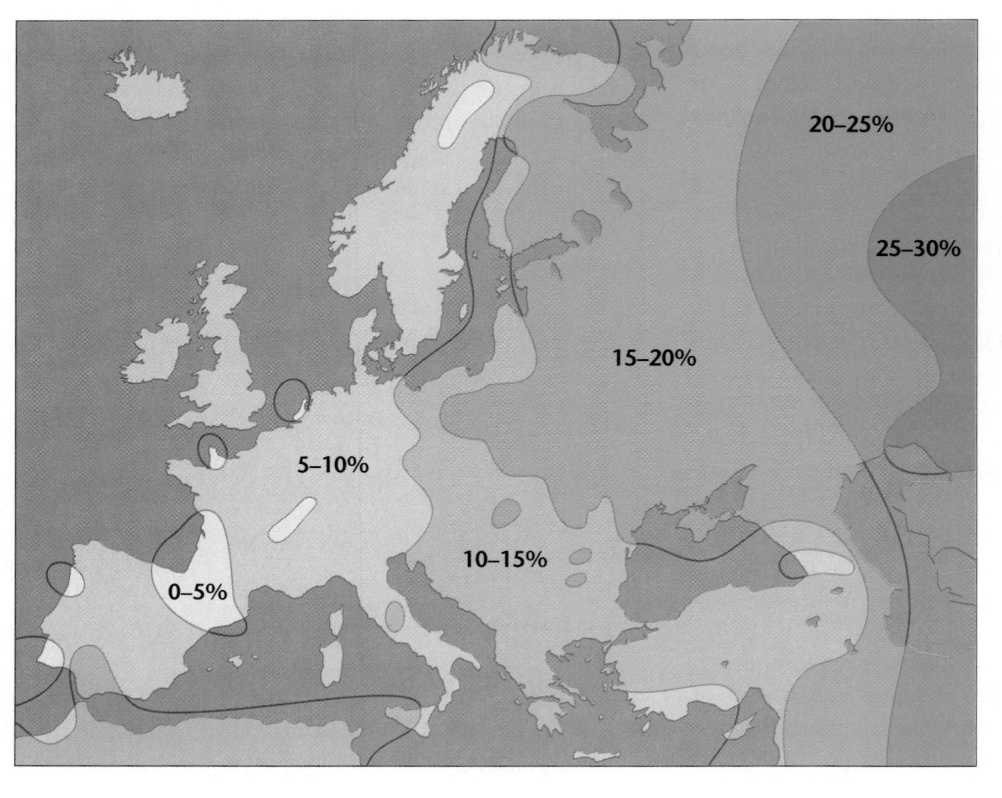

FIGURE 22–19 This map reveals how migration acts as a force in evolution. The *B* allele of the *ABO* locus is present in a gradient from east to west. This allele is at the highest frequency in central Asia and lowest in northeastern Spain. The gradient parallels the waves of Mongol migration into Europe after the fall of the Roman Empire and is a genetic relic of human history.

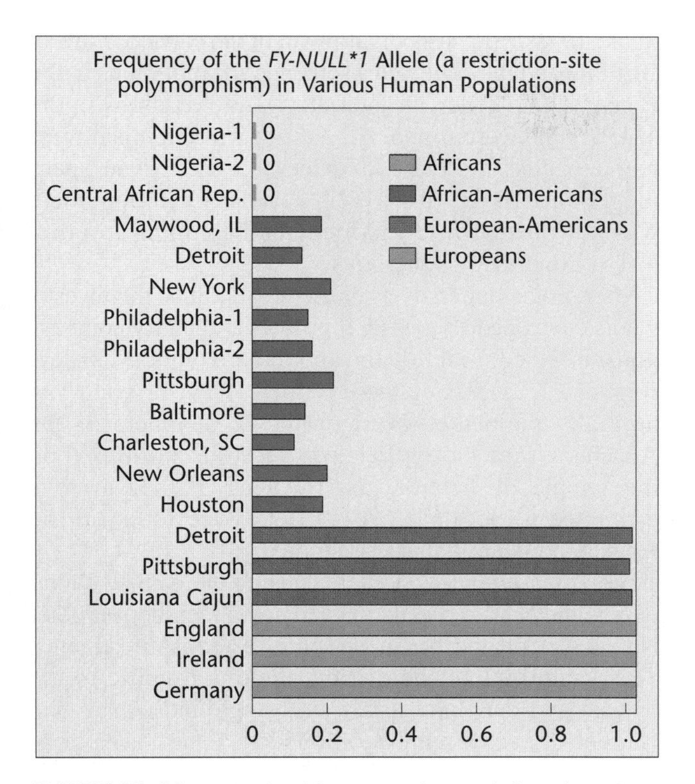

FIGURE 22–20 Frequency histogram of a restriction-site polymorphism allele in several populations. These and other data demonstrate that African-American and European-American populations share ancestry.

Esteban Parra and colleagues measured allele frequencies for several different DNA-sequence polymorphisms in African-American and European-American populations and in African and European populations representative of the ancestral populations from which the two American populations are descended. One locus they studied, a restriction-site polymorphism called *FY-NULL*, has two alleles, *FY-NULL**1* and *FY-NULL**2*. Figure 22–20 shows the frequency of *FY-NULL**1* in each population. The frequency of this allele is 0 in the 3 African populations and 1.0 in the 3 European populations, but it lies between these extremes in all African-American and European-American populations. The simplest explanation for these data is that genes have mixed between American populations with predominantly African ancestry and American populations with predominantly European ancestry. Based on *FY-NULL* and several other loci, researchers estimate that African-American populations derive 11.6–22.5% of their ancestry from Europeans and that European-American populations derive 0.5–1.2% of their ancestry from Africans.

22.9 Genetic Drift

In laboratory crosses, one condition essential to realizing theoretical genetic ratios (e.g., 3 : 1, 1 : 2 : 1, 9 : 3 : 3 : 1) is a fairly large sample size. A large sample size is also important to the study of population genetics as allele and genotype frequencies are examined or predicted. For example, if a population consists of 1000 randomly mating heterozygotes (*Aa*), the next generation will consist of approximately 25% *AA*, 50% *Aa*, and 25% *aa* genotypes. Provided that the initial and subsequent populations are large, only minor deviations from this mathematical ratio will occur and the frequencies of *A* and *a* will remain about equal (at 0.50), thus demonstrating a Hardy-Weinberg relationship.

However, if a population consists of only one set of heterozygous parents and they produce only two offspring, the allele frequency can change drastically. We can also predict allele frequencies and genotypes in the offspring in this case. Table 22.5 shows that in 10 of 16 times such a cross is made, the allele frequencies would be altered. In 2 of 16 crosses, either the *A* or *a* allele would be eliminated in a single generation. In 3 of 16 crosses, the heterozygous genotype would be eliminated in a single generation. This example illustrates two effects of *genetic drift*. The first effect is changes in allele frequency; in an extreme case, allele loss, but more generally, movement of allele frequency toward 1.0 or 0. The second effect is the decrease in heterozygosity; recall from Figure 22–8 that the frequency of heterozygotes decreases as allele frequencies get closer to 1.0 or 0.

Founding a population with only one set of heterozygous parents that produce only two offspring is an extreme example, but it illustrates the point that large interbreeding populations are essential to maintain Hardy-Weinberg equilibrium. In small populations, significant random fluctuations in allele frequencies are possible by chance deviation. The degree of fluctuation increases as the population size decreases, a situation known as **genetic drift**. In addition to small population size, drift can arise through the **founder effect**, which occurs when a population originates from a small number of individuals. Although the population may later increase to a large size, the genes carried by all members are derived from those of the founders (assuming no mutation or migration). Drift can also arise via **genetic bottleneck**. Bottlenecks develop when a large population undergoes a drastic but temporary reduction in numbers. Even though the population recovers, its genetic diversity has been greatly reduced.

TABLE 22.5 All Possible Pairs of Offspring Produced by Two Heterozygous Parents (*Aa* × *Aa*)

Possible Genotype of Two Offspring	*Probability*	*Allele Frequency* *A*	*a*
AA and *AA*	(1/4)(1/4) = 1/16	1.00	0.00
aa and *aa*	(1/4)(1/4) = 1/16	0.00	0.00
AA and *aa*	2(1/4)(1/4) = 2/16	0.50	0.50
Aa and *Aa*	2(2/4)(2/4) = 4/16	0.50	0.50
Aa and *Aa*	2(1/4)(2/4) = 4/16	0.75	0.25
aa and *Aa*	2(1/4)(2/4) = 4/16	0.25	0.75

To study genetic drift in laboratory populations of *Drosophila melanogaster*, Warwick Kerr and Sewall Wright set up over 100 populations, with 4 males and 4 females as the parents for each population. Within each population, the frequency of the sex-linked bristle mutant *forked* (f) and its wild-type allele (f^+) was 0.5. In each generation, four males and four females were chosen at random to produce the next generation. After 16 generations, the complete loss of 1 allele and fixation of the other had occurred in 70 populations—29 in which only the *forked* allele was present and 41 in which the wild-type allele had become fixed. The remaining populations were still segregating the two alleles or had become extinct. If fixation had occurred randomly, then an equal number of populations should have become fixed for each allele. In fact, the experimental results do not differ statistically from the expected ratio of 35 : 35 (70 populations in which fixation of one allele or the other occurred) demonstrating that alleles can spread through a population and other alleles can be eliminated by chance alone.

Founder Effects in Human Populations

Allele frequencies in certain human populations demonstrate the role of genetic drift as an evolutionary force in natural populations. Native Americans living in the southwestern United States have a high frequency of oculocutaneous albinism (OCA). In the Navajo, who live primarily in northeast Arizona, albinism occurs with a frequency of 1 in 1500–2000, compared with whites (1 in 36,000) and African-Americans (1 in 10,000). There are four different forms of OCA (OCA1–4), all with varying degrees of melanin deficiency in the skin, eyes, and hair. The OCA phenotype in the Navajo overlaps those of OCA2 and OCA4. OCA2 is caused by mutations in the *P* gene, which encodes a plasma membrane protein; OCA4 is caused by mutations in a gene called *MAPT*. To investigate the genetic basis of albinism in the Navajo, Murray Brilliant and his colleagues screened for mutations in the *P* and *MAPT* genes. In their study, no mutations in the *MAPT* gene were found. All Navajo with albinism were homozygous for a 122.5 kb deletion in the *P* gene, spanning exons 10–20 (Figure 22–21). This deletion allele was not present in 34 individuals belonging to other Native American populations.

After molecular analysis of the breakpoints, the investigators developed a set of PCR primers to identify homozygous affected individuals and heterozygous carriers (Figure 22–22). Using these primers, they surveyed 134 normally pigmented Navajo and 42 members of the Apache, a tribe closely related to the Navajo. Based on this sample, the heterozygote frequency in the Navajo is estimated to be 4.5%. No carriers were found in the Apache population that was studied.

The 122.5-kb deletion allele causing OCA2 was found only in the Navajo population and not in members of other Native American tribes in the southwestern United States. This suggests that the mutant allele is specific to the Navajo and may have arisen in a single individual who was one of a small number of founders of the Navajo population. Such founder mutations originate on a single chromosome and have a characteristic set of closely linked flanking markers. Over time, mutation and recombination modify and/or replace this set of markers. By knowing the rate of recombination and the rate of mutation in this region of the genome, the origin of the founder mutation can be dated. Using these parameters and studying flanking alleles in deletion and nondeletion chromosomes, the age of the mutation was estimated to be between 400 and 11,000 years. To narrow this range, Brilliant and his colleagues relied on tribal history. Navajo oral tradition indicates that the Navajo and Apache became separate populations between 600 and 1000 years ago. Because the deletion is not found in the Apaches, it probably arose

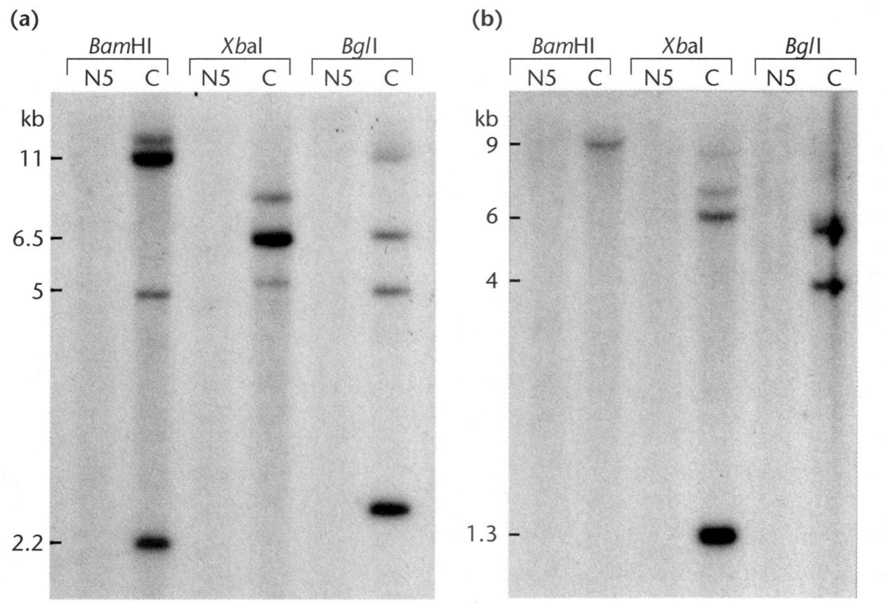

FIGURE 22–21 Genomic DNA digests from a Navajo affected with albinism (N5) and a normally pigmented individual (C). (a) Hybridization with a probe covering exons 11–15 of the *P* gene; there are no hybridizing fragments detected in N5. (b) Hybridization with a probe covering exons 15–20 of the *P* gene; there are no hybridizing fragments detected in N5. This confirms the presence of a deletion in affected individuals. *(Courtesy of Murray Brilliant 2003. A 122.5 kilobase deletion of the P gene underlies the high prevalence of oculocutaneous albinism Type 2 in the Navajo population. From:* American Journal Human Genetics *72: 62–72. fig. 1, parts b and c, p. 65. Published by University of Chicago Press.)*

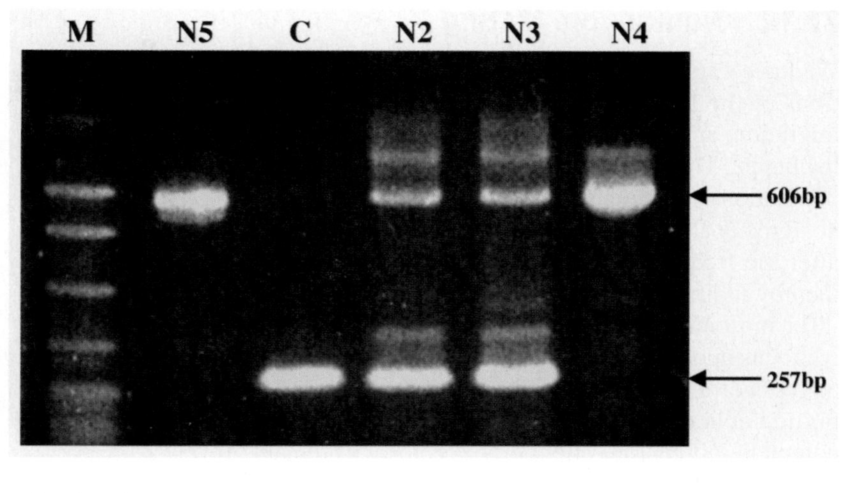

FIGURE 22–22 PCR screens of Navajo affected with albinism (N4 and N5) and the parents of N4 (N2 and N3). Affected individuals (N4, and N5), heterozygous carriers (N2 and N3), and a homozygous normal individual (C) each give a distinctive band pattern, allowing detection of heterozygous carriers in the population. Molecular size markers (M) are in the first lane. *(Courtesy of Murray Brilliant 2003. A 122.5 kilobase deletion of the P gene underlies the high prevalence of oculocutaneous albinism Type 2 in the Navajo population. From:* American Journal Human Genetics *72: 62–72. fig. 3, p. 67. Published by University of Chicago Press.)*

in the Navajo population after the tribes split. On this basis, the deletion is estimated to be 400–1000 years old and probably arose as a founder mutation.

Allele Loss During a Bottleneck

Although many populations have undergone well-documented bottlenecks with presumptive loss of genetic diversity, the amount of genetic variability present in the pre-bottleneck population—and therefore the amount of genetic variation lost in the bottleneck—cannot be measured. In Illinois, the greater prairie chicken population was estimated in the millions in the 1860s. As a result of human activity, the population declined to 25,000 birds in 1933 and to fewer than 50 in 1993. Fitness, as measured by egg hatching, also declined over that time period, from a 93% hatch rate in 1935 to an estimated 56% in 1990. To measure allele variation in pre- and post-bottleneck populations of the prairie chicken, Juan Bouzat and his colleagues used PCR to amplify alleles at six microsatellite loci. Samples were obtained from pre-bottleneck museum specimens and from present-day Illinois populations. In addition, allele frequencies were measured in prairie chicken populations from Kansas, Minnesota, and Nebraska, which have no known history of bottlenecks. A PCR analysis is shown in Figure 22–23, and the results of the allele survey are shown in Table 22.6. For each locus, the alleles in the Illinois population are a subset of those in other populations and in the pre-bottleneck Illinois population. Because all alleles in the present-day Illinois population are shared with the other populations, they were probably present in the Illinois population before the bottleneck. Six alleles at four loci found in museum specimens are missing from the Illinois population, but most of them are common in the other populations. The lost alleles represent a significant loss of genetic diversity caused by the bottleneck. In addition to allele loss, the Illinois population has significantly lower levels of heterozygosity. The outcomes are consequences of genetic drift, in this case caused by a population bottleneck. The small population size during the bottleneck and the resulting inbreeding is probably responsible for the decline in fitness currently observed in the Illinois population.

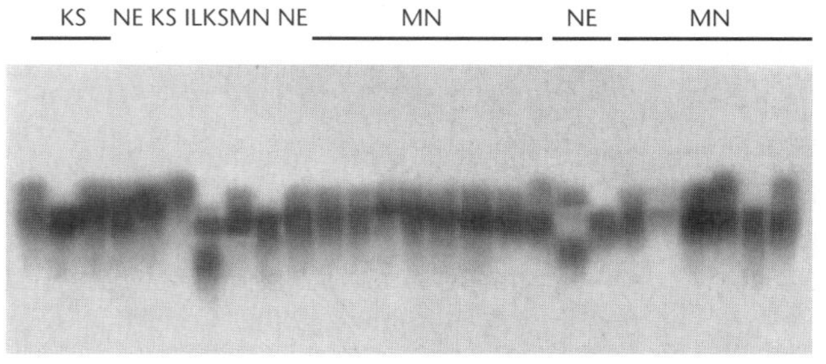

FIGURE 22–23 Alleles of the microsatellite locus ADL42 detected in DNA samples of prairie chickens from Kansas (KS), Nebraska (NE), Illinois (IL), and Minnesota (MN). *(From Bouzat, Juan L. et al. 1998. Genetic evaluation of a demographic bottleneck in the greater prairie chicken.* Conserv. Biol. *12: 836–43. Figure 2a, p. 839.)*

TABLE 22.6 Microsatellite Alleles in Prairie Chicken Populations

	Microsatellite Loci					
Population	*ADL42*	*ADL23*	*ADL44*	*ADL146*	*ADL162*	*ADL230*
Illinois	ABC	ABCD	A D F H	ABC	B E	C EFGHI
Kansas	ABCD	BCDEF	A CDEFGH	ABCDE	ABCDE	ABCDEFGHI
Minnesota	ABCD	ABCD	ABCDEFGH	ABC E	BCDE	BCDEFGHI
Nebraska	ABCD	ABCDE	ABCDEFGH	BC EF	BCDE	BCDEFGHIJK
Museum specimens	AB	BCD**E**	A D F	ABC **E**	B**CD**E*F*G	**B**C**D**EFG I *L*

Note: Bold letters indicate alleles found in the museum specimens which have been lost as a result of the prairie chickens' demographic contraction. Italic letters indicate alleles unique to the Illinois population prior to its demographic contraction.

22.10 Nonrandom Mating

We have explored how violations of first four assumptions of the Hardy-Weinberg law (selection, mutation, migration, and genetic drift) can cause allele frequencies to change. The fifth assumption is that the members of a population will mate randomly. Nonrandom mating does not directly alter the frequencies of alleles. It can, however, alter the frequencies of genotypes in a population and thereby indirectly affect the course of evolution.

The most important form of nonrandom mating, the form we focus upon here, is **inbreeding**—mating between relatives. For a given allele, inbreeding increases the proportion of homozygotes in the population; over time with complete inbreeding, only homozygotes will remain. To demonstrate this concept, let's consider the most extreme form of inbreeding, **self-fertilization**.

Inbreeding

Figure 22–24 shows the results of four generations of self-fertilization, starting with a single individual heterozygous for one pair of alleles. By the fourth generation, only about 6% of the individuals are still heterozygous and 94% of the population is homozygous. Note, however, that alleles A and a remain at 50%.

P₁: Self-fertilization *Aa*

	AA	*Aa*	*aa*
F_1	0.250	0.500	0.250
F_2	0.375	0.250	0.375
F_3	0.437	0.125	0.437
F_4	0.468	0.063	0.468
F_n	$\frac{[1-\frac{1}{2^n}]}{2}$	$\frac{1}{2^n}$	$\frac{[1-\frac{1}{2^n}]}{2}$

FIGURE 22–24 Reduction in heterozygote frequency brought about by self-fertilization. After n generations, the frequencies of the genotypes can be calculated according to the formulas in the bottom row.

In humans, inbreeding (called **consanguineous mating**) is related to population size, mobility, and social customs governing marriages among relatives. To describe the amount of inbreeding in a population, Sewall Wright devised the coefficient of inbreeding. Expressed as F, the **coefficient of inbreeding** is defined as the probability that the two alleles of a given gene in an individual are identical *because they are descended from the same single copy of the allele in an ancestor*. If $F = 1$, all individuals are homozygous and both alleles in every individual are derived from the same ancestral copy. If $F = 0$, no individual has two alleles derived from a common ancestral copy.

Figure 22–25 is the pedigree of a first-cousin marriage. The fourth-generation female (pink) is the daughter of first cousins (purple). If her great-grandmother (green) was a carrier of a recessive lethal allele a, what is the probability that the fourth-generation female will inherit two copies of her great-grandmother's lethal allele? For this to happen, (1) the great-grandmother had to pass a copy of the allele to her son, (2) her son had to pass it to his daughter, and (3) his daughter had to pass it to her daughter (the pink female). In addition, (4) the great-grandmother had to pass a copy of the allele to her daughter, (5) her daughter had to pass it to her son, and (6) her son had to pass it to his daughter. Each of the six necessary events has an individual probability of 1/2, and they *all* have to happen, so the overall probability that the pink female will inherit two copies of her great-grandmother's lethal allele is $(1/2)^6 = 1/64$. This takes us most of the way toward calculating F for a child of a first-cousin marriage. The final step is to note that the fourth-generation female could also inherit two copies of any of the other three alleles that are present in her great-grandparents. Because any of four possibilities would give the pink female two alleles identical by descent from an ancestral copy, $F = 4 \times (1/64) = 1/16$.

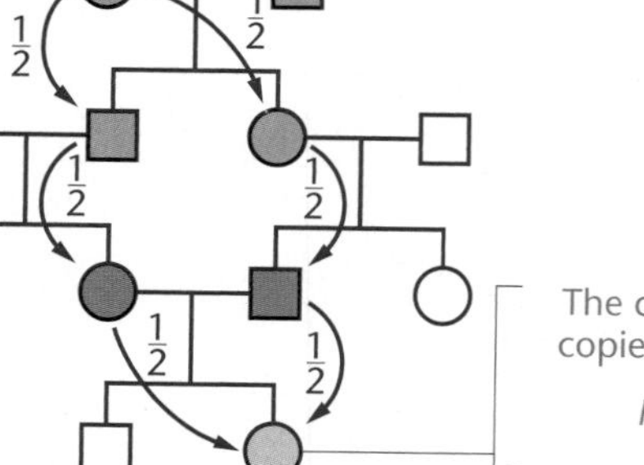

The chance that this female will inherit two copies of her great-grandmother's *a* allele is

$$F = \frac{1}{2} \times \frac{1}{2} \times \frac{1}{2} \times \frac{1}{2} \times \frac{1}{2} \times \frac{1}{2} = \frac{1}{64}$$

Because the female's two alleles could be identical by descent from any of four different alleles,

$$F = 4 \times \frac{1}{64} = \frac{1}{16}$$

FIGURE 22–25 Calculating the coefficient of inbreeding (F) for the offspring of a first-cousin marriage (purple), as illustrated in this pedigree tracing the inberitance of a recessive lethal allele, a.

Genetic Effects of Inbreeding

Inbreeding results in the production of homozygous individuals, including an increase in homozygosity for recessive alleles that were previously concealed in heterozygotes. Because many recessive alleles are deleterious when homozygous, one consequence of inbreeding is an increased chance that an individual will be homozygous for a recessive deleterious allele. Inbred populations often have a lowered mean fitness, and **inbreeding depression** is a measure of this loss of fitness. In domesticated plants and animals, inbreeding and selection have been used for thousands of years, and these organisms have a high degree of homozygosity at many loci. Further inbreeding will usually produce only a small loss of fitness. However, inbreeding among individuals from large, randomly mating populations can produce high levels of inbreeding depression. This effect can be seen by examining mortality rates in the offspring of inbred animals in zoo populations (Table 22.7). To counteract inbreeding depression, many zoos use DNA fingerprinting to estimate the relatedness of their animals and choose the least related animals as parents.

As natural populations of endangered species decrease, concern about inbreeding is an important factor in designing programs to restore these species. One example is a project in Scandinavian zoos to preserve the Fennoscandic wolf. Conservationists established a wolf population that was bred in captivity using four founding individuals. Because the captive-wolf population was founded by such a small number of individuals, it is severely inbred; individuals show inbreeding depression in the form of smaller body weight, reduced reproductive success, and reduced longevity. Furthermore, a number of wolves in the population are blind.

To reduce the level of inbreeding, two Russian wolves were later added to the population, a strategy some conservationists regretted because it introduced alleles from another species into the Fennoscandic gene pool, reducing the number of purebred Fennoscandic wolves. Researchers have concluded that blindness in the captive-wolf line is caused by an autosomal recessive allele at a single locus. The bad news is that all the remaining purebred Fennoscandic wolves (individuals with no genes from the two Russian wolves) have at least a 6% chance of being carriers for blindness—some purebred individuals have a 67% chance of being carriers. The good news is that individuals with greater than 30% chance of being carriers can be removed from the breeding population without reducing the remaining genetic variation in the population by more than 10%. Removing the likely carriers would reduce the frequency of the blindness allele from 14% to 7% and would improve the long-term prospects of maintaining a viable Fennoscandic wolf population.

In humans, inbreeding increases the risk of spontaneous abortions, neonatal deaths, congenital deformities, and recessive genetic disorders. Although less common than in the past, inbreeding occurs in many regions of the world where social customs favor marriage between first cousins. Alan Bittles and James Neel analyzed data from numerous studies on different cultures. They found that the rate of child mortality (i.e., death in the first several years of life) varies dramatically from culture to culture. No matter what the baseline mortality rate is for children of unrelated parents, however, children of first cousins virtually always have a higher death rate—typically by about 4.5 percentage points. Over many generations, inbreeding should eventually reduce the frequency of deleterious recessive alleles if homozygotes die before reproducing. However, some studies indicate that parents who are first cousins tend to have more children to compensate for those lost to genetic disorders. On average, two-thirds of the surviving offspring are heterozygous carriers of the deleterious allele.

It is important to note that inbreeding is not always harmful. Indeed, inbreeding has long been recognized as a useful tool for breeders of domesticated plants and animals. When an inbreeding program is initiated, homozygosity increases, and some breeding stocks become fixed for favorable alleles and others for unfavorable alleles. By selecting the more viable and vigorous plants or animals, the proportion of individuals carrying desirable traits can be increased.

If members of two inbred lines are mated, hybrid offspring are often more vigorous in desirable traits than either of the parental lines. This phenomenon is called

TABLE 22.7 Mortality in Offspring of Inbred Zoo Animals

Species	*Number*	*Noninbred*		*Inbred*
Dorcas gazelle	92	Lived:	36	17
		Died:	14	25
Eld's deer	24	Lived:	13	0
		Died:	4	7
Giraffe	19	Lived:	11	2
		Died:	3	3
Oryx	42	Lived:	35	0
		Died:	2	5
Zebra	32	Lived:	20	3
		Died:	7	2

hybrid vigor. When this approach was used in breeding programs established for maize, crop yields increased tremendously. Unfortunately, as a consequence of segregation, the hybrid vigor extends only through the first generation. Many hybrid lines are sterile, and those that are fertile show subsequent declines in yield. Consequently, the hybrids must be regenerated each time by crossing the original inbred parental lines.

Hybrid vigor has been explained in two ways. The first theory, the **dominance hypothesis**, incorporates the obvious reversal of inbreeding depression, which inevitably must occur in outcrossing. For example, a cross between two strains of maize with genotypes

Strain A		Strain B		F_1
aaBBCCddee	×	*AAbbccDDEE*	→	*AaBbCcDdEe*

generates F_1 hybrids that are heterozygotes at all of the loci shown. The deleterious recessive alleles present in the homozygous form in the parents are masked by the more favorable dominant alleles in the hybrids. Such masking is believed to cause hybrid vigor.

The second theory, **overdominance**, holds that in many cases the heterozygote is superior to either homozygote. This may relate to the fact that in a heterozygote, two forms of a gene product may be present, providing a form of biochemical diversity. Thus, the cumulative effect of heterozygosity at many loci accounts for hybrid vigor. Most likely, hybrid vigor results from a combination of both hypotheses.

We have seen that nonrandom mating can drive the genotype frequencies in a population away from their expected values under the Hardy-Weinberg law. This can indirectly affect the course of evolution. As in stocks purposely inbred by animal and plant breeders, inbreeding in a natural population may increase the frequency of homozygotes for a deleterious recessive allele. With domestic stocks, this increases the efficiency with which selection removes the deleterious allele from the population. Consequently, once deleterious genes are removed, inbreeding no longer causes problems.

GENETICS, TECHNOLOGY, AND SOCIETY

Tracking Our Genetic Footprints out of Africa

Where did we come from? Are we one human family with minor differences, or are we separated into races with profound and ancient roots? For millennia, our efforts to answer these questions invoked legends, mythologies, and the creation stories of our many religions. Over the last century, paleoanthropologists have applied a variety of sophisticated scientific tools to explore our origins and human kinships. The evolving story of our beginnings, based on modern genetics, is as fascinating and controversial as any creation myth.

Based on the physical traits and distribution of hominid fossils, most paleoanthropologists agree that a large-brained, tool-using hominid named *Homo erectus* appeared in east Africa about 2 million years ago. This species used simple stone tools, hunted but did not fish, did not build houses or fireplaces, and lacked ritual burial practices. About 1.7 million years ago, *H. erectus* spread into Eurasia and south Asia. Most scientists also agree that *H. erectus* likely developed into several hominid types including Neanderthals (in Europe) and Peking man or Java man (in Asia). These hominids were anatomically robust, with large heavy skeletons and skulls. Neanderthals and other *H. erectus* groups disappeared 50,000–30,000 years ago—around the same time that anatomically modern humans (*H. sapiens*) appeared all over the world.

It is at this point in our history—when ancient hominids gave way to lighter-skeletoned, anatomically modern humans—that controversy arises.

At present, there are two dominant hypotheses to explain our origins: the out-of-Africa and multiregional hypotheses. The multiregional hypothesis is based primarily on archaeological and fossil evidence. It proposes that *H. sapiens* developed gradually and simultaneously all over the world from existing *H. erectus* groups, including Neanderthals. Interbreeding between these groups eventually made *H. sapiens* a genetically homogeneous species. Natural selection over 1.5 million years then created the regional variants (races) that we see today. In the multiregional view, our genetic makeup should include contributions from Neanderthals and other *H. erectus* groups. In contrast, the out-of-Africa hypothesis, based primarily on genetic analyses of modern human populations, contends that *H. sapiens* evolved from the descendants of *H. erectus* in sub-Saharan Africa about 200,000 to 400,000 years ago. A small band of *H. sapiens* (fewer than 10,000) then left Africa, expanded, and migrated into Europe and Asia around 100,000 years ago. By about 60,000 years ago, populations of *H. sapiens* reached Australia and later migrated into North America. In the out-of-Africa model, *H. sapiens* replaced all the preexisting *H. erectus* types, without interbreeding. In this way, *H. sapiens* became the only species in the genus by about 30,000 years ago.

Although still contentious, most genetic evidence appears to support the out-of-Africa hypothesis. Humans all over the globe are remarkably similar genetically. DNA sequences from any two people chosen at random are 99.9% identical. There is more genetic identity between two persons chosen at random from a human population than there is between two chimpanzees chosen at random from a chimpanzee population. Interestingly, about 90% of the genetic differences that do exist occur between individuals, rather than between populations. This unusually high degree of genetic relatedness in all humans around the world supports the idea that our species arose recently from a small founding group of humans.

Studies of mitochondrial DNA sequences from current human populations reveal that the highest levels of genetic variation occur within African populations. Africans show twice the mitochondrial DNA-sequence diversity of non-Africans. This implies that the earliest branches of *H. sapiens* diverged in Africa and had a longer time to accumulate mitochondrial DNA mutations, which are thought to accumulate at a constant rate over time.

DNA sequences from mitochondrial, Y-chromosome, and chromosome-21 markers support the idea that our roots are in east Africa and the migration out of Africa occurred through Ethiopia, along the coast of the Arabian peninsula, and outward to Eurasia and Southeast Asia. Recent data based on DNA-sequence diversity in nuclear microsatellites further support the notion that humans migrated out of Africa and dispersed throughout the world from a small founding population. Sub-Saharan African populations show the highest levels of microsatellite heterozygosities, followed by those in the Middle East, Europe, East Asia, Oceania, and the Americas—in that order. Native American populations show about 15% less microsatellite heterozygosity than that seen in Africans.

By comparing DNA-sequence differences between populations around the world and by extrapolating back to a time when all sequences would have been the same, paleoanthropologists propose that modern *H. sapiens* developed from a small group in Africa between 200,000 and 400,000 years ago. The time of the out-of-Africa migration is calculated to be 50,000–100,000 years ago.

The recent sequencing of Neanderthal mitochondrial DNA shows that it is so different from ours that Neanderthals were likely a separate species and that Neanderthals and *H. sapiens* diverged about 600,000 years ago. Hence, it appears unlikely that Neanderthals and perhaps other *H. erectus* groups such as Peking man contributed significantly to the *H. sapiens* gene pool.

So, if all people on Earth are so similar genetically, how did we come to have such a range of physical differences, which some describe as racial differences? Many geneticists believe that the genetic changes responsible for these characteristics such as skin color and facial features could accumulate over short periods of time, especially if these characteristics are adaptive to particular climatic and geographic conditions.

As with any explanation of human origins, the out-of-Africa hypothesis is actively debated and may undergo mutation—or even extinction—over time. As methods to sequence DNA from ancient fossils improve, it may be possible to fill the gaps in our genetic pathway leading out of Africa and help us to resolve those age-old questions about our origins.

Reference

Cavalli-Sforza, L. L., and Feldman M. W., 2003. The application of molecular genetic approaches to the study of human evolution. *Nat. Gen. (Suppl.)* 33: 266–75.

Web Sites

Johanson, D. 2001. *Origins of modern humans: multiregional or out of Africa*? [on-line.]
http://www.actionbioscience.org/evolution/johanson.html

Chapter Summary

1. Populations evolve as a result of changes in allele frequency at a number of loci over a period of time. A key insight in understanding evolution was the recognition that changes in the relative abundance of different phenotypes can be tied to changes in the relative abundance of the alleles that influence the phenotypes. This insight led to the development of population genetics, which studies the factors that cause allele frequencies to change.
2. The Hardy-Weinberg law specifies what will happen to allele frequencies at a given locus in a population under a set of simple assumptions. If there is no selection, no mutation, and no migration if the population is large, and if individuals mate at random, then allele frequencies will not change from one generation to the next—the population will not evolve.
3. The Hardy-Weinberg formula can be used to investigate whether a population is in evolutionary equilibrium at a given locus and to estimate the frequency of heterozygotes in a population from the frequency of homozygous recessives.
4. By specifying the conditions under which allele frequencies will not change, the Hardy-Weinberg law identifies the forces that can cause evolution. Selection, mutation, migration, and genetic drift can cause allele frequencies to change and are thus forces of evolution. Nonrandom mating does not alter the frequencies of alleles, but by altering the frequencies of genotypes, nonrandom mating can indirectly affect the course of evolution.
5. Natural selection is the most powerful force that can alter the frequencies of alleles in a population. The rate of evolution under natural selection depends on the initial frequencies of the alleles and on the relative fitness of different genotypes or the relative strength of selection. In small populations, random genetic drift can also be an important force altering the frequencies of alleles.
6. Mutation and migration introduce new alleles into a population, but by themselves mutation and migration usually have little effect on allele frequencies. Instead, the retention or loss of introduced alleles depends on the fitness they confer and the action of selection.
7. Genetic drift is change in allele frequencies as a result of chance events; it drives evolution in small populations.
8. The most important form of nonrandom mating is inbreeding, or mating between relatives. Inbreeding increases the frequency of homozygotes in a population and decreases the frequency of heterozygotes.

Key Terms

INSIGHTS AND SOLUTIONS

1. Tay-Sachs disease is caused by loss-of-function mutations in a gene on chromosome 15 that encodes a lysosomal enzyme and is inherited as an autosomal recessive condition. Among Ashkenazi Jews of central European ancestry, about 1 in 3600 children is born with the disease. What fraction of the individuals in this population are carriers?

Solution: If we let p represent the frequency of the wild-type enzyme allele and q the total frequency of recessive loss-of-function alleles, and if we assume that the population is in Hardy-Weinberg equilibrium, then the frequencies of the genotypes are given by p^2 for homozygous normal individuals, $2pq$ for carriers, and q^2 for individuals with the disease. The frequency of Tay-Sachs alleles is thus

$$q = \sqrt{q^2} = \sqrt{\frac{1}{3600}} = 0.017$$

Since $p + q = 1$, we have

$$p = 1 - q = 1 - 0.017 = 0.983$$

Therefore, we can estimate that the frequency of carriers is

$$2pq = 2(0.983)(0.017) = 0.033$$

or 1 in 30.

2. Eugenics is the term employed for selectively breeding humans to bring about improvements in populations. As a eugenic measure, it has been suggested that individuals suffering from serious genetic disorders should be prevented (sometimes by force of law) from reproducing (by sterilization, if necessary) to reduce the frequency of the disorder in future generations. Suppose that a recessive trait were present in the population at a frequency of 1 in 40,000 and that affected individuals did not reproduce. In 10 generations, or about 250 years, what would be the frequency of the condition? Are the eugenic measures effective in this case?

Solution: Let q represent the frequency of the recessive allele responsible for the disorder. Because the disorder is recessive, we can estimate that

$$q = \sqrt{\frac{1}{40{,}000}} = 0.005$$

If all affected individuals are prevented from reproducing, then in an evolutionary sense, the disorder is lethal: Affected individuals have zero fitness. This means that we can predict the frequency of the recessive allele 10 generations in the future by using the equation

$$q_g = q_0/1 + gq_0$$

Here, $q_0 = 0.005$, and $g = 10$, so we have

$$q_{10} = \frac{(0.005)}{1 + (10 \times 0.005)} = 0.0048$$

If $q_{10} = 0.0048$, then the frequency of homozygous recessive individuals will be roughly

$$(q_{10})^2 = (0.0048)^2 = 0.000023 = \frac{1}{43{,}500}$$

The frequency of the genetic disorder has only been reduced from one in 40,000 to 1 in 43,500 in 10 generations, indicating that this eugenic measure has limited effectiveness.

Problems and Discussion Questions

1. What observations led Wallace and Darwin to formulate the idea that natural selection is the mechanism of adaptive evolution?
2. (a) Define a population. (b) Why is interbreeding a part of this definition?
3. (a) Define a gene pool. (b) Are the genomes of all members of a population part of the gene pool? Why?
4. What are the five assumptions used in the Hardy-Weinberg law, and why is each of them important?
5. Is the Hardy-Weinberg law necessary to calculate allele frequencies for a gene with two codominant alleles? Why?
6. Few, if any, populations in nature meet all the assumptions of the Hardy-Weinberg law. If this is the case, how can the Hardy-Weinberg law be of any use in population genetics?
7. Using microsatellite markers to estimate genetic diversity among invading Argentine ants (*Linepithema humile*), Tsutsui and coworkers (2000. *Proc. Natl. Acad. Sci. (USA)* 97: 5948–53.) found that successfully invasive ants are often less genetically diverse than native populations. In addition, they found that invasive ants are unicolonial and less aggressive than native populations. How might reduced genetic diversity have been achieved in invasive ants, and why might it serve as a selective advantage during an invasion?
8. The ability to taste the compound PTC is controlled by a dominant allele *T*, and individuals homozygous for the recessive allele *t* are unable to taste PTC. In a genetics class of 125 students, 88 can taste PTC and 37 cannot. Calculate the frequency of the *T* and *t* alleles in this population and the frequency of the genotypes.
9. Calculate the frequencies of the *AA*, *Aa*, and *aa* genotypes after one generation when the initial population consists of 0.2 *AA*, 0.6 *Aa*, and 0.2 *aa* genotypes and meets the requirements of the Hardy-Weinberg relationship. What genotypic frequencies will occur after a second generation?
10. *Candida albicans*, a yeast, causes mild infections in individuals with a normal immune system, but can cause life-threatening systemic diseases in immunocompromised individuals. While it is commonly observed that geographical isolation is a major factor causing genetic variation within a species, some studies indicate that extensive geographic isolation can occur without associated genetic variation. In a comparison of 62 *C. albicans* isolates from HIV patients in North Carolina with 64 isolates from similar patients in Brazil, Xu and others (1999. *J. Bacteriol.* 181: 1369–73.) found no statistically significant genetic variation among 16 loci. Coupled with other data, they concluded that the samples, one from North America and the other from South America, were genetically indistinguishable. What factor(s) might account for such genetic homogeneity?
11. Consider rare disorders in a population caused by an autosomal recessive mutation. Calculate the percentage of heterozygous carriers from the following frequencies of the disorder in

a population: (a) 0.0064, (b) 0.000081, (c) 0.09, (d) 0.01, and (e) 0.10.

12. What assumptions must you make to validate the answers in Problem 11?
13. In a population where you know only the total number of individuals with the dominant phenotype, how can you calculate the percentage of carriers and homozygous recessives?
14. Do the two sets of data below represent populations that are in Hardy-Weinberg equilibrium (use chi-square analysis if necessary)?
(a) *CCR5* genotypes: $1/1$, 60%; $1/\Delta 32$, 35.1%; $\Delta 32/\Delta 32$, 4.9%
(b) Sickle-cell hemoglobin: *AA*, 75.6%; *AS*, 24.2%; *SS*, 0.2%
15. If 4% of a population in equilibrium expresses a recessive trait, what is the probability that the offspring of two individuals who do not express the trait will express it?
16. What will be the allele frequencies after one generation if the frequency of allele *A* is $p = 0.7$ and the frequency of allele *a* is $q = 0.3$ and the alleles are codominant for the following fitness values (a) $w_{AA} = 1$, $w_{Aa} = 0.9$, and $w_{aa} = 0.8$ (b) $w_{AA} = 1$, $w_{Aa} = 0.95$, $w_{aa} = 0.9$ (c) $w_{AA} = 1$, $w_{Aa} = 0.99$, $w_{aa} = 0.98$ (d) $w_{AA} = 0.8$, $w_{Aa} = 1$, $w_{aa} = 0.8$
17. If the initial allele frequencies are $p = 0.5$ and $q = 0.5$, and allele *a* is a lethal recessive, what will be the frequencies after 1, 5, 10, 25, 100, and 1000 generations?
18. Determine the frequency of allele *A* in an island population after one generation of migration from the mainland under the conditions in (a) $p_i = 0.6$, $p_m = 0.1$, $m = 0.2$; (b) $p_i = 0.2$, $p_m = 0.7$, $m = 0.3$; and (c) $p_i = 0.1$, $p_m = 0.2$, $m = 0.1$.
19. A large body of evidence indicates that germline mutations in most vertebrates are more common in males. The explanation is that there are more cell divisions before spermatogenesis than oogenesis and a significant number of mutations are replication-dependent. In an examination of germ-line mutations in microsatellite markers (*HrU6* and *HrU10*) in the barn swallow (*Hirundo rustica*) Brohede, et al. (2002, *Nucl. Acids Res.* 30: 1997–2003.) observed the data in the table below. (a) What general conclusions can be drawn from the data? (b) What are the implications of this research in terms of understanding the role of mutation rates in vertebrate evolution?

Marker	*Parental Sex*	*Number of Mutations*
HrU6	Female	10
	Male	2
HrU10	Female	12
	Male	2

Source: Brohede, J. et al. 2002. Heterogeneity in the rate and pattern of germline mutation at individual microsatellite loci. Nucl. Acids Res. *30: 1997-2003.*

20. Assume that an autosomal recessive disorder occurs in 1 of 10,000 individuals (0.0001) in the general population and that in this population about 2% (0.02) of the individuals are carriers for the disorder. Estimate the probability of this disorder occurring in the offspring of a marriage between first cousins, and compare this probability to the population at large.
21. Using samples of scales taken from New Zealand snapper (*Pagrus auratus*) caught in Tasman Bay off the coast of New Zealand, Houser, et al. (2002 *Proc. Natl. Acad. Sci. [USA]* 99: 11,742–47.) assessed allelic frequencies from the commencement of commercial fishing in 1950 through 1998. During this time period, there was an 85% reduction in biomass and a 75% reduction in numbers, reaching an estimated population size of about 3.3 million fish by 1985. The following data indicate a decrease in heterozygosity at a particular microsatellite locus (*GA2B*). (a) What general conclusion can be drawn from the data? (b) What explanation might reasonably account for these data?

Year	*Number of Alleles (GA2B)*
1950	18
1972	14
1981	13
1986	11
1998	9

22. What is the basis of inbreeding depression?
23. Describe how inbreeding can be used in the domestication of plants and animals, and discuss the theories underlying these techniques.
24. Evaluate this statement: Inbreeding increases the frequency of recessive alleles in a population.
25. In a breeding program to improve crop plants, which of the following mating systems should be employed to produce a homozygous line in the shortest possible time? Illustrate your choice with pedigree diagrams.
(a) self-fertilization
(b) brother–sister matings
(c) first-cousin matings
(d) random matings
26. If the recessive trait albinism (*a*) is present in 1 in 10,000 individuals in a population at equilibrium, calculate the frequency of (a) the recessive mutant allele, (b) the normal dominant allele, (c) heterozygotes in the population, and (d) matings between heterozygotes.
27. One of the first Mendelian traits identified in humans was a dominant condition known as brachydactyly, in which there is an abnormal shortening of the fingers or toes (or both). Some scientists thought that the dominant trait would spread until 75% of the population would be affected, because the phenotypic ratio of dominant to recessive is 3 : 1. Show that their reasoning was incorrect.
28. Achondroplasia is a dominant trait that causes a form of dwarfism. In a survey of 50,000 births, 5 infants with achondroplasia were identified. Three of the affected infants had affected parents, while two had normal parents. Calculate the mutation rate for achondroplasia, and express the rate as the number of mutant genes per given number of gametes.
29. A prospective groom, who is normal, has a sister with cystic fibrosis (CF), an autosomal recessive disease; their parents are also normal. He plans to marry a woman who has no history of CF in her family. What is the probability that they will produce a CF child? They are both Caucasian, and the overall frequency of CF in the Caucasian population is 1 affected child per 2500 (assume the population meets the Hardy-Weinberg requirements).
30. A form of dwarfism known as Ellis-van Creveld syndrome was first discovered in the late 1930s, when Richard Ellis and Simon van Creveld shared a train compartment on the way to a pediatrics meeting. In the course of conversation, they discovered that they each had a patient with this syndrome. They published a description of the syndrome in 1940. Affected individuals have a short-limbed form of dwarfism and often have polydactyly (extra fingers) and defects of the lips and teeth. The largest pedigree for the condition was reported in an Old Order Amish population in eastern Pennsylvania by Victor McKusick and his colleagues in 1964. In that community, about 5 chil-

dren per 1000 births are affected, and in the population of 8000, the observed frequency of this disorder is 2 per 1000. All of the affected individuals have unaffected parents, and all of the affected people can trace their ancestry to Samuel King and his wife, who arrived in the area in 1774. It is known that neither King nor his wife was affected with the disorder. There are no cases of the disorder in other Amish communities, such as those in Ohio or Indiana. (a) From this information, derive the most likely mode of inheritance of this disorder. Using the Hardy-Weinberg law and assumptions, calculate the frequency of the mutant allele in the population and the frequency of heterozygotes. (b) What is the most likely explanation for the high frequency of the disorder in the Pennsylvania Amish community and its absence in other Amish communities?

31. The accompanying graph shows the variation in the frequency of allele *A* that occurred over time in two relatively small, independent populations exposed to very similar environmental conditions. A genetics student has analyzed the graph and concluded that the best explanation for these data is that selection for allele *A* is occurring in Population 1. Is the student's conclusion correct? Why, or why not?

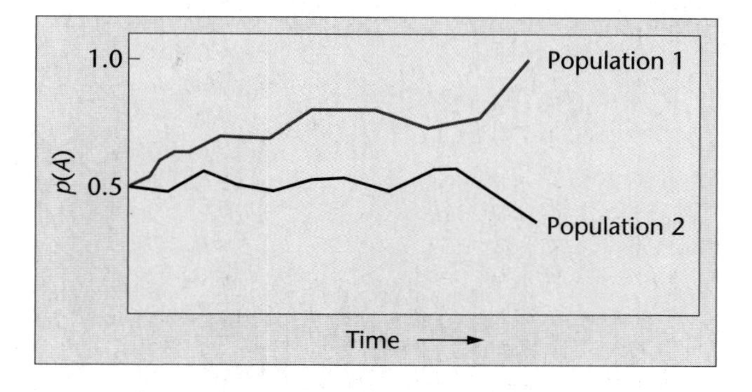

Selected Readings

Ansari-Lari, M. A. et al. 1997. The extent of genetic variation in the *CCR5* gene. *Nat. Genet.* 16: 221–22.

Ballou, J., and Ralls, K. 1982. Inbreeding and juvenile mortality in small populations of ungulates: A detailed analysis. *Biol. Conserv.* 24: 239–72.

Bittles, A. H., and Neel, J. V. 1994. The costs of human inbreeding and their implications for variations at the DNA level. *Nat. Genet.* 8: 117–21.

Bouzat, J. L. et al. 1998. Genetic evaluation of a demographic bottleneck in the greater prairie chicken. *Conserv. Biol.* 12: 836–43.

Carrington, M. et al. 1997. Novel alleles of the chemokine-receptor gene *CCR5*. *Am. J. Hum. Genet.* 61: 1261–67.

Chevillon, C. et al. 1995. Population structure and dynamics of selected genes in the mosquito *Culex pipiens*. *Evol.* 49: 997–1007.

Dudley, J. W. 1977. 76 generations of selection for oil and protein percentage in maize. In *Proceedings of the International Conference on Quantitative Genetics*. (eds.) E. Pollack et al. pp. 459-73. Ames: Iowa State Univ. Press.

Fisher, R. A. 1930. *The Genetical Theory of Natural Selection*. Oxford: Clarendon Press. (Reprinted Dover Press, 1958.)

Freeman, S., and Herron, J. C. 2001. *Evolutionary Analysis*. Upper Saddle River: Prentice-Hall.

Freire-Maia, N. 1990. Five landmarks in inbreeding studies. *Am. J. Med. Genet.* 35: 118–20.

Grant, P. R. 1986. *Ecology and Evolution of Darwin's Finches*. Princeton: Princeton Univ. Press.

Hamblin, M. T., Thompson, E., and Di Rienzo, A. 2002. Complex signatures of natural selection at the Duffy blood group locus. *Am. J. Hum. Genet.* 70: 369–83.

Karn, M. N., and Penrose, L. S. 1951. Birth weight and gestation time in relation to maternal age, parity and infant survival. *Ann. Eugen.* 16: 147–64.

Kerr, W. E., and Wright, S. 1954. Experimental studies of the distribution of gene frequencies in very small populations of *Drosophila melanogaster*. I. *Forked. Evol.* 8: 172–77.

Laikre, L., Ryman, N., and Thompson, E. A. 1993. Hereditary blindness in a captive wolf (*Canis lupus*) population: Frequency reduction of a deleterious allele in relation to gene conservation. *Conserv. Biol.* 7: 592–601.

Leibert, F. et al. 1998. The *DCCR5* mutation conferring protection against HIV-1 in Caucasian populations has a single and recent origin in northeastern Europe. *Hum. Molec. Genet.* 7: 399–406.

Liu, R. et al. 1996. Homozygous defect in HIV-1 coreceptor accounts for resistance of some multiply-exposed individuals to HIV-1 infection. *Cell* 86: 367–77.

Lucotte, G., and Mercier, G. 1998. Distribution of the *CCR5* gene 32-bp deletion in Europe. *J. Acquir Imm. Def. Syndr. Hum. Retrovir.* 19: 174–77.

Markow, T. et al. 1993. HLA polymorphism in the Havasupai: Evidence for balancing selection. *Am. J. Hum. Genet.* 53: 943–52.

Martinson, J. J. et al. 1997. Global distribution of the *CCR5* gene 32-bp deletion. *Nat. Genet.* 16: 100–03.

Parra, E. J. et al. 1998. Estimating African-American admixture proportions by use of population-specific alleles. *Am. J. Hum. Genet.* 63: 1839–51.

Pergams, O. et al. 2003. Rapid changes in mouse mitochondrial DNA. *Nature* 423: 397.

Pier, G. B. et al. 1998. *Salmonella typhi* uses CFTR to enter intestinal epithelial cells. *Nature* 393: 79–82.

Quillent, C. et al. 1998. HIV-1-resistance phenotype conferred by combination of two separate inherited mutations of *CCR5* gene. *Lancet* 351: 14–18.

Samson, M. et al. 1996. Resistance to HIV-1 infection in Caucasian individuals bearing mutant alleles of the *CCR-5* chemokine receptor gene. *Nature* 382: 722–25.

Stephens, J. C. et al. 1998. Dating the origin of the *CCR5-Δ32* AIDS-resistance allele by the coalescence of haplotypes. *Am. J. Hum. Genet.* 62: 1507–15.

Woodworth, C. M., Leng, E. R., and Jugenheimer, R. W. 1952. Fifty generations of selection for protein and oil in corn. *Agron. J.* 44: 60–66.

Yi, Z. et al. 2003. A 122.5 kilobase deletion of the *P* gene underlies the high prevalence of oculocutaneous albinism Type 2 in the Navajo population. *Am. J. Hum. Genet.* 72: 62–72.

CHAPTER 23

Light and dark forms of the pepperred moth, *Biston betularia* on light treebark. *(Breck P. Kent/Animals Animals/Earth Scenes).*

Genetics and Evolution

CHAPTER CONCEPTS

In Chapter 22, we described how populations evolve by changes in allele frequencies. We also outlined forces that cause allele frequencies to change. Mutation, migration, selection, and drift—individually and collectively—bring about evolutionary divergence and species formation. This process depends not only on genetic divergence but also on the presence of environmental or ecological diversity. If a population is spread over a geographic range that contains a number of subenvironments or niches, populations can adapt to the conditions in these niches and become genetically differentiated. Differentiated populations are dynamic: They may continue to exist, become extinct, merge with the parental population, or diverge from the parental population until they become reproductively isolated and form a new species. This process can modify a species over time, transforming it into another species, or it can result in one species splitting into two or more species.

A **species** can be defined as a group of interbreeding or potentially interbreeding organisms that is reproductively isolated in nature from all other such groups. In sexually reproducing organisms, **speciation** divides a single gene pool into two or more separate gene pools. Changes in morphology, physiology, and adaptation to an ecological niche may also occur but are not necessary components of the speciation event. Speciation can take place gradually or within a few generations, but it is difficult to define the exact moment when a new species forms.

Because population and species divergence is accompanied by genetic differentiation, we can use patterns of genetic differences to reconstruct evolutionary history. After exploring the genetic structure of populations, their divergence across space and time, and the process of speciation, we'll discuss how genetic data can be used to answer questions that have an evolutionary context.

HOW DO WE KNOW WHAT WE KNOW?

IN THIS CHAPTER, WE WILL FOCUS ON how forces such as selection and drift act on genetic variation to bring about evolutionary change in populations and the formation of new species. As you study this topic, you should try to answer several fundamental questions:

1. How do we know how much genetic variation is in a population, and how is it measured?
2. How do we know when populations have diverged to the point that they form two different species?
3. How do we know whether the genetic structure of a population is static or dynamic?
4. How do we know how much time is required to form species?
5. How do we know the age of the last common ancestor shared by two species? ■

23.1 Speciation Can Occur by Transformation or Splitting of Gene Pools

Figure 23–1 shows an evolutionary tree, or phylogeny, that describes the history of several hypothetical lizard species. The passage of time is plotted horizontally, and the change in phenotype is plotted vertically. The graph traces the phenotype of the average lizard in each species over time. The history of our hypothetical lizards begins with species 1. For a long time, species 1 experiences evolutionary **stasis**; that is, it does not change. Species 1 then undergoes a period of transformation, also called **phyletic evolution** or **anagenesis**, and becomes species 2. At a later time, species 2 then undergoes **cladogenesis**; it splits into two distinct and reproductively isolated daughter species, which are the only extant species, species 3 and species 4.

In his 1859 book, *On the Origin of Species*, Charles Darwin amassed evidence that all species derive from a single common ancestor by transformation and speciation:

> "... all living things have much in common, in their chemical composition, their germinal vesicles, their cellular structure, and their laws of growth and reproduction.... Therefore I should infer... that probably all the organic beings which have ever lived on this earth have descended from some one primordial form..."

Everything biologists have since learned supports Darwin's conclusion that only one "tree of life" exists. To understand evolution, we must understand the mechanisms responsible for transforming one species into another and for splitting one species into two or more new species.

We have already examined the mechanisms responsible for the transformation of species. Chief among them is natural selection, discovered independently by Alfred Russel Wallace and Darwin. The Wallace-Darwin concept of natural selection is summarized as follows:

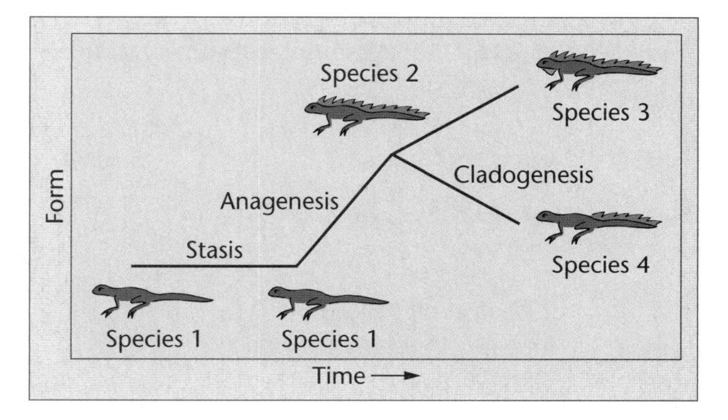

FIGURE 23–1 In anagenesis, one species is transformed over time into another species. At all times, only one species exists. In cladogenesis, one species splits into two or more species.

1. Individuals of a species vary in phenotype, including differences in size, agility, coloration, or ability to obtain food.
2. Many of these variations, even small and seemingly insignificant ones, are heritable and passed on to offspring.
3. Organisms tend to reproduce in an exponential fashion, and more offspring are produced than can survive. This causes members of a species to engage in a struggle for survival, competing with other members of the species for scarce resources.
4. In the struggle for survival, individuals with particular phenotypes will be more successful than individuals with others, allowing them to survive and reproduce at higher rates.

As a consequence of natural selection, populations of a species change over time. The phenotypes that confer improved ability to survive and reproduce become more common, and the phenotypes that confer poor prospects for survival and reproduction disappear.

Although Wallace and Darwin proposed that natural selection explains how evolution occurs, they could not satisfactorily explain how the variations themselves arise nor how such variations are passed from parents to offspring. In the twentieth century, as biologists applied the principles of Mendelian genetics to populations, the source of variation (mutation) and the mechanism of inheritance (segregation of alleles) became apparent. This union of population genetics with natural selection generated a new view of the evolutionary process, called neo-Darwinism. In this view, selection based on phenotype leads to changes in allele frequencies, which is one way evolution occurs. We now define **evolution** as changes in allele frequencies over time in a population or species.

In this chapter, we consider the mechanisms responsible for speciation. As with natural selection, our understanding of speciation builds on key insights about genetic variation.

23.2 Genetic Variation Is Present in Populations and Species

We might naively assume that members of a well-adapted population are genetically homozygous at most or all loci because the most favorable allele at each locus has become fixed. Certainly, examinations of most populations of plants and animals reveal many phenotypic similarities among individuals. However, considerable evidence indicates that most populations contain a high degree of heterozygosity. This built-in genetic diversity is concealed, so to speak, because it is not always apparent in the phenotype. Detecting this concealed genetic variation is not an easy task. Nevertheless, advances in technology make it clear that genetic variation exists at the level of proteins and DNA.

Protein Polymorphisms

Gel electrophoresis separates protein molecules on the basis of differences in size and electrical charge. If a nucleotide variation results in the substitution of a charged amino acid (such as glutamic acid) for an uncharged amino acid (such as glycine), the net electrical charge on the protein will be altered. This difference in charge can be detected as a change in the rate at which proteins migrate through an electrical field. In the mid-1960s, John Hubby and Richard Lewontin used gel electrophoresis to measure protein variation in natural populations of *Drosophila*. Since then, researchers have routinely used gel electrophoresis to study genetic variation in a wide range of organisms (Table 23.1).

Electrophoretically distinct forms of a protein produced by different alleles are called **allozymes**. As shown in Table 23.1, a large percentage of loci from diverse species produce distinct allozymes. Approximately 30 loci per species were examined, and about 30% of the loci are polymorphic, with an average of 10% heterozygosity for each locus.

These values apply only to genetic variation detectable by altered protein migration in an electric field. Elec-

TABLE 23.1 Allozyme Heterozygosity at the Molecular Level

Species	*Populations Studied*	*Loci Examined*	*Polymorphic Loci* per Population (%)*	*Heterozygotes per Locus (%)*
Homo sapiens (human)	1	71	28	6.7
Mus musculus (mouse)	4	41	29	9.1
Drosophila pseudoobscura (fruit fly)	10	24	43	12.8
Limulus polyphemus (horseshoe crab)	4	25	25	6.1

*A polymorphic locus is one for which a population harbors more than one allele.

Source: Lewontin, R. C. 1974. *The Genetic Basis of Evolutionary Change*. New York: Columbia University Press. p. 117.

trophoresis probably detects only about 30% of the actual variation due to amino acid substitutions, because many substitutions do not change the net electric charge on the molecule. Lewontin has estimated that about two-thirds of all loci in a population are polymorphic and that in any individual within that population, about one-third of the loci are heterozygous. The significance of genetic variation detected by electrophoresis is controversial. Some researchers argue that allozymes are functionally equivalent and therefore are unimportant in evolution. We will address this argument later in this section.

Variations in Nucleotide Sequence

The most direct way to estimate genetic variation is by comparing the nucleotide sequence of the genes carried by individuals in a population. With the development of recombinant DNA techniques and the polymerase chain reaction (PCR), nucleotide sequence variations are being cataloged for an increasing number of genes and genomes. In an early study, Alec Jeffreys used differences in restriction-enzyme cutting sites to detect nucleotide sequence differences in 60 unrelated individuals. His results show that within the genes of the β-globin cluster, there is a detectable sequence variation every 1 in 100 bp. If this region is representative of the genome, there may be as many as 3×10^7 nucleotide differences between any two individuals.

In another study of nucleotide sequence variation, Martin Kreitman investigated the *alcohol dehydrogenase* gene (*Adh*) in *Drosophila melanogaster* (Figure 23–2). This locus encodes two allozyme variants: *Adh-f* and *Adh-s*. The proteins differ electrophoretically because of a single amino acid change (Thr versus Lys at codon 192). To determine whether the variation at the amino acid level is reflected in variation at the nucleotide level, Kreitman cloned and sequenced *Adh* genes from five natural populations of *Drosophila*. Sequence analysis of the 11 cloned genes revealed a total of 43 nucleotide variations from the consensus *Adh* sequence of 2721 bp. These variations were distributed throughout the gene: 14 in the exon-coding regions, 18 in the introns, and 11 in the nontranslated and flanking regions. Of the 14 variations in coding regions, only 1 led to an amino acid substitution (Thr for Lys at codon 192), accounting for the 2 allozymes detected by electrophoresis. The other 13 nucleotide substitutions in exons did not result in amino acid substitutions, revealing a high level of concealed genetic diversity.

Is this level of genetic diversity present in human genes? To answer this, let's look at one of the most intensively studied genes in the human genome, the locus encoding the cystic fibrosis transmembrane conductance regulator (CFTR). Recessive loss-of-function mutations in the *CFTR* locus cause cystic fibrosis, a disease with symptoms including salty sweat, excessive amounts of thick mucus in the lungs, and susceptibility to bacterial infections. Geneticists have analyzed *CFTR* genes in some 30,000 chromosomes from individuals with cystic fibrosis and have identified over 500 different mutations that can cause the disease. These include missense mutations, amino acid deletions, nonsense mutations, frameshifts, and splice defects.

Figure 23–3 shows a map of the 27 exons in the *CFTR* locus, with most of the exons labeled by function. The histogram above the map shows the locations of the known disease-causing mutations and the number of copies that have been found of each. A single mutation, a 3-bp deletion in exon 10 called Δ*508*, accounts for 67% of the mutant cystic fibrosis alleles, with several other mutations found in at least 100 chromosomes. In populations of European ancestry, between 1 in 44 and 1 in 20 individuals are heterozygous carriers of cystic fibrosis. Note that Figure 23–3 includes only nucleotide sequence variants that alter the function of the CFTR protein. There are undoubtedly many more *CFTR* alleles with "silent sequence" variants that neither change the structure of the protein nor affect its function. However, even without accounting for these silent variants, the *CFTR* locus shows considerable genetic variation.

Studies of other organisms, including rats and mice, have produced similar estimates of nucleotide diversity. Thus, an enormous reservoir of genetic variability exists within most populations, and at the DNA level, most and perhaps all genes exhibit diversity from individual to individual. But how many genetic differences in allozymes and nucleotide content are the result of natural selection? What is the significance of these differences?

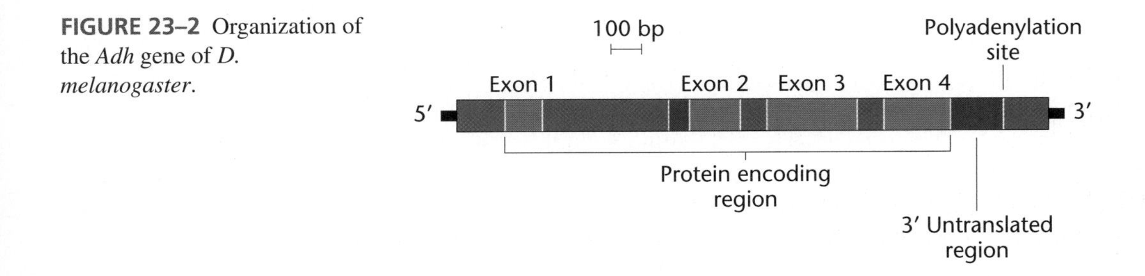

FIGURE 23–2 Organization of the *Adh* gene of *D. melanogaster*.

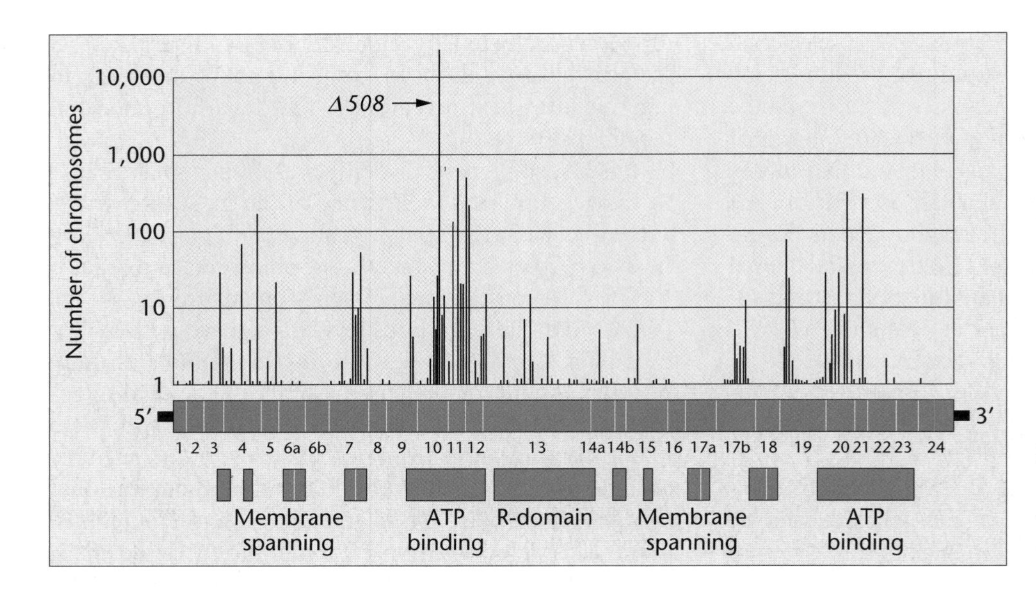

FIGURE 23–3 Disease-causing mutations in the cystic fibrosis gene. The histogram shows the number of copies of each mutation (the vertical axis is on a logarithmic scale). Note the large number of mutations in *Δ508*. The genetic map below the histogram shows the locations and relative sizes of the 27 exons of the *CFTR* locus. The boxes along the bottom indicate the functions of different domains of the CFTR protein. *(Reprinted from* Trends in Genet. *8: 392–98. L. Tsui, The spectrum of cystic fibrosis mutations, 1992, with permission from Elsevier Science.)*

23.3 Explaining High Levels of Genetic Variation in Populations

The finding that populations harbor considerable genetic diversity at the level of amino acid and nucleotide sequences came as a surprise to many evolutionary biologists. Prior to our ability to collect protein and DNA data, the consensus was that selection would favor a single optimal allele at each locus, and as a result, well-adapted populations would be homozygous at most or all loci. This expectation was obviously wrong, and considerable research and argument ensued in an attempt to explain what maintains such high levels of genetic variation.

One view holds that the genetic variation we observe primarily reflects the action of mutation and genetic drift. This idea, proposed by Motoo Kimura and called the **neutral theory of molecular evolution**, argues that mutations leading to amino acid substitutions are rarely favorable. They are sometimes detrimental but most often are neutral or genetically equivalent to the allele that is replaced. The few variations that are favorable or detrimental are preserved or removed from the population, respectively, by natural selection. This process occurs relatively quickly if selection coefficients are large or even moderate. However, most mutations are neutral and are not affected by selection. The frequency of these alleles in a population will be determined by other forces, specifically mutation rates and random genetic drift. Some neutral mutations will drift to fixation in the population; other neutral mutations will be lost. At any given time, a population may contain many neutral alleles at any particular locus, but in general this does not reflect the action of natural selection.

Opposed to the neutral theory are those who favor the role of natural selection in maintaining a high level of variability. **Selectionists** point to examples where enzyme or protein variants are associated with adaptation to certain environmental conditions. The well-known advantage of sickle-cell anemia heterozygotes in resisting infection by malarial parasites exemplifies this type of adaptation. However, because very few examples of heterozygote advantage exist, selectionists postulate that temporal and spatial variation in selective pressure is important.

Selectionists also stress that enzyme variants (like allozymes) often appear to offer no advantage, but exist in such frequency that they cannot be explained as random occurrences. Thus, even though no currently available analytical methods can detect any functional difference, some slight advantage associated with certain amino acid substitutions may exist. Perhaps having two forms of a given protein allows optimum performance under a wider range of cellular conditions, or perhaps natural selection varies in time and space.

The neutralist and selectionist perspectives should not be viewed as mutually exclusive positions. Instead, they are best seen as end points in a spectrum of possibilities. Neutralists do not discount natural selection as a guiding force in evolution; rather, they suggest that most alleles in existence do not confer an adaptive advantage,—they fluctuate randomly because of genetic drift. On the other hand, selectionists certainly don't deny that genetic drift is an important factor in establishing differences in allele frequencies (such as the silent or noncoding nucleotide variations in the *Drosophila Adh* gene). It is difficult to argue against the notion that some genetic variation must be neutral. The difference between the two theories is in their proposed degree of neutrality.

Current data are insufficient to determine what fraction of nucleotide sequence variation is neutral and what fraction is subject to natural selection. The neutral theory nonetheless serves a crucial function. By pointing out that some genetic variation is expected simply as a result of mutation and drift, the neutral theory provides a working hypothesis for studies of molecular evolution. In other words, biologists must find positive evidence that selection is acting on allele frequencies at a particular locus before they can reject the simpler assumption that only mutation and drift are at work.

23.4 The Genetic Structure of Populations Changes Across Space and Time

As geneticists were discovering that most populations harbor considerable genetic diversity, they also found that the genetic structure of populations varies across space and time. To illustrate, let's consider the studies on *Drosophila pseudoobscura* conducted by Theodosius Dobzhansky and his colleagues. This species is found over a wide range of habitats, including the western and southwestern United States. Although flies throughout this range are morphologically similar, Dobzhansky's team discovered that populations from different locations vary in the arrangement of genes on chromosome 3. They found several different inversions in this chromosome that show up as loops in the larval polytene chromosomes of heterozygotes. Each inversion sequence is named after the locale in which it was first discovered (e.g., AR for Arrowhead, British Columbia, and CH for the Chiricahua Mountains). Inversion sequences are compared to one standard sequence, designated ST.

Figure 23–4 shows the frequencies of three arrangements found in populations living at three elevations in the Sierra Nevada mountains in California. The ST arrangement is most common at low elevations, but at 8000 feet, AR is the most common and ST is the least common arrangement. In these populations, the frequency of the CH arrangement gradually increases with elevation. The gradual change in inversion frequencies is probably the result of natural selection and parallels the gradual environmental changes occurring at ascending elevations. Dobzhansky's team also found that if populations are collected at a single site throughout the year, inversion frequencies also change. That is, cyclic variation in chromosome arrangements occurs through the seasons, as shown in Figure 23–5. Such variation was consistently observed over a period of several years. The frequency of ST always declined during the spring, while the frequency of CH always increased in the spring.

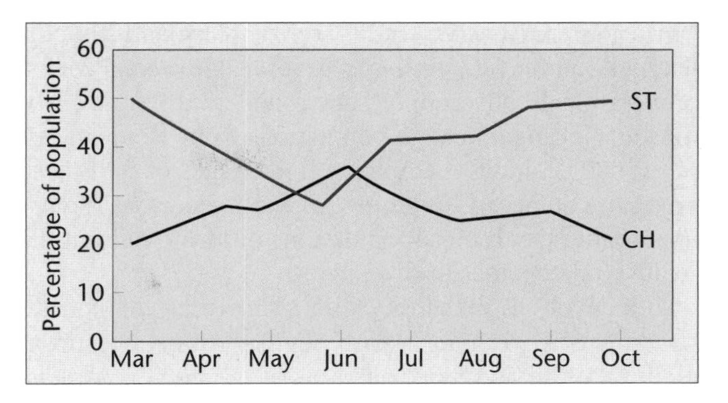

FIGURE 23–5 Changes in the ST and CH arrangements in *D. pseudoobscura* throughout the year.

To test their hypothesis that this cyclic change is a response to natural selection, Dobzhansky's team devised a laboratory experiment. They constructed a large population cage from which samples of *D. pseudoobscura* could be periodically removed and studied. The experiment began with a population having a known inversion frequency: 88% CH and 12% ST. Flies were raised at 25°C and sampled over a one-year period. As shown in Figure 23–6, ST increased gradually until it was present at a level of 70%. At that level, an equilibrium between ST and CH was reached, and no further changes were seen. When the same experiment was performed at 16°C, there was no change in inversion frequency. They concluded that the equilibrium reached at 25°C is a response to the elevated temperature—the only variable in the experiment.

These results indicate that a balance in the frequency of the two inversions and their respective gene arrangements in a population is superior to either inversion by itself. The equilibrium attained presumably represents the highest mean fitness in the population under controlled laboratory conditions. This interpretation of Dobzhansky's experiment suggests that natural selection is the driving force maintaining the diversity and frequency of chromosome 3 inversions.

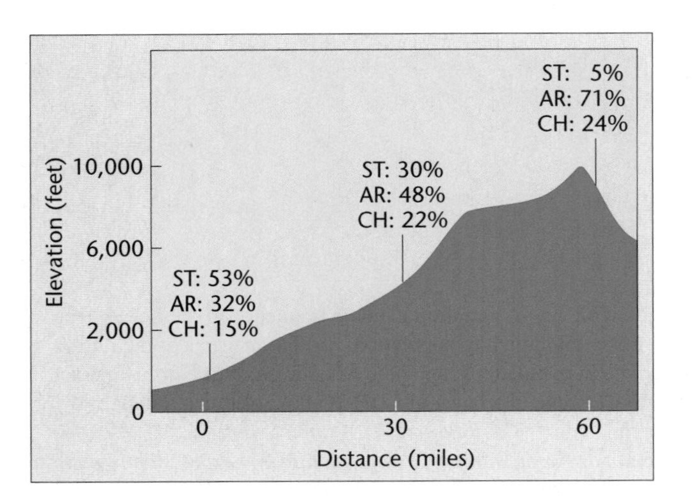

FIGURE 23–4 Inversions in chromosome 3 of *D. pseudoobscura* at different elevations in the Sierra Nevada mountain range near Yosemite National Park.

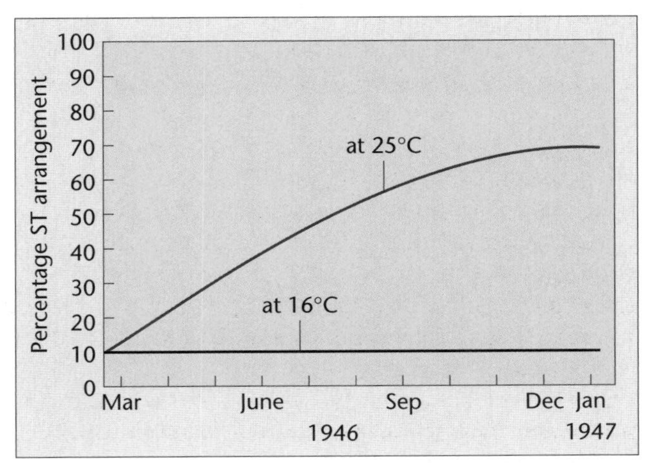

FIGURE 23–6 Graph of the increase in ST arrangement of *D. pseudoobscura* at 25°C versus no change at 16°C in population cages under laboratory conditions.

In a more extensive study, Dobzhansky and his colleagues sampled *D. pseudoobscura* populations over a broader geographic range. They found 22 different chromosome arrangements in populations from 12 locations. Figure 23–7 shows the relative frequencies of 5 of these inversions at these 12 locations. The differences are largely quantitative, with most populations differing only in the relative percentages of inversions.

Collectively, Dobzhansky's data show that the genetic structure of *D. pseudoobscura* populations changes from place to place and from one season to another. At least some of this variation in population genetic structure must be the result of natural selection.

Other studies have provided evidence of how the genetic structure of a species varies among populations. Dennis Powers and Patricia Schulte studied populations of the mummichog (*Fundulus heteroclitus*), a small fish (5–10 cm long) that lives in inlets, bays, and estuaries along the Atlantic coast of North America from Florida to Newfoundland. These workers studied allele frequencies in the gene encoding the enzyme lactate dehydrogenase-B (LDH-B). LDH-B is made in the liver, heart, and red skeletal muscle. It converts lactate to pyruvate and is thus pivotal in both the synthesis of glucose and in aerobic metabolism. There are two allozymes of LDH-B; they differ at two amino acid positions. The alleles encoding the allozymes are *Ldh-B*^{a} and *Ldh-B*^{b}.

The frequencies of the *Ldh-B* alleles vary dramatically among mummichog populations (Figure 23–8a). In northern populations, where the mean water temperature is about 6°C, *Ldh-B*^{b} predominates. In southern populations, where the mean water temperature is about 21°C, *Ldh-B*^{a} predominates. Between these geographic extremes, allele frequencies are intermediate.

To determine whether geographic variation in *Ldh-B* allele frequencies is due to natural selection, Powers and Schulte studied the biochemical properties of the LDH-B allozymes. They found that the enzyme encoded by *Ldh-B*^{b} has high catalytic efficiency at low temperatures, whereas the gene product of *Ldh-B*^{a} is more efficient at

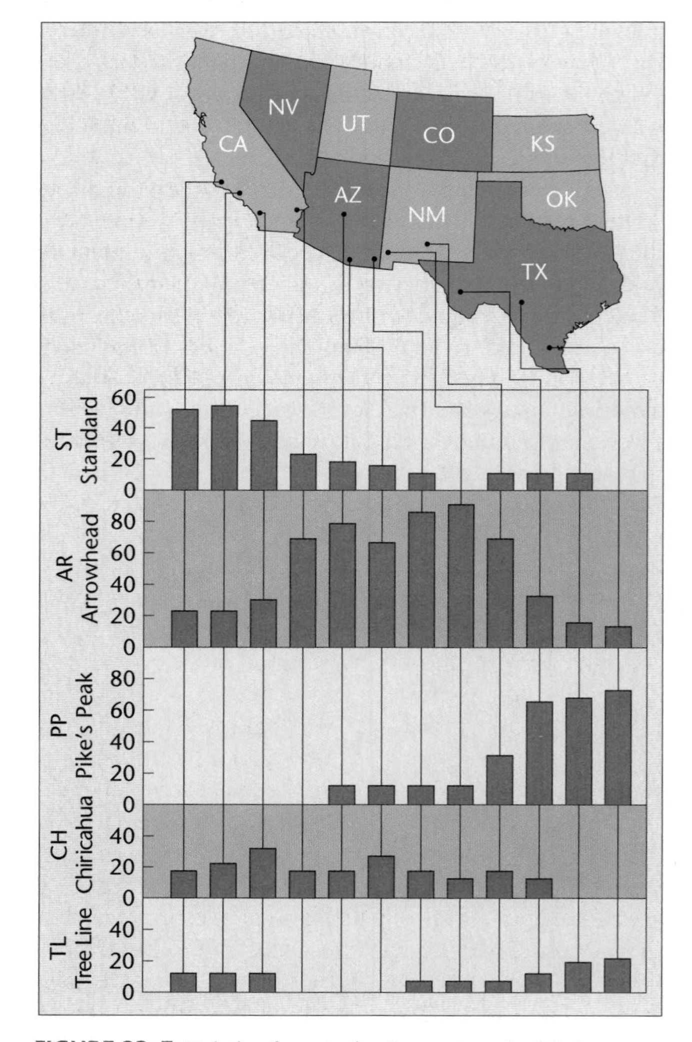

FIGURE 23–7 Relative frequencies (percentages) of 5 chromosomal inversions in *D. pseudoobscura* in 12 different geographic regions.

(a)

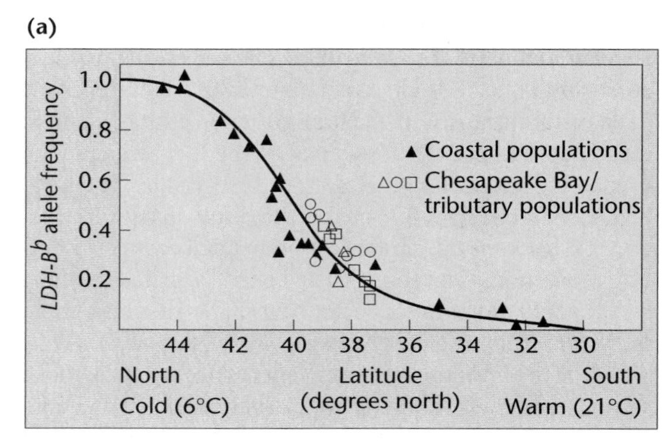

(b)

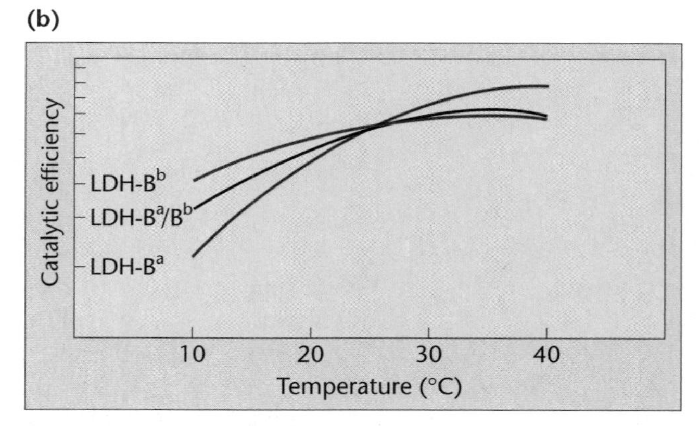

FIGURE 23–8 Variation in allele frequency of the *Ldh-B* gene among mummichog populations. (a) Frequencies of the *Ldh-B*^{b} allele in populations along the Atlantic coast of North America. (b) Catalytic efficiency of LDH-B allozymes as a function of temperature. *(From Powers, D. A., and Schulte, P. M. 1998. Evolutionary adaptations of gene structure and expression in natural populations in relation to a changing environment,* J. Exp. Zool. *Reprinted by permission of Wiley-Liss, Inc., a subsidiary of John Wiley & Sons, Inc; and with permission from* Ann. Rev. Genet. 25. *© 1991 by Annual Reviews.)*

high temperatures (Figure 23–8b). A mixture of the two forms has intermediate efficiency at all temperatures.

In addition to the functional differences between the LDH-B allozymes, the *Ldh-B* alleles have sequence differences in their regulatory regions. The *Ldh-B*^{b} allele's relative rate of transcription is more than twice that of the *Ldh-B*^{a} allele. As a result, northern fish have higher concentrations of the LDH-B enzyme in their cells.

The differences in catalytic efficiency and relative transcription rate associated with the *Ldh-B* allele are consistent with the hypothesis that mummichog populations are adapted to the temperatures at which they live. Mummichogs are ectotherms—their environment determines their body temperature. In general, low body temperatures slow an ectotherm's metabolic rate. The higher transcription rate of the *Ldh-B*^{b} allele and the superior low-temperature catalytic efficiency of its gene product appear to help northern fish compensate for the tendency of their cold environment to reduce their metabolic rate. The superior high-temperature catalytic efficiency and lower transcription rate associated with the *Ldh-B*^{a} allele appear to allow southern fish to economize on the resources devoted to glucose production and aerobic metabolism. Based on this and other evidence, Powers suggests that the differences among mummichog populations in *Ldh-B* allele frequencies are the result of natural selection.

23.5 Reduced Gene Flow, Selection, and Drift Can Form New Species

We have examined several examples illustrating that most populations harbor considerable genetic variation and that different populations within a species may have different alleles and/or allele frequencies at a variety of loci. This genetic divergence between populations within a species can be caused by natural selection, genetic drift, or both. We have seen that the migration of individuals and genes between populations homogenizes allele frequencies. In other words, migration counteracts the tendency of populations to diverge.

When gene flow between two populations is reduced or absent, the populations may diverge to the point where members of one population can no longer successfully interbreed with members of the other. When populations reach the point where they are reproductively isolated from one another, they have become different species. Barriers that prevent interbreeding between populations can be physiological, behavioral, or mechanical. The biological and behavioral factors that prevent or reduce interbreeding are called **reproductive isolating mechanisms** (Table 23.2).

Prezygotic isolating mechanisms prevent individuals from mating. Prezygotic mechanisms prevent individuals from different populations from finding each other at the right time, prevent them from recognizing each other as suitable mates, or impose mechanical barriers that prevent mating.

Postzygotic isolating mechanisms create reproductive isolation even when the members of two populations are willing and able to mate with each other. In these cases, genetic divergence has reached the stage where the viability or fertility of hybrids is reduced. Hybrid zygotes can be formed, but all or most will die before maturity. Alternatively, the hybrids may survive but are sterile or have reduced fertility. In another possibility, the hybrids themselves are fertile, but their progeny have lowered viability or fertility. These postzygotic mechanisms act at or beyond the level of the zygote and are generated by genetic divergence.

Postzygotic isolating mechanisms waste gametes and zygotes and lower the reproductive fitness of hybrid survivors. Selection therefore favors the spread of alleles that reduce or prevent formation of hybrids and lead to the development of prezygotic isolating mechanisms. These mechanisms prevent interbreeding and the formation of hybrid zygotes and offspring. In animal evolution, the most effective prezygotic mechanism is behavioral isolation, involving courtship behavior.

Observing Speciation

As discussed earlier, it is difficult to define exactly when a new species is formed. Reproductive isolation may or may not be accompanied by phenotypic changes or other external clues that speciation has occurred. We will focus on two studies of speciation: one from a laboratory study, the other from a field study.

Diane Dodd and her colleagues studied the evolution of digestive physiology in *D. pseudoobscura*. They collected flies from a wild population and established separate laboratory populations. Some of the laboratory populations were raised on starch-based medium, while others were raised on maltose-based medium. Flies do not use either of these food sources in nature, and both food sources were stressful for the flies. It was only after several months of selection that the populations adapted to their artificial diets and began to thrive.

Dodd's team wanted to know whether the starch-adapted populations and the maltose-adapted populations, which had been diverging under strong selection and in the absence of gene flow, had become different species. Roughly a year after the populations were established, a series of mating trials were performed to see if reproductive isolation was present. For each trial, 48 flies were placed in a mating chamber—12 males and 12 females from a starch-adapted population and 12 males and 12 females from a maltose-adapted population. Then the researchers simply noted which flies mated (Table 23.3).

If populations adapted to different media had speciated, then the flies would prefer to mate with members of their own population. If the populations had not speciated, then the flies would mate at random. About 600 of the 900 matings they observed occurred between males and females from the same population. In other words, the two populations showed some premating isolation. They concluded that the populations had started to speciate but had not yet completed the process.

TABLE 23.2 Reproductive Isolating Mechanisms

Prezygotic Mechanisms
Geographic or ecological
The populations occupy different habitats
Seasonal or temporal
The populations live in the same regions but are sexually mature at different times
Behavioral (only in animals)
The populations are isolated by different and incompatible behavior before mating
Mechanical
Cross-fertilization is prevented or restricted by differences in reproductive structures (genitalia in animals, flowers in plants)
Physiological
Gametes fail to survive in alien reproductive tracts
Postzygotic Mechanisms
Hybrid nonviability or weakness
Developmental hybrid sterility
Hybrids are sterile because gonads develop abnormally or meiosis breaks down before completion
Segregational hybrid sterility
Hybrids are sterile because of abnormal segregation into gametes of whole chromosomes, chromosome segments, or combinations of genes
F_2 breakdown
F_1 hybrids are normal, vigorous, and fertile, but the F_2 contains many weak or sterile individuals

Source: From Stebbins, G. Ledyard. *Processes of Organic Evolution*. 3rd ed. © 1977, p. 143. Reprinted by permission of Prentice Hall, Upper Saddle River, NJ.

TABLE 23.3 Incipient Speciation in Laboratory Populations of *D. pseudoobscura* (number of matings involving each kind of male–female pair)

	Female	
	Starch-Adapted	*Maltose-Adapted*
Male		
Starch-adapted	290	153
Maltose-adapted	149	312

Source: Compiled from Dodd, D. M. B. 1989. Reproductive isolation as a consequence of adaptive divergence in *Drosophila pseudoobscura*. *Evol.* 43: 1308–11.

FIGURE 23–9 A snapping shrimp, genus *Alpheus*. *(Carl C. Hansen/Nancy Knowlton/Smithsonian Institution Photo Services)*

Geological events often separate populations and prevent interbreeding between the isolated populations. One such event is the formation of the Isthmus of Panama, which formed about 3 million years ago, creating a land bridge between North and South America. Formation of this land bridge separated the Caribbean Sea from the Pacific Ocean. Nancy Knowlton and her colleagues took advantage of this natural experiment that isolated populations of several species of snapping shrimp (Figure 23–9). They identified seven Caribbean species and matched each of them with a corresponding Pacific species. Members of each species pair were more phenotypically similar to each other than to any other species in its own ocean, even though they had been separated for several million years. An analysis of allele frequencies and mitochondrial DNA sequences confirmed that the members of each pair were one another's closest genetic relatives. The Knowlton team interpreted these data to mean that prior to the formation of the isthmus, the ancestors of each species pair were a single species. When the isthmus was formed, the seven ancestral species divided into two separate populations: one in the Caribbean, the other in the Pacific.

Meeting in a dish in Knowlton's lab for the first time in 3 million years, would members of Caribbean and Pacific species pair recognize each other as suitable mates? To find out, Knowlton placed males and females together and observed mating behavior. She then calculated the relative inclination of Caribbean/Pacific pairs to mate versus that of Caribbean/Caribbean or Pacific/Pacific pairs. Three of the seven transoceanic pairs did not mate. The other four transoceanic pairs were only 33, 45, 67, and 86% as likely to mate with each other as were same-ocean pairs. For same-ocean pairs that mated, 60% of them produced viable eggs. Of the transoceanic couples that mated, only 1% produced viable eggs. We can conclude from these data that although 3 million years of separation did not produce wide phenotypic divergence, it resulted in complete or nearly complete speciation, involving strong pre- and postzygotic isolating mechanisms for all seven species pairs.

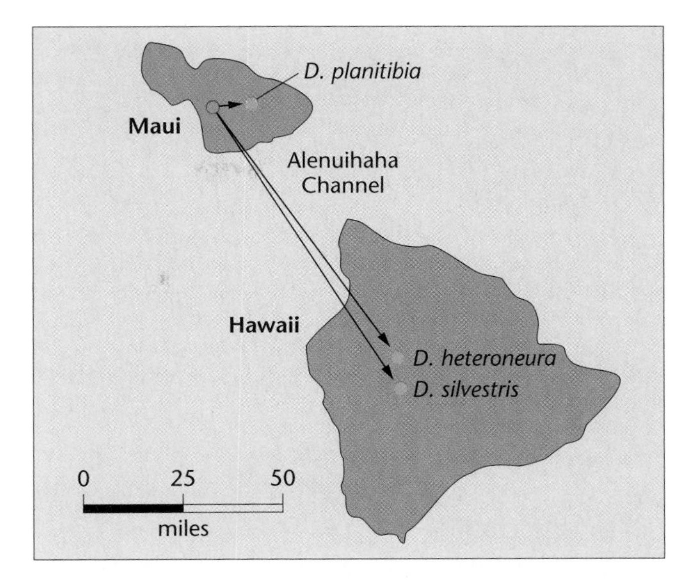

FIGURE 23–10 Proposed pathway for Hawaii's colonization by members of the *D. planitibia* species. The ancestral species colonized Hawaii from Maui. The pink circle represents a population ancestral to the three present-day species.

The Minimum Genetic Divergence Required for Speciation

These studies raise an important question. How much genetic divergence is required between two populations before they become different species? Let's consider two examples, an insect and a plant, which demonstrate that in some cases the answer is, not very much.

Researchers estimate that *Drosophila heteroneura* and *Drosophila silvestris*, found only on the island of Hawaii, diverged from a common ancestral species only about 300,000 years ago. They are thought to be descended from *Drosophila planitibia* colonists from the older island of Maui (Figure 23–10). These two species are separated from each other by different and incompatible courtship and mating behaviors (a prezygotic isolating mechanism), by morphology, and by body and wing pigmentation (Figure 23–11). In spite of their phenotypic and behavioral differences, there are almost no significant differences between these species in chromosomal inversion patterns or protein variants. DNA-hybridization studies using DNA from these species shows the sequence divergence between the two is only about 0.55% (Figure 23–12). Thus, nucleotide sequence diversity precedes the development of protein or chromosomal polymorphisms.

The DNA-hybridization data suggest that there are few genetic differences between these species. Genetic analysis suggests that the differences between *D. heteroneura* and *D. silvestris* are controlled by a relatively small number of genes. For example, as few as 15–19 major loci may be responsible for the morphological differences between the species. In other words, the process of speciation need only involve a small number of genes.

Studies using two closely related species of monkey flowers, a plant that grows in the Sierra Nevada, Rocky Mountain, and Cascade mountain ranges of the western United States, confirm that species can be separated by only a few genetic differences. One species, *Mimulus cardinalis* is fertilized by hummingbirds and does not interbreed with *Mimulus lewisii*, which is fertilized by bumblebees. H. D. Bradshaw and his colleagues studied genetic differences related to reproduction in the two species: flower shape, size, and color, and nectar production. For each trait, a difference in a single gene provided at least 25% of the variation observed among laboratory-created *M. cardinalis* × *lewisii* hybrids. *M. cardinalis* makes 80 times more nectar than *M. lewisii*, and a single

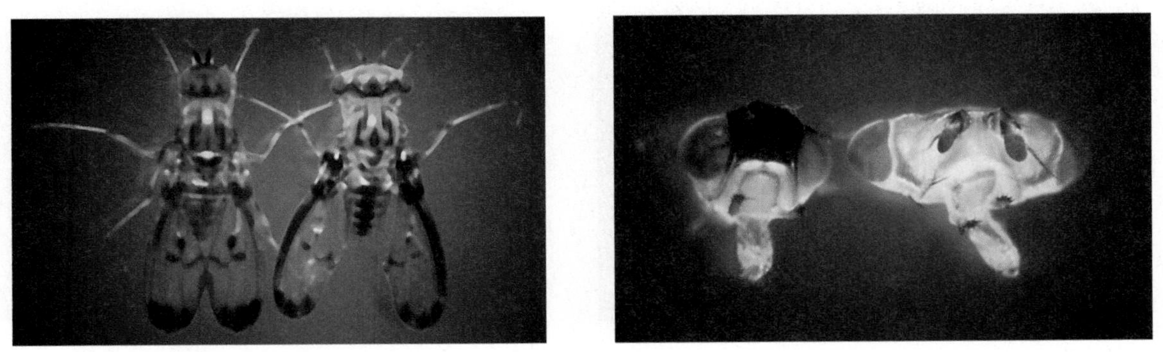

(a) (b)

FIGURE 23–11 (a) Differences in pigmentation patterns in *D. sylvestris* (left) and *D. heteroneura* (right). (b) Head morphology in *D. sylvestris* (left) and *D. heteroneura* (right). *(Kenneth Kaneshiro/University of Hawaii/CCRT)*

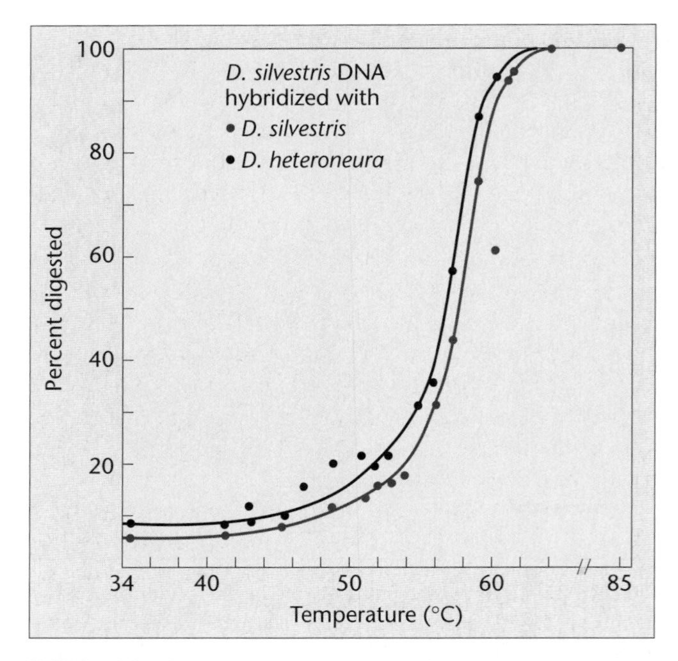

FIGURE 23–12 Nucleotide sequence diversity in the *D. planitibia* species complex. The very slight shift to the left that is observed in the hybrid molecules formed from the DNA of the two species indicates that the degree of nucleotide sequence divergence between the two species is very small.

gene is responsible for at least half the difference. A single gene also controls a large part of the differences in flower color between the two species (Figure 23–13). In this case, as in the Hawaiian *Drosophila*, species differences can be traced to a relatively small number of genes.

The Rate of Speciation

How much time is required for speciation? In many cases, speciation takes place over a long period of time. In other cases, however, speciation is surprisingly rapid. The Great Rift Valley lakes of east Africa (Figure 23–14) support hundreds of species of cichlid fish that are morphologically, behaviorally, and ecologically diverse. Cichlids are highly specialized for different niches (Figure 23–15)—some eat algae floating on the water's surface, others are bottom feeders, insect feeders, mollusk eaters, or predators on other fish species. Genetic analyses using several methods, including recombinant DNA technology, indicate that all of the species in a given lake are more closely related to each other than to species from other lakes. The implication is that most or all of the species in, say, Lake Tanganyika or Lake Victoria are descended from a single common ancestor and that they evolved within their home lake. It is thought that the cichlid species in Lakes Tanganyika, Malawi, and Victoria speciated faster than any other vertebrates in the history of our planet. This species radiation has been intensively studied to identify the biological, geological, and geographical factors involved in speciation.

Lake Tanganyika is the oldest of these lakes (9–12 million years old) and contains the oldest, most diverse, and most complex group of cichlid species. There are almost 200 species in the lake, assigned to 12 species clusters, called tribes. One tribe, the Lamprologini, contains almost 50% of the lake's species. To examine the origin of this and other tribes in Lake Tanganyika, Norihiro Okada and colleagues examined the insertion of a novel family of SINEs (short, interspersed repetitive elements) into the genomes of cichlid species. SINEs are transposable DNA sequences that insert into the genome at random, and integration of a SINE at a site in the genome is an irreversible event. If a SINE is present at the same locus in the genome of all species examined, this is strong evidence that all of these species descend from a common ancestor in which the insertion originally occurred. Using a SINE called AFC, Okada's team screened Lamprologini species and species from three other tribes. The AFC sequence was present only in species belonging to the Lamprologini, was present in all species in this tribe, and was inserted at the same site in all species (Figure 23–16). These results are strong evidence that the 100 or so species belonging to the Lamprologini are descended from a common ancestor (the tribe has a monophyletic origin).

In contrast to Lake Tanganyika, Lake Victoria formed about 250,000–750,000 years ago and contains about 500

FIGURE 23–13 Flowers of two closely related species of monkey flowers, (a) *M. cardinalis* and (b) *M. lewisii,* have very different phenotypes, but only a few genes are responsible for these differences. *(Courtesy of Toby Bradshaw and Doug W. Schemske, University of Washington. Photo by Jordan Rehm.)*

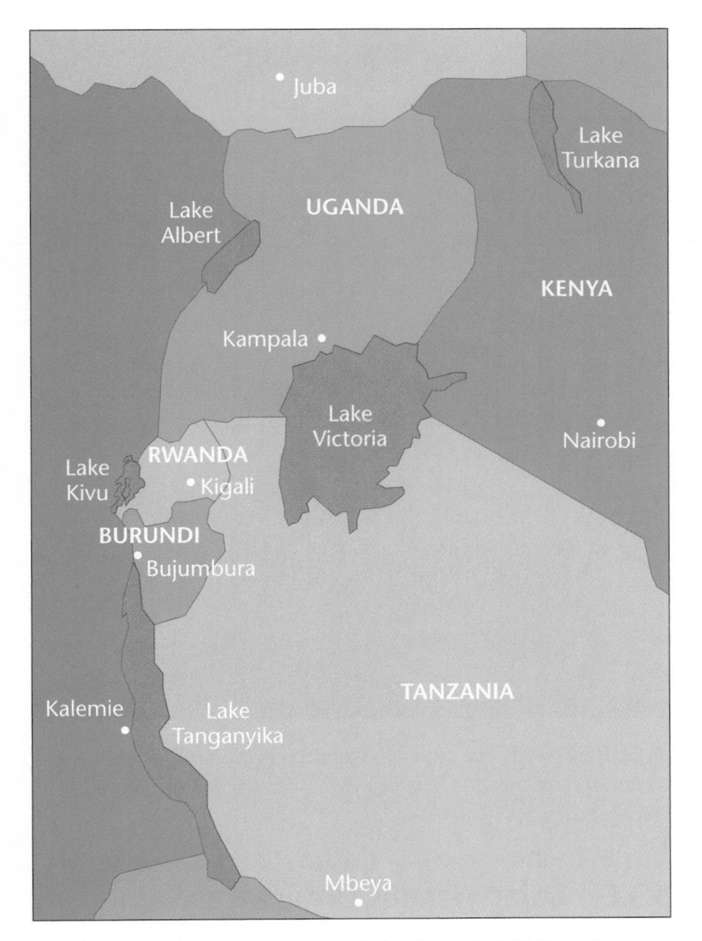

FIGURE 23–14 The great lakes of the Great Rift Valley of east Africa are home to hundreds of species of small fish, the cichlids. The origin of these species has been the subject of intense study.

species of cichlids. Studies using molecular markers suggest that all of these species evolved in Lake Victoria within the last 200,000 years. However, geological evidence indicates that the lake almost completely dried out about 14,000 years ago, raising the hotly debated question of whether the flock of 500 or so species found in the lake today could have evolved in 14,000 years from a single ancestral species.

A recent study by Axel Meyer and his colleagues explored this question using published mitochondrial sequence data and data collected from hundreds of cichlid species from Great Rift Valley lakes and rivers. These results indicate that the species flock in Lake Victoria originated in Lake Kivu, which at one time was connected to it. Two lineages from Lake Kivu, which diverged from a common ancestor about 100,000 years ago, are the ancestors of Lake Victoria cichlids. The split between Lake Kivu and Lake Victoria cichlid species took place about 35,000 years ago, at about the same time that geological changes separated the two lakes. The drying of Lake Victoria, which occurred about 14,000 years ago, led to the extinction of some species. After the lake refilled, there was a great increase in the numbers of individuals—but not a great increase in the number of species. In other words, although cichlid species in the Great Rift Valley lakes have evolved rapidly, the species flock in Lake Victoria is much older than 14,000 years.

Even faster speciation is possible through the mechanism of polyploidy. The formation of animal species by polyploidy is rare, but it has been an important factor in plant evolution. It is estimated that 50% of all flowering plant species have evolved by this mechanism. Polyploidy in ferns is even more common; up to 95% of fern species may have evolved by polyploidy. In these plant groups, species that have evolved by polyploidy are common and successful. They occupy different habitats from their parental species and may be superior in colonizing new niches. Interestingly, most agriculturally important crop plants are polyploid.

There are several types of polyploidy, one of which is allopolyploidy. Allopolyploid species are formed in two steps: Two species of plants interbreed to form a hybrid offspring, which undergoes chromosome doubling (the order of steps may be reversed). For example, if two plant species have chromosome sets designated *RR* and *SS* (where *R* and *S* represent the haploid set of chromosomes in each species), then the F_1 hybrid would have the chromosome constitution *RS*. Normally, such a plant would be sterile because few or no homologous chromosome

FIGURE 23–15 Cichlids occupy a diverse array of niches, and each species is specialized for a distinct food source. *(Dr. Paul V. Loiselle)*

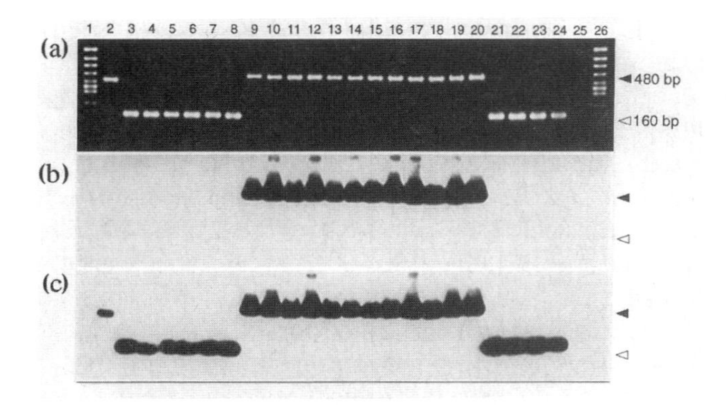

FIGURE 23–16 Tracing species relationships in cichlids from Lake Tanganyika using SINE DNA probes. (a) Photograph of an agarose gel showing PCR fragments generated from primers that flank the AFC family of SINEs. Large fragments containing the SINE are present in all samples of genomic DNA from members of the Lamprologini tribe of cichlids (lanes 9–20). DNA from the species in lane 2 has a similar but shorter fragment, which may represent another repetitive sequence. DNA from species belonging to other tribes (lanes 3–8 and 21–24) produce short fragments that do not contain SINEs. (b) A Southern blot of the gel from part (a) probed with DNA from the AFC family of SINEs. This SINE is present in DNA from all species in the Lamprologini tribe (lanes 9–20) but not in DNA from other species (lanes 3–8 and 21–24). (c) A second Southern blot of the gel in part (a) probed with the genomic sequence at the site of SINE insertion. All of the species examined (lanes 2–24) contain this sequence. The fragments in Lamprologini DNA (lanes 9–20) correspond to those in the previous Southern blot, showing the SINE inserted at the same site in all tribal species. These results are interpreted as showing a common origin for the species of this tribe. *(From Takahashi, K. et al. 1998. A novel family of short interspersed repetitive elements from cichlids: The pattern of insertion of SINEs at orthologous loci support the supposed monophyly of four major groups of cichlid fishes in Lake Tanganyika.* Molec. Biol. Evol. *15(4): 391–407. Figure 5 A, p. 400.)*

pairs exist, and aberrations would arise during meiosis. However, if the hybrid undergoes a spontaneous doubling of chromosome number, a tetraploid *RRSS* plant is produced. Doubling might occur during mitosis in somatic tissue, giving rise to a partially tetraploid plant that produces some fertile tetraploid flowers. Alternatively, aberrant meiotic events may produce *RS* gametes, which when fertilized yield *RRSS* zygotes. The *RRSS* plants are fertile because they possess homologous chromosomes producing viable *RS* gametes. This new, true-breeding tetraploid would have a combination of characters derived from the parental species and would be reproductively isolated from them because the F_1 hybrids are triploids and consequently sterile.

The tobacco plant *Nicotiana tabacum* ($2n = 48$) is the result of the doubling of the chromosome number in the hybrid between *Nicotiana otophora* ($2n = 24$) and *N. silvestris* ($2n = 24$). The origin of *N. tabacum* (Figure 23–17) is an example of virtually instantaneous speciation since it occurred in one generation.

FIGURE 23–17 The cultivated tobacco plant *N. tabacum* is the result of hybridization between two other species. *(C. B. and D. W. Frith/Bruce Coleman, Inc.)*

23.6 Using Genetic Differences to Reconstruct Evolutionary History

Earlier in this chapter, we noted that speciation is associated with changes in genetic structure of populations and genetic divergence. As a result, we can use genetic differences among species to reconstruct their evolutionary histories.

In an important early example of evolutionary reconstruction, W. M. Fitch and E. Margoliash assembled data on the amino acid sequence for **cytochrome c** in a variety of organisms in the 1960s. Cytochrome c is a protein in mitochondria, and its amino acid sequence has evolved very slowly. For example, the amino acid sequence in humans and chimpanzees is identical, and humans and rhesus monkeys show only one amino acid difference. This is remarkable considering that the fossil record indicates that the lines leading to humans and monkeys diverged from a common ancestral species approximately 20 million years ago.

Column (a) in Table 23.4 lists the number of amino acid differences between cytochrome c in humans and a variety of other organisms. The table is consistent with our intuitions about how related we are to these other species. For example, we intuitively know that we are more closely related to other mammals than to insects and that we are more closely related to insects than we are to fungi.

However, more than one nucleotide change may be required to affect a given amino acid, so the measure of amino acid differences may underestimate evolutionary differences. When all of the nucleotide changes necessary for all of the amino acid differences are totaled, it establishes the minimal mutational distance between the genes of any two species. Column (b) in Table 23.4 shows such an analysis for the genes encoding cytochrome c. As expected, these values are larger than the corresponding number of amino acids separating humans from the other nine organisms in the table.

TABLE 23.4 Amino Acid Differences and the Minimal Mutational Distances Between Cytochrome in Humans and Other Organisms

Organism	*(a)* *Amino Acid Differences*	*(b)* *Minimal Mutational Distance*
Human	0	0
Chimpanzee	0	0
Rhesus monkey	1	1
Rabbit	9	12
Pig	10	13
Dog	10	13
Horse	12	17
Penguin	11	18
Moth	24	36
Yeast	38	56

Source: From Fitch W. M., and Margoliash E. 1967. Construction of phylogenetic trees. *Science* 155: 279–84. © 1967 by the American Association for the Advancement of Science.

Fitch and Margoliash used data on the **minimal mutational distances** between the cytochrome c genes of 19 organisms to reconstruct their evolutionary history. The result is an evolutionary tree, or **phylogeny**, that unites the species (Figure 23–18). The black dots at the tips of the branches represent species that are in existence today. These species are connected to the inferred common ancestors, represented by red dots. The ancestral species evolved and diverged to produce modern organisms. The common ancestors are connected to still earlier common ancestors, culminating in a single common ancestor for all of the species on the tree, represented by the red dot on the extreme left of the figure.

Constructing Evolutionary Trees

There are several methods geneticists use to construct phylogenies, but it is beyond the scope of this chapter to review all of them. We will discuss one method, called the *u*nweighted *p*air *g*roup *m*ethod using *a*rithmetic averages or **UPGMA**, to illustrate how genetic differences are used to reconstruct evolutionary history. UPGMA is not the most powerful or commonly used method, but it is, in spite of its name, intuitively straightforward. Furthermore, UPGMA works reasonably well under many circumstances. The underlying assumption in this method is that the more recently two species have shared a common ancestor, the more genetically similar they will be.

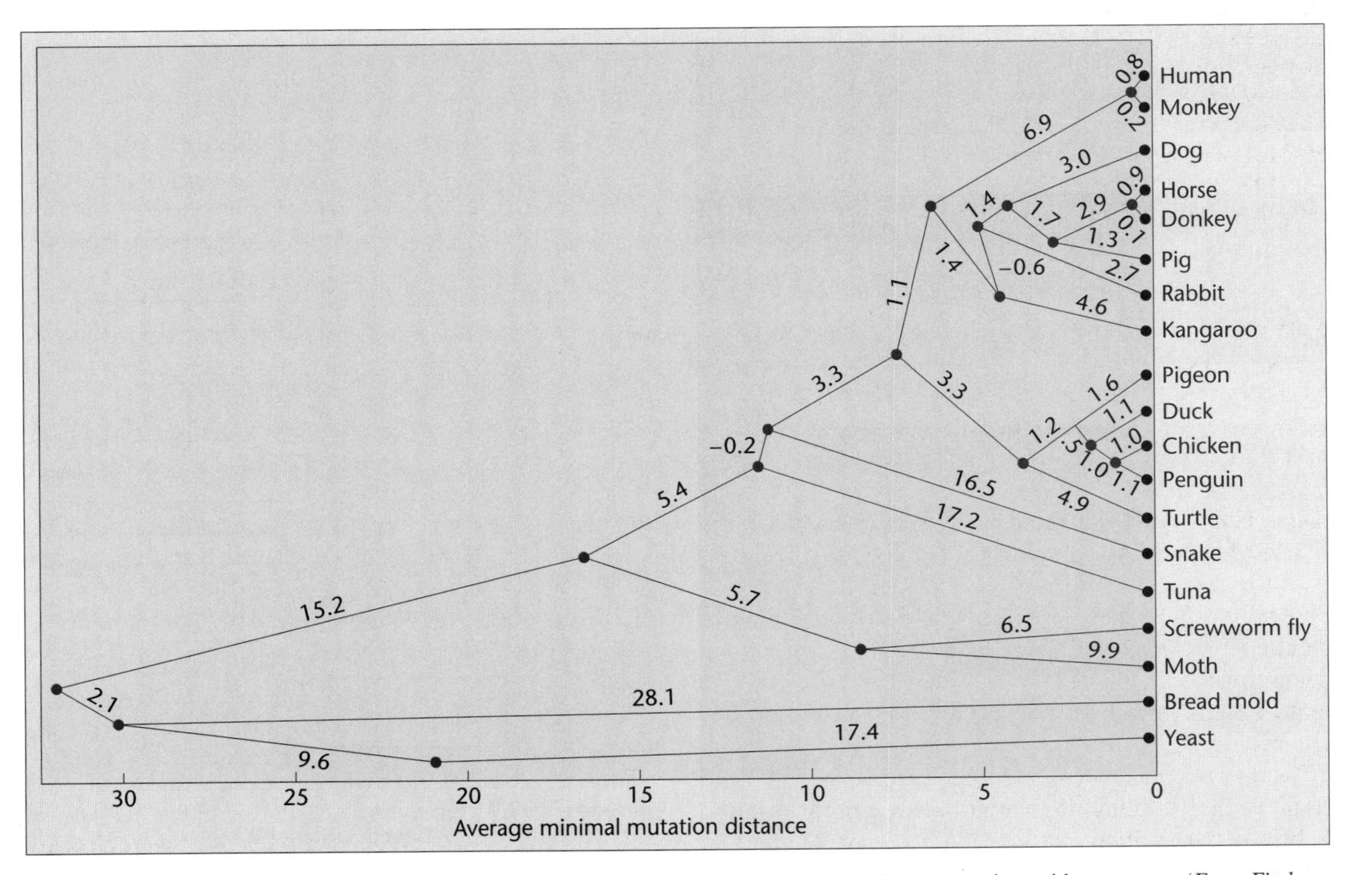

FIGURE 23–18 Phylogeny of eukaryotes constructed by comparing homologies in cytochrome c amino acid sequences. *(From Fitch, W. M., and Margoliash E. 1967. Construction of phylogenetic trees.* Science. *155: 279–84, © 1967 by the American Association for the Advancement of Science.)*

The starting point for UPGMA is a table of genetic distances among a group of species. To demonstrate UPGMA, we'll use data from DNA-hybridization studies by Charles Sibley and Jon Alquist (Figure 23–19). They computed genetic distances among humans and four species of apes: chimpanzee, gorilla, siamang, and gibbon.

The method involves three steps:

1. Search the table for the smallest genetic distance between any pair of species. In Figure 23–19a, this value is 1.628, the distance between human and chimpanzee. Once identified, this species pair is placed on neighboring branches of an evolutionary tree (Figure 23–19b). The length of each branch is half the distance between the species (1.628/2), so the branches connecting human and chimpanzee to their common ancestor are 0.81 genetic distance units.
2. Recalculate the table of genetic distances between this species pair and the other species (Figure 23–19c). The genetic distance between the human–chimpanzee (Hu–Ch) cluster and all other species is the average of the distance between each cluster member and the other species. For example, the genetic distance between the human–chimp cluster and gorilla is the average of the human–gorilla distance and the chimp–gorilla distance. This is $(2.267 + 2.21)/2 = 2.2385$.
3. The smallest distance in the recalculated table is 1.95, the distance between siamang and gibbon (Figure 23–19c). To build the next section of the tree, the siamang and gibbon are placed on neighboring branches, with branch lengths equal to 1/2 of 1.95, or 0.98 (Figure 23–19d).

Repeat steps 1–3 and recalculate the table again. There are two clusters plus gorilla (Figure 23–19e). The genetic distance between the human–chimp cluster and the siamang–gibbon cluster is the average of four distances: human–siamang, human–gibbon, chimp–siamang, and chimp–gibbon. Now the smallest genetic distance is 2.239, the distance between the human–chimp cluster and gorilla. We therefore add a gorilla branch to the tree and connect it to the common ancestor of the human–chimp cluster (Figure 23–19f). The branches are drawn so that the distance between the tips of any two branches in the human–-chimp–gorilla cluster is 2.239.

Redoing the table for the final time (Figure 23–19g), we calculate the genetic distance between the human–chimp–gorilla cluster and the siamang–gibbon cluster as the average

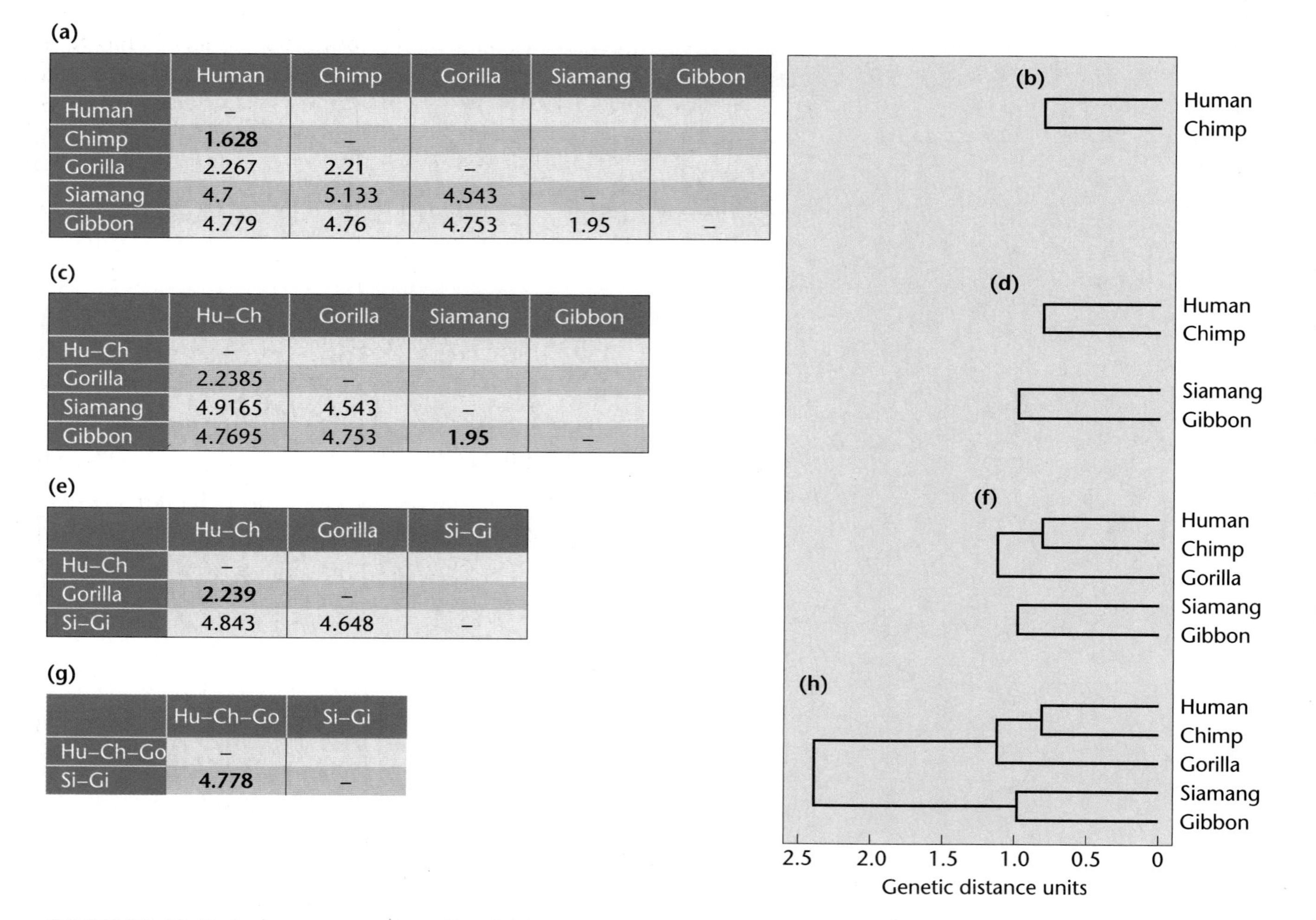

(a)

	Human	Chimp	Gorilla	Siamang	Gibbon
Human	–				
Chimp	**1.628**	–			
Gorilla	2.267	2.21	–		
Siamang	4.7	5.133	4.543	–	
Gibbon	4.779	4.76	4.753	1.95	–

(c)

	Hu–Ch	Gorilla	Siamang	Gibbon
Hu–Ch	–			
Gorilla	2.2385	–		
Siamang	4.9165	4.543	–	
Gibbon	4.7695	4.753	**1.95**	–

(e)

	Hu–Ch	Gorilla	Si–Gi
Hu–Ch	–		
Gorilla	**2.239**	–	
Si–Gi	4.843	4.648	–

(g)

	Hu–Ch–Go	Si–Gi
Hu–Ch–Go	–	
Si–Gi	**4.778**	–

FIGURE 23–19 Phylogeny reconstruction of hominid primates using UPGMA. This phylogeny shows likely evolutionary relationships, but establishes no time scale. *(From Sibley, C.G. and Ahlquist, J.E. 1987. DNA hybridization evidence of hominoid phylogeny: results from an expanded data set.* J. Mol. Evol.*26: 99–121. Table 1, p. 101.)*

of six genetic distances: human–siamang, human–gibbon, chimp–siamang, chimp–gibbon, gorilla–siamang, and gorilla–gibbon. This distance is 4.778, which allows us to complete our evolutionary tree (Figure 23–19h).

Our evolutionary tree indicates that humans and chimpanzees are one another's closest relatives. That is not to say that humans evolved from chimpanzees. Rather, this analysis indicates that humans and chimpanzees share a more recent common ancestor than either share with other species on the tree.

The chief shortcoming of UPGMA is that it does not account for the possibility that the rate of evolution may vary for the different groups of species. For example, how much better does the tree we just created fit our data than a tree that shows chimpanzees and gorillas as closest relatives?

Other methods can be used to search the set of all possible trees connecting a group of species and calculate the relative performance of each. One of these, called **parsimony** analysis, compares trees using the minimum number of evolutionary changes each requires, then selects the simplest possible tree. Another set of tools, called **maximum likelihood** methods, starts with a model of the evolutionary process, calculates how likely it is that evolution will produce each possible tree under the model, and selects the most likely tree.

It is important to keep in mind that a phylogeny produced from data on genetic distances is not necessarily the way species actually evolved, but is instead the most reasonable estimate of their evolutionary history, based on the method that was used. Phylogenies based on combined data sets from several independent loci are usually more reliable than those based on a single locus or protein.

Molecular Clocks

In many cases, we would like to estimate not only which species are most closely related but also when their common ancestors lived. Sometimes we can do so, thanks to **molecular clocks**. Molecular clocks are amino acid sequences or nucleotide sequences in which evolutionary changes accumulate at a constant rate over time.

Research by Fitch and colleagues on the influenza A virus shows how molecular clocks are used. They sequenced part of the hemagglutinin gene from flu viruses isolated at different times over a 20-year period. They calculated the number of nucleotide differences among the copies of the gene from these viruses and constructed an evolutionary tree (Figure 23–20b). Note that most strains of the virus have become extinct; they have no descendents among the more recently isolated strains. Fitch's team then plotted the number of nucleotide substitutions between the first isolate and each subsequent isolate against the year in which the strain was isolated (Figure 23–20a). The points fall very close to a straight line, indicating that nucleotide substitutions in this gene have accumulated at a steady rate—the hemagglutinin gene thus serves as a molecular clock. Molecular clocks are used to compare the sequences of new flu viruses as

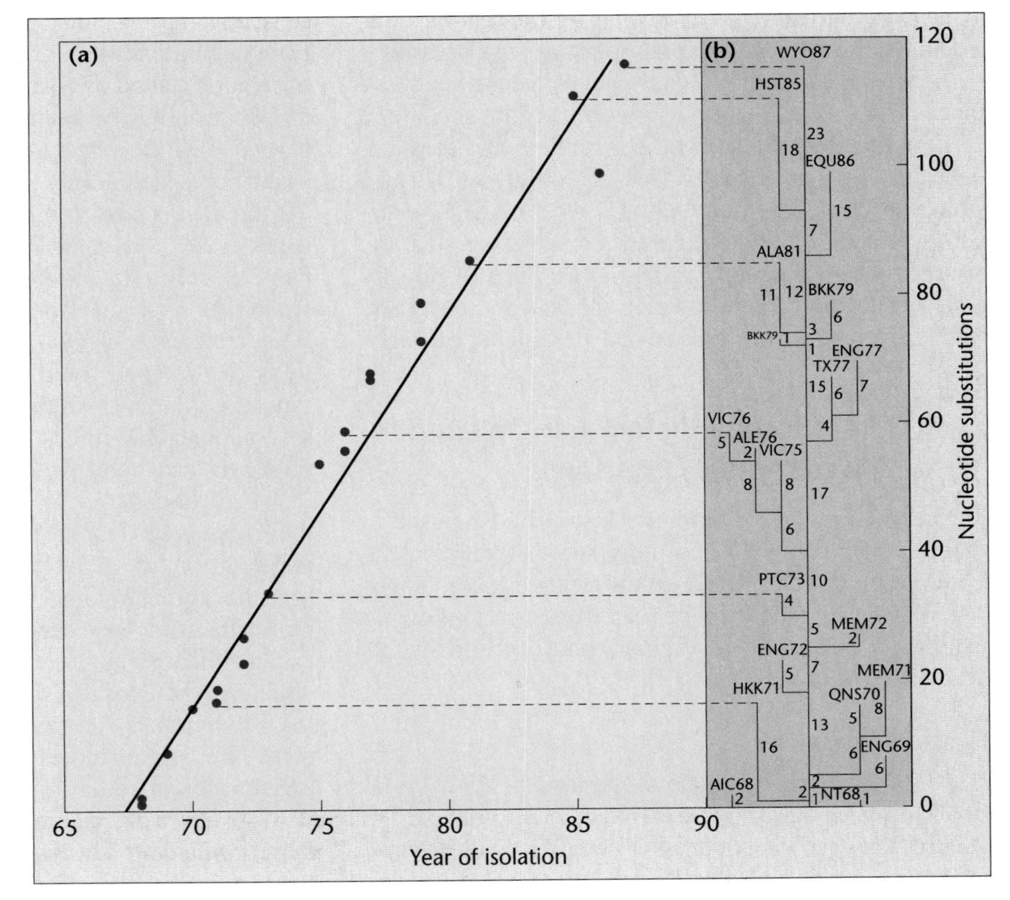

FIGURE 23–20 In this molecular clock of the influenza A hemagglutinin gene, (a) the number of nucleotide differences between the first isolate and each subsequent isolate is shown as a function of the year of isolation. (b) Estimate of the phylogeny of the isolates. *(From Fitch, W. M. et al. 1991. Positive Darwinian evolution in human influenza A viruses.* Proc. Natl. Acad. Sci. [USA] *88: 4270–73.)*

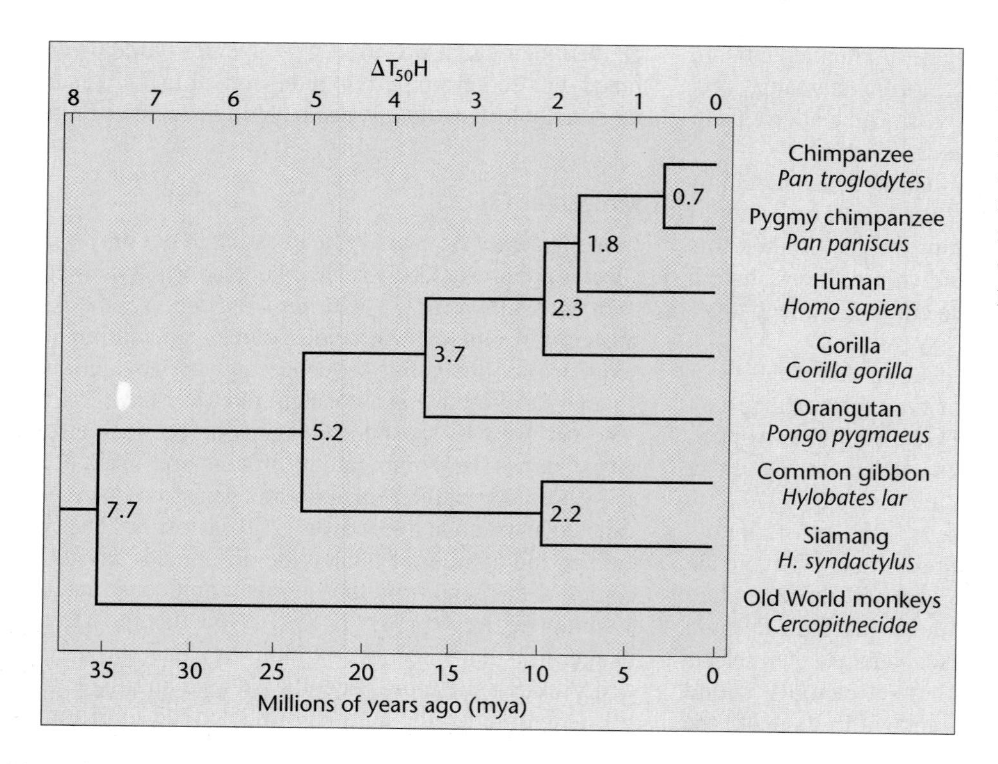

FIGURE 23–21 Phylogeny of hominoid primates and Old World monkeys as estimated by DNA hybridization. The time scale derived from the fossil record along with the degree of nucleotide divergence is used to date the evolutionary branch points.

they appear each year and estimate the time that has passed since each diverged from their common ancestor.

Molecular clocks must be carefully calibrated and used with caution. Fitch's data indicate that strains of influenza A that jump from birds to humans have evolved much more rapidly than strains that remain in birds. Thus, a molecular clock calibrated from human strains of the virus would be highly misleading if applied to bird viruses.

Phylogenies can combine data from molecular and fossil studies. Figure 23–21 shows a phylogeny based on Sibley's data for genetic differences among humans and apes (the same data we used in our UPGMA calculations). Using a molecular clock based on the fossil record, this information is placed on a time scale that estimates ages for the common ancestors of each species cluster. Sibley's tree suggests that the most recent common ancestor of humans and chimps lived between 5 million and 10 million years ago.

23.7 Evolutionary History Can Be Used to Answer Many Questions

Reconstructions of evolutionary history can be used to address a wide range of questions, some of which relate to evolution directly, while others deal with contemporary issues and even criminal activity. In the following sections, we consider several examples of how evolutionary trees can be used.

Transmission of HIV

In late 1986, a Florida dentist tested positive for HIV. Several months later, he was diagnosed with AIDS. He continued to practice dentistry for two more years, until one of his patients, a young woman with no obvious risk factors, discovered that she was infected with HIV. When the dentist publicly urged his other patients to have themselves tested, several more were found to be HIV-positive. How did these patients become infected? Did they get HIV from the dentist, or did the patients become infected by some other means?

At first glance, this situation appears to be a long way from a discussion of evolution; however, the movement of a virus from one individual to another is similar to a new island population started by a small number of migrants. In such a case, gene flow between the ancestral population and the new population is nonexistent, but the genetic history of the island population links it to its ancestral population.

If the dentist passed his HIV infection to his patients, then the HIV strains isolated from the patients should be more closely related to each other and to the dentist's strain than to strains from other individuals living in the same area. Chin-Yih Ou and colleagues sequenced portions of the gene for the envelope protein from HIV viruses collected from the dentist, 9 of his patients, and several other HIV-infected individuals from the area who served as local controls.

An evolutionary tree for the HIV isolates produced from these sequence data is shown in Figure 23–22. Viruses isolated from the study participants are at the tips of the branches; branch points join these to the inferred common ancestors. HIV samples from patients A, B, C, E, and G all share a more recent common ancestor with each other and the dentist's strain than with HIV from other local individuals. The evolutionary relationship among these HIV strains indicates that the dentist did, indeed, transmit his infection to these patients. In contrast, the HIV strains from patients D, F, H, and J are all more closely related to strains from local controls (LC) than they are to the dentist's strain. These patients appear to

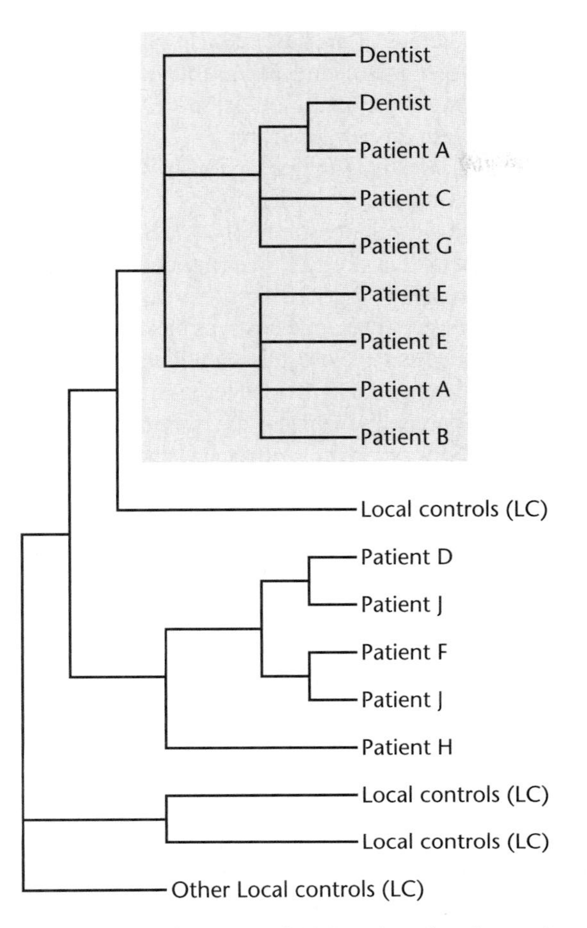

FIGURE 23–22 A phylogeny of HIV strains taken from a dentist, his patients, and several local controls (LC). The viral strains in the shaded portion of the phylogeny are all derived from a common ancestral strain, indicating that the dentist infected some, but not all of his patients. *(Redrawn from Hillis. 1998.* Curr. Biol. *7: R129–R131.)*

have gotten their infection from someone other than the dentist. Patient J, in fact, seems to have acquired his infection from two different sources.

Neanderthals and Modern Humans

Paleontological evidence indicates that the Neanderthals, *Homo neanderthalensis*, lived in Europe and western Asia from some 300,000–30,000 years ago.

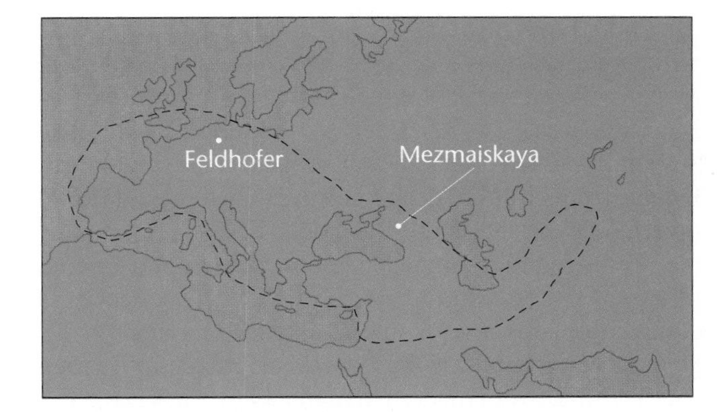

FIGURE 23–23 A map of Europe showing the sites where Neanderthal remains used in phylogenetic analysis were obtained. The dotted line indicates the areas known to have been occupied by Neanderthals. *(Redrawn from Ovchinnikov, I. V. et al. 2000. Molecular analysis of Neanderthal DNA from the northern Caucasus.* Nature *404: 490–93, http://www.nature.com).*

For at least 30,000 years, the Neanderthals coexisted with anatomically modern humans (*H. sapiens*) in several areas. Several questions about Neanderthals and modern humans remain unresolved. (1) Can Neanderthals be regarded as direct ancestors to modern humans? (2) Did *H. neanderthalensis* and *H. sapiens* interbreed, so that the descendants of Neanderthals are alive today? (3) Did the Neanderthals die off and become extinct?

To resolve these questions, in 1997, Svante Pääbo, Matthias Krings, and their colleagues extracted fragments of mitochondrial DNA from a Neanderthal skeleton found in Feldhofer cave near Düsseldorf, Germany (Figure 23–23). After confirming that they had isolated Neanderthal gene fragments, the Neanderthal sequences were placed on a phylogeny with over 2000 modern humans (Figure 23–24a). Based on this analysis, Neanderthals appear to be a distant relative of modern humans. Using a molecular clock calibrated with chimpanzees and humans, the researchers calculated that the last common ancestor between Neanderthals and modern humans lived roughly 600,000 years ago—over 400,000 years before the appearance of modern humans. While this study was important, considerable caution is required when drawing conclusions based on a single Neanderthal specimen.

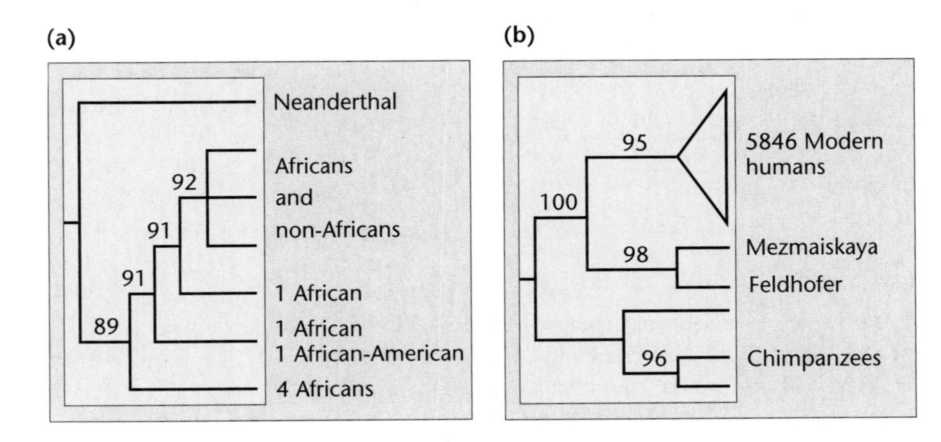

FIGURE 23–24 (a) A phylogeny estimated from mitochondrial DNA sequences of one Neanderthal and over 2000 modern humans. (b) A phylogeny of the Feldhofer and Mezmaiskaya Neanderthal specimens compared with over 5000 modern humans. *(Part (a) from Krings, M. et al. 1997.* Cell *90: 19–30. Figure 7A, p. 26. © 1997 Cell Press; part (b) from Ovchinnikov, I. V. et al. 2000. Molecular analysis of Neanderthal DNA from the northern Caucasus.* Nature *404: 490–93.)*

In 2000, Igor Ovchinnikov and his colleagues analyzed mitochondrial DNA recovered from a set of Neanderthal remains discovered in Mezmaiskaya cave in the Caucasus Mountains east of the Black Sea (see Figure 23–23). Although the two Neanderthal sequences are from individuals from different geographic regions more than 1000 miles apart, they vary by only about 3.5%, suggesting that they belonged to a single gene pool. Further, the amount of variation between the two Neanderthal sequences is comparable to that seen among modern humans. Phylogenetic analysis places the two Neanderthals in a group that is distinct from modern humans (Figure 23–24b). The conclusion here is that although Neanderthals and humans have an ancient common ancestor, Neanderthals were a separate hominid line and did not contribute mitochondrial genes to *H. sapiens*. Taken together, these studies suggest that when the Neanderthals disappeared, their lineage ended with them.

The Origin of Mitochondria

Mitochondria are eukaryotic organelles that carry their own genome as a circular DNA molecule. The human mitochondrial genome encodes 13 proteins required for oxidative phosphorylation and ATP production, 22 tRNA molecules, and 2 rDNA genes. The organization and function of the mitochondrial genome bears many similarities to that of bacterial genomes. Based on this similarity, mitochondria are thought to derive from bacteria. In other words, they are descended from free-living prokaryotic organisms that became intracellular symbionts in host cells. Once it incorporated into a cell with a nucleus, the mitochondrial genome became smaller over time, and the reduced genome we now see in mitochondria is thought to be a product of gene loss as well as the transfer of some genes to the nucleus of the host cell. However, many questions about the origin and evolution of mitochondria remain. Among these is the question of which group of bacteria they are descended from.

Researchers investigating this area have used information derived from nucleotide sequencing of the mitochondrial genes for small-subunit ribosomal RNAs (SSU rRNA). These sequences are under strong functional constraint, and are highly conserved. Phylogenetic analysis of these sequences identified a group of bacteria known as the α-proteobacteria (purple bacteria) as the closest living relatives of mitochondria. More recent work has divided the α-proteobacteria into two subdivisions, one of which is the rickettsial subdivision. Based on the SSU phylogeny, rickettsia have been identified as the bacteria most closely related to mitochondria. Interestingly, rickettsia, like mitochondria, live only inside eukaryotic cells. Unlike mitochondria, however, rickettsia are disease-causing parasites, responsible for typhus, one of the most serious diseases in the history of our species.

Genomic sequencing of one rickettsial species, *Rickettsia prowazekii*, provides additional insight into the relationship between these bacteria and mitochondria. Siv Andersson and colleagues constructed a phylogenetic tree (Figure 23–25) using the genome sequence information for genes involved in ATP synthesis derived from *R. prowazekii*, other bacteria, and mitochondria from several sources. Their tree indicates a close evolutionary relationship between *R. prowazekii* and mitochondria. Evidence from SSU analysis results in a similar tree. In fact, the rickettsia genome has a more recent common ancestor with mitochondrial genomes than with other bacterial genomes. Thus, two lines of proof, from SSUs and proteins involved in ATP synthesis, provide strong evidence that the mitochondria we carry in our cells came from an ancient ancestor of the rickettsia.

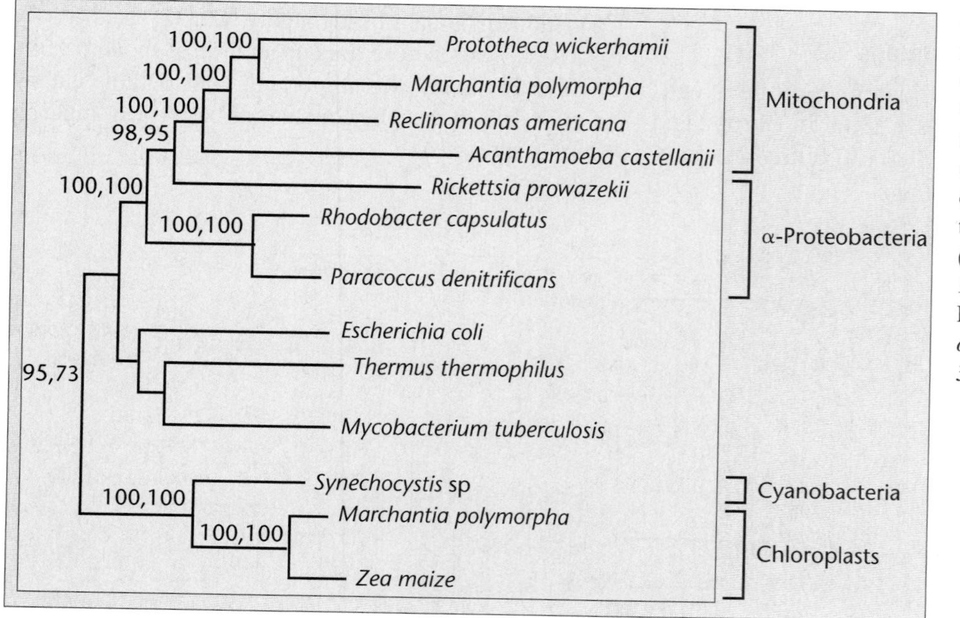

FIGURE 23–25 The evolutionary relationships among mitochondria, α-proteobacteria, cyanobacteria, and chloroplasts. Note that although *R. prowazekii* is an α-proteobacterium, it is more closely related to mitochondria than to other α-proteobacteria. *(From Andersson, S. G. E. et al. 1998. The genome sequence of* Rickettsia prowazekii *and the origin of mitochondria.* Nature *396: 133–43.)*

GENETICS, TECHNOLOGY, AND SOCIETY

What Can We Learn from the Failure of the Eugenics Movement?

The eugenics movement had its origins in the ideas of the English scientist Francis Galton, who became convinced from his study of the appearance of geniuses within families (including his own) that intelligence is inherited. Galton concluded in his 1869 book *Hereditary Genius* that it would be "quite practicable to produce a highly-gifted race of men by judicious marriages during several consecutive generations." The term eugenics, coined by Galton in 1883, refers to the improvement of the human species by such selective mating. Once Mendel's principles were rediscovered in 1900, the *eugenics* movement flourished.

Eugenicists believed that a wide range of human attributes were inherited as Mendelian traits, including many aspects of behavior, intelligence, and moral character. Their overriding concern was that the presumed genetically "feeble–minded" and immoral in the population were reproducing faster than the genetically superior and that this differential birthrate would result in the progressive deterioration of the intellectual capacity and moral fiber of the human race. Several remedies were proposed. *Positive eugenics* called for the encouragement of especially "fit" parents to have more children. However, a central goal of the eugenicists was the *negative eugenics* approach aimed at discouraging the reproduction of the genetically inferior or, better yet, eliminating it altogether.

In the United States, the eugenics movement enjoyed widespread popular support for a time and had a significant impact on public policy. Partially at the urging of prominent eugenicists, 30 states passed laws compelling the sterilization of criminals, epileptics, and inmates in mental institutions; most states enacted laws invalidating marriages between the "feebleminded" and others who were considered to be eugenically unfit. The crowning legislative achievement of the eugenics movement, however, was the passage of the Immigration Restriction Act of 1924, which severely limited the entry of immigrants from eastern and southern Europe because of their perceived mental inferiority.

Throughout the first two decades of the twentieth century, most geneticists passively accepted the views of eugenicists, but by the 1930s critics recognized that the goals of the eugenics movement were determined more by racism, class prejudice, and anti-immigrant sentiment than by sound genetics. Increasingly, prominent geneticists began to speak out against the eugenics movement, among them William Castle, Thomas Hunt Morgan, and Hermann Muller. When the horrific extremes to which the Nazis took eugenics became known, a strong reaction developed that all but ended the eugenics movement.

Paradoxically, the eugenics movement arose at the same time basic Mendelian principles were being developed, principles that eventually undermined the theoretical foundation of eugenics. Today, any student who has completed an introductory course in genetics should be able to identify several fundamental mistakes the eugenicists made.

1. They assumed that complex human traits such as intelligence and personality were strictly inherited, completely disregarding any environmental contribution to the phenotype. Their reasoning was that because certain traits ran in families, the traits must be genetically determined.
2. They assumed that these complex traits were determined by single genes with dominant and recessive alleles. This belief persisted despite research showing that multiple gene pairs contribute in complex ways to many phenotypes.
3. They assumed that a single ideal genotype existed in humans. Presumably, this genotype must be highly homozygous to be sustainable. This precept runs counter to current evidence suggesting that a high level of heterozygosity is present in most populations.
4. They assumed that the frequency of recessively inherited defects in the population could be significantly lowered by preventing homozygotes from reproducing. In fact, for recessive traits that are relatively rare, most of the recessive alleles in the population are carried by asymptomatic heterozygotes who are spared from such selection. Negative eugenic practices, no matter how harsh, are relatively ineffective at eliminating such traits.
5. They assumed that those deemed genetically unfit in the population were outreproducing those thought to be genetically fit. This is the exact reverse of the Darwinian concept of fitness, which equates reproductive success with fitness. (Galton should have understood this since he was Darwin's first cousin!)

More than seven decades have passed since the eugenics movement was in full bloom. We now have a much more sophisticated understanding of genetics, as well as a greater awareness of its potential misuses. But the application of current genetic technologies makes possible a "new eugenics" movement with a scope and power that Francis Galton could not have imagined. In particular, prenatal genetic screening and *in vitro* fertilization enable us to influence the selection of children according to their genotype, a power that will dramatically increase as more and more genes are associated with inherited diseases and perhaps even behaviors.

As we move into this new genetic age, we must not forget the mistakes made by the early eugenicists. We should remember that phenotype is a complex interaction between the genotype and the environment, and we must not lapse into a new hereditarianism that treats a person as only a collection of genes. We need to keep in mind that many genes may contribute to a particular phenotype, whether a disease or a behavior, and that the alleles of these genes may interact in unpredictable ways. We must not fall prey to the assumption that there is an ideal genotype. The success of all populations in nature is correlated with genetic diversity. Most of all, we must not use genetic information to advance ideological goals. We may find that there is a fine line between the legitimate uses of genetic technologies such as having healthy children and other eugenic practices. It will be up to us to decide exactly where the line falls.

References

Allen, G. E., Jacoby, R., and Glauberman, N. (eds). 1995. Eugenics and American social history, 1880–1950. *Genome* 31: 885–89.

Hartl, D. L. 1988. *A Primer of Population Genetics*, 2nd ed. Sunderland: Sinauer.

Kevles, D. J. 1985. *In the Name of Eugenics: Genetics and the Uses of Human Heredity*. Berkeley: Univ. of California Press.

Chapter Summary

1. Today's organisms are the products of an evolutionary history that includes the transformation, splitting, and divergence of species. Alfred Russel Wallace and Charles Darwin formulated the theory of natural selection, which provides a mechanism for the transformation of species. The genetic basis of evolution and the role of natural selection in changing allele frequencies were discovered early in the twentieth century.
2. When geneticists began studying the genetic structure of populations, they discovered that most populations harbor considerable genetic diversity. This diversity is apparent in electrophoretic studies of proteins and at the level of DNA sequences. Whether the genetic diversity of populations is maintained primarily by mutation plus genetic drift or by natural selection is a matter of some debate.
3. The geographic ranges of most species encompass a degree of environmental diversity. As a result of both adaptation to different environments and genetic drift, different populations within a species may have different alleles and/or allele frequencies at many loci.
4. Gene flow among populations homogenizes their genetic composition. When gene flow is reduced, genetic drift and adaptation to different environments can cause populations to diverge. Eventually populations may become so different that the individuals in one population either will not or cannot mate with the individuals in the other. At this point, the divergent populations have become different species.
5. Because speciation is associated with genetic divergence, we can use the genetic differences among species to infer their evolutionary history. By comparing amino acid or nucleotide sequences, we can determine the genetic distances among species, which are then used to reconstruct evolutionary trees. The simplest methods for reconstructing phylogenies assume that the least divergent species are one another's closest relatives.
6. The reconstruction of evolutionary trees is a key technique in answering a surprising number of interesting questions. Examples include tracing the transmission route in diseases, deciphering recent events in human evolution, and determining evolutionary relationships among all organisms.

Key Terms

allozymes, 530
anagenesis 529
cladogenesis, 529
cytochrome, 540
evolution, 530
maximum likelihood, 543
minimal mutational distance, 541
molecular clock, 543
neutral theory of molecular evolution, 532
parsimony, 543
phyletic evolution, 529
phylogeny, 541
postzygotic isolating mechanisms, 535
prezygotic isolating mechanisms, 535
reproductive isolating mechanism, 535
selectionist, 532
speciation, 529
species, 529
stasis, 529
UPGMA, 541

INSIGHTS AND SOLUTIONS

1. Sequence analysis of DNA can be accomplished by several techniques. Protein sequencing, on the other hand, is made more complex by the fact that 20 different subunits must be unambiguously identified and enumerated rather than just 4 nucleotides of DNA. Because of their unique properties, the N-terminal and C-terminal amino acids in a protein are easy to identify, but the array between them offers a difficult challenge since many proteins contain hundreds of amino acids. So how is protein sequencing accomplished?

Solution: The strategy for protein sequencing is the same as for DNA sequencing—divide and conquer. To accomplish this, specific enzymes are used that reproducibly cleave proteins between certain amino acids. These different enzymes produce overlapping fragments. Each fragment is isolated, and its amino acid sequence is determined by chemical means. Sequences from overlapping fragments are then assembled to give the sequence for the entire protein. Alternatively, researchers first sequence the gene that encodes the protein, then infer the protein's amino acid sequence from the genetic code.

2. A single plant twice the size of others in the same population suddenly appears. Normally, plants of this species reproduce by self-fertilization and by cross-fertilization. Is this new giant plant simply a variant or a new species? How would you determine this?

Solution: One of the most widespread mechanisms of speciation in higher plants is polyploidy, the multiplication of entire chromosome sets. The result of polyploidy is usually a larger plant with larger flowers and seeds. There are two ways to test this new variant to determine whether it is a new species. First, the giant plant is crossed with a normal-sized plant to see if it produces viable, fertile offspring. If it does not, the two different types of plants are probably reproductively isolated. Second, the giant plant is cytogenetically screened to examine its chromosome complement. If it has twice as many as its normal-sized neighbors, it is a tetraploid that may have arisen spontaneously. If the chromosome number differs by a factor of two and the new plant is reproductively isolated from its normal-sized neighbors, it is a new species.

Problems and Discussion Questions

1. What is the neo-Darwinian definition of evolution?
2. Define speciation. How does this differ from evolution?
3. Wallace and Darwin identified natural selection as a force acting on phenotypic variation. What two processes related to phenotypic variation were they unable to explain?
4. Two closely related species may have very few phenotypic differences. How can we justify classifying them as two different species instead of one species with a small range of phenotypic variation?
5. Natural selection can lead to speciation. Is this the only way species can form?
6. The maintenance of allozymic diversity in natural populations remains a controversial topic. Nevo (2001. *Proc. Natl. Acad. Sci. [USA]* 98: 6233–40.) examined polymorphisms among 111 aboveground and 132 subterranean mammalian species and found that subterranean mammals are less polymorphic than those living aboveground. Provide a possible explanation.
7. Discuss the rationale behind the statement that inversions in chromosome 3 of *D. pseudoobscura* represent genetic variation.
8. Dobzhansky's studies on populations of *D. pseudoobscura* showed changes in the frequency of the ST and CH chromosome arrangements throughout the year. Why hasn't one of these arrangements been eliminated over a long period of time?
9. Describe how populations with substantial genetic differences can form. What is the role of natural selection?
10. Price et al. (1999. *J. of Bacteriol.* 181: 2358–62.) conducted a genetic study of the toxin transport protein (PA) of *Bacillus anthracis*, the bacterium that causes anthrax in humans. Within the 2294-nucleotide gene in 26 strains they identified 5 point mutations—2 missense and 3 synonyms—among different isolates. Necropsy samples from an anthrax outbreak in 1979 revealed a novel missense mutation and 5 unique nucleotide changes among 10 victims. The authors concluded that these data indicate little or no horizontal transfer between different *B. anthracis* strains. (a) Which types of nucleotide changes (missense or synonyms) cause amino acid changes? (b) What is meant by horizontal transfer? (c) On what basis did the authors conclude that evidence of horizontal transfer is absent from their data?
11. What types of nucleotide substitutions will not be detected by electrophoretic studies of a gene's protein product?
12. A recent study examining the mutation rates of 5669 mammalian genes (17,208 sequences) indicates that, contrary to popular belief, mutation rates among lineages with vastly different generation lengths and physiological attributes are remarkably constant (Kumar, S., and Subramanian S. 2002. *Proc. Natl. Acad. Sci. [USA]* 99: 803–08.). The average rate is estimated at 2.2×10^{-9} per bp per year. What is the significance of this finding in terms of mammalian evolution?
13. Discuss the arguments supporting the neutral theory of molecular evolution. What counterarguments are proposed by the selectionists? Of what value is the debate concerning the neutralist theory?
14. What genetic changes take place during speciation?
15. The mummichog shows *Ldh* allele frequency differences from cold northern waters to warm southern waters. (a) Is there a correlation between allele type and allele frequency, and water temperature? (b) Why are differences in catalytic efficiency *and* differences in transcription rates important in this adaptive strategy?
16. List the barriers that prevent interbreeding, and give an example of each.
17. What are the two groups of reproductive isolating mechanisms? Which of these is regarded as more efficient, and why?
18. In the tobacco plant *N. tabacum*, formed as a hybrid between *N.otophora* and *N. silvestris*, would you expect to find higher levels of heterozygosity in the polyploid species than in the diploid parental species? Is this an advantage or disadvantage?
19. Why are many species of flowering plants polyploid, but only a few animal species are polyploid?
20. Shown below are two homologous lengths of the α and β chains of human hemoglobin. Consult the genetic code dictionary (Figure 13-7) and determine how many amino acid substitutions may have occurred as a result of a single nucleotide substitution. For any that cannot occur as the result of a single change, determine the minimal mutational distance.

α:	Ala	Val	Ala	His	Val	Asp	Asp	Met	Pro
β:	Gly	Leu	Ala	His	Leu	Asp	Asn	Leu	Lys

21. Determine the minimal mutational distances between these amino acid sequences of cytochrome c from various organisms. Compare the distance between humans and each organism.

Human:	Lys	Glu	Glu	Arg	Ala	Asp
Horse:	Lys	Thr	Glu	Arg	Glu	Asp
Pig:	Lys	Gly	Glu	Arg	Glu	Asp
Dog:	Thr	Gly	Glu	Arg	Glu	Asp
Chicken:	Lys	Ser	Glu	Arg	Val	Asp
Bullfrog:	Lys	Gly	Glu	Arg	Glu	Asp
Fungus:	Ala	Lys	Asp	Arg	Asn	Asp

22. The data below are taken from a paper by Heui-Soo Kim and Osamu Takenaka. They are short DNA sequences from five species: human, chimpanzee, gorilla, orangutan, and baboon. The sequences are each 50 bp long. They represent a short piece of a gene for testis-specific protein Y, located on the Y chromosome. The complete human sequence is given; the sequences for the other four species are shown only where they differ from the human sequence. Calculate the genetic difference between each pair of species. (For example, chimps and gorillas differ in 2 out of 50 bases, or 4%; thus, the genetic difference between chimps and gorillas is 0.04.) Then use UPGMA to reconstruct the phylogeny for these five species. Is your phylogeny consistent with those shown in Figures 23–19 and 23–21?

```
Human:       AGAGGTTTTTCAGTGAATGAAGCTATTTTTAAGGGAGTGTGATTGCTGCC
Chimpanzee:                    C
Gorilla:                T      C                             C
Orangutan:    T        GT   C  C                 C          C
Baboon:       C C  G  G T      C         G      GC      C   C G
```

23. The genetic difference between two *Drosophila* species, *D. heteroneura* and *D. sylvestris*, as measured by nucleotide diversity, is about 1.8%. The difference between chimpanzees (*P. troglodytes*) and humans (*H. sapiens*) is about the same, yet the latter species are classified in different genera. In your opinion, is this valid? Why, or why not?
24. In sorting out the complex taxonomic relationships among birds, species with $\Delta T_{50}H$ values of 4.0 are placed in the same genus, even by traditional taxonomy based on morphology. Using the data in Figure 23–21, construct a classification that obeys this rule, using appropriate genus names (*Pongo*, *Pan*, *Homo*).
25. The use of nucleotide sequence data to measure genetic variability is complicated by the fact that the genes of higher eukaryotes are complex in organization and contain 5′ and 3′ flanking regions as well as introns. Researchers have compared the nucleotide sequence of two cloned alleles of the γ-globin gene from a single individual and found a variation of 1%. Those differences include 13 substitutions of one nucleotide for another, and 3 short DNA segments that have been inserted in one allele or deleted in the other. None of the changes takes place in the gene's exons (coding regions). Why do you think this is so, and should it change our concept of genetic variation?
26. In a recent study of cichlid fish inhabiting Lake Victoria in Africa, Nagl et al. (1998. *Proc. Natl. Acad. Sci. [USA]* 95: 14,238–243.) examined suspected neutral sequence polymorphisms in noncoding genomic loci in 12 lake species and their putative riverine ancestors. At all loci, the same polymorphism was found in nearly all of the tested species from Lake Victoria, both lacustrine and riverine. Different polymorphisms at these loci were found in cichlids at other African lakes. (a) Why would you suspect neutral sequences to be located in noncoding genomic regions? (b) What conclusions can be drawn from these polymorphism data in terms of cichlid ancestry in these lakes?
27. Given that there are approximately 400 cichlid species in Lake Victoria and that it dried up almost completely about 14,000 years ago, what evidence indicates that extremely rapid evolutionary adaptation rather than extensive immigration occurred?
28. Some critics have warned that the use of gene therapy to correct genetic disorders will affect the course of human evolution. Evaluate this criticism in light of what you know about population genetics and evolution, distinguishing between somatic gene therapy and germline gene therapy.
29. Comparisons of Neanderthal mitochondrial DNA with that of modern humans indicate they are not related to modern humans and did not contribute to our mitochondrial heritage. However, because Neanderthals and modern humans are separated by at least 25,000 years, this does not rule out some forms of interbreeding causing the modern European gene pool being derived from both Neanderthals and early humans (called Cro-Magnons). To resolve this question, Caramelli et al. (2003. *Proc. Natl. Acad. Sci. [USA]* 100: 6593–97.) analyzed mitochondrial DNA sequences from 25,000 year old Cro-Magnon remains and compared them to four Neanderthal specimens and a large data set derived from modern humans. The results are shown in the graph below: the *x*-axis represents age of the specimens in thousands of years: the *y*-axis represents the average genetic distance. Modern humans are indicated by filled squares; Cro-Magnons, open squares; and Neanderthals, diamonds. (a) What can you conclude about the relationship between Cro-Magnons and modern Europeans? What about the relationship between Cro-Magnons and Neanderthals? (b) From these data, does it seem likely that Neanderthals made any contributions to the Cro-Magnon gene pool or the modern European gene pool?

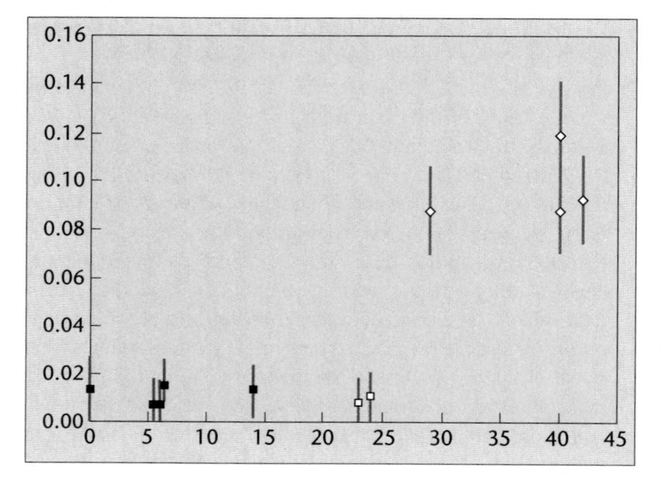

Selected Readings

Andersson, S. G. E. et al. 1998. The genome sequence of *Rickettsia prowazekii* and the origin of mitochondria. *Nature* 396: 133–43.

Avise, J. C. 1990. Flocks of African fishes. *Nature* 347: 512–13.

Ayala, F. J. 1984. Molecular polymorphism: How much is there, and why is there so much? *Dev. Genet.* 4: 379–91.

Bradshaw, H. D., Jr. et al. 1998. Quantitative trait loci affecting differences in floral morphology between two species of monkeyflower (*Mimulus*). *Genetics* 149: 367–82.

Collard, M., and Wood, B. 2000. How reliable are human phylogenetic hypotheses? *Proc. Nat. Acad. Sci. (USA)* 97: 5003–06.

Diamond, J. 1992. *The Third Chimpanzee: The Evolution and Future of the Human Animal.* New York: HarperCollins.

Dobzhansky, T. 1947. Adaptive changes induced by natural selection in wild populations of *Drosophila. Evol.* 1: 1–16.

———. 1948. Genetics of natural populations, XVI. Altitudinal and seasonal changes produced by natural selection in certain populations of *Drosophila pseudoobscura* and *Drosophila persimilis. Genetics* 33: 158–76.

Dobzhansky, T. et al. 1966. Genetics of natural populations: XXXVIII. Continuity and change in populations of *Drosophila pseudoobscura* in western United States. *Evol.* 20: 418–27.

Fitch, W. M. 1973. Aspects of molecular evolution. *Ann. Rev. Genet.* 7: 343–80.

Fitch, W. M., and Margoliash, E. 1967. Construction of phylogenetic trees. *Science* 155: 279–84.

———. 1970. The usefulness of amino acid and nucleotide sequences in evolutionary studies. *Evol. Biol.* 4: 67–109.

Freeman, S., and Herron, J. C. 2001. *Evolutionary Analysis.* Upper Saddle River: Prentice-Hall.

Fryer, G. 1997. Biological implications of a suggested Late Pleistocene desiccation of Lake Victoria. *Hydrobiol.* 354: 177–82.

Gray, M. W. et al. 1999. Mitochondrial evolution. *Science* 283: 1476–81.

Gould, S. J. 1982. Darwinism and the expansion of evolutionary theory. *Science* 216: 380–87.

Hardison, R. 1999. The evolution of hemoglobin. *Amer. Scient.* 87: 126–37.

Hillis, D. M. 1998. Phylogenetic analysis. *Curr. Biol.* 7: R129–31.

Hillis, D. M. et al. 1992. Experimental phylogenetics: Generation of a known phylogeny. *Science* 255: 589–92.

Hillis, D. M., Huelsenbeck, J. P., and Cunningham, C. W. 1994. Application and accuracy of molecular phylogenies. *Science* 264: 671–77.

Höss, M. 2000. Neanderthal population genetics. *Nature* 404: 453–54.

Hunt, J. et al. 1981. Evolution distance in Hawaiian *Drosophila*. *J. Mol. Evol.* 17: 361–67.

Kimura, M. 1979. The neutral theory of molecular evolution. *Sci. Am.* (Nov.) 241: 98–126.

———. 1989. The neutral theory of molecular evolution and the world view of the neutralists. *Genome* 31: 24–31.

Knowlton, N. et al. 1993. Divergence in proteins, mitochondrial DNA, and reproductive compatibility across the Isthmus of Panama. *Science* 260: 1629–32.

Kreitman, M. 1983. Nucleotide polymorphism at the alcohol dehydrogenase locus of *Drosophila melanogaster*. *Nature* 304: 412–17.

Krings, M. et al. 1997. Neandertal DNA sequences and the origin of modern humans. *Cell* 90: 19–30.

Lewontin, R. C., and Hubby, J. L. 1966. A molecular approach to the study of genic heterozygosity in natural populations: II. Amount of variation and degree of heterozygosity in natural populations of *Drosophila pseudoobscura*. *Genetics* 54: 595–609.

Müller, M., and Martin, W. 1999. The genome of *Rickettsia prowazekii* and some thoughts on the origin of mitochondria and hydrogenosomes. *Bio Ess.* 21: 377–81.

Ou, C.-Y. et al., 1992. Molecular epidemiology of HIV transmission in a dental practice. *Science* 256: 1165–71.

Ovchinnikov, I. V. et al. 2000. Molecular analysis of Neanderthal DNA from the northern Caucasus. *Nature* 404: 490–493.

Pääbo, S. 1993. Ancient DNA. *Sci. Am.* (Nov.) 269: 86–92.

Powers, D. A., and Schulte, P. M. 1998. Evolutionary adaptations of gene structure and expression in natural populations in relation to a changing environment: A multidisciplinary approach to address the million-year saga of a small fish. *J. Exper. Zool.* 282: 71–94.

Schemske, D. W., and Bradshaw, H. D., Jr. 1999. Pollinator preference and the evolution of floral traits in monkeyflowers. *Proc. Nat. Acad. Sci. (USA)* 96: 11,910–15.

Seehausen, O. 2002. Patterns in fish radiation are compatible with Pleistocene desiccation of Lake Victoria and 14,600 year history for its cichlid species. *Proc. Roy. Soc. London B* 269: 491–97.

Sibley, C., and Ahlquist, J. 1984. The phylogeny of the hominoid primates, as indicated by DNA–DNA hybridization. *J. Mol. Evol.* 20: 2–15.

Sibley, C. G., Comstock, J. A., and Ahlquist, J. E. 1990. DNA evidence of hominoid phylogeny: A reanalysis of the data. *J. Mol. Evol.* 30: 202–36.

Soltis, P. S., and Soltis, D. E. 2000. The role of genetic and genomic attributes in the success of polyploids. *Proc. Nat. Acad. Sci. (USA)*. 97: 7051–57.

Stiassny, M. L. J., and Meyer, A. 1999. Cichlids of the Rift Lakes. *Sci. Am.* (Feb.) 280: 64–69.

Takahashi, K. et al. 1998. A novel family of short interspersed repetitive elements (SINEs) from cichlids: The pattern of insertion of SINEs at orthologous loci support the proposed monophyly of four major groups of cichlid fishes in Lake Tanganyika. *Mol. Biol. Evol.* 15: 391–407.

Tsui, L.-C. 1992. The spectrum of cystic fibrosis mutations. *Trends Genet.* 8: 392–98.

Val, F. C. 1977. Genetic analysis of the morphological differences between two interfertile species of Hawaiian *Drosophila*. *Evol.* 31: 611–29.

Verhayen, E., Salzberger, W., Snoeks, J., and Meyer, A. 2003. Origin of the superflock of cichlid fishes from Lake Victoria, East Africa. *Science* 300: 325–29.

CHAPTER 24

Cryogenically preserved seeds of rare crop varieties at the U.S. Department of Agriculture National Seed Storage Laboratory. *(Photo by Scott Bauer/ARS Photo Unit/USDA)*

Conservation Genetics

CHAPTER CONCEPTS

As the 21st century progresses, the diversity of life on Earth will be under increasing pressure from the direct and indirect effects of explosive human population growth. Approximately 10 million *Homo sapiens* lived on the planet 10,000 years ago. This number grew to 100 million 2000 years ago and to 2.5 billion by 1950. Within the span of a single lifetime, the world's human population more than doubled to 5.5 billion in 1993 and is projected to reach as high as 19 billion by 2100 (Figure 24–1).

The effect on other species of accelerating human population growth has been dramatic. Data from the World Conservation Union (IUCN) show that, globally, 25% of all mammal species, 11% of birds, 20% of reptiles, 25% of amphibians, and 34% of fish species are vulnerable or endangered. The situation for plants is no better. Based on IUCN surveys, 12% of all vascular plant species worldwide—some 34,000 species—are threatened. Not just wild species are at risk. Genetic diversity in domesticated plants and animals is also being lost as many traditional crop varieties and livestock breeds disappear. The Food and Agriculture Organization (FAO) estimates that since 1900, 75% of the genetic diversity in agricultural crops has been lost. Out of approximately 5000 different breeds of domesticated farm animals worldwide, one-third are at risk of being lost.

Why should we be concerned about losing **biodiversity**—that is, the biological variation represented by these different plants and animals? As the fossil record shows, unrelated to human influences, many different plants and animals that once inhabited the planet have become extinct over millions of years, and others have taken their places. Biologists are concerned, however, at the accelerated rate of species extinctions we are witnessing today, all of which can be ascribed to direct or indirect human impacts. Deliberate hunting or harvesting of plants and animals by humans, habitat destruction through human development activities, and the indirect effects of global climate change are making it increasingly difficult for many species to survive.

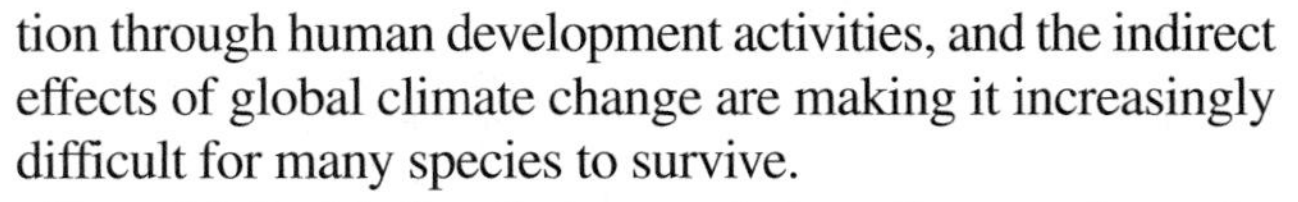

Some biologists fear that **ecosystems**—the complex webs of interdependent, but diverse plants and animals found together in the same environment—may collapse if key sustaining species are lost. This portends possible consequences for our own long-term survival. Other scientists have pointed out that we may be losing the unknown economic potential or other benefits of unexploited plants and animals if we allow them to become extinct. A much-publicized, recent example is the Pacific yew (*Taxus brevifolia*), a rare tree found in coastal forests of the western United States. This slow-growing species was ignored because of its small size and poor timber qualities and was frequently destroyed during logging operations as a "trash tree." In 1991, it was found to be a source of taxol, a compound now widely used as a powerful drug for treating cancer. Quite apart from practical benefits, scientists and nonscientists have made the case that human life will be diminished both spiritually and aesthetically if we do not strive to maintain the fascinating and often beautiful variety of living organisms with which we share the planet.

Conservation biologists work to understand and maintain biodiversity, studying the factors that lead to species decline and the ways in which species can be preserved. The new field of **conservation genetics** has emerged in the last 20 years as scientists have begun to recognize that genetics will be an important tool in maintaining and restoring population viability. The applications of genetics to conservation biology are multifaceted; in this chapter, we will explore but a few of them. Underlying the increasingly important role of genetics in conservation biology is the recognition that biodiversity depends on genetic diversity and that maintaining biodiversity in the long term is unlikely if genetic diversity is lost.

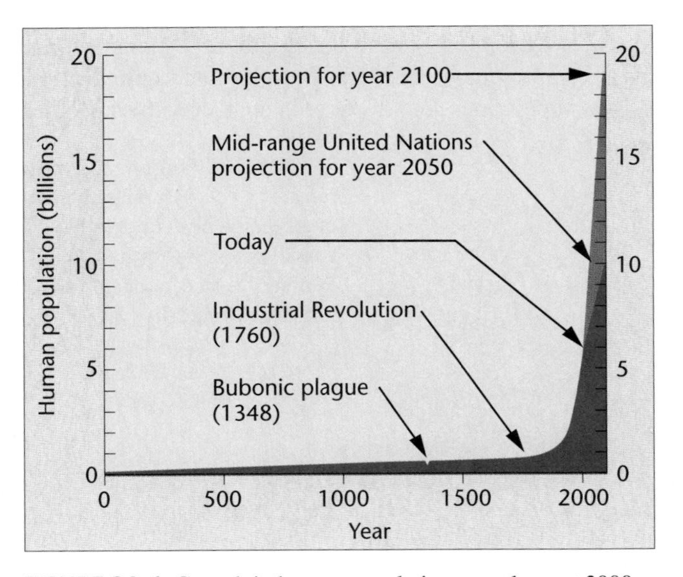

FIGURE 24–1 Growth in human population over the past 2000 years and projected through 2100.

HOW DO WE KNOW WHAT WE KNOW?

IN THIS CHAPTER, WE WILL FOCUS ON the genetic approach that is utilized during conservation efforts. Conservation geneticists assess genetic diversity and work to maintain population numbers for long-term species survival. As you study this topic, you should try to answer several fundamental questions:

1. How do we determine the level of genetic diversity in a species?
2. How do we know that diminished genetic diversity is detrimental to species survival?
3. How do we determine the most effective approach to countering decreased population size?
4. How do we attempt to "conserve" existing genetic diversity? ■

24.1 Genetic Diversity Is at the Heart of Conservation Genetics

While biodiversity encompasses the variation represented by all existing species of plants and animals on this planet at any given time, **genetic diversity** is not as easy to define and study. Genetic diversity can be considered on two levels: **interspecific diversity** and **intraspecific diversity**.

Diversity between species, or interspecific diversity, is reflected in the number of different plant and animal species present in an ecosystem. Some ecosystems have a very high level of species diversity, such as a tropical rain forest in which hundreds of different plant and animal species may be found within a few square meters (Figure 24–2a). Other ecosystems, especially where plants and animals must adapt to a harsh environment, may have much lower levels of interspecific diversity (Figure 24–2b). Lists of the different plant and animal species found in a particular environment are used to compile species inventories. The inventories identify diversity hot spots: geographic areas with especially high levels of interspecific diversity where conservation efforts can be focused. Conservation biologists working at the ecosystem level are interested in preventing species from being lost and in restoring species that were once part of the system but are no longer present. The reestablishment of the gray wolf (*Canis lupus*) in Yellowstone National Park and the release of captive-bred California condors (*Gymnogyps californianus*) into the mountain ranges they previously occupied in the southwest United States are examples of current attempts by conservation biologists to restore missing species to their ecosystems.

Intraspecific diversity—the diversity within a species—is reflected in the level of genetic variation occurring between individuals within a single population of a given species (*intrapopulation diversity*) or between different populations of the same species (*interpopulation diversity*). Genetic variation within populations can be measured as the frequency of individuals in the population that are heterozygous at a given locus or as the number of different alleles at a locus that are present in the population gene pool. When DNA-profiling techniques are used, the percentage of polymorphic loci—those represented by varying bands on the DNA profile in different individuals—can be calculated to indicate the extent of genetic diversity in a population. In outbreeding species, most intraspecific genetic diversity is found at the intrapopulation level.

Significant interpopulation diversity can occur if populations are separated geographically and there is no migration or exchange of gametes between them. On the other hand, predominantly inbreeding species (such as self-fertilizing plants) tend to have greater levels of interpopulation than intrapopulation diversity. There is a limited number of genotypes dominating an individual population, but greater variation between different populations. Understanding the breeding system and the distribution of genetic variation in an endangered species is important to its conservation, not only to ensure continued production of offspring but also to determine the best strategy for maintaining intraspecific diversity. For example, would it be more effective to preserve a few large populations or many small, distinct ones? This information can then be used to guide conservation or restoration efforts.

Loss of Genetic Diversity

Loss of genetic diversity in nondomesticated species is usually associated with a reduction in population size. This may be due to excessive hunting or harvesting. For example, biologists blame commercial overfishing for the collapse in the early 1990s of the deep-sea cod populations off the Newfoundland coast—once one of the most productive fisheries in the world. Habitat loss is also a major cause of population decline. As the global human population increases, more land is developed for housing and transport systems or is put into agricultural production, reducing or eliminating areas that were once home to wild plants and animals. The shrinking available habitat reduces populations of wild species and often also isolates them from each other as individual populations become trapped in pockets of undeveloped land surrounded by areas taken over for agriculture, urban development, or other human uses. This process is known as **population fragmentation**. When populations are no longer in contact with each other,

(a)

(b)

FIGURE 24–2 (a) Tropical rain forest is an ecosystem with high interspecific diversity. (b) A coastal marsh in North Carolina exemplifies an ecosystem with low interspecific diversity. *(Photo [a]: David Austen/Stone; photo [b]: Sarah M. Ward)*

gene flow through migration or gamete exchange between them ceases, and an important mechanism for maintaining genetic variation is lost.

In domesticated species, loss of genetic diversity is not usually the result of habitat loss or collapsing population numbers; there is little risk that cows or corn as species will become extinct any time soon. Reduction in diversity within domesticated species can instead be traced to changes in agricultural practice and consumer demand. Modern farming techniques have greatly increased production levels, but they have also led to greater genetic uniformity. As farmers switch to new crop varieties or improved livestock strains on a large scale, they abandon cultivation of many older local types, which may then disappear if efforts are not made to preserve them.

For example, in 1900, more than 100 different kinds of potatoes were available in the United States. Today, three-quarters of commercial potato production in this country depends on just nine varieties. A single type, Russet Burbank, makes up 43% of the total acreage planted. Modern crop varieties or livestock strains may be better adapted to meet the demands of modern agricultural production, but older types often contain useful genes that can still play a vital role in survival functions, such as resistance to disease, cold, or drought. When the Russian wheat aphid (*Diuraphis noxia*), an insect that causes serious damage to cereal crops, invaded the United States in 1986, plant breeders eventually found genes conferring resistance to this pest—not in modern American wheat lines, but in old traditional varieties from the former Soviet Union, that fortunately had been collected and preserved. The old Russian wheats were crossed with United States wheat plants to transfer the resistance genes, creating new commercial wheat varieties that resist Russian wheat aphid attack and saving wheat growers from crop losses in the millions of dollars.

FIGURE 24–3 Phenotypic variation in seed color and markings in the common bean (*Phaseolus vulgaris*) reveals high levels of intraspecific diversity. *(Sarah M. Ward/Department of Soil and Crop Services/Colorado State University)*

Identifying Genetic Diversity

For many years, population geneticists based estimates of intraspecific diversity on phenotypic differences between individuals, such as different colors of seeds or flowers or variation in markings (Figure 24–3). More recently, techniques have been developed for analyzing intraspecific diversity with greater precision at the molecular level.

A simple molecular method that has been widely used is **allozyme analysis**. Allozymes are multiple versions of a single enzyme occurring within a single species. Allozymes catalyze the same reaction, but differences in the polypeptide chains forming the bulk of the molecule occur because of slight variations in the allelic DNA sequences at the locus coding for the subunits of the enzyme. This results in forms of the protein that differ in size or net charge and can be separated by electrophoresis. The presence of allozyme variation in a population can be used as an indicator of genetic variation because the different versions of the molecule indicate that different alleles are present at a locus.

DNA profiling is a more direct molecular approach used to detect and quantify genetic differences between individuals drawn from a population. This technique has become an important tool for detecting and assessing intra- and interpopulation genetic variation. Nuclear, mitochondrial, and chloroplast DNA can all be analyzed to determine levels of genetic variation. We have already described applications of DNA fingerprinting techniques using either restriction enzymes or PCR. These techniques have also been used with DNA from many different species that are of concern to conservation biologists.

One PCR-based technique that has been widely used to analyze genetic diversity in animal populations focuses on **SSRs** (**simple sequence repeats**), also known as *microsatellites*. SSRs are short (2–5 bases) tandem repeat sequences in the DNA. The number of times a repeat is present at a given SSR locus often varies between individuals, but the nonrepeating DNA sequences on either side of an SSR do not vary. By using PCR primers that match these conserved flanking sequences, the DNA containing the block of SSRs can be selected for amplification. The varying numbers of repeats that are present will produce amplified DNA molecules of different sizes, generating distinctive banding patterns for different individuals when the amplified DNA molecules are separated by gel electrophoresis. Although the SSR regions themselves do not contain expressed genes, conservation biologists can use microsatellite variation among individuals within populations as an indirect estimate of the amount of overall genetic variation present. For example, a study that used microsatellite analysis to compare levels of genetic diversity among North American brown bear populations is described later in this chapter.

A second DNA–profiling technique used to assess genetic diversity in threatened and endangered species uses **amplified fragment-length polymorphisms** (**AFLPs**). This method combines the use of restriction enzymes and PCR. It is especially powerful at detecting very small amounts of genetic diversity in a population, potentially generating different profiles for individuals whose DNA sequence differs by as little as a single base pair.

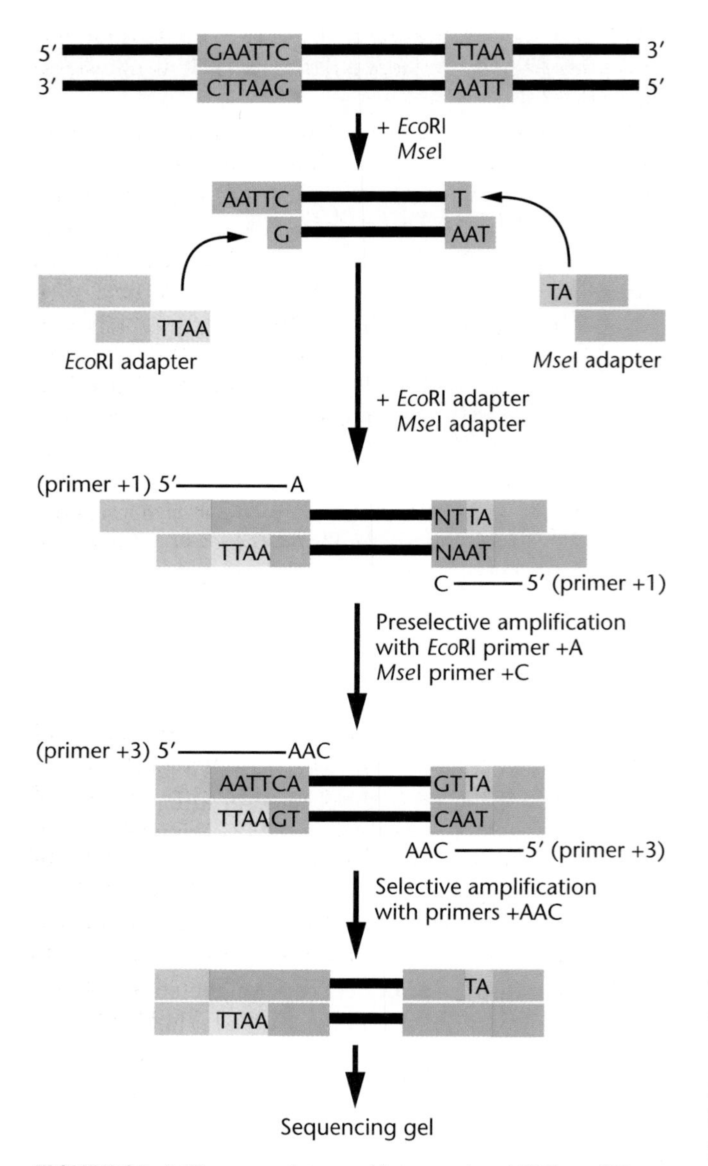

FIGURE 24–4 The procedure used in preparing AFLP-profiling samples.

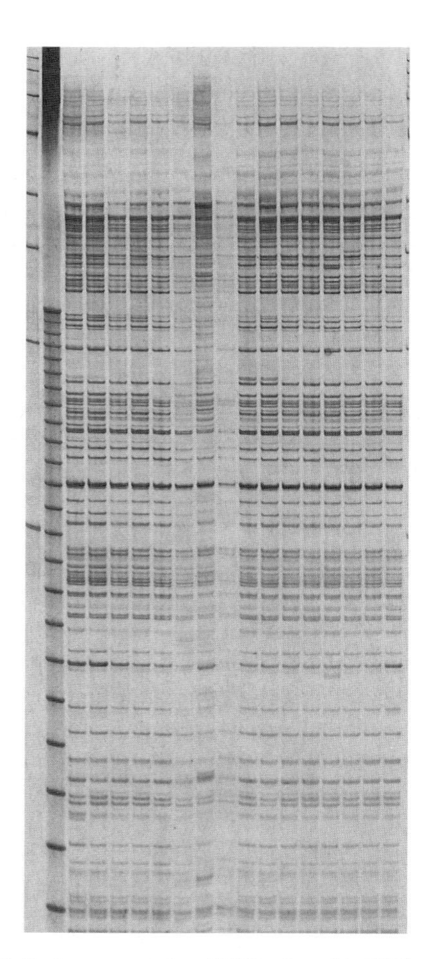

FIGURE 24–5 A representative AFLP gel used in DNA profiling, derived from an analysis of jointed goat grass. *(Gel image courtesy of Todd A. Pester and Sarah M. Ward, Colorado State University)*

The AFLP technique is illustrated in Figure 24–4. DNA is first cut using two restriction enzymes, each with its own target site. This produces a mixture of different DNA fragments of varying sizes, but the last few bases at each end of every fragment are known because they must correspond to one of the two target sites for the restriction enzymes used. The next step is to attach short single-stranded pieces of synthesized DNA called adapters to the cut ends of the fragments. Because we know the sequence of the attached adapters as well as the base sequence for the restriction site at each end of the cut DNA, PCR primers can be designed to match these sequences and amplify the DNA fragments. In practice, not every piece of cut DNA is amplified—it would produce too many PCR products to separate easily on a gel. Instead, a subset of the DNA fragments is amplified with primers that match the adapter and target site and have an extra 1–3 base sequence at the 3′- end. Only DNA fragments with the complementary 1–3 bases at each end are selected for PCR amplification. The AFLP technique generates complex DNA banding profiles (Figure 24–5); differences in AFLP banding patterns arise because of variation in individual DNA sequences affecting the location of the enzyme target sites and the size and number of fragments amplified.

AFLP profiling has been extensively used to analyze genetic variation in crop species and some livestock and is now being used to examine variation and to guide conservation efforts in nondomesticated species. In a recent study, AFLP analysis was performed on the only remaining population of a critically endangered plant species (*Limonium cavanillesii*) growing on the Mediterranean coast of Spain. Researchers found very low levels of diversity but did detect genetically distinct individual plants within the population from which seeds could be collected.

In another recent study, AFLP analysis of the southwestern willow flycatcher, an endangered migratory bird species that nests in southwestern United States, found significant levels of genetic diversity within populations at different breeding sites, but little interpopulation diversity. This indicated to the scientists studying the flycatcher that migration of individual birds between breeding populations is important in maintaining diversity in the species and that conservation efforts should be directed toward preserving nesting sites that allow such migration to continue.

Still a third molecular technique that is useful in conservation is **DNA fingerprinting**. The same DNA-profiling technique commonly used in forensic science can be used in conservation biology to uncover illegal trade in endangered species that are protected by law. For example, since 1986, an international moratorium on commercial whaling has been in place, with only limited hunting of a few species permitted. In 1994, scientists based in New Zealand and Hawaii used PCR to amplify mitochondrial DNA (mtDNA) sequences extracted from meat on sale in markets in Japan, where whale meat is prized as a delicacy. The DNA fingerprints revealed several meat samples from humpback and fin whales, which are protected species. Two meat samples were not from whales at all, but were dolphin meat! This information assisted Japanese customs officials in tracking down sources of illegal whale meat imports.

Unlawful trade in animal products also occurs in the United States. In a recent study by scientists at the University of Florida, amplification of mtDNA sequences from turtle meat sold in Louisiana and Florida revealed that one-quarter of the samples were actually alligator, not turtle. Most of the rest were from small freshwater turtles, indicating that populations of freshwater turtle species were declining through overharvesting and thus were in need of protection. Further refinements to DNA extraction and profiling techniques now allow accurate identification of turtle species from meat already spiced, cooked, and served in restaurants. We now have a tool that hopefully will help to eliminate the illegal harvesting of these endangered reptiles.

24.2 Population Size Has a Major Impact on Species Survival

Some species have never been numerous, especially those that are adapted to survive in unusual habitats. Biologists refer to such species as *naturally rare*. *Newly rare* species, on the other hand, are those whose numbers are in decline because of pressures such as habitat loss. Populations of such species may not only be small in number but also fragmented and isolated from other populations. Both decreased population size and increased isolation have important genetic consequences, leading to an increased risk of loss of diversity.

How small must a population be before it is considered endangered? It varies somewhat with species, but small populations can quickly become vulnerable to genetic phenomena that increase the risk of extinction. In general, a population of fewer than 100 individuals is considered extremely sensitive to these problems, which include genetic drift, inbreeding, and reduction in gene flow. The effects of such problems on species survival are substantial. Studies of bighorn sheep, for example, have shown that populations of fewer than 50 are highly likely to become extinct within 50 years. Projections based on computer models show that for all species, populations of fewer than 10,000 are likely to be limited in adaptive genetic variation and at least 100,000 individuals must be present if a population is to show long-term sustainability.

Determining the number of individuals a population must contain in order to have long-term sustainability is complicated by the fact that not all members of a population are equally likely to produce offspring: Some will be infertile, too young, or too old. The effective population size (N_e) is defined as the number of individuals in a population having an equal probability of contributing gametes to the next generation. N_e is almost always smaller than the absolute population size (N). The effective population size can be calculated in different ways, depending on the factors that are preventing all individuals in a population from contributing equally to the next generation. In a sexually reproducing population that contains different numbers of males and females, for example, the effective population size is calculated as

$$N_e = \frac{4(N_m N_f)}{N_m + N_f}$$

where N_m is the number of males and N_f the number of females in the population. Hence a population of 100 males and 100 females would have an effective size of $4(100 \times 100)/(100 + 100) = 200$. In contrast, if there were 180 males and only 20 females, the effective population size would be $4(180 \times 20)/(180 + 20) = 72$

Effective population size is also influenced by fluctuations in absolute population size from one generation to the next. Here, the effective population size is the harmonic mean of the numbers in each generation, so

$$N_e = 1 \bigg/ \frac{1}{t}\left(\frac{1}{N_1} + \frac{1}{N_2 + \cdots + N_t}\right)$$

where t is the total number of generations being considered. For example, if a population went through a temporary reduction in size in generation 2, so that $N_1 = 100$, $N_2 = 10$, and $N_3 = 100$, then

$$N_e = 1 \bigg/ \frac{1}{3}\left(\frac{1}{100} + \frac{1}{10} + \frac{1}{100}\right) = \frac{1}{0.04} = 25$$

In this case, although the mean actual number of individuals in the population over three generations was 70, the effective population size during that time was only 25. A severe temporary reduction in size such as this is known as a **population bottleneck**. Bottlenecks occur when a population or species is reduced to a few reproducing individuals whose offspring then increase in numbers over subsequent generations to reestablish the population. Although the number of individuals may be restored to healthier levels, genetic diversity in the newly expanded population is often severely reduced because gametes from the handful of surviving individuals functioning as parents do not represent all of the different allelic frequencies present in the original gene pool.

Captive-breeding programs, in which a few surviving individuals from an endangered species are removed from the wild and their offspring raised in a protected environment to rebuild the population, inevitably create population bottlenecks. Bottlenecks also occur naturally when a small number of individuals from one population migrate to establish a new population elsewhere. When a new population derived from a small subset of individuals has significantly less genetic diversity than the original population, it exhibits the **founder effect**. Reduced levels of genetic diversity due to a founder effect can persist for many generations, as shown by studies of two species of Antarctic fur seal (*Arctocephalus gazella* and *Arctocephalus tropicalis*). Seal hunters in the eighteenth and nineteenth centuries had severely reduced populations of these species, eliminating them from parts of their natural range in the Southern Ocean.* Although the number of Antarctic fur seals has now rebounded and the two species have recolonized much of their original habitat, mtDNA fingerprinting reveals that a founder effect can still be detected in *A. gazella*, with reduced genetic variation in current populations descended from a handful of surviving individuals.

The cheetah (*Acinonyx jubatus*), shown in Figure 24–6, is another well-studied example of a species with reduced genetic variation resulting from at least one severe population bottleneck that occurred in its recent history. Allozyme studies of South African cheetah populations have shown levels of genetic variation that are less than 10% of those found in other mammals. The abnormal spermatozoa and poor reproductive rates commonly observed in cheetahs are thought to be linked to the lack of genetic diversity, although when and how the population bottleneck occurred in this species is still unclear.

FIGURE 24–6 The cheetah (*Acinonyx jubatus*), a species with reduced genetic variation following population bottlenecks. *(Johnny Johnson/DRK Photo)*

A serious result of genetic drift in populations with a small effective population size is the loss of genetic variation. Genetic drift is a random process, so both deleterious and advantageous alleles can become fixed within a small population. This means a useful allele can be lost even if it has the potential to increase fitness or long-term adaptability. If we consider a single locus with two alleles, *A* and *a*, genetic drift may result in one of the alleles eventually disappearing, while the other becomes fixed—in other words, it becomes the only version of that gene present in the gene pool of the population. A simple correlation describes the likelihood of the fixation or the loss of an allele. The probability that an allele will be fixed through drift is the same as its initial frequency. So, if $p(A) = 0.8$, the probability of *A*'s becoming fixed is 0.8, or 80%; the probability of *A*'s being lost through drift is $(1 - 0.8) = 0.2$, or 20%. Figure 24–7 is a graph of the effect of genetic drift on a theoretical population.

24.3 Genetic Effects Are More Pronounced in Small, Isolated Populations

Small isolated populations, such as those found in threatened and endangered species, are especially vulnerable to genetic drift, inbreeding, and reduction in gene flow. These phenomena act on the gene pool in different ways, but ultimately have similar effects in that they all can further reduce genetic diversity and long-term species viability.

Genetic Drift

If the number of breeding individuals in a population is small, fewer gametes will form the next generation. The alleles carried by these gametes may not be a representative sample of all those present in the population; purely by chance, some alleles may be underrepresented or not present at all, which will cause changes in allele frequency over time, resulting in **genetic drift**.

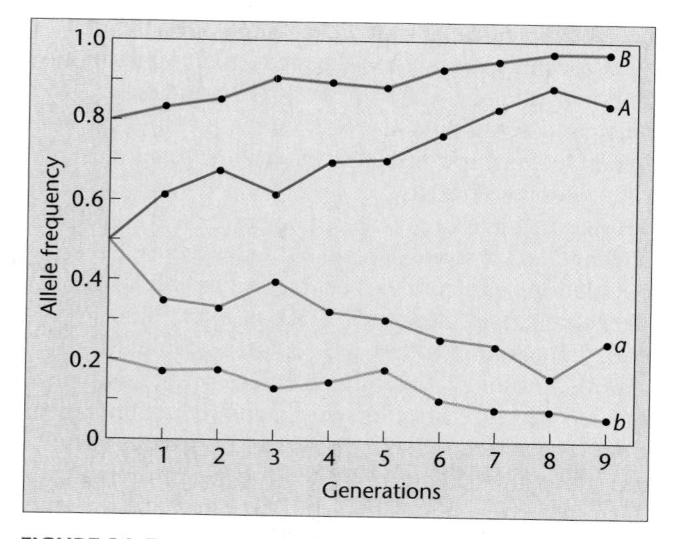

FIGURE 24–7 Change in frequencies over 10 generations for two sets of alleles, *A/a* and *B/b*, in a theoretical population subject to genetic drift.

*In 2000, this fifth world ocean was delimited from the southern portions of the Atlantic, Indian, and Pacific Oceans.

Inbreeding

In small populations, the chance of **inbreeding** (matings between closely related individuals) is greater. As described in Chapter 22, inbreeding increases the proportion of homozygotes in a population, thus increasing the possibility that an individual may be homozygous for a deleterious allele. The **inbreeding coefficient** (*F*) measures the extent of inbreeding occurring in a population; *F* measures the probability that two alleles of a given gene in an individual are derived from a common ancestral allele. The inbreeding coefficient is inversely related to the frequency of heterozygotes in the population, and can be calculated as

$$F = \frac{2pq - H}{2pq}$$

where $2pq$ is the expected frequency of heterozygotes based on the Hardy-Weinberg law and H is the actual frequency of heterozygotes in the population.

In a declining population that has become small enough for drift to occur, heterozygosity (H) will decrease with each generation. The smaller the effective population size, the more rapid the decrease in H and the resulting increase in F, as can be seen from the equation

$$H_t/H_0 = (1 - {}^1\!/_{2N_e})^t$$

where H_0 is the initial frequency of heterozyogtes, H_t is the frequency of heterozygotes after t generations, and N_e is the effective population size. Figure 24–8 compares rates of increase in F for different effective population sizes.

What effect does inbreeding have on the long-term survival of a population? In some cases, inbreeding may not immediately reduce the amount of genetic variation present in the overall gene pool of a species in which numbers of individuals remain high. Self-pollinating plants, for example, often show high levels of homozygosity and relatively little genetic variation within a single population. However, they tend to have considerable variation between different populations, each of which has adapted to slightly different local environmental conditions. On the other hand, in most outbreeding species, including all mammals, inbreeding is associated with reduced fitness and lower survival rates among offspring. This **inbreeding depression** can result from increased homozyosity for deleterious alleles. The number of deleterious alleles present in the gene pool of a population is called the **genetic load** (or genetic burden).

In some species, inbreeding accompanied by selection against less fit individuals homozygous for deleterious alleles has resulted in the elimination of these alleles from the gene pool, a process known as *purging the genetic load*. Species that have successfully purged their genetic load do not show continued reduction in fitness, even after many generations of inbreeding. This is true of numerous domesticated species, especially self-pollinating plants such as wheat. However, computer simulation experiments have shown that it may take 50 generations or more to complete the purging process, during which time inbreeding depression will still occur.

Alternatively, inbreeding depression can result from heterozygous individuals having a higher level of fitness than either of the corresponding homozygotes. In this case, the long-term survival of the population requires that inbreeding be avoided and that the levels of all alleles in the gene pool be maintained. Both of these requirements can be difficult in a species that has already suffered a significant reduction in population size.

The effects of inbreeding depression and loss of genetic variation in a small isolated population have been documented in the case of the Isle Royale gray wolves (Figure 24–9). Around 1950, a pair of gray wolves apparently crossed an ice bridge from the Canadian mainland to Isle Royale in Lake Superior. The island had no other wolves and had an abundance of moose, which became the wolves' main food source. By 1980, the Isle Royale wolf population had increased to over 50 individuals. Over the next decade,

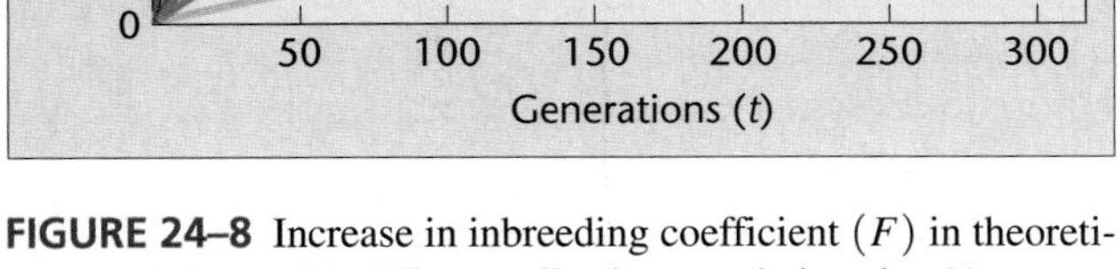

FIGURE 24–8 Increase in inbreeding coefficient (F) in theoretical populations with different effective population size N_e.

FIGURE 24–9 Isle Royale gray wolf (*Canis lupus*). (*Art Wolfe/Stone*)

however, wolf numbers declined to fewer than a dozen with no new litters being born, despite plentiful food and no apparent sign of disease. The genetic variation of the remaining wolves was examined by mtDNA analysis and nuclear DNA fingerprinting. It was found that the Isle Royale wolves had levels of homozygosity that were twice as high as wolves in an adjacent mainland population. Furthermore, the wolves all possessed the same mtDNA genotype, consistent with descent from the same female. Hence, the degree of relatedness between individual Isle Royale wolves was equivalent to that of full siblings, suggesting that the wolves' reproductive failure was due to inbreeding depression, a phenomenon that has also been seen in captive-wolf populations.

Reduction in Gene Flow

Gene flow, the gradual exchange of alleles between two populations, is brought about by the dispersal of gametes or the migration of individuals. It is an important mechanism for introducing new alleles into a gene pool and increasing genetic variation. Migration is the main route for gene flow in animals. In plants, gene flow occurs not through movement of individuals but as a result of cross-pollination between different populations and, to some extent, through seed dispersal. Isolation and fragmentation of populations in rare and declining species significantly reduces gene flow and the potential for maintaining genetic diversity. As we have already discussed, habitat loss is a major threat to species survival. It is not unusual for a threatened or endangered species to be restricted to small separate pockets of the remaining habitat. This isolates and fragments the surviving populations so that movement of individuals can no longer occur between them, thus preventing gene flow.

The term **metapopulation** is used to describe a population consisting of spatially separated subpopulations with limited gene flow, especially if local extinctions and replacements of some of the subpopulations occur over time. One well-studied metapopulation is that of the endangered red-cockaded woodpecker (*Picoides borealis*), which was once common in pinewoods throughout the southeastern United States (Figure 24–10). Habitat loss because of logging has reduced the remaining populations of this bird to small scattered sites isolated from each other, with virtually no migration between them. Studies using allozymes and DNA profiling show that the smallest surviving woodpecker populations (where $N = <100$) have suffered the greatest loss of genetic diversity and are at most risk of inbreeding depression, compared with larger populations where $N = >100$. Management of the red-cockaded woodpecker now includes efforts to increase the genetic diversity of the smallest populations by introducing birds from different larger populations to artificially recreate the gene flow by migration, which would have occurred in the original unfragmented distribution of the species.

FIGURE 24–10 The red-cockaded woodpecker (*Picoides borealis*). (*Tim Thompson/CORBIS BETTMAN*)

Another species in which the effects of gene-flow reduction have been studied is the North American brown bear, *Ursus arctos*. A team of Canadian and U.S. researchers measured heterozygosity in different brown bear populations using amplified microsatellite markers to create DNA profiles for individual bears. The researchers found that levels of heterozygosity in the brown bear population living in and around Yellowstone Park were only two-thirds as high as those in brown bear populations from Canada and mainland Alaska. They concluded that this was due to the isolation and reduced migration of the Yellowstone bears. The habitat of the Canadian and Alaskan bear populations was much less fragmented, allowing migration of individuals and consequent gene flow between populations. The researchers also examined an island population of brown bears on the Kodiak archipelago off the Alaskan coast. Here, heterozygosity was even lower—less than one-half that of the mainland populations—providing further evidence for the effect of restricted migration and gene flow on reduced genetic diversity.

24.4 Genetic Erosion Diminishes Genetic Diversity

The loss of previously existing genetic diversity from a population or species is referred to as **genetic erosion**. Why does it matter if a population loses genetic diversity, especially if the numbers of individuals remain high? Genetic erosion has two important effects on a population. First, it can result in the loss of potentially useful alleles from the gene pool, thus reducing the ability of the population to adapt to changing environmental conditions and increasing its risk of extinction. Several decades before the dawn of modern genetics, Charles Darwin recognized the importance of diversity to long-term species survival and evolutionary success. In *The Origin of Species* (1859), Darwin wrote,

"The more diversified the descendants of any one species . . . by so much will they be better enabled to seize on many widely diversified places in the polity of nature, and so enabled to increase in numbers."

The story of the peppered moth *Biston betularia* in the nineteenth century (see the chapter-opening photograph in Chapter 23) illustrates the importance of maintaining allelic diversity in the gene pool. The allele producing the darker melanic phenotype did not confer any obvious advantage to the species until the moth's environment in parts of Great Britain was significantly altered as a result of increasing industrialization and the accompanying pollution. The continued presence of the melanic allele in the moth's gene pool, however, enabled it to adapt to the new environmental conditions and survive, just as the persistence of the nonmelanic allele in the population now allows the moth to readapt as its environment changes once again. What might have happened to the peppered moth if the melanic allele had been lost from its gene pool before the Industrial Revolution?

The second important effect of genetic erosion is a reduction in levels of heterozygosity. At the population level, reduced heterozygosity will be seen as an increase in the number of individuals homozygous at a given locus. At the individual level, a decrease in the number of heterozygous loci within the genotype of a particular plant or animal will occur. As we have seen, loss of heterozygosity is a common consequence of reduced population size. Alleles may be lost through genetic drift or because individuals carrying them die without reproducing. Smaller populations also increase the likelihood of inbreeding, which inevitably increases homozygosity.

Obviously, once an allele is lost from a gene pool, the potential for heterozygosity is greatly reduced, or it is completely eliminated if there are only two alleles at the locus in question and one is now fixed. The level of homozygosity that can be tolerated varies with species. Studies of populations showing higher-than-normal levels of homozygosity have documented a range of deleterious effects, including reduced sperm viability and reproductive abnormalities in African lions, increased offspring mortality in elephant seals, and reduced nesting success in woodpeckers.

As yet, no evidence has emerged from field studies conclusively linking the extinction of a wild population to genetic erosion. However, laboratory studies using *Drosophila melanogaster* to model evolutionary events show that loss of genetic variation does reduce the ability of a population to adapt to changing environmental conditions. Fruitflies with their small size and rapid generation time of only 10–14 days, are a useful model organism for studying evolutionary events, especially since large populations can be readily developed and maintained through multiple generations.

In one set of experiments carried out by scientists at Macquarie University in Sydney, Australia, *Drosophila* populations were reduced to a single pair for up to three generations to simulate a population bottleneck and were then allowed to increase in number. The capacity of the bottlenecked populations to tolerate increasing levels of sodium chloride was compared with that of normal outbred populations. Researchers found that the bottlenecked populations became extinct at lower salt concentrations (Figure 24–11).

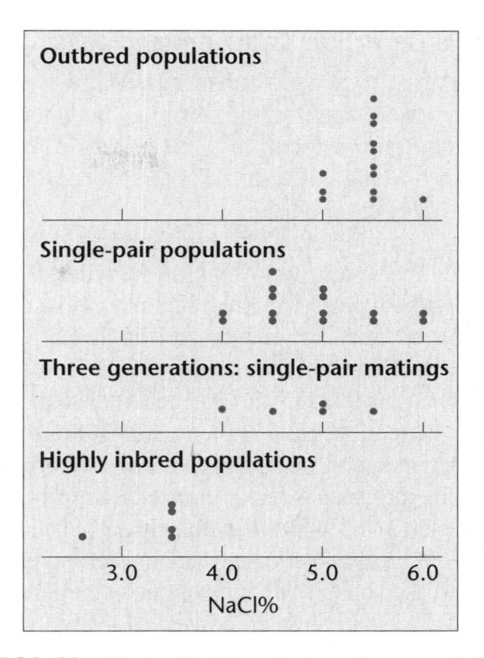

FIGURE 24–11 Effects of bottlenecks in various populations on evolutionary potential in *Drosophila*, as shown by distributions of NaCl concentrations at extinction.

Other experiments comparing inbred and outbred *Drosophila* populations exposed to environmental stresses, such as high temperatures and ethanol presence, have shown that populations with even a low level of inbreeding have a much greater probability of extinction at lower levels of environmental stress than outbred populations. Experimental results such as these indicate that genetic erosion does indeed reduce the long-term viability of a population by reducing its capacity to adapt to changing environmental conditions.

24.5 Conservation of Genetic Diversity Is Essential to Species Survival

Scientists working to maintain biological diversity face several dilemmas. Should they focus on preserving individual populations, or should they take a broader approach by trying to conserve not just one species but all the interdependent plants and animals in an ecosystem? How can genetic diversity be maintained in a species whose numbers are declining? Can genetic diversity lost from a population be restored? Early conservation efforts often focused only on the population size of an endangered species. Biologists now recognize that a complex interplay of different factors must be considered during conservation efforts, including the need to examine the habitat and role of a species within an ecosystem, as well as the importance of genetic variation for long-term survival.

Ex Situ Conservation: Captive Breeding

Ex situ (Latin for off-site) **conservation** involves removing plants or animals from their original habitat to an artificially maintained location such as a zoo or botanic garden. This living collection can then form the basis of a captive-breeding program.

These programs, where a few surviving individuals from an endangered species are removed from the wild and their offspring raised in a protected environment to rebuild the population, have been instrumental in bringing a number of species back from the brink of extinction. However, such programs can have undesirable genetic consequences that potentially jeopardize the long-term survival of the species even after population numbers have been restored. A captive-breeding program is rarely initiated until very few individuals are left in the wild, when the original genetic diversity of the species is already depleted. The breeding program is frequently based on a small number of captured individuals, so genetic diversity is further reduced by the founder effect. Since captive-breeding facilities often accommodate only a small number of individuals in the breeding group, inbreeding is difficult to avoid, especially for animal species that form harems (breeding groups dominated by a single male), so that N_e is considerably smaller than N. Finally, unintended selection for genotypes more suited to captive-breeding conditions over time can reduce the overall capacity of the recovered population to adapt and survive in the wild.

How can this type of program be managed to minimize these genetic effects? As we have already seen, loss of genetic diversity measured as heterozygosity over t generations can be expressed as

$$H_t/H_0 = (1 - 1/2N_e)^t$$

This equation indicates that the loss of genetic diversity H_t/H_0 will be greater with a smaller effective population size N_e and a larger number of generations t. We can expand this equation for a captive-breeding population as follows:

$$H_t/H_0 = [1 - (1/2N_{fo})]\{1 - 1/[2N(N_e/N)]\}^{t-1}$$

where N_{fo} is the effective size of the founding population, N is the mean population size, and N_e is the mean effective population size of the group over t generations. Examination of this equation shows that maximum genetic diversity in a captive-breeding group will be maintained by (1) using the largest possible number of founding individuals to maximize N_{fo}, (2) maximizing N_e/N so that as many individuals as possible produce offspring each generation, and (3) minimizing the number of generations in captivity to reduce t. The importance of maximizing N_{fo} to increase allelic diversity in the founding population can also be seen in Figure 24–12. Successful programs of this sort for endangered species are managed with these three goals in mind. In addition, keeping good pedigree records to avoid matings between relatives and exchanging individuals between different breeding programs when possible will reduce inbreeding.

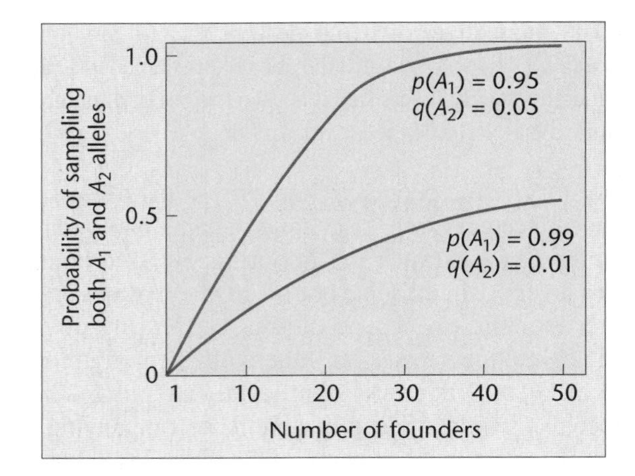

FIGURE 24–12 Effect of captive-population founder number on probability of sampling both A and a alleles at a locus.

Captive Breeding: The Black-Footed Ferret

The black-footed ferret, *Mustela nigripes,* (Figure 24–13) is an excellent example of a species that has been successfully subjected to *captive breeding*. This ferret was once widespread throughout the plains of the western United States, but by the 1970s they were considered to be extinct. Decades of trapping and poisoning of both the ferret and its main prey, the prairie dog, had decimated their populations. However, a small surviving colony of black-footed ferrets was discovered on a ranch near Meeteetse, Wyoming, in 1981. Conservation biologists first tried to conserve this ferret population *in situ*, but in 1985 the colony was infected with canine distemper, which nearly wiped them out. Of the 18 ferrets that were saved and transferred to a captive facility, only 8 were considered to be sufficiently unrelated to become the founders of a breeding program that has produced over 3000 ferrets. The species is now being reintroduced to parts of its original range. The goal is to establish populations of 1500 ferrets in sustainable wild colonies by 2010.

The small founder group of eight individuals caused a severe bottleneck for this species. Additional loss of

FIGURE 24–13 The black-footed ferret (*Mustela nigripes*). *(Jim Brandenburg/Minden Pictures)*

genetic diversity in the captive-breeding program occurred because of drift, limited reproduction from several of the founder females, and a high rate of breeding by one founder male. The risk of inbreeding and further genetic erosion in this type of program is still a problem.

Genetic management strategies, such as using DNA markers to identify the most genetically varied individuals, maintaining careful pedigree records to avoid mating closely related animals, and developing techniques for artificial insemination and sperm cryopreservation have helped to conserve the genetic diversity that remained in the species. Conservation geneticists working on the recovery project estimate that all existing black-footed ferrets share about 12% of their genome. This is roughly the equivalent of being full cousins. The effects of this degree of genetic similarity in the expanding ferret population are unclear. Occasional abnormalities such as webbed feet and kinked or short tails have been observed in the captive population, but without a non-inbred population available for comparison, it is unclear whether this is evidence that inbreeding is increasing the homozygosity for deleterious alleles.

Scientists disagree as to whether the long-term future of the black-footed ferret is jeopardized by the severe bottleneck and subsequent loss of genetic diversity the species has experienced. Some geneticists suggest that because the one surviving Meeteetse population was so isolated, it may have already become sufficiently inbred to purge any deleterious alleles. Other researchers point out that studies of black-footed ferret DNA extracted from museum specimens show that the ferrets alive today have lost significant genetic diversity compared with earlier pre-bottleneck populations, and they suggest that loss of fitness because of inbreeding will inevitably be seen over time.

Ex Situ Conservation: Gene Banks

Another form of ex situ conservation is provided by establishing **gene banks**. In contrast to housing entire animals or plants, these collections instead provide long-term storage and preservation for reproductive components, such as sperm, ova, and frozen embryos in the case of animals, and seed, pollen, and cultured tissue in the case of plants. Many more individual genotypes can be preserved for longer periods in a gene bank than in a living collection. Cryopreserved gametes or seeds can be used to reconstitute lost or endangered animals or plants after many years in storage. Because they are expensive to construct and maintain, most gene banks are used to conserve these components of domesticated species having economic value.

Gene banks have been established in many countries to help preserve genetic material of agricultural importance, such as traditional crop varieties that are no longer grown or old livestock breeds that are becoming rare. One of the most important *ex situ* collections in the United States is the National Center for Genetic Resources Preservation, a Department of Agriculture (USDA) facility in Fort Collins, Colorado, which maintains more than 300,000 different accessions of crop varieties and related wild species. Some of the accessions are stored as seeds and others as cryogenically preserved tissue from which whole plants can be regenerated (see this chapter's opening photograph on page 552). Animal genetic resources, including frozen semen and embryos from endangered livestock breeds, are also preserved at this facility.

Ex situ conservation using gene banks, while often vital, has several disadvantages. A major problem with gene banks is that even large collections cannot contain all the genetic variation that is present in a species. Conservation geneticists attempt to address this problem by identifying a **core collection** for a species. The core collection is a subset of individual genotypes that represents as much as possible of the genetic variation within a species; preserving the core collection takes priority over randomly collecting and preserving large numbers of genotypes. Another disadvantage of *ex situ* conservation is that the artificial conditions under which a species is preserved in a living collection or gene bank often create their own selection pressures. When seeds of a rare plant species are maintained in cold storage, for example, selection may occur for those genotypes better adapted to withstand the lower temperatures, and genotypes may be lost that would actually have greater fitness in the plant's natural environment. Yet another problem posed by *ex situ* conservation is that while the greatest biological diversity in both domesticated and nondomesticated species is frequently found in underdeveloped countries, most *ex situ* collections are situated in developed countries that have the resources to establish and maintain them. This leads to conflict over who owns and has access to the potentially valuable genetic resources maintained in such collections.

In Situ Conservation

In situ (Latin for on-site) **conservation** attempts to preserve the population size and biological diversity of a species while it is maintained in its original habitat. The use of species inventories to identify diversity hot spots is an important tool for determining the best places to establish parks and reserves where plants and animals can be protected from hunting or collecting and where their habitat can be preserved.

For domesticated species, there is increasing interest in "on-farm" preservation, whereby farmers are encouraged with additional resources and financial incentives to maintain traditional crop varieties and livestock breeds. Nondomesticated species with economic potential have also been targeted for *in situ* preservation. In 1998, the USDA established its first *in situ* conservation sites for a wild plant to protect populations of the native rock grape (*Vitis rupestris*) in several eastern states. The rock grape is prized by winegrowers not for its fruit, but for its roots; grape vines grafted onto wild rock grape rootstock are resistant to phylloxera, a serious pest of wine grapes.

The advantage of *in situ* conservation for the rock grape, as with other species, is that larger populations with greater genetic diversity can be maintained. Species conserved *in situ* will also continue to live and reproduce in the environments to which they are adapted, reducing the likelihood that novel selection pressures will produce undesirable changes in allele frequency. However, as the global human population continues to rise, setting aside suitable areas for *in situ* conservation becomes a greater challenge. As we have already seen in the case of the North American brown bear, even in large preserves like Yellowstone National Park, a species' migration and gene flow may be eliminated with the consequent loss of genetic diversity. This problem is even more acute in smaller, more fragmented areas of protected habitat.

Population Augmentation

What genetic considerations should accompany efforts to restore populations or species that are in decline? As we have already seen, populations that go through bottlenecks continue to suffer from low levels of genetic diversity, even after their numbers have recovered. We have also seen that inbreeding and drift contribute to genetic erosion in small populations, and fragmentation interrupts migration and gene flow, further reducing diversity. Captive-breeding programs designed to restore a critically endangered species from a few surviving individuals risk genetic erosion in the renewed population from founder effect and inbreeding, as in the case of the black-footed ferret.

An alternative strategy used by conservation biologists is **population augmentation**—boosting the numbers of a declining population by transplanting and releasing individuals of the same species captured or collected from more numerous populations elsewhere. Attempts to reestablish gene flow in severely fragmented populations of the endangered red-cockaded woodpecker by augmenting the smallest populations were described earlier. Other population augmentation projects in the United States have involved bighorn sheep and grizzly bears in the Rocky Mountains. This method has also been employed with the Florida panther, *Felis coryi* (Figure 24–14), using an isolated population of less than 50 animals confined to the area around the Big Cypress Swamp and Everglades National Park in south Florida. As we will discuss in the Genetics, Technology, and Society essay at the end of this chapter, DNA-profiling patterns show high levels of inbreeding in the Florida panther, with reduced fitness because of severe reproductive abnormalities and increased susceptibility to parasite infections. In this effort, seven unrelated animals from a captive population of South and North American panthers were released into the Everglades in the 1960s and allowed to interbreed with the Florida population, thus producing more genetically diverse family groups. Further population augmentation in this species has been controversial, however, as some biologists argue that the unique features that allow the Florida panther to be classified as a separate subspecies will be lost if mating with other panthers takes place.

Despite such controversies, population augmentation appears to be a valuable restoration tool. This strategy increases population numbers, as well as genetic diversity if the transplanted individuals are unrelated to those in the population to which they are introduced. However, one potential problem with population augmentation is **genetic swamping**. This occurs when the gene pool of the original population is overwhelmed by different genotypes from the transplanted individuals, thus altering their allele frequencies.

FIGURE 24–14 The Florida panther (*Felis coryi*). *(Lynn Stone/Animals Animals/Earth Scenes)*

Another difficulty with population augmentation can be caused by **outbreeding depression**, where reduced fitness occurs in the progeny from matings between genetically diverse individuals. Outbreeding depression occurring in the F_1 generation is thought to be due to the offspring being less well adapted to local environmental conditions than the parents.

This phenomenon has been documented in some plant species in which seeds of the same species—but from a different location—were used to revegetate a damaged area. Outbreeding depression that occurs in the F_2 and later generations is thought to be due to the disruption of **coadapted gene complexes**—groups of alleles that have evolved to work together to produce the best level of fitness in an individual. This type of outbreeding depression has been documented in F_2 hybrid offspring from matings between fish from different salmon populations in Alaska. These studies suggest that restoring the most beneficial type and amount of genetic diversity in a population is more complicated than previously thought, reinforcing the argument that the best long-term strategy for species survival is to prevent the loss of diversity in the first place.

GENETICS, TECHNOLOGY, AND SOCIETY

Gene Pools and Endangered Species: The Plight of the Florida Panther

In the last 400 years, more than 700 of the earth's animal and plant species have become extinct. In the United States alone, at least 30 species have suffered extinction in the last decade. In addition, hundreds of genetically distinct plant and animal species are now endangered. This dramatic loss of biological diversity is a direct consequence of human activity. As humans increase in number and spread over the earth's surface, we harness more and more of the earth's resources for our use. As a consequence of these activities, we have cleared the forests, polluted the water, and permanently altered Earth's natural balance.

Although the destruction continues, we are beginning to understand the implications of our actions and to make efforts to save some of the more endangered life forms. However, despite our best efforts to save threatened species, we often intercede too late—after population numbers have suffered severe declines and after important ecosystems have been permanently compromised. As a result, much of the genetic diversity that existed in these species is lost and the populations must be rebuilt from reduced gene pools. This genetic uniformity can reduce fitness, as well as expose genetic diseases. Reduced fitness may further deplete the numbers of the threatened plant or animal, and the population spirals downward to extinction.

The story of the Florida panther provides a dramatic example of an animal brought to the verge of extinction and the challenges that must be overcome to restore it to a healthy place in the ecosystem. The Florida panther is one of 30 subspecies of cougar and is one of the most endangered mammals in the world. They once roamed the southeastern corner of North America, from South Carolina and Arkansas to the southern tip of Florida. As people settled in the panther's habitats, they were considered to be a threat to livestock and humans. So the panthers were killed by hunting, poisoning, highway collisions, and loss of the habitat that supports their prey—primarily wild deer and hogs. By 1967, Florida panthers were listed as endangered, and only about 30 Florida panthers were left. Population estimates predicted that the Florida panther would be extinct by the year 2055.

Because of about 20 generations of geographical isolation and inbreeding, Florida panthers had the lowest levels of genetic heterozygosity of any subspecies of cougar. The loss of genetic diversity manifested itself in the appearance of some severe genetic defects. For example, almost 80% of panther males born after 1989 in the Big Cypress region showed a rare, heritable (autosomal dominant or sex-linked recessive) condition known as cryptorchidism—failure of one or both testicles to descend. This defect is associated with low testosterone levels and reduced sperm count. Life-threatening congenital heart defects also appeared in the Florida panther population, possibly because of an autosomal dominant gene defect. In addition, some immune deficiencies emerged, making the small panther population more susceptible to diseases and further contributing to the population's decline. Other less serious genetic features appeared, such as a kink in the tail and a whorl of fur on the back.

Over the last two decades, a faint glimmer of hope has appeared for the Florida panther. Federal and state agencies, as well as private individuals, have implemented a Florida Panther Recovery Program. The program's goal is to exceed 500 breeding animals by the year 2010. If successful, the panther species would be granted a 95% probability of survival, while retaining up to 90% of its genetic diversity. The plan includes a captive-breeding program, strict protection, increasing and improving the panther's habitat, and educating the public and private landowners. Wildlife underpasses have been constructed on highways in panther territory, and these have significantly reduced panther highway fatalities (which account for about half of panther deaths). By 2002, Florida panther numbers had increased to about 100 animals.

To retard the detrimental effects of inbreeding, eight wild female Texas panthers (a related subspecies from western Texas) were released into Florida panther territory in 1995. Three of the females died prior to breeding; however, the remaining Texas females gave birth to litters of healthy kittens. It is estimated that 40–70 of the 100 panthers in South Florida are now hybrids of Texas cougars and Florida panthers. None of the hybrids appears to have the kinked tail or other traits characteristic of the inbred Florida panthers. The hybrid cats are more genetically diverse than purebred Florida panthers, with up to 20-30% of their genetic material being contributed by the Texas cougars. This is of concern to some biologists, who worry that the Texas cats may genetically swamp the distinct Florida panther population.

Paradoxically, the success of the restoration program has become a problem. Now that the population has reached about 100 animals, the Florida panther has almost exceeded the capacity of its existing habitat. Biologists estimate that the population must reach at least 250 individuals in order to be self-sustaining. To reach this number, the panthers will need to expand their territory, again putting them in direct competition with human expansion in Florida. Biologists are now evaluating potential new territories for the growing population of Florida panthers—including parts of Louisiana, Arkansas, and South Carolina.

Despite the recent success of the restoration program, the survival of the Florida panther is far from certain. The panther's comeback will require years of monitoring and frequent intervention. In addition, people must be willing to share their land with wild creatures that are dangerous and do not directly further their self-interests. However, public support for the return of the Florida panther has been strong, so there may be hope for this unique, impressive animal.

References

Maehr, D. S., and Lacy, R. C. 2002. Avoiding the lurking pitfalls in Florida panther recovery. *Wildlife Soc. Bull.* 30(3): 971–78.

Mansfield, K. G., and Land, E. D. 2002. Cryptorchidism in Florida panthers: Prevalence, features and influence of genetic restoration. *J. of Wildlife Diseases* 38(4): 693–98.

Web Sites

Derr, M. 2002. *Florida panther's great leap hits a wall.* [on-line]. *The New York Times* [15 October 2002]. ***http://www.nytimes.com/2002/10/15/science/life/15PANT.html***

Florida Panther Net [on-line]. ***http://www.panther.state.fl.us/***

Chapter Summary

1. Biodiversity is lost as increasing numbers of plants and animal species are threatened with extinction. Conservation genetics applies principles of population genetics to the preservation and restoration of threatened species. A major concern of conservation geneticists is the maintenance of genetic diversity.
2. Genetic diversity includes interspecific diversity, which is reflected by the number of different species present in an ecosystem and intraspecific diversity, which is reflected by genetic variation within a population or between different populations of the same species. Genetic diversity can be measured by examining different phenotypes in the population or, at the molecular level, by using allozyme analysis or DNA-profiling techniques.
3. Major declines in a species' population numbers reduce genetic diversity and contribute to their risk of extinction as a result of genetic drift, inbreeding, or loss of gene flow. Populations that suffer severe reductions in effective size and then recover have passed through a population bottleneck and often show reduced genetic diversity.
4. Loss of genetic diversity reduces the capacity of a population to adapt to changing environmental conditions because useful alleles may disappear from the gene pool. Reduced genetic diversity also results in greater levels of homozygosity in a population, often leading to an accumulation of deleterious alleles and inbreeding depression.
5. Conservation of genetic diversity depends on *ex situ* methods, such as living collections, captive-breeding programs, and gene banks, as well as *in situ* approaches, such as the establishment of parks and preserves.
6. Population augmentation, in which individuals are transplanted into a declining population from a more numerous population of the same species located elsewhere, can be used to increase numbers and genetic diversity. However, the risk of outbreeding depression and genetic swamping accompanies this process.

Key Terms

allozyme analysis, 555
amplified fragment-length polymorphisms (AFLPs), 555
biodiversity, 553
coadapted gene complexes, 564
conservation, 563
conservation genetics, 553
core collection, 563
DNA fingerprinting, 557
ecosystems, 553
ex situ conservation, 562
founder effect, 558
gene banks, 563
gene flow, 560
genetic diversity, 554
genetic drift, 558
genetic load, 559
genetic erosion, 560
genetic swamping, 564
inbreeding, 559
inbreeding coefficient, 559
inbreeding depression, 559
in situ conservation, 563
interspecific diversity, 554
intraspecific diversity, 554
metapopulation, 550
outbreeding depression, 564
population augmentation, 564
population bottleneck, 557
population fragmentation, 554
SSRs (simple sequence repeats), 555

INSIGHTS AND SOLUTIONS

1. Is a rare species found as several fragmented subpopulations more vulnerable to extinction than an equally rare species found as one larger population? What factors should be considered when managing fragmented populations of a rare species?

Solution: A rare species in which the remaining individuals are divided among smaller isolated subpopulations can appear to be less vulnerable. If one subpopulation becomes extinct through local causes, such as disease or habitat loss, then the remaining subpopulations may still survive. However, genetic drift will cause smaller populations to experience more rapidly increasing homozygosity over time compared with larger populations. Even with random mating, the change in heterozygosity from one generation to the next because of drift can be calculated as

$$H_1 = H_0(1 - 1/2N)$$

where H_0 is the frequency of heterozygotes in the present generation, H_1 is the frequency of heterozygotes in the next generation, and N is the number of individuals in the population. Thus, in a small population of 50 individuals with an initial heterozygote frequency of 0.5, heterozygosity will decline to $0.5(1 - 1/100) = 0.495$, a loss of 0.5% in just one generation. In a larger population of 500 individuals and the same initial heterozygote frequency, after one generation, heterozygosity will be $0.5(1 - 1/1000) = 0.4995$, a loss of only 0.05%.

Therefore, smaller populations are likely to show the effects of homozygosity for deleterious alleles sooner than larger populations, even with random mating. If populations are fragmented so that movement of individuals or gametes between them is prevented, management options could include transplanting individuals from one subpopulation to another to enable gene flow to occur. Establishment of "wildlife corridors" of undisturbed habitat that connect fragmented populations could be considered. In captive populations, exchange of breeding adults (or their gametes through shipment of preserved semen or pollen) can be undertaken. Management for increased population numbers, however, is vital to prevent further genetic erosion through drift.

Problems and Discussion Questions

1. A wildlife biologist studied four generations of a population of rare Ethiopian jackals. When the study began, there were 47 jackals in the population and analysis of microsatellite loci from these animals showed a heterozygote frequency of 0.55. In the second generation, an outbreak of distemper occurred in the population and only 17 animals survived to adulthood. These jackals produced 20 surviving offspring, which in turn gave rise to 35 progeny in the fourth generation. (a) What was the effective population size for the four generations of this study? (b) Based on this effective population size, what is the heterozygote frequency of the jackal population in generation 4? (c) What is the inbreeding coefficient in generation 4, assuming an inbreeding coefficient of $F = 0$ at the beginning of the study, no change in microsatellite allele frequencies in the gene pool, and random mating in all generations?
2. Chondrodystrophy, a lethal form of dwarfism, has recently been reported in captive populations of the California condor and has killed embryos in 5 out of 169 fertile eggs. Chondrodystrophy in condors appears to be caused by an autosomal recessive allele with an estimated frequency of 0.09 in the gene pool of this species. (a)How do you think California condor populations should be managed in the future to minimize the effect of this lethal allele? (b)What are the advantages and disadvantages of attempting to eliminate it from the gene pool?
3. A geneticist is studying three loci, each with one dominant and one recessive allele, in a small population of rare plants. She estimates the frequencies of the alleles at each of these loci as $A = 0.75$, $a = 0.25$; $B = 0.80$, $b = 0.20$; $C = 0.95$, $c = 0.05$. What is the probability that all the recessive alleles will be lost from the population through genetic drift?
4. How are genetic drift and inbreeding similar in their effects on a population? How are they different?
5. You are the manager of a game park in Africa with a native herd of just 16 black rhinos, an endangered species worldwide. Describe how you would manage this herd to establish a viable population of black rhinos in the park. What genetic factors would you take into consideration in your management plan?
6. Compare the causes and effects of inbreeding depression and outbreeding depression.
7. Cloning, using the techniques similar to those pioneered by the Scottish scientists who produced Dolly the sheep, has been proposed as a way to increase the numbers of some highly endangered mammalian species. Discuss the advantages and disadvantages of using such an approach to aid long-term species survival.
8. In a population of wild poppies found in a remote region of the mountains of eastern Mexico, almost all of the members have pale yellow flowers, but breeding experiments show that pale yellow is recessive to deep orange. Using the tools of the conservation geneticists described in this chapter, how could you experimentally determine whether the prevalence of the recessive phenotype among the eastern Mexican poppy population is due to natural selection or simply due to the effects of genetic drift and/or inbreeding?
9. Contrast *ex situ* conservation techniques with *in situ* conservation techniques.
10. Describe how the captive-breeding *ex situ* conservation approach is applied to a severely endangered species.
11. Explain why a low amount of genetic diversity in a species is a detriment to the survival of that species.
12. Contrast allozyme analysis with AFLP analysis as measures of genetic diversity.
13. A population of endangered lowland gorillas is studied by conservation biologists in the wild. The biologists count 15 gorillas but observe that the population consists of two harems, each dominated by a different single male. One harem contains eight females and the other has five. What is the effective population size, N_e, of the gorilla population?
14. Twenty endangered red pandas are taken into captivity to found a breeding group. (a) What is the probability that at least one of the captured pandas has the genotype B_1/B_2 if $p(B_1) = 0.99$? (b) Careful management keeps the N_e/N ratio for the red panda breeding population at 0.42. If the overall size of the captive population is maintained at 50 individuals, what proportion of the heterozygosity present in the founding population will still be present after 5 generations in captivity?
15. Use your analysis of the red panda in Problem 14 to make suggestions for managing the captive-breeding population of red pandas to maintain as much genetic diversity as possible.
16. Przewalski's horse (*Equus przewalskii*), thought by some biologists to represent the ancestral species from which modern horses were domesticated, is classified as a separate species from the domestic horse although members of the two species can mate and produce fertile offspring. Przewalski's horse was hunted to extinction in its native habitat in the Asian steppes by the 1920s. The few hundred Przewalski's horses alive today are descended from 12 surviving individuals that had been taken into captivity. The founder breeding group also included a domestic mare. Currently, there is interest in reintroducing Przewalski's horse to its original range. (a)What genetic factors associated with the animals available for reintroduction do you think should be considered before a decision is reached? (b)What advice would you give to the conservation biologists managing the reintroduction project?
17. DNA profiles based on different kinds of molecular markers are increasingly used to measure levels of genetic diversity in populations of threatened and endangered species. Do you think these marker-based estimates of genetic diversity are reliable indicators of a population's potential for survival and adaptation in a natural environment? Why, or why not?
18. Seed banks provide managed protection for the conservation of species of economic and noneconomic importance. Schoen and colleagues (*Proc. Natl. Acad. Sci. [USA]*1998. 95: 394-99.) quantified considerable fitness decay in seeds maintained in long-term storage. When germination falls below 65–85%, regeneration (planting and seed collection) of a finite sample of the stored seed type is recommended. What genetic consequences might you expect to accompany the conservation practice of long-term seed storage and regeneration?
19. According to Holmes (2001. *Proc. Natl. Acad. Sci. [USA]* 98: 5072–77.), "One of the first questions a resource manager asks about threatened and endangered species is, 'How bad is it?'" More formally, the question seeks a population viability analysis that includes estimates of extinction risk. What factors would you consider significant in providing an estimate of extinction risk for a species?
20. Microsatellite loci are short (2-to-5 base pair) tandem repeats, which are abundantly and somewhat randomly distributed in all eukaryotic chromosomes. They show high mutability such that new alleles are produced more frequently than in traditional

protein-coding genes. It is likely that most microsatellites are selectively neutral and highly heterozygous in natural populations. Knowing that cheetahs underwent a population bottleneck approximately 12,000 years ago, North American pumas 10,000 years ago, and Gir Forest lions 1,000 years ago, in which of these species would you expect to see the highest degree of microsatellite polymorphism? The lowest?

21. Considering the behavior and evolution of microsatellites described in Problem 20, provide a graph that relates microsatellite polymorphism (variance) with the number of years since a species' bottleneck. Select variances from 1.0 to 9.0, where increasing variance represents increasing polymorphism, and set the range of years since the bottleneck from 10,000 to 50,000.

22. Allozymes are electrophoretically distinct forms of a particular protein, while microsatellites and minisatellites are repetitive DNA sequences that have been found in all eukaryotes studied to date. Such repetitive sequences are rarely associated with coding sequences of DNA. The data shown here represent the percent heterozygosity of allozymes and microsatellite and minisatellite DNAs in nuclear genomes of four species of felines (modified from Driscoll et al. 2002. *Genome Res.* 12: 414–23.): (a) Which species appears to contain the greatest genetic variability? Why might this species be so variable? (b) Which species appears to have the least genetic variability? (c) Why are allozymes less variable than minisatellite or microsatellite DNAs?

	% Heterozygosity			
Marker	*Cheetah*	*Lion*	*Puma*	*Domesticated Cat*
Allozyme	1.4	0.0	1.8	8.2
Minisatellite	43.3	2.9	10.3	44.9
Microsatellite	46.7	7.9	14.7	68.1

23. If we assume that one of the species in Problem 22 had undergone a recent population crisis in which the number of effective breeders reached critically low levels, which species do you think it would be?

24. For many conservation efforts, scientists lack sufficient data to make definitive conservation decisions. Yet the allocation of habitats for conservation cannot be delayed until such data are available. In these cases, conservation efforts are often directed toward three classes of species: flagships (high-profile species), umbrellas (species requiring large areas for habitat), and biodiversity indicators (species representing diverse, especially productive habitats) (Andelman and Fagan, 2000. *Proc. Natl. Acad. Sci. [USA]* 97: 5954–59.). In terms of protecting threatened species, what advantages and disadvantages might accompany investing scarce talent and resources in each of these classes?

Selected Readings

Baker, C. S., and Palumbi, S. R. 1994. Which whales are hunted? A molecular genetic approach to monitoring whaling. *Science* 265: 1538–39.

Bonnell, M. L., and Selander, R. K. 1974. Elephant seals: Genetic variation and near extinction. *Science* 184: 908–09.

Daniels, S. J., and Walters, J. R. 2000. Inbreeding depression and its effects on natal dispersal in red-cockaded woodpeckers. *Condor* 102: 482–91.

Dobson, A., and Lyles, A. 2000. Black-footed ferret recovery. *Science* 288: 985.

Frankham, R. 1995. Conservation genetics. *Ann. Rev. Genet.* 29: 305–27.

Gharrett, A. J., and Smoker, W. W. 1991. Two generations of hybrids between even-year and odd-year pink salmon (*Oncorhynchus gorbuscha*): A test for outbreeding depression? *Canadian J. Fish. Aquat. Sci.* 48: 426–38.

Hedrick, P. W. 2001. Conservation genetics: Where are we now? *Trends Ecol. Evol.* 16: 629–36.

Lacy, R. C. 1997. Importance of genetic variation to the viability of mammalian populations. *J. Mammalogy* 78: 320–35.

Paetkau, D., et. al. 1998. Variation in genetic diversity across the range of North American brown bears. *Conserv. Biol.* 12: 418-29.

Palacios, C., and Gonzalez-Candelas, F. 1999. AFLP analysis of the critically endangered *Limonium cavanillesii*. *J. of Heredity* 90: 485–89.

Ralls, K. et.al. 2000. Genetic management of chondrodystrophy in California condors. *Animal Conserv.* 3: 145–53.

Roman, J., and Bowen, B. W. 2000. The mock turtle syndrome: Genetic identification of turtle meat purchased in the southeastern United States of America. *Animal Conserv.* 3: 61–65.

Wayne, R. K. et. al. 1991. Conservation genetics of the endangered Isle Royale gray wolf. *Conserv. Biol.* 5: 41–51.

Wilson, E. O. (ed.) 1988. *Biodiversity*. Washington, DC: National Academy of Sciences.

Wynen, L. P., et.al. 2000. Postsealing genetic variation and population structure of two species of fur seal. *Mol. Ecol.* 9: 299–314.

Solutions to Problems and Discussion Questions

Chapter 1 An Introduction to Genetics

2. *Epigenesis* refers to the theory that organisms are derived from the assembly and reorganization of substances in the egg, which eventually lead to the development of the adult. *Preformationism* is a seventeenth-century theory stating that the sex cells (eggs or sperm) contain miniature adults, called homunculi, which grow in size to become the adult. Each theory postulates a fundamental difference in the manner in which organisms develop from hereditary determiners.

4. Their theory of natural selection proposed that more offspring are produced than can survive and that in the competition for survival, those with favorable variations survive. Darwin did not understand the nature of heredity and variation.

6. *Transmission* genetics is the most classical approach in which the patterns of inheritance are studied through selective matings or the results of natural matings. *Molecular* and *biochemical* genetic analyses examine the nature of gene expression, regulation, and replication. Recombinant DNA technology has had a significant impact in this area as well as others. In *population* genetics, the interest is in the behavior of genes in groups of organisms (populations), often with an interest in the factors that change gene frequencies in time and space.

8. Norman Borlaug applied Mendelian principles of hybridization and trait selection to the development of superior varieties of wheat. Such varieties are now grown in many countries, including Mexico, and have helped maintain the world supply of food.

10. In the last 40 years, human transmission, cytological, and molecular genetics have provided an understanding of many aspects of both plant and animal biology, including development of pest-resistant crops and identification of hazardous organisms in our food (*E. coli*, for example). In addition, much has been learned about many human diseases. Major medical areas of activity include genetic counseling, gene mapping and identification, disease diagnosis, and genetic engineering.

Chapter 2 Mitosis and Meiosis

2. Chromosomes that are homologous share many properties, including their overall length, position of the centromere (metacentric, submetacentric, acrocentric, and telocentric), banding patterns, type and location of genes, and autoradiographic pattern. *Diploidy* is a term often used in conjunction with the symbol $2n$. It means that both members of a homologous pair of chromosomes are present. *Haploidy* specifically refers to the fact that each haploid cell contains one chromosome of each homologous pair of chromosomes.

6. Notice the different anaphase shapes of chromosomes as they move to the poles: metacentric (a), submetacentric (b), acrocentric (c), telocentric (d).

8. Major divisions of the cell cycle include interphase and mitosis. Interphase is composed of four phases: G1, G0, S, and G2. During the S phase, chromosomal DNA doubles. Karyokinesis involves nuclear division, while cytokinesis involves division of the cytoplasm.

10. The *p53* gene plays a role in regulating the G1 to S transition. It is a tumor-suppressor gene. Mutations cause a lack of control over the cell cycle, and cancer often follows. When conditions exist that a cell which are likely to be harmful in the long term, such as extensive DNA damage, programmed cell death (apoptosis) can be initiated to protect the organism. Apoptosis can be triggered through a number of mechanisms, one of which involves the product of the *p53* gene.

12. Compared with mitosis, which maintains a chromosomal constancy, meiosis provides for a reduction in chromosome number and an opportunity for the exchange of genetic material between homologous chromosomes.

14. Sister chromatids are genetically identical, except where mutations may have occurred during DNA replication. Nonsister chromatids are genetically similar if on homologous chromosomes or genetically dissimilar if on nonhomologous chromosomes. If crossing over occurs, then chromatids attached to the same centromere will no longer be identical.

16. (a) Eight tetrads, (b) eight dyads, (c) eight monads migrating to *each* pole.

18. First, through independent assortment of chromosomes at anaphase I of meiosis, daughter cells (secondary spermatocytes and secondary oocytes) may contain different sets of maternally and paternally derived chromosomes. Second, crossing over, which happens at a much higher frequency in meiotic cells as compared to mitotic cells, allows maternally and paternally derived chromosomes to exchange segments, thereby increasing the likelihood that the daughter cells (that is, secondary spermatocytes and secondary oocytes) are genetically unique.

20. There would be 16 combinations with the addition of another chromosome pair.

22. One-half of each tetrad will have a maternal homolog: $(1/2)^{10}$.

24. In angiosperms, meiosis results in the formation of microspores (male) and megaspores (female), which give rise to the haploid male and female gametophyte stage. Micro- and megagametophytes produce the pollen and the ovules, respectively. Following fertilization, the sporophyte is formed.

26. The folded-fiber model is based on each chromatid's consisting of a single fiber that is wound like a skein of yarn; each fiber consists of DNA and protein. A coiling process occurs during the transition of interphase chromatin to more condensed chromosomes during prophase of mitosis or meiosis. Such condensation leads to a 5000-fold contraction in the length of the DNA within each chromatid. The transition is at the end of interphase and the beginning of prophase when the chromosomes are in the condensation process. This eventually leads to the typically shortened and "fattened" metaphase chromosome.

28. Duplicated chromosomes A^m, A^p, B^m, B^p, C^m, and C^p will align at metaphase. The centromeres divide, and the sister chromatids move to opposite poles at anaphase.

30. As long as you have accounted for eight possible combinations in the previous problem, there would be no new ones added in this problem.

32. The products of nondisjunction of chromosome C at the end of meiosis I are shown here.

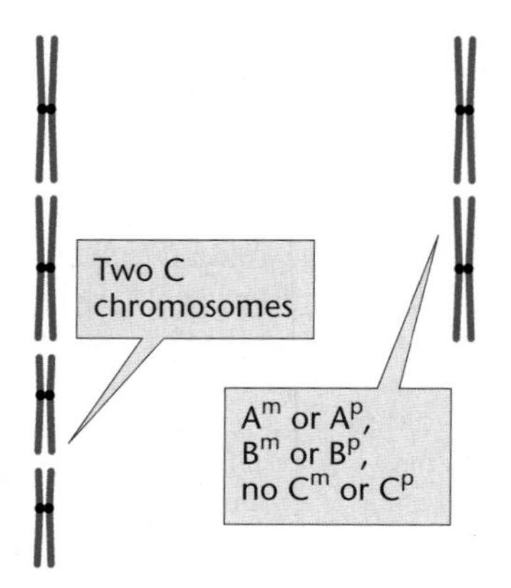

At the end of meiosis II, assuming that the C chromosomes separate as dyads instead of monads during meiosis II, you would have monads for the A and B chromosomes and dyads (from the cell on the left) for both C chromosomes as one possibility. However, another possibility exists for the products of meiosis II, as shown below.

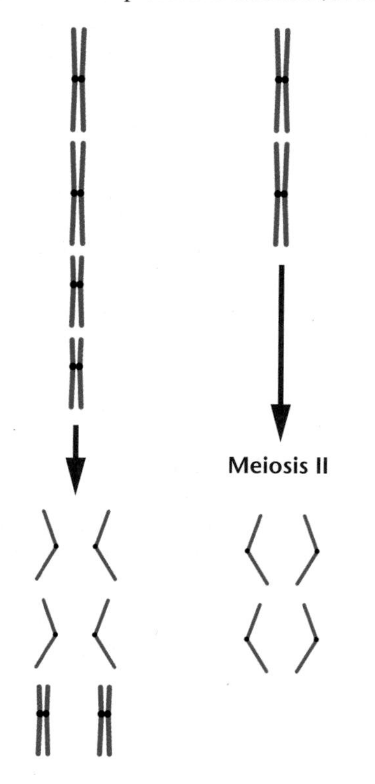

Chapter 3 Mendelian Genetics

2. (a) The parents are both normal, therefore, they could be either *AA* or *Aa*. That they produce an albino child requires that each parent provide an *a* gene to the albino child; thus, the parents must both be heterozygous (*Aa*). (b) The female must be *aa*. Since all the children are normal, you would consider the male to be *AA* instead of *Aa*. However, the male could be *Aa*. Under that circumstance, the likelihood of having six children, all normal, is $1/64$.

6. Symbolism:

w = wrinkled seeds g = green cotyledons
W = round seeds G = yellow cotyledons

$$P_1: WWGG \times wwgg$$

Gametes produced: One member of each gene pair is "segregated" to each gamete.

F_1: $WwGg$
$F_1 \times F_1$: $WwGg \times WwGg$
$9/16$ $W\_G\_$ round seeds, yellow cotyledons
$3/16$ W_gg round seeds, green cotyledons
$3/16$ $wwG\_$ wrinkled seeds, yellow cotyledons
$1/16$ $wwgg$ wrinkled seeds, green cotyledons.

Forked, or branch, diagram:

Seed Shape	*Cotyledon Color*		*Phenotypes*
3/4 round	3/4 yellow	⟶	9/16 round, yellow
	1/4 green	⟶	3/16 round, green
1/4 wrinkled	3/4 yellow	⟶	3/16 wrinkled, yellow
	1/4 green	⟶	1/16 wrinkled, green.

8. In Problem 7, (c) fits this description.

10. Mendel's four postulates are

1. Unit factors occur in pairs.
2. Some genes have dominant and recessive alleles.
3. Alleles segregate from each other during gamete formation. When homologous chromosomes separate from each other at anaphase I, alleles go to opposite poles of the meiotic apparatus.
4. One gene pair separates independently from other gene pairs. Different gene pairs on the same homologous pair of chromosomes (if far apart) or on nonhomologous chromosomes separate independently from each other during meiosis.

12. Homozygosity is a condition where both genes of a pair are the same (i.e., *AA* or *GG* or *hh*), whereas heterozygosity is the condition where members of a gene pair are different (i.e., *Aa* or *Gg* or *Bb*). Homozygotes produce only one type of gamete, whereas heterozygotes will produce 2^n types of gametes, where n = number of heterozygous gene pairs (assuming independent assortment).

14. (a) 4: *AB, Ab, aB, ab*
(b) 2: *AB, aB*
(c) 8: *ABC, ABc, AbC, Abc, aBC, aBc, abC, abc*
(d) 2: *ABc, aBc*

(e) 4: *ABc, Abc, aBc, abc*
(f) $2^5 = 32$

16. Symbols:

Seed Shape	*Seed Color*
W = round	G = yellow
w = wrinkled	g = green

P$_1$: *WWgg* × *wwGG*
F$_1$: *WwGg* cross to *wwgg* (a typical test cross).
The offspring will occur in a typical 1:1:1:1 ratio as

1/4 *WwGg* (round, yellow)
1/4 *Wwgg* (round, green)
1/4 *wwGg* (wrinkled, yellow)
1/4 *wwgg* (wrinkled, green)

18. (a)

Expected Ratio	*Observed (o)*	*Expected (e)*
9/16	315	312.75
3/16	108	104.25
3/16	101	104.25
1/16	32	34.75

$$\chi^2 = 0.47$$

The χ^2 value is associated with a probability greater than 0.90 for 3 degrees of freedom. The observed and expected values do not deviate significantly.

For parts (b) and (c), it is easier to see the observed values for the monohybrid ratios if the phenotypes are listed:

smooth, yellow	315
smooth, green	108
wrinkled, yellow	101
wrinkled, green	32

(b) For the smooth/wrinkled monohybrid component, the smooth types total 423(315 + 108), while the wrinkled types total 133(101 + 32).

Expected Ratio	*Observed (o)*	*Expected (e)*
3/4	423	417
1/4	133	139

The χ^2 value is 0.35 for 1 degree of freedom, giving a p value greater than 0.50 and less than 0.90. We fail to reject the null hypothesis and are confident that the observed values do not differ significantly from the expected values.

(c) For the yellow/green portion of the problem, there are 416 yellow plants (315 + 101) and 140(108 + 32) green plants.

Expected Ratio	*Observed (o)*	*Expected (e)*
3/4	416	417
1/4	140	139

The χ^2 value is 0.01 for 1 degree of freedom, the p value is greater than 0.90. We fail to reject the null hypothesis and are confident that the observed values do not differ significantly from the expected values.

20. Using $p = 0.10$ as the "critical" value for rejecting or failing to reject the null hypothesis instead of $p = 0.05$ would allow more null hypotheses to be rejected. As the χ^2 values increase, there is a higher likelihood that the null hypothesis will be rejected, because the higher values are more likely to be associated with a p value less than 0.05. As the critical p value is increased, it takes a smaller χ^2 value to cause rejection of the null hypothesis. It would take less difference between the expected and observed values to reject the null hypothesis; therefore, the stringency of failing to reject the null hypothesis is increased.

22. The probability of getting *aabbcc* from the *AaBbCC* × *AABbCc* mating is zero.

24. Genes that are recessive can skip generations and exist in a carrier state in parents. For example, notice that II-4 and II-5 produce a female child (III-4) with the affected phenotype. On these criteria alone, the gene must be viewed as being recessive.

I-1 (*Aa*), I-2 (*aa*), I-3 (*Aa*), I-4(*Aa*)
II-1 (*aa*), II-2 (*Aa*), II-3 (*aa*), II-4 (*Aa*), II-5 (*Aa*)
II-6 (*aa*), II-7 (*AA* or *Aa*), II-8 (*AA* or *Aa*)
III-1 (*AA* or *Aa*), III-2 (*AA* or *Aa*), III-3 (*AA* or *Aa*)
III-4 (*aa*), III-5 (probably *AA*), III-6 (*aa*)
IV-1 through IV-7 all *Aa*

26. The gene is inherited as an autosomal recessive.

I-1 (*aa*), I-2 (*Aa* or *AA*), I-3 (*Aa*), I-4 (*Aa*)
II-1 (*Aa*), II-2 (*Aa*), II-3 (*Aa*), II-4 (*Aa*), II-5 (*aa*)
II-6 (*AA* or *Aa*), II-7 (*AA* or *Aa*)
III-1 (*AA* or *Aa*), III-2 (*aa*), III-3 (*AA* or *Aa*)

28. (a) First consider that each parent is homozygous (true-breeding in the question), and since in the F$_1$ only round, axial, violet, and full phenotypes were expressed, they must each be dominant. Because all genes are on nonhomologous chromosomes, independent assortment will occur. (b) Round, axial, violet, and full would be the most frequent phenotypes: 3/4 × 3/4 × 3/4 × 3/4. (c) Wrinkled, terminal, white, and constricted would be the least frequent phenotypes: 1/4 × 1/4 × 1/4 × 1/4. (d) 1/4 × 1/4 × 1/4 × 1/4 × 2. (e) There would be 16 different phenotypes in the test-crossed offspring, just as there are 16 different phenotypes in the F$_2$ generation.

30. (a) Notice in cross #1 that the ratio of straight wings to curled wings is 3:1, and the ratio of short bristles to long bristles is also 3:1. This indicates that straight is dominant to curled, and short is dominant to long. Possible symbols would be (using standard *Drosophila* symbolism)

straight wings $= w^+$ short bristles $= b^+$
curled wings $= w$ long bristles $= b$

(b) Cross 1: $w^+/w; b^+/b \times w^+/w; b^+/b$
Cross 2: $w^+/w; b/b \times w^+/w; b/b$
Cross 3: $w/w; b/b \times w^+/w; b^+/b$
Cross 4: $w^+/w^+; b^+/b \times w^+/w^+; b^+/b$
(one parent could be w^+/w)
Cross 5: $w/w; b^+/b \times w^+/w; b^+/b$.

Chapter 4 Modification of Mendelian Ratios

2. *Incomplete dominance* can be viewed more as a quantitative phenomenon where the heterozygote is intermediate (approximately) between the limits set by the homozygotes. *Codominance* can be viewed in a more qualitative manner where both of the alleles in the heterozygote are expressed. For example in the AB blood group, both the I^A and I^B genes are expressed.

4. *Pp* × *Pp* ↓

1/4 *PP* (lethal)
2/4 *Pp* (platinum)
1/4 *pp* (silver)

The ratio of surviving foxes is 2/3 platinum, 1/3 silver. The *P* allele behaves as a recessive in terms of lethality (seen only in the homozygote), but as a dominant in terms of coat color (seen in the homozygote).

6. Flower color: *RR* = red; *Rr* = pink; *rr* = white
Flower shape: *P* = personate; *p* = peloric
(a) *RRpp* × *rrPP* ⟶ *RrPp*
(b) *RRPP* × *rrpp* ⟶ *RrPp*
(c) *RrPp* × *RRpp* ⟶ { *RRPp*, *RRpp*, *RrPp*, *Rrpp* }
(d) *RrPp* × *rrpp* ⟶ { *rrPp*, *rrpp*, *RrPp*, *Rrpp* }

In the cross of the F_1 of (a) to the F_1 of (b), both of which are double heterozygotes, you would expect

RrPp × *RrPp*

1/4 red
- ╱ 3/4 personate ⟶ 3/16 red, personate
- ╲ 1/4 peloric ⟶ 1/16 red, peloric

2/4 pink
- ╱ 3/4 personate ⟶ 6/16 pink, personate
- ╲ 1/4 peloric ⟶ 2/16 pink, peloric

1/4 white
- ╱ 3/4 personate ⟶ 3/16 white, personate
- ╲ 1/4 peloric ⟶ 1/16 white, peloric

8. This is a case of gene interaction (novel phenotypes), where the yellow and black types (double mutants) interact to give the cream phenotype; and epistasis, where the *cc* genotype produces albino.
(a) *AaBbCc* ⟶ gray (*C* allows pigment).
(b) *A_B_Cc* ⟶ gray (*C* allows pigment).
(c) Use the forked-line method for this portion.

3/4 *A_*
- 3/4 *B_*
  - ╱ 1/2 *Cc* ⟶ 9/32 gray
  - ╲ 1/2 *cc* ⟶ 9/32 albino
- 1/4 *bb*
  - ╱ 1/2 *Cc* ⟶ 3/32 yellow
  - ╲ 1/2 *cc* ⟶ 3/32 albino

1/4 *aa*
- 3/4 *B_*
  - ╱ 1/2 *Cc* ⟶ 3/32 black
  - ╲ 1/2 *cc* ⟶ 3/32 albino
- 1/4 *bb*
  - ╱ 1/2 *Cc* ⟶ 1/32 cream
  - ╲ 1/2 *cc* ⟶ 1/32 albino

Combining the phenotypes gives (always count the proportions to see that they add up to 1.0)

16/32 albino
9/32 gray
3/32 yellow
3/32 black
1/32 cream

10. *RG* = normal vision; *rg* = color-blind
Mother's father: X^{rg}/Y
Father's father: X^{rg}/Y
Mother: $X^{RG}X^{rg}$
Father: X^{RG}/Y
$X^{RG}X^{rg} \times X^{RG}/Y$ ⟶

$X^{RG}X^{RG}$ = 1/4 daughter, normal
$X^{RG}X^{rg}$ = 1/4 daughter, normal
X^{RG}/Y = 1/4 son, normal
X^{rg}/Y = 1/4 son, color-blind

The distribution of offspring will be (a) 1/4, (b) 1/2, (c) 1/4, (d) zero.

12. *RR* = red, *Rr* = red in females
Rr = mahogany in males
rr = mahogany
P_1: female: *RR* (*red*) × male: *rr* (mahogany)
F_1: *Rr* = females, red; males, mahogany
1/2 females, red
1/2 males, mahogany
F_2: 1/4 *RR*; 2/4 *Rr*; 1/4 *rr*
Because half of the offspring are males and half are females, for clarity, you could rewrite the F_2 as

	1/2 females	1/2 males
1/4 *RR*	1/8 red	1/8 red
2/4 *Rr*	2/8 red	2/8 mahogany
1/4 *rr*	1/8 mahogany	1/8 mahogany

14. Symbolism: Normal wing margins = sd^+; scalloped = *sd*.
(a) P_1: $X^{sd}X^{sd} \times X^+/Y$ ⟶
F_1:
1/2 X^+X^{sd} (female, normal)
1/2 X^{sd}/Y (male, scalloped)
F_2:
1/4 X^+X^{sd} (female, normal)
1/4 $X^{sd}X^{sd}$ (female, scalloped)
1/4 X^+/Y (male, normal)
1/4 X^{sd}/Y (male, scalloped)

(b) P_1: $X^+/X^+ \times X^{sd}/Y$ ⟶
F_1:
1/2 X^+X^{sd} (female, normal)
1/2 X^+/Y (male, normal)
F_2:
1/4 X^+X^+ (female, normal)
1/4 X^+X^{sd} (female, normal)
1/4 X^+/Y (male, normal)
1/4 X^{sd}/Y (male, scalloped)

If the *scalloped* gene were not X-linked, then all of the F_1 offspring would be wild (phenotypically) and a 3:1 ratio of normal to scalloped would occur in the F_2.

16. (a) P_1: X^vX^v; +/+ × X^+/Y; b^r/b^r
F_1:
1/2 X^+X^v; $+/b^r$ (female, normal)
1/2 X^vY; $+/b^r$ (male, vermilion)

F_2:

Eye color (X)	Eye color (autosomal)
1/4 females, normal	3/4 normal (3/16)
	1/4 brown (1/16)
1/4 females, vermilion	3/4 normal (3/16)
	1/4 brown (1/16)
1/4 males, normal	3/4 normal (3/16)
	1/4 brown (1/16)

1/4 males, vermilion < 3/4 normal (3/16); 1/4 brown (1/16)

3/16 females, normal
1/16 females, brown
3/16 females, vermilion
1/16 females, white
3/16 males, normal
1/16 males, brown
3/16 males, vermilion
1/16 males, white

(b) P_1: $X^+X^+; b^r/b^r \times X^v/Y; +/+$ ↴

F_1: $1/2\ X^+X^v; +/b^r$ (female, normal)
$1/2\ X^+/Y; +/b^r$ (male, normal)

F_2: 6/16 females, normal
2/16 females, brown
3/16 males, normal
1/16 males, brown
3/16 males, vermilion
1/16 males, white

(c) P_1: $X^vX^v; b^r/b^r \times X^+/Y; +/+$ ↴

F_1: $1/2\ X^+X^v; +/b^r$ (female, normal)
$1/2\ X^v/Y; +/b^r$ (male, vermilion)

F_2: 3/16 females, normal
1/16 females, brown
3/16 females, vermilion
1/16 females, white
3/16 males, normal
1/16 males, brown
3/16 males, vermilion
1/16 males, white

18. (a) Hypothesize that two gene pairs are involved in the inheritance of one trait while one gene pair is involved in the other.
(b) Croaking is due to one (dominant/recessive) gene pair, while eye color is due to two gene pairs. Because there is a 9:4:3 ratio regarding eye color, some gene interaction (epistasis) is indicated.
(c) Symbolism: Croaking: *R_* = rib-it; *rr* = knee-deep.
Eye color: Since the most frequent phenotype is blue eye, let *A_B_* represent the genotypes. For the purple class, a 3/16 group, use the *A_bb* genotypes. The green class, 4/16, would be the *aaB_* and the *aabb* groups.
(d) *AABBrr* × *AAbbRR*, which would produce an F_1 of *AABbRr* that would be blue-eyed and rib-it. The F_2 will follow a pattern of a 9:3:3:1 ratio because of homozygosity for the *A* locus and heterozygosity for both the *B* and *R* loci.

9/16 *AAB_R_* (blue-eyed, rib-it)
3/16 *AAB_rr* (blue-eyed, knee-deep)
3/16 *AAbbR_* (purple-eyed, rib-it)
1/16 *AAbbrr* (purple-eyed, knee-deep)

20. 9/16 *A_B_* (solid white)
3/16 *aaB_* (solid white)
3/16 *A_bb* (black-and-white spotted)
1/16 *aabb* (solid black)

The selection of *bb* as giving the spotted phenotype is arbitrary. You could obtain *AAbb* true-breeding black-and-white spotted cattle.

22. You can envision two pathways leading to the production of green pigment:

⟶ *B* ⟶ blue pigment ↘
green pegment
⟶ *Y* ⟶ yellow pigment ↗

YyBb (green) × *YyBb* (green)

9/16 *Y_B_* (green)
3/16 *yyB_* (blue)
3/16 *Y_bb* (yellow)
1/16 *yybb* (albino)

P_1: *YYBB* (green) × *yybb* (albino)

or

YYbb × *yyBB*
↓
YyBb (green)

Crossing these F_1's gives the observed ratios in the F_2.

24. The test for allelism is made by crossing the various mutant strains. If the resulting offspring are mutant, then the mutations are allelic. If the offspring are wild-type, then the mutations are not allelic and complementation is occurring. In cross 1, all offspring are wild-type, indicating that *r1* and *r2* are complementing and therefore not allelic. In cross 2, all offspring have tan eyes, indicating that the mutations are allelic. Since mutations *r1* and *r3* are in the same gene and *r1* and *r2* are not, the cross *r2* × *r3* should be complementing; that is, *r2* and *r3* are in different genes.

26. (a) Phenotypes: Himalayan × Himalayan ⟶ albino
Genotypes: c^hc^a; c^hc^a; c^ac^a
The Himalayan parents must both be heterozygous to produce an albino offspring.

Phenotypes: full color × albino ⟶ chinchilla
Genotypes: Cc^{ch}; c^ac^a; $c^{ch}c^a$
$c^ac^a \times c^{ch}c^a$ ↴
1/2 chinchilla 1/2 albino

(b) Phenotypes: albino × chinchilla ⟶ albino
Genotypes: c^ac^a; $c^{ch}c^a$; c^ac^a
Phenotypes: full color × albino ⟶ full color
Genotypes: $C\_$; c^ac^a; Cc^a
It is impossible to determine the complete genotype of the full-color parent, but the full-color offspring must be as indicated, Cc^a. Therefore, the cross of the albino with full color would be

$c^ac^a \times Cc^a$ ↴
1/2 full color 1/2 albino

(c) Phenotypes: chinchilla × albino ⟶ Himalayan
Genotypes: $c^{ch}c^h$; c^ac^a; c^hc^a
The chinchilla parent must be heterozygous for Himalayan because of the Himalayan offspring
Phenotypes: full color × albino ⟶ Himalayan
Genotypes: Cc^h; c^ac^a; c^hc^a
Therefore, a cross between the two Himalayan types would produce the following offpsring
$c^hc^a \times c^hc^a$ ↴
3/4 Himalayan 1/4 albino

28. (a) In a cross of *AACC* × *aacc*, the offpsring are all *AaCc* (agouti) because the *C* allele allows pigment to be deposited in the hair; when it is deposited, it will be agouti. F_2 offspring would have "simplified" genotypes with the corresponding phenotypes.

9/16 *A_C_* (agouti)
3/16 *A_cc* (colorless because *cc* is epistatic to *A*)
3/16 *aaC_* (black)

1/16 *aacc* (colorless because *cc* is epistatic to *aa*)
The two colorless classes are phenotyically indistinguishable; therefore, the final ratio is 9:3:4.
(b) The results of crosses
*A_C_* (female, agouti) × *aacc* (male, colorless)
are given in three groups (1) *AACc*, (2) *AaCC*, (3) *AaCc*.

30. The clue to the solution comes from the description of the Dexters as not true-breeding and of low fertility. This indicates that Dexters are heterozygous and the Kerry breed is homozygous recessive. The homozygous dominant type is lethal. Polled is caused by an independently assorting dominant allele, while horned is caused by the recessive allele to polled.

32. Since both of the parents are *Dd*, the parent contributing the eggs must be *Dd*. Therefore, all of the offspring must have the phenotype of the mother's genotype, which is dextral.

34. Developmental phenomena that occur early are more likely to be under maternal influence than those occurring late. Anterior/posterior and dorsal/ventral orientations are among the earliest to be established, and in organisms where their study is experimentally and/or genetically approachable, they often show considerable maternal influence. Maternal-effect genes produce products that are not carried over for more than one generation, as is the case with organelle and infectious heredity. Crosses illustrate the transient nature of a maternal effect; however, depending on particular biochemical/developmental parameters, all crosses may not give these types of patterns:

Aa (female) × *aa* (male) ⟶ all offspring of the *A* phenotype

Take a female *A* phenotype from this cross, and conduct the mating *aa* × *Aa* male. All offspring may be of the *a* phenotype because all of the offspring will reflect the *genotype* of the mother, not her *phenotype*. This cross illustrates that maternal effects last only one generation.

Chapter 5 Sex Determination and Sex Chromosomes

4. Sexual differentiation is the response of cells, tissues, and organs to signals provided by the genetic mechanisms of sex determination. In other words, genes are present that signal developmental pathways whereby the sexes are generated. Sexual differentiation is the complex set of responses to those genetic signals.

6. In *Drosophila*, it is the balance between the number of X chromosomes and the number of haploid sets of autosomes that determines sex. In humans, there is a small region on the Y chromosome that determines maleness.

8. In *primary* nondisjunction, half of the gametes contain two X chromosomes while the complementary gametes contain no X chromosomes. Fertilization of the female gametes with two X chromosomes by a Y-bearing sperm cell would produce the XXY Klinefelter syndrome. Fertilization of the "no-X" female gamete with a normal X-bearing sperm produces Turner syndrome.

10. No. Since the Y chromosome cannot be detected in these crosses, there is no way to distinguish the two modes of sex determination.

12. You would see daughters with the white-eye phenotype and sons with the miniature-wing phenotype.

14. Because synapsis of chromosomes in meiotic tissue is often accompanied by crossing over, it would be detrimental to sex-determining mechanisms to have sex-determining loci on the Y chromosome transferred, through crossing over, to the X chromosome.

16. The simple formula for determining the number of Barr bodies in a given cell is $N - 1$, where N is the number of X chromosomes.

Klinefelter syndrome (XXY)	1
Turner syndrome (XO)	0
47, XYY	0
47, XXX	2
48, XXXX	3

18. Unless other markers (cytological or molecular) are available, you cannot test the Lyon hypothesis with homozygous X-linked genes. The test requires identification of allelic alternatives to see differences in X chromosome activity.

20. Males normally have only one X chromosome; therefore, such mosaicism cannot occur. Females normally have two X chromosomes. However, there are cases of male calico cats that are XXY.

22. In mammals, the scheme of sex determination is dependent upon the presence of a piece of the Y chromosome. If present, a male is produced. In *Bonellia viridis*, the female proboscis produces some substance that triggers a morphological, physiological, and behavioral developmental pattern, that produces males. To elucidate the mechanism, you could attempt to isolate and characterize the active substance by testing different chemical fractions of the proboscis. Mutant analysis usually provides critical approaches into developmental processes. Depending on characteristics of the organism, you could also attempt to isolate mutants that lead to changes in male or female development. Finally, by using micro-tissue transplantations, you could attempt to determine which anatomical "centers" of the embryo respond to the chemical cues of the female.

24. We could account for the significant departures from a 1:1 ratio of males to females by suggesting that at anaphase I of meiosis, the Y chromosome more often goes to the pole that produces the more viable sperm cells. We could also speculate that the Y-bearing sperm has a higher likelihood of surviving in the female reproductive tract or that the egg surface is more receptive to Y-bearing sperm. At this time, the mechanism is unclear.

26.

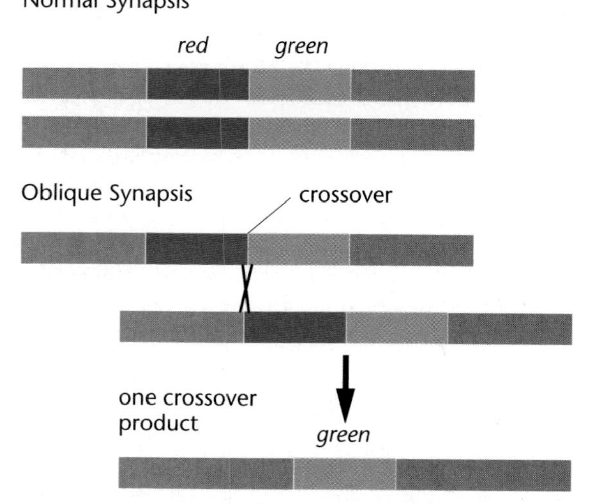

28. (a) Something is missing from the male-determining system of sex determination at the level of the genes, gene products, or receptors, and so forth. (b) The *SOX9* gene or its product is probably involved in male development. Perhaps it is activated by *SRY*. (c) There is probably some evolutionary relationship between the *SOX9* gene and *SRY*. There is considerable evidence that many other genes and pseudogenes are also homologous to *SRY*. (d) Normal female sexual development does not require the *SOX9* gene or gene product(s).

30. In snapping turtles, sex determination is strongly influenced by temperature such that males are favored in the 26–34°C range. Lizards, on the other hand, appear to have their sex determined by factors other than temperature in the 20–40°C range.

32. The white patches of CC are due to an autosomal gene, *S*, for white spotting, which prevents pigment formation in the cell lineages in which it is expressed. Homoygous *SS* cats have more white than heterozygous *Ss* cats, and there is no absolute pattern of patches that are due to the *S* allele; so the distribution of white patches would be expected to be different from Rainbow. In addition, since X chromosome inactivation is random, CC would have a different patch pattern from her genetic mother on the random X inactivation basis alone.

Chapter 6 Quantitative Genetics

4. Any genotype that contains a heterozygote could never act as a true-breeding strain. Those with three uppercase letters, *AABb* for example, weighing 12.4 g, or those with one uppercase letter, *Aabb* for example, weighing 10.8 g.

6. (a) It is possible that two parents of moderate height can produce offspring that are much taller or shorter than either parent, because segregation can produce a variety of gametes and therefore offspring such as

$$\frac{rrSsTtuu}{\text{(moderate)}} \times \frac{RrSsTtUu}{\text{(moderate)}}$$

Offspring from this cross can range from very tall *RrSSTTUu* (12 "tall" units) to very short *rrssttuu* (8 "small" units). (b) If the individual with a minimum height, *rrssttuu*, is married to an individual of intermediate height, *RrSsTtUu*, the offspring can be no taller than the height of the taller parent.

8. (a) The genotypes of the parents would be combinations of alleles that would produce a 6-cm (*aabbcc*) tail and a 30-cm (*AABBCC*) tail, while the 18-cm offspring would have a genotype of *AaBbCc*. (b) For example, a mating of an *AaBbCc* pig with the 6-cm *aabbcc* pig would result in these offspring.

Gametes (18-cm tail)	***Gamete (6-cm tail)***	***Offspring***
ABC		*AaBbCc* (18 cm)
ABc		*AaBbcc* (14 cm)
AbC		*AabbCc* (14 cm)
Abc	*abc*	*Aabbcc* (10 cm)
aBC		*aaBbCc* (14 cm)
aBc		*aaBbcc* (10 cm)
abC		*aabbCc* (10 cm)
abc		*aabbcc* (6 cm)

In this example, a 1:3:3:1 ratio is the result. However, had a different 18-cm-tailed pig been selected, say *AABbcc* × *aabbcc*, a different ratio would occur.

Gametes (18-cm tail)	***Gamete (6-cm tail)***	***Offspring***
ABc	*abc*	*AaBbcc* (14 cm)
Abc		*Aabbcc* (10 cm)

10. The F_2 falls into a normal distribution with 13 classes at 2-inch intervals. We would then conclude inheritance with six gene pairs involved. If the following test cross is involved, *AaBbCcDdEeFf* × *aabbccddeeff*, there would still be 13 classes in the offspring, but they would occur in a 1:1:1:1:1:1:1:1:1:1:1:1:1 ratio rather than a normal distribution.

12. For height, notice that average differences between MZ twins reared together (1.7 cm) and MZ twins reared apart (1.8 cm) are similar (meaning little environmental influence) and considerably less than differences between DZ twins (4.4 cm) or sibs (4.5 cm) reared together. These data indicate that genetics plays a major role in determining height. However, for weight, notice that MZ twins reared together have a much smaller (1.9 kg) difference than MZ twins reared apart, indicating that the environment has a considerable impact on weight. By comparing the weight differences of MZ twins reared apart with DZ twins and sibs reared together, we can conclude that the environment has almost as much an influence on weight as does genetics.

14. (a) Mean = 140 cm

(b) $s^2 = V = n\Sigma f(x^2) - (\Sigma fx)^2/n(n-1)$
$= 374.18$

(c) The standard deviation is the square root of the variance, or 19.34.

(d) The standard error of the mean is the standard deviation divided by the square root of *n*, or about 0.70.

The plot below approximates a normal distribution. Variation is continuous.

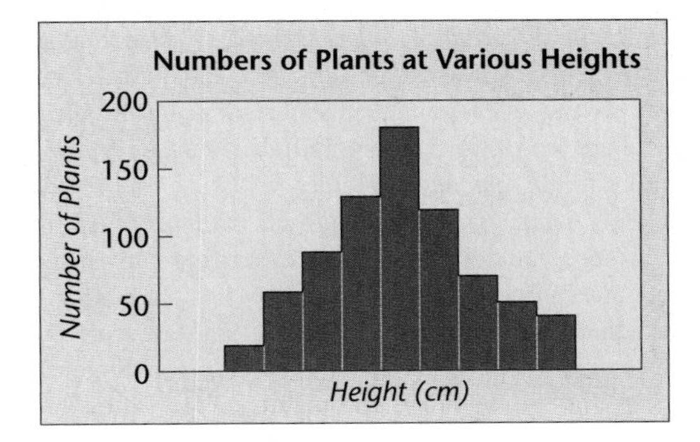

16. For a trait that is quantitatively measured, the relative importance of genetic versus environmental factors may be formally assessed by examining the heritability index (H^2, or broad heritability). In animal and plant breeding, a measure of potential response to selection based on additive variance and dominance variance is termed narrow heritability (h^2). A relatively high narrow heritability is a prediction of the impact selection may have in altering an initial randomly breeding population.

18. The formula for estimating heritability is

$$H^2 = V_G/V_P$$
$$= 0.53$$

This value, when viewed in percentage form, indicates that about 53% of the variation in plant height is due to genetic influences.

20. $h^2 = (7.5 - 8.5/6.0 - 8.5) = 0.4$

Selection will have little relative influence on olfactory learning in *Drosophila*.

22. Chromosome 2 seems to confer considerable resistance to the insecticide, somewhat in the heterozygous state, and more in the homozygous state. Thus, some partial dominance is occurring.

24. (a) Each additive gene contributes about 1.2 mm to the phenotype. (b) The fit to this backcross supports the original hypothesis. (c) These data do not support the simple hypothesis provided in part (a). (d) With these data, there are no distinct phenotypic classes, suggesting that the environment may play a role in eye development or that there are more genes involved.

26. In the case of brachydactyly, there are numerous modifier genes in the genome that can influence brachydactyly expression. Examination of OMIM (*Online Mendelian Inheritance of Man*) through *http://www.ncbi.nlm.nih.gov* will illustrate this point.

28. It is likely that the flies maintained in the *Drosophila* repository are more highly inbred and less heterozygous than those recently obtained from the wild. Response to selection is dependent on genetic variation. The greater the genetic variation in a species, the more likely and dramatic the response to selection. Therefore, we would expect a greater response to selection in the wild population.

30. Breeders attempt to select out this disorder by first maintaining complete and detailed breeding records of afflicted strains. Second, they avoid breeding dogs whose close relatives are afflicted. The molecular-developmental mechanism that causes the "month of birth" effect in canine hip dysplasia is unknown.

Chapter 7 Chromosome Mutations: Variation in Number and Arrangement

4. The fact that there is a significant maternal-age effect associated with Down syndrome indicates that nondisjunction in older females contributes disproportionately to the number of Down syndrome individuals. In addition, certain genetic and cytogenetic marker data indicate the influence of female nondisjunction.

6. Because an allotetraploid has a possibility of producing bivalents at meiosis I, it would be considered the most fertile of the three. Having an even number of chromosomes to match up at the metaphase I plate, autotetraploids would be considered to be more fertile than autotriploids.

8. It is likely that an interspecific hybridization occurred, followed by chromosome doubling. These events probably produced a fertile amphidiploid (allotetraploid).

10. While there is the appearance that crossing over is suppressed in inversion "heterozygotes," the phenomenon extends from the fact that the crossover chromatids end up being abnormal in genetic content. As such, they fail to produce viable (or competitive) gametes or lead to zygotic or embryonic death.

12. In a work entitled *Evolution by Gene Duplication*, Ohno suggests that gene duplication has been essential in the origin of new genes. If gene products serve essential functions, mutation, and therefore evolution, would not be possible unless these gene products could be compensated for by products of duplicated normal genes. The duplicated genes or the original genes would be able to undergo mutational "experimentation" without necessarily threatening the survival of the organism.

14. A Turner syndrome female has the sex chromosome composition of XO. If the father had hemophilia, it is likely that the Turner syndrome individual inherited the X chromosome from the father and no sex chromosome from the mother. If nondisjunction occurred in the mother, either during meiosis I or meiosis II, an egg with no X chromosome can be the result.

16. Given the basic chromosome set of nine unique chromosomes (a haploid complement), other forms with the "n multiples" are forms of polyploidy. Organisms with 27 chromosomes ($3n$) are more likely to be sterile because there are trivalents at meiosis I, which cause a relatively high number of unbalanced gametes to be formed.

18. The cross would be *WWWW* × *wwww*.

F_1: *WWww*

F_2: 35*W* and 1*w*

20. Since two Gl_1 alleles and two ws_3 alleles are present in the triploid, they must have come from the pollen parent. It is implied by the wording of the problem that the pollen parent contributed an unreduced (2n) gamete; however, another explanation is possible: dispermic fertilization. In this case, two $Glws_3$ gametes could have fertilized the ovule.

22. In a paracentric inversion, one recombinant chromatid is dicentric (two centromeres), while the other is acentric (lacks a centromere). In a pericentric inversion, recombinant chromatids have duplications and deletions, however, no acentric or dicentric chromatids are produced.

24. In plants, gametes with aberrant unbalanced genetic complements usually fail to develop normally, leading to aborted pollen or ovules. Therefore, genetic imbalance is revealed prior to fertilization and inviable seeds result. In animals, aberrant or unbalanced gametes tend to function, however, significant embryonic and subsequent developmental abnormalities are likely to occur.

26. (a) Reciprocal translocation.

(b)

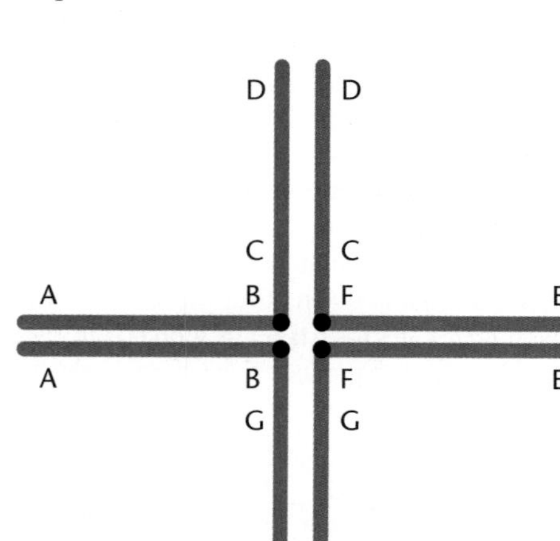

(c) Notice that all chromosomal segments are present and there is no apparent loss of chromosomal material. However, if the breakpoints for the translocation occurred within genes, then an abnormal phenotype may be the result. In addition, a gene's function is sometimes influenced by its position—its neighbors, in other words. If such "position effects" occur, then a different phenotype may result.

28. The symbol t(14;21) indicates that part of chromosome 21 is translocated to chromosome 14. When a gamete containing such a chromosome plus a normal chromosome 21 is fertilized by a standard haploid gamete, the individual has 46 chromosomes but effectively three copies of chromosome 21.

30. (a) In light of this information, meiosis I must have produced the abnormal oocytes with more or less than 24 chromosomes, indicating multiple conditions of nondisjunction. More likely, the oocytes consisted of 22 $^1/_2$ chromosomes, 22 normal dyads, and a single monad. (b) The result will be a monosomic and a normal zygote, assuming that the half chromosome (monad) migrates intact to one pole or the other. (c) In all likelihood, premature division of the centromere (at meiosis I) probably causes the single (nonduplicated) chromosome at meiosis II. (d) These data indicate that some forms of aneuploidy result from premature division of the centromere at meiosis I.

Chapter 8 Linkage and Chromosome Mapping in Eukaryotes

2. First, in order for chromosomes to engage in crossing over, they must be in proximity. It is likely that the side-by-side pairing that occurs during synapsis is the earliest time during the cell cycle that chromosomes achieve the necessary proximity. Second, chiasmata are visible during prophase I of meiosis, and it is likely that these structures are intimately associated with the genetic event of crossing over.

4. Because crossing over occurs at the four-strand stage of the cell cycle (that is, after S phase), notice that each single crossover involves only two of the four chromatids.

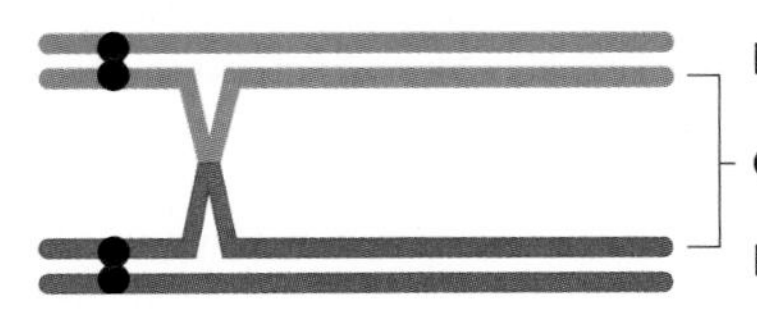

6. Positive interference occurs when a crossover in one region of a chromosome interferes with crossovers in nearby regions. Such interference ranges from 0.0 (no interference) to 1.0 (complete interference). Interference is often explained by a physical rigidity of chromatids such that they are unlikely to make sufficiently sharp bends to allow crossovers to be close together.

8.

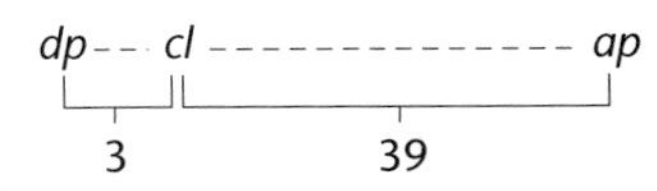

10. The question is whether the arrangement in the parents is *coupled* ($RY/ry \times ry/ry$) or *not coupled* ($Ry/rY \times ry/ry$). Notice that the most frequent phenotypes in the offspring, the parentals, are colored, green (88) and colorless, yellow (92). This indicates that the heterozygous parent in the test cross is coupled ($RY/ry \times ry/ry$). There would be 10 map units between the loci.

12. $PZ/pz \times pz/pz$. Adding the crossover percentages together ($6.9 + 7.1$) gives 14%, which would be the map distance between the two genes.

14.

	female A	*female B*	Frequency
NCO	3, 4	7, 8	first
SCO	1, 2	3, 4	second
SCO	7, 8	5, 6	third
DCO	5, 6	1, 2	fourth

16. (a) $yw+/++ct \times yw+/Y$

(b)

y ---------- w ---------------------- ct

0.0 1.5 20.0

(c) There were $0.185 \times 0.015 \times 1000 = 2.775$ double crossovers expected. (d) Because the cross to the F_1 males included the normal (wild-type) gene for *cut wings,* it would not be possible to unequivocally determine the genotypes from the F_2 phenotypes for all classes.

18. P_1: $+/+$; $p\,e/p\,e$ (females) $\times$ dp/dp; $++/++$ (males)

F_1: $+/dp$; $++/p\,e$ (females) $\times$ dp/dp; $p\,e/p\,e$ (males)

0.20 wild-type
0.05 ebony
0.05 pink
0.20 pink, ebony
0.20 dumpy
0.05 dumpy, ebony
0.05 dumpy, pink
0.20 dumpy, pink, ebony

For the reciprocal cross:

F_1: $+/dp$; $++/p\,e$ (males) $\times$ dp/dp; $p\,e/p\,e$ (females); there would be no crossover classes.

0.25 wild-type
0.25 pink, ebony
0.25 dumpy
0.25 dumpy, pink, ebony.

The results would change because of no crossing over in males.

20. (a, b)

$$a\text{–}b = \frac{32 + 38 + 0 + 0}{1000} \times 100$$
$$= 7 \text{ map units;}$$
$$b\text{–}c = \frac{11 + 9 + 0 + 0}{1000} \times 100$$
$$= 2 \text{ map units}$$

(c) The progeny phenotypes that are missing are $++c$ and $a\,b+$, which, of 1000 offspring, 1.4 would be expected. Perhaps by chance or some other unknown selective factor, they were not observed.

22. These observations as well as the results of other experiments indicate that the synaptonemal complex is required for crossing over.

24. Since the genetic map is more accurate when relatively small distances are covered and when large numbers of offspring are scored, this map would probably not be very accurate with such a small sample size.

26. 33/64 *A_B_*
15/64 *A_bb*
15/64 *aaB_*
1/64 *aabb*

28. By having microscopically visible markers on the chromosomes, Creighton and McClintock were able to show that homologous chromosomal material physically exchanged segments during crossing over.

30.
$a\,b\,c = 168$
$+\,+\,+ = 168$
$a\,+\,+ = 20$
$+\,b\,c = 20$
$+\,+\,c = 10$
$a\,b\,+ = 10$
$+\,b\,+ = 2$
$a\,+\,c = 2$

The map distances would be computed as follows:

$$a\text{–}b = \frac{20 + 20 + 2 + 2}{400} \times 100$$
$$= 11 \text{ map units}$$
$$b\text{–}c = \frac{10 + 10 + 2 + 2}{400} \times 100$$
$$= 6 \text{ map units.}$$

32. B^+ = wild-type eye shape
B = Bar eye shape
m^+ = wild-type wings
m = miniature wings
e^+ = wild-type body color
e = ebony body color

Mapping the distance between B and m:

$$\frac{(57 + 64)}{(226 + 218 + 57 + 64)} \times 100 = \frac{121}{565} \times 100 = 21.4 \text{ map units}$$

We would conclude that the *ebony* locus is either far away from B and m (50 map units or more) or it is on a different chromosome. In fact, *ebony* is on a different chromosome.

Chapter 9 Mapping in Bacteria and Bacteriophages

2. (a) The requirement for physical contact between bacterial cells during conjugation was established by placing a filter in a U-tube such that the medium can be exchanged, but the bacteria cannot come in contact. Under this condition, conjugation does not occur. (b) By treating cells with streptomycin (an antibiotic), it was shown that recombination would not occur if one of the two bacterial strains was inactivated. However, if the other was similarly treated, recombination would occur. Thus, directionality was suggested, with one strain being a donor strain and the other being the recipient. (c) An F^+ bacterium contains a circular, double-stranded, structurally independent DNA molecule, which can direct recombination.

4. Bacteria that are F^+ possess the F factor, while those that are F^- lack the F factor. In Hfr cells, the F factor is integrated into the bacterial chromosome; in F′ bacteria, the F factor is free of the bacterial chromosome yet possesses a piece of the bacterial chromosome.

6. In an Hfr × F^- cross, the F factor is directing the transfer of the donor chromosome. It takes approximately 90 minutes to transfer the entire chromosome. Because the F factor is the last element to be transferred and the conjugation tube is fragile, the likelihood for complete transfer is low.

8. Transformation requires *competence* on the part of the recipient bacterium, meaning that only under certain conditions are bacterial cells capable of being transformed. Transforming DNA must be double-stranded to begin with, yet it is converted to a single-stranded structure upon insertion into the host cell. The most efficient length of the transforming DNA is about 1/200 the size of the host chromosome. Transformation is an energy-requiring process, and the number of sites on the bacterial cell surface is limited.

10. Notice that the incorporation of loci a^+ or b^+ occurs much more frequently than the incorporation of b^+ and c^+ together (210 to 1), and the incorporation of all three genes $a^+b^+c^+$ occurs relatively infrequently. If a and b loci are close together and both are far from locus c, then fewer crossovers would be required to incorporate the two linked loci compared to all three loci. If all three loci were close together, then the frequency of incorporation of all three would be similar to the frequency of incorporation of any two contiguous loci, which is not the case.

14. A single plaque is a clearing of bacteria resulting from the lytic action of millions of bacteriophages.

16. A lytic cycle occurs as bacteriophage enter a bacterial host and form progeny phage after a relatively short period of time. *Lysogeny* is a complex process whereby certain temperate phages can enter a bacterial cell and, instead of following a lytic developmental path, integrate their DNA into the bacterial chromosome.

18. In their experiment, a filter was placed between the two auxotrophic strains, which would not allow contact. F-mediated conjugation requires contact, and without that contact, such conjugation cannot occur. The treatment with DNase showed that the filterable agent was not naked DNA.

20. Cotransduction of genes in generalized transduction allows linkage relationships to be determined because the closer two genes are to each other, the higher the likelihood that they will be physically "linked" together in a single DNA strand during transduction.

22. Viral recombination occurs when there is a sufficiently high number of infecting viruses so that there is a high likelihood that more than one type of phage will infect a given bacterium. Under this condition, phage chromosomes can recombine by crossing over.

24. (a) Greater than 10^5; (b) about 1.4×10^7. (c) Remembering that 0.1 ml is typically used in the plaque assay, the initial concentration of phage is less than 10^7. Coupling this information with the calculations in part (b), it would appear that the initial concentration of phage is around 1×10^7, and the failure to obtain plaques in this portion of the experiment is expected and due to sampling error.

26. Because the frequency of double transformants is quite high (compare the trp^+tyr^+ transformants in A and B experiments), you may conclude that the genes are quite closely linked together.

28. (a) Rifampicin eliminates the donor strain, which is rif^s.
(b)

b a ________________ *c* ______ *F*

(c) An interrupted mating experiment involving a $rif^r\ amp^s$ strain plated on an ampicillin-containing medium would be one strategy. The recombinants must be replated on a rifampicin medium to determine which ones are sensitive.

30. Since g cotransforms with f, it is likely to be in the $b\ c\ f$ "linkage group" and would be expected to cotransform with each. We would not expect transformation with a, d, or e.

32. (a) No, all functional groups do not impact similarly on conjugative transfer of R27. Regions 1, 2, and 4 appear to be least influenced by mutation because transfer is at 100%. (b) Regions 3, 5, 6, 8, 9, 10, 12, 13, and 14 appear to have the most impact on conjugation because when mutant, conjugation is abolished. (c) Regions 7 and 11, when mutant, only partially abolish conjugation; therefore, they probably have less impact on conjugation than those listed in part (b). (d) The data in this problem provide some insight into the complexity of the genetic processes involved in bacterial conjugation. The regions that have the most impact on conjugation fall into three different functional groups. In addition, notice that regions 1, 2, 4, 7, and 11 (which appear to have little if any impact on conjugation) are functionally related as indicated by their shading.

Chapter 10 DNA Structure and Analysis

2. Prior to 1940, most interest in genetics centered on the transmission of similarity and variation from parents to offspring (transmission genetics). While some experiments examined the possible nature of the hereditary material, abundant knowledge of the structural and enzymatic properties of proteins generated a bias that worked to favor proteins as the hereditary substance. In addition, proteins were composed of as many as 20 different subunits (amino acids), thereby providing ample structural and functional variation for the multiple tasks that must be accomplished by the genetic material. The tetranucleotide hypothesis (structure) provided insufficient variability to account for the diverse roles of the genetic material.

4. Transformation is dependent on a macromolecule (DNA), which can be extracted and purified from bacteria. During such purification however, other macromolecular species may contaminate the DNA. Specific degradative enzymes—proteases, RNase, and DNase—were used to selectively eliminate components of the extract, and if transformation is concomitantly eliminated, then the eliminated fraction is the transforming principle. DNase eliminates DNA and transformation; therefore, it must be the transforming principle.

6. Actually, phosphorus is found in approximately equal amounts in DNA and RNA. Therefore labeling with ^{32}P would "tag" both RNA and DNA. However, the T2 phage, in its mature state, contains very little if any RNA; therefore, DNA would be interpreted as being the genetic material in T2 phage.

8. The early evidence would be considered indirect in that at no time was there an experiment—such as transformation in bacteria—in which genetic information in one organism was transferred to another using DNA. Rather, by comparing DNA content in various cell types (sperm and somatic cells) and observing that the *action* and *absorption* spectra of UV light were correlated, DNA was considered to be the genetic material. This suggestion was supported by the fact that DNA was shown to be the genetic material in bacteria and some phages. Direct evidence for DNA's being the genetic material comes from a variety of observations, including gene transfer, which has been facilitated by recombinant DNA techniques.

10. Linkages among the three components require the removal of water (H_2O).

12.

Guanine:	2-amino-6-oxypurine
Cytosine:	2-oxy-4-aminopyrimidine
Thymine:	2,4-dioxy-5-methylpyrimidine
Uracil:	2,4-dioxypyrimidine

16. Because in double-stranded DNA, A=T and G≡C (within limits of experimental error), the data presented would have indicated a lack of pairing of these bases in favor of a single-stranded structure or some other non-hydrogen–bonded structure. Alternatively, from the data, it would appear that A=C and T=G, which would negate the chance for typical hydrogen bonding since opposite charge relationships do not exist. Therefore, it is quite unlikely that a tight helical structure would form at all.

18. Three main differences between RNA and DNA are
1. Uracil in RNA replaces thymine in DNA.
2. Ribose in RNA replaces deoxyribose in DNA.
3. RNA often occurs as both single- and partially double-stranded forms, whereas DNA most often occurs in a double-stranded form.

20. The nitrogenous bases of nucleic acids (nucleosides, nucleotides, and single- and double-stranded polynucleotides) absorb UV light maximally at wavelengths 254–260 nm. Using this phenomenon, we can often determine the presence and concentration of nucleic acids in a mixture. Since proteins absorb UV light maximally at 280 nm, this is a relatively simple way of dealing with mixtures of biologically important molecules. UV absorption is greater in single-stranded molecules (hyperchromic shift) as compared to double-stranded structures; therefore, we can easily determine, by applying denaturing conditions, whether a nucleic acid is in the single- or double-stranded form. In addition, A=T-rich DNA denatures more readily than G≡C-rich DNA. Therefore, we can estimate base content by denaturation kinetics.

22. A *hyperchromic effect* is the increased absorption of UV light as double-stranded DNA (or RNA for that matter) is converted to single-stranded DNA. If you monitor the UV absorption with a spectrophotometer during the melting process, the hyperchromic shift can be observed. The T_m is the point on the temperature profile at which half (50%) of the sample is denatured.

24. For curve A, there is evidence for a rapidly reassociating species (repetitive) and a slowly reassociating species (unique). The fraction that reassociates faster than the *E. coli* DNA is highly repetitive, and the last fraction (with the highest $C_0t_{1/2}$ value) contains primarily unique sequences. Fraction B contains mostly unique, relatively complex DNA.

26. In one sentence of Watson and Crick's paper in *Nature*, they state, "It has not escaped our notice that the specific pairing we have postulated immediately suggests a possible copying mechanism for the genetic material."

28. MS-2 = 200 base pairs
E. coli = 2×10^6 base pairs

30. Left side (a) = right; right side (b) = left.

32. Under this condition, the hydrolyzed 5-methyl cytosine becomes thymine.

34. Heat application would yield a hyprochromic if the DNA is double-stranded. You could also get a rough estimation of the GC content from the kinetics of denaturation, and the degree of sequence complexity from comparative reassociation studies. Determination of base content by hydrolysis and chromatography could be used for comparative purposes and could also provide evidence as to the strandedness of the DNA. Antibodies for Z-DNA could be used to determine the degree of left-handed structures, if present. Sequencing the DNA from both viruses would indicate sequence homology. In addition, through various electronic searches readily available on the Internet (Web site: *blast@ncbi.nlm.nih.gov*, for example), you could determine whether similar sequences exist in other viruses or in other organisms.

Chapter 11 DNA Replication and Synthesis

2. The Meselson-Stahl experiment is conducted by labeling the pool of nitrogenous bases of the DNA of *E. coli* with a heavy isotope, ^{15}N. It would then be possible to follow the "old" DNA.

4. (a) Under a conservative scheme, all of the newly labeled DNA will go to one sister chromatid, while the other sister chromatid will remain unlabeled. In contrast to a semiconservative scheme, the first replicative round would produce one sister chromatid that has labels on both strands of the double helix. (b) Under a dispersive scheme, all of the newly labeled DNA will be interspersed with unlabeled DNA. Because these preparations (metaphase chromosomes) are highly coiled and condensed structures derived from the "spread-out" form at interphase (which includes the S phase), it is impossible to detect the areas where the label is not found. Rather, both sister chromatids would appear as evenly labeled structures.

6. The *in vitro* replication requires a DNA template, a primer to give a double-stranded portion, a divalent cation (Mg^{++}), and all four of the deoxyribonucleoside triphosphates: dATP, dCTP, dTTP, and dGTP. The lowercase d refers to the deoxyribose sugar.

8. Several analytical approaches showed that the products of DNA polymerase I were probably copies of the template DNA. *Base composition* was used initially to compare both templates and products. Within experimental error, those data strongly suggested that the DNA replicated faithfully.

10. Biologically active DNA implies that the DNA is capable of supporting typical metabolic activities of the cell or organism and it is capable of faithful reproduction.

12. None can *initiate* DNA synthesis on a template, but all can *elongate* an existing DNA strand assuming there is a template strand as shown in the figure below. Polymerization of nucleotides occurs in the 5′-to-3′ direction where each 5′ phosphate is added to the 3′ end of the growing polynucleotide.

All three enzymes are large complex proteins with a molecular weight in excess of 100,000 daltons, and each has 3′-to-5′ exonuclease activity.

DNA polymerase I:

- polymerization
- 3′-to-5′ exonuclease activity
- 5′-to-3′ exonuclease activity
- present in large amounts
- relatively stable
- removal of RNA primer

DNA polymerase II:

- polymerization
- 3′-to-5′ exonuclease activity
- possibly involved in repair function

DNA polymerase III:

- polymerization
- 3′-to-5′ exonuclease activity
- essential for replication
- complex molecule

14. (a)

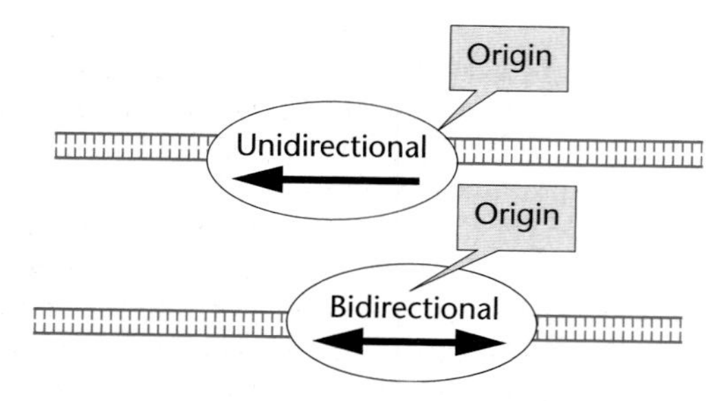

(b) Continuous DNA synthesis occurs in the direction of the replication fork. Discontinuous synthesis occurs in relatively short segments opposite the direction of the replication fork.

16. *Okazaki fragments* are relatively short (1000 to 2000 bases in prokaryotes) DNA fragments that are synthesized in a discontinuous fashion on the lagging strand during DNA replication. *DNA ligase* is required to form phosphodiester linkages in the gaps generated when DNA polymerase I removes RNA primer and meets newly synthesized DNA ahead of it. *Primer RNA* is formed by RNA primase to serve as an initiation point for the production of DNA strands on a DNA template.

18. Eukaryotic DNA is replicated in a manner very similar to that of *E. coli*. Synthesis is bidirectional, continuous on one strand, and discontinuous on the other; and the requirements of synthesis (four deoxyribonucleoside triphosphates, divalent cation, template, and primer) are the same. Okazaki fragments of eukaryotes are about one-tenth the size of those in bacteria. Because there is a much greater amount of DNA to be replicated and DNA replication is slower, there are multiple initiation sites for replication in eukaryotes (and increased DNA polymerase per cell) in contrast to the single replication origin in prokaryotes. Replication occurs at different sites during different intervals of the S phase.

20. (a) About 4,000,000 bp; (b) 1.3 mm.

22. Gene conversion is now considered to be a result of heteroduplex formation accompanied by mismatched bases. When these mismatches are corrected, the conversion occurs.

24. Telomerase activity is present in germline tissue to maintain telomere length from one generation to the next.

26. If replication is conservative, the first autoradiographs would have label distributed on only one side (chromatid) of the metaphase chromosome, as shown below.

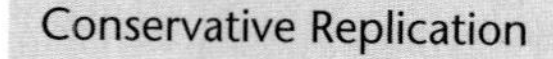

28. If the DNA contained parallel strands in the double helix and the polymerase were able to accommodate such parallel strands, there would be continuous synthesis and no Okazaki fragments. The telomere problem would be at only one end. Several other possibilities exist: If the DNA strands were replicated as complete single strands, the synthesis could begin at the opposite free ends. In addition, if the DNA existed only as a single strand, the same result would occur.

30. Conservative replication can be eliminated.

32. First, given that the label is distributed on both ends of the structure, replication must be bidirectional. Second, since both upper and lower portions are labeled, no restrictions to synthesis are apparent. The distribution of low-density grains in the center would indicate that replication begins in the middle and proceeds to the outward areas in both directions rather evenly.

Chapter 12 Chromosome Structure and DNA Sequence Organization

2. By having a circular chromosome, no free ends present the problem of linear chromosomes, namely, complete replication of terminal sequences.

4. Mitochondrial DNA varies in size from 16 kb in humans to over 350 kb in some plants and codes for ribosomal transfer and messenger RNAs. The protein-synthesizing apparatus and many components of cellular respiration are jointly formed from nuclear and mitochondrial genes. Chloroplast DNA codes for ribosomal RNAs, as well as tRNAs and mRNAs for ribosomal proteins, and photosynthesis. Nuclear genes also contribute to the pool of functional proteins in chloroplasts.

6. Mitochondria and chloroplasts contain proteins more similar to bacterial forms than those found in eukaryotes. This finding supports the endosymbiont hypothesis for the origin of such organelles from bacteria.

8. Since eukaryotic chromosomes are "multirepliconic" in that there are multiple replication forks along their lengths, we would expect to see multiple clusters of radioactivity.

10. Long interspersed elements (LINEs) are repetitive transposable DNA sequences in humans. The most prominent family, designated L1, is about 6.4 kb each and is represented about 100,000 times. LINEs are often referred to as retrotransposons because their mechanism of transposition resembles that used by retroviruses.

12. Because of the diverse cell types of multicellular eukaryotes, a variety of gene products are required, which may be related to the increase in DNA content per cell. In addition, the advantage of diploidy automatically increases DNA content per cell. However, seeing the question in another way, it is likely that a much higher *percentage* of the genome of a prokaryote is actually involved in phenotype production than in a eukaryote. Eukaryotes have evolved the capacity to obtain and maintain what appears to be large amounts of extra, perhaps "junk," DNA. Prokaryotes on the other hand, with their relatively short life cycle, are extremely efficient in their accumulation and use of their genome. In addition, the eukaryotic genome is divided into separate entities (chromosomes) to perhaps facilitate the partitioning process in mitosis and meiosis.

14. Nucleosomes are octomeric structures of two molecules of each histone (H2A, H2B, H3, and H4) except H1. Between the nucleosomes and complexed with linker DNA is histone H1. A 146-base pair sequence of DNA wraps around the nucleosome.

16. *Heterochromatin* is chromosomal material that stains deeply and remains condensed when other parts of chromosomes, euchromatin, are otherwise pale and uncondensed. Heterochromatic regions replicate late in S phase and are relatively inactive in a genetic sense, because there are few genes present, or if they are present, they are repressed. Telomeres and the areas adjacent to centromeres are composed of heterochromatin.

18. Volume of DNA:

$3.14 \times 10\ \text{Å} \times 10\ \text{Å} \times (50 \times 10^4\ \text{Å}) = 1.57 \times 10^8\ \text{Å}^3$

Volume of capsid:

$4/3(3.14 \times 400\ \text{Å} \times 400\ \text{Å} \times 400\ \text{Å}) = 2.67 \times 10^8\ \text{Å}^3$

Because the capsid head has a greater volume than the volume of DNA, the DNA will fit into the capsid.

20. Volume of the nucleus $= 5.23 \times 10^{11}\ \text{nm}^3$

Volume of the chromosome $= 1.9 \times 10^{11}\ \text{nm}^3$

The percentage of the volume of the nucleus occupied by the chromatin is

$$\frac{1.9 \times 10^{11}\ \text{nm}^3}{5.23 \times 10^{11}\ \text{nm}^3} \times 100 = \text{about } 36.3\%$$

22. Chromosomes are not randomly distributed within nuclei. Homologous chromosomes tend to distribute themselves opposite each other and in an antiparallel manner, meaning that their positions are in reverse order on opposite sides of the nucleus. Assuming that such patterns are maintained throughout the entire cell cycle, it is possible that chromosomal positions may influence gene function and/or chromosomal behavior during mitosis and/or meiosis. If gene function is influenced not only by gene position in a chromosome but also by gene position in a nucleus, then an alternative explanation for position effect exists.

24. Nucleosomes follow a *dispersive* pattern with each daughter chromatid containing a mixture of old and original nucleosomes. We could test the distribution of nucleosomes by conducting an autoradiographic experiment similar to Taylor-Woods-Hughes, but instead of labeling the DNA with ^{3}H-thymidine, one would label some or all the histones H2A, H2B, H3, and H4 in nucleosomes.

26. Phage λ is capable of forming a closed, double-stranded circular molecule because of a 12-base pair, single-stranded, complementary "overhanging" sequence at the 5′- end of each single strand.

28. The general frequency and pattern of various trinucleotide repeat motifs are similar in all taxononic groups. Within-gene trinucleotide repeats are the most frequent repeat motif in all taxonomic groups, followed by hexanucleotide repeats. One explanation might be that various microsatellite types (mono, di, tri, etc.) are generated at different rates in different genomic regions (within and between genes). A second possibility is that selection acts differentially depending on the type and location of a repeat. The correlation between the high frequency of tri- and hexanucleotide repeats within genes and a triplet code specifying particular amino acids within genes may not be coincidental.

30. If microsatellites in general are flanked by a conserved sequence, those conserved sequences may be involved in the generation and/or maintenance of the microsatellite. Alternatively, the microsatellite may generate the nonmicrosatellite region. Any hypothesis presented is in need of additional investigation before definitive statements can be made.

Chapter 13 The Genetic Code and Transcription

2. (a) The reason that + + + or − − − restored the reading frame is because the code is triplet. By having the + + + or − − −, the translation system is out of phase until the third + or − is encountered. If the code contained six nucleotides (a sextuplet code), then the translation system is out-of-phase until the sixth + or − is encountered. In this case, the out of phase region would probably be more extensive and likely cause more amino acid alterations; however, the reading frame would eventually be established. (b) Given a sextuplet code, restoration of the reading frames would occur only with the addition or loss of six nucleotides.

4. There are two possibilities for establishing the reading frames: ACA if we start at the first base and CAC if we start at the second base. These would code for two different amino acids (ACA = threonine; CAC = histidine) and would produce repeating polypeptides that would alternate *thr-his-thr-his . . .*, or *his-thr-his-thr . . . .* Given the sequence CUACUACUACUA, notice the different reading frames producing three different sequences, each containing the same amino acid.

Codons:	CUA	CUA	CUA	CUA . . .
Amino acids:	*leu*	*leu*	*leu*	*leu . . .*
	UAC	UAC	UAC	UAC . . .
	tyr	*tyr*	*tyr*	*tyr . . .*
	ACU	ACU	ACU	ACU . . .
	thr	*thr*	*thr*	*thr . . .*

If a tetranucleotide is used, such as ACGUACGUACGU . . . , then

Codons:	ACG	UAC	GUA	CGU	ACG
Amino acids:	*thr*	*tyr*	*val*	*arg*	*thr*
	CGU	ACG	UAC	GUA	CGU
	arg	*thr*	*tyr*	*val*	*arg*
	GUA	CGU	ACG	UAC	GUA
	val	*arg*	*thr*	*tyr*	*val*
	UAC	GUA	CGU	ACG	UAC
	tyr	*val*	*arg*	*thr*	*tyr*

Notice that the sequences are the same except that the starting amino acid changes.

6. From the repeating polymer ACACA . . . , we can say that threonine is either CAC or ACA. From the polymer CAACAA . . . with ACACA . . . , ACA is the only codon in common. Therefore, threonine would have the codon ACA.

8. The basis of the technique is that if a trinucleotide contains bases (a codon) that are complementary to the anticodon of a charged tRNA, a relatively large complex is formed containing the ribosome, the tRNA, and the trinucleotide. This complex is trapped in the filter, whereas the components by themselves are not trapped. If the amino acid on a charged, trapped tRNA is radioactive, then the filter becomes radioactive.

10. Apply the most conservative pathway of change.

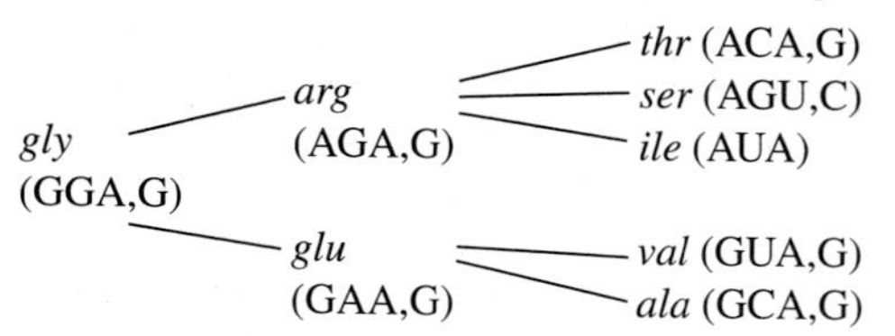

12. Because Poly U is complementary to Poly A, double-stranded structures will be formed. For an RNA to serve as a messenger RNA, it must be single-stranded, thereby exposing the bases for interaction with ribosomal subunits and tRNAs.

14. (a)

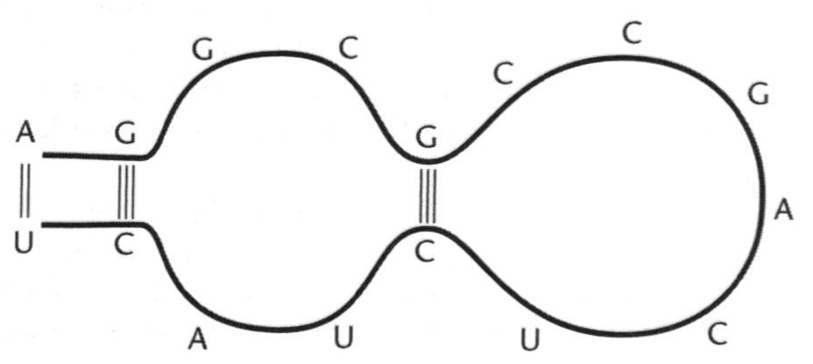

(b) TCCGCGGCTGAGATGA (use complementary bases, substituting T for U). (c) GCU (d) Assuming that the AGG . . . is the 5′-end of the mRNA, the sequence would be *arg-arg-arg-leu-tyr.*

16. (a) Starting from the 5′-end and locating the AUG triplets, we find two initiation sites leading to two sequences:

met-his-thr-tyr-glu-thr-leu-gly
met-arg-pro-leu-asp (or *glu*)

(b) In the shorter of the two reading sequences (the one using the internal AUG triplet), a UGA triplet was introduced at the second codon. While not in the reading frames of the longer polypeptide (using the first AUG codon), the UGA triplet eliminates the product starting at the second initiation codon.

18. The central dogma of molecular genetics and, to some extent, all of biology, states that DNA produces, through transcription, RNA, which is decoded during translation to produce proteins.

20. RNA polymerase from *E. coli* is a complex, large (almost 500,000 Da) molecule composed of subunits $(\alpha, \beta, \beta', \sigma)$ in the proportion $(\alpha_2, \beta, \beta', \sigma)$ for the holoenzyme. The β subunit provides catalytic function, while the sigma subunit is involved in recognition of specific promoters. The core enzyme is the protein without the σ.

22. While some folding (from complementary base pairing) may occur with mRNA molecules, they generally exist as single-stranded structures that are quite labile. Eukaryotic mRNAs are generally processed such that the 5′-end is capped and the 3′-end has a long string of adenine bases. It is thought that these features protect the mRNAs from degradation. Such stability of eukaryotic mRNAs probably evolved with the differentiation of nuclear and cytoplasmic functions. Because prokaryotic cells exist in a more unstable environment (nutritionally and physically, for example) than many cells of multicellular organisms, rapid genetic response to environmental change is likely to be adaptive. To accomplish such rapid responses, a labile gene product (mRNA) is advantageous. For instance, a pancreatic cell, which is developmentally stable and existing in a relatively stable environment, could produce more insulin on stable mRNAs for a given transcriptional rate.

24. First, compute the frequency (percentages would be easiest to compare) for each of the random codons.

For 4/5 C: 1/5 A:

$$\text{CCC} = 4/5 \times 4/5 \times 4/5 = 64/125\ (51.2\%)$$
$$\text{C}_2\text{A} = 3(4/5 \times 4/5 \times 1/5) = 48/125\ (38.4\%)$$
$$\text{CA}_2 = 3(4/5 \times 1/5 \times 1/5) = 12/125\ (9.6\%)$$
$$\text{AAA} = 1/5 \times 1/5 \times 1/5 = 1/125\ (0.8\%)$$

For 4/5 A: 1/5 C:

$$\text{AAA} = 4/5 \times 4/5 \times 4/5 = 64/125\ (51.2\%)$$
$$\text{A}_2\text{C} = 3(4/5 \times 4/5 \times 1/5) = 48/125\ (38.4\%)$$
$$\text{AC}_2 = 3(4/5 \times 1/5 \times 1/5) = 12/125\ (9.6\%)$$
$$\text{CCC} = 1/5 \times 1/5 \times 1/5 = 1/125\ (0.8\%)$$

Proline:	CCC, and one of the C_2A triplets
Histidine:	one of the C_2A triplets
Threonine:	one C_2A triplet, and one A_2C triplet
Glutamine:	one of the A_2C triplets
Asparagine:	one of the A_2C triplets
Lysine:	AAA

26. (a, b) Use the code table to determine the number of triplets that code each amino acid, then construct a graph and plot.

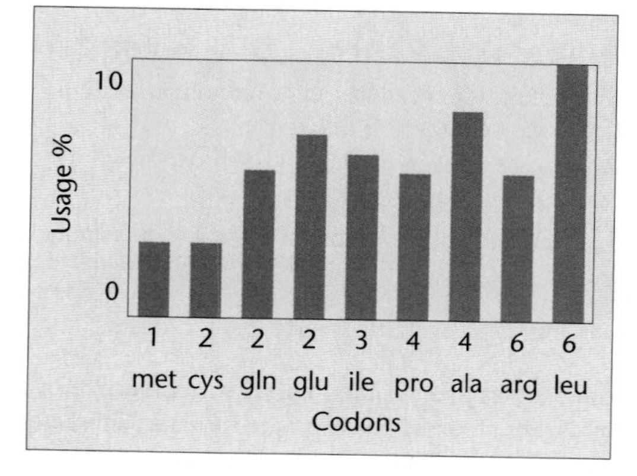

(c) There appears to be a weak correlation between the relative frequency of amino acid usage and the number of triplets for each. (d) To continue to investigate this issue, you could examine additional amino acids in a similar manner. In addition, different phylogenetic groups use code synonyms differently. It may be possible to find situations in which the relationships are more extreme. You could also examine more proteins to determine whether such a weak correlation is stronger with different proteins.

28. (a, b, c) Both nucleic acids include the expected bases; A, T, G, C for DNA and A, U, G, C for RNA. The DNA sample is compatible with a double-stranded structure because the purine/pyrimidine ratio for DNA is 1.0 and is therefore consistent with a Watson-Crick model for DNA. The purine/pyrimidine ratio is not 1.0 for RNA, so the RNA must not be double-stranded. (d) If all the DNA is transcribed as stated in the problem, the (A + U)/(C + G) ratio in RNA should equal the (A + T)/(C + G) ratio in DNA, and it does (1.2). To show that it doesn't matter if one or two strands are copied, draw out a double-stranded DNA with a 1.2 ratio of (A + T)/(C + G). Make RNA copies from one or both strands to see the resulting (A + U)/(C + G) RNA ratios. Notice that if both strands are copied, the purine/pyrimidine ratio (A + G)/(U + C) should be 1.0; it is not. Therefore, it is likely that only one of the two strands is copied and the strand that is copied is richer in Ts and/or Cs compared with As and/or Gs. This would produce more As and Gs in the numerator of the equation, thereby giving the ratio of 1.3.

30. The advantage would be that if sequence homologies can be identified for a variety of HIV isolates, then perhaps a single or a few vaccines could be developed for the multitude of subtypes that infect various parts of the world. In other words, the wider the match of a vaccine to circulating infectives, the more likely the efficacy. On the other hand, the more finely aligned a vaccine is to the target, the more likely it is that new or previously undiscovered variants will escape vaccination attempts.

32. Alternative splicing occurs when pre-mRNAs are spliced in more than one way to yield various combinations of exons in the final mRNA product. Upon translation of a group of alternatively spliced mRNAs, a series of related proteins, called isoforms, is produced. It is likely that alternative splicing evolved to provide a variety of functionally related proteins in a particular tissue from one original source. Some tissues might be more prone to develop alternative splicing if they depend on a number of related protein functions. In addition, if genes found in certain tissues have more exons in their active genes, alternative splicing would be expected. While some information is available concerning the mechanisms of alternative splicing, at this time little is known about the underlying forces that drove the evolution of tissue-specific alternative splicing.

Chapter 14 Translation and Proteins

2. Transfer RNAs are *adaptor* molecules in that they provide a way for amino acids to interact with sequences of bases in nucleic acids. Amino acids are specifically and individually attached to the 3′-end of tRNAs that possess a three-base sequence (the anticodon) to base pair with three bases of mRNA. Messenger RNA, on the other hand, contains a copy of the triplet codes that are stored in DNA. The sequences of bases in mRNA interact, three at a time, with the anticodons of tRNAs. Enzymes involved in transcription include RNA polymerase (*E. coli*) and RNA polymerase I, II, and III (eukaryotes). Those involved in translation include aminoacyl tRNA synthetases, peptidyl transferase, and GTP-dependent release factors.

4. The sequence of base triplets in mRNA constitutes the sequence of codons. A three-base portion of the tRNA constitutes the anticodon.

6. An amino acid in the presence of ATP, Mg^{++}, and a specific aminoacyl synthetase produces an amino acid–AMP enzyme complex ($+PP_i$). This complex interacts with a specific tRNA to produce the aminoacyl tRNA.

8. Phenylalanine is an amino acid that, like other amino acids, is required for protein synthesis. While too much phenylalanine and its derivatives cause PKU in phenylketonurics, too little will restrict protein synthesis.

10. Tyrosine is a precursor to melanin, a skin pigment. Individuals with PKU fail to convert phenyalanine to tyrosine; and even though tyrosine is obtained from the diet, at the population level, individuals with PKU have a tendency for less skin pigmentation.

12.

trp-8 *trp-2* *trp-3* *trp-1*

precursor ––→ AA ––→ IGP ––→ I ––→ TRY

14. The fact that enzymes are a subclass of the general term *protein*, a one-gene:one-protein statement might seem to be more appropriate. However, some proteins are made up of subunits, each different type of subunit (polypeptide chain) being under the control of a different gene. Under this circumstance, *one-gene:one-polypeptide* might be more reasonable. Many functions of cells and organisms are controlled by stretches of DNA that either produce no protein product (operator and promoter regions, for example) or have more than one function, as in the case of overlapping genes and differential mRNA splicing. A simple statement regarding the relationship of a stretch of DNA to its physical product is difficult to justify.

16. Each chain represents a primary structure of amino acids connected by covalent peptide bonds. Secondary structures are determined by hydrogen bonding between components of the peptide bonds; α helices and β-pleated sheets result. Tertiary structures are formed from interactions of the amino acid side chains, while the quaternary level results from the associations of chains.

18. In the late 1940s, Pauling demonstrated a difference in the electrophoretic mobility of HbA and HbS (sickle-cell hemoglobin) and concluded that the difference had a chemical basis. Ingram determined that the chemical change occurs in the primary structure of the globin portion of the molecule, using the fingerprinting technique. He found a change in the sixth amino acid in the β chain.

20. *Colinearity* is the sequential arrangement of subunits, amino acids, and nitrogenous bases in proteins and DNA, respectively. Mutations that occur in certain base sequences will be related to protein structure in a site-specific manner.

22. Since all higher levels of protein structure are dependent upon the sequence of amino acids (primary structure), it is the primary structure that is most influential in determining protein structure and function.

24. Enzymes function to regulate catabolic and anabolic activities of cells. They influence (lower) the *energy of activation*, thus allowing chemical reactions to occur under conditions compatible with living systems.

26. All of the substitutions involve one base change.

28. We can conclude that the amino acid is not involved in recognition of the codon.

30. Because crosses are essentially monohybrid, there would be no difference in the results if crossing over occurred (or did not occur) between the *a* and *b* loci.

32. This cross, *AABBCC* × *aabbcc*, would satisfy the data. Offspring in the F_2:

27	*A_B_C_*	purple
9	*A_B_cc*	pink
9	*A_bbC_*	rose
9	*aaB_C_*	orange

3	*A_bbcc*	pink
3	*aaB_cc*	pink
3	*aabbC_*	rose
1	*aabbcc*	pink

c *b* *a*

pink - - → rose - - → orange - - → purple

This hypothesis could be tested by conducting the backcross *AaBbCc* × *aabbcc*, and should give a 4(pink):2(rose): 1(orange):1(purple) ratio.

34. Evernimicin is likely to bind at the A site of the ribosome.

Chapter 15 Gene Mutation, DNA Repair, and Transposable Elements

2. Mutations are the "windows" through which geneticists look at the normal function of genes, cells, and organisms. When a mutation occurs, it allows the investigator to formulate questions as to the function of the normal allele of that mutation.

4. It is true that *most* mutations are thought to be deleterious to an organism. However, *all* mutations may not be deleterious. Those few, rare beneficial variations will provide a basis for possible differential propagation of the variation.

6. If one unit of output from the normal gene gives the same phenotype as in the normal homozygote, where there are two units of output, the allele is considered recessive.

8. Mutations constantly occur in both somatic and gametic tissues. When in somatic tissue, the mutation will not be passed to the next generation; however, the physiological or structural role of the mutant cell may be compromised or be of little or no impact. Mutations in the gametic tissue may pose problems for future generations.

10. We can conclude that no visible mutations occurred or that those that did occur were not detected. In addition, lethal mutations are not readily detectable by this method, so there may have been lethals induced but missed.

12. Watson and Crick recognized that various tautomeric forms caused by single proton shifts could exist for the nitrogenous bases of DNA. Such shifts could result in mutations by allowing hydrogen bonding of normally noncomplementary bases. Important tautomers involve keto–enol pairs for thymine and guanine, and amino–imino pairs for cytosine and adenine.

14. Frameshift mutations are likely to change more than one amino acid in a protein product because as the reading frame is shifted, new codons are generated. In addition, there is the possibility that a nonsense triplet could be introduced, thus causing premature chain termination. If a single pyrimidine or purine has been substituted, only one amino acid is influenced.

16. In contrast to UV light, X-rays penetrate surface layers of cells and thus can affect gamete-forming tissues in multicellular organisms. In addition, X-rays break chromosomes and a variety of chromosomal aberrations can result. Ions and free radicals are formed in the paths of X-rays, and these interact with components of DNA to cause mutations. UV light generates pyrimidine dimers, primarily thymine, which distort the normal conformation of DNA and inhibit normal function.

18. Because mammography involves the use of X-rays and X-rays are known to be mutagenic, it has been suggested that frequent mammograms may do harm. This subject is presently under considerable debate. At the 2002 World Health Organization conference in Barcelona, Spain, the conclusion was that "mammograms can prevent breast cancer deaths in one in 500 women ages 50 to 69."

20. In the *Ames assay,* the compound to be tested is incubated with a mammalian liver extract to simulate an *in vivo* environment. This solution is then placed on culture plates with an indicator microorganism, *Salmonella typhimurium*, which is defective in its normal repair processes. The frequency of mutations in the test strains is an indication of the mutagenicity of the compound. Most mutagens are carcinogenic because genes are responsible for cell-cycle control.

22. *Xeroderma pigmentosum* is a form of human skin cancer caused by perhaps several rare autosomal genes that interfere with the repair of damaged DNA. The photoreactivation repair enzyme appears to be involved.

24. 1.5×10^{-6}

26. It is likely that the reverse transcriptase, in making DNA, provides a DNA segment capable of integrating into the yeast chromosome, as other types of DNA are known to do.

28. Your study should include examination of the following short-term aspects: immediate assessment of radiation amounts distributed in a matrix of the bomb sites as well as a control area not receiving bomb-induced radiation, radiation exposure as measured by radiation sickness, and evidence of radiation poisoning from tissue samples, abortion rates, birthing rates, and chromosomal studies. Long-term assessment should include sex-ratio distortion (males being more influenced by X-linked recessive lethals than are females), chromosomal studies, birth and abortion rates, cancer frequency and type, and genetic disorders. In each case, data should be compared to the control site to see if changes are bomb-related. In addition, to attempt to determine cause and effect, it is often helpful to show a dose response. Thus, by comparing the location of individuals at the time of exposure to the matrix of radiation amounts, you may be able to determine whether those most exposed to radiation suffer the most physiologically and genetically. If a positive correlation is observed, then statistically significant conclusions may be possible.

30. (a) For those organisms that generate energy by aerobic respiration, a process occurs that involves the reduction of molecular oxygen. Partially reduced species are produced as intermediates and by-products of such molecular action: O_2^-, H_2O_2, and OH^-. These species are potent electrophilic oxidants that escape mitochondria and attack numerous cellular components. Collectively, they are called reactive oxygen species (ROS).

(b)

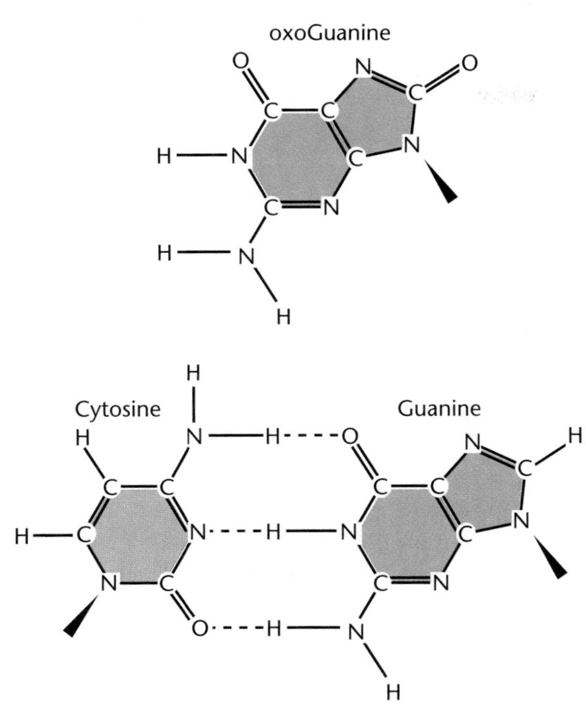

When casually examining the structures in the above diagrams, it is not immediately obvious that oxoG:A pairs should occur. However, hydrogen bonding can occur to any other base, including self pairs. Homopurine (A:A, G:G) and heteropurine (A:G) pairs represent anomalous base-pairing possibilities even with nonaltered bases. While G:C is undoubtedly the most stable, several mispairs are actually stronger than the A=T pair. (c) If you start with a G≡C pair, you end up with an A=T pair. (d) It turns out that transversions are quite commonly found in human cancers and are especially prevalent in the tumor suppressor gene, *p53*. Thus, the cellular defense system has been extensively studied. One component is a triphosphatase that cleanses the nucleotide precursor pool by removing the two outermost phosphates from oxo-dGTP. Another involves a DNA glycosylase that initiates repair of misreplicated oxoG:A by hydrolyzing the glycosidic bond linking the adenine base to the sugar. Another is a DNA gycosylase/lyase system that recognizes oxoG opposite cytosine. Of the three systems, the DNA glycosylases are probably the most effective.

32. Until at least fifteen months of age, there is little or no change in the spontaneous mutation rate for the gene under investigation. However, during the interval from 15 months to 28 months, the mutation rate undergoes an approximately 10-fold increase. At least two factors might be involved. First, stem cells that produce spermatogonial cells last through the majority of life in males, thus allowing a longer exposure to DNA-damage and the accumulation of mutations. Second, male germ cells undergo numerous cycles of DNA replication, thereby allowing for the fixation of mutations during DNA replication.

34. At first glance, we would conclude that the vast majority of mutations in flower color for these two groups are caused by transposons. Only 2 of the 10 mutations are caused by nontransposon activity. You could reasonably argue that transposons are major causes of genetic variation in these plant species and that plants, and perhaps animals too, may have a large proportion of their mutations caused by transposons. Since plants are unable to avoid environmental stressors by moving, perhaps a high level of transposon activity offers them a unique adaptive strategy. However, because human plant domesticators have repeatedly selected for unusual flower phenotypes and interbred those phenotypes for a variety of purposes, primarily esthetic, it is possible that the data presented here represent a biased sample of the sources of mutations.

Chapter 16 Regulation of Gene Expression

2. Under *negative* control, the regulatory molecule interferes with transcription, while in *positive* control, the regulatory molecule stimulates transcription. Negative control is seen in the *lactose* and *tryptophan* systems. The CAP protein exerts positive control.

4. $I^+O^+Z^+$ = inducible
$I^-O^+Z^+$ = constitutive
$I^+O^cZ^+$ = constitutive
$I^-O^+Z^+/F'I^+$ = inducible
$I^+O^cZ^+/F'O^+$ = constitutive
$I^sO^+Z^+$ = repressed
$I^sO^+Z^+/F'I^+$ = repressed

6. *C* codes for the *structural gene*. Because when *B* is mutant, no enzyme is produced, *B* must be the *promoter*. The *A* locus is the *operator,* and the *D* locus is the *repressor* gene.

8. (a) Because of the deletion of a base early in the *lac Z* gene, there will be frameshift of all the reading frames downstream from the deletion. It is likely that either premature chain termination of translation will occur (from the introduction of a nonsense triplet in a reading frame) or the normal chain termination will be ignored. Regardless, a mutant condition for the *Z* gene will be likely. If such a cell is placed on a lactose medium, it will be incapable of growth because β-galactosidase is not available. (b) If the deletion occurs early in the *A* gene, we would expect impaired function of the A gene product, but it will not influence the use of lactose as a carbon source.

10. A single *E. coli* cell contains very few molecules of the *lac* repressor. However, the *lac* I^q mutation causes a 10-fold increase in repressor protein production. With the use of dialysis against a radioactive gratuitous inducer (IPTG), Gilbert and Muller-Hill were able to identify the repressor protein in certain extracts of *lac* I^q cells. The material that bound the labeled IPTG was purified and shown to be heat labile and have other characteristics of protein. Extracts of *lac* I^- cells did not bind the labeled IPTG.

12. Because the deletion of the regulatory gene causes a loss of synthesis of the enzymes, the regulatory gene product can be viewed as one exerting *positive control*. When *tis* is present, no enzymes are made; therefore, *tis* must inactivate the positive regulatory protein. When *tis* is absent, the regulatory protein is free to exert its positive influence on transcription. Mutations in the operator negate the positive action of the regulator. This model of the regulatory system is an example of positive control.

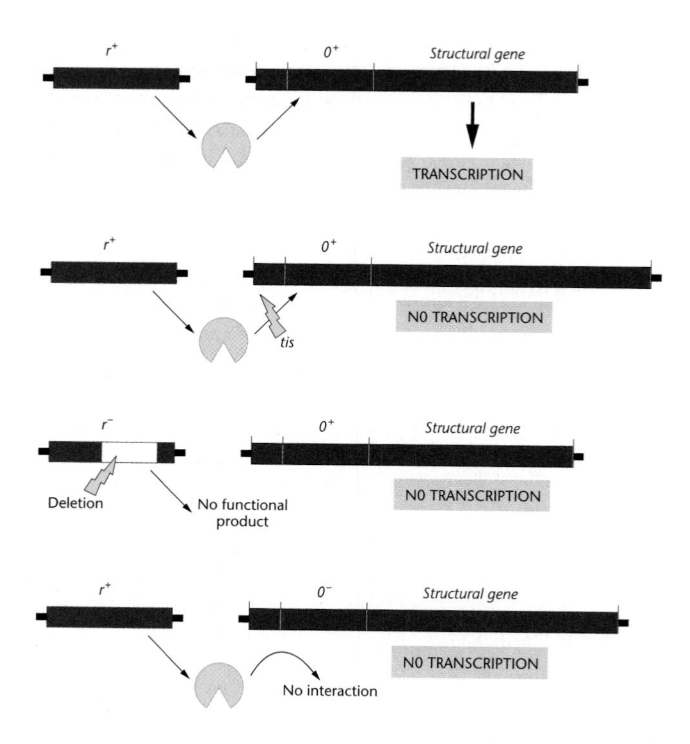

14. If you could develop an assay for the other gene products under SOS control, with a *lexA*⁻ strain, the other gene products should be present at induced levels.

16. In *chromatin remodeling*, changes in DNA/chromosome structure can influence overall gene output. DNA methylation also influences transcription efficiency. Several factors are known to influence *transcription*: promoters, TATA, CAAT, and GC boxes, as well as other upstream regulatory sequences; enhancers, which are cis-acting sequences acting at various locations and orientations; transcription factors with various structural motifs (zinc fingers, homeodomains, and leucine zippers), which bind DNA and influence transcription; and receptor–hormone complexes, which influence transcription. *Processing and transport* types of regulation involve the efficiency of hnRNA maturation as related to capping, polyA-tail addition, intron removal, and mRNA stability. After mRNAs are produced from the processing of hnRNA, they have the potential for *translation*. The stability of the mRNAs appears to be an additional regulatory control point. Certain factors, such as protein subunits may influence a variety of steps in the translational mechanism. For instance, a protein or protein subunit, may activate an RNase, which will degrade certain mRNAs, or a particular regulatory element may cause a ribosome to stall, thus decreasing the speed of translation and increasing the exposure of a mRNA to the action of RNAses.

18. Transcription factors are proteins that are *necessary* for an initiation of transcription. They are not part of the RNA polymerase molecule that initiates and executes the process of transcription; however, they control where, when, and at what rate genes are expressed.

20. Steroid hormones enter target cells and bind to a specific receptor protein in the cytoplasm. The hormone-receptor complex moves into the nucleus and activates transcription of one or more specific genes through interaction with hormone-responsive elements (HREs).

22. Regulation in prokaryotes is based primarily on local control of gene expression through activation or repression of transcription. Such regulation is often in direct response to environmental stimuli. Regulation in eukaryotes involves various levels, from transcriptional through activation and repression, chromatin remodeling, posttranscriptional, translational, and posttranslational levels.

24. (a) There is no place for the TFIID to bind. (b) There is more transcription in the nuclear extracts. Perhaps other factors not present in the purified system are present in the nuclear extracts. (c) There is a region, probably in the −81 to −50 area, that responds to a component in the nuclear extract to bring about high-efficiency transcription.

26. A model might be developed in which certain nuclear domains are more likely to export an RNA than others, thus providing one of the many levels of gene regulation known to exist in eukaryotes. However, results indicating that RNA transport is guided primarily by diffusion rather than directed transport makes such models less appealing. Alternatively, since most of the experiments dealt with polyA RNA, it is possible that pre-polyA RNA is less prone to diffusion and may be directed.

28. If nucleocytoplasmic transport mechanisms can recognize different RNA species and either deny, facilitate, or redirect such species, then posttranscriptional regulation could result. In addition, if certain gene-activating or -repressing molecules are selectively imported or exported from the nucleus, differential gene activity could be the result. If such selective transport exists, then an additional level of genetic regulation will be available to eukaryotes.

30. Methylation of CpGs causes a reduction in luciferase expression that is somewhat proportional to the amount of methylation and patch size. Methylation within the transcription unit more drastically reduces luciferase expression compared to methylation outside the transcription unit. A high degree of methylation outside the transcription unit (593 CpGs) has as great an impact on depressing transcription as the same degree of methylation within the transcription unit.

Chapter 17 Recombinant DNA Technology

2. *Reverse transcriptase* is often used to promote the formation of cDNA (complementary DNA) from an mRNA molecule. Eukaryotic mRNAs typically have a 3′-polyA tail as indicated in the diagram below. The poly-dT segment provides a double-stranded section that serves to prime the production of the complementary strand.

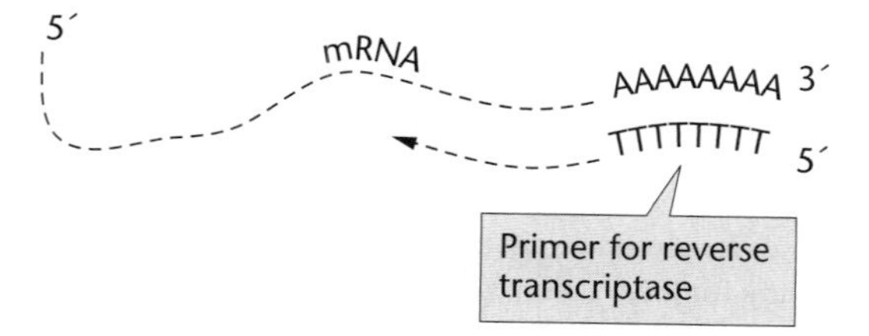

4. It is believed that the protein interacts with the major groove of the DNA helix. This information comes from the structure of the proteins that have been sufficiently well-studied to suggest that the DNA major groove and "fingers," or extensions, of the protein form the basis of interaction.

6. This segment contains the palindromic sequence of GGATCC; which is recognized by the restriction enzyme *Bam*HI. The double-stranded sequence is

CCTAGG
GGATCC

8. Because of their small size, plasmids are relatively easy to separate from the host bacterial chromosome and they have relatively few restriction sites. They can be engineered fairly easily (i.e., polylinkers and reporter genes added). YACs (yeast artificial chromosomes) contain telomeres, an origin of replication, and a centromere and are extensively used to clone DNA in yeast. With selectable markers (TRP1 and URA3) and a cluster of restriction sites, DNA inserts ranging from 100 kb to 1000 kb can be cloned and inserted into yeast. Since yeasts, being eukaryotes, undergo many of the typical RNA- and protein-processing steps of other, more complex eukaryotes, the advantages are numerous when working with eukaryotic genes.

10. This problem can be solved by the following expressions:

*Not*I	4^8
*Hin*fI	$4 \times 4 \times 1 \times 4 \times 4$
*Xho*II	$2 \times 4 \times 4 \times 4 \times 4 \times 2$

12. (a) Because the *Drosophila* DNA has been cloned into the *Bam*H1 site in the ampicillin resistance gene of the plasmid, the gene will be mutated and any bacterium with the recombinant plasmid will be ampicillin-sensitive. The tetracycline resistance gene remains active however. Bacteria that have been transformed with the recombinant plasmid will be resistant to tetracycline, and therefore tetracycline should be added to the medium. (b) Colonies that grow on a tetracycline medium should be tested for growth on an ampicillin medium, either by replica plating or some similar controlled transfer method. Those bacteria that do not grow on the ampicillin medium, probably contain the *Drosophila* DNA insert. (c) Resistance to both antibiotics by a transformed bacterium could be explained in several ways. First, if cleavage with the *Bam*H1 was incomplete, then no change in biological properties of the uncut plasmids would be expected. Or it is possible that the cut ends of the plasmid were ligated together in the original form with no insert.

14. For the antibiotic resistance to be present, the ligation will reform the plasmid into its original form. However, two of the plasmids can join to form a dimer by each rejoining to form a single complex.

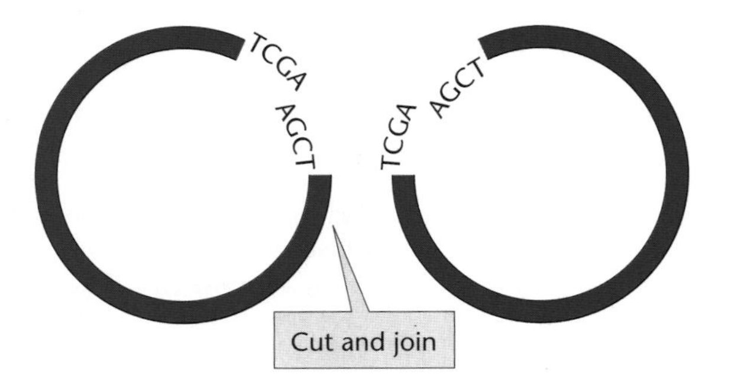

16. Because of complementary base pairing, the 3′- end of the DNA strand often loops back onto itself, thereby providing a primer for DNA polymerase I.

18. A filter is used to bind the DNA from the colonies containing recombinant plasmids. A labeled probe is constructed from the protein sequence of EF1a. Since it is highly conserved, it should show considerable complementation to the human EF-1a cDNA. It is used to detect, through hybridization, the DNA of interest. Cells with the desired clone are then picked from the original plate, and the plasmid is isolated from the cells.

20.

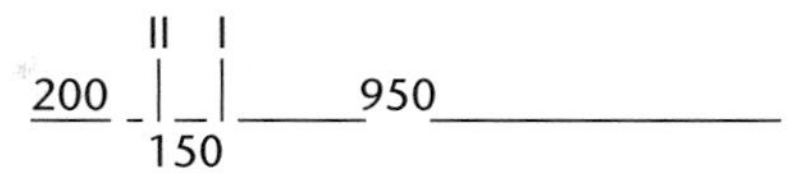

22. Option (b) fits the expectation because the thick band in the offspring probably represents the bands at approximately the same position in both parents. The likelihood of such a match is expected to be low in the general population.

24. (a) By comparing the lanes of the double digests, one can see that there is an A site at 500 bp and a B site at 2500 bp. For the 2000- bp band of the E + B double digest, there are actually two fragments. (b) Notice that the probe hybridizes consistently to the 2000- bp fragment between the A and B restriction sites, so the *rosy* gene is somewhere in that region.

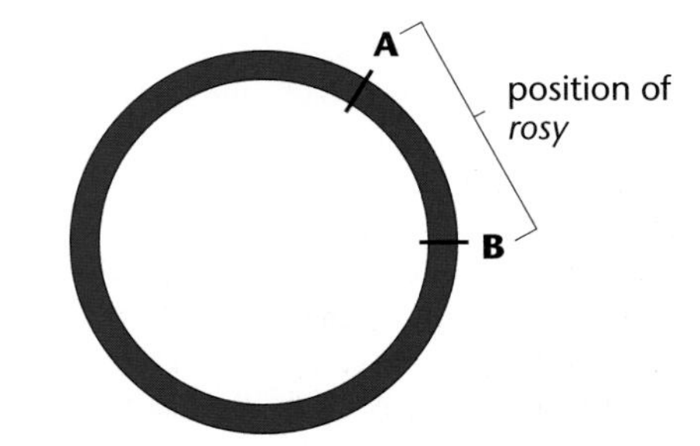

26. With repeated sequences in the genome, chromosome walking is complicated because the clone hybridizes to multiple regions. We can "chromosome jump" over repeated sequences and often block repeats by addition of repetitive DNA.

28. There would be approximately 4096 base pairs between sites. Given that λ DNA contains approximately 48,500 base pairs, there would be about 11.8 sites (48,500/4096).

30. T_m(°C) = about 63°C. Subtracting 5°C gives us a good starting point of about 58°C for PCR with this primer. As the percentage of G≡C and length increase, the T_m(°C) increases. G≡C pairs contain three hydrogen bonds rather than two, as between A═T pairs.

32. ddNTPs are analogs of the "normal" deoxyribonucleotide triphosphates (dNTPs) but they lack a 3′-hydroxyl group. As DNA synthesis occurs, the DNA polymerase occasionally inserts a ddNTP into a growing DNA strand. Since there is no 3′-hydroxyl group, chain elongation cannot take place and resulting fragments are formed that can be separated by electrophoresis. Where the ddNTP was incorporated, the length of each strand, and therefore the position of the particular ddNTP, is established and used to eventually provide the base sequence of the DNA.

Chapter 18 Genomics, Bioinformatics, and Proteomics

2. Knowing the sequence of DNA in an organism is only the beginning. Annotating the DNA is a significant challenge. Even at that, knowing how gene products interact in time and space (proteomics) will take additional rounds of technological advances as yet unconsidered. The work of Haas and others

identified variation in intron/exon splice sites, microexons, and alternative transcription start sites. Correlating various transcriptional and translational schemes with the phenotype will be an interesting adventure.

4. (a) Assuming an average gene size of 5000 base pairs, there would be about 3700 base pairs between genes. (b) 54,934/13,379 = 4.11 exons (c) 48,257/13,379 = 3.61 introns (d) There is a marked increase in the number of genes involved in alternative transcripts. (e) Alternative transcripts are RNAs that are variable in sequence because of different splicing of introns or other processes such as use of alternative promoters (13%) and alternative polyadenylation sites (6%).
6. Gene density in euchromatic regions of the *Drosophila* genome is about one gene per 8730 base pairs, while gene density in heterochromatic regions is one gene per 69,696.9 base pairs (20.7 Mb/297). Clearly, a given region of heterochromatin is much less likely to contain a gene than the same-sized region in euchromatin.
8. Generally one can examine conserved sequences in other organisms to indicate that an ORF is likely a coding region. We can also match a sequence to previously described sequences that are known to code for proteins. The problem is not easily solved, that is, deciding which ORF is actually a gene. The shorter the ORF scan, the more likely the overestimate of genes, because ORFs longer than 200 are less likely to occur by chance.
10. CpG islands are DNA regions containing many copies of this nucleotide doublet. They are found in gene-rich regions of the mammalian genome and are therefore used to help identify coding regions.
12. While archaea are structurally similar to eubacteria, their metabolism more closely resembles eukaryotes. *M. jannaschii* has a circular, double-stranded DNA genome of about 1.7 Mb with about 1738 protein-coding genes. It contains three chromosomes, and most genes (58%) in this organism don't match any other known genes. Genes involved in energy production, cell division, and general metabolism resemble those of eubacteria, while genes involved in RNA synthesis, protein synthesis, and DNA synthesis more closely resemble eukaryotes.
14. Plasmids are capable of carrying both essential and nonessential genes of the host. To complicate the matter, it is likely that all cells contain nonessential genes. An organism's genome will probably come to encompass all genetic elements that can be shown to be stable cellular inhabitants.
16. Because of the diverse cell types of multicellular eukaryotes, a variety of gene products is required, which may be related to the increase in DNA content per cell. In addition, the advantage of diploidy automatically increases DNA content per cell. However, seeing the question in another way, it is likely that a much higher *percentage* of the genome of a prokaryote is actually involved in phenotype production than in a eukaryote. Eukaryotes have evolved the capacity to obtain and maintain what appears to be large amounts of "extra", perhaps "junk," DNA. Prokaryotes on the other hand, with their relatively short life cycle, are extremely efficient in their accumulation and use of their genome. Given the larger amount of DNA per cell in eukaryotes and the requirement that the DNA be partitioned in an orderly fashion to daughter cells during cell division, certain mechanisms and structures (mitosis, nucleosomes, centromeres, etc.) have evolved for *packaging* the DNA. In addition, the genome is divided into separate entities (chromosomes) to perhaps facilitate the partitioning process in mitosis and meiosis.
18. The gene density of *C. elegans* is much lower than in yeast, and about half of the DNA is composed of intergenic spacer DNA. A significant fraction of the genome is highly repetitive, and approximately 25% of the genes are organized into polycistronic transcription units like operons in bacteria. There are more introns in *C. elegans* genes, and many genes are found in the introns of other genes.
20. Increased protein production from approximately 33,000 genes is probably related to alternative splicing and various post-translational processing schemes. In addition, a particular DNA segment may be read in a variety of ways and in two directions.
22. The issue here is whether the organism under consideration is independent and self-reproducing. It appears that the minimum number of genes for a free-living organism is in the range of 250 to 350. Symbionts can have much smaller genomes and exist with fewer genes because of materials supplied by the host cell.
24. Gene duplications can be generated by unequal crossing over and DNA slippage. In addition, chromosomal aberrations such as inversions and translocations can lead to gene duplication. It is possible to duplicate genes through horizontal transfer involving transposons, plasmids, and other transferred DNAs. Gene duplication contributes to evolution by providing mutational opportunities without compromising the functionality of the genome. Gene duplications allow for mutational experimentation and therefore evolution.
26. In general, we would expect certain factors (such as heat or salt) to favor evolution to increase protein stability: distribution of ionic interactions on the surface, density of hydrophobic residues and interactions and number of hydrogen and disulfide bonds. By examining the codon table, a high GC ratio would favor amino acids *Ala*, *Gly*, *Pro*, *Arg*, and *Trp* and minimize the use of *Ile*, *Phe*, *Lys*, *Asn*, and *Tyr*. How codon bias influences actual protein stability is not yet understood.
28. While the β-globin gene family is a relatively large (60 kb) sequence and restriction analyses show that it is composed of six genes, one is a pseudogene and therefore does not produce a product. The five functional genes each contain two similarly sized introns, which when included with noncoding flanking regions (5′ and 3′) and spacer DNA between genes, accounts for the 95% mentioned in the Problem.
30. The number of combinations is determined by a simple multiplication of the number of genes in each class: $V \times D \times J \times C$. Thus, in this case, the answer would be $10\ V \times 30\ D \times 50\ J \times 3\ C = 45{,}000$.
32. Since structural and chemical factors determine the function of a protein, it is likely to have several proteins share a considerable amino acid sequence identity but not be functionally identical. Since the *in vivo* function of such a protein is determined by secondary and tertiary structures as well as local surface chemistries in active or functional sites, the nonidentical sequences may have considerable influence on function.

Chapter 19 Applications and Ethics of Biotechnology

2. Kleter and Peijnenburg used the BLAST tool from the *http://www.ncbi.nlm.nih.gov/BLAST* Web site to conduct a series of alignment comparisons of transgenic sequences with sequences of known allergenic proteins. Of 33 transgenic proteins screened for the identities of at least 6 contiguous amino acids found in allergenic proteins, 22 gave positive results.

4. (a) Both the saline and column extracts of Lkt50 appear to be capable of inducing at least 50% neutralization of toxicity when injected into rabbits. (b) In order for a successful edible vaccine to be developed, numerous hurdles must be overcome. The immunogen must first be stably incorporated into the host plant hereditary material, and the host must express only that immunogen. When ingested, the immunogen must be transported across the intestinal wall unaltered or altered in such a way as to stimulate the desired immune response. There must be guarantees that potentially harmful by-products of transgenesis have not been produced—in other words, broad ecological and environmental issues must be addressed to prevent a transgenic plant from becoming an unintended vector for harm to the environment or any organisms feeding on the plant (directly or indirectly).
6. Enhancement therapy using gene products benefits from the application of modern biotechnology without being burdened by alteration of the genome. Enhancement gene therapy, however, opens the door to a variety of ethical issues. What limits can or should be imposed on individuals or institutions seeking to improve human qualities? What qualities should be open for enhancement? Gene therapy is not without medical and ethical risks.
8. Somatic gene therapy involves attempts to alter the genetic material in non-germline cells. Clinical trials are currently under way. Germline therapy, while certainly being more efficient (although perhaps more difficult technically), alters the germline and is transmitted to offspring. There are considerable ethical problems associated with germ-plasm therapy. It recalls previous attempts of the eugenics movements of past decades, which involved the use of selective breeding to purify the human stock. Some present-day biologists have said publicly that germline gene therapy will *not* be conducted.
10. p53 and pRB are tumor-suppressor proteins and are required by the cell to effectively monitor the cell cycle. Reduction in their activity would diminish normal cell-cycle controls and most likely lead to cancer.
12. (a) A microarray is a solid support containing an orderly arrangement of DNA samples. A typical array contains thousands of DNA spots, which may be small oligonucleotides, cDNAs, or short genomic sequences. Labeled sequences hybridize to the immobilized DNAs by standard base pairing. Such technology allows a method for monitoring the RNA expression levels of thousands of genes in virtually any cell population. (b) Using microarray technology, researchers can observe the overall behavior of the genome in cancer and normal cells, and by comparison, determine which genes are active or inactive under various circumstances. It is possible to identify the set of genes whose expression or lack thereof defines the properties of each tumor type. This application can therefore lead to precise diagnosis and refining possible therapies. In addition, microarray profiling can be used to determine the efficacy of particular therapies.
14. *Drosophila* is a unique experiemental organism in that there is a vast knowledge of its genetics, it is easily cultured and genetically manipulated, and it contains unique polytene chromosomes that allow visual landmarks. Coupled with probe labeling, sequence-tagged sites, visible landmarks (chromomeres), and ease of manipulation, we can actually see where important genes are located in chromosomes. *Drosophila* also contains P elements that allow sequence markers to be inserted into the genome. Microdissection of chromosomes is also useful in developing specific clones for sequencing. In addition, techniques have been developed (*in situ* hybridization) that allow scientists to actually determine the distributions of gene activities in all tissues of the organism.
16. Positional cloning relies on segregation (Mendelian) and linkage analysis. It is a departure from traditional methods of mapping and is used to isolate, map, and clone a gene with no knowledge of its gene product.
18. Positional cloning relies on segregation (Mendelian) and linkage analysis. Given the numerous limitations associated with such analyses in human populations (family and sample size, etc.), it is unlikely that this technique will be successfully applied to genetically complex traits in the near future.
20. With both parents heterozygous, each child born will have a 25% chance of developing CF.
22. In general, bacteria do not process eukaryotic proteins in the same manner as eukaryotes. Transgenic eukaryotes are more likely to correctly process eukaryotic proteins, thus increasing the likelihood of normal biological activity.
24. The answer provided here is based on the condition that individual I-2 is a carrier and the son, II-4, has the disorder. The 3-kb fragment occurs in the normal I-1 father and the normal son II-1. The affected son, II-4, has the 4 kb- fragment. One daughter, II-2, is a carrier, while the other daughter, II-3, is not a carrier.
26. The first problem is the localization of the introduced DNA into the target tissue and target location in the genome. Inappropriate targeting may have serious consequences. In addition, it is often difficult to control the output of introduced DNA. Genetic regulation is complicated and subject to a number of factors including upstream and downstream signals, as well as various posttranscriptional processing schemes. Artificial control of these factors will prove difficult.
28. Using restriction enzyme analysis to detect point mutations in humans is a tedious trial-and-error process. Given the size of the human genome in terms of base sequences and the relatively low number of unique restriction enzymes, the likelihood of matching a specific point mutation, separate from other normal sequence variations, to a desired gene is low.
30. (a) Y-linked excluded, X-linked recessive excluded, autosomal recessive possible but unlikely because the gene is stated as being rare; X-linked dominant possible if heterozygous, autosomal dominant possible. (b) Chromosome 21 with the B1 marker probably contains the mutation. (c) The disease gene is segregating with some certainty with the B1 RFLP marker in the family. Since the mother also has the B3 marker, the offspring could be tested. If the child carries the B3 marker, then he/she does not carry the B1 marker, which has been segregating with the defective gene. However, this prediction is not completely accurate, because a crossover in the mother could put the undesirable gene with the B3 marker. (d) The possibilities would include a crossover between the restriction sites in the father, giving a B1 chromosome, or a mutation eliminating either the B2 or B3 restriction site.

Chapter 20 Genes and Development

2. That nuclei from almost any source remain transcriptionally and translationally active substantiates the fact that the genetic code and the ancillary processes of transcription and translation

are compatible throughout the animal and plant kingdoms. Because the egg represents an isolated "closed" system, which can be mechanically, environmentally, and to some extent biochemically manipulated, various conditions may be developed that allow one to study facets of gene regulation.

4. Imaginal discs are small clusters of cells formed during larval stages. There are 12 bilaterally paired discs composed of eye-antennal, leg, wing, and so forth, and one unpaired genital disc.

6. Genes that control early development are often dependent on the deposition of their products (mRNA, transcription factors, various structural proteins, etc.) in the egg by the mother. Such maternal-effect genes control early events such as defining anterior–posterior polarity. Such products are placed in eggs during oogenesis and are activated immediately after fertilization.

8. Zygotic genes are activated or repressed, depending on their response to maternal-effect gene products. Three subsets of zygotic genes divide the embryo into segments. These segmentation genes are normally transcribed in the developing embryo, and their mutations have embryonic lethal phenotypes. The maternal genotype contains zygotic genes, and these are passed to the embryo as with any other gene.

10. Because the polar cytoplasm contains information to form germ cells, you would expect such a transplantation procedure to generate germ cells in the anterior region. Work done by Illmensee and Mahowald in 1974 verified this expectation.

12. If protein products of a given gene are present in different cell types, we can assume that the responsible gene is being transcribed. If we can actually observe gene activity microscopically, as is the case in some specialized chromosomes (polytene chromosomes), gene activity can be inferred by the presence of localized chromosomal puffs. A more direct and common practice to assess transcription of particular genes is to use labeled probes. If a labeled probe that contains base sequences complementary to the transcribed RNA can be obtained, then such probes will hybridize to that RNA if present in different tissues.

14. A *homeotic mutant* alters the identity of a segment or field within a segment as if it were another segment or field. While homologies do exist between widely diverse groups, some differences in function have evolved. It is likely that functional overlap would not occur; however, only an experiment would answer the question.

16. The gain-of-function *Antp* mutation causes the wild-type *Antennapedia* gene to be expressed in the head, and mutant flies have legs on the head in place of antenna.

18. Because of the regulatory nature of *homeotic* genes in the fundamental cellular activities of determination and differentiation, it would be difficult to ignore their possible impact on oncogenesis. Homeotic genes encode DNA binding domains that influence gene expression, and any factor that influences gene expression may, under some circumstances, influence cell-cycle control.

20. Two coupled approaches might be used. First, you could make transgenic flies that contain a series of deletions spanning all segments of the *bicoid* mRNA, the coding region, and 5′ and 3′ untranslated regions. A comparison of the stabilities of individual, deleted mRNAs with controls would indicate whether a particular segment of the mRNA contains a degradation signal sequence. If a degradation-sensitive region or signal sequence is located by deletion, that same intact region, when ligated to a noninvolved, nondegraded mRNA (like a ribosomal protein or tubulin mRNA), should foster degradation in a manner similar to the *bicoid* mRNA.

22.

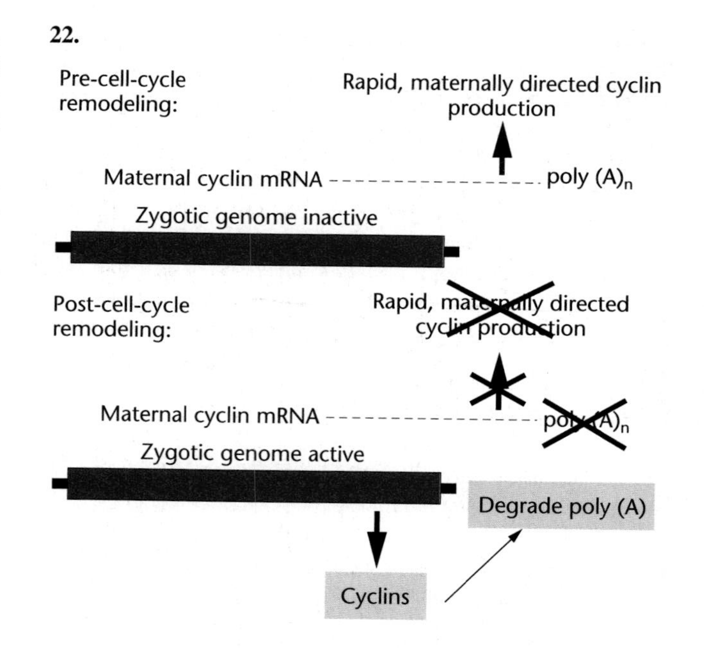

24. It is likely that this gene normally controls the expression of *BX-C* genes in all body segments. The wild-type product of *esc* stored in the egg may be information correctly stored in the egg cortex.

26. Three classes of flower homoeotic genes are known that are activated in an overlapping pattern to specify various floral organs. Class A genes give rise to sepals. Expression of A and B class genes specify petals, B and C genes control stamen formation, and expression of C genes gives rise to carpels.

28. Because signal-receptor interactions depend on membrane-bound structures, the pathway can work only with adjacent cells. The advantage of such a system is that only cells in a certain location will be influenced—those in contact. A disadvantage would occur if large groups of cells are to be induced into a particular developmental pathway or if cells not in contact need to be induced.

30. If the *her*-1^+ product acts as a negative regulator, then when the gene is mutant, suppression over *tra*-1^+ is lost and hermaphroditism would be the result. This hypothesis fits the information provided. The double mutant should be male because even though there is no suppression from *her*-1^-, there is no *tra*-1^+ product to support hermaphrodite development.

Chapter 21 The Genetic Basis of Cancer

2. The G1 stage begins after mitosis and is involved in the synthesis of many cytoplasmic elements. In the S phase, DNA synthesis occurs. G2 is a period of growth and preparation for mitosis. Most cell-cycle time variation is caused by changes in the duration of G1. G0 is the nondividing state.

4. Kinases regulate other proteins by adding phosphate groups. Cyclins bind to the kinases, switching them on and off. CDK4 binds to cyclin D, moving cells from G1 to S. At the G2/mitosis border, a CDK1 (cyclin-dependent kinase) combines with another cyclin (cyclin B). Phosphorylation occurs, bringing about a series of changes in the nuclear membrane via caldesmon, cytoskeleton, and histone H1.

6. When we discuss an inherited predisposition, we usually refer to situations where a particular phenotype is expressed in families in some consistent pattern. However, the phenotype may not always be expressed or may manifest itself in different ways. In retinoblastoma, the gene is inherited as an autosomal dominant and those that inherit the mutant *RB* allele are predisposed to develop eye tumors. However, approximately 10% of the people known to inherit the gene don't actually express it, and in some cases, expression involves only one eye rather than both.

8. There are several ways in which proto-oncogenes are converted to oncogenes: point mutations in which a mutant gene acts as a positive "switch" in the cell cycle, translocations where a hybrid gene might be formed, and overexpression where a gene might acquire a new promoter and/or enhancer.

10. Various kinases can be activated by breaks in DNA. A kinase called ATM and/or a kinase called Chk2 phosphorylate BRCA1 and p53. The activated p53 arrests replication during the S phase to facilitate DNA repair. The activated BRCA1 protein, in conjunction with BRCA2, mRAD51, and other nuclear proteins, is involved in repairing the DNA.

12. The activated p53 protein is a tetramer made up of four p53 subunits. A mutation in any one *p53* allele usually changes the tetramer sufficiently to abolish all p53 activity.

14. Mutations that produce oncogenes alter gene expression either directly or indirectly and act in a dominant capacity. Oncogenes are those that normally function to promote or maintain cell division. In the mutant state, they induce or maintain uncontrolled cell division; that is, there is a gain of function. Generally this gain of function takes the form of increased or abnormally continuous gene output. Loss of function is generally attributed to tumor-suppressor genes that function to halt passage through the cell cycle. When such genes are mutant, they have lost their capacity to halt the cell cycle, and they are generally recessive.

16. Mutations in the *APC* gene are common to both FAP and HNPCC. HNPCC begins with a mutation in DNA-repair genes that produce a mutant *APC*. Individuals with FAP inherit a mutation in the *APC* gene.

18. As with many forms of cancer, a single gene alteration is not the only requirement. The authors (Bose, et al.) state, "only infrequently do the cells acquire the additional changes necessary to produce leukemia in humans." Some studies indicate that variations (often deletions) in the region of the breakpoints may influence expression of CML.

20. Cytogenetic studies using fluorescence *in situ* hybridization suggest the presence of a relatively large region of homology between the two chromosomal regions involved in the BCR–ABL fusion protein. The discovery of a large duplicated region relatively close to the *ABL* and *BCR* genes suggests a possible involvement in the formation of the Philadelphia chromosome translocation. Such a duplication would allow for partial pairing of nonhomologous chromosomes. A crossover in the duplicated region would result in a translocation.

22. It is less expensive, both in terms of human suffering and money, to seek preventive measures for as many diseases as possible. However, having gained some understanding of the mechanisms of disease—in this case, cancer—it must also be stated that no matter what preventive measures are taken, it will be impossible to completely eliminate disease from the human population. It is extremely important, however, that we increase efforts to educate and protect the human population from as many hazardous environmental agents as possible.

24. Any agent that causes damage to DNA is a potential carcinogen, since cell-cycle control is achieved by the gene (DNA) products, proteins. Cigarette smoke is known to contain an agent that changes DNA (in this case, transversions) so numerous modified gene products (including cell-cycle controlling proteins) are likely to be produced. The fact that many cancer patients have such transversions in *p53* strongly suggests that cancer is caused by agents in cigarette smoke.

26. (a, b) Even though there are changes in the *BRAC1* gene, they don't always have physiological consequences. Such neutral polymorphisms make screening difficult in that we can't always be certain that a mutation will cause problems for the patient. (c) The polymorphism in *PM2* is probably a silent mutation because the third base of the codon is involved. (d) The polymorphism in *PM3* is probably a neutral missense mutation because the first base is involved.

28. Surveys often identify unusual cancer rates among certain age groups, occupations, or personal habits. Statistical correlations can provide preliminary support for a cancer-causing circumstance, however, extensive laboratory investigations are required to establish a causal link.

30. No, she will still have the general population risk of about 10%. It addition, it is possible that genetic tests won't detect all breast cancer mutations.

Chapter 22 Population Genetics

2. A population is a group of individuals of the same species that live in the same geographic area that are actually or potentially interbreeding. Interbreeding is a component of this definition because members of a given population must share in a common gene pool.

6. Deviations from the Hardy-Weinberg law provide windows to population structure. Such deviations enable us to determine mutational, migrational, or selectional aspects of a population, which might be available only through direct obsrvation.

8. Because the alleles follow a dominant/recessive mode, we can use the equation $\sqrt{q^2}$ to calculate q from which all other aspects of our answer depend. The frequency of *aa* types is determined by dividing the number of nontasters (37) by the total number of individuals (125):

$$q^2 = 37/125 = 0.296$$
$$q = 0.544$$
$$p = 1 - q$$
$$p = 0.456$$

Frequency of *AA*: $p^2 = (0.456)^2 = 0.208$, or 20.8%

Frequency of *Aa*: $2pq = 2(0.456)(0.544) = 0.496$, or 49.6%

Frequency of *aa*: $q^2 = (0.544)^2 = 0.296$, or 29.6%

10. Genetic variation within and among species is ultimately dependent upon mutation that provides the original source of variation. Selection (directional and disruptive), genetic drift, nonrandom mating, Mendelian assortment, and recombination generate vast new genotypic combinations of the new alleles produced by mutation. Given that such widely separated geographic populations are genetically indistinguishable, we must conclude that something other than those commonly associated with genetic variability exists. Most of the above variation-generating factors are dependent on sexual reproduction. If we remove sexual reproduction as a life process in *C. albicans*, all isolates would be "clonal" in nature. Assuming similar selective forces within each isolate's environment, then genetic uniformity is possible. In fact, even though *C. albicans* is diploid, it employs a primarily clonal mode of reproduction.

12. In order for the Hardy-Weinberg equations to apply, the population must be in equilibrium.

14. (a) Frequency of $1/1$: $p^2 = (0.7755)^2 = 0.6014$, or 60.14%

Frequency of $1/\Delta 32$: $2pq = 2(0.7755)(0.2245) = 0.3482$, or 34.82%

Frequency of $\Delta 32/\Delta 32$: $q^2 = (0.2245)^2 = 0.0504$, or 5.04%

Comparing these equilibrium values with the observed values strongly suggests that the observed values are drawn from a population in equilibrium.

(b) Frequency of *AA*: $p^2 = (0.877)^2 = 0.7691$, or 76.91%

Frequency of *AS*: $2pq = 2(0.877)(0.123) = 0.2157$, or 21.57%

Frequency of *SS*: $q^2 = (0.123)^2 = 0.0151$, or 1.51%

Comparing these equilibrium values with the observed values suggests that the observed values may be drawn from a population that is not in equilibrium. Notice that there are more heterozygotes than predicted and fewer *SS* types. To test for a Hardy–Weinberg equilibrium, apply chi-square analysis.

$$\chi^2 = \frac{\Sigma(o - e)^2}{e} = 1.47$$

Checking the chi-square table, 1 degree of freedom gives a value of 3.84 at the 0.05 probability level. Since the χ^2 value calculated here is smaller, the null hypothesis (the observed values fluctuate from the equilibrium values by chance and chance alone) should not be rejected. Thus, the frequencies of *AA*, *AS*, *SS* sampled a population that is in equilibrium.

16. (a) $q_{g+1} = 0.278$ $p_{g+1} = 0.722$
(b) $q_{g+1} = 0.289$ $p_{g+1} = 0.711$
(c) $q_{g+1} = 0.298$ $p_{g+1} = 0.702$
(d) $q_{g+1} = 0.319$ $p_{g+1} = 0.681$

18. (a) $p_1 = 0.6 + 0.2(0.1 - 0.6) = 0.5$
(b) $p_1 = 0.2 + 0.3(0.7 - 0.2) = 0.35$
(c) $p_1 = 0.1 + 0.1(0.2 - 0.1) = 0.11$

20. Given the frequency of the disorder in the population as 1 in 10,000 individuals (0.0001), then $q^2 = 0.0001$, and $q = 0.01$. The frequency of heterozygosity is $2pq$ or approximately 0.02. The probability for one of the grandparents to be heterozygous would be $0.02 + 0.02$ or 0.04 or $1/25$. If one of the grandparents is a carrier, the probability of the offspring from a first-cousin mating being homozygous for the *recessive* gene is $1/16$. Multiplying the two probabilities together gives $1/16 \times 1/25 = 1/400$. Following the same analysis for the second-cousin mating gives $1/64 \times 1/25 = 1/1600$. The population at large has a frequency of homozygotes of $1/10,000$; therefore, one can see how inbreeding increases the likelihood of homozygosity.

22. *Inbreeding depression* is the reduction in fitness in populations that are inbred. With inbreeding comes an increase in the number of homozygous individuals and a decrease in genetic variability.

24. While inbreeding increases the frequency of homozygous individuals in a population, it does not change the *gene* frequencies. There will be fewer heterozygotes in the population to compensate for the additional homozygotes.

26. (a) q is 0.01
(b) $p = 1 - q$ or 0.99
(c) $2pq = 2(.01)(.99) = 0.0198$ (or about $1/50$)
(d) $2pq \times 2pq = 0.0198 \times 0.0198 = 0.000392$, or about $1/255$

28. Because three of the affected infants had affected parents, only two "new" genes, from mutation, enter into the problem. The gene is dominant; therefore, each new case of achondroplasia arose from a single new mutation. There are 50,000 births, therefore 100,000 gametes (genes) involved. The frequency of mutation is given as follows: $2/100,000$ or 2×10^{-5}.

30. (a) The gene is most likely recessive. For the population, since $q^2 = 0.002$, then $q = 0.045$, $q = 0.955$, and $2(pq) = 0.086$. For the community, since $q^2 = 0.005$, $q = 0.07$, $q = 0.093$, and $2(pq) = 0.13$. (b) The "founder effect" is probably operating here. In small populations, homozygosity is increased as a gene has a higher probability of "meeting itself."

Chapter 23 Genetics and Evolution

2. Speciation is the process that leads to the formation of species; it usually involves some form of reproductive isolation. Evolution is the change in a population over time. Speciation is one of many results of evolution.

4. Organisms may appear to be similar but be reproductively isolated for a sufficient period to justify their species identity. If significant genetic differences occur, they can be considered to be separate species.

6. The subterranean niche is less broad and less dynamic than aboveground. The underground microhabitat consists of a narrower range of climatic changes; thus, genetic polymorphism is not selected for. This conclusion has been supported by additional studies indicating that genetic diversity is positively correlated with niche width.

8. Results from laboratory studies indicated that there was a selective advantage in having the two inversions present rather than either one. Thus, natural selection favored the maintenance of both inversions over the loss of either.

10. (a) Missense mutations cause amino acid changes. (b) Horizontal transfer is the process of passing genetic information from one organism to another without producing offspring. In bacteria, plasmid transfer is an example of horizontal transfer. (c) That none of the isolates shared identical nucleotide changes indicates that there is little genetic exchange among different strains. Each alteration is unique, most likely originating in an ancestral strain and maintained in descendents of that strain only.

12. The approximate similarity of mutation rates among genes and lineages should provide more credible estimates of divergence times of species and allow for broader interpretations of sequence comparisons. It also provides for increased understanding of the mutational processes governing evolution among mammalian genomes. For instance, if the rate of mutation is fairly constant among lineages or cells that have a more rapid turnover, it indicates that replication-related errors do not make a significant contribution to mutation rates.

14. In general, speciation involves the gradual accumulation of genetic changes to a point where reproductive isolation occurs. Depending on environmental or geographic conditions, genetic changes may occur slowly or rapidly. They can involve point or chromosomal changes.

16. Reproductive isolating mechanisms are grouped into prezygotic and postzygotic and include

Prezygotic

geographic or ecological
seasonal or temporal
behavioral
mechanical
physiological

Postzygotic

hybrid inviability or weakness
developmental hybrid sterility
segregational hybrid sterility
F_2 breakdown

18. Polyploid plants resulting from hybridization of two species are expected to be more heterozygous than the diploid parental species because two distinct genomes are combined. Generally, genetic variation is an advantage unless a significant degree of that variation is outside acceptable physiological tolerance.

20. All of the amino acid substitutions (*Ala–Gly*, *Val–Leu*, *Asp–Asn*, *Met–Leu*) require only one nucleotide change. The last change from *Pro* (CC—)–*Lys* (AAA,G) requires two changes (the minimal mutational distance).

22.

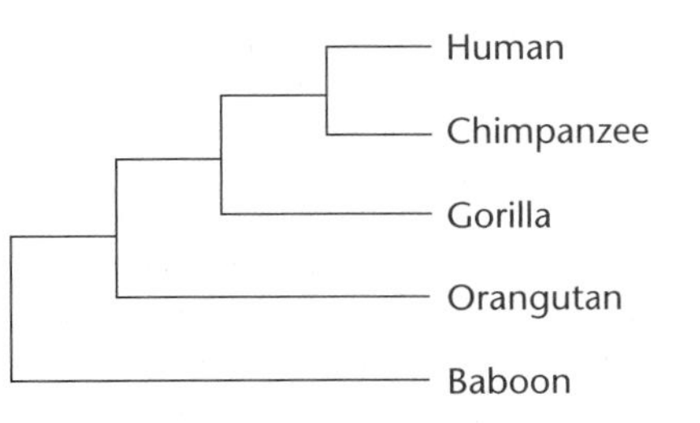

24. In the figure, notice that the $\Delta T_{50}H$ value of 4.0 on the right could be used as a decision point such that any group that diverged above that line would be considered in the same genus, while any group below the line would be in a different genus. Under this rule, you would have the chimpanzee, pygmy chimpanzee, human, gorilla, and orangutan in the same genus. If you assumed that 3.7 is close enough to be considered above 4.0, given considerable experimental error, you could provide a scheme where the orangutan is not included with the chimpanzee, pygmy chimpanzee, human, and gorilla.

26. (a) Since noncoding genomic regions are probably genetically silent, it is likely that they contribute little, if anything, to the phenotype. Selection acts on the phenotype, therefore, such noncoding regions are probably selectively neutral. (b) These polymorphism data indicate that the entire Lake Victoria area (lake and contributing rivers) cichlids are related by recent ancestry, whereas those from neighboring lakes are more distantly related. In addition, since Lake Victoria dried out about 14,000 years ago, it is likely that it was repopulated by a relatively small sample of cichlids.

28. Somatic gene therapy, like any therapy, allows some individuals to live more normal lives than those not receiving therapy. As such, the ability of such individuals to contribute to the gene pool increases the likelihood that less-fit genes will enter and be maintained in the gene pool. This is a normal consequence of therapy, genetic or not; and in the face of disease control and prevention, societies have generally accepted this consequence. Germline therapy could, if successful, lead to limited, isolated, and infrequent removal of a gene from a gene lineage. However, given the present state of the science, its impact on the course of human evolution will be diluted and negated by a host of other factors afflicting mankind.

Chapter 24 Conservation Genetics

2. (a, b) The frequency of the lethal gene in the captive population ($q^2 = 5/169$ and $q = 0.172$) is approximately double that in the gene pool as a whole ($q = 0.09$). Applying the formula

$$q_n = q_o/(1 + nq_o),$$

we can estimate that it would take 10 generations to reduce the lethal gene's frequency to 0.063 in the captive population with no intervention (random mating assumed). Since condors produce very few eggs per year, a more proactive approach seems justified. If detailed records are kept of the breeding partners of the captive birds, knowledge of heterozygotes should be available. Breeding programs could be established to restrict matings between those carrying the lethal gene. Such "kinship management" is often used in captive populations. If kinship records are not available, it is often possible to establish kinship using genetic markers such as DNA microsatellite polymorphisms. Using such markers, we can often identify mating partners and link them to their offspring. By coupling knowledge of mating partners with the likelihood of producing a lethal genetic combination, selective matings can often be used to minimize the influence of a deleterious gene. In addition, such markers can be used to establish matings that optimize genetic mixing, thus reducing inbreeding depression.

4. Both genetic drift and inbreeding tend to drive populations toward homozygosity. Genetic drift is more common when the effective breeding size of the population is low. When this condition prevails, inbreeding is also much more likely. They are different in that inbreeding can occur when certain population structures or behaviors favor matings between relatives, regardless of the effective size of the population. Inbreeding tends to increase the frequency of both homozygous classes at the expense of the heterozygotes. Genetic drift can lead to fixation of one allele or the other, thus producing a single homozygous class.

6. Inbreeding depression, over time, reduces the level of heterozygosity, usually a selectively advantageous quality of a species. When homozygosity increases (through loss of heterozygosity), deleterious alleles are likely to become more of a load on a population. Outbreeding depression occurs when there is a reduction in fitness of progeny from genetically diverse individuals. It is usually attributed to offspring being less well-adapted to the local environmental conditions of the parents. Even though forced outbreeding may be necessary to save a threatened species whose population numbers are low, it significantly and permanently changes the genetic makeup of the species.

8. Often, molecular assays of overall heterozygosity can indicate the degree of inbreeding and/or genetic drift. An allele whose frequency is dictated by inbreeding will not be uniquely influenced. That is, other alleles would decrease heterozygosity as well. So, if the genome has a relatively high degree of heterozygosity, the gene is probably influenced by selection rather than inbreeding or genetic drift.

10. Generally, threatened species are captured and bred in an artificial environment until sufficient population numbers are achieved to ensure species survival. Genetic management strategies are applied to breed individuals in such a way as to increase genetic heterozygosity as much as possible. If plants are involved, seed banks are often used to maintain and facilitate long-term survival.

12. Allozymes are variants of a given allele often detected by electrophoresis. Such variation may or may not impact the fitness of an individual. The greater the allozyme variation, the more genetically heterogeneous the individual. It is generally agreed that such genetic diversity is essential for long-term survival. All other factors being equal, allozyme variation is more likely to reflect physiological variation than AFLP variation because AFLP regions are not necessarily found in the protein-coding regions of the genome. AFLP allows us to detect very small amounts of genetic diversity in a population and is unlikely to encounter an organism that is not in some way variable in terms of AFLP, with respect to other organisms (within and among species).

14. (a) The probability of being a heterozygote is $2pq = 2(0.99)(0.01) = 0.0198$. Multiplying this value by 20 gives the probability of being heterozygous, $0.0198 \times 20 = 0.396$. (b) To determine N_e,

$$N_e/N = 0.42$$
$$N_e = 0.42 \times 50 = 21$$
$$H_t = 0.01755$$
$$H_t/H_0 = 0.01755/0.0198 = 0.886$$

Therefore, there is a loss of approximately 11.4% heterozygosity after five generations.

16. (a, b) It would be necessary to determine whether the native habitat in the Asian steppes of the 1920s is suitable to any introduction. If the original range is supportive of reintroduction, care must be taken to introduce horses with maximum genetic diversity possible. To do so, you might monitor AFLP patterns. Since the founder breeding group included a domestic mare, it may be desirable to select for reintroduction those horses genetically least like the domestic mare. It might also be desirable to release reasonable-sized breeding groups in separate locations within the range to enhance eventual genetic diversity.

18. From a physiological standpoint, cryogenic preservation in liquid nitrogen can allow 100 years or more of seed storage for some species: however, such elaborate storage can be offered to only a small fraction of the world's seeds. Thus, seeds of most species undergo storage loss, which decreases genetic diversity. Seeds of tropical plants are somewhat intolerant to cold storage and must be regenerated frequently, a practice prone to a loss of genetic diversity arising from genetic drift. Only a finite number of seeds can be used in each regeneration procedure, and the restriction of sample size (often fewer than 100 plants) reduces genetic diversity. To somewhat counteract this problem, plants are grown under optimum conditions to reduce selection. Another problem with preserved seeds is the accumulation of deleterious mutations as a result of both seed storage and regeneration. Some studies indicate increased frequencies of chromosomal and mtDNA lesions, chlorophyll deficiency mutations, and decreased DNA polymerase activity associated with long-term seed storage.

20. The longest bottleneck-present interval occurred with cheetahs, and we would expect cheetahs to show the highest degree of microsatellite polymorphism. The shortest bottleneck-present interval occurred with the Gir Forest lions, so it would be expected to have the least polymorphism. Data from Driscoll and others (2002, *Genome Research*, 12:414–23.) include the following estimates of microsatellite polymorphism in the three feline groups that they studied:

cheetahs 84.1%

pumas 42.9%

Gir Forest lions 19.3%

22. (a) The species with the greatest genetic variability, as estimated by these markers, is the domesticated cat. Domesticated cats share an immense and variable gene pool. Their staggering numbers and outbreeding behaviors allow them to maintain a high degree of genetic variability. (b) The lion has the least genetic variability. (c) Since allozymes code for proteins and proteins often provide a significant function in an organism, selection is stronger and mutations are less tolerated. Selection would be expected to be more harsh on DNA segments that are related to function. In addition, by their very nature, microsatellites and minisatellites are more mutable.

24. While flagship species (often large mammals) may make it possible to gather considerable public support and funding, they could reduce support for species that may have a greater impact on a community of species. Primary producers (plants) are a necessary component of a diverse and supportive habitat. If we focus on a flagship species within an area, it is possible that other areas will suffer more dramatically because foundational species are lost. Using umbrella species to protect a large geographic area in hopes of protecting other species in that area is a reasonable approach. However, the size of an area is not necessarily a primary factor in determining species success. Diversity and productivity of a habitat are major contributors to species success. Since land is at a premium, it may be wiser to select umbrella species in diverse and productive habitats rather than on the basis of land size. By selecting sets of species that show considerable biodiversity, we increase the likelihood of protecting a sufficiently rich habitat to support many species. Such habitats are often of considerable economic value, thereby limiting their availability.

GLOSSARY

abortive transduction An event in which transducing DNA fails to be incorporated into the recipient chromosome. See *transduction*.

acentric chromosome Chromosome or chromosome fragment with no centromere.

acquired immunodeficiency syndrome (AIDS) An infectious disease caused by a retrovirus designated as human immunodeficiency virus (HIV). The disease is characterized by a gradual depletion of T lymphocytes, recurring fever, weight loss, multiple opportunistic infections, and rare forms of pneumonia and cancer associated with collapse of the immune system.

acridine dyes A class of organic compounds that bind to DNA and intercalate into the double-stranded structure, producing local disruptions of base-pairing. These disruptions result in additions or deletions in the next round of replication.

acrocentric chromosome Chromosome with the centromere located very close to one end. Human chromosomes 13, 14, 15, 21, and 22 are acrocentric.

active immunity Immunity gained by direct exposure to antigens, followed by antibody production.

active site That portion of a protein, usually an enzyme, whose structural integrity is required for function (e.g., the substrate binding site of an enzyme.

adaptation A heritable component of the phenotype that confers an advantage in survival and reproductive success. The process by which organisms adapt to the current environmental conditions.

additive genes See *polygenic inheritance*.

additive variance Genetic variance that is attributed to the substitution of one allele for another at a given locus. This variance can be used to predict the rate of response to phenotypic selection in quantitative traits.

A-DNA An alternative form of the right-handed double-helical structure of DNA in which the helix is more tightly coiled, with 11 base pairs per full turn of the helix. In this form, the bases in the helix are displaced laterally and tilted in relation to the longitudinal axis. It is not yet clear whether this form has biological significance.

albinism A condition caused by the lack of melanin production in the iris, hair, and skin. It is most often inherited as an autosomal recessive in humans.

aleurone layer In seeds, the outer layer of the endosperm.

alkaptonuria An autosomal recessive condition in humans caused by the lack of the enzyme, homogentisic acid oxidase. Urine of homozygous individuals turns dark upon standing because of oxidation of excreted homogentisic acid. The cartilage of homozygous adults blackens from deposition of a pigment derived from homogentisic acid; affected individuals often develop arthritic conditions.

allele One of the possible mutational states of a gene, distinguished from other alleles by phenotypic effects.

allele frequency Measurement of the proportion of individuals in a population carrying a particular allele.

allele-specific nucleotide (ASO) Synthetic nucleotides, usually 15–20 bp in length that under carefully controlled conditions will hybridize only to a complementary sequence with a perfect match. Under these same conditions, ASOs with a one-nucleotide mismatch will not hybridize.

allelic exclusion In a plasma cell heterozygous for an immunoglobulin gene, the selective action of only one allele.

allelism test See *complementation test*.

allolactose A lactose derivative that acts as the inducer for the *lac* operon.

allopatric speciation Process of speciation associated with geographic isolation.

allopolyploid Polyploid condition formed by the union of two or more distinct chromosome sets with a subsequent doubling of chromosome number.

allosteric effect Conformational change in the active site of a protein brought about by interaction with an effector molecule.

allotetraploid Diploid for two genomes derived from different species.

allozyme An allelic form of a protein that can be distinguished from other forms by electrophoresis.

alpha fetoprotein (AFP) A 70-kDa glycoprotein synthesized during embryonic development by the yolk sac. High levels of this protein in the amniotic fluid are associated with neural tube defects such as spina bifida; lower-than-normal levels may be associated with Down syndrome.

alternative splicing Generation of different protein molecules from the same pre-mRNA by changing the number and order of exons in the mRNA product.

***Alu* sequence** An interspersed DNA sequence of approximately 300 bp found in the genome of primates that is cleaved by the restriction enzyme *Alu*I. These sequences are composed of a head-to-tail dimer. The first monomer is approximately 140 bp and the second, approximately 170 bp. In humans, they are dispersed throughout the genome and are present in 300,000–600,000 copies, constituting some 3–6% of the genome. See short interspersed elements.

amber codon The codon UAG, which does not code for an amino acid but for chain termination.

Ames test An assay developed by Bruce Ames to detect mutagenic and carcinogenic compounds, using reversion to histidine independence in the bacterium *Salmonella typhimurium*.

amino acid Any of the subunit building blocks that are covalently linked to form proteins.

aminoacyl tRNA Covalently linked combination of an amino acid and a tRNA molecule.

amniocentesis A procedure used to test for fetal defects in which fluid and fetal cells are withdrawn from the amniotic layer surrounding the fetus.

amphidiploid See *allotetraploid.*

anabolism The metabolic synthesis of complex molecules from less complex precursors.

analog A chemical compound structurally similar to another, but differing by a single functional group (e.g., 5-bromodeoxyuridine is an analog of thymidine).

anaphase Stage of cell division in which chromosomes begin moving to opposite poles of the cell.

anaphase I The stage in the first meiotic division during which members of homologous pairs of chromosomes separate from one another.

aneuploidy A condition in which the chromosome number is not an exact multiple of the haploid set.

angstrom (Å) Unit of length equal to 10^{-10} meters.

annotation Analysis of genomic nucleotide sequence data to identify the protein-coding genes, the nonprotein-coding genes, their regulatory sequences, and their function(s).

antibody Protein (immunoglobulin) produced in response to an antigenic stimulus with the capacity to bind specifically to the antigen.

anticipation A phenomenon first observed in myotonic dystrophy, where the severity of the symptoms increases from generation to generation and the age of onset decreases from generation to generation. This phenomenon is caused by the expansion of trinucleotide repeats within or near a gene.

anticodon The nucleotide triplet in a tRNA molecule that is complementary to and binds to the codon triplet in an mRNA molecule.

antigen A molecule, often a cell-surface protein, that is capable of eliciting the formation of antibodies.

antiparallel Describing molecules in parallel alignment, but running in opposite directions. Most commonly used to describe the opposite orientations of the two strands of a DNA molecule.

apoptosis A genetically controlled program of cell death, activated as part of normal development or as a result of cell damage.

ascospore A meiotic spore produced in certain fungi.

ascus In fungi, the sac enclosing the four or eight ascospores.

asexual reproduction Production of offspring in the absence of any sexual process.

assortative mating Nonrandom mating between males and females of a species. Selection of mates with the same genotype is positive; selection of mates with opposite genotypes is negative.

ATP Adenosine triphosphate.

attached-X chromosome Two conjoined X chromosomes that share a single centromere.

attenuator A nucleotide sequence between the promoter and the structural gene of some operons that can act to regulate the transit of RNA polymerase, reducing transcription of the related structural gene.

autogamy A process of self-fertilization resulting in homozygosis.

autoimmune disease The production of antibodies that results from an immune response to one's own molecules, cells, or tissues. Such a response results from the inability of the immune system to distinguish self from nonself. Diseases such as arthritis, scleroderma, systemic lupus erythematosus, and juvenile-onset diabetes are examples of autoimmune diseases.

autonomously replicating sequences (ARS) Origins of replication, about 100 nucleotides in length found in yeast chromosomes. ARS elements are also present in organelle DNA.

autopolyploidy Polyploid condition resulting from the replication of one diploid set of chromosomes.

autoradiography Production of a photographic image by radioactive decay. Used to localize radioactively labeled compounds within cells and tissues.

autosomes Chromosomes other than the sex chromosomes. In humans, there are 22 pairs of autosomes.

autotetraploid An autopolyploid condition composed of four similar genomes. In this situation, genes with two alleles (*A* and *a*) can have five genotypic classes: *AAAA* (quadraplex), *AAAa* (triplex), *AAaa* (duplex), *Aaaa* (simplex), and *aaaa* (nulliplex).

auxotroph A mutant microorganism or cell line that requires a substance for growth that can be synthesized by wild-type strains.

backcross A cross involving an F_1 heterozygote and one of the P_1 parents (or an organism with a genotype identical to one of the parents).

bacteriophage A virus that infects bacteria (also, *phage*).

bacteriophage lambda (λ) A member of the lambdoid family of viruses that attach to and infect and replicate within bacterial cells, destroying the host cell in the process. Genetically modified lambda phages are used as vectors in recombinant DNA research.

bacteriophage μ A group of phages whose genetic material behaves as an insertion sequence that can cause inactivation of host genes and rearrangement of host chromosomes.

balanced lethals Recessive, nonallelic lethal genes, each carried on different homologous chromosomes. When organisms carrying balanced lethal genes are interbred, only organisms with genotypes identical to the parents (heterozygotes) survive.

balanced polymorphism Genetic polymorphism maintained in a population by natural selection.

Barr body Densely staining nuclear mass seen in the somatic nuclei of mammalian females. Discovered by Murray Barr, this body is thought to represent an inactivated X chromosome.

base analog See *analog.*

base substitution A single base change in a DNA molecule that produces a mutation. There are two types of substitutions: *transitions*, in which a purine is substituted for a purine or a pyrimidine for a pyrimidine; and *transversions*, in which a purine is substituted for a pyrimidine or vice versa.

β-galactosidase A bacterial enzyme encoded by the *lacZ* gene that converts lactose into galactose and glucose.

bidirectional replication A mechanism of DNA replication in which two replication forks move in opposite directions from a common origin of replication.

biodiversity The genetic diversity present in populations and species of plants and animals.

biometry The application of statistics and statistical methods to biological problems.

biotechnology Commercial and/or industrial processes that utilize biological organisms or products.

bivalents Synapsed homologous chromosomes in the first prophase of meiosis.

Bombay phenotype A rare variant of the ABO system in which affected individuals do not have A or B antigens and thus appear as blood type O, even though their genotype may carry unexpressed alleles for the A and/or B antigens.

bottleneck Fluctuation in allele frequency that occurs when a population undergoes a temporary reduction in size.

BSE (bovine spongiform encephalopathy) A fatal, degenerative brain disease of cattle caused by prion infection. This infection is transmissible to humans and other animals. Also know as mad cow disease.

BrdU (5-bromodeoxyuridine) A mutagenically active analog of thymidine in which the methyl group at the 5′ position in thymine is replaced by bromine; also abbreviated BUdR.

buoyant density A property of particles (and molecules) that depends upon their actual density, as determined by partial specific volume and degree of hydration. Provides the basis for density gradient separation of molecules or particles.

CAAT box A highly conserved DNA sequence found in the untranslated promoter region of eukaryotic genes. This sequence is recognized by transcription factors.

CAP Catabolite activator protein; a protein that binds cAMP and regulates the activation of inducible operons.

carcinogen A physical or chemical agent that causes cancer.

carrier An individual heterozygous for a recessive trait.

catabolism A metabolic reaction in which complex molecules are broken down into simpler forms, often accompanied by the release of energy.

catabolite activator protein See *CAP*.

catabolite repression The selective inactivation of an operon by a metabolic product of the enzymes encoded by the operon.

***cdc* mutation** A class of cell division cycle mutations in yeasts that affect the timing and progression through the cell cycle.

cDNA DNA synthesized from an RNA template by the enzyme reverse transcriptase.

cDNA library A collection of cloned cDNA sequences.

cell cycle Sum of the phases of growth of an individual cell type; divided into G1 (gap 1), S (DNA synthesis), G2 (gap 2), and M (mitosis).

cell-free extract A preparation of the soluble fraction of cells, made by lysing cells and removing the particulate matter, such as nuclei, membranes, and organelles. Often used to carry out the synthesis of proteins by the addition of specific, exogenous mRNA molecules.

CEN In yeasts, fragments of chromosomal DNA, about 120 bp in length, that when inserted into plasmids confer the ability to segregate during mitosis. These segments contain at least three types of sequence elements associated with centromere function.

centimeter (cm) A unit of length equal to 10^{-2} meter.

centimorgan (cM) A unit of distance between genes on chromosomes. One centimorgan represents a value of 1% crossing over between two genes.

central dogma The concept that information flow progresses from DNA to RNA to proteins. Although exceptions are known, this idea is central to an understanding of gene function.

centric fusion See *Robertsonian translocation*.

centriole A cytoplasmic organelle composed of nine groups of microtubules, generally arranged in triplets. Centrioles function in the generation of cilia and flagella and serve as foci for the spindles in cell division.

centromere Specialized region of a chromosome to which sister chromatids remain attached after replication and the site to which spindle fibers attach during cell division. Location of the centromere determines the shape of the chromosome during the anaphase portion of cell division. Also known as the primary constriction.

centrosome Region of the cytoplasm containing the centriole.

chaperone A protein that regulates the folding of a polypeptide into a functional conformation.

character An observable phenotypic attribute of an organism.

charon phages A group of genetically modified lambda phages designed to be used as vectors (carriers) for cloning foreign DNA. Named after the ferryman in Greek mythology who carried the souls of the dead across the River Styx.

chemotaxis Negative or positive response to a chemical gradient.

chiasma (pl., **chiasmata**) The crossed strands of nonsister chromatids seen in diplotene of the first meiotic division. Regarded as the cytological evidence for exchange of chromosomal material, or crossing over.

chi-square(χ^2)analysis Statistical test to determine if an observed set of data fits a theoretical expectation.

chloroplast A cytoplasmic self-replicating organelle containing chlorophyll. The site of photosynthesis.

chorionic villus sampling (CVS) A technique of prenatal diagnosis that intravaginally retrieves chorionic fetal cells and uses them to detect cytogenetic and biochemical defects in the embryo.

chromatid One of the longitudinal subunits of a replicated chromosome; it is joined to its sister chromatid at the centromere.

chromatin The complex of DNA, RNA, histones, and nonhistone proteins that make up uncoiled chromosomes characteristic of the eukaryotic interphase nucleus.

chromatography Technique for the separation of a mixture of solubilized molecules by their differential migration over a substrate.

chromocenter An aggregation of centromeres and heterochromatic elements of polytene chromosomes.

chromomere A coiled, beadlike region of a chromosome most easily visualized during cell division. The aligned chromomeres of polytene chromosomes are responsible for their distinctive banding pattern.

chromosomal aberration Any change resulting in the duplication, deletion, or rearrangement of chromosomal material.

chromosomal mutation See *chromosomal aberration*.

chromosomal polymorphism Alternative structures or arrangements of a chromosome that are carried by members of a population.

chromosome In prokaryotes, an intact DNA molecule containing the genome; in eukaryotes, a DNA molecule complexed with RNA and proteins to form a threadlike structure containing genetic information arranged in a linear sequence and visible during mitosis and meiosis.

chromosome banding Technique for the differential staining of mitotic or meiotic chromosomes to produce a characteristic banding pattern, or selective staining of certain chromosomal regions such as centromeres, the nucleolus organizer regions, and GC- or AT-rich regions. Not to be confused with the banding pattern present in polytene chromosomes, which is produced by the alignment of chromomeres.

chromosome map A diagram showing the location of genes on chromosomes.

chromosome puff A localized uncoiling and swelling in a polytene chromosome, usually regarded as a sign of active transcription.

chromosome theory of inheritance The idea put forward by Walter Sutton and Theodore Boveri that chromosomes are the carriers of genes and the basis for the Mendelian mechanisms of segregation and independent assortment.

chromosome walking A method for analyzing long stretches of DNA, in which the end of a cloned segment of DNA is subcloned and used as a probe to identify other clones that overlap the first clone.

cis configuration The arrangement of two genes or two mutant sites within a gene on the same homolog, such as

$$\frac{a^1 \quad a^2}{+ \quad +}$$

Contrasts with a trans arrangement, where the mutant alleles are located on opposite homologs.

cis-trans test A genetic test to determine whether two mutations are located within the same cistron.

cistron That portion of a DNA molecule coding for a single polypeptide chain; defined by a genetic test as a region within which two mutations cannot complement each other.

cline A gradient of genotype or phenotype distributed over a geographic range.

clonal selection Theory of the immune system that proposes that antibody diversity precedes exposure to the antigen and that the antigen functions to select the cells containing its specific antibody to undergo proliferation.

clone Identical molecules, cells, or organisms derived from a single ancestor by asexual or parasexual methods. For example, a DNA segment that has been enzymatically inserted into a plasmid or chromosome of a phage or a bacterium and replicated to form many copies.

cloned library A collection of cloned DNA molecules representing all or part of an individual's genome.

code See *genetic code*.

codominance Condition in which the phenotypic effects of a gene's alleles are fully and simultaneously expressed in the heterozygote.

codon A triplet of nucleotides that specifies or encodes the information for a single amino acid. Sixty-one codons specify the amino acids used in proteins, and three codons signal termination of growth of the polypeptide chain.

coefficient of coincidence A ratio of the observed number of double crossovers divided by the expected number of such crossovers.

coefficient of inbreeding The probability that two alleles present in a zygote are descended from a common ancestor.

coefficient of selection (*s*) A measurement of the reproductive disadvantage of a given genotype in a population. If for genotype *aa*, only 99 of 100 individuals reproduce, then the selection coefficient is 0.1.

colchicine An alkaloid compound that inhibits spindle formation during cell division. Used in the preparation of karyotypes to collect a large population of cells inhibited at the metaphase stage of mitosis.

colinearity The linear relationship between the nucleotide sequence in a gene (or the RNA transcribed from it) and the order of amino acids in the polypeptide chain specified by the gene.

competence In bacteria, the transient state or condition during which the cell can bind and internalize exogenous DNA molecules, making transformation possible.

complementarity Chemical affinity between nitrogenous bases as a result of hydrogen bonding. Responsible for the base-pairing between the strands of the DNA double helix.

complementation test A genetic test to determine whether two mutations occur within the same gene. If two mutations are introduced into a cell simultaneously and produce a wild-type phenotype (i.e., they complement each other), they are often nonallelic. If a mutant phenotype is produced, the mutations are noncomplementing and are often allelic.

complete linkage A condition in which two genes are located so close to each other that no recombination occurs between them.

complexity The total number of nucleotides or nucleotide pairs in a population of nucleic acid molecules as determined by reassociation kinetics.

complex locus A gene within which a set of functionally related pseudoalleles can be identified by recombinational analysis (e.g., the *bithorax* locus in *Drosophila*).

complex trait A trait whose phenotype is determined by the interaction of multiple genes and environmental factors.

concatemer A chain or linear series of subunits linked together. The process of forming a concatemer is called concatenation (e.g., multiple units of a phage genome produced during replication).

concordance Pairs or groups of individuals identical in their phenotype. In twin studies, a condition in which both twins exhibit or fail to exhibit a trait under investigation.

conditional mutation A mutation that expresses a wild-type phenotype under certain (permissive) conditions and a mutant phenotype under other (restrictive) conditions.

conjugation Temporary fusion of two single-celled organisms for the sexual transfer of genetic material.

consanguineous Related by a common ancestor within the previous few generations.

consensus sequence A basically common, although not necessarily identical, sequence of nucleotides in DNA or amino acids in proteins.

conservation genetics The branch of genetics concerned with the preservation and maintenance of wild species of plants and animals in their natural environments.

continuous variation Phenotype variation exhibited by quantitative traits distributed from one phenotypic extreme to another in an overlapping or continuous fashion.

cosmid A vector designed to allow cloning of large segments of foreign DNA. Cosmids are composed of the *cos* sites of phage λ inserted into a plasmid. In cloning, the recombinant DNA molecules are packaged into phage protein coats, and after infection of bacterial cells, the recombinant molecule replicates and can be maintained as a plasmid.

coupling conformation See *cis configuration*.

covalent bond A nonionic chemical bond formed by the sharing of electrons.

Creutzfeldt-Jakob disease (CJD) A progressive degenerative and fatal disease of the brain and nervous system caused by mutations in the prion protein gene on chromosome 20 that produce aberrant forms of the encoded protein. CDJ is inherited as an autosomal dominant trait.

cri-du-chat syndrome A clinical syndrome in humans produced by a deletion of a portion of the short arm of chromosome 5. Afflicted infants have a distinctive cry that sounds like that of a cat.

crossing over The exchange of chromosomal material (parts of chromosomal arms) between homologous chromosomes by breakage and reunion. The exchange of material between nonsister chromatids during meiosis is the basis of genetic recombination.

cross-reacting material (CRM) Nonfunctional form of an enzyme, produced by a mutant gene, that is recognized by antibodies made against the normal enzyme.

C-terminal amino acid The terminal amino acid in a peptide chain that carries a free carboxyl group.

C terminus The end of a polypeptide that carries a free carboxyl group of the last amino acid. By convention, the structural formula of polypeptides is written with the C terminus at the right.

***C* value** The haploid amount of DNA present in a genome.

***C* value paradox** The apparent paradox that there is no relationship between the size of the genome and the evolutionary complexity of species. For example, the *C* value (haploid genome size) of amphibians varies by a factor of 100.

cyclic adenosine monophosphate (cAMP) An important regulatory molecule in both prokaryotic and eukaryotic organisms.

cyclins A class of proteins found in eukaryotic cells that are synthesized and degraded in synchrony with the cell cycle and regulate passage through stages of the cycle.

cytogenetics A branch of biology in which the techniques of both cytology and genetics are used to study heredity.

cytokinesis The division or separation of the cytoplasm during mitosis or meiosis.

cytological map A diagram showing the location of genes at particular chromosomal sites.

cytoplasmic inheritance Non-Mendelian form of inheritance involving genetic information transmitted by self-replicating cytoplasmic organelles such as mitochondria, chloroplasts, etc.

cytoskeleton An internal array of microtubules, microfilaments, and intermediate filaments that confers shape and the ability to move on a eukaryotic cell.

dalton (Da) A unit of mass equal to that of the hydrogen atom, which is 1.67×10^{-24} gram. A unit used in designating molecular weights.

Darwinian fitness See *fitness*.

deficiency A chromosomal mutation involving the loss or deletion of chromosomal material.

degenerate code The genetic code, where a given amino acid may be represented by more than one codon. For example, some amino acids (leucine) have six codons, others (isoleucine) have three.

deletion See *deficiency*.

deme A local interbreeding population.

denatured DNA DNA molecules that have been separated into single strands.

de novo Newly arising; synthesized from less complex precursors rather than having been produced by modification of an existing molecule.

density gradient centrifugation A method of separating macromolecular mixtures by the use of centrifugal force and solutions of varying density. In buoyant density gradient centrifugation using cesium chloride, the cesium solution establishes a gradient under the influence of the centrifugal field and a mixture of macromolecules such as DNA sediment in the gradient until the density of the cesium chloride solution equals their own, separating them by differences in density.

deoxyribonuclease A class of enzymes that breaks down DNA into oligonucleotide fragments by introducing single-stranded breaks into the double helix.

deoxyribonucleic acid (DNA) A macromolecule usually consisting of antiparallel polynucleotide chains held together by hydrogen bonds, in which the sugar residues are deoxyribose. The primary carrier of genetic information.

deoxyribose The five-carbon sugar associated with the deoxyribonucleotides found in DNA.

dermatoglyphics The study of the surface ridges of the skin, especially of the hands and feet.

determination A regulatory event that establishes a specific pattern of future gene activity and developmental fate for a given cell.

diakinesis The final stage of meiotic prophase I in which the chromosomes become tightly coiled and compacted and move toward the periphery of the nucleus.

dicentric chromosome A chromosome having two centromeres.

dideoxynucleotide A nucleotide containing a dexoyribose sugar lacking a 3′ hydroxyl group. Stops further chain elongation when incorporated into a growing polynucleotide; used in the Sanger method of DNA sequencing.

differentiation The process of complex changes by which cells and tissues attain their adult structure and functional capacity.

dihybrid cross A genetic cross involving two characters in which the parents possess different forms of each character (e.g., tall, round × short, wrinkled peas).

diploid A condition in which each chromosome exists in pairs; having two of each chromosome.

diplotene A stage of meiotic prophase immediately after pachytene. In diplotene, one pair of sister chromatids begins separating from the other, and chiasmata become visible. These overlaps move laterally toward the ends of the chromatids (terminalization).

directional selection A selective force that changes the frequency of an allele in a given direction, either toward fixation or toward elimination.

discontinuous replication of DNA The synthesis of DNA in discontinuous fragments on the lagging strand of the replication fork. The fragments, known as Okazaki fragments, are joined by DNA ligase to form a continuous strand.

discontinuous variation Phenotypic data that fall into two or more distinct, nonoverlapping classes.

discordance In twin studies, a situation where one twin expresses a trait but the other does not.

disjunction The separation of chromosomes at the anaphase stage of cell division.

disruptive selection Simultaneous selection for phenotypic extremes in a population, usually resulting in the production of two phenotypically discontinuous strains.

dizygotic twins Twins produced from separate fertilization events; two ova fertilized independently. Also known as fraternal twins.

DNA See *deoxyribonucleic acid*.

DNA fingerprinting A molecular method for identifying an individual member of a population or species. The pattern of DNA fragments obtained by restriction enzyme digestion, followed by Southern blot hybridization using minisatellite probes. See also *STR sequences*.

DNA footprinting See *footprinting*.

DNA gyrase One of the DNA topoisomerases that functions during DNA replication to reduce molecular tension caused by supercoiling. DNA gyrase produces, then seals double-stranded breaks.

DNA ligase An enzyme that forms a covalent bond between the 5′-end of one polynucleotide chain and the 3′-end of another polynucleotide chain. It is also called polynucleotide-joining enzyme.

DNA polymerase An enzyme that catalyzes the synthesis of DNA from deoxyribonucleotides and a template DNA molecule.

DNase Deoxyribonucleosidase; an enzyme that degrades or breaks down DNA into fragments or constitutive nucleotides.

dominance The expression of a trait in the heterozygous condition.

dominant suppression A form of epistasis in which a dominant allele at one locus suppresses the effect of a dominant allele at another locus, resulting in a 13:3 phenotypic ratio.

dosage compensation A genetic mechanism that regulates the levels of gene products at certain loci on the X chromosome in mammals such that males and females have equal amounts of a gene product. In mammals, this is accomplished by random inactivation of one X chromosome.

double crossover Two separate events of chromosome breakage and exchange occurring within the same tetrad.

double helix The model for DNA structure proposed by James Watson and Francis Crick, involving two antiparallel hydrogen-bonded polynucleotide chains wound into a right-handed helical configuration, with 10 base pairs per full turn of the double helix. Often called B-DNA.

Duchenne muscular dystrophy An X-linked recessive genetic disorder caused by a mutation in the gene for dystrophin, a protein found in muscle cells.

duplication A chromosomal aberration in which a segment of the chromosome is repeated.

dyad The products of tetrad separation or disjunction at the first meiotic prophase. Consists of two sister chromatids joined at the centromere.

dystrophin See *Duchenne muscular dystrophy*.

effective population size The number of individuals in a population that have an equal probability of contributing gametes to the next generation.

effector molecule Small, biologically active molecule that acts to regulate the activity of a protein by binding to a specific receptor site on the protein.

electrophoresis A technique used to separate a mixture of molecules by their differential migration through a stationary medium (such as a gel) in an electrical field.

endocytosis The uptake by a cell of fluids, macromolecules, or particles by pinocytosis, phagocytosis, or receptor-mediated endocytosis.

endomitosis Chromosomal replication that is not accompanied by either nuclear or cytoplasmic division.

endonuclease An enzyme that hydrolyzes internal phosphodiester bonds in a polynucleotide chain or nucleic acid molecule.

endoplasmic reticulum A membranous organelle system in the cytoplasm of eukaryotic cells. The outer surface of the membranes may be ribosome-studded (rough ER) or smooth ER.

endopolyploidy The increase in chromosome sets that results from endomitotic replication within somatic nuclei.

endosymbiont theory The proposal that self-replicating cellular organelles such as mitochondria and chloroplasts were originally free-living organisms that entered into a symbiotic relationship with nucleated cells.

enhancer Originally identified as a 72-bp sequence in the genome of a virus, SV40, that increases the transcriptional activity of nearby structural genes. Similar sequences that enhance transcription have been identified in the genomes of eukaryotic cells. Enhancers can act over a distance of thousands of base pairs and can be located 5′, 3′, or internal to the gene they affect, and thus are different from promoters.

environment The complex of geographic, climatic, and biotic factors within which an organism lives.

enzyme A protein or complex of proteins that catalyzes a specific biochemical reaction.

epigenesis The idea that an organism develops by the appearance and growth of new structures. Opposed to preformationism, which holds that development is the growth of structures already present in the egg.

episome A circular genetic element in bacterial cells that can replicate independently of the bacterial chromosome or integrate and replicate as part of the chromosome.

epistasis Nonreciprocal interaction between genes such that one gene interferes with or prevents the expression of another gene. In *Drosophila* for example, the recessive gene *eyeless*, when homozygous, prevents the expression of eye color genes present in the genome.

epitope That portion of a macromolecule or cell that acts to elicit an antibody response; an antigenic determinant. A complex molecule or cell can contain several such sites.

equational division A division of each chromosome into longitudinal halves that are distributed into two daughter nuclei. Chromosome division in mitosis is an example of equational division.

equatorial plate See *metaphase plate*.

euchromatin Chromatin or chromosomal regions that are lightly staining and are relatively uncoiled during the interphase portion of the cell cycle. Euchromatic regions contain most of the structural genes.

eugenics The improvement of the human species by selective breeding. Positive eugenics refers to the promotion of breeding of people with favorable genes, and negative eugenics refers to the discouragement of breeding among those with undesirable traits.

eukaryotes Those organisms having true nuclei and membranous organelles and whose cells demonstrate mitosis and meiosis.

euphenics Medical or genetic intervention to reduce the impact of defective genotypes.

euploid Polyploid with a chromosome number that is an exact multiple of a basic chromosome set.

evolution The origin of plants and animals from preexisting types. Descent with modifications.

excision repair Removal of damaged DNA segments followed by repair. Excision can include the removal of individual bases (base repair) or a stretch of damaged nucleotides (nucleotide repair). The gap created by excision is filled by polymerase, and the ends are ligated to form an intact molecule.

exon (extron) The DNA segment(s) of a gene that are transcribed and translated into proteins.

exonuclease An enzyme that breaks down nucleic acid molecules by breaking the phosphodiester bonds at the 3′- or 5′- terminal nucleotides.

expressed sequence tags (ESTs) All or part of the nucleotide sequence of cDNA clones. Used as markers in construction of genetic maps.

expression vector Plasmids or phages carrying promoter regions designed to cause expression of inserted DNA sequences.

expressivity The degree or range in which a phenotype for a given trait is expressed.

extranuclear inheritance Transmission of traits by genetic information contained in cytoplasmic organelles such as mitochondria and chloroplasts.

F^- cell A bacterial cell that does not contain a fertility factor. Acts as a recipient in bacterial conjugation.

F^+ cell A bacterial cell having a fertility factor. Acts as a donor in bacterial conjugation.

F factor An episome in bacterial cells that confers the ability to act as a donor in conjugation (also, *fertility factor*).

F′ factor A fertility factor that contains a portion of the bacterial chromosome.

F_1 generation First filial generation; the progeny resulting from the first cross in a series.

F_2 generation Second filial generation; the progeny resulting from a cross of the F_1 generation.

F pilus See *pilus*.

familial trait A trait transmitted through and expressed by members of a family.

fate map A diagram of an embryo showing the location of cells whose development fate is known.

fertility factor See *F factor*.

filial generations See *F_1, F_2 generations*.

fingerprint The unique pattern of ridges and whorls on the tip of a human finger. Also, the pattern obtained by enzymatically cleaving a protein or nucleic acid and subjecting the digest to two-dimensional chromatography or electrophoresis. See also *DNA fingerprinting*.

FISH See *fluorescence* in situ *hybridization*.

fitness A measure of the relative survival and reproductive success of a given individual or genotype.

fixation In population genetics, a condition in which all members of a population are homozygous for a given allele.

fluctuation test A statistical test developed by Salvadore Luria and Max Delbrück to determine whether bacterial mutations arise spontaneously or are produced in response to selective agents.

fluorescence *in situ* hybridization (FISH) A method of *in situ* hybridization that utilizes probes labeled with a fluorescent tag, causing the site of hybridization to fluoresce when viewed in ultraviolet light under a microscope.

flush–crash cycle A period of rapid population growth followed by a drastic reduction in population size.

fmet See *formylmethionine*.

folded-fiber model A model of eukaryotic chromosome organization in which each sister chromatid consists of a single fiber, composed of double-stranded DNA and protein, which is wound like a tightly coiled skein of yarn.

footprinting A technique for identifying a DNA sequence that binds to a particular protein, based on the idea that the phosphodiester bonds in the region covered by the protein are protected from digestion by deoxyribonucleases.

formylmethionine (fmet) A molecule derived from the amino acid methionine by attachment of a formyl group to its terminal amino group. This is the first amino acid inserted in all bacterial polypeptides. Also known as *N*-formyl methionine.

founder effect A form of genetic drift. The establishment of a population by a small number of individuals whose genotypes carry only a fraction of the different kinds of alleles in the parental population.

fragile site A heritable gap or nonstaining region of a chromosome that can be induced to generate chromosome breaks.

fragile X syndrome A genetic disorder caused by the expansion of a CGG trinucleotide repeat and a fragile site at Xq27.3 within the *FMR-1* gene.

frameshift mutation A mutational event leading to the insertion of one or more base pairs in a gene, shifting the codon reading frame in all codons that follow the mutational site.

fraternal twins See *dizygotic twins*.

G1 checkpoint A point in the G1 phase of the cell cycle when a cell becomes committed to initiate DNA synthesis and continue the cycle or withdraw into the G0 resting stage.

G0 A point in the G1 phase where cells withdraw from the cell cycle and enter a nondividing but metabolically active state.

gamete A specialized reproductive cell with a haploid number of chromosomes.

gap genes Genes expressed in contiguous domains along the anterior–posterior axis of the *Drosophila* embryo that regulate the process of segmentation in each domain.

gene The fundamental physical unit of heredity whose existence can be confirmed by allelic variants and which occupies a specific chromosomal locus. A DNA sequence coding for a single polypeptide.

gene amplification The process by which gene sequences are selected and differentially replicated either extrachromosomally or intrachromosomally.

gene conversion The process of nonreciprocal recombination by which one allele in a heterozygote is converted into the corresponding allele.

gene duplication An event in replication leading to the production of a tandem repeat of a gene sequence.

gene flow The gradual exchange of genes between two populations; brought about by the dispersal of gametes or the migration of individuals.

gene frequency The percentage of alleles of a given type in a population.

gene interaction Production of novel phenotypes by the interaction of alleles of different genes.

gene mutation See *point mutation*.

gene pool The total of all alleles possessed by reproductive members of a population.

generalized transduction The transduction of any gene in the bacterial genome by a phage.

genetically modified organism (GMO) A plant or animal that has had a gene from another species transferred to its genome using recombinant DNA technology, and where the gene is expressed to produce a gene product.

genetic anticipation The phenomenon of a progressively earlier age of onset and increasing severity of symptoms in successive generations for a genetic disorder.

genetic background All genes carried in the genome other than the one being studied.

genetic code The nucleotide triplets that code for the 20 amino acids or for chain initiation or termination.

genetic counseling Analysis of risk for genetic defects in a family and the presentation of options available to avoid or ameliorate possible risks.

genetic drift Random variation in allele frequency from generation to generation, most often observed in small populations.

genetic engineering The technique of altering the genetic constitution of cells or individuals by the selective removal, insertion, or modification of individual genes or gene sets.

genetic equilibrium Maintenance of allele frequencies at the same value in successive generations. A condition in which allele frequencies are neither increasing nor decreasing.

genetic erosion The loss of genetic diversity from a population or a species.

genetic fine structure Intragenic recombinational analysis that provides mapping information at the level of individual nucleotides.

genetic load Average number of recessive lethal genes carried in the heterozygous condition by an individual in a population.

genetic polymorphism The stable coexistence of two or more discontinuous genotypes in a population. When the frequencies of two alleles are carried to an equilibrium, the condition is called balanced polymorphism.

genetics The branch of biology that deals with heredity and the expression of inherited traits.

genome The array of genes carried by an individual.

genomic imprinting A condition where the expression of a trait depends on whether the trait has been inherited from a male or a female parent.

genomics The study of genomes, including nucleotide sequence, gene content, organization, and gene number.

genotype The specific allelic or genetic constitution of an organism; often, the allelic composition of one or a limited number of genes under investigation.

germ line An embryonic cell lineage that forms the reproductive cells (eggs and sperm).

germ plasm Hereditary material transmitted from generation to generation.

Goldberg-Hogness box A short nucleotide sequence 20–30 bp upstream from the initiation site of eukaryotic genes to which RNA polymerase II binds. The consensus sequence is TATAAAA. Also known as TATA box.

graft-versus-host disease (GVHD) In transplants, reaction by immunologically competent cells of the donor against the antigens present on the cells of the host. In human bone marrow transplants, often a fatal condition.

green revolution A program that resulted in a 2-3 fold increase in crop yields generated by the development of new varieties of cereal plants with shorter stems and increased disease resistance.

gynandromorph An individual composed of cells with both male and female genotypes.

gyrase One of a class of enzymes known as topoisomerases. Gyrase converts closed circular DNA to a negatively supercoiled form prior to replication, transcription, or recombination.

haploid A cell or organism having a single set of unpaired chromosomes. The gametic chromosome number.

haplotype The set of alleles from closely linked loci carried by an individual and usually inherited as a unit.

Hardy-Weinberg law The principle that both gene and genotype frequencies will remain in equilibrium in an infinitely large population in the absence of mutation, migration, selection, and nonrandom mating.

heat shock A transient response following exposure of cells or organisms to elevated temperatures. The response involves activation of a small number of loci, inactivation of some previously active loci, and selective translation of heat shock mRNA. Appears to be a nearly universal phenomenon observed in organisms ranging from bacteria to humans.

helicase An enzyme that participates in DNA replication by unwinding the double helix near the replication fork.

helix–turn–helix motif The structure of a region of DNA-binding proteins in which a turn of four amino acids holds two α helices at right angles to each other.

hemizygous Conditions where a gene is present in a single dose in an otherwise diploid cell. Usually applied to genes on the X chromosome in heterogametic males.

hemoglobin (Hb) An iron-containing, oxygen-carrying protein occurring chiefly in the red blood cells of vertebrates.

hemophilia An X-linked trait in humans associated with defective blood-clotting mechanisms.

heredity Transmission of traits from one generation to another.

heritability A measure of the degree to which observed phenotypic differences for a trait are genetic.

heterochromatin The heavily staining, late-replicating regions of chromosomes that are condensed in interphase. Thought to be devoid of structural genes.

heteroduplex A double-stranded nucleic acid molecule in which each polynucleotide chain has a different origin. These structures may be produced as intermediates in a recombinational event or by the *in vitro* reannealing of single-stranded, complementary molecules.

heterogametic sex The sex that produces gametes containing unlike sex chromosomes.

heterogeneous nuclear RNA (hnRNA) The collection of RNA transcripts in the nucleus, representing precursors and processing intermediates to rRNA, mRNA, and tRNA. Also represents RNA transcripts that will not be transported to the cytoplasm, such as snRNA.

heterokaryon A somatic cell containing nuclei from two different sources.

heterozygote An individual with different alleles at one or more loci. Such individuals will produce unlike gametes and therefore will not breed true.

Hfr A strain of bacteria exhibiting a high frequency of recombination. These strains have a chromosomally integrated F factor that is able to mobilize and transfer part of the chromosome to a recipient F^- cell.

histocompatibility antigens See *HLA*.

histones Proteins complexed with DNA in the nucleus. They are rich in the basic amino acids arginine and lysine, and they function in coiling DNA to form nucleosomes.

HLA Cell-surface proteins, produced by histocompatibility loci, involved in the acceptance or rejection of tissue and organ grafts and transplants.

hnRNA See *heterogeneous nuclear RNA*.

Holliday structure An intermediate in bidirectional DNA recombination seen in the transmission electron microscope as an X-shaped structure showing four single-stranded DNA regions.

homeobox A sequence of about 180 nucleotides that encodes a sequence of 60 amino acids called a *homeodomain*, which is part of a DNA-binding protein that acts as a transcription factor.

homeotic mutation A mutation that causes a tissue normally determined to form a specific organ or body part to alter its differentiation and form another structure.

homogametic sex The sex that produces gametes that do not differ with respect to sex chromosome content; in mammals, the female is homogametic.

homologous chromosomes Chromosomes that synapse or pair during meiosis. Chromosomes that are identical with respect to their genetic loci and centromere placement.

homozygote An individual with identical alleles at one or more loci. These individuals will produce identical gametes and will therefore breed true.

homunculus The miniature individual imagined by preformationists to be contained within the sperm or egg.

H substance The carbohydrate group present on the surface of red blood cells. When unmodified, it results in blood type O; when modified by the addition of monosaccharides, it results in types A, B, and AB.

human immunodeficiency virus (HIV) A human retrovirus associated with the onset and progression of AIDS.

hybrid An individual produced by crossing two parents of different genotypes.

hybridoma A somatic cell hybrid produced by the fusion of an antibody-producing cell and a cancer cell, specifically, a myeloma. The cancer cell contributes the ability to divide indefinitely, and the antibody cell confers the ability to synthesize large amounts of a single antibody.

hybrid vigor The superiority of a heterozygote over either homozygote for a given trait.

hydrogen bond An electrostatic attraction between a hydrogen atom bonded to a strongly electronegative atom such as oxygen or nitrogen and another atom that is electronegative or contains an unshared electron pair.

hypervariable regions The regions of antibody molecules that attach to antigens. These regions have a high degree of diversity in amino acid content.

identical twins See *monozygotic twins*.

Ig See *immunoglobulin*.

imaginal disc Discrete groups of cells set aside during embryogenesis in holometabolous insects, which are determined to form the external body parts of the adult.

immunoglobulin (Ig) The class of serum proteins having the properties of antibodies.

inborn error of metabolism A biochemical disorder that is genetically controlled; usually an enzyme defect that produces a clinical syndrome.

inbreeding Mating between closely related organisms.

inbreeding depression A decrease in viability, vigor, or growth in progeny after several generations of inbreeding.

incomplete dominance Expression of heterozygous phenotype that is distinct from and often intermediate to that of either parent.

incomplete linkage Occasional separation of two genes on the same chromosome by a recombinational event.

independent assortment The independent behavior of each pair of homologous chromosomes during their segregation in meiosis I. The random distribution of maternal and paternal homologs into gametes.

inducer An effector molecule that activates transcription.

inducible enzyme system An enzyme system under the control of a regulatory molecule, or inducer, which acts to block a repressor and allow transcription.

initiation codon The triplet of nucleotides (AUG) in an mRNA molecule that codes for the insertion of the amino acid methionine as the first amino acid in a polypeptide chain.

insertion sequence See *IS element*.

***in situ* hybridization** A technique for the cytological localization of DNA sequences complementary to a given nucleic acid or polynucleotide.

intercalary deletion A form of chromosome deletion where material is lost from within the chromosome. Deletions that involve the end of the chromosome are called terminal deletions.

intercalating agent A compound that inserts between bases in a DNA molecule, disrupting the alignments and pairing of bases in the complementary strands (e.g., acridine dyes).

interference A measure of the degree to which one crossover affects the incidence of another crossover in an adjacent region of the same chromatid. Negative interference increases the chances of another crossover; positive interference reduces the probability of a second crossover event.

interphase That portion of the cell cycle between divisions.

intervening sequence See *intron*.

intron A portion of DNA between coding regions in a gene that is transcribed but does not appear in the mRNA product.

inversion A chromosomal aberration in which the order of a chromosomal segment has been reversed.

inversion loop The chromosomal configuration resulting from the synapsis of homologous chromosomes, one of which carries an inversion.

in vitro Literally, in glass; outside the living organism; occurring in an artificial environment.

in vivo Literally, in the living; occurring within the living body of an organism.

IS element A mobile DNA segment that is transposable to any of a number of sites in the genome.

isoagglutinogen An antigenic factor or substance present on the surface of cells that is capable of inducing the formation of an antibody.

isochromosome An aberrant chromosome with two identical arms and homologous loci.

isolating mechanism Any barrier to the exchange of genes between different populations of a group of organisms. In general, isolation can be classified as spatial, environmental, or reproductive.

isotopes A form of chemical elements that have the same number of protons and electrons but differ in the number of neutrons contained in the atomic nucleus.

isozyme Any of two or more distinct forms of an enzyme that have identical or nearly identical chemical properties but differ in some property such as net electrical charge, pH optima, number and type of subunits, or substrate concentration.

κ particles DNA-containing cytoplasmic particles found in certain strains of *Paramecium aurelia*. When these self-reproducing particles are transferred into the growth medium, they release a toxin, paramecin, that kills other sensitive strains. A nuclear gene, *K*, is responsible for maintaining kappa particles in the cytoplasm.

karyokinesis The process of nuclear division.

karyotype The chromosome complement of a cell or an individual. Often used to refer to the arrangement of metaphase chromosomes in a sequence according to length and position of the centromere.

kilobase (kb) A unit of length consisting of 1000 nucleotides.

kinetochore A fibrous structure with a size of about 400 nm, located within the centromere. It appears to be the site of microtubule attachment during division.

Klinefelter syndrome A genetic disorder in human males caused by the presence of an extra X chromosome. Klinefelter males are XXY instead of XY. This syndrome is associated with enlarged breasts, small testes, sterility, and, occasionally, mild mental retardation.

knockout mice In producing knockout mice, a cloned normal gene is inactivated by the insertion of a marker, such as an antibiotic resistance gene. The altered gene is transferred to embryonic stem cells, where the altered gene will replace the normal gene (in some cells). These cells are injected into a blastomere embryo, producing a mouse that is bred to yield mice that are homozygous for the mutated gene.

***lac* repressor protein** A protein that binds to the operator in the *lac* operon and blocks transcription.

lagging strand In DNA replication, the strand synthesized in a discontinuous fashion, 5′ to 3′ away from the replication fork. Each short piece of DNA synthesized in this fashion is called an Okazaki fragment.

lampbrush chromosomes Meiotic chromosomes characterized by extended lateral loops, which reach maximum extension during diplotene. Although most intensively studied in amphibians, these structures occur in meiotic cells of organisms ranging from insects to humans.

lariat structure A structure formed by an intron via a 5′- to -2′ bond during processing, and removal of that intron from an mRNA molecule.

leader sequence That portion of an mRNA molecule from the 5′-end to the beginning codon; may contain regulatory or ribosome binding sites.

leading strand During DNA replication, the strand synthesized continuously 5′ to 3′ from the origin of replication toward the replication fork.

leptotene The initial stage of meiotic prophase I, during which the chromosomes become visible and are often arranged in a bouquet configuration, with one or both ends of the chromosomes gathered at one spot on the inner nuclear membrane.

lethal gene A gene whose expression results in death.

leucine zipper A structural motif in a DNA-binding protein that is characterized by a stretch of leucine residues spaced at every seventh amino acid residue, with adjacent regions of positively charged amino acids. Leucine zippers on two polypeptides may interact to form a dimer that binds to DNA.

linkage Condition in which two or more nonallelic genes tend to be inherited together. Linked genes have their loci along the same chromosome; they do not assort independently, but can be separated by crossing over.

linkage group A group of genes that have their loci on the same chromosome.

linking number The number of times that two strands of a closed, circular DNA duplex cross over each other.

locus (pl., **loci**) The site or place on a chromosome where a particular gene is located.

lod score A statistical method used to determine whether two loci are linked or unlinked. A lod (log of the odds) score of 4 indicates that linkage is 10,000 times more likely than nonlinkage. By convention, lod scores of 3–4 are signs of linkage.

long interspersed elements (LINEs) Repetitive sequences found in the genomes of higher organisms, such as the 6-kb L1 sequences found in primate genomes.

long terminal repeat (LTR) Sequence of several hundred base pairs found at the ends of retroviral DNAs.

Lutheran blood group One of a number of blood group systems inherited independently of the ABO, MN, and Rh systems. Alleles of this group determine the presence or absence of antigens on the surface of red blood cells. This gene is located on human chromosome 19.

Lyon hypothesis The random inactivation of the maternal or paternal X chromosome in somatic cells of mammalian females early in development. All daughter cells will have the same X chromosome inactivated, producing a mosaic pattern of expression of genes on the X chromosome.

lysis The disintegration of a cell, brought about by the rupture of its membrane.

lysogenic bacterium A bacterial cell carrying the DNA of a temperate bacteriophage integrated into its chromosome.

lysogeny The process by which the DNA of an infecting phage becomes repressed and integrated into the chromosome of the bacterial cell it infects.

lytic phase The condition in which a temperate bacteriophage loses its integrated status in the host chromosome (becomes induced), replicates, and lyses the bacterial cell.

major histocompatibility (MHC) loci In humans, the HLA complex; and in mice, the H2 complex.

mapping functions Map distance estimates from recombination when the recombination frequency in a region exceeds 15–20%, and double crossovers are undetectable.

map unit A measure of the genetic distance between two genes, corresponding to a recombination frequency of 1%. See *centimorgan*.

maternal effect Phenotypic effects on the offspring produced by the maternal genome. Factors transmitted through the egg cytoplasm that produce a phenotypic effect in the progeny.

maternal influence See *maternal effect*.

maternal inheritance The transmission of traits via cytoplasmic genetic factors such as mitochondria or chloroplasts.

mean The arithmetic average.

median The value in a group of numbers below and above which there is an equal number of data points or measurements.

meiosis The process in gametogenesis or sporogenesis during which one replication of the chromosomes is followed by two nuclear divisions to produce four haploid cells.

melting profile (T_m) The temperature at which a population of double-stranded nucleic acid molecules is half-dissociated into single strands. This is taken to be the melting temperature for that species of nucleic acid.

merozygote A partially diploid bacterial cell containing, in addition to its own chromosome, a chromosome fragment introduced into the cell by transformation, transduction, or conjugation.

messenger RNA See *mRNA*.

metabolism The sum of chemical changes in living organisms by which energy is generated and used.

metacentric chromosome A chromosome with a centrally located centromere, producing chromosome arms of equal lengths.

metafemale In *Drosophila*, a poorly developed female of low viability in which the ratio of X chromosomes to sets of autosomes exceeds 1.0. Previously called a superfemale.

metamale In *Drosophila*, a poorly developed male of low viability in which the ratio of X chromosomes to sets of autosomes is less than 0.5. Previously called a supermale.

metaphase The stage of cell division in which the condensed chromosomes lie in a central plane between the two poles of the cell and in which the chromosomes become attached to the spindle fibers.

metaphase plate The arrangement of mitotic or meiotic chromosomes at the equator of the cell during metaphase.

methylation Enzymatic transfer of methyl groups from S-adenosylmethionine to biological molecules including phospholipids, proteins, RNA, and DNA. Methylation of DNA is associated with reduction in gene expression and with epigenetic phenomena such as imprinting.

MHC See *major histocompatibility loci*.

micrometer (μm) A unit of length equal to 1×10^{-6} meter. Previously called a micron.

micron See *micrometer*.

migration coefficient An expression of the proportion of migrant genes entering the population per generation.

millimeter (mm) A unit of length equal to 1×10^{-3} meter.

minimal medium A medium containing only those nutrients that will support the growth and reproduction of wild-type strains of an organism.

minisatellite Short tandem repeats of 10–100 nucleotides widely dispersed in the genome of eukaryotes. The number of repeats at each locus is variable; these loci are known as variable number tandem repeats (VNTRs). Each variation represents a VNTR allele, and many loci have dozens of alleles. VNTRs are used in DNA fingerprinting. See also *STR sequences*.

mismatch repair A process of excision repair, during which an unpaired base or bases is excised, followed by the synthesis of a new segment, using the complementary strand as a template.

missense mutation A mutation that alters a codon to that of another amino acid, causing an altered translation product to be made.

mitochondrion Found in the cells of eukaryotes, a cytoplasmic, self-reproducing organelle that is the site of ATP synthesis.

mitogen A substance that stimulates mitosis in nondividing cells (e.g., phytohemagglutinin).

mitosis A form of cell division resulting in the production of two cells, each with the same chromosome and genetic complement as the parent cell.

mode In a set of data, the value occurring in the greatest frequency.

monohybrid cross A genetic cross between two individuals involving only one character (e.g., $AA \times aa$).

monosomic An aneuploid condition in which one member of a chromosome pair is missing; having a chromosome number of $2n - 1$.

monozygotic twins Twins produced from a single fertilization event; the first division of the zygote produces two cells, each of which develops into an embryo. Also known as identical twins.

mRNA An RNA molecule transcribed from DNA and translated into the amino acid sequence of a polypeptide.

mtDNA Mitochondrial DNA.

multigene family A gene set descended from a common ancestor by duplication and subsequent divergence from a common ancestor. The globin genes are an example of a multigene family.

multiple alleles Three or more alleles of the same gene.

multiple-factor inheritance See *polygenic inheritance*.

multiple infection Simultaneous infection of a bacterial cell by more than one bacteriophage, often of different genotypes.

μ phage A phage group in which the genetic material behaves like an insertion sequence; capable of insertion, excision, transposition, inactivation of host genes, and induction of chromosomal rearrangements.

mutagen Any agent that causes an increase in the rate of mutation.

mutant A cell or organism carrying an altered or mutant gene

mutation The process that produces an alteration in DNA or chromosome structure; the source of most alleles.

mutation rate The frequency with which mutations take place at a given locus or in a population.

muton The smallest unit of mutation in a gene, corresponding to a single base change.

nanometer (nm) A unit of length equal to 1×10^{-9} meter.

natural selection Differential reproduction of some members of a species resulting from variable fitness conferred by genotypic differences.

neutral mutation A mutation with no immediate adaptive significance or phenotypic effect.

nonautonomous transposon A transposable element that lacks a functional transposase gene.

noncrossover gamete A gamete that contains no chromosomes that have undergone genetic recombination.

nondisjunction An error during cell division in which the homologous chromosomes (in meiosis) or the sister chromatids (in mitosis) fail to separate and migrate to opposite poles; responsible for defects such as monosomy and trisomy.

nonsense codon The nucleotide triplet in an mRNA molecule that signals the termination of translation. Three such codons are known: UGA, UAG, and UAA.

nonsense mutation A mutation that changes an amino acid codon into a termination codon: UAG, UAA, or UGA. Leads to premature termination during translation of mRNA.

NOR See *nucleolar organizer region*.

normal distribution A probability function that approximates the distribution of random variables. The normal curve, also known as a Gaussian or bell-shaped curve, is the graphic display of the normal distribution.

northern blot A technique in which RNA molecules are separated by electrophoresis and transferred by capillary action to a nylon or nitrocellulose membrane. Specific RNA molecules can be identified by hybridization to a labeled nucleic acid probe.

N-terminal amino acid The terminal amino acid in a peptide chain that carries a free amino group.

N terminus The end of a polypeptide that carries a free amino group of the first amino acid. By convention, the structural formula of polypeptides is written with the N terminus at the left.

ν body See *nucleosome*.

nuclease An enzyme that breaks bonds in nucleic acid molecules.

nucleoid The DNA-containing region within the cytoplasm in prokaryotic cells.

nucleolar organizer region (NOR) A chromosomal region containing the genes for rRNA; most often found in physical association with the nucleolus.

nucleolus A nuclear organelle that is the site of ribosome biosynthesis; usually associated with or formed in association with the NOR.

nucleoside A purine or pyrimidine base covalently linked to a ribose or deoxyribose sugar molecule.

nucleosome A complex of four histone molecules, each present in duplicate, wrapped by two turns of a DNA molecule. One of the basic units of eukaryotic chromosome structure. Also known as a ν body.

nucleotide A nucleoside covalently linked to a phosphate group. Nucleotides are the basic building blocks of nucleic acids. The nucleotides commonly found in DNA are deoxyadenylic acid, deoxycytidylic acid, deoxyguanylic acid, and deoxythymidylic acid. The nucleotides in RNA are adenylic acid, cytidylic acid, guanylic acid, and uridylic acid.

nucleotide pair The pair of nucleotides (A and T or G and C) in opposite strands of the DNA molecule that are hydrogen-bonded to each other.

nucleus The membrane-bound cytoplasmic organelle of eukaryotic cells that contains the chromosomes and nucleolus.

null allele A mutant allele that produces no functional gene product. Usually inherited as a recessive trait.

null hypothesis Used in statistical tests, it states that there is no difference between the observed and expected data sets. Statistical methods such as chi-square analysis are used to test the probability of this hypothesis.

nullisomic Describes an individual with a chromosomal aberration in which both members of a chromosome pair are missing.

Okazaki fragment The small, discontinuous strands of DNA produced during DNA synthesis on the lagging strand.

oligonucleotide A linear sequence of about 10–20 nucleotides connected by 5′-to-3′ phosphodiester bonds.

oncogene A gene whose activity promotes uncontrolled proliferation in eukaryotic cells.

open reading frame (ORF) A nucleotide sequence organized as triplets that encodes amino acids. Located between an initiation codon and a termination codon.

operator region A region of a DNA molecule that interacts with a specific repressor protein to control the expression of an adjacent gene or gene set.

operon A genetic unit that consists of one or more structural genes that code for polypeptides, and an adjacent operator gene that controls the transcriptional activity of the structural gene or genes.

origin of replication (ori) Sites along the length of the chromosome where DNA replication begins.

outbreeding depression Reduction in fitness in the offspring produced by mating genetically diverse parents. It is thought to result from a lowered adaptation to local environmental conditions.

overdominance The phenomenon where heterozygotes have a phenotype that is more extreme than either homozygous genotype.

overlapping code A genetic code first proposed by George Gamow in which any given nucleotide is shared by three adjacent codons.

pachytene The stage in meiotic prophase I when the synapsed homologous chromosomes split longitudinally (except at the centromere), producing a group of four chromatids called a tetrad.

pair-rule genes Genes expressed as stripes around the blastoderm embryo during development of the *Drosophila* embryo.

palindrome A word, number, verse, or sentence that reads the same backward or forward (e.g., *able was I ere I saw elba*). In nucleic acids, a sequence in which the base pairs read the same on complementary strands in the 5′-to-3′ direction. For example:

5′-GAATTC-3′
3′-CTTAAG-5′

These often occur as sites for restriction endonuclease recognition and cutting.

pangenesis A discarded theory of development that postulated the existence of pangenes, small particles from all parts of the body that concentrated in the gametes, passing traits from generation to generation, blending the traits of the parents in the offspring.

paracentric inversion A chromosomal inversion that does not include the centromere.

parasexual Condition describing recombination of genes from different individuals that does not involve meiosis, gamete formation, or zygote production. The formation of somatic cell hybrids is an example.

parental gamete See *noncrossover gamete*.

parthenogenesis Development of an egg without fertilization.

partial diploids See *merozygote*.

partial dominance See *incomplete dominance*.

patroclinous inheritance A form of genetic transmission in which the offspring have the phenotype of the father.

pedigree In human genetics, a diagram showing the ancestral relationships and transmission of genetic traits over several generations in a family.

P element Transposable DNA element found in *Drosophila* that is responsible for hybrid dysgenesis.

penetrance The frequency, expressed as a percentage, with which individuals of a given genotype manifest at least some degree of a specific mutant phenotype associated with a trait.

peptide bond The covalent bond between the amino group of one amino acid and the carboxyl group of another amino acid.

pericentric inversion A chromosomal inversion that involves both arms of the chromosome and thus involves the centromere.

phage See *bacteriophage*.

phenocopy An environmentally induced phenotype (nonheritable) that closely resembles the phenotype produced by a known gene.

phenotype The observable properties of an organism that are genetically controlled.

phenylketonuria (PKU) A hereditary condition in humans associated with the inability to metabolize the amino acid phenylalanine. The most common form is caused by the lack of the enzyme phenylalanine hydroxylase.

Philadelphia chromosome The product of a reciprocal translocation that contains the short arm of chromosome 9 carrying the

C-ABL oncogene and the long arm of chromosome 22 carrying the *BCR* gene.

phosphodiester bond In nucleic acids, the covalent bond between a phosphate group and adjacent nucleotides, extending from the 5′ carbon of one pentose (ribose or deoxyribose) to the 3′ carbon of the pentose in the neighboring nucleotide. Phosphodiester bonds form the backbone of nucleic acid molecules.

photoreactivation enzyme (PRE) An exonuclease that catalyzes the light-activated excision of ultraviolet-induced thymine dimers from DNA.

photoreactivation repair Light-induced repair of damage caused by exposure to ultraviolet light. Associated with an intracellular enzyme system.

phyletic evolution The gradual transformation of one species into another over time; vertical evolution.

pilus A filament-like projection from the surface of a bacterial cell. Often associated with cells possessing F factors.

plaque A clear area on an otherwise opaque bacterial lawn, caused by the growth and reproduction of phages.

plasmid An extrachromosomal, circular DNA molecule (often carrying genetic information) that replicates independently of the host chromosome.

pleiotropy Condition in which a single mutation simultaneously affects several characters.

ploidy Term referring to the basic chromosome set or to multiples of that set.

point mutation A mutation that can be mapped to a single locus. At the molecular level, a mutation that results in the substitution of one nucleotide for another.

polar body A cell produced in females at either the first or second meiotic division, which contains almost no cytoplasm as a result of an unequal cytokinesis.

polycistronic mRNA A messenger RNA molecule that encodes the amino acid sequence of two or more polypeptide chains in adjacent structural genes.

polygenic inheritance The transmission of a phenotypic trait whose expression depends on the additive effect of a number of genes.

polylinker A segment of DNA that has been engineered to contain multiple sites for restriction enzyme digestion. Polylinkers are usually found in engineered vectors such as plasmids.

polymerase chain reaction (PCR) A method for amplifying DNA segments that uses cycles of denaturation, annealing to primers, and DNA polymerase-directed DNA synthesis.

polymerases The enzymes that catalyze the formation of DNA and RNA from deoxynucleotides and ribonucleotides, respectively.

polymorphism The existence of two or more discontinuous, segregating phenotypes in a population.

polynucleotide A linear sequence of more than 20 nucleotides, joined by 5′-to-3′ phosphodiester bonds. See also *oligonucleotide*.

polypeptide A molecule made up of amino acids joined by covalent peptide bonds. This term is used to denote the amino acid chain before it assumes its functional three-dimensional configuration.

polyploid A cell or individual having more than two sets of chromosomes.

polyribosome See *polysome*.

polysome A structure composed of two or more ribosomes associated with mRNA, engaged in translation. Formerly called polyribosome.

polytene chromosome A chromosome that has undergone several rounds of DNA replication without separation of the replicated chromosomes, thus forming a giant, thick chromosome with aligned chromomeres producing a characteristic banding pattern.

population A local group of individuals belonging to the same species, which are actually or potentially interbreeding.

position effect Change in expression of a gene associated with a change in the gene's location within the genome.

postzygotic isolation mechanism A factor that prevents or reduces inbreeding by acting after fertilization to produce nonviable, sterile hybrids or hybrids of lowered fitness.

preadaptive mutation A mutational event that later becomes of adaptive significance.

preformationism The discredited idea that an organism develops by growth of structures already present in the egg or sperm.

prezygotic isolation mechanism A factor that reduces inbreeding by preventing courtship, mating, or fertilization.

Pribnow box A 6-bp sequence upstream from the beginning of transcription in prokaryotic genes, to which the σ subunit of RNA polymerase binds. The consensus sequence for this box is TATAAT.

primary protein structure The sequence of amino acids in a polypeptide chain.

primary sex ratio Ratio of males to females at fertilization.

primer In nucleic acids, a short length of RNA or single-stranded DNA that is necessary for the functioning of polymerases.

prion An infectious pathogenic agent devoid of nucleic acid and composed of a protein, PrP, with a molecular weight of 27,000-30,000 Da. Prions are known to cause scrapie, a degenerative neurological disease in sheep, bovine spongiform encephalopathy (BSE or mad cow disease) in cattle, and similar diseases in humans, including kuru and Creutzfeldt-Jakob disease.

probability Ratio of the frequency of a given event to the frequency of all possible events.

proband An individual in whom a genetically determined trait of interest is first detected. Formerly known as a propositus.

probe A macromolecule such as DNA or RNA that has been labeled and can be detected by an assay such as autoradiography or fluorescence microscopy. Probes are used to identify target molecules, genes, or gene products.

product law The law that holds that the probability of two independent events occurring simultaneously is the product of their independent probabilities.

progeny The offspring produced from a mating.

prokaryotes Organisms lacking nuclear membranes, meiosis, and mitosis. Bacteria and blue-green algae are examples of prokaryotic organisms.

promoter Region having a regulatory function and to which RNA polymerase binds prior to the initiation of transcription.

proofreading A molecular mechanism for correcting errors in replication, transcription, or translation. Also known as editing.

prophage A phage genome integrated into a bacterial chromosome. Bacterial cells carrying prophages are said to be lysogenic.

propositus (female, **proposita**) See *proband*.

protein A molecule composed of one or more polypeptides, each composed of amino acids covalently linked together.

proteomics The study of the expressed proteins present in a cell at a given time.

proto-oncogene A gene that normally functions to initiate or maintain cell division. Proto-oncogenes can be converted to oncogenes by alterations in structure or expression.

protoplast A bacterial or plant cell with the cell wall removed. Sometimes called a spheroplast.

prototroph A strain (usually microorganisms) that is capable of growth on a defined, minimal medium. Wild-type strains are usually regarded as prototrophs.

pseudoalleles Genes that behave as alleles to one another by complementation, but can be separated from one another by recombination.

pseudoautosomal inheritance Inheritance of alleles located within the regions of the Y chromosome that are homologous to the X chromosome. Because these alleles are located on both the X and Y chromosome, their pattern of inheritance is indistinguishable from that of autosomal inheritance.

pseudodominance The expression of a recessive allele on one homolog caused by the deletion of the dominant allele on the other homolog.

pseudogene A nonfunctional gene with sequence homology to a known structural gene present elsewhere in the genome. They differ from their functional relatives by insertions or deletions and by the presence of flanking direct repeat sequences of 10–20 nucleotides.

puff *See chromosome puff.*

punctuated equilibrium A pattern in the fossil record of long periods of species stability, punctuated with brief periods of species divergence.

quantitative inheritance See *polygenic inheritance*.

quantitative trait loci (QTLs) Two or more genes that act on a single polygenic trait.

quantum speciation Formation of a new species within a single or a few generations by a combination of selection and drift.

quaternary protein structure Types and modes of interaction between two or more polypeptide chains within a protein molecule.

race A genotypically or geographically distinct subgroup within a species.

rad A unit of absorbed dose of radiation with an energy equal to 100 ergs per gram of irradiated tissue.

radioactive isotope One of the forms of an element, differing in atomic weight and possessing an unstable nucleus that emits ionizing radiation during decay.

random amplified polymorphic DNA (RAPD) A PCR method that uses random primers about 10 nucleotides in length to amplify unknown DNA sequences.

random mating Mating between individuals without regard to genotype.

reading frame Linear sequence of codons in a nucleic acid.

reannealing Formation of double-stranded DNA molecules from dissociated single strands.

recessive A term describing an allele that is not expressed in the heterozygous condition.

reciprocal cross A paired cross in which the genotype of the female in the first cross is present as the genotype of the male in the second cross, and vice versa.

reciprocal translocation A chromosomal aberration in which nonhomologous chromosomes exchange parts.

recombinant DNA A DNA molecule formed by joining two heterologous molecules. Usually applied to DNA molecules produced by *in vitro* ligation of DNA from two different organisms.

recombinant gamete A gamete containing a new combination of genes produced by crossing over during meiosis.

recombination The process that leads to the formation of new gene combinations on chromosomes.

recon A term coined by Seymour Benzer to denote the smallest genetic units between which recombination can occur.

reductional division The chromosome division that halves the diploid chromosome number. The first division of meiosis is a reductional division.

redundant genes Gene sequences present in more than one copy per haploid genome (e.g., ribosomal genes).

regulatory site A DNA sequence that is involved in the control of expression of other genes, usually involving an interaction with another molecule.

rem Radiation equivalent in man; the dosage of radiation that will cause the same biological effect as one roentgen of X-rays.

renaturation The process by which a denatured protein or nucleic acid returns to its normal three-dimensional structure.

repetitive DNA sequences DNA sequences present in many copies in the haploid genome.

replicating form (RF) Double-stranded nucleic acid molecules present as an intermediate during the reproduction of certain viruses.

replication The process of DNA synthesis.

replication fork The Y-shaped region of a chromosome associated with the site of replication.

replicon A chromosomal region or free genetic element containing the DNA sequences necessary for the initiation of DNA replication.

replisome The term used to describe the complex of proteins, including DNA polymerase, that assembles at the bacterial replication fork to synthesize DNA.

repressible enzyme system An enzyme or group of enzymes whose synthesis is regulated by the intracellular concentration of certain metabolites.

repressor A protein that binds to a regulatory sequence adjacent to a gene and blocks transcription of the gene.

reproductive isolation Absence of interbreeding between populations, subspecies, or species. Reproductive isolation can be brought about by extrinsic factors, such as behavior, and intrinsic barriers, such as hybrid inviability.

resistance transfer factor (RTF) A component of R plasmids that confers the ability for cell–cell transfer of the R plasmid by conjugation.

resolution In an optical system, the shortest distance between two points or lines at which they can be perceived to be two points or lines.

restriction endonuclease Nuclease that recognizes specific nucleotide sequences in a DNA molecule and cleaves or nicks the DNA at that site. Derived from a variety of microorganisms, those enzymes that cleave both strands of the DNA are used in the construction of recombinant DNA molecules.

restriction fragment length polymorphism (RFLP) Variation in the length of DNA fragments generated by a restriction endonuclease. These variations are caused by mutations that create or abolish cutting sites for restriction enzymes. RFLPs are inherited in a codominant fashion and can be used as genetic markers.

restrictive transduction See *specialized transduction.*

retrovirus Viruses with RNA as genetic material that utilize the enzyme reverse transcriptase during their life cycle.

reverse transcriptase A polymerase that uses RNA as a template to transcribe a single-stranded DNA molecule as a product.

reversion A mutation that restores the wild-type phenotype.

R factor (R plasmid) Bacterial plasmids that carry antibiotic resistance genes. Most R plasmids have two components: an r-determinant, which carries the antibiotic resistance genes, and the resistance transfer factor (RTF).

RFLP See *restriction fragment length polymorphism.*

Rh factor An antigenic system first described in the rhesus monkey. Recessive r/r individuals produce no Rh antigens and are Rh negative, while R/R and R/r individuals have Rh antigens on the surface of their red blood cells and are classified as Rh positive.

ribonucleic acid (RNA) A nucleic acid characterized by the sugar ribose and the pyrimidine uracil, usually a single-stranded polynucleotide. Several forms are recognized, including ribosomal RNA, messenger RNA, transfer RNA, and heterogeneous nuclear RNA.

ribose The five-carbon sugar associated with the ribonucleotides found in RNA.

ribosomal RNA See *rRNA.*

ribosome A ribonucleoprotein organelle consisting of two subunits, each containing RNA and protein. Ribosomes are the site of translation of mRNA codons into the amino acid sequence of a polypeptide chain.

RNA See *ribonucleic acid.*

RNA editing Alteration of the nucleotide sequence of an mRNA molecule after transcription and before translation. There are two main types of editing: substitution editing, which changes individual nucleotides, and insertion/deletion editing, in which individual nucleotides are added or deleted.

RNA polymerase An enzyme that catalyzes the formation of an RNA polynucleotide strand using the base sequence of a DNA molecule as a template.

RNase A class of enzymes that hydrolyzes RNA.

Robertsonian translocation A form of chromosomal aberration that involves the fusion of the long arms of acrocentric chromosomes at the centromere.

roentgen (R) A unit of measure of the amount of radiation corresponding to the generation of 2.083×10^9 ion pairs in one cubic centimeter of air at 0°C at an atmospheric pressure of 760 mm of mercury.

rolling circle model A model of DNA replication in which the growing point or replication fork rolls around a circular template strand; in each pass around the circle, the newly synthesized strand displaces the strand from the previous replication, producing a series of contiguous copies of the template strand.

R point The point during the G1 stage of the cell cycle when either a commitment is made to DNA synthesis and another cell cycle or the cell withdraws from the cycle and becomes quiescent. Also known as the restriction point.

rRNA The RNA molecules that are the structural components of the ribosomal subunits. In prokaryotes, these are the 16*S*, 23*S*, and 5*S* molecules; in eukaryotes, they are the 18*S*, 28*S*, and 5*S* molecules.

RTF See *resistance transfer factor.*

satellite DNA DNA that forms a minor band when genomic DNA is centrifuged in a cesium salt gradient. This DNA usually consists of short sequences repeated many times in the genome.

SCE See *sister chromatid exchange.*

secondary protein structure The α-helical or β-pleated-sheet form of a protein molecule brought about by the formation of hydrogen bonds between amino acids.

secondary sex ratio The ratio of males to females at birth.

secretor An individual having soluble forms of the blood group antigens A and/or B present in saliva and other body fluids. This condition is caused by a dominant, autosomal gene unlinked to the *ABO* locus (*I* locus).

sedimentation coefficient See *Svedberg coefficient unit.*

segment polarity genes Genes that regulate the spatial pattern of differentiation within each segment of the developing *Drosophila* embryo.

segregation The separation of homologous chromosomes into different gametes during meiosis.

selection The force that brings about changes in the frequency of alleles and genotypes in populations through differential reproduction.

selection coefficient (*s*) A quantitative measure of the relative fitness of one genotype compared with another. See *coefficient of selection.*

selfing In plant genetics, the fertilization of ovules of a plant by pollen produced by the same plant. Reproduction by self-fertilization.

semiconservative replication A model of DNA replication in which a double-stranded molecule replicates in such a way that the daughter molecules are composed of one parental (old) and one newly synthesized strand.

semisterility A condition in which a percentage of all zygotes is inviable.

sex chromatin body See *Barr body.*

sex chromosome A chromosome, such as the X or Y in humans, which is involved in sex determination.

sexduction Transmission of chromosomal genes from a donor bacterium to a recipient cell by the F factor.

sex-influenced inheritance Phenotypic expression that is conditioned by the sex of the individual. A heterozygote may express one phenotype in one sex and the alternate phenotype in the other sex.

sex-limited inheritance A trait that is expressed in only one sex even though the trait may not be X-linked.

sex ratio See *primary* and *secondary sex ratio.*

sexual reproduction Reproduction through the fusion of gametes, which are the haploid products of meiosis.

Shine-Dalgarno sequence The nucleotides AGGAGG present in the leader sequence of prokaryotic genes that serve as a ribosome binding site. The 16*S* RNA of the small ribosomal subunit contains a complementary sequence to which the mRNA binds.

short interspersed elements (SINEs) Repetitive sequences found in the genomes of higher organisms, such as the 300-bp *Alu* sequence.

shotgun experiment The cloning of random fragments of genomic DNA into a vehicle such as a plasmid or phage, usually to produce a library from which clones of specific interest can be selected.

sibling species Species that are morphologically almost identical, but which are reproductively isolated from one another.

sickle-cell anemia A genetic disease in humans caused by an autosomal recessive gene, fatal in the homozygous condition if untreated. Caused by an alteration in the amino acid sequence of the β chain of globin.

sickle-cell trait The phenotype exhibited by individuals heterozygous for the sickle-cell gene.

σ factor A polypeptide subunit of the RNA polymerase that recognizes the binding site for the initiation of transcription.

single-stranded binding proteins (SSBs) In DNA replication, proteins that bind to and stabilize single-stranded regions of DNA that result from the action of unwinding proteins.

sister chromatid exchange (SCE) A crossing over event that can occur in meiotic and mitotic cells; involves the reciprocal exchange of chromosomal material between sister chromatids joined by a common centromere. Such exchanges can be detected cytologically after BrdU incorporation into the replicating chromosomes.

site-directed mutagenesis A process that uses a synthetic oligonucleotide containing a mutant base or sequence as a primer for inducing a mutation at a specific site in a cloned gene.

small nuclear RNA (snRNA) Species of RNA molecules ranging in size from 90 to 400 nucleotides. The abundant snRNAs are present in 1×10^4 to 1×10^6 copies per cell, associated with proteins, and form RNP particles known as snRNPs or *snurps*. Six uridine-rich snRNAs known as U1–U6 are located in the nucleoplasm, and their complete nucleotide sequence is known. snRNAs have been implicated in the processing of pre-mRNA and may have a range of cleavage and ligation functions.

snurps See *small nuclear RNA*.

solenoid structure A level of eukaryotic chromosome structure generated by the supercoiling of nucleosomes.

somatic cell genetics The use of cultured somatic cells to investigate genetic phenomena by parasexual techniques involving the fusion of cells from different organisms.

somatic cells All cells other than the germ cells or gametes in an organism.

somatic mutation A mutational event occurring in a somatic cell and is not heritable.

somatic pairing The pairing of homologous chromosomes in somatic cells.

SOS response The induction of enzymes to repair damaged DNA in *Escherichia coli*. The response involves activation of an enzyme that cleaves a repressor, activating a series of genes involved in DNA repair.

Southern blot A technique developed by Edward Southern in which DNA fragments produced by restriction enzyme digestion are separated by electrophoresis and transferred by capillary action to a nylon or nitrocellulose membrane. Specific DNA fragments can be identified by hybridization to a labeled nucleic acid probe.

spacer DNA DNA sequences found between genes, usually repetitive DNA segments.

specialized transduction Genetic transfer of only specific host genes by transducing phages.

speciation The process by which new species of plants and animals arise.

species A group of actually or potentially interbreeding individuals that is reproductively isolated from other such groups.

spheroplast See *protoplast*.

spindle fibers Cytoplasmic fibrils formed during cell division that are involved with the separation of chromatids at anaphase and their movement toward opposite poles in the cell.

spliceosome The nuclear macromolecule complex within which splicing reactions occur to remove introns from pre-mRNAs.

spontaneous mutation A mutation that is not induced by a mutagenic agent.

spore A unicellular body or cell encased in a protective coat that is produced by some bacteria, plants, and invertebrates; it is capable of survival in unfavorable environmental conditions; and it can give rise to a new individual upon germination. In plants, spores are the haploid products of meiosis.

SRY The sex-determining region of the Y gene found near the pseudoautosomal boundary of the Y chromosome. Accumulated evidence indicates that this gene is the testis-determining factor (TDF).

stabilizing selection Preferential reproduction of those individuals having genotypes close to the mean for the population. A selective elimination of genotypes at both extremes.

standard deviation A quantitative measure of the amount of variation in a sample of measurements from a population.

standard error A quantitative measure of the amount of variation in a sample of measurements from a population.

sterile The condition of being unable to reproduce; free from contaminating microorganisms.

strain A group with common ancestry that has physiological or morphological characteristics of interest for genetic study or domestication.

STR sequences Short tandem repeats of 2–9 base pairs found within minisatellite sequences. These sequences are used in forensic DNA fingerprints, paternity identification, and other applications.

structural gene A gene that encodes the amino acid sequence of a polypeptide chain.

sublethal gene A mutation causing lowered viability, with death before maturity in less than 50% of the individuals carrying the gene.

submetacentric chromosome A chromosome with the centromere placed so that one arm of the chromosome is slightly longer than the other.

subspecies A morphologically or geographically distinct interbreeding population of a species.

sum law The law that holds that the probability of one or the other of two mutually exclusive events occurring is the sum of their individual probabilities.

supercoiled DNA A form of DNA structure in which the helix is coiled upon itself. Such structures can exist in stable forms only when the ends of the DNA are not free, as in a covalently closed circular DNA molecule.

superfemale See *metafemale*.

supermale See *metamale*.

suppressor mutation A mutation that acts to completely or partially restore the function lost by a previous mutation at another site.

Svedberg coefficient unit (*S*) A unit of measure for the rate at which particles (molecules) sediment in a centrifugal field. This unit is a function of several physicochemical properties, including size and shape. A sedimentation value of 1×10^{-13} sec is defined as one Svedberg coefficient unit.

symbiont An organism coexisting in a mutually beneficial relationship with another organism.

sympatric speciation Process of speciation involving populations that inhabit, at least in part, the same geographic range.

synapsis The pairing of homologous chromosomes at meiosis.

synaptonemal complex (SC) An organelle consisting of a tripartite nucleoprotein ribbon that forms between the paired homologous chromosomes in the pachytene stage of the first meiotic division.

syndrome A group of signs or symptoms that occur together and characterize a disease or abnormality.

synkaryon The nucleus of a zygote that results from the fusion of two gametic nuclei. Also used in somatic cell genetics to describe the product of nuclear fusion.

syntenic test In somatic cell genetics, a method for determining whether two genes are on the same chromosome.

S_1 nuclease A deoxyribonuclease that cuts and degrades single-stranded molecules of DNA.

TATA box See *Goldberg-Hogness box.*

tautomeric shift A reversible isomerization in a molecule, brought about by a shift in the localization of a hydrogen atom. In nucleic acids, tautomeric shifts in the bases of nucleotides can cause changes in other bases at replication and are a source of mutations.

TDF (testis-determining factor) The product of the *SRY* gene on the Y chromosome that controls the developmental switch point for the development of the indifferent gonad into a testis.

telocentric chromosome A chromosome in which the centromere is located at the end of the chromosome.

telomerase The enzyme that adds short, tandemly repeated DNA sequences to the ends of eukaryotic chromosomes.

telomere The terminal chromomere of a chromosome.

telophase The stage of cell division in which the daughter chromosomes reach the opposite poles of the cell and re-form nuclei. Telophase ends with the completion of cytokinesis.

telophase I In the first meiotic division, when duplicated chromosomes reach the poles of the dividing cell.

temperate phage A bacteriophage that can become a prophage and confer lysogeny upon the host bacterial cell.

temperature-sensitive mutation A conditional mutation that produces a mutant phenotype at one temperature range and a wild-type phenotype at another temperature range.

template The single-stranded DNA or RNA molecule that specifies the nucleotide sequence of a strand synthesized by a polymerase molecule.

terminalization The movement of chiasmata toward the ends of chromosomes during the diplotene stage of the first meiotic division.

tertiary protein structure The three-dimensional structure of a polypeptide chain brought about by folding upon itself.

test cross A cross between an individual whose genotype at one or more loci may be unknown and an individual who is homozygous recessive for the genes in question.

tetrad The four chromatids that make up paired homologs in the prophase of the first meiotic division. The four haploid cells produced by a single meiotic division.

tetrad analysis Method for the analysis of gene linkage and recombination, using the four haploid cells produced in a single meiotic division.

tetranucleotide hypothesis An early theory of DNA structure proposing that the molecule was composed of repeating units, each consisting of the four nucleotides containing adenine, thymine, cytosine, and guanine.

θ structure An intermediate in the bidirectional replication of circular DNA molecules. At about midway through the cycle of replication, the intermediate resembles the Greek letter theta.

thymine dimer A pair of adjacent thymine bases in a single polynucleotide strand that have chemical bonds formed between carbon atoms 5 and 6. This lesion, usually caused by exposure to ultraviolet light, inhibits DNA replication unless repaired by the appropriate enzymes.

T_m See ***melting profile***.

topoisomerase A class of enzymes that convert DNA from one topological form to another. During DNA replication, these enzymes facilitate the unwinding of the double-helical structure of DNA.

totipotent The ability of a cell or embryo part to give rise to all adult structures. This capacity is usually progressively restricted during development.

trait Any detectable phenotypic variation of a particular inherited character.

trans configuration The arrangement of two mutant sites on opposite homologs, such as

$$\frac{a^1 \quad +}{+ \quad a^2}$$

Contrasts with a cis arrangement, where the sites are located on the same homolog.

transcription Transfer of genetic information from DNA by the synthesis of an RNA molecule copied from a DNA template.

transcriptome The set of mRNA molecules present in a cell at any given time.

transdetermination Change in developmental fate of a cell or group of cells.

transduction Virally mediated genes transfer from one bacterium to another or the transfer of eukaryotic genes mediated by retrovirus.

transfer RNA See *tRNA*.

transformation Heritable change in a cell or an organism brought about by exogenous DNA.

transgenic organism An organism whose genome has been modified by the introduction of external DNA sequences into the germline.

transition A mutational event in which one purine is replaced by another or one pyrimidine is replaced by another.

translation The derivation of the amino acid sequence of a polypeptide from the base sequence of an mRNA molecule in association with a ribosome.

translocation A chromosomal mutation associated with the transfer of a chromosomal segment from one chromosome to another. Also used to denote the movement of mRNA through the ribosome during translation.

transmission genetics The field of genetics concerned with the mechanisms by which genes are transferred from parent to offspring.

transposable element A DNA segment that translocates to other sites in the genome, essentially independent of sequence homology. Usually such elements are flanked by short inverted repeats of 20–40 base pairs at each end. Insertion into a structural gene can produce a mutant phenotype. Insertion and excision of transposable elements depends on two enzymes, transposase and resolvase. Such elements have been identified in both prokaryotes and eukaryotes.

transversion A mutational event in which a purine is replaced by a pyrimidine or a pyrimidine is replaced by a purine.

trinucleotide repeat A tandemly repeated cluster of three nucleotides (such as CTG) in or near a gene, which undergoes an expansion in copy number, resulting in a disease phenotype.

triploidy The condition in which a cell or organism possesses three haploid sets of chromosomes.

trisomy The condition in which a cell or organism possesses two copies of each chromosome, except for one, which is present in three copies. The general form for trisomy is therefore $2n + 1$.

tRNA A small ribonucleic acid molecule that contains a three-base segment (anticodon) that recognizes a codon in mRNA, a binding site for a specific amino acid, and recognition sites for interaction with the ribosomes and the enzyme that links it to its specific amino acid.

tumor-suppressor gene A gene that encodes a gene product that normally functions to suppress cell division. Mutations in tumor-suppressor genes result in the activation of cell division and tumor formation.

Turner syndrome A genetic condition in human females caused by a 45,X genotype. Such individuals are phenotypically female but are sterile because of undeveloped ovaries.

unequal crossing over A crossover between two improperly aligned homologs, producing one homolog with three copies of a region and the other with one copy of that region.

unique DNA DNA sequences that are present only once per genome.

universal code The assumption that the genetic code is used by all life forms. In general, this is true; some exceptions are found in mitochondria, ciliates, and mycoplasmas.

unwinding proteins Nuclear proteins that act during DNA replication to destabilize and unwind the DNA helix ahead of the replicating fork.

variable number tandem repeats (VNTRs) Short, repeated DNA sequences (2–20 nucleotides) present as tandem repeats between two restriction enzyme sites. Variations in the number of repeats creates DNA fragments of differing lengths following restriction enzyme digestion.

variable region Portion of an immunoglobulin molecule that exhibits many amino acid sequence differences between antibodies of differing specificities.

variance A statistical measure of the variation of values from a central value, calculated as the square of the standard deviation.

variegation Patches of differing phenotypes, such as color, in a tissue.

vector In recombinant DNA, an agent such as a phage or plasmid into which a foreign DNA segment will be inserted.

viability The measure of the number of individuals in a given phenotypic class that survive, relative to another class (usually wild type).

virulent phage A bacteriophage that infects and lyses the host bacterial cell.

VNTR See *variable number tandem repeats.*

western blot A technique in which proteins are separated by gel electrophoresis and transferred by capillary action to a nylon membrane or nitrocellulose sheet. A specific protein can be identified through hybridization to a labeled antibody.

wild type The most commonly observed phenotype or genotype, designated as the norm or standard.

wobble hypothesis An idea proposed by Francis Crick, stating that the third base in an anticodon can align in several ways to allow it to recognize more than one base in the codons of mRNA.

writhing number The number of times that the axis of a DNA duplex crosses itself by supercoiling.

W, Z chromosomes Sex chromosomes in species where the female is the heterogametic sex (WZ).

X chromosome Sex chromosome present in species where females are the homogametic sex (XX).

X inactivation In mammalian females, the random cessation of transcriptional activity of one X chromosome. This event, which occurs early in development, is a mechanism of dosage compensation. Molecular basis of inactivation is unknown, but involves a region called the X-inactivation center (XIC) on the proximal end of the p arm. Some loci on the tip of the short arm of the X can escape inactivation. See also *Barr body*, *Lyon hypothesis*.

XIST A locus in the X-chromosome inactivation center that may control inactivation of the X chromosome in mammalian females.

X-linkage The pattern of inheritance resulting from genes located on the X chromosome.

X-ray crystallography A technique to determine the three-dimensional structure of molecules through diffraction patterns produced by X-ray scattering by crystals of the molecule under study.

YAC A cloning vector in the form of a yeast artificial chromosome, constructed using chromosomal elements including telomeres (from a ciliate), centromeres, origin of replication, and marker genes from yeast. YACs are used to clone long stretches of eukaryotic DNA.

Y chromosome Sex chromosome in species where the male is heterogametic (XY).

Y-linkage Mode of inheritance shown by genes located on the Y chromosome.

Z-DNA An alternative structure of DNA in which the two antiparallel polynucleotide chains form a left-handed double helix. Z-DNA has been shown to be present in chromosomes and may have a role in regulation of gene expression.

zein Principal storage protein of maize endosperm, consisting of two major proteins, with molecular weights of 19,000 and 21,000 Da.

zinc finger A DNA-binding domain of a protein that has a characteristic pattern of cysteine and histidine residues that complex with zinc ions, throwing intermediate amino acid residues into a series of loops or fingers.

zygote The diploid cell produced by the fusion of haploid gametic nuclei.

zygotene A stage of meiotic prophase I in which the homologous chromosomes synapse and pair along their entire length, forming bivalents. The synaptonemal complex forms at this stage.

INDEX

F

G

N

O

P

Q

R